**9th Edition
1988**

STANDARDS AND PRACTICES FOR INSTRUMENTATION

**Volume 1
RP2.1-S37.12**

Instrument Society of America

Two-Volume Set
ISBN 1-55617-051-3

INSTRUMENT SOCIETY OF AMERICA
67 Alexander Drive
P.O. Box 12277
Research Triangle Park, North Carolina 27709

ACKNOWLEDGMENTS

The Instrument Society of America wishes to express its deep appreciation of the dedicated work of the many members of the ISA Standards Committees and the other organizations that made this book possible. We also gratefully acknowledge all of the technical societies and associations whose permission to include abstracts of their standards contributed to the success of this project.

ISA is indebted to the following members who have served as the principal officers of the ISA Standards and Practices Department since its inception:

1946-47	F. H. Trapnell	*E. I. du Pont de Nemours & Co., Inc.
1948-49	J. G. Kerley	Shell Oil Company
1950-51	W. H. Fortney	Humble Oil and Refining Company
1952-55	A. V. Novak	E. I. du Pont de Nemours & Co., Inc.
1956	J. W. Percy	United States Steel Corporation
1957-58	E. C. Baran	Standard Oil Company of Ohio
1959-60	G. G. Gallagher	The Fluor Corporation
1961-62	E. A. Adler	United Engineers and Constructors, Inc.
1963-64	J. R. Mahoney	Union Carbide Corporation
1965-66	E. C. Hutchison	Heat Technology Laboratory
1967-68	R. E. Clarridge	International Business Machines, Inc.
1969-70	E. J. Herbster	Mobil Oil Company
1971-72	L. N. Combs	E. I. du Pont de Nemours & Co., Inc.
1973-74	E. J. Byrne	Brown & Root, Inc.
1975-76	W. B. Miller	Moore Products Company
1977-78	A. P. McCauley	SCM Corporation
1979-80	E. C. Magison	Honeywell, Inc.
1981-82	T. J. Harrison	IBM Corporation
1983-84	W. Calder III	The Foxboro Company
1985-86	N. Conger	Fisher Controls International, Inc.
1987-88	D. E. Rapley	Rapley Engineering Services, Inc.

*Note: Company or corporation indicated was the officer's affiliation during his term of office.

The professional enterprise of all these individuals and organizations has helped make ISA Standards and Practices for Instrumentation an authoritative and comprehensive reference for instrumentation technology worldwide.

Glenn F. Harvey
Executive Director
Instrument Society of America

PREFACE

Demands being placed upon the technical standards writing capability of the Instrument Society of America and other standards writing organizations are, perhaps, greater today than at any other time in our history. Legislation being enacted at all levels of government to regulate and structure our modern day world is incorporating the published standard and/or its technical concepts into the primary articles, subparagraphs and far reaching provisos of the laws regulatory guides, and statutes that directly, or indirectly, govern our lives.

The influence and acceptance of our voluntary consensus standards, as produced through the competent efforts of the many individuals, committees and officers of the Standards and Practices Department, places ISA in a position of high esteem — and commensurate responsibility — relative to the demands of a sophisticated society interfacing with a dynamic technology. In keeping with its role as a producer of consensus standards and recommended practices, ISA has been providing these services in the form of single volume editions of *Standards and Practices for Instrumentation* since 1963 when it published the *First Edition*. The *Second Edition* was released in 1966; the *Third Edition* in 1970; the *Fourth Edition* in 1974; the *Fifth Edition* in 1977; the *Sixth Edition* in 1980; the *Seventh Edition* in 1983, and the *Eighth Edition* in 1986.

This, the *Ninth Edition of Standards and Practices for Instrumentation,* provides complete texts of all current ISA Standards and Recommended Practices; titles and abstracts for instrumentation-related standards published by ISA and other U.S. and international organizations; and a subject index, cross-referenced to the titles and abstracts.

Section I is the most comprehensive reference to instrumentation standards presently available. In this section, an attempt was made to present information on representative standards of the organizations that responded. The reader will be able to identify the general area of interest of each organization listed in Section I, and should contact that organization if he can't locate a standard covering a specific area. In addition to the abstracts of standards published by the U.S. Organizations, Section I also contains titles of selected international standards published by the International Organization for Standardization (ISO) and the International Electrotechnical Commission (IEC). The United States is represented in these organizations through its member body, the American National Standards Institute, Inc. (ANSI), formerly, USASI. Complete titles, mailing addresses and ordering information are provided for the organizations listed in Section I. At the end of Section I are the names and addresses of national standardizing organizations of other countries for the readers who may need information on standards activities outside the U.S.

Section II is an index that will help the reader to quickly identify, by subject area, the standards listed in Section I. The subject index was prepared with every consideration to help the reader locate standards in his particular area of interest. All titles of standards contained in Section I were searched for key words and key subject terms. The index is based upon those terms in following format: subject terms are printed in alphabetical order, the titles of standards containing that term or a closely related term are printed under the subject entry along with the acronym of the sponsoring organization and the page within Section I that contains additional information pertaining to each standard.

Section III contains the complete text of each current ISA Standard or Recommended Practice. The ISA Standards and Recommended Practices in this volume, that were not in the *Eighth Edition,* include ten new or revised publications. It is the intent of ISA to publish updated editions of this book at two-year intervals, so the reader should be aware of subsequent editions. Comments and suggestions that will lead to improvement of future editions are always welcome and may be directed to the editor.

TABLE OF CONTENTS

TABLE OF CONTENTS (Continued)

*The Instrument Society of America standards and recommended practices are arranged in numerical order and are individually page numbered. The document titles appear in the upper right corner and document numbers appear in the upper left corner of each odd-numbered page.

**New or revised since the last edition of *Standards and Practices for Instrumentation.*

TABLE OF CONTENTS (Continued)

TABLE OF CONTENTS (Continued)

SECTION I

TITLES AND ABSTRACTS

OF

NATIONAL AND INTERNATIONAL

INSTRUMENTATION STANDARDS

ACOUSTICAL SOCIETY OF AMERICA (ASA)

The following publications are examples of the many acoustical standards available from the Acoustical Society of America, 335 East 45th Street, New York, NY 10017.

ANSI: S1.1-1960 (R 1976), Acoustical Terminology (Including Mechanical Shock and Vibration), 62 pp.

Describes acoustical terminology in areas such as oscillation, vibration and shock, transmission, linear systems, transducers, recording, underwater sound, sonics, architectural acoustics, hearing and speech, and music.

ANSI: S1.4-1983, Specification for Sound Level Meters, 18 pp.

Covers definitions, characteristics, and tolerances, indication, internal noise and extraneous influences, and calibration of general-purpose sound level meters. Includes measurement of transient sound signals and permits use of digital techniques and displays.

ANSI: S1.6-1984, Preferred Frequencies, Frequency Levels, and Band Numbers for Acoustical Measurements, 3 pp.

Provides guide to the design of new acoustical equipment and the selection of frequencies to make comparison of the results of acoustical measurements most convenient.

ANSI: S1.8-1969 (R 1974), Preferred Reference Quantities for Acoustical Levels, 10 pp.

Concerned with the reference quantities and the definitions of some levels for acoustics, electroacoustics and

mechanical vibrations. Purpose is to provide a preferred reference quantity of convenient magnitude for a given kind of acoustical level. (Sponsored by Acoustical Society of America and ASME).

ANSI: S1.10-1966 (R 1976), Method for the Calibration of Microphones, 35 pp.

Describes methods for performing absolute and comparison calibrations of laboratory standard microphones.

ANSI: S1.11-1986 (R 1976), Specification for Octave-Band and Fractional-Octave-Band Analog and Digital Filters

Provides performance requirements for fractional-octave-band bandpass filters, including, in particular, octave-band and one-third-octave-band filters. Basic requirements are given by equations with selected empirical constants to establish limits on the required performance. The requirements are applicable to passive or active analog filters that operate on continuous-time signals, to analog and digital filters that operate on discrete-time signals, and to fractional-octave-band analyses synthesized from narrow-band spectral components. An Appendix is included for reference to terminology used in digital signal processing.

ANSI: S1.12-1967 (R 1977), Specifications for Laboratory Standard Microphones, 11 pp.

Describes types of laboratory microphones suitable for calibration by an absolute method such as the reciprocity technique described in ANSI S1.10-1966 (R 1976), Method for the Calibration of Microphones. These microphones are intended for use as acoustical measurement standards either in a free field or in conjunction with a variety of devices such as artificial voices and couplers for calibrating earphones or microphones.

ANSI: S1.20-1972 (R 1977), Procedures for Calibration of Underwater Electroacoustic Transducers, 40 pp.

Establishes measurement procedures for calibrating underwater electroacoustic transducers and describes forms for presenting the resultant data.

ANSI: S1.23-1976, Method for the Designation of Sound Power Emitted by Machinery and Equipment, 3 pp.

Describes a method for expressing the noise emissons of machinery and equipment in a convenient manner. Applies to all machinery and equipment which is essentially stationary in nature and for which a sound power spectrum may be determined. The designation described in this standard is based on the A-weighted sound power emitted by the source.

ANSI: S1.40-1984, Specification for Acoustical Calibrators

Specifies performance requirements for coupler-type acoustical calibrators. For each microphone type that may be used with the calibrator, requirements include the sound pressure level in the coupler, the frequency of the sound, and the determination of the influence of atmospheric pressure, temperature, humidity, and magnetic fields on the pressure level and frequency of the sound produced by the calibrator. Specifications are to be met within stated tolerances at each frequency and sound pressure level of operation.

ANSI: S1.42-1986, Design Response of Weighting Networks for Acoustical Measurements

Provides the design criteria for both the frequency domain response (amplitude and phase) and time domain response of the A-, B-, and C-weighting networks used in acoustically related measurements. The poles and zeros for each weighting network are given, along with equations for computing the amplitude and phase responses as functions of frequency and the impulse and step responses as functions of time. In the Appendix, similar information is provided for the D- and E-weighting networks.

ANSI: S2.4-1976 (R 1982), Method for Specifying the Characteristics of Auxiliary Analog Equipment for Shock and Vibration Measurements, 10 pp.

Provides a uniform technology and format for the presentation of the performance and other characteristics of auxiliary analog equipment for shock and vibration measurements.

ANSI: S2.9-1976 (R 1982), Nomenclature for Specifying Damping Properties of Materials, 8 pp.

Presents the preferred nomenclature (parameters, symbols, and definitions) for specifying the damping properties of uniform materials and uniform specimens where "uniform" implies homogeneity on a macroscopic scale.

ANSI: S2.11-1969 (R 1978), Selection of Calibration and Tests for Electrical Transducers Used for Measuring Shock and Vibration, 19 pp.

Includes considerations relevant to commonly employed electromechanical shock and vibration measurement transducers, but not to those transducers primarily designed for measurement of acoustic or pressure phenomena.

ANSI: S2.31-1979, Method for the Experimental Determination of Mechanical Mobility. Part I: Basic Definitions and Transducers, 17 pp.

Covers the experimental determination of mechanical mobility of structures by a variety of methods appropriate for different test situations. Part I covers basic concepts and definitions and serves as a guide for the selection, calibration and evaluation of the transducers and instruments used in mobility measurements. The material in Part I is common to most mobility measurement tasks. This document supersedes ANSI S2.6-1963 (R1976).

AIR-CONDITIONING AND REFRIGERATION INSTITUTE (ARI)

The following are examples of many HVAC standards that are available from the Air-Conditioning and Refrigeration Institute, 1501 Wilson Blvd., 6th Floor, Arlington, VA 22209.

ARI: STD. 130-82, Graphic Electrical Symbols for Air-Conditioning and Refrigerating Equipment, 1982.

Contains over 100 electrical symbols commonly used for air-conditioning and refrigeration equipment.

ARI: STD. 575-87, Method of Measuring Machinery Sound Within Equipment Spaces, 1987, 12 pp.

Establishes a uniform method of measuring, recording and determining the sound pressure level of machinery installed in mechanical equipment spaces.

ARI: STD. 720-76, *Standard for Refrigerant Access Valves and Hose Connectors, 1976.*

Applies to 1/4 inch SAE flare refrigerant access valves and hose connectors. It has definitions of components and establishes location requirements of the valve core pin in the access valve body and the location of the hose connector gasket and core pin depressor. It includes drawings, dimensions and gaging procedures.

ARI: STD. 750-87, *Thermostatic Refrigerant Expansion Valves, 1987, 4 pp.*

Definitions; testing and rating requirements; a specification for minimum published data; recommended standard maximum operating pressures for pressure-limiting type valves; recommended refrigerant designation and color coding; and recommended standard connection sizes.

ARI: STD. 760-87 *Solenoid Valves for Use with Volatile Refrigerants and Water, 1987, 19 pp.*

Definitions and classifications; testing and rating requirements; specifications for minimum published data; performance requirements; electrical specifications; recommended line connection sizes; design, construction and assembly recommendations for safety; nameplate data; and conformance conditions.

ARI: STD. 770-84, *Refrigerant Pressure Regulating Valves, 1984, 11 pp.*

Establish, for refrigerant pressure regulating valves: definitions; requirements for testing and rating; specifications, literature and advertising requirements; requirements for marking; recommendations for safety; and conformance conditions.

ARI: STD. 780-86, *Definite Purpose Magnetic Contactors, 1986, 5 pp.*

Establish, for definite purpose magnetic contactors within its scope: definitions; requirements for testing; performance requirements; safety recommendations; and conformance conditions.

ARI: STD. 790-86, *Definite Purpose Magnetic Contactors for Limited Duty, 1986, 4 pp.*

Limited to magnetic contactors for repeatedly establishing and interrupting an electrical power circuit or primary safety circuit (refrigeration or air conditioning).

AIR MOVEMENT AND CONTROL ASSOCIATION (AMCA)

The following standards are available from the Air Movement and Control Association, 30 W. University Drive, Arlington Heights, IL 60004.

AMCA: *Standard 300-85: Reverberant Room Method for Sound Testing of Fans, 1985, 25 pp.*

Detailed specifications of test setups, instrumentation, procedures, and calculations for determining sound power output of fans.

AMCA: *Standard 210-85; Laboratory Methods of Testing Fans for Rating, 1985, 60 pp.*

Detailed specifications of test setups, instrumentation, and procedures for measuring air performance of fans.

AMCA: *Standard 500-83, Test Method for Louvers, Dampers, and Shutters, 1975, 28 pp.*

Detailed specifications of instrumentation, calibration, test procedures and apparatus for determining pressure drop, leakage, and water penetration across ventilating louvers and dampers.

AMCA: *Interim Standard 220-82, Test Methods for Air Curtain Units, 1982.*

Establishes uniform methods for laboratory testing of Air Curtain units to determine performance in terms of flow rate, outlet velocity uniformity, power consumption, and velocity profile.

ALUMINUM ASSOCIATION (AA)

The following publications are available from the Aluminum Association, 900 19th St. NW, Washington, DC 20006.

ASD-1, *Aluminum Standards and Data, 1984, 212 pp.*

Issued biennially, this reference book contains data on chemical compositions, mechanical, physical and other properties, tolerances and aluminum products in general use. It also includes separate sections on sheet and plate, rolled rod and bar, extruded rod, bar and shapes, drawn and extruded tube, forgings, electrical conductors and other aluminum forms and shapes.

ASD-1-M, Aluminum Standards and Data Metric SI, 1986, 212 pp.

Metric version of Aluminum Standards and Data which is divided into three sections: general information, including typical physical and mechanical data; nomenclature unique to the aluminum industry; and chemical compositions, mechanical properties and tolerances.

AMERICAN ASSOCIATION OF TEXTILE CHEMISTS AND COLORISTS (AATCC)

The following annual publication is available from the American Association of Textile Chemists and Colorists, P.O. Box 12215, Research Triangle Park, NC 22709.

Technical Manual of the AATCC, 1983.

An annual publication which contains AATCC Test Methods, committee rosters and reports plus related ASTM and Federal Government testing procedures. The listing of the methods is divided into four sections: 1) Identification and Analysis, 2) Colorfastness, 3) Physical Properties, 4) Biological Properties.

AMERICAN BOILER MANUFACTURERS ASSOCIATION (ABMA)

The following are available from the American Boiler Manufacturers Association, Suite 160, 950 No. Glebe Road, Arlington, VA 22203.

Recommended Design Guidelines for Stoker Firing of Bituminous Coals (1st Edition)

Provides for the application of stokers firing bituminous coal. Contains coal sizing and carbon loss curves, which have been developed as an aid in boiler and stoker design. They provide empirical data covering projected stoker operation and carbon loss expectations.

Fluidized Bed Combustion Guidelines (1987, 1st Edition)

Provides architects, engineers, fuel producers and users with general information about fluid bed combustion systems for steam generation. Intended as a supplement to other sources of information furnished by the equipment manufacturers.

Guidelines for Industrial Boiler Performance Improvement

Provides general guidelines for use by industrial boiler operators to reduce stack emissions on nitrogen oxides and improve boiler operating efficiency. Deals primarily with boiler adjustments that are typically within the control of boiler operators and plant engineering personnel.

Operation and Maintenance Safety Manual

Prepared by the ABMA Product Safety Committee for the purpose of alerting boiler operators and maintenance personnel to some of the hazards of operating and maintaining boiler systems. Applies to utility, industrial and commercial boilers.

Procedure for the Measurement of Sound from Field-Erected Stationary Steam Generators (1973, 3rd Edition)

Sets forth a method for the measurement and recording on data sheets, the sound pressure levels of field-erected stationary steam generators.

Procedure for the Measurement of Sound from Boiler Units, Bottom Supported, Shop or Field-Erected (1973, 3rd Edition)

Provides a standard test method for the measurement of airborne sound from bottom supported steam or hot water generators (boilers), using water or other fluids and from liquid phase heaters.

Matrix of Recommended Quality Control Requirements

This recommended matrix of the ABMA concerning items to be considered for quality control/assurance in connection with boiler manufacturer's procured items was prepared by the ABMA Committee on Quality

Assurance/Control and represents those items a number of experts believe should be considered when qualifying procured materials.

Thermal Shock Damage to Hot Water Boilers as a Result of Energy Conservation Measures

Warns users about problems that can arise from incorrect application of energy management systems to boilers and related systems.

AMERICAN CHEMICAL SOCIETY (ACS)

The following publication is available from the American Chemical Society, Books and Journals' Division, 1155 Sixteenth Street NW, Washington, DC 20036.

Reagent Chemicals, ACS Specifications, 6th Edition, 1981, 685 pp.

ACS Specifications for 320 reagent chemicals; includes 50 pages of definitions, tests, and reagent solutions. Features flame and flameless atomic absorption methods; new polarographic and chromatographic procedures; and new colorimetric test for arsenic.

AMERICAN CONFERENCE OF GOVERNMENTAL INDUSTRIAL HYGIENISTS (ACGIH)

The following publications are available from the American Conference of Governmental Industrial Hygienists, Committee on Industrial Ventilation, P.O. Box 16153, Lansing, MI 48901.

Industrial Ventilation — A Manual of Recommended Practice, 18th Edition, 1984.

Eleven sections, from Section 1, "Principles of Air-Flow," to Section XI, "Air Cleaning Devices," discuss basic ventilation principles and provide useful information.

Ventilation System Testing

From Industrial Ventilation — A Manual of Recommended Practice.

AMERICAN GAS ASSOCIATION (AGA)

The following publications are available from Order and Billing Department, American Gas Association, 1515 Wilson Boulevard, Arlington, VA 22209.

Z21.2b ANSI Z21.2-1983, Gas Hose Connectors for Portable Indoor Gas-Fired Equipment

Details test and examination criteria for gas hose connectors for use indoors with laboratory, shop or ironing

equipment that requires mobility during operation. Such connectors may have end fittings of the slip end type or fittings provided with taper pipe threads; are limited to a maximum nominal inside diameter of 3/8 inch and a maximum nominal length of 6 feet; are intended for use only in unconcealed indoor locations and where they will not be likely to be subject to excessive temperatures (above 125 F); are for use only with natural, manufactured and mixed gases having a specific gravity less than 1.0; and are for use on gas piping systems having fuel gas pressures not in excess of 1/2 psig. This standard does not apply to gas appliance connectors covered under ANSI Z21.24, Z21.45, Z21.54 or Z21.69.

Z21.13 Gas-Fired Low-Pressure Steam and Hot Water Boilers (revision of Z21.13-1982 and Z21.13a-1983)

Details test and examination criteria for low-pressure steam and hot water boilers for use with natural, manufactured and mixed gases, liquefied petroleum gases, and LP gas-air mixtures. A low-pressure boiler is defined in the standard as a boiler operating at or below the following pressures or temperatures: steam heating boiler — 15 psig steam pressure; hot water heating or supply boiler — 160 psig water pressure, 250 F water temperature.

Z21.15 ANSI Z21.15-1979, Manually Operated Gas Valves

Details test and examination criteria for manually operated gas valves which are substantially of the plug and body, or rotating disc type, and to valves of other types which will provide equivalent performance. The standard presents minimum levels for the substantial and durable construction, safe operation and acceptable performance of such valves.

Z21.21 Automatic Valves for Gas Appliances (revision of ANSI Z21.21-1974, Z21.21a-1977 and Z21.21b-1981).

Details test and examination criteria for individual automatic valves, valves utilized as parts of automatic gas ignition systems, or the automatic valve functions of combination controls, which have maximum operating pressure ratings of 1/2 psi, 2 psi, 5 psi, or higher than 5 psi in 5 psi increments up to and including a maximum operating pressure of 60 psi.

Z21.22 Relief Valves and Automatic Gas Shutoff Devices for Hot Water Supply Systems (revision of ANSI Z21.22-1979, Z21.22a-1983 and Z21.22b-1984)

Details test and examination criteria for: (1) temperature relief valves and combination temperature and pressure relief valves for use on storage tanks of hot water supply systems without heater input limitation; (2) valves having only pressure relief features for use on storage tanks of hot water supply systems with inputs up to and including 200,000 Btu per hour; (3) automatic gas shutoff valves and devices; and (4) vacuum relief valves.

Z21.24 Metal Connectors for Gas Appliances (Revisions to Z21.24-1981, Z21.24a-1983 and Z21.24b-1985)

Details test and examination criteria for gas appliance connectors comprised of semi-rigid metal tubing and having a fitting at each end provided with taper pipe threads for connection to a gas appliance and to house piping, or consisting of corrugated tubing depending on all-metal construction for gastightness. Such connectors are suitable for connecting gas-fired appliances to fixed gas supply lines at pressure not in excess of 1/2 psig. These connectors are limited to a maximum nominal length of 6 feet and are not intended for continuous movement.

Z21.41 ANSI Z21.41-1978, Quick-Disconnect Devices for Use With Gas Fuel

Details test and examination criteria for hand-operated devices which provide means for connecting and disconnecting gas-fired appliances or gas appliance connectors to gas supplies and which are for use under indoor or outdoor applications. These devices are equipped with automatic means to shut off gas flow when disconnected.

ANSI: Z21.45-1985, Flexible Connectors of Other Than All-Metal Construction for Gas Appliances.

Details test and examination criteria for gas appliance connectors consisting of flexible tubing dependent on other than all-metal construction for gas-tightness. Such connectors are suitable for connecting gas-burning appliances to fixed gas supply lines containing fuel gases at pressures not in excess of 1/2 psig. These connectors are limited to a maximum nominal inside diameter of 1 inch and a nominal length of 6 feet and are not intended for continuous movement.

Z21.45a Addenda to ANSI Z21.45-1985, Flexible Connectors of Other Than All-Metal Construction for Gas Appliances

Details test and examination criteria for gas appliance connectors consisting of flexible tubing dependent on other than all-metal construction for gastightness. Such connectors are suitable for connecting gas-fired appliances to fixed gas supply lines containing natural, manufactured or mixed gases, liquefied petroleum gases or LP gas-air mixtures at pressures not in excess of 1/2 psig. These connectors are limited to a maximum nominal inside diameter of 1 inch and a nominal length of 6 feet and are not intended for continuous movement. This standard does not apply to gas appliance connectors covered under ANSI Z21.2, Z21.24, Z21.54 or Z21.69.

Z21.70 Earthquake Actuated Automatic Gas Shutoff Systems (new standard)

Details test and examination criteria for automatic gas shutoff systems consisting of, (1) a seismic sensing means and, (2) an actuating means designed to automatically actuate a companion gas shutoff means installed in gas piping. Such systems are designed to automatically shut off the gas supply downstream of the gas shutoff means in the event of a seismic disturbance. The system may consist of separable components or may incorporate all functions in a single body.

Z83.3-1971, Z83.3a-1972, Z83.3b-1986 Gas Utilization Equipment in Large Boilers (reaffirmation)

Details criteria for the installation of gas utilization equipment in boilers having inputs over 400,000 Btu per hour per combustion chamber, except water-tube boilers having outputs of 10,000 pounds of steam per hour or more.

ANSI: Z223.1-1974, National Fuel Gas Code, 161 pp.

A safety code for gas piping systems on consumers' premises, the installation of gas utilization equipment and

6

accessories for use with fuel gases. Piping systems covered by this code are limited to a maximum operating pressure of 60 psig.

Z223.1a Addenda to ANSI Z223.1-1984, National Fuel Gas Code

A safety code which applies to the installation of fuel gas piping systems, fuel gas utilization equipment and related accessories.

AMERICAN LEATHER CHEMISTS ASSOCIATION (ALCA)

ALCA is working with ASTM in developing uniform test procedures for leather processing. ALCA has accepted Standard Methods for the Examination of Water and Wastewater, 14th Edition, published by the American Wastewater Association, WPCF Association, and the American Public Health Association, 1015 18th Street, NW., Washington, DC 20036. This publication replaces ACLA's former M-series of methods. Other information on leather processing is available from the American Leather Chemists Association, c/o Campus Station, Location 14, Cincinnati, OH 45221.

AMERICAN NATIONAL STANDARDS INSTITUTE (ANSI)

The American National Standards Institute (ANSI) does not develop standards but rather approves standards developed by other organizations after ANSI has verified that the requirements for due process and consensus have been met. In addition ANSI coordinates the private sector standards activity within the United States and manages US participation in nongovernmental, international standards developing organizations, primarily the International Organization for Standardization (ISO) and the International Electrotechnical Commission (IEC). Copies of ANSI-approved standards may be purchased either from the standards developer or ANSI. ANSI sells IEC and ISO standards and serves as a clearinghouse for information on other international standards organizations. Contact: American National Standards Institute, 1430 Broadway, New York, New York 10018.

AMERICAN NUCLEAR SOCIETY (ANS)

The following ANS and ANSI standards are available from the American Nuclear Society, 555 North Kensington Avenue, La Grange Park, IL 60521.

ANS: 10.2-1982, Recommended Programming Practices to Facilitate the Portability of Scientific Computer Programs, 1982.

> This standard recommends programming practices to facilitate the portability of computer programs prepared for scientific and engineering computations.

ANS: 2.2-1978, Earthquake Instrumentation Criteria for Nuclear Power Plants, 1978.

> This ANSI standard specified earthquake instrumentation for the site, structures, equipment, and piping. It is intended for use at water-cooled nuclear power plants, and may be used for guidance at other types of nuclear power plants.

ANS: 2.10-1979, Guidelines for Retrieval, Review, Processing and Evaluation of Records Obtained from Seismic Instrumentation, 1979.

Provides instructions for treatment of data from a variety of seismic instruments used in light water-cooled, land-based nuclear power plants. Defines the type and timing of plant owner activities, required in the event of an earthquake and includes specific procedures for the evaluation of records obtained from seismic instrumentation specified in ANS-2.2-1978.

ANS: 3.1-1987 Selection, Qualification and Training of Personnel for Nuclear Power Plants (Revision of ANSI/ANS:3.1-1981), 1987.

This standard provides criteria for the selection, qualification, and training of personnel for stationery nuclear power plants. Qualifications, responsibilities, and training of personnel in operating and support organizations appropriate for the safe and efficient operation of nuclear power plants are addressed.

ANS: 4.5-1986, Criteria for Accident Monitoring Functions in Light-Water-Cooled Reactors, 1986.

Criteria are provided for determining the variable to be monitored by the control room operator of a light water reactor, as required for safety, during the course of an accident, including long-term stable shutdown. Also included are criteria for determining the requirements for the equipment used to monitor those variables.

ANS: 6.6.1-1979, Calculaton and Measurement of Direct and Scattered Gamma Radiation from LWR Nuclear Power Plants, 1979.

This standard defines calculational requirements and discusses measurement techniques for estimates of dose rates near nuclear power plants due to direct and scattered gamma-rays from contained sources on site. It describes the considerations necessary to computer dose rates, including component self-shielding, shielding afforded by walls and structures, and scattered radiation.

ANS: 8.3-1986, Criticality Accident Alarm System, (Revision of ANSI/ANS: 8.3-1979), 1986.

This standard applies to all operations with plutonium, 233U, 235U, and other fissionable materials in which inadvertent criticality may occur and cause the exposure of personnel to unacceptable amounts of radiation. It is directed principally toward gamma-radiation rate-sensing systems. This revision incorporates relevant features of ANSI/ANS-8.3-1979 and ANS-N2.3-1979.

ANS: 10.3-1986, Guidelines for the Documentation of Digital Computer Programs (Revision of ANSI/N413-1974), 1986.

This standard presents guidelines for the documentation of digital computer programs prepared for scientific and engineering computations. The guidelines are designed to facilitate effective usage, transfer, conversion, and modification of computer programs.

ANS: 59.3-1984, Safety Criteria for Control Air Systems, 1984.

This standard provides nuclear safety criteria for the control air system that furnishes compressed air to safety-related components in nuclear power plants.
It applies only to the air supply system and does not apply to air-operated devices.

AMERICAN PETROLEUM INSTITUTE (API)

The following API Publications, Recommended Practices, Standards, and Bulletins are available from the American Petroleum Institute, 1220 L Street, NW, Washington, DC 20005.

API: Bull 5T1, Bulletin on Nondestructive Testing Terminology, Seventh Edition, 1985.

Provides definitions in English, French, German, Italian, Japanese, and Spanish for a number of defects which commonly occur in steel pipe.

API: Spec. 6D, Specification for Pipeline Valves, End Closures, Connectors and Swivels, Eighteenth Edition, 1982.

Covers materials, dimensions, and pressure ratings for flanged and welding-end pipeline valves.

Guide for Inspection of Refinery Equipment:

Chap. IV, Inspection Tools, Third Edition, 1983.

This chapter describes and illustrates both homemade and purchasable refinery inspection tools, such as hammer, calipers, gages, electronic and mechanical devices, and special tools. It also covers the use of tools to solve specific inspection problems.

Chap. XV, Instruments and Control Equipment, Third Edition, 1981.

This chapter is a guide for instrument inspection. The inspection procedures given cover the majority of types of standard commercial instruments and associated equipment used in modern refineries.

API: RP 500A, Classification of Areas for Electrical Installations in Petroleum Refineries, Fourth Edition, 1982.

A guide that applies to refinery areas when flammable vapors and liquids are processed, stored, loaded, unloaded, or otherwise handled. It is intended to serve as a supplement to the *National Electrical Code.* 19 pages.

API- RP 500B, Recommended Practice for Classification of Areas for Electrical Installations at Drilling Rigs and Production Facilities on Land and on Marine Fixed and Mobile Platforms, Second Edition, 1973.

Classifies areas surrounding drilling rigs and production facilities on land and on marine fixed and mobile platforms for the safe installation of electrical equipment. 14 pages.

API: RP 500C, Classification of Areas for Electrical Installation, Petroleum, Second Edition, 1984.

Classified areas include pump stations, compressor stations, storage facilities, loading racks, and manifold and pipeline right-of-way areas where flammable liquids and gases are handled.

API: RP 520, Design and Installation of Pressure-Relieving Systems in Refineries, Parts I and II.

Part I — Design, Fourth Edition, 1976.

Applies to relieving devices and their discharge systems on refinery pressure vessels and equipment designed for maximum allowable working pressure of more than 15 psig.

Part II — Installation, Second Edition, 1963, (Reaffirmed 1973).

Applies to installation of pressure relief valves in gas, vapor, and liquid service. It includes information on inlet and discharge piping, valve location and position, valve setting, and handling and testing.

API: RP 521, Guide for Pressure Relief and Depressuring Systems, Second Edition, 1982.

Supplements the material set forth in API RP 520, Parts I and II. Guidelines are provided for examining the principal causes of over-pressure; for determining individual relieving rates; and for selecting and designing disposal systems.

API: Std. 526, (ANSI/API Std. 526-1984), Flanged Steel Safety Valves, Third Edition, 1984.

Specifies dimensions of carbon and alloy steel safety relief valves for the purpose of promoting interchangeability. Basic requirements are given for orifice designation and area, valve size and rating, materials, pressure-temperature limits, and center-to-face dimensions.

API: Std. 527, (ANSI/API Std. 527-1978), Commercial Seat Tightness of Safety Relief Valves with Metal-to-Metal Seats, Second Edition, 1978.

Describes a method of determining seat tightness of safety relief valves as covered in API Std. 526.

API: RP 550, Manual on Installation of Refinery Instruments and Control Systems, Parts I, II, III, and IV.

Part I — Process Instrumentation and Control.

Section 1 — Flow, Third Edition, 1977
Section 2 — Level, Fourth Edition, 1980 (Reaffirmed 1983)
Section 3 — Temperature, Fourth Edition, 1985
Section 4 — Pressure, Fourth Edition, 1980 (Reaffirmed 1983)
Section 5 — Controllers and Control Systems, Fourth Edition, 1985
Section 6 — Control Valves and Accessories, Fourth Edition, 1985
Section 7 — Transmission Systems, Third Edition, 1974
Section 8 — Seals, Purges, and Winterizing, Fourth Edition, 1980
Section 9 — Air Supply Systems, Fourth Edition, 1980
Section 10 — Hydraulic Systems, Fourth Edition, 1981
Section 11 — Electrical Power Supply, Third Edition, 1981
Section 12 — Control Centers, Third Edition, 1977
Section 13 — Alarms and Protective Devices, Fourth Edition, 1985
Section 14 — Process Computer Systems, First Edition, 1982

Part II — Process Stream Analyzers

Section 1 — Analyzers, Fourth Edition, 1985
Section 2 — Process Chromatographs, Fourth Edition, 1981
Section 4 — Moisture Analyzers, Fourth Edition, 1983
Section 5 — Oxygen Analyzers, Fourth Edition, 1983
Section 6 — Analyzers for the Measurement of Sulfur and Its Compounds, Fourth Edition, 1983
Section 7 — Electrochemical Liquid Analyzers, Fourth Edition, 1984
Section 9 — Water Quality Analyzers, Fourth Edition, 1984
Section 10 — Area Safety Monitors, Fourth Edition, 1983

Part III — Fired Heaters and Inert Gas Generators, Third Edition, 1985

Part IV — Steam Generators, Second Edition, 1984

API: Std. 598, Valve Inspection and Test, Fifth Edition, 1982.

Covers inspection and pressure test requirements for both resilient seated and metal-to-metal seated valves of the gate, globe, plug, check, ball, and butterfly types.

Bull. 2509B, Shop Testing of Automatic Liquid-Level Gages, 1961.

This publication is not included in the manual. 49 pages.

API: Manual of Petroleum Measurement Standards (Complete Set)

This manual, which includes all subject matter found in API measurement publications, is an ongoing project.

Chapter 4, Proving Systems, First Edition, 1978 (ANSI/API MPMS 4-1978)

Serves as a guide for the design, installation, calibration, and operation of meter proving systems.

Chapter 5, Metering

Covers the dynamic measurement of liquid hydrocarbons, or metering. It is divided into subchapters.

Chapter 5.2, Measurement of Liquid Hydrocarbons by Displacement Meter Systems, First Edition, 1977. (ANSI/API MPMS 5.2-1977)

Describes and illustrates methods and practices that may be used to obtain optimum measurement of liquid hydrocarbons and maximum service life when using displacement meters.

Chapter 5.3, Turbine Meters, First Edition, 1976.

Specifies the characteristics of turbine meters and gives rules for applying appropriate considerations to the nature of the liquids to be measured. It also covers the installation of metering systems that use a turbine meter, and their performance, operation, and maintenance in liquid hydrocarbon service.

Chapter 5.4, Instrumentation or Accessory Equipment for Liquid Hydrocarbon Metering Systems, First Edition, 1976. (ANSI/API MPMS 5.4-1976)

Specifies the characteristics of available and necessary equipment that can be used to attain desired purposes when used in conjunction with volumetric hydrocarbon meters.

Chapter 5.5, Fidelity and Security of Flow Measurement Pulsed-Data Transmission Systems, First Edition, June 1982.

Provides a guide to the selection, operation, and maintenance of pulsed-data, cabled transmission systems for fluid metering systems to provide the desired level of fidelity and security of transmitted data.

Chapter 6, Metering Assemblies

Discusses the design, installation and operation of metering systems for coping with special situations in hydrocarbon measurement. Portions of Chapter 6 are in preparation.

Chapter 7, Temperature Determination

Covered the sampling, reading, averaging, and rounding of the temperature of liquid hydrocarbons in both the static and dynamic modes of measurement for volumetric purposes. Portions of Chapter 7 are in preparation.

Chapter 8, Sampling

Covers standardized procedures for sampling crude oil or its products.

Chapter 9, Density Determination

Describes the standard methods and apparatus used to determine the specific gravity of crude petroleum products normally handled as liquids. It is divided into subchapters as follows.

Chapter 10, Sediment and Water

Describes methods for determining the amount of sediment and water, either together or separately. Laboratory and field methods are covered as follows.

Chapter 12, Calculation of Petroleum Quantities

Describes the standard procedures for calculating net standard volumes, including the application of correction factors and the importance of significant figures. The purpose of standardizing the calculation procedure is to achieve the same result regardless of what person or computer does the calculating.

Chapter 13, Statistical Aspects of Measuring and Sampling

Covers the application of statistical methods to petroleum measurement and sampling. Chapter 13 is in preparation.

Chapter 14, Natural Gas Fluids Measurement

Standardizes practices for measuring, sampling, and testing natural gas fluids. Chapter 14 is in preparation.

Chapter 16, Measurement of Petroleum by Weight

Provides references to model regulations promulgated by NCWM regarding commercial weighing, tolerances, and other technical requirements and to the recognized practices of the petroleum industry when products are handled on a weight basis. Chapter 16 is in preparation.

AMERICAN SOCIETY FOR QUALITY CONTROL (ASQC)

The following ASQC and ANSI standards are available from the American Society of Quality Control, 310 West Wisconsin Avenue, Milwaukee, WI 53203.
ANSI/ASQC: A1-1978, Definitions, Symbols, Formulas and Tables for Control Charts.

A standardization of the symbols, concepts, terms, and procedures relation to Shewhart control charts, control charts with warning limits, moving averages and ranges, exponentially smoothed averages and ranges, exponentially smoothed averages, Cusum charts, multivariate control, trend control, and the acceptance control chart.

ANSI/ASQC: A2-1978, Terms, Symbols and Definitions for Acceptance Sampling.

Covers the major forms of acceptance sampling schemes for both attributes and variables measures, including extensive comments, explanations, and comparison of the various acceptance sampling approaches.

ANSI/ASQC: A3-1978, Quality Systems Terminology.

Presents basic definitions dealing with quality assurance, quality control, quality programs, and quality systems for general use within U.S. commerce and industry.

ANSI/ASQC: C1-1985 (ANSI Z1.8-1971), Specifications of General Requirements for a Quality Program.

This standard concerns the establishment and maintenance of a quality program by a contractor to assure compliance requirements in the areas of quality management, design information, procurement, manufacture, acceptance, and documentation.

ANSI/ASQC: Z1.15-1979, Generic Guidelines for Quality Systems.

Describes the significant elements that should be considered in the quality system of a manufactured product, including quality policy, design assurance, purchased materials control, quality control at various production stages, field performance, and product liability.

ANSI/ASQC: Z1.4-1980, Sampling Procedures and Tables for Inspection by Attributes.

This standard, which corresponds to MIL-STD-105D, establishes sampling plans and procedures for inspection by attributes. Its tables and procedures are completely compatible with MIL-STD-105D. It is also compatible and interchangeable with ANSI/ASQC Z1.9-1980 for variables inspection.

ANSI/ASQC: Z1.9-1980, Sampling Procedures and Tables for Inspection by Variables for Percent Nonconforming.

This standard, establishing sampling plans and procedures for inspection by variables, corresponds to the military standard MIL-STD-414 and is interchangeable with ISO/DIS3951. It contains tables and procedures of MIL-STD-414, suitably modified to achieve correspondence with ISO/DIS 3951 and matching with MIL-STD-105D and ANSI/ASQC Z1.4-1980.

ANSI/IEEE: 730-1981, Software Quality Assurance Plans.

This standard assists in the preparation of quality assurance plans for the development and maintenance of critical software. It provides developers, users, and the public with criteria against which such plans can be prepared and assessed.

ANSI/ASQC: E2-1984, Guide to Inspection Planning.

This standard describes the significant elements that should be considered in the development of inspection activities. It provides generic guidelines for planning and applying a product/process inspection system for construction, manufacturing, operating, or service functions.

ANSI/ASQC B1, B2 and B3-1985 (ANSI Z1.1-1958, Z1.2-1958 and Z1.3-1958 Revised 1975) Guide for Quality Control, Control Chart Method of Analyzing Data, and Control Chart Method of Controlling Quality During Production.

Contains three standards; a guide for handling problems concerning the economic control of quality of materials and manufactured products with particular reference to methods of collecting, arranging, and analyzing inspection and test records to detect lack of uniformity of quality; a guide to the control chart method of analyzing a collection of data, with particular reference to quality data resulting from inspections and tests of materials and manufactured products; and a guide outlining the control chart method of identifying and eliminating causes of trouble in repetitive production processes, in order to reduce variations in the quality of manufactured product and materials.

ANSI/ASQC Q1-1986, Generic Guidelines for Auditing of Quality Systems.

Describes the significant elements that should be considered in the planning and execution of audits. The standard is intended to provide generic guidelines for internal and external Audits of Quality Systems. Not intended to be applied in all specific situations, but is to be used in developing and describing criteria for effective and efficient auditing.

ANSI/ASQC S1-1987, An Attribute Skip-Lot Sampling Program.

Provides a procedure for reducing the inspection effort on products submitted by those suppliers who have demonstrated their ability to control, in an effective manner, all facets of product quality and consistently produce superior quality material. Shall not be applied to the inspection of product characteristics which involve the safety of personnel.

AMERICAN SOCIETY FOR TESTING AND MATERIALS (ASTM)

The following ASTM Standards are available from the American Society for Testing and Materials, 1916 Race Street, Philadelphia, PA 19103. Joint ASTM-API Standards are available from either organization.

ASTM: Index to Standards; The 1985 Annual Book of ASTM Standards, Volume 00.01, 1985.

This book is a key reference volume. It lists by number designation, and cross-indexed title every one of over 8000 ASTM Standards. The index refers the reader to the exact volume of the 66-volume Annual Book of ASTM Standards where a particular standard may be found.

ASTM: A 105/A 105M-86a, Standard Specification for Forgings, Carbon Steel, for Piping Components, 7 pp.

Included are flanges, fittings, valves, and similar parts to specified dimensions or to dimensional standards such as ANSI and API specifications.

ASTM: A 181/A 181M-85a, Standard Specification for Forgings, Carbon Steel for General Purpose Piping, 5 pp.

Two Grades of material are covered, designated as grades I and II, respectively, and are classified in accordance with their chemical and physical properties.

ASTM: A 182/A 182M-86a, Standard Specification for Forged or Rolled Alloy-Steel Pipe Flanges, Forged Fittings, and Valves and Parts for High-Temperature Service, 14 pp.

Twenty-five grades are covered including eleven ferritic steels and fourteen austenitic steels. Selection will depend upon design and service conditions, mechanical properties, and the high-temperature characteristics.

ASTM: A 522/A 522M-86, Standard Specification for Forged or Rolled 8 and 9 Percent Nickel Alloy Steel Flanges, Fittings, Valves, and Parts for Low-Temperature Service, 5 pp.

The specification is applicable to forgings with maximum section thickness of 33 in. (76.2 mm) in the double normalized and tempered condition and 5 in. (127.0 mm) in the quenched and tempered condition. Forgings to this specification are intended for service at operating temperatures not lower than -320 F (-196 C) or higher than 250 F (121 C).

ASTM: B 106-84, Standard Test Methods for Flexivity of Thermostat Metals, 8 pp.

Intended for determining the flexure-temperature characteristics of thermostat metals tested in the form of flat strips 0.012 in. or over in thickness and in the form of spiral coils less than 0.012 in. in thickness.

ASTM: B 223-85, Standard Test Method for Modulus of Elasticity of Thermostat Metals (Cantilever Beam Method), 7 pp.

Covers the procedures for determining the modulus of elasticity of thermostat metals at any temperature between -300 and -1000°F by mounting the specimen as a cantilever beam and measuring the deflection when subjected to a mechanical load.

ASTM: B 244-79, Standard Method for Measurement of Thickness of Anodic Coatings on Aluminum and of Other Nonconductive Coatings on Nonmagnetic Basis Metals with Eddy-Current Instruments, 1 pp.

Describes procedures for measuring by non-destructive means the thickness of anodic coatings on aluminum using eddy-current instruments.

ASTM- B305-56 (1978),[e1] Standard Test Method for Maximum Loading Stress at Temperature of Thermostat Metals (Cantilever Beam Method), 5 pp.

Intended for the evaluation of the maximum stress at temperatures that can be applied to thermostat metals by a static or dead load before the combination of thermal and mechanical stresses causes displacement of the metal in excess of the elastic limit of the material. In this test the specimen is mounted as a cantilever beam.

ASTM: B 362-81, Standard Test Method for Mechanical Torque Rate of Spiral Coils of Thermostat Metal, 3 pp.

Covers the principles of determining the mechanical torque rate of spiral coils of thermostat metal.

ASTM: B 388-86, Standard Specification for Thermostat Metal Sheet and Strip, 1974, 9 pp.

Covers thermostat metals in the form of sheet or strip which are used for the temperature-sensitive elements of devices for controlling, compensating, or indicating temperature and is intended to supply acceptance requirements to purchasers ordering this material by type designation.

ASTM: B 389-81 (1986), Standard Test Method for Thermal Deflection Rate of Spiral and Helical Coils of Thermostat Metal, 6 pp.

Covers a procedure for determining thermal deflection rate of spiral and helical coils for thermostat metal.

ASTM: B 430-84, Standard Test Method for Particle Size Distribution of Refractory Metal-Type Powders by Turbidimetry. 6 pp.

Covers the procedure for the determination of particle size distribution of refractory metal powder with a turbidimeter.

ASTM: B 462-82, Standard Specification for Forged or Rolled Chromium-Nickel-Iron-Molybdenum-Copper-Columbium Stabilized Alloy (UNS N08020) Pipe Flanges, Forged Fittings, and Valves and Parts for Corrosive High-Temperature Service, 1970, 4 pp.

ASTM: C 115-86, Standard Test Method for Fineness of Portland Cement by the Turbidimeter, 10 pp.

Describes the Wagner turbidimeter apparatus and procedure for determining the fineness of portland cement. Includes details for construction, calibration, and application of the turbidimeter.

ASTM: C 518-85, Standard Test Method for Steady-State Heat Flux Measurements and Thermal Transmission Properties by Means of the Heat Flow Meter Apparatus, 32 pp.

Covers the determination of, by means of a heat flow meter, the thermal conductivity of homogeneous insulating, building, and other materials whose thermal conductivities do not exceed 2.0 Btu/h; $\times$ ft $2 \times$ °F (1.13 mW/cm2x°C).

ASTM: C 604-86, Standard Test Method for True Specific Gravity of Refractory Materials by Gas-Comparison Pycnometer, 3 pp.

Covers the determination of the true specific gravity of solid materials, and is particularly useful for easily hydrateable materials which are not suitable for test with ASTM Method C 135, Test for True Specific Gravity of Refractory Materials.

ASTM: D 240-85, Standard Test Method for Heat of Combustion of Liquid Hydrocarbon Fuels by Bomb Calorimeter, 10 pp.

Describes procedures for determining heat of combustion. It is applicable to a variety of substances but particularly to liquid hydrocarbon fuels of both low and high volatility.

ASTM: D 287-82 - (API: STD 2544), Standard Test Method for API Gravity of Crude Petroleum and Petroleum Products (Hydrometer Method), 4 pp.

Covers the determination by means of a glass hydrometer of the API gravity of crude petroleum and petroleum products normally handled as liquids and having a Reid vapor pressure of 26 lb. or less.

ASTM: D 941-83, Standard Test Method for Density and Relative Density (Specific Gravity) of Liquids by Lipkin Bicapillary Pycnometer, 6 pp.

Intended for the measurement of the density of any hydrocarbon material that can be handled in a normal fashion as a liquid at the specified test temperatures of 20 and 25°C. Its application is restricted to liquids having vapor pressures less than 600 mm Hg (approximately 0.8 atm) and having viscosities less than 15 cSt at 20°C.

ASTM: D 1071-83, Standard Method for Measurement of Gaseous Fuel Samples, 19 pp.

Are applicable to the measuring of gaseous fuel samples, including, liquefied petroleum gases, in the gaseous state at normal temperature and pressures.

ASTM: D 1085-65 (1984) - (API: STD 2545), Standard

Method of Gaging Petroleum and Petroleum Products, 39 pp. Published as reprint only.

Describes the procedure for gaging crude petroleum and its liquid products in various types of tanks, containers, and carriers.

ASTM: D 1142-86, Standard Test Method for Water Vapor Content of Gaseous Fuels by Measurement of Dew-Point Temperature. 13 pp.

Covers the determination of the water vapor content of gaseous fuels by measurement of the dew-point temperature and the calculation therefrom of the water vapor content.

ASTM: D 1145-80, Standard Method of Sampling Natural Gas, 14 pp.

Covers the procedures for the sampling of (1) natural gases containing primarily hydrocarbons and nitrogen, (2) natural gases containing hydrogen, (3) natural gases containing hydrogen sulfide, or organic sulfur compounds, or other sulfur contaminants, (4) natural gas containing carbon dioxide, (5) natural gas containing gasoline and condensables.

ASTM: D 1186-81, Standard Method for Nondestructive Measurement of Dry Film Thickness of Nonmagnetic Coatings Applied to a Ferrous Base, 5 pp.

Covers the measurement of film thickness of nonmagnetic dried films of paint, varnish, lacquer, and related products applied over a magnetic base material.

ASTM: D 1217-81, Standard Test Method for Density and Relative Density (Specific Gravity) of Liquids by Bingham Pycnometer, 8 pp.

Intended for the measurement of the density of pure hydrocarbons or petroleum distillates boiling between 194 and 230°F (90 and 110°C) that can be handled in a normal fashion as a liquid at the specified test temperatures of 68 and 77°F (20 and 25°C). The method was developed especially for the reference fuels n-heptane and iso-octane and is designed to provide values having an accuracy of 0.00003 g/mL.

ASTM: D 1238-86, Standard Test Method for Measuring Flow Rates of Thermoplastics by Extrusion Plastometer, 13 pp.

Covers measurement of the rate of extrusion of molten resins through an orifice of a specified length and diameter under prescribed conditions of temperature and pressure.

ASTM: D 1247-80, Standard Method of Sampling Manufactured Gas, 11 pp.

Covers the procedures for securing representative samples of manufactured gas, and correlates the size or type of sample with the analysis to be done subsequently on that sample.

ASTM: D 1298-85, Standard Test Method for Density,

Relative Density, (Specific Gravity), or API Gravity of Crude Petroleum and Liquid Petroleum Products by Hydrometer Method, 7 pp.

Covers the laboratory determination, using a glass hydrometer, of the density, specific gravity, or API gravity of crude petroleum, petroleum products, or mixtures of petroleum and non-petroleum products normally handled as liquids, and having a Reid vapor pressure [ASTM Method D 323, Test for Vapor Pressure of Petroleum Products (Reid Method) or IP 69] of 26 lb. or less.

ASTM: D 1356-73(1979), Standard Definitions of Terms Relating to Atmospheric Sampling and Analysis, 6 pp.

Includes definitions relating to atmospheric conditions, sampling devices, and methods of analysis.

ASTM: D 1408-65 (1984), Standard Methods for Measurement and Calibration of Spherical and Spheroidal Tanks, 1 pg. Available as separate reprint only.

Describes the procedures for calibrating spherical and spheroidal tanks which are used as liquid containers.

ASTM: D 1410-65 (1984), Standard Method for Measurement and Calibration of Stationary Horizontal Tanks, 1 pg. Available as separate reprint only.

Describes external measurement procedures for calibrating horizontal aboveground stationary tanks larger than a barrel or drum.

ASTM: D 1480-81, Standard Test Method for Density and Relative Density (Specific Gravity) of Viscous Materials by Bingham Pycnometer, 9 pp.

Describes two procedures for the measurement of the density of materials which are fluid at the desired test temperature. Its application is restricted to liquids of vapor pressures below 600 mm Hg and viscosities below about 400 cSt at the test temperature.

ASTM: D 1481-81, Standard Test Method for Density and Relative Density (Specific Gravity) of Viscous Materials by Lipkin Bicapillary Pycnometer, 9 pp.

Intended for determining the density of oils more viscous than 15 cSt at 20°C (68°F), and of viscous oils and melted waxes at elevated temperatures, but not at temperatures at which the sample would have a vapor pressure of 100 mm Hg or above.

ASTM: D 1605-60 (1979), Standard Recommended Practices for Sampling Atmospheres for Analysis of Gases and Vapors, 22 pp.

Covers two types of sampling methods for the sampling of atmospheres for analysis of gases and vapors. Includes a description of procedures, apparatus, and methods for the determination of performance.

ASTM: D 1657-83, Standard Test Method for Density or Relative Density of Light Hydrocarbons by Pressure Hydrometer, 5 pp.

Covers a procedure for determining the specific gravity 60/60°F of light hydrocarbons including liquefied petroleum hydrocarbons, LPG, and butadiene.

ASTM: D 1671-72, Standard Test Method for Absorbed Gamma Radiation Dose in the Fricke Dosimeter, 4 pp. Discontinued 10/26/84. Replaced by E 1026.

The Fricke dosimeter method describes a procedure for measurement of gamma radiation in the range of 0.2×10^4 to 4×10^4 rads of oxidation of ferrous ammonium sulfate and determination of ferric ion concentration spectro-photometrically.

ASTM: D 1717-65 (1975), Standard Method for Analysis of Commercial Butane-Butene Mixtures and Isolutylene by Gas Chromatography, 4 pp. Discontinued 10/26/84. Replaced by D 4424.

Applies to the analysis by gas chromatography of: (1) commercial butane-butene mixtures (b-b mix), and (2) commercial isobutylene of at least 98% purity. Describes the apparatus, preparation of apparatus, procedure, and calculations.

ASTM: D 1826-83, Standard Test Method for Calorific Value of Gases in Natural Gas Range by Continuous Recording Calorimeter, 15 pp.

Covers a procedure for determining with the continuous recording calorimeter the total calorific value of fuel gas produced or sold in the natural gas range of 900 to 1200 Btu per standard cubic foot. Includes definitions, description of apparatus, methods of installation, and operation of the calorimeter.

ASTM: D 1890-81, Standard Test Method for Beta Particle Radioactivity of Water, 7 pp.

Covers the measurement of beta particle activity of water by means of several types of instruments composed of a detecting device and combined amplifier, power supply, and scaler.

ASTM: D 1943-81, Standard Test Method for Alpha Particle Radioactivity of Water, 5 pp.

Covers the measurement of alpha particle activity of water. It is applicable to alpha emitters having energies above 3.9 MeV and at activity levels above 9.5 pCi/mL of radioactivity homogeneous water. The method is not applicable to samples containing alpha-emitting radio-elements that are volatile under conditions of the analysis.

ASTM: D 1945-81, Standard Method for Analysis of Natural Gas by Gas Chromatography, 15 pp.

Describes the method determination of the complete chemical composition of reformed gases and similar gaseous mixtures containing the following components: hydrogen, oxygen, nitrogen, carbon monoxide, carbon dioxide, methane, ethane, and ethylene. Includes definitions, measurement variables, description of apparatus, sampling, standardization and calibration, and calculations.

ASTM: D 1946-82, Standard Method for Analysis of Reformed Gas by Gas Chromatography, 6 pp.

Describes the method determination of the complete chemical composition of reformed gases and similar gaseous mixtures containing the following components: hydrogen, oxygen, nitrogen, carbon monoxide, carbon dioxide, methane, ethane, and ethylene. Includes definitions, measurement variables, description of apparatus, sampling, standardization and calibration, and calculations.

ASTM: D 2009-65 (1979),[ε1] Recommended Practice for Collection by Filtration and Determination of Mass, Number, and Optical Sizing of Atmospheric Particulates, 8 pp.

Covers the collection and measurement of mass particle size and particle size distribution of atmospheric material. Applies to both solid and liquid particles.

ASTM: D 2124-70 (1984),[ε1] Standard Method for Analysis of Components in Poly (Vinyl Chloride) Compounds Using an Infrared Spectrophotometric Technique, 7 pp.

Provides for the infrared identification of resins, plasticizers, stabilizers, and fillers in poly (vinyl chloride) (PVC) compounds.

ASTM: D 2162-86, Standard Method for Basic Calibration of Master Viscometers and Viscosity Oil Standards, 9 pp.

Covers procedures for calibrating master viscometer and viscosity oil standards both of which may be used to calibrate routine viscometers as described in ASTM Method D 445, Test for Kinematic Viscosity of Transparent and Opaque Liquids (and the Calculation of Dynamic Viscosity).

ASTM: D 2163-82, Standard Method for Analysis of Liquefied Petroleum (LP) Gases and Propane Concentrates by Gas Chromatography, 6 pp.

Covers the determination of the composition of liquefied petroleum (LP) gases. It is applicable to analysis of propane, propylene, and butane in all concentration ranges 0.1% and above.

ASTM: D 2168-80, Standard Methods for Calibration of Laboratory Mechanical-Rammer Soil Compactors, 7 pp.

This method of calibration is intended for use only in the calibration of mechanical compactors equipped with rammers striking directly the surface of the soil.

ASTM: D 2186-84, Standard Test Methods for Deposit-Forming Impurities in Steam, 9 pp.

Covers the determination of the amount of deposit-forming impurities in steam by the evaporative electrical conductivity and sodium tracer methods. Special techniques for silica and certain metal oxides are also presented.

ASTM: D2389-83, Standard Test Method for Minimum Pressure for Vapor Phase Ignition of Monopropellants, 15 pp.

Covers the determination of the minimum pressure for the ignition of monopropellants in the vapor phase. This measure can lead to the determination of minimum ignition energy.

ASTM: D2597-83, Standard Method for Analysis of Natural Gas-Liquid Mixtures by Gas Chromatography, 7 pp.

Covers the analysis of wide-range natural gas-liquid (NGL) mixtures, such as commercial de-ethanized and depropanized natural gasoline mixtures, that cannot readily be entered into the chromatograph as a liquid by syringe or as a vapor at atmospheric pressure because of both highly volatile and heavy-end components.

ASTM: D 2600-82, Standard Test Method for Aromatic Traces in Light Saturated Hydrocarbons by Gas Chromatography, 11 pp.

Covers the determination of benzene, toluene, and C8 aromatics in light saturate hydrocarbon samples. The method is limited by aromatic selectivity of the stationary liquid to samples containing n-decane as the highest boiling compound.

ASTM: D2650-83, Standard Test Method for Chemical Composition of Gases by Mass Spectrometry, 10 pp.

Covers the quantitative analysis of gases containing specific combinations of the following components; hydrogen, hydrocarbons with up to six carbon atoms per molecule; carbon monoxide; carbon dioxide, mercaptans with one or two carbon atoms per molecule; hydrogen sulfide; and air (nitrogen, oxygen, and argon).

ASTM: E1-85, Standard Specification for ASTM Thermometers, 61 pp.

Covers specifications for 184 etched-stem liquid-in-glass thermometers graduated in Celsius or Fahrenheit degrees.

ASTM: E70-77 (1986),[ε1] Standard Test Method for pH of Aqueous Solutions with the Glass Electrode, 7 pp.

Covers the definition of pH and the apparatus and procedures for the electrometric measurement of pH values of aqueous solutions or extracts with the glass electrode.

ASTM: E74-83, Standard Practice of Calibration of Force Measuring Instruments for Verifying the Load Indication of Testing Machines, 9 pp.

Covers procedures for the verification of calibration devices suitable for calibrating testing machines in accordance with the requirements of the Methods of Load Verification of Testing Machines (ASTM Designation: E4).

ASTM: E77-84,[ε1] Standard Method for Verification and Calibration of Liquid-In-Glass Thermometers, 18 pp.

Describes the principles, apparatus, and procedures, for visual and dimensional insection, test for permanency of pigment, test for bulb stability, and test for scale accuracy to be used in the verification and calibration of etched-stem liquid-in-glass thermometers.

ASTM: E83-85, Standard Practice for Verification and Classification of Extensometers, 8 pp.

Covers procedures for the verification and classification of extensometers. The method applies only to instruments that indicate or record values which are proportional to changes in length. Extensometers are classifed on the basis of the magnitude of their errors.

ASTM: E94-84a, Standard Guide for Radiographic Testing, 13 pp.

Provides a guide for satisfactory radiographic testing. Statements about preferred practices are given without discussion of the technical reasons leading to the preference.

ASTM: E116-81 (1986), Standard Practice for Photographic Photometry in Spectrochemical Analysis, 31 pp.

Provides a practical guide to the preparation and use of emulsion calibration curves for determining spectral line intensity ratios in spectrochemical analysis.

ASTM: E131-84, Standard Definitions of Terms and Symbols Relating to Molecular Spectroscopy, 9 pp.

ASTM: E135-86a, Standard Definitions of Terms and Symbols Relating to Emission Spectroscopy, 7 pp.

ASTM: E137-82, Standard Practice for Evaluation of Mass Spectrometers for Chemical Analysis, 4 pp.

Provides means for evaluation of the suitability of mass spectrometers for use in ASTM mass spectrometric methods of chemical analysis. Also includes discussion of tests that are generally helpful in evaluating the performance of a particular mass spectrometer as used in a particular ASTM method of analysis.

ASTM: E168-67 (1977), Standard Recommended Practices for General Techniques of Infrared Quantitative Analysis, 9 pp.

Provides general information on the various techniques most often used in infrared quantitative analysis. Includes definitions and symbols, theory, apparatus, calculation methods, and special techniques.

ASTM: E170-84b, Terminology Relating to Radiation Measurements and Dosimetry, 9 pp.

Wherever possible, these definitions are the same as, or similar to, those recommended by the International Commission on Radiological Units and Measurements (ICRU) as presented in the National Bureau of Standards Handbook 62.

ASTM: E172-85, Standard Practice for Describing and Specifying the Excitation Source in Emission Spectrochemical Analysis, 5 pp.

Provides general recommendations for the description of the various types of sources used in spectrographic analysis and for the specification of source parameters.

ASTM: E177-86, Standard Practice for Use of the Terms Precision and Bias in ASTM Test Methods, 16 pp.

The purpose of this recommended practice is to outline some general concepts regarding the terms "precision" and "accuracy", to provide some standard usages for ASTM committees in reference to precision and accuracy, and to illustrate some important features of the experimental determination of precision.

ASTM: E179-81, Standard Practice for Selection of Geometric Conditions for Measurement of Reflectance and Transmittance, 8 pp.

Intended for use in selecting terminology, measurement scales, and instrumentation for describing or evaluating such appearance characteristics as glossiness, opacity, lightness, transparency, and haziness, in terms of reflected or transmitted light.

ASTM: E189-63 (1975),$^{\epsilon 1}$ Standard Recommended Practice for Determining Temperature-Electrical Resistance Characteristics (EMF) of Metallic Materials, 5 pp.

Covers procedures determining the temperature versus electrical resistance of emf characteristics of metallic materials.

ASTM: E230-83, Standard Temperature Electromotive Force (EMF) Table for Standardized Thermocouples, 100 pp.

Consists of temperature-emf tables for thermocouple types B, E, J, K, R, S, and T; standard and special limits of error and upper temperature limits All intervals are 1 degree.

ASTM: E235-82, Standard Specification for Thermocouples, Sheathed, Type K, for Nuclear or Other High-Reliability Applications, 7 pp.

Presents the material, operating, and environmental requirements two-wire thermocouples intended for nuclear service. Provisions for temperatures up to 900 degrees C (1650 degrees F) are covered.

ASTM: E261-77, Standard Method for Determining Neutron Flux, Fluence, and Spectra by Radioactivation Techniques, 12 pp.

Covers the determination of neutron flux in a radiation field from the radioactivity that is induced in a detector specimen. The description is directed toward the need for characterization of the magnitude and energy distribution of neutron flux in connection with radiation effects on materials.

ASTM: E275-83, Standard Practice for Describing and Measuring Performance of Ultraviolet, Visible, and Near Infrared Spectrophotometers, 16 pp.

Covers the description of requirements of spectrophotometric performance especially for ASTM methods, and the testing of the adequacy of available equipment for a specific method.

ASTM: E304-81, Standard Practice for Use and Evaluation of Spark Source Mass Spectrometers for the Analysis of Solids, 6 pp.

Provides guidelines for evaluation of suitability of solids mass spectrometers for use in ASTM methods for analysis of solids which specify the use of such apparatus. This practice is restricted to those instruments in which ions are produced in an electrical discharge directly from the solid.

ASTM: E306-71 (1976),[e1] Standard Method for Absolute Calibration of Reflectance Standards, 5 pp.

Describes the requirements for an auxiliary-integrating sphere to be used with a spectrophotometer already equipped with a sphere to measure diffusely-reflected flux, and describes the calibration of an instrument standard on a scale on which the perfectly-reflecting, perfectly-diffusing specimen is assigned a value of unity.

ASTM: E317-85, Standard Practice for Evaluating Performance Characteristics of Ultrasonic Pulse-Echo Testing Systems Without the Use of Electronic Measurement Instruments, 17 pp.

Describes procedures for determining some important performance characteristics that establish the capabilities of pulse-echo ultrasonic testing systems in which test results are displayed on an A-scan cathode-ray tube screen.

ASTM: E334-81, Standard Practices for General Techniques in Infrared Microanalysis, 8 pp.

These recommended practices cover general information on techniques that are of general use in securing and subsequently analyzing (by infrared spectrophotometric techniques) microgram quantities of solid or liquid samples.

ASTM: E337-84 — (ANSI: L14/19-1963), Standard Test Method for Measuring Humidity With a Psychrometer (The Measurement of Wet-Bulb and Dry-Bulb Temperatures), 24 pp.

Covers the procedure for determining relative humidity of atmospheric air by means of wet- and dry-bulb temperature readings.

ASTM: E355-77 (1983), Standard Recommended Practice for Gas Chromatography Terms and Relationships, 9 pp.

List of operating parameters and relationships that applies in most cases only to steady-gas elution chromatography.

ASTM: E380-86, Standard Metric Practice (Complete in Vol. 14.02 Only; Excerpts in Related Material Section of All Other Volumes), 42 pp.

Gives guidance for application of the International System of Units (SI). Includes information on SI, a limited list of non-SI units recognized for use with SI, a list of conversion factors from non-SI to SI units, and general guidance on proper style and usage.

ASTM: E386-78 (1984), Standard Definitions of Terms, Symbols, Conventions, and References Relating to High-Resolution Nuclear Magnetic Resonance (NMR) Spectroscopy, 12 pp.

ASTM: E425-85, Standard Definitions of Terms Relating to Leak Testing, 8 pp.

ASTM: E432-71 (1984),[e1] Standard Guide for the Selection of a Leak Testing Method, 4 pp.

Intended as a guide for the selection of a leak testing method. The type of item to be tested or the test system and the method considered for either leak measurement or location are related in order of increasing sensitivity.

ASTM: F305-70 (1980), Standard Practice for Sampling Particulates from Reservoir-Type Pressure-Sensing Instruments by Fluid Flushing, 2 pp.

Covers the sampling of reservoir-type pressure-sensing instruments which enclose a volume that has dubious drainage capabilities.

AMERICAN SOCIETY OF AGRICULTURAL ENGINEERS (ASAE)

The following publications are available from the American Society of Agricultural Engineers, 2950 Niles Rd., Joseph, MI 49085.

ASAE: S313.2, Soil Cone Penetrometer, revised 1985, 1 p.

Described is the soil cone penetrometer, recommended as a measuring device to provide a standard uniform method characterizing the penetration resistance of soils.

ASAE: D271.2, Psychrometric Data, reconfirmed 1983, 5 charts.

Convenient reference psychrometric charts that yield data for a dry bulb temperature range of -35 degrees to 600 degrees F.

ASAE: D272.2, Resistance to Airflow of Grains, Seeds, Other Agricultural Products, and Perforated Metal Sheets, 5 pp.

Provides an estimate of airflow resistance that can be used as the basis for the design of systems to aerate grain and seed.

ASAE: D293.1 Dielectric Properties of Grain and Seeds, revised 1984, 8 pp.

These data are intended to provide a basis for design of equipment and application of radio-frequency energy for treatment of grain and seed, or possible electrical measurement of moisture content, and development of capacitive type of RF-energy absorption sensing devices.

ASAE: S368.1 Compression Test of Food Materials of Convex Shape, reconfirmed 1984.

This standard is intended for use in determining mechanical attributes of food texture, resistance of mechanical injury and force deformation behavior of food materials of convex shape, such as fruits and vegetables, seeds and grains, and manufactured food materials.

ASAE: D309, Wet-Bulb Temperatures and Wet-Bulb Depressions, reconfirmed 1981, 4 pp.

Explanation and accompanying set of maps show mean wet-bulb temperatures, mean wet-bulb depressions, and their standard deviations.

AMERICAN SOCIETY OF HEATING, REFRIGERATING, AND AIR-CONDITIONING ENGINEERS, INC. (ASHRAE)

The following ASHRAE Standards may be obtained from the American Society of Heating, Refrigerating, and Air-Conditioning Engineers, Inc., 1791 Tullie Circle N.E., Atlanta, GA 30329.

ASHRAE: Std. 12-75, (ANSI: B53.1-1974), Refrigeration Terms and Definitions, 1975, 33 pp.

Intended to provide authoritative definitions of words and terms employed in all phases of activity connected with refrigeration and air-conditioning.

ASHRAE: Std. 28-78, Methods of Testing Flow Capacity of Refrigerant Capillary Tubes, 1978, 4 pp.

Covers air flow capacity tests for tubes used in refrigerant metering.

ASHRAE: Std. 41.1-74, Part 1, Standard Measurements Guide: Section on Temperature Measurements, 1974, 18 pp.

Provides methods for accurate temperature measurement for the particular needs of heating, refrigeration, and air conditioning. The rates of heat flow, both to and from moving volatile and non-volatile fluids, in the range of -40 to 400 degrees F are covered. The use of thermometers, thermocouples, and thermistors and the effect of changes in enthalpy are discussed.

ANSI/ASHRAE: 41.4-1984, Standard Method of Measurement of Proportion of Oil in Liquid Refrigerant, 4 pp.

Intended to apply only where it is known that the sample is from a single-phase solution of oil in liquid refrigerant. Does not apply to measurement of oil concentrations so low that the criterion of 6.1 is not met.

ASHRAE: 41.5-1975, Standard Measurement Guide: Engineering Analysis of Experimental Data, 15 pp.

To provide recommended practices for reporting of uncertainty in results for data obtained from an experiment.

Defines terms and sets forth procedures for applying statistical methods to experimental data.

ANSI/ASHRAE: 41.6-1982, Standard Method for Measurement of Moist Air Properties, 24 pp.

Sets forth recommended practices and procedures for the measurement and calculation of moist air properties in order to promote accurate measurement methods for specific use in the preparation of other ASHRAE standards.

ASHRAE: 41.7-1984, (reaffirmed, supersedes 41.7-78), Standard Method for Measurement of Flow of Gas, 7 pp.

Provide recommended practices for the measurement of the flow of dry gases for use in the preparation of ASHRAE Standards.

ASHRAE: 41.8-78, Standard Methods of Measurement of Flow of Fluids — Liquids, 14 pp.

Establish recommended practices for the measurement of flow of fluids as liquids. It shall also establish the standard technique to be used for the calibration of other instruments more convenient to use. This standard is not intended to be used as a replacement for the calibration of flow meters by facilities traceable to NBS nor restrict the use of such facilities that do not incorporate the methods outlined below.

ASHRAE: Std. 74-73, Method of Measuring Solar-Optical Properties of Materials, 1973, 7 pp.

The purpose of this standard is to develop a standard method of measuring and reporting the solar-optical properties of materials.

ANSI/ASHRAE: 86-1983, Methods of Testing Floc Point of Refrigeration Grade Oils, 3 pp.

The test for floc point is intended to determine the waxing tendency of refrigeration grade oils at low temperatures and is based on evaluation of wax precipitation tendency of a mixture of 90% Refrigerant 12 and 10% of oil being tested, the results of which can be used to compare several different oils.

ANSI/ASHRAE: 97-1983, Sealed Glass Tube Method to Test the Chemical Stability of Material for Use within Refrigerant Systems, 10 pp.

Establishes a procedure using sealed glass tubes for the evaluation of materials to be used in refrigerant systems. Detailed safety precautions are included.

ANSI/ASHRAE: 101-1981, Application of Infrared Sensing Devices to the Assessment of Building Heat Loss Characteristics, 27 pp.

Describe acceptable procedures and specifications for the applied use of infrared radiation sensing devices for assessment of building heat loss characteristics and interpretation of data resulting therefrom.

AMERICAN SOCIETY OF MECHANICAL ENGINEERS (ASME)

The following ASME and ANSI Standards may be obtained from The American Society of Mechanical Engineers, 345 East 47th Street, New York, NY 10017.

ASME: PTC 6A, Appendix A to Test Code for Steam Turbines, 1982, 56 pp. Bk. No. C00029.

This appendix provides numerical examples of various turbine test calculations.

ASME: PTC 6R, Guidance for Evaluation of Measurement Uncertainty in Performance Tests of Steam Turbines, 1969, 32 pp. Bk. No. D00041.

This report provides guidance to establish the degree of uncertainty with test results when there are deviations from the requirements of Performance Test Code No. 6 on Steam Turbines.

ASME: PTC 6S, Simplified Procedures for Routine Performance Tests of Steam Turbines, 1970, 4 pp. Bk. No. D00042.

Provides for the testing of steam turbines operating predominantly within the moisture region for the purpose of determining the level of performance with minimum uncertainty.

ASME: PTC 9, Displacement Compressor, Vacuum Pumps, and Blowers, 1970, 41 pp. Bk. No. C00009.

Applies to tests for determining the performance of positive displacement, compressors, blowers, and vacuum pumps whether reciprocating or rotating.

ASME: PTC 19.2, Instruments and Apparatus: Pressure Measurement, 1964, 58 pp. Bk. No. D00029.

Discusses the technology of pressure measurement: general considerations and definitions, pressure connections, liquid-level gages, deadweight gages and testers, elastic gages, and low-pressure measurement.

ASME: PTC 19.3, Instruments and Apparatus: Temperature Measurement, (R 1979), 118 pp., Bk. No. C00035.

Presents a revision, expansion, and consolidation of all earlier pamphlets on temperature measurement instruments, with particular emphasis on basic sources of errors and means of coping with them.

ASME: PTC 19.5, Interim Supplement on Instruments and Apparatus: Application, Part II of Fluid Meters. Sixth Edition, 1972, 140 pp., Bk. No. G00018.

Presents the recommended conditions, procedures and data for measuring the flow of fluids, particularly with the three principal differential pressure meters: the orifice, the flow nozzle, and the venturi tube.

ASME: PTC 19.5.1, Instruments and Apparatus: Weighing Scales, 1964, 17 pp., Bk. No. D00028.

Discusses weighing scales suitable for quantity measurement of materials in connection with tests of power equipment.

ASME: PTC 19.6, Electrical Measurements in Power Circuits, 1965, 40 pp., Bk. No. D00007.

The methods given include measurements made with either indicating or integrating instruments of power, voltage and current in direct-current and alternating-current single-phase and poly-phase rotating machinery, transformers induction apparatus, arc and resistance heating equipment, and mercury are rectifiers.

ASME: PTC 19.7, Measurement of Shaft Horsepower, 1980, 29 pp., Bk. No. D00009.

Shows how measurement of shaft horsepower of rotating machines can be accomplished either by the direct method of utilizing dynamometers or the indirect method of using calibrated motors or generators, heat balance, or heat exchangers.

ASME: PTC 19.8, Measurement of Indicated Horsepower, 1970, 29 pp., Bk. No. D00008.

This supplement of the Performance Test Codes treats the direct measurement of indicated power of piston engines and compressors by use of the engine indicator.

ASME: PTC 19.10, Flue and Exhaust Gas Analyses, Instruments and Apparatus — V Part 10, 1981, 32 pp., Bk. No. C00031.

This supplement provides descriptions of methods, apparatus, and calculations which are used in conjunction with Performance Test Codes to determine quantitatively the constituents of the gases resulting from combustion of carbonaceous or hydrocarbon fuel in the solid, liquid, or gaseous form.

ASME: PTC 19.11, Water and Steam in the Power Cycle (Purity and Quality, Leak Detection, and Measurement), 1970, 98 pp., Bk. No. D00011.

Specifies and discusses the methods of instrumentation for testing boiler feedwater, steam, and condensate in relation to performance testing in the power cycle and for leak detection and leakage measurement for surface condensers and other cycle components.

ASME: PTC 19.12, Measurement of Time, 1958, 12 pp., Bk. No. D00012.

General purpose clocks, chronometers, clocks or regulators for indicating time to the nearest second, astronomical clocks, watches, stop watches, timers, chronographs, and oscillographs are the types of timekeepers described.

ASME: PTC 19.13, Measurement of Rotary Speed, 1961, 17 pp., Bk. No. D00013.

Covers commonly used instruments and methods, and discusses characteristics and limitations of commercially available instruments used for testing rotating machinery, turbines, blowers, or electric motors.

ASME: PTC 19.14, Linear Measurements, 1958, 14 pp., Bk. No. D00014.

Coverage includes tapes, rules and scales, calipers, and dividers, slide calipers, depth gages, vernier calipers, vernier depth gages, micrometer calipers, micrometer depth gages, internal micrometers, telescoping and small hole gages, dial indicators, dial bore gages, dial caliper gages, thickness gages, and gage blocks.

ASME: PTC 19.16, Density Determinations of Solids and Liquids, 1965, 12 pp., Bk. No. D00016.

Considers available methods for determining specific gravity.

ASME: PTC 19.17, Determination of the Viscosity of Liquids, 1965, 16 pp., Bk. No. D00017.

Gives procedures for and information on various types of viscometers and their applications.

ASME: PTC 32.1, Nuclear Steam Supply Systems, 1969, 34 pp., Bk. No. C00012.

Contains instructions for the performance testing of nuclear steam supply systems. It establishes procedures for conducting tests to determine the thermal performance of a nuclear steam supply system as a unit.

The following Standards, identified by their ANSI number, are ANSI approved, but published by ASME. They are available from either ASME or ANSI.

ANSI: B40.1-1980, Gauges — Pressure and Vacuum, Indicating Dial Type — Elastic Element, 1980, 15 pp., Bk. No. K00015.

The scope is confined to dial type indicating gauges which indicate pressure or vacuum by means of a pointer and a graduated scale, utilizing an elastic element for measuring the pressure or vacuum. It does not include dead weight types, mercury-floated piston types, or other special constructions which do not utilize an elastic element.

ANSI: MC88.1-1972 (R 1978), Guide for Dynamic Calibration of Pressure Transducers, 29 pp., Bk. No. L00042.

Provides calibration techniques and methods for use with pressure transducers.

ANSI: Y1.1-1972, Abbreviations for Use on Drawings and in Text, Bk. No. J00003.

Provides list of basic abbreviations.

ANSI: Y14.1-1980, Drawing Sheet Size and Format, Bk. No. N00001.

Established standard drawing sheet sizes.

ANSI: Y14.26M-1981, Engineering Drawing and Related Practices — Digital Representation for Communication of Product Definition Data, Bk. No. N00099.

Establishes data required to describe and communicate the essential engineering characteristics of physical objects as manufactured products. Establishes information structures to be used for the digital representation and communication of products definition data.

ANSI: Y32.10-1967 (R 1979), Graphic Symbols for Fluid Power Diagrams, 1967, 22 pp., Bk. No. N00022.

Presents a system of graphic symbols for fluid power diagrams. Elementary forms of symbols are: Circles, Squares, Rectangles, Triangles, Arcs, Arrows, Lines, Dots, and Crosses.

ANSI: Y32.11-1961, Graphical Symbols for Process Flow Diagrams in Petroleum and Chemical Industries, Bk. No. K00040.

A preliminary set of standard symbols, developed for use on the basic process flow diagrams in order to represent the major items of equipment used by the petroleum and chemical industries.

ANSI: Z32.2.3-1949 (R 1953), *Graphical Symbols for Pipe Fittings, Valves and Piping, 12 pp., Bk. No. K00006.*

ANSI: Z32.2.4-1949 (R 1953), *Graphical Symbols for Heating, Ventilating, and Air Conditioning, 16 pp., Bk. No. K00005.*

ANSI: Z32.2.6-1950 (R 1956), *Graphical Symbols for Heat-Power Apparatus, 8 pp., Bk. No. K00004.*

ANSI: B88.2-1974 (R 1981), *Procedure for Bench Calibration of Tank Level Gauging Tapes and Sounding Rules, (R 1981), 4 pp., Bk. No. L00043.*

Procedure applies to any gauging tape or sounding rule using a graduated scale to determine level of liquid in tanks. Procedures for both linear and non-linear scales are provided.

ANSI: B89.1.9-1973 (R 1980), *Precision Inch Gauge Blanks for Length Measurement (Thru 20 Inches), (R 1980), 16 pp., Bk. No. L00044.*

Covers specifications for gauge blocks up to and including 20 inches in length, including physical properties, general dimensions, tolerance grades, flatness, parallelism and surface texture requirements.

ANSI: B89.3.1-1972, (R 1979), *Measurement of Out-Of-Roundness, 1972, 27 pp., Bk. No. L00020.*

Covers the specification and measurement of out-of-roundness of a surface of revolution. Deals primarily with precision spindle instruments for out-of-roundness measurement and polar chart presentation.

ANSI: B89.6.2-1973 (R 1979), *Temperature and Humidity Environment for Dimensional Measurement, 1973, 31 pp., Bk. No. L00047.*

Covers methods of describing and testing temperature-controlled environments for dimensional measurement and ensuring adequate temperature control for the calibration of measuring equipment as well as the manufacture and acceptance of work-pieces.

ANSI: C85.1-1981, *Terminology for Automatic Control, 1981, Bk. No. N00036.*

Terminology pertaining to systems such as: automatic process control, feedback control, regulating, and other related systems not requiring human intervention as a part of the regulating procedure.

AMERICAN VACUUM SOCIETY (AVS)

The following tentative Standards are available from the American Vacuum Society, 335 East 45th Street, New York, NY 10017.

AVS: 2.1, *Calibration of Leak Detectors of the Mass Spectrometer Type, 1973, 10 pp.*

Prescribes procedures to be used for calibrating leak detectors of the mass spectrometer type, that is, for determining a sensitivity figure for such leak detectors.

AVS: 2.2, *Method for Vacuum Leak Calibration, 1968, 4 pp.*

This standard describes an apparatus for measuring the leak rate of vacuum leaks, in the range of 10^{-5} to 10^{-3} atm cm^3/s, and a procedure for using the apparatus to determine such leak rates.

AVS: 2.3-1972, *Procedure for Calibrating Gas Analyzers of the Mass Spectrometer Type, 1972, 15 pp.*

Concerned with calibration procedures for determining the minimum detectable partial pressure of gas analyzers of the mass spectrometer type. Procedures are also given for evaluating the resolution of the analyzer.

AVS: 5.3, *Method for Measuring Pumping Speed of Mechanical Vacuum Pumps for Permanent Gases, 1967, 3 pp.*

Describes procedures for determining the pumping speed for permanent gases of all positive displacement mechanical vacuum pumps.

AVS: 6.2-1969, *Procedure for Calibrating Vacuum Gages of the Thermal Conductivity Type, 1969, 4 pp.*

Procedures are given and apparatus described for calibrating vacuum gages of the thermal conductivity type of direct comparison with measurements made with an absolute reference instruments such as the McLeod gage. The pressure range considered is of the order of 10^{-4} to several Torr.

AVS: 6.4-1969, *Procedure for Calibrating Hot Filament Ionization Gauges Against a Reference Manometer in the Range of 10^{-2} — 10^{-5} Torr, 1969, 5 pp.*

Procedures are given for the calibration of hot cathode ionization gauges and gauge tube by direct comparison against a McLeod gauge or other absolute manometer in the pressure range of 10^{-2} — 10^{-5} Torr.

AVS: 6.5-1971, *Procedures for the Calibration of Hot Filament Ionization Gauge Controls, 1971, 7 pp.*

Guidelines and methods are provided for the electrical calibration of hot cathode ionization gauge controls.

AVS: 7.1, Graphic Symbols in Vacuum Technology, 1966, 4 pp.

A uniform system of graphic symbols to be used in vacuum technology.

AMERICAN WATER WORKS ASSOCIATION (AWWA)

The following AWWA Publications are available from the American Water Works Association, 6666 West Quincy Avenue, Denver, CO 80235.

AWWA: C700-77, Cold-Water Meters — Displacement Type, 1977, 16 pp.

Covers the various types and classes of cold-water displacement meters in sizes 5/8 in. through 6 in., and the materials and workmanship employed in their fabrication.

AWWA: C701-78, Cold-Water Meters — Turbine Type for Customer Service, 1978, 20 pp.

Covers the various types and classes of cold-water turbine meters in sizes 1-1/2 in. — 12 in. for water works customer service, and the material and workmanship employed in their fabrication.

AWWA: C702-86, Cold-Water Meters — Compound Type, 1986.

Covers the various types and classes of cold-water compound meters in sizes 2 in. — 10 in. and covers the materials and workmanship employed in their fabrication.

AWWA: C703-86, Cold-Water Meters — Fire Service Type, 1986.

Covers the various types and classes of cold-water fire service type meters in sizes 3 in. through 10 in., and the materials and workmanship employed in their fabrication.

AWWA: C704-70 (R 84), Cold-Water Meters — Propeller Type for Main Line Application, 1970, Reaffirmed without revision, 1984, 12 pp.

Covers the various types and classes of cold-water propeller meters in sizes 2 in. to 36 in., for main line applications, and the materials and worksmanship employed in their fabrication.

AWWA: C706-86, Direct-Reading Remote Registration Systems for Cold-Water Meters, 1986, 12 pp.

Covers direct-reading remote registration systems for use on cold-water meters for water utility customer service, and the materials and workmanship employed in their fabrication and assembly.

AWWA: C707-82, Encoder-Type Remote Registration Systems for Cold-Water Meters, 1982, 16 pp.

Covers encoder-type remote-registration systems for use on cold-water meters for water-utility customer service and the materials and workmanship employed in their fabrication and assembly.

AWWA: C708-82, Cold-Water Meters Multi-Jet Type for Customer Service, 1982, 16 pp.

Covers the various types and classes of cold-water multi-jet meters in sizes 5/8 in. through 2 in., and the materials and workmanship employed in their fabrication.

AWWA: C500-86, Gate Valves — 3 in. through 48 in. — for Water and Other Liquids, 1986.

Covers iron-body, bronze-mounted non-rising stem gate valves, 3 in. through 48 in. in diameter with either double disc gates having parallel or inclined seats, or solid-wedge gates.

AWWA: C501-80, Sluice Gates, 1980, 28 pp.

Covers wall thimble, vertically-mounted sluice gates designed for either seating head or unseating head or both, in ordinary water supply or wastewater service.

AWWA: C504-80, Rubber-Seated Butterfly Valves, 1980, 24 pp.

Covers cast-iron and ductile-iron body, rubber-seated tight-closure butterfly valves, 3 — 72 in. in size with four body types for fresh water having a pH greater than 6 and temperatures generally less than 125 degrees F.

AWWA: C506-78 (R 83), Backflow Prevention Devices — Reduced Pressure Principle and Double Check Valve Types, 1978, 20 pp.

Covers two types of backflow prevention devices designed for operation on cold-water lines (maximum 110 degrees F) at 150 psi operating pressure.

AWWA: C507-85, Ball Valves, Shaft- or Trunion-Mounted — 6 in. through 48 in. — for Water Pressures up to 300 psi, 1985, 16 pp.

Covers cast-iron, ductile-iron, and cast-steel flanged end, tight shut-off, shaft- or trunion-mounted, full port, double- and single-seated ball valves for use in fresh water with a pH greater than 6 and temperature generally less than 125 degrees F.

AWWA: C508-82, Swing-Check Valves for Waterworks Service, 2 in. through 24 in. NPS, 1982, 16 pp.

Covers iron-body, bronze-mounted swing-check valves, 2-24 in. NPS, for normal horizontal installation in water systems.

AWWA: C509-87, Resilient-Seated Gate Valves, 3 through 12 NPS, for Water and Sewage Systems, 1987, 20 pp.

Covers iron-body resilient-seated gate valves with non-

rising stems (NRS) and outside screw-and-yoke (OS&Y) rising stems for installation in water and sewage systems.

AWWA: Standard Method for the Examination of Water and Wastewater, 1984, 1332 pp.

"Standard Methods" presents the best current practice of American water analysis in connection with the ordinary water purification, sewage disposal and sanitary investigations.

AMERICAN WELDING SOCIETY (AWS)

The following publications are available from the American Welding Society, P.O. Box 351040, Miami, FL 33135.

AWS: A4.2-86, Standard Procedures for Calibrating Magnetic Instruments to Measure the Delta Ferrite Content of Austenitic Stainless Steel Weld Metal, 1974, 16 pp.

Prescribes procedures for the calibration and maintenance of calibration of magnetic instruments for measuring the delta ferrite content of austenitic stainless steel weld metals in terms of their Ferrite Number.

AWS: F6.1-78, Method for Sound Level Measurement of Manual Arc Welding and Cutting Processes, 1978, 8 pp.

Describes the equipment and procedure to be used in measuring sound levels of manual arc welding and cutting processes. The procedure described allows the user to measure the sound level associated with specific processes in a reproducible manner that permits comparison with other selected processes. This method is not applicable to the determination of operator exposure to process sound.

ANTI-FRICTION BEARING MANUFACTURERS ASSOCIATION, INC. (AFBMA)

The following document is available from the Anti-Friction Bearing Manufacturers Association, Inc., 1101 Connecticut Ave., NW, Washington, DC 20036.

ANSI/AFBMA: Std. 12.1-1985, Instrument Ball Bearings, Metric Design, 1985, 43 pp.

ANSI/AFBMA: Std. 12.2-1985, Instrument Ball Bearings, Inch Design, 1985, 43 pp.

These Standards for Instrument Ball Bearings have been established by The Anti-Friction Bearing Manufacturers Association, Inc., for the purpose of defining the characteristics of these bearings, such as boundary dimensions, tolerances, classification used for selective assembly, radial internal clearance values, recommended gaging practices, mounting practices and starting torque values.

ASSOCIATION FOR THE ADVANCEMENT OF MEDICAL INSTRUMENTATION (AAMI)

The following publication and other related standards are available from Association for the Advancement of Medical

Instrumentation, Suite 602, 1901 N. Fort Myer Drive, Arlington, VA 22209.

AAMI: ANSI/AAMI ES1-1985, Safe Current Limits for Electromedical Apparatus, 1978, 20 pp.

Provides designers and consumers with limits and measuring techniques for risk currents of electromedical apparatus as a function of frequency, the characteristics of the apparatus, and the nature of the intentional contact with the patient.

AAMI: ANSI/AAMI DF2 5/81, Standard for Cardiac Defibrillator Devices, 1982, 20 pp.

Provides minimum labeling, performance and safety requirements for cardiac defibrillator devices. Also included are referee test methods by which compliance can be verified.

AAMI: ANSI/AAMI RD 5 6/81, Standard for Hemodialysis Systems, 1982, 35 pp.

Establishes requirements for materials, components, monitors, accessories, maintenance and labeling for hemodialysis systems. The quality of water and hemodialysis bath concentrate used in the system is also defined. Additionally, a guideline for the user of the device, with particular emphasis on water purity assurance and monitoring, is appended to this standard.

AAMI: ANSI/AAMI AT6 6/81, Standard for Autotransfusion Devices, 1982, 12 pp.

Describes labeling requirements sufficient to assure than adequate information is available to the clinician to choose the appropriate device for his or her particular application, requirements for safety and performance of the device and test methods to verify that the labeling and performance requirements of the standard have been met.

AAMI: ANSI/AAMI EC11—1982, American National Standard for Diagnostic Electrocardiographic Devices, 1983, 36 pp.

Establishes minimum safety and performance requirements for ECG systems with direct writing devices which are intended for use in EGC contour analysis for diagnostic purposes.

AAMI: ANSI/AAMI EC12—1983, American National Standard for Pregelled EGC Disposable Electrodes, 1984, 16 pp.

Contains minimum labeling and electrical performance requirements, test methods, and terminology for any pregelled electrocardiographic disposable electrode in which the electrolyte has been placed in contact with the sensing element by the manufacturer.

AAMI: ANSI/AAMI EC13—1983, American National Standard for Cardiac Monitors, Heart Rate Meters and Alarms, 1984, 40 pp.

Establishes minimum safety and performance requirements for cardiac monitors, heart rate meters, and alarms which are used to acquire and/or display ECG signals with the primary purpose of continuous detection of cardiac rhythm.

AAMI: ANSI/AAMI NS14—1984, American National Standard for Implantable Spinal Cord Stimulators, 1984, 16 pp.

Establishes minimum labeling, safety, and performance requirements for implantable spinal cord stimulators intended for use in the relief of chronic, severe pain.

AAMI: ANSI/AAMI NS15—1984, American National Standard for Implantable Peripheral Nerve Stimulators, 1984, 16 pp.

Establishes minimum labeling, safety, and performance requirements for implantable peripheral nerve stimulators intended for use in the relief of chronic, severe pain.

AAMI: ECGC—5/83, Standard for ECG Connectors, 1983, 16 pp.

Establishes a preferred connector set for cable and panel and certain cable requirements to achieve universal defibrillation protection and in order to allow cable and apparatus interchangeability of ECG monitoring systems with isolated patient connections.

AAMI: ANSI/AAMI BP22—1986, Blood Pressure Transducers, General, 1986, 20 pp.

Establishes safety and performance requirements for isolated transducers, including cables, designed for blood pressure measurements through an indwelling catheter or direct puncture. Also included are disclosure requirements to permit the user to determine compatibility between the transducer and the blood pressure monitor.

AAMI: ANSI/AAMI BP23—1986, Blood Pressure Transducers, Interchangeability and Performance of Resistive Bridge Type, 1986, 20 pp.

Establishes interchangeability and performance requirements for resistive bridge-type blood pressure transducers, including the connector interface to the blood pressure monitor. The standard addresses compatibility between transducers and monitors regardless of the manufacturer.

AAMI: ANSI/AAMI HF18—1986, Electrosurgical Devices, 1986, 42 pp.

Provides minimum labeling, performance and safety requirements to help ensure safe and effective use of electrosurgical devices. Devices included are electrosurgical high frequency generators, directly related accessories, including active electrodes and cables, dispersive electrodes and cables, and footswitches and other operator-controlled mechanisms for activating the generator output.

AAMI: ANSI/AAMI NS4—1985, Transcutaneous Electrical Nerve Stimulators, 1986, 18 pp.

Establishes certain requirements for portable, battery-powered, transcutaneous electrical nerve stimulators that are used in the treatment of pain syndromes, that are intended for use on intact skin and mucuous membranes, and that do not require surgical intervention or violation of the skin surface.

AAMI: ANSI/AAMI SP9—1985, Sphygmomanometers, Non-Automated, 1986, 20 pp.

Provides safety and performance requirements for pneumatic or other non-automated sphygmomanometers that are used with an occluding cuff for the indirect determination of blood pressure. Aneroid and mercury gravity sphygmomanometers are included, used in conjunction with a stethoscope or other manual methods for detecting Korotkoff sounds.

ASSOCIATION OF OFFICIAL ANALYTICAL CHEMISTS (AOAC)

The following publication is available from the Association of Official Analytical Chemists, 1111 N. 19th Street, Arlington, VA 22209.

Official Methods of Analysis of the AOAC, 1984, 1141 pp.

Methods of analysis, adopted after collaborative studies to demonstrate that they are reliable, have convenient practical application, and give reproducible results in the hands of professional analytical chemists. These standardized methods are used by regulatory agencies of federal, state, and municipal governments and by the regulated industries; in specifications to government, private, and research workers in agriculture and public health.

THE CHLORINE INSTITUTE (CI)

The following publications are available from The Chlorine Institute, Inc., 2001 L Street, NW, Washington, DC 20036.

CI: No. 6, Piping Systems for Dry Chlorine (11th Ed.), 1985, 14 pp.

Contains information on selected pipe, valves and fittings suitable for use with dry chlorine (gas or liquid) within the range — 20 F and 300 F.

CI: No. 39, Maintenance Instructions for Chlorine Institute Standard Safety Valves, Type 1-1/2 JQ (8th Ed.), 1982, 32 pp.

Illustrated instructions for maintenance of Crosby style 1-1/2JQ chlorine safety valves.

CI: No. 40, Maintenance Instructions for Chlorine Institute Standard Angle Valve (4th Ed.), 1982, 19 pp.

Illustrated instructions on maintenance of standard chlorine angle valve.

CI: No. 41, Maintenance Instructions for Chlorine Institute Standard Safety Valves, Type 4 JQ (3rd Ed.), 1975, 20 pp.

Illustrated instructions on maintenance of Crosby style 4 JQ chlorine safety valves.

CI: No. 42, Maintenance Instructions for Chlorine Institute Standard Excess Flow Valves (3rd Ed.), 1981, 16 pp.

Illustrated instructions on maintenance of standard chlorine excess flow valves.

COOLING TOWER INSTITUTE (CTI)

The following documents are available from the Cooling Tower Institute, P.O. Box 73383, Houston, TX 77273.

CTI: ATC-105, Acceptance Test Code for Industrial Water-Cooling Towers, 1982, 43 pp (Addendum 1986).

This Code covers the determination of the capability of water-cooling towers. The purpose is to describe instrumentation and test procedures which will yield accuracy and uniformity in calculated tower capacity.

CTI: Gear Speed Reducers, 1979.

Rating practice and operating considerations for use with propellor-type fans: includes AGMA-approved service factors; recommendations for lubrication, alignment and protection during shutdown.

CTI: ACT-128, Code for Measurement of Sound from Water Cooling Towers, 1981, 3 pp.

Applies to mechanical and natural draft towers. Test and measurement procedures, operating conditions, and instrumentation are specified.

CTI: STD-201, Certification Standard for Commercial Water-Cooling Towers, 1981.

This standard sets forth procedures whereby the Cooling Tower Institute may certify that a line of water cooling towers being offered for sale by a specific manufacturer, will perform in accordance with the manufacturers' published ratings. The certification thus provided will assure customers of specified water cooling tower capacity without costly field testing on an individual tower basis.

DAIRY AND FOOD, 3-A SANITARY STANDARDS COMMITTEES (DFSSC)

The following 3-A Sanitary Standards were formulated by the cooperative effort of industry and regulatory groups as represented by the Dairy Industry Committee, International Association of Milk, Food, and Environmental Sanitarians, U.S. Public Health Service, U.S. Department of Agriculture, Dairy and Food Industries Supply Association, and Poultry and Egg Institute of America. 3-A and E-3-A Standards and Practices are published in Dairy and Food Sanitation, P.O. Box 701, Ames, IA 50010.

3-A: 09-07, Sanitary Standards for Instrument Fittings and Connections Used on Milk and Milk Products Equipment. Part One: Text; Part Two: Drawings.

These standards cover the sanitary aspects of instrument fittings and connections for milk and milk product equip-

ment and on lines which hold or convey milk or milk products.

3-A: 28-00, Sanitary Standards for Flow Meters for Milk and Liquid Milk Products, 1972, 3 pp.

Covers the sanitary aspects of flow meters for milk and liquid milk products, and includes that portion of any device integral with the meter such as strainers, temperature sensors and density sensors, which is in contact with the flowing product.

3-A: 37-00, Sanitary Standards for Pressure and Level Sensing Devices, 1978, 3 pp.

Covers the sanitary aspects of elements used on milk and milk products equipment for sensing pressure and/or product level.

ELECTRONIC INDUSTRIES ASSOCIATION (EIA)

The following EIA Standards are available from the Electronics Industries Association, Standards Sales Office, 2001 Eye Street, NW, Washington, DC 20006. In addition, EIA, NEMA and EEI have jointly prepared several standards. The joint standards may also be listed under EEI or NEMA and may be ordered from any of the three.

EIA: RS-156-B, (R 1978), Battery Socket Patterns.

This standard is intended to provide clearance dimensions on the mating plates used in battery sockets.

EIA: RS-186-E, Standard Test Methods for Passive Electronic Component Parts, 1978.

Establishes uniform methods for testing electronic component parts. The methods provide a number of test conditions of varying degrees of severity so that appropriate test conditions may be selected for any component. Included in this base document are definitions and general instructions, as well as an index of the test methods which are available individually.

EIA: RS:186-1E, Method 1: Humidity (Steady State).

This test is intended to evaluate the effect of absorption and diffusion of moisture and moisture vapor on component parts.

EIA: RS-186-2E, Method 2: Moisture Resistance (Cycling).

This test is intended to evaluate in an accelerated manner, the resistance of component parts to deterioration resulting from high humidity and heat conditions typical of tropical environments.

EIA: RS-186-3E, Method 3: Humidity (Steady State-Sealed Container).

This test method provides a means for performing humidity testing without the need for specialized humidity test equipment. It is more severe than Method I.

EIA: RS-186-4E, Method 4: Dielectric Test (Withstanding Voltage).

This dielectric test is performed for the purpose of determining the ability of component parts to withstand a potential at sea level or at a specified altitude.

EIA: RS-186-5E, Method 5: Salt Spray (Corrosion).

The salt spray test is performed for the purpose of determining the adequacy of protective coatings or finishes, and has been widely used to evaluate the resistance of metals to corrosion in marine service or in exposed shore locations.

EIA: RS186-6E, Method 6: Mechanical Robustness of Terminals.

These tests are provided to cover various significant characteristics and are intended to determine the ability of terminals to withstand the usual stresses which may be applied during assembly or disassembly operations.

EIA: RS-186-7E, Method 7: Vibration Fatigue Test (Low Frequency, 10 to 55 Hertz).

This vibration fatigue test is performed for the purpose of determining the ability of component parts and their mountings to withstand vibration in the low frequency range of 10 to 55 Hertz.

EIA: RS-186-8E: Method 8: Vibration, High Frequency.

This sinusoidal high frequency vibration test is performed for the purpose of determining the effect on component parts of vibration in the frequency range of 10 to 500 Hz or 10 to 2000 Hz as may be encountered in aircraft, missiles, space or automotive vehicles.

EIA: RS-186-9E, Method 9: Solderability.

The purpose of this test standard is to determine the solderability of solid lead wires, terminals, and other terminations which are normally joined by means of soft solder.

EIA: RS:186-10E, Method 10: Effect of Soldering.

This test is performed to determine the effect of normal soldering operations on component parts.

EIA: RS-186-11E, Method 11: Thermal Shock in Air.

This test is conducted for the purpose of determining the resistance of a component part to exposures at extremes of high and low temperatures in air, and to the shock of alternate exposures to these extremes.

EIA: RS-186-12E, Method 12: Heat-Life Test.

This test is performed to determine the effect of storing or operating component parts at elevated temperatures for various time periods.

EIA: RS-186-13E, Method 13: Insulation Resistance Test.

This test is to measure the resistance offered by the insulating members of a component part to an impressed direct voltage tending to produce a leakage of current through or on the surface of these members.

EIA: RS-186-14E, Method 14: Panel Seal Test.

This test is intended to determine the effectiveness of panel seals on electronic components which are intended for mounting through holes in panels or enclosures. The panel seals are exposed to water under pressure and observed for leakage.

EIA: RS-232-C, Interface Between Data Terminal Equipment and Data Communication Equipment Employing Serial Binary Data Interchange, 1981.

Applicable to the interconnection of data terminal equipment (DTE) and data communication equipment (DCE) employing serial binary data interchange. It defines Electrical Signal Characteristics, Interface Mechanical Characteristics, Function Description of Interchange Circuits and Solid Interfaces for Selected Communication System Configurations. Included are thirteen specific interface configurations intended to meet the needs of fifteen defined system applications. (A companion document to RS-232-C is Industrial Electronics Bulletin No. 9. It provides the application notes to RS-232-C.)

EIA: RS-236-B (Also NEMA SK 502-1968), Color Coding of Semiconductor Devices, 1979.

Describes a color code system for marking semiconductor devices with their JEDEC assigned type numbers by means of color bands.

EIA: RS-267-B (ANSI/EIA-RS-267-B-83), Axis and Motion Nomenclature for Numerically Controlled Machines, 1983.

Intended to simplify programming, to simplify the training of programmers, and to facilitate the interchangeability of control tapes; it applies to all numerically controlled machines.

EIA: RS-272. Definition and Measurement of Voltage Jump — for Voltage Regulator and Reference Tubes, 1979.

Provides a definition of voltage jump and establishes a general test method including the necessary circuit constants.

EIA: RS-275-A (ANSI: C83.68-1972), Thermistor Definitions and Test Methods, 1981.

Cover definitions of terms and test methods for measurement of the performance characteristics of thermistors. The following are defined: Zero Power Temperature Coefficient, Maximum Operating Temperature, Dissipation Constant, Zero Power Resistance Temperature Characteristic, Temperature Wattage Characteristic, Current-time Characteristic, Resistance Ratio, Beta, Stability, and Maximum Power.

EIA: RS-281-B, Electrical and Construction Standards for Numerical Machine Control.

The provisions of this standard apply when equipment of one of the following categories is furnished as part of a numerically controlled machine system: (1) electronic control; (2) electro-mechanical low power logic and switching equipment (small relays, stepping switches, etc.); (3) electro-mechanical transducers (synchros, tachometers, position sensors); (4) other servo components.

EIA: RS-309 (ANSI: C83.27-1968), (R 1977), General Specifications for Thermistors, Insulated and Non-Insulated.

Covers insulated and non-insulated thermistor disks and rods with leads.

EIA: RS-310-C, (ANSI/EIA) RS-310-C-1977), Racks, Panels and Associated Equipment.

Establishes those dimensions which are critical in ensuring compatibility between racks (open and enclosed), panels, and the equipment or apparatus installed thereon. It is intended as a guide to equipment manufacturers and designers. Three cabinet and rack widths to accommodate each of the three standard panel widths 19″, 24″ and 30″ are covered.

EIA: RS-313-B, Thermal Resistance Measurements of Conduction Cooled Power Transistors.

Described are conditions and methods for thermal resistance measurements on conduction cooled power transistors.

EIA: RS-337, (ANSI: C83.28-1968, (R 1977), General Specification for Glass Coated Thermistor Beads and Thermistor Beads in Glass Probes and Glass Rods (Negative Temperature Coefficient).

Covers definitions and test methods, characteristics and requirements of glass coated thermistor beads and thermistor beads in glass probes and glass rods.

EIA: RS-359 (ANSI/EIA RS-359-A-84), EIA Standard Colors for Color Identification and Coding, 1984.

Supersedes EIA GEN-101-A, "Color Coding for Numerical Values," and associated color chips. All of the nominal and limit colors of GEN-101-A fall within this new standard with the exception of dark limits of yellow and orange which could easily have been confused with orange or brown respectively. The colors defined in this standard are intended to be applied to the marking of electronic components, wires, terminals, and circuit functions, based on the visual color attributes of hue, value, and chroma of the Munsell System of color notation.

EIS: RS-364 (ANSI: C83.63-1971), Standard Test Procedures for Low Frequency (Below 3 MHz) Electrical Connectors.

Provides uniform procedures for testing a wide range of electrical and mechanical parameters of electrical connectors at frequencies below 3 MHz.

EIA: RS-364-1, Addendum No. 1 to RS-364 (ANSI C83.63a-1972), 1970 (R 1976).

EIA: RS-364-2, Addendum No. 2 to RS-364 (ANSI C83.63b-1972), 1971 (R 1976).

EIA: RS-364-3, Addendum No. 3 to RS-364 (ANSI C83.63c-1973), 1974.

EIA: RS-364-4, Addendum No. 4 to RS-364 (ANSI C83.63d-1975), 1975.

EIA: RS-364-7, Addendum No. 7 (ANSI/EIARS-364-7-1978), 1978.

EIA: RS-394 (ANSI: C83.71-1972), Recorded Tape Formats for 7, 14 and 21 Tracks on 1/2 Inch Magnetic Tape and 14, 28 and 42 Tracks on 1 Inch Magnetic Tape for Instrumentation Recording.

EIA: RS-405, (ANSI C83.99-1973), Flutter Measurement of Instrumentation Magnetic Tape Recorder/Reproducers — Recommended Test Method.

Covers acceptable instrumentation and procedures for the measurement of flutter in instrumentation magnetic recording equipment. The purpose is to promote interchangeability and to eliminate misunderstandings between manufacturers and users by specifying standardized and reproducible flutter measurement techniques. In addition, it is intended to help the user ascertain the suitability of magnetic recording equipment for his requirements.

EIA:RS-408, Interface Numerical Control Equipment and Data Terminal Equipment Employing Parallel Binary Data Interchange, 1973.

This standard applies to the interconnection of data terminal equipment and numerical control equipment at the tape reader interface. The data terminal would typically be connected to a remote data source/sink such as a computer.

EIA: RS-413 (ANSI: C83.94-1973), Recommended Test Method — Timing Error Measurements of Instrumentation Magnetic-Tape Recorder/Reproducers.

This Standard covers acceptable instrumentation and procedures for the measurement of Time Base Error, Intertrack Time Displacement Error, Composite Time Base Error, and Pulse-to-Pulse Jitter in instrumentation magnetic-recording equipment.

EIA: RS-414-A, Simulated Shipping Test for Consumer Electronic Products.

This standard covers all consumer electronic products, as packaged for shipping. It recommends test procedures and minimum performance requirements.

EIA: RS-422-A, Electrical Characteristics of Balanced Voltage Digital Interface Circuits.

Specifies the electrical characteristics of the balanced voltage digital interface circuit normally implemented in integrated circuit technology.

EIA: RS-423-A, Electrical Characteristics of Unbalanced Voltage Digital Interface Circuits.

Specifies the electrical characteristics of the unbalanced digital interface circuit normally implemented in integrated circuit technology.

JEDEC Standard 77, Recommendations for Letter Symbols, Abbreviations, Terms and Definitions for Semi-conductor Device Data Sheets and Specifications.

The purpose of this publication is to provide supplemental information which will facilitate the use of Joint Electron Device Engineering Council (JEDEC) registration formats. Registration formats contain a number of technical terms and symbols for which definitions should be of assistance to the writers and users of specifications.

FACTORY MUTUAL SYSTEM (FM)

The following Approval Standards and Data Sheets, although prepared for the information of manufacturers and Factory Mutual engineers and policyholders, are available for help on specific problems from Factory Mutual Research Corporation, 1151 Boston-Providence Turnpike, Norwood, Massachusetts 02062.

FM: Data Sheet 5-0/14-1, Graphical Symbols for Electrical Diagrams — Device Numbers — Functions, 1977, 12 pp.

The symbols shown in the bulletin, based on American Standard Y32.2, are those which might be encountered when studying electrical power or control diagrams.

FM: Data Sheet 5-1, Electrical Equipment in Hazardous Locations, 1976, 10 pp.

Discusses the classification of hazardous (classified) locations for electrical installations, and the types of electrical equipment that should be provided in Class I (flammable gases or vapors) and Class II (combustible dusts) hazardous locations.

FM: Data Sheet 5-10/14-10, Protective Grounding of Electrical Power Systems and Equipment, 1984, 19 pp.

Describes the various methods used for grounding electrical systems and the non-current carrying metal parts of electrical wiring systems and equipment. The advantages and disadvantages of the different grounding methods, and the means employed to safeguard property from arc damage and fire are also discussed.

FM: Data Sheet 5-32, Electronic Computer Systems, 1978, 11 pp.

Covers protection of computer equipment, records and tapes for protection of electrical equipment, occupancy and location for electronic computer systems. Also covers protection under raised floors for wiring.

FM: Data Sheet 14-6, Insulation-Resistance Tests, April 1980, 10 pp.

Presented are insulation-resistance tests which are of considerable value in detecting grounds, damp windings, carbonized or damaged insulation, foreign deposits, current leakage to ground and other conditions that cause or contribute to electrical breakdowns.

FM: Data Sheet 17-1, Nondestructive Flaw-Detection Methods, 1978, 11 pp.

Gives nondestructive flaw detection methods, readily adaptable to Code procedure and loss prevention including: magnetic particle flaw detection; liquid penetrant inspection; eddy-current, ultrasonic and radiographic testing.

FM: Approval Standard, Intrinsically Safe Apparatus and Associated Apparatus for use in Class I, II and III Division I Hazardous Locations, 1979, 59 pp.

Covers requirements for approval of intrinsically safe apparatus and associated apparatus. These requirements are based to a large extent on NFPA 493 and previous work done by the Instrument Society of America and the International Electrotechnical Commission.

FM: Approval Standard, Explosionproof Electrical Equipment, 1979, 21 pp.

Covers safety requirements for approval of explosion-proof (flameproof) electrical equipment.

FM: Approval Standard, Electrical Utilization Equipment, 1979, 35 pp.

Covers safety requirements for approval of ordinary location electrical equipment. States construction criteria and test requirements for protection against shock and fire and for electrical equipment. Requirements are based to a large extent on ANSI C39.5 and IEC 348.

FM: Approval Standard, Fire Safe Valves, 1981, 5 pp.

Covers requirements for approval of valves designed for flammable liquid service.

FM: Approval Standard, Less Flammable Transformer Fluids, 1979, 3 pp.

Covers fire point, fluid properties and convection and radiative heat release rates for approval of less flammable fluid. Tests to a large degree are based on ASTM procedures.

FM: Approval Standard, Flame Radiation Detectors for Automatic Fire Alarm Signaling, 1977, 4 pp.

Covers requirements for approval of flame radiation detectors used to sense accidental fire and to actuate alarm or extinguishing systems.

FM: Approval Standard, Smoke Actuated Detectors for Automatic Fire Alarm Signaling, 1976, 5 pp.

Covers requirements for approval of detectors sensitive to airborne products of combustion responding to particulates or gaseous products.

FM: Approval Standard, Electrostatic Finishing Equipment, 1974, 6 pp.

Covers safety requirements for approval of electrostatic paint or powder finishing equipment.

FM: Approval Standard, Electric Interlocking Fuel Gas and Fuel Oil Crocks, 1973, 4 pp.

Covers requirements for approval of electric interlocking cocks which are manually operated to shut off the fuel supply to gas or oil fired burner equipment.

FM: Approval Standard, Airflow Interlocking Switches and Pressure Supervisory Switches for Fuel Oil, Fuel Gas and Ventilation or Combustion Air, 1974, 4 pp.

Covers requirements for approval of switches that electrically interlock air or fuel pressure with other combustion control equipment.

FM: Approval Standard, Nonprogramming and Programming Single or Multi-Burner Combustion Safeguards of the Industrial Gas and/or Fuel Oil Flame Supervising Types, 1970, 6 pp.

Covers requirements for approval of combustion safeguards for oil or gas fired industrial heating equipment. Units sense the presence of flame and cause the fuel to be shut off in the event of accidental flame failure.

FM: Approval Standard, Fuel Gas and Oil Safety Shutoff Valves, 1976, 3 pp.

Covers requirements for approval of shutoff valves installed in the fuel supply of industrial heating equipment.

FM: Approval Standard, Combustible Gas Detectors, 1982, 20 pp.

Covers proposed requirements for approval of fixed and portable combustible gas detectors. Requirements are based to a large extent on work done by the Instrument Society of America and the Canadian Standards Association.

FM: Approval Standard, Electrical Equipment for Use in Class I, Division 2, Class II, Division 2 and Class III, Division 1 and 2 Hazardous Locations.

Covers requirements for approval of electrical equipment having Nonincendive circuits, field wiring and/or components. These requirements are based to a large extent on ISA-S12.12.

FLUID CONTROLS INSTITUTE, INC. (FCI)

The following FCI publications are available from the Fluid Controls Institute, Inc., P.O. Box 9036, 31 South Street, Morristown, NJ 07960.

FCI: 65-3, Standards for Determining Industrial Steam Trap Capacity Rating, 1965, 16 pp.

A brief description of the operating principles of various types of traps and the temperature depression below saturation on which capacities are based. This information provides an authoritative source of reference and makes possible a comparison of the relative capacities of traps of different types and manufacture.

FCI: 68-1, Procedure in Rating Flow and Pressure Characteristics of Solenoid Valves for Gas Service, 1968, 12 pp.

A recognized procedure under which solenoid valves intended for gas service may be rated for flow and pressure characteristics.

FCI: 68-2, Procedure in Rating Flow and Pressure Characteristics of Solenoid Valves for Liquid Service, 1968, 12 pp.

A recognized procedure for rating flow and pressure characteristics of solenoid valves for liquid service.

ANSI/FCI 69-1-1977, Pressure Rating Standard for Steam Traps, 1977, 7 pp.

Provides the minimum requirements for the design, fabrication, pressure rating, marking and testing of pressure-containing housing for steam traps.

FCI: 70-1, Standard Terminology and Definition for Filled Thermal Systems of Remote Sensing Temperature Regulators, 1970, 10 pp.

The purpose is to establish uniform terminology and definition for filled thermal system of remote sensing temperature regulators.

ANSI B16.104-1976 (FCI-70-2), Quality Control Standard for Control Valve Seat Leakage, 1976, 7 pp.

Establishes a series of seat leakage classes for control valves and defines the test procedures.

FCI: 71-1, Standard Terminology for Regulators, 1971, 6 pp.

The terminology in this standard is intended to identify by preferred current usage the different parts in a regulator.

FCI: 73-1, Pressure Rating Standard for "Y" Type Strainers, 1973, 6 pp.

Provides the minimum requirements for the design, fabrication, pressure rating, marking and testing of pressure containing housings for FCI approved "Y" Type Strainers for use with pipe conforming to dimensions specified in ANSI B36.10 and ANSI B36.19.

ANSI/FCI 74-1-1979, Silent Check Valve Standard, 1979.

A standard establishing uniform methods for design, rating and testing silent check valves.

ANSI/FCI 75-1-1979, Test Conditions and Procedures for Measuring Electrical Characteristics of Solenoid Valves, 1979.

A guide for manufacturers, users or others interested in test conditions and procedures for measuring electrical characteristics of solenoid valves.

FCI: 78-1, Pressure Rating Standard for Pipeline Strainers Other than "Y" Type, 1978.

A standard defining uniform methods for design, rating and testing of pipeline strainers other than "Y" Type.

FCI: 79-1, Standard for Proof of Pressure Ratings for Pressure Reducing Regulators, 1979.

Establishes guidelines for proof-of-design testing of pressure reducing regulators.

FCI: 81-1, Standard for Proof of Pressure Rating for Temperature Regulators, 1981.

Establishes guidelines for proof-of-design testing of temperature regulators.

FCI: 82-1, Recommended Methods for Testing and Classifying the Water Hammer Characteristics of Electrically Operated Valves.

A design guide to establish test procedures and to set performance criteria for valves which cause a shock wave in water media on shut-off.

FCI: 84-1, Metric Definition of the Valve Flow Coefficient Cv.

A standard defining the valve flow coefficient Cv commonly used for the purpose of calculating valve flow capacity or leakage in metric terminology.

FCI: 85-1, Standard for Production and Performance Tests for Steam Traps.

A standard to assist manufacturers, users and specifiers to the product to comply with production and performance characteristics of automatic steam traps.

FCI: 86-2, Regulator Terminology

The terminology in this standard is intended to establish a common and preferred usage of terms as they apply to regulators.

GAS PROCESSORS ASSOCIATION (GPA)

The following are available from the Gas Processors Association, 1812 First National Bank Building, Tulsa, OK 74103.

GPA: Standard 2140-86, Liquefied Petroleum Gas Specifications and Test Methods, 1986, 52 pp.

Covers definitions and specifications for liquefied petroleum gases, vapor pressure, test, specific gravity test (pycnometer method), specific gravity test (hydrometer method), corrosion test (copper strip method), volatile sulphur test (lamp method), dryness test — commercial propane (dew point method) residue test — commercial propane (mercury freeze method), weathering test, residue test (end point index method), and sampling methods.

GPA: Standard 2261-86, Method for Natural Gas Analysis by Gas Chromatography, 1986, 13 pp.

Covers the chromatographic analysis of natural gas to oxygen, nitrogen, helium, carbon dioxide, and hydrocarbons, methane through hexanes, and heavier. This method is applicable to natural gases and similar mixtures.

GPA: Publication 2165-75, Method for Analysis of Natural Gas Liquid Mixtures by Gas Chromatography, 1975, 6 pp.

This method is intended for the analysis of wide-range NGL mixtures, such as commercial de-ethanized and de-propanized natural gasoline mixtures, that cannot readily be entered into the chromatograph as a liquid by syringe or as a vapor at atmospheric pressure because of the presence of both highly volatile and heavy-end components. This method is intended to determine ethane and heavier hydrocarbons. Hydrogen sulfide, carbon dioxide, air, and methane are not determined.

GPA: Publication 2166-86, Methods for Obtaining Natural Gas Samples for Analysis by Gas Chromatography, 1986, 9 pp.

Describes procedures for obtaining representative ''spot'' samples of natural gas from vacuum or pressure sources in containers which are suitable for transporting the gas to a laboratory. Comparison of the degree of accuracy of eight methods.

GPA: Publication 2174-83, Method for Obtaining Hydrocarbon Fluid Samples Using A Floating Piston Cylinder, 1983, 5 pp.

Describes an alternate procedure for obtaining a representative sample of a hydrocarbon fluid and the subsequent preparation of that sample for laboratory analysis.

GPA: Standard 2186-86, Tentative Method for the Extended Analysis of Hydrocarbon Liquid Mixtures Containing Nitrogen and Carbon Dioxide by Temperature Programmed Gas Chromatography, 1986, 20 pp.

Intended for the compositional analysis of natural gas liquid streams where precise physical property data of the hexanes and heavier fraction are required.

GPA: Standard 2286-86, Tentative Method of Extended Analysis for Natural Gas and Similar Gaseous Mixtures by Temperature Programmed Gas Chromatography, 1986, 17 pp.

Covers determination of the chemical composition of natural gas and similar gaseous mixtures. Intended for use with rich gas systems and in situations where the heptanes plus compositional breakdown is desired.

INDUSTRIAL FASTENERS INSTITUTE (IFI)

The following standards are available from the Industrial Fasteners Institute, 1505 East Ohio Building, Cleveland, OH 44114.

IFI: Metric Standards, 1983, 500 pp.

Covers screw threads, materials, bolts, screws, studs, slotted and recessed screws, nuts, non-threaded fasteners. Provides data in metric only.

IFI: Fastener Standards, Fifth Edition, 1970, 500 pp.

Provides standards on all inch fasteners. Information is arranged by product group and contains over 100 pages of supporting technical data.

IFI: Capability Guide, 1986, 100 pp.

Covers fastener types, specifications, materials, bolt nut compatibility, platings, bolted joint design and tightening.

INDUSTRIAL RISK INSURERS (IRI)

The following are available from the Publications Department, Industrial Risk Insurers, 85 Woodland St., Hartford, CT 06102.

IRI: Loss Prevention and Protection for Chemical and Petrochemical Plants, 2nd edition, 28 pp.

Highlights the problems of insuring chemical and petro-chemical plants and outlines prevention and protection guidelines aimed at preventing or minimizing losses in these properties.

IRI: Recommended Good Practice for the Protection of Electronic Data Processing and Computer-Controlled Industrial Processes, 2nd edition, Revised 1983, 62 pp.

Data processing and automatic control equipment, even though not directly involved in a fire, can be severely damaged or rendered inoperative by even a moderate exposure to heat, smoke, or corrosive fumes. Discussion or preferred fire prevention and protection measures to reduce potential loss or associated effects.

Pulp and Paper Mills: Loss Prevention and Protection, 2nd edition, 1983, 36 pp.

Discusses loss experience within the pulp and paper industry with recommendations for reducing and controlling losses.

INSTITUTE OF ELECTRICAL AND ELECTRONICS ENGINEERS (IEEE)

Publications of the former American Institute of Electrical Engineers (AIEE) and Institute of Radio Engineers (IRE) are identified by the IEEE number. All of these publications are available from the Institute of Electrical and Electronics Engineers, 345 East 47th Street, New York, 10017.

IEEE: STD 74-1958 (REAFF 1974), Test Code for Industrial Control (600 Volts or Less), 20 pp.

Establishes procedures for tests on a representative sample of an industrial control device or apparatus in order to substantiate conformance of that type of device or apparatus with a recognized standard of performance.

ANSI/IEEE: STD 81-1983, Guide for Measuring Earth Resistivity, Ground Impedence, and Earth Surface Potentials of a Ground System, 1983, 42 pp.

Describes and discusses the present state of the technique of measuring ground impedance, earth resistivity, potential gradients from currents in the earth, and the prediction of the magnitudes of ground resistance and potential gradients from scale model tests.

IEEE: STD 85-1973, (REAFF 1986), Test Procedure for Airborne Noise Measurements on Rotating Electric Machinery, 34 pp.

Outlines practical techniques and procedures which can be followed for conducting and reporting tests on rotating and reporting tests on rotating electrical machines of all sizes to determine the airborne noise characteristics under steady state conditions.

ANSI/IEEE: STD 91-1984, Graphic Symbols for Logic Diagrams (Two-State Devices), 1984, 146 pp.

Sets forth principles governing the formation of graphical symbols for logic diagrams in which connections between symbols are generally shown with lines. Definitions of logic functions, the graphical representations of these functions, and examples of their application are given.

IEEE: STD 94-1970, Definition of Terms for Automatic Generation Control on Electric Power Systems, 8 pp.

Covers terms applicable to power system operation, governors, automatic control elements and action, tele-metering, and load economics, includes nonstandard and alphabetical cross-indexes.

ANSI/IEEE: STD 100-1984, Dictionary of Electrical and Electronics Terms, 1984, 1173 pp.

IEEE: STD 108-1955, Proposed Recommended Guide for Specification of Electronic Voltmeters, 6 pp.

Provides a common basis of comparison among electronic voltmeters offered for general purpose applications.

IEEE: STD 118-1978, Standard Master Test Code for Resistance Measurements, 31 pp.

Provides instructions for those measurements of electric resistance which are commonly needed in determining the performance characteristics of electric machinery and equipment.

IEEE: STD 120-1955 (REAFF 1980), Master Test Code for Electrical Measurements in Power Circuits, 40 pp.

Gives instructions for those measurements of electrical quantities which are commonly needed in determining the performance characteristics of electric machinery and equipment.

IEEE: STD 139-1953, Recommended Practice for Measurement of Field Intensity Above 300 MHz from Radio-Frequency Industrial, Scientific and Medical Equipments, 14 pp.

Includes information on methods of measurement, associated measuring equipment and measurement precautions and factors affecting accuracy of measurement.

ANSI/IEEE: STD 141-1986, Recommended Practice for Electric Power Distribution for Industrial Plants, 608 pp.

Includes chapters on: System Planning; Voltage Considerations; Surge Voltage Protection; Application and Coordination of System Protective Devices; Fault Calculations; Grounding; Power Factors and Related Considerations; Power Switching; Transformation, and Motor-Control Apparatus; Instruments and Meters; Cable Systems; Busways; Electrical Energy Conservation; and Cost Estimating of Industrial Power Systems.

ANSI/IEEE: STD 142-1982, Recommended Practice for Grounding of Industrial and Commercial Power Systems, 136 pp.

Includes chapters on: Systems Grounding; Equipment Grounding; Static and Lightning Protection Grounding; and Connection to Earth.

ANSI/IEEE: STD 146-1980, Definitions of Fundamental Waveguide Terms, 1980, 10 pp.

Gives standard definitions of fundamental waveguide terms placing emphasis on waveguides (hollow uniconductor transmission lines).

ANSI/IEEE: STD 147-1979, Definitions for Waveguide Components, 1979, 7 pp.

Contains definitions for 41 of the more general, basic, and established terms for waveguide components in which the waveguide is treated as a generic term covering both transmission line and uniconductor waveguide.

IEEE: STD 148-1959, Waveguide and Waveguide Component Measurements, 1959 (REAFF 1971), 16 pp.

Provides general techniques for measurement of quantities which characterize a waveguide. References are included to cover specific procedures and equipment. Contents are limited to linear, reciprocal systems.

ANSI/IEEE: STD 162-1963, Definitions of Terms for Electronic Digital Computers, 2nd edition, 1963, (REAFF 1984), 12 pp.

Contains nearly all important terms related to electronic digital computers. Analog and programming terms are not included because they are covered in other standards.

ANSI/IEEE: STD 165-1977, (REAFF 1984), Definitions of Terms for Analog Computers, 1977, 12 pp.

Contains definitions for 172 terms common to analog computing techniques and associated hardware.

ANSI/IEEE: STD 166-1977, (REAFF 1984), Definitions of Terms for Hybrid Computer Linkage Components, 1977.

ANSI/IEEE: STD 181-1977, Pulse Measurement and Analysis by Objective Techniques, 16 pp.

Provides definitions and descriptions of the techniques and procedures for time domain pulse measurements.

IEEE: STD 182-1961, Definitions of Terms for Radio Transmitters, 6 pp.

Defines 36 terms basic to the design of radio transmitters.

ANSI/IEEE: STD 186-1948 (REAFF 1976) Standard Methods of Testing Amplitude — Modulation Broadcast Receivers and ANSI/IEEE STD 189-1955 (REAFF 1976) Standard Method of Testing Receivers Employing Ferrite Core Loop Antennas, 28 pp.

Describes test assemblies and adjustments of input and output, operating condition, and radio receiver adjustments as applied to any type of receiver. Recommends procedures for measuring sensitivity, selectivity, and fidelity and other characteristics.

ANSI/IEEE: STD 206-1960, REAFF 1978), Television: Measurement of Differential Gain and Differential Phase, 12 pp.

Describes a method of measuring differential gain and differential phase in video transmission systems.

IEEE: STD 216-1960, Definitions of Semiconductor Terms, 1960, (REAFF 1980), 4 pp.

Provides definitions of important terms relating to the physical aspects of semiconductor of materials and basic components developed from them.

ANSI/IEEE: STD 242-1986, Recommended Practice for Protection and Coordination of Industrial and Commercial Power Systems, 588 pp.

Includes chapters on: Calculation of Short-Circuit Currents; Instrument Transformers; Selection and Application of Protective Relays; Fuses; Low-Voltage Circuit Breakers; Ground Fault Protection; Conductor Protection; Motor Protection; Transformer Protection; Generator Protection; Bus and Switchgear Protection; Service Supply Line Protection; Overcurrent Coordination; and Maintenance Testing and Calibration.

ANSI/IEEE: STD 251-194, Proposed Heat Procedure for Direct-Current Tachometer Generators, 15 pp.

Covers instructions for conducting and reporting the more generally applicable and acceptable tests to determine the performance characteristics of direct-current tachometer generators.

IEEE: STD 261-1965, Letter Symbols for Thermoelectric Devices, 1965, 3 pp.

Presents standard letter symbols for quantities used in the field of thermoelectric devices.

ANSI/IEEE: STD 268-1982, Standard for Metric Practice, 48 pp.

Provides guidance for application of the modernized metric system and includes sections on: SI units and symbols; Application of the metric system; Recommendations concerning units; style and usage; Rules for conversion and rounding; and terminology.

IEEE: STD 284-1968, Standards Report on State-of-the-Art of Measuring Field Strength, Continous Wave, Sinusoidal, 1968, 7 pp.

Report on the state-of-the-art of measuring field strength

of radio-frequency electromagnetic waves, with respect to available and desirable accuracies.

ANSI/IEEE: STD 290-180, Standard for Electric Couplings, 19 pp.

Covers the more generally applicable performance characteristics and conducting and reporting of the test for determining the performance characteristics of electric couplings.

ANSI/IEEE: STD 300-1982, Standard Test Procedure for Semiconductor Charge Particle Detectors, 28 pp.

Establishes standard test procedures for semiconductor charged-particle detectors. These detectors are in widespread use for the detection and high-resolution spectroscopy of charged particles.

ANSI/IEEE: STD 301-1976, Standard Test Procedures for Amplifiers and Preamplifiers for Semiconductor Radiation Detectors for Ionizing Radiation, 28 pp.

Includes sections on: Noise Linewidth Measurements, Preamplifier Noise; Pulse-Height Linearity; Count Rate Effects; Overload Effects; Pulse-Height Dependence on Rise Time; Pulse-Height Stability; and Crossover Walk.

IEEE: STD 302-1969, (REAFF 1981), Methods for Measuring (Below 1000 MHz) Electromagnetic Field Strength for Frequencies Below 1000 MHz in Radiowave Propagation, 1969, 15 pp.

Most measurements with which radio wave propagation are concerned involve the measurement of field strength. Standard methods for the measurement of this fundamental quantity for frequencies below 1000 megahertz are described.

IEEE: STD 306-1969, (REAFF 1981), Test Procedure for Charging Inductors, 1969, 57 pp.

This document pertains to the methods of measurement of the electrical characteristics of charging inductors used in radar transmitters, linear particle accelerators; and similar equipment.

ANSI/IEEE: STD 308-1980, Standard Criteria for Class IE Power Systems for Nuclear Power Generating Stations, 1980, 24 pp.

Applies to the Class 1E portions of the following systems and equipment in single unit and multi-unit nuclear power generating stations: vital instrumentation and control power systems; alternating current power systems; and direct current power systems.

ANSI/IEEE: STD 309-1970, (REAFF 1984), Test Procedure for Geiger-Muller Counters, 8 pp.

Presents standard test procedures for Geiger-Muller counter radiation detectors.

IEEE: STD 311-1970, Standard Specification of General-Purpose Laboratory Cathode-Ray Oscilloscopes, 16 pp.

This standard documents the minimum information that users of general-purpose laboratory cathode-ray oscilloscopes typically need; provides potential purchasers and others with a common means for making comparisons between instruments; and provides uniformity of information from manufacturers.

IEEE: STD 314-1971, Standard Report on State of the Art of Measuring Unbalanced Transmission Line Impedance, 8 pp.

Applies to reporting the state-of-the-art of measuring impedance in distributed parameter coaxial waveguide systems, propagating a TEM wave.

IEEE: STD 316-1971, Requirements for Direct Current Instrument Shunts, 7 pp.

Applies to shunts for use in direct current circuits to extend the range of instruments or other measuring devices.

ANSI/IEEE: STD 325-1971, (REAFF 1982), Test Procedures for Germanium Gamma-Ray Detectors, 19 pp.

Presents test procedure for germanium gamma-ray detectors used for the detection and analysis of gamma-radiation.

ANSI/IEEE: STD 336-1985, Installation, Inspection, and Testing Requirements for Class 1E Instrumentation and Electric Equipment at Nuclear Power Generating Stations, 1980, 12 pp.

Sets forth the requirements for installation, inspection, and testing of Class 1E electric power, instrumentation, and control equipment and systems during the construction phase of a nuclear power generating station.

IEEE: STD 337-1972, Specifications Format Guide and Test Procedure for Linear, Single-Axis, Pendulous, Analog Torque Balance Accelerometer, 1972, (REAFF 1978), 47 pp.

Defines the requirements for a linear, single-axis, pendulous, analog torque balance accelerometer. The instrument is equipped with a permanent magnet torquer and is used as a sensing element to provide an electrical signal proportional to acceleration.

ANSI/IEEE: STD 338-1977, (REAFF 1984), Criteria for Periodic Testing of Nuclear Power Generating Station Safety Systems, 1977, 15 pp.

Establishes specific criteria for the periodic testing required to ensure operational availability of protection systems utilizing the capability called for in sections of the Criteria for Nuclear Power Generating Station Protection Systems, IEEE Std. 279-1971.

ANSI/IEEE: STD 344-1975, (REAFF 1980), Guide for Seismic Qualification of Class 1E Equipment for Nuclear Power Generating Stations, 1975, 24 pp.

Provides direction for establishing procedures that will yield data which verify that the Class I electric equipment can meet its performance requirements during and following a design basis earthquake.

ANSI/IEEE: STD 376-1975, (REAFF 1986), Measurement of Impulse Strength and Impulse Bandwidth, 16 pp.

Provides basic information relating to the use of the impulse generator and interpretation of measurements made using instruments based on it.

ANSI/IEEE: STD 381-1977, (REAFF 1984), Criteria for Type Tests of Class 1E Modules, Used in Nuclear Power Generating Stations, 27 pp.

Describes the basic requirements of a type test program with the objective of verifying that a module used as Class IE Equipment in a nuclear power generating station meets or exceeds its design specifications.

IEEE: Std 389-1979, Recommended Practice for Testing Electronic Transformers and Inductors. 60 pp.

Presents a number of tests for use in determining the significant parameters and performance characteristics of electronic transformers and inductors. Included are tests for: electric strength, DC resistance, power losses, balance, equivalent impedances, terminated impedances, and other transformer properties.

ANSI/IEEE: Std 398-1972, Standard Test Procedures for Photomultipliers for Scintillation Counting and Glossary for Scintillation Counting Field. 28 pp.

Includes sections on: Photomultiplier characteristics, testing of photomultiplier characteristics, test conditions for photomultipliers, test instrumentation, and glossary for scintillation counting field.

ANSI/IEEE: Std 399-1980, Recommended Practice for Industrial and Commercial Power System Analysis. 224 pp.

Chapters include: Applications of Power System Analysis; Analytical Procedures; System Modeling; Load Flow Studies; Motor Starting Studies; Harmonic Studies; Switching Transients Studies; Reliability Studies; Grounding Mat Studies; and Computer Services.

ANSI/IEEE: Std 415-1986 Guide for Planning of Pre-operational Testing Programs for Class 1E Power Systems for Nuclear Power Generating Stations. 15 pp.

Provides direction for establishing an acceptable pre-operational testing program for Class 1E Power Systems.

ANSI/IEEE: Std 416-1984, ATLAS Test Language, 1984, 448 pp.

Defines the Abbreviated Test Language for All Systems (ATLAS). The term "all" was substituted for "avionics" in recognition of the wider application of the language. Used in preparation and documentation of test procedures which can be implemented either manually or with automatic or semi-automatic test equipment.

ANSI/IEEE: STD 446-1987, Emergency and Standby Power Systems for Industrial and Commercial Applications, 1987, 272 pp.

Provides commercial facility designers, operators and owners with guidelines for assuring uninterrupted power.

ANSI/IEEE: Std 449-1984, Standard for Ferroresonant Voltage Regulators. 26 pp.

Pertains to ferroresonant voltage regulators which operate at relatively constant frequencies and provide substantially constant ouput voltages in spite of relatively large changes in input voltage, and to controlled ferroresonant regulators which maintain substantially constant output voltages regardless of variations, within limits, of input voltage, temperature, frequency, and output load.

IEEE: STD 457-1982, Definitions of Terms for Nonlinear, Active, and Nonreciprocal Waveguide Components, 1982.

IEEE: STD 467-1980, Quality Assurance Program Requirements for the Design and Manufacture of Class 1E Instrumentation and Electric Equipment for Nuclear Power Generating Stations, 1980, 12 pp.

Sets forth the quality assurance program requirements for the design and manufacture of Class 1E instrumentation and electric equipment in a nuclear power generating station.

IEEE: STD 470-1972, Application Guide for Bolometric Power Meters, 26 pp.

Applies to bolometric power meters as complete instruments and to their constituent parts: bolometric detectors, bolometer units, and bolometer elements.

ANSI/IEEE: STD 474-1973, (REAFF 1982), Specifications and Test Methods for Fixed and Variable Attenuators, DC to 40 GHz, 1973, 27 pp.

Covers absorptive and reflective attenuators, both fixed as well as continuously variable or variable in fixed steps, both manual and programmable types. It does not cover electronic or solid-state-type attenuators.

ANSI/IEEE: Std 475-1983, Standard Measurement Procedure for Field-Disturbance Sensors (rf Intrusion Alarms). 16 pp.

Defines a test procedure for field-disturbance sensors to measure radio-frequency (rf), radiated field strength of the fundamental frequency, including second and third harmonics, and of any nonharmonic spurious emission within the frequency range from 0.3 GHz to 40.0 GHz.

ANSI/IEEE: STD 488-1978, Standard Digital Interface for Programmable Instrumentation, 83 pp.

Applies to interface systems used to interconnect both programmable and nonprogrammable electronic measuring apparatus with other apparatus and accessories necessary to assemble instrumentation systems.

ANSI/IEEE: STD 497-1981, Criteria for Accident Monitor-

ing Instrumentation for Nuclear Power Generating Stations, 1981, 14 pp.

Establishes the minimum design criteria for accident monitoring instrumentation.

ANSI/IEEE: STD 498-1985, Supplementary Requirements for the Calibration and Control of Measuring and Test Equipment Used in the Construction and Maintenance of Nuclear Power Generating Stations, 8 pp.

Establishes the requirements for a calibration program to control and verify the accuracy of M&TE (measuring and test equipment) which is used to assure that important parts of nuclear power generating stations are in conformance with prescribed technical requirements. This standard is intended to be used in conjunction with ANSI N45.2-1971.

ANSI/IEEE: STD 500-1984, Reliability Data Manual, 1984, 1424 pp.

Contains comprehensive and up-to-date collection of reliability data including maintenance rates as well as failure rates, failure rate ranges, failure modes and environmental factor information on generic components actually or potentially in use in nuclear power generating stations.

IEEE: STD 510-1983, Recommended Practices for Safety in High-Voltage and High-Power Testing, 1983, 19 pp.

Provide guidance for those who are establishing or revising their safety rules. Cites some of the hazards present in making high-voltage and high-power measurements, and some of the procedures and equipment which can be used to reduce personnel hazards.

ANSI/IEEE: STD 518-1982, Guide for the Installation of Electrical Equipment to Minimize Electrical Noise Inputs to Controllers from External Sources, 160 pp.

Develops a guide for the installation and operation of industrial controllers to minimize the disturbing effects of electrical noise on these controllers.

IEEE: STD 544-1975, Electrothermic Power Meters, 1975, 26 pp.

Provides a common means for making comparisons between instruments and to provide a uniformity of information among manufacturers. Also documents the minimum information that users of electrothermic power meters may need.

ANSI/IEEE: STD 583-1982, Modular Instrumentation and Digital Interface Systems (CAMAC), 1982, 81 pp.

Serves as a basis for a range of modular instrumentation capable of interfacing transducers and other devices to digital controllers for data and control. Specifies a data bus (dataway) by means of which instruments can communicate with each other with peripherals, computers and other external controllers.

ANSI/IEEE: STD 595-1982, Serial Highway Interface Systems (CAMAC), 1982, 85 pp.

Defines a serial highway system using byte-organized messages, and configured as a unidirectional loop to which are connected a system controller and up to sixty-two CAMAC crate assemblies or other control devices.

ANSI/IEEE: STD 596-1982, Parallel Highway Interface System (CAMAC), 1982, 42 pp.

Defines the CAMAC parallel highway interface system for interconnecting up to seven CAMAC crates (or other devices) and a system controller.

ANSI/IEEE: Std 622A-1984, Recommended Practice for the Design and Installation of Electric Pipe Heating Control and Alarm Systems for Power Generating Stations. 24 pp.

Provides recommended practices for designing and installing electric pipe heating control and alarm systems as applied to mechanical piping systems that require heat. Includes selection of control and alarm systems, accuracy considerations, local control usage, centralized control usage, qualification criteria of controls and alarms, and calibration and testing of controls and alarms.

IEEE: Std 625-1979, Recommended Practices to Improve Electrical Maintenance and Safety in the Cement Industry. 24 pp.

Recommendations apply to all electrical equipment such as, substations, power transformers, motor controls, generators, distribution systems, instruments and storage batteries commonly used in cement plants.

ANSI/IEEE: STD 645-1977, (REAFF 1982), Test Procedures for High-Purity Germanium Detectors for Ionizing Radiation, 11 pp.

Standard is a supplement to ANSI/IEEE STD 325-1971 (REAFF 1977) and provides additional test procedures required for high-purity germanium detectors.

ANSI/IEEE: STD 649-1980, Qualifying Class 1 E Motor Control Centers for Nuclear Power Generating Stations, 1980, 22 pp.

Describes the basic principles, requirements, and methods for qualifying Class 1E motor control centers for outside containment applications in nuclear power generating stations.

ANSI/IEEE: STD 675-1982, Multiple Controllers in a CAMAC Crate, 1982, 32 pp.

Provides for the use of auxiliary controllers in a CAMAC crate to extend the capabilities and fields of application of the CAMAC modular instrumentation and interface system.

ANSI/IEEE: STD 683-1976 (REAFF 1987), Recommended Practice for Block Transfers in CAMAC Systems, 1976, 19 pp.

Presents recommended algorithms to encourage uniformity in design of CAMAC modules and controllers with resulting increased compatibility.

ANSI/IEEE: STD 690-1984, Standard for the Design and Installation of Cable Systems for Class 1E Circuits in Nuclear Power Generating Stations, 30 pp.

Provides direction for the design and installation of safety related electrical cable systems, including associated circuits. Also provided is guidance for the design and installation of those nonsafety related cable systems that may affect the function of safety related systems.

IEEE: 696-1983, 696 Interface Devices, 1983, 40 pp.

Defines a general-purpose interface system for designers of new computer system components that will ensure their compatibility with present and future IEEE Std. 696 computer systems.

ANSI/IEEE: STD 716-1985, C/ATLAS Test Language, 1985, 438 pp.

Describes the language to be used for the writing of test programs for Units Under Test (UUTs), so that these programs can operate on various makes and models of Automatic Test Equipment (ATE).

ANSI/IEEE: STD 726-1982, Real-Time BASIC for CAMAC, 1982, 20 pp.

Defines the declarations and real-time statements for use with CAMAC hardware.

ANSI/IEEE: STD 728-1982, Recommended Practice for Code and Format Conventions, 1982, 52 pp.

Elaborates a family of codes and formats to be generated, processed, and interpreted by the device functions of apparatus operating in concert with interface functions.

ANSI/IEEE: STD 729-1983, Glossary of Software Engineering Terminology, 1983, 38 pp.

Defines more than 450 terms in general use in the software engineering field.

ANSI/IEEE: STD 730-1984, Software Quality Assurance Plans, 1984, 12 pp.

Provides uniform, minimum acceptable requirements for preparation and content of Software Quality Assurance Plans and provides a standard against which such plans can be assessed.

ANSI/IEEE: STD 739-1984, Bronze Book, 1984, 160 pp.

Energy conservation and cost-effective planning are addressed in the areas of engineering, design, applications, utilization, and to some extent the operation and maintenance of electric power systems to provide for the optimal use of electrical energy.

ANSI/IEEE: STD 746-1984, Standard for Performance Measurements of A/D and D/A Converters for PCM Television Video Circuits, 28 pp.

Describes methods for measuring the performance of uniformly coded analog-to-digital (A/D) converters and digital-to-analog (D/A) converters for pulse code modulation (PCM) television video signals.

IEEE: STD 748-1979, Spectrum Analyzers, 1979, 20 pp.

Provides a reference for specification comparison allowing a user to evaluate different instruments from a common base.

ANSI/IEEE: STD 758-1979 (REAFF 1987), Subroutines for CAMAC, 1979, 24 pp.

Presents a recommended set of software subroutines for use with CAMAC modular instrumentation and interface system.

ANSI/IEEE: STD 759-1984, Standard Test Procedures for Seminconductor X-Ray Energy Spectrometers, 39 pp.

Presents standard test procedures for semiconductor X-Ray energy spectrometers. Such systems consist of a semiconductor radiation detector assembly and signal processing electronics interfaced to a pulse height analyzer/computer.

ANSI/IEEE: STD 771-1984, Guide to the Use of ATLAS Test Language, 1984, 318 pp.

Defines ATLAS, Abbreviated Test Language for All Systems, which is standardized test language for expressing test specifications and test procedures.

ANSI/IEEE: STD 796-1983, Microcomputer System Bus, 1983, 46 pp.

Prepared for those users who evaluate or design products that will be compatible with the IEEE Std 796 system bus structure. Deals only with the interface characteristics of microcomputer devices, not with design specification, performance requirements, and safety requirements of modules.

ANSI/IEEE: STD 802.2-1985, Logical Link Control, 1985, 111 pp.

Provides a description of the peer-to-peer protocol procedures that are defined for the transfer of information and control between any pair of Data Link Layer service access points on a local area network.

ANSI/IEEE: STD 802.3-1985, Carrier Sense Multiple Access with Collision Detection, 1985, 143 pp.

Encompasses several media types and techniques for signal rates from 1 Mb/s to 20 Mb/s. Provides the necessary specifications and related parameter values for a 10 Mb/s baseband implementation.

ANSI/IEEE: STD 802.4-1985, Token-Passing Bus Access Method and Physical Layer Specifications, 1985, 238 pp.

Deals with all elements of the token-passing bus access method and its associated physical signalling and media technologies. Deals exclusively with the broadcast type.

ANSI/IEEE: STD 802.5-1985, Local Area Networks: Token

Ring Access Method and Physical Layer Specifications, 89 pp.

Defines: the frame format and introduces medium access control frames, timers and priority stacks; medium access control protocol; physical layer functions of symbol encoding and decoding, symbol timing, and latency buffering; and the 1 and 4 Mb/s, shielded twisted pair attachment of the station to the medium including the definition of the medium interface connector.

ANSI/IEEE: STD 803-1983, Principles and Definitions, 1983.

Presents principles, definitions, and a procedure whereby systems/structures and component functions of power plant projects and related facilities can be uniquely identified.

ANSI/IEEE: STD 803A-1983, Component Function Identifiers, 1983, 20 pp.

Provides component function identifiers which have been selected considering the functional levels within a power plant which are significant to design, procurement, construction, operation, and maintenance applications.

ANSI/IEEE: STD 804-1983, Implementation, 1983, 15 pp.

Provides instruction for the implementation of the EIIS and includes information for establishing codes, code structure, and identification codes on equipment and design documents, and for reporting.

ANSI/IEEE: STD 805-1984, System Identification in Nuclear Power Plants, 1984, 177 pp.

Source of nuclear power plant system description concentrating on system function and includes internal detail where necessary to clearly support the system function description. System descriptions and diagrams represent typical systems.

ANSI/IEEE: STD 828-1983, Software Configuration Management Plans, 1983, 11 pp.

Provides the minimum requirements for preparation and content of Sofware Configuration Management (SCM) Plans and is applicable to the entire life cycle of critical software. Identifies essential items that appear in SCM plans, and provides examples to enhance clarity and promote understanding.

ANSI/IEEE: STD 829-1983, Software Test Documentation, 1983, 48 pp.

Describes a set of basic test documents associated with the dynamic aspects of software testing and defines the purpose, outline and content of each basic document.

ANSI/IEEE: STD 830-1984, Software Requirements Specifications, 1984, 24 pp.

Describes approaches to good practice in the specification of software requirements.

ANSI/IEEE: STD 960-1986, Standard FASTBUS Modular High-Speed Data Acquisition and Control System, 232 pp.

Standardized modular data-bus system for data acquisition, data processing and control. The system consists of multiple bus segments that can operate independently, but link together for passing data and other information.

ANSI/IEEE/ANS: STD 7432-1982, Application Criteria for Programmable Digital Computer Systems in Safety Systems of Nuclear Power Generating Stations, 1982, 5 pp.

Establishes application criteria for programmable digital computer systems of nuclear power generating stations.

ANSI/IEEE: C57.13-1978, Standard Requirements for Instrument Transformers, 62 pp.

Covers certain electrical, dimensional and mechanical characteristics, and takes into consideration certain safety features of current and inductively coupled voltage transformers of types generally used in the measurement of electricity and the control equipment associated with the generation, transmission and distribution of alternating current.

ANSI: C63.2-1980, Specifications for Electromagnetic Noise and Field Strength Instrumentation, 10 kHz to 1 GHz, 24 pp.

Describes requirements for instruments measuring quasi-peak, peak, rms, and average values.

ANSI: C95.3-1973 (Reaff 1979) Standard Techniques and Instrumentation for the Measurement of Potentially Hazardous Electromagnetic Radiation at Microwave Frequencies, 36 pp.

Sections included on: electromagnetic environment, theoretical calculations, methods of measurement and instrumentation requirements and calibration.

ANSI: C95.5-1981, Recommended Practice for the Measurement of Hazardous Electromagnetic Fields — RF and Microwave, 35 pp.

Specifies techniques and instrumentation for measurement of potentially hazardous electromagnetic fields both in the near field and far field of the electromagnetic source. Management techniques and instruments also apply to fields in the neighborhood of flammable materials and explosive devices.

ANSI: N13.4-1971 (Reaff 1983) Standard for the Specification of Portable X- or Gamma-Radiation Survey Instruments, 10 pp.

Provides the means of stating or describing performance as applicable to portable X- or gamma-radiation survey instruments, which includes a detector, and visual, analog and/or digital type readout.

ANSI: N42.4-1971 (Reaff 1985) Standard for High Voltage Connectors for Nuclear Instruments, 8 pp.

Applicable to coaxial high voltage connectors on nuclear

instruments for dc applications up to 3500 volts rms at 60 Hz. The connectors are of the ''safe'' type in that the pin and socket contacts are well and securely recessed in the connector housing.

ANSI: N42.14-1978, (REAFF 1985), Calibration and Usage of Germanium Detectors for Measurement of Gamma-Ray Emission of Radionuclides, 1978, 13 pp.

Covers the energy and full energy peak efficiency calibration as well as the determination of gamma-ray energies in the 0.06 to 2 MeV energy region and is designed to yield gamma-ray emission rates with an uncertainty of plus/minus 3 percent.

ANSI: N42.18-1980 (REAFF 1985), Specification and Performance of On-Site Instrumentation for Continuously Monitoring Radioactivity in Effluents, 1980, 16 pp.

Applies to continuous monitors that measure normal releases, detect inadvertent releases, show general trends, and annunciate radiation levels that have exceeded predetermined values.

ANSI: N317-1980 (Reaff 1985), Standard Performance Criteria for Instrumentation Used for Inplant Plutonium Monitoring, 16 pp.

Presents performance criteria for radiation protection instrumentation essential to inplant plutonium monitoring. Plutonium radiations are also characterized. Criteria

is limited to instruments capable of measuring: photon radiations within the energy range of 0.010 to 1.25 MeV; neutron radiations within the energy range from thermal to 10 MeV; and alpha radiations within the emitted energy range of 4.5 to 7.5 MeV.

ANSI: N320-1979 (Reaff 1985), Standard Performance Specifications for Reactor Emergency Radiological Monitoring Instrumentation, 15 pp.

Defines the essential performance parameters, and general placement of emergency instrumentation, radiological instrumentation systems, systems for monitoring conditions within the reactor facility, instrumentation systems for detecting and quantifying the release to the environs, installed systems for monitoring conditions in the environs, and portable instrumentation for monitoring the release of radionuclides associated with a postulated serious accident at a reactor facility.

ANSI: N323-1978 (Reaff 1983), Standard Radiation Protection Instrumentation Test and Calibration, 23 pp.

Establishes calibration methods for portable radiation protection instruments used for detection and measurement of levels of ionizing radiation fields or levels of radioactive surface contamination, and includes conditions, equipment, and techniques for calibration as well as the degree of precision and accuracy required.

INSTITUTE OF ENVIRONMENTAL SCIENCES (IES)

IES: RP-CC-001-86, Recommended Practice for HEPA Filters, 1986, 10 pp.

Covers basic requirements for HEPA (high efficiency particulate air) filter units as a basis for agreement between buyer and seller.

IES: RP-CC-002-86, Recommended Practice for Laminar Flow Clean Air Devices, 1986, 10 pp.

Covers definitions, procedures for evaluating performance, and major requirements of laminar flow clean air devices.

IES: RP-CC-006-84-T, Recommended Practice for Testing Clean Rooms, 1984, 16 pp.

Covers testing methods for characterizing the performance of clean rooms.

IES: RP-CC-008-84, Recommended Practice for Gas-Phase Adsorber Cells, 1984, 8 pp.

Covers the design and testing of modular gas-phase adsorber cells for use where high efficiency removal of gaseous contaminants is required.

IES: CC-009-84, Compendium of Standards, Practices, Methods, Relating to Contamination Control, 1984, 53 pp.

Lists standards, practices, methods, technical orders, specifications, and similar documents developed by government, industry, and technical societies in the United States and other countries, which are related to the field of contamination control.

IES: RP-CC-013-86-T, Recommended Practice for Equipment Calibration or Validation Procedures.

Covers definitions and procedures for calibrating instruments used for testing clean rooms and clean air devices, and for determining intervals of calibration.

INSTITUTE FOR INTERCONNECTING AND PACKAGING ELECTRONIC CIRCUITS (IPC)

The following documents are available from the Institute for Interconnecting and Packaging Electronic Circuits, 7380 N. Lincoln Ave., Lincolnwood, IL 60646.

Each of the manuals listed is constantly being expanded and updated. Your initial purchase entitles you to receive all new information developed for these publications for two full years, plus the balance of the months during the year the publication is purchased.

IPC: Assembly-Joining Handbook.

The new Assembly and Joining Handbook contains practical and useful information regarding various approaches and techniques for the interconnection of electronic components. This material, developed by experts in the field of assembly and joining techniques and approved by the IPC committee on assembly and joining, is divided into eight sections. These sections contain information on: *Introductory Material; Printed Wiring Boards; Component/Lead types; Joining Materials; Component Mounting; Joining Techniques; Packaging; Quality Assurance and Testing.*

It is recognized that ideas and techniques discussed in the initial release will be subject to improvement and change. As changes occur in the technology the material will be revised and the changes or new information sent to subscribers of this handbook.

IPC: TM-650, Printed Circuit Test Methods Manual.

Provides a comprehensive compendium of pertinent information on test methods that will be useful to manufacturers and users of printed circuit boards. Designed to provide specific information on test methods it does not attempt to establish acceptability levels for performance. Most of the test methods have been taken from previously approved standards and specfications of the IPC. Other test methods have been taken from either government standards or from other standards that have extensive use in the printed circuit industry. Contains over 100 test methods, and is divided into five major sections. A spearate Table of Contents is shown at the beginning of each section.

IPC: Printed Wiring Design Guide.

Contains a compilation of valuable and useful data pertinent to the problems encountered in the application of printed wiring design principles. Sections include: General Printed Wiring Information, Single-Sided Printed Wiring (Rigid), Double-Sided Printed Wiring (Rigid), Multilaver Printed Wiring (Rigid), Flexible Printed Wiring, Printed Wiring Assemblies, Documentation, Quality Assurance and Testing, Computer Aided Design, plus Appendix. Update supplements available.

IPC: Technical Manual.

Divided into three volumes, it includes copies of all IPC (more than 50) Specifications and Standards.

INSTRUMENT SOCIETY OF AMERICA (ISA)

The following ISA Standards and Recommended Practices are available from the Instrument Society of America, 67 Alexander Drive, P.O. Box 12277, Research Triangle Park, NC 27709. A complete copy of each Standard is in Section III of this book.

ISA:RP2.1, Manometer Tables, Reaffirmed 1978, 31 pp.

Presents abbreviations and fundamental conversion factors commonly used in manometry, recommended definitions of pressure in terms of a column of mercury and water, and for a large number of liquids, tables of pressures indicated by, or equivalent to, heights of columns at various temperatures.

ISA-S5.1, Instrumentation Symbols and Identification (Formerly ANSI Y32.20), (ANSI/ISA-1984), 1984, 52 pp.

Establishes a uniform means of designating instruments and instrumentation systems used for measurement and control. The differing established procedural needs of various organizations are recognized (where not inconsistent with the objectives of the standard) by providing alternative symbolism methods. A number of options are provided for adding information or simplifying the symbolism, if desired. Includes additional information on symbolism for function blocks, function designations, computer functions, and programmable logic control.

ISA-S5.2, (ANSI/ISA-1976, R1981), Binary Logic Diagrams for Process Operations, R1981, 19 pp.

Provides symbols, both basic and non-basic, for binary operating functions. Intended to symbolize the binary operating functions of a system in a manner that can be applied to any class of hardware, whether electronic, electrical, fluidic, pneumatic, hydraulic, mechanical, manual, optical, or other.

ISA-S5.3, Graphic Symbols for Distributed Control/Shared Display Instrumentation, Logic and Computer Systems, 1982, 14 pp.

Establishes documentation for that class of instrumentation consisting of computers, programmable controllers, minicomputers and micro-processor based systems that have shared control, shared display or other interface features. Symbols are provided for interfacing field instrumentation, control room instrumentation and other hardware to the above.

ISA-S5.4, (ANSI/ISA-1976, R1981), Instrument Loop Diagrams, 1981, 11 pp.

Provides a method and practice for the preparation and use of instrument loop diagrams in the design, construction, checkout, startup, operation, maintenance, and reconstruction of instrument systems in industrial plants.

ISA-S5.5, Graphic Symbols for Process Displays, (ANSI/ISA-1985), 1986, 40 pp.

Provides a system of graphic symbols for conveying information on visual display units (VDUs) used for process monitoring and control. Intended to ensure compatibility of symbols on process VDUs with related symbols used in other disciplines. The standard applies to computers, distributed control systems, etc., and covers both color and monochromatic displays. It supplements ISA-S5.1 and ISA-S5.3.

ISA-RP7.1 Pneumatic Control Circuit Pressure Test, 1956, 6 pp.

Intended to provide a satisfactory procedure for the testing of pneumatic control circuits for leaks together with reasonable criteria for acceptance of work done and suitable aids for performance.

ISA-S7.3 (ANSI/ISA-1975, R1981), Quality Standard for Instrument Air, R1981, 6 pp.

Establishes a maximum allowable moisture content at which the instruments will function satisfactorily; a maximum entrained particle size which will avoid plugging and wear/erosion of air passages and orifices; a maximum allowable oil content which will avoid malfunction due to clogging and war of the components; an awareness of a possible source of corrosive or toxic contamination entering the air system; through the compressor suction, plant air system cross connections, or instrument air connections directly connected to processes.

ISA-S7.4, (ANSI/ISA-1981), Air Pressures for Pneumatic Controllers, Transmitters, and Transmission Systems, 1981, 4 pp.

Purpose is to establish standard operating pressure ranges for pneumatic intelligence transmission systems; and standard air supply pressures (with limit values) for oper-

ation of pneumatic controllers and pneumatic intelligence transmission systems.

ISA-RP7.7 Recommended Practice for Producing Quality Instrument Air, 1984, 16 pp.

Establishes general equipment guidelines for producing instrument air of the quality defined in ANSI/ISA-S7.3-1975 (R1981). This document enumerates equipment characteristics to include types, range, and performance of the components necessary to meet air quality requirements of ANSI/ISA-S7.3. This recommended practice lists tests of system, and components where applicable, to check performance of instrument air supply system to ANSI/ISA-S7.3 requirements.

ISA-RP12.1, Electrical Instruments in Hazardous Atmospheres, 1960, 7 pp.

Provides general guidance for the safe installation of electrical instruments using appropriate means to prevent ignition of flammable gases and vapors.

ISA-S12.4, Instrument Purging for Reduction of Hazardous Area Classification, 1970, 14 pp.

Covers a technique for reducing the hazard classification by the continuous addition of an air or inert gas within a general purpose enclosure. Refers only to hazards created by gases or vapors, and is concerned only with those system design criteria related to electrical ignition of a hazardous gas or vapor.

ISA-RP12.6 (ANSI/ISA-1977), Installation of Intrinsically Safe Instrument Systems in Class I Hazardous Locations, 10 pp.

Provides guidance for the design and installation of field installed wiring in non-hazardous locations and for the layout and wiring of panels which contain intrinsically safe wiring. Intended for use in conjuction with nationally recognized codes covering wiring practices, such as NEC Code NFPA 70 (ANSI C1) and Canadian Electrical Code, Part 1.

ISA-S12.10, Area Classification in Hazardous Dust Locations, 1973, 24 pp.

Evaluates the degree of dust hazard in locations made hazardous by the presence of a cloud or blanket of dust. Is in conformance with (and attempts to expand and clarify) the National Electrical Code and the Canadian Electrical Code.

ISA-S12.11, Electrical Instruments in Hazardous Dust Locations, 1973, 12 pp.

Summarizes the requirements for safe and economical installation of electrical instruments in locations made hazardous by a cloud or blanket of dust. (See ISA S12.10 for classification of such areas.) Is in conformance with (and attempts to expand and clarify) the National Electrical Code and the Canadian Electrical Code.

ISA-S12.12 Electrical Equipment for Use in Class I, Divi-

sion 2 Hazardous (Classified) Locations, (ANSI/ISA-1984), 1984, 24 pp.

Provides requirements for the design, construction, and marking of electrical equipment, or parts of such equipment used in Class I, Division 2 locations. This document establishes uniformity in test methods for determining the suitability of the equipment and associated circuits and components as they are related to their ability to ignite a specified flammable gas or vapor-in-air mixture. This standard applies only to equipment, circuits, or components designed and assessed specifically for use in Class I, Division 2, hazardous locations as defined by the National Electrical Code NFPA No. 70, Articles 500 and 501, or the Canadian Electrical Code (Part I), C22.1, Section 18.

ISA-S12.13, Part I, Performance Requirements, Combustible Gas Detectors, (ANSI/ISA-S12.13, Part I-1986), 1986, 24 pp.

Improves the level of electrical safety and safety-oriented performance of combustible gas detection instruments used in hazardous (classified) locations. Covers the details of construction, performance, and test for portable, mobile, and stationary electrical instruments for sensing the presence of combustible gas or vapor concentrations in ambient air; parts of these instruments may be installed or used in Class I hazardous locations and gaseous mines in accordance with codes specified by authorities having jurisdiction. This standard does *not* cover gas detection instruments of the laboratory- or scientific-type used for analysis or measurement, instruments used for process control and process monitoring purposes, or instruments used for residential purposes.

ISA-RP12.13, Part II, Installation, Operation, and Maintenance of Combustible Gas Detection Instruments, 1987, 152 pp.

Establishes user criteria for the installation, operation, and maintenance of combustible gas detection instruments. Covers storage, user record-keeping, maintenance, checkout procedures, calibration, external power supply systems, etc. Provides a substantial list of references.

ISA-RP16.1, 2, 3, Terminology, Dimensions, and Safety Practices for Indicating Variable Area Meters (Rotameters, Glass Tube, Metal Tube, Extension Type Glass Tube), 1959, 6 pp.

Combined RP16.1, 16.2 and 16.3 — intended to (a) establish uniformity of connection dimensions to permit interchangeability of one manufacturer's meters with another manufacturer's meters of the same size; (b) provide a common ground of understanding of the terminology, use, and component parts and accuracies of these meters; and (c) to provide a reference for the safe working pressures of these meters.

ISA-RP16.4, Nomenclature and Terminology for Extension Type Variable Area Meters (Rotameters), 1960, 3 pp.

Defines the nomenclature and terminology of various types of extensions applicable to 5 in. (125 mm) glass and metal tube variable area meters (rotameters) covered in ISA-RP16.1, 2, 3.

ISA-RP16.5, Installation, Operation, Maintenance Instructions for Glass Tube Variable Area Meters (Rotameters), 1961, 6 pp.

Covers the general considerations, important to the installation, operation and maintenance of meters to obtain the most reliable results.

ISA-RP16.6, Methods and Equipment for Calibration of Variable Area Meters (Rotameters), 1961, 7 pp.

Describes the methods and equipment used for calibrating the glass and metal metering tube area meters (rotameters) covered in RP16.1, 2, 3.

ISA-S18.1 (ANSI/ISA-1979, R 1985), Annunciator Sequences and Specifications, 36 pp.

Covers electrical annunciators that call attention to abnormal process conditions by the use of individual illuminated visual displays and audible devices. Sequence designations provided can be used to describe basic annunciator sequences and also many sequence variations.

ISA-S20, Specification Forms for Process Measurement and Control Instruments, Primary Elements and Control Valves, R1981, 72 pp.

These forms are intended to assist the specification writer to present the basic information. In this sense they are "short-form" specifications or "check sheets" and may not include all necessary engineering data or definitions of application requirements. While the types of instruments described by these forms are more common to the process industries the forms should also prove useful in the other areas if special requirements are defined elsewhere.

ISA-S26 (ANSI MC4.1-1975), Dynamic Response Testing of Process Control Instrumentation, 25 pp.

Incorporating four revised ISA recommended practices, the standard establishes the basis for dynamic response testing of measurement and control equipment with pneumatic output and electric output, and for closed loop actuators for externally actuated control valves and other final control elements. Pulse testing techniques as well as methods for sine wave, step, and pulse-type signals are included.

ISA-RP31.1 (ANSI/ISA RP31.1-1977), Specification, Installation, and Calibration of Turbine Flowmeters, 21 pp.

Establishes minimum ordering information, recommended acceptance and qualification test methods including calibration techniques, uniform terminology and drawing symbols, and recommended installation techniques for volumetric turbine flow transducers having an electrical output.

ISA-S37.1 (ANSI/ISA-1975, R1982), Electrical Transducer Nomenclature and Terminology, R1982, 15 pp.

Establishes uniform nomenclature for transducers and uniform simplified terminology for transducer characteristics.

ISA-RP37.2, Guide for Specifications and Tests for Piezoelectric Acceleration Transducers for Aerospace Testing, R1982, 19 pp.

Covers piezoelectric acceleration transducers, primarily those used in aerospace test instrumentation. Terminology used in this document follows ISAS37.1, "Electrical Transducer Nomenclature and Terminology," except that additional terms considered applicable to piezoelectric vibration transducers are defined.

ISA-S37.3 (ANSI/ISA-1975, R, 1982), Specifications and Tests for Strain Gage Pressure Transducers, R 1982, 22 pp.

Establishes for strain gage pressure transducers; uniform minimum specifications for design and performance characteristics; uniform acceptance and qualification test methods, including calibration techniques; uniform presentation of minimum test data; and a drawing symbol for use in electrical schematics.

ISA-S37.5 (ANSI/ISA-1975, R 1982), Specifications and Tests for Strain Gage Linear Acceleration Transducers, R1982, 18 pp.

Establishes uniform minimum specifications for design and performance characteristics, uniform acceptance and qualification test methods including calibration techniques, uniform presentation of minimum test data, and a drawing symbol for use in electrical schematics for strain gage linear acceleration transducers.

ISA-S37.6 (ANSI/ISA-1976, R1982), Specifications and Tests of Potentiometric Pressure Transducers, R1982, 27 pp.

Establishes for potentiometric pressure transducers; uniform minimum specifications for design and performance characteristics; uniform acceptance and qualification test methods, including calibration techniques; uniform presentation of minimum test data; and a drawing symbol for use in electrical schematics.

ISA-S37.8 (ANSI/ISA-1977, R1982), Specifications and Tests for Strain Gage Force Transducers, R1982, 16 pp.

Outlines uniform general specifications, acceptance and qualification methods, methods for data presentation, and includes a drawing symbol used in electrical schematics for tension, compression and combination tension/compression transducers.

ISA-S37.10 (ANSI/ISA-1975, R1982), Specifications and Tests for Piezoelectric Pressure and Sound-Pressure Transducers, R1982, 22 pp.

Establishes uniform specifications for describing design and performance characteristics, acceptance and qualification test methods and calibration techniques, and procedures for presenting test data for piezoelectric (including ferro-electric) pressure and sound-pressure transducers.

ISA-S37.12 (ANSI/ISA-1977, R1982), Specification and Test for Potentiometric Displacement Transducers, R1982, 21 pp.

Covers potentiometric displacement transducers, primarily those used in measuring systems. The specifications are not intended to cover transducers used in hazardous locations as specified in the National Electrical Code nor are all requirements covered for transducers used in nuclear power plants.

ISA-RP42.1 Nomenclature for Instrument Tube Fittings, 1982, 8 pp.

Defines nomenclature for tubing fittings most commonly used in instrumentation. It is not intended as a substitute for manufacturer's catalog numbers, nor does it apply to special fittings. It is intended to apply to a mechanical fitting rather than a sweat fitting.

ISA-S50.1 (ANSI/ISA-1975, R1982), Compatibility of Analog Signals for Electronic Industrial Process Instruments, R1982, 11 pp.

This standard applies to analog dc signals used in process control and monitoring systems to transmit information between subsystems or separated elements of systems. Its purpose is to provide for compatibility between the several subsystems or separated elements of given systems.

ISA-S51.1 (ANSI/ISA 1979), Process Instrumentation Terminology, 1979, 41 pp.

Intended to include all specialized terms used to describe the use and performance of the instrumentation and instrument systems used for measurement, control or both in the process industries.

ISA-RP52.1, Recommended Environments for Standard Laboratories, 1975, 18 pp.

Recommendations for three levels of standardization are presented — from the more general National Bureau of Standards, through commercial, industrial and government laboratories. Requirements for nine environmental factors are discussed.

ISA-RP55.1 (ANSI: MC8.1-1975), Hardware Testing of Digital Process Computers, R1983, 54 pp.

Establishes a basis for evaluating functional hardware performance of digital process computers. Covers general recommendations applicable to all hardware performance testing, specific tests for pertinent subsystems and system parameters. Includes a brief glossary of terms used.

ISA-RP60.3, Human Engineering for Control Centers, 16 pp.

Assists the design engineer in establishing concepts which accommodate physical and mental capabilities of the operator while recognizing the operator's limitations. This recommended practice is limited to those aspects of human engineering that will affect the layout of and equipment selection for the control center. It is recognized that some of the human factors discussed in this document are also used in the design and manufacture of instruments.

ISA-RP60.6, Nameplates, Labels and Tags for Control Centers, 1984, 24 pp.

Assists the designer or engineer in choosing and specifying the method of identifying items mounted on a control center or associated with a control center facility. This recommended practice summarizes identification methods and suggests the use of nameplates, labels, and tags. Examples are included for guidance in preparing drawings and specifications. This recommended practice also covers functional definitions associated with nameplates, labels, and tags.

ISA-RP60.8, Electrical Guide for Control Centers, 1978, 6 pp.

Assists the design engineer in establishing the electrical requirements of a control center; it is also intended to comply with the provisions of the NEC. Special considerations which may apply to particular devices or circuits are not taken into account in this recommended practice.

ISA-RP60.9, Piping Guide for Control Centers, 1981, 11 pp.

Assists the design engineer in defining the piping requirements for pneumatic signals and supplies in control centers.

ISA-S61.1 (ANSI/ISA-S61.1-1977), Industrial Computer System FORTRAN Procedures for Executive Functions, Process Input-Output and Bit Manipulation, 1977, 11 pp.

Presents external procedure references for use in industrial computer control systems. These external procedure references permit interface with executive programs, process input and output functions and allow manipulation of bit strings. The FORTRAN statements described in this standard conform to the ANSI:X3.9-1966 Standard FORTRAN. No changes to standard FORTRAN syntax are intended.

ISA-S61.2 (ANSI/ISA-S61.2-1978), Industrial Computer System FORTRAN Procedures for File Access and the Control of File Contention, 1978, 7 pp.

Presents external procedure references for use in industrial computer control systems. These external procedure references provide means for accessing files, and also provide means for resolving problems of file access contention in a multiprogramming multiprocessing environment.

ISA-S67.01, (ANSI/ISA-S67.01-1981, R1986), Transducer and Transmitter Installation for Nuclear Safety Applications, R1986, 14 pp.

Covers the installation of transducers for nuclear-safety-related applications, excepting those for measurands of liquid metals. It establishes requirements and recommendations for the installation of transducers and auxiliary

equipment for nuclear power plant applications outside of the main reactor vessel.

ISA-S67.02, Nuclear-Safety-Related Instrument Sensing Line Piping and Tubing Standards for Use in Nuclear Power Plants, (ANSI/ISA-1980), 1981, 14 pp.

Covers design, protection and installation of nuclear-safety-related instrument sensing lines for light water cooled nuclear power plants. The standard covers the pressure boundary requirements for piping, capillary tubing, and tubing lines up to and including one inch (25.4 mm) outside diameter or three-quarter inch nominal pipe.

ISA-S67.03, Standard for Light Water Reactor Coolant Pressure Boundary Leak Detection, (ANSI/ISA-1982), 1982.

Defines design criteria that are intended to insure that adequate Reactor Coolant Pressure Boundary leak detection capabilities are provided to the nuclear plant operator and to meet the Code of Federal Regulations.

ISA-S67.04, Setpoints for Nuclear Safety-Related Instrumentation Used in Nuclear Power Plants, 1982, 15 pp.

Develops a basis for establishing setpoints for actions determined by the design basis for protection systems and to account for instrument errors and drift in the channel from the sensor through and including the bistable trip device.

ISA-S67.06, Response Time Testing of Nuclear-Safety-Related Instrument Channels in Nuclear Power Plants, (ANSI/ISA-1984), 1984, 20 pp.

Delineates requirements and methods for determining the response time characteristics of nuclear-safety-related instrument channels. The standard applies only to those instrument channels whose primary sensors measure pressure, temperature, or neutron flux. This document provides the nuclear power industry with requirements and acceptable methods for response time testing nuclear-safety-related instrument channels.

ISA-S67.10, Sample-Line Piping and Tubing Standard for Use in Nuclear Power Plants, (ANSI/ISA-1986), 1986, 24 pp.

Covers design, protection, and installation of sample lines connecting nuclear-safety related power plant processes with sampling instrumentation. The standard applies to light-water-cooled nuclear power plants, covering the pressure boundary requirements for piping and tubing. It applies to the areas from the process tap to the upstream side of the sample panel, bulkhead fitting, or analyzer shut-off valve, and it includes in-line sample probes.

ISA-S67.14, Qualifications and Certification of Instrumentation and Control Technicians in Nuclear Power Plants, (ANSI/ISA-1983), 1983, 16 pp.

Identifies the criteria for certification of instrumentation and control technicians in nuclear power plants. These criteria address qualifications based on education, experience, training, and job performance.

ISA-S71.01, Environmental Conditions for Process Measurement and Control Systems: Temperature and Humidity, (ANSI/ISA-1985), 1985, 18 pp.

Establishes uniform classifications of the environmental conditions of temperature and humidity as they relate to industrial process measurement and control equipment. The standard is compatible with IEC Publication 654-1, 1979, *Operating Conditions for Industrial-Process Measurement and Control Equipment, Part 1: Temperature, Humidity and Barometric Pressure.*

ISA-S71.04, Environmental Conditions for Process Measurement and Control Systems: Airborne Contaminants, (ANSI/ISA-1985), 1985, 16 pp.

Classifies airborne contaminants that may affect process measurement and control instruments. This classification system provides a means of specifying the type and concentration of airborne contaminants to which a specified instrument may be exposed. This standard is limited to airborne contaminants and biological influences only, covering contamination influences that affect industrial process measurement and control systems.

ISA-S72.01, PROWAY-LAN Industrial Data Highway, (ANSI/ISA-1985), 1985, 200 pp.

Specifies those elements which are required for compatible interconnection of stations by way of a Local Area Network (LAN) using the Token Bus access method in an industrial environment. The standard is compatible with (but more restrictive than) the IEEE 802.2 and 802.4 standards for general LANS.

ISA-RP74.01, Application and Installation of Continuous-Belt Weighbridge Scales, 1984, 28 pp.

Furnishes design criteria inducive to simplified specifications and provides recommendations for installation, calibration, and maintenance of continuous-belt, weighbridge type scales. This recommended practice provides an effective base of comparison of scale suppliers, establishes minimum values, and ensures that a scale specification and purchase incorporates the essentials to satisfy a particular weighing job. It permits early belt conveyor design, with the full knowledge of the weight scale configuration, regardless of the manufacturer.

ISA-S75.01, Flow Equations for Sizing Control Valves, (ANSI/ISA-1985), 1985, 34 pp.

Establishes equations for predicting the flow of compressible and incompressible fluids through control valves. The equations are not intended for use when mixed-phase fluids, dense slurries, dry solids, or non-Newtonian liquids are encountered. The prediction of cavitation, noise, or other effects is not a part of this standard.
The equations are not, however, intended for use with mixed phases.

ISA-S75.02 (ANSI/ISA-S75.02-1982, Formerly S39.2 and S39.4), Control Valve Capacity Test Procedure, 1982, 16 pp.

Utilizes the mathematical equations outlined in ANSI/ISA-S75.01 in providing a test procedure for obtaining the valve sizing coefficient, liquid pressure recovery factor, Reynolds number factor, liquid critical pressure ratio factor, piping geometry factor, and pressure drop factors.

ISA-S75.03, Face-to-Face Dimensions for Flanged Globe-Style Control Valve Bodies (Formerly ISA-S4.01.1), (ANSI/ISA-1985), 1984, 12 pp.

Applies to flanged globe-style control valves, sizes ½ inch through 16 inches, having top, top and bottom, port, or cage guiding. This standard aids users in the piping design by providing ANSI Class 125, flat face, and ANSI Classes 150, 250, 300 and 600, raised face, flanged control valve dimensions, without giving special consideration to the equipment manufacturer to be used.

ISA-S75.04, Face-to-Face Dimensions for Flangeless Control Valves (Formerly ISA-S4.01.2), (ANSI/ISA-1985), 1984, 8 pp.

Applies to flangeless control valves, sizes ¾ inch through 16 inches for ANSI Classes 150 through 600. This standard aids users in their piping designs for flangeless control valves without giving special consideration to the equipment manufacturer to be used. This standard applies to flangeless ball control valves utilizing a full ball or a segment of a ball and other rotary-stem or sliding-stem flangeless control valves. It does not apply to weld-end valves, butterfly valves, or other rotary-stem valves that may be covered by other standards.

ISA-S75.05, Control Valve Terminology, (ANSI/ISA-1983), 1983, 33 pp.

Provides terminology for control valves of seven different types and also for common types of actuators used with these valves. This standard names individual valve parts, defines assemblies of parts, and provides terminology for part and assembly functions.

ISA-RP75.06, (Formerly ISA-RP4.2), Control Valve Manifold Designs, 1981, 20 pp.

Presents six control valve manifold types with space estimates for various sizes. Each of these six types consists of a straight through globe control valve, isolation upstream and downstream block valves and bypass piping with a manually activated valve.

ISA-S75.07, Laboratory Measurement of Aerodynamic Noise Generated by Control Valves, 1987, 16 pp.

Defines equipment, methods, and procedures for the laboratory testing and measurement of airborne sound radiated by a compressible fluid flowing through a control valve and its associated piping, including fixed-flow restrictions. The test may be conducted under any conditions agreed upon by the user and the manufacturer. Although this standard is designed for measurement of noise radiated from the piping downstream of the valve, other test variations are optional, including the use of insulation and nonstandard piping. Applications of this standard to control valves discharging directly to atmosphere are excluded.

ISA-S75.08, Installed Face-to-Face Dimensions for Flanged Clamp or Pinch Valves, (ANSI/ISA-S75.08-1985), 1985, 12 pp.

Applies to clamp or pinch valves sizes 1 inch through 8 inches. The purpose of this standard is to aid users in their piping design by providing installed face-to-face dimensions for control valves, incorporating clamp or pinch elements, which have flanges that mate with ANSI B16.1 Class 125 (PN20) and/or ANSI B16.5 Class 125 (PN20) flanges, without giving special consideration to the manufacturer of the equipment to be used.

ISA-S75.11, Inherent Flow Characteristic and Rangeability of Control Valves, (ANSI/ISA-1985), 1984, 16 pp.

Defines the statement of typical control valve inherent flow characteristics and inherent rangeabilities and establishes criteria for adherence to manufacturer-specified flow characteristics. This standard uses the basic definitions from ISA-S75.05 and also defines specific terms related to flow characteristic and rangeability. A table listing inherent flow characteristic deviations and sample plots of relative flow coefficient versus relative travel are also given.

ISA-S75.12, Face-to-Face Dimensions for Socket Weld-End and Screwed-End Globe-Style Control Valves, (ANSI classes 150, 300, 600, 900, 1500, and 2500), (ANSI/ISA-S75.12-1987), 1986, 12 pp.

Applies to socket weld-end globe-style control valves, sizes ½ inch through 4 inches, and screwed-end globe-style control valves, size ½ inch through 2½ inches, having top, top and bottom, port, or cage guiding. This standard aids users in their piping designs by providing ANSI Classes 150 through 2500 socket weld-end control valve dimensions and ANSI Classes 150 through 600 screwed-end control valve dimensions, without giving special considerations to the equipment manufacturer to be used.

ISA-S75.14, Face-to-Face Dimensions for Buttweld-End Globe-Style Control Valves, (ANSI/ISA-1984), 1984, 8 pp.

Applies to buttweld-end globe-style control valves, sizes ½ inch through 8 inches, having top and cage guiding. This standard aids users in their piping designs by providing ANSI Class 4500 buttweld-end control valve dimensions, without giving special consideration to the equipment manufacturer to be used.

ISA-S75.15, Face-to-Face Dimensions for Buttweld-End Globe-Style Control Valves (ANSI Classes 150, 300, 600, 900, 1500, and 2500), (ANSI/ISA-S75.15-1987), 1986, 12 pp.

Applies to buttweld-end globe-style control valves, sizes ½ inch through 18 inches, for ANSI Classes 150 through 2500, having top, top and bottom, port, or cage guiding.

This standard aids users in their piping designs by providing buttweld-end control valve dimensions, without giving special consideration to the equipment manufacturer to be used.

ISA-S75.16, Face-to-Face Dimensions for Flanged Globe-Style Control Valve Bodies (ANSI Classes 900, 1500, and 2500), (ANSI/ISA-S75.16-1987), 1986, 12 pp.

Applies to flanged globe-style control valves, sizes ½ inch through 18 inches, having top, top and bottom, port, or cage guiding. This standard aids users in their piping designs by providing ANSI Classes 900, 1500, and 2500 raised-face, flanged control valve dimensions, without giving special consideration to the equipment manufacturer to be used.

ISA-S77.42, Fossil-Fuel Plant Feedwater Control System— Drum-Type, 1987, 32 pp.

Establishes minimum criteria for the control of levels, pressures, and flow for the safe and reliable operation of drum-type feedwater systems in fossil power plants. Aids in the development of design specifications covering the measurement and control of feedwater systems. The following requirements are defined for minimum system design: (1) process measurement requirements; (2) control and logic requirements; (3) final control device requirements; (4) system reliability and availability; (5) alarm requirements; and (6) operator interface. The safe physical containment of the feedwater shall be in accordance with applicable piping codes and standards and is beyond the scope of this standard.

The following standard, identified by its ANSI number, was sponsored and published by ISA, approved by ANSI, and is available from either ISA or ANSI.

MC96.1, American National Standard for Temperature Measurement Thermocouples, 1982, 48 pp.

Covers coding of thermocouple and extension wire; coding of insulated duplex thermocouple extension wires; terminology, limits of error and wire sizes for thermocouples and thermocouple extension wires; temperature EMF tables for thermocouples; plus appendices that cover fabrication, checking procedures, selection, and installation.

ANSI C100.6-3, American National Standard for Voltage or Current Reference Devices; Solid State Devices, 1984, 12 pp.

Applies to physical devices used to maintain the unit of dc voltage or current having uncertainties of 100 ppm of output or less. This standard treats these devices from the standpoint of performance characteristics, but does not specify design or construction details or techniques. This part of the standard, C100.6-3, applies to solid state devices used to maintain the unit of dc voltage or current having uncertainties of 100 ppm of output or less.

INSULATED CABLE ENGINEERS ASSOCIATION, INC. (ICEA)

The following publications are available from the Insulated Cable Engineers Association, Inc. (Prior to March 7, 1979, the Insulated Power Cable Engineers Association), P.O. Box 9, South Yarmouth, MA 02664. Standards developed by NEMA-ICEA are available from NEMA, and standards developed by IEEE-ICEA are available from IEEE.

IPCEA: T-22-294, (R 1983), Test Procedures for Extended Time-Testing of Wire and Cable Insulation for Service in Wet Locations, 1983, 3 pp.

Describes procedures for extended time testing of extruded wire and cable insulations for service in wet locations.

IPCEA: T-24-380, Guide for Partial-Discharge Test Procedures, 1980, 6 pp.

Applies to the detection and measurement of partial discharges occurring in the insulation of single-conductor shielded cables and assemblies thereof and multiple-conductor cables with individually shielded conductors.

IPCEA: T-25-425, Guide for Establishing Stability of Volume Resistivity for Conducting Polymeric Components of Power Cables, 1981, 4 pp.

Applies to testing of extruded conducting polymeric components of power cables with extruded insulation. It describes a method of demonstrating the stability over a period of time of the volume resistivity (calculated from longitudinal resistance) of these components at temperatures up to the emergency operating temperature of the cable.

IPCEA: T-28-562, (R 1983), Test Method for Measurement of Hot Creep of Polymeric Insulations, 1983, 3 pp.

Provides procedure for determining the relative degree of crosslinking of polymeric electrical cable insulations.

INTERNATIONAL ASSOCIATION OF PLUMBING AND MECHANICAL OFFICIALS (IAPMO)

The following documents are available from the International Association of Plumbing and Mechanical Officials, 5032 Alhambra Avenue, Los Angeles, CA 90032.

IAPMO: PS 8-77, 1977, 2 pp.

Implements Section 209 of the Uniform Plumbing Code published by the International Association of Plumbing and Mechanical Officials (IAPMO) for backwater valves.

IAPMO: PS 10-84, Globe-Type Loglighter Valves Angle or Straight Pattern, 1984, 4 pp.

Establishes a generally acceptable standard for globe-type loglighter valves, angle or straight pattern.

IAPMO: PS 15-77, Pressure Reducing and Regulating Valves for Installation on Domestic Water Supply Lines, 1977, 3 pp.

Serves to supplement the provisions of the Uniform Plumbing Code, Section 1007 for pressure reducing and regulating valves as required on domestic water supply lines; to prescribe minimum standards for materials in the construction of such valves and providing for test standards.

IAPMO: PS 31-77, Backflow Prevention Devices, 1977, 20 pp.

Covers material requirements, dimensions, and design and other specific properties, in addition to general description of materials.

Uniform Solar Energy Code, 1985, 74 pp.

Provisions apply to the erection, installation, alteration, addition, repair, relocation, replacement, maintenance or use of any solar system except as otherwise provided for in this Code.

INTERNATIONAL CONFERENCE OF BUILDING OFFICIALS (ICBO)

The following handbook is available from the International Conference of Building Officials, 5360 South Workman Mill Road, Whittier, CA 90601.

Uniform Building Code Standards, 1985, 1340 pp.

A collection of building code standards consisting of specifications developed by ANSI, ASTM, UL and other organizations. Some materials testing standards are included.

MANUFACTURERS STANDARDIZATION SOCIETY OF THE VALVE AND FITTINGS INDUSTRY (MSS)

The following MSS Standards publications can be obtained from the Manufacturers Standardization Society of the Valve and Fitting Industry, 127 Park St., NE, Vienna, VA 22180. No further information was available.

MMS: SP-6-1985, Standard Finishes for Contact Faces of Pipe Flanges and Connecting-End Flanges of Valves and Fittings.

MMS: SP-9-1984, Spot Facing for Bronze, Iron and Steel Flanges.

MMS: SP-25-1978 (R 1983), Standard Marking System for Valves, Fittings, Flanges and Unions.

MMS: SP-42-1985, Class 150 Corrosion Resistant Gate, Globe, Angle and Check Valves with Flanged and Butt-Weld Ends.

MMS: SP-43-1982 (R 1986), Wrought Stainless Steel Butt-Welding Fittings.

MMS: SP-44-1985, Steel Pipeline Flanges.

MMS: SP-45-1982, By-pass and Drain Connection Standard.

MMS: SP-51-1986, Class 150LW Corrosion Resistant Cast Flanges and Flanged Fittings.

MMS: SP-53-1985, Quality Standard for Steel Castings and Forgings for Valves, Flanges, and Fittings and Other Piping Components — Magnetic Particle Examination Method.

MMS: SP-54-1985, Quality Standard for Steel Castings for Valves, Flanges and Fittings and Other Piping Components — Radiographic Examination Method.

MMS: SP-55-1985, Quality Standard for Steel Castings for Valves, Flanges and Fittings and Other Piping Components — Visual Method.

ANSI/MMS: SP-58-1983, Pipe Hangers and Supports — Materials, Design and Manufacture.

ANSI/MMS: SP-60-1982 (R 1986), Connecting Flange Joint Between Tapping Sleeves and Tapping Valves.

ANSI/MMS: SP-61-1985, Pressure Testing of Steel Valves.

ANSI/MMS: SP-65-1983, High Pressure Chemical Industry Flanges and Threaded Stubs for Use with Lens Gaskets.

ANSI/MMS: SP-67-1983, Butterfly Valves.

ANSI/MMS: SP-68-1984, High Pressure-Offset Seat Butterfly Valves.

ANSI/MMS: SP-69-1983, Pipe Hangers and Supports — Selection and Application.

ANSI/MMS: SP-70-1984, Cast Iron Gate Valves, Flanged and Threaded Ends.

ANSI/MMS: SP-71-1984, Cast Iron Swing Check Valves, Flanged and Threaded Ends.

ANSI/MMS: SP-72-1970, Ball Valves with Flanged or Butt-Welding Ends for General Service.

ANSI/MMS: SP-73-1986, Brazing Joints for Wrought and Cast Copper Alloy Solder Joint Pressure Fittings.

ANSI/MMS: SP-75-1983, Specification for High Test Wrought Butt-Welding Fittings.

ANSI/MMS: SP-77-1984, Guidelines for Pipe Support Contractual Relationships.

ANSI/MMS: SP-78-1977, Cast Iron Plug Valves, Flanged and Threaded Ends.

ANSI/MMS: SP-79-1980, Socket-Welding Reducer Inserts.

ANSI/MMS: SP-80-1979, Bronze Gate, Globe, Angle and Check Valves.

ANSI/MMS: SP-81-1981 (R 1986), Stainless Steel, Bonnetless, Flanged, Knife Gate Valves.

ANSI/MMS: SP-82-1976 (R 1981, 1986), Valve Pressure Testing Methods.

ANSI/MMS: SP-83-1976, Carbon Steel Pipe Unions, Socket-Welding and Threaded.

ANSI/MMS: SP-84-1985, Steel Valves — Socket Welding and Threaded Ends.

ANSI/MMS: SP-85-1985, Cast Iron Globe & Angle Valves, Flanged and Threaded Ends.

ANSI/MMS: SP-86-1981, Guidelines for Metric Data in Standards for Valves, Flanges, and Fittings.

ANSI/MMS: SP-87-1982 (R 1986), Factory-Made Butt-Welding Fittings for Class 1 Nuclear Piping Applications.

ANSI/MMS: SP-88-1983, Diaphragm Type Valves.

ANSI/MMS: SP-89-1985, Pipe Hangers and Supports — Fabrication and Installation Practices.

ANSI/MMS: SP-90-1986, Guidelines on Terminology for Pipe Hangers and Supports.

ANSI/MMS: SP-91-1984, Guidelines for Manual Operation of Valves.

ANSI/MMS: SP-92-1980, MSS Valve User Guide.

ANSI/MMS: SP-93-1982, Quality Standards for Steel Castings and Forgings for Valves, Flanges, and Fittings and Other Piping Components — Liquid Penetrant Examination Method.

ANSI/MMS: SP-94-1982, Quality Standards for Ferritic and Martensitic Steel Castings and Forgings for Valves, Flanges, and Fittings and Other Piping Components — Ultrasonic Examination Method.

ANSI/MMS: SP-95-1986, Swage(d) Nipples and Bull Plugs.

ANSI/MMS: SP-96-1986, Guidelines on Terminology for Valves and Fittings.

METAL POWDER INDUSTRIES FEDERATION (MPIF)

The following publication is available from the Metal Powder Industries Federation, 105 College Rd. East, Princeton, NJ 08540.

Standard Test Methods for Metal Powders and Powder Metallurgy Products, 1985-86 edition, 95 pp.

Bound edition of MPIF standards covering five categories: P/M Nomenclature, Powder Testing Standards, P/M Material Standards, Material Testing Standards, and Safety Standards.

Materials Standards for P/M Structural Parts, 1978-1988 Edition, 20 pp.

Materials Standards for P/M Self-Lubricating Bearings, 1986-1987 Edition, 14 pp.

NATIONAL ASSOCIATION OF PIPE COATING APPLICATORS (NAPCA)

The following publications are available from NAPCA, 717 Commercial National Bank Building, Shreveport, LA 71101.

1-65-83, Recommended Specification Designations for Enamel Coatings (Rev. April 1983).

2-66-83, Standard Applied Pipe Coating Weights for NAPCA Coating Specifications (Rev. April 1983).

3-67-83, External Application Procedures for Hot Applied Coal Tar and Asphalt Enamel Coatings to Steel Pipe (Rev. April 1983).

5-69-83, NAPCA Specifications Pipeline Felts (Rev. April 1983).

6-69-83-1, Suggested Procedures to Hand Wrap Field Joints Using Hot Enamel (Rev. April 1983).

6-69-83-2, Suggested Procedures for Coating of Girth Welds With Fusion Bonded Epoxy (Rev. April 1983).

6-69-83-3, Suggested Procedures for Coating Field Joints, Fittings, Connections and Pre-Fabricated Sections Using Tape Coatings (Rev. April 1983).

6-69-83-4, Suggested Procedures for Field Joint Application Using Mastic Mix and Field Mold (Rev. April 1983).

6-69-83-5, Suggested Application Procedures for Coating Field Joints Using Heat Shrinkable Materials (Rev. April 1983).

12-78-83, Application Specifications Mill Applied Fusion Bonded Epoxy Coatings (Rev. April 1983).

13-79-83, Application Specifications for Coal Tar Epoxy Protective Coatings (Rev. April 1983).

14-83, Application Specifications for Polyolefin Pipe Coating Applied by the Cross Head Extrusion Method or the Side Extrusion Method (Rev. April 1983).

15-83, Plant Applied Tape Coating Application Specification for the Exterior of Steel Pipe (Rev. April 1983).

Pocket Edition of National Association of Pipe Coating Applicators Specifications and Plant Coating Guide (1-65-83 through 15-83, Rev. April 1983).

NATIONAL ASSOCIATION OF RELAY MANUFACTURERS (NARM)

The following publications are available from the National Association of Relay Manufacturers, P.O. Box 1505, Elkhart, IN 46515.

Engineer's Relay Handbook, 3rd Edition, 1980.

Purpose is to bring together information that simplifies and clarifies the specifying and obtaining of correct relays. Information is directed at individuals responsible for specifying the correct type of relays for a given application; it isn't intended for designers and manufacturers or relays.

Definitions of Relay Terms, 1980.

A glossary of words and terms in common use by relay users and manufacturers, includes symbols.

NATIONAL BOARD OF BOILER AND PRESSURE VESSEL INSPECTORS (NBBI)

The following publications are available from NBBI, 1055 Crupper Avenue, Columbus, OH 43229.

NB-23, National Board Inspection Code, Revision 5, 1985, 227 pp.

This manual for inspectors, owners and users of boilers and pressure vessels presents rules and guidelines for inspection after installation, repairs, alterations and re-rating; thereby, helping to ensure that these objects may continue to be safely used.

NB-18, Pressure Relief Device Certifications, Revision 3, 1987, 328 pp., plus addenda.

Lists safety and safety relief valves by manufacturer and model number and provides the relieving capacities of these valves as determined by test and certified by the National Board. Also lists names of companies which have been authorized by the National Board to repair safety and safety relief valves.

NB-169, Making It With Metric, 1st Edition, 1985, 89 pp.

Provides a basis for metric communication in a coordinated and orderly fashion.

NATIONAL BUREAU OF STANDARDS (NBS)

The Standards Code and Information (SCI) Program of the NBS Office of Product Standards Policy maintains the National Center for Standards and Certification Information (NCSCI), the central repository of standards-related information in the United States. NCSCI provides access to more than 240,000 titles of standards and related documents published by U.S. professional and technical organizations, U.S. Federal and State Government groups, and foreign national and international organizations. For further information concerning the Standards Code and Information Program contact SCI Program, A629 Administration, National Bureau of Standards, Gaithersburg, MD 20899.

NBS: SP-681, Standards Activities of Organizations in the United States.

Lists over 750 U.S. organizations and describes their standards-related activities. The standards activities covered include those of the Federal government, the private sector, and state procurement offices. The directory also lists sources of standards information and documents.

NBS: SP-649, Directory of International and Regional Organizations Conducting Standards-Related Activities.

Lists 272 international and regional organizations involved in standardization, certification, laboratory accreditation or other standards-related activities. Describes the scope of each organization and their activities related to standards. Provides information concerning U.S. participants and availability of any standards documents in English.

KWIC Index (Computer Output Microform (COM) produced).

The KWIC Index contains the titles of more than 25,000 U.S. voluntary product and engineering standards. A standard can be located by means of any significant or key word in the title. Key words are arranged alphabetically.

NATIONAL CABLE TELEVISION ASSOCIATION (NCTA)

The following documents are available from NCTA, 1724 Massachusetts Avenue, NW, Washington, DC 20036.

NCTA: Standard Graphic Symbols for Cable Television Systems, 16 pp.

This standard provides a list of graphic symbols for the designation of electrical, electronic, and pole line devices for layout drawings of cable television systems.

NCTA: Recommended Practices for Measurements on Cable Television Systems, 1983, 120 pp. (supplements issued 1985).

Provides informative, readily updated descriptions of good engineering practices required for making test measurements to the head end and distribution system; current satellite transmission practices, NTC no. 7.

NATIONAL COUNCIL OF RADIATION PROTECTION AND MEASUREMENTS (NCRP)

The following, and other publications on radiation monitoring and protection, are available from the National Council of Radiation Protection and Measurements, 7910 Woodmont Avenue, Suite 1016, Bethesda, MD 20814.

NCRP: Report No. 23, Measurement of Neutron Flux and Spectra for Physical and Biological Applications, 1960, 92 pp.

NCRP Report No. 23 presents a discussion of neutron flux and spectra, compares various methods of neutron source calibration and presents the results of the inter-comparisons. There is a discussion of methods of measurement of the emission rate of neutron sources, thermal neutron flux, intermediate neutron flux, fast neutron flux, and neutron spectra. Neutron radiation instruments for area survey and personnel monitoring involving flux and spectrum measurements are discussed also. Typical spectra of various neutron sources are shown. (Published in 1960 as National Bureau of Standards Handbook 72).

NCRP: Report No. 25, Measurement of Absorbed Dose of Neutrons and of Mixtures of Neutrons and Gamma Rays, 1961, 86 pp.

NCRP Report No. 25 represents a summary of methods for determining energy absorption in matter as a result of its interaction with neutrons. Since neutrons are almost invariably accompanied by gamma radiation, mixtures of gamma radiation and neutrons are discussed. Discussions are general wherever possible; however, most of the detailed examples have been drawn from the fields of health physics and radiobiology.

NCRP: Report No. 33, Medical X-Ray and Gamma-Ray Protection for Energies up to 10 MeV — Equipment Design and Use, 1968, 66 pp.

NCRP Report No. 33 is concerned with radiation protection in connection with the medical use of x and gamma rays of energies up to 10 MeV. This report presents recommendations pertaining to equipment design, use and operating conditions, and to radiation protection surveys and personnel monitoring. It includes sections for the guidance of the physician and his associates; the equipment designer and manufacturer; and the radiological physicist concerned with calibration procedures, equipment inspection and survey measurements.

NCRP: Report No. 47, Tritium Measurement Techniques, 1976, 97 pp.

Describes and discusses methods for the measurement of tritium in a variety of media. Included are most of the important methods for the measurement of tritium and information on their advantages and disadvantages. Step-by-step procedures and detailed descriptions of equipment are not included.

NCRP: Report No. 50, Environmental Radiation Measurements, 1976, 246 pp.

Presents information on the properties of widely-distributed radionuclides and of typical radiation fields in the environment and on methods for their measurement. Emphasis is placed on the role of measurements in the realistic assessment of dose to man. Techniques applicable to routine monitoring programs during normal operation of nuclear facilities are described.

NCRP: Report No. 51, Radiation Protection Design Guidelines for 0.1-100 MeV Particle Accelerator Facilities, 1977, 159 pp.

Provides design guidelines for radiation protection in particle-accelerator facilities and describes one or more methods by which this protection may be achieved.

NCRP: Report No. 57, Instrumentation and Monitoring Methods for Radiation Protection, 1978, 177 pp.

Describes techniques, instruments, and practices applicable to all types of institutions concerned with radiation or radioactive materials. The first section presents information of a general character related to radiation surveys and instrumentation. Subsequent sections contain discussions of specific installations and types of measurement.

NCRP: Rpt. No. 58, A Handbook of Radioactivity Measurements Procedures, 2nd edition, 1985.

Updates progress made in the field of radionuclide metrology. Includes material dealing with liquid scintillation counting and the latest data from the Oak Ridge data banks.

NCRP: Rpt. No. 68, Radiation Protection in Pediatric Radiology, 1981.

Offers practical information on how to conduct the radiological examinations of children to reduce the radiation dose to the children and those responsible for their care.

NCRP: Rpt. No. 69, Dosimetry of X-Ray and Gamma-Ray Beams for Radiation Therapy in the Energy Range 10 keV to 50 MeV, 1982.

Describes a dosimetric process that will allow the delivery of a prescribed absorbed dose from x-ray and gamma-ray sources to a uniform phantom within the accuracy needed for radiation therapy.

NCRP: Rpt. No. 72, Radiation Protection and Measurements for Low Voltage Neutron Generators, 1983.

Provides information on the radiation protection problems in the use of generators that operate at voltages below a few hundred kilovolts and produce neutrons chiefly by the T(d,n) reaction.

NCRP: Rpt. No. 73, Protection in Nuclear Medicine and Ultrasound Diagnostic Procedures in Children, 1983.

Provides information on the manner of conducting nuclear medicine studies in children to reduce the dose to these patients and those responsible for their care. Also addresses the application of ultrasound to children and discusses the factors that need to be considered to insure continued safe use in clinical practice.

NCRP: Rpt. No. 79, Neutron Contamination From Medical Electron Accelerators, 1985.

Reviews the source of neutrons generated from medical electron accelerators and provides an examination of the transport of the neutrons in the protective housing of the accelerator, as well as in structural shielding barriers when equipment is operated at energies above 10 MeV.

NATIONAL ELECTRICAL MANUFACTURERS ASSOCIATION (NEMA)

The following NEMA Standards publications are available from the National Electrical Manufacturers Association, 2101 L St. NW, Washington, DC 20037.

NEMA: DC 2-1982, Quick-Connect Terminals.

Specifies dimensional construction of flat quick-connect

terminals and performance characteristics of female connectors.

NEMA: DC 4-1986, Warm Air Limit and Fan Controls.

Describes the characteristics of temperature actuated electric devices intended to control the temperature of air through air-heating equipment intended primarily for residential use.

NEMA: DC 10-1983, Temperature Limit Controls for Electric Baseboard Heaters.

Defines basic construction standards and performance characteristics of temperature limit controls and control systems for use with electric baseboard heaters.

NEMA: DC 12-1985, Hot-Water Immersion Controls.

Defines basic construction standards and performance characteristics of electric-switch-type hot-water immersion controls intended primarily for use with hot-water boilers and heaters used in residential heating.

NEMA: DC 13-1979 (R 1985), Line-Voltage Integrally-Mounted Thermostats for Electric Heaters.

Defines basic construction standards and performance characteristics of integrally-mounted thermostats.

NEMA: DC 22-1977 (R 1982), Load Control for Use on Central Electric Heating Systems.

Defines classifications, ratings, and other characteristics of load controls for use on central electric heating systems intended primarily for residential use.

NEMA: CC 1-1984, Electrical Power Connectors for Substations.

Defines manufacturing, rating, and testing standards.

ANSI/NEMA: CC 3-1973 (R 1978, 1983), Connectors for Use Between Aluminum or Aluminum-Copper Overhead Conductors.

Defines performance requirements and testing procedures.

NEMA: CC 4-1986, 8.3 kV and 8.3/14.4 kV Probe for Separable Insulated Loadbreak Connectors.

Establishes dimensions, design tests, test conditions, and interchangeable construction features for probes of loadbreak separable insulated connectors rated 8.3 kV and 8.3/14.4 kV AC, 200 Amperes.

ANSI/NEMA: ICS 1-1983, General Standards for Industrial Control and Systems.

Provides practical general information concerning ratings, construction, testing, performance, and manufacture of industrial control and systems equipment, terminal blocks, and resistance welding controls. This publication is recommended for use in conjunction with other NEMA ICS publications.

NEMA: ICS 1.1-1984, Safety Guidelines for the Application Installation, Maintenance of Solid State Control.

NEMA: ICS 1.3-1986, Preventive Maintenance of Industrial and Systems Equipment.

Covers fundamental principles, safety precautions and common guidelines for preventive maintenance of equipment within the scope of the NEMA Industrial Control and Systems Section. (Also published as Part 1-115 of ICS 1.)

NEMA: ICS 2-1983, Industrial Control Devices, Controllers and Assemblies.

Provides practical information concerning ratings, construction, testing, performance, and manufacture of industrial control devices and equipment.

NEMA: ICS 2.2-1983, Maintenance of Motor Controllers After a Fault Condition.

Covers the procedures to be followed in order to return to service a motor controller which has been subjected to a short circuit or ground fault.

NEMA: ICS 2.3-1983, Instructions for the Handling, Installation, Operation, and Maintenance of Motor Control Centers.

Guide to practical information containing instructions for the handling, installation, operation, and maintenance of motor control centers rated 600 volts or less.

ANSI/NEMA: ICS 3-1983, Industrial Systems.

Provides practical information concerning ratings, construction, testing, performance, and manufacture of industrial systems equipment.

NEMA: ICS 3.1-1983, Safety Standards for Construction and Guide for Selection, Installation and Operation of Adjustable-Speed Drive Systems.

These standards apply to all industrial equipment electrical components and wiring which are part of the electrical drive system, commencing at the point of input power.

ANSI/NEMA: ICS 4-1983, Terminal Blocks for Industrial Use.

Covers terminal blocks with screw type or screwless clamping units intended for industrial use, to provide electrical and mechanical connection for conductors having a cross section of 24 AWG to 2000 MCM.

NEMA: ICS 5-1983, Resistance Welding Control.

Defines construction standards, performance requirements, and safety standards for resistance welding control equipment.

ANSI/NEMA: ICS 6-1983, Enclosures for Industrial Controls and Systems.

This publication is recommended for use in conjunction with other NEMA ICS publications. It should also be used in conjunction with NEMA Standards Publication No. 250. (Unless otherwise specified, ICS 6 is shipped with a copy of NEMA Standards Publication No. 250.)

ACR/NEMA 300-1985, Digital Imaging and Communications Standard.

Specifies a standard method for communicating between digital diagnostic imaging devices and associated equipment.

NEMA: PB 1-1984, Panelboards.

Covers single panelboards or groups of panel units suitable for assembly in the form of single panelboards, including buses, and with or without switches or automatic overload protective devices (fuses or circuit breakers), or both. These units are used in the distribution of electricity for light, heat, and power at: 600 volts and less, 1600-ampere mains and less, and 1200-ampere branch circuits and less.

NEMA: SM 23-1985, Steam Turbines for Mechanical Drive Service.

Defines construction standards and testing and performance requirements for single and multistage turbines intended to drive equipment such as pumps, fans, compressors, and generators.

NEMA: WC 55-1986, Instrumentation Cables and Thermocouple Wire (ICEA S-82-552).

Applies to materials, constructions and testing of multi-conductor instrumentation cables including thermocouple extension cables.

NATIONAL ENVIRONMENTAL BALANCING BUREAU (NEBB)

The following manuals are available from NEBB, 8224 Old Courthouse Road, Vienna, VA 22180.

NEBB: Procedural Standards for Testing, Adjusting, and Balancing of Environmental Systems, 4th edition, 1983, 136 pp.

Contains a comprehensive reference on instruments and measurement accuracies, including a description of each TAB instrument required by NEBB for certification, its recommended uses, limitations, precision of readings and calibration required, U.S. unit and metric unit equations, and HUAC system air and hydronic design tables.

NEBB: Procedural Standards for Measuring Sound and Vibration, 1st edition, 1977, 96 pp.

Covers NEBB specifications, instruments and accuracy, conditions required for sound tests, sound measurement procedures, vibration isolation devices and systems, field inspection and measurement of vibration, sound and vibration table, charts and equations, reduced copies of NEBB certified report forms, and other topics.

NEBB: Environmental Systems Technology, 1st Edition, 1984.

Incorporating HVAC system history and fundamentals, engineering principles, system design, equipment, components and installation testing and balancing, controls, acoustics, and an extensive glossary and set of engineering tables.

NEBB: Testing, Adjusting, Balancing Manual for Technicians, 1st Edition, 1986.

A basic educational text on testing, adjusting and balancing work, as well as a comprehensive reference manual.

NATIONAL FIRE PROTECTION ASSOCIATION (NFPA)

The following NFPA standards and codes are not instrumentation standards as such; however, their content is important to anyone planning measurement and automatic control systems. They are special applications of the long-accepted guides for protective systems. Attention is drawn to them, but abstracts are not included. Copies of them may be obtained from the National Fire Protection Association, Batterymarch Park, Quincy, MA 02269.

NFPA: 70-84, National Electrical Code, 1984.

Contains most widely adopted set of electrical safety requirements for electricians, inspectors, contractors, electrical manufacturers, architects, builders, and consulting engineers.

NFPA: 70A, Electrical Code for One- and Two-Family Dwellings, 1984.

All you need to know to meet Code regulations when installing electrical services in dwellings. Excerpted and edited from the 1984 National Electrical Code.

NFPA: 70B, Electrical Equipment Maintenance, 1983.

A recommended practice confined to preventive maintenance for industrial-type electrical systems and equipment to reduce hazards that result from their failure.

NFPA: 70E, Electrical Safety Requirements for Employee Work Places, 1983.

Standard includes installation safety requirements necessary to provide practical and safe working areas for employees.

NFPA: 71, Central Station Signaling Systems, 1977.

Provides a standard for a system, or group of systems, maintained and supervised from an approved central station controlled and operated by a person, firm, or corporation whose principal business is furnishing and maintaining a supervised signaling service.

NFPA: 72A, Local Protective Signaling Systems, 1979.

Describes fire alarm or supervisory signals within the protected premises primarily for the protection of life and secondarily for the protection of property.

NFPA: 72B, Auxiliary Protective Signaling Systems, 1979.

A standard on protection of an individual occupancy or building or group of buildings of a single occupancy where the municipal fire alarm facilities are utilized to transmit an alarm to the fire department.

NFPA: 72C, Remote Station Protective Signaling Systems, 1982.

Provides a standard for employing a direct circuit connection between signaling devices at protected premises and signal receiving equipment in a remote station.

NFPA: 72D, Proprietary Protective Signaling Systems, 1981.

Provides a standard for systems having their operation under the control or domination of the owner or others interested in the property to be protected.

NFPA: 72E, Automatic Fire Detectors, 1984.

Covers minimum performance, location, mounting, testing, and maintenance requirements of automatic fire detectors.

NFPA: 72H, Guide for Testing Procedures for Protective Signaling Systems, 1984.

Procedures for acceptance and periodic testing of installed protective signaling systems.

NFPA: 75, Protection of Electronic Computer/Data Processing Equipment, 1981.

State requirements for installations needing fire protection or special building construction, rooms, areas or operating environment.

NFPA: 77, Recommended Practice on Static Electricity, 1977.

Assists in reducing the fire hazard of static by discussing its nature and origin, methods of mitigation, and dissipation.

NFPA: 493, Intrinsically Safe Apparatus in Division 1 Hazardous Locations, 1978.

Provides standard for the construction and evaluation of equipment of limited energy for use in hazardous locations.

NFPA: 496, Purged and Pressurized Enclosures for Electrical Equipment, 1982.

Provides standard for the design of equipment to eliminate or reduce a hazardous atmosphere.

NFPA: 497, Classification of Class 1 Hazardous Locations for Electrical Installations in Chemical Plants, 1975.

Recommended practice for classifying the zones of potential hazards from flammable atmospheres in chemical plants.

NFPA: 803, Standard for Fire Protection for Nuclear Power Plants, 1978.

Covers the protection of nuclear electric generating facilities from the consequences of fire, including safety to life, protection of property, and continuity of production.

NATIONAL FLUID POWER ASSOCIATION (NFLDP)

The following are available from the National Fluid Power Association, 3333 N. Mayfair Rd., Milwaukee, WI 53222.

National Fluid Power Standards, 10 Volume Edition, 1987, over 2500 pp.

This bound, 10-volume set is the most complete compilation of NFPA and ANSI Standards relating to fluid power. Contains over 130 fluid power standards. The 10-volume set includes the following: A - Communications Standards; B - Pressure Rating Standards; C - Pump, Motor, Power Unit and Reservoir Standards; D - Filtration and Contamination Standards; E - Conductor and Associated Component Standards; F - Control Products Standards; G - Cylinder and Accumulator Standards; H - Fluid, Lubricant and Sealing Device Standards; I - Testing Standards; and J - Bibliographies Standards. Volumes may be purchased individually.

Volume A — Communications

ANSI/B93.2-1986, American National Standard Fluid Power Systems and Products — Glossary.

NFPA/T2.10.1M-1978, Metric Units for Fluid Power Applications.

NFPA/T2.10.2M-1977, Survey of Metric Language Usage by the U.S. Fluid Power Industry.

NFPA/T3.9.13-1982, Hydraulic Fluid Power — Pumps and Motors — Glossary.

ISO 1000-1981, SI Units and Recommendations for the Use of Their Multiples and of Certain Other Units.

ISO 1219-1976, Fluid Power Systems and Components — Graphic Symbols.

ISO 2944-1974, Fluid Power Systems and Components — Nominal Pressures.

Volume B — Pressure Rating

ANSI/B93.2-1986, Excerpts Extracted From American National Standard Fluid Power Systems and Products — Glossary.

NFPA/T1.21.1-1978, Procedures for Self-Certification by Fluid Power Manufacturers.

NPFA/T2.6.12-1974 (R1982), Method for Verifying the Fatigue and Static Pressure Ratings of the Pressure Containing Envelope of a Metal Fluid Power Component.

NFPA/T3.4.7M-1975 (R1980), Method for Establishing and Verifying the Fatigue and Static Pressure Ratings and Conducting Production Testing of the Pressure Containing Envelope of a Metal Fluid Power Accumulator.

NFPA/T2.6.1M S9 (T3.5.26M)-1977 (R1982), Hydraulic Valve Pressure Rating Supplement No. 9 to NFPA Recommended Standard for Verifying the Fatigue and Static Pressure Ratings of the Pressure Containing Envelope of a Metal Fluid Power Component.

ANSI/B93.10M-1969 (R1982), American National Standard Static Pressure Rating Methods of Square Head Fluid Power Cylinders.

NFPA/T3.6.29M-1976 (R1981), Method for Establishing and Verifying the Fatigue and Static Pressure Ratings of the

Metal Pressure Containing Envelope of a Tie Rod or Bolted Fluid Power Cylinder.

NFPA/T2.6.1M S7 (T3.6.31M)-1976 (R1981), Telescopic Cylinders and Cylinder of Non-Bolted End Construction Pressure Rating Supplement No. 7 to NFPA Recommended Standard for Verifying the Fatigue and Static Pressure Ratings of the Pressure Containing Envelope of Metal Fluid Power Component.

NFPA/T3.9.22-1982, Pump/Motor Pressure Rating Supplement No. 6 to NFPA Recommended Standard Method for Verifying the Fatigue and Static Pressure Ratings of the Pressure Containing Envelope of a Metal Fluid Power Component.

NFPA/T2.6.1M S1 (T3.10.5.1M)-1976 (R1981), Hydraulic Filter/Separator Housing Pressure Rating Supplement No. 1 to NFPA Recommended Standard for Verifying the Fatigue and Static Pressure Ratings of the Pressure Containing Envelope of a Metal Fluid Power Component.

NFPA/T2.6.1M S4 (T3.12.10M)-1976 (R1981), Air Line Filter, Regulator and/or Lubricator Pressure Rating Supplement No. 4 to NFPA Recommended Standard for Verifying the Fatigue and Static Pressure Ratings of the Pressure Containing Envelope of a Metal Fluid Power Components.

NFPA/T2.6.1M S8 (T3.16.8M)-1975 (R1982), Hydraulic Reservoir Pressure Rating Supplement No. 8 to NFPA Recommended Standard for Verifying the Fatigue and Static Pressure Ratings of the Pressure Containing Envelope of a Metal Fluid Power Component — Part 1 — Static Ratings.

NFPA/T2.6.1M S5 (T3.20.8M-1975) (R1981), Quick Action Couplings Pressure Rating Supplement No. 5 to NFPA Recommended Standard for Verifying the Fatigue and Static Pressure Ratings of the Pressure Containing Envelope of a Metal Fluid Power Component.

NFPA/T2.6.1M S2 (T3.231.4M)-1977 (R1982), Pneumatic Valve Pressure Rating Supplement No. 2 to NFPA Recommended Standard for Verifying the Fatigue and Static Pressure Ratings of the Pressure Containing Envelope of a Metal Fluid Power Component.

NFPA/T2.6.1M S3 (T3.29.2M)-1976 (R1982), Pressure Switch Pressure Rating Supplement No. 3 to NFPA Recommended Standard for Verifying the Fatigue and Static Pressure Ratings of the Pressure Containing Envelope of a Metal Fluid Power Component.

Volume C — Pumps, Motors, Power Units and Reservoirs

ANSI/B93.2-1986, Excerpts Extracted from American National Standard Fluid Power Systems and Products — Glossary.

NFPA/T1.21.1-1978 (R1983), Procedure for Self-Certification by Fluid Power Manufacturers.

NFPA/T2.6.1-1974 (R1982), Method for Verifying the Fatigue and Static Pressure Ratings of the Pressure Containing Envelope of a Metal Fluid Power Component.

NFPA/T2.6.1M S8 (T3.16.8M)-1975, Hydraulic Reservoir Pressure Rating — Supplement No. 8 to NFPA Recommended Standard for Verifying the Fatigue and Static Pres-

sure Ratings of the Pressure Containing Envelope of a Metal Fluid Power Component.

ANSI/B93.71M-1986, American National Standard Hydraulic Fluid Power — Pumps — Test Code for the Determination of Airborne Noise Levels.

ANSI/B93.73M-1986, American National Standard Hydraulic Fluid Power — Motors — Test Code for the Determination of Airborne Noise Levels.

ANSI/B93.6M-1972 (R1981), American National Standard Dimensions and Identification Code for Mounting Flanges and Shafts for Positive Displacement Hydraulic Fluid Power Pumps and Motors.

NFPA/T3.9.13-1982, Hydraulic Fluid Power — Pumps and Motors — Glossary.

ANSI/B93.27M-1973 (R1979), American National Standard Method of Testing and Presenting Basic Performance Data for Positive Displacement Hydraulic Fluid Power, Pumps and Motors.

NFPA/T3.9.18M R1-1978, Method of Establishing the Flow Degradation of Fixed Displacement Hydraulic Fluid Power Pumps When Exposed to Particulate Contaminant.

NFPA/T3.9.22-1982, Pump/Motor Pressure Rating Supplement No. 6 to NFPA Recommended Standard Method for Verifying the Fatigue and Static Pressure Ratings of the Pressure Containing Envelope of a Metal Fluid Power Component.

NFPA/T3.9.25M-1977 (R1982), Method of Establishing the Speed Degradation of Hydraulic Fluid Power Motors When Exposed to Particulate Contaminant.

ANSI/B93.57M-1982, American National Standard Hydraulic Fluid Power — Pumps and Motors — Geometric Displacement.

ANSI/B93.18M-1973 (R1980), American National Standard Non-Integral Industrial Fluid Power Hydraulic Reservoirs.

ANSI/B93.41M-1976 (R1982), American National Standard Requirements of Non-Integral Industrial Fluid Power Hydraulic Power Units.

ANSI/B93.12M-1971 (R1977), American National Standard Method of Rating for Mechanical Vacuum Pumps.

NFPA/T3.9.21-1978, Bibliography of Fluid Power Pump/Motor Standards.

NFPA/T3.16.9-1977 (R1982), Bibliography of Hydraulic Fluid Power Reservoirs and Power Units Standards.

Volume D — Filtration and Contamination

ANSI/B93.2, Excerpts Extracted from American National Standard Fluid Power Systems and Products — Glossary.

NFPA/T1.21.1-1978 (R1983), Procedure for Self-Certification by Fluid Power Manufacturers.

NFPA/T2.6.1-1974 (R1982), Method for Verifying the Fatigue and Static Pressure Ratings of the Pressure Containing Envelope of a Metal Fluid Power Component.

NFPA/T2.6.1M S1 (T3.10.5.1M)-1976 (R1981), Hydraulic Filter/Separator Housing Pressure Rating Supplement No. 1

to NFPA Recommended Standard for Verifying the Fatigue and Static Pressure Ratings of the Pressure Containing Envelope of a Metal Fluid Power Component.

ANSI/B93.19M-1972 (R1980), American National Standard Method for Extracting Fluid Samples from the Lines of an Operating Hydraulic Fluid Power System (for Particulate Contamination Analysis).

ANSI/B93.20M-1972 (R1980), American National Standard Procedure for Qualifying and Controlling Cleaning Methods for Hydraulic Fluid Power Fluid Sample Containers.

ANSI/B93.30M-1980, American National Standard — Hydraulic Fluid Power — Contamination Analysis Data — Reporting Method.

NFPA/T2.9.5M-1976 (R1980), Hydraulic Fluid Power — Calibration Method — To Count and Measure Computer Assisted Image Analysis Systems — Particles in the 1-10 μm Range.

ANSI/B93.28M-1973 (R1980), American National Standard Hydraulic Fluid Power — Calibration of Liquid Automatic Particle-Count Instruments — Method Using Air Cleaner Fine Test Dust Contaminant.

ANSI/B93.54M-1981, American National Standard Hydraulic Fluid Power — Assembled Systems — Method for Achieving Roll-off Cleanliness.

ANSI/B93.44M-1978 (R1986), American National Standard Method for Extracting Fluid Samples from a Reservoir of an Operating Hyrdaulic Fluid Power System.

ANSI/B93.73M-1986, American National Standard Hydraulic fluid power — In-line Liquid Automatic Particle Counting Systems — Method of Validation.

NFPA/T3.10.4M-1968 (R1980), Graphic Symbols for Hydraulic Fluid Power Filters and Separators.

ANSI/B93.21M-1972 (R1980), American National Standard End Load Test Method for a Hydraulic Fluid Power Filter Element.

ANSI/B93.22M-1972 (R1979), American National Standard Hydraulic Fluid Power — Filter Elements — Determination of Fabrication Integrity.

ANSI/B93.25M-1972 (R1980), American National Standard Hydraulic Fluid Power — Filter Elements — Verification of Collapse/Burst Resistance.

ANSI/B93.23M-1972 (R1980), American National Standard Hydraulic Fluid Power — Filter Elements — Verification of Material Compatibility Fluids.

ANSI/B93.24M-1972 (R1980), American National Standard Method for Verifying the Flow Fatigue Characteristics of a Hydraulic Fluid Power Filter Element.

ANSI/B93.31M-1973 (R1981), American National Standard Multi-Pass Method for Evaluating the Filtration Performance of a Fine Hydraulic Fluid Power Filter Element.

ANSI/B93.46M-1978 (R1986), American National Standard Method for Determining the Pore Size of a Cleanable Surface Type Hydraulic Fluid Power Filter Element.

NFPA/T3.10.8.18M-1977 (R1982), Multi-Pass Method for

Evaluating the Filtration Performance of a Coarse Hydraulic Fluid Power Filter Element.

NFPA/T3.10.12 R1-1983, NFPA Information Report — Bibliography of Existing Standards Relating to Filtration and Contamination.

Volume E — Conductors and Associated Products

ANSI/B93.2-1986, Excerpts extracted from American National Standard Fluid Power Systems and Products — Glossary.

NFPA/T1.21.1-1978 (R1983), Procedure for Self-Certification by Fluid Power Manufacturers.

NFPA/T2.6.1-1974 (R1982), Method for Verifying the Fatigue and Static Pressure Ratings of the Pressure Containing Envelope of a Metal Fluid Power Component.

ANSI/B93.48M-1979, American National Standard Pneumatic Fluid Power Applications — Metal Separable Tube Fittings — Qualifications Test.

NFPA/T3.8.11-1977, A Bibliography of Fluid Power Tube Fittings and Conductors Standards.

ANSI/B93.59M-1982, American National Standard Fluid Power Systems and Products — Connectors and Associated Components — Outside Diameters of Tubes and Inside Diameters of Hoses.

ANSI/B93.60M-1982, American National Standard Fluid Power Systems and Products — Connectors and Associated Components — Nominal Pressures.

ANSI/B93.4M-1981, American National Standard Hydraulic Fluid Power — Line Tubing — Electric Resistance Welded, Mandrel Drawn.

ANSI/B93.11M-1981, American National Standard Hydraulic Fluid Power — Line Tubing — Seamless Low Carbon Steel.

ANSI/B93.42M-1977 (R1983), American National Standard Method for Testing Hydraulic Fluid Power Quick Action Couplings.

ANSI/B93.51M-1980, American National Standard Pneumatic Fluid Power — Quick Action Couplings — Test Conditions and Procedures.

NFPA/T3.20.1-1973 (R1981), Glossary For Fluid Power Quick Action Couplings.

NFPA/T3.20.7 R1-1983, A Bibliography of Fluid Power Quick Action Coupling Standards.

NFPA/T2.6.1M S5 (T3.20.8M)-1975 (R1981), Quick Action Couplings Pressure Rating Supplement No. 5 to NFPA Recommended Standard for Verifying the Fatigue and Static Pressure Ratings of the Pressure Containing Envelope of a Metal Fluid Power Component.

ANSI/B93-68M-1983, American National Standard Hydraulic Fluid Power — Quick Action Couplings — Surge Flow Test (Short Duration Flow).

ANSI/B93.69M-1983, American National Standard Hydraulic Fluid Power — Quick Action Couplings — Surge Flow Test (Long Duration Flow).

NFPA/T3.26.1 R1-1977, A Bibliography of Fluid Power Hose, Hose Fittings and Hose Assemblies.

Volume F — Control Products

ANSI/B93.2-1986, Excerpts extracted from American National Standard Fluid Power Systems and Products — Glossary.

NFPA/T3.27.1-1972 (R1981), Glossary of Terms for Compressed Air Dryers.

NFPA/T1.21.1-1978 (R1983), Procedure for Self-Certification by Fluid Power Manufacturers.

NFPA/T2.6.1-1974 (R1982), Method of Verifying the Fatigue and Static Pressure Ratings of the Pressure Containing Envelope of a Metal Fluid Power Component.

NFPA/T2.6.1M S2 (T3.21.4M)-1977 (R1982), Pneumatic Valve Pressure Rating Supplement No. 2 to NFPA Recommended Standard for Verifying the Fatigue and Static Pressure Ratings of the Pressure Containing Envelope of a Metal Fluid Power Component.

NFPA/T2.6.1M S3 (T3.29.2M)-1976 (R1982), Pressure Switch Pressure Rating Supplement No. 3 to NFPA Recommended Standard for Verifying the Fatigue and Static Pressure Ratings of the Pressure Containing Envelope of a Metal Fluid Power Component.

NFPA/T2.6.1M S4 (T3.12.10M)-1976 (R1981), Air Line Filter, Regulator and/or Lubricator Pressure Rating Supplement No. 4 to NFPA Recommended Standard for Verifying the Fatigue and Static Pressure Ratings of the Pressure Containing Envelope of a Metal Fluid Power Component.

NFPA/T2.6.1M S9 (T3.5.26)-1977 (R1982), Hydraulic Valve Pressure Rating Supplement No. 9 to NFPA Recommended Standard for Verifying the Fatigue and Static Pressure Ratings of the Pressure Containing Envelope of a Metal Fluid Power Component.

ANSI/B93.7M-1986, American National Standard Hydraulic Fluid Power — Valves — Mounting interfaces.

ANSI/B93.9M-1969 (R1981), American National Standard Symbols for Marking Electrical Leads and Ports on Fluid Power Valves.

ANSI/B93.40M-1976 (R1982), American National Standard Series of Mounting Interfaces for 4567 Maximum psi (315 bar) Four-Port Hydraulic Fluid Power Directional Valves.

ANSI/B93.66M-1983, American National Standard Hydraulic Fluid Power — Directional Control Valve — Method for Determining the Metering Characteristics.

ANSI/B93.49-1980, American National Standard Hydraulic Fluid Power — Valves — Pressure Differential-flow Characteristic — Method of Measuring and Reporting.

ANSI/B93.55M-1981, American National Standard Hydraulic Fluid Power — Solenoid-piloted Industrial Valves — Interface Dimensions for Electrical Connectors.

NFPA/T3.5.33M-1985, Hydraulic fluid power — Cylinder actuator mounted valves — Standard dimensions for mounting surfaces.

ANSI/B93.65M-1983, American National Standard Hydraulic Fluid Power — Code for Identification of Valve Mounting Surfaces.

NFPA/T3.7.2M-1968 (R1980), Graphic Symbols for Fluidic Devices and Circuits.

ANSI/B93.14M-1971 (R1979), American National Standard Methods for Presenting Basic Performance Data for Fluidic Devices.

ANSI/B93.39M-1978 (R1986), American National Standard Requirements for Presentation of Catalog Data, Fluid Compatibility, Cleaning Media, Markings and Dimensional Identification Codes and Pressure Drop Characteristics for Fluid Power Air Line Filters.

ANSI/B93.13M-1981, American National Standard Pneumatic Fluid Power — Pressure Regulators — Industrial Type.

ANSI/B93.33M-1974 (R1981), American National Standard Interfaces for 4-Way General Purpose Industrial Pneumatic Directional Control Valves.

NFPA/T3.21.7M-1976 (R1981), Defining Interface Surfaces for each Pneumatic Valve Interface in NFPA Recommended Standard T3.21.1-1973.

NFPA/T3.21.9M-1976 (R1981), Definition of Port Communication for the Fluid Power Pneumatic Valve Interface to NFPA Recommended Standard T3.21.1 with the valve in Position in Response to a Remote Pilot Signal or Electrical Energization.

ANSI/B93.67M-1983, American National Standard Pneumatic Fluid Power — Five-Port Directional Control Valves — Mounting Surfaces — Optional Electrical Connector — Dimensions and Requirements.

ANSI/B93.45M-1982, American National Standard Pneumatic Fluid Power — Compressed Air Dryers — Methods for Rating and Testing.

ANSI/B93.38-1976 (R1981), American National Standard Method of Diagramming for Moving Parts Fluid Controls.

NFPA/T3.5.27-1976 (R1982), A Bibliography of Hydraulic Valve Standards and Test Procedures.

NFPA/T3.12.9-1977, A Bibliography of Fluid Power Pneumatic FRL Standards.

NFPA/T3.21.5-1978, A Bibliography of Fluid Power Pneumatic Valve Standards.

NFPA/T3.27.4-1979, A Bibliography of Compressed Air Dryers Standards.

NFPA/T3.28.11-1982, A Bibliography of Fluid Logic Devices.

Volume G — Cylinders and Accumulators

ANSI/B93.2-1986, Excerpts extracted from American National Standard Fluid Power Systems and Products — Glossary.

NFPA/T1.21.1-1978 (R1983), Procedure for Self-Certification by Fluid Power Manufacturers.

NFPA/T2.6.1-1974 (R1982), Method for Verifying the Fatigue and Static Pressure Ratings of the Pressure Containing Envelope of a Metal Fluid Power Component.

NFPA/T2.6.1M S7 (T3.6.31M)-1976 (R1981), Telescopic Cylinders and Cylinders of Non-Bolted End Construction Pressure Rating Supplement No. 7 to NFPA Recommended Standard for Verifying the Fatigue and Static Pressure Ratings of the Pressure Containing Envelope of a Metal Fluid Power Component.

NFPA/T3.4.7M-1975 (R1980), Method for Establishing and Verifying the Fatigue and Static Pressure Ratings and Conducting Production Testing of the Pressure Containing Envelope of a Metal Fluid Power Accumulator.

ANSI/B93.3-1984, American National Standard Fluid Power Systems and Products — Cylinder Bores and Piston Rod Diameters — Inch Series.

ANSI/B93.1M-1964 (R1982), American National Standard Dimension Identification Code for Fluid Power Cylinders.

ANSI/B93.8M-1968 (R1986), American National Standard Bore and Rod Size Combinations and Rod End Configurations for Cataloged Square Head Industrial Fluid Power Cylinders.

ANSI/B93.10-1969 (R1982), American National Standard Static Pressure Rating Methods of Square Head Fluid Power Cylinders.

ANSI/B93.15-1981, American National Standard Fluid Power Systems and Products — Square Head Industrial Cylinders — Mounting Dimensions.

ANSI/B93.29M-1986, American National Standard Fluid Power Systems — Cylinders — Dimensions for Accessories for Catalogued square Head.

ANSI/B93.34M-1973 (R1979), American National Standard Bore and Rod Size Combinations, Rod End Configurations, Dimensional Identification Code, and Mounting Dimensions for 3/4, 1 and 1 1/8 inch Bore Cataloged Square Head Tie Rod Type Industrial Fluid Power Cylinders.

NFPA/T3.6.17M-1971 (R1980), Port Nominal Pipe Sizes for Merged Inch and Metric Series Cataloged Square Head Industrial Pneumatic Fluid Power Cylinders.

NFPA/T3.6.29M-1976 (R1981), Method for Establishing and Verifying the Fatigue and Static Pressure Ratings of the Metal Pressure Containing Envelope of a Tie Rod or Bolted Fluid Power Cylinder.

ANSI/B93.52M-1981, American National Standard Fluid Power Systems and Products — Cylinder Bores and Piston Rod Diameters — Metric Series.

NFPA/T3.6.34-1979, Fluid Power Systems and Components — Cylinder Bores and Piston Rod Diameters — Inch Series.

ANSI/B93.53M-1981, American National Standard Fluid Power Systems and Products — Cylinder — Nominal Pressures.

ANSI/B93.56M-1982, American National Standard Fluid Power Systems and Products — Cylinders — Basic Series of Piston Strokes.

ANSI/B93.61M-1982, American National Standard Fluid

Power Systems and Products — Cylinders — Piston Rod and Thread Dimensions and Types.

ANSI/B93.62M-1982, American National Standard Hydraulic Fluid Power — Reciprocating Dynamic Sealing Devices in Linear Actuators — Method of Testing, Measuring and Reporting Leakage.

NFPA/T3.6.54M-1986, Hydraulic Fluid Power — Cylinder Ports — SAE Straight Thread O-ring and 4-bolt Flange Ports — Heavy Duty and Light Duty Cylinders.

NFPA/T3.6.36-1978 (R1984), A Bibliography of Fluid Power Cylinder Standards.

Volume H — Fluid, Lubricant and Sealing Devices

ANSI/B93.2-1986, Excerpts extracted from American National Standard Fluid Power Systems and Products — Glossary.

NFPA/T1.21.1-1978 (R1983), Procedure for Self-Certification by Fluid Power Manufacturers.

ANSI/B93.50M-1979, American National Standard Pneumatic Fluid Power — Use of Synthetic Lubricants — Guidelines.

ANSI/B93.5M-1979, American National Standard Practice for the Use of Fire Resistant Fluids in Industrial Hydraulic Fluid Power Systems.

NFPA/T2.13.2 R2-1980, Hydraulic Fluid Power — Fire Resistant Fluids — Information Report on Company Trade Names.

NFPA/T2.13.3-1979, Index of Non-proprietary Hydraulic Fluid Specifications and Selected Recommended Practices.

ANSI/B93.63M-1984, American National Standard Hydraulic Fluid Power — Petroleum Fluids — Prediction of Bulk Moduli.

NFPA/T3.19.4M R1-1985, Hydraulic Fluid Power — Seal Housings — Dimensions and Tolerances — Cylinder Rod and Piston Seals for Reciprocating Applications — Nominal Series.

ANSI/B93.17M-1979 (R1986), American National Standard Fluid Power Systems And Components — Multiple Lip Packing Sets — Methods for Measuring Stack Heights.

ANSI/B93.35M-1978 (R1986), American National Standard Cavity Dimensions For Fluid Power Exclusion Devices (Inch Series).

ANSI/B93.36M-1973 (R1986), American National Standard Grove Dimensions For Floating Type Metallic and Non-Metallic Fluid Power Piston Rings.

ANSI/B93.62M-1982, American National Standard Hydraulic Fluid Power — Reciprocating Dynamic Sealing Devices in Linear Actuators — Method of Testing, Measuring, and Reporting Leakage.

ANSI/B93.32M-1973 (R1986), American National Standard Groove Dimensions for Fluid Power Radial Compression Type Piston Rings.

ANSI/B93.58M-1982, American National Standard Fluid

Systems — O-rings — Inside Diameters, Cross Sections, Tolerances and Size Identification Code.

NFPA/T3.19.22-1982, A Bibliography of Fluid Power Sealing Devices Standards.

Volume I — Testing

ANSI/B93.2-1986, Excerpts extracted from American National Standard Fluid Power Systems and Products — Glossary.

NFPA/T1.21.1-1978 (R1983), Procedure for Self-Certification by Fluid Power Manufacturers.

NFPA/T2.6.1-1974 (R1982), Method of Verifying the Fatigue and Static Pressure Ratings of the Pressure Containing Envelope of a Metal Fluid Power Component.

NFPA/T2.12.5-1983, National Fluid Power Association Information Report Fluid Power Laboratory Guidelines.

ANSI/B93.49M-1980, American National Standard Hydraulic Fluid Power — Valves — Pressure Differential-Flow Characteristic — Method of Measuring and Reporting.

ANSI/B93.10M-1969 (R1982), American National Standard Static Pressure Rating Methods of Square Head Fluid Power Cylinders.

ANSI/B93.48M-1979, American National Standard Pneumatic Fluid Power, Applications — Metal Separable Tube Fittings — Qualifications Test.

ANSI/B93.71M-1986, American National Standard Hydraulic Fluid Power — Pumps — Test Code for the Determination of Airborne Noise Levels.

ANSI/B93.72M-1986, American National Standard Hydraulic Fluid Power — Motors — Test Code for the Determination of Airborne Noise Levels.

ANSI/B93.27M-1973 (R1979), American National Standard Method of Testing and Presenting Basic Performance Data for Positive Displacement Hydraulic Fluid Power Pumps and Motors.

NFPA/T3.9.18M R1-1978, Method of Establishing the Flow Degradation of Fixed Displacement Hydraulic Fluid Power Pumps When Exposed to Particulate Contaminant.

NFPA/T3.9.25M-1977 (R1982), Method of Establishing the Speed Degradation of Hydraulic Fluid Power Motors When Exposed to Particulate Contaminant.

ANSI/B93.21M-1972 (R1980), American National Standard End Load Test Method for a Hydraulic Fluid Power Filter Element.

ANSI/B93.22M-1972 (R1979), American National Standard Method for Determining the Fabrication Integrity of a Hydraulic Fluid Power Filter Element.

ANSI/B93.25M-1972 (R1980), American National Standard Hydraulic Fluid Power — Filter Elements Verification of Collapse/Burst Resistance.

ANSI/B93.23M-1972 (R1980), American National Standard Hydraulic Fluid Power — Filter Elements — Verification of Material Compatibility with Fluids.

ANSI/B93.24M-1972 (R1980), American National Standard

Method for Verifying the Flow Fatigue Characteristics of a Hydraulic Fluid Power Filter Element.

ANSI/B93.31M-1973 (R1981), American National Standard Multi-Pass Method for Evaluating the Filtration Performance of a Fine Hydraulic Fluid Power Filter Element.

ANSI/B93.46M-1978 (R1986), American National Standard Method of Determining the Pore Size of a Cleanable Surface Type Hydraulic Fluid Power Filter Element.

NFPA/T3.10.8.18M-1977 (R1982), Multi-Pass Method for Evaluating the Filtration Performance of a Coarse Hydraulic Fluid Power Filter Element.

ANSI/B93.62M-1982, American National Standard Hydraulic Fluid Power — Reciprocating Dynamic Sealing Devices

in Linear Actuators — Method of Testing, Measuring and Reporting Leakage.

ANSI/B93.42M-1977 (R1983), American National Standard Method for Testing Hydraulic Fluid Power Quick Disconnect Couplings.

ANSI/B93.51M-1980, American National Standard Pneumatic Fluid Power — Quick Action Couplings — Test Conditions and Procedures.

ANSI/B93.12-1971 (R1977), American National Standard Method of Rating for Mechanical Vacuum Pumps.

ANSI/B93.45M-1982, American National Standard Pneumatic Fluid Power — Compressed Air Dryers — Methods for Rating and Testing.

PIPE FABRICATION INSTITUTE (PFI)

The following standards are available from PFI, P.O. Box 173, Springdale, PA 15144.

PFI: ES-1, Internal Machining and Solid Machined Backing Rings for Circumferential Butt Welds, 1983.

PFI: ES-2, Method of Dimensioning Piping Assemblies, Revised 1984.

PFI: ES-3, Fabricating Tolerances, Reaffirmed 1984.

PFI: ES-4, Hydrostatic Testing of Fabricated Piping, Revised 1985.

PFI: ES-5, Cleaning of Fabricated Piping, Revised 1984.

PFI: ES-7, Minimum Length and Spacing for Welded Nozzles, 1984.

PFI: ES-11, Permanent Marking of Piping Materials, 1974 (R 1984), 2 pp.

Covers recommended identification of piping materials welder's symbols or other data. Methods of marking only are involved.

PFI: ES-16, Access Holes and Plugs for Radiographic Inspection of Pipe Welds, Revised 1985.

PFI: ES-20, Wall Thickness Measurement by Ultrasonic Examination, Revised 1985.

PFI: ES-21, Internal Machining and Fit-up of GTAW Root Pass Circumferential Butt Welds, 1983.

PFI: ES-22, Recommended Practice for Color Coding of Piping Materials, 1974 (R 1984), 2 pp.

Provides identification of piping materials by a general material classification. It is not intended that the color coding will distinguish between the various grades of a particular material (A106B vs. A106A) or between speci-

fications of the same materials (as ASTM A53, A106 or A155 pipe), or representing different manufacturing methods (such as seamless and seam welded).

PFI: ES-24, Pipe Bending Tolerances — Minimum Bending Radii — Minimum Tangents, Revised 1984.

PFI: ES-25, Random Radiography of Pressure Retaining Girth Butt Welds, Revised 1985.

PFI: ES-26, Welded Load Bearing Attachments to Pressure Retaining Piping Materials, 1984.

PFI: ES-27, Visual Examination — The Purpose, Meaning and Limitation of the Term, Reaffirmed 1984.

PFI: ES-29, Abrasive Blast Cleaning of Ferritic Piping Materials, Reaffirmed 1984).

PFI: ES-30, Random Ultrasonic Examination of Butt Welds, 1979.

PFI: ES-31, Standard for Protection of Ends of Fabricated Piping Assemblies, Revised 1985.

PFI: ES-32, Tool Calibration, Reaffirmed 1985.

PFI: ES-33, Circumferential Butt Welds in The Arc of Pipe Bends, Reaffirmed 1985.

PFI: ES-34, Painting of Fabricated Piping, 1983.

PFI: ES-35, Nonsymmetrical Bevels and Joint Configurations for Butt Welds, 1984.

PFI: TB1-1974, Pressure Temperature Ratings of Seamless Pipe Used in Power Plant Piping Systems, Reaffirmed 1984, 40 pp.

The information is derived from formulae and stress values contained in Section 1 Power Boiler and Pressure Vessel Code, and the Power Piping Section of the USA Standard Code for Pressure Piping ANSI: B31.1.0 Usage of these as maximum values is mandatory for piping systems within the jurisdiction of either of these Codes.

PFI: TB2-1983, Reinforcement Tables for 45 Degrees and 90 Degrees Branch Connections of Seamless Steel Pipe.

PFI: TB3-Revised 1985, Guidelines Clarifying Relationships and Design Engineering Responsibilities Between Purchasers' Engineers & Pipe Fabricator or Pipe Fabricator Erector.

PLUMBING AND DRAINAGE INSTITUTE (PDI)

The following publication is available from the Plumbing and Drainage Institute, 5342 Boulevard Place, Indianapolis, IN 46208.

PDI: WH201, Water Hammer Arresters, 1977, 24 pp.

Covers certification, sizing, placement and reference data on water hammer arresters used for reduction of noise, vibration and destruction in piping systems, valves, meters, etc.

PDI: G101, Testing and Rating Procedure for Grease Interceptors, with Appendix of Sizing and Installation Data, 1981, 12 pp.

This project includes the design and construction of the testing equipment, preliminary research and testing, the development of a certification test procedure and the development of a standard method of rating the flow capacities and grease capacities of grease interceptors.

PDI: Code Guide 302, Glossary of Industry Terms, 1979, 17 pp.

Describes plumbing devices which are normally used in plumbing and drainage systems and should be included when plumbing codes are written and adopted.

RADIO TECHNICAL COMMISSION FOR AERONAUTICS (RTCA)

The following publications are available from the RTCA Secretary, 1425 K Street, NW, Suite 500, Washington, DC 20005.

RTCA: DO-52, Calibration Procedures for Signal Generators Used in the Testing of VOR and ILS Receivers, 1953.

Recommends procedures for testing and calibrating signal generators used in the servicing of airborne VOR and ILS receivers. The accuracy of the components of simulated VOR and ILS signals is stated for signal generators calibrated as described.

RTCA: DO-56, VOR Test Signals, 1954.

Describes methods for determining, in an aircraft, the accuracy of VOR bearing indications. The causes of VOR bearing error due to VOR receiver malfunctioning are analyzed.

RTCA: DO-62, Calibration Procedures — Test Standard Omni-Bearing Selector Test Sets, 1954.

Recommends procedures to aid operators of aircraft radio service stations in the Calibration of Test Standard Omni-Bearing Selectors and Omni-Bearing Selector Test Sets used in testing and adjusting VOR receivers and their associated Omni-Bearing Selectors.

RTCA: DO-88, Altimetry, 1959.

Reports on studies of the problems associated with the measurement of aircraft altitude. States requirement that would permit all aircraft to maintain assigned heights within specific limits as related to terrain clearance and the safe vertical separation of aircraft in flight. Appendix I reports on Meteorological Aspects of Pressure Altimetry.

RTCA: DO-117, Standard Adjustment Criteria for Airborne Localizer and Glide Slope Receivers, 1963.

Recommends procedures for adjustment of Airborne Glide Slope and Localizer Receivers.

RTCA: DO-119, Interference to Aircraft Electronic Equipment from Devices Carried Aboard, 1963.

Recommends limits of permissible radiation of RF energy from portable electronic equipment used aboard aircraft in flight, including test procedures for the measurement thereof. Also recommends regulatory actions relating to the operation and identification of passenger-operated devices.

RTCA: DO-127, Standard Procedure for the Measurement of the Radio-Frequency Radiation from Aviation Radio Receivers Operating Within the Radio-Frequency Range of 30-890 Megacycles, 1965.

Recommends standards and test procedures for use by manufacturers of aviation receivers in making necessary radiation measurements using the Far-Field method. In addition, the report discusses the alternative methods of performing such measurement using the Near-Field method.

RTCA: DO-136, Universal Air-Ground Digital Communication System Standard, 1968.

Recommends universal digital standards for linking aircraft into the ground communications and data processing environment of the air traffic control system and airlines and military management information systems.

RTCA: DO-143, Minimum Performance Standards — Airborne Radio Marker Receiving Equipment Operating on 75 Megahertz, 1970.

Recommends standards and test procedures for Airborne Radio Marker Receiving Equipment. Coordinated with EUROCAE.

RTCA: DO-144, Minimum Operational Characteristics — Airborne ATC Transponder Systems, 1970.

Part I defines the concepts, philosophy and development of MOCs for airborne systems, and Part II covers the MOCs for Airborne ATC Transponder Systems, including system characteristics; provides information for demonstration of compliance, and guidance material.

RTCA: DO-148, A New Guidance System for Approach and Landing, 1970.

Defines a system concept and technical description (signal format) for a new precision instrument approach and landing guidance system (LGS) intended to satisfy the varied operational needs of different classes of aviation users, civil and military, in the United States and abroad. Volume 1 is an 80-page summary of findings and recommendations. Volume II is a 400-page compilation of the milestone Special Committee Reports, including the Tentative Operational Requirements, the Report of the Techniques Assessment Team and the Report of the Signal Format Development Team.

RTCA: DO-152, Minimum Operational Characteristics — Vertical Guidance Equipment Used in Airborne Volumetric Navigational Systems, 1922/Appendix 1974.

Part I defines the concepts, philosophy and development of MOCs for airborne systems, and Part II covers the

MOCs for vertical guidance equipment used in airborne volumetric navigation systems, including system characteristics; provide information for demonstration of compliance; and guidance accuracy analysis. Appendix D provides a VOR/DME/Altimeter vertical guidance analysis.

RTCA: DO-154, Recommended Basic Characteristics for Airborne Radio Homing and Alerting Equipment for Use with Emergency Locator Transmitters (ELT), 1973.

Recommends basic system characteristics and provides test and guidance material for Airborne Radio Homing and Alerting Equipment.

RTCA: DO-155, Minimum Performance Standards — Airborne Low-Range Radar Altimeters, 1974.

Recommends standards and test procedures for those characteristics of an Airborne Low-Range Altimeter which are essential for its operation in application to provide measured height above terrain for obstruction clearance and landing. Coordinated with EUROCAE.

RTCA: DO-158, Minimum Performance Standards — Airborne Doppler Radar Navigation Equipment, 1975.

Recommends standards and test procedures for Airborne Doppler Radar Navigation Equipment. Appendices include conditions of testing and detailed test procedures. Coordinated with EUROCAE.

RTCA: DO-160B, Environmental Conditions and Test Procedures for Airborne Equipment, 1984.

Standard procedures and environmental test criteria for testing airborne equipment for the whole spectrum of aircraft from light general aviation aircraft and helicopters through the ''Jumbo Jets'' and SST categories of aircraft. Coordinated with EUROCAE, RTCA DO-160B and EUROCAE/ED-14B are identically worded. Endorsed by the International Organization for Standardization (ISO) as de facto international standard ISO 7137.

RTCA: DO-161A, Minimum Performance Standards — Airborne Ground Proximity Warning Equipment, 1976.

Recommends standards and test procedures for Ground Proximity Warning Equipment. This is a revision of DO-161 and includes changes (1 & 2) to that document and other improvements suggested by operating experience. Appendices include envelopes of conditions for warning, conditions for testing, and detailed test procedures.

RTCA: DO-162, Report on Air-Ground Communications — Operational Considerations for 1980 and Beyond, 1975.

Provides an analysis of air-ground communications requirements anticipated for the post-1980 time frame. Includes definitions of U.S. aviation system and future trends, requirements, systems concepts, and recommendations. This is a companion report to RTCA Paper No. 128-72/EC-671 issued 8-18-72, entitled Proposed U.S. National Aviation Standard for the VHF A/G Communications System.

RTCA: DO-163, Minimum Performance Standards — Airborne HF Radio Communications Transmitting and Receiving Equipment Operating Within the Radio-Frequency Range of 1.5 to 3.0 Megahertz, 1976/Errata.

Recommends standards and test procedures for HF/SSB receivers and transmitters designed to operate in a 3 kHz channel environment. Also includes standards for the provision of AM equivalent mode of operation. Appendices include conditions for testing and detailed test procedures.

RTCA: DO-164A, Minimum Performance Standards — Airborne Omega Receiving Equipment, 1979.

Recommends standards and test procedures for Airborne Omega Navigation Receivers, Systems Sensors, and Navigation Systems. Also included are operational characteristics. Appendices include conditions for testing, detailed test procedures, and a description of Omega Error Mechanisms.

RTCA: DO-165, Initial Report of Civil Aviation Frequency Spectrum Requirements — 1980–2000, 1976.

Provides a comprehensive report on civil aviation's frequency requirements. Appendices recommend revisions to the ITU Table of Allocations and the footnotes thereto; provides justification for stated operational requirements; provides an aviation forecast.

RTCA: DO-166, Microwave Landing System (MLS) Implementation, 1977.

Reports on a study to develop user recommendations for a national implementation policy for MLS as the primary landing system in service by the year 2000. Volume I provides recommendations on how best to transition from ILS to MLS; recommends implementation strategy and a national implementation policy, which are summarized in a findings and recommendation chapter. Volume II includes six appendices which are the reports of working groups in special categories such as Benefits, Airborne Systems Operational Capabilities, and Civil System Costs.

RTCA: DO-167, Airborne Electronics and Electrical Equipment Reliability, 1977.

Provides a tutorial discussion of reliability related to aircraft accidents. Discusses airborne electronic equipment failures and means of reducing failures.

RTCA: DO-169, VHF Air-Ground Communication Technology and Spectrum Utilization, 1979.

Reports on VHF (118–136 MHz) spectrum utilization including the investigation of modulation techniques and reduced channel separation. Identifies problem areas and recommends, among other things, use of reduced channel spacing on a selective basis.

RTCA: DO-170, Audio Systems Characteristics and Minimum Performance Standards — Aircraft Microphones (Ex-

cept Carbon), Aircraft Headsets and Speakers, Aircraft Audio Selector Panels and Amplifiers, 1980.

Part I discusses audio systems response characteristics affecting the intelligibility of air-ground voice communications and recommends means for improvement by users and designers of communication equipment. Part II recommend standards and test procedures for microphones (except carbon), headsets and speakers, and audio and interphone amplifiers for use in aircraft. Coordinated with EUROCAE.

RTCA: DO-171, Recommendations on Policies and Procedures for Off-the-Shelf Electronic Test Equipment Acquisition and Support, 1980.

Provides rationale and recommendations for various conditions and procedures that could provide major benefits to those responsible for drafting legislation, policies, procedures and guidelines for the acquisition and support of electronic test equipment.

RTCA: DO-172, Minimum Operational Performance Standards for Airborne Radar Approach and Beacon Systems for Helicopters, 1980.

Postulates operational goals and applications, and recommends standards and test procedures for Airborne Radar Approach (ARA) systems for helicopters, particularly when operating under IFR, IMC conditions or at night, including standards for the ground-based radar beacon. Includes test conditions and procedures for installed equipment performance, and operational characteristics with test procedures.

RTCA: DO-173, Minimum Operational Performance Standards for Airborne Weather and Ground Mapping Pulsed Radars, 1980.

Postulates operational goals and applications, and recommends standards and test procedures for airborne weather and ground mapping pulsed radars. It takes into account new radar technology and is applicable to both large aircraft and general aviation aircraft systems. Includes test conditions and procedures for installed equipment performance, and operational characteristics with test procedures.

RTCA: DO-174, Minimum Operational Performance Standards for Optional Equipment Which Displays Non-Radar Derived Data on Weather and Ground Mapping Radar Indicators, 1981.

Postulates operational goals and applications, and recommends standards and test procedures for use of weather and ground mapping radar indicators for display of non-radar graphic and/or alphanumeric data. Includes test conditions and procedures for installed equipment performance and operational characteristics with test procedures.

RTCA: DO-175, Minimum Operational Performance Standards for Ground-Based Automated Weather Observation Equipment, 1981.

Postulates operational goals and applications, and recommends standards and test procedures for ground-based automated weather observation equipment. Provides system characteristics for users, designers, manufacturers and installers of such equipment — of interest to various users, including airfield operators, meteorological services, aviation administrations, airplane and helicopter operators, etc. Includes test conditions and procedures for installed system performance, and operational characteristics with test procedures.

RTCA: DO-176, FM Broadcast Interference Related to Airborne ILS, VOR and VHF Communications, 1981.

Reviews the various aspects of the problem of commercial FM broadcast stations contributing to the interference of airborne systems. Recommends improved intragovernmental coordination procedures; and recommends steps to limit growth of the problem, to reduce the problem with installed receivers, and to minimize the problem with new receivers or installations.

RTCA: DO-177, Minimum Operational Performance Standards for Microwave Landing System (MLS) Airborne Receiving Equipment, 1981.

Postulates operational goals and applications, and recommends standards and test procedures for use of Microwave Landing Systems (MLS) airborne receiving equipment. Includes test conditions and procedures for installed equipment performance and operational characteristics with test procedures. Coordinated with EUROCAE.

RTCA: DO-178A, Software Considerations in Airborne Systems and Equipment Certification, 1981.

Describes techniques and methods that may be used for the orderly development and management of software for airborne digital computer-based equipment and systems. Provides guidance to both industry and regulators for use in the certification process. Coordinated with EUROCAE. EUROCAE/ED-12A is identically worded.

RTCA: DO-179, Minimum Operational Performance Standards for Automatic Direction Finding (ADF) Equipment, 1982.

Postulates operational goals and applications, and recommends standards and test procedures for airborne automatic direction finding equipment. Includes test conditions and procedures for installed equipment performance and operational characteristics with test procedures. Coordinated with EUROCAE.

RTCA: DO-180, Minimum Operational Performance Standards for Airborne Area Navigation Equipment Using VOR/DME Reference Facility Sensor Inputs, 1982.

Postulates operational goals and applications, and recommends standards and test procedures for airborne area navigation equipment (2S and 3D) using VOR/DME reference facility sensor inputs. Includes test conditions and procedures for installed equipment performance and

operational characteristics with test procedures. Coordinated with EUROCAE.

RTCA: DO-181, Minimum Operational Performance Standards for Air Traffic Control Radar Beacon System/Mode Select (ATCRBS/Mode S) Airborne Equipment, 1983.

Postulates operational goals and applications, and recommends standards and test procedures for ATCRBS and Mode S airborne equipment. Includes test conditions and procedures for installed equipment performance and operational characteristics with test procedures. Coordinated with EUROCAE.

RTCA: DO-182, Emergency Locator Transmitter (ELT) Equipment Installation and Performance, 1982.

Provides analyses of ELT performance in regard to false alarms and activations in crash environments; provides criteria and guidelines for placement and installation of ELTs in aircraft; reports on ELT system performance in a variety of typical installations; and provides specific recommendations on all of the above standards.

RTCA: DO-183, Minimum Operational Performance Standards for Emergency Locator Transmitters — Automatic Fixed — ELT (AF), Automatic Portable — ELT (AP), Automatic Deployable — ELT (AD), Survival — ELT(s) Operating on 121.5 and 243.0 Megahertz, 1983.

Postulates operational goals and applications, and recommends standards and test procedures for emergency locator transmitters. Includes test conditions and procedures for installed equipment performance and operational characteristics with test procedures.

RTCA: DO-184, Traffic Alert and Collision Avoidance Systems (TCAS) I Functional Guidelines, 1983.

Sets forth minimum requirements and describes the various elements of TCAS I. Discusses both passive and active TCAS I applications. Provides the minimum performance requirements for electromagnetic compatibility for an active TCAS I, and test procedures for both active and passive systems. Appendix A addresses cross-link advisories.

RTCA: DO-185, Minimum Operational Performance Standards for Traffic Alert and Collision Avoidance Systems (TCAS) Airborne Equipment, 1984.

Volume I sets forth operational goals and applications and recommends standards and test procedures for airborne traffic alert and collision avoidance equipment intended primarily for use on transport category aircraft. Includes test conditions and procedures for ensuring installed equipment performance. This baseline TCAS is described as having omnidirectional interrogation/reception capability. Additional features are described which will provide sectorized interrogations for operations in higher density airspace and, separately, will provide bearing estimation measurements. When all features are included, the system will meet FAA Minimum TCAS II requirements for collision avoidance. Volume II contains the required Collision Avoidance Algorithms. The algorithms are presented as high-level pseudocode to convey functional design, and as low-level pseudocode to serve as detailed specification. Coordinated with EUROCAE.

RTCA: DO-186, Minimum Operational Performance Standards for Airborne Radio Communications Equipment Operating Within the Radio Frequency Range 117.975-137.000 MHZ, 1984.

Postulates operational goals and applications, and recommends standards and test procedures for airborne VHF communications receivers and transmitters. Includes test conditions and procedures for installed equipment performance, and operational characteristics with test procedures. Coordinated with EUROCAE.

RTCA: DO-187, Minimum Operational Performance Standards for Airborne Area Navigation Equipment Using Multi-Sensor Inputs.

Postulates operational goals and applications, and rec-

ommends standards and test procedures for airborne area navigation equipment (2D and 3D) using multisensor inputs. Includes test conditions and procedures for installed equipment performance and operational characteristics with test procedures. Coordinated with EUROCAE.

RTCA: DO-189, Minimum Operational Performance Standards for Airborne Distance Measuring Equipment (DME) Operating Within the Radio Frequency Range of 960-1215 Megahertz.

Postulates operational goals and applications, and recommends standards and test procedures for airborne distance measuring equipment (DME). It updates the former DME operational characteristics and performance standards for airborne equipment that operate with conventional DME (DME/N) ground facilities and establishes standards for airborne equipment which will operate with both DME/N and precision DME (DME/P) ground facilities. Coordinated with EUROCAE.

RANGE COMMANDERS COUNCIL (RCC)

The following Interrange Instrumentation Group (IRIG) and Range Commanders Council (RCC) Standards have been prepared and published by the RCC for use by government agencies and industries under contract to them who have an interest in missiles, rockets and associated equipment. Limited copies of these publications are available to authorized government agencies and contractors with active government contracts from the Secretariat, Range Commanders Council, STEWS-SA-R, U.S. Army White Sands Missile Range, NM 88002. Others may obtain copies from the Defense Logistics Agency, Defense Technical Information Center, ATTN: FDRA, Cameron Station, Alexandria, VA 22304-7633. Use the AD/AO numbers that follow each listing below when ordering from DTIC.

IRIG: 106-86, Telemetry Standards, 1986, 132 pp.

Provides development and coordination agencies with the necessary criteria on which to base equipment designs and modification. The standards are intended to ensure efficient spectrum and interference-free operation of the radio link for telemetry systems at the RCC member ranges.

RCC: 118-79, Test Methods for Telemetry Systems and Subsystems, Volume I, End-to-End Methods for Telemetry Systems, 1979, 97 pp.

Addresses the methodology employed to accomplish solar calibrations and transducer-based system calibrations.

Volume II, Test Methods for Telemetry RF Subsystems, 1979, 183 pp.

Contains test procedures for telemetry antenna systems, RF preamps, multicouplers, receivers, and diversity combiners.

Volume III, Test Methods for Recorder/Reproducer Systems and Magnetic Tape, 1979, 136 pp.

Provides test methodologies for tape heads, tape speed, speed variation and timing error, direct record systems, FM systems, and serial high density digital systems.

Volume IV, Test Methods for Data Multiplex Equipment, 1979, 164 pp.

Contains test procedures for frequency division multiplexing, time division multiplex systems, subcarrier oscillators, and bit synchronizers.

RCC: 118-82, Test Methods for Telemetry Systems and Subsystems, Volume V, Test Methods for Vehicle Telemetry Systems, 120 pp.

Addresses test methodologies which apply to transducers, charge amplifiers, differential dc amplifiers, power supplies, and telemetry transmitters.

IRIG: 152-83, IRIG Standard for Distributing Vector Acquisition Data, 1983, 14 pp.

Defines a standard procedure for distributing interrange vector acquisition data to remote tracking instruments.

IRIG: 154-71, IRIG Standards for Distributing Raw Radar Antenna Data, 1971, 22 pp.

Defines a standard procedure for distributing raw antenna data from remote tracking sites via teletype.

IRIG: 161-65, IRIG Standard Data Format for Interrange Transmission of Tracking Data from Computer to Computer (2400 BPS Synchronous EFG Format O), 1985, 12 pp.

Provides a standard format for interrange transmission of tracking data from a computer on one range to a computer at another range.

IRIG: 200-70, IRIG Standard Time Formats, 1970, 26 pp.

Describes standard instrumentation timing formats suitable for recording on magnetic tape, recording oscillographs, film, and real-time transmission, which meet both manual and automatic data reduction requirements. Definitions of terms relating to the standard formats are included.

RCC: 203-64, Standard Format for Interrange Distribution of Visual Count Status Information, 1964, 18 pp.

Specifies a format to be used for the transfer of visual count status information between elements of a global range.

IRIG: 205-77, IRIG Standard Parallel Binary Time Code Formats, 1977, 12 pp.

Presents the ''ground rules'' for attaining maximum compatibility between present and future time generation equipment and the user interface. The selected combination codes included contain most, if not all, of the various parallel binary time codes.

IRIG: 206-77, IRIG Standard for VHF Time Transmission, 1977, 5 pp.

Provides guidance and criteria for procurement and evaluation of projected systems which transmit signals via VHF at various Department of Defense installations.

IRIG: 208-85, IRIG Standard for UHF Command Systems, 1985, 16 pp.

Serves as a guide for the orderly implementation and application of UHF command systems for test ranges. Provides development and coordination agencies with the necessary criteria on which to base equipment designs and modifications.

IRIG: 251-80, IRIG Standard for Pulse Repetition Frequencies and Reference Oscillator Frequency for C-Band Radars, 1980, 3 pp.

Contains standards established to accommodate existing C-band instrumentation radars.

IRIG: 252-74, IRIG Tracking Radar Compatibility and Design Standards for G-Band (4 to 6 GHz) Radars, 1974, 18 pp.

Outlines the minimum noncoherent G-band radar com-patibility standards for interrange use with emphasis on transmitter and receiver characteristics. Also provides guidelines for the design of new G-band noncoherent instrumentation radars.

IRIG: 253-65, IRIG Standard Coordinate System and Data Formats for Antenna patterns, 1965, 100 pp.

Defines a vehicle antenna coordinate system and antenna pattern formats recommended for use at the National and Service Ranges. Also addresses antenna polarization considerations and includes a number of definitions associated with vehicle antenna pattern representations.

RCC: 257-86, Coherent C-Band Transponder Standards, 1986, 25 pp.

Defines minimum transponder parameters in such a manner that any C-band instrumentation radar on any test range may use the transponder.

IRIG: 303-64, Standardized Test Procedures for UHF Flight Termination Receivers, revised 1968, 45 pp.

Provides standardized interrange test procedures for the UHF Flight Termination Receivers for use in determining their adequacy for range safety applications and their compatibility with existing ground instrumentation systems.

IRIG: 352-72, IRIG Standards for Range Meteorological Data Reduction Part I — Rawinsonde, 1972, 129 pp.

Establishes a standard method for reducing rawinsonde data and defines all formulae and computation routines needed to carry out this data reduction process.

IRIG: 352-85, IRIG Standards for Range Meteorological Data Reduction Part I — Rocketsonde, 1985, 437 pp.

Contains computer program documentation for two different computer systems used for the reduction of range meteorological data. Section A addresses a program used by a large central computer to process data from either non-transponder or transponder rocketsondes. Section B deals with the program for the minicomputer (NOVA-3/12) used in the Meteorological Sounding System to process data from transponder rocketsondes.

RCC: 452-86, Video Standards and Formats, 1986, 50 pp.

Provides users of standard and high resolution monochrome and standard color closed circuit television equipment with the criteria essential for the interchange and compatibility of equipment, tape recordings and live signals. Applies to locally generated signals as well as metric video applications involving video tape recorders, video disc recorders and data insertion equipment.

RCC: 503-82, Standard for Data Labels and Data Annotation Procedures, 1982, 31 pp.

Establishes a standard data labeling process to simplify and expedite the annotation of selected data (items) generated from test operations conducted by the various RCC member and associate member ranges and facilities.

RESISTANCE WELDER MANUFACTURERS ASSOCIATION (RWMA)

The following publication and others dealing with special welding applications (piping, exotic materials, etc.) are available from the Resistance Welder Manufacturers Association, 1900 Arch Street, Philadelphia, PA 19103.

RWMA: Bulletin No. 16, Resitance Welding Equipment Standards, 1984, 105 pp.

Contains 10 standards including such topics as nomenclature and definitions, butt welding, electrical standards and fluid power standards.

SCIENTIFIC APPARATUS MAKERS ASSOCIATION (SAMA)

The following publications are available from SAMA, 1101 16th Street, NW, Suite 300, Washington, DC 20036.

SAMA: AI 1.1, Recommended Test Procedures for Glass pH and Reference Electrodes, 1978, 6 pp.

Provides a uniform means for users of glass pH electrodes to evaluate and calibrate these sensors. Procedures for percent theoretical slope, sodium error, zero potential pH, isopotential point, and time response are provided.

SAMA: AI 2.2, Safety Practices for Atomic Absorption Spectrophotometers, 1973, 4 pp.

Describes recommended safety practices for the installation and use of atomic absorption spectrophotometers. Information which manufacturers should provide is listed, and codes about which users should be informed are noted.

SAMA: AI 2.3, Instrumental Specifications for Atomic Absorption Spectrophotometers, 1973, 6 pp.

Provides a set of performance specifications which users of atomic absorption spectrophotometers can expect to obtain from manufacturers. These specifications are classified by function and relative importance.

SAMA: AI 2.1, Guidelines for Purity and Handling of Gases Used in Atomic Absorption Spectroscopy, 1978, 6 pp.

Describes the purity requirements for gases used in atomic absorption spectroscopy. Contains information on handling and gas volume requirements for acetylene, hydrogen, nitrous oxide, argon, and nitrogen.

SAMS: AI 2.4B, Terminology for Atomic Absorption Spectrophotometers, 1978, 6 pp.

Provides readily usable definitions for 34 terms significant in atomic absorption spectroscopy. Usage is consistent with ASTM where applicable.

SAMA: MTI 1 and 2, Load Cell Terminology and Recommended Test Procedures, second edition, 1964, 16 pp.

Provides recommended terminology and definitions for hydraulic, pneumatic, and mechanical load cells used for measurement of weight and force. Also provides recommended general purpose test procedures for qualification and acceptance testing of these types of load cells.

SAMA: RC 2-5, Air Pressures for Pneumatic Controllers and Transmission Systems, 1967, 2 pp.

Establishes standard operating pressure ranges for pneumatic intelligence transmission systems and standard air supply pressures (with limit valves) for operation of pneumatic controllers and pneumatic intelligence transmission systems.

SAMA: PMC 4-1-1962, Bimetallic Thermometers.

SAMA: PMC 5-10-1963, Resistance Thermometers.

SAMA: PMC 6-10-1963, Filled System Thermometers.

SAMA: PMC 8-10-1963, Thermocouple Thermometers (Pyrometers).

SAMA: PMC 17-10-1963, Bushings and Wells for Temperature Sensing Elements.

SAMA: PMC 18-10, Markings for Adjustable Means in Automatic Controllers, 2nd edition, 1965, 2 pp.

Establishes standard terms and units for the marking of adjustment means of controllers capable of producing one or more of the following control actions — proportional action, reset or integral action, and rate of derivative action.

SAMA: PMC 19-10, Tubing Connection Markings for Pneumatic Instruments, 1963, 1 pp.

Establishes a uniform system of marking tubing connection to simplify the inter-connection of pneumatic instruments in their application to industrial processes.

SAMA: PMC 20.1-1973, *Process Measurement and Control Terminology, 1973, 44 pp.*

Applies to terminology associated with industrial process instrumentation used in industries such as chemical, petroleum, metallurgical, power, food, textile and paper. It includes terms relating to measurement and control, and the static and dynamic performance of indicators, recorders, controllers, indicating controllers, recording controllers, transmitters and transducers.

SAMA: PMC 21.4-1966, *Temperature-Resistance Valves for Resistance Thermometer Elements for Platinum, Nickel and Copper.*

SAMA: PMC 22.1-1981, *Functional Diagramming of Instrument and Control Systems, 1981, 20 pp.*

Presents both symbols and diagramming format for use in representing measuring, controlling and computing systems as sued in industrial practice. The purpose is to establish uniformity of symbols and practices in diagramming such systems in their basic functional form, exclusive of their operating media or specific equipment detail.

SAMA: PMC 23-2-1971, *Hydrostatic Testing of Control Valves, 1971, 6 pp.*

This standard applies to ferrous, including stainless steel, control valves. The purpose of the hydrostatic test is to prove the structural integrity and liquid tightness of the valve.

SAMA: PMC 27.1-1980, *Pressure Safety for Pressure and Differential Pressure Process Control Devices, 1980.*

SAMA: PMC 28.1-1973, *Dimensions of Wide Chart Recorders, 1973, 3 pp.*

Covers mounting of wide chart recorders in panels, racks, and rack panels. The purpose is to establish mounting dimensions so that panel mounting configurations and rack configurations shall be compatible.

SAMA: PMC 28.2-1976, *Dimensions for Panel and Rack Mounted Industrial Process Measurement and Control Instruments.*

Covers dimensions for panel and rack mounted industrial process measurement and control instruments.

SAMA: PMC 31.1-1980, *Generic Test Methods for the Testing and Evaluation of Process Control Instrumentation, 1980.*

Describes the conditions and procedures for the testing under static and dynamic conditions of process control instrumentation with analog input and output signals, using the manufacturer's specifications and instructions for installation and operation.

SAMA: 31.2-1983, *Guidelines for Presenting Specifications of Analog Process Measurement and Control Instruments.*

SAMA: PMC 32.0-1981, *Process Instrumentation Reliability Terminology Handbook.*

Defines process instrumentation and process instrumentation reliability terms.

SAMA: PMC 32.1-1976, *Process Instrumentation Reliability Terminology.*

Establishes terminology for use in process instrumentation reliability.

SAMA: PMC 33.1-1978, *Electromagnetic Susceptibility of Process Control Instrumentation.*

Discusses electromagnetic susceptibility as it relates to process control instrumentation.

SAMA-ABMA: *Recommended Standard Instrument Connections Manual, 1981 Edition, 148 pp.*

The recommendations represent the continuation of joint efforts between the Process Measurement and Control Section (PMC) SAMA and members of the American Boiler Manufacturers Association (ABMA) for the purpose of standardizing identification and location of instrument and control connections for water-tube boilers. No attempt has been made to incorporate connections used specifically for environmental monitoring control.

SOCIETY OF AUTOMOTIVE ENGINEERS, INC. (SAE)

The following and numerous other SAE publications are available from the Society of Automotive Engineers, 400 Commonwealth Drive, Warrendale, PA 15096.

SAE/J 254, *Instrumentation and Techniques for Exhaust Gas Emissions Measurement, 1984.*

Establishes uniform laboratory techniques for the continuous and bag-sample measurement of various constituents in the exhaust gas of the gasoline engines installed in passenger cars and light-duty trucks. Concentrates on the measurement of the following components in exhaust gas: hydrocarbons (HC), carbon monoxide (CO), carbon dioxide (CO_2), oxygen (O_2), and nitrogen oxides (NO_x).

SAE/J 247, Instrumentation for Measuring Acoustic Impulses Within Vehicles, 1980.

Provides guidelines for selection and application of instrumentation for proper measurement of acoustic impulses within vehicles, as typified by those generated during the deployment of a passive restraint system. The objective is to achieve uniformity in instrumentation practice and reporting of test measurements. Use should provide a basis for meaningful comparisons of test results from different sources.

SAE/J 211, Instrumentation for Impact Tests, 1980.

Provides guidelines for instrumentation used in automotive safety impact tests. The aim is to achieve uniformity in instrumentation practice and in reporting test results, without imposing undue restrictions on the performance characteristics of the individual elements in an instrumentation or data analysis system. Provides a basis for meaningful comparisons of test results from different sources.

SAE/AS 942, Pressure Altimeter System — Minimum Safe Performance Standard.

Specifies the requirements for minimum safe performance of an altimeter system in its normal mode of operation on subsonic aircraft. The instrument system specified shall accept an input of the static pressure and in some equipment other inputs that contribute altitude information to provide a visual indication of pressure altitude. If equipped with an automatic correction mechanism, it shall indicate by a positive means when the automatic correction mechanism is not in use. If the static source pressure error compensating mechanism is operational it shall be functional throughout the required operating envelope of the particular aircraft. Each aircraft type has its own static source error data which shall be obtained from the airframe manufacturer's certified data. When a Central Air Data Computer is used in the altimeter system, the CADC shall be certified to its own governing document and the altimeter system (CADC and display) shall comply with the requirements of this document. NOTE: The instrument system specified herein does not include the aircraft pressure lines and pressure sources.

SAE/AS 793, Total Temperature Measuring Instruments (Turbine Powered Subsonic Aircraft).

Establishes essential minimum safe performance requirements for total temperature measuring instruments, primarily for use with turbine-powered subsonic transport aircraft, the operation of which may subject the instruments to the environmental conditions specified in this report. Covers three basic types of total temperature measuring instruments used as a means of determining the total temperature developed by adiabatic heating of the air due to motion of the aircraft through the air.

SAE/J 1045, Instrumentation and Techniques for Vehicle Refueling Emissions Measurement.

Describes a procedure for measuring the hydrocarbon emissions occurring during the refueling of passenger cars and light trucks. It can be used as a method for investigating the effects of temperatures, fuel characteristics, etc., on refueling emissions in the laboratory. It also can be used for determining the reduction in emissions achieved with emission control hardware. For this latter use, standard temperatures, fuel volatility, and fuel quantities are specified.

SAE/AS 407B, Fuel Flowmeters.

To specify minimum requirements for Fuel Flowmeters for use primarily in reciprocating engine powered civil transport aircraft, the operation of which may subject the instruments to the environmental conditions specified in Section 3.3. This Aeronautical Standard covers two basic types of instruments, or combinations thereof, intended for use in indicating fuel consumption of aircraft engines as follows: TYPE I — Measure rate of flow of fuel used. TYPE II — Totalize amount of fuel consumed or remaining.

SAE/AS 406, Flight Directors (Turbine-Powered Subsonic Aircraft).

This standard establishes the essential minimum safe performance requirements for flight director instruments, primarily for use with turbine-powered transport aircraft, the operation of which may subject the instruments to the environmental conditions specified in paragraph 3.3. This standard covers flight directors for use on aircraft to indicate to the pilot, by visual means, the correct control application for the operation of an aircraft in accordance with a preselected flight plan.

SAE/AS 404B, Electric Tachometer: Magnetic Drag (Indicator and Generator).

To specify minimum requirements for Electric Tachometers primarily for use in reciprocating engine powered civil transport aircraft, the operation of which may subject the instruments to the environmental conditions specified in Section 3.3. This Aeronautical Standard covers magnetic drag tachometers with or without built-in synchroscopes.

SAE/AS 394A, Rate of Climb Indicator, Pressure Actuated (Vertical Speed Indicator).

To specify minimum requirements for pressure, actuated Climb Indicators for use in aircraft, the operation of which may subject the instruments to the environmental conditions specified in paragraph 3.3. This Aeronautical Standard covers four (4) basic types of direct indicating instruments as follows: TYPE I — Range 0-2000 feet per minute climb and descent, TYPE II — Range 0-3000 feet per minute climb and descent, TYPE III — Range 0-4000 feet per minute climb and descent, TYPE IV — Range 0-4000 feet per minute climb and descent.

SAE/AS 392C, Altimeter, Pressure Actuated Sensitive Type.

To specify minimum requirements for Pressure Actuated Sensitive Altimeters for use in aircraft, the operation of which may subject the instrument to the environmental conditions specified in paragraph 3.3.

SAE/AS 391C, Airspeed Indicator (Pitot Static) (Reciprocating Engine Powered Aircraft).

To establish the essential minimum safe performance standards for pitot static pressure type of airspeed indicators, primarily for use with reciprocating engine power transport aircraft, the operation of which may subject the instruments to the conditions specified in Section 3.3.

SAE/AS 1104, Specification for Single-Degree-of-Freedom Spring-Restrained Rate Gyros.

This specification covers that gyroscopic instrument normally defined as a "subminiature rate gyro." The rate gyro, when subjected to an angular rate abut its input axis, provides an AC output voltage proportional to the angular rate. The subminiature size category generally includes gyro instruments of one (1) inch diameter or less and three and one-half (3½) inches length or less. This specification defines the requirements for a subminiature spring-restrained, single-degree-of-freedom rate gyro for aircraft, missile, and spacecraft applications.

SAE/ARP 416, Directional Indicating System (Turbine Powered).

To recommend the essential minimum safe performance requirements for gyroscopically stabilized Directional Indicating System, primarily for use with turbine powered subsonic transport aircraft, the operation of which may subject the instruments to the environmental conditions specified in paragraph 3.3. This recommended practice covers the requirements for gyroscopically stabilized Directional Indicating Systems, which will operate as a 1 degree/hour latitude corrected, free directional gyro or as a slaved gyro, magnetic compass with ½ degree accuracy.

SAE/ARP 1278, Oscilloscopic Method of Measuring Spark Energy.

This report provides specific information on instrumentation and procedure for the measurement of capacitance discharge spark energy using an oscilloscope. This report describes basic method for measurement of spark energy on all types of capacitance discharge exciters. Reference is made to other methods which may be used if limitations are observed.

SAE/ARP 1267, Electromagnetic Interference Measurement Impluse Generators; Standard Calibration Requirements and Techniques.

This Aerospace Recommended Practice (ARP) describes a standard method and means for measuring or calibrating the "Spectrum Amplitude" output of an impulse generator. This ARP also outlines the method for the measurement of EMI instruments impulse bandwidth.

SAE/ARP 1217, Instrumentation Requirements for Turboshaft Engine Performance Measurements.

This Aerospace Recommended Practice (ARP) defines the measurement parameters that may be used by a pilot or operator to monitor the thermodynamic health of a turboshaft engine in a helicopter and the measurement system accuracies desired.

SAE/ARP 24B, Determination of Hydraulic Pressure Drop.

This ARP is intended to serve as an instrument to determine hydraulic pressure drop, utilizing the best known practices for accessories in the hydraulic, fuel, oil, and coolant systems to aerospace vehicles.

SAE/AIR 1255, Spectrum Analyzers for Electromagnetic Interference Measurements.

This AIR was prepared to inform the aerospace industry about the electromagnetic interference measurement capability of spectrum analyzers. The spectrum analyzers considered are of the wide dispersion type which are electronically tuned over an octave or wider frequency range. The reason for limiting the AIR to this type of spectrum analyzer is that several manufacturers produce them as general-purpose instruments, and their use for EMI measurement will give significant time and cost savings. The objective of the AIR is to give a description of the spectrum analyzers, consider the analyzer parameters, and describe how the analyzers are usable for collection of EMI data. The operator of a spectrum analyzer should be thoroughly familiar with the analyzer and the technical concepts reviewed in the AIR before performing EMI measurements.

SAE/AIR 1092, High Tension Exciter Output Voltage Measurement Using Cathode-Ray Oscilloscope.

The purpose of this report is to provide specific information on instrumentation and procedure for the measurement of high tension exciter output voltage. This report describes a method of voltage measurement using a cathode-ray oscilloscope and high-voltage probe, with emphasis on calibration.

ASTM/B, Test Method for Indentation Hardness of Aluminum Alloys by Means of a Newage Portable Non-Caliper-Type Instrument.

Aluminum alloys — indentation hardness — portable non-caliper-type hardness instrument, test; Hardness (indentation) — aluminum alloys — portable non-caliper-type hardness instrument, test; Portable non-caliper-type hardness method — indentation hardness (of aluminum alloys), by portable non-caliper-type hardness instrument, test.

SAE/J 209, Instrument Face Design and Location for Construction and Industrial Equipment, 1980.

The instrument design criteria and grouping described are recommended to manufacturers of construction and industrial equipment for all new designs. Adherence to these recommendations will promote improved performance and ease of machine and operation, protect machine and operator, and simplify instrument design and production.

SAE/AE 8021, Minimum Performance Standards for Direction Instrument, Non-Magnetic (Gyroscopically Stabilized).

This Aerospace Standard (AS) defines minimum performance requirements under standard and environmental conditions for Gyroscopically Stabilized Non-Magnetic Direction Instruments for use in aircraft. This document establishes the minimum requirements for design and qualification of equipment identified as Gyroscopically Stabilized Non-Magnetic Direction Instruments.

SAE/AS 8016, Vertical Velocity Instrument (Rate of Climb).

This Aerospace Standard (AS) establishes the minimum performance standards for vertical velocity instruments for aircraft use.

SAE/AS 8013, Minimum Performance Standard for Direction Instrument, Magnetic (Gyroscopically Stabilized).

This Aerospace Standard (AS) defines minimum performance requirements under standard and environmental conditions for Gyroscopically Stabilized Magnetic Direction Instruments for use in aircraft. This document establishes the minimum requirements for design and qualification of equipment identified as Gyroscopically Stabilized Magnetic Direction Instruments.

SAE/AS 8004, Minimum Performance Standard for Turn and Slip Instrument.

This standard establishes the minimum performance standards for turn and slip instruments for aircraft use.

SAE/AS 439, Stall Warning Instrument (Turbine Powered Subsonic Aircraft).

This Aerospace Standard establishes the essential minimum safe performance standards for stall warning instruments primarily for use with turbine powered subsonic transport aircraft, the operation of which may subject the instruments to the environmental conditions specified in paragraph 3.4. This standard covers stall warning instruments to provide positive warning to the pilot of an impending stall. Stall, as defined for the purpose of this standard, is the minimum steady flight speed at which the airplane in controllable.

SAE/AS 403A, Stall Warning Instrument.

To specify minimum requirements for stall warning instruments for use in aircraft, the operation of which may subject the instrument to environmental conditions specified in Section 3.3.

SAE/AS 399A, Direction Instrument, Magnetic, (Stabilized Type).

To specify minimum requirements for gyroscopically stabilized i Magnetic Direction Instruments for use in aircraft, the operation of which may subject the instruments to the environmental conditions specified in Paragraph 3.3 This Aeronautical Standard covers minimum requirements for gyroscopically stabilized Magnetic Direction Instruments for use in aircraft.

SAE/AS 398A, Direction Instrument, Magnetic, Non-Stabilized Type (Magnetic Compass).

To specify minimum requirements for non-stabilized magnetic direction instruments for use in aircraft, the operation of which may subject the instrument to the environmental conditions specified in Paragraph 3.3 This Aeronautical Standard covers two basic types of instruments: Type I — Direct Reading, Type II — Remote Indicating.

SAE/AS 397A, Direction Instrument, Non-Magnetic, Stabilized Type (Directional Gyro).

To specify minimum requirements for non-magnetic gyroscopically stabilized direction indicators for use in aircraft, the operation of which may subject the instruments to the environmental conditions specified in Paragraph 3.3. This Aeronautical Standard covers two basic types: Type I — Air Operated, Type II — Electrically Operated.

SAE/AIR 818C, Aircraft Instrument Standards: Wording, Terminology, Phraseology, and Environmental and Design Standards.

Provides the sponsors of Aerospace Standards, (AS), with standard wording, formatting, and minimum environment and design requirements for use in the preparation of their document.

SAE/AS 802B, Powerplant Fire Detection Instruments, Thermal & Flame Contact Types (Reciprocating and Turbine Engine Powered Aircraft).

This Standard establishes minimum requirements for powerplant fire detection instruments primarily for use in reciprocating and turbine engine powered aircraft.

SAE/AS 8019, Airspeed Instruments.

This standard establishes minimum performance standards for total and static pressure actuated airspeed instruments.

SAE/AS 8005, Minimum Performance Standard — Temperature Instruments.

This Aerospace Standard (AS) establishes the essential minimum performance requirements for electrical type temperature instruments primarily for use on aircraft which may subject the instruments to environmental conditions specified herein.

SAE/AS 431A, True Mass Fuel Flow Instruments.

Establishes the essential minimum safe performance standards for True Mass Fuel Flow Instruments primarily for use with turbine powered, subsonic transport aircraft, the operation of which may subject the instruments to the environmental conditions specified in Section 3.3. This Aerospace Standard covers three basic types of true mass flow indicating instruments. Each may consist of an indicator, transmitter and other auxiliary means such as a power supply or amplifier as required.

SAE/AS 428, Exhaust Gas Temperature Instruments.

This standard establishes the essential minimum safe performance standards for exhaust gas temperature instruments primarily for use with turbine powered, subsonic aircraft, the operation of which may subject the instruments to the environmental conditions specified in paragraph 3.3 et seq. The exhaust gas temperature instruments covered by this standard are of the electrical servonull balance type, actuated by varying emf output or one or more parallel connected Chromel-Alumel thermocouples.

SAE/AS 414A, Temperature Instruments (Turbine Powered Subsonic Aircraft).

Establishes the essential minimum safe performance standards for electrical type temperature instruments primarily for use with turbine powered subsonic transport aircraft, the operation of which may subject the instruments to the environmental conditions specified in Section 3.4. Covers basic types of temperature instruments: TYPE I: Ratiometer type, actuated by changes in electrical resistance of a temperature sensing electrical resistance element; TYPE II: Millivoltmeter type, operated and actuated by varying EMF input to the instrument being obtained by temperature changes of the temperature sensing thermocouple.

SAE/AS 413B, Temperature Indicator.

Establishes the minimum sage performance standards for electrical type temperature instruments primarily for use with reciprocating engine powered transport aircraft, the operation of which may subject the instruments to the environmental conditions specified in Section 3.4. Covers two basic types of temperature instruments: TYPE I: Ratiometer type, actuated by changes in electrical resistance of a temperature sensing electrical resistance element; TYPE II: Millivoltmeter type, operated and actuated by varying E.M.F. output of a thermocouple.

SAE/AS 412A, Carbon Monoxide Detector Instruments.

To specify minimum requirements for carbon monoxide detector instruments for use in aircraft, the operation of which may subject the instrument to the environmental conditions specified in Paragraph 3.3. Not intended to cover fire detectors. Covers the basic type of carbon monoxide detector instrument used to determine toxic concentrations of carbon monoxide by the measurement of heat changes through catalytic oxidation.

SAE/AS 411A, Manifold Pressure Indicating Instruments.

Establishes the essential minimum safe performance standards for manifold pressure instruments primarily for use with reciprocating engine powered transport aircraft, the operation of which may subject the instruments to the environmental conditions specified in Section 3.3. Covers two basic types of manifold pressure instruments: TYPE I — Direct Indicating; TYPE II — Remote Indicating.

SAE/AS 408B, Pressure Instruments — Fuel, Oil, and Hydraulic (Reciprocating Engine Powered Aircraft).

Establishes the essential minimum safe performance standards for fuel, oil and hydraulic pressure instruments primarily for use with reciprocating engine powered transport aircraft, the operation of which may subject the instruments to the environmental conditions specified in Section 3.3. Covers two basic types of fuel, oil and hydraulic pressure instruments: TYPE I — Direct Indicating; TYPE II — Remote Indicating.

SAE/AS 405B, Fuel and Oil Quantity Instruments.

To specify minimum requirements for Fuel and Oil Quantity Instruments for use in aircraft, the operation of which may subject the instruments to the environmental conditions specified in Paragraph 3.3. Covers two basic types of instruments: Type I — Float Instruments; Type II — Capacitance Instruments.

SAE/ARP 427, Pressure Ratio Instruments.

To recommend requirements for electrical Pressure Ratio Indicating Instruments for use in aircraft, the operation of which may subject the instruments to the environmental conditions specified in Para. 3.3. Covers two types of two unit Pressure Ratio Instruments each of which consists of a Transducer and an Indicator.

SAE/ARP 1254, Fluidics Test Methods and Instrumentation.

Establishes acceptable methods, procedures and instrumentation required for testing fluidic devices. The tests described include only those necessary to predict the performance of a device when used in a system of circuit. The term "instrumentation" is understood to include all laboratory instrumentation (electrical, fluidic or mechanical, etc.).

THE SOCIETY OF NAVAL ARCHITECTS AND MARINE ENGINEERS (SNAME)

The following publications are available from SNAME, One World Trade Center, Suite 1369, New York, NY 10048.

SNAME: Code C-1, Code for Shipboard Vibration Measurements, 1975.

Establishes standard procedures for gathering and interpreting data on hull vibrations in single-screw commercial ships. These data are needed to compare the vibration characteristics of different ships of a given class, to establish vibration reference levels, and to provide a basis for the improvement of individual ships.

SNAME: Code C-2, Code for Sea Trials, 1973.

Includes a section which describes the instruments and apparatus commonly used for making measurements of the performance of various items of machinery in ship's trials.

SNAME: 3-8, Code on Installation and Shop Tests, 1960, 48 pp.

Contains general outlines and practices to be followed in conducting shop and installation tests and dock trials of merchant vessel machinery and equipment. Operating tests are only included.

SNAME: Code C-4, Local Shipboard Structures and Machinery Vibration Measurements, 1976, 28 pp.

Establishes standard procedures for gathering and presenting data on vibrations measured on structural elements of ships.

SNAME: Code C-5, Acceptable Vibration of Marine Steam

and Gas Turbine Noise and Auxiliary Machinery Plants, 1976, 16 pp.

Provides criteria for mechanical vibration and serves as a refence standard in establishing ships' specifications and procurement documents for new marine equipment.

SNAME: T&R Bulletin 3-23, Guide for Centralized Control and Automation of Ship's Steam Propulsion Plant, 1970, 60 pp.

Gives technical guidance to establishing the desired degree and methods for employing centralized control and for automating a ship's steam propulsion plant.

SNAME: T&R Bulletin 3-29, Guide to Centralized Control and Automation of Ship's Gas Turbine Propulsion Plant, 1978, 55 pp.

Provides technical guidance in the development of centralized control and automation of geared or electric drive gas turbine ship propulsion plants.

SNAME: T&R Bulletin 4-18, Propulsion Monitoring Instrumentation for Shipboard Energy Conservation, 1984, 24 pp.

Reviews state of the art machinery instrumentation and provides guidance to ship operators on the selection of appropriate hardware to optimize propulsion plant performance.

SPRING MANUFACTURERS INSTITUTE, INC. (SMI)

The following handbook is available from the Spring Manufacturer's Institute, Inc., 1211 West 22nd Street, Oak Brook, IL 60521.

SMI: Handbook of Spring Design, 1977, 34 pp.

Gives specifications of material and various test procedures and standards for commercial springs including compression, extension, torsion, flat and hot wound. Contains glossary of spring and related testing terminology.

TECHNICAL ASSOCIATION OF THE PULP AND PAPER INDUSTRY (TAPPI)

The following and other TAPPI publications are available from TAPPI, Technology Park/Atlanta, P.O. Box 105113, Atlanta, GA 30348.

TAPPI: T 210 hm-86, Weighing, Sampling and Testing Pulp for Moisture, 1986, 4 pp.

The methods selected for moisture testing are specified for each given kind of pulp.

TAPPI: T 656 hm-83, Measuring, Sampling, and Analyzing White Waters, 1983, (Historical Method), 4 pp.

Presents methods of white water evaluation so that different mills may use substantially the same procedures and thus establish a common basis of comparison.

TAPPI: T 1206 rp-86, Precision Statement for Test Methods, 1986, 5 pp.

This recommended practice defines terms and describes how to estimate precision from available data.

TAPPI: T 1209 rp-87, Identification of Instrumental Methods of Color or Color Difference Measurement, 1987, 4 pp.

Provides brief, yet specific, recommendations for identification of instrumentation methods for the measurement of color or color difference.

TAPPI: TIS 0414-01, Instrument Symbols and Nomenclature, 1981, 7 pp.

Provides for the pulp and paper industry modifications to ISA Standard 5.1 and recommendations for symbolizing recent innovations in instrument hardware.

TAPPI: TIS 0414-02, Instrument Air Tubing and Piping Materials Recommends, 1982, 3 pp.

Assists in the proper specification and application of instrument air tubing in the pulp and paper mill.

TAPPI: TIS 0804-06, Photometric Linearity of Optical Properties Instruments, 1981, 2 pp.

Describes a test for linearity of optical properties instruments used in several TAPPI optical test methods.

ULTRASONIC INDUSTRY ASSOCIATION, INC. (UIA)
(Formerly, Ultrasonic Manufacturers Association, Inc., UMA)

The following publication is available from the Ultrasonic Industry Association, Inc., P.O. Box 5126, Old Bridge, NJ 08857.

UIA: Recommended Standard Rating for Electric Generators, 1965, 1 pg.

Provides ratings for ultrasonic electrical generators on the basis of the average power output developed into a pure resistance or simulated load. The effects of peak pulse power and application of intermittent loads are also discussed.

UNDERWRITERS' LABORATORIES, INC. (UL)

The following "Standards for Safety" are available from Underwriters' Laboratories, Inc., Publications Stock, 333 Pfingsten Road, Northbrook, IL 60062. Offices and testing stations also located in Melville, LI, NY; Santa Clara, CA; and Tampa, FL. Standards approved by ANSI have ANSI number in parentheses and are available from either UL or ANSI.

UL 25 (ANSI/UL25-1979), Standard for Meters for Flammable and Combustile Liquids and LP-Gas 1979, 11 pp.

These requirements cover meters for measuring flammable and combustible liquids such as gasoline, kerosene, fuel oil, and similar petroleum products, and liquefied-petroleum gas in the liquid state. Liquid-measuring meters of the designs covered by this standard are commonly used in the assembly of dispensing equipment, tank trucks, and other low- or medium-pressure applications where there is a need for measurement of the product transferred.

UL 132, Standard for Relief Valves for Anhydrous Ammonia and LP-Gas, 1973, 10 pp.

These requirements cover safety valves and hydrostatic relief valves for anhydrous ammonia and liquefied petroleum gas (LP-Gas) for use in nonrefrigerated systems in facilities covered by the following American National and other Standards: ANSI K61.1, ANSI Z106.1, NFPA No. 58, and NFPA No. 59.

UL 1144 (ANSI/UL 144-1977), Standard for Pressure Regulators for LP-Gas, 1978, 14 pp.

These requirements cover pressure regulators for use with LP-Gas equipment other than in automotive and marine applications or gas-welding and cutting operations. They are also not intended to cover regulators for use in chemical, petro-chemical, petroleum, or utility power plants; nor pipeline or marine terminals; nor related storage facilities at such plants or terminals.

UL 268, (ANSI/UL 268-1981), Standard for Smoke Detectors for Fire-Protective Signaling Systems, 1981, 83 pp.

These requirements cover smoke detectors to be employed in ordinary indoor locations in accordance with the following Standards of the National Fire Protection Association NPFA No. 72E-1978 and 74-1980.

UL 180 (ANSI/UL1.80-1980), Standard for Liquid-Level Indicating Gauges and Tank-Filling Signals for Petroluem Products, 1980, 11 pp.

These requirements cover liquid-level indicating gauges and tank filling signals for use with vented tanks for the storage of petroleum products, such as gasoline, kerosene, fuel, oil, and other similar petroleum products.

UL 252, Standard for Compressed Gas Regulators, 1979, 12 pp.

These requirements cover pressure regulators which reduce the storage cylinder or line gas pressure to the use pressure. Regulators covered by these requirements are intended for use with air, inert gases, and fuel gases.

UL 404, Standard for Indicating Pressure Gauges for Compressed Gas Service, 1979, 7 pp.

These requirements cover indicating pressure gauges of the elastic element type usually employed in the high-pressure side of regulators or reducing valves used on compressed gas containers or cylinders of oxygen, hydrogen, nitrogen, and other gases. Such gauges usually have pressure ranges of 0-1500, 0-2000, 0-3000, or 0-4000 psi.

UL 187 (ANSI: C33.67-1974), Standard for X-Ray Equipment, 1974, 25 pp.

These requirements cover X-ray equipment for medical commercial, and industrial use, to be employed in accordance with the National Electrical Code.

UL 429, Standard for Electrically Operated Valves, 1982, 45 pp.

These requirements cover electrically general purpose and safety operated valves for the control of fluids such as air, gases, oils, refrigerants, steam, water, etc. Electrically operated valves covered by these requirements are intended to be employed in ordinary locations in accordance with the National Electrical Code.

UL 466 (ANSI: C 33.16-1976), Standard for Electrically Illuminated Scales, 1976, 8 pp.

These requirements cover portable, electrically illuminated counter scales rated at 250 volts or less and ordinarily of the computing type, intended for the measurement of weight and to be employed in accordance with the National Electrical Code.

UL 508, (ANSI: C33.76-1976), Standard for Industrial Control Equipment, 1977, 63 pp.

These requirements cover industrial control equipment for use in ordinary locations in accordance with the National Electrical Code. Includes apparatus and the devices immediately accessory thereto for starting, stopping, regulating, controlling, or protecting electric motors.

UL 521, (ANSI/UL 521-1980), Standard for Fire-Detection Thermostats, 1978, 23 pp.

These requirements cover heat detectors for fire protective signaling systems intended to be installed in ordinary indoor and outdoor locations in accordance with the Standard for Automatic Fire Detectors, NFPA No. 72E.

UL 565, Standard for Liquid-Level Gages and Indicators for Anhydrous Ammonia and LP-Gas, 1973, 7 pp.

These requirements cover liquid-level gages and indicators for anhydrous ammonia and liquefied petroleum gas (LP-Gas) for use with pressure vessels in non-refrigerated systems in installations covered by the following American National and other Standards: ANSI: K61.1, ANSI: Z106.1, NFPA No. 58, and NFPA No. 59.

UL 632 (ANSI/UL 632-1980), Standard for Electrically Actuated Transmitters, 1980, 31 pp.

These requirements cover electrically actuated transmitters intended for permanent installation and use in ordinary indoor locations. They do not cover manually actuated signaling boxes.

UL698 (ANSI: C33.30-1973), Standard for Industrial Control Equipment for Use in Hazardous Locations, Class I, Groups A, B, C, and D and Class II, Groups, E, F, and G, 1973, 28 pp.

These requirements cover industrial control equipment for installation and use in hazardous locations, Class I, Groups, A, B, C, and D, and Class II, Groups, E, F, and G, in accordance with the National Electrical Code. They do not cover intrinsically safe electrical circuits of industrial control equipment for use in hazardous locations.

UL 873 (ANSI/UL 873-1981), Standard for Temperature-Indicating and Regulating Equipment, 1979, 59 pp.

These requirements cover general-use field-installation equipment and controls intended to be factory installed on or in certain appliances as safety, limiting, or operating controls. These controls respond directly or indirectly to changes in temperature, humidity, or pressure to effect control for equipment or appliance operation, etc.

UL 894-1972 (ANSI/UL 894-1977), Standard for Switches for Use in Hazardous Locations Class I, Groups, A, B, C, and D: Class II, Groups, E, F, and G, 1977, 18 pp.

These requirements cover snap and similar type switches rated at 60 amperes or less at 250 volts or less; switches 30 amperes or less at 600 volts or less; switches rated at 2 horsepower or less at 600 volts or less; Class I, Groups A,

B, C, and D, and Class II, Groups, E, F, and G as defined in Article 500 of the National Electrical Code.

UL 913, (ANSI/UL/NFPA 4913-1979), Standard for Intrinsically Safe Apparatus and Associated Apparatus for Use in Class I, II, III, Division I, Hazardous Locations, 1979, 42 pp.

These requirements shall apply to: Apparatus or parts of apparatus in Class I, II, or III; Division 1 locations. Those parts of apparatus located outside of the Class I, II, or III; Division 1 location whose design and construction may influence the intrinsic safety of an electrical circuit within the Class I, II, or III; Division 1 location.

UL 1002 (ANSI/UL 1002-1977), Standard for Electrically Operated Valves, 1977, 18 pp.

These requirements cover electrically operated valves for installation and use in hazardous locations, Class I, Division I, Groups A, B, C, and D, and Class II, Division 1, Groups E, F, and G, in accordance with the National Electrical Code. They are not intended to cover valves for a fluid power system, which is a system that transmits and controls power through use of a pressurized fluid within an enclosed circuit.

UL 1244, Electrical and Electronic Measuring and Testing Equipment, 1980.

Applies to electrical, electronic, or electromechanical measuring or testing equipment designed to measure or observe and indicate quantities of electrical, or electronic phenomena (measuring equipment).

UL 1262, Laboratory Equipment, 1976.

These requirements cover cord-connected equipment rated 250 volts or less and permanently connected equipment rated 600 volts or less intended for use on interior wiring systems in accordance with the National Electrical Code.

UL 1437, Electrical Analog Instruments — Panel Board Types, 1979.

These requirements cover electrical and electrically operated indicating and recording instruments of the analog type that are powered only from the measured parameter and are intended for ordinary use in panel boards and the like.

INTERNATIONAL STANDARDIZATION

Standards of two international organizations are available from the American National Standards Institute (ANSI), 1430 Broadway, New York, NY 10018. The number of instrumentation-related international standards is so extensive that only representative titles are listed in this book. In addition the listing has been limited to reference number and title. Additional information on individual standards or a complete listing of international standards available from the following organizations may be obtained from ANSI.

INTERNATIONAL ORGANIZATION FOR STANDARDIZATION (ISO) — The American National Standards Institute is the Member Body representing the United States in the International Organization for Standardization (ISO). Sixtynine national standards bodies comprise the world membership and cooperate in formulating the technical program in which each member maintains a status as a participant or observer in accordance with the interest of the member in the specific standard under consideration.

INTERNATIONAL ELECTROTECHNICAL COMMISSION (IEC)— The American National Standards Institute has administrative and technical affiliation with the U.S. National Committee of the IEC. This committee, in turn, represents the U.S. in the IEC. The IEC is composed of 43 National Committees that collectively represent about 80 percent of the world's population that produces and consumes 95 percent of all electrical energy. The IEC holds the international responsibility for the coordination and unification of all national electrotechnical standards and it is affiliated with the ISO. It also acts as the coordinating body for the activities of other international organizations whose responsibilities relate to or overlap the electrotechnical field.

INTERNATIONAL ORGANIZATION FOR STANDARDIZATION (ISO)

The following standards are available from ANSI, 1430 Broadway, New York, NY 10018.

ISO: 1-1975, Standard Reference Temperature for Industrial Length Measurements.

ISO: 31/I-1978, Quantities and Units of Space and Time.

ISO: 31/II-1978, Quantities of Units of Periodic and Related Phenomena.

ISO: 31/IV-1978, Quantities and Units of Heat.

ISO: R31/V-1965, Quantities and Units of Electricity and Magnetism.

ISO: 31/VI-1973, Quantities and Units of Light and Related Electromagnetic Radiations.

ISO: 31/VII-1978, Quantities and Units of Acoustics.

ISO: 31/IX-1973, Quantities and Units of Atomic and Nuclear Physics.

ISO: 31/X-1973, Quantities and Units of Nuclear Reactions and Ionizing Radiations.

ISO: R91-1970, Petroleum Measurement Tables including Amendments.

ISO: R128-1959, Engineering Drawing, Principles of Presentation.

ISO: 140/I-1978, Part I: Requirements for Laboratories.

ISO: 140/2-1978, Part II: Statement of Precision Requirements.

ISO: 140/3-1978, Part III: Laboratory Measurements of Airborne Sound Insulation of Building Elements.

ISO: 228/I-1978, Pipe Threads where Pressure-Tight Joints are not made on Threads Part I: Designation Dimension and Tolerances.

ISO: 261-1973, ISO General Purpose Metric Screw Threads, General Plan.

ISO: 386-1977, Liquid-in-Glass Laboratory Thermometers — Principles of Design, Construction, and Use.

ISO: 454-1975, Acoustics Relation Between Sound Pressure Levels of Narrow Bands of Noise in a Diffuse Field and in a Frontally-Incident Free Field of Equal Loudness.

ISO: R495-1966, General Requirements for the Preparation of Test Codes for Measuring the Noise Emitted by Machines.

ISO: R508-1966, Identification Colors for Pipes Conveying Fluids in Liquid or Gaseous Condition in Land Installations and on Board Ships.

ISO: 532-1975, Acoustics — Method for Calculating Loudness Level.

ISO: R541-1967, Measurement of Fluid Flow by Means of Orifice Plates and Nozzles.

ISO: 554-1967, Standard Atmospheres for Conditioning and/or Testing Specifications.

ISO: 555, Liquid Flow Measurements in Open Channels-Dilution Method for Measurements of Steady Flow.

ISO: 605-1977, Pulses: Methods of Test.

ISO: 651-1975, Solid-Stem Calorimeter Thermometers.

ISO: 652-1975, Enclosed-Scale Calorimeter Thermometers.

ISO: 748-1973, Liquid Flow Measurement in Open Channels by Velocity Area Methods.

ISO: 772-1978, Liquid Flow Measurement in Open Channels — Vocabulary and Symbols, Bilingual Edition.

ISO: 921-1972, (E/F/R) Nuclear Energy Glossary.

ISO: 1028-1973, Information Processing Flowchart Symbols.

ISO: 1070-1973, Liquid Flow Measurement in Open Channels by Slope Area Method.

ISO: R1087-1969, Vocabulary of Terminology.

ISO: 1100-1973, Liquid Flow Measurement in Open Channels: Establishment and Operation of a Gauging Station and Determination of the Stage-Discharge Relation.

ISO: R1607-1970, Methods of Measurements of the Performance Characteristics of Positive-Displacement Vacuum Pumps, Part 1: Measurement of the Volume Rate of Flow (Pumping Speed).

ISO: R1608-1970, Methods of Measurement of the Performance Characteristics of Vapor Vacuum Pumps, Part I: Measurement of the Volume Rate of Flow (Pumping Speed).

ISO: R1660-1971, Technical Drawings — Tolerances of form and of Position — Part III: Dimensioning and Tolerancing of Profiles.

ISO: R1661-1971, Technical Drawings — Tolerances of Form and of Position — Part IV: Practical Examples of Indications on Drawings.

ISO: 1709-1975, Nuclear-Energy Fissile Materials — Principles of Criticality Safety in Handling and Processing.

ISO: 1999-1975, Acoustics — Assessments of Occupational Noise Exposure for Hearing Conservation Purposes.

ISO: 2186-1973, Fluid Flow in Closed Conduits — Connections for Pressure Signal Transmissions between Primary and Secondary Elements.

ISO: 2636-1973, Information Processing-Conventions for Incorporating Flowchart Symbols in Flowcharts.

ISO: 2975, *Measurement of Water Flow in Closed Circuits — Tracer Method.*

ISO: 2975/1-1974, *Part I: General.*

ISO: 2975/2-1975, *Part II: Constant Rate Injection Method Using Non-Radioactive Tracers.*

ISO: 2975/3-1976, *Part III: Constant Injection Method Using Radioactive Tracers.*

ISO: 2975/6-1977, *Part VI: Transit Time Method Using Non-Radioactive Tracers.*

ISO: 2975/7-1977, *Part VII: Transit Time Method Using Radioactive Tracers.*

ISO: 3313-1974, *Measurement of Pulsating Fluid Flow in a Pipe by Means of Orifice Plates, Nozzles, or Venturi Tubes, in Particular in the Case of Sinusoidal or Square Wave Intermittent Periodic-Type Fluctuations.*

ISO: 3354-1975, *Measurement of Clean Water Flow in Closed Conduits — Velocity Area Method Using Current-Meters.*

ISO: 3966-1977, *Measurement of Fluid Flow in Closed Conduits — Velocity Area Method Using Pitot Static Tubes.*

ISO: 4006-1977, *Measurement of Fluid Flow in Closed Conduits — Vocabulary and Symbols, Bilingual Edition.*

INTERNATIONAL ELECTROTECHNICAL COMMISSION (IEC)

The following standards are available from ANSI, 1430 Broadway, New York, NY 10018.

IEC: 27, *Letter Symbols to be Used in Electrical Technology:*

IEC: 27-1 (1971), *Part 1: General, incorporating Amendments No. 1 (1974) and No. 2 (1977) and including Supplement 27-1A (1976): Time-Dependent Quantities.*

IEC: 27-2 (1972), *Part 2: Telecommunications and Electronics, including Supplement 27-2A (1975).*

IEC: 27-3 (1974), *Part 3: Logarithmic Quantities and Units.*

IEC: 38 (1975), *IEC Standard Voltages, including Amendment No. 1 (1977).*

IEC: 50, *International Electrotechnical Vocabulary.*

A glossary of the terms, with their definitions in English and French, used in electrical engineering. The equivalent terms only are given in Dutch, German, Italian, Polish, Swedish and Spanish. A separate index is given for each of the eight languages. The vocabulary is issued in the form of separate booklets, each dealing with a specific field.

IEC: 50(00) (1975), *International Electrotechnical Vocabulary, General Index,* $63.00.

IEC: 50(08) (1960), *Electro-Acoustics.*

IEC: 50(12) (1955), *Transductors.*

IEC: 68, *Basic Environmental Testing Procedures:*

Describes a standard general procedure for climatic and mechanical robustness tests, designed to assess the durability, under various conditions of use, transport and storage, of components used in equipment for radio-communication and in electronic equipment employing similar techniques.

IEC: 68-1 (1978), *Part 1: General.*

IEC: 68-2, *Part 2: Tests.*

This part describes the different tests in detail. Each test is identified by a letter of the alphabet and is issued in the form of a separate booklet.

IEC: 79, *Electrical Apparatus for Explosive Gas Atmospheres:*

IEC: 79-0 (1971), *Part O: General Introduction.*

IEC: 79-1 (1971), *Part 1: Construction and Test of Flameproof Enclosures of Electrical Apparatus, including Supplement 79-1a (1975): Appendix D: Method of Test for Ascertainment of Maximum Experimental Safe Gap.*

IEC: 79-2 (1975), *Part 2: Pressurized Enclosures.*

IEC: 79-3 (1972), *Part 3: Spark Test Apparatus for intrinsically-Safe Circuits.*

IEC: 79-4 (1975), *Part 4: Method of Test for Ignition Temperature, including Supplement 79-4A (1970).*

IEC: 79-5 (1967), *Part 5: Sand-Filled Apparatus.*

IEC: 79-6 (1968), *Part 6: Oil-Immersed Apparatus.*

IEC: 79-7 (1969), *Part 7: Construction and Test of Electrical Apparatus, Type of Protection "e".*

IEC: 79-8 (1969), *Part 8: Classification of Maximum Surface Temperatures.*

IEC: 79-9 (1970), *Part 9: Marking.*

IEC: 79-10 (1972), Part 10: Classification of Hazardous Area.

IEC: 113, Diagrams, Charts, Tables.

IEC: 113-1 (1971), Part 1: Definitions and Classification.

IEC: 113-2 (1971), Part 2: Item Designation.

IEC: 117, Recommended Graphical Symbols; Graphical Symbols.

IEC: 117-0 (1973), Part O: General Index.

IEC: 117-1 (1960), Part 1: Kind of Current, Distribution Systems, Methods of Connection and Circuit Elements, incorporating Amendments No. 1 (1966), No. 2 (1967) and No. 3 (1973) and including Supplementl 117-1A (1976).

IEC: 117-2 (1960), Part 2: Machines, Transformers, Primary Cells and Accumulators, Transductors and Magnetic Amplifiers, Inductors, incorporating Amendments No. 1 (1966), No. 2 (1971), No. 3 (1973) and first supplement (1974).

IEC: 117-3 (1977), Part 3: Switching and Protective Devices, superseding first edition (1963), Amendments No. 1 (1966), No. 2 (1972), No. 3 (1973), No. 4 (1974), and Supplement 117-3A (1970) and 117-3B (1972).

IEC: 117-4 (1963), Part 4: Measuring Instruments and Electric Clocks, incorporating Amendments No. 1 (1971) and including Amendments No. 2 (1973) and No. 3 (1974) and Supplement 117-4A (1974).

IEC: 150 (1963), Testing and Calibration of Ultrasonic Therapeutic Equipment.

IEC: 271 (1974), List of Basic Terms, Definitions and Related Mathematics for Reliability, including Supplement 271A (1978).

IEC: 272 (1968), Preliminary Reliability Considerations.

IEC: 278 (1968), Documentation to be Supplied with Electronic Measuring Apparatus, including Supplement 278A (1974).

IEC: 284 (1968), Rules of Behavior with Respect to Possible Hazards when Dealing with Electronic Equipment and Equipment Employing Similar Techniques, including Amendment No. 1 (1972).

IEC: 319 (1978), Presentation of Reliablity Data on Electronic Components (or parts).

IEC: 351, Expression of the Properties of Cathrode-Ray Oscilloscopes.

IEC: 351-1 (1976), Part 1: General, superseding 351 (1971).

IEC: 351-2 (1976), Part 2: Storage Oscilloscopes.

IEC: 359 (1971), Expression of the Functional Performance of Electronic Measuring Equipment.

IEC: 393, Potentiometers.

IEC: 393-1 (1973), Part 1: Terms and Methods of Test, including Supplements 393-1A (1977) and 393-1B (1978).

IEC: 393-2 (1976), Part 2: Sectional Specification: Lead-Screw Actuated Preset Potentiometers. Selection of Methods of Test and General Requirements.

IEC: 393-3 (1977), Part 3: Sectional Specification: Single-Turn Rotary Wirewound and Non-Wirewound Potentiometers. Selection of Methods of Test and General Requirements.

IEC: 393-4 (1978), Part 4: Sectional Specifications: Single-Turn Rotary Power Potentiometers. Selection of Methods of Test and General Requirements.

IEC: 393-5 (1978), Part 5: Sectional Specification: Single-Turn Rotary Low-Power Wirewound and Non-Wirewound Potentiometers. Selection of Methods of Test and General Requirements.

IEC: 405 (1972), Nuclear Instruments: Constructional Requirements to Afford Personal Protection Against Ionizing Radiation.

IEC: 414 (1973), Safety Requirements for Indicating and Recording Electrical Measuring Instruments and their Accessories.

IEC: 416 (1972), General Principles for the Formulation of Graphical Symbols, including Amendment No. 1 (1978).

IEC: 473 (1974), Dimensions for Panel-Mounted Indicating and Recording Electrical Measuring Instruments.

IEC: 477 (1974), Laboratory D.C. Resistors.

IEC: 482 (1975), Dimensions of Electronic Instrument Modules (for Nuclear Electronic Instruments).

IEC: 484 (1974), Indirect Acting Electrical Measuring Instruments.

IEC: 485 (1974), Digital Electronic D.C. Voltmeters and D.C. Electronic Analogue-to-Digital Converters.

IEC: 529 (1976), Classification of Degrees of Protection Provided by Enclosures, including Amendment No. 1 (1978).

IEC: 536 (1976), Classification of Electrical and Electronic Equipment with Regard to Protection Against Electric Shock.

IEC: 539 (1976), Directly Heated Negative Temperature Coefficient Thermistors.

IEC: 540 (1976), Test Methods for Insulation and Sheaths of Electric Cables and Cords (elastomerica and thermoplastic compounds), superseding 330 (1970).

IEC: 544-1 (1977), Guide for Determining the Effects of Ionizing Radiation on Insulating Materials Part 1: Radiation Interaction.

IEC: 546 (1976), Methods of Evaluating the Performance of Controllers with Analogue Signals for Use in Industrial Process Control.

IEC: 552 (1977), CAMAC — Organization of Multi-Crate Systems, Specification of the Branch-Highway and CAMAC Crate Controller Type A1.

IEC: 561 (1976), Electro-Acoustical Measuring Equipment for Aircraft Noise Certification.

IEC: 601-1 (1977), Safety of Medical Electrical Equipment Part 1: General Requirements.

NATIONAL STANDARDIZING ORGANIZATIONS
(Outside of the United States)

ISA recognizes the extensive standardizing activities taking place in other countries; however, it wasn't practical to provide complete coverage in this volume of all instrumentation-related standards published by organizations within these countries. Readers engaged in international trade should consult the standards organizations of the country involved. To this end, addresses are provided for the following partial list of organizations.

Algeria: INAPI, Institut Algerien de Normalisation et de Propriete Industrielle, 5 Rue Abou Hamou Moussa, B.P. 1021, Centre de Tri, Alger

Argentina: IRAM, IRAM, Instituto Argentino de Racionalizacion de Materiales, Chile 1192, Buenos Aires

Austrailia: SAA, AS, Standards Association of Australia, Standards House, 80–86 Arthur Street, North Sidney, N.S.W. 2060

Austria: ON, ONORM, Oesterreichisches Normungsinstitut, Leopoldsgasse 4, Postfach 130, A 1020 Wien 2

Bangladesh: BDSI, Bangladesh Standards Institution, 3-DIT (Extension) Avenue, Motijheel Commercial Area, Dacca-2

Barbados: BNSI, Barbados National Standards Institution, "Flodden" Culloden Road, St. Michael

Belgium: IBN, NBN, *Institut Belge de Normalisation, av. de la Brabanconne, 29, B-1040 Bruxelles

Bolivia: Direccion General de Normas y Tecnologia, Ministerio de Industria, Comercio y Turismo, Casilla 4430, Piso 9, La Paz

Brazil: ABNT, NB, EB, *Associacao Brasileira de Normas Tecnicas, 13 Av. Treze de Maio, Andar 28, Caixa Postal 1680, CEP 20,000, Rio de Janeiro

Bulgaria: DKC, State Committee for Standardization at the Council of Ministers, 21, 6th September Street, Sofia

Cameroon: Service de Normalisation, Direction de l'Industrie, Ministere de l'Economie et du Plan, B.P. 1604, Yaounde

Canada: SCC, CAN, Standards Council of Canada, International Standardization Branch, Meadowvale Corporate Centre, 2000 Argentia Road, Suite 2-401, Mississauga, Ontario L5N 1V8

Chile: INN, Instituto Nacional de Normalizacion, Matias Cousino 64, Piso 6, Casilla 995, Correo 1, Santiago

China: CNS, CNS, National Bureau of Standards, Ministry of Economic Affairs, 5th Floor, Hsin Kuang Life Insurance Bldg., Taipei, Taiwan 104, Republic of China

Columbia: ICONTEC, Instituto Colombiano de Normas Tecnicas, Carrera 37 No. 52–95, P.O. Box 14237, Bogota

Costa Rica: Instituto Centroamericano de Investigaciones y Tecnologia Industrial, 4a Calle y Avenida la Reforma, Zona 10, Guatemala City, Guatemala

Cyprus: Cyprus Organization for Standards and Control of Quality, Ministry of Commerce and Industry, Nicosia

Czechoslovakia: CSN, ON, Urad pro Normalizaci Mereni, Vaclavske Namesti 19, 113 47 Praha 1

Denmark: DS, DS, Dansk Standardiseringsraad, Aurehojvej 12 and 15, Postbox 77, DK-2900 Hellerup

Ecuador: INEN, Instituto Ecuatoriano de Normalizacion, Casilla 3999, Av, Universitaria 784, Quito

Egypt: EOS, Egyptian Organization for Standardization, 2 Latin America Street, Garden City, Cairo-Egypt

El Salvador: Instituto Centroamericano de Investigaciones y Tecnologia Industrial, 4a Calle y Avenida la Reforma, Zona 10, Guatemala City, Guatemala

Ethiopia: ESI, Ethiopian Standards Institution, P.O. Box 2310, Addis Ababa

Finland: SFS, SFS, Suomen Standardisoimisliitto r.y., P.O. Box 205, SF-00121 Helsinki 12

France: AFNOR, NF, Association Francaise de Normalisation, Tour Europe, Cedex 7, 92080 Paris La Defense

Germany: DIN, DNA, DIN Deutsches Institut fur Normung, Burggrafenstrasse 4–10, Postfach 1107, D-1000 Berlin 30

Ghana: GSB, Ghana Standards Board, P.O. Box M.245, Accra

Greece: ELOT, Hellenic Organization for Standardization, Didotou 15, Athens 144

Guatemala: ICAITI, Instituto Centroamericano de Investigaciones y Tecnologia Industrial, 4a Calle y Avenida la Reforma, Zona 10, Apartado Postal 1552, Guatemala City

Honduras: Instituto Centroamericano de Investigaciones y Tecnologia Industrial, 4a Calle y Avenida la Reforma, Zona 10, Guatemala City, Guatemala

Hong Kong: Hong Kong Standards and Testing Centre, Eldex Industrial Bldg. 12th Floor, Unit-A, 21 Ma Tau Wei Road, Hung Hom, Kowloon

Hungary: MSZH, Magyar Szabvanyugyi Hivatel, Postafiok 24, 1450 Budapest 9

Iceland: Industrial Development Institute, Skipholt 37, Reykjavik

India: ISI, IS, Indian Standards Institution, Manak Bhavan, 9 Bahadur Shah Zafar Marg, New Delhi 110001

Indonesia: YDNI, NI, Yayasan Dana Normalisasi Indonesia, Indonesian Institute of Sciences, Jalan Teuku Chik Ditro No. 43, P.O. Box 250, Jakarta

Iran: ISIRI, ISIRI, Institute of Standards and Industrial Research of Iran, Ministry of Industries and Mines, P.O. Box 2937, Teheran

Iraq: IOS, Iraqi Organization for Standards, Planning Board, P.O. Box 11185, Baghdad

Ireland: IIRS, IS, Institute for Industrial Research and Standards, Ballymun Road, Dublin-9

Israel: SII, Standards Institution of Israel, 42 University Street, Tel Aviv 69977.

Italy: UNI, Ente Nazionale Italiano de Unificazione, Piazza Armando Diaz 2, 1 20123 Milano

Ivory Coast: BIN, Bureau Ivoirien de Normalisation, B.P. 1318, Abidjan

Jamaica: JBS, Bureau of Standards, 6 Winchester Road, P.O. Box 113, Kingston 10

Japan: JISC, JIS, Japanese Industrial Standards Committee, Ministry of International Trade and Industry, 1-3-1 Kasumigaseki Chiyodaku, Tokyo

Jordan: Directorate of Standards, Ministry of Industry and Trade, P.O. Box 2019, Amman

Kenya: KEBS, Kenya Bureau of Standards, NHC House, Harambee Avenue, P.O. Box 54974, Nairobi

Korea: Korean Standards Association, C.P.O. Box 783, Seoul, Republic of Korea

Kuwait: KSS, Standards and Metrology Department, Ministry of Commerce and Industry, Post Box No. 2944, Kuwait

Lebanon: LIBNOR, Institut Libanais de Normalisation, B.P. 195144, Beyrouth

Liberia: Ministry of Commerce, Industry and Transportation, Division of Standards, Monrovia

Libya: Standards and Specifications Section, Department of Industrial Organization, Secretariat of Industry, Tripoli

Madagascar: Ministere du Developpement Rural et de la Reforme Agraire, Direction l'agriculture, Service du Controle des Qualites et du Conditionnement, B.P. 1.316, Antananarivo

Malawi: Malawi Bureau of Standards, P.O. Box 946, Blantyre

Malaysia: SIRIM, Standards and Industrial Research Institute of Malaysia, Lot 10810, Phase 3, Federal Highway, P.O. Box 35, Shah Alam, Selangor

Mauritius: Mauritius Standards Bureau, Ministry of Commerce and Industry, Reduit

Mexico: DGN, DGN, Direccion General de Normas, Tuxpan No. 2, Mexico 7, D.F.

Morocco: SNIMA, Service de Normalisation Industrielle Marocaine, Direction de l'Industrie, Ministere du Commerce, de l'Industrie, des Mines et de la Marine Marchande, Rabat

Netherlands: NNI, NEN, Nederlands normalisatie-Institut, Polakweg 5, P.O. Box 5810, 2280 HV Rijswijk ZH

New Zealand: SANZ, Standards Association of New Zealand, Private Bag, Wellington

Nicaragua: Instituto Centroamericano de Investigaciones y Tecnologia Industrial, 4a Calle y Avenida la Reforma, Zona 10, Guatemala City, Guatemala

Nigeria: NSO, Nigerian Standards Organization, Federal Ministry of Industries, 11 Kofo Abayomi Road, Victoria Island, Lagos

Norway: NSF, NS, Norges Standardiseringsforbund, Haakon VII's gt. 2, N-Oslo 1

Oman: Directorate General for Specifications and Measurements, Ministry of Commerce and Industry, P.O. Box 550, Muscat

Pakistan: PSI, PS, Pakistan Standards Institution, 39 Garden Road, Saddar, Karachi-3

Panama: Instituto Centroamericano de Investigaciones y Tecnologia Industrial, 4a Calle y Avenida la Reforma, Zona 10, Guatemala City, Guatemala

Peru: ITINTEC, Instituto de Investigacion Tecnologia, Industrial y de Normas Tecnicas, Jr. Morelli—2da Cuadra, Urbanizacion San Borja—Surquillo, Lima 34

Philippines: PS, Philippines Bureau of Standards, TML Commercial Bldg, 100 Quezon Avenue, Quezon City, P.O. Box 3719, Manila

Poland: PKNiM, Polski Komitet Normalizacji i Miar, UI, Electoraina 2, 00–139 Warszawa

Portugal: IGPAI, NP, Reparticao de Normalizacao, Avenida de Berna 1, Lisboa-1

Republic of South Africa: SABS, SABS, South African Bureau of Standards, Private Bag X191, Pretoria 0001

Rhodesia: Standards Association of Central Africa, Coventry Road, Workington P.O. Box 2259, Salisbury 4

Romania: IRS, STAS, Institutul Roman de Standardizare, Casuta Postala 6214, Bucarest 1

Saudi Arabia: SASO, Saudi Arabian Standards Organization, Airport Street, P.O. Box 3437, Riyadh

Singapore: SISIR, Singapore Institute of Standards and Industrial Research, 179 River Valley Road, P.O. Box 2611, Singapore 6

Spain: IRANOR, Instituto Nacional de Racionalizacion y Normalizacion, Serrano 150, Madrid 6

Sri Lanka: BCS, Bureau of Ceylon Standards, 53 Dharmapala Mawatha, Colombo 3

Sweden: SIS, SIS—Standardiseringskommissionen i Sverige, Tegnergatan 11, Box 3 295, S-103 66 Stockholm

Switzerland: SNV, SNV, *Association Suisse de Normalisation, Kirchenweg 4, Postfach, 8032 Zurich

Syria: Industrial Testing and Research Centre (Syrian Standards Organization), P.O. Box 845, Damascus

Thailand: TISI, Thai Industrial Standards Institute, Department of Science, Ministry of Industry, Rama VI Street, Bangkok 4

Trinidad and Tobago: Trinidad and Tobago Bureau of Standards, Room 318, Salvatori Bldg, Frederick Street, P.O. Box 288, Port of Spain

Tunisia: Ministere de l'Economie Nationale, Tunis

Turkey: TSE, TS, Turk Standardlari Enstitusu, Necatibey Caddesi 112, Bakanliklar, Ankara

United Kingdom: BSI, BS, British Standards Institution, 2 Park Street, London W1A 2BS, England

USSR: GOST, GOST, Gosudarstvennyj Komitet Standartov, Soveta Ministrov S.S.S.R, Leninsky Prospekt 9, Moskva 117049

Venezuela: COVENIN, Comision Venezolana de Normas industriales, Av. Boyaca (Cota Mil), Edf. Fundacion La Salle Piso 5, Caracas 105

Yugoslavia: JZS, Jugoslovenski zavod za Standardizaciju, Slobodana Penezica-Krcuna br. 35, Post Pregr. 933, 11000 Beograd

Zambia: Zambia Standards Institute, P.O. Box RW 259, Lusaka

SECTION II

SUBJECT INDEX TO SECTION I STANDARDS

This index is provided to help the reader locate standards that apply to a particular area. The index is based on key words or subject terms that are reflected within the titles of the standards contained in Section I.

Each key word is followed by the titles of standards that are related to the entry. The acronym and page number that follow each title refer to the Section I page that contains abstracts and ordering information.

The reader, when seeking standards dealing with a subject area, should search the index under all possible terms that may be related to the subject. Standards tend to have narrow and concise titles and the index terms are not grouped by general subject areas; therefore, related standards may be listed under a number of entries.

A

C

Calorimetry

Certification

Chromatography

Connectors

Contamination

D

E

Electromagnetic Radiation

Electronic Components

Electronic Hazards

Flow Measurement

Fluid Power

G

Gages

Gases

H

Hazardous Locations

Heat Measurements

Humidity Measurements

Metrification

Modulation

Moisture Measurements

N

Nuclear

Precision

Pressure Control & Measurement

134

Stress Measurement

Surface Texture

Symbols

Temperature Measurement & Control

144

146

Velocity Data

Ventilating Systems

Vibration Measurements

Viscometry

SECTION III

STANDARDS AND RECOMMENDED PRACTICES

ISSUED BY THE INSTRUMENT SOCIETY OF AMERICA

INTRODUCTION

The Standards and Practices Department has been an integral part of the Instrument Society of America for forty-two years. To accomplish its primary goal of uniformity in the field of instrumentation, the Standards Department relies on the work of volunteer committee members whose efforts are coordinated and directed by volunteer Committee Chairmen and Standards and practices Department Directors. These volunteer members provide the time and technical expertise needed to develop and update standards in the field of instrumentation and control.

The Society also participates in many national and international standards activities. On March 26, 1976, the Instrument Society of America was approved as an Accredited Standards Writing organization by the American National Standards Institute (ANSI). Through strict adherence to our ANSI approved procedures, ISA has maintained its ANSI accreditation and continues to submit new and revised ISA standards to ANSI for approval as American National Standards. ISA is currently represented on three ANSI Standards Management Boards and is actively involved with the work of ISO and IEC. Through ANSI, ISA holds the Secretariat of ISO/TC 10/SC 3, ''Graphical Symbols for Instrumentation,'' and IEC/TC SC65B, ''Industrial-Process Measurement and Control: Elements of Systems.'' In addition twelve ISA standards committees serve as the advisory groups for U.S. representation to ISO and IEC committees and working groups.

The products of the ISA Standards program are well-surveyed, carefully written standards and recommended practices that outline accepted procedures in instrumentation throughout U.S. industry. To maintain their value, these standards and practices are reviewed and modified where necessary by the committee responsible for them, at intervals after publication. Toward the goal of improving its standards program, ISA welcomes criticism of its work. Correspondence should be directed to the attention of Standards Board Secretary, Instrument Society of America, 67 Alexander Drive, Research Triangle Park, North Carolina 27709.

Lois M. Ferson
Manager of Standards Services

SECTION III

TABLE OF CONTENTS

*The Instrument Society of America standards and recommended practices are arranged in numerical order and are individually page numbered. The document titles appear in the upper right corner and document numbers appear in the upper left corner of each odd-numbered page.

**New or revised since the last edition of *Standards and Practices for Instrumentation.*

The following standards and recommended practices have been withdrawn by the Instrument
Society of America:

ISA-RP2.1-1978

Recommended Practice

Manometer Tables

Instrument Society of America

ISBN 0-87664-325-X

ISA-RP2.1   Manometer Tables

INSTRUMENT SOCIETY OF AMERICA
67 Alexander Drive
P.O. Box 12277
Research Triangle Park, North Carolina 27709

FOREWORD

This Recommended Practice has been prepared as a part of the service of the Instrument Society of America toward a goal of uniformity in the field of instrumentation. To be of real value this report should not be static, but should be subjected to periodic review. Toward this end the Society welcomes all comments and criticisms, and asks that they be addressed to the Standards and Practices Board Secretary, Instrument Society of America, 67 Alexander Drive, P.O. Box 12277, Research Triangle Park, North Carolina 27709.

This report was prepared and revised by the 8D-RP2 Committee of the Production Processes Standards Division on Manometer Tables, chaired by W. G. Brombacher, Consultant, National Bureau of Standards, Washington, D.C.

The assistance of those who aided in the preparation of this Recommended Practice by answering questionnaires, offering suggestions, and in other ways is gratefully acknowledged.

The following reviewed the original practice and served as a Board of Review:

E. A. Adler - United Engineers & Constructors, Inc.
C. W. Bates - Humble Oil & Refining Co.
W. E. Boyle - Shell Oil Company
R. E. Clarridge - General Electric Co.
J. R. Connell - Imperial Oil Co., Ltd.
W. A. Crawford - E. I. duPont deNemours & Co., Inc.
Howard Ecker - Minnesota Mining & Mfg. Co.
G. G. Gallagher - The Fluor Corporation
R. L. Galley - Convair Astronautics
W. A. Hagerbaumer - Socony-Vacuum Oil Co.
D. E. Hostedler - Foster Wheeler Corp.

P. R. Hoyt - Shell Development Company
J. G. Kerley - Shell Oil Company
A. E. Krogh - Minneapolis-Honeywell Reg. Co.
J. L. Lopez - Lago Oil & Transport Co., Ltd.
H. M. McCarthy - Standard Oil Development Co.
E. S. Mehnert - Colgate-Palmolive-Peet Co.
A. V. Novak - E. I. duPont deNemours & Co., Inc.
J. W. Percy - U. S. Steel Corporation
R. R. Proctor - The Pure Oil Company
J. E. Read - E. I. duPont deNemours & Co., Inc.
A. F. Sperry - Consultant
F. H. Trapnell - E. I. duPont deNemours & Co., Inc.
J. E. White - Black, Sivalls & Bryson, Inc.

This Recommended Practice was approved for publication by the Recommended Practices Committee on September 6, 1952, and approved for revision by the Standards & Practices Board in 1962.

E. A. Adler - United Engineers & Constructors, Inc.
   Standards & Practices Department Vice President
F. E. Bryan - Douglas Aircraft Company
   Aero-Space Standards Division Director
R. E. Clarridge - General Electric Company
   Intersociety Standards Division Director
J. P. Green - Arthur G. McKee Company
   Metals Standards Division Director
E. C. Hutchison - Union Carbide Nuclear Co.
   Production Processes Standards Division Director
W. J. Ladniak - Minneapolis-Honeywell Reg. Co.
   Nuclear Standards Division Director
E. J. Minnar - ISA Headquarters Staff
   Standards & Practices Board Secretary

This Revision, under the direction of W. G. Brombacher, 8D-RP2 Committee Chairman, has been prepared to include changes by the National Bureau of Standards since the date of last publication (1952).

SCOPE

1.1 This report presents abbreviations and fundamental conversion factors commonly used in manometry, recommended definitions of pressure in terms of a column of mercury and water, and for a large number of liquids, tables of pressures indicated by, or equivalent to, heights of columns at various temperatures. These data have the object of facilitating and standardizing the use of manometers and U-tubes as direct pressure indicating instruments or in the calibration of pressure recorders and controllers.

1.2 A discussion is included of the more frequent or more important sources of error in manometric measurements, together with correction tables for mercury and water columns. To conform to general practice in this country, English units of measurements are largely used (i.e., pounds, inches, etc.) and with several exceptions, decimals are used to denote parts of units, rather than octonary fractions.

1.3 In particular, it will be seen that conversion factors for kerosine, gage oil, alcohol, dibromobenzene, dibromoethane, and acetylene tetrabromide are given only to obtain equivalent water columns at 60 F. Further, no conversion data are presented for benzol, butyl collosolve, carbitol, nbutyl phthalate, or halowax oil. The filling in of these omissions will have to await demonstrated need.

MANOMETER TABLES

DISCUSSION OF STANDARDS OF MANOMETERS

For a more complete discussion on mercury manometers, see "Mercury barometers and manometers," Nat. Bur. of Standards Monograph No. 8, 1960.

Standards. The pressure in terms of height of a column of liquid is given by the fundamental relation

$$P = Dhg \qquad (1)$$

where P is the pressure in absolute units, as dynes/sq cm or poundals/sq in.

D is the density of the liquid in grams/cu cm or lb/cu in.

h is the height of the column from horizontal surface to horizontal surface of the menisci in cm or in.

g is the acceleration of gravity in cm/sec^2 or in./sec^2

P/g is the pressure in grams/sq cm or psi

If the height of the column of a liquid is used as a unit of pressure,

$$h = \frac{P}{Dg} \qquad (2)$$

from which it is evident that some value of the density of the liquid and of gravity must be selected as standard for greater convenience, preferably the same by everyone. The standard density will be considered only for water and mercury, and is more conveniently set in terms of the temperature of the liquid, rather than in terms of density.

At its meeting of September 7, 1947, the Recommended Practices Committee approved as a recommended practice the following standards:

(a) Standard conditions for a mercury column used as unit of pressure:

Gravity: 980.665 cm/sec^2 = 32.1740 ft/sec^2
Temperature: 0°C (32°F)
760 mm of mercury (=29.9213 in. of mercury)
$\qquad$ = a pressure of one atmosphere

(b) Standard conditions for a water column used as a unit of pressure:

Gravity: 980.655 cm/sec^2 = 32.1740 ft/sec^2
Temperature: 20°C (68°F)
407.513 in. of water (= 1035.08 cm of water)
$\qquad$ = a pressure of one atmosphere

The value of standard gravity given above is an international standard accepted in engineering and physics and therefore is applicable for all columns of all liquids in addition to water and mercury if used as a pressure unit. It should be noted that meteorologists in 1953 adopted internationally this value, previously having used the value 980.62 cm/sec^2.

The pressure of one atmosphere internationally defined for fixed points on the International Temperature scale is 1013.250 millibars. This value has been rather universally accepted. This pressure is equivalent to that exerted by a column of mercury 760 mm high, having a density of 13.5951 grams/cm^3 and subjected to a gravitational acceleration of 980.665 cm/sec^2. A mm or cm or inch of mercury of differential or absolute pressure is defined as a submultiple of the atmosphere above defined. However, at high pressures correction for compressibility may have to be applied.

For convenience, mercury manometers are often calibrated to indicate the height of the mercury column in terms of mercury at 0°C when the manometer is at another temperature.

The definition of a unit height of a water column at 20°C as a unit of pressure is a recommended practice, made in an effort to eventually obtain national standardization. It is obviously worth while to standardize on a single pressure unit, but it is realized that changes from other units now used may cause serious inconvenience. Use of the proposed unit is advocated in fields where no standard has been defined. Three manometer temperatures, 15°C (59°F), 60°F (15.56°C), and 20°C (68°F), are now used on a national scale to define the unit of pressure as the height of a water column. In aeronautics, 15°C is used; The American Gas Association, and perhaps other societies, uses 60°F; and in orifice flowmeter work, 20°C is commonly used. The temperature used to define the pressure does not affect other conditions desired as standard, as for example, volume of gas flow measured in terms of the gas at 60°F and 29.92 inches of mercury. Some have suggested the maximum density of water (3.98°C) as the unit. This temperature is far from normal laboratory temperatures and will require the application of temperature corrections in many cases where none would be necessary with 20°C as the standard. The proposal is attractive, however.

It is intended that the definition apply to distilled water, free from absorbed gases. It is easier to eliminate absorbed gases than to determine the amount present. In mercury, absorbed gases give no trouble. Lack of purity fouls the mercury meniscus long before there is a significant change in density.

No action on standards for liquids other than water and mercury was taken, since these other liquids are not often used as standards of pressure.

ERRORS OF LIQUID COLUMN MANOMETERS

3.1 The errors in the indications of manometers are (a) scale errors, (b) temperature errors, (c) gravity errors, (d) capillary errors, (e) compressibility, and (f) effect of absorbed gases. Of these, corrections are ordinarily applied only for the scale, temperature, and gravity errors. The other errors are corrected for only precise work or under special conditions of use.

3.2 *Scale Errors.* After a manometer reading has been corrected for temperature and gravity error, and for such of the other errors as are warranted, there usually remains a residual error which may be called the scale error. This can be determined only by calibration against a standard instrument. If corrections are made only for the temperature and gravity errors, the scale error will include (a) errors in graduating the scale, and (b) the effect of capillarity, and perhaps of compressibility and of absorbed gases. Under (a) will be included to a high degree of accuracy the error introduced when the scale is graduated true at one temperature and is assumed correct at another temperature. Further, when the effects under (b) are included under the scale error, only the variation in these effects need to be accounted for.

In general, the procedure of calibrating manometers is such that correction for scale error should be applied first in applying corrections to a manometer reading.

3.3 *Temperature Errors.* Since the densities of liquids vary with temperature, any deviation in manometer temperature from that selected as standard for the pressure unit will introduce an error in the manometer indication. Further, since the scale expands or contracts with changes in its temperature, an additional error is introduced. An expansion of the scale reduces

the reading of the manometer held at constant pressure and conversely an expansion of the liquid increases the reading; the two expansions tend to balance one another, but the effect of the expansion of the liquid is usually much larger.

Temperature corrections are given by the following relations:

$$H_o = H_t + C \tag{3}$$

$$C = \frac{s(t-t_s)-m(t-t_o)}{1 + m(t-t_o)} H_t \tag{4}$$

where H_o is the height of the liquid column at the standard temperature

H_t is the indicated height of the liquid column at temperature t, corrected for scale errors

C is the temperature correction

s is the coefficient of linear expansion of the scale

m is the coefficient of cubical expansion of the liquid

t is the manometer temperature

t_s is the temperature at which the scale indicates the true height

t_o is the standard temperature at which the height of the liquid column is in terms of a pressure unit.

When the correction is negative in sign, the correction is subtracted from H_t as indicated by equation (3).

The coefficient of expansion is usually available with reference to 0°C or 32°F, but no significant error is ordinarily introduced by using this value if t_o differs from 0°C. If warranted, the proper value of s can be computed.

Both the cubical and linear thermal expansions of most manometer liquids and scales are hyperbolic in character and therefore in general cannot be accurately represented by the single terms s and m used in equation (4). For the small temperature ranges over which manometers are generally used, the errors introduced are not usually significant. Consideration of this point is essential in very precise measurements.

Exceptions to the above are mercury, for which m requires no modification over the temperature range of interest, and water, for which m varies with temperature considerably and significantly for most measurements at all temperatures ordinarily of interest. Therefore, for water, equation (4) does not apply, and corrections must be applied in two parts, first for the temperature error for the scale $\{=s(t-t_s)\}$ and then for the error for the liquid by a multiplier obtained from Table 12.

If t_o be substituted in equation (4) for the temperature t_s at which the scale is calibrated true and the resulting error be incorporated into the scale correction, considerable simplification in constructing tables is obtained. The error thus introduced is usually not significant. On this basis, equation (4) becomes

$$C = \frac{s(t-t_o)-m(t-t_o)}{1 + m(t-t_o)} H_t = C_o H_t \tag{5}$$

For cistern manometers where the level of the liquid in the cistern shifts with change in the pressure while the scale remains fixed, equations (4) or (5) do not apply. Usually the temperature correction can be put in the form

$$C_1 = (H_t + h)C_o \qquad (6)$$

where C_1 is the temperature correction, C_o is defined by equation (5), and h is an addition to the indicated column height H_t which in general must be determined by calibration tests at various temperatures of a manometer of any given design.

3.4 *Gravity Errors.* The effect of variation of gravity from the standard value on manometer indications is as follows:

$$P = \frac{g}{g_s} H_o \qquad (7)$$

$$\text{or } P = H_o + H_o \frac{(g-g_s)}{g_s} = H_o + C \qquad (8)$$

where P is the height of the liquid column subjected to standard gravity. This is the pressure if, as is common practice, no further corrections are applied.

H_o is the indicated height of the liquid column at the standard temperature

C is the correction to be applied to H_o

g is the ambient value of gravity

g_s is the standard value of gravity, 980.665 cm/sec²

The ambient value of gravity at any location can be obtained as a function of latitude from the Smithsonian Meteorological (or Physical) Tables and more accurate values for many locations from the U. S. Coast & Geodetic Survey, (Washington 25, D. C.).

In some cases the value of gravity must be obtained from the accepted relation for gravity as a function of latitude. This relation holds only for sea level. To obtain the value of gravity at a station h feet above sea level:

$$g = g_o - 0.000094h \qquad (9)$$

where g is the value of gravity at the station, g_o is the value of gravity at sea level, both in cm/sec², and h is the altitude in feet. Equation (9) holds only for free air, but is sufficiently accurate for moderate elevations of terrain above sea level.

3.5 *Capillary Errors.* If a liquid wets the walls of the manometer tube or cistern, the center of the liquid column is elevated above the level which would obtain if the surface were infinite in extent. Conversely if the liquid does not wet the surface, of which mercury is a conspicuous example, the column level is depressed. The amount of the depression tends to vary with the direction of the change in pressure and greatly with local differences in the condition of the surface in contact with the liquid.

Correction for the capillary effect, if made at all, is normally made only for the depression of the mercury in the tube. Part of the table given by Gould and Vickers in the Journal Scientific Instruments 29, 85, 1952, is reproduced, based on a surface tension of mercury of 450 dynes per centimeter.

DEPRESSION OF MERCURY COLUMN IN MM

Bore of tube, mm	Height of meniscus in mm				
	0.6	0.8	1.0	1.5	2.0
6	0.73	0.94	1.12	1.43	–
8	0.37	0.48	0.58	0.78	0.89
10	0.20	0.26	0.32	0.44	0.52
12	0.12	0.15	0.19	0.26	0.31
15	0.05	0.07	0.09	0.12	0.15
20	0.015	0.02	0.024	0.034	0.042

To reduce the variable effects of surface tension, certain practices may be recommended, assuming that the scale correction includes the capillary effect.

(a) In the design of the manometer, large bore tubes greatly reduce the change in capillary depression (or elevation) with meniscus height. For mercury manometers the minimum bore should be about 10 mm (0.4 in.); there is little overall advantage in bores exceeding 15 to 20 mm (0.6 to 0.8 in.). The above recommended minimum and maximum bores apply equally well to water manometers and probably also to those with other liquids. Equal size of bore is obviously indicated for U-tube manometers.

(b) Cistern manometers should also have the recommended tube bores. The capillary effect in the cistern is minimized by making the area of the cistern large. In mercury manometers of this type especially, it is necessary to tap or vibrate the cistern vigorously in order to reduce scatter in the readings at a constant pressure. It also helps to vibrate the tube, but not to the same degree.

(c) The gradual accumulation of corrosion products and dirt at mercury surfaces changes the capillary correction. This change is larger in its effect on the reading the smaller the bore of the tube or of the cistern. Corrosion seems easier to prevent with other liquids.

(d) After a pressure change, drainage of liquids which wet the surface of tubes is a source of error, particularly in manometers in which the measured pressure depends upon the indication of one surface level only, as in the case of most cistern manometers. In effect a time lag in reading occurs while drainage takes place. The effect reduces with increase in the bore of the manometer tubes.

(e) The addition of a few drops of a wetting agent, such as Aerosol OT100, or Dreen, to the water helps greatly in obtaining a symmetrical meniscus in water manometers. This decreases the difficulty in making readings and also eliminates in large measure the effect of films of foreign material on the glass.

3.6 *Compressibility Effects.* The absolute pressure acting on a liquid compresses it and thereby changes its density. It is of importance for manometers only in the cases where the difference between two high pressures is measured.

The compressibility coefficient C, or change in volume per-unit pressure, is defined as follows:

$$C = \frac{1}{V_o}\frac{dV}{dP} \tag{10}$$

where V_o is the volume at pressure P or more practically is the volume at the lowest pressure

dV/dP is the rate of change of volume with pressure normally negative since the volume usually decreases with the application of pressure

A few values of the coefficient are given in the table for water, mercury, and ethyl alcohol. The pressures are absolute values. The values for water were obtained from "Properties of Ordinary Water Substance" by E. N. Dorsey; those on mercury from the "Smithsonian Physical Tables"; and those on ethyl alcohol from the "Handbook of Chemistry and Physics."

COMPRESSIBILITY OF LIQUIDS

Liquid	Temp. deg C	*Coefficient	Pressure Range, Atmospheres
Water	10	48.0	1-25
	20	47.2	1-25
	10	49.2	25-50
	20	47.6	25-50
	10	46.0	100-200
	20	44.2	100-200
	50	42.8	100-200
	20	37.4	500-1000
	20	32.6	1000-1500
	20	29.2	1500-2000
	20	25.9	2000-2500
	20	23.6	2500-3000
Mercury	20	3.95	300
	22	3.97	500
	22	3.91	1000
Ethyl Alcohol	20	112	1-50
	28	86	150-200
	28	81	150-400

Petroleum oils with a specific gravity 60/60°F between 0.80 and 0.90 have a compressibility coefficient of approximately 70×10^{-6} at 20°C (68°F) for a gage pressure change from 0 to 50 atmospheres. Reference: R. S. Jessup, BS Jnl. of Research, Vol. 5, 1930, p985; RP 244. Most organic liquids have compressibilities of the same order as that of oil.

As a consequence of compressibility it requires 4×10^{-6} more pressure to obtain a change in indication of one mm of mercury at an absolute pressure of 760 mm than it does at one mm of mercury. This difference is 12 times as great for water and 38 times as

great for ethyl alcohol. Thus the unit of pressure defined in terms of a liquid column is also a function of pressure, but the effect in manometers as ordinarily used to measure low gage pressures is less than 0.001 per cent.

3.7 *Effect of Absorbed Gases.* Air absorbed in water decreases the density, contrary to what might be expected. At 100°C the decrease of water density presumably fully saturated with air is 1.6, and at 20°C, 0.4 parts per million. The weight of absorbed air is 27 at 10°C and 22 parts per million at 20°C. The effect is quite variable and requires consideration for each gas in contact with a particular liquid.

The "Handbook of Chemistry and Physics" and the "International Critical Tables" contain data on the solubility of many gases in a number of liquids. These references, however, contain relatively little information on the effect of solubility on the density of those liquids that are used in manometers.

Mercury is an exception in that small amounts of gases go into solution. No data were found, but no sensible outgassing is evident when mercury is subject to a sudden decrease in pressure, as is the case with most other manometric liquids.

Since there is considerable time lag in both outgassing and in the dissolving of gases in liquids, considerable uncertainty usually exists as to the precise amount of gas in solution. For this reason, densities used in computing pressures should be for gas-free liquids if available, and when necessary, corrections made for the change in density due to the gas in solution. The correction is rarely made except in precise work or under unusual conditions of measurement.

BASIS, DESCRIPTION AND USE OF TABLES

4.1 Table 1. *Abbreviations.* Abbreviations used are those proposed by the American Standards Association (Z10.1-1941).

4.2 Table 2. *Fundamental Constants and Common Factors.* As indicated in the table the values are largely taken from "Units of Weight and Measure," NBS M233, although the number of digits in the values is far in excess of that needed in manometry. The conversion factors are primarily from metric to English units and vice versa.

4.3)

4.4) – Tables 3, 4, and 5. No comment is required.

4.5)

4.6 Table 6. *Density of Mercury.* The values of the density of mercury given here are based on the mean value of the thermal coefficient of volume expansion. Accurate (not mean) values of the coefficient of cubical thermal expansion of mercury are given by Beattie, Blaisdell, Kaye, Gerry and Johnson, Proc. Am. Acad. Arts and Sciences, 74, 370, 1941. The two values of the density do not differ more than one part in 100000. At 0°C, the density is generally considered accurate to one part in 100,000 for virgin natural mercury. There is no great difficulty in purifying mercury to the degree necessary to realize the densities given; even minute impurities of the base metals make the mercury unusable in a manometer.

The density at 0°C is the standard used at present in defining pressure in terms of the height of a mercury column.

4.7 Table 7. *Density of Water.* The values given here differ slightly from those given by various experimenters. The data for temperatures above 40°C (104°F) are less reliable than those for lower temperatures, which is indicated by dropping one digit from the values. As pointed out under "Discussion of Standards and Errors of Manometers", the density is affected somewhat by the content of absorbed air. Also, the dissolved salts will affect the density, which effect is avoided by the use of distilled water.

The unit height of a water column at either 15°C, 60°F, or 68°F is used in various fields of engineering to define a pressure. The ISA has adopted as a recommended practice 68°F (20°C) as the definition.

4.8 Tables 8A and 8B. *Gravity Corrections at Sea Level at Various Latitudes or at Various Values of Gravity.* In Table 8A are given the gravity corrections at sea level for a column of any liquid in any unit of height, mm, cm, in., or ft, against latitude. The values of gravity corresponding to latitude are from the formulas and tables given in the Smithsonian Meteorological Tables, 1958 edition.

Table 8B presents the gravity corrections, again for any liquid or in any unit of height, against evenly divided values of gravity. This table is convenient to use if the value of gravity for the station is known. It is preferable to base corrections on the actual value of gravity of the station, if available. The U. S. Coast and Geodetic Survey, (Washington, D. C.) has determined and can furnish values of gravity for hundreds of locations in the United States.

When Table 8A is used, an additional correction is required for the elevation of the station above sea level. The altitude correction to be applied to the value of gravity, which is strictly speaking only for free air and applies only approximately to large elevations found in mountainous areas, is as follows:

$$\text{Correction (cm/sec}^2) = -94 \times 10^{-6} h$$

where h is the elevation in feet of the manometer above sea level.

As an example of the use of the tables, assume a liquid column height of 72 inches (after application of scale correction, before or after application of the temperature correction) at latitude 38 degrees and elevation above sea level of 150 feet.

From Table 8A the value of gravity at sea level at 38 degrees latitude is 979.997 cm/sec^2.

The altitude correction is $-94 \times 150 \times 10^{-6}$ or -0.014.

The value of gravity is 979.983 cm/sec^2.

By interpolation in Table 8B, the gravity correction is -0.048 in.

The column height corrected for gravity is 72 -0.048 or 71.952 inches.

Neglecting the altitude correction and using Table 8A to obtain the correction, there is obtained 0.049 inch. Correcting gravity for the altitude effect introduces negligible error in this case.

4.9 Table 9. *Conversion Factors for Various Pressure Units.* This table was computed based largely on the data given in Tables 2, 6, and 7. The number of digits of the factors is greater than required for most manometer use. The factors were used insofar as applicable in preparing the other tables.

The factors apply only to manometer readings after all corrections have been made to obtain the pressures in the units stated.

4.10 Table 10. *Properties of Manometer Liquids.* This table is not complete, either in the properties of the liquids listed or in the best liquids which might be used for a particular application. Members of the ISA who have additional data are urged to furnish it so that this table can be perfected.

Water is probably the most widely used manometric liquid, being not only low in cost, readily available, and stable in nature, but also having a low factor of thermal expansion. Its major disadvantages are its limited range of utility — usually from 0.05 to 2.0 psi, and only 40°F to 100°F — and its corrosive effect on ferrous metals. One of the most desirable methods of increasing its visibility in glass tubing is by the addition of a few drops of fluorescein solution; coloring by inks or dyes is not satisfactory.

Mercury is the second most widely used manometric liquid, being quite stable and reasonably available, immiscible with other liquids, and having a low thermal expansion and low vapor pressure. It can be used over a wide range of temperature conditions — from −35°F to over 250°F, and normally in the pressure range from 0.2 psi to 20-30 psi. Many metals, such as copper, tin, silver, and zinc are soluble in mercury, so that instrument parts in contact with it should be glass, carbon steel, stainless steel or good grade iron.

The major use of kerosine as a manometric liquid is a substitute for water in locations of low temperatures; for many years steel mills and coke oven plants have used "mm of oil" as a manometric unit on exposed mains. Normal practice specifies the use of 40-41 degrees API kerosine, having a specific gravity of 0.8200-0.8204 at 60°F. It has some advantages over water in that it can be used at low temperatures, will not become discolored by rusting of iron fittings with which it is in contact, and provides slightly more accurate measurements because of its lower density. Its use provides a fire hazard, and it cannot be used at temperatures above 75-85°F because of vapor phase and meniscus effects.

Ellison gage oil is a special petroleum fraction, similar to kerosine, having a specific gravity of 0.830-0.834 designed for use in inclined gages measuring low differentials in pressure.

Alcohol, and mixtures of alcohol and ethylene glycol, are quite frequently used as substitutes for water in locations in which temperatures below 35°F are encountered. Ethyl alcohol, having a specific gravity of only 0.794, may provide a slightly more sensitive measurement of pressures, but has the drawbacks of being inconstant in density due to absorption of water vapor, becoming discolored by rusting, promoting a fire hazard, and being undesirable above 90-100°F. A mixture of 36 per cent by volume of ethyl alcohol, and 64 per cent of ethylene glycol (or Prestone) has a specific gravity at 60°F comparable to that of water, and is therefore frequently used in water manometers for below freezing temperatures. However, both alcohol and glycol have relatively large thermal expansion coefficients which, if not taken into account when these substances are used at low temperatures, may lead to serious calibration errors.

Glycerine finds some use as a manometric liquid at temperatures above those at which water or kerosine may be used from 100°F to 250°F. At ordinary temperatures it is relatively viscous and slow draining; deliquescence alters its density.

Both dibromobenzene and dibromoethane have densities approximately twice as great as that of water, and they are therefore used to obtain more accurate measurements within the range 10 to 150 in H_2O than can be obtained by use of mercury. Both have the serious drawback of being hydrolyzed by water and becoming slightly acidic; they should be used only in glass tubes or instruments with special brass chambers and fittings.

Acetylene tetrabromide, or tetrabromoethane, is another high density liquid which is used for measuring pressures greater than those that can be conveniently reached by use of water columns, i.e., equivalents of 5 to 250 in H_2O or 0.2 to 10.0 psi. In spite of a greater degree of bromination, tetrabromoethane appears to be somewhat more stable and less reactive than dibromoethane, but not sufficiently so to permit use in contact with appreciable concentrations of water vapor, nor in instruments that are not provided with special brass chambers or fittings.

Butyl phthalate has low vapor pressure and forms a satisfactory meniscus. It has a relatively low freezing point. It absorbs air readily to a degree which may be troublesome, but no quantitative data were found.

Liquid mixtures such as alcohol-water, alcohol-glycol, and liquids which absorb water readily, such as glycol, alcohol, and glycerin do not hold their initial densities in a manometer open to the air and therefore must be used with caution.

Many organic liquids, such as the petroleum derivatives, Halowax oil and others, have densities which vary from lot to lot. The petroleum derivatives are mixtures, while the organic liquids in the technical grades contain impurities to a degree sufficient to affect the density. The densities given are therefore approximate and should be determined for each lot if pressures are to be measured accurately. Comparison of manometers containing liquids of unknown density

against a water column is a convenient method of determining their density.

4.11 Table 11. *Expansion Coefficients of Scale Materials.* The length of the scale is determined from the coefficients of linear expansion of scale materials by the relation

$$L = L_o(1+at) \qquad \text{for t in deg C}$$
$$L = L_o\{1+b(t-32)\} \qquad \text{for t in deg F}$$

where L is the length at temperature t, L_o is the length at 0°C or 32°F, and a and b are the expansion coefficients per C and per F respectively.

The above relation is not exact, but the error is tolerable for the small temperature ranges ordinarily experienced by manometers. In fact, if the scale length is known at some other temperature than 0°C, the above relations can be written without serious loss in accuracy as follows:

$$L = L_1\{1+a(t-t_1)\} \qquad \text{for t in deg C}$$
$$L = L_1\{1+b(t-t_1)\} \qquad \text{for t in deg F}$$

where L_1 is the length measured at t_1 and the other terms are defined as above.

It is obvious for precise measurements, that the variations in the coefficients for a given material are such that the coefficient of expansion of the scale material must be measured. Knowledge of the composition of the scale material, and of the degree of cold work for metals, may often serve to pick up a reasonably accurate value from the literature on the subject.

4.12 Table 12. *Conversion Factors for Water Columns at Various Temperatures.* For a number of temperatures of the water column, the equivalent psi is given for one inch of water, and the equivalent height of a mercury column at 32°F and of water columns at 60°F, 68°F (20°C), and 77°F (25°C) are given for unit height of a water column in any unit as cm, in., etc. The table must be entered in all cases with the true height, that is, the indicated height must be corrected for scale errors and for the effect of temperature on the scale, and any other errors, if significant.

If psi corresponding to one cm of water are desired, multiply the given values in psi by 0.3937.

As an example, find the equivalent psi, in. of mercury and in. of water at 68°F of 35 in. of water at 75°F. Entering the table at 75°F there is obtained

$$35 \times 0.036031 = 1.261 \text{ psi}$$
$$35 \times 0.073359 = 2.568 \text{ in. of mercury at 0°C}$$
$$35 \times 0.9991 = 34.97 \text{ in. of water at 68°F}$$

4.13 Tables 13 and 14. *Temperature Corrections for Water Manometers with Brass Scales to Obtain Reading at 68°F (20°C) and at 60°F.* These tables were constructed from the following relation:

$$\text{Correction} = \left[\overline{\frac{D-D_o}{D_o}} + s(t-t_o) \right] h$$

where D is density of water at temperature t; D_o the density at the reference temperature t_o which is 68°F in Table 13 and 60°F in Table 14; s is the linear coefficient of expansion of brass; and h is the indicated height of the water column at temperature t after correction for scale error. The temperatures are in degrees Fahrenheit.

This form of the equation has to be used rather than equation (4) of the section on "Temperature Errors", because the coefficient of expansion of water varies too much with temperature to be assumed constant.

Table 11 shows that the coefficient s varies for brass. The value was taken as 10.2×10^{-6} per degree F following the usual practice in barometry. A deviation in s of 10 per cent from the value chosen will on the average affect the corrections in the order of 1 per cent. The density of water was obtained from Table 7.

Either table may be entered with the height of the column in inches or mm, or any other unit, to obtain the temperature correction in the same unit. Also, it has been assumed that the scales are graduated to indicate correctly at 68°F in Table 13, and 60°F in Table 14. The tables can be used nevertheless if the effect of deviations from this assumption can be lumped in with the scale error.

As an example, assume a manometer with a brass scale to read 720 mm at a manometer temperature of 82°F, to obtain the height of the water column at 68°F. Assume the scale correction to be −1 mm when calibrated at 68°F. Applying this correction gives a reading of 719 mm. Enter Table 13 successively with 700, 100 and 900 mm, and interpolate between 80 and 85°F. The temperature correction is then, properly locating the decimal point, 1.24 + 0.02 + 0.02 or −1.28 mm, and the reading corrected for temperature is 717.7 mm.

4.14 Tables 15A and 15B. *Equivalents in psi of Water Columns at Various Temperatures in Inches and in Feet and Inches.* These tables were constructed from the conversion factors given in Table 9. The tables are entered with the true height of the water column at the designated temperature, that is, entered with manometer readings corrected for all errors.

As an example of the use of Table 15B, 8 ft 11½ in. of water at 95°F = 3.447 + 0.395 + 0.018 = 3.860 psi.

4.15 Table 16. *Conversion Factors for Mercury Columns at Various Temperatures.* The table presents the equivalent of one inch of mercury in psi at the designated temperature, and the factor for converting the height of a mercury column at the designated temperature to the height of a mercury column at 32°F (0°C) and to the height of a water column at any one of the three temperatures, 60, 68, or 77°F. The conversion factor can be used to convert the mercury column in any unit of height to obtain the height in the same unit of the desired liquid column. The table should be entered with the true height of the mercury column at the designated temperature.

The computations are based on the data given in Tables 6, 7, and 9.

As an example, convert 25.00 mm of mercury at 85°F to psi, to a mercury column at 32°F, and to a water column at 68°F.

(a) 25 x 0.48854 = 12.21 psi
(b) 25 x 0.9947 = 24.87 mm of mercury at 32°F
(c) 25 x 13.547 = 338.7 mm of water at 68°F

To obtain psi for 1 mm of mercury, multiply the psi factor given in the table by 0.03937.

4.16 Tables 17 and 18. *Temperature Corrections for Mercury Manometers and Barometers with Brass Scales to Obtain Readings at 0°C (32°F).* The two tables are alike except that one is for deg C and the other is for deg F. The indicated column heights given in the tables were selected for convenience: if temperatures are measured in deg C, the manometer is more likely to be graduated in millimeters; in deg F, in inches. However, either table may be entered with the liquid height either in millimeters or inches, or any other unit, to obtain the temperature correction in the same unit.

The corrections were computed from equation (4) of the section on "Temperature Errors".

$$\text{Correction} = \frac{s(t-t_o) - m(t-t_o)}{1 + m(t-t_o)} H_t$$

where H_t = reading of the manometer at temperature t corrected for scale error.

  t = temperature of the manometer in deg C for Table 17, in deg F for Table 18.

  t_o = 0°C for Table 17, 32°F for Table 18.

  m = coefficient of expansion of mercury, 0.0001818 per deg C for Table 17, 0.0001010 for Table 18.

  s = linear coefficient of expansion for the brass scale, 0.0000184 for deg C for Table 17, 0.0000102 per deg F for Table 18.

It is obvious that the correction is linear with the reading and very nearly so with temperature, sufficiently for the purposes of interpolation in the tables.

Brass scales will not necessarily have the value given above for the coefficient of expansion, which is that commonly used in barometry. A deviation of as much as 10 per cent from the value used will introduce an error of about 1 per cent in the correction.

It will be noted that the brass scale has been assumed accurate at 32°F or 0°C. Such is not generally the case, but the corrections will still apply with high accuracy if the error due to this assumption is merged with the scale error. This procedure is facilitated by the common practice of calibrating a manometer against a "standard" manometer.

Table 17 for deg C is identical with the barometer temperature correction table for deg C in the Smithsonian Meteorological Tables (SMT). Table 18 for deg F differs from that in the SMT in that the brass scale is assumed calibrated correctly at 62°F (t_o in the term $s(t-t_o)$ in the equation above equals 62°F) in SMT, instead of 32°F as here assumed. Since neither temperature is in correspondence with normal practice, it seems more practical and convenient to use 32°F.

The corrections apply only to manometers and barometers in which the scale is read at both liquid levels, or the zero of the scale and the lower liquid level are brought to coincidence. In instruments of the design, including the cistern type, where the scale is calibrated to take care of the rise or fall of one of the liquid surfaces, the corrections given in the tables do not apply. The effect on the reading of the expansion of the mercury in the cistern, or below the lowest surface of the mercury, is not covered by the equation given above. However, the corrections can be used for cistern type instruments if the reading with which the table is entered be increased roughly by the height of the mercury in the cistern at zero differential pressure. The exact amount of this increase must be determined by tests at two temperatures.

As an example, assume that a U-tube manometer with a brass scale at 75°F reads 22.21 inches of mercury after correction for scale error. From Table 18 the correction is 0.078 + 0.42 x 0.019 or 0.086 inch. The corrected reading is 22.21 − 0.09 or 22.12 inches of mercury.

4.17 Table 19. *Equivalents in psi of Mercury Columns in Inches of Mercury at Various Temperatures.* This table was computed from the data given in Tables 6 and 9. The heights of the mercury column are in true inches of mercury at the temperature designated, and should be in terms of the height under the condition of standard gravity.

4.18 Tables 20A and 20B. *Equivalents of Mercury Columns at Various Temperatures to Water Columns at 20°C (68°F) and at 60°F.* These tables are computed from the data given in Tables 6 and 7. They can be entered with true inches of mercury at the designated temperature, either before or after correcting the height for gravity error. If the mercury height is uncorrected for gravity error, the equivalent water column will be also.

4.19 Table 21. *Density of Kerosine and Equivalent Pressures in psi of Kerosine Columns at Various Temperatures.* This table is for kerosine of specific gravity 0.8200 at 60/60 (41.06 API degrees Baume). The densities at various temperatures are based on the specific gravity data given in the National Standard Petroleum Tables, National Bureau of Standards Circular 410, 1936.

In practice, it is necessary to select the kerosine to have a specific gravity of 0.8200 at 60/60 in order for the table to apply. Evaporation of any components of the kerosine will affect the density, but data on this possibility are not available.

Where the height, one inch or 100 mm, is specified, the heights are in true inches at the temperature designated, corrected for gravity error for which Tables 8A and 8B are applicable.

As an example, a kerosine column of 52.08 true mm at 75°F equals

(a) 52.08 x 0.00157 = 0.06026 psi
(b) 52.08 x 0.8134 = 42.36 g/sq cm

4.20 Table 22. *Equivalents in psi of Columns of Kerosine in Inches at Various Temperatures.* The definition of the kerosine and basic data for the table are as indicated in Table 21.

The heights of the kerosine column are in true inches at the temperature designated, corrected for gravity error, for which Tables 8A and 8B are applicable.

As an example, a kerosine column of 45.28 inches at 75°F, with all corrections applied, equals

1.176 + 0.147 + 0.006 + 0.002 = 1.331 psi

4.21 Table 23. *Equivalents in Inches of Water at 60°F of Kerosine Columns in mm at Various Temperatures.* This table is for kerosine of the same density as for Tables 21 and 22. The table was computed with the aid of Table 3 of NBS Circular 410, National Standard Petroleum Tables.

The table must be entered with heights in true mm and inches. Conversion may be made before or after applying correction for gravity, as convenient.

As an example, 115 mm of kerosine at 75°F equals 3.205 + 0.320 + 0.161 or 3.686 inches of water at 60°F.

4.22 Table 24. *Equivalents in Inches of Water at 60°F of Gage Oil Columns at Various Temperatures.* This oil is a kerosine known commercially as "Ellison Gage Oil". The data given are based on a coefficient of volume expansion of the oil about 4 per cent less than for an oil of the same specific gravity 60/60 given in NBS Handbook C410. The spread in the densities with temperature is therefore about 4 per cent greater in the table.

The table serves equally well to convert oil columns in any unit of height to water columns in the same unit of height, as mm of oil to mm of water.

The table must be entered only with true inches or other units of height at the designated temperature. Gravity corrections can be applied before or after conversion.

As an example, 18.72 inches (or mm) of oil at 75°F equals 8.282 + 6.625 + 0.584 + 0.017 or 15.51 inches (or mm) of water at 60°F.

4.23 Table 25. *Equivalents in Inches of Water at 60°F of Alcohol Columns at Various Temperatures.* This table is for 100 per cent ethyl alcohol which, unfortunately, is not easily obtained or maintained in manometers. For 95 per cent alcohol by weight, the remainder water, the density increases about 2 per cent and correspondingly increases the equivalent column height in inches of water.

The table can be used equally well to convert alcohol columns in any unit of height to water columns in the same unit of height, as mm of alcohol to mm of water.

The table applies only to true inches or other units of height at the designated temperatures.

As an example, 20.52 inches (or mm) of alcohol at 75°F equals 15.74 + 0.39 + 0.02 or 16.15 inches (or mm) of water.

4.24 Table 26. *Equivalents in Inches of Water at 60°F of Alcohol-Glycol Columns at Various Temperatures.* This mixture is so chosen that the specific gravity is the same as that of water at 60°F, but as indicated in the table differs significantly from the specific gravity of water at other temperatures. It is useful when the manometer temperature is below 32°F. At these temperatures, the use of carbitol or butyl phthalate, or perhaps alcohol alone, may be preferable. See Table 10.

The application of the table to other units of height and the restriction to true heights are as stated under Table 25.

As an example, 29.22 inches (or mm) of alcohol-glycol at 75°F equals 19.87 + 8.94 + 0.20 + 0.02 or 29.03 inches (or mm) of water at 60°F.

4.25 Table 27. *Equivalents in Inches of Water at 60°F of Dibromobenzene Columns at Various Temperatures.* The Handbook of Physics and Chemistry gives the specific gravity of o-dibromobenzene as $1.9557 \frac{20.5}{4}$ which equals $1.9576 \frac{69}{60}$. It has the serious drawback of being hydrolyzed by water and becoming slightly acidic. It should be used only in glass tubes or instruments with brass chambers and fittings.

The application of the table to other units of height and the restriction to true heights are as stated under Table 25.

As an example, 29.22 inches (or mm) of dibromobenzene at 75°F equals 39.06 + 17.58 + 0.39 + 0.04 or 57.07 inches (or mm) of water.

4.26 Table 28. *Equivalents in psi of Dibromobenzene Columns at Various Temperatures.* The height of the column should be entered with true inches at the designated temperature with corrections applied for all errors.

As an example, 29.22 inches of dibromobenzene at 75°F equals 1.410 + 0.635 + 0.014 + 0.001 or 2.060 psi.

4.27 Table 29. *Equivalents in Inches of Water at 60°F of Dibromoethane (also known as ethylidene bromide) Columns at Various Temperatures.* The Handbook of Chemistry and Physics gives $2.089 \frac{20.5}{4}$ (equals $2.09 \frac{69}{60}$ for the specific gravity and 108-110°C (226-230°F) for the boiling point. In Table 29 the specific gravity is $2.045 \frac{70}{60}$ which is 2.2 per cent less than the above value. The lower value apparently applies to the technical grade of the liquid.

Dibromoethane like dibromobenzene is hydrolyzed by water and becomes slightly acidic. The same restrictions apply as to the use of glass tubes, or brass, in manometers.

The application of the table to other units of height and the restriction to true heights are as stated under Table 25.

As an example, 29.22 inches (or mm) of dibromoethane at 75°F equals 40.79 + 18.36 + 0.41 + 0.04 or 59.60 inches (or mm) of water at 60°F.

4.28 Table 30. *Equivalents in psi of Dibromoethane Columns at Various Temperatures.* The table should be entered with true inches at the designated temperature with corrections applied for all errors.

As an example, 29.22 inches of dibromoethane at 75°F equals 1.472 + 0.663 + 0.015 + 0.002 or 2.152 psi.

4.29 Table 31. *Equivalents in Inches of Water at 60°F of Columns of Acetylene Tetrabromide at Various Temperatures.* The Handbook of Chemistry and Physics gives the specific gravity of 1, 1, 2, 2-tetrabromoethane (or acetylene tetrabromide) as $2.9638 \frac{20}{4}$, equivalent to $2.9667 \frac{68}{60}$. In Table 31 it is $2.9614 \frac{68}{60}$, which is 0.17 per cent lower than for the presumably pure chemical.

The application of the table to other units of height and the restriction to true heights are stated under Table 25.

As an example, 29.22 inches (or mm) of acetylene tetrabromide at 75°F equals 59.04 + 26.57 + 0.59 + 0.06 or 86.26 inches (or mm) of water at 60°F.

4.30 Table 32. *Equivalents in psi of Acetylene Tetrabromide at Various Temperatures.* The table should be entered with true inches at the designated temperature with corrections applied for all errors.

In spite of a greater degree of bromination, tetrabromoethane appears to be somewhat more stable and less reactive than dibromoethane, but not sufficiently so to permit use in contact with appreciable concentrations of water vapor, nor in instruments that are not provided with special brass chambers or fittings.

As an example 29.22 inches of acetylene tetrabromide at 75°F equals 2.131 + 0.959 + 0.021 + 0.002 or 3.113 psi.

Table 1

ABBREVIATIONS

Term	Abbreviations
absolute	abs
atmosphere	atm
boiling point	bp
centimeter	cm
cubic	cu
cubic centimeter	cu cm or cm³
cubic foot	cu ft
cubic inch	cu in.
cubic meter	cu m or m³
foot	ft
freezing point	fp
gram	g
grams per cubic centimeter	g/cu cm or g/cm³
grams per square centimeter	g/sq cm or g/cm²
inch	in.
inches of mercury	in. Hg
inches of water	in. H₂O
kilogram	kg
kilograms per square meter	kg/sq m or kg/m²
melting point	mp
millibars	mb

Term	Abbreviations
micron (0.001 mm)	μ or mu
milliliter	ml
millimeter	mm
millimeters of mercury	mm Hg
millimeters of water	mm H₂O
ounce	oz
ounce avoirdupois	oz avdp
ounces per square inch	oz/sq in. or oz/in.²
pound	lb
pounds per cubic inch	lb per cu in.
pounds per cubic foot	lb per cu ft
pounds per square foot	psf
pounds per square inch	psi
pounds per square inch gage	psig
pounds per square inch absolute	psia
specific gravity	sp gr
square centimeter	sq cm or cm²
square foot	sq ft
square inch	sq in.
square meter	sq m or m²
square millimeter	sq mm or mm²
temperature	temp

Table 2

FUNDAMENTAL CONSTANTS AND CONVERSION FACTORS

Largely from "Units of Weight and Measure"
National Bureau of Standards Misc. Publ. No. M233

Fundamental constants are underlined

1 cm = 0.3937 inches
1 meter = 39.37 inches
1 meter = 3.280840 feet

1 inch = 2.54 cm

1 foot = 30.48 cm

1 sq cm = 0.1550003 sq inch
1 sq meter = 10.76391 sq feet

1 sq inch = 6.4516 sq cm
1 sq foot = 929.0304 sq cm

1 cu cm = 0.06102374 cu inch
1 cu meter = 35.31467 cu feet

1 cu inch = 16.387064 cu cm
1 cu foot = 0.028316847 cu meter

1 kilogram = 2.204623 pounds
1 gram = 0.03527397 ounce

1 pound = 453.59237 grams
1 ounce = 28.349523 grams

1 liter = 1000.028 cu cm
1 liter = 61.02545 cu inches
1 cu cm = 0.9999720 liter

1 cu inch = 0.01638661 liter

1 gram/cu cm = 62.4280 lb/cu foot
1 gram/cu cm = 0.0361273 lb/cu inch
1 gram/ml = 0.9999730 gram/cu cm

1 lb/cu foot = 0.0160185 gram/cu cm
1 lb/cu inch = 27.6799 gram/cu cm
1 gram/cu cm = 1.000028 grams/ml

Standard acceleration of gravity = 980.665 cm/sec² = 32.1740 ft/sec²
(International value in physics and engineering)

Table 3
DECIMAL EQUIVALENTS OF FRACTIONS OF ONE INCH

1-64	0.015625	17-64	0.265625	33-64	0.515625	49-64	0.765625
1-32	0.031250	9-32	0.281250	17-32	0.531250	25-32	0.781250
3-64	0.046875	19-64	0.296875	35-64	0.546875	51-64	0.796875
1-16	0.062500	5-16	0.312500	9-16	0.562500	13-16	0.812500
5-64	0.078125	21-64	0.328125	37-64	0.578125	53-64	0.828125
3-32	0.093750	11-32	0.343750	19-32	0.593750	27-32	0.843750
7-64	0.109375	23-64	0.359375	39-64	0.609375	55-64	0.859375
1-8	0.125000	3-8	0.375000	5-8	0.625000	7-8	0.875000
9-64	0.140625	25-64	0.390625	41-64	0.640625	57-64	0.890625
5-32	0.156250	13-32	0.406250	21-32	0.656250	29-32	0.906250
11-64	0.171875	27-64	0.421875	43-64	0.671875	59-64	0.921875
3-16	0.187500	7-16	0.437500	11-16	0.687500	15-16	0.937500
13-64	0.203125	29-64	0.453125	45-64	0.703125	61-64	0.953125
7-32	0.218750	15-32	0.468750	23-32	0.718750	31-32	0.968750
15-64	0.234375	31-64	0.484375	47-64	0.734375	63-64	0.984375
1-4	0.250000	1-2	0.500000	3-4	0.750000	–	–

Table 4
MILLIMETER EQUIVALENTS OF INCHES
Millimeters

Inches	0	0.1	0.2	0.3	0.4	0.5	0.6	0.7	0.8	0.9
0	0.00	2.54	5.08	7.62	10.16	12.70	15.24	17.78	20.32	22.86
1	25.40	27.94	30.48	33.02	35.56	38.10	40.64	43.18	45.72	48.26
2	50.80	53.34	55.88	58.42	60.96	63.50	66.04	68.58	71.12	73.66
3	76.20	78.74	81.28	83.82	86.36	88.90	91.44	93.98	96.52	99.06
4	101.60	104.14	106.68	109.22	111.76	114.30	116.84	119.38	121.92	124.46
5	127.00	129.54	132.08	134.62	137.16	139.70	142.24	144.78	147.32	149.86
6	152.40	154.94	157.48	160.02	162.56	165.10	167.64	170.18	172.72	175.26
7	177.80	180.34	182.88	185.42	187.96	190.50	193.04	195.58	198.12	200.66
8	203.20	205.74	208.28	210.82	213.36	215.90	218.44	220.98	223.52	226.06
9	228.60	231.14	233.68	236.22	238.76	241.30	243.84	246.38	248.92	251.46
10	254.00	256.54	259.08	261.62	264.16	266.70	269.24	271.78	274.32	276.86

Table 5
INCH EQUIVALENTS OF MILLIMETERS
Inches

mm	0	1	2	3	4	5	6	7	8	9
0	0.0000	0.0394	0.0787	0.1181	0.1575	0.1969	0.2362	0.2756	0.3150	0.3543
10	0.3937	0.4331	0.4724	0.5118	0.5512	0.5906	0.6299	0.6693	0.7087	0.7480
20	0.7874	0.8268	0.8661	0.9055	0.9449	0.9842	1.0236	1.0630	1.0124	1.1417
30	1.1811	1.2205	1.2598	1.2992	1.3386	1.3780	1.4173	1.4567	1.4961	1.5354
40	1.5748	1.6142	1.6535	1.6929	1.7323	1.7717	1.8110	1.8504	1.8898	1.9291
50	1.9685	2.0079	2.0472	2.0866	2.1260	2.1654	2.2047	2.2441	2.2835	2.3228
60	2.3622	2.4016	2.4409	2.4803	2.5197	2.5590	2.5984	2.6378	2.6772	2.7165
70	2.7559	2.7953	2.8346	2.8740	2.9134	2.9528	2.9921	3.0315	3.0709	3.1102
80	3.1496	3.1890	3.2283	3.2677	3.3071	3.3464	3.3858	3.4252	3.4646	3.5039
90	3.5433	3.5827	3.6220	3.6614	3.7008	3.7402	3.7795	3.8189	3.8583	3.8976
100	3.9370	3.9764	4.0157	4.0551	4.0945	4.1338	4.1732	4.2126	4.2520	4.2913

Table 6

DENSITY OF MERCURY

At 0 C (32 F) = 13.5951 grams per cu cm
= 0.491157 lb weight per cu in.
= 848.719 lb weight per cu ft

At other temperatures the density can be computed from the following:

$V = V_o (1 + 0.0001818\ t)$ where t is in deg C (1)

$= V_o \left[1 + 0.0001010\ (t - 32) \right]$ where t is in deg F (2)

$D = 1/V$ (3)

where V = specific volume at temperature t
V_o = specific volume at 0 C or 32 F in the same unit as V
D = the mass per unit volume

Density of mercury from Smithsonian Physical Tables. These values agree with those computed from formulas (1) and (3) above.

Temperature deg C	deg F	Density g per cu cm
−20	−4	13.6446
−10	14	13.6198
0	32	13.5951
5	41	13.5827
10	50	13.5704
15	59	13.5581
20	68	13.5458
25	77	13.5336
30	86	13.5214
35	95	13.5091
40	104	13.4970
45	113	13.4847
50	122	13.4725
60	140	13.4482
70	158	13.4240
80	176	13.3998
90	194	13.3755
100	212	13.3515

Temperature deg F	Density lb per cu in.	g per cu cm
0	0.49275	13.6391
10	0.49225	13.6253
20	0.49175	13.6116
32	0.491157	13.5951
40	0.49076	13.5841
50	0.49026	13.5704
60	0.48977	13.5568
70	0.48928	13.5431
80	0.48879	13.5295
90	0.48830	13.5159
100	0.48780	13.5023
110	0.48732	13.4888
120	0.48683	13.4753
130	0.48634	13.4617
140	0.48585	13.4482
150	0.48536	13.4347
160	0.48488	13.4213
170	0.48439	13.4079
180	0.48391	13.3944
190	0.48342	13.3809
200	0.48293	13.3675

Table 7

DENSITY OF WATER

Distilled and free from air

From NBS Jnl. of Research, v. 18, Feb. 1937, RP 971 and from Table 93, volume on "Water-Substance" by N. E. Dorsey, 1940.

Temperature deg C	Density g per cu cm	lb per cu in.
0	0.999841	0.0361218
3.98	0.999973	0.0361265
5	0.999965	0.0361262
10	0.999701	0.0361167
15	0.999102	0.0360951
20	0.998207	0.0360627
25	0.997048	0.0360209
30	0.995651	0.0359704
35	0.994037	0.0359121
40	0.992221	0.0358465
45	0.99021	0.035774
50	0.98804	0.035695
60	0.98324	0.035522

Temperature deg F	Density lb per cu in.	g per cu cm
35	0.0361232	0.999882
40	0.0361265	0.999971
45	0.0361236	0.999892
50	0.0361167	0.999701
55	0.0361060	0.999406
60	0.0360919	0.999015
65	0.0360746	0.998536
68	0.0360627	0.998207
70	0.0360542	0.997971
75	0.0360309	0.997327
80	0.0360050	0.996608
85	0.0359764	0.995819
90	0.0359454	0.994960
95	0.0359121	0.994037
100	0.0358764	0.993051
105	0.0358387	0.992006
110	0.035799	0.99090
115	0.035757	0.98974
120	0.035713	0.98854
130	0.035622	0.98600
140	0.035522	0.98324
150	0.035414	0.98025
200	0.034792	0.96304

Table 8A

Gravity Corrections at Sea Level at Various Latitudes for Manometers filled with any Liquid.

Corrections in the same unit as the height of the column.

Latitude degrees	Gravity cm/sec²	Height of liquid column in any unit 20	40	60	80	100
0	978.036	-0.054	-0.107	-0.161	-0.214	-0.268
10	978.191	-0.050	-0.101	-0.151	-0.201	-0.252
20	978.638	-0.041	-0.083	-0.124	-0.165	-0.207
25	978.955	-0.035	-0.070	-0.104	-0.139	-0.174
30	979.324	-0.027	-0.055	-0.082	-0.110	-0.137
35	979.731	-0.019	-0.038	-0.057	-0.076	-0.095
40	980.167	-0.010	-0.020	-0.030	-0.040	-0.051
45	980.616	-0.001	-0.002	-0.003	-0.004	-0.005
—	980.665	0	0	0	0	0
50	981.065	+0.008	+0.017	+0.025	+0.033	+0.041
55	981.500	+0.017	+0.034	+0.052	+0.069	+0.086
60	981.911	+0.025	+0.051	+0.076	+0.102	+0.127
65	982.281	+0.033	+0.066	+0.099	+0.132	+0.166
70	982.601	+0.040	+0.079	+0.119	+0.158	+0.197
80	983.051	+0.049	+0.097	+0.146	+0.195	+0.243
90	983.208	+0.052	+0.104	+0.156	+0.207	+0.259

Table 8B

GRAVITY CORRECTIONS FOR MANOMETERS FILLED WITH ANY LIQUID

Corrections in the same unit as the height of the column.

Gravity cm/sec²	Height of liquid column in any unit 20	40	60	80	100
978.0	-0.054	-0.109	-0.162	-0.217	-0.272
978.5	-0.044	-0.088	-0.132	-0.177	-0.221
979.0	-0.034	-0.068	-0.102	-0.136	-0.170
979.5	-0.024	-0.048	-0.071	-0.095	-0.119
980.0	-0.014	-0.027	-0.041	-0.054	-0.068
980.5	-0.003	-0.007	-0.010	-0.013	-0.017
980.665	0	0	0	0	0
981.0	+0.007	+0.014	+0.021	+0.027	+0.034
981.5	+0.017	+0.034	+0.051	+0.068	+0.085
982.0	+0.027	+0.054	+0.082	+0.109	+0.136
982.5	+0.037	+0.075	+0.112	+0.150	+0.187
983.0	+0.048	+0.095	+0.143	+0.191	+0.238
983.5	+0.058	+0.116	+0.174	+0.231	+0.289

Table 9

CONVERSION FACTORS FOR VARIOUS PRESSURE UNITS

Equivalent Value in Various Units

Pressure Unit Value	mm mercury 0°C	inches mercury 0°C	millibars	pounds per sq in.	pounds per sq ft	ounces per sq in.	grams per sq cm	cm water 60°F	in. water 60°F	cm water 20°C	in. water 20°C	cm water 25°C	in. water 25°C	atmosphere
mm mercury 0°C	1	0.03937	1.3332	0.019337	2.7845	0.30939	1.3595	1.3609	0.53577	1.3619	0.53620	1.3635	0.53682	1.31579×10^{-3}
in. mercury 0°C	25.400	1	33.864	0.49115	70.726	7.8585	34.531	34.566	13.609	34.593	13.619	34.634	13.635	0.0334210
millibars	0.75006	0.029530	1	0.014504	2.0885	0.23206	1.0197	1.0207	0.40186	1.0215	0.40218	1.0227	0.40265	9.86923×10^{-4}
lb per sq in.	51.715	2.0360	68.948	1	144	16	70.307	70.376	27.707	70.433	27.730	70.515	27.762	0.0680460
lb per sq ft	0.35913	0.014139	0.47880	0.0069444	1	0.11111	0.48824	0.48872	0.19241	0.48912	0.19257	0.48969	0.19279	4.72540×10^{-4}
oz per sq in.	3.2322	0.12725	4.3092	0.0625	9	1	4.3942	4.3985	1.7317	4.4021	1.7331	4.4072	1.7351	4.25286×10^{-3}
grams per sq cm	0.73556	0.028959	0.98067	0.014223	2.0482	0.22757	1	1.0010	0.39409	1.0018	0.39441	1.0030	0.39487	9.67842×10^{-4}
cm water 60°F	0.73483	0.028930	0.97970	0.014209	2.0461	0.22735	0.99901	1	0.3937	1.0008	0.39402	1.0020	0.39448	9.66887×10^{-4}
in. water 60°F	1.8665	0.073483	2.4884	0.036092	5.1972	0.57747	2.5375	2.5400	1	2.5421	1.0008	2.5450	1.0020	2.45589×10^{-3}
cm water 20°C	0.73424	0.028907	0.97891	0.014198	2.0444	0.22717	0.99821	0.99919	0.39338	1	0.3937	1.0012	0.39416	9.66105×10^{-4}
in. water 20°C	1.8650	0.073424	2.4864	0.036063	5.1930	0.57700	2.5354	2.5380	0.99919	2.5400	1	2.5430	1.0012	2.45392×10^{-3}
cm water 25°C	0.73339	0.028873	0.97777	0.014181	2.0421	0.22690	0.99705	0.99803	0.39292	0.99884	0.39324	1	0.3937	9.64984×10^{-4}
in. water 25°C	1.8628	0.073339	2.4835	0.036021	5.1870	0.57633	2.5325	2.5350	0.99803	2.5371	0.99884	2.5400	1	2.45106×10^{-3}
atmosphere	760	29.9213	1013.250	14.69595	2116.22	235.136	1033.227	1034.25	407.184	1035.08	407.511	1036.29	407.987	1

1 millibar = 1000 dynes per sq cm
1 kg per sq meter = 0.1 gram per sq cm
1 kg per sq cm = 1000 grams per sq cm
1 gram per sq meter = 0.0001 gram per sq cm.

Table 10

PROPERTIES OF MANOMETRIC LIQUIDS

Liquid	Specific Gravity 20/20	Action with Water Vapor	Vapor Pressure at 68 F (mm Hg)	Coefficient of Thermal Expansion per deg F $\times 10^6$	per deg C $\times 10^6$	Range deg F	Melting Point deg F	Boiling Point deg F	Flash Point deg F
1. Ethyl Alcohol, C_2H_6O	0.7939	absorbs	43.9	600	1080	50-86	-179	173	55
2. Kerosine, 41 API at 60 F	0.8200 60/60	negligible	-	480	864	30-100	-20	300+	120
3. Ellison Gage Oil	0.8340 60/60	negligible	-	466	839	30-100	-	300+	140
4. Benzene (Benzol), C_6H_6	0.8794	negligible	74.7	687	1237	68	42	176	12
5. Butyl Cellosolve $C_6H_{14}O_2$ (Ethylene Glycol Monobutyl Ether)	0.9019	absorbs	0.85	-	-	-	-100	340	165
6. Water	1.000	-	17.5	115	207	68	32	212	non-inflam.
7. Alcohol Glycol	1.000	absorbs	-	427	769	30-100	-60	173	70
8. Carbitol, $C_6H_{14}O_3$	1.024-30	absorbs	-	-	-	-	-76	202	210
9. n-Butyl Phthalate, $C_{16}H_{22}O_4$ (Diethylene Glycol Mono-ethyl Ether)	1.0477	negligible	10^{-4}	433	780	-	-31	644	340
10. Ethylene Glycol (Glycol), $C_2H_6O_2$	1.1155 20/4	absorbs slowly	0.09	354	638	68	+0.8	387	241
11. Halowax Oil	1.19-1.25	absorbs	0.3-50 C	367	660	-	-24 - -42	554	203
12. Glycerine (Glycerol), $C_3H_8O_3$	1.260 20/4	absorbs	low	281	505	68	64	430	320
13. o-Dibromobenzene, $C_6H_4Br_2$	1.956 20/4	negligible	-	432	778	30-100	35.2	230	150+
14. 1, 1-Dibromoethane, $C_2H_4Br_2$	2.089 20/4	negligible	34.7	532	958	30-100	40	-	75+
15. Acetylene tetrabromide (Tetrabromoethane), $C_2H_2Br_4$	2.964 20/4	absorbs slightly	-	370	660	-	-4	-	non-inflam.
16. Mercury	13.570	negligible	0.0012	101	181.8	-20 to 250	-38	679	non-inflam.

Table 11

EXPANSION COEFFICIENTS OF SCALE MATERIALS

Material	Temperature Range deg C	Coefficient of Linear Expansion Per Degree C $\times 10^{-6}$	Coefficient of Linear Expansion Per Degree F $\times 10^{-6}$	Observer or Source
Slate	20 to 100	6.3 to 8.8	3.5 to 4.9	Griffiths
	0 to 100	9.4	5.2	Watertown Arsenal
	Ordinary temp.	10.4	5.8	Adie
Porcelain	20 to 200	1.6 to 19.6	0.9 to 10.9	Souder and Hidnert
Wood				
Along Grain	a	1. to 11.	0.6 to 6.	Various observers
Across grain	a	32. to 73.	18. to 41.	Various observers
Glass	0 to 300	0.8 to 12.8	0.4 to 7.1	Corning Glass Works
Quartz, fused	20 to 60	0.40	0.22	Souder and Hidnert
Iron	25 to 100	12.0	6.7	Souder and Hidnert
Wrought Iron	20 to 100	12.1	6.7	Circular NBS C447
Cast Iron	20 to 100	8.7 to 11.1	4.8 to 6.2	Circular NBS C447
Steel (carbon)	20 to 100	9.4 to 12.5	5.2 to 6.9	Metals Handbook (1939)
Stainless Steel				
12 Cr	20 to 100	10.0	5.6	Hidnert
18 Cr, 9 Ni	20 to 60	16.1	8.9	Hidnert
Brass	20 to 100	16.9 to 19.4	9.4 to 10.8	Hidnert
Bronze	20 to 100	16.8 to 19.1	9.3 to 10.6	NBS Research Paper RP 1518
Bakelite	20 to 60	21. to 33.	12. to 18.	Souder and Hidnert
Aluminum	0 to 60	23.2	12.9	Hidnert
Hard Rubber b	c	50. to 84.	28. to 47.	Proc. Rubber Technology Conference, London, England (1938)

a. Various temperature ranges between 2 and 100 C
b. Includes terms "ebonite" and "vulcanite"
c. Various temperature ranges between 0 and 100 C

Table 12

CONVERSION FACTORS FOR WATER COLUMNS AT VARIOUS TEMPERATURES

Observed Temperature of Water deg F	deg C	Equiv. Psi of 1 in. of water	Factor for Converting Water Column Height at Various Temperatures to Mercury Column at 32 F	Water Column at 60 F	Water Column at 68 F	Water Column at 77 F
32	0.00	0.036122	0.073544	1.0008	1.0016	1.0028
35	1.67	0.036123	0.073547	1.0009	1.0017	1.0028
39.2	3.98	0.036126	0.073554	1.0010	1.0018	1.0029
45	7.22	0.036124	0.073548	1.0009	1.0017	1.0028
50	10.00	0.036117	0.073534	1.0007	1.0015	1.0027
55	12.78	0.036106	0.073512	1.0004	1.0012	1.0024
60	15.56	0.036092	0.073483	1.0000	1.0008	1.0020
65	18.33	0.036075	0.073448	0.9995	1.0003	1.0015
68	20.00	0.036063	0.073424	0.9992	1.0000	1.0012
70	21.11	0.036054	0.073407	0.9990	0.9998	1.0009
75	23.89	0.036031	0.073359	0.9983	0.9991	1.0003
77	25.00	0.036021	0.073339	0.9980	0.9988	1.0000
80	26.67	0.036005	0.073306	0.9976	0.9984	0.9996
85	29.44	0.035976	0.073248	0.9968	0.9976	0.9988
90	32.22	0.035945	0.073185	0.9959	0.9967	0.9979
95	35.00	0.035912	0.073117	0.9950	0.9958	0.9970
100	37.78	0.035876	0.073045	0.9940	0.9948	0.9960
110	43.33	0.035799	0.072887	0.9919	0.9927	0.9938
120	48.89	0.035713	0.072713	0.9895	0.9903	0.9915
130	54.44	0.035622	0.072526	0.9870	0.9878	0.9889
140	60.00	0.035522	0.072323	0.9842	0.9850	0.9862
150	65.56	0.035414	0.072103	0.9812	0.9820	0.9832
200	93.33	0.034792	0.070837	0.9640	0.9648	0.9659

Table 13

TEMPERATURE CORRECTIONS FOR WATER MANOMETERS WITH BRASS SCALES TO OBTAIN READING AT 68 F

Plus corrections are to be added to, minus corrections to be subtracted from reading.

Enter with indicated column height (mm or inches) and its temperature.

Temperature deg F	deg C	100	150	200	250	300	350	400	450	500	550	600	650	700	750	800	850	900	950	1000
32	0.00	+0.13	+0.19	+0.25	+0.32	+0.38	+0.44	+0.51	+0.57	+0.63	+0.70	+0.76	+0.82	+0.89	+0.95	+1.02	+1.08	+1.14	+1.21	+1.27
35	1.67	0.13	0.20	0.27	0.34	0.40	0.47	0.54	0.60	0.67	0.74	0.80	0.87	0.94	1.01	1.07	1.14	1.21	1.27	1.34
40	4.44	0.15	0.22	0.30	0.37	0.44	0.52	0.59	0.67	0.74	0.81	0.89	0.96	1.04	1.11	1.18	1.26	1.33	1.41	1.48
45	7.22	0.15	0.22	0.29	0.36	0.44	0.51	0.58	0.65	0.73	0.80	0.87	0.94	1.02	1.09	1.16	1.24	1.31	1.38	1.46
50	10.00	0.13	0.20	0.26	0.33	0.39	0.46	0.53	0.59	0.66	0.72	0.79	0.85	0.92	0.98	1.05	1.12	1.18	1.25	1.32
55	12.78	0.11	0.16	0.21	0.27	0.32	0.37	0.43	0.48	0.53	0.59	0.64	0.69	0.75	0.80	0.85	0.91	0.96	1.01	1.07
60	15.56	0.07	0.11	0.15	0.18	0.22	0.25	0.29	0.33	0.36	0.40	0.44	0.47	0.51	0.55	0.58	0.62	0.65	0.69	0.73
65	18.33	+0.03	+0.04	+0.06	+0.07	+0.09	+0.10	+0.12	+0.13	+0.15	+0.16	+0.18	+0.19	+0.21	+0.22	+0.24	+0.25	+0.27	+0.28	+0.30
68	20.00	0.00	0.00	0.00	0.00	0.00	0.00	0.00	0.00	0.00	0.00	0.00	0.00	0.00	0.00	0.00	0.00	0.00	0.00	0.00
70	21.11	-0.02	-0.03	-0.04	-0.05	-0.06	-0.08	-0.09	-0.10	-0.11	-0.12	-0.13	-0.14	-0.15	-0.16	-0.17	-0.18	-0.19	-0.21	-0.22
75	23.89	0.08	0.12	0.16	0.20	0.24	0.28	0.32	0.36	0.41	0.45	0.49	0.53	0.57	0.61	0.65	0.69	0.73	0.77	0.81
77	25.00	0.11	0.16	0.21	0.27	0.32	0.37	0.43	0.48	0.53	0.59	0.64	0.69	0.75	0.80	0.86	0.91	0.96	1.02	1.07
80	26.67	0.15	0.22	0.30	0.37	0.44	0.52	0.59	0.67	0.74	0.81	0.89	0.96	1.04	1.11	1.18	1.26	1.33	1.41	1.48
85	29.44	0.22	0.33	0.44	0.55	0.67	0.78	0.89	1.00	1.11	1.22	1.33	1.44	1.55	1.66	1.78	1.89	2.00	2.11	2.22
90	32.22	0.30	0.45	0.61	0.76	0.91	1.06	1.21	1.36	1.51	1.67	1.82	1.97	2.12	2.27	2.42	2.57	2.73	2.88	3.03
95	35.00	0.39	0.59	0.78	0.98	1.17	1.37	1.56	1.76	1.95	2.15	2.34	2.54	2.73	2.93	3.12	3.32	3.51	3.71	3.90
100	37.78	0.48	0.73	0.97	1.21	1.45	1.69	1.94	2.18	2.42	2.66	2.90	3.15	3.39	3.63	3.87	4.11	4.36	4.60	4.84
105	40.56	0.58	0.88	1.17	1.46	1.75	2.04	2.33	2.63	2.92	3.21	3.50	3.79	4.08	4.38	4.67	4.96	5.25	5.54	5.83
110	43.33	0.69	1.03	1.38	1.72	2.07	2.41	2.76	3.10	3.45	3.79	4.14	4.48	4.82	5.17	5.51	5.86	6.20	6.55	6.89
115	46.11	0.80	1.20	1.60	2.00	2.40	2.80	3.20	3.60	4.00	4.40	4.80	5.20	5.60	6.00	6.40	6.80	7.20	7.60	8.00
120	48.89	-0.92	-1.37	-1.83	-2.29	-2.75	-3.20	-3.66	-4.12	-4.58	-5.03	-5.49	-5.95	-6.41	-6.87	-7.32	-7.78	-8.24	-8.70	-9.15

Table 14

TEMPERATURE CORRECTIONS FOR WATER MANOMETERS WITH BRASS SCALES TO OBTAIN READING AT 60 F

Plus corrections are to be added to, minus corrections are to be subtracted from reading.

Enter with indicated column height (mm or inches) and its temperature

Temperature deg F	deg C	100	150	200	250	300	350	400	450	500	550	600	650	700	750	800	850	900	950	1000
32	0.00	+0.06	+0.08	+0.11	+0.14	+0.17	+0.19	+0.22	+0.25	+0.28	+0.30	+0.33	+0.36	+0.39	+0.41	+0.44	+0.47	+0.50	+0.52	+0.54
35	1.67	0.06	0.09	0.12	0.15	0.18	0.21	0.25	0.28	0.31	0.34	0.37	0.40	0.43	0.46	0.49	0.52	0.55	0.58	0.61
40	4.44	0.08	0.11	0.15	0.19	0.23	0.26	0.30	0.34	0.38	0.41	0.45	0.49	0.53	0.56	0.60	0.64	0.68	0.72	0.75
45	7.22	0.07	0.11	0.14	0.18	0.22	0.25	0.29	0.33	0.36	0.40	0.43	0.47	0.51	0.54	0.58	0.62	0.65	0.69	0.72
50	10.00	0.06	0.09	0.12	0.15	0.18	0.20	0.23	0.26	0.29	0.32	0.35	0.38	0.41	0.44	0.47	0.50	0.53	0.56	0.58
55	12.78	+0.03	+0.05	+0.07	+0.08	+0.10	+0.12	+0.14	+0.15	+0.17	+0.19	+0.20	+0.22	+0.24	+0.26	+0.27	+0.29	+0.31	+0.32	+0.34
60	15.56	0.00	0.00	0.00	0.00	0.00	0.00	0.00	0.00	0.00	0.00	0.00	0.00	0.00	0.00	0.00	0.00	0.00	0.00	0.00
65	18.33	-0.04	-0.06	-0.09	-0.11	-0.13	-0.15	-0.17	-0.19	-0.21	-0.24	-0.26	-0.28	-0.30	-0.32	-0.34	-0.36	-0.39	-0.41	-0.43
68	20.00	0.07	0.11	0.15	0.18	0.22	0.25	0.29	0.33	0.36	0.40	0.44	0.47	0.51	0.55	0.58	0.62	0.65	0.69	0.73
70	21.11	0.09	0.14	0.19	0.24	0.28	0.33	0.38	0.42	0.47	0.52	0.57	0.61	0.66	0.71	0.75	0.80	0.85	0.90	0.94
75	23.89	0.15	0.23	0.31	0.38	0.46	0.54	0.61	0.69	0.77	0.85	0.92	1.00	1.08	1.15	1.23	1.31	1.38	1.46	1.54
77	25.00	0.18	0.27	0.36	0.45	0.54	0.63	0.72	0.81	0.90	0.99	1.08	1.17	1.26	1.35	1.44	1.53	1.62	1.71	1.80
80	26.67	0.22	0.33	0.44	0.55	0.66	0.77	0.88	0.99	1.10	1.21	1.32	1.43	1.54	1.65	1.76	1.87	1.98	2.09	2.21
85	29.44	0.29	0.44	0.59	0.74	0.88	1.03	1.18	1.32	1.47	1.62	1.77	1.91	2.06	2.21	2.36	2.50	2.65	2.80	2.94
90	32.22	0.38	0.56	0.75	0.94	1.13	1.31	1.50	1.69	1.88	2.06	2.25	2.44	2.63	2.81	3.00	3.19	3.38	3.57	3.75
95	35.00	0.46	0.69	0.93	1.16	1.39	1.62	1.85	2.08	2.31	2.54	2.78	3.01	3.24	3.47	3.70	3.93	4.16	4.39	4.63
100	37.78	0.56	0.83	1.11	1.39	1.67	1.95	2.22	2.50	2.78	3.06	3.34	3.62	3.89	4.17	4.45	4.73	5.01	5.28	5.56
105	40.56	0.66	0.98	1.31	1.64	1.97	2.29	2.62	2.95	3.28	3.61	3.93	4.26	4.59	4.92	5.25	5.57	5.90	6.23	6.56
110	43.33	0.76	1.14	1.52	1.90	2.28	2.66	3.05	3.43	3.81	4.19	4.57	4.95	5.33	5.71	6.09	6.47	6.85	7.23	7.61
115	46.11	0.87	1.31	1.74	2.18	2.62	3.05	3.49	3.93	4.36	4.80	5.23	5.67	6.11	6.54	6.98	7.41	7.85	8.29	8.72
120	48.89	-0.99	-1.48	-1.97	-2.47	-2.96	-3.46	-3.95	-4.44	-4.94	-5.43	-5.92	-6.42	-6.91	-7.40	-7.90	-8.39	-8.89	-9.38	-9.87

Table 15A
EQUIVALENTS IN PSI, OF WATER COLUMNS IN INCHES AT VARIOUS TEMPERATURES
Temperature of Water in deg F

Inches Water	40	50	60	68	70	80	90	100	150	200
	psi	psi	psi	psi	psi	psi	psi	psi	psi	psi
0.1	0.0036	0.0036	0.0036	0.0036	0.0036	0.0036	0.0036	0.0036	0.0035	0.0035
0.2	0.0072	0.0072	0.0072	0.0072	0.0072	0.0072	0.0072	0.0072	0.0071	0.0070
0.3	0.0108	0.0108	0.0108	0.0108	0.0108	0.0108	0.0108	0.0108	0.0106	0.0104
0.4	0.0145	0.0144	0.0144	0.0144	0.0144	0.0144	0.0144	0.0144	0.0142	0.0139
0.5	0.0181	0.0181	0.0180	0.0180	0.0180	0.0180	0.0180	0.0180	0.0179	0.0177	0.0174
0.6	0.0217	0.0217	0.0217	0.0216	0.0216	0.0216	0.0216	0.0215	0.0213	0.0209
0.7	0.0253	0.0253	0.0253	0.0252	0.0252	0.0252	0.0252	0.0251	0.0248	0.0244
0.8	0.0289	0.0289	0.0289	0.0288	0.0288	0.0288	0.0288	0.0287	0.0283	0.0278
0.9	0.0325	0.0325	0.0325	0.0325	0.0324	0.0324	0.0324	0.0323	0.0319	0.0313
1.0	0.0361	0.0361	0.0361	0.0361	0.0361	0.0360	0.0359	0.0359	0.0354	0.0348
2.	0.0723	0.0722	0.0722	0.0721	0.0721	0.0720	0.0719	0.0718	0.0708	0.0696
3.	0.1084	0.1084	0.1083	0.1082	0.1082	0.1080	0.1078	0.1076	0.1062	0.1044
4.	0.1445	0.1445	0.1444	0.1444	0.1443	0.1442	0.1440	0.1438	0.1435	0.1417	0.1392
5.	0.1806	0.1806	0.1805	0.1803	0.1803	0.1800	0.1797	0.1794	0.1771	0.1740
6.	0.2168	0.2167	0.2166	0.2164	0.2163	0.2160	0.2157	0.2153	0.2125	0.2088
7.	0.2529	0.2528	0.2526	0.2524	0.2524	0.2520	0.2516	0.2511	0.2479	0.2435
8.	0.2890	0.2889	0.2887	0.2885	0.2884	0.2880	0.2876	0.2870	0.2833	0.2783
9.	0.3251	0.3251	0.3248	0.3246	0.3245	0.3240	0.3235	0.3229	0.3187	0.3131
10.	0.3613	0.3612	0.3609	0.3606	0.3605	0.3600	0.3595	0.3588	0.3541	0.3479
20.	0.7225	0.7223	0.7218	0.7213	0.7211	0.7201	0.7189	0.7175	0.7083	0.6958
30.	1.0838	1.0835	1.0828	1.0819	1.0816	1.0802	1.0784	1.0763	1.0624	1.0438
40.	1.4451	1.4447	1.4437	1.4425	1.4422	1.4402	1.4378	1.4351	1.4166	1.3917
50.	1.8063	1.8058	1.8046	1.8031	1.8027	1.8002	1.7973	1.7938	1.7707	1.7396
60.	2.1676	2.1670	2.1655	2.1638	2.1632	2.1603	2.1567	2.1526	2.1248	2.0875
70.	2.5289	2.5282	2.5264	2.5244	2.5238	2.5204	2.5162	2.5113	2.4790	2.4354
80.	2.8901	2.8893	2.8874	2.8850	2.8843	2.8804	2.8756	2.8701	2.8331	2.7834
90.	3.2514	3.2505	3.2483	3.2456	3.2449	3.2404	3.2351	3.2289	3.1873	3.1313
100.	3.6126	3.6117	3.6092	3.6063	3.6054	3.6005	3.5945	3.5876	3.5414	3.4792

Table 15B
EQUIVALENTS IN PSI OF WATER COLUMNS IN FEET AND INCHES AT VARIOUS TEMPERATURES
Temperature of water columns in deg F

Water Column Feet In.	40	50	60	68	70	80	90	100	150	200
	psi	psi	psi	psi	psi	psi	psi	psi	psi	psi
1/8	0.0045	0.0045	0.0045	0.0045	0.0045	0.0045	0.0045	0.0045	0.0044	0.0043
1/4	0.0090	0.0090	0.0090	0.0090	0.0090	0.0090	0.0090	0.0090	0.0089	0.0087
3/8	0.0135	0.0135	0.0135	0.0135	0.0135	0.0135	0.0135	0.0135	0.0133	0.0130
1/2	0.0181	0.0181	0.0180	0.0180	0.0180	0.0180	0.0180	0.0179	0.0177	0.0174
5/8	0.0226	0.0226	0.0226	0.0225	0.0225	0.0225	0.0225	0.0224	0.0221	0.0217
3/4	0.0271	0.0271	0.0271	0.0270	0.0270	0.0270	0.0270	0.0269	0.0266	0.0261
7/8	0.0316	0.0316	0.0316	0.0316	0.0315	0.0315	0.0315	0.0314	0.0310	0.0304
1	0.0361	0.0361	0.0361	0.0361	0.0361	0.0360	0.0359	0.0359	0.0354	0.0348
2	0.0723	0.0722	0.0722	0.0721	0.0721	0.0720	0.0719	0.0718	0.0708	0.0696
3	0.1084	0.1084	0.1083	0.1082	0.1082	0.1080	0.1078	0.1076	0.1062	0.1044
4	0.1445	0.1445	0.1444	0.1443	0.1442	0.1440	0.1438	0.1435	0.1417	0.1392
5	0.1806	0.1806	0.1805	0.1803	0.1803	0.1800	0.1797	0.1794	0.1771	0.1740
6	0.2168	0.2167	0.2166	0.2164	0.2163	0.2160	0.2157	0.2153	0.2125	0.2088
7	0.2529	0.2528	0.2526	0.2524	0.2524	0.2520	0.2516	0.2511	0.2479	0.2435
8	0.2890	0.2889	0.2887	0.2885	0.2884	0.2880	0.2876	0.2870	0.2833	0.2783
9	0.3251	0.3251	0.3248	0.3246	0.3245	0.3240	0.3235	0.3229	0.3187	0.3131
10	0.3613	0.3612	0.3609	0.3606	0.3605	0.3600	0.3595	0.3588	0.3541	0.3479
11	0.3974	0.3973	0.3970	0.3967	0.3966	0.3961	0.3954	0.3946	0.3896	0.3827
12	0.4335	0.4334	0.4331	0.4328	0.4326	0.4321	0.4313	0.4305	0.4250	0.4175
1	0.4335	0.4334	0.4331	0.4328	0.4326	0.4321	0.4313	0.4305	0.4250	0.4175
2	0.8670	0.8668	0.8662	0.8655	0.8653	0.8641	0.8627	0.8610	0.8499	0.8350
3	1.3005	1.3002	1.2993	1.2983	1.2979	1.2962	1.2940	1.2915	1.2749	1.2525
4	1.7340	1.7336	1.7324	1.7310	1.7306	1.7282	1.7254	1.7220	1.6999	1.6700
5	2.1676	2.1670	2.1655	2.1638	2.1632	2.1603	2.1567	2.1526	2.1248	2.0875
6	2.6011	2.6004	2.5968	2.5965	2.5959	2.5924	2.5880	2.5831	2.5498	2.5050
7	3.0346	3.0338	3.0317	3.0293	3.0285	3.0244	3.0194	3.0136	2.9748	2.9225
8	3.4681	3.4672	3.4648	3.4620	3.4612	3.4565	3.4507	3.4441	3.3997	3.3400
9	3.9016	3.9006	3.8979	3.8948	3.8938	3.8885	3.8821	3.8746	3.8248	3.7575
10	4.3351	4.3340	4.3310	4.3276	4.3265	4.3206	4.3134	4.3051	4.2497	4.1750
11	4.7686	4.7674	4.7641	4.7603	4.7591	4.7527	4.7447	4.7356	4.6746	4.5925
12	5.2021	5.2008	5.1972	5.1931	5.1918	5.1847	5.1761	5.1661	5.0996	5.0100

Table 16

CONVERSION FACTORS FOR MERCURY COLUMNS AT VARIOUS TEMPERATURES

Mercury temperature		Equiv. psi of 1 in. Hg	Mercury column at 32 F	Factor for converting mercury column height at various temperatures to:		
				at 60 F	Water column 68 F	77 F
deg F	deg C					
0	−17.78	0.49275	1.0032	13.652	13.663	13.680
10	−12.22	0.49225	1.0022	13.638	13.649	13.665
20	− 6.67	0.49175	1.0012	13.625	13.636	13.652
30	− 1.11	0.49126	1.0002	13.611	13.622	13.638
32	0.00	0.49116	1.0000	13.609	13.620	13.635
35	1.67	0.49101	0.9997	13.604	13.615	13.631
40	4.44	0.49076	0.9992	13.598	13.609	13.624
45	7.22	0.49051	0.9987	13.591	13.602	13.618
50	10.00	0.49026	0.9982	13.584	13.595	13.611
55	12.78	0.49002	0.9977	13.577	13.588	13.604
60	15.56	0.48977	0.9972	13.570	13.581	13.597
65	18.33	0.48952	0.9967	13.564	13.575	13.590
68	20.00	0.48938	0.9964	13.560	13.570	13.586
70	21.11	0.48928	0.9962	13.557	13.568	13.584
75	23.89	0.48904	0.9957	13.550	13.561	13.577
77	25.00	0.48894	0.9955	13.547	13.558	13.574
80	26.67	0.48879	0.9952	13.543	13.554	13.570
85	29.44	0.48854	0.9947	13.536	13.547	13.563
90	32.22	0.48830	0.9942	13.530	13.541	13.556
95	35.00	0.48805	0.9937	13.523	13.534	13.549
100	37.78	0.48780	0.9932	13.516	13.527	13.543
110	43.33	0.48732	0.9922	13.502	13.513	13.529
120	48.89	0.48683	0.9912	13.489	13.500	13.515
130	54.44	0.48634	0.9902	13.475	13.486	13.502
140	60.00	0.48585	0.9892	13.462	13.472	13.488
150	65.56	0.48536	0.9882	13.448	13.459	13.475
160	71.11	0.48488	0.9872	13.434	13.445	13.461
170	76.67	0.48439	0.9862	13.421	13.432	13.447
180	82.22	0.48391	0.9852	13.407	13.418	13.434
190	87.78	0.48342	0.9842	13.393	13.404	13.420
200	93.33	0.48293	0.9832	13.380	13.391	13.406

Table 17

TEMPERATURE CORRECTIONS FOR MERCURY MANOMETERS AND BAROMETERS WITH BRASS SCALES TO OBTAIN READING AT 0 C

All corrections are to be subtracted.

Enter with indicated column height (mm, in., or mb) and its temperature:

Temperature deg C	100	150	200	250	300	350	400	450	500	550	600	650	700	750	800	850	900	950	1000	1050	1100	1150	1200
0	0.00	0.00	0.00	0.00	0.00	0.00	0.00	0.00	0.00	0.00	0.00	0.00	0.00	0.00	0.00	0.00	0.00	0.00	0.00	0.00	0.00	0.00	0.00
5	0.08	0.12	0.16	0.20	0.24	0.29	0.33	0.37	0.41	0.45	0.49	0.53	0.57	0.61	0.65	0.69	0.73	0.78	0.82	0.86	0.90	0.94	0.98
10	0.16	0.24	0.33	0.41	0.49	0.57	0.65	0.73	0.82	0.90	0.98	1.06	1.14	1.22	1.30	1.39	1.47	1.55	1.63	1.71	1.79	1.88	1.96
11	0.18	0.27	0.36	0.45	0.54	0.63	0.72	0.81	0.90	0.99	1.08	1.17	1.26	1.35	1.44	1.52	1.61	1.70	1.79	1.88	1.97	2.06	2.15
12	0.20	0.29	0.39	0.49	0.59	0.68	0.78	0.88	0.98	1.08	1.17	1.27	1.37	1.47	1.57	1.66	1.76	1.86	1.96	2.05	2.15	2.25	2.35
13	0.21	0.32	0.42	0.53	0.64	0.74	0.85	0.95	1.06	1.17	1.27	1.38	1.48	1.59	1.70	1.80	1.91	2.01	2.12	2.23	2.33	2.44	2.54
14	0.23	0.34	0.46	0.57	0.68	0.80	0.91	1.03	1.14	1.25	1.37	1.48	1.60	1.71	1.83	1.94	2.05	2.17	2.28	2.40	2.51	2.62	2.74
15	0.24	0.37	0.49	0.61	0.73	0.86	0.98	1.10	1.22	1.34	1.47	1.59	1.71	1.83	1.96	2.08	2.20	2.32	2.44	2.57	2.69	2.81	2.93
16	0.26	0.39	0.52	0.65	0.78	0.91	1.04	1.17	1.30	1.43	1.56	1.69	1.82	1.96	2.09	2.22	2.35	2.48	2.61	2.74	2.87	3.00	3.13
17	0.28	0.42	0.55	0.69	0.83	0.97	1.11	1.25	1.38	1.52	1.66	1.80	1.94	2.08	2.22	2.35	2.49	2.63	2.77	2.91	3.05	3.18	3.32
18	0.29	0.44	0.59	0.73	0.88	1.03	1.17	1.32	1.47	1.61	1.78	1.91	2.05	2.20	2.35	2.49	2.64	2.79	2.93	3.08	3.22	3.37	3.52
19	0.31	0.46	0.62	0.77	0.93	1.08	1.24	1.39	1.55	1.70	1.86	2.01	2.17	2.32	2.48	2.63	2.78	2.94	3.09	3.25	3.40	3.56	3.71
20	0.33	0.49	0.65	0.81	0.98	1.14	1.30	1.47	1.63	1.79	1.95	2.12	2.28	2.44	2.60	2.77	2.93	3.09	3.26	3.42	3.58	3.74	3.91
21	0.34	0.51	0.68	0.85	1.03	1.20	1.37	1.54	1.71	1.88	2.05	2.22	2.39	2.56	2.73	2.91	3.08	3.25	3.42	3.59	3.76	3.93	4.10
22	0.36	0.54	0.72	0.90	1.07	1.25	1.43	1.61	1.79	1.97	2.15	2.33	2.51	2.69	2.86	3.04	3.22	3.40	3.58	3.76	3.94	4.12	4.30
23	0.37	0.56	0.75	0.94	1.12	1.31	1.50	1.68	1.87	2.06	2.25	2.43	2.62	2.81	2.99	3.18	3.37	3.56	3.74	3.93	4.12	4.30	4.49
24	0.39	0.59	0.78	0.98	1.17	1.37	1.56	1.76	1.95	2.15	2.34	2.54	2.73	2.93	3.12	3.32	3.51	3.71	3.90	4.10	4.30	4.49	4.69
25	0.41	0.61	0.81	1.02	1.22	1.42	1.63	1.83	2.03	2.24	2.44	2.64	2.85	3.05	3.25	3.46	3.66	3.86	4.07	4.27	4.47	4.68	4.88
26	0.42	0.63	0.85	1.06	1.27	1.48	1.69	1.90	2.11	2.33	2.54	2.75	2.96	3.17	3.38	3.59	3.81	4.02	4.23	4.44	4.65	4.86	5.07
27	0.44	0.66	0.88	1.10	1.32	1.54	1.76	1.98	2.20	2.41	2.63	2.85	3.07	3.29	3.51	3.73	3.95	4.17	4.39	4.61	4.83	5.05	5.27
28	0.46	0.68	0.91	1.14	1.37	1.59	1.82	2.05	2.28	2.50	2.73	2.96	3.19	3.41	3.64	3.87	4.10	4.32	4.55	4.78	5.01	5.23	5.46
29	0.47	0.71	0.94	1.18	1.41	1.65	1.89	2.12	2.36	2.59	2.83	3.06	3.30	3.54	3.77	4.01	4.24	4.48	4.71	4.95	5.19	5.42	5.66
30	0.49	0.73	0.98	1.22	1.46	1.71	1.95	2.19	2.44	2.68	2.93	3.17	3.41	3.66	3.90	4.14	4.39	4.63	4.88	5.12	5.36	5.61	5.85
31	0.50	0.76	1.01	1.26	1.51	1.76	2.01	2.27	2.52	2.77	3.02	3.27	3.53	3.78	4.03	4.28	4.53	4.79	5.04	5.29	5.54	5.79	6.04
32	0.52	0.78	1.04	1.30	1.56	1.82	2.08	2.34	2.60	2.86	3.12	3.38	3.64	3.90	4.16	4.42	4.68	4.94	5.20	5.46	5.72	5.98	6.24
33	0.54	0.80	1.07	1.34	1.61	1.88	2.14	2.41	2.68	2.95	3.22	3.48	3.75	4.02	4.29	4.56	4.82	5.09	5.36	5.63	5.90	6.16	6.43
34	0.55	0.83	1.10	1.38	1.66	1.93	2.21	2.48	2.76	3.04	3.31	3.59	3.87	4.14	4.42	4.69	4.97	5.25	5.52	5.80	6.07	6.35	6.63
35	0.57	0.85	1.14	1.42	1.70	1.99	2.27	2.56	2.84	3.13	3.41	3.69	3.98	4.26	4.55	4.83	5.11	5.40	5.68	5.97	6.25	6.54	6.82
36	0.58	0.88	1.17	1.46	1.75	2.05	2.34	2.63	2.92	3.21	3.51	3.80	4.09	4.38	4.68	4.97	5.26	5.55	5.84	6.14	6.43	6.72	7.01
37	0.60	0.90	1.20	1.50	1.80	2.10	2.40	2.70	3.00	3.30	3.60	3.90	4.20	4.50	4.80	5.10	5.40	5.71	6.01	6.31	6.61	6.91	7.21
38	0.62	0.92	1.23	1.54	1.85	2.16	2.47	2.77	3.08	3.39	3.70	4.01	4.32	4.62	4.93	5.24	5.55	5.86	6.17	6.47	6.78	7.09	7.40
39	0.63	0.95	1.27	1.58	1.90	2.21	2.53	2.85	3.16	3.48	3.80	4.11	4.43	4.75	5.06	5.38	5.69	6.01	6.33	6.64	6.96	7.28	7.59
40	0.65	0.97	1.30	1.62	1.95	2.27	2.60	2.92	3.24	3.57	3.89	4.22	4.54	4.87	5.19	5.52	5.84	6.16	6.49	6.81	7.14	7.46	7.79
45	0.73	1.09	1.46	1.82	2.19	2.55	2.92	3.28	3.65	4.01	4.38	4.74	5.11	5.47	5.83	6.20	6.56	6.93	7.29	7.66	8.02	8.39	8.75
50	0.81	1.21	1.62	2.02	2.43	2.83	3.24	3.64	4.05	4.45	4.86	5.26	5.67	6.07	6.48	6.88	7.29	7.69	8.10	8.50	8.91	9.31	9.73

Table 18

TEMPERATURE CORRECTIONS FOR MERCURY MANOMETERS AND BAROMETERS WITH BRASS SCALES TO OBTAIN READINGS AT 32 F

All corrections are to be subtracted

Enter with indicated column height (mm, in. or mb) and its temperature:

Temperature deg F	10	15	20	25	30	35	40	45	50	55	60	70	80	90	100
32	0.000	0.000	0.000	0.000	0.000	0.000	0.000	0.000	0.000	0.000	0.000	0.000	0.000	0.000	0.000
35	0.003	0.004	0.005	0.007	0.008	0.009	0.011	0.012	0.014	0.015	0.016	0.019	0.022	0.024	0.027
40	0.007	0.011	0.014	0.018	0.022	0.025	0.029	0.033	0.036	0.040	0.044	0.051	0.058	0.065	0.072
45	0.012	0.018	0.024	0.029	0.035	0.041	0.047	0.053	0.059	0.065	0.071	0.082	0.094	0.106	0.118
50	0.016	0.024	0.033	0.041	0.049	0.057	0.065	0.073	0.081	0.090	0.098	0.114	0.130	0.147	0.163
51	0.017	0.026	0.034	0.043	0.052	0.060	0.069	0.077	0.086	0.095	0.103	0.120	0.138	0.155	0.172
52	0.018	0.027	0.036	0.045	0.054	0.063	0.072	0.081	0.090	0.100	0.109	0.127	0.145	0.163	0.181
53	0.019	0.028	0.038	0.048	0.057	0.066	0.076	0.086	0.095	0.104	0.114	0.133	0.152	0.171	0.190
54	0.020	0.030	0.040	0.050	0.060	0.070	0.080	0.090	0.099	0.109	0.119	0.139	0.159	0.179	0.199
55	0.021	0.031	0.042	0.052	0.062	0.073	0.083	0.094	0.104	0.114	0.125	0.146	0.166	0.187	0.208
56	0.022	0.033	0.043	0.054	0.065	0.076	0.087	0.098	0.108	0.119	0.130	0.152	0.174	0.196	0.217
57	0.023	0.034	0.045	0.056	0.068	0.079	0.090	0.102	0.113	0.124	0.136	0.158	0.181	0.204	0.226
58	0.024	0.035	0.047	0.059	0.070	0.082	0.094	0.106	0.117	0.129	0.141	0.164	0.188	0.212	0.235
59	0.024	0.037	0.049	0.061	0.073	0.086	0.098	0.110	0.122	0.134	0.147	0.171	0.196	0.220	0.244
60	0.025	0.038	0.051	0.064	0.076	0.089	0.102	0.114	0.127	0.139	0.152	0.178	0.203	0.228	0.254
61	0.026	0.039	0.053	0.066	0.079	0.092	0.105	0.118	0.131	0.144	0.158	0.184	0.210	0.236	0.263
62	0.027	0.041	0.054	0.068	0.082	0.095	0.109	0.122	0.135	0.149	0.163	0.190	0.218	0.244	0.272
63	0.028	0.042	0.056	0.070	0.084	0.098	0.112	0.126	0.140	0.154	0.169	0.197	0.225	0.252	0.281
64	0.029	0.044	0.058	0.072	0.087	0.102	0.116	0.130	0.145	0.159	0.174	0.203	0.232	0.260	0.290
65	0.030	0.045	0.060	0.075	0.090	0.105	0.120	0.134	0.149	0.164	0.179	0.209	0.239	0.268	0.299
66	0.031	0.046	0.062	0.077	0.092	0.108	0.123	0.139	0.154	0.169	0.185	0.216	0.246	0.277	0.308
67	0.032	0.048	0.063	0.079	0.095	0.111	0.127	0.143	0.158	0.174	0.190	0.222	0.253	0.285	0.317
68	0.033	0.049	0.065	0.081	0.098	0.114	0.130	0.147	0.163	0.179	0.195	0.228	0.260	0.293	0.326
69	0.034	0.050	0.067	0.084	0.100	0.117	0.134	0.151	0.167	0.184	0.201	0.234	0.268	0.301	0.335
70	0.034	0.052	0.069	0.086	0.103	0.120	0.138	0.155	0.172	0.189	0.206	0.241	0.275	0.310	0.344
71	0.035	0.053	0.071	0.088	0.106	0.124	0.141	0.159	0.176	0.194	0.212	0.247	0.282	0.318	0.353
72	0.036	0.054	0.072	0.090	0.109	0.127	0.145	0.163	0.181	0.199	0.217	0.253	0.290	0.326	0.362
73	0.037	0.056	0.074	0.093	0.111	0.130	0.148	0.167	0.185	0.204	0.223	0.260	0.297	0.334	0.371
74	0.038	0.057	0.076	0.095	0.114	0.133	0.152	0.171	0.190	0.209	0.228	0.266	0.304	0.341	0.380
75	0.039	0.059	0.078	0.097	0.117	0.136	0.156	0.175	0.194	0.214	0.233	0.272	0.311	0.349	0.389
76	0.040	0.060	0.080	0.100	0.119	0.139	0.159	0.179	0.199	0.219	0.239	0.279	0.318	0.358	0.398
77	0.041	0.061	0.081	0.102	0.122	0.142	0.163	0.183	0.203	0.224	0.244	0.285	0.325	0.366	0.407
78	0.042	0.062	0.083	0.104	0.125	0.146	0.166	0.187	0.208	0.229	0.250	0.291	0.333	0.374	0.416
79	0.042	0.064	0.085	0.106	0.128	0.149	0.170	0.191	0.212	0.234	0.255	0.298	0.340	0.383	0.425
80	0.043	0.065	0.087	0.108	0.130	0.152	0.174	0.195	0.217	0.239	0.260	0.304	0.347	0.391	0.434
81	0.044	0.066	0.089	0.111	0.133	0.155	0.177	0.199	0.221	0.244	0.266	0.310	0.354	0.399	0.443
82	0.045	0.068	0.090	0.113	0.136	0.158	0.181	0.203	0.226	0.249	0.271	0.316	0.362	0.407	0.452
83	0.046	0.069	0.092	0.115	0.138	0.161	0.184	0.207	0.230	0.254	0.277	0.323	0.369	0.415	0.461
84	0.047	0.070	0.094	0.118	0.141	0.164	0.188	0.211	0.235	0.258	0.282	0.329	0.376	0.423	0.470
85	0.048	0.072	0.096	0.120	0.144	0.168	0.192	0.215	0.239	0.263	0.287	0.335	0.383	0.431	0.479
86	0.049	0.073	0.098	0.122	0.146	0.171	0.195	0.219	0.244	0.268	0.293	0.341	0.390	0.439	0.488
87	0.050	0.075	0.099	0.124	0.149	0.174	0.199	0.224	0.248	0.273	0.298	0.348	0.397	0.447	0.497
88	0.050	0.076	0.101	0.126	0.152	0.177	0.202	0.227	0.252	0.278	0.303	0.354	0.404	0.455	0.505
89	0.051	0.077	0.103	0.128	0.154	0.180	0.206	0.231	0.257	0.283	0.308	0.360	0.411	0.463	0.514
90	0.052	0.078	0.105	0.131	0.157	0.183	0.209	0.235	0.261	0.288	0.314	0.366	0.418	0.471	0.523
91	0.053	0.080	0.106	0.133	0.160	0.186	0.213	0.239	0.266	0.293	0.319	0.372	0.426	0.479	0.532
92	0.054	0.081	0.108	0.135	0.162	0.189	0.216	0.243	0.270	0.298	0.324	0.379	0.433	0.487	0.541
93	0.055	0.082	0.110	0.138	0.165	0.192	0.220	0.248	0.275	0.302	0.330	0.385	0.440	0.495	0.550
94	0.056	0.084	0.112	0.140	0.168	0.196	0.224	0.252	0.279	0.307	0.335	0.391	0.447	0.503	0.559
95	0.057	0.085	0.114	0.142	0.170	0.199	0.227	0.256	0.284	0.312	0.341	0.398	0.455	0.511	0.568
96	0.058	0.087	0.115	0.144	0.173	0.202	0.231	0.260	0.288	0.317	0.346	0.404	0.462	0.519	0.577
97	0.059	0.088	0.117	0.146	0.176	0.205	0.234	0.264	0.293	0.322	0.352	0.410	0.469	0.527	0.586
98	0.060	0.089	0.119	0.149	0.178	0.208	0.238	0.268	0.297	0.327	0.357	0.416	0.476	0.535	0.595
99	0.060	0.091	0.121	0.151	0.181	0.211	0.242	0.272	0.302	0.332	0.362	0.423	0.483	0.544	0.604
100	0.061	0.092	0.123	0.153	0.184	0.215	0.245	0.276	0.306	0.337	0.368	0.429	0.490	0.552	0.613
101	0.062	0.093	0.124	0.156	0.187	0.218	0.249	0.280	0.311	0.342	0.373	0.435	0.497	0.560	0.622
102	0.063	0.095	0.126	0.158	0.189	0.221	0.252	0.284	0.315	0.347	0.379	0.442	0.505	0.568	0.631
103	0.064	0.096	0.128	0.160	0.192	0.224	0.256	0.288	0.320	0.352	0.384	0.448	0.512	0.576	0.640
104	0.065	0.097	0.130	0.162	0.195	0.227	0.260	0.292	0.324	0.357	0.389	0.454	0.519	0.584	0.649
105	0.066	0.099	0.132	0.164	0.197	0.230	0.263	0.296	0.329	0.362	0.395	0.461	0.526	0.592	0.658
106	0.067	0.100	0.133	0.167	0.200	0.233	0.266	0.300	0.333	0.366	0.400	0.466	0.533	0.599	0.666
107	0.068	0.101	0.135	0.169	0.203	0.236	0.270	0.304	0.338	0.371	0.405	0.472	0.540	0.607	0.675
108	0.068	0.103	0.137	0.171	0.205	0.239	0.274	0.308	0.342	0.376	0.410	0.479	0.547	0.616	0.684
109	0.069	0.104	0.139	0.173	0.208	0.243	0.277	0.312	0.347	0.381	0.416	0.485	0.554	0.624	0.693
110	0.070	0.105	0.140	0.176	0.211	0.246	0.281	0.316	0.351	0.386	0.421	0.491	0.562	0.632	0.702
111	0.071	0.107	0.142	0.178	0.213	0.249	0.284	0.320	0.356	0.391	0.427	0.498	0.569	0.640	0.711
112	0.072	0.108	0.144	0.180	0.216	0.252	0.288	0.324	0.360	0.396	0.432	0.504	0.576	0.648	0.720
113	0.073	0.109	0.146	0.182	0.219	0.255	0.292	0.328	0.365	0.401	0.438	0.511	0.583	0.656	0.729
114	0.074	0.111	0.148	0.184	0.221	0.258	0.295	0.332	0.369	0.406	0.443	0.517	0.590	0.664	0.738
115	0.075	0.112	0.149	0.187	0.224	0.261	0.299	0.336	0.374	0.411	0.448	0.523	0.598	0.672	0.747
116	0.076	0.113	0.151	0.189	0.227	0.265	0.302	0.340	0.378	0.416	0.454	0.529	0.605	0.680	0.756
117	0.076	0.115	0.153	0.191	0.230	0.268	0.306	0.344	0.383	0.421	0.459	0.536	0.612	0.689	0.765
118	0.077	0.116	0.155	0.194	0.232	0.271	0.310	0.348	0.387	0.426	0.464	0.542	0.619	0.697	0.774
119	0.078	0.117	0.157	0.196	0.235	0.274	0.313	0.352	0.392	0.431	0.470	0.548	0.626	0.705	0.783
120	0.079	0.119	0.158	0.198	0.238	0.277	0.317	0.356	0.396	0.436	0.475	0.554	0.634	0.713	0.792

Table 19
EQUIVALENTS IN PSI OF MERCURY COLUMNS IN INCHES OF MERCURY AT VARIOUS TEMPERATURES

Mercury	Mercury Temperature in deg F									
Inches	20	30	32	40	50	60	70	80	90	100
	psi	psi	psi	psi	psi	psi	psi	psi	psi	psi
0.1	0.0492	0.0491	0.0491	0.0491	0.0490	0.0490	0.0489	0.0489	0.0488	0.0488
0.2	0.0984	0.0983	0.0982	0.0982	0.0981	0.0980	0.0979	0.0978	0.0977	0.0976
0.3	0.1475	0.1474	0.1474	0.1472	0.1471	0.1469	0.1468	0.1468	0.1465	0.1463
0.4	0.1967	0.1965	0.1965	0.1963	0.1961	0.1959	0.1957	0.1955	0.1953	0.1951
0.5	0.2459	0.2456	0.2456	0.2454	0.2451	0.2449	0.2446	0.2444	0.2442	0.2439
0.6	0.2951	0.2948	0.2947	0.2945	0.2942	0.2939	0.2936	0.2933	0.2930	0.2927
0.7	0.3442	0.3439	0.3438	0.3435	0.3432	0.3428	0.3425	0.3422	0.3418	0.3415
0.8	0.3934	0.3930	0.3929	0.3926	0.3922	0.3918	0.3914	0.3910	0.3906	0.3902
0.9	0.4426	0.4421	0.4420	0.4417	0.4412	0.4408	0.4404	0.4399	0.4395	0.4390
1.0	0.4918	0.4913	0.4912	0.4908	0.4903	0.4898	0.4893	0.4888	0.4883	0.4878
2	0.9835	0.9825	0.9823	0.9815	0.9805	0.9795	0.9786	0.9776	0.9766	0.9756
3	1.4753	1.4738	1.4735	1.4723	1.4708	1.4693	1.4678	1.4664	1.4649	1.4634
4	1.9670	1.9650	1.9646	1.9630	1.9611	1.9591	1.9571	1.9551	1.9532	1.9512
5	2.4588	2.4563	2.4558	2.4538	2.4513	2.4489	2.4464	2.4439	2.4415	2.4390
6	2.9505	2.9475	2.9469	2.9446	2.9416	2.9386	2.9357	2.9327	2.9298	2.9268
7	3.4423	3.4388	3.4381	3.4353	3.4319	3.4284	3.4249	3.4215	3.4181	3.4146
8	3.9340	3.9300	3.9293	3.9261	3.9221	3.9182	3.9142	3.9103	3.9064	3.9024
9	4.4258	4.4213	4.4204	4.4168	4.4124	4.4080	4.4035	4.3991	4.3947	4.3902
10	4.9175	4.9126	4.9116	4.9076	4.9026	4.8977	4.8928	4.8879	4.8830	4.8780
20	9.8351	9.8251	9.8231	9.8152	9.8053	9.7955	9.7856	9.7757	9.7659	9.7561
30	14.753	14.738	14.735	14.723	14.708	14.693	14.678	14.664	14.649	14.634
40	19.670	19.650	19.646	19.630	19.611	19.591	19.571	19.551	19.532	19.512
50	24.588	24.563	24.558	24.538	24.513	24.489	24.464	24.439	24.415	24.390
60	29.505	29.475	29.469	29.446	29.416	29.386	29.357	29.327	29.298	29.268
70	34.423	34.388	34.381	34.353	34.319	34.284	34.249	34.215	34.181	34.146
80	39.340	39.300	39.293	39.261	39.221	39.182	39.142	39.103	39.064	39.024
90	44.258	44.213	44.204	44.168	44.124	44.080	44.035	43.991	43.947	43.902
100	49.175	49.126	49.116	49.076	49.026	48.977	48.928	48.879	48.830	48.780

Example: 48.36 in. of mercury at 85 F = (19.542 + 3.908 + 0.146 + 0.029) = 23.625 psi.

Table 20A
EQUIVALENTS OF MERCURY COLUMNS AT VARIOUS TEMPERATURES TO WATER COLUMNS AT 20 C (68 F)
HEIGHT OF WATER COLUMN

Mercury	Temperature of mercury column deg F							
Height	32	40	50	60	70	80	90	100
0.1	1.362	1.361	1.359	1.358	1.357	1.355	1.354	1.353
0.2	2.724	2.722	2.719	2.716	2.713	2.711	2.708	2.705
0.3	4.086	4.083	4.078	4.074	4.070	4.066	4.062	4.058
0.4	5.448	5.443	5.438	5.432	5.427	5.422	5.416	5.411
0.5	6.810	6.804	6.797	6.791	6.784	6.777	6.770	6.763
0.6	8.172	8.165	8.157	8.149	8.140	8.132	8.124	8.116
0.7	9.534	9.526	9.516	9.507	9.497	9.488	9.478	9.469
0.8	10.896	10.887	10.876	10.865	10.854	10.843	10.832	10.821
0.9	12.258	12.248	12.235	12.223	12.211	12.198	12.186	12.174
1	13.620	13.608	13.595	13.581	13.567	13.554	13.540	13.527
2	27.239	27.217	27.190	27.162	27.135	27.108	27.080	27.053
3	40.859	40.826	40.784	40.743	40.702	40.661	40.621	40.580
4	54.478	54.434	54.379	54.325	54.270	54.215	54.161	54.106
5	68.098	68.042	67.974	67.906	67.837	67.769	67.701	67.633
6	81.717	81.651	81.569	81.487	81.405	81.323	81.241	81.159
7	95.337	95.260	95.163	95.068	94.972	94.877	94.781	94.686
8	108.956	108.868	108.758	108.649	108.539	108.430	108.321	108.212
9	122.576	122.476	122.353	122.230	122.107	121.984	121.862	121.739
10	136.195	136.085	135.948	135.812	135.674	135.538	135.402	135.266
20	272.390	272.170	271.896	271.623	271.349	271.076	270.804	270.531
30	408.586	408.255	407.843	407.434	407.023	406.614	406.205	405.797
40	544.781	544.340	543.791	543.246	542.697	542.152	541.607	541.062
50	680.976	680.425	679.739	679.058	678.372	677.690	677.009	676.328
60	817.171	816.510	815.687	814.869	814.046	813.228	812.411	811.593
70	953.366	952.595	951.635	950.680	949.720	948.766	947.813	946.859
80	1089.56	1088.68	1087.58	1086.49	1085.39	1084.30	1083.21	1082.12
90	1225.76	1224.76	1223.53	1222.30	1221.07	1219.84	1218.62	1217.39
100	1361.95	1360.85	1359.48	1358.12	1356.74	1355.38	1354.02	1352.66

The table can be used for any consistent unit of column height.
Example: 150 in. or mm of mercury at 70 F = 1356.74 + 678.37 = 2035.11 in. or mm or water at 68 F.

Table 20B

EQUIVALENTS OF MERCURY COLUMNS AT VARIOUS TEMPERATURES TO WATER COLUMNS AT 60 F

HEIGHT OF WATER COLUMN

Mercury Height	Temperature of mercury column deg F							
	32	40	50	60	70	80	90	100
0.1	1.361	1.360	1.358	1.357	1.356	1.354	1.353	1.352
0.2	2.722	2.719	2.717	2.714	2.711	2.709	2.706	2.703
0.3	4.083	4.079	4.075	4.071	4.067	4.063	4.059	4.055
0.4	5.443	5.439	5.434	5.428	5.423	5.417	5.412	5.406
0.5	6.804	6.799	6.792	6.785	6.778	6.771	6.765	6.758
0.6	8.165	8.158	8.150	8.142	8.134	8.126	8.118	8.109
0.7	9.526	9.518	9.509	9.499	9.490	9.480	9.470	9.461
0.8	10.887	10.878	10.867	10.856	10.845	10.834	10.823	10.812
0.9	12.248	12.238	12.225	12.213	12.201	12.189	12.176	12.164
1	13.609	13.597	13.584	13.570	13.556	13.543	13.529	13.516
2	27.217	27.195	27.168	27.140	27.113	27.086	27.058	27.031
3	40.826	40.792	40.751	40.711	40.669	40.629	40.588	40.547
4	54.434	54.390	54.335	54.281	54.226	54.171	54.117	54.062
5	68.043	67.987	67.919	67.851	67.782	67.714	67.646	67.578
6	81.651	81.585	81.503	81.421	81.339	81.257	81.175	81.094
7	95.260	95.182	95.086	94.991	94.895	94.800	94.705	94.609
8	108.868	108.780	108.670	108.561	108.452	108.343	108.234	108.125
9	122.477	122.377	122.254	122.132	122.008	121.886	121.763	121.640
10	136.085	135.975	135.838	135.702	135.565	135.428	135.292	135.156
20	272.170	271.950	271.676	271.403	271.129	270.857	270.585	270.312
30	408.255	407.925	407.513	407.105	406.694	406.285	405.877	405.468
40	544.340	543.900	543.351	542.807	542.258	541.714	541.169	540.624
50	680.425	679.874	679.189	678.508	677.823	677.142	676.461	675.781
60	816.510	815.849	815.027	814.210	813.387	812.570	811.754	810.937
70	952.595	951.824	950.865	949.912	948.952	947.999	947.046	946.093
80	1088.68	1087.80	1086.70	1085.61	1084.52	1083.43	1082.34	1081.25
90	1224.77	1223.77	1222.54	1221.32	1220.08	1218.86	1217.63	1216.40
100	1360.85	1359.75	1358.38	1357.02	1355.65	1354.28	1352.92	1351.56

The table can be used for any consistant unit of column height.

Example: 150 in. or mm of mercury at 70 F = 1355.65 + 677.82 = 2033.47 in. or mm of water at 60 F.

Table 21

DENSITY OF KEROSINE AND EQUIVALENT PRESSURES IN PSI OF KEROSINE COLUMNS AT VARIOUS TEMPERATURES*

Kerosine Temp. deg F	Density lb per cu ft	Density** g per cm³	Specific gravity***	Equiv. psi of one inch	Equiv. psi of 100 mm
−20	53.10	0.8505	0.8513	0.03073	0.1210
−15	52.98	0.8486	0.8494	0.03066	0.1207
−10	52.85	0.8466	0.8474	0.03058	0.1204
− 5	52.73	0.8447	0.8455	0.03052	0.1201
0	52.61	0.8427	0.8435	0.03044	0.1199
5	52.49	0.8408	0.8416	0.03038	0.1196
10	52.37	0.8389	0.8397	0.03031	0.1193
15	52.25	0.8369	0.8377	0.03023	0.1190
20	52.12	0.8348	0.8356	0.03016	0.1187
25	52.00	0.8330	0.8338	0.03009	0.1185
30	51.88	0.8310	0.8318	0.03002	0.1182
35	51.75	0.8290	0.8298	0.02995	0.1179
40	51.63	0.8271	0.8279	0.02988	0.1176
45	51.51	0.8251	0.8259	0.02981	0.1174
50	51.38	0.8231	0.8239	0.02974	0.1171
55	51.27	0.8212	0.8220	0.02967	0.1168
60	51.14	0.8192	0.8200	0.02960	0.1165
65	51.02	0.8172	0.8180	0.02952	0.1162
70	50.90	0.8153	0.8161	0.02945	0.1160
75	50.78	0.8134	0.8142	0.02939	0.1157
80	50.65	0.8113	0.8121	0.02931	0.1154
85	50.53	0.8094	0.8102	0.02924	0.1151
90	50.40	0.8074	0.8082	0.02917	0.1148
95	50.28	0.8054	0.8062	0.02910	0.1146
100	50.17	0.8036	0.8044	0.02903	0.1143

 * Kerosine specific gravity 0.820 at 60/60; or 41.06° API.
 ** Pressure equivalent in g per sq cm of 1 cm column of kerosine.
 *** Specific gravity referred to water at 60 F.

Table 22

EQUIVALENTS IN PSI OF COLUMNS OF KEROSINE IN INCHES AT VARIOUS TEMPERATURES

Inches Kerosine	Temperature of Kerosine Column in deg F										
	0	10	20	30	40	50	60	70	80	90	100
	psi	psi	psi	psi	psi	psi	psi	psi	psi	psi	psi
0.1	0.003	0.003	0.003	0.003	0:003	0.003	0.003	0.003	0.003	0.003	0.003
0.2	0.006	0.006	0.006	0.006	0.006	0.006	0.006	0.006	0.006	0.006	0.006
0.3	0.009	0.009	0.009	0.009	0.009	0.009	0.009	0.009	0.009	0.009	0.009
0.4	0.012	0.012	0.012	0.012	0.012	0.012	0.012	0.012	0.012	0.012	0.012
0.5	0.015	0.015	0.015	0.015	0.015	0.015	0.015	0.015	0.015	0.015	0.015
0.6	0.018	0.018	0.018	0.018	0.018	0.018	0.018	0.018	0.018	0.018	0.017
0.7	0.021	0.021	0.021	0.021	0.021	0.021	0.021	0.021	0.021	0.020	0.020
0.8	0.024	0.024	0.024	0.024	0.024	0.024	0.024	0.024	0.024	0.023	0.023
0.9	0.027	0.027	0.027	0.027	0.027	0.027	0.027	0.027	0.026	0.026	0.026
1.0	0.030	0.030	0.030	0.030	0.030	0.030	0.030	0.029	0.029	0.029	0.029
2	0.061	0.061	0.060	0.060	0.060	0.060	0.059	0.059	0.059	0.058	0.058
3	0.091	0.091	0.090	0.090	0.090	0.089	0.089	0.088	0.088	0.088	0.087
4	0.122	0.121	0.121	0.120	0.120	0.119	0.118	0.118	0.117	0.117	0.116
5	0.152	0.152	0.151	0.150	0.149	0.149	0.148	0.147	0.147	0.146	0.145
6	0.183	0.182	0.181	0.180	0.179	0.178	0.178	0.177	0.176	0.175	0.174
7	0.213	0.212	0.211	0.210	0.209	0.208	0.207	0.206	0.205	0.204	0.203
8	0.244	0.242	0.241	0.240	0.239	0.238	0.237	0.236	0.235	0.233	0.232
9	0.274	0.273	0.271	0.270	0.269	0.268	0.266	0.265	0.264	0.263	0.261
10	0.304	0.303	0.302	0.300	0.299	0.297	0.296	0.295	0.293	0.292	0.290
20	0.609	0.606	0.603	0.600	0.598	0.595	0.592	0.589	0.586	0.584	0.581
30	0.913	0.909	0.905	0.901	0.896	0.892	0.888	0.884	0.879	0.875	0.871
40	1.218	1.212	1.206	1.201	1.195	1.190	1.184	1.178	1.173	1.167	1.161
50	1.522	1.516	1.508	1.501	1.494	1.487	1.480	1.473	1.466	1.459	1.452
60	1.826	1.819	1.810	1.801	1.793	1.784	1.776	1.767	1.759	1.750	1.742
70	2.131	2.122	2.111	2.101	2.092	2.082	2.072	2.062	2.052	2.042	2.032
80	2.435	2.425	2.413	2.402	2.391	2.379	2.368	2.356	2.345	2.334	2.322
90	2.740	2.728	2.714	2.702	2.689	2.676	2.664	2.651	2.638	2.626	2.613
100	3.044	3.031	3.016	3.002	2.988	2.974	2.960	2.945	2.931	2.917	2.903

Table 23

EQUIVALENTS IN INCHES OF WATER AT 60 F OF KEROSINE COLUMNS IN MM AT VARIOUS TEMPERATURES

MM Kerosine	Temperature of Kerosine Column in deg F										
	0	10	20	30	40	50	60	70	80	90	100
	Inches of water										
1	0.033	0.033	0.033	0.033	0.033	0.032	0.032	0.032	0.032	0.032	0.032
2	0.066	0.066	0.066	0.065	0.065	0.065	0.065	0.064	0.064	0.064	0.063
3	0.100	0.099	0.099	0.098	0.098	0.097	0.097	0.096	0.096	0.095	0.095
4	0.133	0.132	0.132	0.131	0.130	0.130	0.129	0.129	0.128	0.127	0.127
5	0.166	0.165	0.164	0.164	0.163	0.162	0.161	0.161	0.160	0.159	0.158
6	0.199	0.198	0.197	0.196	0.196	0.195	0.194	0.193	0.192	0.191	0.190
7	0.232	0.231	0.230	0.229	0.228	0.227	0.226	0.225	0.224	0.223	0.222
8	0.266	0.264	0.263	0.262	0.261	0.259	0.258	0.257	0.256	0.255	0.253
9	0.299	0.298	0.296	0.295	0.293	0.292	0.291	0.289	0.288	0.286	0.285
10	0.332	0.331	0.329	0.327	0.326	0.324	0.323	0.321	0.320	0.318	0.317
20	0.664	0.661	0.658	0.655	0.652	0.649	0.646	0.643	0.639	0.636	0.633
30	0.996	0.992	0.987	0.982	0.978	0.973	0.968	0.964	0.959	0.955	0.950
40	1.328	1.322	1.316	1.310	1.304	1.297	1.291	1.285	1.279	1.273	1.267
50	1.660	1.653	1.645	1.637	1.630	1.622	1.614	1.606	1.599	1.591	1.583
60	1.993	1.984	1.974	1.965	1.956	1.946	1.937	1.928	1.918	1.909	1.900
70	2.325	2.314	2.303	2.292	2.282	2.271	2.260	2.249	2.238	2.227	2.217
80	2.657	2.645	2.632	2.620	2.608	2.595	2.583	2.570	2.558	2.546	2.534
90	2.989	2.975	2.961	2.947	2.933	2.919	2.905	2.892	2.877	2.864	2.850
100	3.321	3.306	3.290	3.275	3.259	3.244	3.228	3.213	3.197	3.182	3.167
200	6.642	6.612	6.580	6.550	6.519	6.487	6.457	6.426	6.394	6.364	6.334
300	9.963	9.918	9.869	9.824	9.778	9.731	9.685	9.639	9.592	9.546	9.507
400	13.284	13.224	13.159	13.099	13.038	12.975	12.913	12.852	12.789	12.728	12.668
500	16.604	16.530	16.449	16.374	16.297	16.218	16.142	16.065	15.986	15.910	15.834
600	19.925	19.835	19.739	19.649	19.556	19.462	19.370	19.278	19.183	19.091	19.001
700	23.246	23.141	23.029	22.924	22.816	22.706	22.598	22.491	22.380	22.273	22.168
800	26.567	26.447	26.318	26.198	26.075	25.950	25.826	25.704	25.578	25.455	25.335
900	29.888	29.753	29.608	29.473	29.335	29.193	29.055	28.917	28.775	28.637	28.502
000	33.209	33.059	32.898	32.748	32.594	32.437	32.283	32.130	31.972	31.819	31.669

Table 24

EQUIVALENTS IN INCHES OF WATER AT 60 F, OF GAGE OIL* COLUMNS AT VARIOUS TEMPERATURES

Inches Oil	Temperature of Gage Oil Column in deg F												
	−20	−10	0	10	20	30	40	50	60	70	80	90	100
	Inches of Water												
0.01	0.009	0.009	0.009	0.009	0.009	0.009	0.009	0.008	0.008	0.008	0.008	0.008	0.008
0.02	0.017	0.017	0.017	0.017	0.017	0.017	0.017	0.017	0.017	0.017	0.017	0.017	0.016
0.03	0.026	0.026	0.026	0.026	0.026	0.025	0.025	0.025	0.025	0.025	0.025	0.025	0.025
0.04	0.035	0.035	0.034	0.034	0.034	0.034	0.034	0.034	0.033	0.033	0.033	0.033	0.033
0.05	0.043	0.043	0.043	0.043	0.043	0.043	0.042	0.042	0.042	0.042	0.042	0.041	0.041	0.041
0.06	0.052	0.052	0.051	0.051	0.051	0.051	0.051	0.050	0.050	0.050	0.050	0.049	0.049
0.07	0.061	0.060	0.060	0.060	0.060	0.059	0.059	0.059	0.058	0.058	0.058	0.058	0.057
0.08	0.069	0.069	0.069	0.068	0.068	0.068	0.067	0.067	0.067	0.066	0.066	0.066	0.066
0.09	0.078	0.078	0.077	0.077	0.077	0.076	0.076	0.075	0.075	0.075	0.074	0.074	0.074
0.1	0.087	0.086	0.086	0.085	0.085	0.085	0.084	0.084	0.083	0.083	0.083	0.082	0.082
0.2	0.173	0.172	0.172	0.171	0.170	0.169	0.168	0.168	0.167	0.166	0.165	0.165	0.164
0.3	0.260	0.258	0.257	0.256	0.255	0.254	0.253	0.251	0.250	0.249	0.248	0.247	0.246
0.4	0.346	0.345	0.343	0.341	0.340	0.338	0.337	0.335	0.334	0.332	0.331	0.329	0.327
0.5	0.433	0.431	0.429	0.427	0.425	0.423	0.421	0.419	0.417	0.415	0.413	0.411	0.409
0.6	0.519	0.517	0.514	0.512	0.510	0.507	0.505	0.503	0.500	0.498	0.496	0.493	0.491
0.7	0.606	0.603	0.600	0.597	0.595	0.592	0.589	0.587	0.584	0.581	0.578	0.576	0.573
0.8	0.692	0.689	0.686	0.683	0.680	0.677	0.673	0.670	0.667	0.664	0.661	0.659	0.655
0.9	0.779	0.775	0.772	0.768	0.765	0.761	0.758	0.754	0.751	0.747	0.744	0.740	0.737
1.	0.865	0.861	0.857	0.853	0.850	0.846	0.842	0.839	0.834	0.830	0.826	0.822	0.818
2.	1.730	1.722	1.715	1.707	1.699	1.691	1.684	1.676	1.668	1.660	1.653	1.645	1.637
3.	2.595	2.584	2.572	2.560	2.549	2.537	2.525	2.514	2.502	2.490	2.479	2.467	2.455
4.	3.460	3.445	3.429	3.414	3.398	3.383	3.367	3.352	3.336	3.320	3.305	3.289	3.274
5.	4.325	4.306	4.287	4.267	4.248	4.228	4.209	4.190	4.170	4.151	4.131	4.112	4.092
6.	5.191	5.167	5.144	5.121	5.097	5.074	5.051	5.027	5.004	4.981	4.957	4.934	4.911
7.	6.056	6.028	6.001	5.974	5.947	5.920	5.892	5.865	5.838	5.811	5.784	5.756	5.729
8.	6.921	6.890	6.859	6.827	6.796	6.765	6.734	6.703	6.672	6.641	6.610	6.579	6.548
9.	7.786	7.751	7.716	7.681	7.646	7.611	7.576	7.541	7.506	7.471	7.436	7.401	7.366
10.	8.651	8.612	8.573	8.534	8.495	8.457	8.418	8.379	8.340	8.301	8.262	8.223	8.185

* Specifically, "Ellison Gage Oil" for inclined gages: spec. grav. 60/60 = 0.834

Table 25

EQUIVALENTS IN INCHES OF WATER AT 60 F, OF ALCOHOL COLUMNS AT VARIOUS TEMPERATURES
(100% ethyl alcohol)

Inches Alcohol	Temperature of Alcohol Column in deg F												
	−30	−20	−10	0	10	20	30	40	50	60	70	80	90
	Inches of water												
0.01	0.008	0.008	0.008	0.008	0.008	0.008	0.008	0.008	0.008	0.008	0.008	0.008	0.008
0.02	0.017	0.017	0.017	0.016	0.016	0.016	0.016	0.016	0.016	0.016	0.016	0.016	0.016
0.03	0.025	0.025	0.025	0.025	0.025	0.024	0.024	0.024	0.024	0.024	0.024	0.024	0.023
0.04	0.034	0.033	0.033	0.033	0.033	0.033	0.032	0.032	0.032	0.032	0.032	0.031	0.031
0.05	0.042	0.042	0.041	0.041	0.041	0.041	0.040	0.040	0.040	0.040	0.040	0.039	0.039
0.06	0.050	0.050	0.050	0.049	0.049	0.049	0.049	0.048	0.048	0.048	0.047	0.047	0.047
0.07	0.059	0.058	0.058	0.058	0.057	0.057	0.057	0.056	0.056	0.056	0.055	0.055	0.055
0.08	0.067	0.067	0.066	0.066	0.065	0.065	0.065	0.064	0.064	0.064	0.063	0.063	0.062
0.09	0.075	0.075	0.074	0.074	0.074	0.073	0.073	0.072	0.072	0.072	0.071	0.071	0.070
0.1	0.084	0.083	0.083	0.082	0.082	0.081	0.081	0.080	0.080	0.079	0.079	0.079	0.078
0.2	0.167	0.166	0.165	0.164	0.164	0.163	0.162	0.161	0.160	0.159	0.159	0.157	0.156
0.3	0.251	0.250	0.248	0.247	0.245	0.244	0.242	0.241	0.240	0.238	0.237	0.235	0.234
0.4	0.335	0.333	0.331	0.329	0.327	0.325	0.323	0.321	0.319	0.318	0.316	0.314	0.312
0.5	0.418	0.416	0.414	0.411	0.409	0.406	0.404	0.402	0.399	0.397	0.395	0.392	0.390
0.6	0.502	0.499	0.496	0.493	0.491	0.488	0.485	0.482	0.479	0.476	0.474	0.471	0.468
0.7	0.585	0.582	0.579	0.576	0.572	0.569	0.566	0.562	0.559	0.556	0.552	0.549	0.546
0.8	0.669	0.665	0.662	0.658	0.654	0.650	0.646	0.643	0.639	0.635	0.631	0.628	0.624
0.9	0.753	0.749	0.744	0.740	0.736	0.732	0.727	0.723	0.719	0.715	0.710	0.706	0.702
1.	0.836	0.832	0.827	0.822	0.818	0.813	0.808	0.803	0.799	0.794	0.789	0.785	0.780
2.	1.673	1.663	1.654	1.644	1.635	1.626	1.616	1.607	1.597	1.588	1.578	1.569	1.560
3.	2.509	2.495	2.481	2.467	2.452	2.438	2.424	2.410	2.396	2.382	2.368	2.353	2.339
4.	3.345	3.327	3.308	3.289	3.270	3.251	3.232	3.213	3.194	3.176	3.157	3.138	3.119
5.	4.182	4.158	4.135	4.111	4.087	4.064	4.040	4.017	3.993	3.970	3.946	3.922	3.899
6.	5.018	4.990	4.961	4.933	4.905	4.877	4.848	4.820	4.792	4.763	4.735	4.707	4.679
7.	5.854	5.821	5.788	5.755	5.722	5.689	5.656	5.623	5.590	5.557	5.524	5.491	5.458
8.	6.691	6.653	6.615	6.578	6.540	6.502	6.464	6.427	6.389	6.351	6.313	6.276	6.238
9.	7.527	7.485	7.442	7.400	7.357	7.315	7.272	7.230	7.188	7.145	7.103	7.060	7.018
10.	8.363	8.316	8.269	8.222	8.175	8.128	8.081	8.033	7.986	7.939	7.892	7.945	7.798

Table 26

EQUIVALENTS IN INCHES OF WATER AT 60 F, OF ALCOHOL-GLYCOL* COLUMNS AT VARIOUS TEMPERATURES

Temperature of Alcohol-Glycol Column in deg F

Inches Alcohol-Glycol	−40	−30	−20	−10	0	10	20	30	40	50	60	70	80	100
0.1	0.104	0.104	0.103	0.103	0.103	0.102	0.102	0.101	0.101	0.100	0.100	0.100	0.099	0.098
0.2	0.209	0.208	0.207	0.206	0.205	0.204	0.203	0.203	0.202	0.201	0.200	0.199	0.198	0.197
0.3	0.313	0.312	0.310	0.309	0.308	0.306	0.305	0.304	0.303	0.301	0.300	0.299	0.297	0.295
0.4	0.417	0.415	0.414	0.412	0.410	0.409	0.407	0.405	0.403	0.402	0.400	0.398	0.397	0.393
0.5	0.521	0.519	0.517	0.515	0.513	0.511	0.509	0.506	0.504	0.502	0.500	0.498	0.496	0.492
0.6	0.626	0.623	0.621	0.618	0.615	0.613	0.610	0.608	0.605	0.603	0.600	0.597	0.595	0.590
0.7	0.730	0.727	0.724	0.721	0.718	0.715	0.712	0.709	0.706	0.703	0.700	0.697	0.694	0.688
0.8	0.834	0.831	0.827	0.824	0.821	0.817	0.814	0.810	0.807	0.803	0.800	0.797	0.793	0.786
0.9	0.938	0.935	0.931	0.927	0.923	0.919	0.916	0.912	0.908	0.904	0.900	0.896	0.892	0.885
1.0	1.043	1.038	1.034	1.030	1.026	1.021	1.017	1.013	1.009	1.004	1.000	0.996	0.992	0.983
2	2.085	2.077	2.068	2.060	2.052	2.043	2.034	2.026	2.017	2.009	2.000	1.992	1.983	1.966
3	3.128	3.115	3.103	3.090	3.077	3.064	3.051	3.038	3.026	3.013	3.000	2.987	2.974	2.949
4	4.171	4.154	4.137	4.120	4.103	4.085	4.068	4.051	4.034	4.017	4.000	3.983	3.966	3.932
5	5.214	5.192	5.171	5.150	5.128	5.107	5.085	5.064	5.043	5.021	5.000	4.979	4.957	4.915
6	6.256	6.231	6.205	6.179	6.154	6.128	6.103	6.077	6.051	6.026	6.000	5.974	5.949	5.898
7	7.299	7.269	7.239	7.209	7.179	7.150	7.120	7.090	7.060	7.030	7.000	6.970	6.940	6.880
8	8.342	8.307	8.273	8.239	8.205	8.171	8.137	8.103	8.068	8.034	8.000	7.966	7.932	7.863
9	9.384	9.346	9.307	9.269	9.231	9.192	9.154	9.115	9.077	9.038	9.000	8.962	8.923	8.846
10	10.427	10.384	10.342	10.299	10.256	10.214	10.171	10.128	10.085	10.043	10.000	9.957	9.915	9.829
20	20.854	20.769	20.683	20.598	20.512	20.427	20.342	20.256	20.171	20.085	20.000	19.915	19.829	19.658
30	31.281	31.153	31.025	30.897	30.769	30.641	30.512	30.384	30.256	30.128	30.000	29.872	29.744	29.488
40	41.708	41.537	41.366	41.196	41.025	40.854	40.683	40.512	40.342	40.171	40.000	39.830	39.658	39.317
50	52.135	51.922	51.708	51.495	51.281	51.068	50.854	50.641	50.427	50.214	50.000	49.787	49.573	49.146
60	62.562	62.306	62.050	61.793	61.537	61.281	61.025	60.769	60.512	60.256	60.000	59.744	59.488	58.975
70	72.989	72.690	72.391	72.092	71.793	71.495	71.196	70.897	70.598	70.299	70.000	69.701	69.402	68.804
80	83.416	83.074	82.733	82.391	82.050	81.708	81.366	81.025	80.683	80.342	80.000	79.658	79.317	78.633
90	93.843	93.456	93.074	92.690	92.306	91.922	91.537	91.153	90.769	90.384	90.000	89.616	89.231	88.463
100	104.27	103.84	103.42	102.99	102.56	102.14	101.71	101.28	100.85	100.43	100.00	99.573	99.146	98.292

* 36% by vol. ethyl alcohol; 64% ethylene glycol; spec. gravity at 60 F = 1.000

Table 27

EQUIVALENTS IN INCHES OF WATER AT 60 F, OF DIBROMOBENZENE* COLUMNS AT VARIOUS TEMPERATURES

Temperature of Dibromobenzene columns in deg F

Inches of Water

Inches Dibromobenzene	30	40	50	60	70	80	90	100	110
0.01	0.020	0.020	0.020	0.020	0.020	0.019	0.019	0.019	0.019
0.02	0.040	0.040	0.039	0.039	0.039	0.039	0.039	0.039	0.038
0.03	0.060	0.059	0.059	0.059	0.059	0.058	0.058	0.058	0.058
0.04	0.080	0.079	0.079	0.079	0.079	0.078	0.078	0.077	0.077
0.05	0.100	0.099	0.099	0.098	0.098	0.097	0.097	0.097	0.096
0.06	0.120	0.119	0.118	0.118	0.117	0.117	0.116	0.116	0.115
0.07	0.139	0.139	0.138	0.138	0.137	0.136	0.136	0.135	0.135
0.08	0.159	0.159	0.158	0.157	0.157	0.156	0.155	0.155	0.154
0.09	0.179	0.178	0.178	0.177	0.176	0.175	0.175	0.174	0.173
0.1	0.199	0.198	0.197	0.197	0.196	0.195	0.194	0.193	0.192
0.2	0.398	0.397	0.395	0.393	0.392	0.390	0.388	0.386	0.385
0.3	0.597	0.595	0.592	0.590	0.587	0.585	0.582	0.580	0.577
0.4	0.797	0.793	0.790	0.786	0.783	0.780	0.776	0.773	0.769
0.5	0.996	0.992	0.987	0.983	0.979	0.975	0.970	0.966	0.962
0.6	1.195	1.190	1.185	1.180	1.175	1.169	1.164	1.159	1.154
0.7	1.394	1.388	1.382	1.376	1.370	1.364	1.358	1.352	1.347
0.8	1.593	1.586	1.580	1.573	1.566	1.559	1.552	1.546	1.539
0.9	1.792	1.785	1.777	1.769	1.762	1.754	1.747	1.739	1.731
1	1.992	1.983	1.974	1.966	1.958	1.949	1.941	1.932	1.924
2	3.983	3.966	3.949	3.932	3.915	3.898	3.881	3.864	3.847
3	5.974	5.949	5.923	5.898	5.873	5.847	5.822	5.796	5.771
4	7.965	7.932	7.898	7.864	7.830	7.796	7.762	7.728	7.694
5	9.957	9.915	9.872	9.830	9.788	9.745	9.703	9.660	9.618
6	11.949	11.898	11.847	11.796	11.745	11.694	11.643	11.592	11.541
7	13.940	13.881	13.821	13.762	13.703	13.643	13.584	13.524	13.465
8	15.932	15.864	15.796	15.728	15.660	15.592	15.524	15.456	15.388
9	17.923	17.847	17.770	17.694	17.618	17.541	17.465	17.388	17.312
10	19.915	19.830	19.745	19.660	19.575	19.490	19.405	19.320	19.235

* ortho-dibromobenzene: spec. gravity at 60/60 = 1.9660; m. p. = 30 F.

Table 28

EQUIVALENTS IN PSI, OF DIBROMOBENZENE COLUMNS AT VARIOUS TEMPERATURES

Temperature of dibromobenzene column in deg F

Inches Dibromobenzene	30	40	50	60	70	80	90	100	110
					Psi				
0.1	0.007	0.007	0.007	0.007	0.007	0.007	0.007	0.007	0.007
0.2	0.014	0.014	0.014	0.014	0.014	0.014	0.014	0.014	0.014
0.3	0.022	0.022	0.021	0.021	0.021	0.021	0.021	0.021	0.021
0.4	0.029	0.029	0.029	0.028	0.028	0.028	0.028	0.028	0.028
0.5	0.036	0.036	0.036	0.036	0.035	0.035	0.035	0.035	0.035
0.6	0.043	0.043	0.043	0.043	0.042	0.042	0.042	0.042	0.042
0.7	0.050	0.050	0.050	0.050	0.050	0.050	0.049	0.049	0.049
0.8	0.058	0.057	0.057	0.057	0.057	0.056	0.056	0.056	0.056
0.9	0.065	0.064	0.064	0.064	0.064	0.063	0.063	0.063	0.063
1.0	0.072	0.072	0.071	0.071	0.071	0.070	0.070	0.070	0.069
2	0.144	0.143	0.143	0.142	0.141	0.141	0.140	0.140	0.139
3	0.216	0.215	0.214	0.213	0.212	0.211	0.210	0.209	0.208
4	0.288	0.286	0.285	0.284	0.283	0.281	0.280	0.279	0.278
5	0.359	0.358	0.356	0.355	0.353	0.352	0.350	0.349	0.347
6	0.431	0.429	0.428	0.426	0.424	0.422	0.420	0.419	0.417
7	0.503	0.501	0.499	0.497	0.495	0.492	0.490	0.488	0.486
8	0.575	0.573	0.570	0.568	0.565	0.563	0.560	0.558	0.555
9	0.647	0.644	0.641	0.639	0.636	0.633	0.630	0.628	0.625
10	0.719	0.716	0.713	0.710	0.706	0.703	0.700	0.697	0.694
20	1.437	1.431	1.425	1.419	1.413	1.407	1.400	1.395	1.388
30	2.156	2.147	2.138	2.129	2.119	2.110	2.101	2.092	2.083
40	2.875	2.863	2.850	2.838	2.826	2.814	2.801	2.789	2.777
50	3.594	3.578	3.563	3.548	3.532	3.517	3.502	3.486	3.471
60	4.312	4.294	4.276	4.257	4.239	4.220	4.202	4.184	4.165
70	5.031	5.010	4.988	4.967	4.945	4.924	4.902	4.881	4.859
80	5.750	5.725	5.701	5.676	5.652	5.627	5.603	5.578	5.554
90	6.469	6.441	6.413	6.386	6.358	6.331	6.303	6.275	6.248
100	7.187	7.157	7.126	7.095	7.065	7.034	7.003	6.972	6.942

Table 29

EQUIVALENTS IN INCHES OF WATER AT 60 F, OF DIBROMOETHANE* COLUMNS AT VARIOUS TEMPERATURES

Temperature of Dibromoethane Column in deg F

Inches Dibromoethane	10	20	30	40	50	60	70	80	90	100
					Inches of Water					
0.01	0.021	0.021	0.021	0.021	0.021	0.021	0.021	0.020	0.020	0.020
0.02	0.042	0.042	0.042	0.042	0.041	0.041	0.041	0.041	0.041	0.041
0.03	0.063	0.063	0.063	0.062	0.062	0.062	0.062	0.061	0.061	0.060
0.04	0.084	0.084	0.084	0.083	0.083	0.082	0.082	0.081	0.081	0.081
0.05	0.105	0.105	0.104	0.104	0.103	0.103	0.102	0.102	0.101	0.101
0.06	0.127	0.126	0.126	0.125	0.124	0.123	0.123	0.122	0.121	0.121
0.07	0.148	0.147	0.146	0.146	0.145	0.144	0.143	0.142	0.142	0.141
0.08	0.169	0.168	0.167	0.166	0.165	0.165	0.164	0.163	0.162	0.161
0.09	0.190	0.189	0.188	0.187	0.186	0.185	0.184	0.183	0.182	0.181
0.1	0.211	0.210	0.209	0.208	0.207	0.206	0.205	0.203	0.202	0.201
0.2	0.422	0.420	0.418	0.416	0.413	0.411	0.409	0.407	0.405	0.403
0.3	0.633	0.630	0.627	0.623	0.620	0.617	0.614	0.610	0.607	0.604
0.4	0.844	0.840	0.836	0.831	0.827	0.822	0.818	0.814	0.809	0.805
0.5	1.055	1.050	1.044	1.039	1.034	1.028	1.023	1.017	1.012	1.006
0.6	1.266	1.260	1.253	1.247	1.240	1.234	1.227	1.221	1.214	1.207
0.7	1.478	1.470	1.462	1.455	1.447	1.439	1.436	1.424	1.416	1.409
0.8	1.689	1.680	1.671	1.662	1.654	1.645	1.636	1.627	1.619	1.610
0.9	1.900	1.890	1.880	1.870	1.860	1.850	1.841	1.831	1.821	1.811
1.0	2.111	2.100	2.088	2.078	2.067	2.056	2.045	2.034	2.023	2.012
2	4.221	4.200	4.178	4.156	4.134	4.112	4.090	4.068	4.046	4.025
3	6.332	6.299	6.266	6.234	6.201	6.168	6.135	6.102	6.070	6.037
4	8.443	8.399	8.355	8.312	8.268	8.224	8.180	8.136	8.093	8.049
5	10.553	10.499	10.444	10.389	10.335	10.280	10.225	10.171	10.116	10.061
6	12.664	12.599	12.533	12.467	12.402	12.336	12.270	12.205	12.139	12.074
7	14.775	14.698	14.622	14.545	14.469	14.392	14.315	14.239	14.162	14.086
8	16.886	16.798	16.710	16.623	16.536	16.448	16.360	16.273	16.186	16.098
9	18.996	18.898	18.799	18.701	18.602	18.504	18.406	18.307	18.209	18.110
10	21.107	20.998	20.888	20.779	20.669	20.560	20.451	20.341	20.232	20.123

* 1, 1 — Dibromoethane: spec gravity, 2.056 at 60/60.

Table 30

EQUIVALENTS IN PSI, OF DIBROMOETHANE COLUMNS AT VARIOUS TEMPERATURES

| Inches Dibromoethane | Temperature of Dibromoethane Column in deg F | | | | | | | | | |
	10	20	30	40	50	60	70	80	90	100
	psi									
0.1	0.007	0.007	0.007	0.007	0.007	0.007	0.007	0.007	0.007	0.007
0.2	0.015	0.015	0.015	0.015	0.015	0.015	0.015	0.015	0.015	0.015
0.3	0.023	0.023	0.023	0.023	0.022	0.022	0.022	0.022	0.022	0.022
0.4	0.031	0.030	0.030	0.030	0.030	0.030	0.030	0.029	0.029	0.029
0.5	0.038	0.038	0.038	0.038	0.037	0.037	0.037	0.037	0.037	0.036
0.6	0.046	0.046	0.045	0.045	0.045	0.045	0.044	0.044	0.044	0.044
0.7	0.053	0.053	0.053	0.053	0.052	0.052	0.052	0.051	0.051	0.051
0.8	0.061	0.061	0.060	0.060	0.060	0.059	0.059	0.059	0.058	0.058
0.9	0.069	0.068	0.068	0.068	0.067	0.067	0.066	0.066	0.066	0.065
1.0	0.076	0.076	0.075	0.075	0.075	0.074	0.074	0.073	0.073	0.073
2	0.152	0.152	0.151	0.150	0.149	0.148	0.148	0.147	0.146	0.145
3	0.229	0.227	0.226	0.225	0.224	0.223	0.221	0.220	0.219	0.218
4	0.305	0.303	0.302	0.300	0.298	0.297	0.295	0.294	0.292	0.291
5	0.381	0.379	0.377	0.375	0.373	0.371	0.369	0.367	0.365	0.363
6	0.457	0.455	0.452	0.450	0.448	0.445	0.443	0.441	0.438	0.436
7	0.533	0.531	0.528	0.525	0.522	0.519	0.517	0.514	0.511	0.508
8	0.609	0.606	0.603	0.600	0.597	0.594	0.591	0.587	0.584	0.581
9	0.686	0.682	0.679	0.675	0.671	0.668	0.664	0.661	0.657	0.654
10	0.762	0.758	0.754	0.750	0.746	0.742	0.738	0.734	0.730	0.726
20	1.524	1.516	1.508	1.500	1.492	1.484	1.476	1.468	1.460	1.452
30	2.285	2.273	2.262	2.250	2.238	2.226	2.214	2.202	2.191	2.179
40	3.047	3.031	3.015	3.000	2.984	2.968	2.952	2.936	2.921	2.905
50	3.809	3.789	3.769	3.750	3.730	3.710	3.690	3.671	3.651	3.631
60	4.571	4.547	4.523	4.499	4.476	4.452	4.428	4.405	4.381	4.357
70	5.332	5.305	5.277	5.249	5.222	5.194	5.166	5.139	5.111	5.084
80	6.094	6.062	6.031	5.999	5.968	5.936	5.905	5.873	5.841	5.810
90	6.856	6.820	6.785	6.749	6.714	6.678	6.643	6.607	6.572	6.536
100	7.618	7.578	7.539	7.499	7.460	7.420	7.381	7.341	7.302	7.262

Table 31

EQUIVALENTS IN INCHES OF WATER AT 60 F, OF COLUMNS OF ACETYLENE TETRABROMIDE* AT VARIOUS TEMPERATURES

| Inches Acetylene Tetrabromide | Temperature of Acetylene Tetrabromide Column in deg F | | | | | | | | | | |
	0	10	20	30	40	50	60	70	80	90	100
	Inches of Water										
0.01	0.031	0.030	0.030	0.030	0.030	0.030	0.030	0.030	0.030	0.029	0.029
0.02	0.061	0.061	0.061	0.060	0.060	0.060	0.059	0.059	0.059	0.059	0.058
0.03	0.091	0.091	0.091	0.090	0.090	0.090	0.089	0.089	0.088	0.088	0.088
0.04	0.122	0.121	0.121	0.121	0.120	0.119	0.119	0.118	0.118	0.117	0.117
0.05	0.152	0.152	0.151	0.151	0.150	0.149	0.149	0.148	0.147	0.147	0.146
0.06	0.183	0.182	0.182	0.181	0.180	0.179	0.178	0.178	0.177	0.176	0.175
0.07	0.213	0.212	0.212	0.211	0.210	0.209	0.208	0.207	0.206	0.205	0.204
0.08	0.244	0.243	0.242	0.241	0.240	0.239	0.238	0.237	0.236	0.235	0.234
0.09	0.274	0.273	0.272	0.271	0.270	0.269	0.268	0.266	0.265	0.264	0.263
0.1	0.305	0.304	0.303	0.301	0.300	0.299	0.297	0.296	0.295	0.293	0.292
0.2	0.610	0.607	0.605	0.602	0.600	0.597	0.594	0.592	0.589	0.587	0.584
0.3	0.914	0.911	0.907	0.904	0.900	0.896	0.892	0.888	0.884	0.880	0.876
0.4	1.221	1.216	1.210	1.205	1.200	1.194	1.189	1.184	1.178	1.173	1.168
0.5	1.526	1.519	1.512	1.506	1.500	1.493	1.486	1.479	1.473	1.466	1.460
0.6	1.830	1.822	1.815	1.807	1.799	1.791	1.783	1.775	1.767	1.760	1.752
0.7	2.136	2.126	2.117	2.108	2.099	2.090	2.080	2.071	2.062	2.053	2.044
0.8	2.441	2.431	2.420	2.409	2.399	2.388	2.378	2.367	2.357	2.346	2.335
0.9	2.746	2.734	2.722	2.710	2.699	2.687	2.675	2.663	2.651	2.639	2.527
1	3.051	3.038	3.025	3.012	2.998	2.985	2.972	2.959	2.946	2.932	2.919
2	6.102	6.076	6.049	6.023	5.997	5.970	5.944	5.918	5.891	5.865	5.839
3	9.153	9.114	9.074	9.035	8.995	8.956	8.916	8.876	8.837	8.797	8.758
4	12.204	12.151	12.099	12.046	11.993	11.941	11.888	11.835	11.783	11.730	11.677
5	15.255	15.189	15.123	15.058	14.992	14.926	14.860	14.794	14.728	14.662	14.597
6	18.306	18.227	18.148	18.069	17.990	17.911	17.832	17.753	17.674	17.595	17.516
7	21.357	21.265	21.173	21.081	20.988	20.896	20.804	20.712	20.619	20.527	20.435
8	24.408	24.303	24.197	24.092	23.987	23.881	23.776	23.670	23.565	23.460	23.354
9	27.459	27.341	27.222	27.104	26.985	26.867	26.748	26.629	26.511	26.392	26.274
10	30.510	30.379	30.247	30.115	29.983	29.852	29.720	29.588	29.456	29.325	29.193

* 1, 1, 2, 2 — Tetrabromoethane: spec. grav. 2.972 at 60/60.

Table 32

EQUIVALENTS IN PSI, OF COLUMNS OF ACETYLENE TETRABROMIDE AT VARIOUS TEMPERATURES

Temperature of Acetylene Tetrabromide Column in deg F

Inches Acetylene Tetrabromide	0	10	20	30	40	50	60	70	80	90	100
					psi						
0.1	0.011	0.011	0.011	0.011	0.011	0.011	0.011	0.011	0.011	0.011	0.011
0.2	0.022	0.022	0.022	0.022	0.022	0.022	0.021	0.021	0.021	0.021	0.021
0.3	0.033	0.033	0.033	0.033	0.032	0.032	0.032	0.032	0.032	0.032	0.032
0.4	0.044	0.044	0.044	0.043	0.043	0.043	0.043	0.043	0.043	0.042	0.042
0.5	0.055	0.055	0.055	0.054	0.054	0.054	0.054	0.053	0.053	0.053	0.053
0.6	0.066	0.066	0.066	0.065	0.065	0.065	0.064	0.064	0.064	0.064	0.063
0.7	0.077	0.077	0.076	0.076	0.076	0.075	0.075	0.075	0.074	0.074	0.074
0.8	0.088	0.088	0.087	0.087	0.087	0.086	0.086	0.085	0.085	0.085	0.084
0.9	0.099	0.099	0.098	0.098	0.097	0.097	0.097	0.096	0.096	0.095	0.095
1.0	0.110	0.110	0.109	0.109	0.108	0.108	0.107	0.107	0.106	0.106	0.105
2.	0.220	0.219	0.218	0.217	0.216	0.215	0.215	0.214	0.213	0.212	0.211
3.	0.330	0.329	0.327	0.326	0.325	0.323	0.322	0.320	0.319	0.318	0.316
4.	0.440	0.438	0.437	0.435	0.433	0.431	0.429	0.427	0.425	0.423	0.421
5.	0.550	0.548	0.546	0.543	0.541	0.539	0.536	0.534	0.532	0.529	0.527
6.	0.660	0.657	0.655	0.652	0.649	0.646	0.644	0.641	0.638	0.635	0.632
7.	0.770	0.767	0.764	0.761	0.757	0.754	0.751	0.747	0.744	0.741	0.738
8.	0.880	0.876	0.873	0.869	0.866	0.862	0.858	0.854	0.850	0.847	0.843
9.	0.990	0.986	0.982	0.978	0.974	0.970	0.965	0.961	0.957	0.953	0.948
10	1.101	1.096	1.092	1.087	1.082	1.077	1.073	1.068	1.063	1.058	1.054
20	2.202	2.193	2.183	2.174	2.164	2.155	2.145	2.136	2.126	2.117	2.107
30	3.303	3.289	3.275	3.261	3.246	3.232	3.218	3.204	3.189	3.175	3.161
40	4.404	4.385	4.366	4.347	4.328	4.309	4.290	4.271	4.252	4.233	4.214
50	5.506	5.482	5.458	5.434	5.411	5.387	5.363	5.339	5.315	5.292	5.268
60	6.607	6.578	6.550	6.521	6.493	6.464	6.436	6.407	6.379	6.350	6.322
70	7.708	7.675	7.641	7.608	7.575	7.542	7.508	7.475	7.442	7.408	7.375
80	8.809	8.771	8.733	8.695	8.657	8.619	8.581	8.543	8.505	8.467	8.429
90	9.910	9.867	9.824	9.782	9.739	9.696	9.653	9.611	9.568	9.525	9.482
100	11.011	10.964	10.916	10.869	10.821	10.774	10.726	10.678	10.631	10.583	10.536

INSTRUMENT SOCIETY of AMERICA
Research Triangle Park, North Carolina

ANSI/ISA-S5.1-1984
Approved November 5, 1986

American National Standard

Instrumentation Symbols and Identification

Instrument Society of America

ISBN 0-87664-844-8

ISA-S5.1 Instrumentation Symbols and Identification

INSTRUMENT SOCIETY OF AMERICA
67 Alexander Drive
P.O. Box 12277
Research Triangle Park, North Carolina 27709

PREFACE

This preface is included for information and is not a part of ISA-S5.1.

This standard has been prepared as part of the service of the Instrument Society of America (ISA) toward a goal of uniformity in the field of instrumentation. To be of real value, this document should not be static, but should be subject to periodic review. Toward this end, the Society welcomes all comments and criticisms, and asks that they be addressed to the Secretary, Standards and Practices Board, Instrument Society of America, 67 Alexander Drive, P.O. Box 12277, Research Triangle Park, NC 27709, Telephone (919) 549-8411.

The ISA Standards and Practices Department is aware of the growing need for attention to the metric system of units in general, and the International System of Units (SI) in particular, in the preparation of instrumentation standards. The Department is further aware of the benefits to U.S.A. users of ISA standards of incorporating suitable references to the SI (and the metric system) in their business and professional dealings with other countries. Toward this end, this Department will endeavor to introduce SI-acceptable metric units in all new and revised standards to the greatest extent possible. The Metric Practice Guide, which has been published by the Institute of Electrical and Electronics Engineers as ANSI/IEEE Std. 268-1982, and future revisions will be the reference guide for definitions, symbols, abbreviations, and conversion factors.

It is the policy of the Instrument Society of America to encourage and welcome the participation of all concerned individuals and interests in the development of ISA standards. Participation in the ISA standards-making process by an individual in no way constitutes endorsement by the employer of that individual, of the Instrument Society of America, or of any of the standards that ISA develops.

The information contained in the preface, footnotes, and appendices is included for information only and is not a part of the standard.

The instrumentation symbolism and identification techniques described in the standard accommodate the advances in technology and reflect the collective industrial experience gained since the publication of Recommended Practice RP5.1 in 1949.

This revision attempts to strengthen the standard in its role as a tool of communication in the process industries. Communication presupposes a common language; or, at the very least, it is facilitated by one. The standard offers the foundation for that common language.

When integrated into a system, the symbols and designations presented here form a concise, dedicated language which communicates concepts, facts, intent, instructions, and knowledge about measurement and control systems in the process industries.

This document is a consensus standard rather than a mandatory one. As such, it has many of the strengths and the weaknesses of consensus standards. Its primary strength is that it can be used in widespread, interdisciplinary ways. Its weakness is generally that of not being specific enough to satisfy the special requirements of particular interest groups.

The symbols and identification contained in ISA-S5.1 have evolved by the consensus method and are intended for wide application throughout the process industries. The symbols and designations are used as conceptualizing aids, as design tools, as teaching devices, and as a concise and specific means of communication on all types and kinds of technical, engineering, procurement, construction, and maintenance documents.

In the past, the standard has been flexible enough to serve all of the uses just described. In the future, it must continue to do so. To this end, this revision offers symbols, identification, and definitions for concepts that were not previously described; for example, shared display/control, distributed control, and programmable control. Definitions were broadened to accommodate the fact that, although similar functions are being performed by the new control systems, these functions are frequently not related to a uniquely identifiable instrument; yet they still must be conceptualized and identified. The excellent SAMA (Scientific Apparatus Makers Association) method of functional diagramming was used to describe function blocks and function designators. To help the batch processing industries, where binary (on-off) symbolism is extremely useful, new binary line symbols were introduced and first-letter *Y* was selected to represent an initiating variable which could be categorized as an event, presence, or state. In general, breadth of application as opposed to narrowness has been emphasized.

3

The ISA Standards Committee on Instrumentation Symbols and Identification operates within the ISA Standards and Practices Department, with William Calder III as vice president. The persons listed below served as members of or advisors to the SP5.1 committee. The SP5.1 committee is deeply appreciative of the work of previous SP5.1 committees and has tried to treat their work with the respect it deserves. In addition, this committee would like to acknowledge the work of the SP5.3 committee in developing ISA-S5.3, ''Graphic Symbols for Distributed Control/ Shared Display Instrumentation, Logic and Computer Systems.'' The key elements of ISA-S5.3 have been incorporated into ISA-S5.1, and it is the Society's intent to withdraw ISA-S5.3 after publication of this revision of ISA-S5.1.

The following people served as members of ISA Committee SP5.1, which prepared this standard:

NAME	COMPANY
R. Mulley, Chairman	Fluor Engineers, Inc.
E. J. Blahut	Blahut Engineering, Inc.
P. R. Boubel	TXE Inc.
J. P. Carew	Stone and Webster Engineering Corporation
N. Dogra	ANK Engineers
J. E. Doyle	Tweedcrest Limited
C. R. Gross	EXXON Company U.S.A.
T. E. Hamler	Owens Corning Fiberglas Corporation
F. Horn	Allied Chemical Company
A. A. Iverson	ARCO Chemical Company
A. Langelier	Polaroid Coroporation
W. E. Mapes	Eastman Kodak Company
T. C. McAvinew	Vertech Treatment Systems
W. L. Mostia	AMOCO Chemicals
G. K. Pace	Phelps Dodge Corporation
*G. Platt, Past Chairman	Bechtel Power Corporation
A. W. Reeve	AWR Controls (Canada) Ltd.
S. Sankaran	McDermott Engineering
R. M. Shah	Olin Chemicals Corporation
D. G. Turnbull	Sandwell and Company, Limited
R. von Brecht	The M. W. Kellogg Company
G. Wilbanks	The Rust Engineering Company

*Member Emeritus

The following people served as members of ISA Committee SP5:

NAME	COMPANY
D. E. Rapley, Chairman	Stearns Catalytic Corporation
R. C. Greer	Bailey Controls Company
D. G. Kempfer	Standard Oil Company of Ohio
R. H. Kind	El Paso Natural Gas Company
R. Mulley	Fluor Engineers, Inc.
T. J. Myron	The Foxboro Company

This standard was approved for publication by the ISA Standards and Practices Board in September 1984.

TABLE OF CONTENTS

LIST OF FIGURES

1 PURPOSE

The purpose of this standard is to establish a uniform means of designating instruments and instrumentation systems used for measurement and control. To this end, a designation system that includes symbols and an identification code is presented.

2 SCOPE

2.1 General

2.1.1 The procedural needs of various users are different. The standard recognizes these needs, when they are consistent with the objectives of the standard, by providing alternative symbolism methods. A number of examples are provided for adding information or simplifying the symbolism, as desired.

2.1.2 Process equipment symbols are not part of this standard, but are included only to illustrate applications of instrumentation symbols.

2.2 Application to Industries

2.2.1 The standard is suitable for use in the chemical, petroleum, power generation, air conditioning, metal refining, and numerous other, process industries.

2.2.2 Certain fields, such as astronomy, navigation, and medicine, use very specialized instruments that are different from the conventional industrial process instruments. No specific effort was made to have the standard meet the requirements of those fields. However, it is expected that the standard will be flexible enough to meet many of the needs of special fields.

2.3 Application to Work Activities

2.3.1 The standard is suitable for use whenever any reference to an instrument or to a control system function is required for the purposes of symbolization and identification. Such references may be required for the following uses, as well as others:

> Design sketches
> Teaching examples
> Technical papers, literature, and discussions
> Instrumentation system diagrams, loop diagrams, logic diagrams
> Functional descriptions
> Flow diagrams: Process, Mechanical, Engineering, Systems, Piping (Process) and Instrumentation
> Construction drawings
> Specifications, purchase orders, manifests, and other lists
> Identification (tagging) of instruments and control functions
> Installation, operating and maintenance instructions, drawings, and records

2.3.2 The standard is intended to provide sufficient information to enable anyone reviewing any document depicting process measurement and control (who has a reasonable amount of process knowledge) to understand the means of measurement and control of the process. The detailed knowledge of a specialist in instrumentation is not a prerequisite to this understanding.

2.4 Application to Classes of Instrumentation and to Instrument Functions

The symbolism and identification methods provided in this standard are applicable to all classes of process measurement and control instrumentation. They can be used not only to describe discrete instruments and their functions, but also to describe the analogous functions of systems which are variously termed ''shared display,'' ''shared control,'' ''distributed control,'' and ''computer control.''

2.5 Extent of Functional Identification

The standard provides for the identification and symbolization of the key functions of an instrument. Additional details of the instrument are better described in a suitable specification, data sheet, or other document intended for those requiring such details.

2.6 Extent of Loop Identification

The standard covers the identification of an instrument and all other instruments or control functions associated with it in a loop. The user is free to apply additional identification—by serial number, unit number, area number, plant number, or by other means.

3 DEFINITIONS

For the purpose of understanding this standard, the following definitions apply. For a more complete treatment, see ISA-S51.1 and the ISA-S75 series of standards. Terms italicized in a definition are also defined in this section.

Accessible - A term applied to a device or *function* that can be used or be seen by an operator for the purpose of performing control actions, *e.g., set point* changes, auto-manual transfer, or on-off actions.

Alarm - A device or *function* that signals the existence of an abnormal condition by means of an audible or visible discrete change, or both, intended to attract attention.

It is not recommended that the term *alarm switch* or *alarm* be used to designate a device whose operation is simply to close or open a circuit that may or may not be used for normal or abnormal interlock, start-up, shutdown, actuation of a *pilot light* or an *alarm* device, or the like. The first device is properly designated as a level *switch*, a flow *switch, etc.,* because ''switching'' is what the device does. The device may be designated as an *alarm* only if the device itself contains the *alarm function*. [*See also* Table 1, note (13).]

Assignable - A term applied to a feature permitting the channeling (or directing) of a signal from one device to another without the need for switching, patching, or changes in wiring.

Auto-Manual Station - Synonym for *control station*.

Balloon - Synonym for *bubble*.

Behind the Panel - A term applied to a location that is within an area that contains (1) the *instrument panel*, (2) its associated rack-mounted hardware, or (3) is enclosed within the *panel*. *Behind the panel* devices are not *accessible* for the operator's normal use, and are not designated as *local* or front-of-*panel*-mounted. In a very broad sense, *"behind the panel"* is equivalent to "not normally *accessible* to the operator."

Binary - A term applied to a signal or device that has only two discrete positions or states. When used in its simplest form, as in *"binary* signal" (as opposed to "analog signal"), the term denotes an "on-off" or "high-low" state, *i.e.*, one which does not represent continuously varying quantities.

Board - Synonym for *panel*.

Bubble - The circular symbol used to denote and identify the purpose of an *instrument* or *function*. It may contain a tag number. Synonym for *balloon*.

Computing Device - A device or *function* that performs one or more calculations or logic operations, or both, and transmits one or more resultant output signals. A *computing device* is sometimes called a computing *relay*.

Configurable - A term applied to a device or system whose functional characteristics can be selected or rearranged through programming or other methods. The concept excludes rewiring as a means of altering the configuration.

Controller - A device having an output that varies to regulate a controlled variable in a specified manner. A *controller* may be a self-contained analog or *digital instrument*, or it may be the equivalent of such an *instrument* in a shared-control system.

An automatic *controller* varies its output automatically in response to a direct or indirect input of a measured *process variable*. A manual *controller* is a *manual loading station*, and its output is not dependent on a measured *process variable* but can be varied only by manual adjustment.

A *controller* may be integral with other functional elements of a control *loop*.

Control Station - A *manual loading station* that also provides switching between manual and automatic control modes of a control *loop*. It is also known as an *auto-manual station*. In addition, the operator interface of a *distributed control system* may be regarded as a *control station*.

Control Valve - A device, other than a common, hand-actuated ON-OFF valve or self-actuated check valve, that directly manipulates the flow of one or more fluid process streams.

It is expected that use of the designation "hand *control valve*" will be limited to hand-actuated valves that (1) are used for process throttling, or (2) require *identification* as an *instrument*.

Converter - A device that receives information in one form of an instrument signal and transmits an output signal in another form.

An *instrument* which changes a sensor's output to a standard signal is properly designated as a *transmitter*, not a *converter*. Typically, a temperature element (*TE*) may connect to a *transmitter* (*TT*), not to a *converter* (*TY*).

A converter is also referred to as a *transducer*; however, *"transducer"* is a completely general term, and its use specifically for signal conversion is not recommended.

Digital - A term applied to a signal or device that uses *binary* digits to represent continuous values or discrete states.

Distributed Control System - A system which, while being functionally integrated, consists of subsystems which may be physically separate and remotely located from one another.

Final Control Element - The device that directly controls the value of the manipulated variable of a control *loop*. Often the *final control element* is a *control valve*.

Function - The purpose of, or an action performed by, a device.

Identification - The sequence of letters or digits, or both, used to designate an individual *instrument* or *loop*.

Instrument - A device used directly or indirectly to measure and/or control a variable. The term includes *primary elements, final control elements, computing devices,* and electrical devices such as annunciators, *switches,* and pushbuttons. The term does not apply to parts (*e.g.*, a receiver bellows or a resistor) that are internal components of an *instrument*.

Instrumentation - A collection of *instruments* or their application for the purpose of observation, *measurement*, control, or any combination of these.

Local - The location of an *instrument* that is neither in nor on a *panel* or console, nor is it mounted in a control room. *Local instruments* are commonly in the vicinity of a *primary element* or a *final control element*. The word "field" is often used synonymously with *local*.

Local Panel - A *panel* that is not a central or main *panel*. *Local panels* are commonly in the vicinity of plant subsystems or sub-areas. The term *"local panel instrument"* should not be confused with *"local instrument."*

Loop - A combination of two or more *instruments* or control *functions* arranged so that signals pass from one to another for the purpose of *measurement* and/or control of a *process variable*.

Manual Loading Station - A device or *function* having a manually adjustable output that is used to actuate one or more remote devices. The station does not provide switching between manual and automatic control modes of a control *loop (see controller* and *control station)*. The station may have integral indicators, lights, or other features. It is also known as a manual station or a manual loader.

Measurement - The determination of the existence or the magnitude of a variable.

Monitor - A general term for an *instrument* or *instrument* system used to measure or sense the status or magnitude of one or more variables for the purpose of deriving useful information. The term *monitor* is very unspecific—sometimes meaning analyzer, indicator, or alarm. *Monitor* can also be used as a verb.

Monitor Light - Synonym for *pilot light*.

Panel - A structure that has a group of *instruments* mounted on it, houses the operator-process interface, and is chosen to have a unique designation. The *panel* may consist of one or more sections, cubicles, consoles, or desks. Synonym for *board*.

Panel-Mounted - A term applied to an *instrument* that is mounted on a *panel* or console and is *accessible* for an operator's normal use. A *function* that is normally *accessible* to an operator in a *shared-display* system is the equivalent of a discrete *panel-mounted* device.

Pilot Light - A light that indicates which of a number of normal conditions of a system or device exists. It is unlike an *alarm* light, which indicates an abnormal condition. The *pilot light* is also known as a *monitor light*.

Primary Element - Synonym for *sensor*.

Process - Any operation or sequence of operations involving a change of energy, state, composition, dimension, or other properties that may be defined with respect to a datum.

Process Variable - Any variable property of a *process*. The term *process variable* is used in this standard to apply to all variables other than *instrument* signals.

Program - A repeatable sequence of actions that defines the status of outputs as a fixed relationship to a set of inputs.

Programmable Logic Controller - A *controller*, usually with multiple inputs and outputs, that contains an alterable *program*.

Relay - A device whose *function* is to pass on information in an unchanged form or in some modified form. *Relay* is often used to mean *computing device*. The latter term is preferred.

The term *"relay"* also is applied specifically to an electric, pneumatic, or hydraulic *switch* that is actuated by a signal. The term also is applied to *functions* performed by a *relay*.

Scan - To sample, in a predetermined manner, each of a number of variables intermittently. The *function* of a scanning device is often to ascertain the state or value of a variable. The device may be associated with other *functions* such as recording and alarming.

Sensor - That part of a *loop* or *instrument* that first senses the value of a process variable, and that assumes a corresponding, predetermined, and intelligible state or output. The *sensor* may be separate from or integral with another functional element of a *loop*. The *sensor* is also known as a detector or *primary element*.

Set Point - An input variable that sets the desired value of the controlled variable. The *set point* may be manually set, automatically set, or programmed. Its value is expressed in the same units as the controlled variable.

Shared Controller - A *controller*, containing preprogrammed algorithms that are usually *accessible*, *configurable*, and *assignable*. It permits a number of *process variables* to be controlled by a single device.

Shared Display - The operator interface device (usually a video screen) used to display *process* control information from a number of sources at the command of the operator.

Switch - A device that connects, disconnects, selects, or transfers one or more circuits and is not designated as a *controller*, a *relay*, or a *control valve*. As a verb, the term is also applied to the *functions* performed by *switches*.

Test Point - A *process* connection to which no *instrument* is permanently connected, but which is intended for the temporary or intermittent connection of an *instrument*.

Transducer - A general term for a device that receives information in the form of one or more physical quantities, modifies the information and/or its form, if required, and produces a resultant output signal. Depending on the application, the *transducer* can be a *primary element, transmitter, relay, converter* or other device. Because the term *"transducer"* is not specific, its use for specific applications is not recommended.

Transmitter - A device that senses a *process variable* through the medium of a sensor and has an output whose steady-state value varies only as a predetermined *function* of the *process variable*. The *sensor* may or may not be integral with the *transmitter*.

4 OUTLINE OF THE IDENTIFICATION SYSTEM

4.1 General

4.1.1 Each instrument or function to be identified is designated by an alphanumeric code or tag number as shown in Figure 1. The loop identification part of the tag number generally is common to all instruments or functions of the loop. A suffix or prefix may be added to complete the identification. Typical identification is shown in Figure 1.

```
                 TYPICAL  TAG  NUMBER

TIC 103.  -  Instrument Identification or Tag Number
T    103  -  Loop Identification
     103  -  Loop Number
TIC       -  Functional Identification
T         -  First-letter
  IC      -  Succeeding-Letters

              EXPANDED  TAG  NUMBER

10-PAH-5A -  Tag Number
10        -  Optional Prefix
       A  -  Optional Suffix

Note: Hyphens  are  optional  as  separators.
```

Figure 1. Tag Numbers

4.1.2 The instrument loop number may include coded information, such as plant area designation. It is also possible to set aside specific series of numbers to designate special functions; for instance, the series 900 to 999 could be used for loops whose primary function is safety-related.

4.1.3 Each instrument may be represented on diagrams by a symbol. The symbol may be accompanied by a tag number.

4.2 Functional Identification

4.2.1 The functional identification of an instrument or its functional equivalent consists of letters from Table 1 and includes one first-letter (designating the measured or initiating variable) and one or more succeeding-letters (identifying the functions performed).

4.2.2 The functional identification of an instrument is made according to the function and not according to the construction. Thus, a differential-pressure recorder used for flow measurement is identified by *FR;* a pressure indicator and a pressure-actuated switch connected to the output of a pneumatic level transmitter are identified by *LI* and *LS,* respectively.

4.2.3 In an instrument loop, the first-letter of the functional identification is selected according to the measured or initiating variable, and not according to the manipulated variable. Thus, a control valve varying flow according to the dictates of a level controller is an *LV,* not an *FV.*

4.2.4 The succeeding-letters of the functional identification designate one or more readout or passive functions and/or output functions. A modifying-letter may be used, if required, in addition to one or more other succeeding-letters. Modifying-letters may modify either a first-letter or succeeding-letters, as applicable. Thus, *TDAL* contains two modifiers. The letter *D* changes the measured variable *T* into a new variable, "differential temperature." The letter *L* restricts the readout function *A,* alarm, to represent a low alarm only.

4.2.5 The sequence of identification letters begins with a first-letter selected according to Table 1. Readout or passive functional letters follow in any order, and output functional letters follow these in any sequence, except that output letter *C* (control) precedes output letter *V* (valve), *e.g., PCV,* a self-actuated control valve. However, modifying-letters, if used, are interposed so that they are placed immediately following the letters they modify.

4.2.6 A multiple function device may be symbolized on a diagram by as many bubbles as there are measured variables, outputs, and/or functions. Thus, a temperature controller with a switch may be identified by two tangent bubbles—one inscribed *TIC-3* and one inscribed *TSH-3.* The instrument would be designated *TIC/TSH-3* for all uses in writing or reference. If desired, however, the abbreviation *TIC-3* may serve for general identification or for purchasing, while *TSH-3* may be used for electric circuit diagrams.

4.2.7 The number of functional letters grouped for one instrument should be kept to a minimum according to the judgment of the user. The total number of letters within one group should not exceed four. The number within a group may be kept to a minimum by:

(1) Arranging the functional letters into subgroups. This practice is described in Section 4.2.6 for instruments having more than one measured variable or input, but it may also be used for other instruments.

(2) Omitting the *I* (indicate) if an instrument both indicates and records the same measured variable.

4.2.8 All letters of the functional identification are uppercase.

4.3 Loop Identification

4.3.1 The loop identification consists of a first-letter and a number. Each instrument within a loop has assigned to it the same loop number and, in the case of parallel numbering, the same first-letter. Each instrument loop has a unique loop identification. An instrument common to two or more loops should carry the identification of the loop which is considered predominant.

4.3.1 Loop numbering may be parallel or serial. Parallel numbering involves starting a numerical sequence for each new first-letter, *e.g., TIC-100, FRC-100, LIC-100, AI-100, etc.* Serial numbering involves using a single sequence of numbers for a project or for large sections of a project, regardless of the first-letter of the loop identification, *e.g., TIC-100, FRC-101, LIC-102, AI-103, etc.* A loop numbering sequence may begin with *1* or any other convenient number, such as *001, 301* or *1201.* The number may incorporate coded information; however, simplicity is recommended.

4.3.3 If a given loop has more than one instrument with the same functional identification, a suffix may be appended to the loop number, *e.g., FV-2A, FV-2B, FV-2C, etc.,* or *TE-25-1, TE-25-2, etc.* However, it may be more convenient or logical in a given instance to designate a pair of flow transmitters, for example, as *FT-2* and *FT-3* instead of *FT-2A* and *FT-2B.* The suffixes may be applied according to the following guidelines:

(1) An uppercase suffix letter should be used, *i.e., A, B, C, etc.*

(2) For an instrument such as a multipoint temperature recorder that prints numbers for point identification, the primary elements may be numbered *TE-25-1, TE-25-2, TE-25-3, etc.,* corresponding to the point identification number.

(3) Further subdivisions of a loop may be designated by serially alternating suffix letters and numbers. (*See* Section 6.9R(3).)

4.3.4 An instrument that performs two or more functions may be designated by all of its functions. For example, a flow recorder *FR-2* with a pressure pen *PR-4* may be designated *FR-2/PR-4.* A two-pen pressure recorder may be *PR-7/8,* and a common annunciator window for high and low temperature alarms may be *TAHL-21.* Note that the slash is not necessary when distinctly separate devices are not present.

4.3.5 Instrument accessories such as purge meters, air sets, and seal pots that are not explicitly shown on a diagram but that need a designation for other purposes should be tagged individually according to their functions and should use the same loop identification as the instrument they directly serve. Application of such a designation does not imply that the accessory must be shown on the diagram. Alternatively, the accessories may use the identical tag number as that of their associated instrument, but with clarifying words added. Thus an orifice flange union associated with orifice plate *FE-7* should be tagged *FX-7,* but may be designated *FE-7 FLANGES.* A purge meter associated with pressure gauge *PI-8* may be tagged *PI-8 PURGE.* A thermowell used with thermometer *TI-9* should be tagged *TW-9,* but may be tagged *TI-9 THERMOWELL.*

The rules for loop identification need not be applied to instruments and accessories that are purchased in bulk quantities if it is the user's practice to identify these items by other means.

4.4 Symbols

4.4.1 The examples in this standard illustrate the symbols that are intended to depict instrumentation on diagrams and drawings. Methods of symbolization and identification are demonstrated. The examples show identification that is typical for the pictured instrument or functional interrelationships. The symbols indicating the various instruments or functions have been applied in typical ways in the illustrations. This usage does not imply, however, that the applications or designations of the instruments or functions are restricted in any way. No inference should be drawn that the choice of any of the schemes for illustration constitutes a recommendation for the illustrated methods of measurement or control. Where alternative symbols are shown without a statement of preference, the relative sequence of symbols does not imply a preference.

4.4.2 The bubble may be used to tag distinctive symbols, such as those for control valves, when such tagging is desired. In such instances, the line connecting the bubble to the instrument symbol is drawn close to, but not touching, the symbol. In other instances, the bubble serves to represent the instrument proper.

4.4.3 A distinctive symbol whose relationship to the remainder of the loop is easily apparent from a diagram need not be individually tagged on the diagram. For example, an orifice flange or a control valve that is part of a larger system need not be shown with a tag number on a diagram. Also, where there is a primary element connected to another instrument on a diagram, use of a symbol to represent the primary element on the diagram is optional.

4.4.4 A brief explanatory notation may be added adjacent to a symbol or line to clarify the function of an item. For instance, the notations *3-9 psig* and *9-15 psig* adjacent to the signal lines to two valves operating in split range, taken together with the symbols for the failure modes, allow complete understanding of the intent. Similarly, when two valves are operated in a diverting or mixing mode from a common signal, the notations *3-15 psig* and *15-3 psig,* together with the failure modes, allow understanding of the function.

4.4.5 The sizes of the tagging bubbles and the miscellaneous symbols shown in the examples are the sizes generally recommended; however, the optimum sizes may vary depending on whether or not the finished diagram is to be reduced in size and depending on the number of characters that are expected in the instrument tagging designation. The sizes of the other symbols may be selected as appropriate to accompany the symbols of other equipment on a diagram.

4.4.6 Aside from the general drafting requirements for neatness and legibility, symbols may be drawn with any orientation. Likewise, signal lines may be drawn on a diagram entering or leaving the appropriate part of a symbol at any angle. However, the function block designators of Table 3 and the tag numbers should always be drawn with a horizontal orientation. Directional arrowheads should be added to signal lines when needed to clarify the direction of flow of information. The judicious use of such arrowheads, especially on complex drawings, will often facilitate understanding of the system.

4.4.7 The electrical, pneumatic, or other power supply to an instrument is not expected to be shown unless it is essential to an understanding of the operation of the instrument or the loop.

4.4.8 In general, one signal line will suffice to represent the interconnections between two instruments on flow diagrams even though they may be connected physically by more than one line.

4.4.9 The sequence in which the instruments or functions of a loop are connected on a diagram should reflect the functional logic or information flow, although this arrangement will not necessarily correspond to the signal connection sequence. Thus, an electronic loop using analog voltage signals requires parallel wiring, while a loop using analog current signals requires series interconnections. However, the diagram in both instances should be drawn as though all the wiring were parallel, to show the functional interrelationships clearly while keeping the presentation independent of the type of instrumentation finally installed. The correct interconnections are expected to be shown on a suitable diagram.

4.4.10 The degree of detail to be applied to each document or sketch is entirely at the discretion of the user of the standard. The symbols and designations in this standard can depict both hardware and function. Sketches and technical papers will usually contain highly simplified symbolism and identification. Process flow diagrams will usually be less detailed than engineering flow diagrams. Engineering flow diagrams may show all in-line components, but may differ from user to user in the amount of off-line detail shown. In any case, consistency should be established for each application. The terms *simplified, conceptual,* and *detailed* as applied to the diagrams of 6.12 were chosen to represent a cross section of typical usage. Each user must establish the degree of detail that fulfills the purposes of the specific document or sketch being generated.

4.4.11 It is common practice for engineering flow diagrams to omit the symbols of interlock-hardware components that are actually necessary for a working system, particularly when symbolizing electric interlock systems. For example, a level switch may be shown as tripping a pump, or separate flow and pressure switches may be shown as actuating a solenoid valve or other interlock devices. In both instances, auxiliary electrical relays and other components may be considered details to be shown elsewhere. By the same token, a current transformer sometimes will be omitted and its receiver shown connected directly to the process—in this case the electric motor.

4.4.12 Because the distinctions between shared display/shared control and computer functions are sometimes blurred, in choosing symbols to represent them the user must rely on manufacturers' definitions, usage in a particular industry, and personal judgment.

5 TABLES

The purpose of Section 5, Tables, is to define certain of the building blocks of the identification and symbolic representation system used in this standard in a concise, easily-referenced manner.

Table 1, Identification Letters, together with the Notes for Table 1, define and explain the individual letter designators used as functional identifiers in accordance with the rules of Section 4.2, Functional Identification.

Table 2, Typical Letter Combinations, attempts to facilitate the task of choosing acceptable combinations of identifying letters.

Table 3, Function Blocks - Function Designations, is an adaptation of the SAMA (Scientific Apparatus Manufacturers Association) method of functional diagramming. Two basic uses are found for these symbols: as standalone function blocks on conceptual diagrams, or as flags which designate functions performed by bubbles on more detailed drawings. A third use is a combination of the first two and is found in shared control systems where, for instance, the measured variable signal line enters a square root function block that is drawn adjacent to a shared controller.

Two omissions will be noted: The SAMA symbol for *Transfer* and that for an *Analog Signal Generator*. Since the ultimate use of ISA-S5.1 symbolism usually requires identification to be associated with a symbol, it is advisable to use the *HIC* (manual loader) bubble for an analog signal generator and an *HS* (hand switch) with or without a relay bubble for a transfer function.

5.1 Notes for Table 1

(1) A "user's choice" letter is intended to cover unlisted meanings that will be used repetitively in a particular project. If used, the letter may have one meaning as a first-letter and another meaning as a succeeding-letter. The meanings need to be defined only once in a legend, or other place, for that project. For example, the letter *N* may be defined as "modulus of elasticity" as a first-letter and "oscilloscope" as a succeeding-letter.

(2) The unclassified letter *X* is intended to cover unlisted meanings that will be used only once or used to a limited extent. If used, the letter may have any number of meanings as a first-letter and any number of meanings as a succeeding-letter. Except for its use with distinctive symbols, it is expected that the meanings will be defined outside a tagging bubble on a flow diagram. For example, *XR-2* may be a stress recorder and *XX-4* may be a stress oscilloscope.

(3) The grammatical form of the succeeding-letter meanings may be modified as required. For example, "indicate" may be applied as "indicator" or "indicating," "transmit" as "transmitter" or "transmitting," *etc*.

(4) Any first-letter, if used in combination with modifying letters *D* (differential), *F* (ratio), *M* (momentary), *K* (time rate of change), *Q* (integrate or totalize), or any combination of these is intended to represent a new and separate measured variable, and the combination is treated as a first-letter entity. Thus, instruments *TDI* and *TI* indicate two different variables, namely, differential-temperature and temperature. Modifying letters are used when applicable.

(5) First-letter *A* (analysis) covers all analyses not described by a "user's choice" letter. It is expected that the type of analysis will be defined outside a tagging bubble.

(6) Use of first-letter *U* for "multivariable" in lieu of a combination of first-letters is optional. It is recommended that nonspecific variable designators such as *U* be used sparingly.

(7) The use of modifying terms "high," "low," "middle" or "intermediate," and "scan" is optional.

(8) The term "safety" applies to emergency protective primary elements and emergency protective final control elements only. Thus, a self-actuated valve that prevents operation of a fluid system at a higher-than-desired pressure by bleeding fluid from the system is a back-pressure-type *PCV*, even if the valve is not intended to be used normally. However, this valve is designated as a *PSV* if

it is intended to protect against emergency conditions, *i.e.*, conditions that are hazardous to personnel and/or equipment and that are not expected to arise normally.

The designation *PSV* applies to all valves intended to protect against emergency pressure conditions regardless of whether the valve construction and mode of operation place them in the category of the safety valve, relief valve, or safety relief valve. A rupture disc is designated *PSE*.

(9) The passive function *G* applies to instruments or devices that provide an uncalibrated view, such as sight glasses and television monitors.

(10) "Indicate" normally applies to the readout—analog or digital—of an actual measurement. In the case of a manual loader, it may be used for the dial or setting indication, *i.e.*, for the value of the initiating variable.

(11) A pilot light that is part of an instrument loop should be designated by a first-letter followed by the succeeding-letter *L*. For example, a pilot light that indicates an expired time period should be tagged *KQL*. If it is desired to tag a pilot light that is not part of an instrument loop, the light is designated in the same way. For example, a running light for an electric motor may be tagged *EL*, assuming voltage to be the appropriate measured variable, or *YL*, assuming the operating status is being monitored. The unclassified variable *X* should be used only for applications which are limited in extent. The designation *XL* should not be used for motor running lights, as these are commonly numerous. It is permissible to use the user's choice letters *M*, *N* or *O* for a motor running light when the meaning is previously defined. If *M* is used, it must be clear that the letter does not stand for the word "motor," but for a monitored state.

(12) Use of a succeeding-letter *U* for "multifunction" instead of a combination of other functional letters is optional. This nonspecific function designator should be used sparingly.

(13) A device that connects, disconnects, or transfers one or more circuits may be either a switch, a relay, an ON-OFF controller, or a control valve, depending on the application.

If the device manipulates a fluid process stream and is not a hand-actuated ON-OFF block valve, it is designated as a control valve. It is incorrect to use the succeeding-letters *CV* for anything other than a self-actuated control valve. For all applications other than fluid process streams, the device is designated as follows:

A switch, if it is actuated by hand.

A switch or an ON-OFF controller, if it is automatic and is the first such device in a loop. The term "switch" is generally used if the device is used for alarm, pilot light, selection, interlock, or safety.

The term "controller" is generally used if the device is used for normal operating control.

A relay, if it is automatic and is not the first such device in a loop, *i.e.*, it is actuated by a switch or an ON-OFF controller.

(14) It is expected that the functions associated with the use of succeeding-letter *Y* will be defined outside a bubble on a diagram when further definition is considered necessary. This definition need not be made when the function is self-evident, as for a solenoid valve in a fluid signal line.

(15) The modifying terms "high," and "low," and "middle" or "intermediate" correspond to values of the measured variable, not to values of the signal, unless otherwise noted. For example, a high-level alarm derived from a reverse-acting level transmitter signal should be an *LAH*, even though the alarm is actuated when the signal falls to a low value. The terms may be used in combinations as appropriate. (*See* Section 6.9A.)

(16) The terms "high" and "low," when applied to positions of valves and other open-close devices, are defined as follows: "high" denotes that the valve is in or approaching the fully open position, and "low" denotes that it is in or approaching the fully closed position.

(17) The word "record" applies to any form of permanent storage of information that permits retrieval by any means.

(18) For use of the term "transmitter" versus "converter," see the definitions in Section 3.

(19) First-letter *V*, "vibration or mechanical analysis," is intended to perform the duties in machinery monitoring that the letter *A* performs in more general analyses. Except for vibration, it is expected that the variable of interest will be defined outside the tagging bubble.

(20) First-letter *Y* is intended for use when control or monitoring responses are event-driven as opposed to time- or time schedule-driven. The letter *Y,* in this position, can also signify presence or state.

(21) Modifying-letter *K,* in combination with a first-letter such as *L, T,* or *W,* signifies a time rate of change of the measured or initiating variable. The variable *WKIC,* for instance, may represent a rate-of-weight-loss controller.

(22) Succeeding-letter *K* is a user's option for designating a control station, while the succeeding-letter *C* is used for describing automatic or manual controllers. (*See* Section 3, Definitions.)

TABLE 1
IDENTIFICATION LETTERS

	FIRST-LETTER (4)		SUCCEEDING-LETTERS (3)		
	MEASURED OR INITIATING VARIABLE	MODIFIER	READOUT OR PASSIVE FUNCTION	OUTPUT FUNCTION	MODIFIER
A	Analysis(5,19)		Alarm		
B	Burner, Combustion		User's Choice(1)	User's Choice(1)	User's Choice(1)
C	User's Choice(1)			Control(13)	
D	User's Choice(1)	Differential(4)			
E	Voltage		Sensor (Primary Element)		
F	Flow Rate	Ratio (Fraction)(4)			
G	User's Choice(1)		Glass, Viewing Device(9)		
H	Hand				High(7,15,16)
I	Current (Electrical)		Indicate(10)		
J	Power	Scan(7)			
K	Time, Time Schedule	Time Rate of Change(4,21)		Control Station (22)	
L	Level		Light(11)		Low(7,15,16)
M	User's Choice(1)	Momentary(4)			Middle, Intermediate(7,15)
N	User's Choice(1)		User's Choice(1)	User's Choice(1)	User's Choice(1)
O	User's Choice(1)		Orifice, Restriction		
P	Pressure, Vacuum		Point (Test) Connection		
Q	Quantity	Integrate, Totalize(4)			
R	Radiation		Record(17)		
S	Speed, Frequency	Safety(8)		Switch(13)	
T	Temperature			Transmit(18)	
U	Multivariable(6)		Multifunction(12)	Multifunction(12)	Multifunction(12)
V	Vibration, Mechanical Analysis(19)			Valve, Damper, Louver(13)	
W	Weight, Force		Well		
X	Unclassified(2)	X Axis	Unclassified(2)	Unclassified(2)	Unclassified(2)
Y	Event, State or Presence(20)	Y Axis		Relay, Compute, Convert(13,14,18)	
Z	Position, Dimension	Z Axis		Driver, Actuator, Unclassified Final Control Element	

Note: Numbers in parentheses refer to specific explanatory notes on pages 15 and 16.

S5.1

TABLE 2
TYPICAL LETTER COMBINATIONS

First-Letters	Initiating or Measured Variable	Controllers Recording	Controllers Indicating	Controllers Blind	Self-Actuated Control Valves	Readout Recording	Readout Indicating	Switches/Alarm High**	Switches/Alarm Low	Switches/Alarm Comb	Transmitters Recording	Transmitters Indicating	Transmitters Blind	Solenoids, Relays, Computing Devices	Primary Element	Test Point	Well or Probe	Viewing Device, Glass	Safety Device	Final Element
A	Analysis	ARC	AIC	AC		AR	AI	ASH	ASL	ASHL	ART	AIT	AT	AY	AE	AP	AW			AV
B	Burner/Combustion	BRC	BIC	BC		BR	BI	BSH	BSL	BSHL	BRT	BIT	BT	BY	BE		BW	BG		BZ
C	User's Choice																			
D	User's Choice																			
E	Voltage	ERC	EIC	EC		ER	EI	ESH	ESL	ESHL	ERT	EIT	ET	EY	EE					EZ
F	Flow Rate	FRC	FIC	FC	FCV, FICV	FR	FI	FSH	FSL	FSHL	FRT	FIT	FT	FY	FE	FP		FG		FV
FQ	Flow Quantity	FQRC	FQIC			FQR	FQI	FQSH	FQSL			FQIT	FQT	FQY	FQE, FE					FQV
FF	Flow Ratio	FFRC	FFIC	FFC		FFR	FFI	FFSH	FFSL											FFV
G	User's Choice																			
H	Hand		HIC	HC						HS										HV
I	Current	IRC	IIC	IC		IR	II	ISH	ISL	ISHL	IRT	IIT	IT	IY	IE					IZ
J	Power	JRC	JIC	JC		JR	JI	JSH	JSL	JSHL	JRT	JIT	JT	JY	JE					JV
K	Time	KRC	KIC	KC	KCV	KR	KI	KSH	KSL	KSHL	KRT	KIT	KT	KY	KE					KV
L	Level	LRC	LIC	LC	LCV	LR	LI	LSH	LSL	LSHL	LRT	LIT	LT	LY	LE		LW	LG		LV
M	User's Choice																			
N	User's Choice																			
O	User's Choice																			
P	Pressure/Vacuum	PRC	PIC	PC	PCV	PR	PI	PSH	PSL	PSHL	PRT	PIT	PT	PY	PE	PP			PSV, PSE	PV
PD	Pressure/Vacuum, Differential	PDRC	PDIC	PDC	PDCV	PDR	PDI	PDSH	PDSL	PDSHL	PDRT	PDIT	PDT	PDY	PE	PP				PDV
Q	Quantity	QRC	QIC	QC		QR	QI	QSH	QSL	QSHL	QRT	QIT	QT	QY	QE					QZ
R	Radiation	RRC	RIC	RC		RR	RI	RSH	RSL	RSHL	RRT	RIT	RT	RY	RE		RW			RZ
S	Speed/Frequency	SRC	SIC	SC	SCV	SR	SI	SSH	SSL	SSHL	SRT	SIT	ST	SY	SE					SV
T	Temperature	TRC	TIC	TC	TCV	TR	TI	TSH	TSL	TSHL	TRT	TIT	TT	TY	TE	TP	TW		TSE	TV
TD	Temperature, Differential	TDRC	TDIC	TDC	TDCV	TDR	TDI	TDSH	TDSL	TDSHL	TDRT	TDIT	TDT	TDY	TE	TP	TW			TDV
U	Multivariable					UR	UI							UY						UV
V	Vibration/Machinery Analysis	VRC	VIC	VC	VCV	VR	VI	VSH	VSL	VSHL	VRT	VIT	VT	VY	VE					VZ
W	Weight/Force	WRC	WIC	WC	WCV	WR	WI	WSH	WSL	WSHL	WRT	WIT	WT	WY	WE					WZ
WD	Weight/Force, Differential	WDRC	WDIC	WDC	WDCV	WDR	WDI	WDSH	WDSL	WDSHL	WDRT	WDIT	WDT	WDY	WE					WDZ
X	Unclassified																			
Y	Event/State/Presence		YIC	YC		YR	YI	YSH	YSL	YSHL		YIT	YT	YY	YE					YZ
Z	Position/Dimension	ZRC	ZIC	ZC	ZCV	ZR	ZI	ZSH	ZSL	ZSHL	ZRT	ZIT	ZT	ZY	ZE					ZV
ZD	Gauging/Deviation	ZDRC	ZDIC	ZDC	ZDCV	ZDR	ZDI	ZDSH	ZDSL	ZDSHL	ZDRT	ZDIT	ZDT	ZDY	ZDE					ZDV

Note: This table is not all-inclusive.

*A, alarm, the annunciating device, may be used in the same fashion as S, switch, the actuating device.

**The letters H and L may be omitted in the undefined case.

Other Possible Combinations:

FO	(Restriction Orifice)
FRK, HIK	(Control Stations)
FX	(Accessories)
TJR	(Scanning Recorder)
LLH	(Pilot Light)

PFR	(Ratio)
KQI	(Running Time Indicator)
QQI	(Indicating Counter)
WKIC	(Rate-of-Weight-Loss Controller)
HMS	(Hand Momentary Switch)

TABLE 3
FUNCTION BLOCKS-FUNCTION DESIGNATIONS

THE FUNCTION DESIGNATIONS ASSOCIATED WITH CONTROLLERS, COMPUTING DEVICES, CONVERTERS AND RELAYS MAY BE USED INDIVIDUALLY OR IN COMBINATION (ALSO, SEE TABLE 1, NOTE 14.). THE USE OF A BOX AVOIDS CONFUSION BY SETTING OFF THE SYMBOL FROM OTHER MARKINGS ON A DIAGRAM AND PERMITS THE FUNCTION TO BE USED AS A STAND-ALONE BLOCK ON CONCEPTUAL DESIGNS.

S5.1

5.4

NO	FUNCTION	SYMBOL	MATH EQUATION	GRAPHIC REPRESENTATION	DEFINITION
1	SUMMING	$\boxed{\Sigma}$	$M = X_1 + X_2 + \ldots + X_n$		THE OUTPUT EQUALS THE ALGEBRAIC SUM OF THE INPUTS. (THE INPUTS MAY BE LABELED WITH POSITIVE OR NEGATIVE SIGNS).
2	AVERAGING	$\boxed{\Sigma/n}$	$M = \dfrac{X_1 + X_2 + \ldots + X_n}{n}$		THE OUTPUT EQUALS THE ALGEBRAIC SUM OF THE INPUTS DIVIDED BY THE NUMBER OF INPUTS.
3	DIFFERENCE	$\boxed{\Delta}$	$M = X_1 - X_2$		THE OUTPUT EQUALS THE ALGEBRAIC DIFFERENCE OF THE TWO INPUTS.
4	PROPORTIONAL	$\boxed{K}$ $\boxed{1:1}$ $\boxed{2:1}$	$M = KX$		THE OUTPUT IS DIRECTLY PROPORTIONAL TO THE INPUT. IN THE CASE OF A VOLUME BOOSTER, "K" MAY BE REPLACED BY 1:1. FOR INTEGER GAINS, 2:1, 3:1, ETC., MAY BE SUBSTITUTED FOR K.
5	INTEGRAL	$\boxed{\int}$	$M = \dfrac{1}{T_I}\int X dt$		THE OUTPUT VARIES IN ACCORDANCE WITH BOTH MAGNITUDE AND DURATION OF THE INPUT. THE OUTPUT IS PROPORTIONAL TO THE TIME INTEGRAL OF THE INPUT.
6	DERIVATIVE	$\boxed{d/dt}$	$M = T_D \dfrac{dX}{dt}$		THE OUTPUT IS PROPORTIONAL TO THE RATE OF CHANGE (DERIVATIVE) OF THE INPUT.

5.4   TABLE 3 - CONTINUED

S5.1

NO	FUNCTION	SYMBOL	MATH EQUATION	GRAPHIC REPRESENTATION	DEFINITION
7	MULTIPLYING	$\boxed{\times}$	$M = X_1 X_2$		THE OUTPUT EQUALS THE PRODUCT OF THE TWO INPUTS.
8	DIVIDING	$\boxed{\div}$	$M = \dfrac{X_1}{X_2}$		THE OUTPUT EQUALS THE QUOTIENT OF THE TWO INPUTS.
9	ROOT EXTRACTION	$\boxed{\sqrt[n]{\;}}$	$M = \sqrt[n]{X}$		THE OUTPUT EQUALS THE ROOT (I.E., CUBE ROOT, FOURTH ROOT, 3/2 ROOT, ETC.) OF THE INPUT. IF n IS OMITTED, A SQUARE ROOT IS ASSUMED.
10	EXPONENTIAL	$\boxed{X^n}$	$M = X^n$		THE OUTPUT EQUALS THE INPUT RAISED TO A POWER (I.E., SECOND, THIRD, FOURTH, ETC.).
11	NONLINEAR OR UNSPECIFIED FUNCTION	$\boxed{f(x)}$	$M = f(x)$		THE OUTPUT EQUALS SOME NONLINEAR OR UNSPECIFIED FUNCTION OF THE INPUT.
12	TIME FUNCTION	$\boxed{f(t)}$	$M = X f(t)$ $M = f(t)$		THE OUTPUT EQUALS THE INPUT TIMES SOME FUNCTION OF TIME OR EQUALS SOME FUNCTION OF TIME ALONE.
13	HIGH SELECTING	$\boxed{\wedge}$	$M = \begin{cases} X_1 \text{ FOR } X_1 \geq X_2 \\ X_2 \text{ FOR } X_1 \leq X_2 \end{cases}$		THE OUTPUT IS EQUAL TO THE GREATER OF THE INPUTS.

5.4 TABLE 3 - CONTINUED

S5.1

NO	FUNCTION	SYMBOL	MATH EQUATION	GRAPHIC REPRESENTATION	DEFINITION
14	LOW SELECTING	(symbol)	$M = \begin{cases} X_1 \text{ FOR } X_1 \leq X_2 \\ X_2 \text{ FOR } X_1 \geq X_2 \end{cases}$	(graphic)	THE OUTPUT IS EQUAL TO THE LESSER OF THE INPUTS.
15	HIGH LIMITING	(symbol)	$M = \begin{cases} X \text{ FOR } X \leq H \\ H \text{ FOR } X \geq H \end{cases}$	(graphic)	THE OUTPUT EQUALS THE INPUT OR THE HIGH LIMIT VALUE WHICHEVER IS LOWER.
16	LOW LIMITING	(symbol)	$M = \begin{cases} X \text{ FOR } X \geq L \\ L \text{ FOR } X \leq L \end{cases}$	(graphic)	THE OUTPUT EQUALS THE INPUT OR THE LOW LIMIT VALUE WHICHEVER IS HIGHER.
17	REVERSE PROPORTIONAL	-K	$M = -KX$	(graphic)	THE OUTPUT IS REVERSELY PROPORTIONAL TO THE INPUT.
18	VELOCITY LIMITER	(symbol)	$\dfrac{dM}{dt} = \dfrac{dX}{dt} \left\{ \dfrac{dX}{dt} \leq H \text{ AND } M = X \right\}$ $\dfrac{dM}{dt} = H \left\{ \dfrac{dX}{dt} \geq H \text{ OR } M \neq X \right\}$	(graphic)	THE OUTPUT EQUALS THE INPUT AS LONG AS THE RATE OF CHANGE OF THE INPUT DOES NOT EXCEED A LIMIT VALUE. THE OUTPUT WILL CHANGE AT THE RATE ESTABLISHED BY THIS LIMIT UNTIL THE OUTPUT AGAIN EQUALS THE INPUT.
19	BIAS	+ - +|	$M = X \pm b$	(graphic)	THE OUTPUT EQUALS THE INPUT PLUS (OR MINUS) SOME ARBITRARY VALUE (BIAS).
20	CONVERT	*/*	OUTPUT = f(INPUT)	NONE	THE FORM OF THE OUTPUT SIGNAL IS DIFFERENT FROM THAT OF THE INPUT. * E - VOLTAGE H - HYDRAULIC I - CURRENT O - ELECTROMAGNETIC, SONIC P - PNEUMATIC R - RESISTANCE(ELECT.) A - ANALOG D - DIGITAL B - BINARY

S5.1

5.4 TABLE 3 - CONTINUED

NO	FUNCTION	SYMBOL	MATH EQUATION	GRAPHIC REPRESENTATION	DEFINITION
21	SIGNAL MONITOR	**H	STATE 1 $X \leq H$ STATE 2 $X > H$ (ENERGIZED OR ALARM STATE) STATE 1 $X < L$ (ENERGIZED OR ALARM STATE) STATE 2 $X \geq L$ STATE 1 (FIRST OUTPUT M_1 ENERGIZED $X < L$ OR ALARM STATE) STATE 2 (BOTH OUTPUTS INACTIVE $L \leq X \leq H$ OR DE-ENERGIZED) STATE 3 (SECOND OUTPUT M_2 ENERGIZED $X > H$ OR ALARM STATE)		THE OUTPUT HAS DISCRETE STATES WHICH ARE DEPENDENT ON THE VALUE OF THE INPUT. WHEN THE INPUT EXCEEDS (OR BECOMES LESS THAN) AN ARBITRARY LIMIT VALUE THE OUTPUT CHANGES STATE.

Symbols in the SYMBOL column: **H , **L , **HL

THE VARIABLES USED IN THE TABLE ARE:

b - ANALOG BIAS VALUE.

$\dfrac{d}{dt}$ - DERIVATIVE WITH RESPECT TO TIME.

H - AN ARBITRARY ANALOG HIGH LIMIT VALUE.

$\dfrac{1}{T_I}$ - INTEGRATING RATE.

L - AN ARBITRARY ANALOG LOW LIMIT VALUE.

M - ANALOG OUTPUT VARIABLE.

n - NUMBER OF ANALOG INPUTS OR VALUE OF EXPONENT.

t - TIME.

T_D - DERIVATIVE TIME.

X - ANALOG INPUT VARIABLE.

$X_1, X_2, X_3, \ldots, X_n$ - ANALOG INPUT VARIABLE (1 TO N IN NUMBER).

* - TABLE 1 LETTER DESIGNATORS.

NOTE: THE SQUARE MAY BE USED AS A FLAG

[1-0] ON-OFF

[REV] REVERSE ACTION

THIS TABLE HAS BEEN MODIFIED FROM SAMA PMC 22-11-1981 WITH PERMISSION OF THE COPYRIGHT HOLDER, SCIENTIFIC APPARATUS MAKERS ASSOCIATION. COPIES OF PMC 22-11-1981 ARE AVAILABLE FROM SAMA, 1101 16TH STREET NW, WASHINGTON, D.C., 20036.

6 DRAWINGS

6.1 Cautionary Notes

(1) If a given drawing, or set of drawings, uses graphic symbols that are similar or identical in shape or configuration and that have different meanings because they are taken from different standards, then adequate steps must be taken to avoid misinterpretation of the symbols used. These steps may be to use caution notes, reference notes, comparison charts that illustrate and define the conflicting symbols, or other suitable means. This requirement is especially critical in cases where symbols taken from different disciplines are intermixed and their misinterpretation might cause danger to personnel or damage to equipment.

(2) The titles *Simplified Diagrams, Conceptual Diagrams* and *Detailed Diagrams* of Section 6.12 were chosen to represent a cross section of symbol usage, not any particular generic document. (*See* 4.4.10 for a more complete discussion.)

(3) The line symbols of Section 6.2 offer ''user's choice'' alternative electrical symbols and optional binary symbols. The subsequent examples use one consistent set of these alternatives and apply the binary options. This was done for consistency of appearance of the standard.

It is recommended that the user choose either the dashed line electrical symbol or the triple cross hatch symbol and apply it consistently. The optional binary (on-off) symbols are available for those applications where the user finds it necessary to distinguish between analog and binary signals. If, in the user's judgment, the application does not require such differentiation, the reverse slash may be omitted from on-off signal line symbols. Consistency is recommended on a given set of documents.

6.2 INSTRUMENT LINE SYMBOLS
ALL LINES TO BE FINE IN RELATION TO PROCESS PIPING LINES.

(1) INSTRUMENT SUPPLY *
     OR CONNECTION TO PROCESS

(2) UNDEFINED SIGNAL

(3) PNEUMATIC SIGNAL **

(4) ELECTRIC SIGNAL

(5) HYDRAULIC SIGNAL

(6) CAPILLARY TUBE

(7) ELECTROMAGNETIC OR SONIC SIGNAL ***
     (GUIDED)

(8) ELECTROMAGNETIC OR SONIC SIGNAL ***
     (NOT GUIDED)

(9) INTERNAL SYSTEM LINK
     (SOFTWARE OR DATA LINK)

(10) MECHANICAL LINK

OPTIONAL BINARY (ON-OFF) SYMBOLS

(11) PNEUMATIC BINARY SIGNAL

(12) ELECTRIC BINARY SIGNAL

NOTE: 'Or' means user's choice. Consistency is recommended.

* The following abbreviations are suggested to denote the types of power
  supply. These designations may also be applied to purge fluid supplies.

  AS - Air Supply   HS - Hydraulic Supply
     IA - Instrument Air } Options   NS - Nitrogen Supply
     PA - Plant Air   SS - Steam Supply
  ES - Electric Supply   WS - Water Supply
  GS - Gas Supply

  The supply level may be added to the instrument supply line, e.g., AS-100,
  a 100-psig air supply; ES-24DC, a 24-volt direct current power supply.

** The pneumatic signal symbol applies to a signal using any gas as the
   signal medium. If a gas other than air is used, the gas may be
   identified by a note on the signal symbol or otherwise.

*** Electromagnetic phenomena include heat, radio waves, nuclear radiation,
    and light.

6.3 GENERAL INSTRUMENT OR FUNCTION SYMBOLS

	PRIMARY LOCATION *** NORMALLY ACCESSIBLE TO OPERATOR	FIELD MOUNTED	AUXILIARY LOCATION *** NORMALLY ACCESSIBLE TO OPERATOR
DISCRETE INSTRUMENTS	1 * ◯ IPI**	2 ◯	3 ⊖
SHARED DISPLAY, SHARED CONTROL	4 ⊡	5 ▣	6 ⊟
COMPUTER FUNCTION	7 ⬡	8 ⬡	9 ⬡
PROGRAMMABLE LOGIC CONTROL	10 ◈	11 ◇	12 ◈

* Symbol size may vary according to the user's needs and the type of
  document. A suggested square and circle size for large diagrams
  is shown above. Consistency is recommended.

** Abbreviations of the user's choice such as IPI (Instrument
   Panel #1), IC2 (Instrument Console #2), CC3 (Computer Console #3), etc.,
   may be used when it is necessary to specify instrument or function
   location.

*** Normally inaccessible or behind-the-panel devices or functions
    may be depicted by using the same symbols but with dashed horizontal
    bars, i.e.

6.3 GENERAL INSTRUMENT OR FUNCTION SYMBOLS (Contd.)

13	14	15
	(6TE 2584-23) INSTRUMENT WITH LONG TAG NUMBER	⭘⭘ INSTRUMENTS SHARING COMMON HOUSING *
16 PILOT LIGHT	17 C 12 PANEL MOUNTED PATCHBOARD POINT 12	18 ** P PURGE OR FLUSHING DEVICE
19 ** R RESET FOR LATCH-TYPE ACTUATOR	20 DIAPHRAGM SEAL	21 ** *** I UNDEFINED INTERLOCK LOGIC

   * It is not mandatory to show a common housing.

  ** These diamonds are approximately half the size of the larger ones.

*** For specific logic symbols, see ANSI/ISA Standard S5.2.

6.4 CONTROL VALVE BODY SYMBOLS, DAMPER SYMBOLS

1	2	3	4
GENERAL SYMBOL	ANGLE	BUTTERFLY	ROTARY VALVE
5	6	7	8
THREE-WAY	FOUR-WAY	GLOBE	
9	10	11	12
DIAPHRAGM	DAMPER OR LOUVER		

Further information may be added adjacent to the body symbol either by note or code number.

6.5 ACTUATOR SYMBOLS

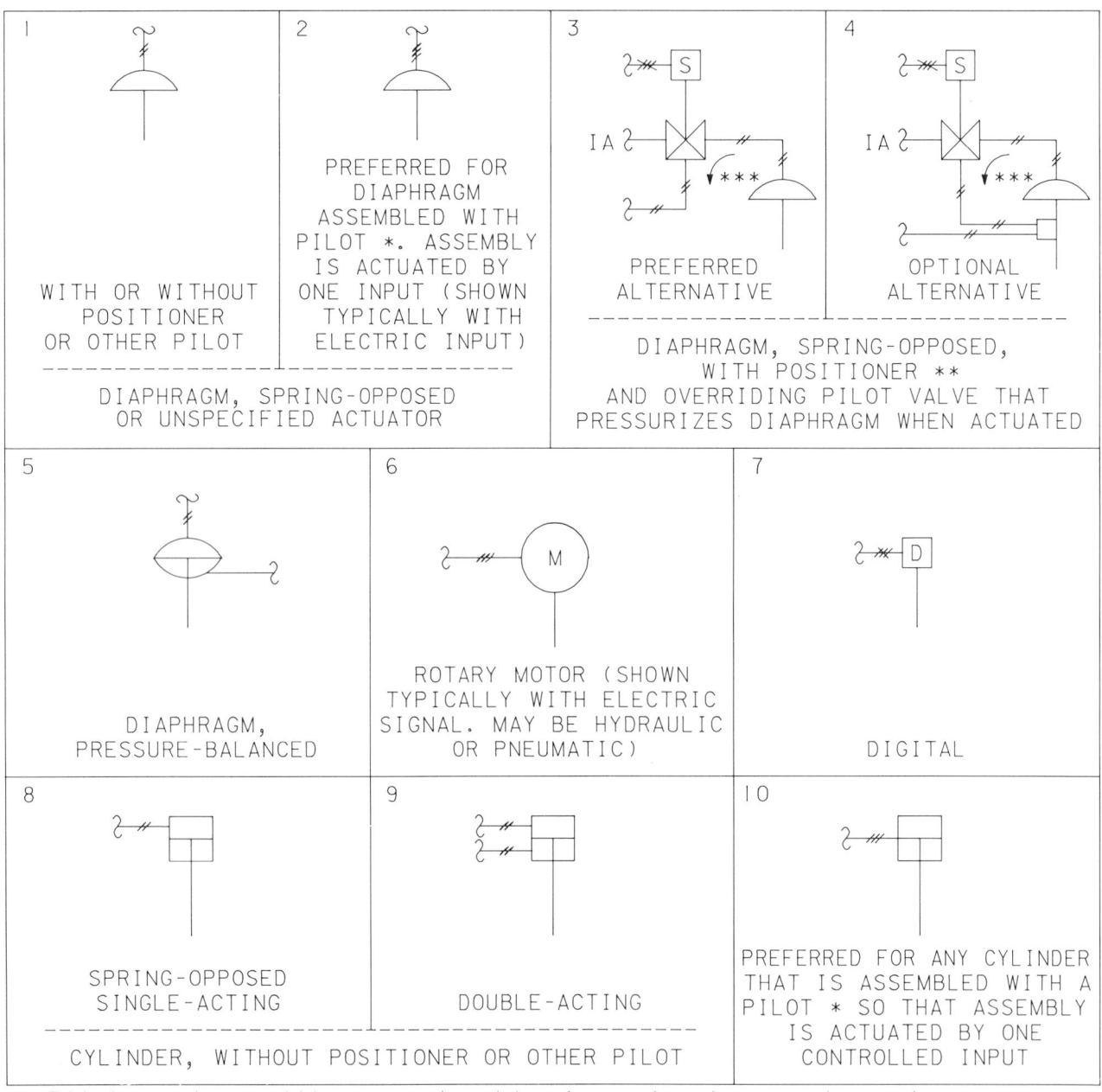

1 WITH OR WITHOUT POSITIONER OR OTHER PILOT	2 PREFERRED FOR DIAPHRAGM ASSEMBLED WITH PILOT *. ASSEMBLY IS ACTUATED BY ONE INPUT (SHOWN TYPICALLY WITH ELECTRIC INPUT)

DIAPHRAGM, SPRING-OPPOSED
OR UNSPECIFIED ACTUATOR

3 PREFERRED ALTERNATIVE | 4 OPTIONAL ALTERNATIVE

DIAPHRAGM, SPRING-OPPOSED,
WITH POSITIONER **
AND OVERRIDING PILOT VALVE THAT
PRESSURIZES DIAPHRAGM WHEN ACTUATED

5 DIAPHRAGM, PRESSURE-BALANCED

6 ROTARY MOTOR (SHOWN TYPICALLY WITH ELECTRIC SIGNAL. MAY BE HYDRAULIC OR PNEUMATIC)

7 DIGITAL

8 SPRING-OPPOSED SINGLE-ACTING

9 DOUBLE-ACTING

CYLINDER, WITHOUT POSITIONER OR OTHER PILOT

10 PREFERRED FOR ANY CYLINDER THAT IS ASSEMBLED WITH A PILOT * SO THAT ASSEMBLY IS ACTUATED BY ONE CONTROLLED INPUT

* Pilot may be positioner, solenoid valve, signal converter, etc.

** The positioner need not be shown unless an intermediate device is on its output. The positioner tagging, ZC, need not be used even if the positioner is shown. The positioner symbol, a box drawn on the actuator shaft, is the same for all types of actuators. When the symbol is used, the type of instrument signal, i.e., pneumatic, electric, etc., is drawn as appropriate. If the positioner symbol is used and there is no intermediate device on its output, then the positioner output signal need not be shown.

*** The arrow represents the path from a common to a fail open port. It does not correspond necessarily to the direction of fluid flow.

6.5 ACTUATOR SYMBOLS (Contd.)

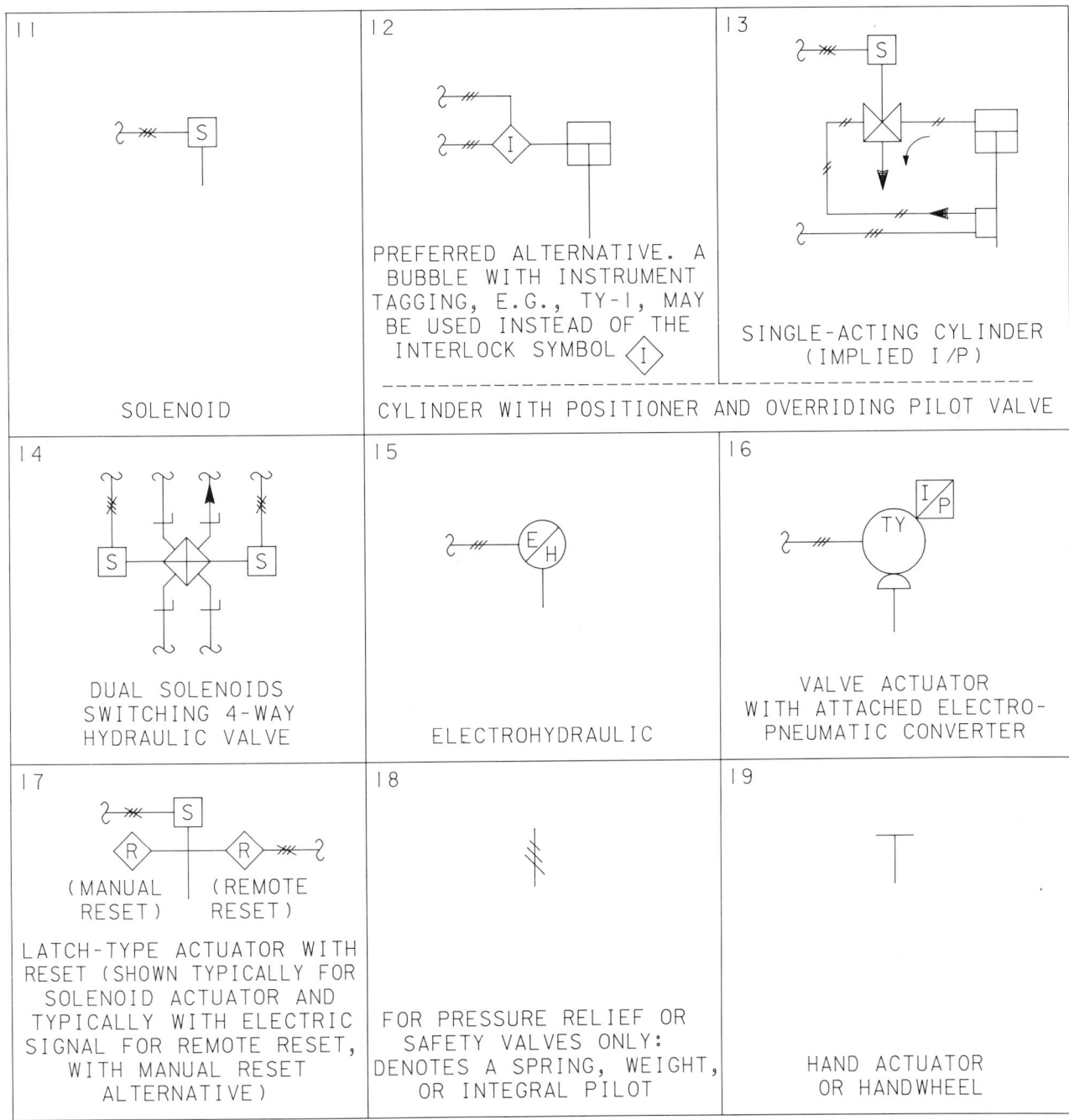

11	12	13
SOLENOID	PREFERRED ALTERNATIVE. A BUBBLE WITH INSTRUMENT TAGGING, E.G., TY-1, MAY BE USED INSTEAD OF THE INTERLOCK SYMBOL	SINGLE-ACTING CYLINDER (IMPLIED I/P)
	CYLINDER WITH POSITIONER AND OVERRIDING PILOT VALVE	
14	15	16
DUAL SOLENOIDS SWITCHING 4-WAY HYDRAULIC VALVE	ELECTROHYDRAULIC	VALVE ACTUATOR WITH ATTACHED ELECTRO-PNEUMATIC CONVERTER
17	18	19
(MANUAL RESET) (REMOTE RESET) LATCH-TYPE ACTUATOR WITH RESET (SHOWN TYPICALLY FOR SOLENOID ACTUATOR AND TYPICALLY WITH ELECTRIC SIGNAL FOR REMOTE RESET, WITH MANUAL RESET ALTERNATIVE)	FOR PRESSURE RELIEF OR SAFETY VALVES ONLY: DENOTES A SPRING, WEIGHT, OR INTEGRAL PILOT	HAND ACTUATOR OR HANDWHEEL

6.6 SYMBOLS FOR SELF-ACTUATED REGULATORS, VALVES, AND OTHER DEVICES

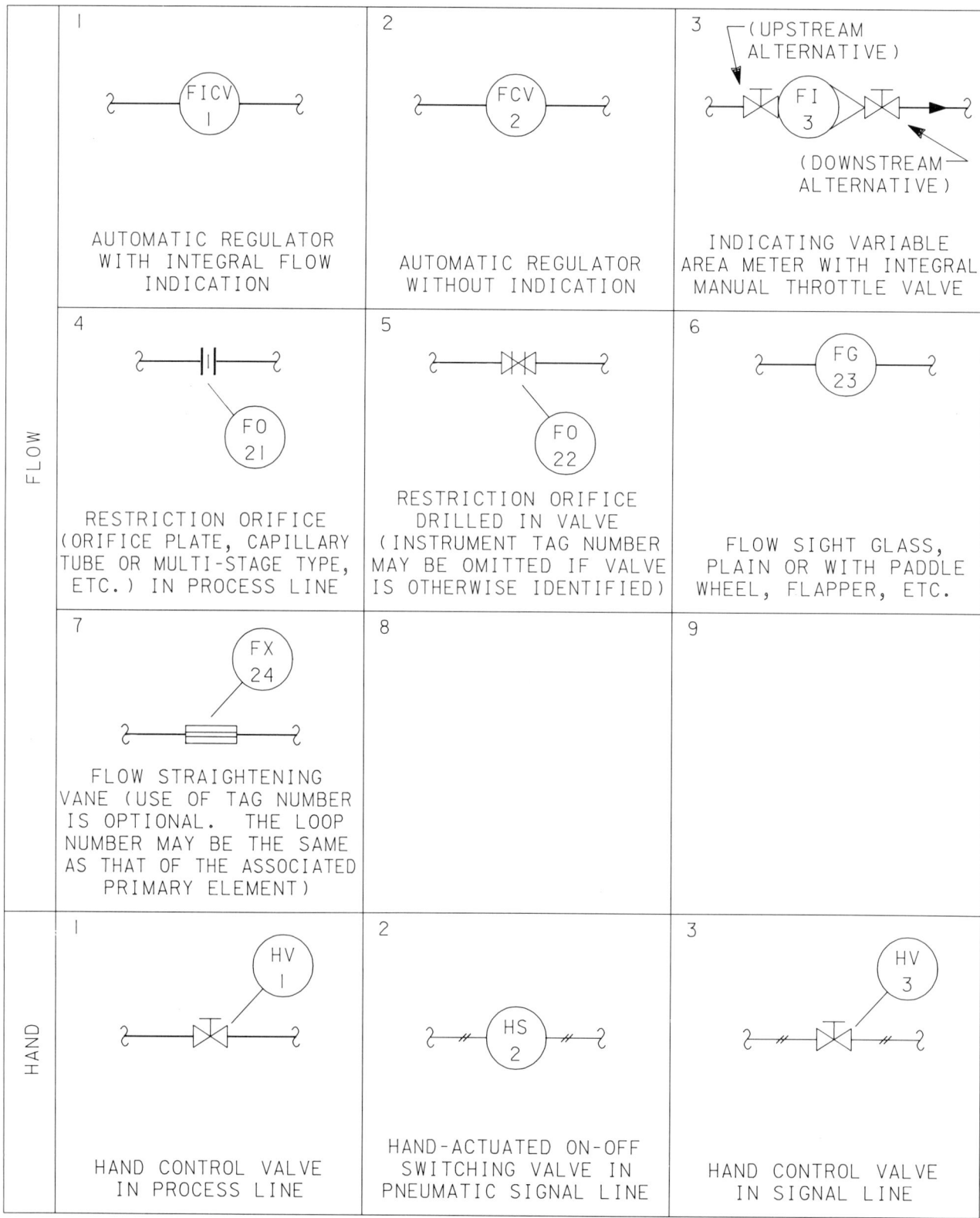

FLOW

1 AUTOMATIC REGULATOR WITH INTEGRAL FLOW INDICATION

2 AUTOMATIC REGULATOR WITHOUT INDICATION

3 (UPSTREAM ALTERNATIVE) (DOWNSTREAM ALTERNATIVE) INDICATING VARIABLE AREA METER WITH INTEGRAL MANUAL THROTTLE VALVE

4 RESTRICTION ORIFICE (ORIFICE PLATE, CAPILLARY TUBE OR MULTI-STAGE TYPE, ETC.) IN PROCESS LINE

5 RESTRICTION ORIFICE DRILLED IN VALVE (INSTRUMENT TAG NUMBER MAY BE OMITTED IF VALVE IS OTHERWISE IDENTIFIED)

6 FLOW SIGHT GLASS, PLAIN OR WITH PADDLE WHEEL, FLAPPER, ETC.

7 FLOW STRAIGHTENING VANE (USE OF TAG NUMBER IS OPTIONAL. THE LOOP NUMBER MAY BE THE SAME AS THAT OF THE ASSOCIATED PRIMARY ELEMENT)

8

9

HAND

1 HAND CONTROL VALVE IN PROCESS LINE

2 HAND-ACTUATED ON-OFF SWITCHING VALVE IN PNEUMATIC SIGNAL LINE

3 HAND CONTROL VALVE IN SIGNAL LINE

6.6 SYMBOLS FOR SELF-ACTUATED REGULATORS, VALVES, AND OTHER DEVICES (Contd.)

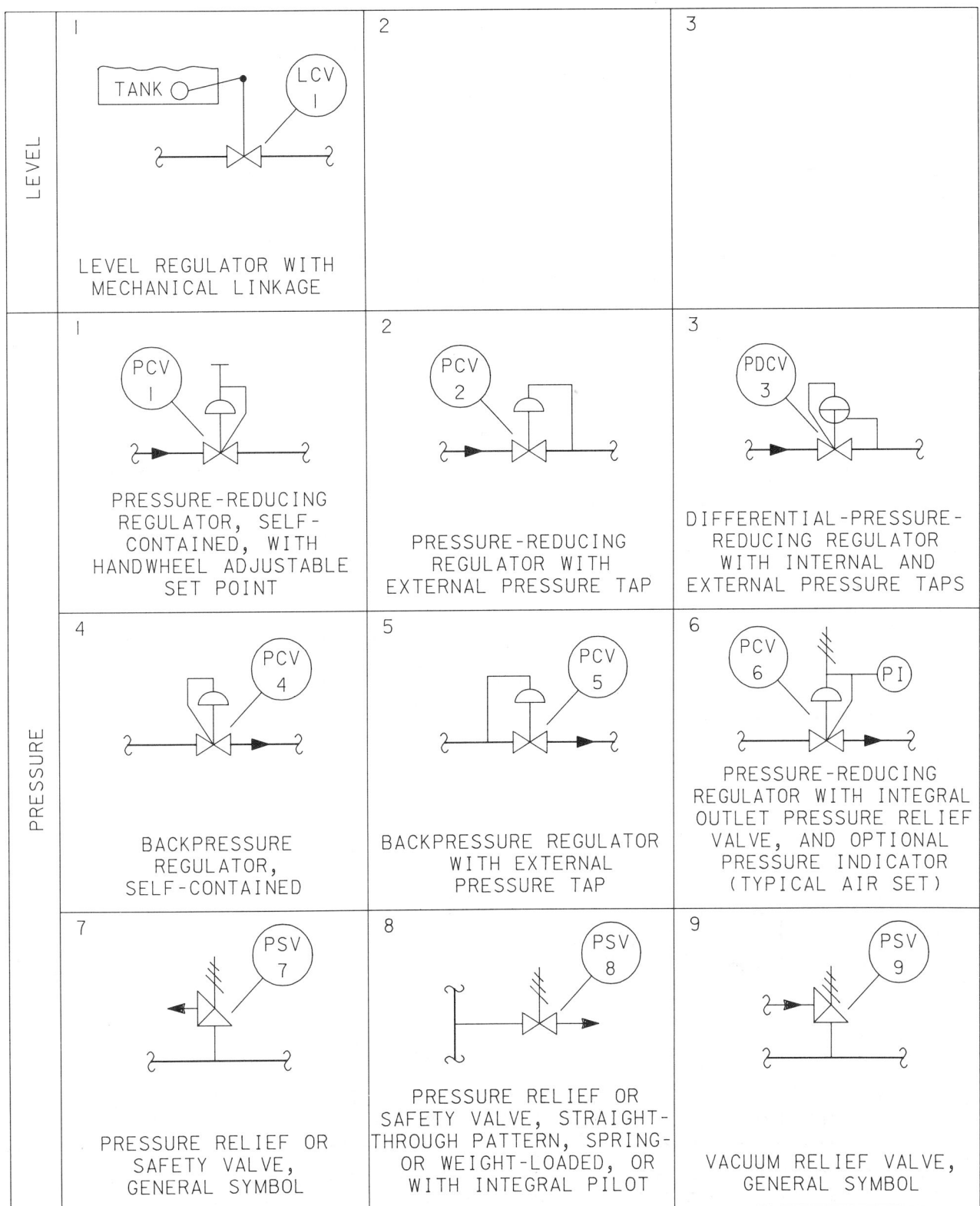

	1	2	3
LEVEL	LCV 1 — TANK ◯ LEVEL REGULATOR WITH MECHANICAL LINKAGE		
PRESSURE	1 PCV 1 PRESSURE-REDUCING REGULATOR, SELF-CONTAINED, WITH HANDWHEEL ADJUSTABLE SET POINT	2 PCV 2 PRESSURE-REDUCING REGULATOR WITH EXTERNAL PRESSURE TAP	3 PDCV 3 DIFFERENTIAL-PRESSURE-REDUCING REGULATOR WITH INTERNAL AND EXTERNAL PRESSURE TAPS
	4 PCV 4 BACKPRESSURE REGULATOR, SELF-CONTAINED	5 PCV 5 BACKPRESSURE REGULATOR WITH EXTERNAL PRESSURE TAP	6 PCV 6 — PI PRESSURE-REDUCING REGULATOR WITH INTEGRAL OUTLET PRESSURE RELIEF VALVE, AND OPTIONAL PRESSURE INDICATOR (TYPICAL AIR SET)
	7 PSV 7 PRESSURE RELIEF OR SAFETY VALVE, GENERAL SYMBOL	8 PSV 8 PRESSURE RELIEF OR SAFETY VALVE, STRAIGHT-THROUGH PATTERN, SPRING- OR WEIGHT-LOADED, OR WITH INTEGRAL PILOT	9 PSV 9 VACUUM RELIEF VALVE, GENERAL SYMBOL

6.6 SYMBOLS FOR SELF-ACTUATED REGULATORS, VALVES, AND OTHER DEVICES (Contd.)

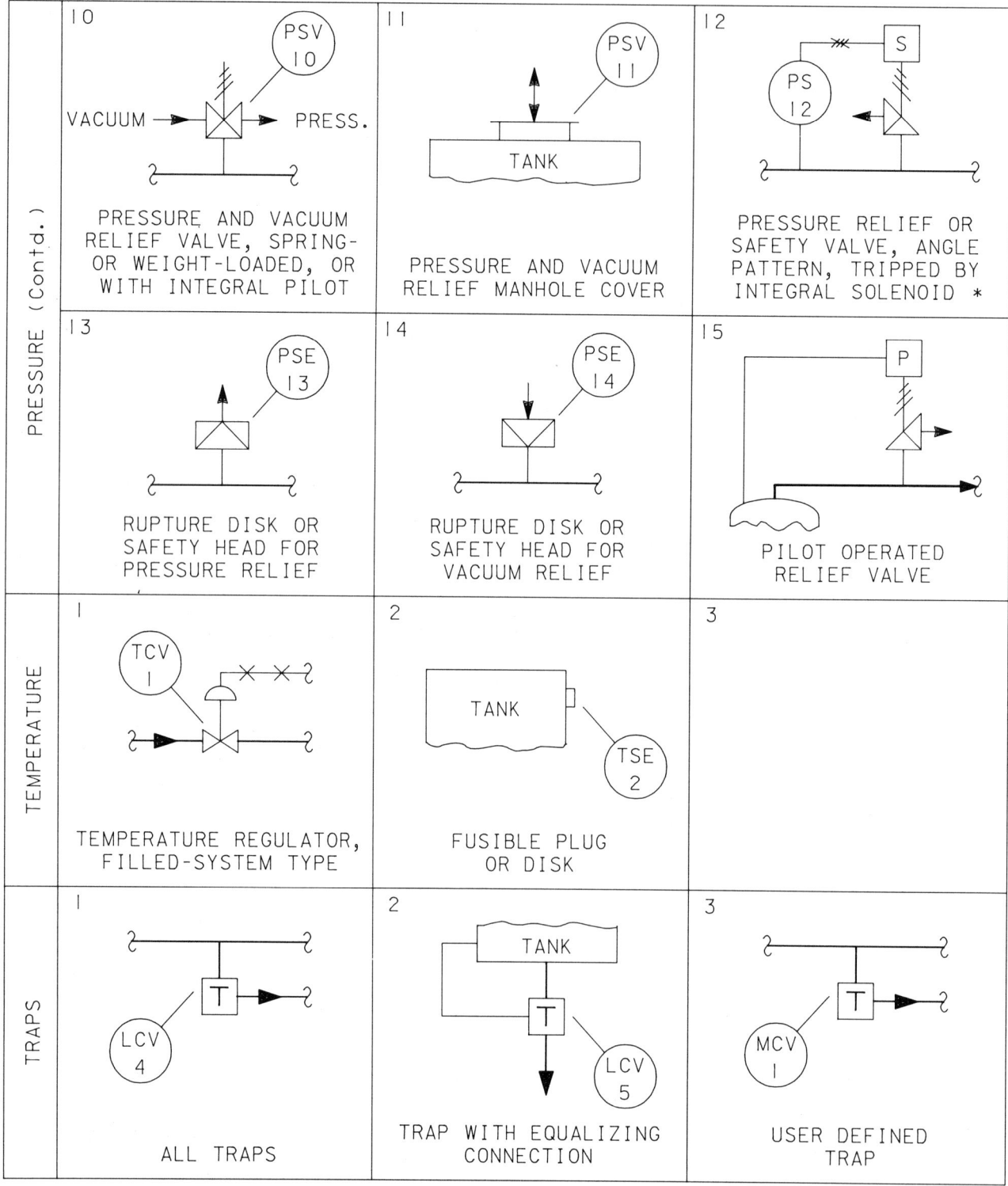

* The solenoid-tripped pressure relief valve is one of the class of power-actuated relief valves and is grouped with the other types of relief valves even though it is not entirely a self-actuated device.

6.7 SYMBOLS FOR ACTUATOR ACTION IN EVENT OF ACTUATOR POWER FAILURE. (SHOWN TYPICALLY FOR DIAPHRAGM-ACTUATED CONTROL VALVE)

1 TWO-WAY VALVE, FAIL OPEN	2 TWO-WAY VALVE, FAIL CLOSED	3 THREE-WAY VALVE, FAIL OPEN TO PATH A-C
4 FOUR-WAY VALVE, FAIL OPEN TO PATHS A-C AND D-B	5 ANY VALVE, FAIL LOCKED (POSITION DOES NOT CHANGE)	6 ANY VALVE, FAIL INDETERMINATE

The failure modes indicated are those commonly defined by the term, "shelf-position". As an alternative to the arrows and bars, the following abbreviations may be employed:

FO - Fail Open
FC - Fail Closed
FL - Fail Locked (last position)
FI - Fail Indeterminate

6.8 PRIMARY ELEMENT SYMBOLS

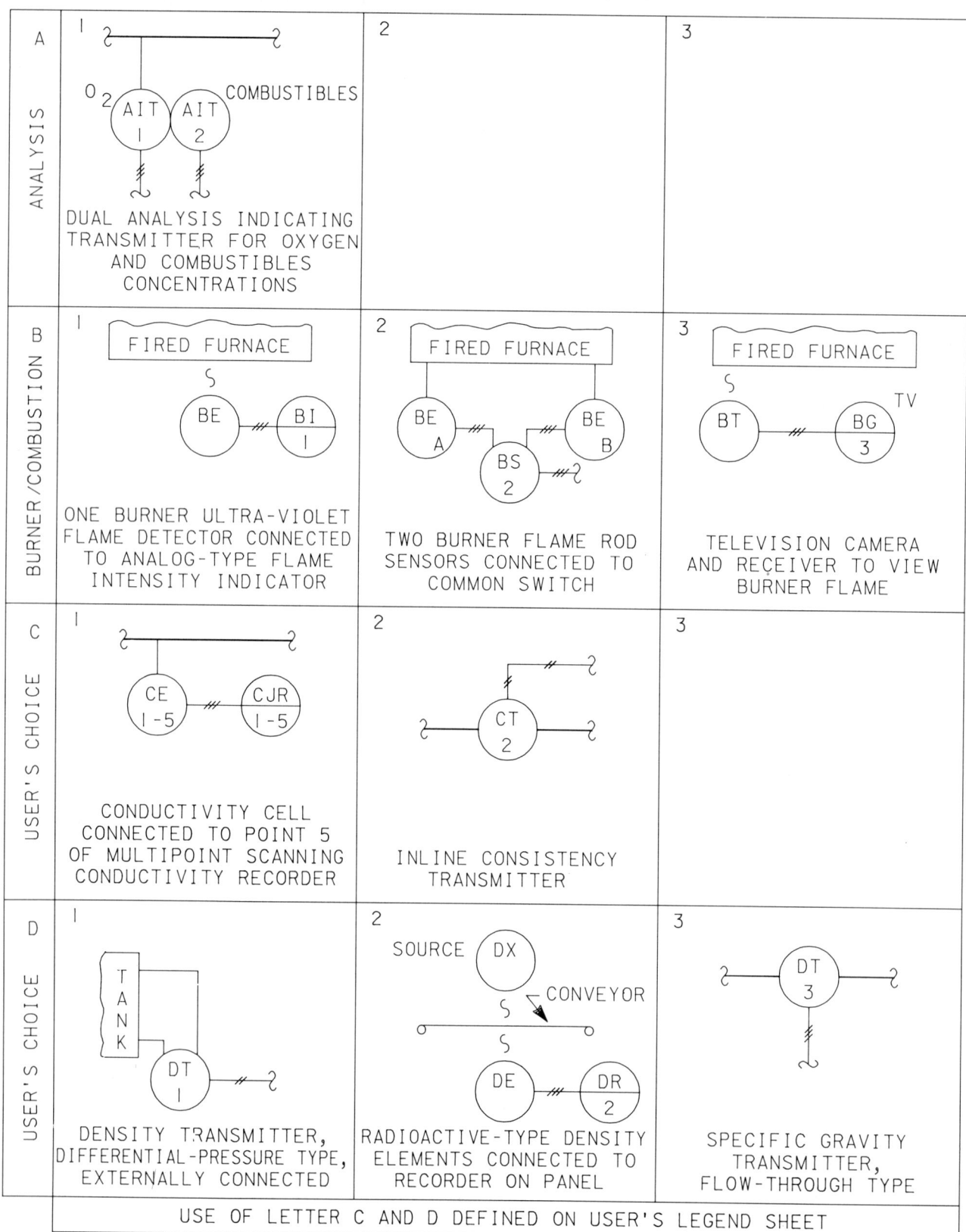

USE OF LETTER C AND D DEFINED ON USER'S LEGEND SHEET

6.8 PRIMARY ELEMENT SYMBOLS (Contd.)

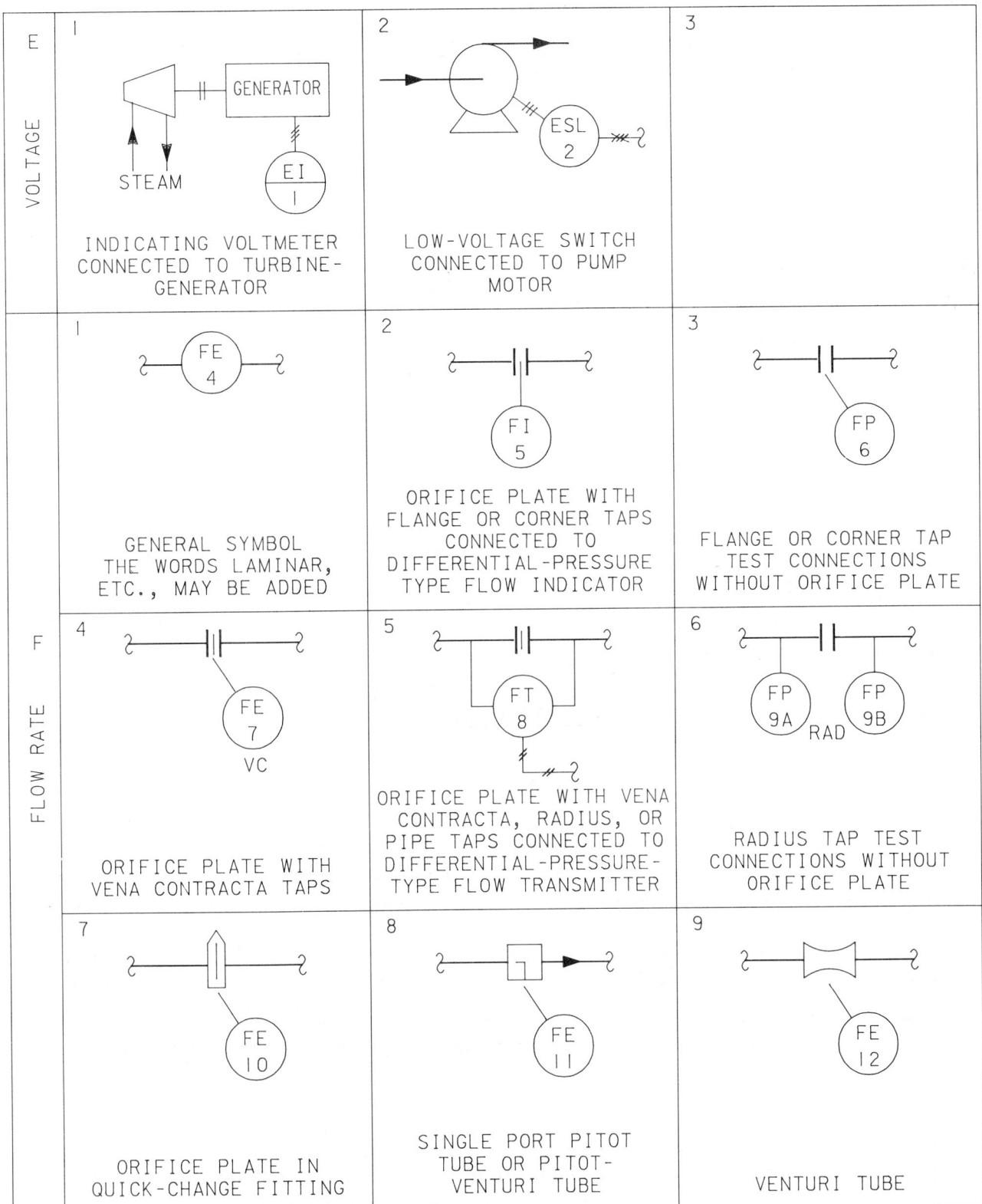

	1	2	3
VOLTAGE	INDICATING VOLTMETER CONNECTED TO TURBINE-GENERATOR	LOW-VOLTAGE SWITCH CONNECTED TO PUMP MOTOR	
FLOW RATE	GENERAL SYMBOL THE WORDS LAMINAR, ETC., MAY BE ADDED	ORIFICE PLATE WITH FLANGE OR CORNER TAPS CONNECTED TO DIFFERENTIAL-PRESSURE TYPE FLOW INDICATOR	FLANGE OR CORNER TAP TEST CONNECTIONS WITHOUT ORIFICE PLATE
	ORIFICE PLATE WITH VENA CONTRACTA TAPS	ORIFICE PLATE WITH VENA CONTRACTA, RADIUS, OR PIPE TAPS CONNECTED TO DIFFERENTIAL-PRESSURE-TYPE FLOW TRANSMITTER	RADIUS TAP TEST CONNECTIONS WITHOUT ORIFICE PLATE
	ORIFICE PLATE IN QUICK-CHANGE FITTING	SINGLE PORT PITOT TUBE OR PITOT-VENTURI TUBE	VENTURI TUBE

6.8 PRIMARY ELEMENT SYMBOLS (Contd.)

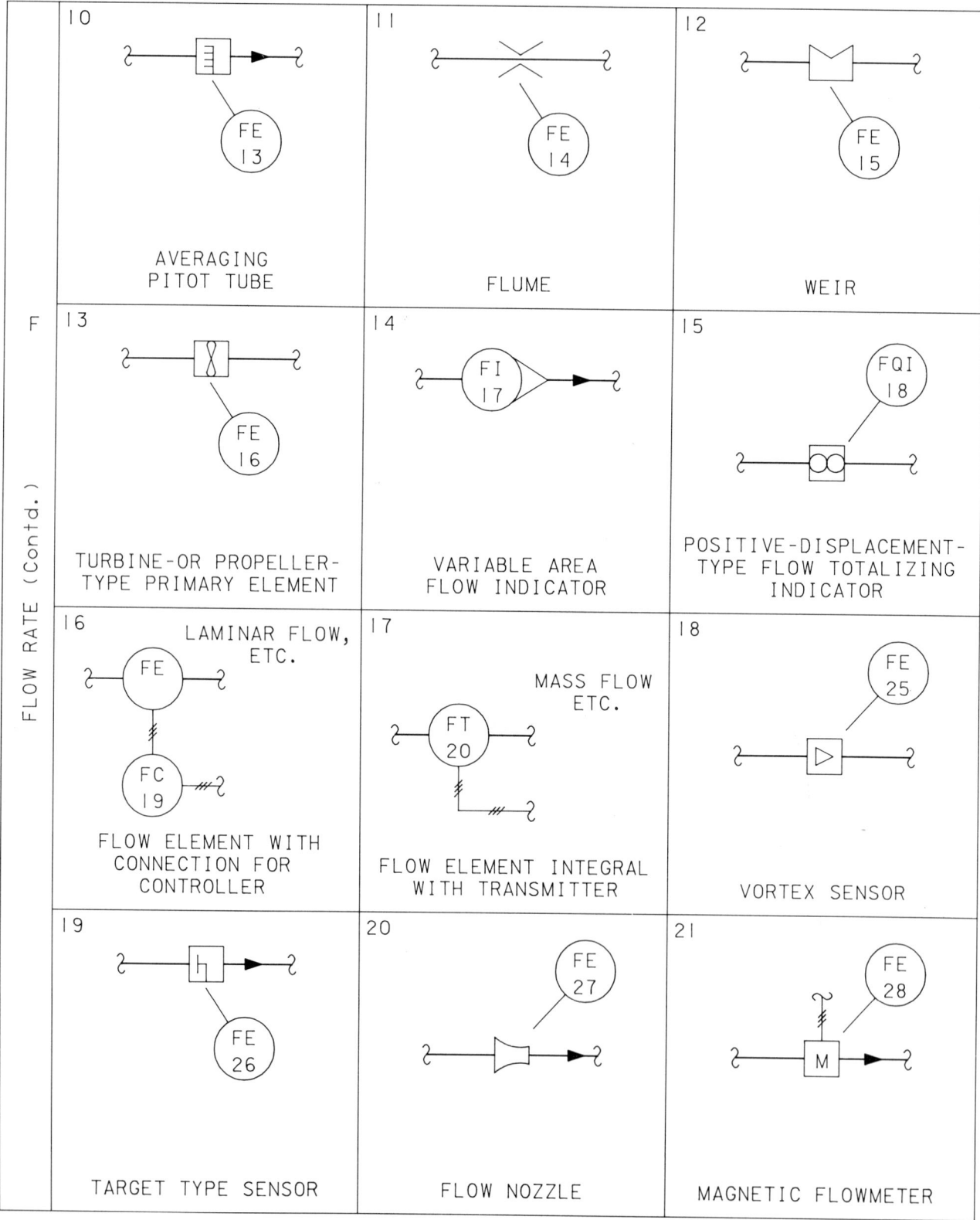

	10	11	12
	FE 13 AVERAGING PITOT TUBE	FE 14 FLUME	FE 15 WEIR
F FLOW RATE (Contd.)	13 FE 16 TURBINE-OR PROPELLER-TYPE PRIMARY ELEMENT	14 FI 17 VARIABLE AREA FLOW INDICATOR	15 FQI 18 POSITIVE-DISPLACEMENT-TYPE FLOW TOTALIZING INDICATOR
	16 LAMINAR FLOW, ETC. FE FC 19 FLOW ELEMENT WITH CONNECTION FOR CONTROLLER	17 MASS FLOW ETC. FT 20 FLOW ELEMENT INTEGRAL WITH TRANSMITTER	18 FE 25 VORTEX SENSOR
	19 FE 26 TARGET TYPE SENSOR	20 FE 27 FLOW NOZZLE	21 FE 28 M MAGNETIC FLOWMETER

6.8 PRIMARY ELEMENT SYMBOLS (Contd.)

	22	23	24
F **FLOW RATE (Contd.)**	FT 29 M → MAGNETIC FLOWMETER WITH INTEGRAL TRANSMITTER	FE 30 ~ → SONIC FLOWMETER "DOPPLER" OR "TRANSIT TIME" MAY BE ADDED	
	1	2	3
I **CURRENT**	IE 1 CURRENT TRANSFORMER MEASURING CURRENT OF ELECTRIC MOTOR		
	1	2	3
J **POWER**	JI 1 INDICATING WATTMETER CONNECTED TO PUMP MOTOR		
	1	2	3
K **TIME OR TIME-SCHEDULE**	KI 1 CLOCK	KIS 2-7 → MULTIPOINT ON-OFF TIME SEQUENCING PROGRAMMER POINT 7	KC 3 → SP → TIC 4 TIME-SCHEDULE CONTROLLER, ANALOG TYPE, OR SELF-CONTAINED FUNCTION GENERATOR

6.8 PRIMARY ELEMENT SYMBOLS (Contd.)

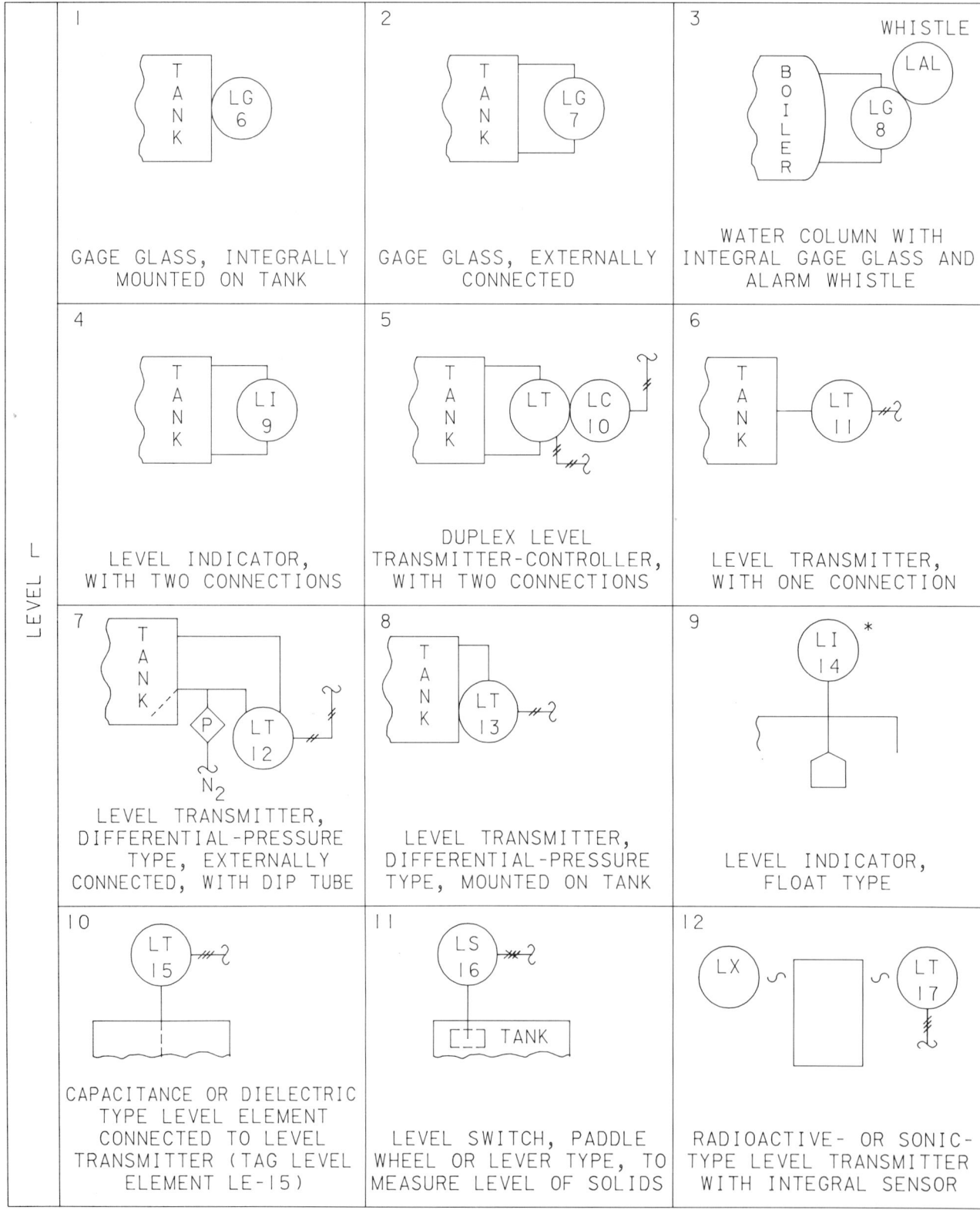

LEVEL L

1 — GAGE GLASS, INTEGRALLY MOUNTED ON TANK

2 — GAGE GLASS, EXTERNALLY CONNECTED

3 — WATER COLUMN WITH INTEGRAL GAGE GLASS AND ALARM WHISTLE

4 — LEVEL INDICATOR, WITH TWO CONNECTIONS

5 — DUPLEX LEVEL TRANSMITTER-CONTROLLER, WITH TWO CONNECTIONS

6 — LEVEL TRANSMITTER, WITH ONE CONNECTION

7 — LEVEL TRANSMITTER, DIFFERENTIAL-PRESSURE TYPE, EXTERNALLY CONNECTED, WITH DIP TUBE

8 — LEVEL TRANSMITTER, DIFFERENTIAL-PRESSURE TYPE, MOUNTED ON TANK

9 — LEVEL INDICATOR, FLOAT TYPE

10 — CAPACITANCE OR DIELECTRIC TYPE LEVEL ELEMENT CONNECTED TO LEVEL TRANSMITTER (TAG LEVEL ELEMENT LE-15)

11 — LEVEL SWITCH, PADDLE WHEEL OR LEVER TYPE, TO MEASURE LEVEL OF SOLIDS

12 — RADIOACTIVE- OR SONIC-TYPE LEVEL TRANSMITTER WITH INTEGRAL SENSOR

* Notations such as "mounted at grade" may be added.

6.8 PRIMARY ELEMENT SYMBOLS (Contd.)

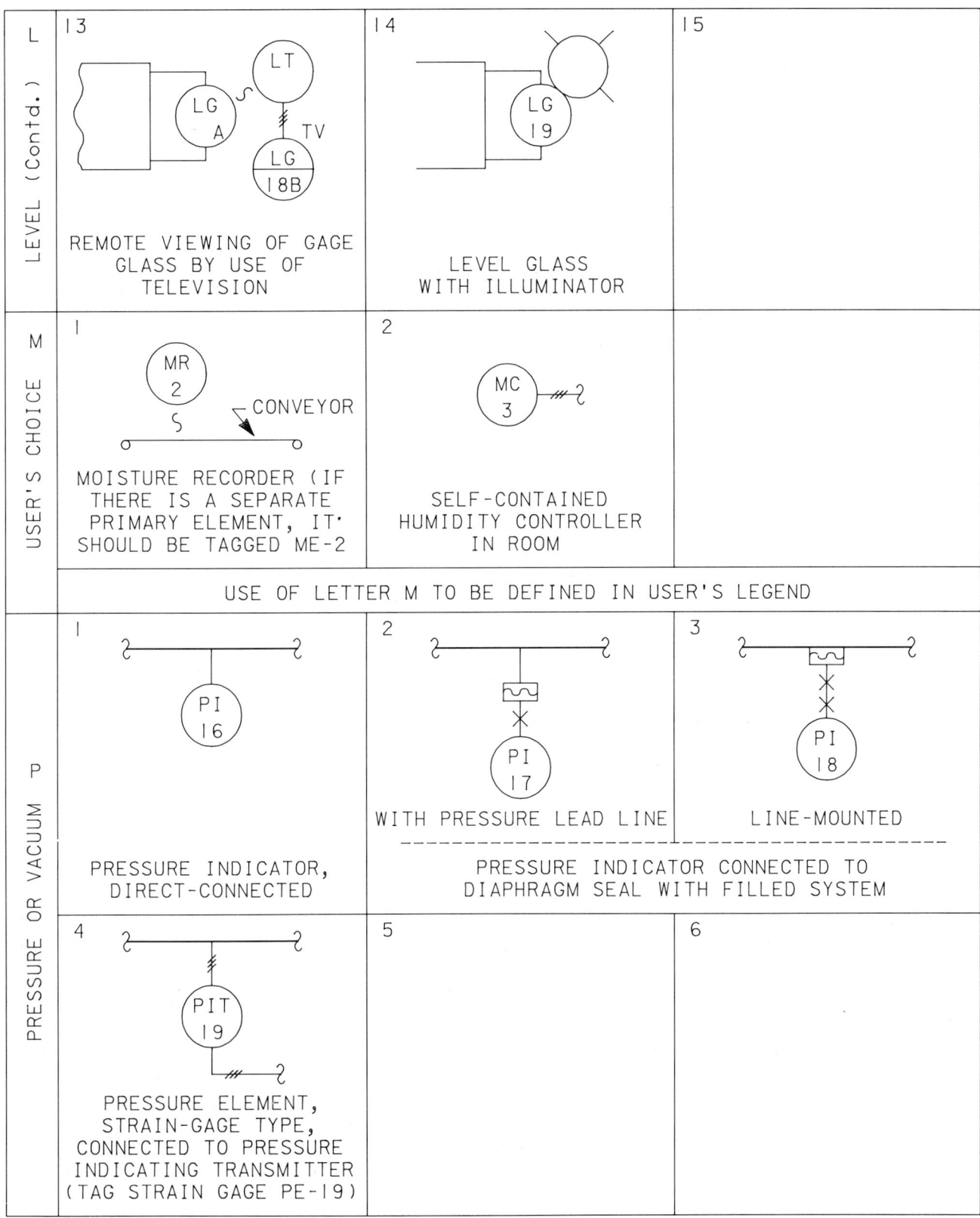

L / LEVEL (Contd.)	13 — REMOTE VIEWING OF GAGE GLASS BY USE OF TELEVISION	14 — LEVEL GLASS WITH ILLUMINATOR	15

M / USER'S CHOICE

1 — MOISTURE RECORDER (IF THERE IS A SEPARATE PRIMARY ELEMENT, IT· SHOULD BE TAGGED ME-2

2 — SELF-CONTAINED HUMIDITY CONTROLLER IN ROOM

USE OF LETTER M TO BE DEFINED IN USER'S LEGEND

P / PRESSURE OR VACUUM

1 — PRESSURE INDICATOR, DIRECT-CONNECTED

2 — WITH PRESSURE LEAD LINE

3 — LINE-MOUNTED

PRESSURE INDICATOR CONNECTED TO DIAPHRAGM SEAL WITH FILLED SYSTEM

4 — PRESSURE ELEMENT, STRAIN-GAGE TYPE, CONNECTED TO PRESSURE INDICATING TRANSMITTER (TAG STRAIN GAGE PE-19)

5

6

6.8 PRIMARY ELEMENT SYMBOLS (Contd.)

Q QUANTITY	1 LIGHT SOURCE — QX — CONVEYOR — QS 1 COUNTING SWITCH, PHOTO-ELECTRIC TYPE, WITH SWITCH ACTION FOR EACH EVENT	2 LIGHT SOURCE — QX — CONVEYOR — QQS 2 COUNTING SWITCH, PHOTO-ELECTRIC TYPE, WITH SWITCH ACTION BASED ON CUMULATIVE TOTAL	3 CONVEYOR — QQI 3 INDICATING COUNTER, MECHANICAL TYPE
R RADIATION	1 RI 1 RADIATION INDICATOR	2 RE 2 — RT 2 RADIATION MEASURING ELEMENT AND TRANSMITTER	3
S SPEED OR FREQUENCY	1 ROTATING MACHINE — ST 1 SPEED TRANSMITTER	2	3
T TEMPERATURE	1 TW 4 TEMPERATURE CONNECTION WITH WELL	2 TP 5 TEMPERATURE TEST CONNECTION WITHOUT WELL	3 TE 6 TEMPERATURE ELEMENT WITHOUT WELL (ELEMENT NOT CONNECTED TO SECONDARY INSTRUMENT)

6.8 PRIMARY ELEMENT SYMBOLS (Contd.)

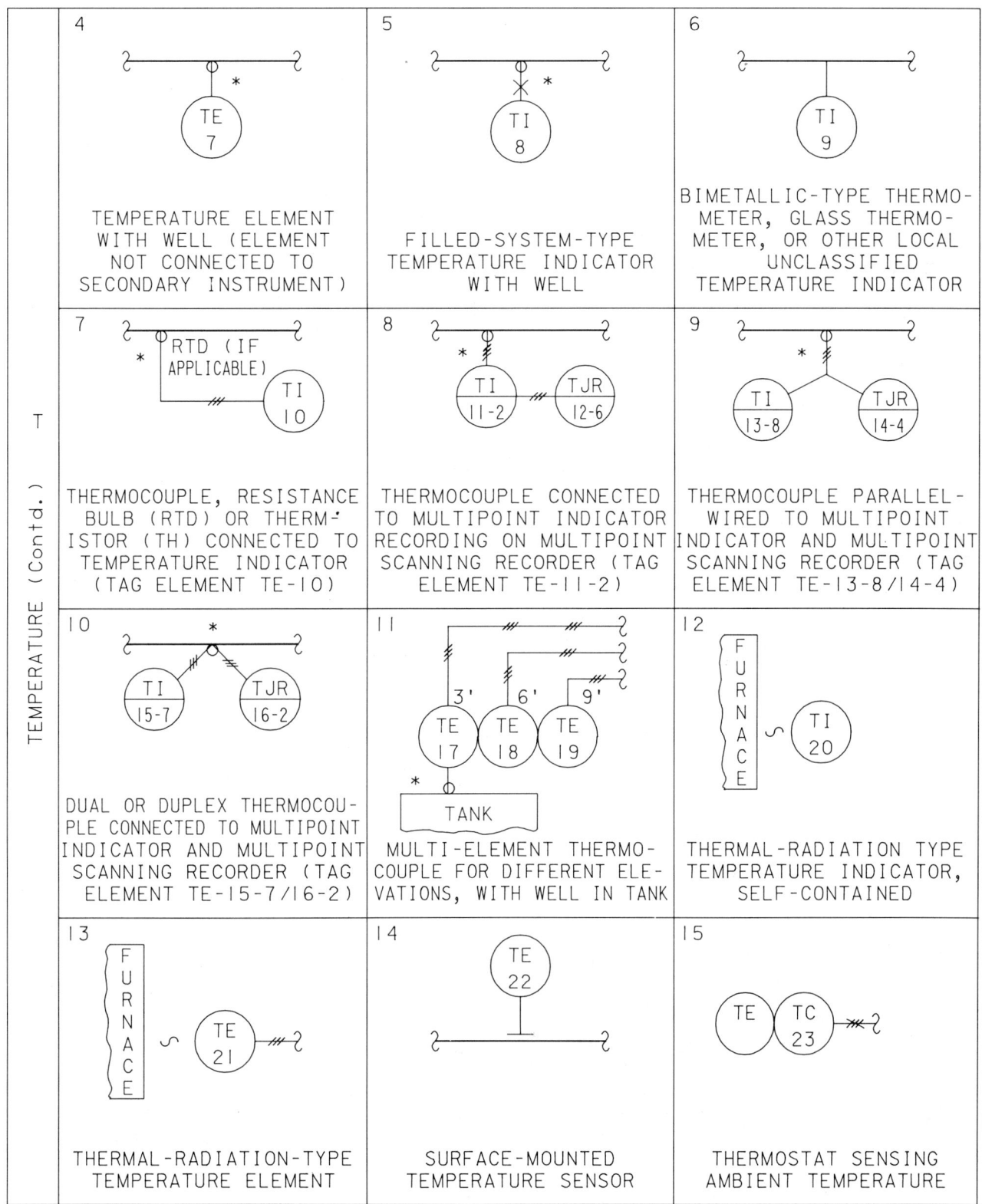

4	5	6
TE 7	TI 8	TI 9
TEMPERATURE ELEMENT WITH WELL (ELEMENT NOT CONNECTED TO SECONDARY INSTRUMENT)	FILLED-SYSTEM-TYPE TEMPERATURE INDICATOR WITH WELL	BIMETALLIC-TYPE THERMO-METER, GLASS THERMO-METER, OR OTHER LOCAL UNCLASSIFIED TEMPERATURE INDICATOR
7	8	9
RTD (IF APPLICABLE) TI 10	TI 11-2 TJR 12-6	TI 13-8 TJR 14-4
THERMOCOUPLE, RESISTANCE BULB (RTD) OR THERM-ISTOR (TH) CONNECTED TO TEMPERATURE INDICATOR (TAG ELEMENT TE-10)	THERMOCOUPLE CONNECTED TO MULTIPOINT INDICATOR RECORDING ON MULTIPOINT SCANNING RECORDER (TAG ELEMENT TE-11-2)	THERMOCOUPLE PARALLEL-WIRED TO MULTIPOINT INDICATOR AND MULTIPOINT SCANNING RECORDER (TAG ELEMENT TE-13-8/14-4)
10	11	12
TI 15-7 TJR 16-2	TE 17 TE 18 TE 19 3' 6' 9' TANK	FURNACE TI 20
DUAL OR DUPLEX THERMOCOU-PLE CONNECTED TO MULTIPOINT INDICATOR AND MULTIPOINT SCANNING RECORDER (TAG ELEMENT TE-15-7/16-2)	MULTI-ELEMENT THERMO-COUPLE FOR DIFFERENT ELE-VATIONS, WITH WELL IN TANK	THERMAL-RADIATION TYPE TEMPERATURE INDICATOR, SELF-CONTAINED
13	14	15
FURNACE TE 21	TE 22	TE TC 23
THERMAL-RADIATION-TYPE TEMPERATURE ELEMENT	SURFACE-MOUNTED TEMPERATURE SENSOR	THERMOSTAT SENSING AMBIENT TEMPERATURE

T   TEMPERATURE (Contd.)

* Use of the thermowell symbol is optional. However, use or omission of the symbol should be consistent throughout a project.

6.8 PRIMARY ELEMENT SYMBOLS (Contd.)

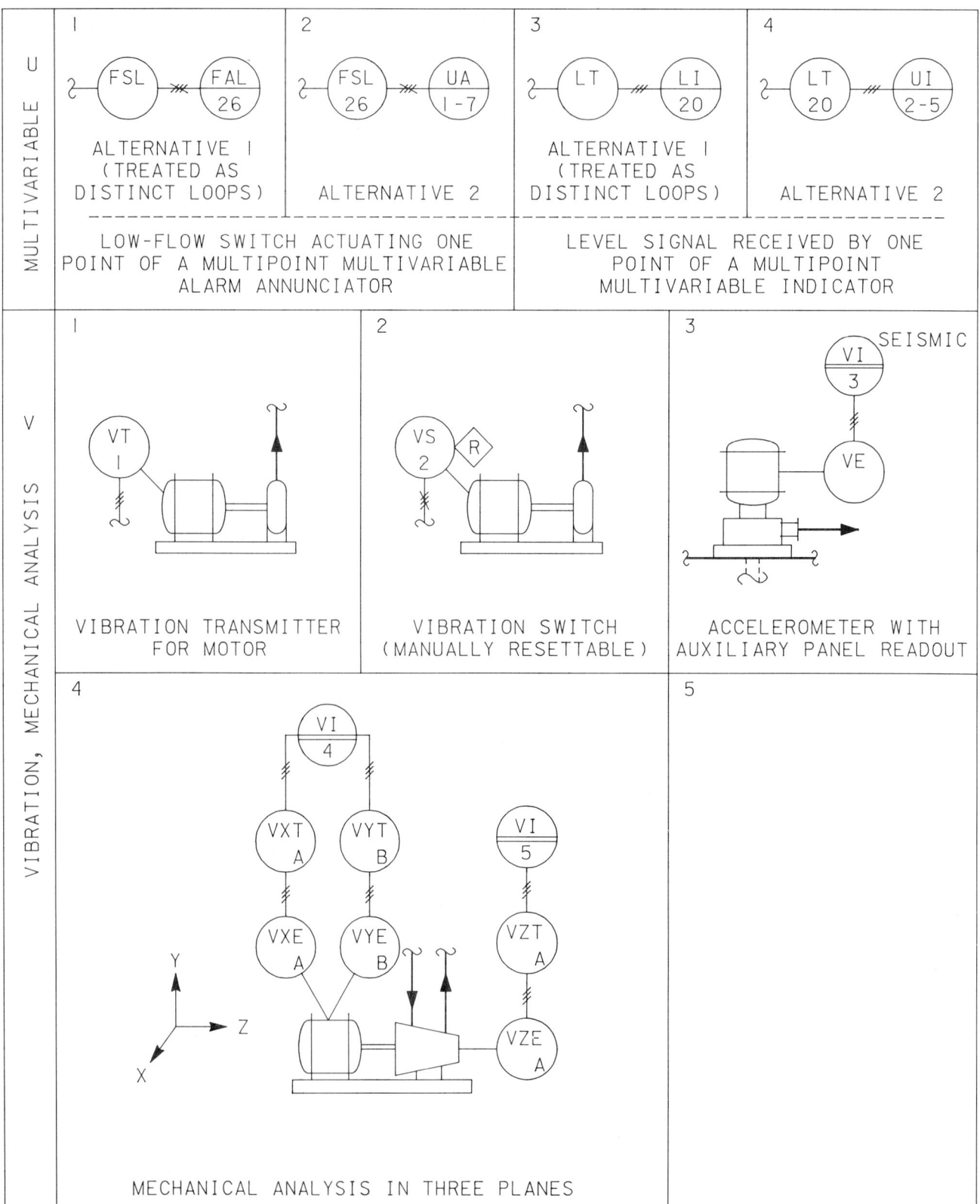

MULTIVARIABLE U

1. ALTERNATIVE I (TREATED AS DISTINCT LOOPS)
2. ALTERNATIVE 2

LOW-FLOW SWITCH ACTUATING ONE POINT OF A MULTIPOINT MULTIVARIABLE ALARM ANNUNCIATOR

3. ALTERNATIVE I (TREATED AS DISTINCT LOOPS)
4. ALTERNATIVE 2

LEVEL SIGNAL RECEIVED BY ONE POINT OF A MULTIPOINT MULTIVARIABLE INDICATOR

VIBRATION, MECHANICAL ANALYSIS V

1. VIBRATION TRANSMITTER FOR MOTOR
2. VIBRATION SWITCH (MANUALLY RESETTABLE)
3. ACCELEROMETER WITH AUXILIARY PANEL READOUT

4. MECHANICAL ANALYSIS IN THREE PLANES

5.

6.8 PRIMARY ELEMENT SYMBOLS (Contd.)

	1	2	3
W WEIGHT OR FORCE	**WT 1** TANK WEIGHT TRANSMITTER, DIRECT-CONNECTED	**WT 2** TANK STRAIN GAGE CONNECTED TO SEPARATE WEIGHT TRANSMITTER (TAG STRAIN GAGE WE-2)	**WT 3** CONVEYOR WEIGH-BELT SCALE TRANSMITTER
Z POSITION, DIMENSION	**ZT 1** CONVEYOR ROLL-THICKNESS TRANSMITTER	SOURCE **ZDX** CONVEYOR **ZDS 2** THICKNESS SWITCH, RADIOACTIVE TYPE	**TC 24** **ZSL 3** LIMIT SWITCH THAT IS ACTUATED WHEN VALVE IS CLOSED TO A PRE-DETERMINED POSITION
	4	5	6
	DRIVEN MACHINE **ZDT 4** TURBINE SHELL/ROTOR DIFFERENTIAL-EXPANSION TRANSMITTER (TAG PRIMARY ELEMENT ZDE-4)		

6.9 EXAMPLES-FUNCTIONS

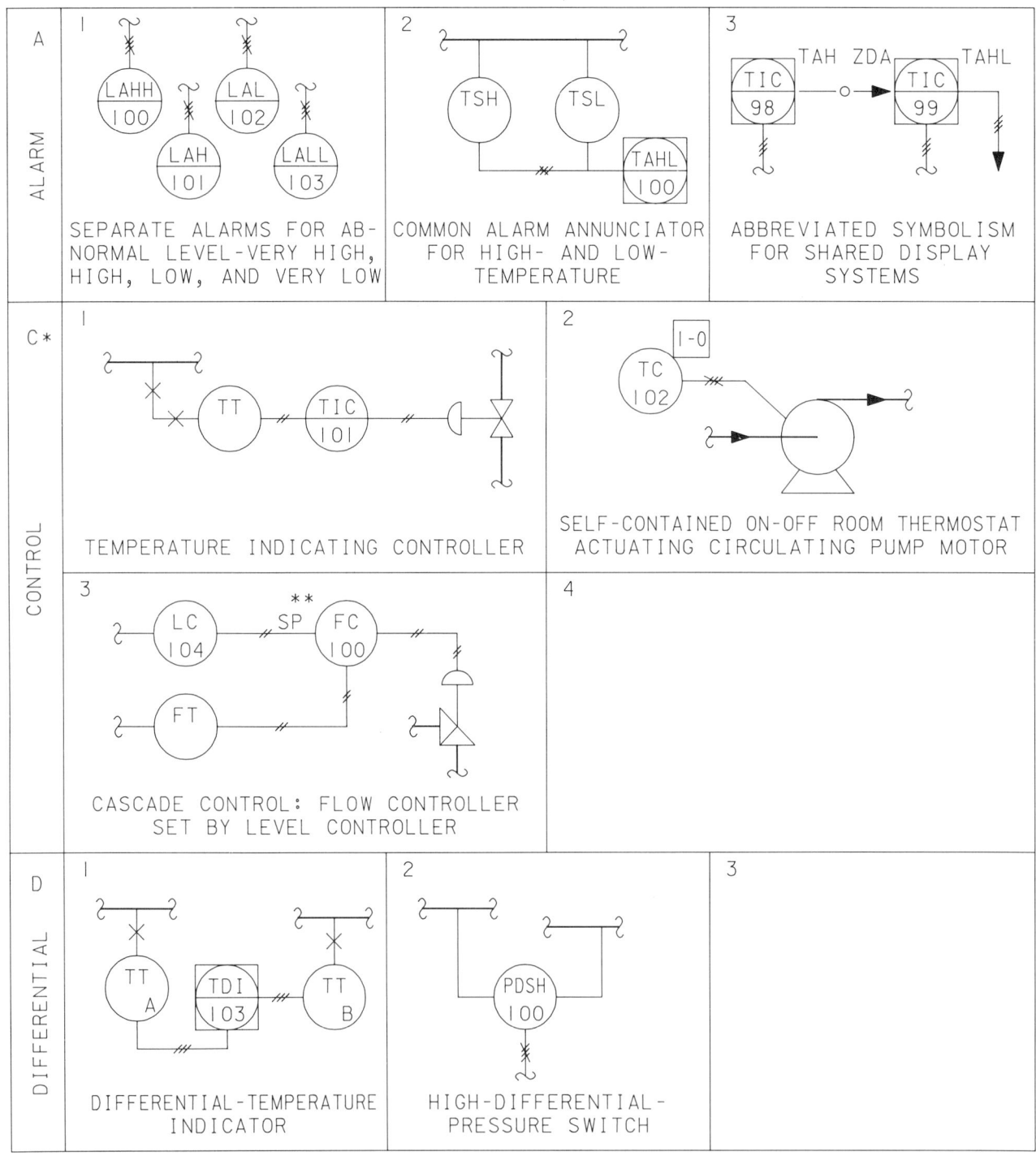

A ALARM

1 SEPARATE ALARMS FOR AB-NORMAL LEVEL-VERY HIGH, HIGH, LOW, AND VERY LOW

2 COMMON ALARM ANNUNCIATOR FOR HIGH- AND LOW-TEMPERATURE

3 ABBREVIATED SYMBOLISM FOR SHARED DISPLAY SYSTEMS

C* CONTROL

1 TEMPERATURE INDICATING CONTROLLER

2 SELF-CONTAINED ON-OFF ROOM THERMOSTAT ACTUATING CIRCULATING PUMP MOTOR

3 CASCADE CONTROL: FLOW CONTROLLER SET BY LEVEL CONTROLLER

4

D DIFFERENTIAL

1 DIFFERENTIAL-TEMPERATURE INDICATOR

2 HIGH-DIFFERENTIAL-PRESSURE SWITCH

3

* It is expected that control modes will not be designated on a diagram. However, designations may be used outside the controller symbol, if desired, in combinations such as %, ∫, I-0.

** A controller is understood to have integral manual set-point adjustment unless means of remote adjustment is indicated. The remote set-point designation is SP.

6.9 EXAMPLES-FUNCTIONS (Contd.)

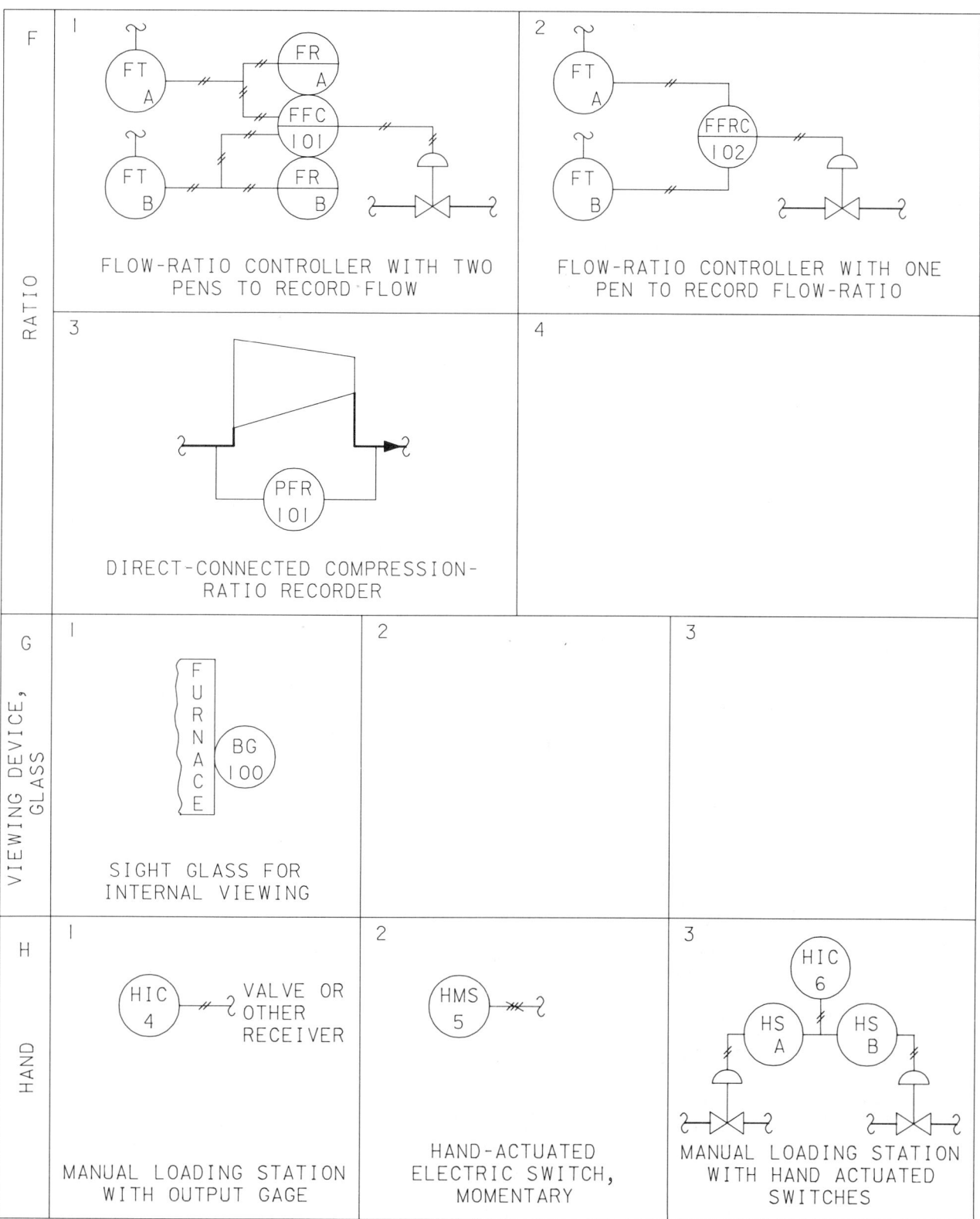

F / RATIO

1. FLOW-RATIO CONTROLLER WITH TWO PENS TO RECORD FLOW

2. FLOW-RATIO CONTROLLER WITH ONE PEN TO RECORD FLOW-RATIO

3. DIRECT-CONNECTED COMPRESSION-RATIO RECORDER

4.

G / VIEWING DEVICE, GLASS

1. SIGHT GLASS FOR INTERNAL VIEWING

2.

3.

H / HAND

1. MANUAL LOADING STATION WITH OUTPUT GAGE

2. HAND-ACTUATED ELECTRIC SWITCH, MOMENTARY

3. MANUAL LOADING STATION WITH HAND ACTUATED SWITCHES

6.9 EXAMPLES-FUNCTIONS (Contd.)

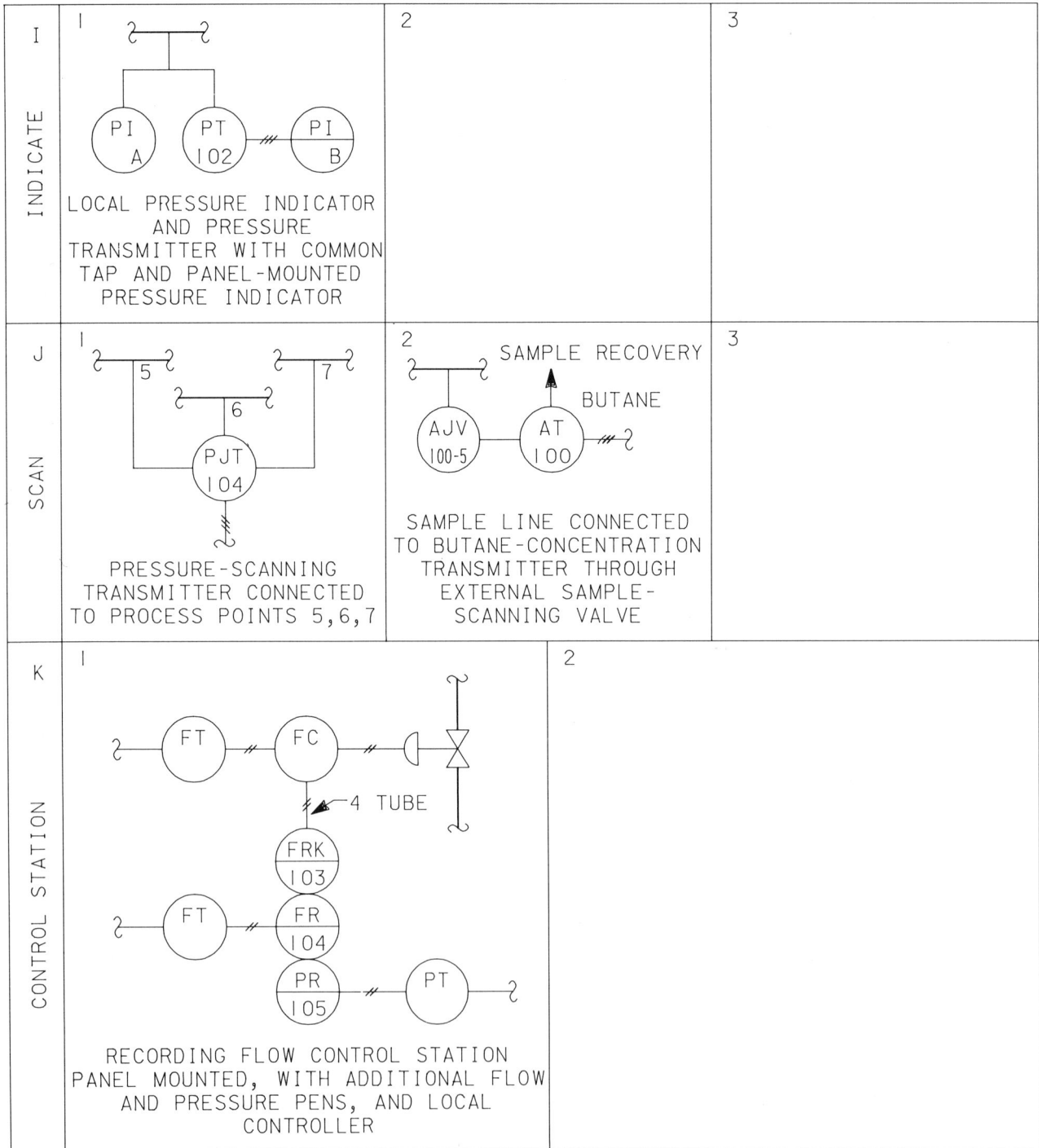

	1	2	3
INDICATE	PI A — PT 102 — PI B LOCAL PRESSURE INDICATOR AND PRESSURE TRANSMITTER WITH COMMON TAP AND PANEL-MOUNTED PRESSURE INDICATOR		
SCAN	5 6 7 — PJT 104 PRESSURE-SCANNING TRANSMITTER CONNECTED TO PROCESS POINTS 5,6,7	SAMPLE RECOVERY BUTANE AJV 100-5 — AT 100 SAMPLE LINE CONNECTED TO BUTANE-CONCENTRATION TRANSMITTER THROUGH EXTERNAL SAMPLE-SCANNING VALVE	
CONTROL STATION	FT — FC 4 TUBE FRK 103 FT — FR 104 PR 105 — PT RECORDING FLOW CONTROL STATION PANEL MOUNTED, WITH ADDITIONAL FLOW AND PRESSURE PENS, AND LOCAL CONTROLLER		

6.9 EXAMPLES-FUNCTIONS (Contd.)

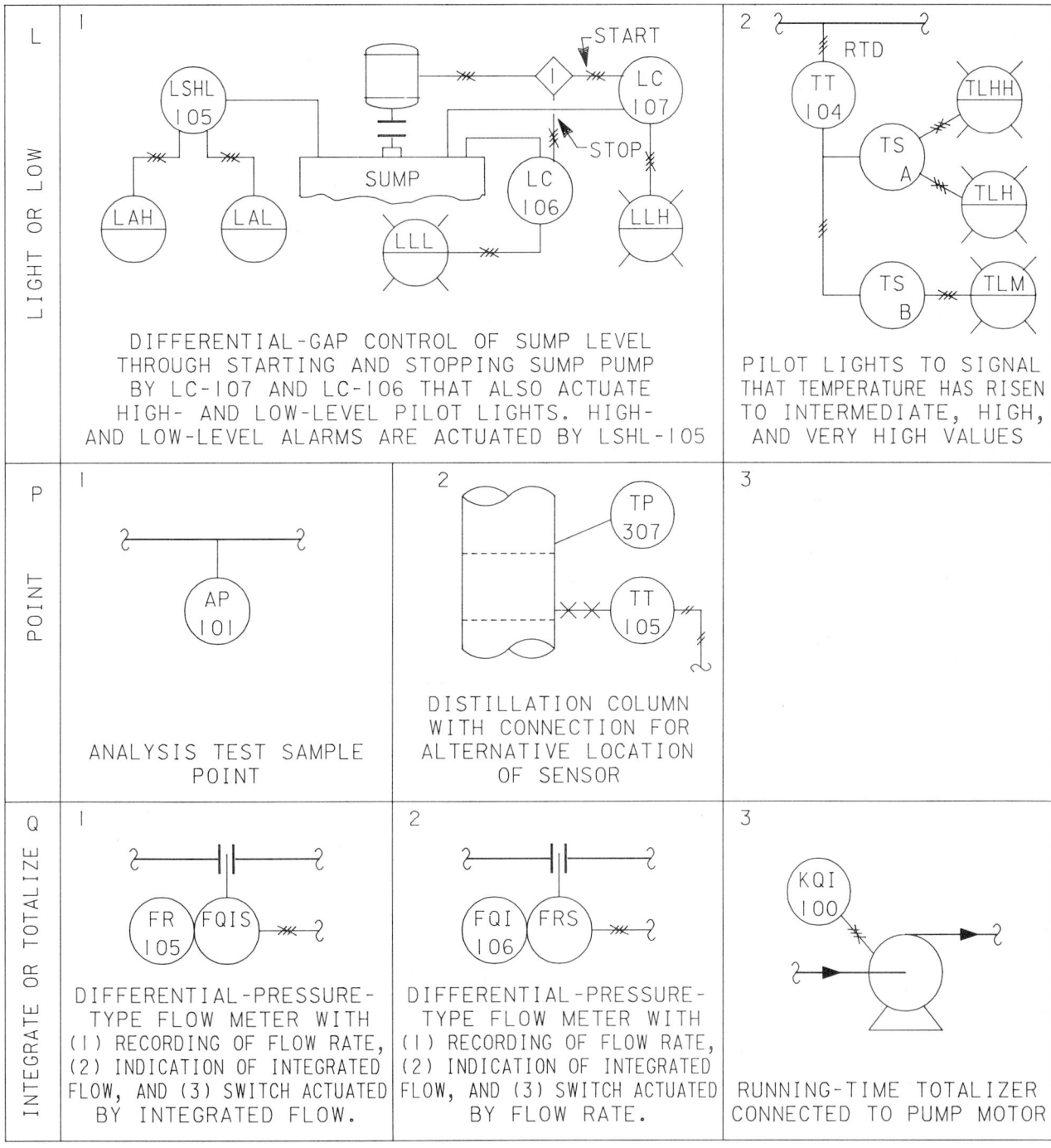

L — LIGHT OR LOW

1. DIFFERENTIAL-GAP CONTROL OF SUMP LEVEL
THROUGH STARTING AND STOPPING SUMP PUMP
BY LC-107 AND LC-106 THAT ALSO ACTUATE
HIGH- AND LOW-LEVEL PILOT LIGHTS. HIGH-
AND LOW-LEVEL ALARMS ARE ACTUATED BY LSHL-105

2. PILOT LIGHTS TO SIGNAL
THAT TEMPERATURE HAS RISEN
TO INTERMEDIATE, HIGH,
AND VERY HIGH VALUES

P — POINT

1. ANALYSIS TEST SAMPLE
POINT

2. DISTILLATION COLUMN
WITH CONNECTION FOR
ALTERNATIVE LOCATION
OF SENSOR

3.

Q — INTEGRATE OR TOTALIZE

1. DIFFERENTIAL-PRESSURE-
TYPE FLOW METER WITH
(1) RECORDING OF FLOW RATE,
(2) INDICATION OF INTEGRATED
FLOW, AND (3) SWITCH ACTUATED
BY INTEGRATED FLOW.

2. DIFFERENTIAL-PRESSURE-
TYPE FLOW METER WITH
(1) RECORDING OF FLOW RATE,
(2) INDICATION OF INTEGRATED
FLOW, AND (3) SWITCH ACTUATED
BY FLOW RATE.

3. RUNNING-TIME TOTALIZER
CONNECTED TO PUMP MOTOR

6.9 EXAMPLES-FUNCTIONS (Contd.)

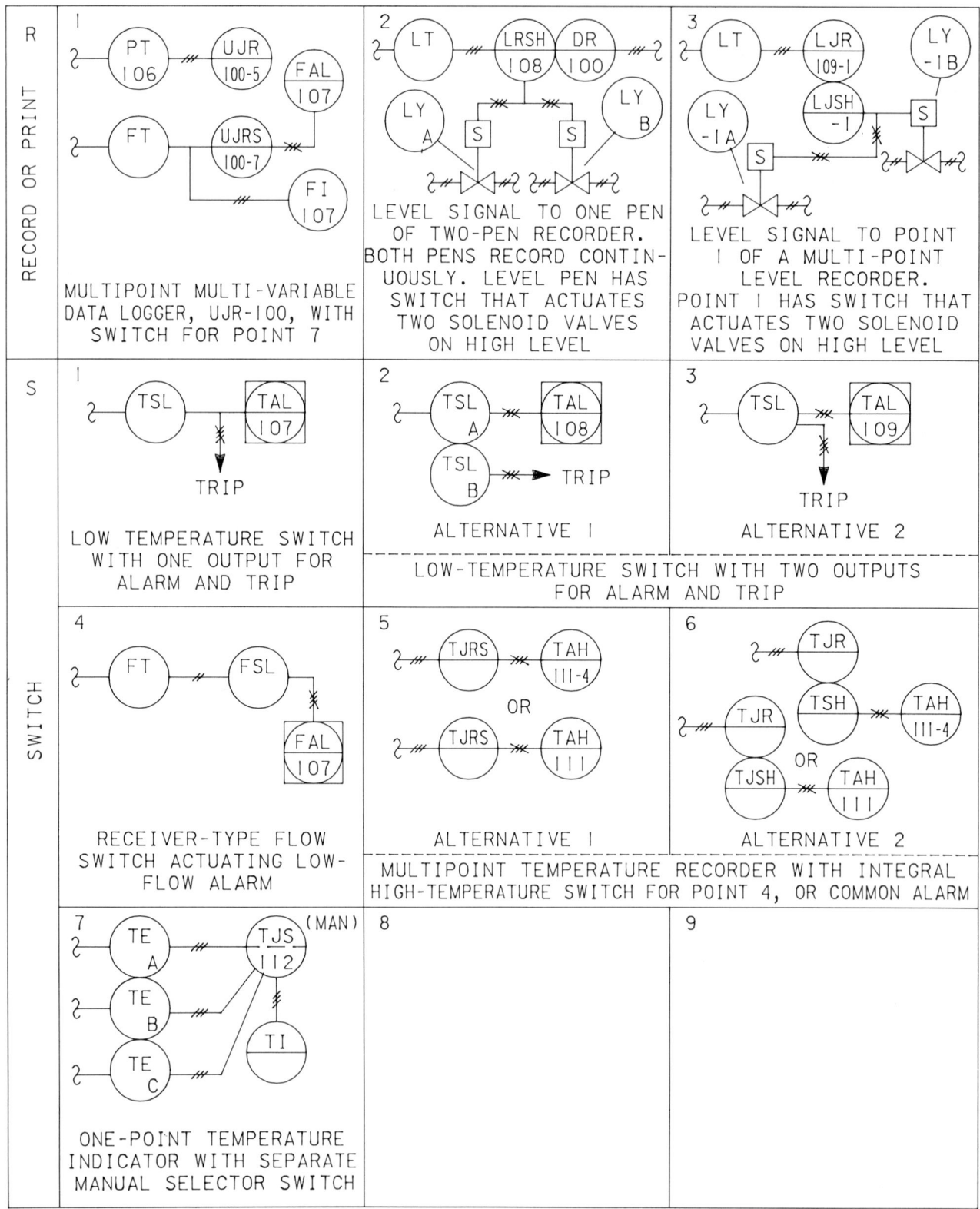

RECORD OR PRINT

R-1 MULTIPOINT MULTI-VARIABLE DATA LOGGER, UJR-100, WITH SWITCH FOR POINT 7

R-2 LEVEL SIGNAL TO ONE PEN OF TWO-PEN RECORDER. BOTH PENS RECORD CONTINUOUSLY. LEVEL PEN HAS SWITCH THAT ACTUATES TWO SOLENOID VALVES ON HIGH LEVEL

R-3 LEVEL SIGNAL TO POINT 1 OF A MULTI-POINT LEVEL RECORDER. POINT 1 HAS SWITCH THAT ACTUATES TWO SOLENOID VALVES ON HIGH LEVEL

SWITCH

S-1 LOW TEMPERATURE SWITCH WITH ONE OUTPUT FOR ALARM AND TRIP

S-2 ALTERNATIVE 1

S-3 ALTERNATIVE 2

LOW-TEMPERATURE SWITCH WITH TWO OUTPUTS FOR ALARM AND TRIP

S-4 RECEIVER-TYPE FLOW SWITCH ACTUATING LOW-FLOW ALARM

S-5 ALTERNATIVE 1

S-6 ALTERNATIVE 2

MULTIPOINT TEMPERATURE RECORDER WITH INTEGRAL HIGH-TEMPERATURE SWITCH FOR POINT 4, OR COMMON ALARM

S-7 ONE-POINT TEMPERATURE INDICATOR WITH SEPARATE MANUAL SELECTOR SWITCH

6.9 EXAMPLES-FUNCTIONS (Contd.)

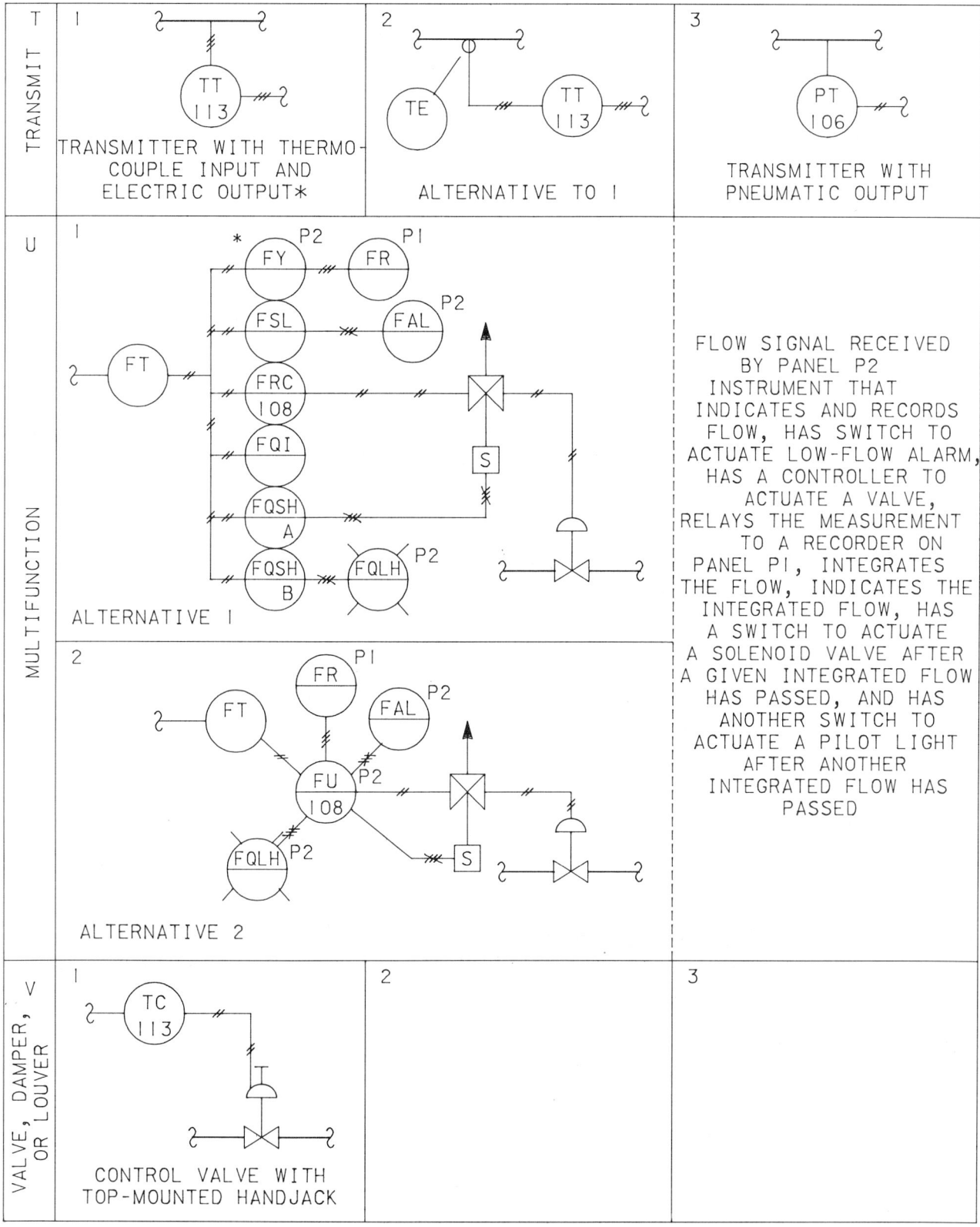

* See definition of converter versus transmitter.

6.9 EXAMPLES-FUNCTIONS (Contd.)

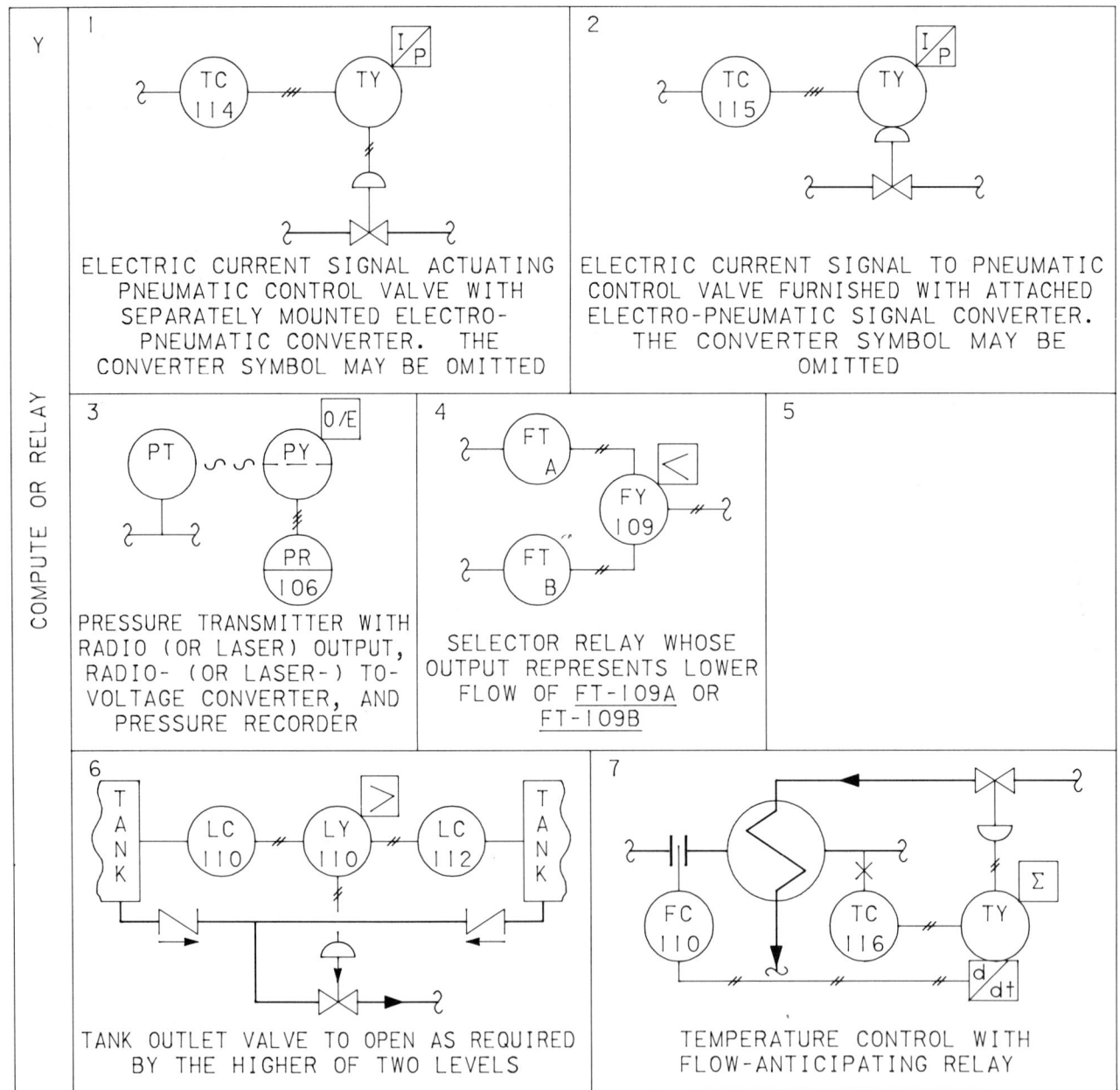

Y — COMPUTE OR RELAY

1. ELECTRIC CURRENT SIGNAL ACTUATING PNEUMATIC CONTROL VALVE WITH SEPARATELY MOUNTED ELECTRO-PNEUMATIC CONVERTER. THE CONVERTER SYMBOL MAY BE OMITTED

2. ELECTRIC CURRENT SIGNAL TO PNEUMATIC CONTROL VALVE FURNISHED WITH ATTACHED ELECTRO-PNEUMATIC SIGNAL CONVERTER. THE CONVERTER SYMBOL MAY BE OMITTED

3. PRESSURE TRANSMITTER WITH RADIO (OR LASER) OUTPUT, RADIO- (OR LASER-) TO-VOLTAGE CONVERTER, AND PRESSURE RECORDER

4. SELECTOR RELAY WHOSE OUTPUT REPRESENTS LOWER FLOW OF FT-109A OR FT-109B

5.

6. TANK OUTLET VALVE TO OPEN AS REQUIRED BY THE HIGHER OF TWO LEVELS

7. TEMPERATURE CONTROL WITH FLOW-ANTICIPATING RELAY

6.9 EXAMPLES-FUNCTIONS (Contd.)

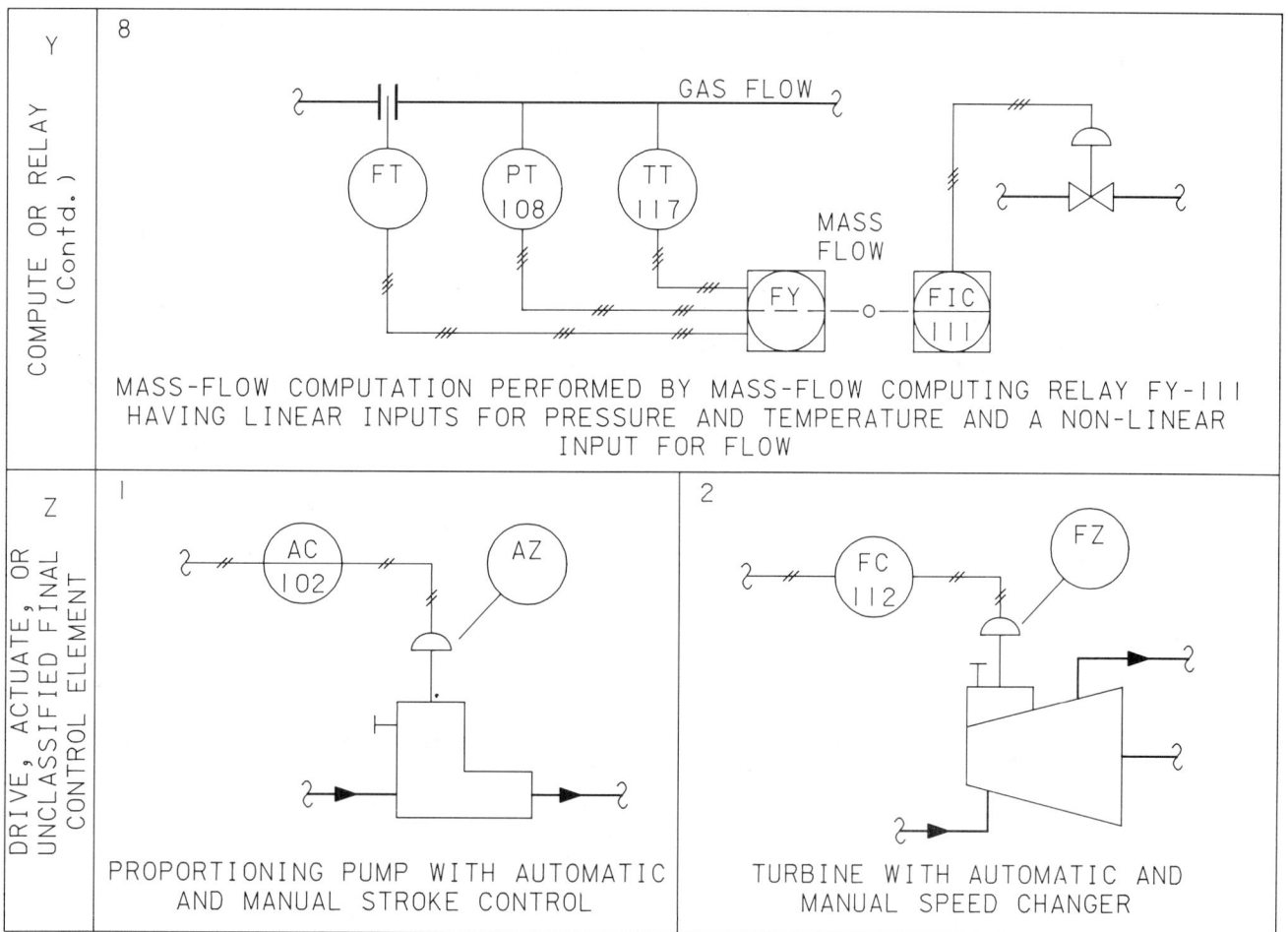

Y	8	
COMPUTE OR RELAY (Contd.)		MASS-FLOW COMPUTATION PERFORMED BY MASS-FLOW COMPUTING RELAY FY-111 HAVING LINEAR INPUTS FOR PRESSURE AND TEMPERATURE AND A NON-LINEAR INPUT FOR FLOW

MASS-FLOW COMPUTATION PERFORMED BY MASS-FLOW COMPUTING RELAY FY-111
HAVING LINEAR INPUTS FOR PRESSURE AND TEMPERATURE AND A NON-LINEAR
INPUT FOR FLOW

Z — DRIVE, ACTUATE, OR UNCLASSIFIED FINAL CONTROL ELEMENT

1

PROPORTIONING PUMP WITH AUTOMATIC
AND MANUAL STROKE CONTROL

2

TURBINE WITH AUTOMATIC AND
MANUAL SPEED CHANGER

6.10 EXAMPLES-MISCELLANEOUS COMBINATIONS

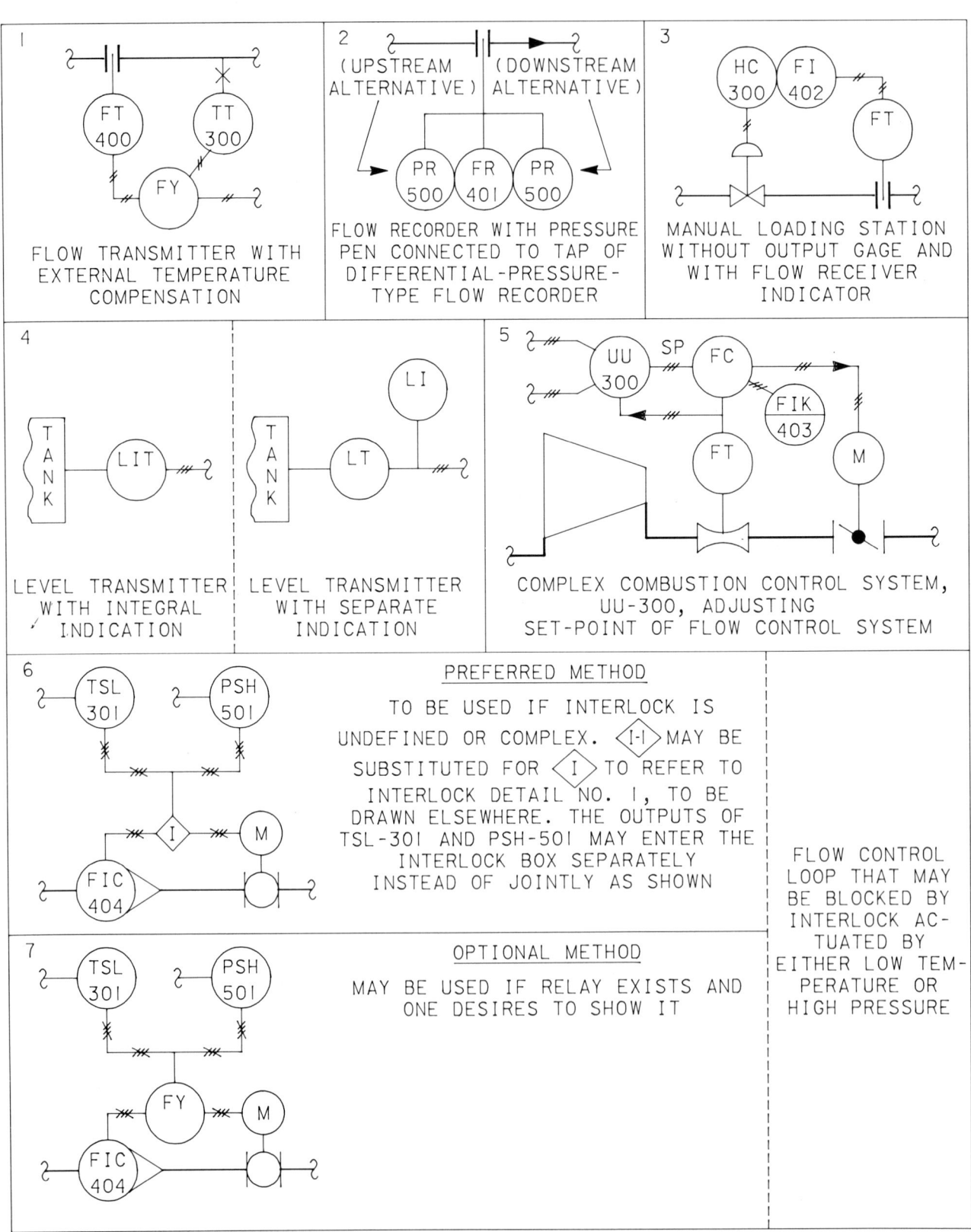

1. FLOW TRANSMITTER WITH EXTERNAL TEMPERATURE COMPENSATION

2. (UPSTREAM ALTERNATIVE) (DOWNSTREAM ALTERNATIVE)
FLOW RECORDER WITH PRESSURE PEN CONNECTED TO TAP OF DIFFERENTIAL-PRESSURE-TYPE FLOW RECORDER

3. MANUAL LOADING STATION WITHOUT OUTPUT GAGE AND WITH FLOW RECEIVER INDICATOR

4. LEVEL TRANSMITTER WITH INTEGRAL INDICATION

LEVEL TRANSMITTER WITH SEPARATE INDICATION

5. COMPLEX COMBUSTION CONTROL SYSTEM, UU-300, ADJUSTING SET-POINT OF FLOW CONTROL SYSTEM

6. PREFERRED METHOD

TO BE USED IF INTERLOCK IS UNDEFINED OR COMPLEX. ⟨I-I⟩ MAY BE SUBSTITUTED FOR ⟨I⟩ TO REFER TO INTERLOCK DETAIL NO. I, TO BE DRAWN ELSEWHERE. THE OUTPUTS OF TSL-301 AND PSH-501 MAY ENTER THE INTERLOCK BOX SEPARATELY INSTEAD OF JOINTLY AS SHOWN

7. OPTIONAL METHOD

MAY BE USED IF RELAY EXISTS AND ONE DESIRES TO SHOW IT

FLOW CONTROL LOOP THAT MAY BE BLOCKED BY INTERLOCK AC-TUATED BY EITHER LOW TEM-PERATURE OR HIGH PRESSURE

6.10 EXAMPLES-MISCELLANEOUS COMBINATIONS (Contd.)

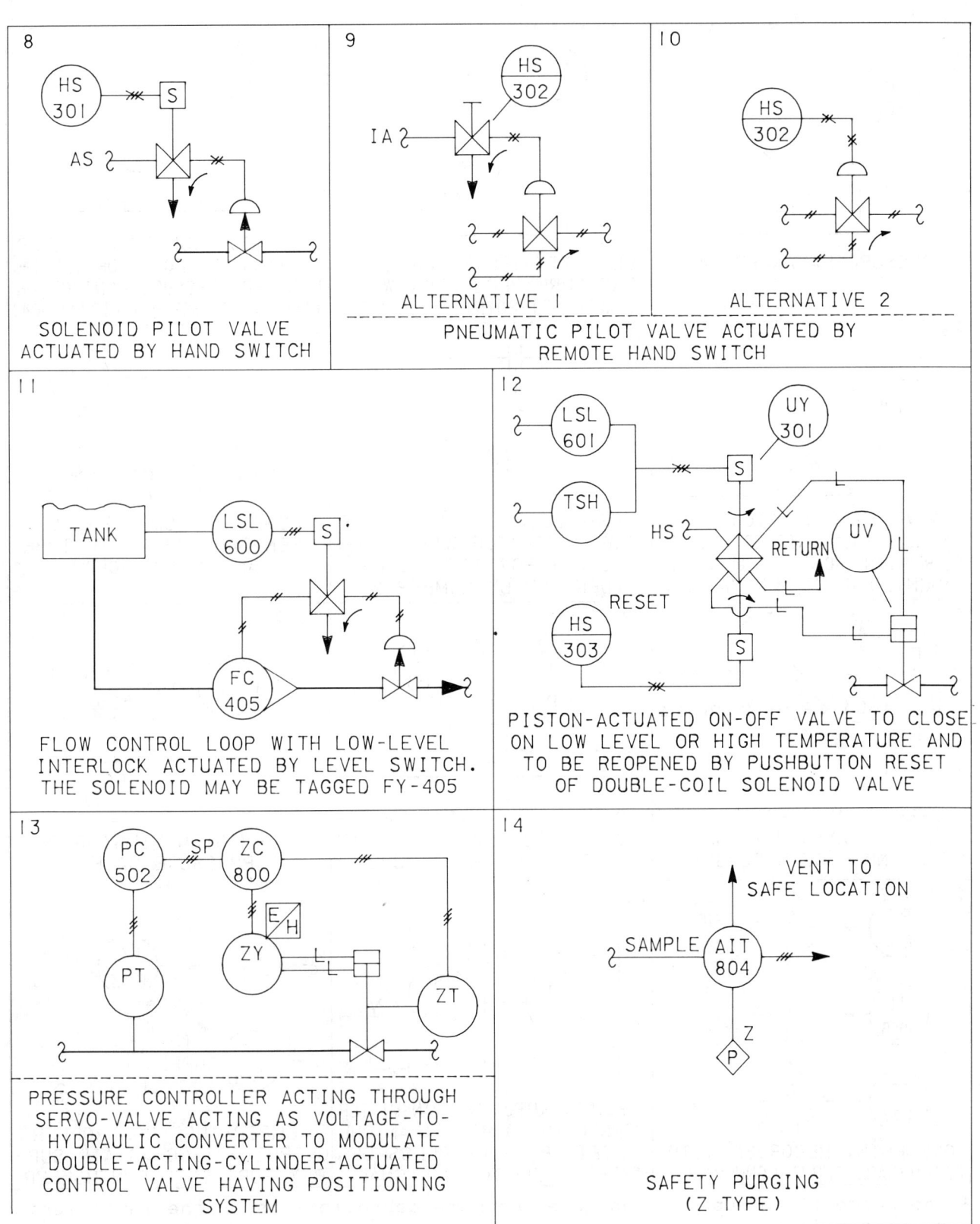

8

SOLENOID PILOT VALVE
ACTUATED BY HAND SWITCH

9

ALTERNATIVE 1

10

ALTERNATIVE 2

PNEUMATIC PILOT VALVE ACTUATED BY
REMOTE HAND SWITCH

11

FLOW CONTROL LOOP WITH LOW-LEVEL
INTERLOCK ACTUATED BY LEVEL SWITCH.
THE SOLENOID MAY BE TAGGED FY-405

12

PISTON-ACTUATED ON-OFF VALVE TO CLOSE
ON LOW LEVEL OR HIGH TEMPERATURE AND
TO BE REOPENED BY PUSHBUTTON RESET
OF DOUBLE-COIL SOLENOID VALVE

13

PRESSURE CONTROLLER ACTING THROUGH
SERVO-VALVE ACTING AS VOLTAGE-TO-
HYDRAULIC CONVERTER TO MODULATE
DOUBLE-ACTING-CYLINDER-ACTUATED
CONTROL VALVE HAVING POSITIONING
SYSTEM

14

SAFETY PURGING
(Z TYPE)

6.10 EXAMPLES-MISCELLANEOUS COMBINATIONS (Contd.)

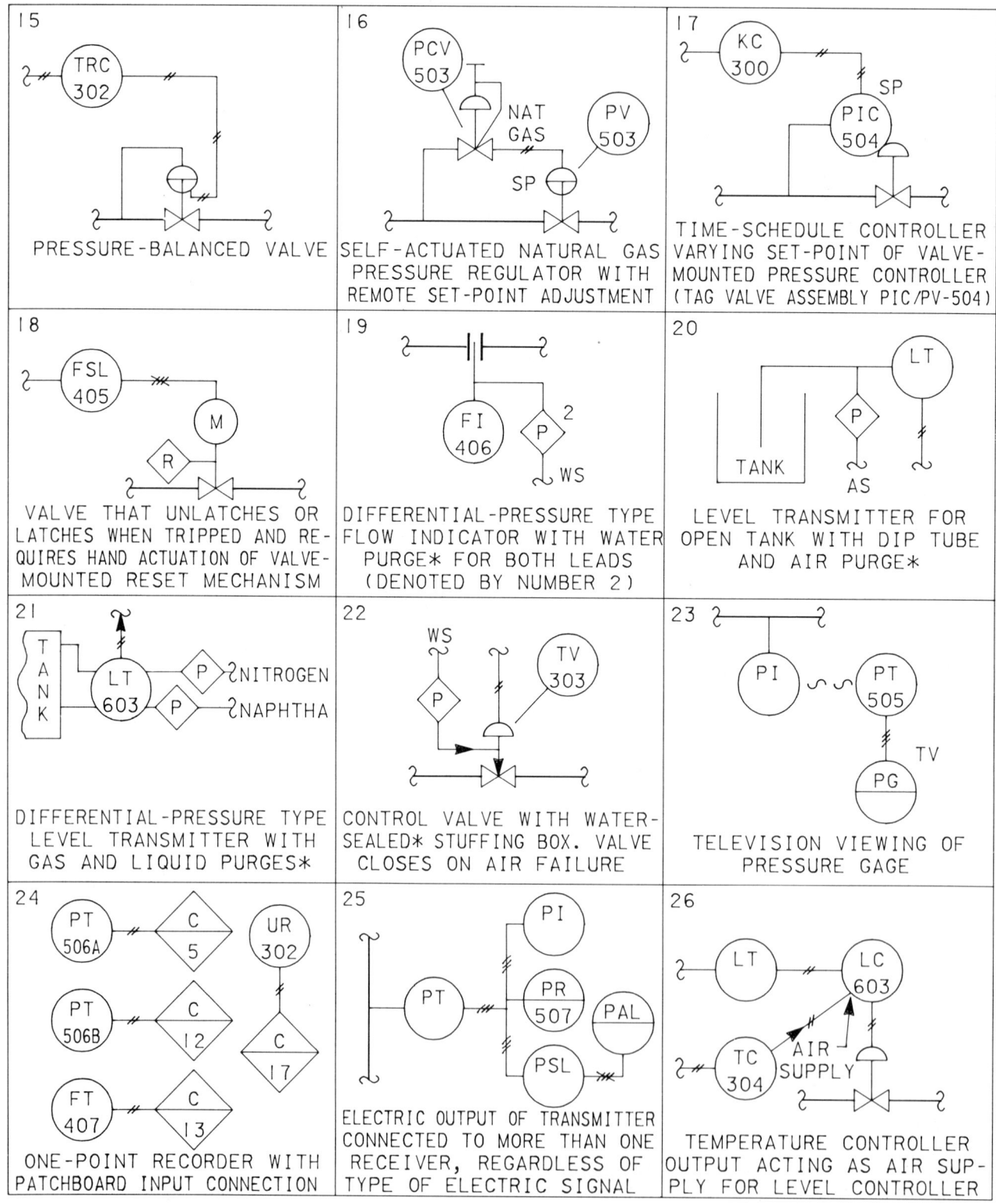

15 PRESSURE-BALANCED VALVE

16 SELF-ACTUATED NATURAL GAS PRESSURE REGULATOR WITH REMOTE SET-POINT ADJUSTMENT

17 TIME-SCHEDULE CONTROLLER VARYING SET-POINT OF VALVE-MOUNTED PRESSURE CONTROLLER (TAG VALVE ASSEMBLY PIC/PV-504)

18 VALVE THAT UNLATCHES OR LATCHES WHEN TRIPPED AND REQUIRES HAND ACTUATION OF VALVE-MOUNTED RESET MECHANISM

19 DIFFERENTIAL-PRESSURE TYPE FLOW INDICATOR WITH WATER PURGE* FOR BOTH LEADS (DENOTED BY NUMBER 2)

20 LEVEL TRANSMITTER FOR OPEN TANK WITH DIP TUBE AND AIR PURGE*

21 DIFFERENTIAL-PRESSURE TYPE LEVEL TRANSMITTER WITH GAS AND LIQUID PURGES*

22 CONTROL VALVE WITH WATER-SEALED* STUFFING BOX. VALVE CLOSES ON AIR FAILURE

23 TELEVISION VIEWING OF PRESSURE GAGE

24 ONE-POINT RECORDER WITH PATCHBOARD INPUT CONNECTION

25 ELECTRIC OUTPUT OF TRANSMITTER CONNECTED TO MORE THAN ONE RECEIVER, REGARDLESS OF TYPE OF ELECTRIC SIGNAL

26 TEMPERATURE CONTROLLER OUTPUT ACTING AS AIR SUPPLY FOR LEVEL CONTROLLER

* The purge fluid supplies may use the same abbreviations as the instrument power supplies.

6.10 EXAMPLES-MISCELLANEOUS COMBINATIONS (Contd.)

27

ALTERNATIVE 1

28

ALTERNATIVE 2

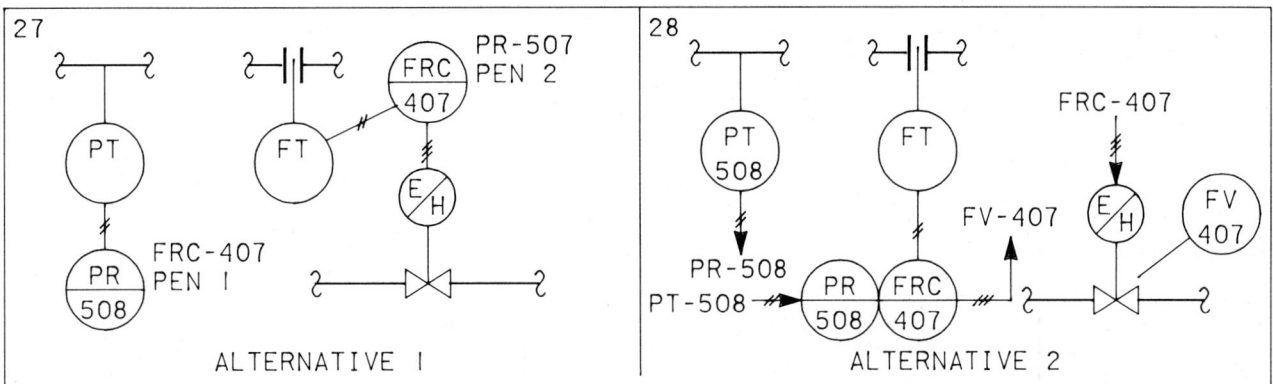

INTERRELATED INSTRUMENTS WHOSE SYMBOLS ARE SCATTERED ON DIAGRAM. (A MULTI-POINT INSTRUMENT, SUCH AS A DATA LOGGER, THAT IS DESIGNATED WITH POINT NUMBERS ON A DIAGRAM IS NOT EXPECTED TO HAVE THE SYMBOLS FOR THE VARIOUS POINTS TIED OR REFERENCED TOGETHER). PEN ASSIGNMENTS NEED NOT BE SHOWN ON A DIAGRAM IF IT IS THE USER'S PREFERENCE TO SHOW THEM IN AN INDEX

29

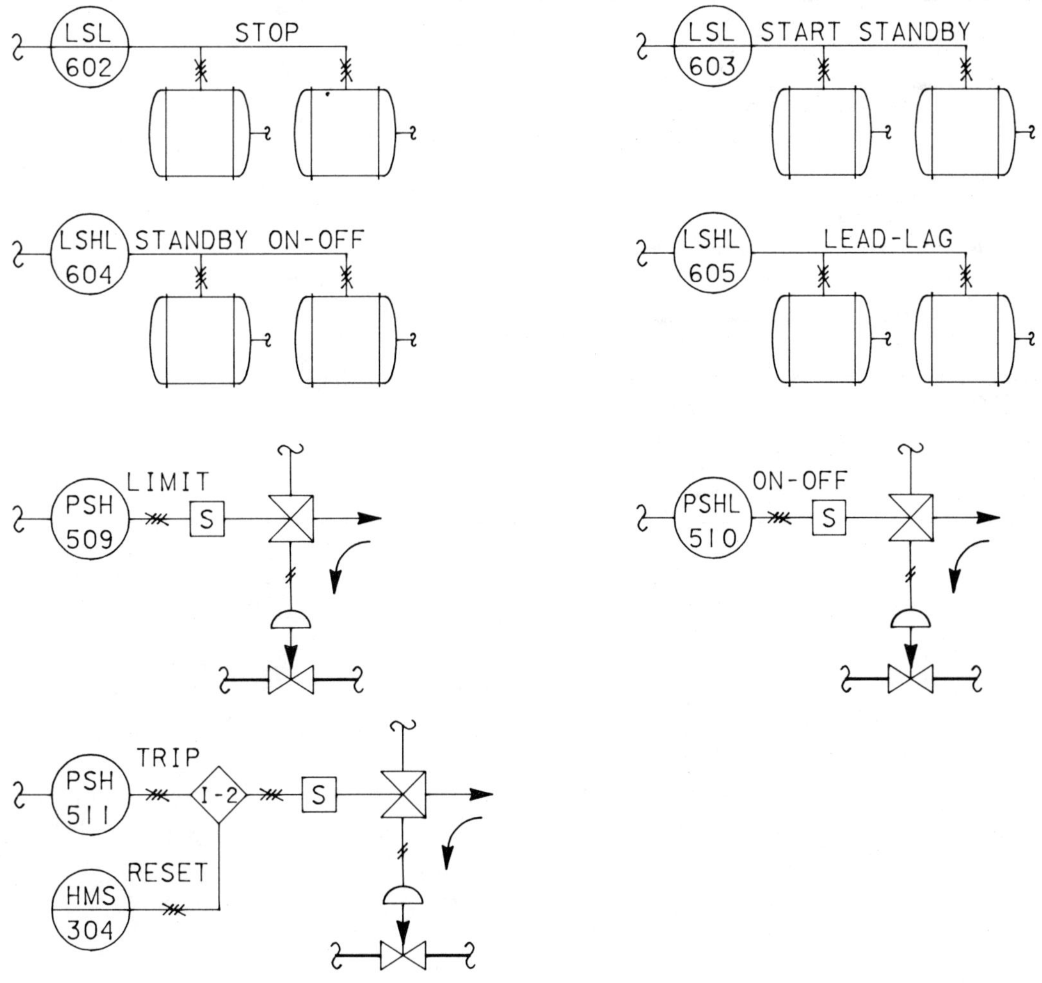

THE JUDICIOUS USE OF WORDS CAN CLARIFY DESIGN INTENT

6.10 EXAMPLES-MISCELLANEOUS COMBINATIONS (Contd.)

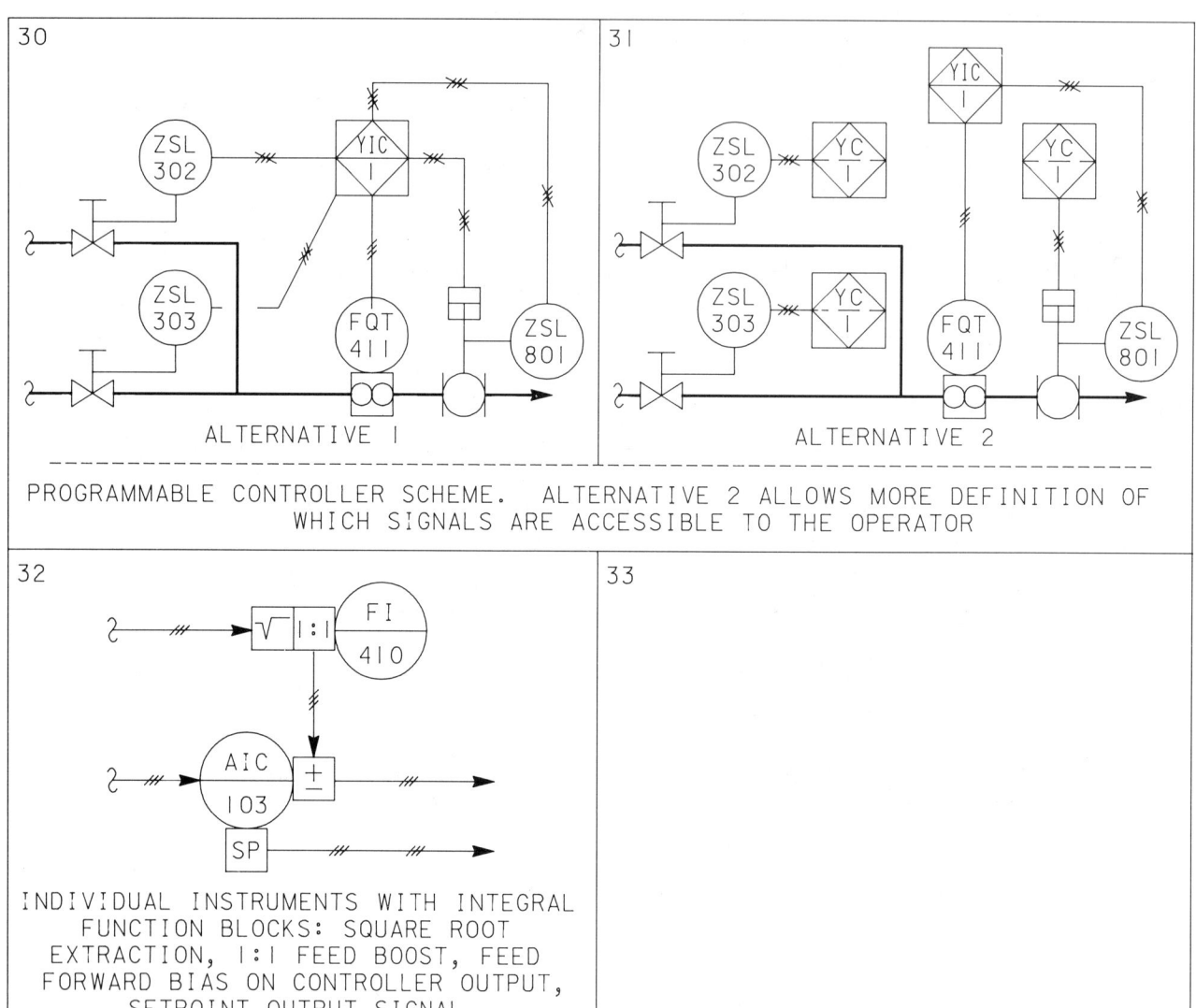

30 ALTERNATIVE 1

31 ALTERNATIVE 2

PROGRAMMABLE CONTROLLER SCHEME.   ALTERNATIVE 2 ALLOWS MORE DEFINITION OF WHICH SIGNALS ARE ACCESSIBLE TO THE OPERATOR

32

INDIVIDUAL INSTRUMENTS WITH INTEGRAL FUNCTION BLOCKS: SQUARE ROOT EXTRACTION, 1:1 FEED BOOST, FEED FORWARD BIAS ON CONTROLLER OUTPUT, SETPOINT OUTPUT SIGNAL

33

6.11 EXAMPLE - COMPLEX COMBINATIONS

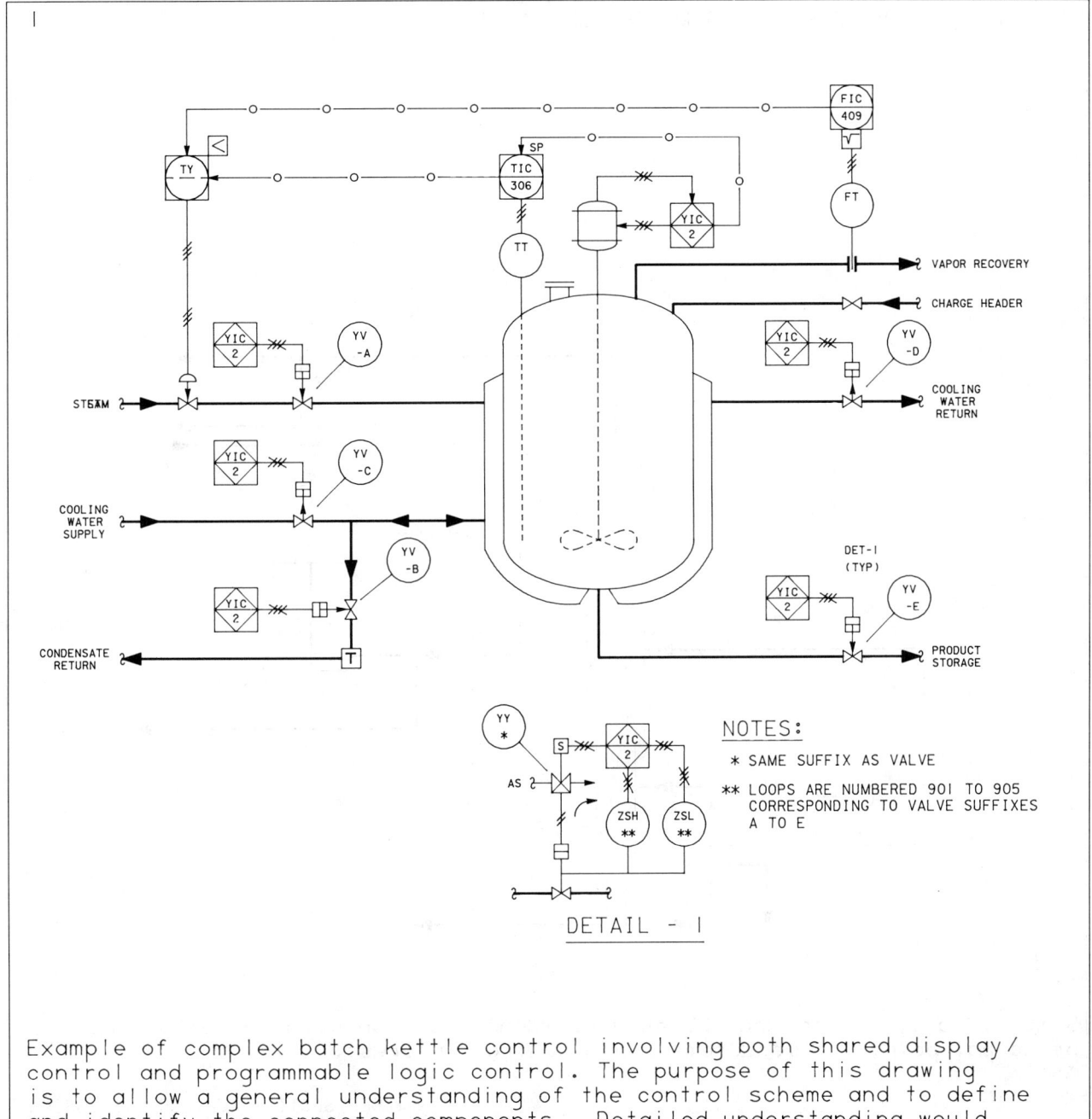

NOTES:

* SAME SUFFIX AS VALVE

** LOOPS ARE NUMBERED 901 TO 905
   CORRESPONDING TO VALVE SUFFIXES
   A TO E

DETAIL - 1

Example of complex batch kettle control involving both shared display/
control and programmable logic control. The purpose of this drawing
is to allow a general understanding of the control scheme and to define
and identify the connected components. Detailed understanding would
be obtained from the study of other documents.

6.12 EXAMPLE - DEGREE OF DETAIL *

I TYPICAL SYMBOLISM FOR SIMPLIFIED DIAGRAMS

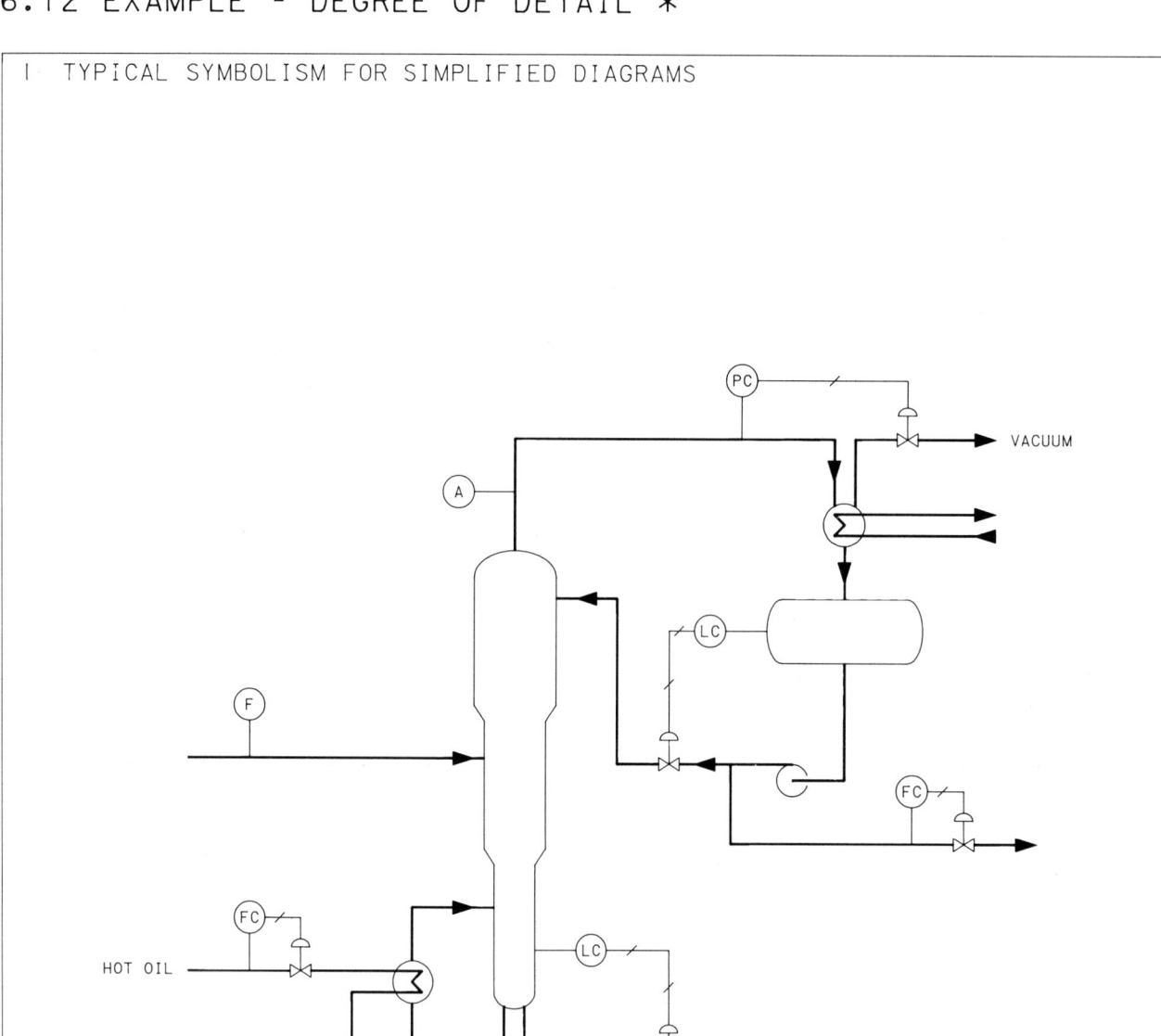

Simplified symbolism and abbreviated identification used to define the principal points of measurement and control interest.

* SEE SECTION 4.4 FOR DISCUSSION

6.12 EXAMPLE - DEGREE OF DETAIL *

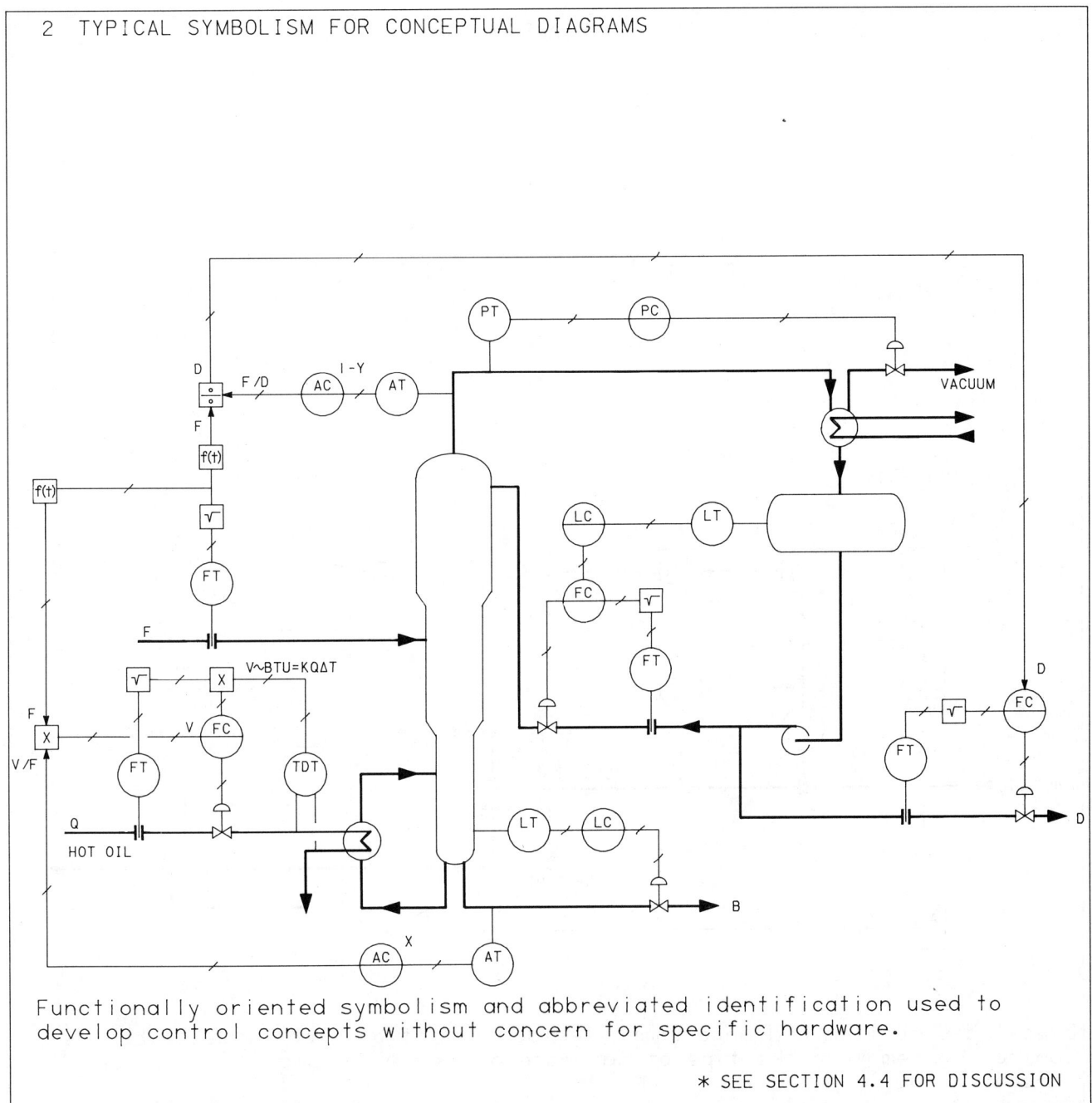

2   TYPICAL SYMBOLISM FOR CONCEPTUAL DIAGRAMS

Functionally oriented symbolism and abbreviated identification used to develop control concepts without concern for specific hardware.

* SEE SECTION 4.4 FOR DISCUSSION

6.12 EXAMPLE - DEGREE OF DETAIL *

3   TYPICAL SYMBOLISM FOR DETAILED DIAGRAMS

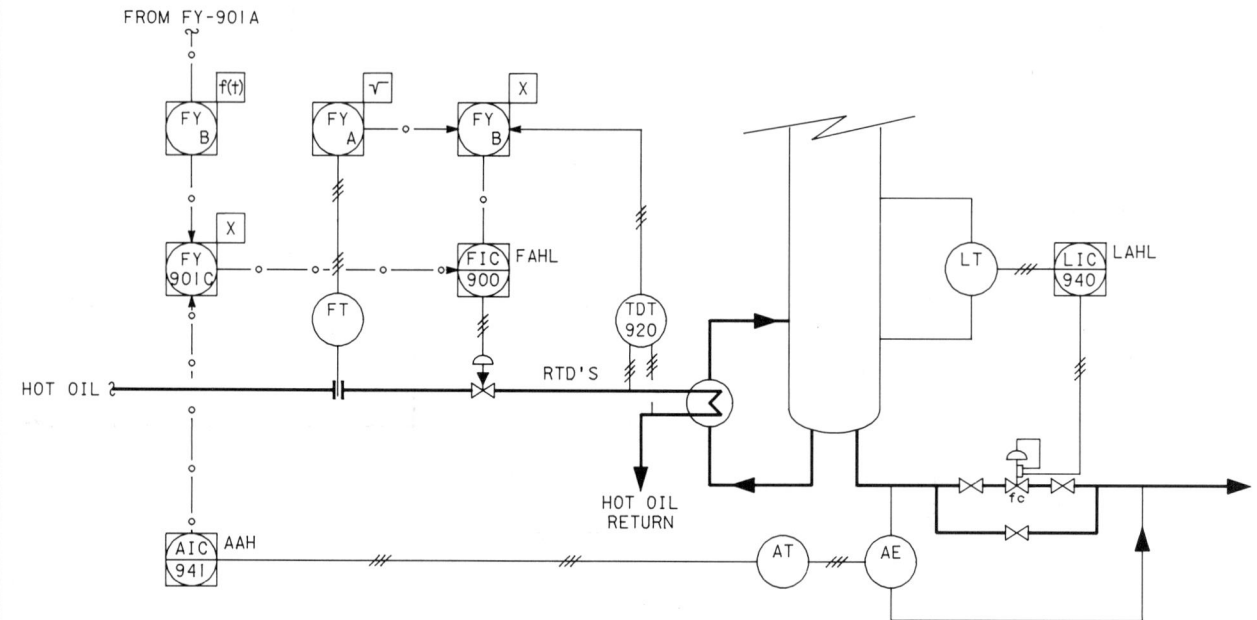

Detailed symbolism and more complete identification used to describe the control system when the type of hardware and kinds of signals have been chosen.

* SEE SECTION 4.4 FOR DISCUSSION

INDEX

INSTRUMENT SOCIETY of AMERICA
Research Triangle Park, North Carolina

ANSI/ISA-S5.2-1976
(R1981)
Approved October 9, 1981

American National Standard

Binary Logic Diagrams For Process Operations

Instrument Society of America

ISBN 0-87664-331-4

ISA-S5.2   Binary Logic Diagrams for
            Process Operations

INSTRUMENT SOCIETY OF AMERICA
67 Alexander Drive
P.O. Box 12277
Research Triangle Park, North Carolina 27709

PREFACE

This preface is included for informational purposes and is not part of Standard ISA-S5.2.

This Standard has been prepared as a part of the service of the Instrument Society of America toward a goal of uniformity in the field of instrumentation. To be of real value, this document should not be static, but should be subject to periodic review. Toward this end, the Society welcomes all comments and criticisms, and asks that they be addressed to the Secretary, Standards and Practices Board, Instrument Society of America, 67 Alexander Drive, P.O. Box 12277, Research Triangle Park, NC 27709, Telephone (919) 549-8411.

The ISA Standards and Practices Department is aware of the growing need for attention to the metric system of units in general, and the International System of Units (SI) in particular, in the preparation of instrumentation standards. The Department is further aware of the benefits to USA users of ISA Standards of incorporating suitable references to the SI (and the metric system) in their business and professional dealings with other countries. Toward this end this Department will endeavor to introduce SI — acceptable metric units in all new and revised standards to the greatest extent possible. The Metric Practice Guide, which has been published by the American Society for Testing and Materials as ANSI designation Z210.1 (ASTM E380-76, IEEE Std. 286-1975), and further revisions, will be the reference guide for definitions, symbols, abbreviation, and conversion factors.

It is the Policy of the Instrument Society of America to encourage and welcome the participation of all concerned individuals and interests in the development of ISA Standards. Participation in the ISA Standards making process by an individual in no way constitutes endorsement by the employer of that individual of the Instrument Society of America or any of the Standards which ISA develops.

The system described in this Standard is intended to meet the needs of people who are concerned with the operation of process systems. The guide for the Standard was American National Standards Institute (ANSI) Standard Y32.14.1973, Graphic Symbols for Logic Diagrams, which the committee attempted to follow so far as practical for the intended users of the ISA Standard.

The Committee also referred to National Electric Manufacturers Association Standards ICS 1-102, Graphic Symbols for Logic Diagrams, whose symbols bear resemblance to those of the ANSI Standard, and ICS 1-103, Static Switching Control Devices, which may eventually be supplanted by ICS 1-102. Reference was also made to National Fluid Power Association Recommended Standard T.3.7.68.2, Graphic Symbols for Fluidic Devices and Circuits. In addition, numerous other industrial standards were reviewed.

The following people served on the 1976 SP5.2 Committee:

NAME	COMPANY
George Platt, Chairman	Bechtel Power Company
Edward J. Blahut	Procon Incorporated, Pacific Operations
Sanford Chalfin	Fluor Corporation
Louis Costea	Hunt-Wesson Foods, Incorporated
Russell C. Greer	Bailey Meter Company
Roy Lazar	Carnation Company
Frank Mehle (deceased)	Procon Incorporated, Pacific Operation
Gary L. Pierce	Shell Oil Company
Chuck Simms	Fisher Controls Company
John Vance	United Process Control Systems
Robert Woo	Los Angeles Department of Water & Power

The following people served on the 1981 (reaffirmation) SP5.2 Committee:

NAME	COMPANY
Russell C. Greer, Chairman	Bailey Controls Company
Edward J. Blahut	Blahut Engineering, Inc.
Sanford Chalfin	Fluor Corporation (retired)
Louis Costea	Hunt-Wesson Foods, Inc.
Roy Lazar	Carnation Company
Gary L. Pierce	Shell Oil Company
George Platt	Bechtel Power Corperation
Charles Simms	Fisher Controls Company
John Vance	Comsysco
Robert Woo	Los Angeles Department of Water & Power

The 1976 edition of this Standard was approved by the ISA Standards and Practices Board in November, 1975.

NAME	COMPANY
W. B. Miller, Chairman	Moore Products Company
P. Bliss	Pratt & Whitney Aircraft Company
E. J. Byrne	Brown & Root, Inc.
W. Calder	The Foxboro Company
B. A. Christensen	Continental Oil Company
L. N. Combs	Retired from E. I. duPont deNemours & Company
R. L. Galley	Bechtel Power Corporation
R. G. Hand, Secretary	Instrument Society of America
T. J. Harrison	IBM Corporation
T. S. Imsland	Fisher Controls Company
P. S. Lederer	National Bureau of Standards
O. P. Lovett, Jr.	E. I. duPont deNemours & Company
E. C. Magison	Honeywell, Inc.
R. L. Martin	Tex-A-Mation Engineering Inc.
A. P. McCauley	Glidden-Durkee Div. SCM Corporation
T. A. Murphy	The Fluor Corporation, Ltd.
R. L. Nickens	Reynolds Metals Company
G. Platt	Bechtel Power Corporation
A. T. Upfold	Polysar Ltd.
K. A. Whitman	Allied Chemical Corporation

This standard was reaffirmed by the ISA Standards and Practices Board in March, 1981.

NAME	COMPANY
T. J. Harrison, Chairman	IBM Corporation
P. Bliss	Consultant
W. Calder	The Foxboro Company
B. A. Christensen	Continental Oil Company
M. R. Gorden-Clark	Scott Paper Company
R. T. Jones	Philadelphia Electric Company
R. Keller	Boeing Company
O. P. Lovett, Jr.	Jorden Valve
E. C. Magison	Honeywell, Inc.
A. P. McCauley	Diamond Shamrock Corporation
E. M. Nesvig	ERDCO Engineering Corporation
R. L. Nickens	Reynolds Metals Company
G. Platt	Bechtel Power Corporation
R. Prescott	Moore Products Company
R. W. Signor	General Electric Company
W. C. Weidman	Gilbert Associates, Inc.
K. A. Whitman	Allied Chemical Corporation
*L. N. Combs	
*R. L. Galley	
*R. G. Marvin	
*W. B. Miller	Moore Products Company
*J. R. Williams	Stearns-Rogers, Inc.

*Director Emeritus

CONTENTS

1. PURPOSE

1.1 The purpose of this Standard is to provide a method of logic diagramming of binary interlock and sequencing systems for the startup, operation, alarm, and shutdown of equipment and processes in the chemical, petroleum, power generation, air conditioning, metal refining, and numerous other industries.

1.2 The Standard is intended to facilitate the understanding of the operation of binary systems, and to improve communications among technical, management, design, operating, and maintenance personnel concerned with the systems.

2. SCOPE

2.1 The Standard provides symbols, both basic and non-basic, for binary operating functions. The use of symbols in typical systems is illustrated in appendices.

2.2 The Standard is intended to symbolize the binary operating functions of a system in a manner that can be applied to any class of hardware, whether it be electronic, electrical, fluidic, pneumatic, hydraulic, mechanical, manual, optical, or other.

3. USE OF SYMBOLS

3.1 By using the symbols designated as "basic," logic systems may be described with the use of only the most fundamental logic building blocks. The remaining symbols, not basic, are more comprehensive and enable logic systems to be diagrammed more concisely. Use of the non-basic symbols is optional.

3.2 A logic diagram may be more or less detailed depending on its intended use. The amount of detail in a logic diagram depends on the degree of refinement of the logic and on whether auxiliary, essentially non-logic, information is included.

As an example of refinement of detail: A logic system may have two opposing inputs, e.g., a command to open and a command to close, which do not normally exist simultaneously; the logic diagram may or may not go so far as to specify the outcome if both the commands were to exist at the same time. In addition, explanatory notes may be added to the diagram to record the logic rationale.

Non-logic information may also be added, if desired, e.g., reference document identification, tag numbers, terminal markings, etc.

In these ways, the diagram may provide the level of detail appropriate, for example, for communication between a designer of pneumatic circuits and a designer of electric circuits, or may provide a broad-view system-description for a plant manager.

3.3 The existence of a logic signal may correspond physically to either the existence or the non-existence of an instrument signal, depending on the particular type of hardware system and the circuit design philosophy that are selected.* For example, a high-flow alarm may be chosen to be actuated by an electric switch whose contacts open on high flow; on the other hand, the high-flow alarm may be designed to be actuated by an electric switch whose contacts close on high flow. Thus, the high-flow condition may be represented physically by the absence of an electric signal or by the presence of the electric signal. The Standard does not attempt to relate the logic signal to an instrument signal of any specific kind.

3.4 A logic symbol that is shown in Section 4 with three inputs — A, B, and C — is typical for the logic function having any number of two or more inputs.

3.5 The flow of intelligence is represented by lines that interconnect logic statements. The normal direction of flow is from left to right, or top to bottom. Arrowheads may be added to the flow lines wherever needed for clarity, and shall be added to lines whose flow is not in a normal direction.

3.6 A summary of the status of an operating system may be put in the diagram wherever it is deemed useful as a reference point or landmark in the sequence.

3.7 There may be misunderstanding of binary logic statements involving devices that are not recognizable as inherently having only two specific alternative states. For example, if it is stated that a valve is not closed, this could mean either (a) that the valve is open fully, or (b) that the valve is simply not closed, namely, that it may be in any position from almost closed to wide open. To aid accurate communication between writer and reader of the logic diagram, the diagram should be interpreted literally. Therefore, possibility (b) is the correct one.

If a valve is an open-close valve, then, to avoid misunderstanding, it is necessary to do one of the following:

1. Develop the logic diagram in such a way that it says exactly what is intended. If the valve is intended to be open, then it should be so stated and not be stated as being not closed.

2. Have a separate note specifying that the valve always assumes either the closed or the open position.

By contrast, a device such as a motor-driven pump is either operating or stopped, barring some special situations. To say that the pump is not operating usually clearly denotes that it has stopped.

---

*In process operations, binary instrument signals are commonly either *ON* or *OFF*. However, as a more general case, logic systems exist that make use of binary hardware having signals with two alternate real values, e.g., +5 volts and −3 volts. In *positive logic*, the more positive signal, +5 volts, represents the existence of a logic condition, e.g., *pump stopped*. In *negative logic*, the less positive signal, −3 volts, represents the existence of a logic condition of *pump stopped*.

The following definitions apply to devices that have open, closed, or intermediate positions. The positions stated are nominal to the extent that there are differential-gap and dead band in the instrument that senses the position of the device.

Open position: a position that is 100-percent open.

Not-open position: a position that is less than 100-percent open. A device that is not open may or may not be closed.

Closed position: a position that is zero-percent open.

Not-closed position: a position that is more than zero-percent open. A device that is not closed may or may not be open.

Intermediate position: a SPECIFIED position that is greater than zero- and less than 100-percent open.

Not-at-intermediate position: a position that is either above or below the SPECIFIED intermediate position.

For a logic system having an input statement that is derived inferentially or indirectly, a condition may arise that will lead to an erroneous conclusion. For example, an assumption that flow exists because a pump motor is energized may be false because of a closed valve, a broken shaft, or other mishap. Factual statements, that is, statements based on positive measurements that a certain condition specifically exists or does not exist, are generally more reliable.

3.8 A process operation may be affected by loss of the power supply* to memories and to other logic elements. In order to take such operating eventualities into account, it may therefore be necessary to consider the effect of loss of power to any logic component or to the entire logic system. In such cases, it may be necessary to

---

*The term *power supply* covers the energizing medium, whether it be electric, pneumatic, or other.

enter power supply or loss of power supply as logic inputs to a system or to individual logic elements. For memories, the consideration of power supply may be handled in this manner or as shown in Sections 4.7b, c, and d.

By the same token, it may be necessary to consider the effect of restoration of power supply.

Logic diagrams do not necessarily have to cover the effect of logic power supplies on process systems but may do so for thoroughness.

3.9 It is recommended, for clarity, that a single time-function symbol, as appropriate, be used to represent each time function in its entirety. Though not incorrect, the representation of a complex or uncommon time function by using a time-function symbol in immediate sequence with a second time-function symbol or with a *NOT* symbol should be avoided (see Section 4.8).

3.10 Process instrument symbols and designations follow ISA Standard S5.1-1973 (American National Standards Institute Standard Y32.20-1975), "Instrumentation Symbols and Designations." However, these symbols are included for illustrative purposes, only, and are not part of Standard S5.2.

3.11 If a drawing, or set of drawings, uses graphic symbols that are similar or identical to one another in shape or configuration and that have different meanings because they are taken from different standards, then adequate steps shall be taken to avoid misinterpretation of the symbols used. These steps may be to use caution notes or reference notes, comparison charts that illustrate and define the conflicting symbols, or other suitable means. This requirement is especially critical if the graphic symbols used, being from different disciplines, represent devices, conductors, flow lines, or signals whose symbols, if misinterpreted, may result in danger to personnel or damage to equipment.

4 Symbols

The symbols for diagramming binary logic are defined as follows:

FUNCTION	SYMBOL	DEFINITION	EXAMPLE
4.1 INPUT	Statement of Input —| Alternatively: Statement of Input — ⊙ Initiating instrument or device number, if known	An input to the logic sequence	The start position of a hand switch HS-1, is actuated to provide an input to start a conveyor. Alternative diagrams: a) HS-1 Start Conveyor Manually —| b) ⊙ HS 1 Start Conveyor Manually —|
4.2 OUTPUT	|— Statement of Output Alternatively: |— Statement of Output ⊙ Operated instrument or device number, if known	An output from the logic sequence.	An output from the logic sequence commands valve HV-2 to open. Alternative diagrams: a) |— Open Valve HV-2 b) |— ⊙ HV 2 Open Valve
4.3 AND **BASIC**	A —| B —| A | C —| |— D	Logic output D exists if and only if all logic inputs A, B, and C exist.	Operate pump if suction tank level is high and discharge valve is open. Tank Level High —| A |— Operate Valve Open —| Pump
4.4 OR **BASIC**	A —| B —| OR | C —| |— D	Logic output D exists if and only if one or more of logic inputs A, B, and C exist.	Stop compressor if cooling water pressure is low or bearing temperature is high. Water Pressure Low —| OR |— Stop Bearing Temperature High —| Compressor

FUNCTION	SYMBOL	DEFINITION	EXAMPLE
4.5 QUALIFIED OR	A — B — * — D C — *Internal details represent numerical quantities (see "Definition").	Logic output D exists if and only if a specified number of logic inputs A, B, and C exist. Mathematical symbols, including the following, shall be used, as appropriate, in specifying the number: a. $=$ equal to b. $\neq$ not equal to c. $<$ less than d. $>$ greater than e. $\not<$ not less than f. $\not>$ not greater than g. $\leq$ less than or equal to [equivalent to f] h. $\geq$ greater than or equal to [equivalent to e]	a) Operate mixer if two, and only two, bins are in service. Red Bin In Service — Blue Bin In Service — = 2 — Operate Mixer White Bin In Service — Yellow Bin In Service — b) Stop reaction if at least two safety devices call for stop. Device #1 Actuated — Device #2 Actuated — Device #3 Actuated — $\not< 2$ — Stop Reaction Device #4 Actuated — Device #5 Actuated — c) Operate materials feeder if at least one and no more than two mills are in service. Mill #1 In Service — >1 Mill #2 In Service — and — Operate Feeder Mill #3 In Service — $\not> 2$
4.6 NOT BASIC	A — ◯ — B **The *NOT* symbol may be drawn tangent to an adjacent logic symbol.**	Logic output B exists if and only if logic input A does not exist.	Shut off fuel gas if burners no. 1 and no. 2 are not on. Burner No. 1 On — ◯ Burner No. 2 On — ◯ — A — Shut Off Fuel Gas Some Alternatives: Burner No. 1 On — ◯ — A — Shut Off Fuel Gas Burner No. 2 On — ◯ — Shut Off Fuel Gas Burner No. 1 On — Burner No. 2 On — OR — ◯ — Shut Off Fuel Gas

FUNCTION	SYMBOL	DEFINITION	EXAMPLE

4.7 MEMORY (Flip-Flop) **BASIC**	a) A ─── S ─── C B ─── R ─── D*
	
	 *Output D shall not be shown if it is not used.
		S represents *set memory* and R represents *reset memory*.
		
		    Logic output C exists as soon as logic input A exists. C continues to exist, regardless of the subsequent state of A, until the memory is reset, i.e., terminated by logic input B existing. C remains terminated regardless of the subsequent state of B, until A causes the memory to be set.
		
		    Logic output D, if used, exists when C does not exist, and D does not exist when C exists.
		
		 Input-Override Option
		    If inputs A and B exist simultaneously, and if it is desired to have A override B, then S should be encircled, i.e., Ⓢ ; if B is to override A, then R should be encircled, i.e., Ⓡ.
		
		 Loss-Of-Power-Supply Option
		    The unmodified letter S denotes that *no consideration* has been given to the action of the memory on loss of the logic power supply. See paragraphs 4.7 b, c, and d, below, and 3.8.
			If tank pressure becomes high, vent tank and continue venting, regardless of pressure, until venting is stopped by manual actuation of hand switch, HS-1, provided that the pressure is not high. If the venting is stopped, a compressor may be started.

	b) A ─── LS ─── C B ─── R ─── D (See Appendix C)	Similar to definition of symbol (a) except that the memory shall be *lost* in the event of loss of the logic power supply.	If feed begins to flow, the cooler shall operate until the feed tank is empty. In the event of loss of the logic power supply, the cooler shall not operate
	c) A ─── MS ─── C B ─── R ─── D	Similar to definition of symbol (a) except that the memory shall be *maintained* in the event of loss of the logic power supply.	If standby pump operation is initiated, the pump shall operate, even on loss of the logic power supply, until the process sequence is terminated. The pump shall operate if start and stop commands exist simultaneously.

Examples (diagrams):

Tank Pressure High ─── Ⓢ / R ─── Vent Tank

HS-1 ─── (to R) ─── Permit Compressor Start

Feed Flowing ─── LS / R ─── Operate Cooler
Feed Tank Empty ───

Standby Pump Operation Initiated ─── Ⓜ︎Ⓢ / R ─── Operate Standby Pump
Process Sequence Terminated ───

(cont'd)

FUNCTION	SYMBOL	DEFINITION	EXAMPLE
4.7 (cont'd)	d) A —[NS]— C, B —[R]— D	Similar to definition of symbol (a) except that *after consideration* it is deemed *not significant*, so far as the process is concerned, whether the memory is maintained or lost in the event of loss of power supply.	If reservoir level is low, operate fill pump until either level is high or water quality is unsatisfactory. It is not significant to the process what happens to the pump on loss of the logic supply. If start and stop commands are simultaneous, the pump shall stop. Reservoir Level Low —[NS]—(R)— Operate Fill Pump; Reservoir Level High, Water Quality Unsatisfactory —[OR]
4.8 TIME ELEMENT	a) A —[*]— B *For functional details, see the following (also see Section 3.9):	Logic output B exists with a time relationship to logic input A as specified.	
BASIC	b) A —[DI t]— B (Delay Initiation of output)	The continuous existence of logic input A for time t causes logic output B to exist when t expires. B terminates when A terminates.	If reactor temperature exceeds a high limit continuously for 10 seconds, block catalyst flow. Resume flow when temperature does not exceed the limit. Reactor Temp. High —[DI 10 s]— Block Catalyst Flow
BASIC	c) A —[DT t]— B (Delay Termination of output)	The existence of logic input A causes logic output B to exist immediately. B terminates when A has terminated and has not again existed for time t.	If system pressure falls below a low limit, operate compressor at once. Stop the compressor when pressure is not low continuously for one minute. System Press. Low —[DT 1 min.]— Operate Compressor
BASIC	d) A —[PO t]— B (Pulse Output)	The existence of logic input A, regardless of its subsequent state, causes logic output B to exist immediately. B exists for time t and then terminates.	If vessel purge fails for any period of time, operate evacuation pump for 3 minutes and then stop the pump. Vessel Purge Fails —[PO 3 min.]— Operate Evacuation Pump

(cont'd)

FUNCTION	SYMBOL	DEFINITION	EXAMPLE
4.8 (cont'd)		A generalized method for diagramming all time functions is outlined as follows. The symbols that are defined are intended to be illustrative but are not all-inclusive.	
	e1)	Input logic state exists. Input logic state does not exist. Output logic state exists. Output logic state does not exist. The time at which the logic input A is initiated is represented by the left-hand edge of the box. Passage of time is from left to right and is usually shown unscaled. The logic output B always begins and ends in the same state within the time-element box. More than one output may be shown, if required.	
	e2)	The timing of logic may be applied to either the existence state or the non-existence state, as applicable. Output logic state exists. Output logic state does not exist.	
	f1)	The continuous existence of logic input A for time t_1 causes logic output B to exist when t_1 expires. B terminates when A terminates.	Avoid nuisance alarms on high level by actuating alarm only if level remains high continuously for 0.5 second. The alarm signal terminates when there is no high level.
(cont'd)	f2)	The continuous existence of logic input A for time t_1 causes logic output B to exist when t_1 expires. B terminates when A has been terminated continuously for time t_2.	Purge immediately with inert gas when combustibles concentration is high. Stop the purge when concentration is not high continuously for 5 minutes.

FUNCTION	SYMBOL	DEFINITION	EXAMPLE
4.8 (cont'd)	f3) A t_3 t_4 B	The termination of logic input A and its continuous non-existence for time t_3 cause logic output B to exist when t_3 expires. B terminates when either (1) B has existed for time t_4, or (2) A again exists, whichever occurs first.	Steam is turned on for 15 minutes beginning 6 minutes after agitator has stopped except that the steam shall be turned off if the agitator restarts. Agitator Operating | 6 min. | 15 min. | Steam On
	f4) A t_1 t_4 B	The existence of logic input A, regardless of its subsequent state, causes logic output B to exist when time t_1 expires. B exists for time t_4 and then terminates.*	If pressure dips to low value momentarily, block modulating control of turbine immediately, maintain for 1½ minutes, then release turbine to modulating control. Pressure Low | 0 | 1½ min. | Turbine Modulating Control Blocked
	f5) A t_1 t_4 B	The continuous existence of logic input A for time t_1 causes logic output B to exist when t_1 expires. B exists for time t_4, regardless of the state of A, and then terminates.*	If pH is low continuously for ½ minute, add caustic for 3 minutes. pH Low | ½ min. | 3 min. | Add Caustic
	f6) A t_1 t_4 B	The continuous existence of logic input A for time t_1 causes logic output B to exist when t_1 expires. B terminates when either (1) B has existed for time t_4, or (2) A terminates, whichever occurs first.*	If temperature is normal continuously for 5 minutes, add reagent for 2 minutes except that reagent shall not be added if temperature is abnormal. Temperature Normal | 5 min. | 2 min. | Add Reagant

*For symbols f4, f5, and f6, the action of logic output B depends on how long logic input A is in continuous existence, up to the line break for A. Beyond the break in A, the state of A is not significant to the completion of the B sequence. If it is desired to have a B time segment, e.g., t_1, go to completion only if A exists continuously, then A must be drawn beyond that segment. If A is drawn past the beginning but not beyond the end of a time segment, then the segment will be initiated and go to completion regardless of whether A exists only momentarily or longer.

FUNCTION	SYMBOL	DEFINITION	EXAMPLE
4.9 SPECIAL	A B Statement of Special Requirements	Logic output B exists with a relationship to logic input A as specified in the statement of special requirements. The statement may cover a logic function not otherwise specified in this standard or a logic system that is further defined elsewhere.	

12

5. BIBLIOGRAPHY

American National Standards Institute Standard Y32.14-1973, Graphic Symbols for Logic Diagrams (Two-State Devices).

American National Standards Institute Standard X3.5-1970, Flowchart Symbols and Their Usage in Information Processing.

International Electrotechnical Commission Recommendation, Publication 117-15, 1972, Binary Logic Elements.

National Electric Manufacturers Association Standard ICS 1-102, Graphic Symbols for Logic Diagrams.

National Electric Manufacturers Association Standard ICS 1-103, Static Switching Control Devices.

National Fluid Power Association Standard T.3.7.68.2, Graphic Symbols for Fluidic Devices and Circuits.

APPENDIX "A"

GENERAL APPLICATION EXAMPLE

1. INTRODUCTION

This example uses a representative process whose instruments are denoted by the symbols of ISA-S5.1-1973, (ANSI Y32.20-1975.) The process equipment symbols are included only to illustrate applications of instrumentation symbols. The example is not a part of Standard S5.2.

2. SIMPLIFIED FLOW DIAGRAM

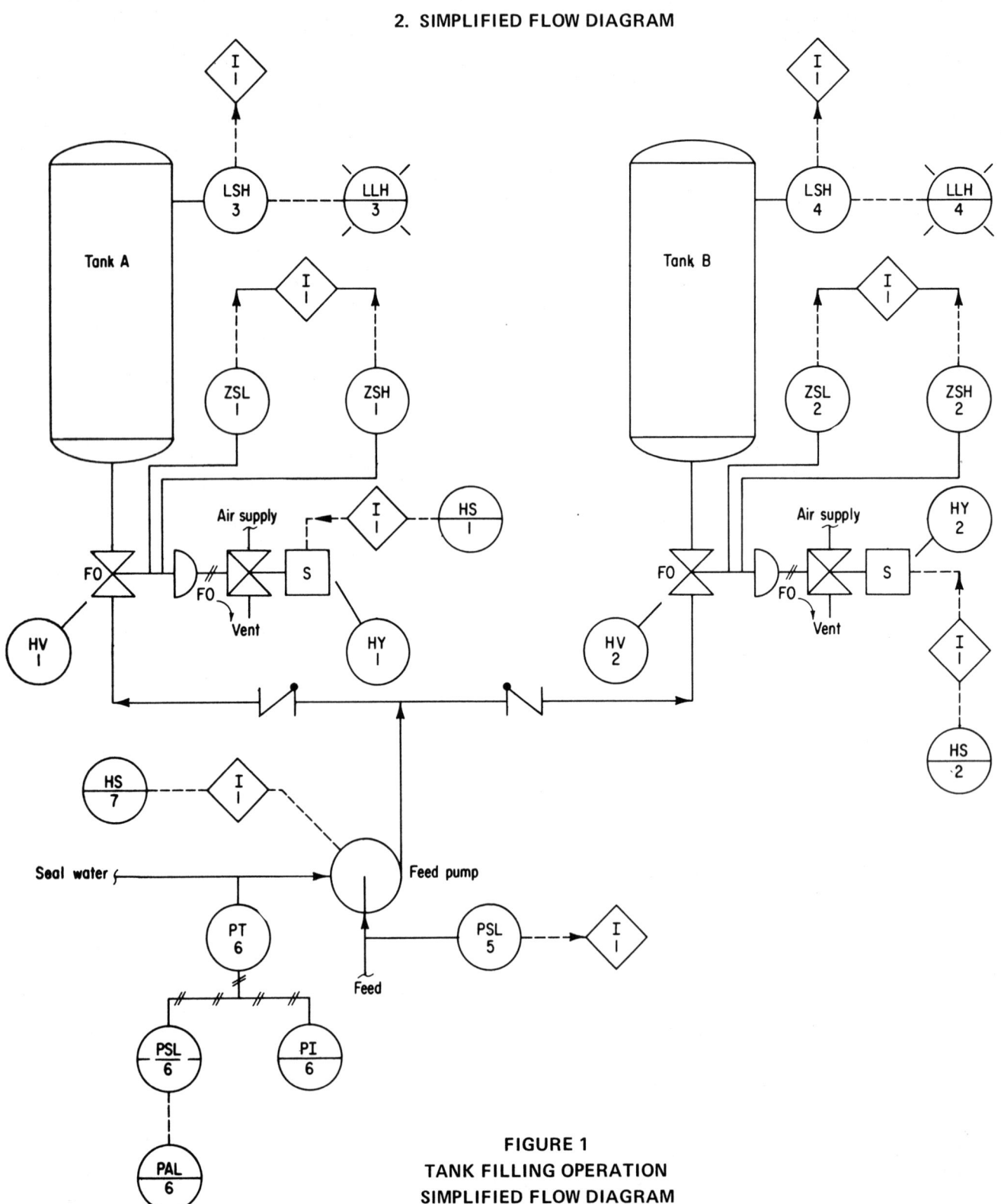

FIGURE 1
TANK FILLING OPERATION
SIMPLIFIED FLOW DIAGRAM

3. WORD DESCRIPTION

3.1 Pump Start

Feed is pumped into either tank *A* or tank *B*. The pump may be operated manually or automatically, as selected manually on a local maintained-output selector switch, *HS-7*, which has three positions: *ON*, *OFF*, and *AUTO*. When the pump is operating, red pilot light *L-8A* is on; when not operating, green pilot light *L-8B* is on. Once started, the pump continues to operate until a stopping command exists or until the control power supply is lost.

The pump may be operated manually at any time provided that no trouble condition exists: The suction pressure must not be low; the seal water pressure must not be low; and the pump motor must not be overloaded and its starter must be reset.

In order to operate the pump automatically, all the following conditions must be met:

3.1.1 Board-mounted electric momentary-contact hand switches, *HS-1* and *HS-2*, start the filling operation for tanks *A* and *B*, respectively. Each switch has two positions, *START* and *STOP*. *START* de-energizes the associated solenoid valves, *HY-1* and *HY-2*. De-energizing a solenoid valve causes it to go to the fail-safe position, i.e., to vent. This depressurizes the penumatic actuator of the associated control valves, *HV-1* and *HV-2*. Depressurizing a control valve causes it to go to the fail-safe position, i.e., to open. The control valves have associated open-position switches, *ZSH-1* and *ZSH-2*, and closed-position switches, *ZSL-1* and *ZSL-2*.

The *STOP* position of switches *HS-1* and *HS-2* causes the opposite actions to occur so that the solenoid valves are energized, the control valve actuators are pressurized, and the control valves close.

If starting circuit power is lost, the starting memory is lost and the filling operation stops. The command to stop filling can override the command to start filling.

To start the pump automatically, either control valve *HV-1* or *HV-2* must be open and the other control valve must be closed, depending on whether tank *A* or tank *B* is to be filled.

3.1.2 The pump suction pressure must be above a given value, as signalled by pressure switch *PSL-5*.

3.1.3 If valve *HV-1* is open to permit pumping into tank *A*, the tank level must be below a given value, as signalled by level switch *LSH-3*, which also actuates a board-mounted high-level pilot light, *LLH-3*. Similarly, high-level switch, *LSH-4*, permits pumping into tank *B*, if not actuated, and actuates pilot light *LLH-4*, if actuated.

3.1.4 Pump seal water pressure must be adequate, as indicated on board-mounted receiver gage, PI-6. This is a non-interlocked requirement that depends on the operator's attention before he starts the operation. Pressure switch, *PSL-6*, behind the board, actuates board-mounted low-pressure alarm, *PAL-6*.

3.1.5 The pump drive motor must not be overloaded and its starter must be reset.

3.2 Pump Stop

The pump stops if any of the following conditions exists:

3.2.1 While pumping into a tank, its control valve leaves the fully-open position, or the valve of the other tank leaves its fully-closed position, provided that the pump is on automatic control.

3.2.2 The tank selected for filling becomes full, provided that the pump is on automatic control.

3.2.3 The pump suction pressure is continuously low for 5 seconds.

3.2.4 The pump drive motor is overloaded. It is immaterial to the process logic whether or not the memory of the pump motor overload is retained on loss of power in this system because the maintained memory that operates the pump is defined as losing memory on loss of power, and this by itself will cause the pump to stop. However, an existing motor-overload condition prevents the motor starter from being reset.

3.2.5 The sequence is stopped manually through *HS-1* or *HS-2*. If stop and start commands for pump operation exist simultaneously, then the stop command overrides the operate command.

3.2.6 The pump is stopped manually by HS-7.

3.2.7 The pump seal water pressure is low. This condition is not interlocked, and requires manual intervention to stop the pump.

4. LOGIC DIAGRAM

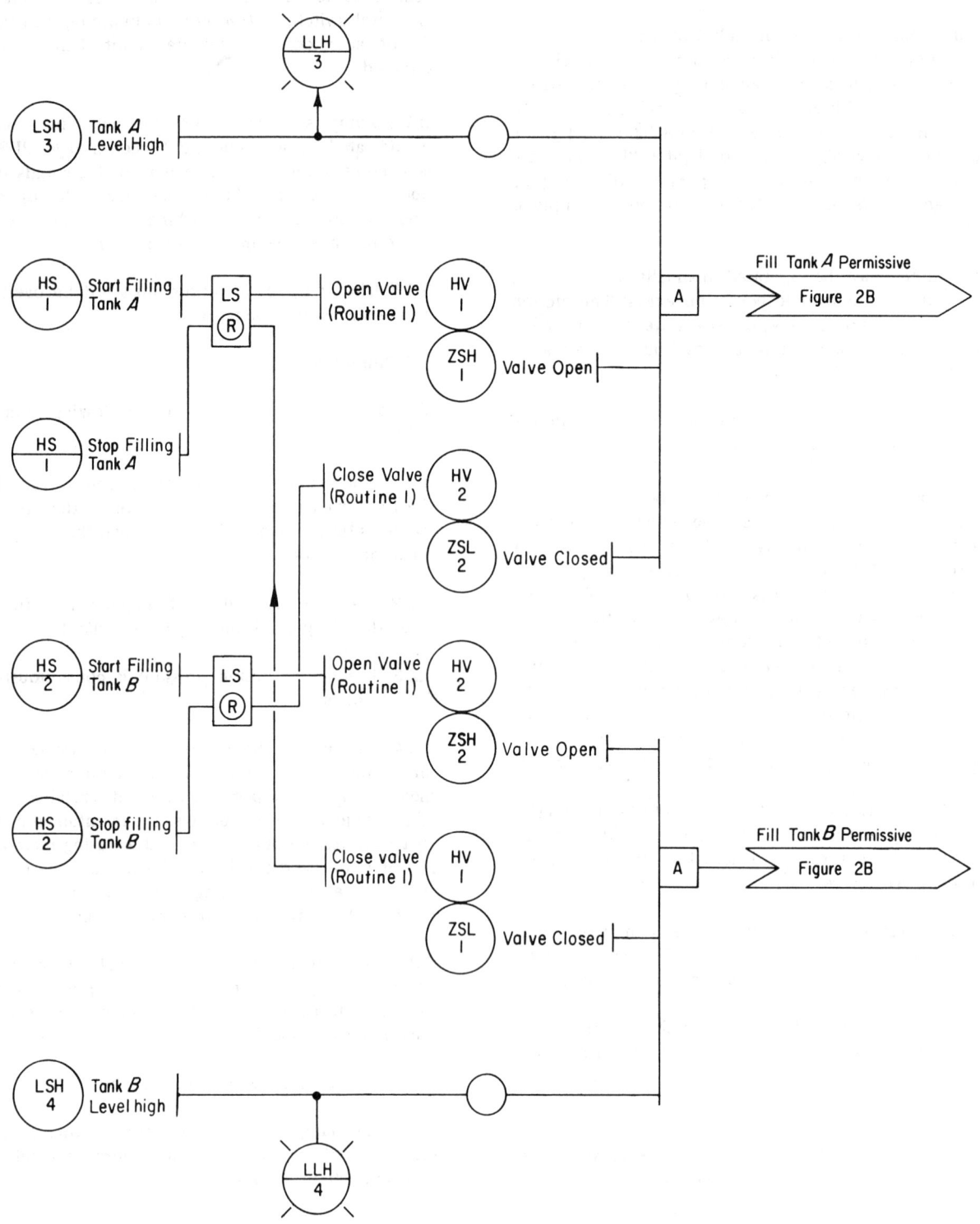

FIGURE 2A
TANK FILLING OPERATION
INTERLOCK 1 LOGIC DIAGRAM — PART I

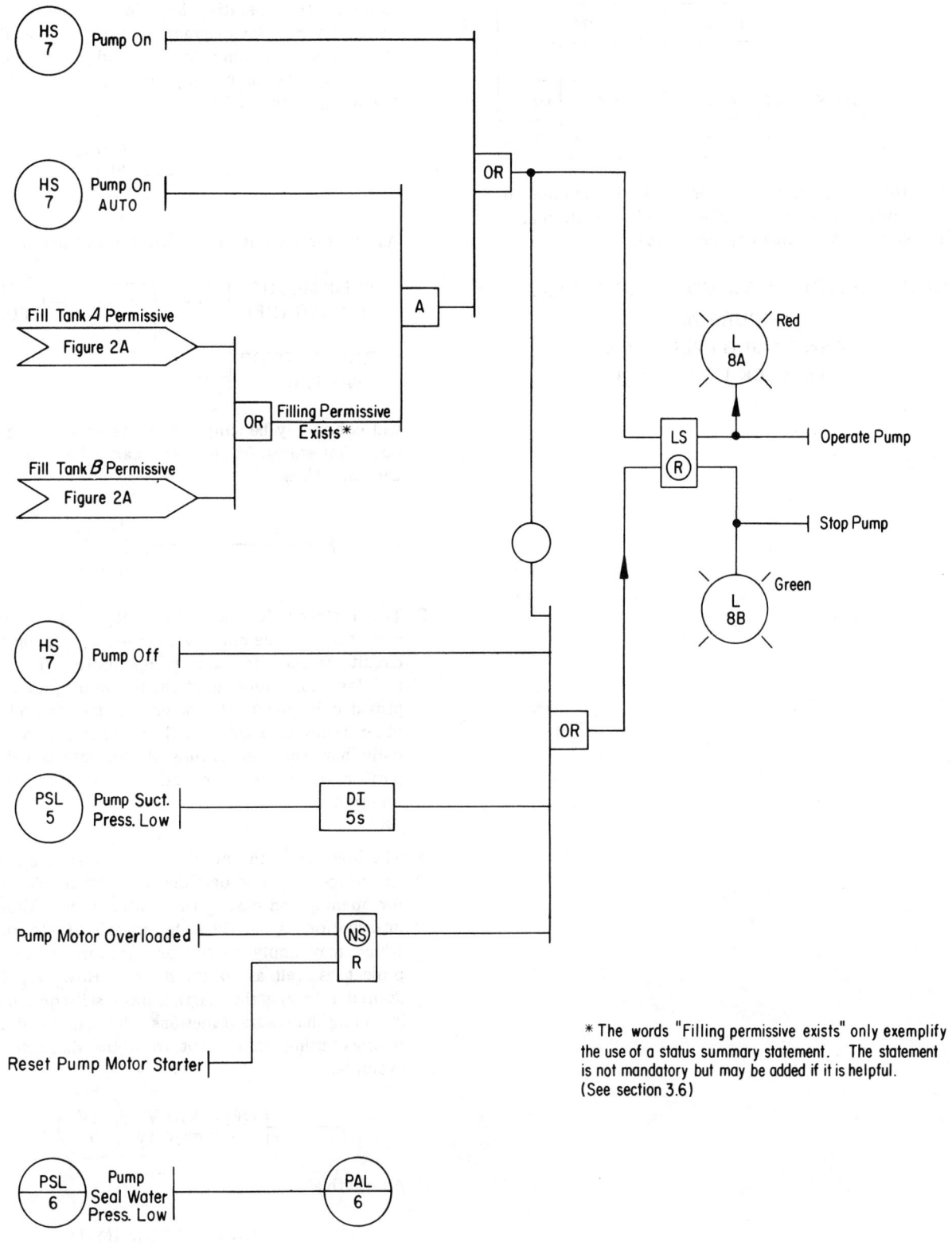

FIGURE 2B
TANK FILLING OPERATION
INTERLOCK 1 LOGIC DIAGRAM — PART II

* The words "Filling permissive exists" only exemplify the use of a status summary statement. The statement is not mandatory but may be added if it is helpful. (See section 3.6)

Solenoid Valve		Control Valve	
HY-1		HV-1	
HY-2		HV-2	
		Actuator	Port
Operation	Open Valve	De-Energized	Vented	Open
	Close Valve	Energized	Pressurized	Closed

The information stated in this figure is required if detailed design work is to be done. The information may be presented in any other convenient form.

DESCRIPTION OF VALVE ACTUATION SCHEME
FIGURE 2C
TANK FILLING OPERATION
INTERLOCK 1 ROUTINE 1

Comments on the logic diagram for Interlock 1:

1. The diagram may be simplified by using general notes (GN) for a project, especially for repetitive items. For example, the operating light for the pump may be omitted from the diagram by using a general note that states: "All pumps have red and green pilot lights to denote that the pump motors are operating or not operating, respectively," thus,

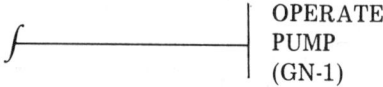

As another example, the motor lockout detail

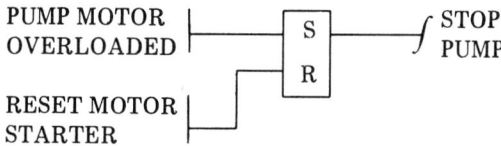

will commonly be simplified by referring to a general note that states: "The motor starter locks out when tripped," thus:

$$\text{STOP PUMP (GN-2)}$$

2. The memory function that keeps the pumps in operation may be but is not necessarily provided by a circuit breaker for the pump motor. The other maintained-memory functions in the diagram may be provided by pneumatic or electric latching relays or other types of hardware. This illustrates the essentially hardware-free nature of the operational logic portion of the diagram and the emphasis on logic function.

3. The logic diagram emphasizes the operating logic of the process by not detailing the system mechanism for opening and closing the control valves. Thus, this information is provided by means of Routine 1, which may apply to similar hardware of an entire project as well as to Interlock 1. However, if it is desired to make the diagram more self-contained by including hardware functions, this can be done as follows, using an excerpt from the diagram as an example:

OPEN VALVE (ROUTINE 1) HV 1

Alternative:

(DEENERGIZE HY-1)
(VENT HV-1)
OPEN VALVE HV 1

APPENDIX "B"
COMPLEX TIME-ELEMENT EXAMPLE

1. WORD DESCRIPTION

Assume a process operation, as follows:

If air flow becomes high and is so sustained for 4 seconds, then open vent, actuate alarm, and initiate heating by east and west heaters. If heating by east heater is initiated, the heater goes on for 2 seconds, off for one second, and on again for 4 seconds, regardless of whether the air flow remains high while this is occurring. If heating by west heater is initiated, then heater goes on for 30 seconds, off for 18 seconds, and on for 40 seconds, but only if the air flow remains high while this is occurring.

If high flow of air is sustained for 10 seconds, stop the auxiliary blower if it is running.

When air flow is no longer high, close the vent, permit the auxiliary blower to be restarted and the alarm to be reset.

2. LOGIC DIAGRAM

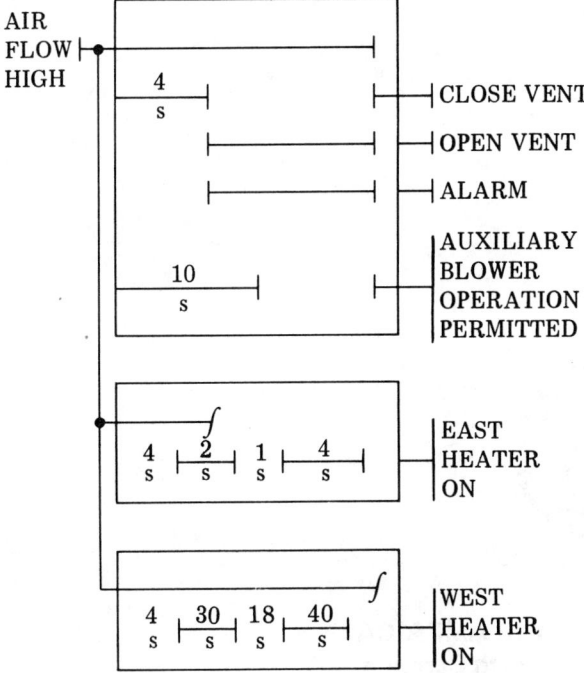

APPENDIX "C"
LOSS OF POWER SUPPLY FOR MEMORY

Section 4.7b indicates how to symbolize memories that are lost in the event of loss of power supply. The use of a logic feedback to symbolize a memory is deprecated. Thus, the following symbolisms shall not be used:

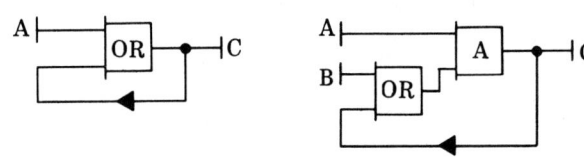

INSTRUMENT SOCIETY of AMERICA
Research Triangle Park, North Carolina

ISA-S5.3-1983

Standard

Graphic Symbols for Distributed Control/Shared Display Instrumentation, Logic and Computer Systems

Instrument Society of America

ISBN 0-87664-707-7

ISA-S5.3 Graphic Symbols for
Distributed Control/Shared Display
Instrumentation, Logic and Computer Systems

INSTRUMENT SOCIETY OF AMERICA
67 Alexander Drive
P.O. Box 12277
Research Triangle Park, North Carolina 27709

PREFACE

This preface is included for informational purposes and is not part of ISA-S5.3.

This Standard has been prepared as a part of the service of the Instrument Society of America toward a goal of uniformity in the field of instrumentation. To be of real value, this document should not be static but should be subject to periodic review. Towards this end, the Society welcomes all comments and criticisms and asks that they be addressed to the Secretary, Standards and Practices Board, Instrument Society of America, 67 Alexander Drive, P.O. Box 12277, Research Triangle Park, North Carolina 27709.

The ISA Standards and Practices Department is aware of the growing need for attention to the metric system of units in general, and the International System of Units (SI) in particular, in the preparation of instrumentation standards. The Department is further aware of the benefits to USA users of ISA Standards of incorporating suitable references to the SI (and the metric system) in their business and professional dealings with other countries. Towards this end this Department will endeavor to introduce SI and SI-acceptable metric units in all new and revised standards to the greatest extent possible. The Metric Practice Guide, which has been published by the American Society for Testing and Materials as ANSI designation Z210.1 (ASTM E380-76, IEEE Std. 268-1975), and future revisions, will be the reference guide for definitions, symbols, abbreviations, and conversion factors.

The systems referenced in this Standard are based on advances in control systems technology since the publication of ISA-S5.1, ''Instrumentation Symbols and Identification.'' During recent years, technology has evolved in terms of microprocessor-based systems presently manufactured by many companies as ''Distributed Control Systems.''

These systems may include components identified as ''computers'' as distinct from the integral processor, which derives the various functions of the system. The computer component may be integrated into the overall system, via the communication link, or it may be a stand-alone computer.

In attempting to implement these systems, the need for supplementary symbolism has become apparent.

The symbols defined in ISA-S5.3 are intended to complement those of ISA-S5.1, ''Instrumentation Symbols and Identification,'' for use on flow diagrams. In this way, the integration of distributed controllers and process computers into the more traditional instrument systems—analog, binary, and digital—can be depicted clearly on flow diagrams and other documents to give an overall and comprehensive picture of how process variables are measured and controlled.

Distributed control systems appear to be similar to each other; however, they are so diverse in philosophy that there must be a generic way to document their application.

The ISA Standards Committee on Graphic Symbols for Distributed Control/Shared Display Instrumentation, Logic, and Computer Systems, SP5.3, operates within the ISA Standards and Practices Department, Dr. Thomas J. Harrison, Vice President. The persons listed below served as members of the SP5.3 Committee.

NAME	COMPANY
D. E. Rapley, Chairman	Stearns-Roger Engineering Corporation
A. Bohnenberger, Secretary (deceased)	Johns-Manville Sales Corporation
R. Barber	Ralph M. Parsons Company
G. V. Barta	Dow Corning Corporation
G. Bennett	Bechtel Power Corporation
J. Biggs	Dow Chemical USA
R. V. Bins	Toledo Edison Company
M. A. Blaschke	Weyerhaeuser Company
C. T. Carroll	Foster Wheeler
W. Cohen	M. W. Kellogg Co.
W. M. Dillow	Gulf Science & Technology
R. E. Dragoo	Honeywell, Inc.
S. E. Gaertner	Bechtel Inc.
L. K. Haberman	Bailey Meter Company
E. Harrison	Exxon Coal USA
T. Herrera	Stone & Webster Engineering Corporation
C. Kenyon	Amoco Oil
S. Keown	Diamond Shamrock Corp.
T. H. King	Ralph M. Parsons Company
P. Kramer	Ralph M. Parsons Company
G. K. Kullberg	Dow Corning Corporation

D. G. Leonard	Aramco Services
G. Lind	Sybron Taylor
W. M. Lydecker	Carnation Co
R. E. Lynch	Bechtel Petroleum
J. E. Macko	Westinghouse
A. F. Marks	Bechtel Petroleum
R. G. Martin	A. E. Staley Manufacturing
T. C. McAvinew	International Coal Refining Co.
J. M. McHenry	Brown & Root
T. J. Myron, Jr.	Foxboro Company
J. Nevelus	Bechtel Power Corporation
R. L. Nicholson	Sohio
H. C. Prendergast	Coors Container Co.
W. A. Rock	International Engineering
F. Sandt	Pennsylvania Power & Light
R. Shearer	Stone & Webster Engineering Corp
J. W. Stuckey	Georgia Pacific Corporation
W. Su	Stearns-Roger Engineering Corporation
R. C. von Brecht	M. W. Kellogg Co.
S. Whitaker	Tex-A-Mation
J. H. Young	Bechtel Power Corporation

This Standard was approved for publication by the ISA Standards and Practices Board in June 1982.

NAME	COMPANY
T. J. Harrison, Chairman	IBM Corporation
P. Bliss	Consultant
W. Calder	The Foxboro Company
N. Conger	Continental Oil Co.
B. Feikle	Bailey Controls Co.
R. T. Jones	Philadelphia Electric Co.
R. Keller	Boeing Company
O. P. Lovett, Jr.	Isis Corp.
E. C. Magison	Honeywell, Inc.
A. P. McCauley	Diamond Shamrock Corp.
J. W. Mock	EG&G Idaho, Inc.
E. M. Nesvig	ERDCO Engineering Corp.
G. Platt	Bechtel Power Corp.
R. Prescott	Moore Products Company
W. C. Weidman	Gilbert Associates
K. A. Whitman	Allied Chemical Corp.
J. R. Williams	Stearns-Roger, Inc.
B. A. Christensen*	
L. N. Combs*	
R. L. Galley*	
R. G. Marvin*	
W. B. Miller*	Moore Products Company
R. L. Nickens*	

*Director Emeritus

TABLE OF CONTENTS

1 PURPOSE

The purpose of this standard is to establish documentation for that class of instrumentation consisting of computers, programmable controllers, minicomputers and micro-processor based systems that have shared control, shared display or other interface features. Symbols are provided for interfacing field instrumentation, control room instrumentation and other hardware to the above. Terminology is defined in the broadest generic form to describe the various categories of these devices.

It is not the intent of this standard to mandate the use of each type symbol for each occurrence of a generic device within the overall control system. Such usage could result in undue complexity in the case of a Piping and Instrument Drawing (P&ID). If, for example, a computer component is an integral part of a distributed control system, the use of the computer symbol would normally be an undesirable redundancy. If, however, a separate general purpose computer is interfaced with the system, the inclusion of the computer symbol may provide the degree of clarity needed for control system understanding.

This standard attempts to provide the users with defined symbolism and rules for usage, which may be applied as needed to provide sufficient clarity of intent. The extent to which these symbols are applied to various types of drawings remains with the users. The symbols may be as simple or complex as needed to define the process.

2 SCOPE

This standard satisfies the requirements for symbolically representing the functions of distributed control/shared display instrumentation, logic, and computer systems. The instrumentation is generally composed of field hardware communication networks and control room operator devices. This standard is applicable to all industries using process control and instrumentation systems.

No effort will be made on the flow diagram to explain the internal construction, configuration, or method of operation of this type of instrumentation, logic and computer systems. Personnel needing to understand flow diagrams must have a basic understanding of the total system in order to correctly interpret the diagram. The type of computation or the use of the process variable within a program is not indicated except in those cases where the process variable is an integral part of the control strategy. In applications where all instrument system data base information is available to the computer via the communication link, the depiction of the computer interconnections is optional in order to conserve space on flow diagrams.

2.1 Application to Work Activities

This standard is intended for use whenever any reference to an instrument is required. Such references may be required for the following uses as well as others:

Flow diagrams, process and mechanical;
Instrumentation system diagrams;
Specifications, purchase orders, manifests, and other lists;
Construction drawings;
Technical papers, literature, and discussions;
Tagging of instruments; and
Installation, operation, and maintenance instructions, drawings, and records.

2.2 Relationship to Other ISA Standards

This standard complements ISA-S5.1, "Instrumentation Symbols and Identification," for symbols and formats representing functional identification codes. For clarification of examples, a limited amount of ISA-S5.1 symbology has been included in this document.

2.3 Relationship to Other Standards

Where applicable, definitions not included in Section 3 are in accordance with ANSI X3/TR-1-77, "American National Dictionary for Information Processing," and/or ISA-S5.1.

3 DEFINITIONS AND ABBREVIATIONS

Accessible - A system feature that is viewable by and interactive with the operator, and allows the operator to perform user permissible control actions, e.g. set point changes, auto-manual transfers, or on-off actions.

Assignable - A system feature that permits an operator to channel (or direct) a signal from one device to another, without the need for changes in wiring, either by means of switches or via keyboard commands to the system.

Communication Link - Computer control is a device in which control and/or display actions are generated for use by other system devices. When used with other control devices on the communication link the computer normally performs or functions in a hierarchical relationship to the other control devices.

Computer Control System - A system in which all control action takes place within the control computer. Single or redundant computers may be used.

Configurable - A system feature that permits selection through entry of keyboard commands of the basic structure and characteristics of a device or system, such as control algorithms, display formats, or input/output terminations.

Distributed Control System - That class of instrumentation (input/output devices, control devices and operator interface devices) which in addition to executing the stated control functions also permits transmission of control, measurement, and operating information to and from a single or a plurality of user specifiable locations, connected by a communication link.

I/O - Input/Output

Shared Controller - A control device that contains a plurality of pre-programmed algorithms which are user retrievable, configurable, and connectable, and allows user defined control strategies or functions to be implemented. Control of multiple process variables can be implemented by sharing the capabilities of a single device of this kind.

Shared Display - The operator interface device used to display signals and/or data on a time shared basis. The signals and/or data, i.e. alphanumeric and/or graphic, reside in a data base from where selective accessibility for display is at the command of a user.

Software - Digital programs, procedures, rules, and associated documentation required for the operation and/or maintenance of a digital system.

Software Link - The interconnection of system components or functions via software or keyboard instruction.

Supervisory Set Point Control System - The generation of set point and/or other control information by a computer control system for use by shared control, shared display or other regulatory control devices.

4 SYMBOLS

4.1 General

Standard instrumentation symbols as shown in ISA-S5.1 are retained as much as possible for flow diagram use, but are supplemented as necessary by the new symbols in Sections 4.2 through 4.6. Symbol size should be consistent with ISA-S5.1, Section 3. The symbol descriptions listed to the right of each symbol are intended as guidelines for applications, and are not intended to be all inclusive. The symbol may be used if one or more of the descriptions apply. Shared signal lines can be expressed by the symbol for a system link (See Section 4.6.1.).

4.2 Distributed Control/Shared Display Symbols

Advances in control systems brought about by microprocessor based instrumentation permit shared functions such as display, control and signal lines. Therefore, the symbology defined here should be "Shared Instruments," which means shared display and/or shared control. The square portion of this symbol, as shown in paragraphs 4.2.1 through 4.2.3 has the meaning of shared type instrument.

4.2.1 Normally Accessible to Operator

Indicator/Controller/Recorder or Alarm Points—usually used to indicate video display.

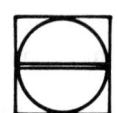

(1) Shared display.
(2) Shared display and shared control.
(3) Access limited to communication link.
(4) Operator Interface on communication link.

4.2.2 Auxiliary Operator's Interface Device

(1) Panel mounted-normally having an analog faceplate—not normally mounted on main operator console.
(2) Can be a backup controller or manual station.

(3) Access may be limited to communication link.
(4) Operator interface via the communication link.

4.2.3 Not Normally Accessible to Operator

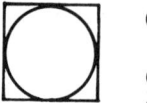

(1) Shared blind controller.
(2) Shared display installed in field.
(3) Computation, signal conditioning in shared controller.
(4) May be on communication link.
(5) Normally blind operation.
(6) May be altered by configuration

4.3 Computer Symbols

The following symbols should be used where systems include components identified as "computers," as distinct from an integral processor, which drive the various functions of a "distributed control system." The computer component may be integrated with the system via the data link, or it may be a stand-alone computer.

4.3.1 Normally Accessible to Operator

Indicator/Controller/Recorder or Alarm Point—usually used to indicate video display.

4.3.2 Not Normally Accessible to Operator

(1) Input/Output interface.
(2) Computation/Signal conditioning within a computer.
(3) May be used as a blind controller or a software calculation module.

4.4 Logic and Sequential Control Symbols

4.4.1 General Symbol - For undefined complex interconnecting logic or sequence control. (Also see ISA-S5.1).

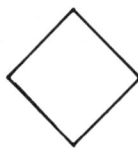

4.4.2 Distributed control interconnecting logic controller with binary or sequential logic functions.

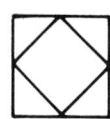

(1) Packaged programmable logic controller, or digital logic controls integral to the distributed control equipment.
(2) Not normally accessible to the operator.

4.4.3 Distributed control interconnecting logic controller with binary or sequential logic functions.

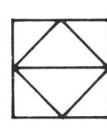

(1) Packaged programmable logic controller, or digital logic controls integral to the distributed control equipment.

(2) Normally accessible to the operator.

4.5 Internal System Function Symbols

4.5.1 Computation/Signal Conditioning

(1) For block identification refer to ISA-S5.1, Table 2 "Function Designations for Relays."

(2) For extensive computational requirements, use designation "C". Explain on supplementary documentation.

(3) Used in conjunction with function relay bubbles per ISA-S5.1.

4.6 Common Symbols

4.6.1 System Link

(1) Used to indicate either a software link or manufacturer's system supplied connections between functions.

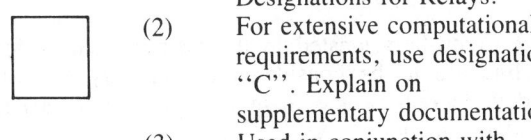

(2) Alternatively, link can be implicitly shown by contiguous symbols.

(3) May be used to indicate a communication link at the user's option.

4.7 Recorders and Other Historical Data Retention

4.7.1 Conventional hard-wired recording devices such as strip chart recorders shall be shown in accordance with ISA-S5.1. (Refer to Appendix A.2.2. of this standard.)

4.7.2 For assignable recording devices use Symbol 4.2.1.

4.7.3 Long term/mass storage of a process variable by digital memory means such as tape, disc, etc., shall be depicted in accordance with 4.2 or 4.3 of this standard, depending on the location of the device.

5 IDENTIFICATION

For purposes of this standard, identification codes shall be consistent with ISA-S5.1, with the following additions.

5.1 Software Alarms

Software alarms may be identified by placing ISA-S5.1, Table 1, letter designators on the input or output signal lines of the controls, or other specific integral system component. See Section 6 Alarms of this standard.

5.2 Contiguity of Symbols

Two or more symbols can adjoin to express the following means in addition to those shown in ISA-S5.1:

(1) Communication among the associated instruments, e.g.

- Hard wiring
- Internal system link
- Backup

(2) Instrument integrated with multiple functions, e.g.

- Multipoint recorder
- Control valve with integrally mounted controller.

The application of contiguous symbols is a user option.

If the intent is not absolutely clear, contiguous symbols should *not* be used.

6 ALARMS

6.1 General

All hard-wired standard devices and alarms, as distinct from those devices and alarms specifically covered in this standard, shall be shown in accordance with ISA-S5.1, Table 1.

The examples in paragraph 6.2 illustrate principles of the methods of symbolization and identification. Additional applications that adhere to these principles may be devised as required. The location of the alarm identifiers is left to the discretion and convenience of the user.

6.2 Instrument System Alarms

6.2.1 Multiple alarm capability is provided in most systems. Alarms covered by this standard should be identified as shown by the examples in 6.2.2 and 6.2.3.

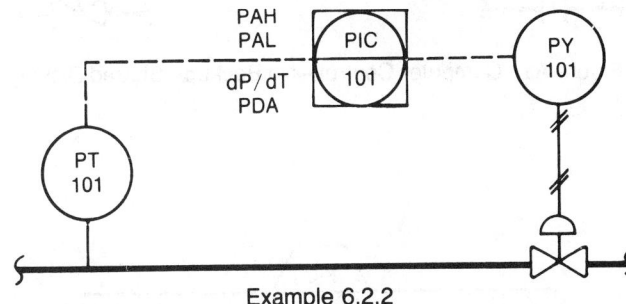

Example 6.2.2

6.2.2 Alarms on measured variables shall include the variable identifiers, i.e.:

Pressure:	PAH	(High)
	PAL	(Low)
	dP/dt	(Rate of change)
	PDA	(deviation from set point)

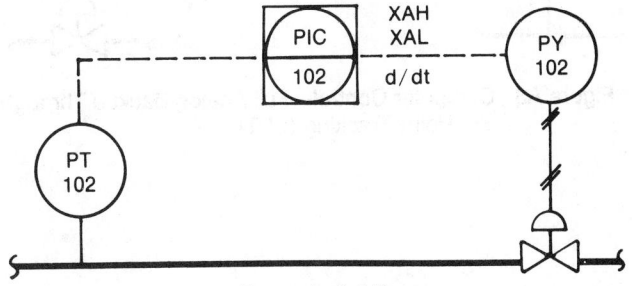

Example 6.2.3

6.2.3 Alarms on controller output shall use the undefined variable identifier X, i.e.:

XAH	(High)
XAL	(Low)
d/dt	(Rate of change)

APPENDIX A
EXAMPLES

A.1 Examples of Use

A.1.1 The following figures illustrate some of the various combinations of symbols presented in this standard and ISA-S5.1. These symbols may be combined as necessary to fulfill the needs of the user.

A.1.2 Controllers located in the diagram main information line are to be considered the primary controllers. All devices outside the main line provide a backup or secondary function.

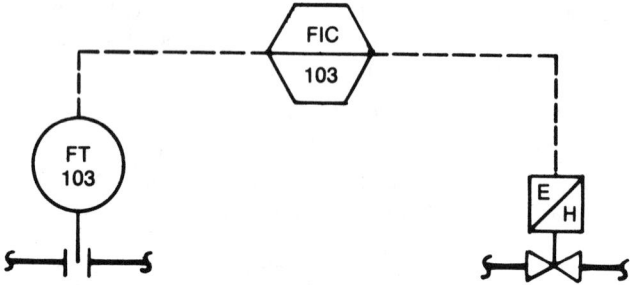

Figure A1. Computer Control—No Backup - Shared Display

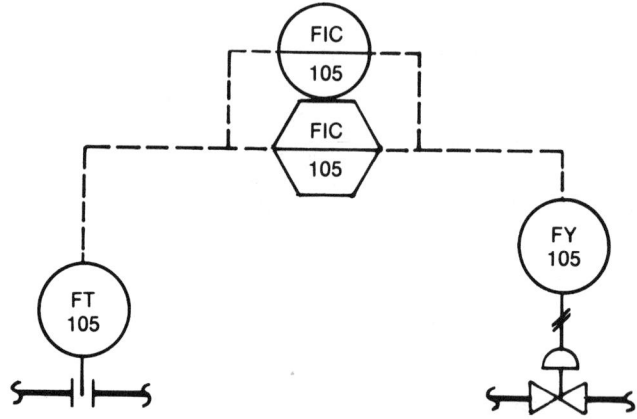

Figure A2. Computer Control—With Analog Backup

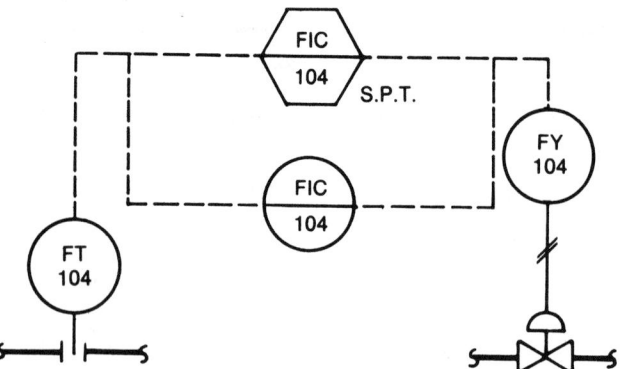

Figure A3. Computer Control—Full Analog Backup Through Set Point Tracking (SPT)

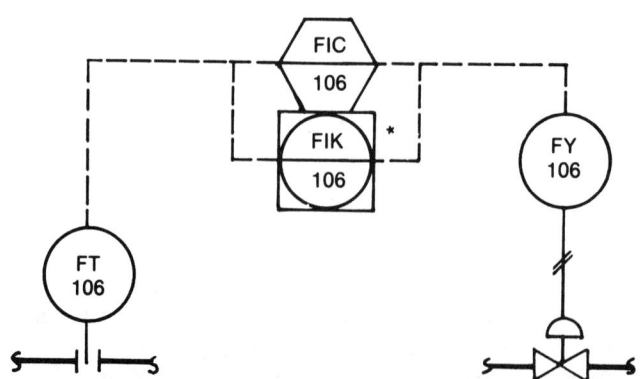

Figure A4. Computer Control-Full Backup from Distributed Control Instrumentation. Computer Uses Instrument System Communication Link

*Usage of suffix (K) is optional.

10

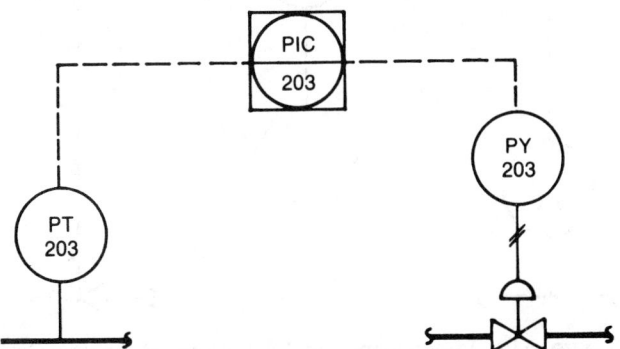

Figure A5. Shared Display/Shared Control—No Backup

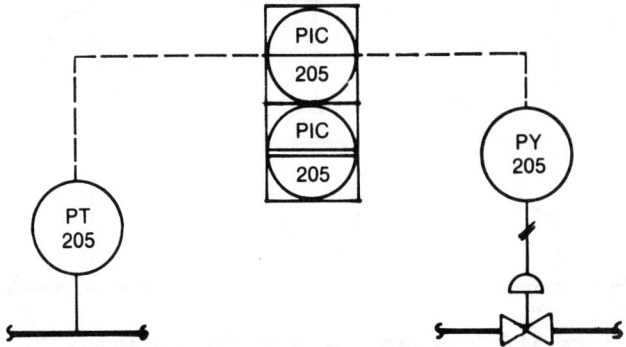

Figure A6. Shared Display/Shared Control—With Auxiliary
Operator's Interface Device

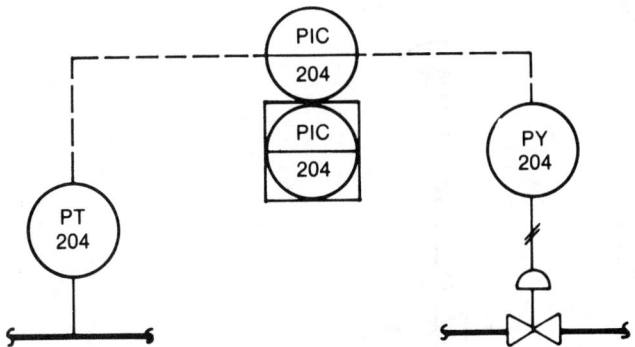

Figure A7. Analog Control—Interfaced with Shared Display.
Shared Control Backup

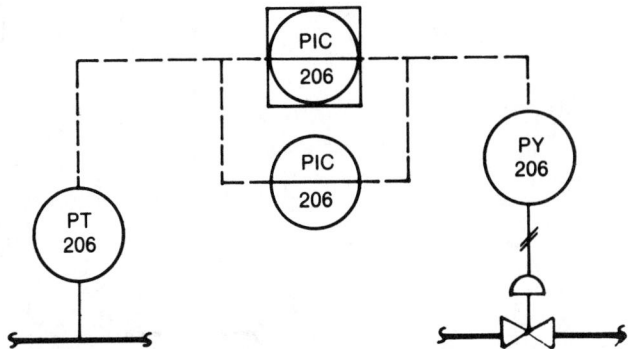

Figure A8. Shared Display/Shared Control—With Analog
Controller Backup

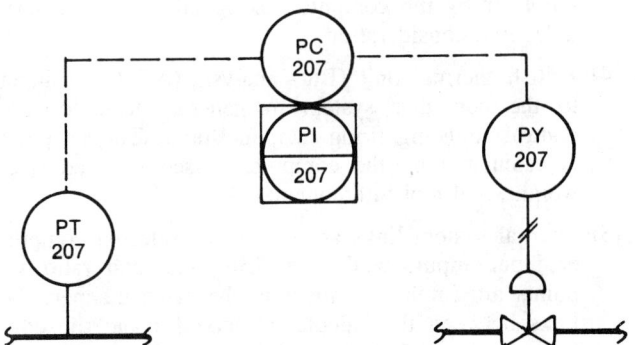

Figure A9. Analog Control—Blind Controller. Shared Display

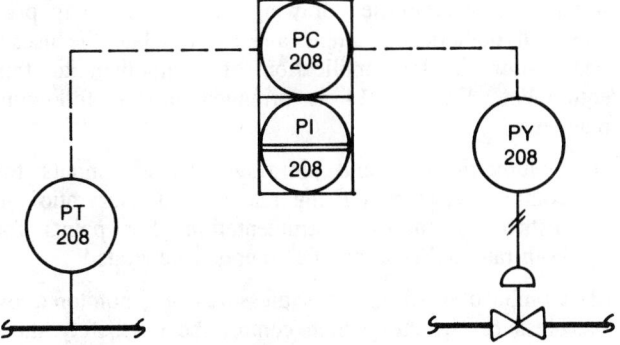

Figure A10. Blind Shared Control—With Auxiliary Operators
Interface Backup

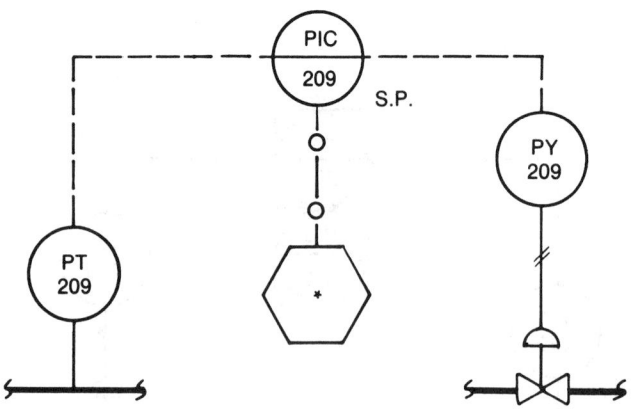

Figure A11. Supervisory Set Point Control—Analog Controller with Conventional Faceplate. Computer Supervisory Set Point via Communication Link

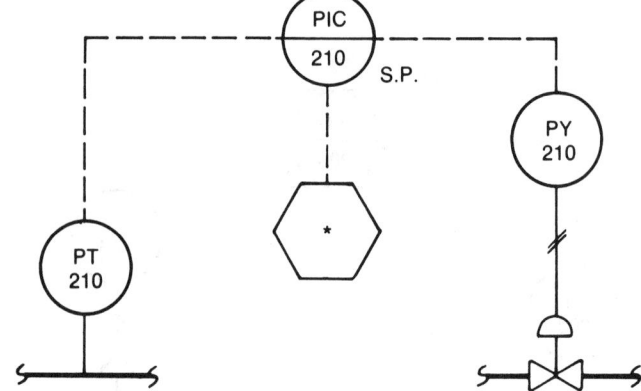

Figure A12. Supervisory Set Point Control—Analog Controller Complete with Conventional Faceplate. Computer Supervisory Set Point Hardwired.

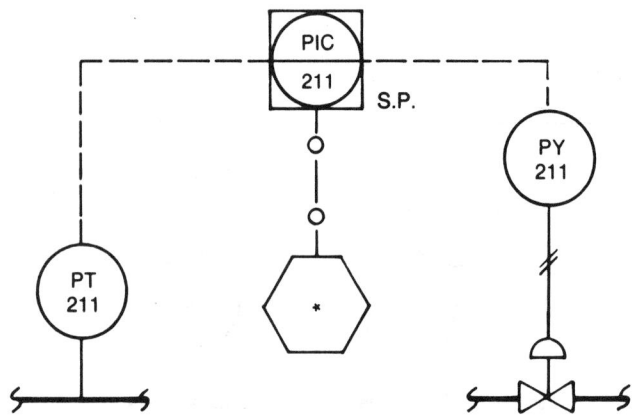

Figure A13. Supervisory Set Point Control—Shared Display Shared Control with Full Computer Access via the Communication Link

*User identification is optional

A.2 Typical Flow Diagrams

A.2.1 Figure A14 combines the basic symbols of this standard in a simplified drawing. It is intended to provide a hypothetical example and to stimulate the user's imagination in the application of symbolism to this equipment. Figure A14 is arranged in the following manner:

(1) Volumetric fuel and air flows provide inputs for combustion system firing rate and fuel air ratio via distributed control instrumentation. Set points for both rate and ratio can be computer generated.

(2) Combustion air and gas pressures are monitored by pressure switches which control the gas safety shut-off valve via UC-600 "distributed control interconnecting logic."

(3) Material moisture content is measured, dry weight of the input material is calculated, and feed rate is controlled by MT-300 and WC-301. Discharged material moisture content is read by MT-302. At this point, firing rate and/or feed rate could be controlled by the Distributed Control System (DCS) instrumentation or by the computer taking other process variables into consideration.

(4) British thermal unit (Btu) analysis (AT-97) is input to the computer system to generate feed forward control adjusting firing rate, in Btu/hr. The set point is calculated by the computer, based on feed rate, weight, and moisture content.

(5) Internal system links are shown for selected computer input/output, while the firing rate and ratio set points are implied. Shown in the same manner, the links between the calculation modules and the controllers are implied by contiguous symbols, while the wild flow to the ratio control is shown in the system link symbol.

A.2.2. Figure A15 combines the symbols to depict a cascade loop with alarms. Notes are added on the diagram itself for clarification purposes only.

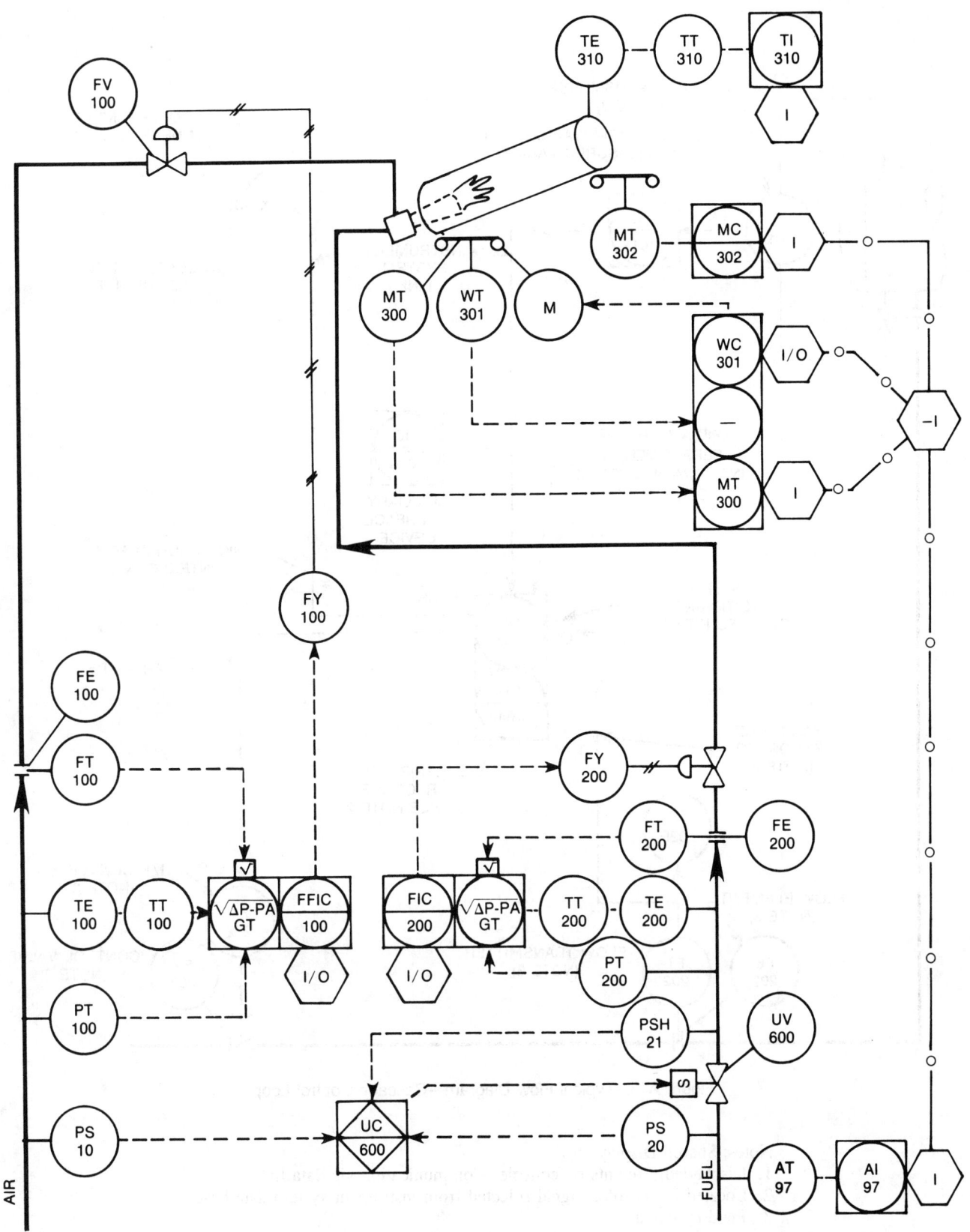

Figure A14. Example—Simplified Drawing

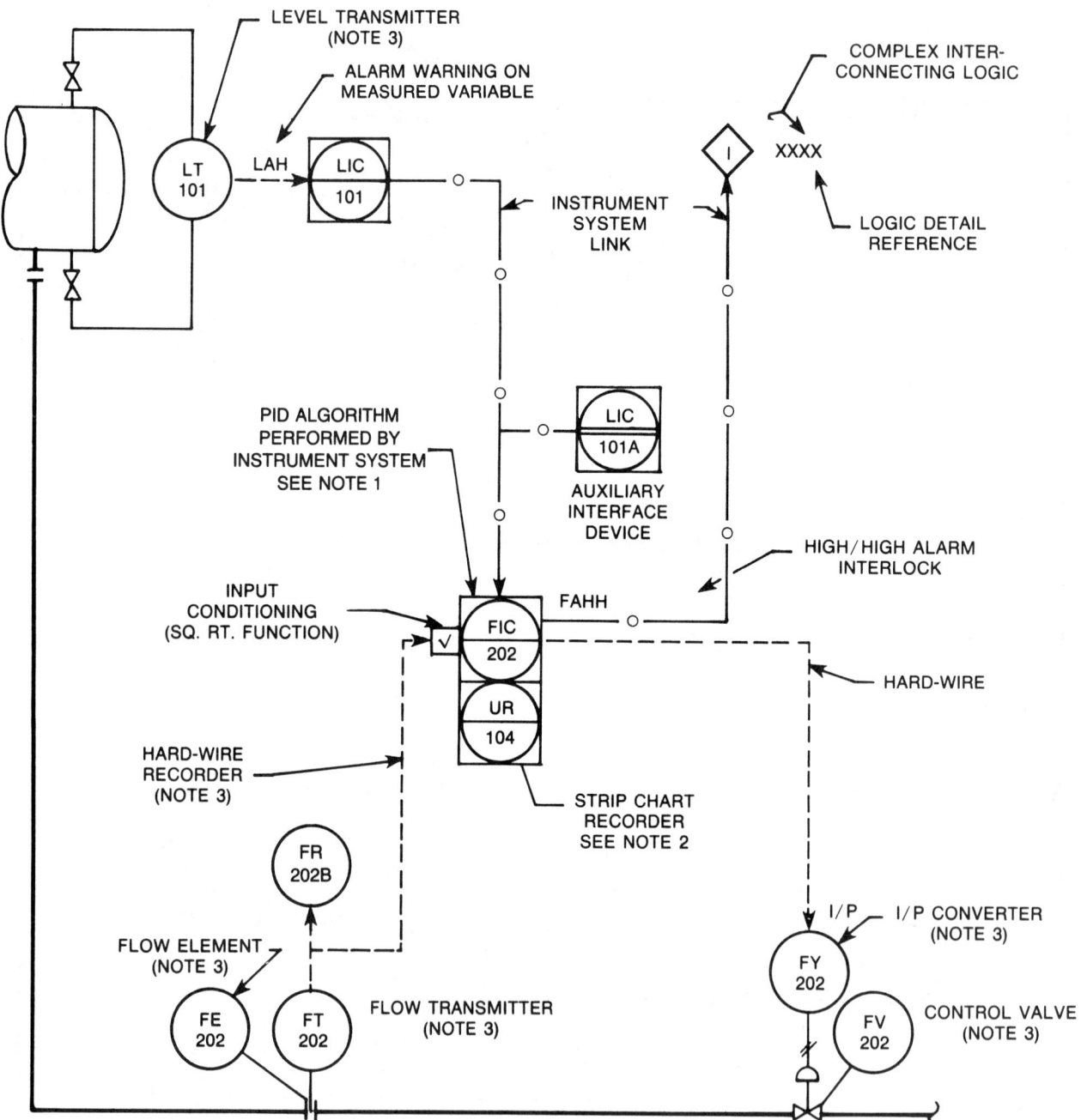

Figure A15. Typical Flow Diagram—Cascade Control Loop

Notes: Shared Display
1. Display/adjustments on console. Communication via data link.
2. Located in console. Signal selected from instrument system data base.
3. Field mounted.

INSTRUMENT SOCIETY of AMERICA
Research Triangle Park, North Carolina

ANSI/ISA-S5.4-1976
(R1981)
Approved October 9, 1981

American National Standard

Instrument Loop Diagrams

Instrument Society of America

ISBN 0-87664-332-2

ISA-S5.4   Instrument Loop Diagrams

INSTRUMENT SOCIETY OF AMERICA
67 Alexander Drive
P.O. Box 12277
Research Triangle Park, North Carolina 27709

PREFACE

This preface is included for informational purposes and is not part of Standard ISA-S5.4.

This Standard has been prepared as a part of the service of the Instrument Society of America toward a goal of uniformity in the field of instrumentation. To be of real value, this document should not be static, but should be subject to periodic review. Toward this end, the Society welcomes all comments and criticisms, and asks that they be addressed to the Secretary, Standards and Practices Board, Instrument Society of America, 67 Alexander Drive, P.O. Box 12277, Research Triangle Park, NC 27709, Telephone (919) 549-8411.

The ISA Standards and Practices Department is aware of the growing need for attention to the metric system of units in general, and the International System of Units (SI) in particular, in the preparation of instrumentation standards. The Department is further aware of the benefits to USA users of ISA Standards of incorporating suitable references to the SI (and the metric system) in their business and professional dealings with other countries. Toward this end this Department will endeavor to introduce SI — acceptable metric units in all new and revised standards to the greatest extent possible. The Metric Practice Guide, which has been published by the American Society for Testing and Materials as ANSI designation Z210.1 (ASTM E380-76, IEEE Std. 286-1975), and further revisions, will be the reference guide for definitions, symbols, abbreviation, and conversion factors.

It is the Policy of the Instrument Society of America to encourage and welcome the participation of all concerned individuals and interests in the development of ISA Standards. Participation in the ISA Standards making process by an individual in no way constitutes endorsement by the employer of that individual of the Instrument Society of America or any of the Standards which ISA develops.

Committee ISA-SP5.4, formed in 1971, adopted this standard on April 16, 1975. Loop diagrams are intended for general use throughout industry. It is important in considering the use of loop diagrams to review their proper value during their use for design, construction, checkout, startup, operation, maintenance, rearrangement and reconstruction. Personnel experienced in their use report benefits such as reduction in engineering costs, improved loop integrity and purchasing accuracy plus an extremely useful maintenance troubleshooting tool.

Loop diagrams offer the outstanding advantage of presenting on one conveniently sized sheet all the information needed for installation, checkout, startup and maintenance. When loop diagrams are not used, that information is often spread among many other documents. Updating of loop diagrams to "as built" is more easily achieved than updating the variety of documents frequently used.

The committee recognizes that loop diagrams can be used effectively on any size project from one or two loops up to large and complex installations.

The following people served on the 1975 SP5.4 Committee:

NAME	COMPANY
Larkin A. Spence, Chairman	Union Carbide Corporation
Robert T. Coupal, Representing the Scientific Apparatus Makers Association (SAMA)	Taylor Instrument Process Control Div., Sybron Corporation
Wm. G. Deutsch	Fluor Pioneer Incorporated
James W. Fay, Jr.	Bechtel Power Corporation
John R. Lavigne, Representing Technical Association of Pulp and Paper Industry (TAPPI)	The Foxboro Company
John Lorenz, Representing American Petroleum Institute (API)	Leeds and Northrup Company
Julius Slavin	E. I. duPont deNemours and Company
Richard E. Terhune	Exxon Company, U.S.A.

The following people served on the 1981 (reaffirmation) SP5.4 Committee:

NAME	COMPANY
Larkin A. Spence, Chairman	Matthew Hall Engineering, Inc.
Robert T. Coupal	Taylor Instruments

This Standard was approved by the ISA Standards and Practices Board in April, 1976.

This standard was reaffirmed by the ISA Standards and Practices Board in March, 1981.

TABLE OF CONTENTS

LIST OF FIGURES

1 PURPOSE

1.1 The purpose of this Standard is to provide a method and practice for the preparation and use of instrument loop diagrams in the design, construction, checkout, startup, operation, maintenance, rearrangement and reconstruction of instrument systems in industrial plants.

1.2 The Standard is intended to facilitate the understanding of loop diagrams and to improve communications among technical, non-technical, management, design, operating, and maintenance personnel concerned with instrument systems.

1.3 The Piping and Instrument Drawing (P&ID)/Flow Diagram, see ANSI Y32.20-1975 (ISA S5.1) "Instrumentation Symbols and Identification" (revised 1973), applies to the whole process loop, while the loop diagram gives further information on the control loop of an individual parameter.

2 SCOPE

This Standard presents guidelines to encourage general use and acceptance of loop diagrams throughout industry. This Standard covers the following:

      Uses
      Content
      Format
      Symbols
      Illustrations

3 USES OF AN INSTRUMENT LOOP DIAGRAM

Some present uses of loop diagrams are listed in the chronology of project development. Selection of desirable uses can be made to fulfill individual needs.

3.1 Engineering

a. As a design tool

   When prepared early, maximum benefits can be realized from use of loop diagrams to express control philosophy.

b. As an extension of Piping and Instrument Drawings (P&ID's).

   The loop diagram must show the *components and accessories* of the instrument loop, highlighting special safety and other requirements.

c. As a tool for the specification of instrument hardware items and auxiliaries.

d. As definition of instrumentation scope on a project, loop diagrams can supplement P&ID's as official documents.

e. As a means of communicating desires to potential vendors.

f. As a verification as to the completeness of submitted data.

3.2 Construction

a. Initial construction of panels.

b. Field installation of instrumentation, including panels.

c. Instrument interconnection.

d. Instrument loop checkout.

e. Inspection and documentation.

3.3 Commissioning and Startup

a. Pre-startup checking and calibration.

b. Extension of P&ID's for commissioning and startup personnel, highlighting safety and other requirements.

c. As an extension of P&ID's, loop diagrams can be used as training tools or aids.

3.4 Maintenance Work

a. Scheduled and non-scheduled maintenance.

b. Modification.

c. Reconstruction.

3.5 Operation

a. Communication medium between Operations, Maintenance and Engineering personnel.

b. Operations training device.

4 CONTENT OF AN INSTRUMENT LOOP DIAGRAM

A loop diagram must contain the information needed to accommodate the uses selected from Section 3 above. *Categorized below are minimum, desirable, and optional requirements which can be combined to match the desired uses. To accomplish minimum needs, a loop diagram shall contain at least the information covered in 4.1 through 4.6.*

4.1 All devices or items with clear labeling and identification.

4.2 Identification of the loop and each component of the loop, including connections to such things as trend or multipoint recorders and computers. All tagging or numbering must agree with the P&ID.

4.3 Word description of loop functions. The title should be adequate, but if not, a supplemental note can be added. A description of special features or functions which are not obvious or implied in the title, especially safety and shutdown circuits, is required. The identification of safety and shutdown circuits is especially important.

4.4 All interconnections with identifying numbers for electrical cables, conductor pairs, pneumatic multi tubes and individual pneumatic and hydraulic tubing. This identification of connections includes junction boxes, terminals, bulkheads, ports, computer input/output (I/O connections, grounding systems, grounding connections and signal levels).

4.5 Location of devices, such as, but not limited to: field, panel front, panel rear, auxiliary equipment, rack, termination cabinet, cable spreading room, and computer I/O cabinet.

4.6 Energy sources: electrical power, air supply and hydraulic fluid supply; designating voltage, pressure and other applicable requirements.

It is desirable that loop diagrams contain information described in 4.7 through 4.11.

4.7 Enough process lines and equipment to describe the process side of the loop and provide clarity of control action. This should include what is being measured, what is being controlled, and other information required to complete the process loop.

4.8 Reference to supplementary records and drawings. This would include inter-relation to other control loops (including overrides, interlocks, cascaded setpoint, shutdowns).

4.9 Controller action, control valve action, control valve fail safe action (electronic and/or pneumatic failure) and solenoid valve action.

4.10 Installation standards and details for all devices (including process connections, purge or blowback, heat trace, cooling, thermal insulation, seals, vents) shall be defined and source of energy or fluid designated.

4.11 Exact location of the device including the elevation should be designated.

Information described in 4.12 through 4.15 is optional for loop diagrams.

4.12 Purchase specifications, usually in abbreviation form and supplemented elsewhere, or reference made to purchase specification or data sheets.

4.13 Manufacturer's model numbers of devices for quick identification by project or maintenance personnel, when specifications are not included.

4.14 Calibration information, including set point values for alarm and shutdown devices, in consistent units.

4.15 Identifying numbers for equipment, including racks, panels, and junction boxes.

5 FORMAT

It is recommended that a consistent pattern be developed by the user.

5.1 Size of Drawing

A loop diagram should be on a small drawing which is easy to handle, preferably 8-1/2 in. x 11 in. or 11 in. x 17 in. One recommended method is to produce the original on larger drawings and reduce for field use. Attention must be given to size of letters and details to keep them legible in reduced sizes.

5.2 Content

A loop diagram should normally contain only one loop (See 4.8 above). Judgment must be used to accommodate individual situations. Care must be used to prevent overcrowding and space should be provided for future additions and loop data.

5.3 General Layout

It is recommended that a consistent pattern (horizontal or vertical) be developed by the user for each project. A suggested layout is to divide the drawing into sections for relative location of devices as discussed in Section 4.5.

6 SYMBOLS

6.1 The symbols in ANSI Y32.20-1975 (ISA S5.1) are intended for P&ID's and are suitable for loop diagrams. However, these symbols must be expanded as follows to include connection points, power source (electrical, air, hydraulic) and instrument action and range, to clarify certain connection and operation details required on loop diagrams.

LOOP DIAGRAM SYMBOLS

A. Terminal Block Symbol

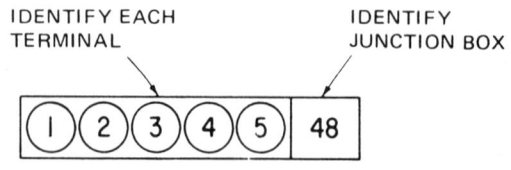

B. Pneumatic Bulkhead Symbol

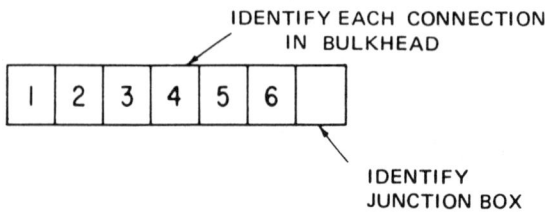

C. Instrument Terminals and/or Ports

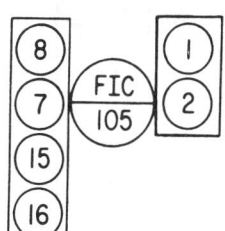

Instrument terminal or port with identifying letter and/or number. (It is suggested that identifying number or letters be the manufacturer's designation.)

Instrument symbol per ISA Standard ANSI Y32.20-1975 (ISA S5.1).

NOTE: These terminals or ports are not intended to be pictorial.

D. Instrument System Energy Supply
(See ANSI Y32.20-1975 (ISA S5.1) page 17)

1. Identify electrical power supply followed by the appropriate voltage, etc.

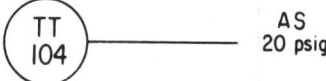

2. Identify air supply followed by air supply pressure.

3. Identify hydraulic fluid followed by the fluid supply pressure.

E. Controller Action

1. Controller, direct acting
   A controller in which the value of the output signal increases as input (measured variable) increases is represented by an arrow pointing up (↑).

2. Controller, reverse acting
   A controller in which the value of the output signal decreases as the value of the input (measured variable) increases is represented by an arrow pointing down (↓).

NOTE:
    (1) The arrow can be located on either side of the circle, but must be vertical.
    (2) A similar arrow may be used to designate a transmitter with reverse output.

F. Set Point and Calibration

1. Identify set points by a diamond adjacent to the instrument symbol with the set point listed in the diamond.

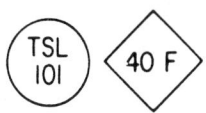

2. Identify calibration information by a rectangle adjacent to the instrument symbol with the data listed inside the rectangle.

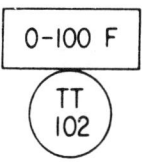

6.2 Illustration 7.1 and 7.2 indicate the use of these symbols to describe all the mandatory items on a loop diagram. Illustrations 7.3 and 7.4 indicate the desirable and optional items as well as the minimum required loop diagram components.

7 ILLUSTRATIONS

Loop diagrams have been prepared to illustrate the use of these symbols when describing a relatively simple feedback flow control loop.

7.1 Loop diagram, pneumatic control, minimum required items.

7.2 Loop diagram, electronic control, minimum required items.

7.3 Loop diagram, pneumatic control, minimum required items plus desirable and optional items.

7.4 Loop diagram, electronic control, minimum required items plus desirable and optional items.

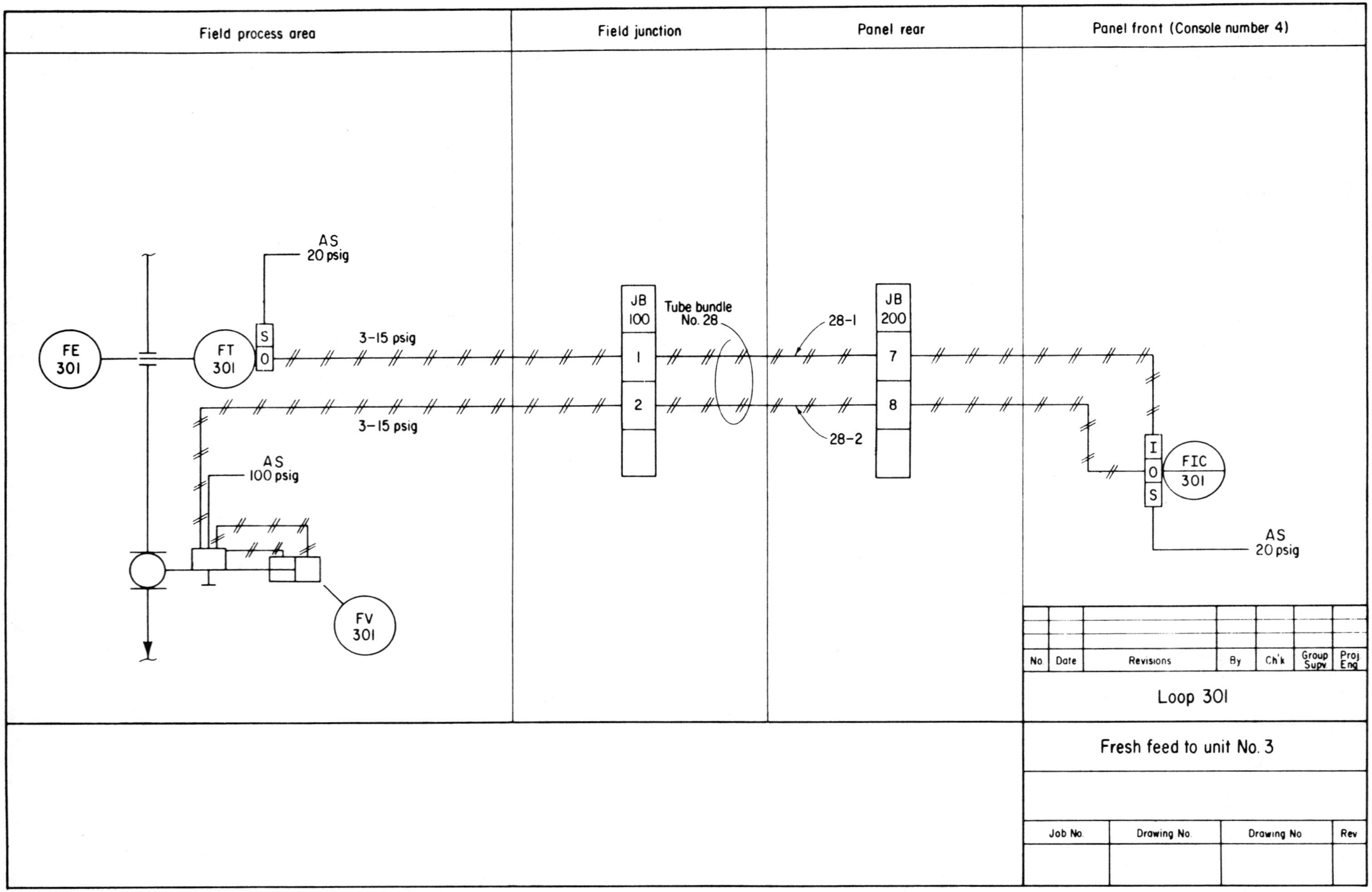

Illustration 7.1
Loop diagram, pneumatic control, minimum required items.

8

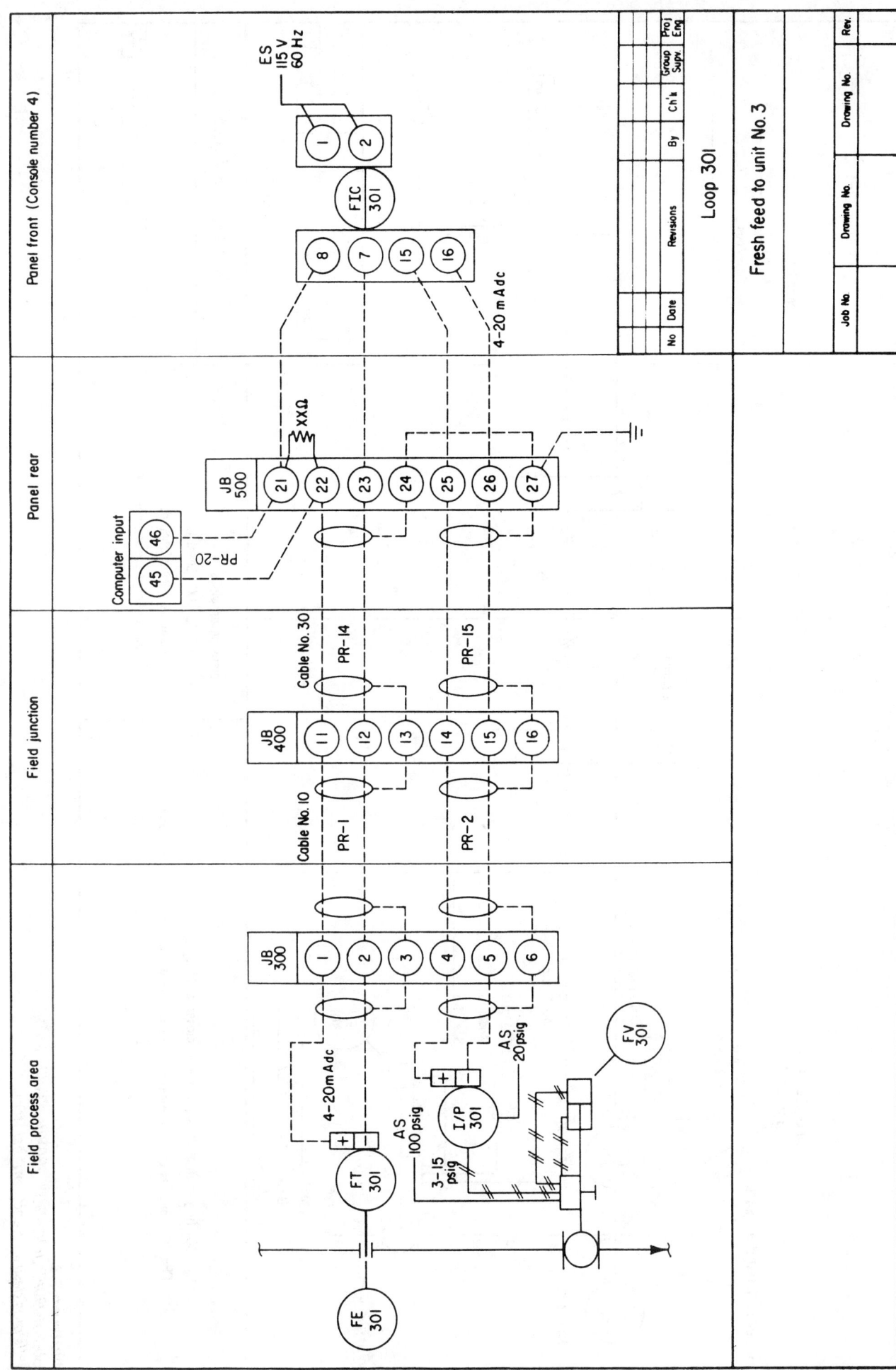

Illustration 7.2
Loop diagram, electronic control, minimum required items.

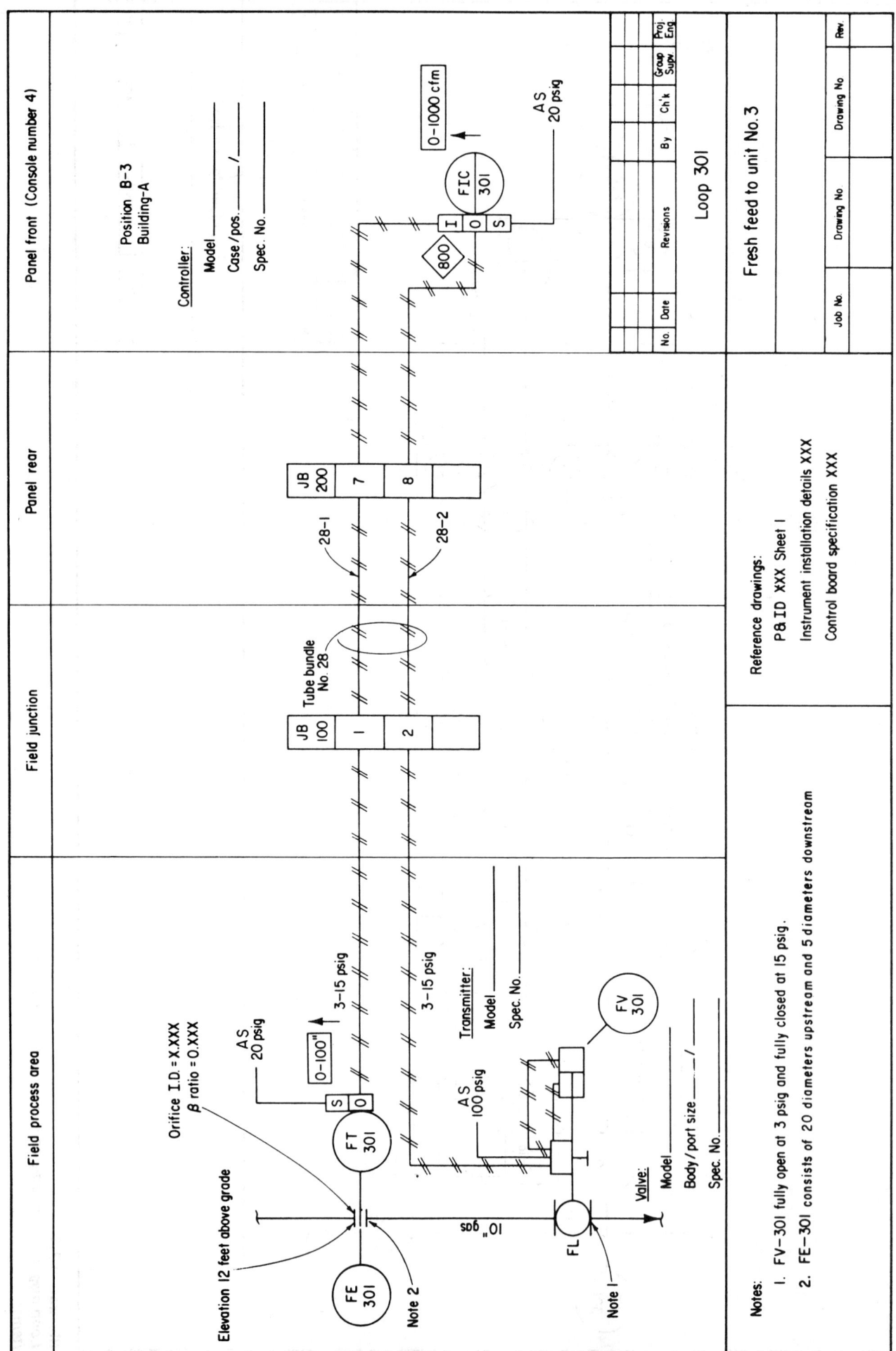

Illustration 7.3
*Loop diagram, pneumatic control, minimum required
items plus desirable and optional items.*

10

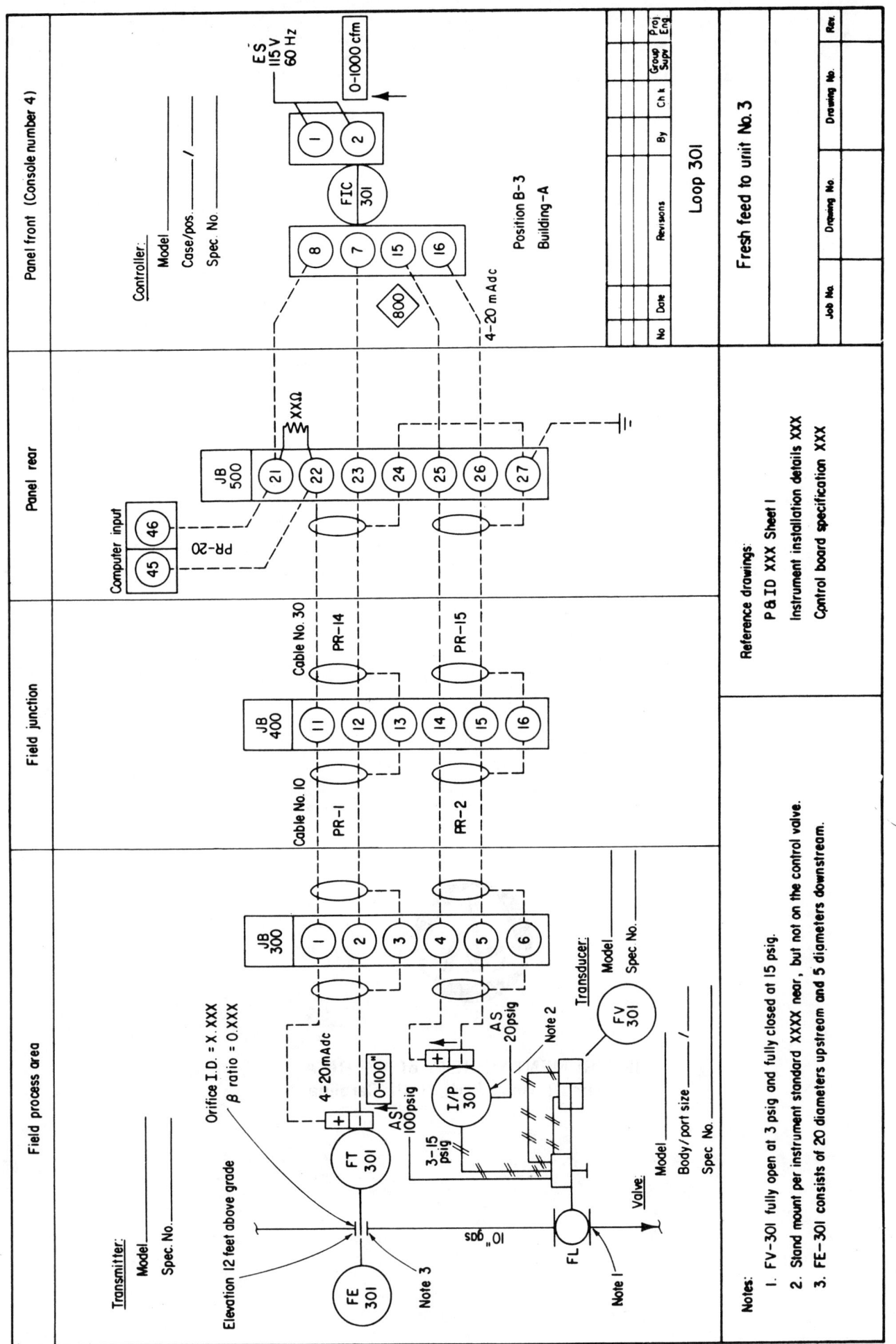

Illustration 7.4.
*Loop diagram, electronic control, minimum required
items plus desirable and optional items.*

INSTRUMENT SOCIETY of AMERICA
Research Triangle Park, North Carolina

ANSI/ISA-S5.5-1985
Approved: February 3, 1986

American National Standard

Graphic Symbols
for Process Displays

Instrument Society of America

ISBN 0-87664-935-5

ISA-S5.5   Graphic Symbols for Process Displays

INSTRUMENT SOCIETY OF AMERICA
67 Alexander Drive
P.O. Box 12277
Research Triangle Park, NC 27709

PREFACE

This preface is included for informational purposes and is not a part of Standard S5.5.

This Standard has been prepared as a part of the service of the Instrument Society of America (ISA) toward a goal of uniformity in the field of instrumentation. To be of real value, this document should not be static, but should be subject to periodic review. Toward this end, the Society welcomes all comments and criticisms and asks that they be addressed to the Secretary, Standards and Practices Board, Instrument Society of America, 67 Alexander Drive, P.O. Box 12277, Research Triangle Park, North Carolina 27709, Telephone (919) 549-8411.

The ISA Standards and Practices Department is aware of the growing need for attention to the metric system of units in general and the International System of Units (SI) in particular, in the preparation of instrumentation standards. The Department is further aware of the benefits to U.S.A. users of ISA Standards of incorporating suitable references to the SI (and the metric system) in their business and professional dealings with other countries. Toward this end, this Department will endeavor to introduce SI-acceptable metric units in all new and revised standards to the greatest extent possible. *The Metric Practice Guide*, which has been published by the Institute of Electrical and Electronics Engineers as ANSI/IEEE Std. 268-1982, and future revisions will be the reference guide for definitions, symbols, abbreviations, and conversion factors.

It is the policy of the Instrument Society of America to encourage and welcome the participation of all concerned individuals and interests in the development of ISA Standards. Participation in the ISA Standards-making process by an individual in no way constitutes endorsement by the employer of that individual of the Instrument Society of America or any of the standards which ISA develops.

The information contained in this preface, in the footnotes, and in the appendices is included for information only and is not part of the standard.

The original draft of this document resulted from the committee work of the International Purdue Workshop on Industrial Computer Systems, the Man/Machine Communication Committee TC-6.

The use of graphic symbols representing entities and characteristics of processes has evolved rapidly during the course of the last decade. Technology has allowed the presentation of a physical process to be represented and controlled by the use of computers and advanced electronic systems. These systems use video-display technologies such as CRTs, plasma screens, and other media to present to the user a graphic representation of his process. It is through these devices and the symbology used to represent the process in question that the user monitors and controls the particular operation.

Process displays convey information to the user in the form of both text and graphic symbols. Text information is based on the use of numeric data and the alphabet to construct the words necessary to convey the meaning of the information. This text information is structured around the use of written language and is highly ordered and understood by users. On the other hand, the use of graphic symbols for process and information presentation is highly dependent upon the manufacturer and the user of the product. These graphic symbols are generally customized to the particular application at hand.

Standard graphic symbols provide a more logical and uniformly understandable mechanism for modern control processes. For example, a control system may be constructed of several control systems and a central control system. In cases such as this, the operator often finds that he must become familiar with the graphic symbology of several different systems, although they may represent common elements.

It is the intent of this document that both the manufacturers and users of process displays use these graphic symbols in their systems whenever applicable. It is recognized that techology is rapidly changing in the types of devices available for process display use. The graphic symbols suggested in this standard should provide a foundation for all display systems that are used to display and control processes. The graphic symbols that are represented in this standard are divided into 13 major groups. Attributes associated with the various types of symbols such as color usage, blink, orientation, etc., are addressed in the document.

The symbols defined in ISA-S5.5 are intended to supplement those of ISA-S5.1 and ISA-S5.3 to provide a cohesive integration of graphic symbology and common industry usage of flow diagrams. ISA-S5.1 and ISA-S5.3 are drafting standards which govern the depiction of process and instrumentation symbols for drawings and other printed documents. The ISA-S5.5 symbols were developed for use on video devices that represent both character display and pixel addressable displays. Use of the symbols also applies to both color and monochromatic video displays as well as other media. Therefore, the symbols that are represented in this standard may differ from those in the other standards because of the nature of the physical devices used to display the symbols. The principle users of these symbols are operators and other personnel who use information concerning process operations.

The main intent of the graphic symbols is to provide to the user an easily understandable representation of his process on a display device. Computers, distributed control systems, stand-alone microprocessor-based systems, etc., can appear to be similar or to perform similar functions; however, they are diverse in philosophy and graphic presentation. Therefore, it is essential that a common set of symbols be used to convey process information to the users of such devices.

The symbols presented in this standard are by no means all that were suggested or that may be required; however, by adopting these as a standard, the majority of present processes may be adequately represented. When it becomes necessary to develop special symbols for equipment not included in the standard, simplicity of form is considered of paramount importance.

The ISA Standards Committee on Graphic Symbols for Process Displays SP5.5 operates within the ISA Standards and Practices Department, Norman Conger, Vice President. The persons listed on the following page served as members of the ISA Committee SP5.5.

The persons listed below served as members of ISA Committee SP5.5, which prepared this standard:

NAME	COMPANY
D. G. Kempfer, Chairman 1982–85	Standard Oil Company of Ohio
A. T. Bonina	Industrial Data Terminals
R. F. Carroll, Chairman 1981	Setpoint, Inc.
A. S. Fortunak	Inland Steel Company
W. K. Greene	Union Carbide
F. W. Magalski	Industrial Data Terminals
R. F. Sapita, Chairman 1979–80	The Foxboro Company
B. J. Selb	Rosemount
J. A. Shaw	Taylor Instrument Company
J. Ventresca	AccuRay Corporation
D. Winward	Aydin Controls

The persons listed below served as members of ISA Committee SP5, which approved this standard:

NAME	COMPANY
D. E. Rapley, Chairman	Rapley Engineering Services
R. C. Greer	Bailey Controls Company
D. G. Kempfer	Standard Oil Company of Ohio
R. H. Kind	El Paso Natural Gas Company
R. Mulley	S. F. Braun
T. J. Myron	The Foxboro Company

This standard was approved for publication by the ISA Standards and Practices Board in December 1985.

NAME	COMPANY
N. Conger, Chairman	Fisher Controls Company
P. V. Bhat	Monsanto Company
W. Calder III	The Foxboro Company
R. S. Crowder	Ship Star Associates
H. S. Hopkins	Westinghouse Electric Company
J. L. Howard	Boeing Aerospace Company
R. T. Jones	Philadelphia Electric Company
R. Keller	The Boeing Company
O. P. Lovett, Jr.	ISIS Corporation
E. C. Magison	Honeywell, Inc.
A. P. McCauley	Chagrin Valley Controls, Inc.
J. W. Mock	Bechtel Corporation
E. M. Nesvig	ERDCO Engineering Corporation
R. Prescott	Moore Products Company
D. E. Rapley	Rapley Engineering Services
C. W. Reimann	National Bureau of Standards
J. Rennie	Factory Mutual Research Corporation
W. C. Weidman	Gilbert/Commonwealth, Inc.
K. Whitman	Consultant
*P. Bliss	Consultant
*B. A. Christensen	Continental Oil Company
*L. N. Combs	Retired
*R. L. Galley	Consultant
*T. J. Harrison	IBM Corporation
*R. G. Marvin	Roy G. Marvin Company
*W. B. Miller	Moore Products Company
*G. Platt	Consultant
*J. R. Williams	Stearns Catalytic Corporation

*Director Emeritus

TABLE OF CONTENTS

1.0 PURPOSE

The purpose of this standard is to establish a system of graphic symbols for process displays that are used by plant operators, engineers, etc., for process monitoring and control. The system is intended to facilitate rapid comprehension by the users of the information that is conveyed through displays, and to establish uniformity of practice throughout the process industries.

Resulting benefits are intended to be as follows:

A. A decrease in operator errors

B. A shortening of operator training

C. Better communication of the intent of the control system designer to the system users

An objective of the standard is to insure maximum compatibility of symbols on process visual display units (VDUs) with related symbols used in other disciplines.

The symbols in this standard are intended to depict processes and process equipment. The symbols are suitable for use on visual display units (VDUs), such as Cathode Ray Tubes (CRTs).

2.0 SCOPE

The standard is suitable for use in the chemical, petroleum, power generation, air conditioning, metal refining, and numerous other industries.

Though the standard may make use of standard symbols now used for piping and instrument diagrams, logic diagrams, loop diagrams, and other documents, the symbols of the standard are generally expected to be used in ways complementing existing types of engineering documents.

The symbolism is intended to be independent of type or brand of hardware or computer software.

2.1 Application to Work Activities

This standard is suitable for use whenever any reference to process equipment on VDUs is required. Such references may be required for the following uses as well as others:

A. Process displays on CRTs

B. Process displays on other visual media such as plasma displays, liquid crystal displays, etc.

2.2 Relationship to Other ISA Standards

This standard complements, whenever possible, ISA Standards S5.1 "Instrumentation Symbols and Identification," S5.3 "Flow Diagram Graphic Symbols for Distributed Control/Shared Display Instrumentation Logic and Computer Systems," RP60.05 "Graphic Displays for Control Centers," and ANSI/ISA S51.1 "Process Instrumentation Terminology."

2.3 Relationship to Other Symbol Standards

This document complements the ANSI Standard for process flow sheets, ANSI Y32.11M — "Graphic Symbols for Process Flow Diagrams in the Petroleum and Chemical Industries" and ANSI/NEMA Standard ICS 1-1978 "General Standards for Industrial Control and Systems" whenever possible and practical.

2.4 Definitions

Aspect Ratio — The ratio of a symbol's height to its width.

Background — The field that information is displayed upon for contrast.

Blinking — A periodic change of hue, saturation, or intensity of a video display unit pixel, character, or graphic symbol.

Character — A term used to refer to a predefined group of pixels.

Chromaticity — The color quality of light, which is characterized by its dominant wavelength and purity.

Color Coding — The use of different background and foreground colors to symbolically represent processes and process equipment attributes, such as status, quality, magnitude, identification, configuration, etc.

Foreground — The information element on a background field.

Graphic Symbol — An easily recognized pictorial representation.

Highlighting — A term encompassing various attention-getting techniques, such as blinking, intensifying, underscoring, and color coding.

Intensity — The lumination level (i.e., brightness) of the pixels of a VDU.

Pixel — The smallest controllable display element on a VDU. Also referred to as picture element (PEL).

Process Visual Display — A dynamic display intended for operators and others engaged in process monitoring and control.

Reverse Video — The interchange of foreground and background attributes, such as intensity, color, etc.

Task/Surround Lumination Ratio — The luminance ratio between the keyboard and screen (TASK) and workplace (SURROUND) within the operator's field of view.

Visual Display Unit (VDU) — A generic term used for display units based on technologies such as Cathode Ray Tubes (CRTs), Plasma Discharge Panels (PDPs), Electroluminescent Devices (ELs), Liquid Crystal Displays (LCDs), etc.

3.0 SYMBOLS

3.1 Symbol Usage

3.1.1 General

1. The graphic symbols in this standard are intended for use on VDUs.

2. Because size variations of symbols representing the various pieces of equipment are anticipated, no scale is indicated on the graphic symbol sketches. The integrity of the defined symbols should be preserved by maintaining the aspect ratio depicted.

3. Color coding to improve the perception of information and ease of interpretation of the displayed image is anticipated.

4. Graphic symbols should be arranged to depict spatial relationships, energy, material and data flows in a consistent manner (e.g., left to right, top to bottom, etc.). Equipment outlines and piping lines may be differentiated by color, intensity, or width.

5. Symbols may be rotated in any orientation on a VDU in order to represent the process in the most effective manner.

6. Arrows may be used on process lines to indicate direction of flow.

7. Symbols should be shown only when they are important to understanding the operation or are an integral part of the process depicted. Symbol qualities, such as luminance, size, color, fill, and contrast should be considered collectively and judiciously in order to avoid any psychophysiological masking of adjacent display targets, such as measurement values, alarm messages, labels, etc.

8. Numeric values and text may be included to enhance comprehension. The values may be either static or dynamic.

9. Graphic displays may contain both static and dynamic symbols and data. The symbol set, while intended for color displays, is also usable on monochromatic displays.

10. Special characteristics of displays should be used to enhance the understanding of process symbols. These characteristics may be used to indicate the status of process devices:

- Reverse video
- Blinking
- Intensity variation
- Color coding

These characteristics can be used for both static and dynamic symbol applications.

11. The use of outline and solid (filled) forms to indicate status is as follows:

- An outline symbol form indicates an off, stopped, or nonactive state.

- A solid (filled) symbol form indicates an on, running, or active state.

Status designation by use of solid or outline forms are particularly applicable to the Rotating Equipment and Valves and Actuators groups of symbols. Prudence in judgement should be used when adhering to this practice as some symbols should not change from their outline form. In depicting valve position, use solid to show open (material flowing or active) and outline to show closed (material stopped or nonactive). Another usage is solid/outline to represent a pump running/stopped as the generally accepted practice. Some industries, such as the power industry, use solid/outline to show closed (active or unit energized)/open (nonactive or unit deenergized). In these special cases, the explicit uses of these conventions are to be made clear to the operator and noted in operation manuals.

12. A symbol may be partially filled or shaded to represent the characteristic of the contents of a vessel, e.g., level, temperature, etc.

13. Properties of physical or chemical states, as measured by primary elements or instruments, can be represented on a VDU by symbolic characters. It is not normal to display these characters on a process display, but they are available if required. Appendix B contains the recommended designated characters and an example of their usage. This list has been derived from character designations based on the ISA Standard S5.1, "Instrumentation Symbols and Identification". It has been modified for use on VDU displays. An excerpt of the S5.1 document explaining the identification-letter usage is also included in Appendix B.

3.1.2 Color

Color is an effective coding technique used either singularly or redundantly with symbol, shape, and alphanumeric coding. Although this standard pertains exclusively to the definition and configuration of display symbols, certain color application guidelines have, nevertheless, been included for the convenience of the display designer. They are as follows:

1. Information-bearing color schemes should be simple, consistent, and unambiguous.

2. The most common color technology is the CRT using the raster display scheme and an additive color generation technique based on the three primaries: red, blue, and green. The number of selectable colors can range from six plus black and white to the thousands. The number of colors in one display should be limited to the minimum necessary to satisfy the process interface objectives of the display. Color is an effective coding technique for dynamic identification and classification of display elements. Used judiciously, it can improve operator performance, e.g., reduce search time, improve element identification, etc. Conversely, irrelevant color can act as visual noise and negate the positive effects of color coding. Typically, four colors can accommodate the dynamic coding requirements of process displays.

3. Large background areas should be black. In situations where the black background results in a high task/surround lumination ratio, a brighter background may be used, preferably blue or brown.

4. Compatible color combinations, i.e., those with high chromaticity contrast, should be used. Some good combinations include: black-on-yellow, red-on-white, blue-on-white, and green-on-white. Combinations to avoid include: yellow-on-white, yellow-on-green, red-on-magenta, and cyan-on-green. In each case, the weight or size of the foreground element must also be considered. Certain combinations like blue-on-black can be acceptable only when the blue element is sufficiently large. These generalizations neglect the effects of lumination levels and ambient lighting. Each pair should be evaluated on a per-case basis.

5. Use color as a redundant indicator along with text, symbol, shape, size, reverse video, blinking, and intensity coding to preserve communications of critical process state and quality information with individuals having limited color perception.

6. To insure fast operator reponse, use highly saturated colors such as red or yellow.

7. Colors should not be used to indicate quantitative value.

8. The display designer should establish a project-related set of generic color meanings before developing a list of specific color-to-display-element associations. This generic set should be based on applicable plant, industry, and agency (OSHA, NRC, ANSI, etc.) conventions. Each project may have its unique set of generic definitions; e.g., Project A uses red to indicate closed or inactive states, while Project B uses green. In some special cases, such as the power industry, red may indicate closed and active or unit energized. This is suitable as long as the color meanings are defined as such for the particular project. Listed below is an example of a unique project-related color plan:

COLOR PLAN EXAMPLE

Color	Generic Meaning	Element Association
Black	Background	
Red	Emergency	A) Stop B) Highest Priority Alarm C) Closed D) Off
Yellow	Caution	A) Abnormal Condition B) Second Priority Alarm
Green	Safe	A) Normal Operation B) Start C) Open D) On
Cyan (Light Blue)	Static & Significant	A) Process Equipment in Service B) Major Labels
Blue	Nonessential	A) Standby Process Equipment B) Labels, Tags, etc.
Magenta (Purple)	Radiation	A) Radiation Alarms B) Questionable Values
White	Dynamic Data	A) Measurements & State Information B) System Messages C) Trend D) Active Sequential Step

3.2 Grouping of Symbols

The graphic symbols for process displays have been divided into related groups. There are 13 groups and their contents are as follows:

Group	Symbol	Page
Connectors		13
Containers and Vessels		14
Process	Distillation Tower	14
	Jacketed Vessel	14
	Reactor	14
	Vessel	14

The symbols are presented in Section 3.3, Structure of Symbols. The symbols are categorized into their respective groups and are presented in alphabetical order. Each symbol is described with the following information:

Group	An associated classification of similar symbols
Subgroup	Represents further division within a group
Symbol Name	The name of the process symbol
Symbol Mnemonic	A four-character name given to the symbol to be used as its reference name in a computer system
Description	A brief description of what the symbol represents
Symbol Drawing	The actual drawing of the symbol itself. Although no specific aspect ratio is given, the shape that is drawn should be depicted as closely as possible. Process connections and flow directions have been included with some symbols for functional clarity. These may be arranged as necessary.

Heads shown on containers and vessels are those most frequently encountered for that specific type. However, dished, elliptical, hemispherical, conical, or flat heads may be substituted where appropriate to match the actual configuration of the device.

3.3 STRUCTURE OF SYMBOLS

3.3.1 Group: Connectors

Subgroup: N/A
Symbol Name: N/A
Symbol Mnemonic: N/A

Description: For the purpose of this document, the various possible connectors have been excluded. In the majority of cases, pipe connections are not required to be detailed. A recommended practice to avoid any confusion on the video display is to use line breaks to indicate that the lines do not join. The most important lines should be kept solid with the secondary lines being broken. If all lines are of equal importance, a usual convention is to break the vertical line.

3.3 STRUCTURE OF SYMBOLS

3.3.2 Group: Containers and Vessels

Subgroup: Process
Symbol Name: Distillation Tower
Symbol Mnemonic: DTWR

Description: A packed or trayed distillation tower used for separation. Packing or trays may be shown to indicate type of distillation tower.

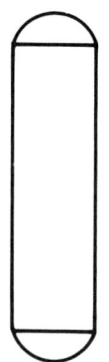

Subgroup: Process
Symbol Name Jacketed Vessel
Symbol Mnemonic: JVSL

Description: A vessel with a heating or cooling jacket. Jacket may be on straight shell, on bottom head, on top head, or any combination, as required to match the actual process vessel.

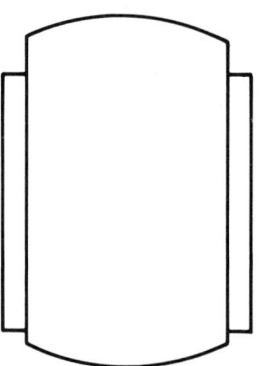

Subgroup: Process
Symbol Name: Reactor
Symbol Mnemonic: RCTR

Description: A chemical reactor. Internal details may be shown to indicate type of reactor.

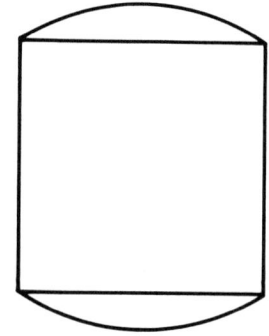

Subgroup: Process
Symbol Name: Vessel
Symbol Mnemonic: VSSL

Description: A vessel or separator. Internal details may be shown to indicate type of vessel. Can also be used as a pressurized vessel in either a vertical or horizontal arrangement.

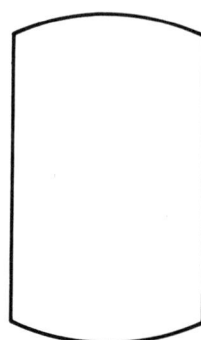

3.3 STRUCTURE OF SYMBOLS

3.3.2 Group: Containers and Vessels (Cont'd)

Subgroup: Storage
Symbol Name: Atmospheric Tank
Symbol Mnemonic: ATNK

Description: A tank for material stored under atmospheric pressure.

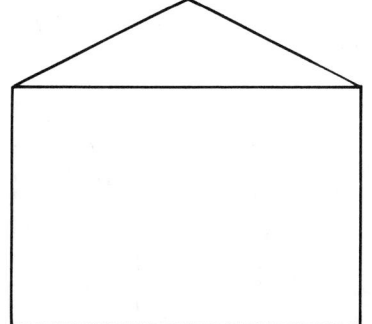

Subgroup: Storage
Symbol Name Bin
Symbol Mnemonic: BINN

Description: A container used to store solid or granular material that is discharged from the bottom.

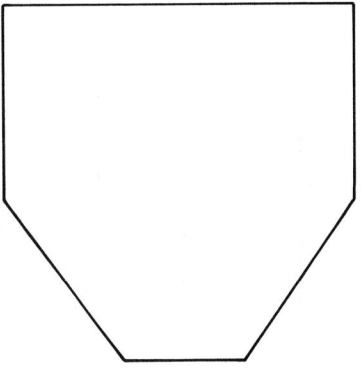

Subgroup: Storage
Symbol Name: Floating Roof Tank
Symbol Mnemonic: FTNK

Description: A tank for liquids with roof of vessel moving up and down with a change in stored volume.

Subgroup: Storage
Symbol Name: Gas Holder
Symbol Mnemonic: GHDR

Description: A tank with for gases roof of vessel moving up and down with a change in stored volume.

3.3 STRUCTURE OF SYMBOLS

3.3.2 Group: Containers and Vessels (Cont'd)

Subgroup: Storage
Symbol Name: Pressure Storage Vessel
Symbol Mnemonic: PVSL

Description: A pressurized spherical vessel for storage of gases and liquids.

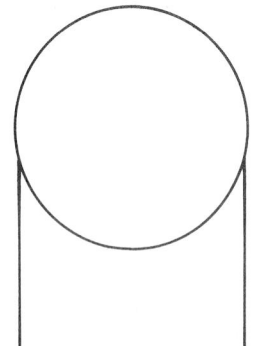

Subgroup: Storage
Symbol Name Weigh Hopper
Symbol Mnemonic: WHPR

Description: A vessel used for weighing material.

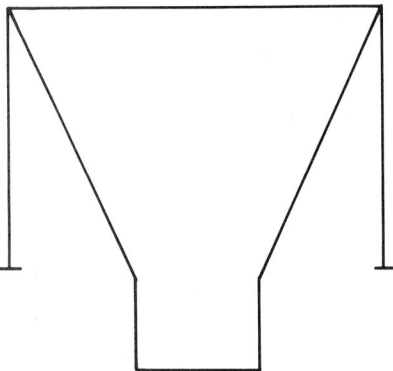

3.3 STRUCTURE OF SYMBOLS

3.3.3 Group: Electrical

Subgroup: N/A
Symbol Name: Circuit Breaker
Symbol Mnemonic: CBRK

Description: Representation of a circuit breaker for electrical systems. See STATE INDICATOR symbol for alternative use.

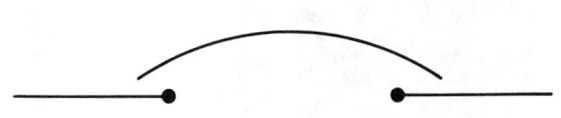

Subgroup: N/A
Symbol Name Manual Contactor
Symbol Mnemonic: MCTR

Description: A power distribution switch used for device isolation.

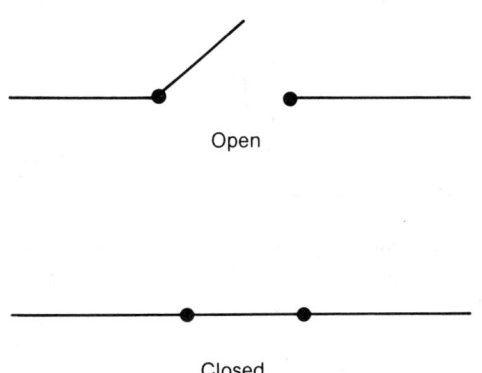

Open

Closed

Subgroup: N/A
Symbol Name: Delta Connection
Symbol Mnemonic: DLTA

Description: Representation of a 3-phase delta connection.

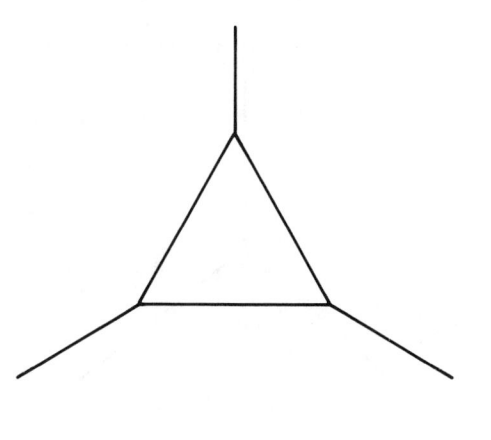

Subgroup: N/A
Symbol Name: Fuse
Symbol Mnemonic: FUSE

Description: Representation of a fuse as an over-current protection device.

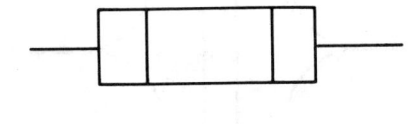

3.3 STRUCTURE OF SYMBOLS

3.3.3 Group: Electrical (Cont'd)

Subgroup: N/A
Symbol Name: Motor
Symbol Mnemonic: MOTR

Description: An ac or dc motor.

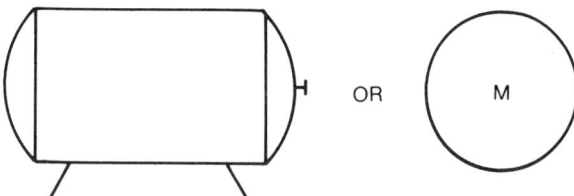

This is the preferred symbol for process diagrams (base optional).

This is the preferred symbol for electrical diagrams.

Subgroup: N/A
Symbol Name State Indicator
Symbol Mnemonic: STAT

Description: Used to represent binary states. For example: Circuit Closed/Circuit Open, etc.

Circuit Closed   Circuit Open

Subgroup: N/A
Symbol Name: Transformer
Symbol Mnemonic: XFMR

Description: A universal transformer.

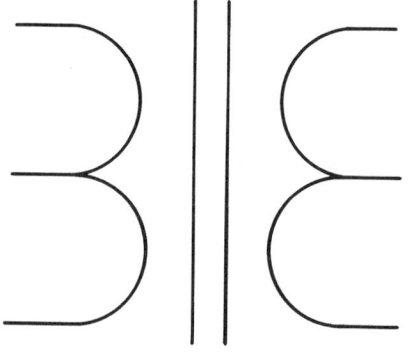

Subgroup: N/A
Symbol Name: WYE Connection
Symbol Mnemonic: WYEC

Description: Representation of a 3-phase wye (star) connection.

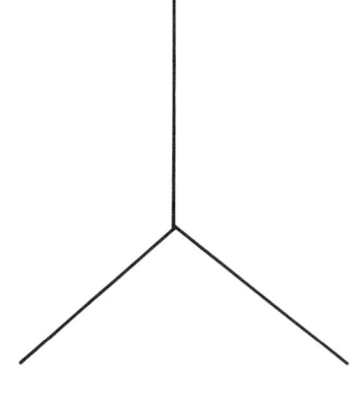

3.3 STRUCTURE OF SYMBOLS

3.3.4 Group: Filters

Subgroup: N/A
Symbol Name: Liquid Filter
Symbol Mnemonic: LFLT

Description: A liquid filter.

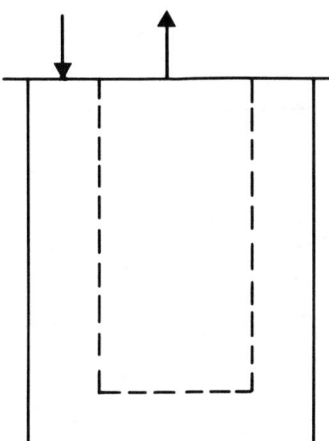

Subgroup: N/A
Symbol Name Vacuum Filter
Symbol Mnemonic: VFLT

Description: A vacuum-assisted filtration device.

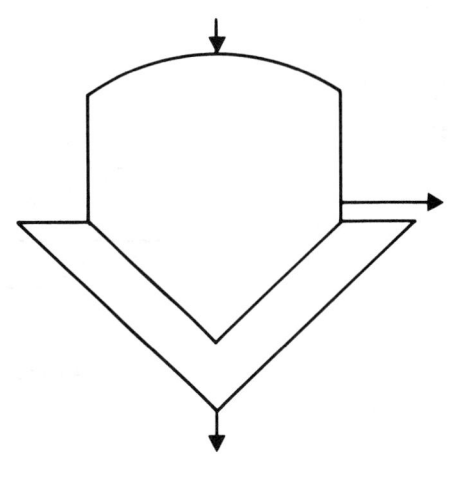

3.3 STRUCTURE OF SYMBOLS

3.3.5 Group: Heat Transfer Devices

Subgroup: N/A
Symbol Name: Exchanger
Symbol Mnemonic: XCHG

Description: Heat transferral equipment. An alternative symbol is depicted.

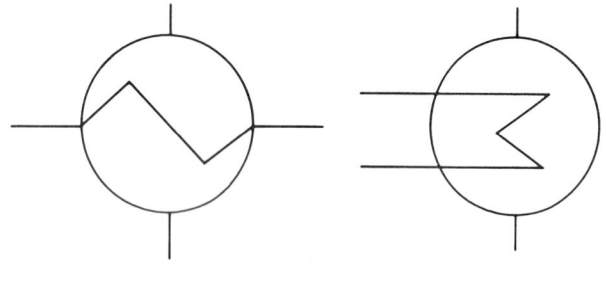

Alternative

Subgroup: N/A
Symbol Name Forced Air Exchanger
Symbol Mnemonic: FAXR

Description: A forced-air heat exchanger.

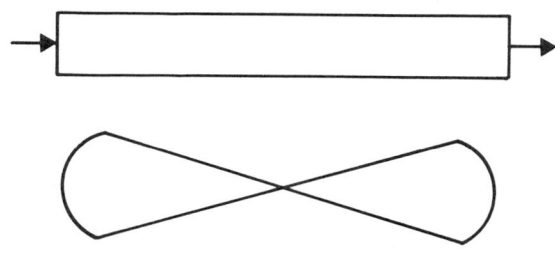

Subgroup: N/A
Symbol Name: Furnace
Symbol Mnemonic: FURN

Description: Process heater or furnace. Internal details may be shown as needed.

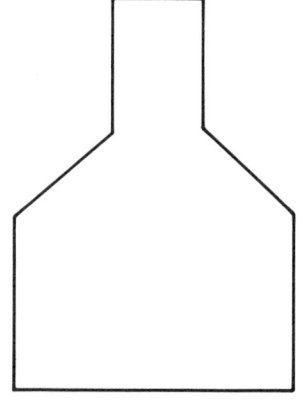

Subgroup: N/A
Symbol Name: Rotary Kiln
Symbol Mnemonic: KILN

Description: Typical gas, oil, coal or coke-fired kiln.

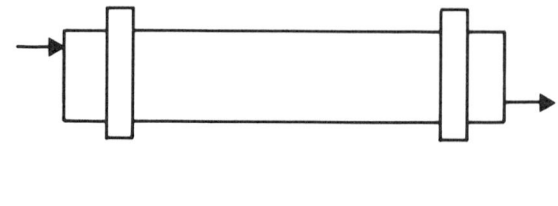

3.3 STRUCTURE OF SYMBOLS

3.3.6 Group: HVAC (Heating Ventilation & Air Conditioning)

Subgroup: N/A
Symbol Name: Cooling Tower
Symbol Mnemonic: CTWR

Description: A device for use in HVAC or other processes indicating the atmospheric cooling of water by forced evaporation.

Subgroup: N/A
Symbol Name Evaporator
Symbol Mnemonic: EVPR

Description: An HVAC device used to represent the exchange of heat between a liquid or gas and a refrigerant.

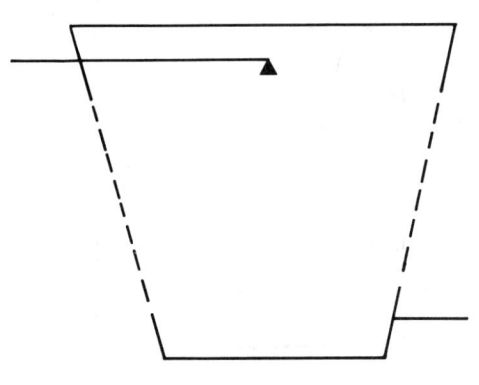

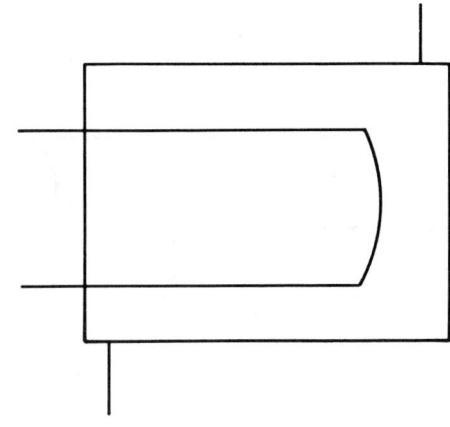

Subgroup: N/A
Symbol Name: Finned Exchanger
Symbol Mnemonic: FNXR

Description: A high surface transfer device used to exchange heat between a liquid or gas and air.

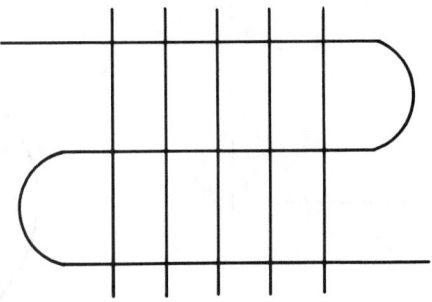

3.3 STRUCTURE OF SYMBOLS

3.3.7 Group: Material Handling

Subgroup: N/A
Symbol Name: Conveyor
Symbol Mnemonic: CNVR

Description: Belt conveyors, chain conveyors, and roller conveyors used in association with other symbols to represent more complex equipment such as a paper machine.

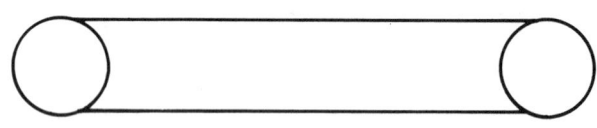

Subgroup: N/A
Symbol Name Mill
Symbol Mnemonic: MILL

Description: Rotating rod, ball, autogenous, or semiautogenous mill used for size reduction of solids.

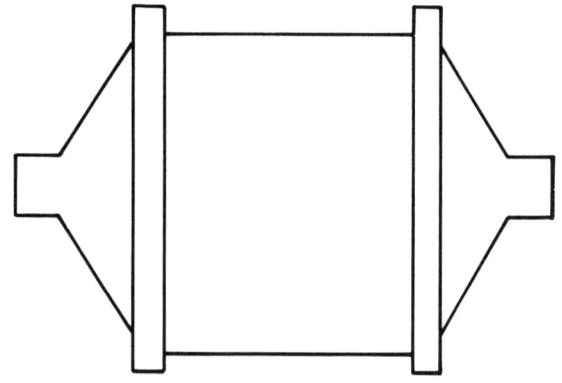

Subgroup: N/A
Symbol Name: Roll Stand
Symbol Mnemonic: RSTD

Description: Roll stand used in metal, paper, rubber, plastic, and glass industries.

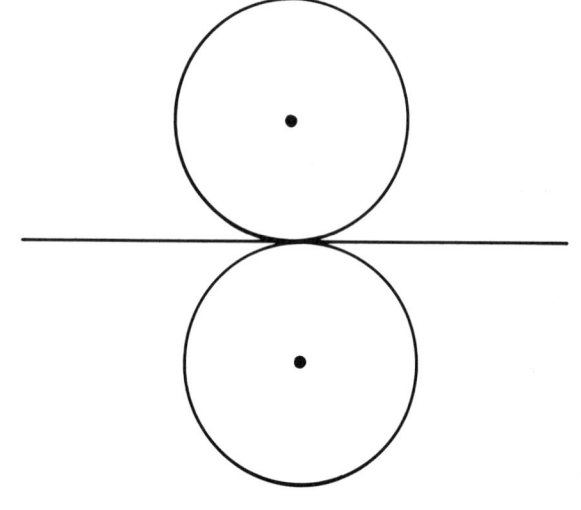

Subgroup: N/A
Symbol Name: Rotary Feeder
Symbol Mnemonic: RFDR

Description: A rotary feeder used to convey material in dry powder form from one location to another.

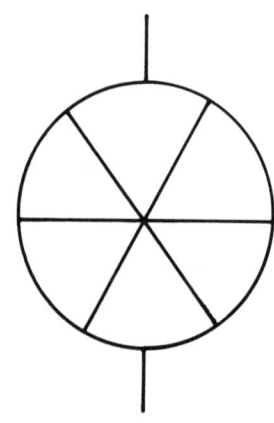

3.3 STRUCTURE OF SYMBOLS

3.3.7 Group: Material Handling (Cont'd)

Subgroup: N/A
Symbol Name: Screw Conveyor
Symbol Mnemonic: SCNV

Description: A typical screw conveyor or screw pump.

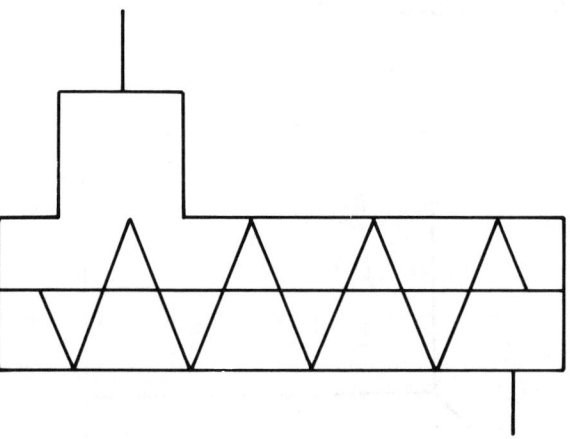

3.3.8 Group: Mixing

Subgroup: N/A
Symbol Name Agitator
Symbol Mnemonic: AGIT

Description: A blade, propeller, or paddle-type agitator.

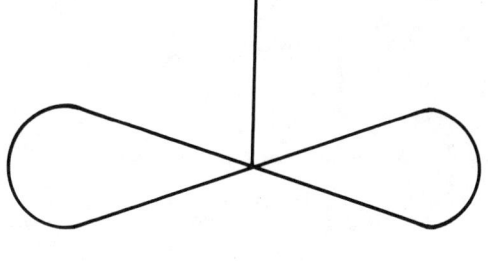

Subgroup: N/A
Symbol Name Inline Mixer
Symbol Mnemonic: IMIX

Description: A mixing device used to continuously blend materials.

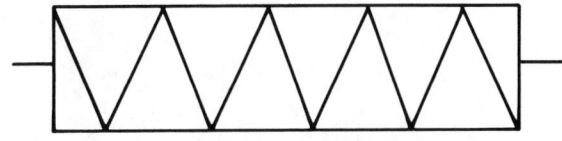

3.3 STRUCTURE OF SYMBOLS

3.3.9 Group: Reciprocating Equipment

Subgroup: N/A
Symbol Name: Reciprocating Compressor
Symbol Mnemonic: RECP

Description: A reciprocating compressor or pump represents that class of equipment used to transport slurries or liquids by reciprocating action. Examples are pistons, diaphragms, plungers, etc.

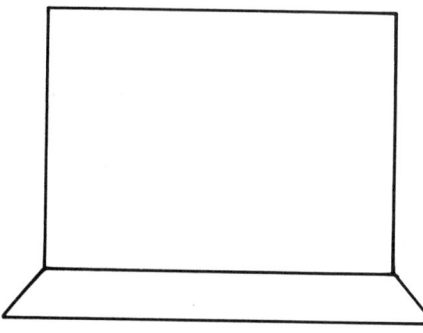

3.3 STRUCTURE OF SYMBOLS

3.3.10 Group: Rotating Equipment

Subgroup: N/A
Symbol Name: Blower
Symbol Mnemonic: BLWR

Description: A device used to convey a gas under slight pressure.

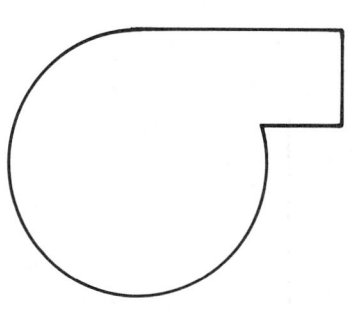

Subgroup: N/A
Symbol Name Compressor
Symbol Mnemonic: CMPR

Description: A device used to convey a gas under high pressure.

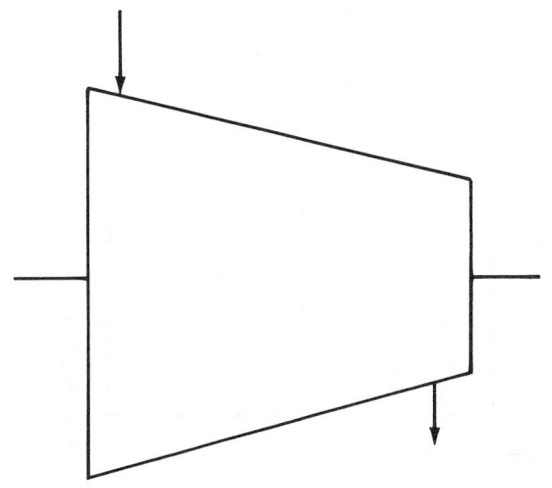

Subgroup: N/A
Symbol Name: Pump
Symbol Mnemonic: PUMP

Description: Represents that class of equipment used to transport slurries or liquids by internal rotatory action. Examples are centrifugal, gear, lobe, etc.

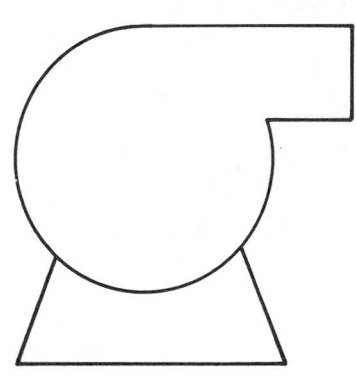

Subgroup: N/A
Symbol Name: Turbine
Symbol Mnemonic: TURB

Description: A device using the force of expanding gas to propel rotating equipment.

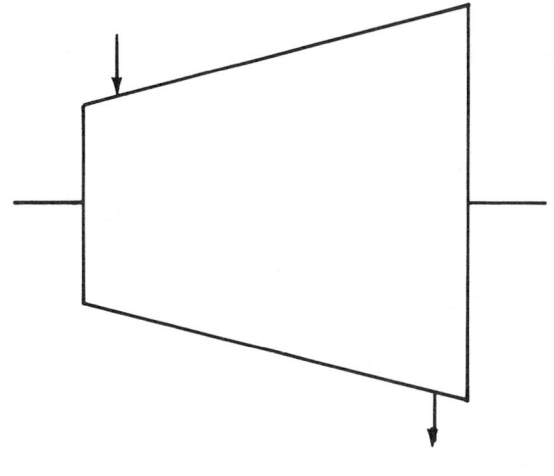

3.3 STRUCTURE OF SYMBOLS

3.3.11 Group: Scrubber and Precipitators

Subgroup: N/A
Symbol Name: Electrostatic Precipitator
Symbol Mnemonic: EPCP

Description: A device used to separate solid particles from a gas (e.g., in a smoke stack) by means of an electrostatically charged grid.

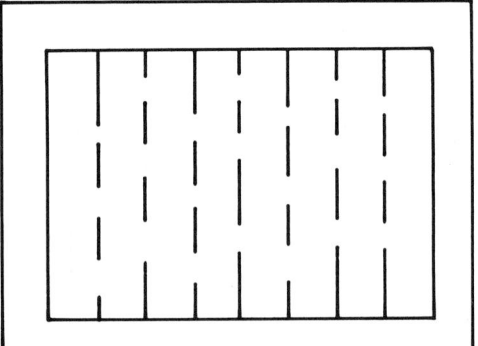

Subgroup: N/A
Symbol Name Scrubber
Symbol Mnemonic: SCBR

Description: A device that uses a liquid spray to scrub gas.

3.3 STRUCTURE OF SYMBOLS

3.3.12 Group: Separators

Subgroup: N/A
Symbol Name: Cyclone Separator
Symbol Mnemonic: CSEP

Description: A device used for solid, liquid, or vapor separation.

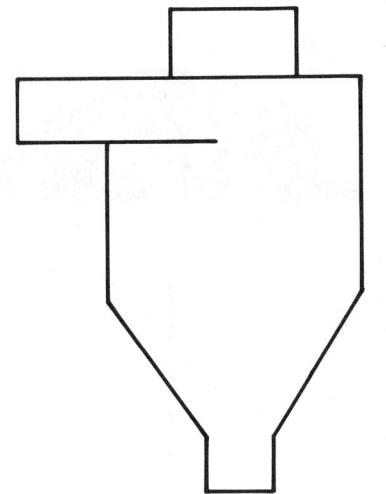

Subgroup: N/A
Symbol Name Rotary Separator
Symbol Mnemonic: RSEP

Description: A rotary device for separating solids from liquids.

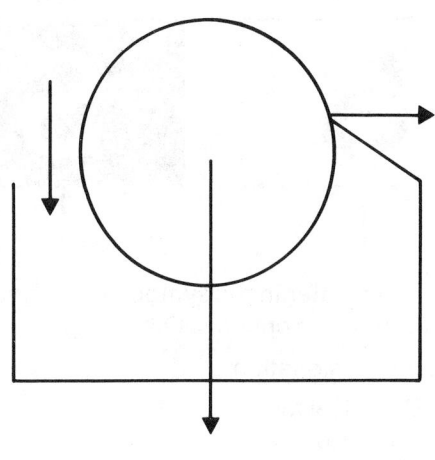

Subgroup: N/A
Symbol Name: Spray Dryer
Symbol Mnemonic: SDRY

Description: A device used for evaporation of liquids from mixtures of solids and liquids.

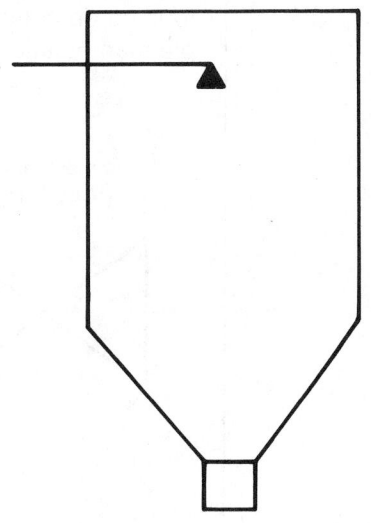

3.3 STRUCTURE OF SYMBOLS

3.3.13 Group: Valves and Actuators

Subgroup: Actuators
Symbol Name: Actuator
Symbol Mnemonic: ACTR

Description: Reprsents the final control element that determines the state of a two-state device.

Desired Device State is CLOSED   Desired Device State is OPEN

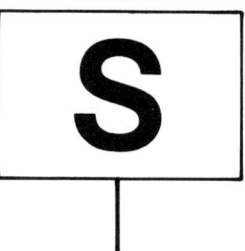 

The use of a letter in the symbol to designate the type of actuator is optional. Other choices include:

Character Designation

M = Electrical Motor
S = Solenoid
H = Hydraulic
A = Air Motor

Subgroup: Actuators
Symbol Name Throttling Actuator
Symbol Mnemonic: TACT

Description: Represents a diaphragm actuator that can affect multiple positions of the controlled device.

Subgroup: Actuators
Symbol Name: Manual Actuator
Symbol Mnemonic: MATR

Description: Represents a manually-operated valve actuator.

Subgroup: Valves
Symbol Name: Valve
Symbol Mnemonic: VLVE

Description: Represents GLOBE, GATE, BALL, and NEEDLE valves used to regulate fluid flow through piping systems. Can be used with various combinations of actuators to convey multiple manipulation schemes.

Actual State is CLOSED   Actual State is OPEN

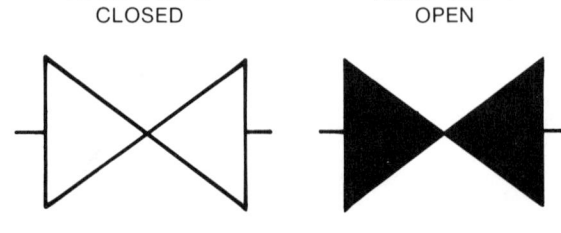

3.3 STRUCTURE OF SYMBOLS

3.3.13 Group: Valves and Actuators (Cont'd)

Subgroup: Valves
Symbol Name: 3-Way Valve
Symbol Mnemonic: VLV3

Description: Represents a valve used in piping systems to select flow paths or regulate between flow paths. Can be used with various combinations of actuators to convey multiple manipulation schemes.

THROTTLING

SELECTING
(Pathway open
only between
Ports 2 & 3)

SELECTING
(Pathway open
only between
Ports 1 & 3)

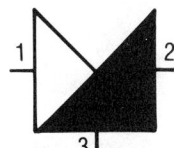
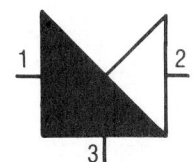

Note: Port numbers are not part of symbol.

Subgroup: Valves
Symbol Name Butterfly Valve
Symbol Mnemonic: BVLV

Description: Represents a butterfly valve, damper, or vane used to throttle (modulate) fluid flow through a pipe, duct, or stack.

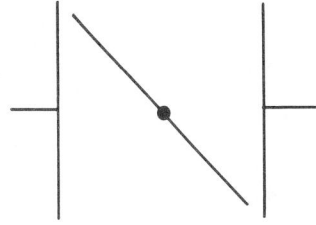

Subgroup: Valves
Symbol Name: Check Valve
Symbol Mnemonic: CVLV

Description: Represents a device that mechanically limits fluid flow to only one direction in a piping system — typically a check valve or back-draft damper.

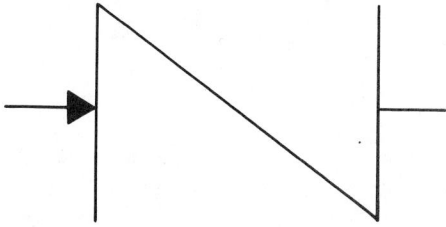

Arrow shows direction of allowable
flow and is part of the symbol.

Subgroup: Valves
Symbol Name: Relief Valve
Symbol Mnemonic: RVLV

Description: Represents a one-way mechanically actuated pressure relief valve. While these valves are normally closed, two symbols are shown to accommodate those situations where feedback signals are provided to indicate actual status.

Normally closed valve
that is actually
CLOSED

Normally closed valve
that is actually
OPEN

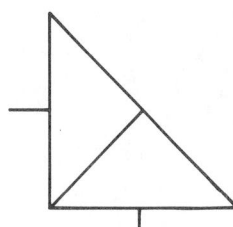
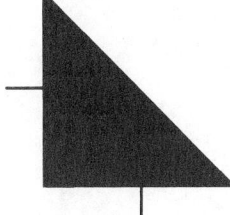

APPENDIX A

EXAMPLES OF USE

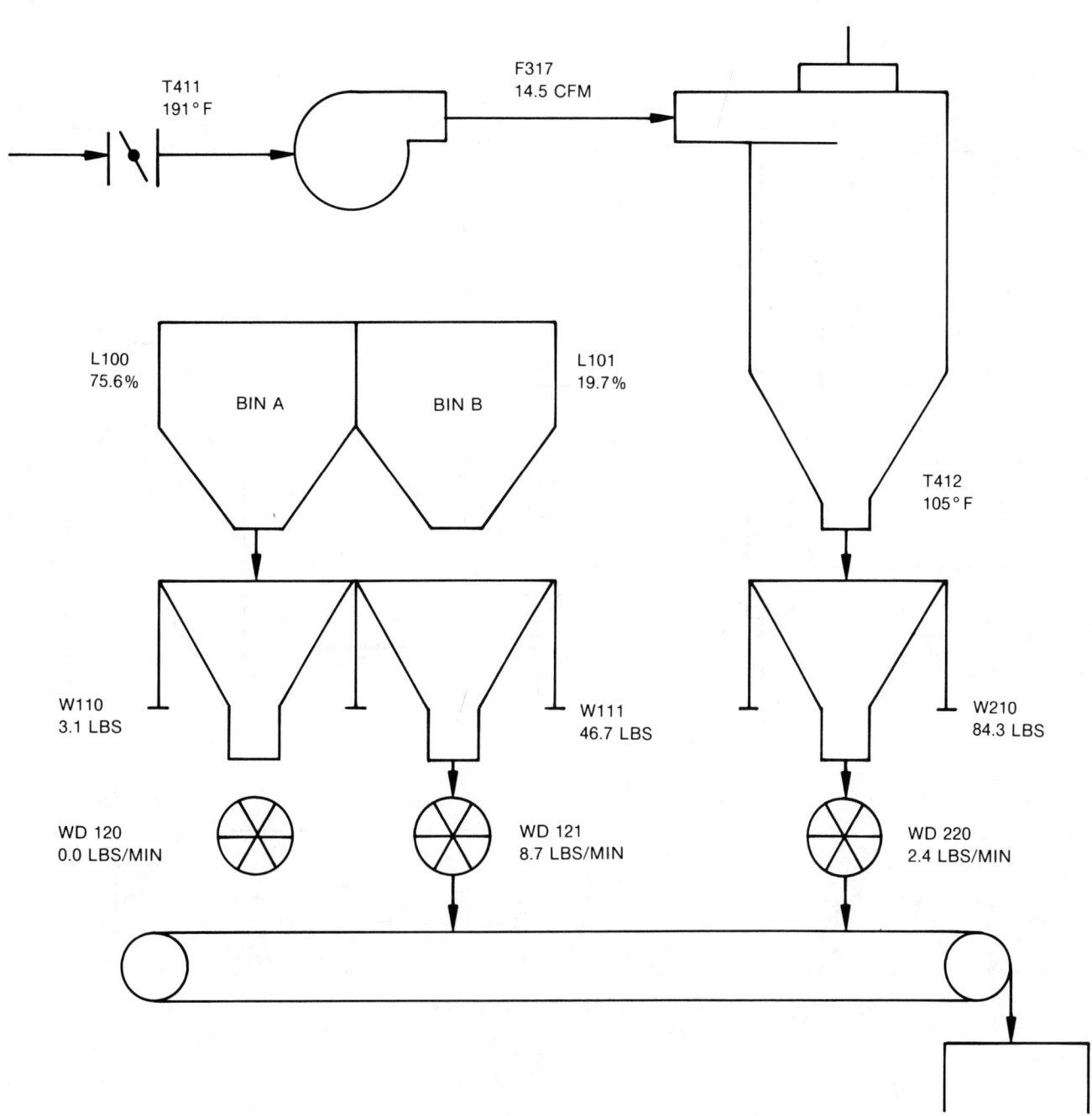

Figure A-1.   Gas Cleaning and Particle Collection

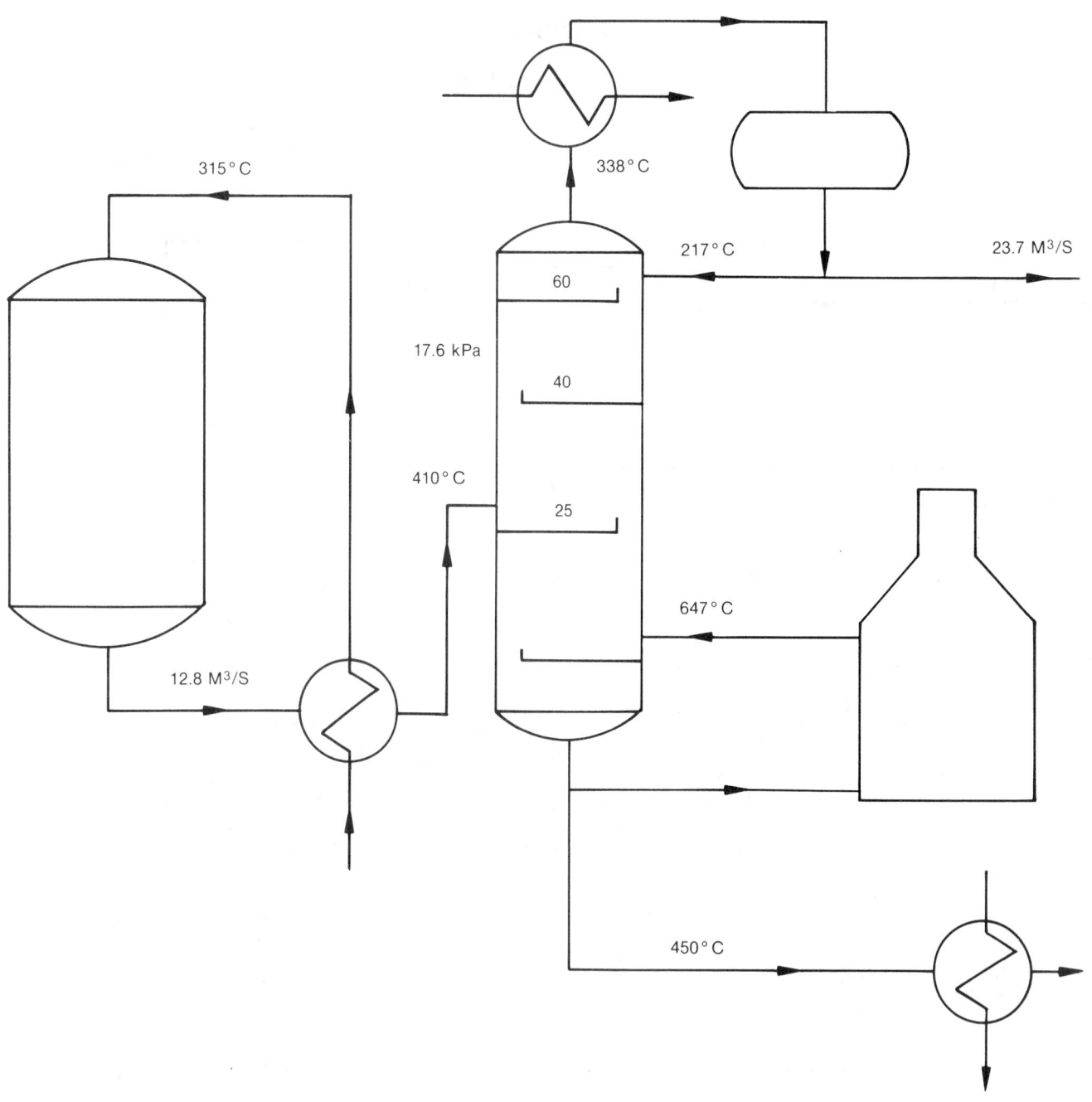

Figure A-2. Chemical Process

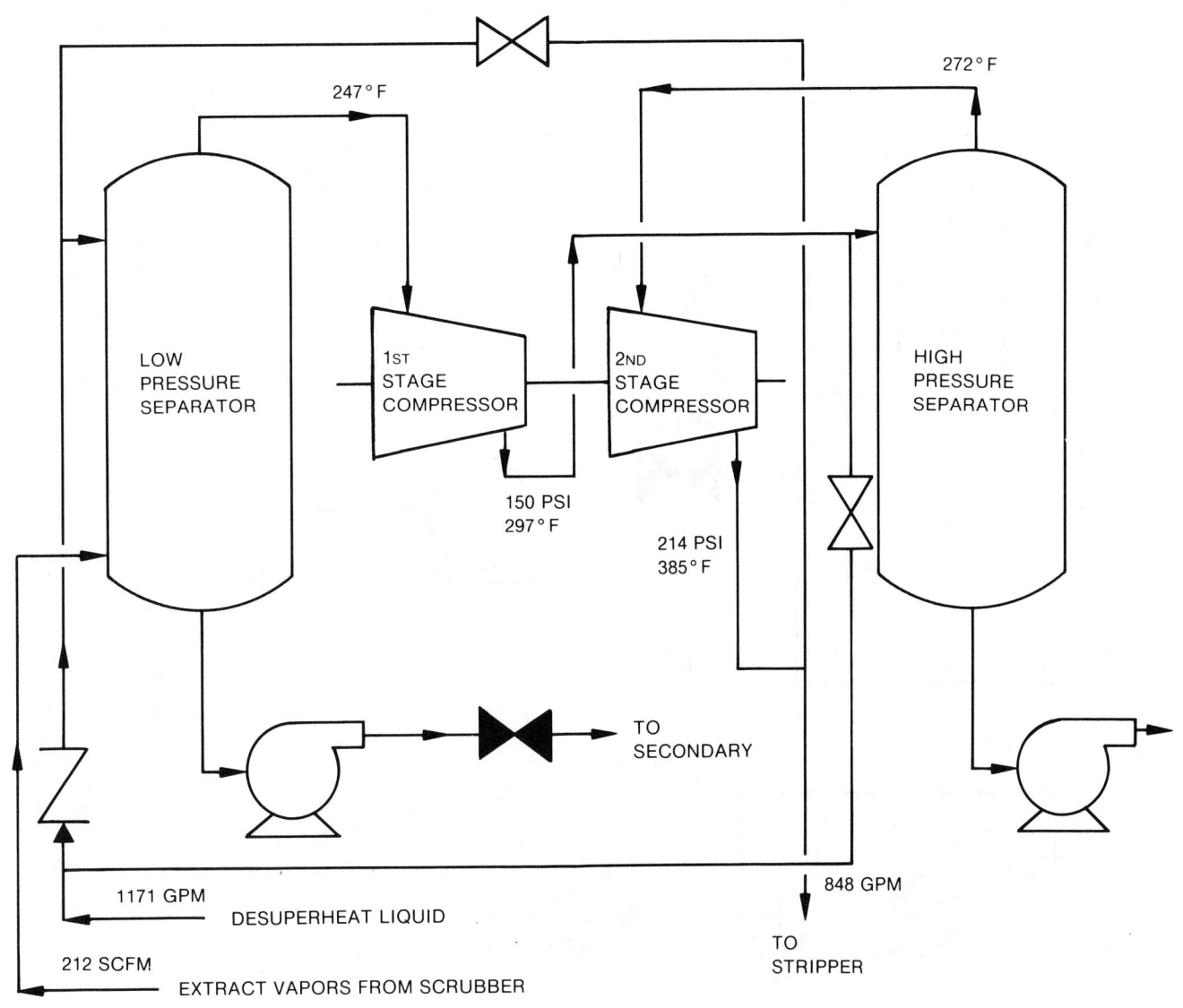

Figure A-3.   Heat Pump System

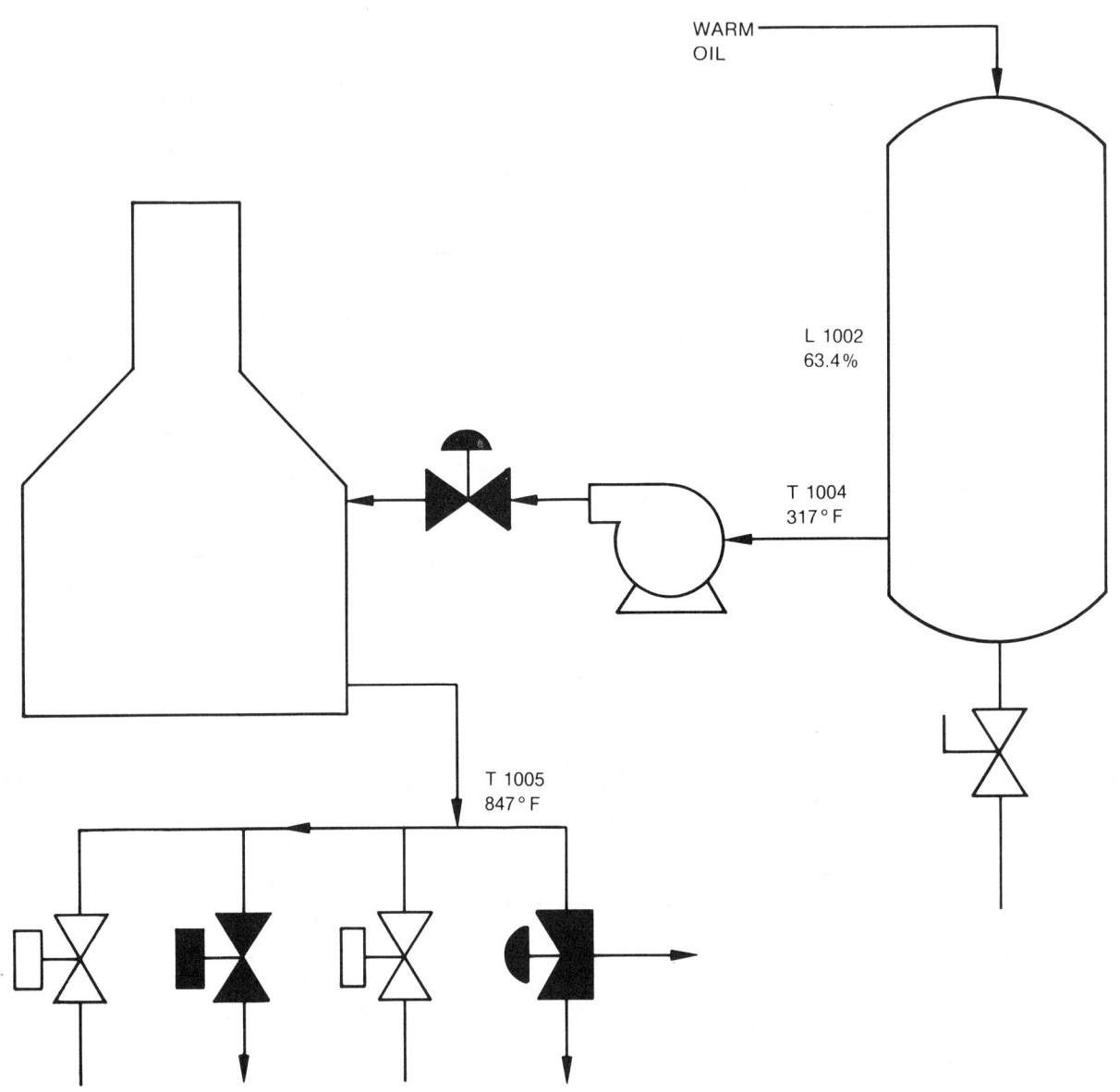

Figure A-4. Hot Oil System

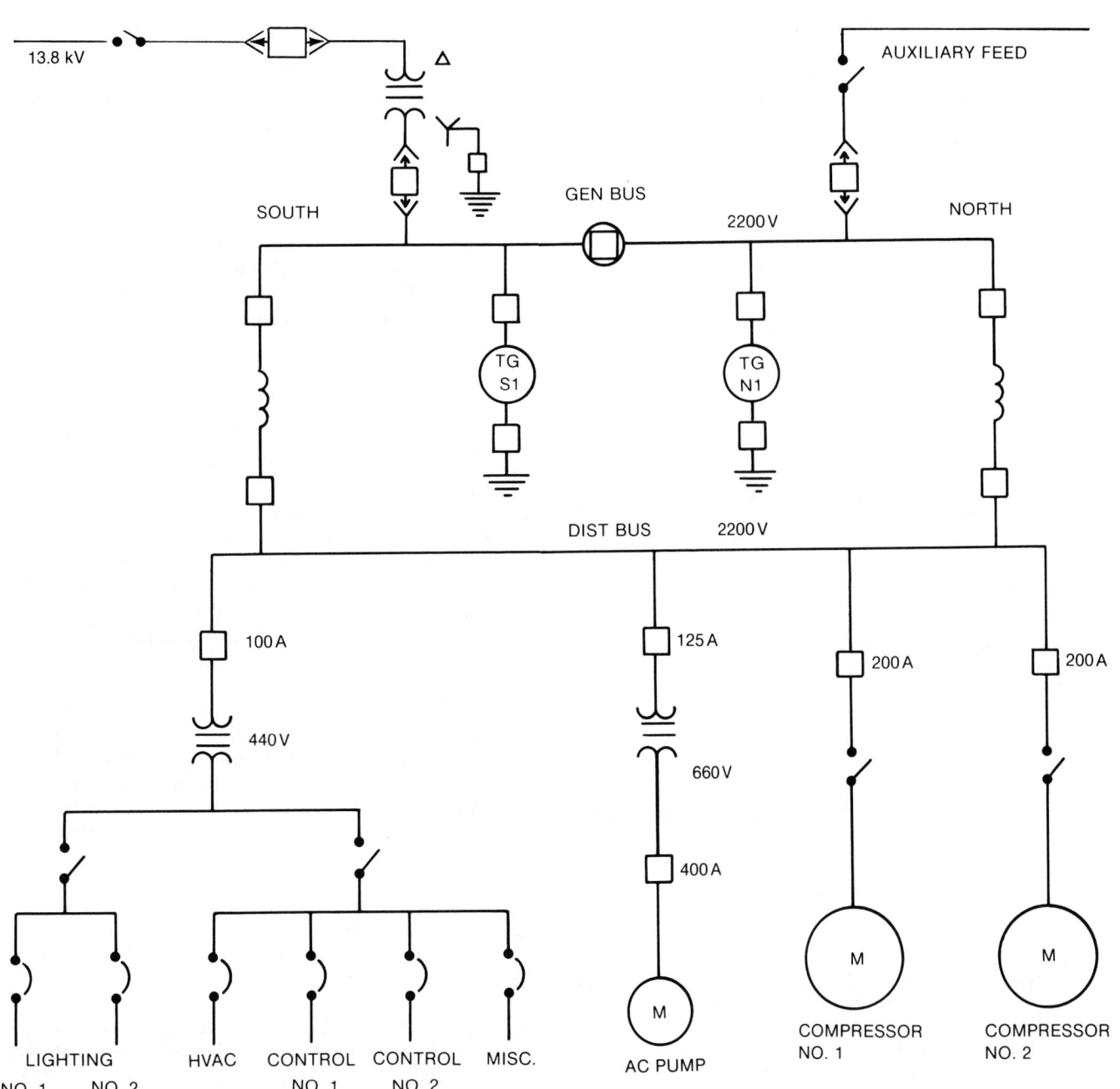

Figure A-5. Electrical Power System

APPENDIX B

PRIMARY MEASUREMENT RECOMMENDED USAGE

Primary elements or instruments can be depicted on a VDU by a character(s). The characters that are recommended for use are:

First Character	Type of Measurement	Notes
A	Analysis	4
B	Burner, Combustion	
C	User's Choice	1
D	User's Choice	1
E	Voltage (EMF)	
F	Flow Rate	
G	User's Choice	1
H	Hand (Manual)	
I	Current (Electric)	
J	Power	
K	Time	
L	Level	
M	User's Choice	1
N	User's Choice	1
O	User's Choice	1
P	Pressure/Vacuum	
Q	Quantity	
R	Radiation	
S	Speed, Frequency	
T	Temperature	
U	Multivariable	5
V	Vibration, Mechanical Analysis	6
W	Weight, Force	
X	Unclassified	2
Y	Event, State or Presence	7
Z	Position, Dimension	

First Modifier*	Type of Measurement	Notes
D	Differential	
F	Ratio	
K	Time Rate of Change	8
Q	Integrate or Totalize	

*(See Note 3)

The above character designations are based on ISA Standard S5.1, "Instrumentation Symbols and Identification".

NOTES:

1. A "USER'S CHOICE" letter is intended to cover unlisted meanings for primary measurements that will be used repetitively in a particular project. If used, the letter will have one meaning as a first letter and a different meaning for the second letter. The meanings need be defined only once in the beginning of the project. For example, the letter 'M' may be defined as "MOISTURE" in one project, but as "MASS" in another.

2. The unclassified letter 'X' is intended to cover unlisted meanings that will be used only once or to a limited extent. If used, the letter may have any number of meanings as a first letter and any number of meanings as a succeeding letter.

   Except for its use with distinctive measurements, it is expected that the meaning will be defined outside the symbol. For example, 'X' may be a stress measurement at one point and a volume measurement at another point.

   The units of the quantity measured will assist in determining the actual usage of the letter 'X'.

3. Any first letter, if used in combination with modifying letters 'D' (differential), 'F' (ratio), 'K' (time rate of change), or 'Q' (integrate or totalize), or any combination of them, shall be construed to represent a new and separate measured variable, and the combination shall be treated as a first-letter entity. Thus, instrument measurements 'T' and 'TD' measure two different variables, namely, temperature and differential temperature. These modifying letters shall be used when applicable.

4. First letter 'A' for analysis covers all analyses not described by a "USER'S CHOICE" letter. It is expected that the type of analysis will be defined outside the symbol. The units of the quantity measured will assist in determining the actual type of analysis occurring. Additional information can be added as text to the visual display unit.

5. Use of the first letter 'U' for "Multivariable" in lieu of a combination of first letters is optional. It is recommended that nonspecific designators such as 'U' be used sparingly.

6. First letter 'V', "Vibration or Mechanical Analysis", is intended to perform the duties in machinery monitoring that the letter 'A' performs in more general analyses. Except for vibration, it is expected that the variable of interest will be defined outside the actual symbol. This definition can occur as a result of units of the quantity measured or as additional text shown on the visual display unit.

7. First letter 'Y' is intended for use when control or monitoring responses are event-driven as opposed to time or time-schedule driven. It can also signify presence or state.

8. Second letter 'K', in combination with a first letter such as 'L', 'T', or 'W', signifies a time rate of change of the primary measurement. As an example, 'WK' may represent "Rate of Weight Loss or Gain".

The following are Identification Letters and their usage from ISA Standard S5.1, "Instrumentation Symbols and Identification," Revision 4.

TABLE B-1
Identification Letters

First Letter (4)		Succeeding Letters (3)		
Measured or Initiating Variable	Modifier	Readout or Passive Function	Output Function	Modifier
A Analysis (5, 19)		Alarm		
B Burner, Combustion		User's Choice (1)	User's Choice (1)	User's Choice (1)
C User's Choice (1)			Control (13)	
D User's Choice (1)	Differential (4)			
E Voltage		Sensor (Primary Element)		
F Flow Rate	Ratio (Fraction) (4)			
G User's Choice (1)		Glass, Viewing Device (9)		
H Hand				High (7, 15, 16)
I Current (Electrical)		Indicate (10)		
J Power	Scan (7)			
K Time, Time Schedule	Time Rate of Change (4, 21)		Control Station (22)	
L Level		Light (11)		Low (7, 15, 16)
M User's Choice (1)	Momentary (4)			Middle, Intermediate (7, 15)
N User's Choice (1)		User's Choice (1)	User's Choice (1)	User's Choice (1)
O User's Choice (1)		Orifice, Restriction		
P Pressure, Vacuum		Point (Test) Connection		
Q Quantity	Integrate, Totalize (4)			
R Radiation		Record (17)		
S Speed, Frequency	Safety (8)		Switch (13)	
T Temperature			Transmit (18)	
U Multivariable (6)		Multifunction (12)	Multifunction (12)	Multifunction (12)
V Vibration, Mechanical Analysis			Valve, Damper, Louver (13)	

TABLE B-1
Identification Letters (Cont'd)

First Letter (4)		Succeeding Letters (3)		
Measured or Initiating Variable	Modifier	Readout or Passive Function	Output Function	Modifier
W Weight, Force		Well		
X Unclassified (2)	X Axis	Unclassified (2)	Unclassified (2)	Unclassified (2)
Y Event, State or Presence (20)	Y Axis		Relay, Compute, Convert (13, 14, 18)	
Z Position Dimension	Z Axis		Driver, Actuator, Unclassified Final Control Element	

Notes for Table B-1:

1. A "USER'S CHOICE" letter is intended to cover unlisted meanings that will be used repetitively in a particular project. If used, the letter may have one meaning as a first letter and another meaning as a succeeding letter. The meanings need to be defined only once in a legend, or otherwise, for that project. For example, the letter 'N' may be defined as "MODULUS OF ELASTICITY" as a first letter and "OSCILLOSCOPE" as a succeeding letter.

2. The unclassified letter 'X' is intended to cover unlisted meanings that will be used only once or to a limited extent. If used, the letter may have any number of meanings as a first letter and any number of meanings as a succeeding letter. Except for its use with distinctive symbols, it is expected that the meanings will be defined outside a tagging bubble on a flow diagram. For example, XR-2 may be a stress recorder and XX-4 may be a stress oscilloscope.

3. The grammatical form of the succeeding letter meanings may be modified as required. For example, "indicate" may be applied as "indicator" or "indicating," "transmit" as "transmitter" or "transmitting," etc.

4. Any first letter, if used in combination with modifying letters 'D' (differential), 'F' (ratio), 'M' (momentary), 'K' (time rate of change), 'Q' (integrate or totalize), or any combination of these is intended to represent a new and separate measured variable, and the combination is treated as a first-letter entity. Thus, instruments 'TDI' and 'TI' indicate two different variables, namely, differential temperature and temperature. Modifying letters are used when applicable.

5. First letter 'A,' "Analysis," covers all analyses not described by a "USER'S CHOICE" letter. It is expected that the type of analysis will be defined outside a tagging bubble.

6. Use of first letter 'U' for "Multivariable" in lieu of a combination of first letters is optional. It is recommended that nonspecific designators such as 'U' be used sparingly.

7. The use of modifying terms "high," "low," "middle" or "intermediate," and "scan" is optional.

8. The term "safety" applies to emergency protective primary elements and emergency protective final control elements only. Thus, a self-actuated valve that prevents operation of a fluid system at a higher than desired pressure by bleeding fluid from the system is a backpressure-type PCV, even if the valve is not intended to be used normally. However, this valve is designated as a PSV if it is intended to protect against emergency conditions, i.e., conditions that are hazardous to personnel and/or equipment and that are not expected to arise normally.

The designation 'PSV' applies to all valves intended to protect against emergency pressure conditions regardless of whether the valve construction and mode of operation place them in the category of the safety valve, relief valve, or safety relief valve. A rupture disc is designated 'PSE.'

9. The passive function 'G' applies to instruments or devices that provide an uncalibrated view such as sight glasses and television monitors.

10. "Indicate" normally applies to the readout, analog or digital, of an actual measurement. In the case of a manual loader, it may be used for the dial or setting indication, i.e., for the value of the initiating variable.

11. A pilot light that is part of an instrument loop should be designated by a first letter followed by the succeeding letter 'L.' For example, a pilot light that indicates an expired time period should be tagged 'KQL.' If it is desired to tag a pilot light that is not part of an instrument loop, the light is designated in the same way. For example, a running light for an electric motor may be tagged 'EL,' assuming voltage to be the appropriate measured variable, or 'YL,' assuming the operating status is being monitored. The unclassified variable 'X' should be used only for applications that are limited in extent. 'XL' should not be used for motor running lights as these are commonly numerous. It is permissible to use the "USER'S CHOICE" letters 'M,' 'N,' or 'O' for a motor running light when the meaning is previously defined. If 'M' is used, it must be clear that the letter does not stand for the word "Motor," but for a monitored state.

12. Use of a succeeding letter 'U' for "Multifunction" instead of a combination of other functional letters is optional. This nonspecific variable designator should be used sparingly.

13. A device that connects, disconnects, or transfers one or more circuits may be either a switch, a relay, an ON-OFF controller, or a control valve, depending on the application.

    If the device manipulates a fluid process stream and is not a hand-actuated ON-OFF block valve, it is designated as a control valve. It is incorrect to use the succeeding letters 'CV' for anything other than a self-actuated control valve. For all applications, other than fluid process streams, the device is designated as follows:

    A *switch*, if it is actuated by hand.

    A switch or an ON-OFF controller, if it is automatic and is the first such device in a loop. The term "Switch" is generally used if the device is used for alarm, pilot light, selection, interlock, or safety. The term "Controller" is generally used if the device is used for normal operating control.

    A relay, if it is automatic and is not the first such device in a loop, i.e., it is actuated by a switch or an ON-OFF controller.

14. It is expected that the functions associated with the use of succeeding letter 'Y' will be defined outside a bubble on a diagram when further definition is considered necessary. This definition need not be made when the function is self-evident, as for a solenoid valve in a fluid signal line.

15. The modifying terms "high," "low," and "middle" or "intermediate" correspond to values of the measured variable, not of the signal, unless otherwise noted. For example, a high-level alarm derived from a reverse-acting level transmitter signal shall be an 'LAH,' even though the alarm is actuated when the signal falls to a low value. The terms may be used in combinations as appropriate (see Section 6.9A ISA-S5.1).

16. The terms "high" and "low," when applied to positions of valves and other open-close devices, are defined as follows: "high" denotes that the valve is in or approaching the fully open position, and "low" denotes it is in or approaching the fully closed position.

17. The word "record" applies to any form of permanent storage of information that permits retrieval by any means.

18. For use of the term "transmitter" versus "converter," see the definitions in Section 3, ISA-S5.1.

19. First letter 'V,' "Vibration or Mechanical Analysis," is intended to perform the duties in machinery monitoring that the letter 'A' performs in more general analyses. Except for vibration, it is expected that the variable of interest will be defined outside the tagging bubble.

20. First letter 'Y' is intended for use when control or monitoring responses are event-driven as opposed to time- or time-schedule-driven. 'Y,' in this position, can also signify presence or state.

21. Modifying letter 'K,' in combination with a first letter, such as 'L,' 'T,' or 'W,' signifies a time rate of change of the measured or initiating variable. 'WKIC,' for instance, may represent a rate-of-weight-loss controller.

22. Succeeding letter 'K' is a user's option for designating a *control station*, while the succeeding letter 'C' is used for describing automatic or manual *controllers*. See Definitions, ISA S5.1.

TABLE B-2

Typical Letter Combinations

First Letters	Initiating or Measured Variable	Controllers Recording	Controllers Indicating	Controllers Blind	Self-Actuated Control Valves	Readout Recording	Readout Indicating	Switches High	Switches Low†	Switches Comb	Transmitters Recording	Transmitters Indicating	Transmitters Blind	Solenoids Relays Computing Devices	Primary Element	Test Point	Well or Probe	Viewing Device Glass	Safety Device	Element
A	Analysis	ARC	AIC	AC		AR	AI	ASH	ASL	ASHL	ART	AIT	AT	AY	AE	AP	AW			AV
B	Burn./Comb.	BRC	BIC	BC		BR	BI	BSH	BSL	BSHL	BRT	BIT	BT	BY	BE		BW	BG		BZ
C	User's Choice																			
D	User's Choice																			
E	Voltage	ERC	EIC	EC		ER	EI	ESH	ESL	ESHL	ERT	EIT	ET	EY	EE					EZ
F	Flow Rate	FRC	FIC	FC	FCV, FICV	FR	FI	FSH	FSL	FSHL	FRT	FIT	FT	FY	FE	FP		FG		FV
FQ	Flow Quantity	FQRC	FQIC			FQR	FQI	FQSH	FQSL			FQIT	FQT	FQY	FQE					FQV
FF	Flow Ratio	FFRC	FFIC	FFC		FFR	FFO	FFSH	FFSL						FE					FV
G	User's Choice																			
H	Hand		HIC	HC																
I	Current	IRC	IIC			IR	II	ISH	ISL	ISHL	IRT	IIT	IT	IY	IE					IZ
J	Power	JRC	JIC			JR	JI	JSH	JSL	JSHL	JRT	JIT	JT	JY	JE					JV
K	Time	KRC	KIC	KC	KCV	KR	KI	KSH	KSL	KSHL	KRT	KIT	KT	KY	KE					KV
L	Level	LRC	LIC	LC	LCV	LR	LI	LSH	LSL	LSHL	LRT	LIT	LT	LY	LE		LW	LG		LV
M	User's Choice																			
N	User's Choice																			
O	User's Choice																			
P	Press./Vacuum	PRC	PIC	PC	PCV	PR	PI	PSH	PSL	PSHL	PRT	PIT	PT	PY	PE	PP			PSV, PSE	PV
PD	Press./Diff.	PDRC	PDIC	PDC	PDCV	PDR	PDI	PDSH	PDSL		PDRT	PDIT	PDT	PDY	PDE	PDP				PDV
Q	Quantity	QRC	QIC			QR	QI	QSH	QSL	QSHL	QRT	QIT	QT	QY	QE					QZ
R	Radiation	RRC	RIC	RC		RR	RI	RSH	RSL	RSHL	RRT	RIT	RT	RY	RE		RW			RZ
S	Speed/Frequency	SRC	SIC	SC	SCV	SR	SI	SSH	SSL	SSHL	SRT	SIT	ST	SY	SE					SV
T	Temperature	TRC	TIC	TC	TCV	TR	TI	TSH	TSL	TSHL	TRT	TIT	TT	TY	TE	TP	TW			TV
TD	Temperature/Diff.	TDRC	TDIC	TDC	TDCV	TDR	TDI	TDSH	TDSL	TDSHL	TDRT	TDIT	TDT	TDY	TE	TP	TW		TSE	TDV
U	Multivariable					UR	UI							UY						UV
V	Vibration Machinery Analysis					VR	VI	VSH	VSL	VSHL	VRT	VIT	VT	VY	VE					VZ
W	Weight/Force	WRC	WIC	WC	WCV	WR	WI	WSH	WSL	WSHL	WRT	WIT	WT	WY	WE					WZ
WD	Weight/Force/Diff.	WDRC	WDIC	WDC	WDCV	WDR	WDI	WDSH	WDSL	WDSHL	WDRT	WDIT	WDT	WDY	WDE					WDZ
X	Unclassified																			
Y	Event, State Presence			YC		YR	YI	YSH	YSL				YT	YY	YE					YZ
Z	Pos./Dimen.	ZRC	ZIC	ZC	ZCV	ZR	ZI	ZSH	ZSL	ZSHL	ZRT	ZIT	ZT	ZY	ZE					ZV
ZD	Gaug/Devia.	ZDRC	ZDIC	ZDC	ZDCV	ZDR	ZDI	ZDSH	ZDSL	ZDSHL	ZDRT	ZDIT	ZDT	ZDY	ZDE					ZDV

NOTE: This table is not all inclusive.

* A. alarm, the annunciating device. may be used in the same fashion as S. switch. the actuating device.

† The letters H and L may be omitted in the undefined case.

Other Possible Combinations:

FO	(Restriction Orifice)	PFR	(Ratio)
FRK, HIK	(Control Stations)	KQI	(Running Time Indicator)
FX	(Accessories)	QQI	(Indicating Counter)
TJR	(Scanning Recorder)	WKIC	(Rate-of-Weight Loss Controller)
LLH	(Pilot Light)	HMS	(Hand Momentary Switch)

Recommended Practice

Pneumatic Control Circuit Pressure Test

Instrument Society of America

ISBN 0-87664-333-0

ISA-S7.1   Pneumatic Control Circuit Pressure Test

INSTRUMENT SOCIETY OF AMERICA
67 Alexander Drive
P.O. Box 12277
Research Triangle Park, North Carolina 27709

FOREWORD

This Tentative Recommended Practice (Issued March 21, 1955, and Revised August 1, 1956) has been prepared as a part of the service of the Instrument Society of America toward a goal of uniformity in the field of instrumentation. Toward this end the Society welcomes all comments and criticisms, and asks that they be addressed to the Standards and Practices Board Secretary, Instrument Society of America, 67 Alexander Drive, P.O. Box 12277, Research Triangle Park, North Carolina 27709.

This revision introduces an improved test assembly and incorporates editorial changes.

This revision was prepared by the Subcommittee on Pneumatic Circuits (RP-7):

R. U. Stanley, Chairman 1956 - Standard Oil Company of California

N. A. Austin - Standard Oil Company of California

A. Carpenter - Fischer & Porter Company

E. C. Cottingham - Shell Development Company

G. W. Jacobs - Shell Development Company

J. W. Jones - Jensen Instrument Company

J. T. Kirkland - Bushnell Controls & Equipment Co., Inc.

E. E. Larson - G. R. Friederich & Company

J. Power - Swanson Engineering & Manufacturing Company

J. F. Rogers - Samuel Moore & Company

N. S. Waner - Hallikainen Instruments, Inc.

Approved for Tentative Publication by Main Recommended Practices Committee on July 1, 1956.

A. V. Novak, Chairman — E. I. du Pont de Nemours & Co., Inc.
C. W. Bates — Humble Oil & Refining Co.
R. E. Clarridge — Taylor Instrument Co.
W. A. Crawford — E. I. du Pont de Nemours & Co., Inc.
E. S. Day — General Electric Company
G. G. Gallagher — Fluor Corporation

T. B. McCraray — Dow Chemical Company
R. L. Patton — Gulf Oil Corporation
J. W. Percy — United States Steel Corporation
E. D. Phillips — Aro, Inc.
C. A. Prior — Diamond Alkali Company
J. E. Read — E. I. du Pont de Nemours & Co., Inc.

The following have reviewed the report and served as a Board of Review:

H. D. Bishop — West V. Pulp & Paper Co.
D. M. Boyd, Jr. — Universal Oil Products Co.
W. J. Claffie, Jr. — Winthrop-Stearns Co.
Hart U. Fisher — Niagara Blower Company
Henry C. Frost — Corn Products Refining Co.
Louis Gess — Minneapolis-Honeywell Regulator Company
Homer C. Givens — La Gloria Corporation
H. H. Gorie — Bailey Meter Company
R. W. Groendycke — Cities Service Oil Co.
C. R. Horton — Roy Horton Engineering

N. I. Isenhour — Carbide & Carbon Chemical Company
A. H. Keyser — Minneapolis-Honeywell Regulator Company
D. J. Lane — Leeds & Northrup Company
George A. Larsen — The Texas Company
Dale E. Mattix — Cities Service Oil Co.
R. Robert Proctor — Pure Oil Company
Fred S. Rich — Arabian-American Oil Co.
F. L. Spies — J. F. Pritchard & Company
Robert N. Wilson — Kelly Springfield Tire Co.

1. PURPOSE

1.1--This Recommended Practice is made to provide a satisfactory procedure for the testing of pneumatic control circuits for leaks together with reasonable criteria for acceptance of work done and suitable aids for performance.

2. INTRODUCTION

2.1--The procedures described are a consolidation of the best ideas now being used.

2.2--The permissible leakage tolerance in a pneumatic control circuit cannot be critically defined. These circuits, including instrument pilots, vary in characteristics to the extent that a serious leak in a given control circuit may not be serious in another.

2.3--Widely different methods of testing are in current use varying in:

2.3.1--Test pressures at maximum pilot pressures or at some higher pressure. This report recommends test at maximum pilot pressure to avoid the danger of damaging instrument elements by over-pressure.

2.3.2--Static or cycling pressures.

2.3.3--Time duration of holding test pressures.

2.3.4--Incremental time of holding test pressure as a function of tubing size and length and terminal capacity of controlled element.

2.4--Designs of pneumatic control circuits should be scrutinized closely to minimize the number of probable sources of leakage toward obtaining a dependable pneumatic circuit. The fewer fittings, valves, etc., the less chance for leakage. Infrequently used test connections sometimes installed for operator or mechanic convenience should be minimized.

3. SCOPE OF APPLICATION

3.1--Pneumatic Control Circuit Pressure Test RP-7.1 may be used in part or in whole for the following:

3.1.1--To specify testing to be done by contractor or operators at plant start-up.

3.1.2--To guide trouble-shooting activities by instrument men.

4. TESTING AND INSPECTIONS REQUIRED PRIOR TO PRESSURE TESTING

4.1--Establish identity of each component of a given control loop and its conformance to design specification. This commonly includes, but is not limited to, air supply valves, transmitter, controller; recorder or indicator, control panel bulkhead connections; field cable terminal connections, control valve or device, etc.

4.2--Blow air lines clean.

4.3--Identify positively the air circuit from air supply valve through instrument to air-operated element by setting instrument to deliver air and observing air leakage at the disconnected air line at the air-operated element.

5. PRESSURE TEST

5.1--Disconnect air-consuming pilots, if any, from the section of tubing to be tested.

5.2--Connect the test assembly shown in Figure RP-7.1.1 to the tubing to be tested and to the air supply.

5.3--Operation of the test assembly for a 3-15 psig system:

5.3.1--Open the toggle valve and adjust the pressure regulator until the red pointer reads 15 psi.

5.3.2--Adjust precision relay until black pointer reads 15 psi.

5.3.3--Close toggle valve. A drop in pressure as indicated by the black pointer will show a tubing leak.

6. PERMISSIBLE LEAK TOLERANCE

6.1--Tubing leaks which cause the black pointer to decrease at the rate of 1 psi per hundred feet of 1/4-inch tubing per 5 seconds (1 psi per 200 feet of 1/4-inch tubing per 10 seconds) are usually not acceptable. Instrument air lines terminated in control valve topworks must be corrected by converting the topworks capacity to equivalent feet of tubing by means of the following table:

Conventional Control Valve Topworks	Approximate Equivalent Feet of 1/4-inch Copper Tubing
Valves with positioners	0
11" OD topworks	104
15" OD topworks	385
17" OD topworks	770
20" OD topworks	1050

7. PROBABLE SOURCES OF LEAKS

7.1--Suggested items are:

7.1.1--Valves--packed or packless. Faulty packing or seals. Porous castings.

7.1.2--Valve operators--diaphragm seal, packing on reverse acting topworks.

7.1.3--Fittings--screwed. Joint not made up tight, improper selection or use of compound, poor threads, body porosity.

7.1.4--Fittings--flared. Incorrect flare, cracked flare, foreign material under flare.

7.1.5--Fittings--compression. Improperly made joint, poor ferrule, ferrule reversed, no bite.

7.1.6--Fittings--brazed or soldered. Improperly made joint.

7.1.7--Metallic tubing--at points where work hardened by bending or vibration, chemical or atmospheric corrosion, galvanic corrosion.

7.1.8--Plastic tubing--at sharp edged supports, overheated, chemical attack or deterioration.

7.1.9--Cable tubing--at cable ends (damage when cutting sheath).

7.1.10--Instruments--seldom are leaks found in the instrument or control element.

7.1.11--Welded bulkhead connections--pin holes at welds.

8. LOCATION OF LEAKS

8.1--Paint suspected leaking parts with any suitable commercial bubble fluid or soap solution.

9. EQUIPMENT LIST

9.1--Source of clean dry air at 25 to 100 psi.

9.2--Test assembly per Figure 7.1.1.

9.3--Soap solution.

9.4--Usual tubing working and pipe fittings, tools and supplies.

9.5--Watch with sweep second hand. (Stop-watch is desirable.)

9.6--(Desirable on construction work.) Sound-power telephone set with ring circuit.

10. ALTERNATIVES

10.1--Higher test pressures may be required to locate tubing leaks for pilots having higher maximum operating pressures.

10.1.1--A portable test unit consisting of a supply air valve, a bleed valve, an instrument air regulator and a pressure gage connected with 1/4-inch pipe and equipped with suitable flexible end connections may be used.

10.1.2--Tubing leaks which cause the pressure gage to decrease more than 1 psi during the test period are usually not acceptable. The test period is 2 minutes for each 100 feet of 1/4-inch tubing. Use the table in Paragraph 6.1 for control valve topworks volume conversion to equivalent feet of 1/4-inch copper tubing.

10.2--Contaminate Gas Leakage Detectors.

10.2.1--Several reliable instruments are on the market for the detection of Freon or other halide gases. These electronic instruments are of specific value for the detection of leaks in instrument tubing work. A trace of the halide gas can be injected into the pneumatic system. The detector is then brought into the suspected leak area. Due to the high sensitivity of these detectors, very small leaks can be located. This method is particularly useful in testing a large number of joints such as on the rear of panelboards where many fittings are located within a limited area.

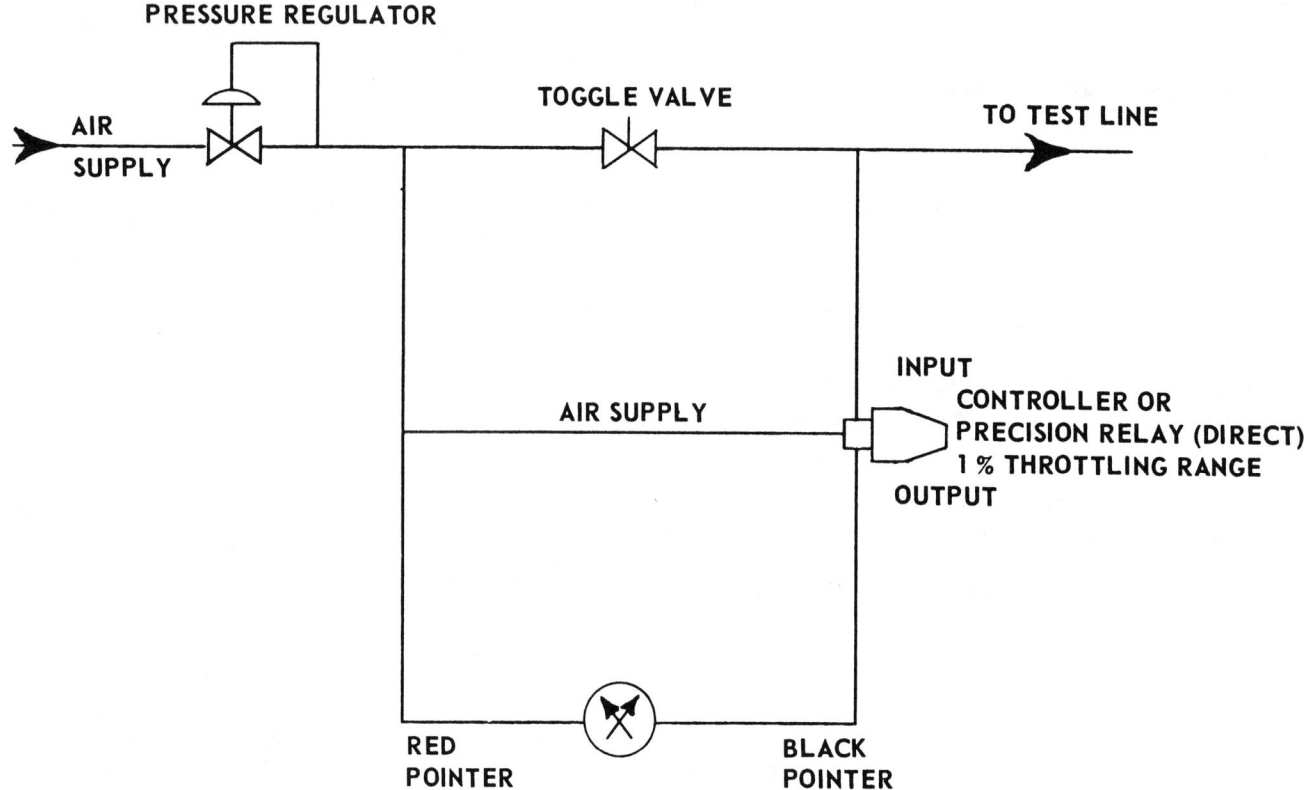

PRESSURE REGULATOR

AIR
SUPPLY

TOGGLE VALVE

TO TEST LINE

AIR SUPPLY

INPUT
CONTROLLER OR
PRECISION RELAY (DIRECT)
1 % THROTTLING RANGE
OUTPUT

RED
POINTER

BLACK
POINTER

DUPLEX PRESSURE GAGE

TEST ASSEMBLY
FIG. RP 7.1.1

INSTRUMENT SOCIETY of AMERICA
Research Triangle Park, North Carolina

ANSI/ISA-S7.3-1975
(R1981)
Approved November 16, 1981
(Formerly ANSI MC 11.1-1976

American National Standard

Quality Standard
for Instrument Air

Instrument Society of America

ISBN 0-87664-335-7

ISA-S7.3 Quality Standard for
Instrument Air

INSTRUMENT SOCIETY of AMERICA
67 Alexander Drive
P.O. Box 12277
Research Triangle Park, North Carolina 27709

PREFACE

This preface is included for information purposes and is not part of Standard ISA-S7.3.

This Standard has been prepared as a part of the service of the Instrument Society of America toward a goal of uniformity in the field of instrumentation. To be of real value, this document should not be static but should be subject to periodic review. Towards this end, the Society welcomes all comments and criticisms and asks that they be addressed to the Secretary, Standards and Practices Board, Instrument Society of America, 67 Alexander Drive, P.O. Box 12277, Research Triangle Park, NC 27709.

The ISA Standards and Practices Department is aware of the growing need for attention to the metric system of units in general, and the International System of Units (SI) in particular in the preparation of instrumentation standards. The Department is further aware of the benefits to USA users of ISA Standards of incorporating suitable references to the SI (and the metric system) in their business and professional dealings with other countries. Towards this end this Department will endeavor to introduce SI and SI-acceptable metric units in all new and revised standards to the greatest extent possible. The Metric Practice Guide, which has been published by the American Society for Testing and Materials as ANSI designation Z210.1 (ASTM E380-79, IEEE Std. 286-1979), and further revisions, will be the reference guide for definitions, symbols, abbreviation, and conversion factors.

It is the policy of the Instrument Society of America to encourage and welcome the participation of all concerned individuals and interests in the development of ISA Standards. Participation in the ISA Standards making process by an individual in no way constitutes endorsement by the employer of that individual of the Instrument Society of America or of any of the Standards which ISA develops.

The following individuals served on the 1975 SP7.3 committee:

NAME	COMPANY
W. F. Kayes, Chairman	Inland Steel Company
H. N. Cook	Enpro Inc
R. H. Kearns	Lectrodryer Company
D. M. Rowe, Secretary	Taylor Instrument Company
A. L. Weiner	Pall Trinity Micro Company

The following individuals served on the 1981 SP7.3 committee:

NAME	COMPANY
J. H. Higgins, Chairman	Pall Corporation
H. N. Cook	Enpro Inc
D. M. Rowe	Taylor/Synbron
C. R. Surprise	Johnson Controls, Inc.

The following individuals served on the 1981 SP7 committee:

NAME	COMPANY
D. M. Rowe, Chairman	Taylor/Synbron
A.J. Bolten	Monsanto
M. Bradner	The Foxboro Company
R. E. Brelin	Detroit Edison Co
H. D. Grymes	Process Systems, Inc
G. Hagerty	Stone & Webster Company
J. H. Higgins	Pall Corporation
J. J. Holtgrefe	The Badger Company, Inc
G. J. House	Amoco
D. C. Hughes	
M. Radford	Measurements & Control Ltd
H. E. Ressinger	Tek Sales, Inc
J. J. Schetselaar	Badger B. V.
J. J. Soos	The Foxboro Company
S. W. Thrift	Catalytic Enterprises

This standard was approved for publication by the ISA Standards and Practices Board in May, 1981.

1 PURPOSE

The purpose of this Standard is to establish instrument air quality values.

2 SCOPE

2.1 Establish a maximum allowable moisture content at which the instruments will function satisfactorily.

2.2 Establish a maximum entrained particle size which will avoid plugging and wear / erosion of air passages and orifices.

2.3 Establish a maximum allowable oil content which will avoid malfunction due to clogging and wear of the components.

2.4 Establish an awareness of a possible source of corrosive or toxic contamination entering the air system; through the compressor suction, plant air system cross connections, or instrument air connections directly connected to processes.

3 DEFINITIONS

3.1 Dew Point, Temperature.

The temperature, referred to a specific pressure, at which water vapors condense.

3.2 Dew Point, (at line pressure) (for the purpose of this Standard)

The dew point value of the air at line pressure of the compressed air system (usually measured at the outlet of the dryer system, or at any instrument air supply source, prior to pressure reduction). When presenting or referencing dew point, the value shall be given in terms of the line pressure; e.g., -40°C (-40°F) dew point at 100 psig.

3.3 Micrometre.

A metric measure with a value of 10^{-6} metres or 0.000001 metre, (previously referred to as "micron").

3.4 Parts per Million (ppm).

Represents parts per million and should be given on a weight basis. The abbreviation shall be ppm (w/w). If inconvenient to present data on a weight basis (w/w) it may be given in a volume basis; (v/v) must be stated after the term ppm, e.g., 5ppm (v/v) or 7 ppm (w/w).

3.5 Ambient Temperature (For the purpose of this Standard)

The temperature of the atmosphere encompassing the area of the entire instrument air system installation, including the compressor, piping, dryer, and the instruments.

3.6 Relative Humidity.

The ratio of the amount of water vapor contained in the air at a given temperature and pressure to the maximum amount it could contain at the same temperature and pressure under saturated conditions.

4 INSTRUMENT AIR, QUALITY STANDARD.

This Standard establishes four elements for the quality of instrument air for use in pneumatic instruments.

4.1 Dew Point, (at line pressure)

4.1.1 Outdoor installations (where any part of the instrument air system is exposed to the outdoor atmosphere).

The dew point at line pressure shall be at least 10°C (18°F) below the minimum local recorded ambient temperature at the plant site.

4.1.2 Indoor Installations (Where the entire instrument air system is installed indoors).

The dew point at line pressure shall be at least 10°C (18°F) below the minimum temperature to which any part of the instrument air system is exposed at any season of the year. In no case should dew point at line pressure exceed 2°C (approximately 35°F).

4.2 Particle Size.

The maximum particle size in the air stream at the instrument shall be three (3) micrometres.

4.3 Oil Content

The maximum total oil or hydrocarbon content, exclusive of noncondensables, shall be as close to zero (0) w/ w or v/v as possible; and under no circumstances shall it exceed one (1) ppm w/w or v/v under normal operating conditions.

NOTE: The ANSI spelling of "micrometre" for dimension and "micrometer" for tool is used in this standard.

NOTE: All pneumatic devices may not require this quality of air while others (fluidics) may require a higher quality of air. Revisions or additional standards may be desired later for special pneumatic systems.

4.4 Contaminants

The instrument air shall be free of all corrosive contaminants and hazardous gases, flammable or toxic, which may be drawn into the instrument air stream. If contamination exists in the compressor intake area, the air should be taken from an elevation or remote location free from contamination or processed to remove such contamination. Any cross connections or process connections to the instrument air piping shall be isolated to preclude contamination of the air system. A regular periodic check should be made to assure high quality instrument air.

REFERENCES

1. Eberly, G.L.: "Conditioning the Instrument Air Supply"; Handbook of Applied Instrumentation: McGraw-Hill.

2. Hankison, Paul M.: "Theory and Filtering Technique for Compressed Air": Instruments. Nov. 1953.

3. Hehn, A.H.: "Can Component Failures in Air and Oil Systems be Predicted?" Hydraulics and Pneumatics. July 1971.

4. Lapple, C.E.: "Characteristics of Particles and Particle Dispersoids": Stanford Research Institute Journal. 1961.

5. Queer, Elmer R. and E. R. McLaughlin: "Desiccation of Air for Air Control Instruments": State College, Pa.

6. Weiner, Arnold L.: "How to Clean and Dry Compressed Air": Hydrocarbon Processing. February 1966.

7. NFPA%T3.27.1-1972: "Glossary of Terms for Compressed Air Dryers".

8. SAE Aerospace Recommended Practice. ARP-1156: "Requisites for Design Specifications for Absorptive Systems".

9. ASHRAE: "Handbook of Fundamentals" Chapter 16.

10. Handbook of Chemistry and Physics 51st Edition: Pages F193-F199: CRC: Chemical Rubber Company.

INSTRUMENT SOCIETY of AMERICA
Research Triangle Park, North Carolina

ANSI/ISA-S7.4-1981
Approved April 6, 1983

American National Standard

Air Pressures for Pneumatic Controllers, Transmitters, and Transmission Systems

Instrument Society of America

ISBN 0-87664-336-5

ISA-S7.4 Air Pressures for Pneumatic Controllers,
Transmitters, and Transmission Systems

INSTRUMENT SOCIETY OF AMERICA
67 Alexander Drive
P.O. Box 12277
Research Triangle Park, North Carolina 27709

PREFACE

This preface is included for informational purposes and is not part of Standard ISA-S7.4.

This Standard has been prepared as a part of the service of the Instrument Society of America toward a goal of uniformity in the field of instrumentation. To be of real value, this document should not be static but should be subject to periodic review. Towards this end, the Society welcomes all comments and criticisms and asks that they be addressed to the Secretary, Standards and Practices Board, Instrument Society of America, 67 Alexander Drive, P.O. Box 12277, Research Triangle Park, NC 27709.

The ISA Standards and Practices Department is aware of the growing need for attention to the metric system of units in general, and the International System of Units (SI) in particular, in the preparation of instrumentation standards. The Department is further aware of the benefits to USA users of ISA Standards of incorporating suitable references to the SI (and the metric system) in their business and professional dealings with other countries. Towards this end this Department will endeavor to introduce SI and SI - acceptable metric units in all new and revised standards to the greatest extent possible. The Metric Practice Guide, which has been published by the American Society for Testing and Materials as ANSI designation Z210.1 (ASTM E380-79, IEEE Std. 286-1979), and further revisions, will be the reference guide for definitions, symbols, abbreviation, and conversion factors.

It is the policy of the Instrument Society of America to encourage and welcome the participation of all concerned individuals and interests in the development of ISA Standards. Participation in the ISA Standards making process by an individual in no way constitutes endorsement by the employer of that individual of the Instrument Society of America or of any of the Standards which ISA develops.

ISA-S7.4 was approved by the Standards and Practices Board in September 1970 as an adoption without change of SAMA Standard RC 2-5-1967, "Air Pressures for Pneumatic Controllers and Transmission Systems." This revision is an adaptation of SAMA RC 2-5-1975 which consists of additional definitions, metrication, and changes in supply pressure limits.

The ISA Standards Committee on Pneumatic Circuts, SP7, operates within the ISA Standards and Practices Department, T. J. Harrison, Vice President. Because of the common interests of SP 7.4 and the Process Measurement and Control Section of SAMA a joint committee was formed for drafting this revised standard. The men listed below served as members of this joint committee.

NAME	COMPANY
A. R. Catheron, Chairman (Deceased)	Consultant, Concord, MA
M. Bradner, Secretary and Chairman representing the Process Measurement & Control Section of the Scientific Apparatus Makers Association	The Foxboro Company
R. G. Beach	Sybron/Taylor
W. F. Kibble	Powers Regulator Company
E. C. Magison	Honeywell, Inc.
J. S. McChesney	ACCO, Bristol Division
P. Wing	Masoneilan International, Inc.

This standard was approved for publication by the ISA Standards and Practices Board in May, 1981.

NAME	COMPANY
T. J. Harrison, Chairman	IBM Corporation
P. Bliss	Consultant
B. A. Christensen	Continental Oil Company
M. R. Gorden-Clark	Scott Paper Company

1 SCOPE AND PURPOSE

1.1 This standard specifies the pneumatic transmission signals used in industrial process measurement and control systems to transmit information between elements of systems. It does not apply to signals used entirely within an element. It does apply to:

a) *Pneumatic Controllers*
b) *Pneumatic Transmitters and Information Transmission Systems*
c) *Current-to-Pressure Transducers*

1.2 The purpose of this standard is to establish:

a) Standard operating pressure ranges for pneumatic information transmission systems.
b) Standard air supply pressures (with limit values) for operation of pneumatic controllers and transmitters, pneumatic information transmission systems, current-to-pressure transducers, and similar devices.

NOTE:

For pneumatic controllers and transmitters, the standards for air supply pressures apply directly. However, the output of a controller is frequently (and that of a current-to-pressure transducer may be) connected to a valve actuator or other means of positioning a final control element. For proper positioning of the latter in such service, the air pressure may vary from near zero to near the supply pressure. This extra range is often necessary for proper final control element performance. Linear response is assumed limited to the nominal operating range, and it is common practice to refer to a particular controller as (say) a 20 to 100 kPa (3-15 psi)* device.

2 DEFINITIONS

2.1 Elements of industrial process measurement and control systems - Functional units or integrated combinations thereof which ensure the transducing, transmitting or processing of measured values, control quantities or variables, and reference variables. A valve actuator in combination with a current to pressure transducer, valve positioner, or a booster relay is considered an element which receives the standard pneumatic transmission signal or standard electric current transmission signal.

2.2 Pneumatic controller - A pneumatic controller is a device which compares the value of a variable quantity or condition to a selected reference and operates by pneumatic means to correct or limit the deviation.

2.3 Pneumatic information transmission system - A pneumatic information system is a system for conveying information comprising (1) a transmitting mechanism converting input information into a corresponding air pressure, (2) interconnecting tubing and (3) a receiving element responsive to air pressure which develops an output directly corresponding to the input information.

2.4 Current-to-pressure transducer - A device which receives an analog electrical signal and converts it to a corresponding air pressure.

2.5 Pneumatic transmission signal - A signal used for information transmission which varies in a continuous manner.

2.6 Measured value of a pneumatic transmission signal - The indicated value during a stated duration.

2.7 Range of a pneumatic transmission signal - The range determined by the lower limit and the upper limit of the signal pressure.

2.8 Span of a pneumatic transmission signal - The difference between the stated high and low pneumatic pressure values of a transmission range.

2.9 Lower limit - The pneumatic signal corresponding to the minimum value of the transmitted input.

2.10 Upper limit - The pneumatic signal corresponding to the maximum value of the transmitted input.

2.11 Supply pressure - The pneumatic pressure supply which enables the system element to generate the pneumatic transmission signals specified in this standard or to provide the valve or final element with required operational force.

3 SPECIFIED VALUES

3.1 Ranges of pneumatic pressure transmission signals.

*See Note 1

3.1.1 80 kPa (12 psi) span (preferred). The operating pressure range for the 80 kPa operating pressure span shall be 20 kPa (3 psi) to 100 kPa (15 psi).

3.1.2 160 kPa (24 psi) span. The operating pressure range for the 160 kPa operating pressure span shall be 40 kPa (6 psi) to 200 kPa (30 psi).

4 SUPPLY PRESSURES

4.1 Supply pressure limits

4.1.1 80 kPa (12 psi) span. A minimum of 130 kPa (19 psi) and a maximum of 150 kPa (22 psi).

4.1.2 160 kPa (24 psi) span. A minimum of 260 kPa (38 psi) and a maximum of 300 kPa (44 psi).

Note 1. Except for span, all expressions of pressure are given in gauge pressure rather than absolute pressure. The unit for pressure in Systems International d'Unities (SI) is the pascal (Pa), defined as one newton per square meter. The conversion factor for Pa to psi is 1000 Pa = 1 kPa = 0.145038 psi.

Note 2. The bar (1 bar = 10^5 Pa) is a customary unit in some countries. It is *not* an SI unit, but "may be used" according to ISO 1000, International Standards Organization (ISO) standard number 1000.

Note 3. The psi values given in this standard represent current American practice and the kPa values are the same pressures as used in Europe; previously expressed as bar, or kgf/cm² or kp/cm².

INSTRUMENT SOCIETY OF AMERICA
Research Triangle Park, North Carolina

ISA-RP7.7-1984

Recommended Practice

Recommended Practice for Producing Quality Instrument Air

Instrument Society of America

ISBN 0-87664-845-6

ISA-RP7.7 Recommended Practice for Producing
Quality Instrument Air

INSTRUMENT SOCIETY OF AMERICA
67 Alexander Drive
P.O. Box 12277
Research Triangle Park, North Carolina 27709

PREFACE

This preface is included for information purposes and is not part of ISA-RP7.7.

This recommended practice has been prepared as part of the service of the Instrument Society of America (ISA) toward a goal of uniformity in the field of instrumentation. To be of real value, this document should not be static, but should be subject to periodic review. Toward this end, the Society welcomes all comments and criticisms, and asks that they be addressed to the Secretary, Standards and Practices Board, Instrument Society of America, 67 Alexander Drive, P.O. Box 12277, Research Triangle Park, NC 27709, Telephone (919) 549-8411.

The ISA Standards and Practices Department is aware of the growing need for attention to the metric system of units in general, and the International System of Units (SI) in particular, in the preparation of instrumentation standards. The Department is further aware of the benefits to U.S.A. users of ISA standards of incorporating suitable references to the SI (and the metric system) in their business and professional dealings with other countries. Toward this end, this Department will endeavor to introduce SI-acceptable metric units in all new and revised standards to the greatest extent possible. The Metric Practice Guide, which has been published by the Institute of Electrical and Electronics Engineers as ANSI/IEEE Std. 268-1982, and future revisions will be the reference guide for definitions, symbols, abbreviations, and conversion factors.

It is the policy of the Instrument Society of America to encourage and welcome the participation of all concerned individuals and interests in the development of ISA standards. Participation in the ISA standards-making process by an individual in no way constitutes endorsement by the employer of that individual, of the Instrument Society of America, or of any of the standards that ISA develops.

The information contained in the preface, footnotes, and appendices is included for information only and is not a part of the recommended practice.

The persons listed below served as members of ISA subcommittee SP7.3, which prepared this recommended practice.

NAME	COMPANY
J.H. Higgins, 1983 Chairman	Systems & Engineered Equipment Company
A.R. Catheron	Deceased
G. Kirkland	Pall Pneumatics Products Corporation
F.C. Quinn	Hygrodynamics
D. Rowe	Taylor Instrument Company
R.D. Wall	C.M. Kemp Mfg. Company

The persons listed below served as members of ISA Committee SP7, which approved this recommended practice.

NAME	COMPANY
M. Bradner	The Foxboro Company
B.A. Christensen	
N.L. Conger	Conoco
H.D. Grymes	Earl and Wright
G. Hagerty	Stone & Webster Engineering Corporation
J.H. Higgins	Consultant
D.C. Hughes	
A.F. Marks	Bechtel Petroleum, Inc.
P.C. Menmuir	Eaton Corporation
P.A. Papish	Pall Pneumatic Products Corporation
F.C. Quinn	Hygrodynamics
C. Seyffert	Brown & Root, Inc.
J.S. Soos	Fischer & Porter Company
W. Valday	Mercury Company
R.D. Wall	C.M. Kemp Mfg. Company
J.D. Warnock	Moore Products Company

This recommended practice was approved for publication by the ISA Standards and Practices Board in September 1984.

TABLE OF CONTENTS

LIST OF ILLUSTRATIONS

LIST OF TABLES

1 PURPOSE

The purpose of this recommended practice is to establish general equipment guidelines for producing instrument air of the quality defined in ANSI/ISA-S7.3-1975 (R 1981), "Quality Standard for Instrument Air."

2 SCOPE

The scope of this recommended practice is to:

(a) Enumerate equipment characteristics to include types, range, and performance of the components necessary to meet air quality requirements of ANSI/ISA-S7.7-1975 (R 1981).*

(b) Conduct tests of system—and components where applicable—to check performance of instrument air supply systems to meet ANSI/ISA-S7.3-1975 (R 1981) requirements.

(c) Provide information useful for maintaining air quality from instrument air systems.

NOTE

This recommended practice is restricted to major components necessary to assure a supply of air which meets the standards of ANSI/ISA-S7.3-1975 (R 1981). It is not intended to encompass minor components such as pressure relief valves or other portions of the air system—such as piping, valves, controls, and related instrumentation.

The units used in this recommended practice are based on the International System of Units (SI). Metric and English units are enclosed in parentheses following SI units.

UNITS USED: SI (METRIC, ENGLISH)

TEMPERATURE

SI UNITS	K	degrees Kelvin
METRIC	C	degrees Celsius
ENGLISH	F	degrees Fahrenheit

FLOW

SI UNITS	m^3/s	cubic meter per second (standard conditions)
METRIC	m^3/h	cubic meter per hour (standard conditions)
ENGLISH	SCFM	standard cubic feet per minute

PRESSURE

SI UNITS	kPa	kilopascal
METRIC	bar	bar 100 kPa
ENGLISH	psia	pound per square inch absolute
	psig	pound per square inch gage
	in. Hg	inch of mercury

ABBREVIATIONS USED:

| ppm | part per million |
| w/w | weight per weight |

TABLE 1
VALUES FOR CONSTANT A--
DEVIATION FROM IDEAL GAS
LAWS AT ELEVATED PRESSURES*

Pressure (kPa)	(Absolute) psi	Constant A
0	0	1
103	15	0.9971
207	30	0.9943
310	45	0.9915
414	60	0.9986
552	80	0.9848
690	100	0.9811
1030	150	0.9717
1380	200	0.9625

*A.W. Diniak and E.R. Weaver, "A New Computer for Calculating the Water Content of Gases," *The Journal of Research of the National Bureau of Standards,* 56 (No. 5, 1956).

3 INSTRUMENT AIR SUPPLY AND QUALITY CONDITIONING SYSTEM

3.1 Introduction

An instrument air supply and conditioning system consists of components required to assure an adequate supply of compressed air of the proper quality (clean and dry) and at the desired pressure.

---

*ANSI/ISA-S7.3-1975 (R 1981), "Quality Standard for Instrument Air," Instrument Society of America, Research Triangle Park, NC, 1981.

TABLE 2
SATURATION VAPOR PRESSURE OVER WATER
(Smithsonian Meteorological Tables)*

METRIC UNITS				ENGLISH UNITS			
Temperature	Vapor pressure	Temperature	Vapor pressure	Temperature	Vapor pressure	Temperature	Vapor pressure
°C	10^{-3} bar	°C	10^{-3} bar	°F	10^{-3} in. Hg	°F	in. Hg
−50	0.06356	0	6.1078	−50	3.089	0	0.04477
−49	0.07124	1	6.5662	−49	3.281	1	0.04691
−48	0.07975	2	7.0547	−48	3.488	2	0.04915
−47	0.08918	3	7.5753	−47	3.703	3	0.05149
−46	0.09961	4	8.1294	−46	3.930	4	0.05392
−45	0.1111	5	8.7192	−45	4.170	5	0.05646
−44	0.1239	6	9.3465	−44	4.924	6	0.05910
−43	0.1379	7	10.013	−43	4.692	7	0.06185
−42	0.1534	8	10.722	−42	4.973	8	0.06471
−41	0.1704	9	11.474	−41	5.271	9	0.06769
−40	0.1891	10	12.272	−40	5.584	10	0.07080
−39	0.2097	11	13.119	−39	5.915	11	0.07403
−38	0.2323	12	14.017	−38	6.263	12	0.07740
−37	0.2571	13	14.969	−37	6.630	13	0.08089
−36	0.2842	14	15.977	−36	7.016	14	0.08454
−35	0.3139	15	17.044	−35	7.424	15	0.08832
−34	0.3463	16	18.173	−34	7.849	16	0.09226
−33	0.3818	17	19.367	−33	8.298	17	0.09634
−32	0.4205	18	20.630	−32	8.770	18	0.10060
−31	0.4628	19	21.964	−31	9.270	19	0.10501
−30	0.5088	20	23.373	−30	9.789	20	0.10960
−29	0.5589	21	24.861	−29	10.34	21	0.11437
−28	0.6134	22	26.430	−28	10.91	22	0.11933
−27	0.6727	23	28.086	−27	11.52	23	0.12446
−26	0.7371	24	29.831	−26	12.15	24	0.12980
−25	0.8070	25	31.671	−25	12.82	25	0.13534
−24	0.8827	26	33.608	−24	13.52	26	0.14109
−23	0.9649	27	35.649	−23	14.25	27	0.14705
−22	1.0538	28	37.796	−22	15.02	28	0.15324
−21	1.1500	29	40.055	−21	15.83	29	0.15966
−20	1.2540	30	42.430	−20	16.68	30	0.16631
−19	1.3664	31	44.927	−19	17.56	31	0.17321
−18	1.4877	32	47.551	−18	18.49	32	0.18036
−17	1.6186	33	50.307	−17	19.46	33	0.18778
−16	1.7597	34	53.200	−16	20.48	34	0.19546
−15	1.9118	35	56.236	−15	21.55	35	0.20342
−14	2.0755	36	59.422	−14	22.66	36	0.21168
−13	2.2515	37	62.762	−13	23.83	37	0.22020
−12	2.4409	38	66.264	−12	25.05	38	0.22904
−11	2.6443	39	69.934	−11	26.33	39	0.23819
−10	2.8627	40	73.777	−10	27.66	40	0.24767
− 9	3.0971	41	77.802	− 9	29.06	41	0.25748
− 8	3.3484	42	82.015	− 8	30.52	42	0.26763
− 7	3.6177	43	86.423	− 7	32.04	43	0.27813
− 6	3.9061	44	91.034	− 6	33.63	44	0.28899
− 5	4.2148	45	95.855	− 5	35.29	45	0.30023
− 4	4.5451	46	100.89	− 4	37.03	46	0.31185
− 3	4.8981	47	106.16	− 3	38.84	47	0.32387
− 2	5.2753	48	111.66	− 2	40.73	48	0.33629
− 1	5.6780	49	117.40	− 1	42.71	49	0.34913
− 0	6.1078	50	123.40	− 0	44.77	50	0.36240

*By permission of the Smithsonian Institution Press from *Smithsonian Meteorological Tables, Smithsonian Miscellaneous Collections*, Volume 114 (Sixth Revised Edition), prepared by Robert J. List. "Saturation Vapor Pressure Over Water," Table 94, pp. 351-359. Smithsonian Institution, Washington, D.C., 1951.

3.2 Air Supply System

Typical components of the air supply system (see Figure 1) include:

Filters
Compressors
Aftercoolers and condensate separators
Air receivers and drain traps

3.2.1 Compressor

Compressors must be sized to deliver air at the specified pressure under maximum conditions. The compressor intake should be located outside of regions subject to process spills and should be provided with an adequate filter.

Various types of compressors are available, including:

Reciprocating oiled piston
Reciprocating oilless piston
Rotary vane
Rotary liquid ring
Diaphragm
Rotary screw
Centrifugal

Compressors of the nonlubricated type are commonly used to prevent problems with oil or lubricant contamination. Where lubricated compressors are used, provision must be made to adequately remove these lubricants from the air supply.

3.2.2 Aftercooler and Moisture Separator

The aftercooler is a heat exchanger utilizing water, air, or other coolant to cool the hot compressor discharge air. Cooling compressed air below its dew point temperature causes moisture to condense. The condensate is collected in a mechanical separator and drained away.

Water-cooled aftercoolers are usually sized to cool outlet air to with 5.5 K to 8.5 K (5.5°C to 8.5°C; 10°F to 15°F) of the inlet cooling water temperature.

Air-cooled aftercoolers are usually sized to cool outlet air to within 14 K to 17 K (14°C to 17°C; 25°F to 30°F) of the ambient air temperature.

3.2.3 Air Receiver and Drain Trap

The air receiver must be sized to store enough air to handle system demand surges, to allow time for moisture separation, and to provide a reserve for orderly (or emergency) shutdown or for temporary operation. The receiver pressure is a control for loading and unloading the compressor to balance compressor and demand cycle variations.

The receiver ambient temperature is typically lower than the dew point temperature of the air entering the receiver. This causes moisture to condense inside the receiver. A liquid drain device (an automatic drain recommended for all but the smallest systems) must always be furnished on an air receiver to dispose of the condensate.

3.3 Air Quality Conditioning System

The air quality conditioning system consists of components to remove moisture, oil, and particulate matter from the compressed air.

The typical air quality conditioning system consists of the following components:

Prefilter
Air dryer
Afterfilter

3.3.1 Prefilter

A coalescing prefilter is required to prevent liquids, oil, and water in aerosol form from entering the air dryer.

3.3.2 Air Dryer

The air drying equipment must meet the dew point requirements of ANSI/ISA-S7.3-1975 (R 1981).

Various types of dryers are available to remove moisture from compressed air. (See Table 3.) Selection of the proper type and size of dryer must be based upon the actual inlet and flow conditions under which the dryer is expected to perform and upon the quality of air to be produced.

The essential information required for dryer selection is as follows:

Maximum flow: m^3/s (m^3/h, SCFM)

Maximum inlet temperature: K (°C, °F)

Maximum percentage moisture saturation of inlet (if unknown, assume inlet temperature as pressure dew point)

Minimum inlet pressure: kPa (bar, psig)

Maximum inlet pressure: kPa (bar, psig)

Maximum allowable outlet dew point temperature at dryer outlet pressure: K (°C, °F)

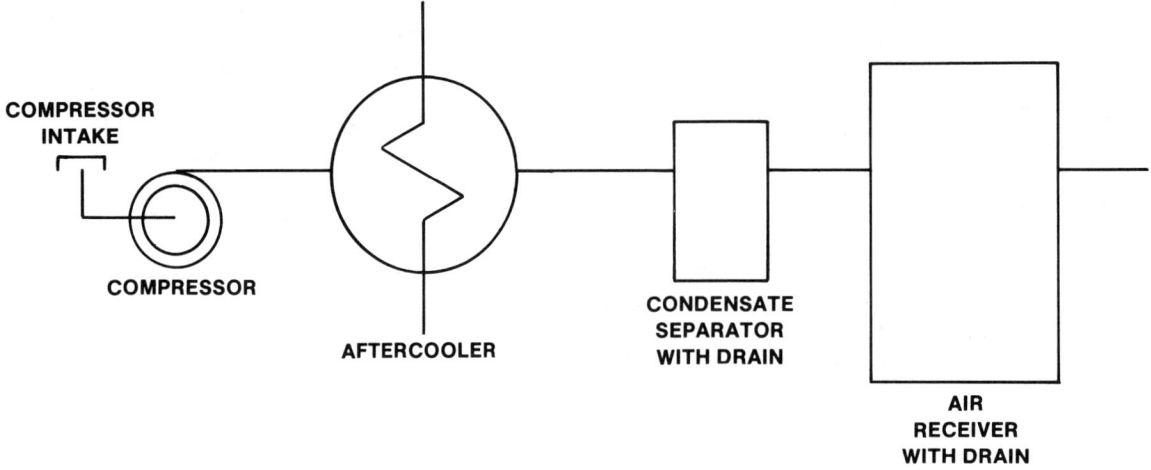

AIR SUPPLY SYSTEM

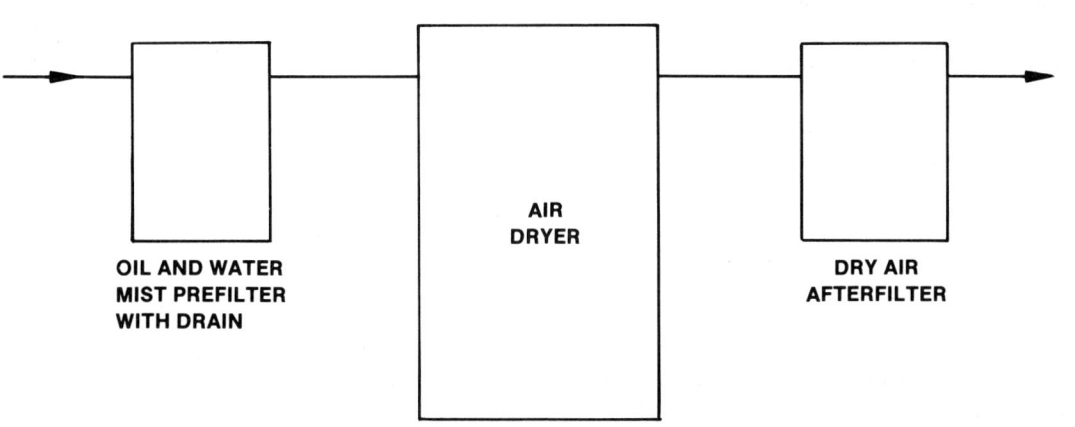

AIR QUALITY CONDITIONING SYSTEM

Figure 1. Instrument Air System — Typical Schematic

TABLE 3
COMPRESSED AIR DRYER TYPES

I. Deliquescent

Flow Range:[1]   0–8500 m^3/h @ 311 K & 690 kPa (gage)
                 (0–5000 SCFM @ 100 °F & 100 psig)

Outlet Dew Point at Line Pressure: 272 K & higher[2]
                                     (+30 °F & higher)

Utility Requirements: None

II. Refrigeration

Flow Range:[1]   0-16 990 m^3/h @ 311 K & 690 kPa (gage)
                 (0-10 000 SCFM @ 100 °F & 100 psig)

Outlet Dew Point at Line Pressure: 274 K & higher[2]
                                     (35 °F & higher)

Utility Requirements: Electric

III. Regenerative Desiccant Dryers

A. Regeneration with Heaters

Flow Range:[1]   0-16 990 m^3/h @ 311 K & 690 kPa (gage)
                 (0-10 000 SCFM @ 100 °F & 100 psig)

Outlet Dew Point at Line Pressure: 233 K higher & lower[3]
                                     (–40 °F higher & lower)

Utility Requirements: Electric or Steam

B. Regeneration without Heaters

Flow Range:[1]   0-16 990 m^3/h @ 311 K & 690 kPa (gage)
                 (0-10 000 SCFM @ 100 °F & 100 psig)

Outlet Dew Point at Line Pressure: 233 K higher & lower[3]
                                     (–40 °F higher & lower)

Utility Requirements: Air

(1) Flow values are given at standard conditions (60 °F and 14.7 psig, for example).

(2) This dew point may be inadequate for many instrument air system applications. Refer to ANSI/ISA-7.3-1975 (R 1981) for dew point requirements.

(3) Traditionally, regenerative desiccant dryers for instrument air systems are sized to provide –40 °F dew point air at pressure. However, in extremely cold climates, instrument air applications may require dew points as low as –100 °F at operating pressure.

Required accessories: pressure gages, relief valves, thermometers, timers, safety switches, etc.

Other pertinent information, such as: contaminants that may be present (oil, liquid, etc.)

Utilities available, such as: electricity, steam, water, etc.

Area electrical classification where equipment is to be installed

Information on whether filters are to be provided, and

If filters are to be provided, the degree of filtration required must be adequate to meet the requirements of ANSI/ISA-S7.3-1975 (R 1981).

3.3.3 Afterfilter

The afterfilter provides final cleaning of the compressed airstream by removing particulate matter from the dryer discharge.

Afterfilters are required on all instrument air systems, particularly on desiccant-type dryers to prevent desiccant dust from passing downstream.

4 AIR QUALITY CONSIDERATIONS

4.1 Dew Point (Air/Water Quality)

ANSI/ISA-S7.3-1975 (R 1981) establishes a maximum allowable line dew point for instrument air systems. This maximum protects instruments from problems related to the presence of liquid water or ice. (Preventing formation of harmful materials from contaminants of the ambient air supply or system materials, e.g., lubricants and scale, is not necessarily assured by meeting the dew point requirements of ANSI/ISA-S7.3-1975 (R 1981).

Compression and cooling stages in an instrument air system can promote the formation of liquid water or ice. Compression increases the partial pressure of the water vapor present. If the water vapor partial pressure is increased to the saturation water vapor pressure, liquid water or ice results. Cooling reduces the saturation water vapor pressure, a temperature-dependent variable. If the saturation water vapor pressure is reduced to the partial pressure of the water vapor present, liquid water or ice results. Therefore, moisture removal is a major consideration of instrument air quality conditioning systems.

The most common methods of moisture removal are compression cooling, adsorption, chemical methods, and combinations of these methods (e.g., use of a refrigerant dryer followed by a desiccant dryer).

ANSI/ISA-S7.3-1975 (R 1981) establishes the maximum value in terms of dew point temperature at line pressure. The instrument air system dew point should always reference the line pressure. If the determination of this value is made at other than line pressure, both pressures (determination and reference or line) must be used in any computation.

A change in the total pressure of air causes a proportional change in the partial pressure of all gases or vapors, including water vapor; the increase in partial pressure is limited to the saturation partial pressure for condensables such as water.

$$\frac{e_1}{e_2} = \frac{P_1}{P_2} \times \frac{A_1}{A_2} \qquad \text{Equation (1)}$$

where: e is the water vapor (partial pressure)
P is the total pressure (absolute pressure)
A is a factor for deviation from ideal gas laws at elevated pressures

Note: Table 1 lists values of A applying to instrument air.

The above equation can be used to calculate the change in water vapor partial pressure with change in total pressure. The resultant partial pressure then may be converted to the dew point temperature using psychrometric tables such as Table 2.

Example: Assume a dew point of 253 K (−20°C; 14°F) at two atmospheres absolute pressure (203 kPa absolute; 2.03 bar absolute; 29.4 psia)

To determine atmospheric dew point using Equation 1:

From Saturation Vapor Pressure Tables (Table 2), e_2, water vapor pressure is 1.254×10^{-3} bar (0.037 in. Hg)

From Table 1, constant A_2 is 0.9944 and A_1 is 0.9972

Solving Equation 1, e_1 is 0.6287×10^{-3} bar (0.01838 in. Hg)

From Saturation Vapor Pressure Tables (Table 2), a water vapor pressure of 0.6224×10^{-3} bar (0.01838 in. Hg) is an atmospheric dew point of approximately 245 K (−28°C; −18.4°F)

The change in dew point with pressure can also be determined using the curves of Figures 2, 3, and 4. (The curves in Figures 2, 3, and 4 were prepared by the committee using the data in Table 1 and the equations in Reference 1.)

4.2 Oil Content

ANSI/ISA-S7.3-1975 (R 1981) establishes a maximum allowable oil content for instrument air. Oil as liquid or vapor may be present in compressed air when an oil-lubricated compressor is used.

4.2.1 Liquid Oil

Oil problems develop when oil adheres to air piping and instrument passages and contaminates filtering equipment and desiccants. The cumulative effects of oil in the airstream are instrument inaccuracy and downtime.

Oil aerosols generally enter an air system from lubricated compressors. Oil vapor from the compressor condenses during cooling. Oil quantities can range from 3 ppm to 30 ppm (w/w) depending upon the age, type, and mechanical condition of the compressor.

Aerosols are extremely small sized liquid droplets (0.01 to 2 μm). They can be removed from compressed air only with efficient coalescing filters.

4.2.2 Oil Vapor

4.2.2.1 Oil vapor will be present in compressed air in small quantities. Actual amounts depend upon the compressor and the properties of the lubrication oil used. Oil vapor usually does not cause instrumentation and control problems if it remains in the vapor state. However, it may condense from cooling during sharp air expansion, and the resultant liquid oil can cause problems.

4.2.2.2 Special adsorption or collection equipment is needed to remove oil vapor—equipment such as heaterless desiccant dryers, activated carbon adsorbers, etc.

4.3 Particulate Matter (Particle Size)

ANSI/ISA-S7.3-1975 (R 1981) establishes a maximum allowable particle size for instrument air.

Particulate matter causes equipment malfunction by clogging and eroding small orifices and working parts in pneumatic instruments and controls.

Particulate matter can be introduced into an instrument air system via a variety of sources: from the ambient air through the intake filter; from the formation of rust and scale, which can develop from moisture and chemical corrosion in pipelines and components; and from desiccant dust, which can be carried over from the air drying equipment.

A dry element afterfilter can be used to remove particulate matter from the instrument air to meet the requirements of ANSI/ISA-S7.3-1975 (R 1981).

4.4 Other Contaminants

ANSI/ISA-S7.3-1975 (R 1981) indicates that contaminants other than particulate matter, water, or oil carryover from compressors can cause operating problems in instrument air systems. These sections refer to chemical (gas or vapor) compounds which may be toxic, corrosive, or both.

Unless the air intake can be located in an area which is free of contaminants, an appropriate scrubber or absorber may be required for the protection of the pneumatic components. The range of possible contaminants is so wide that every installation must be considered separately. The kind and concentration of contaminant, the air dryness, and the amount of compression are all factors for consideration.

Contaminants can also originate from the system components, such as the generation of corrosive vapors from the phosphate esters used for fireproofing synthetic lubricants for compressors. Materials such as seals and diaphragms used in pneumatic components should be compatible with the synthetic lubricant; or, an appropriate scrubber should be used in the air system to remove contaminants.

5 TESTS

To assure continuous delivery of instrument air in compliance with the standard ANSI/ISA-S7.3-1975 (R 1981), it is necessary to verify the quality of the system air. Tests may be necessary for oil content, dew point, particles, and other contaminants. Tests or analysis must be conducted on initial start-up and routinely at intervals determined by experience.

It is necessary to check the performance of individual components of the system when improper usage or malfunctioning can adversely affect the performance of other parts. For example, an oil and water mist prefilter could malfunction or be improperly sized or maintained. This could result in coating the desiccant with oil, thus destroying the effectiveness of the dryer.

Finite values need not be verified for acceptability. Testing is necessary only to show that air quality meets ANSI/ISA-S7.3-1975 (R 1981). For example, if the dew point requirement, per ANSI/ISA-S7.3-1975 (R 1981), is 253 K (–20°C or –4°F), a low offscale reading on a dew point instrument starting at 248 K (–25°C or –13°F) would indicate sufficient dryness.

Tests need not be performed when component design obviates testing. For example, ANSI/ISA-S7.3-1975 (R 1981) limits particle size to 3 μm. The use of a selected absolute rated filter in operating condition assures adequate filtration; however, the filter must be replaced on a maintenance schedule.

5.1 Dew Point Tests

The proper dew point temperature should be established based on the requirements of ANSI/ISA-S7.3-1975 (R 1981). Periodic checks must be scheduled to assure the proper performance of the instrument air system with regard to moisture.

Various methods are available for determining the moisture in air. These include dew point instruments (dewcup, mirror, cloud chamber, hygroscopic salts), electrical hygrometers, capacitance, spectroscopy and thermal conductivity.

The dew point temperature value must be expressed at a specific line pressure. If the determination is made at other than line pressure, the measured value and the pressure of measurement should also be noted. (See Section 4.1.)

5.2 Oil Content Tests

When oil contamination is possible (use of oil-lubricated compressor), periodic checks are required to assure air quality.

Methods which can be used to determine oil content in air include gravimetric techniques, particle counters, microscopic techniques, infrared spectrometry and ultraviolet molecular emission for liquid oil. Gas chromatographs can be used for oil vapor.

5.3 Particle Size Tests

Periodic checks for particulate matter are recommended if operating problems are prevalent.

Microscopic techniques are normally required for determining particle size.

5.4 Other Contamination Tests

Periodic analysis should be performed to check the effectiveness of scrubbers or other equipment used to remove chemical contaminants.

6 CHECK LIST

An air dryer and appropriate filters are needed when using air-operated instruments.

Where oil-lubricated compressors are used, some means for removing the oil ahead of desiccant drying equipment are necessary.

Summer humidity loads as well as winter temperatures must be taken into design considerations. If it is necessary to install an instrument air line or header along the inside surface of an exterior wall, check that the air line or header is not exposed to any undue chilling of the air stream. Use anticipated minimum temperature at this location to determine dew point limits of ANSI/ISA-S7.3-1975 (R 1981).

Reliability of equipment and safeguards must be considered with each application.

Instrument air system maintenance must include: investigtion for pipeline corrosion, leaks, freezing of lines or valves; servicing of compressors, dryers, and filters; and service and replacement of all related equipment.

Consideration must be given to standby facilities, duplicate equipment, or bypass piping which may be needed to eliminate instrument shutdown. This equipment must likewise comply with the conditions of ANSI/ISA-S7.3-1975 (R 1981).

For any instrument air system normal maintenance procedures should include checks on proper operation of all system components (e.g., compressors, aftercoolers, drain traps, air dryers, and filters, etc.)

7 REFERENCES

Diniak, A.W., and Weaver, E.R. "A New Computer for Calculating the Water Content of Gases." *Journal of Research of the National Bureau of Standards* 56 (No. 5, May 1956).

Landsbaum, E.M.; Dodds, W.S.; and Stutzman, L.F. "Humidity of Compressed Air." *Industrial and Engineering Chemistry* 47 (No. 1, January 1955).

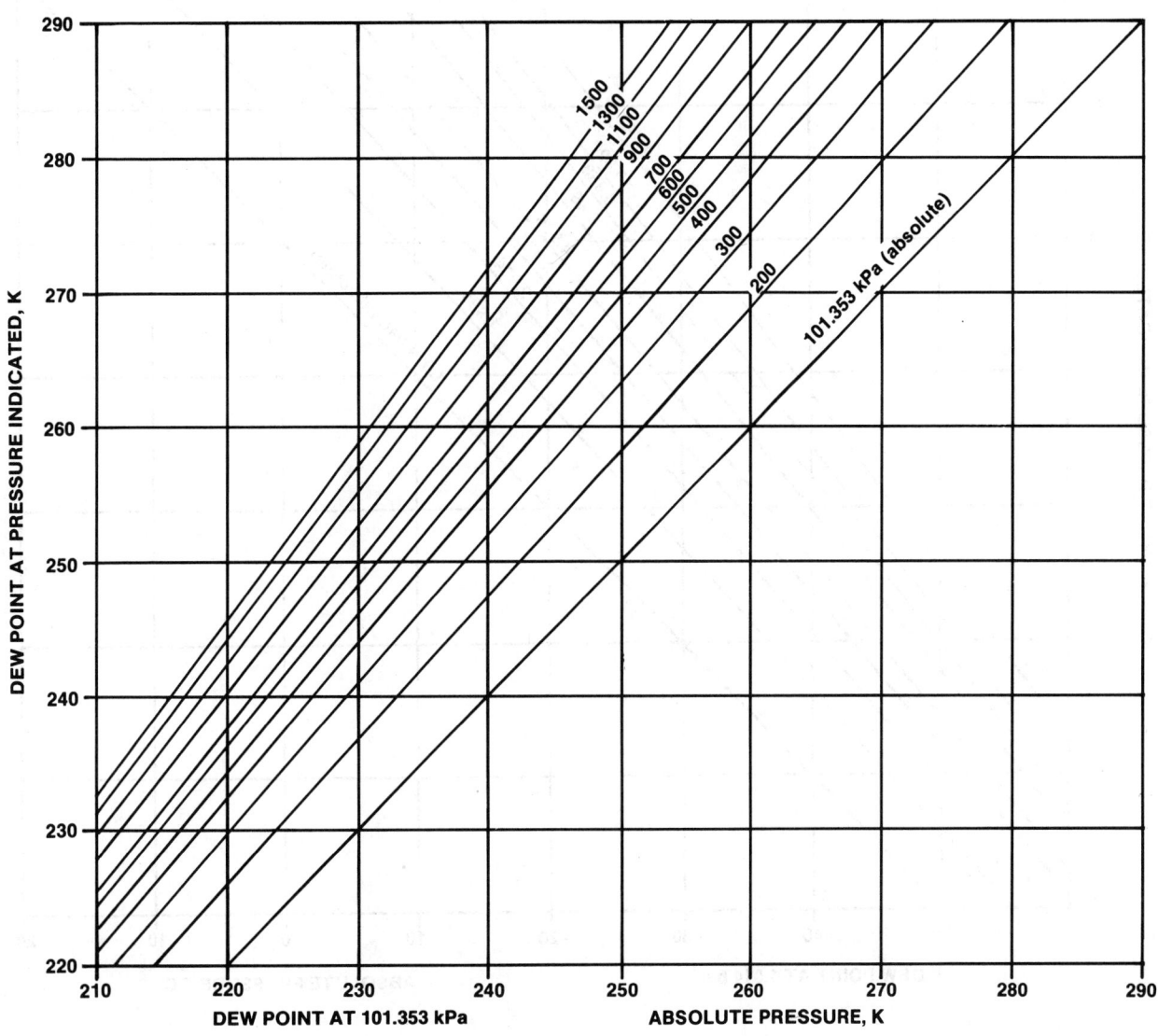

Figure 2. Change in Dew Point — Conversion Chart (SI Units)
(Prepared by Committee using data from Table 1 and equations in Reference 1)

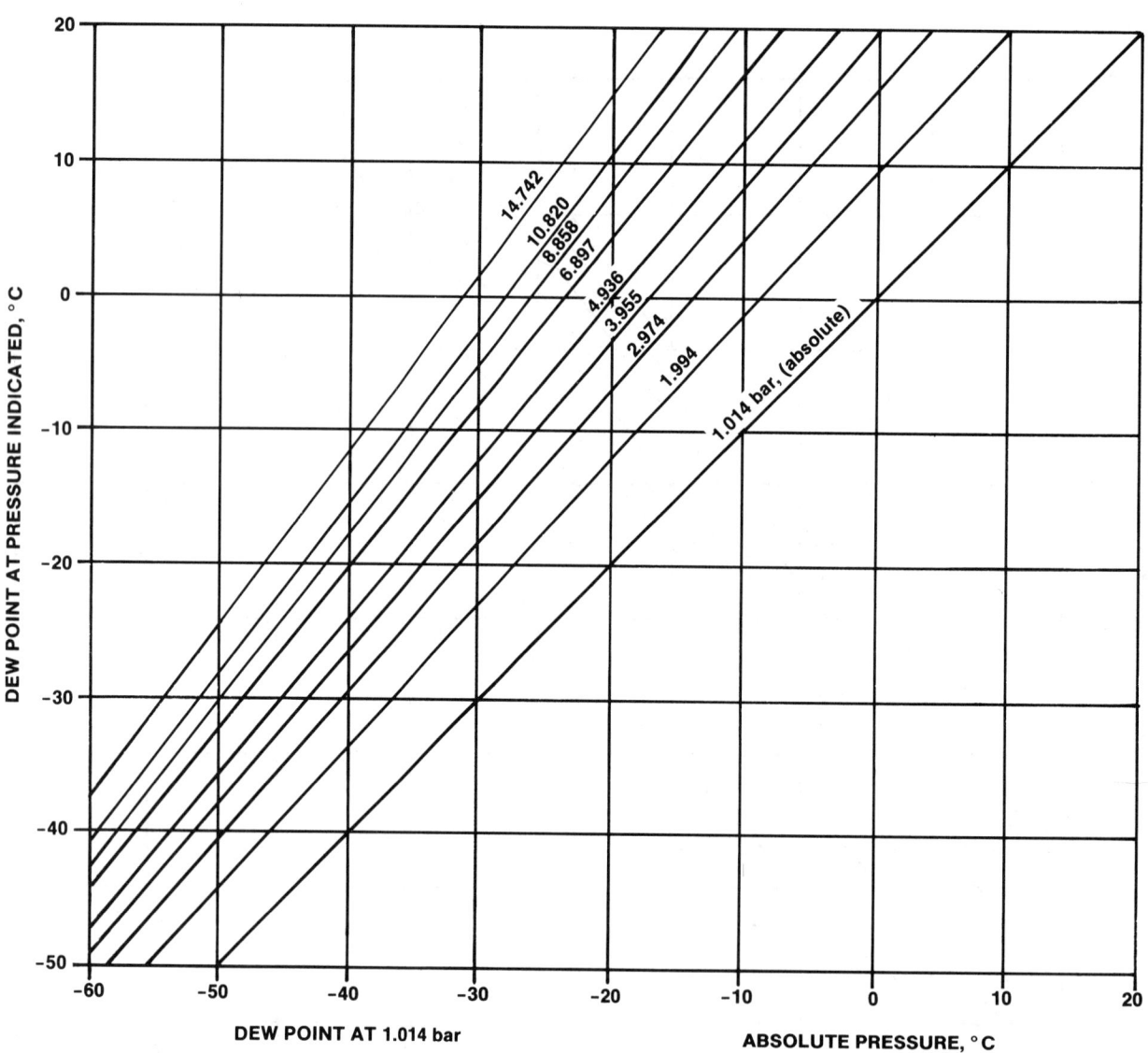

Figure 3. Change in Dew Point — Conversion Chart (Metric Units)
(Prepared by the Committee using data from Table 1 and equations in Reference 1)

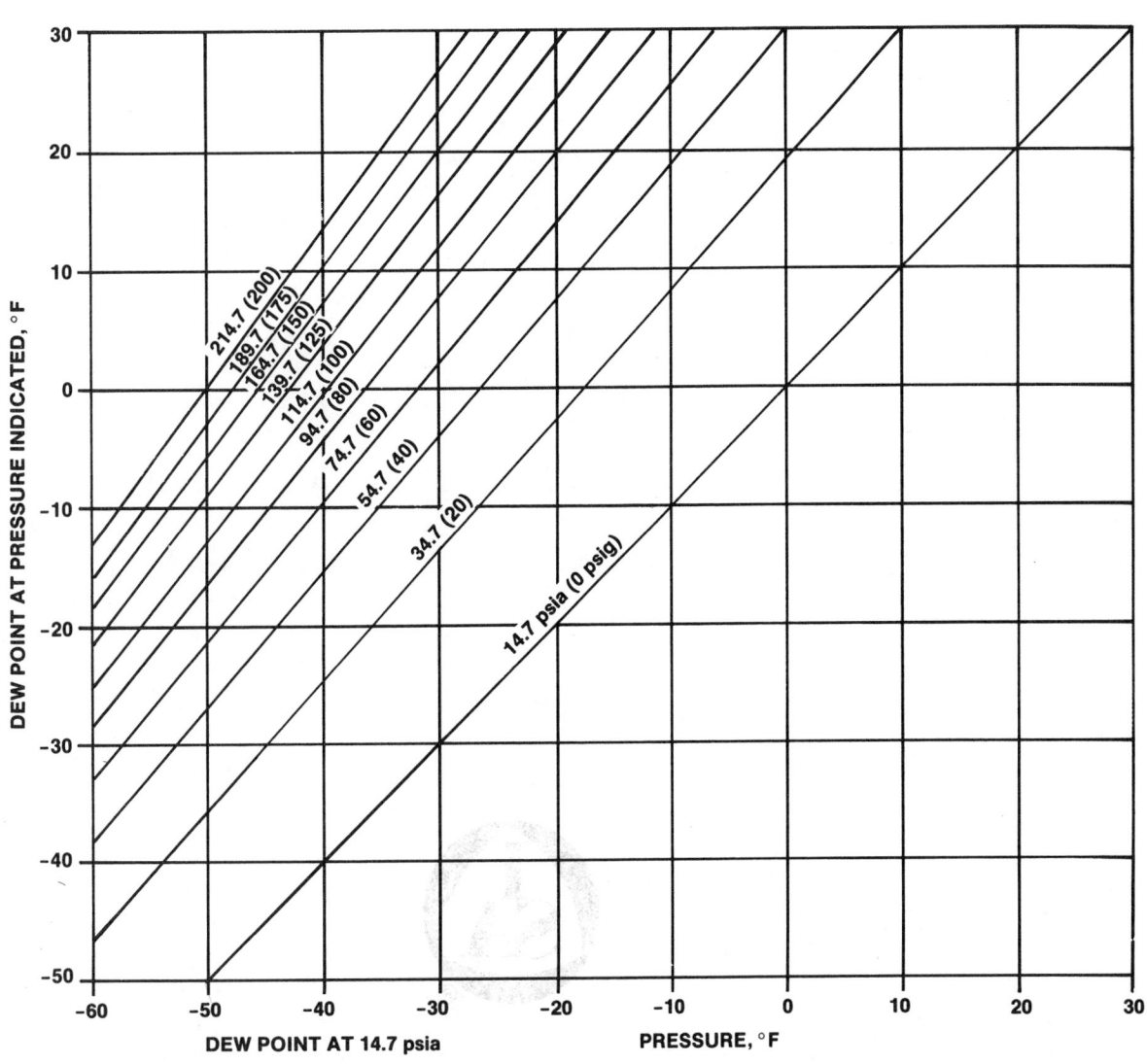

Figure 4. Change in Dew Point — Conversion Chart (English Units)
(Prepared by the Committee using data from Table 1 and equations in Reference 1)

INSTRUMENT SOCIETY OF AMERICA
Research Triangle Park, North Carolina 27709

ISA-RP12.1-1960

Recommended Practice

Electrical Instruments
in Hazardous Atmospheres

Instrument Society of America

ISBN 0-87664-337-3

ISA-RP12.1   Electrical Instruments in
             Hazardous Atmospheres

INSTRUMENT SOCIETY OF AMERICA
67 Alexander Drive
P.O. Box 12277
Research Triangle Park, North Carolina 27709

TABLE OF CONTENTS

FOREWORD

This Recommended Practice has been prepared as a part of the service of the Instrument Society of America toward a goal of uniformity in the field of instrumentation. To be of real value, this report should not be static, but should be subject to periodic review. Toward this end, the Society welcomes all comments and criticisms, and asks that they be addressed to the Standards and Practices Board Secretary, Instrument Society of America, 67 Alexander Drive, P.O. Box 12277, Research Triangle Park, North Carolina 27709.

This report was prepared by the Committee on Instrumentation for Hazardous Areas (8D-RP12).

Committee Members

F. L. Maltby (Chairman) - Drexelbrook Engineering Co.
L. E. Cuckler - Robertshaw-Fulton Co.
C. F. Kisselstein - Crouse-Hinds Co.
Robert McCarron - Leeds & Northrup Co.
W. F. Hickes - The Foxboro Co.
E. C. Magison - Minneapolis-Honeywell Regulator Co.
A. H. McKinney - E. I. du Pont de Nemours & Co., Inc.
Walter Jacobson - Bristol Co.
Marvin Lorig - Panellit, Inc.
D. L. Olson - Minnesota Mining & Manufacturing Co.

Sub-Committee A

Marvin Lorig (Chairman) - Panellit, Inc.
T. L. Cliff - Standard Oil Company of Ohio
Reese Littlefield - E. I. du Pont de Nemours & Co., Inc.
Lee Eddy - Universal Oil Products, Inc.
R. D. Guimond - The Foxboro Co.
Robert Tuck - Cities Service Co.

Review Board

D. J. Beckwith - Dow Chemical Co.
H. W. Crouch - Eastman Kodak Co.
H. L. Knight - E. I. du Pont de Nemours & Co., Inc.
D. E. Ridgley - General Electric Co.
C. M. Buehler - Taylor Instrument Cos.
A. F. Du Fresne - Consolidated Electrodynamics Corp.
R. H. Lee - E. I. du Pont de Nemours & Co., Inc.
Eugene W. Cray - Factory Mutual
Howard Ecker - Minnesota Mining & Manufacturing Co.
Dr. Bernard Lewis - Combustion & Explosives Research, Inc.
L. M. Goldsmith - Consultant
J. A. Schoke - Zeleny Industrial Systems
E. I. Thomas - Union Carbide Chemical Co.
Hamilton Lewis - E. I. du Pont de Nemours & Co., Inc.
E. P. Shoub - U. S. Bureau of Mines

F. W. Velguth - Corn Products Co.
V. A. Riggio - U. S. Industrial Chemical Co.
A. F. Sperry - Panellit Div., Information Systems, Inc.
C. W. Wilkins - Minneapolis-Honeywell Regulator Co.

Approved for publication by the Standards and Practices Board on April 11, 1960.

G. G. Gallagher - The Fluor Corporation
  Standards & Practices Department Vice President
F. E. Bryan - Douglas Aircraft Co.
  Aeronautical Standards Division Director
R. E. Clarridge - The General Electric Co.
  Intersociety Standards Division Director
F. H. Winterkamp - E. I. du Pont de Nemours & Co., Inc.
  Production Processes Standards Division Director
H. S. Kindler - ISA Director
  Technical & Educational Services, Secretary

1. PURPOSE

1.0 This practice describes recommended methods for the installation of instruments in hazardous atmospheres.

1.1 This and other Recommended Practices published by the 8D-RP12 Committee have as their objectives safe and practical instrumentation in hazardous areas.

1.2 Each Recommended Practice has been written based on evaluations of the physical situations involved. Specifications given allow the use of instruments without significantly increasing the existing risk. Recommendations are based on analyses of each situation avoiding arbitrary decisions wherever possible.

2. SCOPE

2.0 This Recommended Practice, RP-12.1 provides general guidance for the safe installation of electrical instruments using appropriate means to prevent ignition of flammable gasses and vapors. Specific safe practices for instruments are given in companion Recommended Practices.

Designation	Subject
RP 12.2	Intrinsic Safety
RP 12.3	Explosion-Proof
RP 12.4	Purging
RP 12.5	Sealing and Immersion
RP 12.6	Wiring Practices

2.1 This Recommended Practice gives recommendations on maximum fuse sizes to prevent the external surface of a housing from reaching specific ignition temperatures. This consideration applies to equipment in Division I areas protected by purging, potting, sealing or by being in explosion-proof housings.

2.2 This Recommended Practice refers only to hazards created by gasses or vapors. These are classified in the National Electrical Code as Class I, Groups A, B, C, and D. The recommendations given do not apply to protection against corrosive atmospheres or Class II or III, hazards due to flammable dusts, fibers, or metal powders.

2.3 These Recommended Practices are for the guidance of engineers in the design and installation of instrument systems and to assist local inspection authorities in approving such installations. This Recommended Practice embodies the conclusions of the ISA Committee 8D-RP 12 on Instrumentation for Hazardous Areas on what constitutes safe and economical instrumentation practice. In any particular case it may or may not be recognized by local Code enforcing authorities.

2.4 This Recommended Practice is being submitted to the Electrical Code Committee of the National Fire Protection Association for future recognition in the National Electrical Code.

3. DEFINITIONS

3.0 A hazardous atmosphere for the purpose of this Recommended Practice, is a combustible mixture of gases and/or vapors.

3.1 For the purpose of this Recommended Practice, instruments are measuring, recording, controlling, and similar apparatus requiring the use of small to moderate amounts of electrical energy in normal operation.

    3.1.1 Intrinsically safe equipment and wiring is equipment and wiring which is incapable of releasing sufficient electrical energy under normal or abnormal conditions to cause ignition of a specific hazardous atmospheric mixture. (See NEC Code 5001 for detailed definition.)

    3.1.2 Non-incendive equipment and wiring is equipment and wiring which in its normal operating condition would not ignite a specific hazardous atmosphere. The cir-cuits may include sliding or make-and-break contacts releasing insufficient energy to cause ignition. Circuits not containing sliding or make-and-break contacts may operate at energy levels potentially capable of causing ignition. Equipment and wiring having exposed surface temperatures above 80% of the ignition temperature in °C of the specific hazardous material shall not be classed as non-incendive.

    3.1.3 Ignition-capable equipment and wiring is equipment and wiring which in its normal operating condition releases sufficient electrical or thermal energy to cause ignition of a specific hazardous atmosphere, under normal operating conditions.

4. CLASS I AREA CLASSIFICATION

4.0 Areas are classified as Division 1 or Division 2 locations, or not classified, in accordance with the expectations of the presence of explosive or ignitible mixtures.

4.1 Division 1 locations may contain hazardous mixtures under normal operating conditions. (See NEC Code 5004A for detailed definition.)

4.2 Division 2 locations are those in which the atmosphere is normally non-hazardous but may become hazardous due to equipment failures, failure of ventilating systems, etc. (See NEC Code 5004B for detailed definitions.) (Appendix 8.1.)

4.3 Locations not classified as Division 1 or Division 2 may be called non-hazardous areas.

4.4 Some of the methods given in the ISA Recommended Practices are based on classifying the location within a suitable enclosure as being different from the classification of the area external to the container. The following table indicates the classification of the areas within purged and non-purged general purpose housings directly connected through sealed conduits to sources of hazardous materials.

INSTRUMENT HOUSINGS WITH CONNECTIONS
TO HAZARDOUS PROCESS MATERIALS

Classification of Area External to Housing	Number of Seals in Protecting Conduit	Classification of Area inside Housing			
		Purging Type (1)			
		X	Y	Z	Not Purged
Div. 1	2 (2)	None	Div. 2	(3)	Div. 1
Div. 2	2 (2)	None	None	None	Div. 2
None	2 (2)	None	None	None	None
Div. 1	1 (4)	Div. 2	Div. 2	(3)	Div. 1
Div. 2	1 (4)	Div. 2	Div. 2	Div. 2	Div. 2
None	1 (4)	Div. 2	Div. 2	Div. 2	Div. 2

(1) Types of purging indicated in Section 5.4
(2) Two seals, with adequate vent between, see 4.4.1
(3) Type Z purging not permitted in Div. 1 area
(4) See Sections 4.4.3.1. and 4.4.3.2. for exceptions

None in the above table means no area classification, or non-hazardous. The following Sections 4.4.2 and 4.4.3 give the basis for this table.

4.4.1 Enclosures which are connected to process lines, vessels, or sources of hazardous materials should be protected by two seals with an adequate vent between these seals. (Appendix 8.2)

4.4.2 Enclosures which are connected by singly-sealed conduits to sources of hazardous materials and which are not purged shall be classified as Division 1 if the outside area is Division 1, and as Division 2 if the outside area is not Division 1.

4.4.3 Enclosures which are connected by singly-sealed conduits to sources of hazardous materials and which are air-purged shall be classified as Division 2 locations, except as noted in sections 4.4.3.1 and 4.4.3.2. (Appendix 8.3).

    4.4.3.1 The provision of Section 4.4.3 does not apply if the pressure of the air purge exceeds the pressure of the hazardous materials.

    4.4.3.2 The provision of Section 4.4.3 does not apply if the purge flow is sufficiently large to prevent explosive concentrations of the hazardous material from accumulating within the enclosure in case of failure of the single seal.

5. SAFE PRACTICES

5.0 Protection is required for different types of electrical equipment according to the following table:

Type of Equipment or Wiring	Protection Requirements for	
	Division I Area	Division II Area
Intrinsically Safe	Not Required	Not Required
Non-Incendive	Required	Not Required
Ignition Capable	Required	Required

Where protection is required, it may be accomplished by purged or explosion-proof housings, potting or sealing, as indicated in the following sections:

5.1 Intrinsically safe equipment and wiring may be installed in either Division 1 or Division 2 areas in general purpose housings.

    5.1.1 ISA Recommended Practice (RP-12.2, Recommended Practice for Intrinsically Safe and Non-Incendive Equipment) describes the considerations applying to the classification of equipment as intrinsically safe.

5.2 Non-Incendive equipment and wiring may be installed in Division 2 areas in general purpose housings.

    5.2.1 ISA Recommended Practice (RP-12.2, Recommended Practice for Intrinsically Safe and Non-Incendive Equipment) describes the considerations applying to the classification of equipment as non-incendive and possible circuit modifications whereby the equipment may be classed as intrinsically safe.

5.3 Ignition-capable equipment and wiring may be installed in explosion-proof housings and located in Division 1 or Division 2 areas. Nonincendive equipment and wiring may be installed in explosion-proof housing and located in Division 1 areas. A discussion of proper installation and maintenance procedures for explosion-proof housings is given in ISA Recommended Practice (RP-12.3, Maintenance of Explosion-Proof and Dust Ignition-Proof Enclosures for Electrical Equipment).

5.4 Ignition-capable equipment and wiring may be installed in purged housings and located in Division 1 or Division 2 areas. Non-incendive equipment and wiring may be installed in purged housings and located in Division 1 areas. Definitions of suitable housings, purge rates, and power shut-off requirements when the purge fails, are given in ISA Recommended Practice (RP-12.4, Recommended Practice for Instrument Purging for Reduction of Hazardous Area Classification).

5.5 Ignition-capable equipment and wiring may be installed in isolation containers and located in Division 1 and Division 2 areas. Non-incendive equipment and wiring may be installed in isolation containers and located in Division 1 areas. ISA Recommended Practice (RP-12.5, Recommended Practice for the Use of Hermetic Seals, Fluid Immersion, and Potting in Hazardous Areas) describes the design requirements for isolation containers in Division 1 and Division 2 areas.

6. SURFACE HOT-SPOT LIMIT

6.0 A hazard exists when instruments are installed in Division 1 areas if sufficient power is available to heat or burn through the housing. An abnormal event, such as a short circuit or insulation failure, could raise an external portion of the case to the ignition temperature of the surrounding atmosphere.

6.1 Instrument housings should be inspected for adequate mechanical strength and for resistance to damage or overheating in case of any failure or short of the electrical components contained. Information is given for steel and aluminum housings in the following sections. Other materials of construction should be equally appropriate for instrument housings if the equivalent mechanical ruggedness and resistance to external over-heating and burn-through are preserved.

6.2 Currents in circuits contained in suitable housings that are located in Division 1 atmospheres should be limited (as with fuses) consistent with section 6.3 if the equipment is protected by any of the following arrangements: (Appendix 8.4).

(a) Explosion-proof housings
(b) Purging
(c) Hermetic seals
(d) Potting or fluid immersion

6.3 The unit shall be fused such that the release at a spot inside the case of 20% of the available energy will not burn through the case or raise the external surface temperature to 80% of the ignition temperature in °C. of the gas or vapor involved as determined by ASTM test procedure designation D-286-58T.

The conformance of aluminum or steel cases with this requirement can be determined by reference to Figure 12.1.1.

In the event that double wall or box within box construction is used and the purge is connected to the inside box, purging both, the fuse size for circuits within the inner box may be increased by a factor of 10. The minimum spacing between the two walls must be at least ten times the wall thickness.

7. WIRING

7.0 ISA Recommended Practice (RP-12.6, Recommended Practices for Wiring in Hazardous Areas) provides general design information on standards for the installation and interconnection of of instruments in Division 1 and Division 2 areas.

8. APPENDIX

This appendix to RP-12.1 provides additional information and explanations for portions of the Recommended Practice.

8.1 Refers to 4.2. Section 5013-2 of the National Electrical Code permits the use of resistors, resistance devices and rectifiers used in or in connection with meters, instruments, and relays if no make-and-break contacts exist, and if the exposed surfaces do not exceed 80% of the ignition temperature of the hazardous atmosphere involved.

8.2 Refers to 4.4.1. Examples of enclosures connected to hazardous atmospheres are found in pressure, temperature, flow connections and sampling lines to instruments. These may be easily recognized piping connections, as thermocouple conduit, or less obvious channels between stranded wires in flexible covers.

Seals consist of thermocouple wells, diaphragms, bourdon tubes, walls of analysis cells, etc. Some seals, such as required in electrical conduits in Division 1 areas, should more properly be considered as controlled leaks.

Double seals should have an adequate vent between them, sufficient to allow the escape of hazardous vapors if both seals fail.

8.3 Refers to 4.4.3. A doubly sealed system can be arranged for analyzers and pressure gauges using box-within-box construction with adequate vent between.

The inside of housings connected to hazardous materials through singly-sealed systems should be classified as Division 2 locations, even if the area outside of the housing is non-hazardous.

8.4 Refers to 6.2. The temperatures existing on the outside surface of a metal housing due to the liberation of heat energy on the inside surface have been calculated based on the following assumptions:

(a) Ambient temperature outside the housing is 30 °C.

(b) Temperature inside the housing is dependent upon the energy released by electrical short (no sparking assumed) and upon the rate of heat dissipation from the case.

(c) A specific housing has been assumed having the dimensions 18″ x 14″ x 12″.

(d) The point of electrical contact is assumed to be a 14 A.W.G. wire.

(e) The temperature distribution around the point of contact is in conformity with established laws of convection, conduction and radiation. Maximum temperature is achieved under steady state conditions.

(f) The combined radiation and convection coefficient for metal to air heat flow was assumed to be 2 p.c.u./(Hr.) (ft.^{2}) (°C.).

(NOTE: At the ignition temperature for Group IV gases the heat transfer coefficient from clean metal surfaces would be more nearly in the region of 4-8 p.c.u./hr. The lower assumed value of 2 p.c.u./hr. allows for a substantial dust loading on the outside of the instrument case.)

(g) The inside of the box was assumed to be covered 50% with perfectly insulating material.

8.4.1 The following changes in the actual conditions will increase the rate of energy release required to raise the external housing temperature to the values as indicated in the chart:

(a) Liberation of the energy over an area larger than an 0.064 inch diameter circle or at a location not directly contacting the internal surface.

(b) Liberation of 20% of the energy continuously available for an insufficient interval of time to reach the final temperature.

(c) Improvement of the expected heat transfer to the atmosphere by:

(i) Applying any appreciable external air velocity.

(ii) Removing some of the expected external dust loading.

(iii) Removing any of the equipment mounting inside the housing assumed to insulate 50% of the internal surface.

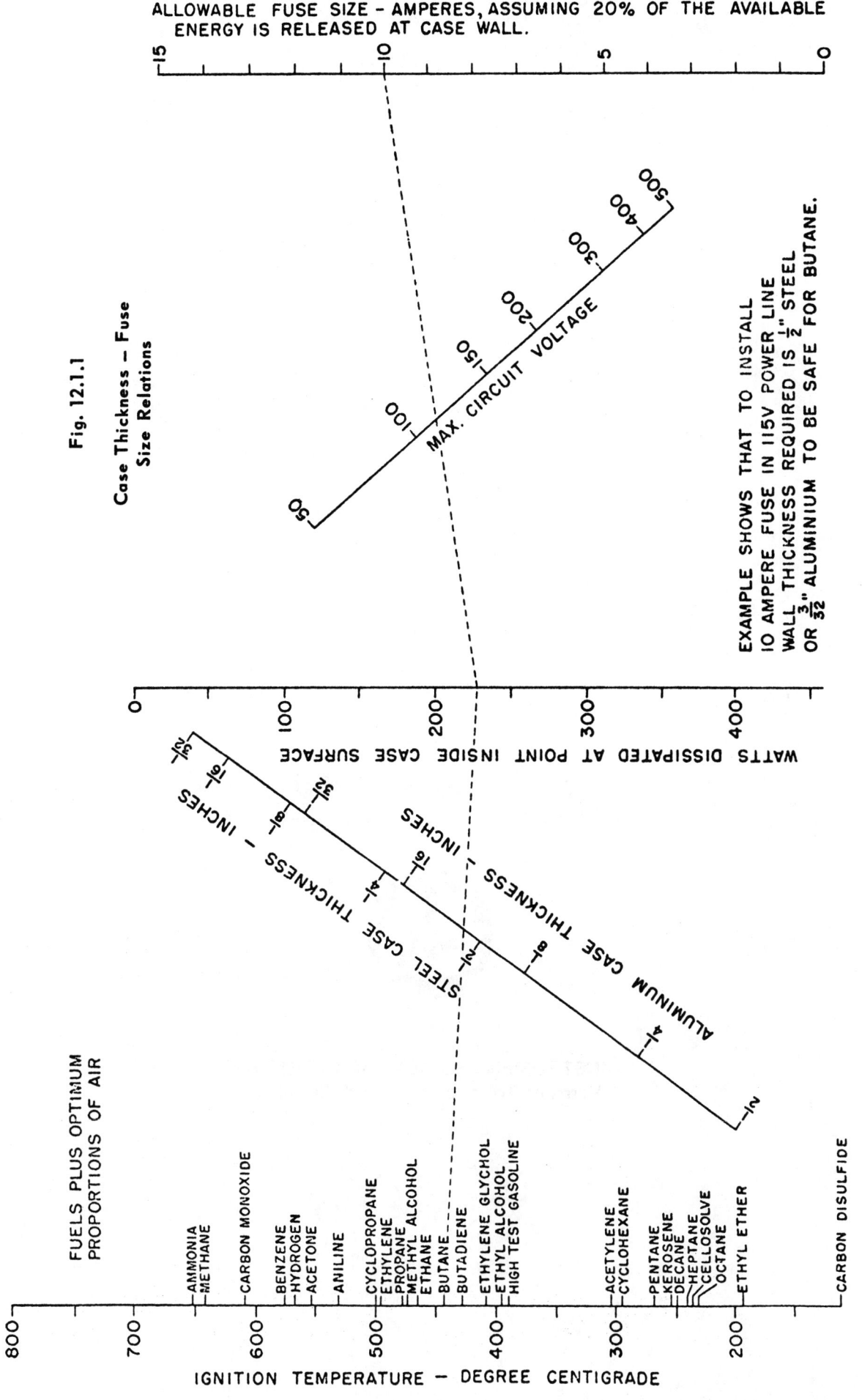

Fig. 12.1.1

Case Thickness – Fuse Size Relations

INSTRUMENT SOCIETY of AMERICA
Research Triangle Park, North Carolina

ISA-S12.4 -1970

Standard

Instrument Purging for Reduction of Hazardous Area Classification

Instrument Society of America

ISBN 0-87664-112-5

ISA-S12.4   Instrument Purging for Reduction of
              Hazardous Area Classification

INSTRUMENT SOCIETY OF AMERICA
67 Alexander Drive
P.O. Box 12277
Research Triangle Park, North Carolina 27709

FOREWORD

This Standard has been prepared as a part of the service of the Instrument Society of America toward a goal of uniformity in the field of instrumentation. To be of real value this report should not be static, but should be subjected to periodic review. Towards this end, the Society welcomes all comments and criticisms, and asks that they be addressed to the Standards and Practices Board Secretary, Instrument Society of America, 67 Alexander Drive, P.O. Box 12277, Research Triangle Park, North Carolina 27709.

This Standard was originally prepared by the 8D-SP12 Committee on Instrumentation for Hazardous Areas. It was, after considerable time and effort had been expended on its preparation, issued in 1960 as Tentative Standard ISA-RP12.4.

In 1966 a subcommittee undertook the revision of the standard to bring it up to date and to remove the tentative status.

Subcommittee members directly responsible for the current standard are:

Chairman, Robert McCarron	Leeds & Northrup Company
Samuel P. Axe	Atlantic Richfield Company
Francis S. Becker	Sun Oil Company
A. W. Jacobson	The Bristol Company
A. P. Rodriguez	Honeywell, Inc.
George H. Robinson	. E. I. duPont deNemours & Company

Members of the main 8D-SP12 Committee who have reviewed the standard are:

Chairman, F. L. Maltby	Drexelbrook Engineering
Francis Becker	Sun Oil Company
J. A. Bossert	CSA Testing Labs
Eugene W. Gray	Factory Mutual
W. F. Hickes	Foxboro Company
G. Hunt	Monsanto Company
A. W. Jacobson	Bristol Company
C. F. Kisselstein	Crouse-Hinds Company
E. C. Magison	Honeywell, Inc.
Robert McCarron	Leeds & Northrup Company
A. H. McKinney	E. I. duPont deNemours & Company
T. W. Moodie	Pillsbury Company

CORRESPONDING MEMBERS

James Beardsley	Upjohn Company
C. M. Buehler	Taylor Instrument Companies
R. D. Coffee	Eastman Kodak Company
Reese Littlefield	E. I. duPont deNemours & Company
D. L. Olson	Minnesota Mining & Manufacturing Co.
D. E. Ridgley	General Electric Company
R. A. Robinson	Consulting Engineer
E. P. Shoub	U. S. Bureal of Mines

The following people have served as a volunteer Board of Review for this standard:

T. L. Clift	Standard Oil Company of Ohio
J. G. Converse	Sun Oil Company
H. Ecker	Minnesota Mining & Manufacturing Co.
G. R. Griffith	Sinclair Refining Company
R. D. Guimond	Foxboro Company
W. J. Knatt	Monsanto Company
R. H. Lee	E. I. duPont deNemours & Company
Bernard Lewis	Combustion and Explosion Research, Inc.
Hamilton Lewis	E. I. duPont deNemours & Company
R. C. MacMullan	Esso Research and Engineering
R. W. Moore	Socony-Mobil Oil Company
V. A. Riggio	U.S. Industrial Chemicals Company
J. Senyk	Gulf Research & Development Company
C. H. St. Onge	Esso Research & Engineering Company
E. I. Thomas	Union Carbide Chemicals Company
J. V. Walsh	Consolidated Systems Corporation

INSTRUMENT PURGING FOR REDUCTION OF HAZARDOUS AREA CLASSIFICATION

CONTENTS

1 PURPOSE

1.1 This Standard is one of a series recommending safe and economical procedures for installing electrical instruments in hazardous atmospheres.

2 SCOPE

2.1 This Standard covers one specific phase of the above, namely, a technique for reducing the hazard classification by the continuous addition of an air or inert gas within a general purpose enclosure. In addition to the requirements specified in this Standard (12.4), the general requirements specified in RP-12.1 "Electrical Instruments in Hazardous Areas" must also be observed, with particular attention paid to the requirements specified when enclosures are directly connected to sources of flammable gases or vapors.

2.2 This principle is accepted in the National Electrical Code; Chapter 5, Article 500, Paragraph 500.1:

"In some cases hazards may be reduced or hazardous areas limited or eliminated by adequate positive pressure ventilation from a source of clean air in conjunction with effective safeguards against ventilation failure."

2.3 This Standard embodies the conclusions of the ISA-SP-8D Committee on Instrumentation for Hazardous Areas of what constitutes safe and economical instrumentation practices. In any particular case it may or may not be recognized by local Code-enforcing authorities.

2.4 The concept of this Standard has been accepted by the Sectional Committee on Electrical Equipment in Chemical Atmospheres of the NFPA Committee on Chemicals and Explosives and has been issued essentially as presented here as a part of NFPA Standard 496.

2.5 This Standard refers only to hazards created by gases or vapors. These are classified in the National Electrical Code as Class 1, Groups A, B, C, and D.

2.6 This Standard is concerned only with those system design criteria related to electrical ignition of a hazardous gas or vapor. It is recognized that hazards to personnel, corrosion problems or other environmental conditions must also be considered as a part of a complete system design. These problems are not, however, within the scope of this Standard.

3 DEFINITIONS

3.1 Area Classification

3.1.1 Division 1 (Hazardous). Where concentrations of flammable gases or vapors exist (1) continously or periodically during normal operations; (2) frequently during repair or maintenance or because of leakage; or (3) due to equipment breakdown or faulty operation which could cause simultaneous failure of electrical equipment. (See National Electrical Code, Paragraph 500-4(a) for detailed definition.)

3.1.2 Division 2 (Normally Non-Hazardous). Locations in which the atmosphere is normally non-hazardous and may become hazardous only through the failure of the ventilating system, opening of pipe lines, or other unusual situations. (See National Electrical Code, Paragraph 500-4(b) for detailed definition.)

3.1.3 Non-Hazardous. Areas not classified as Division 1 or Division 2 are considered non-hazardous.

NOTE: It is safe to have open flames or other continuous sources of ignition in non-hazardous areas.

3.2 Purging

The addition of air or inert gas (such as nitrogen) into the enclosure around the electrical equipment at sufficient flow to remove any hazardous vapors present and sufficient pressure to prevent their re-entry.

3.3 Purging Classifications

3.3.1 Type Z Purging. Covers purging requirements adequate to reduce the classification of the area within an enclosure from Division 2 (normally non-hazardous) to non-hazardous.

3.3.2 Type Y Purging. Covers purging requirements adequate to reduce the classification of the area within an enclosure from Division 1 (hazardous) to Division 2 (normally non-hazardous).

3.3.3 Type X Purging. Covers purging requirements adequate to reduce the classification of the area within an enclosure from Division 1 (hazardous) to non-hazardous.

3.4 Non-Incendive

Non-incendive equipment is equipment which in its normal operating condition would not ignite a specific hazardous atmosphere in its most easily ignited concentration. (The circuits may include sliding or make-and-break contacts releasing insufficient energy to cause ignition. Circuits not containing sliding or make-and-break contacts may operate at energy levels potentially capable of causing ignition.)

NOTE: Equipment having exposed surface temperatures above 80% of the ignition temperature in °C. of the specific hazardous atmosphere shall not be classed as non-incendive.

4 GENERAL REQUIREMENTS

4.1 Instrument Enclosure

4.1.1 This Standard applies to instrument enclosures not exceeding 10 cubic feet. (Appendix 6.1).

4.1.2 In like manner for purposes of this Standard the ratio of the maximum internal dimension to the minimum shall not exceed 10 to 1. (Appendix 6.2).

4.1.3 An internal enclosure or an adjacent enclosure that is being considered as part of and purged with the main instrument enclosure must have nonrestricted top and bottom vents, common to the purged main enclosure, having a minimum size for each vent of 1 square inch per 400 cubic inches of the volume of the internal or adjacent enclosure. (Appendix 6.3).

4.1.4 Enclosure shall be of non-combustible material and construction not likely to be broken under conditions to which it may be subjected.

4.1.5 Any window in a purged enclosure shall be of a material that is resistant to breakage such as 1/4 in. tempered glass or equivalent.

4.2 If hazardous gases or vapors have collected within the enclosure, either because the door has been opened or the purge has failed, then enclosure must be purged before power is applied. Once purged of hazardous concentration, it is not obligatory to maintain any given flow rate. It is only necessary that positive pressure be maintained within the case.

4.3 Since the intent is to purge an enclosure to reduce the concentration of hazardous gases or vapors to an acceptably safe level, enclosures within the instrument or adjacent enclosures connected to the instrument must be considered separately.

4.3.1 They may be adequately vented to the main enclosure.

4.3.2 They may be separately purged.

4.3.3 Equipment contained therein may be protected by other approved means.

4.4 If the enclosure is opened or if a failure occurs within the purging system, the purging system pressure may not be adequate to exclude the entrance of flammable gases or vapors. Suitable precautions such as indicators, interlocks, etc., must therefore be provided to safeguard the installation.

4.5 The purging supply shall be essentially clean and free of dust and liquids. It shall contain no more than trace amounts of flammable vapors or gases.

4.5.1 Air of instrument quality is acceptable as is other suitable supply such as inert gas.

NOTE: Ordinary plant compressed air is usually not suitable.

4.5.2 The compressor intake must be located in a non-hazardous area. The compressor suction line should preferably not pass through any area having hazardous atmospheres. (Appendix 6.4).

5 SPECIFIC REQUIREMENTS

5.1 Type Z - Requirements for the Use of Purging

to Reduce the Classification of the Area within an Instrument from Division 2 (Normally Non-hazardous) to Non-hazardous (Appendix 6.5).

5.1.1 Before power is turned on, at least four enclosure volumes of purge gas must have passed through the enclosure while maintaining an internal enclosure pressure of at least 0.1 in. of water. (Appendix 6.6).

Exception: Power may be turned on immediately if a pressure of at least 0.1 in. of water exists and if the atmosphere in the enclosure is known to be non-hazardous.

5.1.2 The enclosure must be maintained under a positive pressure of not less than 0.1 in. water when the power is on. (Appendix 6.7).

5.1.3 Under normal operation and with 125 per cent of rated voltage applied to the instrument, the external enclosure temperature or the temperature of the egress air shall not exceed 80 per cent of the ignition temperature (in °C) of the vapor or gas involved as determined by Method of Test for Autogenous Ignition Temperatures of Petroleum Products, ASTM D2155. (Appendix 6.8).

5.1.4 Safety interlocks to remove power upon failure of purging supply are not required.

5.1.5 Acceptable installations are shown in Figure 12.4.1, 12.4.2, and 12.4.3.

5.1.6 An alarm or indication of purge system failure must be provided. The device may be mechanical, pneumatic or electric; audible or visual.

5.1.6.1 If electrical, it must meet the requirements for its location.

5.1.6.2 To avoid plugging when a pneumatic device is used, any restrictions between the pneumatic device and the enclosure shall have passages no smaller than the smallest passage before the pneumatic device.

5.1.6.3 No valve between the alarm or indicator and the enclosure shall be permitted.

5.1.6.4 The pressure or flow device must be capable of indicating (or actuating an alarm) when the purging pressure or flow is inadequate to maintain a static pressure within the enclosure of 0.1 in. water. (Appendix 6.7).

5.1.7 A red warning nameplate must be mounted on the instrument. The nameplate shall be mounted in a prominent location and be visible before the enclosure is opened. It shall state:

5.1.7.1 "Enclosure shall not be opened unless area is known to be non-hazardous or unless the power has been removed from all devices within the enclosure."

5.1.7.2 "Power shall not be restored after enclosure has been opened until enclosure has been purged for 'X' minutes."

NOTE: Manufacturer to recommend purge conditions and flow rate, based upon his enclosure design, necessary to purge with four purge volumes in the stated time.

5.1.8 The maximum operating temperature of any internal surface exposed to the atmosphere within the enclosure shall not exceed 80 per cent of the ignition temperature (°C) of the gases or vapors involved under normal operating conditions and at 125 per cent of rated voltage. If any temperature exists over 80 per cent of the ignition temperature of the gases or vapors involved, then:

5.1.8.1 The warning nameplate shall contain a statement that such conditions exist and that power must be removed for "X" minutes (period to be determined and specified by the manufacturer to be sufficient to permit

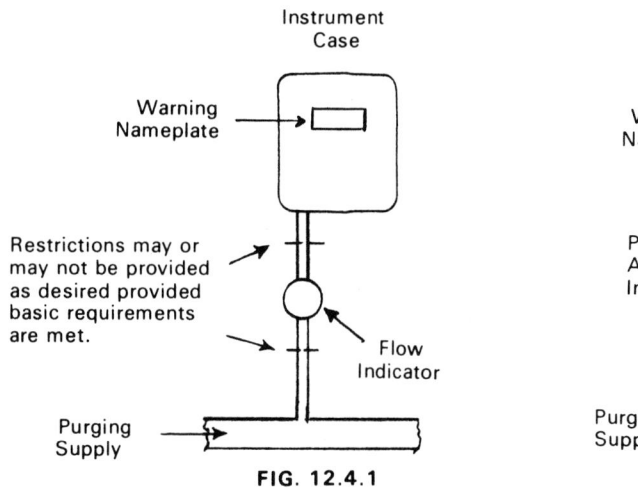

FIG. 12.4.1

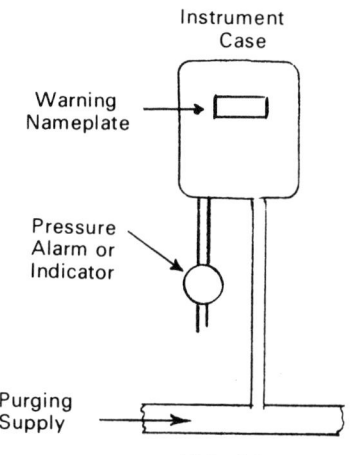

FIG. 12.4.2

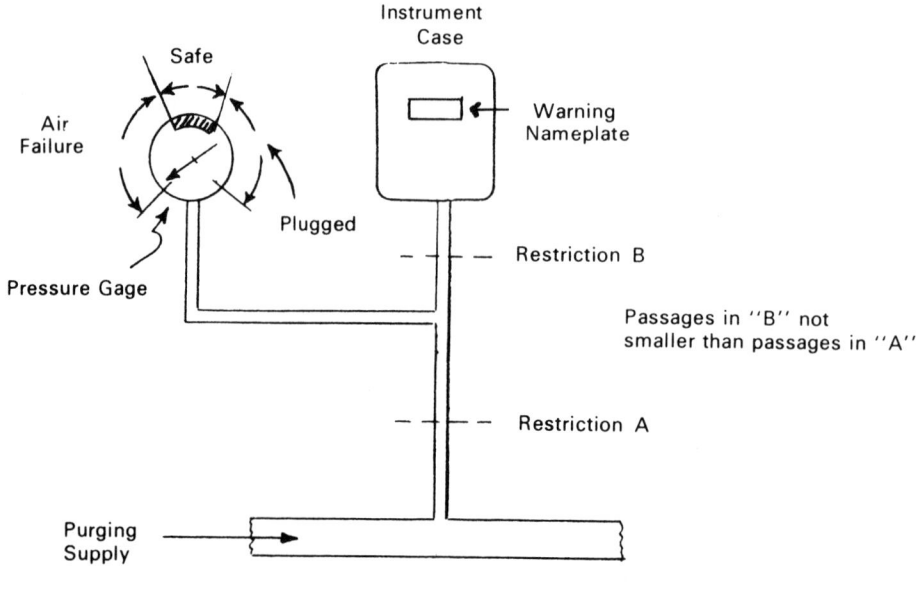

FIG. 12.4.3

ACCEPTABLE INSTALLATIONS FOR TYPES Y & Z PURGING

unit to cool to safe limit) before the door is opened unless the area is demonstrated to be non-hazardous at the time, or

5.1.8.2 the hot component may be separately housed so that the temperature of its housing is below safe limits. This housing shall be purged or sealed and provided with a warning name-plate that its cover may not be removed for "X" minutes (period to be determined and specified by manufacturer to be sufficient to permit unit to cool to safe limit) unless the area is demonstrated to be non-hazardous at that time.

5.2 Type Y-Requirements for the Use of Purging to Reduce the Classification of the Area within an Instrument from Division 1 (Hazardous) to Division 2 (Normally Non-hazardous). (Appendix 6.9).

5.2.1 All requirements 5.1.1 to and including 5.1.7 must be met.

5.2.2 Precautions must be taken to insure that a malfunction (short circuit) between the power wiring and the enclosure walls shall not burn through the enclosure or otherwise raise the external surface temperature to 80 per cent of the ignition temperature in °C of the gas or vapor involved. (Appendix 6.10) Precautions involve a combination of fuse type (i.e. quick blow and medium blow), fuse rating and thickness of case wall. Conformance of aluminum or steel cases with this requirement can be determined by reference to Figure 12.4.7. Other materials meeting the requirements will be equally acceptable.

5.2.3 Equipment shall be non-incendive (i.e., shall conform to the requirements for Divison 2 locations). (Appendix 6.11). The specific requirements are:

5.2.3.1 Make-and-break or sliding contacts shall be either immersed in oil, enclosed within a chamber hermetically sealed against the entrance of gases or vapors, or in circuits which under normal conditions do not release sufficient energy to ignite a specific hazardous atmosphere mixture. (Appendix 6.12).

NOTE: For a standard on Intrinsic Safety, see ISA S12.2 (Intrinsically Safe and Non-Incendive Electrical Instruments).

5.2.3.2 The maximum operating temperature of any internal surface exposed to the atmosphere within the enclosure shall not exceed 80 per cent of the ignition temperature in degrees Celsius of the gases or vapors involved under normal operating condi-

tions and at 125 per cent of rated voltage. If any temperature exists over 80 per cent of the ignition temperature in °C of the gases or vapors involved, the surface having this temperature shall be enclosed within a chamber hermetically sealed against the entrance of gases or vapors. (Appendix 6.13).

NOTE: If the conditions specified under 5.2.3 are met, the equipment can be located in a Division 2 location in a general purpose enclosure without purging.

5.3 Type X-Requirements for the Use of Purging to Reduce the Classification of the Area within an Instrument from Division 1 (Hazardous) to Non-hazardous. (Appendix 6.14).

5.3.1 A timing device must be incorporated to prevent power being applied until after the elapse of a time sufficient to permit at least four enclosure volumes of purge gas to have passed through the enclosure while maintaining an internal pressure of at least 0.1 in. of water. Timing device must meet the requirements of its location.

NOTE: Manufacturer to recommend purge condition and flow rate, based on his enclosure design necessary to purge with four volumes in a stated time.

5.3.2 The enclosure must be maintained under a positive pressure of not less than 0.1 in. of water when the power is on.

5.3.3 A device must be incorporated to automatically remove potential from all circuits or equipment within the enclosure not suitable for Division 1, upon failure of the purging supply.

5.3.4 A door switch must be provided to remove potential automatically from all circuits, within the enclosure not suitable for Division 1, if the enclosure can be readily opened without the use of a key or tools. The door switch, even though located within the enclosure must be suitable for Division 1 locations. (Appendix 6.15).

5.3.5 The maximum operating temperature of any surface exposed to the atmosphere within the enclosure shall not exceed 80 per cent of the ignition temperature in °C of the gases or vapors involved under normal operating conditions and at 125 per cent of rated voltage.

If any temperature exists over 80 per cent of the ignition temperature in °C of the gases or vapors involved, the surface having this temperature shall be enclosed within a chamber hermetically sealed against the entrance of gases or vapors. (Appendix 6.16).

5.3.6 Precautions must be taken to insure that a malfunction (short circuit) between the power wiring and the enclosure walls shall not burn through the enclosure or otherwise raise the external surface temperature to 80 per cent of the ignition temperature in ° C of the gas or vapor involved. (Appendix 6.10) Precautions involve a combination of fuse type (i.e. quick blow, and medium blow) fuse rating and thickness of case wall. Conformance of aluminum or steel cases with this requirement can be determined by reference to Figure 12.4.7. Other materials meeting the requirements will be equally acceptable.

5.3.7 Acceptable installations are shown in Figures 12.4.4, 12.4.5 and 12.4.6.

5.3.8 The power cut-off switch provided to remove power upon failure of the purging system shall be either flow or pressure actuated.

5.3.8.1 It must conform to the requirements of its location.

5.3.8.2 The pressure or flow device must be capable of cutting off power when the purging pressure flow is inadequate to maintain a static pressure within the enclosure of 0.1 in. water. (Appendix 6.7).
If a pressure device (Fig. 12.4.6) is used, it must be capable of cutting off power if pressure exceeds predetermined safe limits.

5.3.8.3 To avoid plugging when a pneumatic device is used, any restrictions between the device and the enclosure shall have passages no smaller than the smallest passage before the device.

5.3.8.4 No valve between the alarm or indicator and the enclosure shall be permitted.

5.3.9 A red warning nameplate must be mounted on the instrument. The nameplate shall be mounted in a prominent location and be visible before enclosure is opened. It shall state:

5.3.9.1 "Enclosure shall not be opened or any cover removed unless area is known to be non-hazardous or unless the power has been removed from all devices within the enclosure."

5.3.9.2 "Power shall not be restored after enclosure has been opened until enclosure has been purged for 'X' minutes."

NOTE: Manufacturer to recommend purge conditions and flow rate necessary to pass at least four enclosure volumes in the stated time "X".

6 APPENDIX

This Appendix is intended to clarify Standard ISA-SP12.4 and to explain the philosophy on which it

is based. Wherever possible, this Standard is based upon an analysis of the physical conditions involved with a minimum reliance upon arbitrary rules.

6.1 Refers to 4.1.1 - This Standard is intended to cover instrument enclosures but not cubicles or rooms. It is considered that the stated volume of 10 cubic feet will make this distinction. For information concerning requirements for larger enclosures refer to NFPA 496 Standard for Purged Enclosures for Electrical Equipment in Hazardous Locations.

6.2 Refers to 4.1.2 - The 10 to 1 limitation on the maximum internal dimension to the minimum internal dimension is considered to describe normal enclosure practice. Further, this limitation permits the insertion of purged gas at at a single point to purge the entire enclosure volume. This would not necessarily be true in elongated shapes outside of this limitation.

6.3 Refers to 4.1.3 - In order for any internal or adjacent enclosure to be automatically purged as the main enclosure is purged, adequate vents must be provided to permit air circulation between it and the main enclosure. If top and bottom vents are not practical, vents otherwise suitably separated may be used. The area required to provide adequate venting will obviously depend upon the internal or adjacent enclosure. It is considered that meeting this requirement will prevent the formation of unpurged pockets of gas within the enclosure. It is not intended to imply that internal or adjacent enclosures not meeting these venting requirements are prohibited but that such enclosures must be provided with their own purging connections.

6.4 Refers to 4.5.2 - It is recognized that if the compressor suction line passes through any area in which hazardous atmospheres may exist, there is a possibility that flammable gases or vapors may be drawn into the system in sufficiently high concentrations that the purging gas itself becomes hazardous. Consequently, the best practice is not to permit the suction line to pass through any hazardous area. It is recognized, however, that this is not always practical. In the event that it is essential that the compressor suction line pass through a hazardous area, it must be of non-combustible material and suitably protected against mechanical damage and leaks.

6.5 Refers to 5.1 - A hazard is created under Type Z conditions only if the purge should fail at the same time the area which is normally non-hazardous becomes hazardous. Because of this, it is not considered essential to remove power from the equipment upon failure of the purge but only that adequate warnings must be provided to prevent continuing operation without purge protection.

6.6 Refers to 5.1.1 - Any time the enclosure has been opened or the purging gas removed, there exists the possibility that explosive gases or vapors have accumulated within the enclosure. Passing four enclosure volumes of

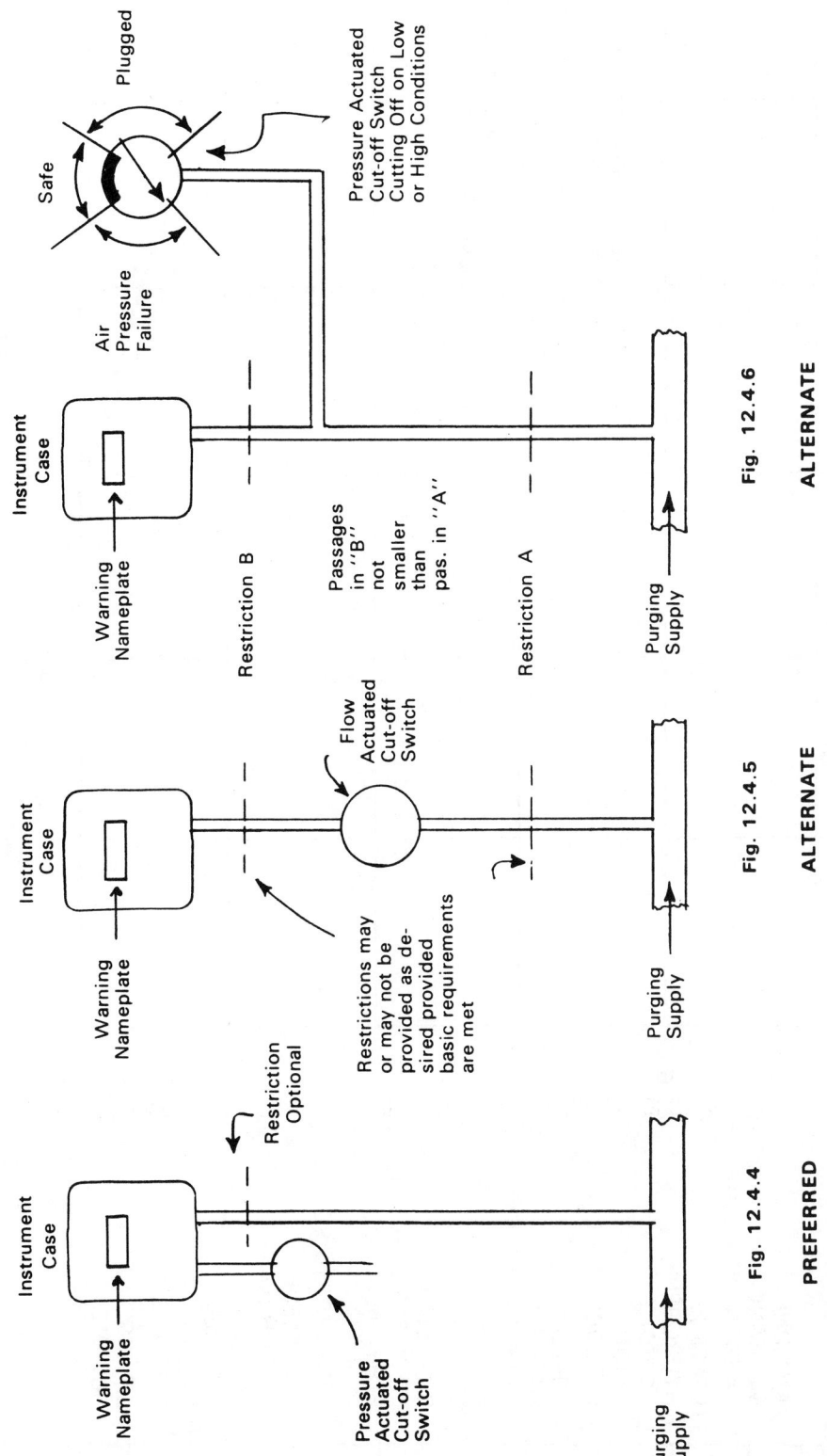

ACCEPTABLE INSTALLATIONS FOR TYPE X PURGING

Fig. 12.4.6 ALTERNATE

Fig. 12.4.5 ALTERNATE

Fig. 12.4.4 PREFERRED

9

EXTERNAL CASE SPOT TEMPERATURES VS CASE THICKNESS & FUSE TYPE & RATING

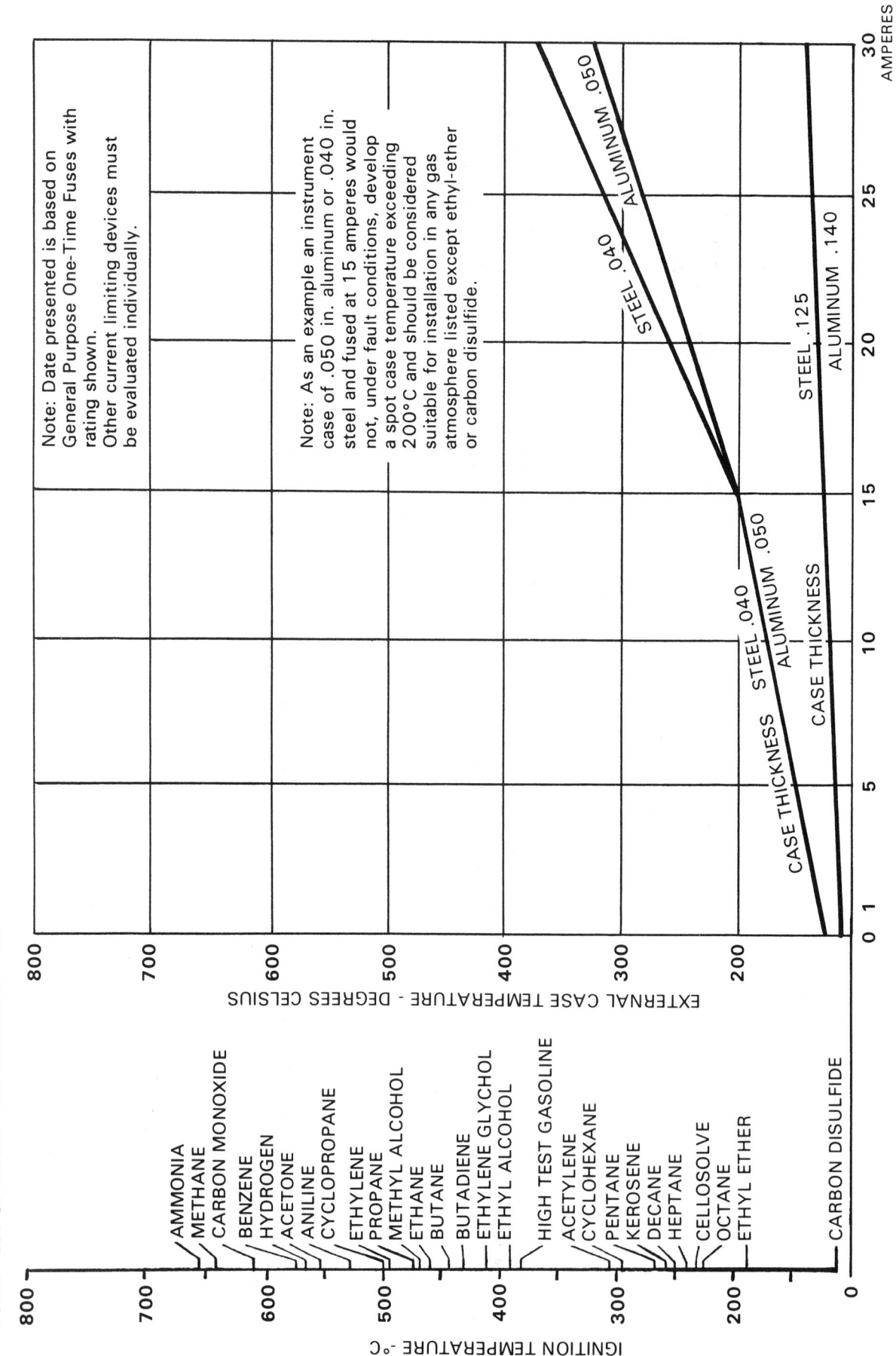

Note: Date presented is based on General Purpose One-Time Fuses with rating shown.
Other current limiting devices must be evaluated individually.

Note: As an example an instrument case of .050 in. aluminum or .040 in. steel and fused at 15 amperes would not, under fault conditions, develop a spot case temperature exceeding 200°C and should be considered suitable for installation in any gas atmosphere listed except ethyl-ether or carbon disulfide.

FUELS PLUS OPTIMUM PROPORTIONS OF AIR

EXTERNAL CASE TEMPERATURE - DEGREES CELSIUS

FUSE RATING-GENERAL PURPOSE FUSE (CASE SHAWMUT TYPE OT)
FIG. 12.4.7

Revised
7-22-69

IGNITION TEMPERATURE -°C

AMMONIA
METHANE
CARBON MONOXIDE
BENZENE
HYDROGEN
ACETONE
ANILINE
CYCLOPROPANE
ETHYLENE
PROPANE
METHYL ALCOHOL
ETHANE
BUTANE
BUTADIENE
ETHYLENE GLYCHOL
ETHYL ALCOHOL
HIGH TEST GASOLINE
ACETYLENE
CYCLOHEXANE
PENTANE
KEROSENE
DECANE
HEPTANE
CELLOSOLVE
OCTANE
ETHYL ETHER
CARBON DISULFIDE

STEEL .040
ALUMINUM .050
CASE THICKNESS

STEEL .040
ALUMINUM .050
CASE THICKNESS

ALUMINUM .050
STEEL .040

STEEL .125
ALUMINUM .140

AMPERES

the purging gas through the enclosure is adequate to dilute even a 100% hydrogen atmosphere to below the lower explosive limit.

6.7   Refers to 5.1.2 and 5.1.6.4 and 5.3.8.2 - The reason for requiring that a positive pressure be maintained is to prevent flammable gases or vapors from being forced into the enclosure by external air velocities. It is considered that the velocity of a hazardous mixture encountered in the plant will not exceed 15 miles per hour. The pressure limit of 0.1 in. of water is equal to static pressure corresponding to 15 miles per hour velocity.

NOTE: This pressure (0.1 in. water) is the minimum acceptable level. In order to insure operation at this figure the system setting should be at a level sufficiently above this figure to compensate for any system uncertainties to be expected in the specific device employed.

In establishing this pressure level due consideration has also been given to the possible effect of diffusion. Studies have determined that even under extremely adverse conditions i.e. a sustained high concentration of combustibles and a case approaching a hermetically sealed condition diffusion would not raise the interior atmosphere to explosive limits. When it is further considered that the purging requirements apply to instrument cases of typical industrial construction that are not well sealed and that have purge flow rates on the order of .1 to 1 standard cubic feet per minute per inch of water purge pressure it is unlikely that a hazard due to diffusion can occur.

6.8   Refers to 5.1.3 - In the event that an external enclosure temperature in excess of the ignition temperature of the gas or vapor involved existed, it is obvious the purging cannot prevent an explosion; therefore, it is essential that excessive surface temperatures be prevented.

6.9   Refers to 5.2 - The equipment that can be included within the enclosure under Type Y conditions must be suitable for Division 2. This requires that it does not normally contain a source of ignition. Thus, a hazard is created within the enclosure only upon failure of the purging system simultaneously with a failure of the internal equipment causing it to produce a source of ignition. Therefore, it is not considered essential that on failure of the purging system the power be automatically removed from the equipment but that a warning be provided to prevent continuing operation without purge protection.

6.10   Refers to 5.2.2 & 5.3.6 - Because the exterior atmosphere may have a high probability of being hazardous at any particular time it is essential that the probability of an internal failure raising the temperature of the external enclosure surface to a temperature capable of igniting the gas or vapor involved negligible. This requirement can be met by a correct selection of fuse size and type and a suitable thickness of the case wall.

Tests have determined that at the instant a short circuit occurs, for a fraction of a cycle, a tremendous surge current, in the order of 1000 Amperes, develops depending on the

capacity of the power source and the fuse type and rating.

External case temperatures depicted on Figure 12.4.7 are based on actual tests made on metal plates of the material and thickness listed and using general purpose fuse of the ratings indicated. Fuses were listed as Chase Shawmut OT or One Time general purpose type. Quick Blow fuses of the Chase Shawmut Amp-Trap current limiting type sufficiently reduced the current surge that the external case temperatures were generally in the safe region despite the fuse size.

6.11   Refers to 5.2.3 - The requirement for equipment to be used in Division 2 locations is that it should not normally contain a source of ignition. Such sources of ignition can be provided by either make-and-break contacts or by high surface temperatures in contact with the atmosphere which may become hazardous.

6.12   Refers to 5.2.3.1 - Examples of contacts normally operating at energy levels which would not cause ignition are: slide-wire and switching contacts in thermo-couple, resistance thermometer, strain gauge and pH electrode, etc. circuits.

6.13   Refers to 5.2.3.2 - Internal temperatures above the ignition temperature of the gas or vapor involved, such as vacuum tube filaments, are hermetically sealed to prevent them from normally coming in contact with the atmosphere which may become hazardous. It is essential, of course that in such enclosure the surface of the glass envelope, which does come in contact with the atmosphere, not have a temperature in excess of 80% of the ignition temperature of the gas or vapor involved.

6.14   Refers to 5.3 - Because the probability of a hazardous concentration of gas or vapor external to the enclosure is high and the enclosure normally contains a source of ignition, it is essential that any disruption of the purging will result in the removal of power from the equipment. It is also essential that the power within the enclosure be limited to minimize the possibility that an internal fault can raise the external surface temperature to a hazardous degree.

6.15   Refers to 5.3.4 - It is considered essential that any door or other opening which can be opened by untrained people without tools be protected with door interlock switches. Consistent with the practice which has been established with explosion-proof enclosures, it is considered that the commonly displayed warning nameplate is adequate protection for the enclosure that can only be opened by the use of suitable tools.

6.16   Refers to 5.3.5 - Because the source of ignition caused by high temperature is not immediately removed by cutting off power to the equipment, it is considered essential that no surface temperature approaching the ignition temperature of the gas or vapor involved should be permitted to come in contact with the internal enclosure atmosphere.

INSTRUMENT SOCIETY of AMERICA
Research Triangle Park, North Carolina

ANSI/ISA-RP12.6-1976
Approved May 26, 1977

Recommended Practice

Installation of Intrinsically Safe Instrument Systems in Class I Hazardous Locations

Instrument Society of America

ISBN 0-87664-339-4

ISA-RP12.6   Installation of Intrinsically Safe Instrument
              Systems in Class I Hazardous Locations

INSTRUMENT SOCIETY OF AMERICA
67 Alexander Drive
P.O. Box 12277
Research Triangle Park, North Carolina 27709

PREFACE

(This Preface is included for informational purposes and is not a part of ISA RP12.6 — 1976).

This Recommended Practice has been prepared as a part of the service of the Instrument Society of America toward a goal of uniformity in the field of instrumentation. To be of real value this Recommended Practice should not be static, but should be subjected to periodic review. Toward this end, the Society welcomes all comments and criticisms, and asks that they be addressed to the Standards and Practices Board Secretary, Instrument Society of America, 67 Alexander Drive, P.O. Box 12277, Research Triangle Park, North Carolina 27709.

The ISA Standards and Practices Department is aware of the growing need for attention to the metric system of units in general, and the International System of Units (SI) in particular, in the preparation of instrumentation standards. The Department is further aware of the benefits to USA users of ISA Standards of incorporating suitable references to the SI (and the metric system) in their business and professional dealings with other countries. Towards this end this Department will endeavor to introduce SI and SI-acceptable metric units in all new and revised standards to the greatest extent possible. The Metric Practice Guide, which has been published by the American Society for Testing and Materials as ANSI Z210.1 (ASTM E380-76; IEEE 268-1975), and future revisions, will be the reference guide for definitions, symbols, abbreviations, and conversion factors.

This practice was prepared by the Committee on Installation Practices (RP12.6). This Committee was established in 1973 to formulate standard practices for the installation of instrument systems in hazardous locations. In addition to the standard practices the Committee has included, through footnotes and appendices, many recommendations which are considered as preferred engineering practices for installing instrument systems. Subsequent parts will include recommendations for the installation of non-incendive, explosion-proof and dust ignition-proof systems. Committee members represent Suppliers, Contractors, Users, and Testing Labs. The Committee has explored from all sides the questions that arise when installing instrument systems in hazardous locations.

In preparing this Recommended Practice, the Committee considered the following:
   a. Intrinsically safe wiring is not required to comply with the requirements in national codes for non-intrinsically safe wiring in hazardous locations.
   b. Intrinsically safe writing must comply with the applicable requirements in national codes for non-hazardous location wiring.
   c. Intrusion of unsafe energy into intrinsically safe wiring must be prevented.

The ISA Standards Committee SP12.6 operates as a subcommittee of ISA Standards Committee SP12 within the ISA Standards and Practices Department, W. B. Miller, Vice President. Committee SP12 approved this Recommended Practice in May, 1976.

Those listed below serve as members of Committee SP12.6:

John R. Williams, Chairman	Stearns Roger, Inc.
John M. Bacon	Mobay Chemical Company
John Bossert	Canadian Explosive Atmospheres Lab
Richard Buschart	Monsanto Company
William M. Dillow	Gulf Oil Company
Wallace Engard	Eastman Kodak Company
Michael J. Morgan	Taylor Instrument Company
Daniel J. Oswald	Taylor Instrument Company
Gerald Posner	Fischer & Porter
Peter J. Schram	Underwriters Laboratories Inc.
Robert West	Monsanto Company
Edward C. Yeaton	Eastman Kodak Company

The following people have aided in editing this document and have served as a Board of Review:

R. B. Adams	Dow Chemical Company
J. E. Altimier	USI Chemicals Company
P. Bliss	Pratt & Whitney Aircraft Division, United Technologies Corporation
W. Calder	Foxboro Company
C. M. Crain	Monsanto
R. D. Coffee	Eastman Kodak Company
R. H. Daniel	Shell Oil Company
S. L. Davis	E. I. duPont deNemours & Company
T. Diebold	General Mills Inc.
W. Dreier	
G. F. Erk	Sun Oil Company
R. Evans	St. Regis Paper Company
H. F. Fabisch	Fluor Corporation
B. J. Feikle	Bailey Meter Company

RP12.6

Installation of Intrinsically Safe Instrument Systems
In Class I Hazardous Locations.

CONTENTS

RP12.6

Installation of Intrinsically Safe Instrument Systems
In Class I Hazardous Locations.

1. PURPOSE

1.1 This Recommended Practice has been formulated to provide safe installation practices for intrinsically safe instrument systems in Class I hazardous locations.

1.2 This Recommended Practice is intended to promote uniformity among specialists. It is intended that it be applied only by those who have familiarized themselves with fundamental principles of safety in instrument systems. It is not intended as an instruction manual for untrained persons, nor as a standard for use by inspection personnel who are not fully acquainted with these principles.

1.3 Many national codes covering electrical installation do not currently provide adequate guidance for the installation of intrinsically safe systems. For example the 1975 National Electrical Code Section 500-1 states "Equipment and associated wiring approved as intrinsically safe shall be permitted in any hazardous (classified) location for which it is approved, and the provisions of Articles 500 through 517 shall not be considered applicable to such installations. Means shall be provided to prevent the passage of gases and vapors." This Recommended Practice is intended to provide guidance for wiring of intrinsically safe systems.

2. SCOPE

2.1 This Recommended Practice is for the guidance of those involved in the design and installation of field installed instrument wiring in hazardous locations. (See Appendix A.1.1)

2.2 Of necessity, this Recommended Practice also provides guidance for the design and installation of field installed wiring in non-hazardous locations.

2.3 This Recommended Practice provides guidance for the layout and wiring of panels which contain intrinsically safe wiring. (For information concerning wiring practices for central control panels, see ISA S60.8 "Electrical Guide for Control Centers" which is in draft form at this writing.)

2.4 This Recommended Practice is intended to be used in conjunction with nationally recognized codes covering wiring practices, such as the National Electrical Code NFPA 70 (ANSI C1) and Canadian Electrical Code, Part 1.

2.5 This Recommended Practice is not intended to include guidance for equipment design.

2.6 This Recommended Practice is not intended to include guidance for maintenance and testing.

2.7 This Recommended Practice is not intended to be applied to the utilization of intrinsically safe portable equipment in hazardous locations.

3. DEFINITIONS

3.1 *Intrinsically safe equipment and wiring* is equipment and wiring which is incapable of releasing sufficient electrical or thermal energy under normal or abnormal conditions to cause ignition of a specific hazardous atmospheric mixture in its most easily ignited concentration.

3.2 *Different intrinsically safe systems* are defined as intrinsically safe systems: (1) operating at different voltage levels or polarities, or (2) having different signal ground reference points, or (3) approved for different hazardous location groups. An intrinsically safe system may include more than one intrinsically safe circuit. (See Appendix A.2.1)

3.3 An intrinsic safety barrier is a device or system of devices that will, when properly installed, render any circuit intrinsically safe provided:

3.3.1 There is no intrusion of outside power sources exceeding the barrier rating (usually 250 Vac) on the non-hazardous location side of the barrier. (See Appendix A.2.2)

3.3.2 There is no intrusion of outside power sources including other intrinsically safe circuits on the hazardous location side of the barrier.

3.3.3 There is no energy storage system (capacitive or inductive) in excess of the maximum permitted by the barrier design on the hazardous location side of the barrier. (See Appendix A.2.3)

3.4 *Signal ground reference point* is a single grounding point to which all grounding conductors of any single intrinsically safe system are connected. It may be used as the ground reference for more than one intrinsically safe system.

4. NON-HAZARDOUS LOCATION WIRING OF INTRINSICALLY SAFE CIRCUITS

4.1 *Precautions shall be taken to insure against intrusion of unsafe energy from other circuits.* Since wiring rules for non-hazardous locations are less stringent than for hazardous locations, the likelihood for intrusion of unsafe energy into intrinsically safe circuits is greater in non-hazardous locations than in hazardous locations.

4.2 External to panels, the intrinsically safe wiring shall be separated from non-intrinsically safe wiring in enclosures, cables, raceways or cable trays which are identified as containing intrinsically safe wiring. The identification shall be of sufficient durability to withstand the environment involved. A portion of enclosure or cable tray compartmented by an adequate insulating or grounded metal partition may be considered a separate enclosure or cable tray. Alternatively, intrinsically safe and non-intrinsically safe wiring may occupy the same enclosure, raceway or cable tray if the intrinsically safe and non-intrinsically safe wiring are spaced at least 50mm (2 inches) apart and separately tied down.

4.2.1 Connections to terminals shall be in accordance with Section 6.

4.3 Inside panels, the field wiring terminals for intrinsically safe circuits shall be adequately segregated or separated from non-intrinsically safe terminals. This may be accomplished by locating intrinsically safe and non-intrinsically safe terminals in separate enclosures, by use of an insulating or grounded metal partition between terminals, or by adequate separation between terminals. When a partition is used to segregate terminals, it shall extend close enough to the enclosure walls to effectively separate the terminals. When spacing is used, the distance between shall be at least 50mm (2 inches). Additional precautions such as wire tie-downs or special wiring methods may be necessary to provide adequate segregation of circuits. This is especially true when terminals are arranged one above the other. Spacing alone will not usually provide adequate separation in such cases. Care shall be taken in the layout of terminals and wiring method used to prevent contact between intrinsically safe and non-intrinsically safe circuits should a wire become disconnected. (See Appendix A.3.2)

4.3.1 Inside panels, non-intrinsically safe wiring shall be segregated from all intrinsically safe wiring by positive means, such as installation in different raceways. (See Appendix A.3.2)

4.3.2 Inside panels, cables and wiring of intrinsically safe circuits shall be identified as intrinsically safe. The identification shall be of sufficient durability to withstand the environment involved.

4.3.2.1 Color coding alone may be used for identification of intrinsically safe wiring, if the color used is bright blue and no other wiring is color coded bright blue. If other wiring is color coded bright blue, such as thermocouple wiring in non-intrinsically safe circuits, the intrinsically safe wiring shall be identified by other than or in addition to color coding.

4.4 Wiring of intrinsically safe circuits and systems shall be in accordance with 6.7.

5. GROUNDING AND BONDING OF INTRINSICALLY SAFE CIRCUITS

5.1 The signal ground system shall consist of conductors, separate from the power grounding system, connected to the signal ground reference point. This system shall have a single connection to the plant equipment grounding system. (See Appendix A.4.1)

5.2 The maximum resistance allowable between the farthest point on the intrinsic safety barrier ground bus and the signal ground reference point shall be in accordance with the barrier manufacturer's recommendation. At the present time the general recommendation is one ohm or less. (See Appendix A.4.2)

5.3 All metal enclosures for intrinsically safe circuits shall be bonded to the signal ground reference point, through a metallic path. (See Appendix A.4.3)

5.3.1 A metal enclosure for intrinsically safe circuits need not be bonded to the signal ground reference point if the potential of the enclosure cannot be raised above ground potential, taking into consideration faults in other equipment. (See Appendix A.4.4)

6. HAZARDOUS LOCATION WIRING OF INTRINSICALLY SAFE CIRCUITS

6.1 *Precautions shall be taken to insure against intrusion of unsafe energy from other circuits.*

6.2 All intrinsically safe wiring shall be kept separate from non-intrinsically safe wiring. The intrinsically safe wiring may be installed using any wiring method suitable for ordinary locations as though the hazardous location were not present, provided the installation complies with all other requirements in this Recommended Practice.

6.3 Conductor insulation thickness should be no less than .25 mm (0.010-inch).[1]

6.4 In enclosures containing different intrinsically safe circuits, terminal strips shall be the type with physical partitions between terminals, i.e. "barrier-type". Terminals for different intrinsically safe systems shall be separated by insulating or grounded metal partitions or in accordance with 4.3.

6.5 All wiring at terminal strips shall be arranged so that it is unlikely for any conductor, upon coming loose from its termination, to contact a terminal of a different intrinsically safe circuit or non-intrinsically safe circuit.

6.6 Grounding and bonding conductors shall be insulated where practicable. Such conductors shall be insulated from ground except at points where they are intentionally connected to grounding electrodes, to prevent sparking in a hazardous location.

6.7 Different intrinsically safe systems shall not be run in the same multiconductor cable. Different intrinsically safe circuits of the same intrinsically safe system shall not be run in the same cable unless at least .25 mm (0.010-inch) thickness insulation is used on each conductor, or unless no hazard can result from interconnection.

6.8 Where conduit or other raceway is used to enclose intrinsically safe wiring, the conduit or raceway shall be sealed and/or vented in such a manner that the conduit or raceway will not serve to transmit the flammable atmosphere from one division of a hazardous location to another or from a hazardous to a non-hazardous location.

6.9 Where intrinsically safe circuits are contained in multiconductor cable, and flammable mixtures can be transmitted through the cable core, the cable core shall be sealed and/or vented in such a manner that the cable will not serve to transmit the flammable atmosphere from one division of a hazardous location to another or from a hazardous to a non-hazardous location. (See Appendix A.4.5)

6.10 Cables and wiring of intrinsically safe circuits shall be identified in accordance with 4.3.2.

7. GENERAL

7.1 The manufacturers' instructions with regard to electrical parameters (e.g. inductance, L/R ratio, capacitance) and permissible system configurations shall be followed.

---

[1] Conductor insulation is not always required for intrinsic safety. However, good engineering practice usually requires insulation of this quality to insure reliability.

RP12.6

Installation of Intrinsically Safe Instrument Systems
In Class I Hazardous Locations.

7.2 Where intrinsically safe circuits may be exposed to disturbing magnetic or electric fields, suitable attention shall be given to twisting or shielding to ensure that these fields do not adversely affect the intrinsic safety of the circuit.[2]

7.3 Unless different intrinsically safe circuits run in the same multiconductor cable are electrically isolated from each other so as to preclude summing of the available current in each circuit should the circuits become shorted together, the cable shall where practicable be constructed so that each circuit consists of a twisted pair of conductors.

APPENDIX A

A.1 INSTRUMENT WIRING IN HAZARDOUS LOCATIONS

A.1.1 The table "Wiring in Class I Locations" includes the concept of a Division 0 location, although Division 0 is not included in the 1975 National Electrical Code or the 1972 Canadian Electrical Code as a separate location. Division 0 (called Zone 0) is defined in International Electrotechnical Commission Publication 79-10, 1972 (Electrical Apparatus for Explosive Gas Atmospheres. Part 10 Classification of Hazardous Areas.), as an area in which an explosive atmosphere is continuously present or present for long periods. This is part of the definition of Division 1 in the 1975 National Electrical Code and 1972 Canadian Electrical Code. The concept is separated here for the guidance of those who wish to recognize the increased degree of hazard in that small percentage of Division 1 locations and provide a level of safety over and above that required by the national codes. The table "Wiring in Class II Locations" is included for additional information.

A.2 DEFINITIONS

A.2.1 Different intrinsically safe systems approved for different hazardous location groups may be considered a single intrinsically safe system if approved for use in the most severe location in which any part of the system is used and the system otherwise complies with 3.2.

A.2.2 Intrinsic safety barrier terminals are identified as hazardous location and non-hazardous location, or equal. The terminals identified for non-hazardous location wiring are those which may be at unsafe energy levels. These terminals, however, may be physically in Division 2 hazardous locations.

A.2.3 Energy storage systems include both field installed wiring and field installed devices. Some approved field devices have reactive components that store ignition capable energy which cannot be released into the intrinsically safe circuit.

A.3 NON-HAZARDOUS LOCATION WIRING OF INTRINSICALLY SAFE CIRCUITS

A.3.1 If a conductor contacts a terminal of another intrinsically safe loop in the same intrinsically safe system, the safe energy level is not likely to be exceeded. However, if the conductor simultaneously contacts two or more intrinsically safe loops in the same intrinsically safe system, it is possible the safe energy level will be exceeded.

A.3.2 When intrinsic safety barriers are used, some acceptable methods of separation of circuits are shown by separate wireways in the following drawings.

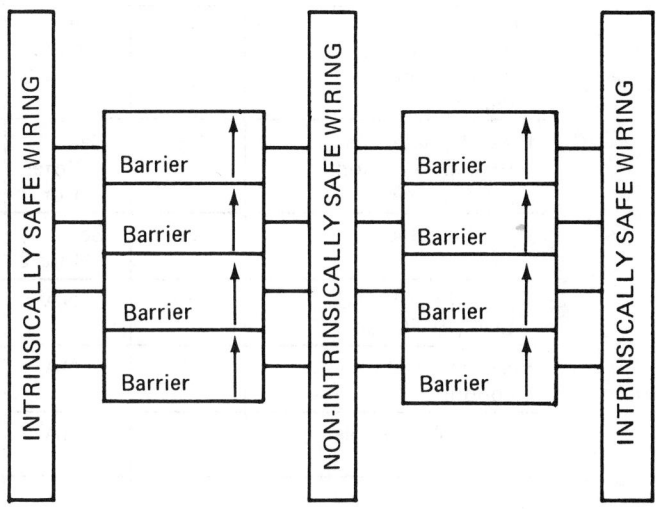

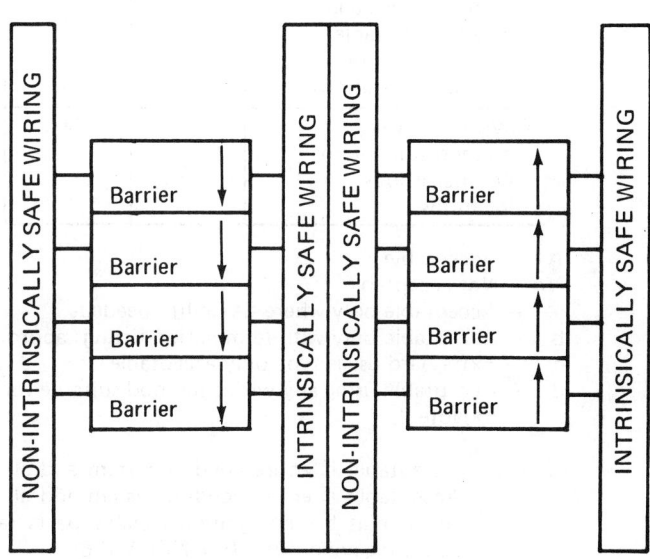

Raceway should be used. Wire lacing, wire ties, or equal are also acceptable methods to separate wiring, but when used, consideration should be given to the possibility of lacing or ties not being replaced after removal, such as for servicing.

---

[2] Suitable attention to transposition or shielding may also be necessary to permit proper operation of the equipment.

WIRING IN CLASS I LOCATIONS

Note: See National Electrical Code for description and use of wiring systems. Division O wiring not presently required by National Electrical Code. Divisions 1 and 2 wiring per 1975 National Electrical Code. Division O requirements are provisional recommendations only and do not represent a proposed standard.

Wiring System	Division O		Division 1		Division 2	
	Intrin-sically Safe	Not Intrin-sically Safe	Intrin-sically Safe	Not Intrin-sically Safe	Intrin-sically Safe	Not Intrin-sically Safe
Threaded rigid metal conduit: explosionproof wiring	A	See Notes 1 or 2	A	A	A	A
Flexible metal fitting: explosion-proof type[a]	A	See Notes 1 or 2	A	A	A	A
Type MI cable	A	See Note 2[b]	A	A[b]	A	A
Type ALS, CS, MC, SNM, & TC cable	A	NA	A	NA	A	A
Flexible metal conduit	A	NA	A	NA	A	A[a]
Liquidtight flexible metal conduit	A	NA	A	NA	A	A[a]
Electrical metallic tubing	A	NA	A	NA	A	See Note 2
Intermediate metal conduit	A	NA	A	NA	A	See Note 2
Flexible cord	A	NA	A	See Note 3[c]	A	A[a,c]
Other recognized rigid raceways and wireways suitable for wiring in non-hazardous locations	A	NA	A	NA	A	NA[d]
Any other wiring method suitable for non-hazardous locations	A	NA	A	NA	A	NA[d]

A — Acceptable.
NA — Not acceptable.
a — Acceptable only where flexibility needed.
b — Acceptable only with termination fittings approved for Class I, Division 1 locations of the proper group.
c — Extra hard usage type only acceptable.
d — Acceptable using any wiring method suitable for wiring in non-hazardous locations if circuit conforms to Note 2.

Note 1 — Acceptable if entire conduit system and all enclosures purged and pressurized, Type X purging. Acceptable if entire conduit system and all enclosures purged and pressurized, Type Y purging, and circuit has no ignition capable parts (arcing, sparking or high temperature) under normal operating conditions. (See NFPA 496)

Note 2 — Acceptable if circuit, under normal operating conditions, cannot release sufficient energy to ignite hazardous atmospheric mixture when any conductor opened, shorted to ground, or shorted to any other conductor in same cable or raceway.

Note 3 — Acceptable only on approved portable equipment where provisions made for cord replacement.

WIRING IN CLASS II LOCATIONS

Note: See National Electrical Code for description and use of wiring system.

Wiring System	Division 1		Division 2	
	Intrinsically Safe	Not Intrinsically Safe	Intrinsically Safe	Not Intrinsically Safe
Threaded rigid metal conduit: dust-ignition-proof wiring	A	A	A	A
Flexible metal fittings: dust-ignition-proof type[a]	A	A	A	A
Type MI cable	A	A[b]	A	A
Type MC cable	A[c]	NA	A	A
Types ALS, CS, SNM and TC cables	A	NA	A	A[d]
Flexible metal conduit	A[c]	NA	A	A[a]
Liquidtight flexible metal conduit	A	A[a]	A	A[a]
Intermediate metal conduit	A	NA	A	NA
Flexible cord	A[f]	A[a,e,f]	A	A[a,e]
Dust-tight wireways and raceways	A	NA	A	A
Any other wiring method suitable for non-hazardous locations	A[f]	NA	A	NA

A — Acceptable.
NA — Not acceptable.
a — Acceptable only where flexibility needed.
b — Acceptable only with termination fittings approved for Class II, Division 1 locations of the proper groups.
c — Not acceptable when electrically conductive dusts will be present.
d — Type TC cable not acceptable.
e — Extra hard usage type only acceptable.
f — Acceptable only with dust-tight seals at both ends when electrically conductive dusts will be present.

A.4 GROUNDING AND BONDING OF INTRINSICALLY SAFE CIRCUITS

A.4.1 Earth grounding systems will vary with the condition of the soil. Good engineering practices and local codes should be used to design the most effective type of grounding system. A ground reference is not required on some intrinsically safe systems. In general, a low resistance ground connection is required when energy limiting is dependent upon shunting to ground in order to prevent a large difference in potential from developing between circuit ground and the ground in hazardous locations.

A.4.2 Consideration should also be given to the mechanical strength or physical protection provided for the grounding conductor, to assure that it is not likely to be broken. At least a #12 AWG conductor is recommended.

A.4.3 It is important that metal enclosures for intrinsically safe circuits except as noted in 5.4.1 be bonded to the signal ground reference point in order to obtain equalization of potential. Several methods of accomplishing this are as follows:

a. Bonding of the enclosure to a building's structural steel which in turn is bonded to the signal ground reference point.

b. Bonding the enclosure to a system designed to serve as a plant system ground grid which, in turn, is bonded to the signal ground reference point.

c. Bonding to the intrinsic safety barrier ground bus or directly to the signal ground reference point by means of a separate conductor.

It is recommended that where intrinsically safe and non-intrinsically safe circuits are in the same enclosure, the intrinsic safety barrier ground bus and other intrinsic safety grounds be connected to the signal ground reference point by a conductor separate from the enclosure grounding conductor, and the two grounding conductors be insulated from each other.

Consideration should also be given to the mechanical strength or physical protection provided for the bonding conductor, to assure that it is not likely to be broken. At least a #12 AWG conductor is recommended.

A.4.4 Grounding may be necessary to drain electrostatic charges or to prevent shock hazard if the enclosure houses high voltage intrinsically safe circuits.

A.4.5 Cables which will not transmit gases or vapors through the core in excess of that permitted for sealing fittings as determined by ANSI Standard C33.27 — 1974 (.007 cubic ft./hr. of air at a pressure differential of 6 inches of water) need not be sealed or vented. The length of the cable run shall not be less than the length which limits gas or vapor flow to the rate permitted for seal fittings. Both ends of the cable are at atmospheric pressure.

INSTRUMENT SOCIETY of AMERICA
Research Triangle Park, North Carolina

ISA-S12.10-1973

Standard

Area Classification in Hazardous Dust Locations

Instrument Society of America

ISBN 0-87664-340-3

ISA-S12.10   Area Classifications in
             Hazardous Dust Locations

INSTRUMENT SOCIETY OF AMERICA
67 Alexander Drive
P.O. Box 12277
Research Triangle Park, North Carolina 27709

PREFACE

This Standard has been prepared as a part of the service of the Instrument Society of America toward a goal of uniformity in the field of instrumentation. To be of real value, this document should not be static, but should be subject to periodic review. Toward this end, the Society welcomes all comments and criticisms, and asks that they be addressed to the Standards and Practices Board Secretary, Instrument Society of America, 67 Alexander Drive, P.O. Box 12277, Research Triangle Park, North Carolina 27709.

The development of this Standard was initiated as a result of a survey of the flour milling industry of need for additional electrical standards. Members of S12 had been working in gas and vapor standardization and recognized the need for others to deal with the similar problems in dusty locations. The Electrical Safety in Hazardous Dust Locations SP12 Committee was formed under the Committee SP12 Instrumentation in Hazardous Locations SP12 Committee. The dust committee has exposed its views and findings to numerous forums of representatives of government, industry, and labor. A Canadian representative has made significant contributions. The Standard was passed through several mail-review-revision cycles until a concensus of reviewers was reached.

The ISA Standards and Practices Department is aware of the growing need for attention to the metric system of units in general, and the International System of Units (SI) in particular, in the preparation of instrumentation standards. The Department is further aware of the benefits to USA users of ISA Standards of incorporating suitable references to the SI (and the metric system) in their business and professional dealings with other countries. Towards this end this Department will endeavor to introduce SI and SI-acceptable metric units as optional alternatives to English units in all new and revised standards to the greatest extent possible. The Metric Practice Guide, which has been published by the American National Standards Institute as Z210.1-71 and future revisions, will be the reference guide for definitions, symbols, abbreviations, and conversion factors.

The assistance of those aided in the preparation of this document by answering questionnaires, offering suggestions, and in other ways, is gratefully acknowledged.

SP12 INSTRUMENTATION FOR HAZARDOUS LOCATIONS S12

F. L. Maltby, Chairman	Drexelbrook Engineering Co.
J. A. Bossert	Canadian Explosive Atmospheres Laboratory
E. N. Calder	The Foxboro Co.
H. G. Conner	E. I. DuPont deNemours & Co.
E. J. Cranch	Leeds & Northrup Co.
W. F. Hickes	The Foxboro Co.
George Hunt	Monsanto Co.
W. Jacobson	Bristol Co.
C. F. Kisselstein	Crouse Hinds Co.
E. C. Magison	Honeywell, Inc.
L. E. Miller	Factory Mutual Research Corp.
T. W. Moodie	The Pillsbury Co.
M. Morgan	Taylor Instruments Co.
P. J. Schram	Underwriters' Laboratories, Inc.
R. L. Swift	Mine Safety Appliances
J. R. Williams	Stearns-Roger Inc.
E. Yeaton	Eastman Kodak Co.

ISA ELECTRICAL SAFETY IN HAZARDOUS DUST LOCATIONS S12.7

T. Moodie, Chairman	The Pillsbury Co.
J. Anderson, Secretary	International Multifoods, Inc.
C. Russel Backes	General Mills, Inc.
Warren Carlson	CPC International, Inc.

The Committee SP12 served to review several times the work of the Committee SP12.7 in review revision cycles. Also in one of these cycles was a Review Board composed of:

J. Beardsley	Up-John Co.
C. M. Buehler	Taylor Instruments Co.
R. D. Coffee	Eastman Kodak
R. Littlefield	E. K. DuPont deNemours & Co.
D. L. Olson	Minnesota Mining & Mfg. Co.
R. A. Robinson	Consultant
E. P. Shoub	Public Health Service
G. V. Oswald	Boeing Co.
J. Abbott	Factory-Mutual Research Corp.
D. E. Ridgley	General Electric Corp.

The ISA Committee SP12, Electrical Instruments for Hazardous Locations operates within the ISA Standards and Practices Department. This Standard was approved by the ISA Standards and Practices Board in October, 1973.

E. J. Byrne, Vice President	Brown & Root, Inc.
J. A. Berger	Instrument Society of America
P. Bliss	Pratt & Whitney Aircraft Co.
L. N. Combs	Retired from E. I. duPont deNemours & Co.
N. Conger	Continental Oil Co.
G. G. Gallagher	The Fluor Corp., Ltd.
R. L. Galley	Bechtel Corp.
T. J. Harrison	IBM Corp.
P. S. Lederer	National Bureau of Standards
E. C. Magison	Honeywell, Inc.
J. R. Mahoney	IBM Corp.
F. L. Maltby	Drexelbrook Engineering Co.
R. M. Marvin	Dow Chemical Co.
A. P. McCauley	The Glidden Co.
W. B. Miller	Moore Products
H. N. Norton	Jet Propulsion Laboratory
G. Platt	Bechtel Corp.
C. E. Ryker	Cummins Engine Co.
K. A. Whitman	Allied Chemical Corp.

TABLE OF CONTENTS

1 PURPOSE

This Standard is one of a series recommending safe and economical procedures for installing electrical instruments in locations made hazardous by the presence of a cloud or blanket of dust. It is in conformance with the NEC (National Electrical Code) and the CEC (Canadian Electrical Code) and attempts to expand and clarify the NEC and CEC.

It evaluates the degree of the dust hazard of a specific location. To this end the physical principles involved and the techniques for the use of acceptable instruments are explained.

It presents guides for the classification of specific locations consistent with the degree of hazard determined.

2 SCOPE

This Standard refers only to hazards created by combustible dusts - agricultural plastics, chemicals, and metal dusts. These are classified in the NEC-Article 500-2 as Class II, Groups E, F and G materials.

It relates the classification of a location to dust cloud concentration, to accumulated dust layer thickness, and to dust bulk resistivity.

It contains references and data on the explosibility of common dusts as well as references to laboratory equipment and test procedures for evaluating the explosibility of dusts.

It is for the guidance of persons traned in the design and installation of instrument systems and to assist inspection authorities in approving such installations. It embodies the conclusions of the ISA committee SP12 on "Area Classification in Hazardous Locations" on what constitutes safe and economical classification practice.

3 DEFINITIONS

Classification of a location—the assignment of a rating such as Division 1, Division 2, or non-hazardous.

Conductive dust—a dust whose resistivity is less than 100 ohm-cm or which breaks down with 1000 volts per cm. applied across the bulk sample when tested in accordance with methods outlined in the appendix. Such dusts are defined as Group E in the NEC.

Division 1—the classification assigned to a location where either there is a high probability of a dust hazardous atmosphere occurring frequently, or regularly, or the dust is electrically conductive. (See examples in 5.2, Divisions).

Division 2—the classification assigned to a location where there is a low probability of a dust hazardous atmosphere occurring and/or a high probability of the presence of a hazardous dust layer. (See examples in 5.2 Divisions).

Dust-ignition-proof enclosure—one which excludes ignitible amounts of dusts or amounts which might affect performance or rating and which, when installation and protection are in conformance with the NEC, will not permit arcs, sparks or heat otherwise generated or liberated inside of the enclosure to cause ignition of exterior accumulations or of atmospheric suspensions of a specified dust on or in the vicinity of the enclosure. D.I.P. enclosures are used in Division 1 locations.

Dust tight enclosure—an enclosure of substantial mechanical construction provided with gaskets or otherwise designed to exclude dust. It has no open through holes and no knock outs. Conduit entrance is by tapped threads with a minimum of three threads engaged or by a gasketed, bonded conduit hub. It has a substantial door or cover made dust tight to the enclosure by a securely fastened gasket or by width and closeness of fit of the mating flanges. Door or cover fasteners are of substantial construction and are permanently captive. The door or cover is permanently captive to the enclosure. Threaded-hub conduit fittings are made dust tight by welding or gasketing. Threaded hub conduit fittings are solidly bonded to the enclosure by welding or bonding through proper fittings. NEMA 3, 3S, 4, 4X, 6, 12 or 12 enclosures are such enclosures. (See Reference 9.2).

Hazardous atmosphere—for the purpose of this Standard, an explosible mixture of dust in air.

Hazardous dust layer—any accumulation of combustible dust that will propagate or cause a fire.

Hazardous location—a space of limited and definable extent in which may be found a hazardous atmosphere or a hazardous dust layer.

Infrequently—as referring to the frequency of a hazardous event, connotes "not normally occurring" and "not likely during normal operations of the facility."

Ignition capable—equipment and wiring is equipment or wiring which in its normal operating condition releases sufficient electrical or thermal energy to cause ignition of a specific hazardous atmosphere or hazardous dust layer.

Instruments—are measuring, indicating, recording, computing, controlling, and similar apparatus requiring the use of small to moderate amounts of electrical energy in normal operation.

Intrinsically safe—equipment and wiring which is incapable of releasing sufficient electrical or thermal energy under normal or abnormal conditions to cause ignition of a specific hazardous atmospheric mixture or hazardous layer.

Minimum cloud ignition temperature—the minimum

temperature at which a hazardous atmosphere will ignite. (See Reference 9.1).

Minimum dust layer ignition temperature—the minimum temperature of a surface which will ignite dust laying on it after a long (theoretically infinite) time.[1] (See Reference 9.1). In most dusts, free moisture has been vaporized before ignition.

Minimum Explosible Concentration—the minimum concentration of combustible dust that when ignited produces light explosive force.[1] (See Reference 9.1).

Non-hazardous location—a location where neither a hazardous atmosphere nor a hazardous dust layer is to be expected.

Non-incendive equipment and wiring—equipment and wiring which in its normal operating conditions is incapable of igniting a specific hazardous atmosphere or hazardous dust layer. Equipment and wiring having exposed blanketed surface temperatures above 80 percent of the ignition temperature in C degrees of the specific hazardous dust layer shall *not* be classed as non-incendive. The blanketed surface temperature shall be determined at the outside surface of the enclosure beneath the surface of a dust accumulation 0.2 or more thickness.

Pressurized enclosure—maintained at a pressure higher than the surrounding area and the areas communicated with by conduit runs. The pressurizing medium shall be clean, dry air or an inert gas. In Division 1 locations, the pressure shall be supervised by a suitable pressure switch, to deenergize supply conductors in case of pressure failure. See footnote 2 with Table 1 in ISA-S12.11 for more detail.

TABLE 1
COMPARISON OF IGNITION PROPERTIES
CLASS I & II

	CLASS I - GASES AND VAPORS				CLASS II DUST CLOUDS			
	METHANE	PROPANE	HYDROGEN	ACETYLENE	WHEAT STARCH	RICE	SAFFLOWER MEAL	SUGAR
MINIMUM IGNITION ENERGY-[2] MJ.	0.30	0.25	0.019	0.017	20	40	20	30
EXPLOSION PRESSURE[3]-PSI (kg/cm^2) IPSI = 0.0703 (kg/cm^2)	102 (7.17)	122 (8.58)	105 (7.39)	146 (10.3)	105 (7.38)	93 (6.54)	84 (5.90)	91 (6.40)
TIME TO PEAK PRESSURE[3]-SEC	70	46	7	14				
EXPLOSION PRESSURE RATE OF RISE - psi/s (kg/cm^2/s)					8500 (597.6)	3600 (253.1)	2900 (203.9)	5000 361.6
CL I-L.E.L. %[3] VOLUME LOWER EXPLOSIVE LIMIT = L.E.L.	7.5	4.3	- - -	- - -				
%[4] VOLUME	5.3	2.2	4.1	2.5				
CL II-MIN. IGNITION CONCENTRATION oz/kcf. 1OZ/KCF = 0.9989 mg/1. KCF = 1000 CUBIC FEET					25	45	55	35
CL I AUTO IGNITION TEMPERATURE[4] °C	435	198	470	272				
CLASS II LAYER SURFACE IGNITION TEMPERATURE °C					- - -	220	210	400
CL II - CLOUD-SURFACE IGNITION TEMPERATURE °C					420	440	460	370

1 - REFERENCE #4 - ALL CLASS II DATA
2   #11
3   13
4   10

4 DUST EXPLOSION PARAMETERS AND VARIABLES

4.1 The Nature of the Dust Explosion

A dust explosion is burning in a cloud of dispersed dust. The ignited cloud releases heat energy in a pressure wave. When unrelieved, the pressure wave causes property damage. The flame rapidly traverses the dust cloud, igniting other combustibles.

The initial explosion often dislodges dust from the buiding and machinery. This dust may be ignited by glowing residue from the initial explosion and a secondary explosion follows. The quantity of settled dust thus redispersed may have a greater effect on total property damage than the initial dust cloud. Housekeeping in a dust hazardous area is vitally important. Good housekeeping keeps accumulations on the structures and machinery to a minimum.

The triggering of the explosion may be just a small puff or fire. Dust accumulated as a layer will not ordinarily explode. It will melt, char, or burn, however, and this burning can generate heat and air turbulance that may disperse some dust which may then explode. A significant pile of dust or other combustible matter is necessary for such turbulance.

4.2 Dust Dispersion Mechanics, Dispersion Control and Their Relations to Electrical Equipment Protection

Dust can be seen. Dusts are defined,[1] (See Reference 9.1) to be particles smaller than 74 microns (0.0029 in.) (through 200 mesh USS). In diameter one micron is one millionth meter.

Dust disperses horizontally from its source. The extent of dispersion depends on initial horizontal velocity, release height and the particle settling time. The Settling time can be estimated from Strokes Law for spherical particles:

$$V_s = \frac{h}{t} = \frac{g \rho D^2 10^{-8}}{18 \lambda}$$

V_s = still air settling velocity, cm/sec
h = release height, cm
t = settling time, sec
g = gravitational constant, 980 cm/sec^2
ρ = particle density, gm/cm^3 (1.44 for dry flour)
D = particle size, microns
λ = air viscosity, poise

$\lambda = 180 \times 10^{-6}$ poise at normal temperature and pressure.

The above equation holds for particles of 1 to 100 micron size in free fall, with about 10 percent error at lower concentrations. The larger particles will fall nearby. The small will float some distance from the release point.

Well designed process equipment and buildings have easily cleaned, smooth surfaces, without ledges or inaccessible pockets where dust can accumulate. Dust should be mechanically sealed into the process. Where the dusty material must be handled in the open, dust collection should be applied.

Frequent cleaning up of settled dust will materially reduce the real hazard of a location.

4.3 Dust Cloud Ignition

The factors usually considered in dust cloud ignition are:

> Minimum explosible concentrates
> Minimum cloud ignition temperature
> Minimum ignition energy
> Maximum explosion pressure
> Maximum rate of explosion pressure rise

These are influenced by:

> Composition
> Particle size
> Moisture content

Table 1 shows a comparison of ignition properties of Class I with Class II materials. Minimum ignition energies, are roughly two to three orders of magnitude higher for Class II than Class I materials. Laboratory explosion pressures are about the same, but detonation and pressure piling[12] (See Reference 9.2) in Class I materials can drive that pressure much higher. Ability to do so is a function of geometry of the Class I enclosure. Lower explosive limit is a measure of Class I explosive concentration and they are shown for two sources and are a function of geometry of test apparatus. Minimum explosible concentration is a similar measure for Class II materials. Contrary to Class I materials, surface ignition temperature for dusts must be expressed both for layer and cloud form. The two are not usually equal.

The dust ignition explosibility factors are related to changes in dust physical properties. For example, see Figures 2 through 10 increasing cornstarch[3] (See Reference 9.2) moisture content increases its ignition energy, ignition temperature, and minimum explosible concentration, and decreases explosion pressure and the rate of rise of pressure. Increasing cornstarch particle size increases the ignition energy, ignition temperature, minimum explosible concentration, and decreases the explosion pressure and the rate of pressure rise.

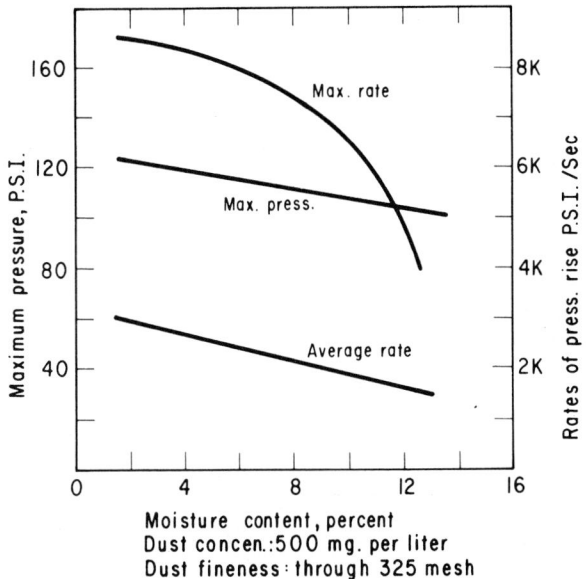

Fig. 1. Pressure and Rate of Rise (Cornstarch Explosion) vs. percent of Moisture.

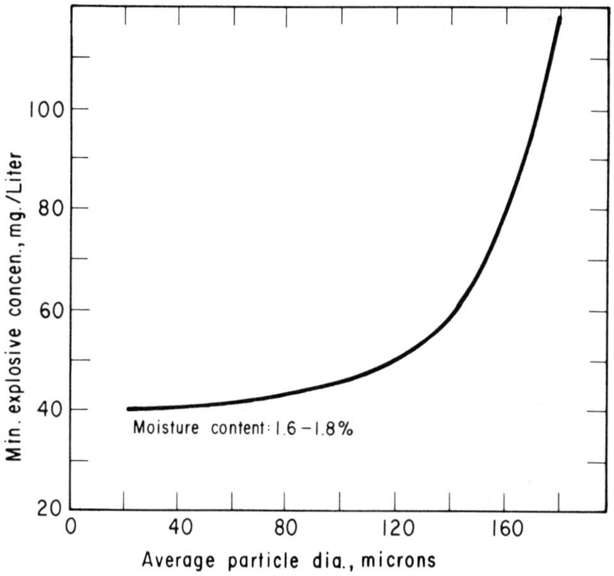

Fig. 2. Minimum Explosion Concentration (Cornstarch vs. Particle Diameter).

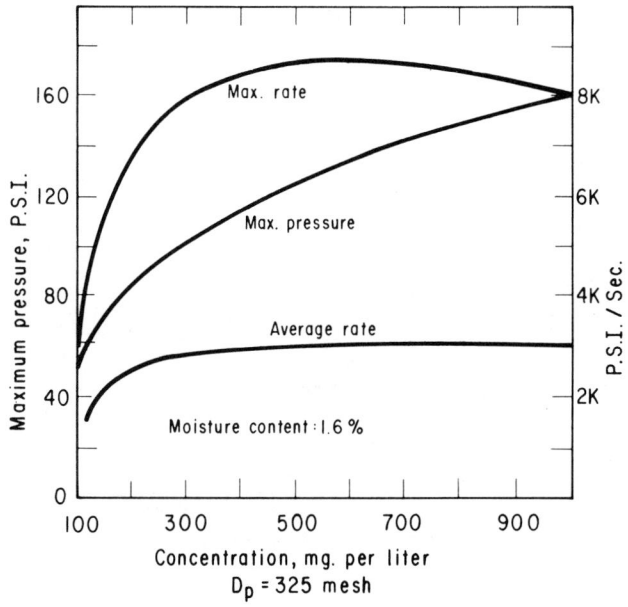

Fig. 3. Pressure and Rate of Rise (Cornstarch Explosion) vs. Concentration.

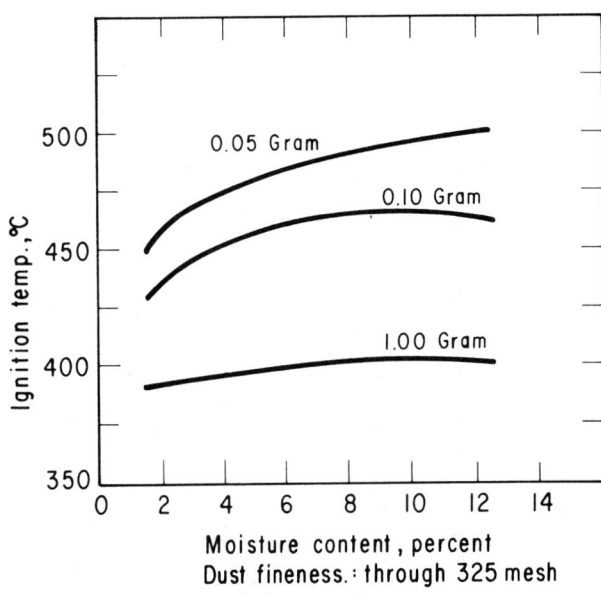

Fig. 4. Ignition Temperature vs. percent of Moisture Cornstarch).

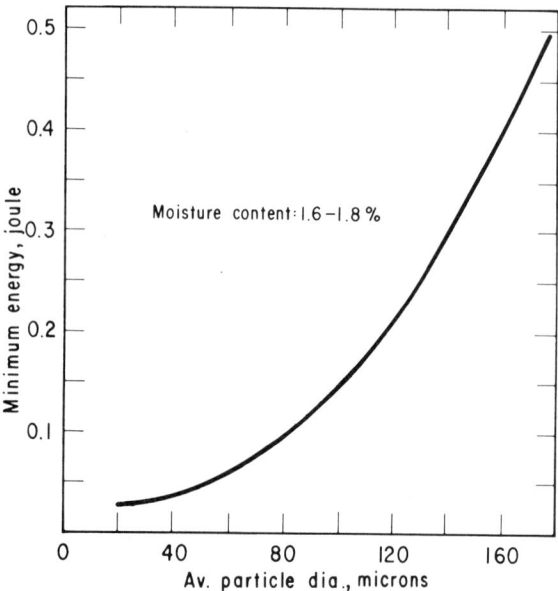

Fig. 5. Minimum (Ignition) Energy - (Cornstarch Explosion vs. Particle Diameter).

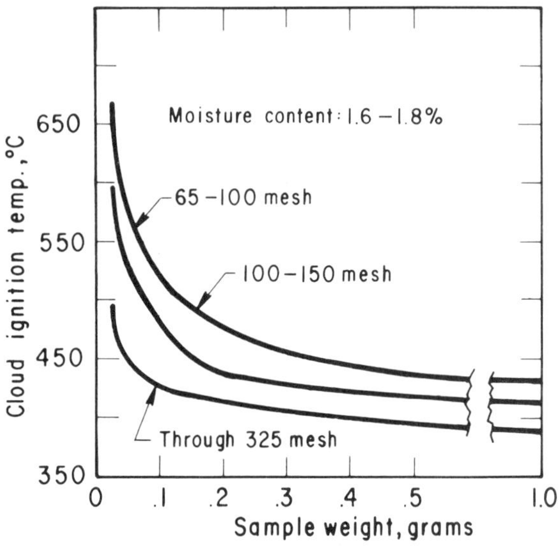

Fig. 6. Ignition Temperature (Cornstarch Cloud Explosion) vs. Sample Diameter Weight.

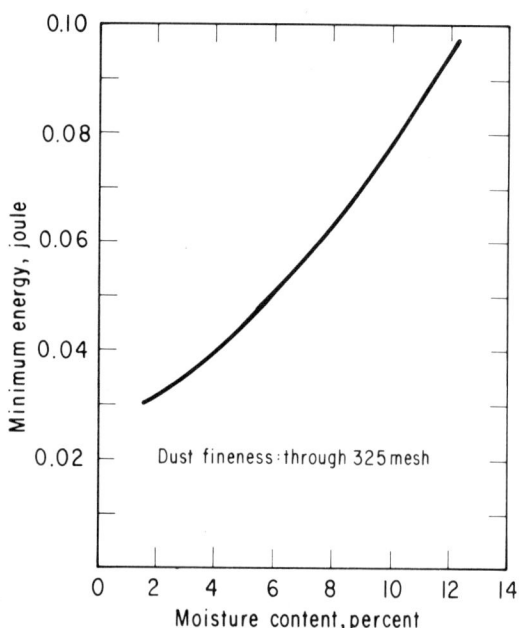

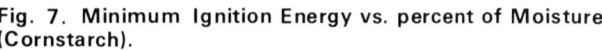

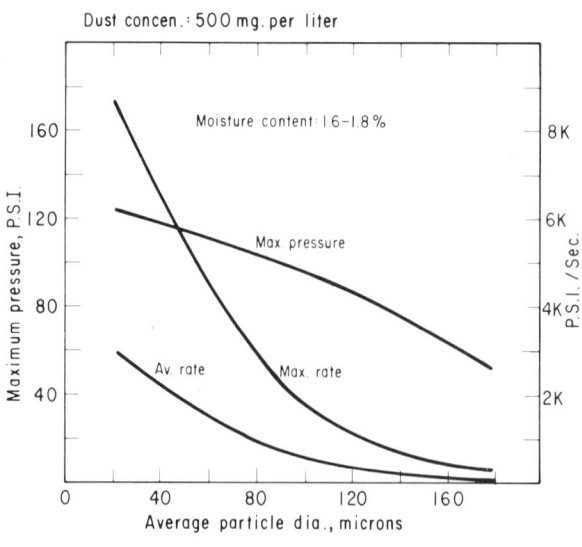

Fig. 7. Minimum Ignition Energy vs. percent of Moisture (Cornstarch).

Fig. 8. Minimum (Explosion) Pressure vs. Particle Diameter (Cornstarch).

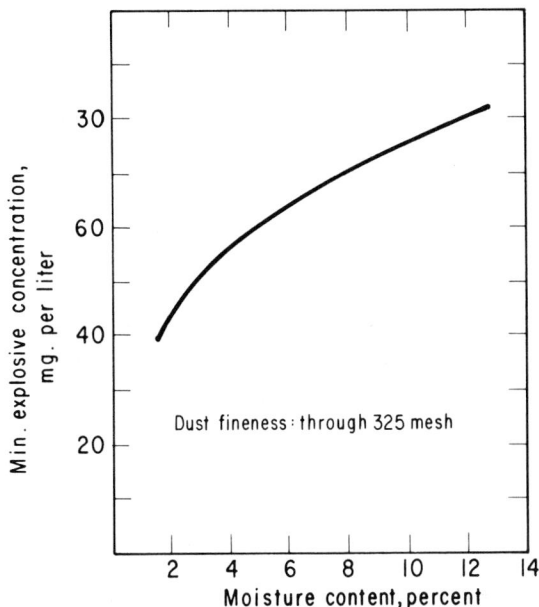

Fig. 9. Minimum Explosive Concentration vs. percent of Moisture (Cornstarch).

4.4 Ignition of Dust Layers

Factors considered in dust layer ignition are layer ignition temperature and minimum combustible layer thickness.

The layer ignition temperature of many materials differ greatly from their cloud ignition temperatures. Others are nearly the same. A few are compared in Table 1.

A prolonged elevated temperature may decrease minimum layer ignition temperature. For cornstarch, the layer ignition temperature changes from 500°C for 0.3 hours to 150°C for 70 hours.

A second important factor in dust layer ignition is minimum hazardous dust layer thickness. It is the thickness which must be exceeded for a fire or glow to propagate in the settled dust. The method for determining the minimum hazardous layer thickness is given in Appendix 1.2.

5. AREA CLASSIFICATION FOR COMBUSTIBLE DUST LOCATIONS

5.1 Groups

In Article 500-2 of the NEC, Class II materials are subdivided in Groups E, F and G. New definitions of the groups based on resistivity are given below.

Group E dusts are those having resistivities lower than 100 ohm-cm or which break down when subjected to 1000 volts/cm across a bulk sample when tested in accordance with Appendix 8.6, Dust Layer Thickness.
(This group includes the dusts of metals, generally regarded as conductors.)

Group F dusts are those having resistivities between 1– ohm-cm and 100 meg-ohm-cm and which do not break down when subjected to 1000 volts/cm across the bulk sample when tested in accordance with Appendix 8.6, Dust Layer Thickness). *(This group includes the carbonaceous dusts, generally regarded as semiconductors.)*

Group G dusts are those having resistivities greater than 100 meg-ohm-cm and which do not break down when subjected to 1000 volts/cm across the bulk sample when tested in accordance with Appendix 8.6, Dust Layer Thickness.
(This group includes the agricultural and plastic dusts generally regarded as insulators.)

5.2 Divisions

5.2.1 Division 1 locations are places where:

(1) The dust cloud concentration is above the minimum explosible concentration continuously, intermittently, or periodically.

(2) The explosive dust cloud is produced by a malfunction of the process or handling machinery and this malfunction causes a simultaneous source of ignition.

(3) The dust is Group E.

5.2.2 Division 2 locations are places where:

(1) The dust cloud concentration rises above the minimum explosible concentration only infrequently or when an abnormal condition exists at that location. Examples:

(a.) Sack dump with dust collector in operation, but malfunctioning.
(b.) Malfunction of equipment - removal for maintenance or cleanout.
(c.) When the process is making dust and the dust collection equipment is not operating, or malfunctioning.

(2) A hazardous dust layer exists.

5.3 Non-Hazardous Locations

A hazardous atmosphere location is classified as non-hazardous if there is no significant likelihood that either (2) a hazardous explosive cloud or (b) hazardous dust layer will exist, i.e., it is not Division 1 or 2.

Dust cloud concentration is prevented from rising above the minimum explosible concentration by proper equipment design and by effective housekeeping. Housekeeping must be frequent enough to prevent hazardous dust layers from accumulating.

6 PROCEDURE FOR CLASSIFICATION OF THE VARIOUS AREAS OF EXISTING PLANTS

(See Measurement Procedures and Examples in Appendix.)

6.1   Division 1

Locations should be classified Division 1 if:
6.1.1 The combustible dust cloud is frequently found to be above the minimum explosible concentration. Examples are found inside machinery:

Mills	Pneumatic conveying systems
Fans	Screw conveyors
Sifters	Drag conveyors
Purifers	Dust Collection systems

or where:

Air pressure jets and wind currents disperse settled dust.

Dust clouds occur at open machinery, where the dusty granular material encounters a sudden change in speed or direction, without dust collection. Examples are:

at side discharges, on trippers, on belt conveyors and
at transfer points between belt conveyors.

bag dumps
inside of bags being filled
spouts or scoops discharging bulk product

6.1.2 A process equipment failure that releases a dust cloud above the minimum explosible concentration and releases sufficient energy to cause ignition.

6.1.3 The hazardous dust has a resistivity of less than 100 ohm-cm (conductivity more than 0.01 ohm/cm) or the dust breaks down on test with 1000 volts per cm applied across a bulk sample, (see Appendix).

DIVISION 1 is the proper classification for these locations because when the circuit is energized with a layer of such dust across the insulation between live parts, a short circuit with high heat release can be expected. Frequent tripping can also be over-current devices. The resultant heat can be expected to ignite a dust accumulation on the enclosed equipment.

Means of defining the extent of areas to be classified Division 1 are:

Measurement - see Appendix.
Appraisal based on experience with measurement of similar or less dusty areas.

6.2   Division 2

Locations should be classified Division 2 for the following conditions:

6.2.1 Dust clouds reach minimum explosible concentrations infrequently. Examples:

Dust bearing materials are handled entirely within tightly enclosed machinery, conveyors or containers with leakage rate below that for generation of a Hazardous Atmosphere. Handling of dust bearing

materials in the open is under the control of an effective dust collection system.

The only process is the pipe of a conveying or dust control system. This is one exception to paragraph 6.1 where low resistivity (Group E) dust is being handled, the area outside of the pipe is classified DIVISION 2. Divison 2, not NON-HAZARDOUS because the low resistivity (metal) dusts are abrasive and eventually wear through the pipe.

Space surrounding the machinery or pipes are made dusty, reaching the minimum explosible concentrations only at such times as the equipment or pipe develops a severe leak or the dust collection is broken or ineffective.

5.2.2 Combustible dust layers exceed the hazardous layer thickness. Examples are:

Machinery, conveyor or containers where slow leakage occurs, but dust is removed by careful methods so a Hazardous Atmosphere does not occur frequently. Leakage rate must not normally reach the Minimum Explosible Concentration.

Means of defining the extent of areas to be classified Division 2 are:

Measurement - see Appendix
Appraisal based on experience with measurement of more dusty areas.

6.3   Non-Hazardous

Locations should be classified NON-HAZARDOUS for the following conditions:

6.3.1 The dust cloud concentration does not exceed the Minimum explosible concentration.

6.3.2 Combustible dust layers exceed the Hazardous Layer thickness only infrequently. Examples:

There is excellent housekeeping adjacent to DIVISION 2 locations.

The dust layer thickness has diminished to less than the Hazardous Layer thickness at points further than a certain distance from a Division 2 location, i.e., ten feet beyond the machinery in a Division 2 location.

Means of defining the extent of the areas to be classified NON-HAZARDOUS are:

Measurement - see Appendix
Appraisal - based on experience with measurement of more dusty areas.

7 PROCEDURE FOR CLASSIFICATION OF THE VARIOUS AREAS OF NEW PLANTS DURING DESIGN

Classification goals in the design stage are set for each location in a building by a team of knowledgeable, responsible people. This team includes those responsible for:

(1) Process and electrical design
(2) Corporate Insurance
(3) Enforcement of local fire and electrical codes.

They prepare and approve a confirming reference drawing (or written description) delineating the agreed limits of Division 1, Division 2 and Non-Hazardous areas.

The team establishes the classification of each location on the basis of:

(1) experience and data from similar installations.
(2) evaluation of the dusting characteristics of the proposed facilities.
(3) layout of the facilities
(4) housekeeping standards

When the area classifications have been assigned, design criteria are specified to insure construction of a facility meeting the requirements for the classified area.

The owner designates those that are knowledgeable and responsible to accomplish the work. These individuals see that the installation meets the classification goals.

Those knowledgeable and responsible shall insure that the facilities installation and all subsequent installations are compatible with the classification goals.

It shall be mandatory that those knowledgeable and responsible shall maintain proper housekeeping practices so that the established area classifications are not violated.

8 APPENDIX

This Appendix is intended to clarify Standard ISA-S12.10 and to explain the philosophy on which it is based. Wherever possible, this Standard is based upon an analysis of the physical conditions involved with a minimum reliance upon arbitrary rules.

8.1 Procedure for Classifying Areas in the Field as Division 1 or Division 2

The following test procedures supplement procedures described in sections 6 and 7.

8.2 Procedure for Determining Dust Cloud Weight Concentration

The maximum dust cloud concentration at the proposed electrical installation is to be determined. The most dense dust concentration is found by observing present conditions. When a plant operation disperses dust into the air as a normal routine, these conditions should be duplicated for the purposes of this test.

8.2.1 Apparatus

(1) For Disc Sampling:

(a) Vacuum pump 45 LPM free air capacity up to 20 in. Vacuum, carbon-vane type with 115 V.A.C. Motor.
(b) Rotameter 0-35 LPM.
(c) 25 in. of 3/8 in. O.D. Rubber hose.
(d) 2 - 2 in. diameter filter disc holders.
(e) Laboratory balance.
(f) Supply of filter discs. 2 in. Glass Fibre, submicron pore size, low pressure drop at 35 LPM. flow.
(g) Business envelopes
(h) Stop watch or watch with second hand.

(2) For Tape Sampling:

(a) Tape Sampler consisting of carbon-vane vacuum pump, adjustable timer, and mechanism for indexing the tape from "Sampling Reading" position. Photo densitometer for "reading" the tape darkening by dust.
(b) Recording Milliameter for Photo densitometer output recording.

(3) Extension Cords

(4) Face Masks

8.2.2 Procedure

Long time samples are taken with the 45 LPM Vacuum Pump for approximately five-minute periods, using a 2 in. glass fibre filter disc as the collector. Filter holders are connected to the vacuum pump unit through a 3/8 in. tubing. The tubing connection permits free movement to select by visual observation the most dusty location to sample. Tubing also permits vertical down flow into the filter for the most conservative orientation of collector. Once collected, the loaded collector is carefully covered with the paired filter disc. The pair is carefully placed in a small business envelope to preserve the total sample. Net weight gain is determined by analytical balance, as the difference in weight of the two filters after and before sampling. Air quantity is found from readings of a Rotameter on the vacuum pump and a stop watch. The disc sampler concentration in ounces/cu. ft. can then be calculated.

Short time samples are taken using the tape spot sampler with photo densitometer. This sampler automatically collects timed samples through the inlet tube for an adjustable sampling period of 1 to 120 seconds. Spots of dust on the tape are analyzed for percent light transmission in the photo densitometer. The purpose of using the tape sampler is to learn the maximum short-time concentrations rather than the average over five minutes - as by the disc sampler. Sample periods of from 7 to 60 seconds are used. The optimum sampling period lies between too short a period, when insufficient staining occurs, and too long a period, when a dust load thick enough to choke the sampling slot accumulates. When the choke occurs, the tape tears. The best sampling period is found by trial.

The samplers are correlated by taking concurrent samples from the same area. Densitometer readings are converted to Coefficient to Haze (C.O.H.) numbers by means of the following formula:

$$COH/1000 \ L.F. = \frac{Area \ 10^5}{Volume} \ log_{10} \ \frac{100}{(\% \ Transmittance)}$$

COH determination is facilitated by reading it from the graph - Figure 1. COH values for each test are averaged and a ratio of maximum to average is calculated. This ratio multiplied by the calculated disc sampler concentration is called the short time average concentration. Area and volume constants are used as a ratio and, therefore, cancel out.

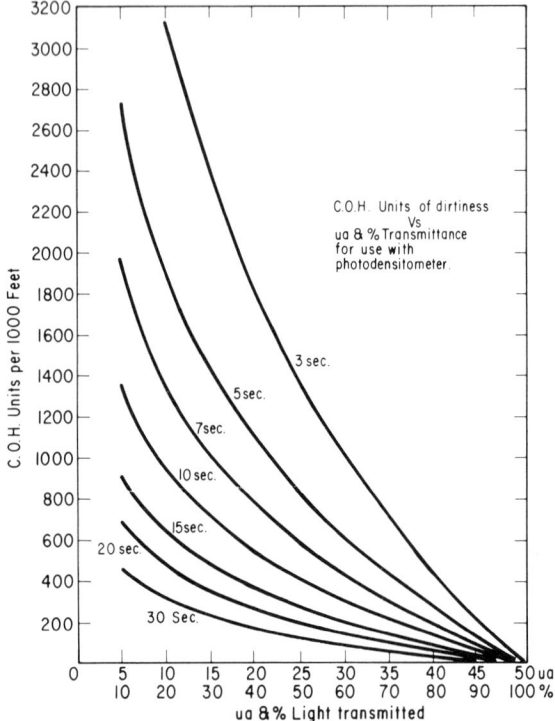

Fig. 10. "COH Units of Dirtiness vs. U.A. and percent of Transmittance."

Short time concentrations (STC) can then be compared to the Bureau of Mines Minimum Explosive Concentration for the product involved. See References 4, 5, 6, 7, 8 or 9. A safety factor is then applied to the measured data to account for momentary peaks of concentration higher than the measured data, see paragraph below. By this comparison (with the safety factor applied to the S.T.C.) and an estimate of the frequency of the sampled condition a decision is reached on the Classification of the location.

The Bureau of Mines records explosion pressure and rate of pressure rise. From their recordings, a dust explosion starts within 10 to 20 milliseconds of ignition. The peak pressure will pass with 50 to 100 milliseconds. Minimum tape sampling period per spot is 1 to 20 seconds. Since a peak concentration lasting only a portion of this period could result in an explosion, an additional safety factor is applied to the measurements.

Since instruments are usually mounted within five feet of the floor, samples are also taken within five feet of the floor. Agitate settled dust to insure that a maximum dust concentration is present when sampling. This adds settled material to any dust cloud normally present during operations. Of all possible locations in the room for sampling, the operator chooses by sight that location having the densest dust cloud after agitating settled dust. This method probably puts more dust into the air than most abnormal operating conditions at this location, an inherent safety factor.

8.2.3 Safety Factor Considerations

Safety factors are added to account for:

(1) Probability that the concentration for ten milliseconds at the test location exceeds the average during five seconds measuring period.
(2) Errors of apparatus and technique.

Momentary concentrations in 1b exceeds the measured concentrations and require an additional safety factor. The tape sampler is limited to five to ten second minimum sampling period, and dust explosions can begin within 20 milliseconds after supplied with ignition energy. Until such time as speedier instruments are introduced and proven, an additional safety factor of two is recommended to allow for possible high consistency processing.

Erros of apparatus and technique, stem largely from two sources, errors in reading the stop watch and the rotameter. (Errors in the balances are usually negligible when weighing over one gram on a $\pm 1/10$ mg. accuracy balance.) Because of poor lighting in a dust cloud, two sec error in measuring a 30 second liters/min error in 35 liters/min error are reasonable. These periods occur in the worst combination, and a safety factor of $1/.78 = 1.28$. The error is 22 percent.

Accumulated additional safety factor is 2 x 1.22 = 2.56. Rounded off, the required additional safety factor is three.

8.3 Procedure for Determining Dust Layer Thickness

Measured of the thickness of a uniform layer is best done by sweeping a small (about two feet square) measured area and noting the volume swept. Thickness is calculated from volume and area measured. Care should be taken to allow time for settlement of swept up material in the measurer before reading volume.

The area chosen for sweeping should be close to or on the electrical instrument location in question. If the measured thickness exceeds the hazardous layer thickness, the area is rated Division 2 if less, Non-Hazardous. Refer to Appendix 8.6.

8.4 Laboratory Methods

8.5 Dust Cloud Concentration

See Reference 9.1.

8.6 Dust Layer Thickness

Maximum dust thickness that will not support combustion is determined.

8.6.1 Apparatus:

Settling Chamber with dispersing pan at top.
Volume Measuring Containers.
Dispersing Blower.
Heating wire or electric charcoal fire starter.

Dust is poured into the dispersing pan; then dispersed and to produce a layer. The required volume of dust can be approximated from:

Volume	Area	Typical Layer
1 pt.	9 ft.^{2}	0.022 in.
2 pt.	9 ft.	0.044 in.
3 pt.	9 ft.	0.066 in.

^{2} See Reference 9.2

The heater, heated to dull red color, is carefully laid into the layer. For a layer or more, charring of only the dust in contact with the heater is observed, no propogation. A point is reached when the dust will continue to burn away from the heat source. This is the hazardous layer thickness.

Additions smaller than one pint will improve the estimate of hazardous layer thickness.

8.6.2 Discussion

For continuous charring to take place, a thickness must be reached where an excess of heat or combustion is available. This explanation is supported by the upturn of ignition temperature vs. thickness as thickness is decreased in Reference 9.14. The excess heat condition is achieved when the heat of combustion more than equals the total heat capacity of the materials to be burned and losses to surroundings.

For certain materials (Wheat elevator dust) a large percentage (67 percent of the heat of combustion is required to heat the non-combustible[15] (See Reference 9.15) water and ash part of the fuel.

8.7 Laboratory Resistivity Measurement of Dusts

8.7.1 Objective

To measure the resistivity of a dust, to facilitate classifications of resistivity as high, medium, or low.

The test provides:

(1) Values of resistivity
(2) Evidence of pass or failure of a voltage withstand test.
(3) A sample cell and measuring means, see Figure 11, in which a sample can be tested and on which various environmental influences can be imposed. Such influences, when present in the area to be classified, should include:

  (a) ambient temperature, including heating by electric apparatus.
  (b) wetting by liquids.
  (c) packing (not pressurized).
  (d) vibrating.
  (e) washing by non-hazarous gases or vapors.
  (f) aging with voltage applied.

The test voltage for passage of the withstand test shall exceed by a significant safety factor the allowable voltage per distance of the class of equipment in use. For reference[16] the NEC uses 1/2, 3/4, 1, and 2 in. clearance so that bare conductors have voltage to spacing relationships from 166 to 600 volts per inch (66 to 236 volts per cm.)

8.7.2 Theory

Resistivity, Conductivity and Resistance are all related. Conductivity is the inverse of Resistivity; accordingly, Conductivity = 1/Resistivity. The relationship between resistance and resistivity is shown in Equation 1.

$$R_s = \rho \frac{L}{a} = \rho \frac{L}{hw} \qquad (1)$$

R_s = sample resistance, ohms (see Figure 11)
L = resistance path length, cm. (1.26 cm. used)
a = area of the resistance path, h x w, where w and h are in cm. (9.92 and 1.42 cm. used)
ρ = resistivity of the path, ohm-cm . . . . (empty cell air capacitive reactance is calculated to be 3000 meg ohms at 60 HZ, so it is neglected).

The sample cell is shown in Figure 11. Empty cell resistance influenced the sample resistance determination. The sample cell serves to:

(1) Conform the bulk material to simple dimensions that will fit Equation 1.
(2) Provide electrodes between which the bulk material resistance is measured. The empty cell resistance is assumed to form a parallel resistance path with the cell samples in a different form in Equation 3.

$$\frac{1}{R_c} = \frac{1}{R_a} + \frac{1}{R_s} \qquad (2)$$

R_a = empty cell resistance
R_c = filled cell restiance
R_s = sample resistance in the cell - Measured Resistance

$$R_s = \frac{R_c \; R_a}{R_a - R_c} \qquad (3)$$

8.7.3 Apparatus and Measuring Equations

A circuit as shown in Figure 11 is provided for measuring voltage and current of the sample cell and for protecting the instruments in case of sudden sample failure. Resistor R_1 is sized to protect the transformer in case of cell flashover. Shunt switch is opened to take a reading at voltmeter eight. By measuring the voltage drop across shunt resistor R_2, cell current can be calculated from ohms law. A vacuum tube voltmeter with input resistance over one meg ohm is used. Depending on the loaded cell resistance, the shunt resistor is assigned values of 10,000 to 160,000 ohms. The empty cell resistance is approximately 1000 meg ohms. The cell voltage is one of the terms, IR_c, in Kirkhoff's Law written for voltage drops around the transformer secondary circuit, Equation 4. Refer to Figure 11 when considering Equation 4.

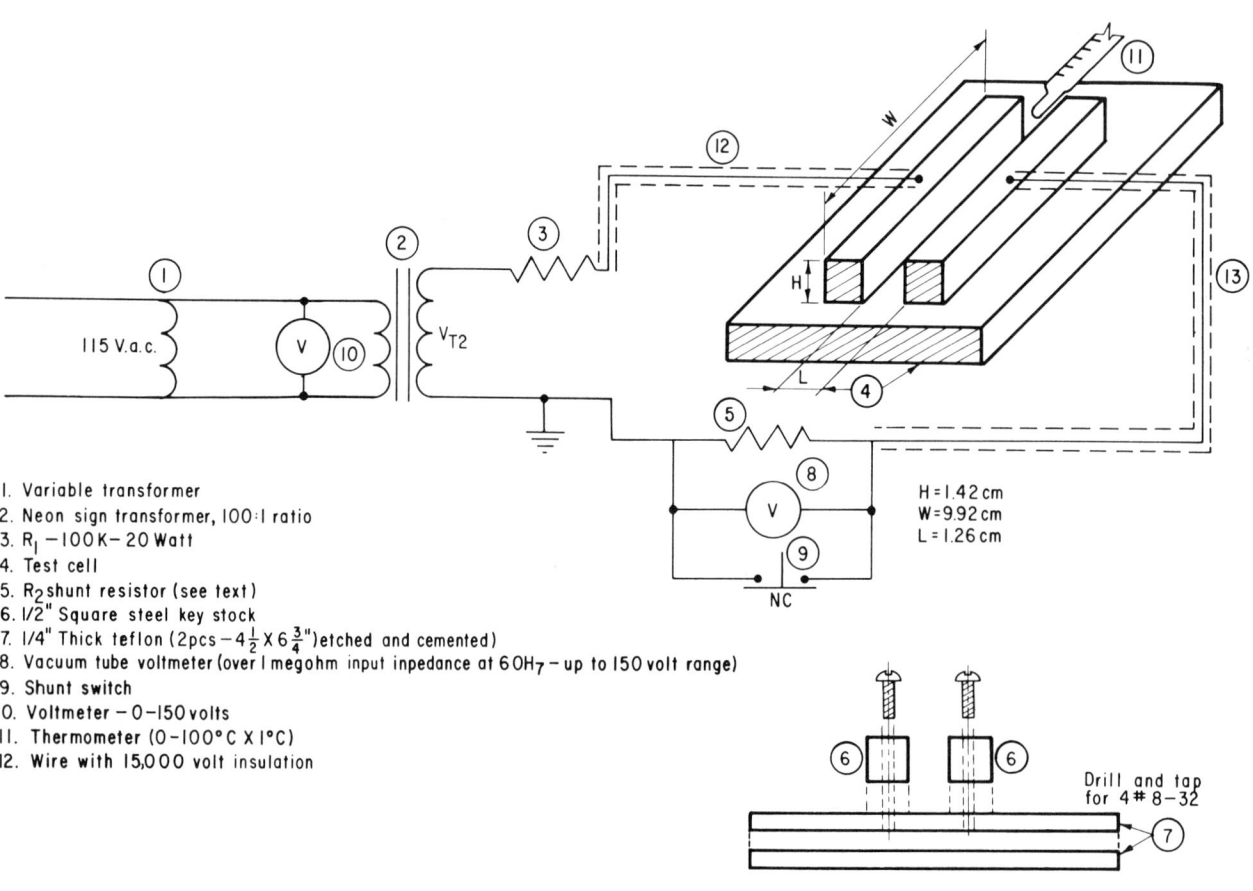

1. Variable transformer
2. Neon sign transformer, 100:1 ratio
3. R_1 — 100K — 20 Watt
4. Test cell
5. R_2 shunt resistor (see text)
6. 1/2" Square steel key stock
7. 1/4" Thick teflon (2pcs — $4\frac{1}{2}$ X $6\frac{3}{4}$")etched and cemented)
8. Vacuum tube voltmeter (over I megohm input inpedance at 6 OH₇ — up to 150 volt range)
9. Shunt switch
10. Voltmeter — 0-150 volts
11. Thermometer (0-100°C X I°C)
12. Wire with 15,000 volt insulation

H = 1.42 cm
W = 9.92 cm
L = 1.26 cm

Drill and tap for 4 # 8-32

Test cell detail

Fig. 11. Non-Pressurized Resistivity Test Apparatus

$$V_2 = IR_1 + IR_c + IR_2 \qquad (4)$$

Values for the various terms in Equation 4 and their accuracy are considered. When current through the voltmeter eight is neglected, cell current is given by Equation 5. When shunt resistance is less than ten percent of voltmeter eight input resistance, and when ten percent error is acceptable in the sample resistance calculation, current through the voltmeter eight is negligible. With a 100 to 1 ratio neon sign transformer, secondary voltage V_2 is 100 times the primary voltage V_1 read on voltmeter ten.

$$I = \frac{V_8}{R_2} \qquad (5)$$

Into Equation 4, substitutions from Equation 5 for I and $100 \times V_1$ for V_2 are made. Equation 4 is then solved for R_c. R_c is then substituted into Equation 3. The result is Equation 6.

$$R_s = R_a \frac{\left[\frac{(100V_1}{V_8} - 1)R_2 - R_1\right]}{R_a - (100V_1}{V_8} - 1)R_2 + R_1)} \qquad (6)$$

Sample resistance in the cell is calculated from a relationship of circuit constants and measured quantities.

Equation 1 is solved for Resistivity, ρ ; R_s, resistance of sample is substituted. The result is Resistivity given in Equation 7.

$$\rho = \frac{(hw)}{L} \frac{R_a \left[\frac{(100V_1}{V_8} -1)R_2 - R_1\right]}{\left[R_a - \frac{(100V_1}{V_8} -1)R_2 + R_1\right]} = \frac{hw}{L} R_s \qquad (7)$$

Resistivity is expressed entirely in terms of circuit constants, cell dimensions and measured electrical quantities.

8.7.4 Sample Calculations

A sample calculation for activated carbon follows:

$V_1 = 60$ volts, $V_8 = 75.3$ volts, $R_1 = 100,000$ ohms, $R_2 = 6,000$ ohms, $R_a = 2,400 \times 10^6$ ohms by equation 6

(Sample)

$$R_s = 2,400,000,000 \frac{\left[\frac{(100 \times 60}{75.3} -1)6,000 - 100,000\right]}{2,400,000,000 + 100,000 - \frac{(100 \times 60}{75.3} -1) 60,000}$$

$$= 1(80 - 1)6,000 - 100,000) = 4,730,000 \text{ ohms}$$

Cell dimensions are:

h = 1.42 cm., w = 9.92 cm, L = 1.26 cm

By equation 7a

$$\rho = \frac{(1.42 \times 9.92)}{1.26} 4.73 \times 10^6 \text{ ohm-cm} \qquad \text{(7a Sample)}$$

$$= 54.4 \text{ meg-ohm cm.}$$

Cell voltage is V_2 minus the voltage across $R_1 + R_2$

$$V_2 = 100V_1 = 6,000 \text{ volts} \qquad (8)$$

across R_2, $V_8 = 75.3$ volts

across R_1, Voltage $= IR_1$, by Equation 5

$$IR_1 = \frac{75.3}{6,000}(100,000) = 1550 \text{ volts}$$

Cell voltage is 6,000 - 75.3 - 1550 = 4,374.7 volts

Electrode spacing is 1.26 cm.

Voltage stress across bulk sample is

$$\frac{4374.7}{1.26} = 3470 \frac{\text{Volts}}{\text{cm}}$$

8.7.5 Procedure

Steps in test procedures follow:

(1) Prepare circuit and sample cell shown in Figure 11.
(2) Clean cell, metal and teflon, using electrical equipment cleaning solvent.
(3) Read meters with the applied voltage increased in five steps so as to apply up to 60 volts at V_1 (6,000 volts at VT2).
(4) Using Equations 3, 4 and 5 calculate empty cell resistance, R_a (assume R_s is infinite) then $R_a - R_c$.
(5) Return applied voltage to zero. Load cell sample chamber, scraping bulk material off even with the top and ends of chamber.
(6) Subject bulk material to the necessary modifying environmental influences for test.

8.7.6 Wet Test

(1) With a face tissue (as a filter) lining a paper cup, fill tissue with bulk powder and then add liquid encountered in the plant.
(2) Lift tissue and bulk material carefully out of cup, wrap up, and squeeze out the excess liquid by hand.

(3) Carefully open tissue, spoon wet material into sample cell, tapping down until air voids are eliminated. Scrape off excess over top and ends.

(4) Read meters with applied voltage increased in five steps so as to apply up to 60 volts at $\dot{V}10$. (6,000 volts at VT2).

(5) Calculate sample resistance and resistivity by Equations 6 and 7.

8.7.7 Vibration Test

(6) Using dry or wet sample, spoon material into sample cell, tapping down to fill air voids, scrape off excess.

(7) Clamp cell to table of jig saw or other vibrating shop tool.

(8) Repeat applied voltage steps as in 4, 5 and 6.

8.7.8 Aging Test

(9) Repeat above steps as desired for wet, vibrated, or dry samples - recording meter readings.

(10) Insert a precision thermometer into end of sample cavity, see Figure 11. Read and record initial temperature after a steady value is reached.

(11) Read and record other significant environmental data, like wet and dry bulbs temperatures.

(12) Leave cell under high applied voltage for desired period and check periodically for changes in shunt voltage drop, V8, and sample temperature. Read and record values from meters and thermometers.

(13) Make a complete run of five applied voltage levels at end of aging period and calculate R_S and ρ by Equations 6 and 7.

8.7.9 Dry Test

Repeat steps 3, 4 and 5 without wetting sample.

8.7.10 Collecting and Aging of Samples

Collect samples where material is likely to leak into equipment enclosures, where dust might cause tracing or insulation flashover. Keep sample (wet, dry, hot, greasy, or otherwise) and test as soon as possible. Otherwise measure temperature and other variables. Reproduce conditions on test.

8.7.11 Personnel

Tests and calculations shall be under the direction of a technically qualified person. He shall be qualified in the usual laboratory skills of observation, deduction and mathematics through algebra.

8.7.12 Breakdown

On above tests, when the resistance decreases as the voltage applied increases, watch for avalanche breakdown; as evidenced by sudden rise of voltmeter reading.

8.7.13 Results

Some materials were tested by these methods. The results are given in Figures 12 through 18.

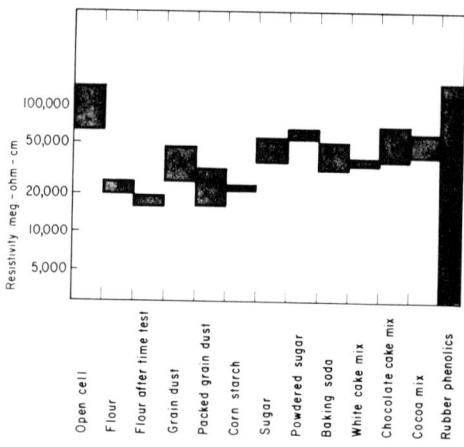

Fig. 12. Resistivity Range at Food Powders.

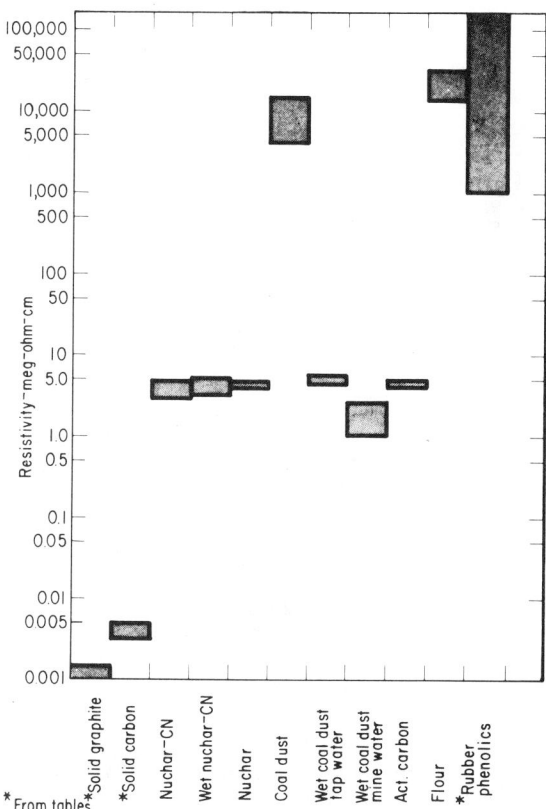

Fig. 13. Resistivity Range of Carbon Powders.

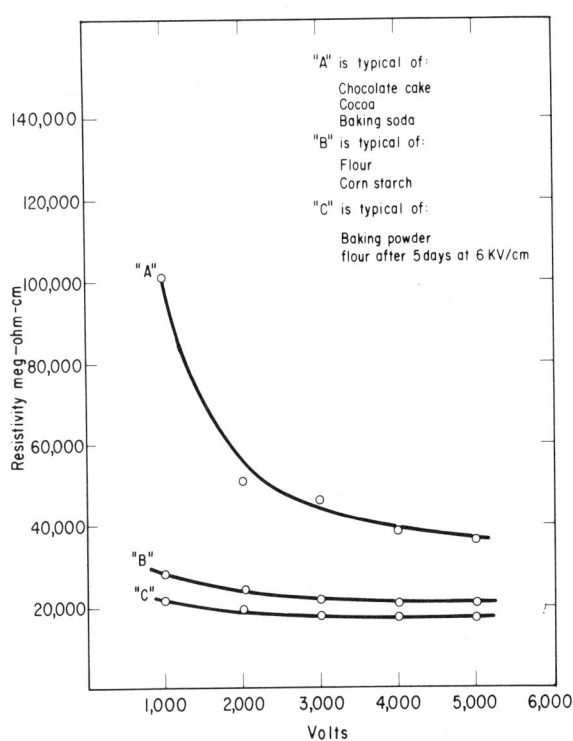

Fig. 14. Resistivity Test of Insulator Powders - to 5000 V/CM.

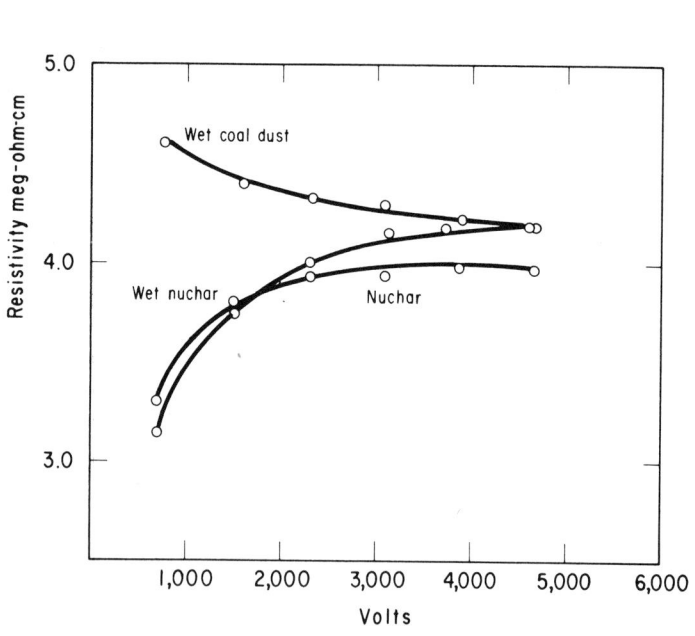

Fig. 15. Resistivity Test of Carbon Powders - to 5000 V/CM wet and dry.

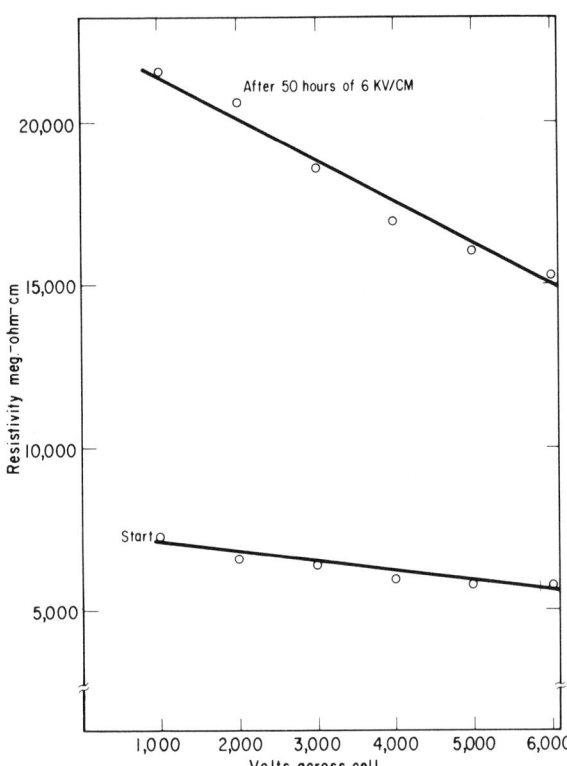

Fig. 16. Resistivity Test of Coal - to 5000 V/CM (Aged and Starting).

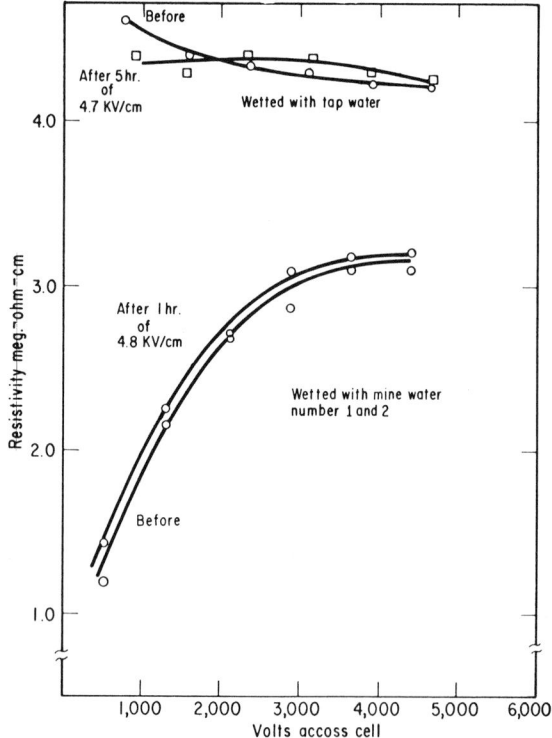

Fig. 17. Resistivity of Wet Coal - to 5000 V/CM (Tap and Mine Water).

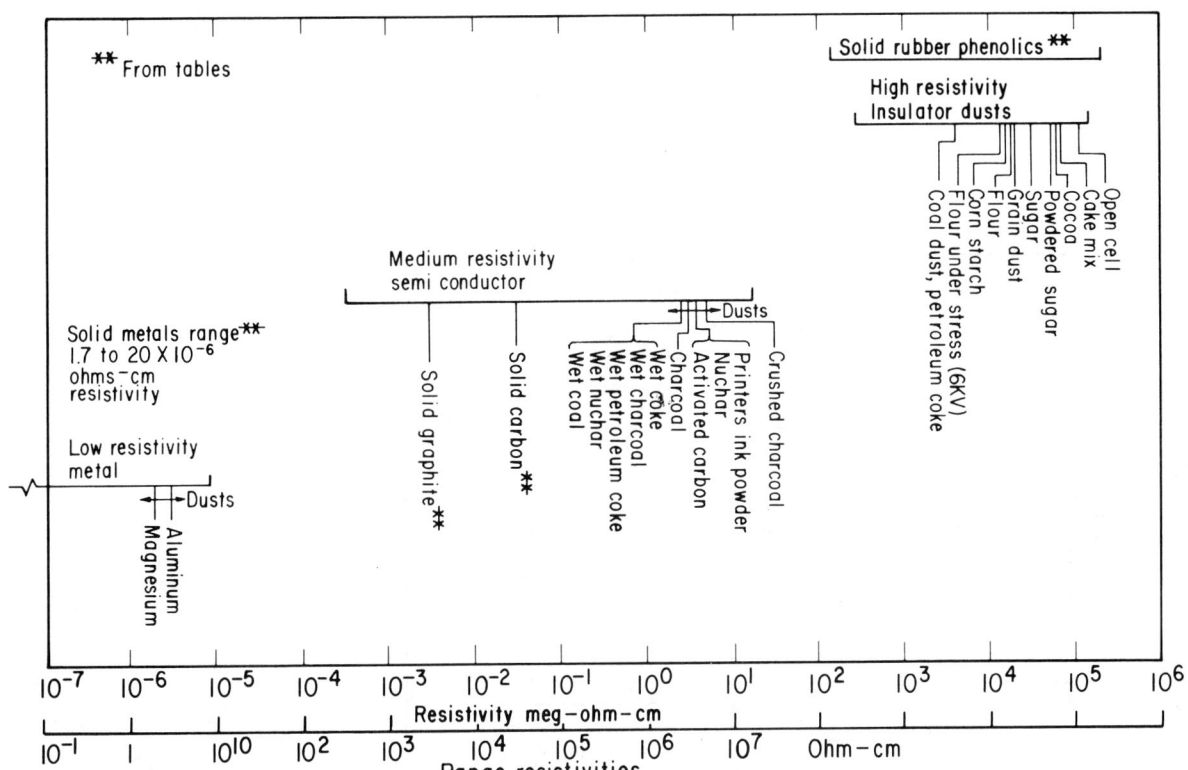

Fig. 18. Ranges of Resitivities.

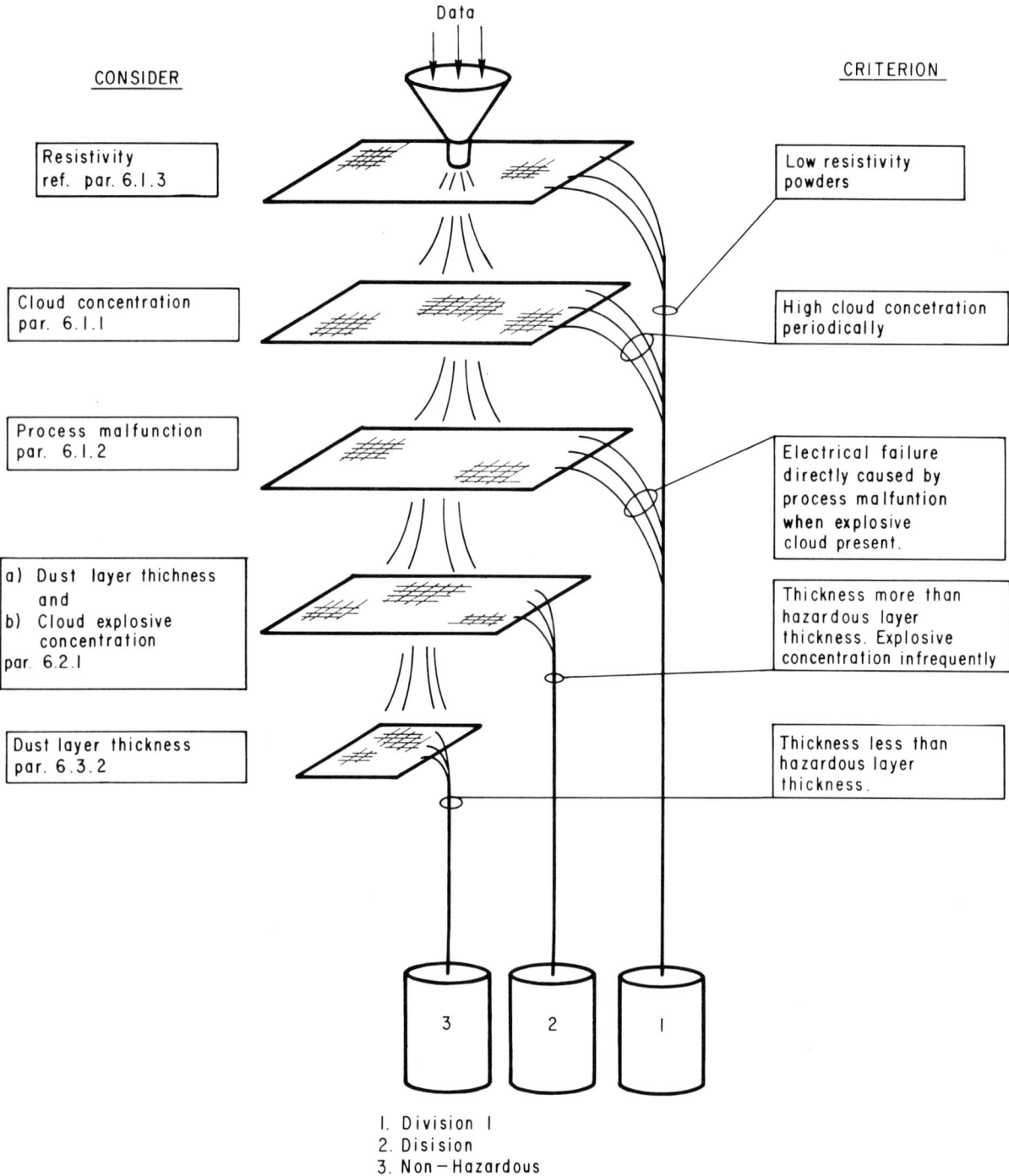

Fig. 19. Classification Sieves.

All materials are placed into the cell, completely filling it without packing. Materials in the semiconductor range (see Figure 15) were held in a field of four to six KV/CM for a period of time to detect resistivity change. Selected materials in the high resistivity range and wetted carbon materials were also aged four to six KV/CM.

For most of the tests, temperature of the materials and temperature and relative humidity of the room were measured. Ambient temperature was between 74-78 F and relative humidity 25 to 35 percent.

The tests did not produce linear curves of resistivity to volts per cm. over a range of voltages, see Figures 14, 15, 16 and 17. Tables 2 and 3 give mean and minimum resistivity for insulating and semiconductor dusts. Activated carbon, coal, petroleum coke and Xerox toner (Table 5) should not be grouped with the other semiconductor dusts, but rather with the insulators of Table 2.

TABLE 2
RESISTIVITY OF DUSTS
INSULATORS

MATERIAL	AVG. RESISTIVITY MEGOHM-CM*	MIN. RESISTIVITY MEGOHM-CM*
Sugar	55000	35300
Powdered Sugar	76700	54200
Chocolate Cake Mix	70000	35300
Nestle's Cocoa Mix	54700	37700
Baking Soda	51600	28000
Corn Starch	23700	21000
Flour (Before Test)	23300	20000
Flour (After five days of 6KV-CM)	19100	17500
Sulfer	56200	27700
Grain Dust	41600	22000
Grain Dust (Packed)	26100	17200
Gorilla Milk	35100	33000
White Cake Mix	35000	32700
Baking Powder	19300	15600
Xerox Toner	63500	22300
Coal	5430	5040
Crushed Petroleum Coke	5160	3590

*1 Megohm - cm = 10^6 ohm - cm

TABLE 3
RESISTIVITY OF DUSTS
SEMI CONDUCTORS (NON-METALLIC-CONDUCTORS)

MATERIAL	LENGTH OF TEST	BEGIN RES. AVG. MEG-OHM-CM*	END. RES. AVG. MEG-OHM-CM*	MIN. RES. MEG-CM*	TEMPERATURE RISE FROM AMBIENT - °F.
Flour**	5 Days	23300.00	19100.00	17500.00	2
Coal Dust	4 Days	5430.00	15600.00	5040.00	3
Crushed Petroleum Coke	- -	5160.00		3590.00	- -
Xerox^{TM} Toner		63500.00		22300.00	
Act. Carbon	15 Hrs.	4.08	4.15	3.98	9
Nuchar	3 Days	4.37	SAME	4.31	11
Nuchar - CN	3 Days	3.89	3.90	3.09	17
Wet Nuchar-CN	2 Days	3.79	SAME	3.29	4
Wet Coal Dust Tap Water	5 Hrs.	4.34	4.36	4.19	9
Mine Water #1	1 Hr.	2.58	2.60	1.30	- -
Mine Water #2	1 Hr.	2.59	2.59	1.21	- -
Printers Ink Powder	- -	4.12	- -	4.02	- -
Crushed Coal Coke	- -	4.09	- -	3.99	- -
Crushed Charcoal after vibrating	- -	6.86 5.56	- - - -	4.50 4.50	- - - -
Wet Charcoal, Coal coke and Petroelum coke	- -	4.07	- -	3.99	- -

* 1 Meghohm - cm = 10^6 ohm - cm.

** For Reference

Figures 12 and 13 show maximum and minimum values measured. Figure 18 places resistivities of all dusts sampled on the same chart. It groups them by resistivity range into insulators, semiconductors, and conductors according to high, medium and low resistivity of sample.

(1.) In Figure 12, food powders and grain dust resistivities were in the range of the poorer insulators, 104 to 106 meg-ohm-cm.

(2.) In Figure 13, dry carbon dust resistivities were 3 to 6 orders of magnitude greater than solid carbon.

(3.) In Figures 13 and 17, mine water wetted coal dust resistivity was half the tap water wetted resistivity.

(4.) In Figure 13, tap water decreased coal dust resistivity by three orders of magnitude.

(5.) In Figure 14 B and C, flour resistivity diminished slightly after aging at 6 KV/CM for five days.

(6.) In Figure 14, chocolate cake, cocoa and baking soda resistivity dropped to under 40 percent of the first reading as volts per cm. (stress) increased to 5 KV/CM.

(7.) Comparing Figures 15 and 17, nuchar, wet nuchar, and mine water wetted coal dust resistivities increased as the voltage per cm. increased.

(8.) In Figure 15, water wetting did not significantly change nuchar resistivity.

(9.) In Figures 15, 16, 17 WET coal dust resistivity decreased as voltage per cm. increased.

(10.) In Figure 17, mine water wetted coal dust resistivity increased by three times the first value as stress increased.

(11.) In Figure 17, initial and aged tap water wetted samples had nearly the same resistivity as the volts per cm. applied was increased.

8.7.14 Conclusions

When a sample has withstood, without breakdown 1,000 volts per cm. and has not shown a temperature rise of over ten percent in C during prolonged test at high voltage, it has passed the breakdown test. Breakdown is defined as the resistivity decreasing to under 100 ohm-cm. as applied cell voltage is increased and/or other environmental tests are applied. Passing means that the dust shall be considered non-conducting. When the sample does not withstand the breakdown or temperature rise test, the material fails.

Passing the test **does not** mean that the dust will not interfere with sensitive instruments if quantities of dust seep into the instrument enclosure. Treatment of instrument reliability in conducting combustible dust locations is beyond the scope of this Standard.

When the sample is a metal powder; it probably will, as on the committee's tests, act like a dead short at low applied voltage. This can be detected with a sample ohmmeter. Such samples are unquestionably conductors.

9 REFERENCES

9.1 Dorsett, Henry etal; LABORATORY EQUIPMENT AND TEST PROCEDURES FOR EVALUATION OF EXPLOSIBILITY OF DUSTS; Report of Investigation 5624; Bureau of Mines; Pittsburgh; 1960.

9.2 ICI INDUSTRIAL CONTROL; National Electrical Manufacturer's Association, New York 1959 (Rev 1972).

9.3 Hartmann, Irving etal; EXPLOSIBILITY OF CORNSTARCH; RI 4725; Bureau of Mines, Pittsburgh; 1950.

9.4 Jacobson, Murray etal; EXPLOSIBILITY OF AGRICULTURAL DUSTS; RI 5753, Bureau of Mines; Pittsburgh; 1961.

9.5 Jacobson, Murray etal; EXPLOSIBILITY OF DUSTS USED IN THE PLASTICS INDUSTRY; RI 5971; Bureau of Mines; Pittsburgh; 1962.

9.6 Jacobson, Murray etal; INFLAMMABILITY AND EXPLOSIBILITY OF METAL POWDERS; RI 6516; Bureau of Mines; Pittsburgh; 1964.

9.7 Nagy, John etal; EXPLOSIBILITY OF CARBONACEOUS DUSTS; RI 6597; Bureau of Mines; Pittsburgh; 1965.

9.8 Dorsett, Henry etal; DUST EXPLOSIBILITY OF CHEMCALS, DRUGS, DYES AND PESTICIDES; RI 7132; Bureau of Mines; Pittsburgh; 1968.

9.9 Nagy, John etal; EXPLOSIBILITY OF MISCELLANEOUS DUSTS; RI 7208; Bureau of Mines; Pittsburgh; 1968.

9.10 Staff of Factory Mutual Engineering Corp.; HANDBOOK OF INDUSTRIAL LOSS PREVENTION; 2nd Ed.; McGraw Hill; New York; 1967.

9.11 Magison, E. C.; ELECTRICAL INSTRUMENTS IN HAZARDOUS LOCATIONS: 2nd Ed.; Instrument Society of America; Pittsburgh; 1972.

9.12 Zabetakis, Michael; FLAMMABILITY CHARACTERISTICS OF COMBUSTIBLE GASES AND VAPORS; Bulletin 627; Bureau of Mines; Pittsburgh; 1965.

9.13 IEC 79 RECOMMENDATIONS FOR THE CONSTRUCTION OF FLAMEPROOF ENCLOSURES OF ELECTRICAL APPARATUS; Central Office of the International Electrotechnical Commission; Geneva; 1957.

9.14 VDEO165 BESTIMMUNGEN FUR DIE ERRICHTURN ELECTRISCHER ANLAGEN IN EXPLOSIONGEFAHRDETEN VETRIEBSSTATTEN; 0165/8.69; VDE Verlag Gmbh; Berlin; 1969 (Figurel); 1969.

9.15 Trostel, L. J.; "The Lower Limits of Concentration For Explosions of Dusts in Air"; Chemical and Metallurgical Engineering Vol. 30; Jan. 28, 1924.

9.16 NFPA No. 70 NATIONAL ELECTRICAL CODE; National Fire Protection Association; Boston; 1971.

INSTRUMENT SOCIETY of AMERICA
Research Triangle Park, North Carolina

ISA-S12.11-1973

Standard

Electrical Instruments in Hazardous Dust Locations

Instrument Society of America

ISBN 0-87664-341-1

ISA-S12.11   Electrical Instruments in
             Hazardous Dust Locations

INSTRUMENT SOCIETY OF AMERICA
67 Alexander Drive
P.O. Box 12277
Research Triangle Park, North Carolina 27709

PREFACE

This Standard has been prepared as a part of the service of the Instrument Society of America toward a goal of uniformity in the field of instrumentation. To be of real value, this document should not be static, but should be subject to periodic review. Toward this end, the Society welcomes all comments and criticisms, and asks that they be addressed to the Standards and Practices Board Secretary, Instrument Society of America, 67 Alexander Drive, P.O. Box 12277, Research Triangle Park, North Carolina 27709.

The development of this Standard was initiated as a result of a survey of the flour milling industry of need for additional electrical standards. Members of S12 had been working in gas and vapor standardization and recognized the need for others to deal with the similar problems in dusty locations. The Committee for Instrumentation for Hazardous Dust Locations, SP12 was formed under the Instrumentation for Hazardous Locations S12 Committee, SP12. The dust committee has exposed its views and findings to numerous forums of representatives of government, industry, and labor. A Canadian representative has made significant contributions. The Standard was passed through several mail-review-revision cycles until a consensus of reviewers was reached.

The ISA Standards and Practices Department is aware of the growing need for attention to the metric system of units in general, and the International System of Units (SI) in particular, in the preparation of instrumentaton standards. The Department is further aware of the benefits to USA users of ISA Standards of incorporating suitable references to the SI (and the metric system) in their business and professional dealings with other countries. Towards this end this Department will endeavor to introduce SI and SI-acceptable metric units as optional alternatives to English units in all new and revised standards to the greatest extent possible. The Metric Practice Guide, which has been published by the American National Standards Institute as Z210.1-71 and future revisions, will be the reference guide for definitions, symbols, abbreviations, and conversion factors.

The assistance of those aided in the preparation of this document by answering questionnaires, offering suggestions, and in otherways, is gratefully acknowledged.

ISA SP12 INSTRUMENTATION FOR HAZARDOUS LOCATIONS

F. L. Maltby, Chairman	Drexelbrook Engineering Co.
J. A. Bossert	Canadian Explosive Atmospheres Laboratory
W. Calder	The Foxboro Co.
H. G. Conner	E. I. DuPont deNemours & Co.
E. J. Cranch	Leeds & Northrup Co.
W. F. Hickes	The Foxboro Co.
George Hunt	Monsanto Co.
W. Jacobson	Bristol Co.
C. F. Kisselstein	Crouse Hinds Co.
E. C. Magison	Honeywell, Inc.
L. E. Miller	Factory Mutual Research Corp.
T. W. Moodie	The Pillsbury Co.
M. Morgan	Taylor Instruments Co.
P. J. Schram	Underwriters' Laboratories, Inc.
R. L. Swift	Mine Safety Appliances
J. R. Williams	Stearns-Roger Inc.
E. Yeaton	Eastman Kodak Co.

ISA SP12.7 ELECTRICAL SAFETY IN HAZARDOUS DUST LOCATIONS

T. Moodie, Chairman	The Pillsbury Co.
J. Anderson, Secretary	International Multifoods, Inc.
C. Russell Backes	General Mills, Inc.
Warren Carlson	CPC International, Inc.

atorcalal

The SP12 Committee served to review several times the work of the SP12 Committee in review revision cycles. Also in one of these cycles was a Review Board composed of:

J. Beardsley	Up-John Co.
C. M. Buehler	Taylor Instrument Co.
R. D. Coffee	Eastman Kodak
R. Littlefield	E. I. DuPont deNemours & Co.
D. L. Olson	Minnesota Mining & Mfg. Co.
R. A. Robinson	Consultant
E. P. Shoub	Public Health Service
G. V. Oswald	Boeing Co.
J. Abbott	Factory Mutual Research Corp.
D. E. Ridgley	General Electric Co.

The ISA Committee SP12, Electrical Instruments for Hazardous Locations operates within the ISA Standards and Practices Department. This Standard was approved by the ISA Standards and Practices Board in October, 1973.

E. J. Byrne, Vice President	Brown & Root, Inc.
J. A. Berger	Instrument Society of America
P. Bliss	Pratt & Whitney Aircraft Co.
L. N. Combs	Retired from E. I. duPont deNemours & Co.
N. Conger	Continental Oil Co.
G. G. Gallagher	The Fluor Corp., Ltd.
R. L. Galley	Bechtel Corp.
T. J. Harrison	IBM Corp.
P. S. Lederer	National Bureau of Standards
E. C. Magison	Honeywell, Inc.
J. R. Mahoney	IBM Corp.
F. L. Maltby	Drexelbrook Engineering Co.
R. M. Marvin	Dow Chemical Co.
A. P. McCauley	The Glidden Co.
W. B. Miller	Moore Products
H. N. Norton	Jet Propulsion Laboratory
G. Platt	Bechtel Corp.
C. E. Ryker	Cummins Engine Co.
K. A. Whitman	Allied Chemical Corp.

TABLE OF CONTENTS

1 PURPOSE

This Standard is one of a series recommending safe and economical procedures for installing electrical instruments in locations made hazardous by a presence of a cloud or blanket of combustible dust. It is in conformance with the NEC (National Electrical Code) and CEC (Canadian Electrical Code) and attempts to expand and clarify the NEC and CEC.

This Standard summarizes the requirements for safe installation of electrical instruments in Hazardous Locations. The requirements depend on the hazard classification of the location.

2 SCOPE

It provides guidelines for safe installation of electrical instruments using appropriate means to prevent electrical ignition of combustible dusts.

It refers only to hazards created by combustible dusts - agricultural dusts, plastics, chemicals, and metal dusts. These are classified in the NEC - Article 500-2 as Class II, Groups E, F and G, materials.

It is for the guidance of persons trained in the design and installation of instrument systems and to installation of instrument systems and to assist inspection authorities in approving such installations. It embodies the conclusions of the ISA committee SP12 on "Instrumentation for Hazardous Locations" on what constitutes safe and economical practices.

3 DEFINITIONS

Classification—the assignment of a hazard rating such as Division 1, Divison 2 or non-hazardous.

Conductive dust—A dust whose resistivity is less than 100 ohm-cm or which breaks down with 1000 volts per cm. applied across the bulk sample when tested in accordance with methods outlined in ISA-S12.10. Such dust is denoted Group E in the NEC.

Division 1—the classification assigned to a location where either there is a high probability of a dust hazardous atmosphere occurring frequently, or regularly, or where the dust is electrically conductive. (See examples in 5.2, Divisions)

Division 2—the classification assigned to a location where there is a low probability of a dust hazard-ouse atmosphere occurring and/or a high probability of the presence of a hazardous dust layer. (See examples in 5.2 Divisions)

Dust-ignition-proof enclosure—one which excludes ignitible amounts of dusts or amounts which might affect performance or rating and which, when installation and protection are in conformance with the NEC will not permit arcs, sparks or heat otherwise generated or liberated inside of the enclosure, to cause ignition of exterior accumulations or atmospheric suspensions of a specified dust on or in the vicinity of the enclosure. Underwriter's laboratories specifications for a D.I.P. enclosure are summarized in Appendix 1. D.I.P. enclosures may be used in Division' location.

Dust tight enclosure—an enclosure of substantial mechanical construction provided with gaskets or otherwise designed to exclude dust. It has no open through holes and no knock outs. Conduit entrance is by tapped threads with a minimum of 3-1/2 threads engaged or by a gasketed, bonded conduit hub. It has a substantial door or cover made dust tight to the enclosure by a securely fastened gasket or by width and closeness of fit of the mating flanges. Door or cover fasteners are of substantial construction and are permanently captive. The door or cover is permanently captive to the enclosure. Threaded-hub conduit connections are made dust tight by welding or gasketing. Threaded hub conduit connections are solidly bonded to the enclosure by welding or bonding through proper fittings. Such enclosures are NEMA 3, 3X, 4, 4X, 6, 12 or 13 enclosures. D. T. enclosures may be used where permitted by S12.10 and Section 502 of the NEC.

Hazardous atmosphere—for the purpose of this Standard an explosible mixture of dust in air.

Hazardous dust layer—any accumulation of combustible dust that will propagate or cause a fire.

Hazardous location—a space of limited and definable extent in which may be found a hazardous atmosphere or a hazardous dust layer.

Infrequently—Referring to the frequency of a hazardous event connotes "not normally occurring" and "not be likely during normal operations of the facility."

Ignition capable—equipment or wiring which in its normal operating condition releases sufficient electrical or thermal energy to cause ignition of a specific hazardous atmosphere or hazardous dust layer.

Intrinsically safe—equipment and wiring which is incapable of releasing sufficient electrical or thermal energy under normal or abnormal conditions to cause ignition of a specific hazardous atmospheric mixture or hazardous layer.

Minimum cloud ignition temperature—the minimum temperature at which a hazardous atmosphere will ignite. 1 (See Reference 1)

Minimum dust layer ignition temperature—the minimum temperature of a surface which will ignite dust laying on it after a long (theoretically infinite) time. (See Reference 1). In most dusts, free moisture has been vaporized before ignition.

Minimum Explosible Concentration—the minimum concentration of combustible dust that when ignited produces light explosive force. 1 (See Reference 1).

Non-hazardous location—a location where neither a hazardous atmosphere nor a hazardous dust layer is to be expected.

Non-incendive—equipment and wiring which in its normal operating condition is incapable of igniting a specific hazardous atmosphere or hazardous dust layer. Equipment and wiring having exposed blanketed surface temperatures above 80 percent of the ignition temperature in degrees centigrade of the specific hazardous dust layer shall *not* be classed as non-incendive. The blanketed surface temperature shall be determined at the outside surface of the enclosure beneath the surface of a dust accumulation 0.2 inch or more thickness.

Instruments—measuring, indicating, recording, computing, controlling, and similar apparatus requiring the use of small to moderate amounts of electrical energy in normal operation.

Pressurized enclosure—maintained at a pressure higher than the surrounding area and the areas communicated with by conduit runs. The pressurizing medium shall be clean, dry air or an inert gas. In Division 1 locations, the pressure shall be supervised by a suitable pressure switch, to de-energize supply conductors in case of pressure failure. (See footnote 2 with Table 1.)

4 PLANNED REDUCTION OF DUST HAZARD

Well designed process equipment and buildings have easily cleaned, smooth surfaces, without ledges or inaccessible pockets where dust can accumulate. Dust should be mechanically sealed into the process. Where the dusty material must be handled in the open, dust collection should be applied.

Because dusts have discrete particles they can be effectively excluded from electrical enclosures. Dusts are defined in most Bureau of Mines tests 1 (See Reference 8.1) as smaller than 74 microns (through 200 mesh USS) 2.96 x 10.4 in. Devices to exclude dust are commercially available.

Frequent removal of settled dust will materially reduce the real hazard of a location.

Planned location of installation to an area of lower classification effectively reduces hazard and cost.

5 AREA CLASSIFICATION FOR COMBUSTIBLE DUST LOCATIONS

5.1 Groups

In Article 500-2 of the NEC, Class 11 materials are subdivided in Groups E, F and G. New definitions of the NEC groups based on resistivities are given below.

Group E dusts are those having resistivities lower than 100 ohm-cm or which break down when subjected to 1000 volts/cm across a bulk sample when tested in accordance with ISA-S12.10 Appendix 8.7 Laboratory Resistivity Measurement of Dusts.

(This group includes the dusts of metals, generally regarded as conductors.)

Group F dusts are those having resistivities between 100 ohm-cm and 100 meg-ohm-cm and which do not break down when subjected to 1000 volts/em across the bulk sample when tested in accordance with ISA S12.10 Appendix 8.7 Laboratory Resistivity Measurement of Dusts.

(This group includes the carbonaceous dusts, generally regarded as semiconductors.)

Group G dusts are those having resistivities greater than 100 meg-ohm-cm and which do not break down when subjected to 1000 volts/cm across the bulk sample when tested in accordance with ISA-S12.10. Appendix 8.7 Laboratory Resistivity Measurement of Dusts.

(This group includes the agricultural and plastic dusts generally regarded as insulators.)

5.2 Divisions

5.2.1 Division 1 locations are places where:

(1) The dust cloud concentration is above the minimum explosible concentration continuously, intermittently, or periodically.

(2) The explosive dust cloud is produced by a malfunction of the process or handling machinery and this malfunction causes a simultaneous source of ignition.

(3) The dust is Group E.

5.2.2 Division 2 locations are places where:

(1) The dust cloud concentration rises above the explosive concentration only when

an abnormal condition exists at that location. Examples:

(a) Sack dumps with dust collector in operation but malfunctioning.
(b) Malfunction of equipment - removal for maintenance or cleanout.
(c) When the process is making dust and the dust collection equipment is not operating, or malfunctioning.

(2) A hazardous dust layer exists.

5.3 Non-Hazardous Locations

A location is classified as non-hazardous if there is no significant likelihood that either (a) a hazardous atmosphere or (b) hazardous dust layer will exist, i.e., it is not Division 1 or 2.

Dust cloud concentration is prevented from rising above the explosible concentration by proper equipment design and by effective housekeeping. Housekeeping must be frequent enough to prevent hazardous dust layers from accumulating.

6 SAFE PRACTICES

No special protection is required when there is an extremely low probability that fuel is available in either cloud or layer form. This condition is met in a non-hazardous location and no special protection of electrical equipment is needed.

As the available fuel increases, areas classified Division 2 or Division 1 will be encountered. For locations classified Division 2, one level of protection must be present to safeguard against a dust explosion. For Division 1, two levels are required. The meaning of safety level will become apparent in Table 1 and the definitions. In Table 1, safety level is equivalent to sum of ratings. A review of DEFINITIONS involving Intrinsically Safe 3.10, Non-Incendive and Ignition Capable is recommended.

Planned location of installation to an area of lower classification reduces hazard and cost.

This Standard relates only to hazards created by combustible dusts. Protection against corrosion, personnel shock injury, mechanical injury, and influences causing faulty operation are beyond the scope of this Standard.

7 APPENDIX

This Appendix is intended to clarify Standard ISA-S12.11 and to explain the philosophy on which it is based. Wherever possible, this Standard is based upon an analysis of the physical principles involved, with a minimum reliance upon arbitrary rules.

Appendices 7.1 and 7.2 contain material referenced within this standard.

7.1 Summary Of Underwriters' Laboratories Standards For Class II/Division 1 Equipment

Standards reviewed were:

UL 694A (1964)   Electric Motors and Generators for Use in Hazardous Locations, Class II, E, F or G
UL 698 (1973)   Industrial Control Equipment for Use in Hazardous Locations
UL 844 (1972)   Electrical Lighting Fixtures for Use in Hazardous Locations
UL 886 (1969)   Outlet Boxes and Fittings for Use in Hazardous Locations
UL 894 (1972)   Switches for Use in Hazardous Locations
UL 1002 (1972)   Electrically Operated Values for Use in Hazardous Locations
UL 1010 (1973)   Receptacle Plug Combinations for Use in Hazardous Locations

Flanges/Gaps-

Flange Width	Gap
3/16 in. min (4.75 mm)	0.002 in. (0.031 mm.)
1/4 in. (6.9 mm.)	0.003 in. (0.076 mm.)
3/8 in. min. (9.5 min.)	Gasketted

Threaded Joint - 3 threads min., not more than 32 threads per inch.

Gaskets - Polyletrafluorethylene woven asbestos or plant fiber sheet. No rubber is allowed. No attachment by adhesive or cement is allowed. Attachment mechanical. Except: for enclosures also intended to be raintight, a gasket, min. 3/8 inch (9.5 mm) along joint, passing specified aging test.

Housing Thickness - Table states by maximum dimension, area of a side, and composition (same as for Class I).

Housing Material - Protected against corrosion; sheet steel, cast iron and aluminum; cast brass, bronze, copper or malleable iron, polymeric material. Polymeric material shall pass tests of equivalency to metal enclosures. Zinc, manganese, or their alloys and certain non-metals are excluded.

Fastenings or Hinges - spacing or strength not stated.

Supports - Spacing or strength not stated.

Shaft Clearances, (Controls, Motors or Generators) Groups F & G (Group E less Clearance) - 1/2 in. (12.7 mm.) minimum path length, diametral clearance 0.005 in. (0.127 mm) for 1/2 in. (12.7 mm) path length, 0.008 in. (0.203 mm) maximum for 1 in. (25.4 mm) path length, 0.01 in. (0.254 mm) maximum for 1-1/2 in. (3.81 mm) path length; intermediate values proportional.

Shaft Clearances, Rotating or In-Out Motions (Controls) Groups F & G (Group E less clearance) - 1/2 in. (12.7 mm) minimum path length, diametral clearance 0.010 in. (0.254 mm) for 1/2 in. (12.7 mm) path length, 0.016 in. (0.406 mm) for 1 in. path length, 0.022 in. (0.56 mm) 1-1/2 in. (38.1 mm) path length, intermediate values proportional.

Hubs - With conduit stop or internal space for bushing without stop, 3/4 in. taper per foot thread, 1/4 in. (6.35 mm) minimum thickness 5 full threads when enclosure is a lighting fixture; 3 threads minimum in control enclosures, except 5 threads if device is to be supported.

Lenses or Windows on lighting fixtures or meter and instrument enclosures - glass or similar rigid material. Table of thicknesses given.

Internal Clearances - 1/4 in. (6.35 mm) to ground minimum.

Test - Expose boxes or fittings to circulating dust in atmosphere for at least 30 hours. Times for other items are unspecified. Loads shall be applied at full rating, continuously and intermittently.

Test Materials - for test of boxes or fittings and valves to be classified for use in Class II, Group E, F and G areas the dust shall have the following characteristics:

Group E - Mangnesium Dust

Mesh	Opening	% Through
60	0.0097 in. (0.246 mm)	100
100	0.0058 in. (0.147 mm)	66
200	0.0029 in. (0.074 mm)	22

Groups F + G - Wheat and/or corn

Mesh	Opening	% Through
100	0.0058 in. (0.147 mm)	100 (no breakdown of fractions) wheat and/or corn

Dusts for testing other items unspecified.

Criteria - No entrance of dust into enclosure. Temperatures of external surface shall not exceed safe limits for materials of consideration. No charring or burning is acceptable.

Surface Temperature, when blanketed with dust

Group	Maximum Temperature (40 C ambient)	
	°C	°F
E	200	392
F	200	392
G	165	329

Vibration Test - Unspecified vibration tests shall be conducted on lighting fixtures. Flexible connections to boxes, fittings, or fixtures shall withstand 35 hours of 200 cpm vibration at 1/16 in. (0.159 mm) peak to peak amplitude without loosening or damage.

7.2 NEMA enclosures for Indoor Nonhazardous Locations (Reference 8.2)

	Type of Enclosure							
Provides Protection Against	1*	2*	4	4X	6	11	12*	13
Accidental contact with enclosed equipment	yes	yes	yes	yes	yes	yes	yes	yes
Falling dirt	yes	yes	yes	yes	yes	yes	yes	yes
Falling liquids and light splashing		yes	yes	yes	yes	yes	yes	yes
Dust, lint, fibers and flyings			yes	yes	yes	yes	yes	yes
Hosedown and splashing water			yes	yes	yes			
Oil and coolant seepage							yes	yes
Oil and coolant spraying and splashing								yes
Corrosvie agents				yes		yes		
Occasional submersion					yes			

7.3 NEMA Enclosures for Outdoor Nonhazardous Locations (Reference 8.2)

Provides Protection Against	Type 3	Type 3R*	Type 3S	Type 4	Type 4X	Type 6
Accidental contact with enclosed equipment	Yes	Yes	Yes	Yes	Yes	Yes
Rain, snow and sleet-	Yes	Yes	Yes	Yes	Yes	Yes
Sleet while external operating mechanisms remain operable	. . .	. . .	Yes	. . .	. . .	. . .
Windblown dust	Yes	. . .	Yes	Yes	Yes	Yes
Hosedown	. . .	. . .	. . .	Yes	Yes	Yes
Corrosive Agents	. . .	. . .	. . .	. . .	Yes	Yes
Occasional submersion	. . .	. . .	. . .	. . .	. . .	Yes

* These enclosures may be ventilated, except that Type 12 will not be dusttight if ventilated. Consult manufacturer. See ICS 1-110.07
External operating mechanisms are not required to be operable when the enclosure is ice covered.
External operating mechanisms are operable when the enclosure is ice covered.
Subparagraph and Tables approved as Authorized Engineering Information 7-16-1969, Table of suffixes approved as NEMA Standard 7-16-1969.

7.4 NEMA Enclosures for Indoor Hazardous Locations (Reference 8.2)

If the installation is outdoors and/or additional protection is required a combination-type enclosure is required.

Provides Protection Against Atmospheres Containing	Class†	Group†	Type of Enclosure					
			7A or 8A	7B or 8B	9E*	9F*	9G†	10
Metal dust	II	E	. . .	. . .	yes	. . .	. . .	. . .
Carbon black, coal dust, coke dust . .	II	F	. . .	. . .	. . .	yes	. . .	. . .
Flour, starch, grain dust	II	G	. . .	. . .	. . .	yes	yes	. . .
Methane with or without coal dust. . . . . Bureau of Mines			. . .	. . .	. . .	. . .	. . .	yes

* These enclosures may be ventilated, except that Type 9 will not be dusttight if ventilated.
† As described in Article 5(X) of the National Electrical Code.

8 REFERENCES

8.1. Dorsett, Henry etal; LABORATORY EQUIPMENT AND TEST PROCEDURES FOR EVALUATION OF EXPLOSIBILITY OF DUSTS; Report of Investigation 5624; Pittsburgh; Bureau of Mines; 1960

8.2. NEMA ICES (1970); INDUSTRIAL CONTROL; National Electrical Manufacturer's Association, New York; 1970 (Rev. 1971)

8.3. NFPA 496; PURGED AND PRESSURIZED ENCLOSURE FOR ELECTRICAL EQUIPMENT IN HAZARDOUS LOCATIONS; National Fire Protection Association; Boston; 1971

TABLE 1 - PROTECTION RELATED TO AREA CLASSIFICATION

SPECIAL PROTECTION BY:

RATING CIRCUIT	+ RATING DEVICE	+ RATING ENCLOSURE	NEMA[3] SUM	= RATING SAFE FOR[1]
(0) Ignition Capable	- (0) Open	- (0) General Purpose[5]	1 -	(0) Non-Hazardous
(0) Ignition Capable	- (0) Open	- (1) Industrial Use	13 -	(1) Division 2[4]
(0) Ignition Capable	- (0) Open	- (2) Dust Ignition Proof	9 -	(2) Division 1
(0) Ignition Capable	- (1) Potted, Sealed, Oil Immersed	- (0) General Purpose[5,6]	1 -	(1) Division 2[4]
(0) Ignition Capable	- (1) Potted, Sealed, Oil Immersed	- (1) Industrial Use[6]	13 -	(2) Division 1
(0) Ignition Capable	- (0) Open	- (1) Pressurized[2], General Purpose[5]	1 -	(1) Division 2[4]
(0) Ignition Capable	- (0) Open	- (2) Industrial Use, Pressurized[2]	13 -	(2) Division 1
(1) Non-Incendive	- (0) Open	- (0) General Purpose[5,6]	1 -	(1) Division 2[4]
(1) Non-Incendive	- (0) Open	- (1) Industrial Use[6]	13 -	(2) Division 1
(1) Non-Incendive	- (0) Open	- (0) General Purpose[5], Pressurized[2]	1 -	(2) Division 1
(1) Non-Incendive	- (1) Potted, Sealed, Oil Immersed	- (0) General Purpose[5]	1 -	(2) Division 1
(2) Intrinsically Safe	- (0) Open	- (0) General Purpose[5]	1 -	(2) Division 1

[1] When the sum of the special protection ratings equal 0, the combination may safely be located in a Non-Hazardous location. When the sum equals 1, a Division 2 location is acceptable. When the sum equals 2, a Division 1 location is acceptable. This means that in a Division 1 location, two improbable failures must occur to make a combination established above unsafe. In Division 2 locations, one highly probable failure must occur to make the installation unsafe. A lower value is reasonable for Division 2 because the hazardous atmosphere itself is improbable of occurring.

[2] The following requirements apply to enclosures of 10 cubic feet or less volume. The enclosure shall be pressurized with alarm or indication of pressurizing system failure. The device may be mechanical, pneumatic, or electrical and the signal may be audible or visual. Pressure shall be set at 0.1" w.g. min. and 0.5" w.g. min. for dust of specific particle density 130 or less and greater than 130#/c.f. respectively. In Division 1 locations, power supply shall be interlocked with the door switch. There shall be an appropriate caution sign, for cases where the door switch cannot be used, warning against opening the enclosure in the presence of a hazardous atmosphere. The enclosure temperature under a dust layer shall not exceed 80% of the layer ignition temperature of the combustible dusts encountered. These and other requirements for pressurized enclosures are in NFPA Standard 496, Reference 8.3.

[3] NEMA - National Electrical Manufacturer's Association. The type numbers are further defined in Reference 2. Part 2.68 defines Type 1 as General Purpose, Type 9 as for Hazardous Locations, Class II, Group E, F or G and Type 13 as Industrial Use (to exclude dust, lint, fibers, flyings, oil or coolant seepage), also acceptable with Industrial Use are NEMA 3, 3S, 4, 4X, 6 or 12 if not ventilated because there types are dust tight enclosed. (See Definitions.) See Appendix 2, Environments protected against for NEMA types of enclosures are shown in Appendixes 7.2, 7.3, and 7.4.

[4] No Division 2 allowed for Group E (metal) dusts.

[5] General purpose enclosures shall have six full sides and have all unused openings closed.

[6] Temperature limits of footnote [2] must be met.

INSTRUMENT SOCIETY of AMERICA
Research Triangle Park, North Carolina

ANSI/ISA-S12.12-1984
Approved October 29, 1986

American National Standard

Electrical Equipment for Use in Class I, Division 2 Hazardous (Classified) Locations

Instrument Society of America

ISBN 0-87664-846-4

ISA-S12.12 Electrical Equipment for Use in Class I, Division 2 Hazardous (Classified) Locations

INSTRUMENT SOCIETY OF AMERICA
67 Alexander Drive
P.O. Box 12277
Research Triangle Park, North Carolina 27709

PREFACE

This preface is included for information purposes and is not a part of ISA-S12.12.

This standard has been prepared as part of the service of the Instrument Society of America (ISA) toward a goal of uniformity in the field of instrumentation. To be of real value, this document should not be static, but should be subject to periodic review. Toward this end, the Society welcomes all comments and criticisms, and asks that they be addressed to the Secretary, Standards and Practices Board, Instrument Society of America, 67 Alexander Drive, P.O. Box 12277, Research Triangle Park, NC 27709, Telephone (919) 549-8411.

The ISA Standards and Practices Department is aware of the growing need for attention to the metric system of units in general, and the International System of Units (SI) in particular, in the preparation of instrumentation standards. The Department is further aware of the benefits to U.S.A. users of ISA standards of incorporating suitable references to the SI (and the metric system) in their business and professional dealings with other countries. Toward this end, this Department will endeavor to introduce SI-acceptable metric units in all new and revised standards to the greatest extent possible. The Metric Practice Guide, which has been published by the Institute of Electrical and Electronics Engineers as ANSI/IEEE Std. 268-1982, and future revisions will be the reference guide for definitions, symbols, abbreviations, and conversion factors.

It is the policy of the Instrument Society of America to encourage and welcome the participation of all concerned individuals and interests in the development of ISA standards. Participation in the ISA standards-making process by an individual in no way constitutes endorsement by the employer of that individual, of the Instrument Society of America, or of any of the standards that ISA develops.

The information contained in the preface, footnotes, and appendices is included for information only and is not a part of the standard.

The following people served as members of ISA Committee SP12.12:

NAME	COMPANY
F. J. McGowan, Chairman	The Foxboro Company
A. A. Bartkus	Underwriters Laboratories, Inc.
J. A. Bossert	Energy Research Laboratories
R. Lelievre	Factory Mutual Research Corporation
R. Masek	Bailey Controls, Inc.
A. E. Turner	Westinghouse Electric Corporation
R. K. Weinzler	Taylor Instrument Company
H. Heinaman (deceased)	Honeywell, Inc.
M. Joshi	Honeywell, Inc.
J. B. Simpson	Beckman Industrial Corporation

The following people served as members of ISA Committee SP12:

NAME	COMPANY
E. M. Nesvig, Chairman	ERDCO Engineering Corporation
A. A. Bartkus	Underwriters Laboratories, Inc.
J. A. Bossert	Energy Research Laboratories
R. Buschart	Monsanto Company
B. Colbacchini	Motorola, Inc.
K. M. Collins	Canadian Standards Association
H. C. Conner	E. I. DuPont de Nemours & Company
J. A. Davenport	Industrial Risk Insurers
J. R. Dolphin (Alternate)	Underwriters Laboratories, Inc.
W. J. Engard	Eastman Kodak Company
L. G. Heine	Leeds & Northrup Company
B. Jackson	United States Coast Guard
M. N. Joshi	Honeywell, Inc.
F. L. Maltby	Drexelbrook Engineering
H. H. Marshall, Jr.	Mine Safety Appliances Company
F. J. McGowan	The Foxboro Company
E. W. Olson	Minnesota Mining & Manufacturing Company
D. L. Plahn	Fisher Controls International Company
G. Posner	Fischer & Porter Company
J. Rennie	Factory Mutual Research Corporation
D. E. Ridgley	General Electric Company
R. L. Swift	Mine Safety Appliances Company
R. K. Weinzler	Sybron Corporation
D. C. Whitten	Bristol Babcock, Inc.
Z. Zborovszky	U. S. Bureau of Mines

This standard was approved for publication by the ISA Standards and Practices Board in September 1984.

NAME	COMPANY
W. Calder III, Chairman	The Foxboro Company
P. V. Bhat	Monsanto Company
N. L. Conger	Conoco
B. Feikle	Bailey Controls Company
H. S. Hopkins	Westinghouse Electric Company
J. L. Howard	Boeing Aerospace Company
R. T. Jones	Philadelphia Electric Company
R. Keller	The Boeing Company
O. P. Lovett, Jr.	ISIS Corporation
E. C. Magison	Honeywell, Inc.
A. P. McCauley	Chagrin Valley Controls, Inc.
J. W. Mock	Bechtel Corporation
E. M. Nesvig	ERDCO Engineering Corporation
R. Prescott	Moore Products Company
D. Rapley	Stearns Catalytic Corporation
W. C. Weidman	Gilbert Commonwealth, Inc.
K. A. Whitman	Consultant
*P. Bliss	Consultant
*B. A. Christensen	Continental Oil Company
*L. N. Combs	Retired
*R. L. Galley	Consultant
*T. J. Harrison	IBM Corporation
*R. G. Marvin	Roy G. Marvin Company
*W. B. Miller	Moore Products Company
*G. Platt	Bechtel Power Corporation
*J. R. Williams	Stearns Catalytic Corporation

*Director Emeritus

TABLE OF CONTENTS

APPENDICES

LIST OF ILLUSTRATIONS

LIST OF TABLES

1 PURPOSE

1.1* The purpose of this standard is to provide requirements for the design, construction, and marking of electrical equipment or parts of such equipment used in Class I, Division 2 locations and which, in normal operation, is incapable of causing ignition under the conditions prescribed in this standard. In addition, it is the intent of this document to establish uniformity in test methods for determining the suitability of the equipment and associated circuits and components as they relate to their ability to ignite a specified flammable gas or vapor-in-air mixture.

2 SCOPE

2.1* This standard applies only to equipment, circuits, or components designed and assessed specifically for use in Class I, Division 2 hazardous locations as defined by the National Codes.[1] It is recognized that general purpose equipment may also be suitable for Class I, Division 2 in accordance with National Codes.

2.2 This standard is primarily intended to provide requirements for process measurement and control equipment; however, the principles may be applied to similar types of electrical equipment.

2.3 This standard is concerned only with equipment construction and test criteria related to electrical or thermal ignition of a specified flammable gas or vapor-in-air mixture in its most easily ignitible concentration. Equipment shall comply with the ordinary location requirements for the particular types of equipment[2] except as amended herein.

2.4 This standard does not cover equipment for use in Class I, Division 1 locations, such as equipment constructed to be intrinsically safe or explosionproof; however, such equipment is suitable for use in Class I, Division 2 locations in the same Group for which it is suitable in Division 1.

2.5 This standard does not cover equipment utilizing purged or pressurized enclosures; however, such equipment is considered suitable for use in Class I, Division 2 locations.[3]

2.6 This standard does not cover mechanisms of ignition from external sources, such as static electricity or lightning, which are not related to the electrical characteristics of the equipment.

2.7 This standard is not an instructional manual for untrained persons. It is intended to promote uniformity of practice among those skilled in the art.

2.8* The requirements of this standard are based on consideration of ignition in locations made hazardous by the presence of flammable gases or vapor-in-air mixtures under atmospheric conditions. For the purposes of this standard, conditions are generally considered to be:

(1) An ambient temperature of 40°C (104°F)

(2) An oxygen concentration of 21 percent by volume, and

(3) A pressure of 86 to 106 kPa (12.5 to 15.4 psia)

NOTE: Equipment specified for atmospheric conditions beyond the above are subject to special investigation.

3 DEFINITIONS

3.1 *Hermetically Sealed Device* is a device which is sealed against the entrance of an external atmosphere and in which the seal is made by fusion, e.g., soldering, brazing, welding, or the fusion of glass to metal.

3.2 *Maintenance*

3.2.1 *Corrective Maintenance* is an activity that is not normal in the operation of the equipment and which requires access to the interior. Such activities are expected to be performed by qualified personnel who are aware of the hazards involved.

Such activities typically include locating causes of faulty performance, replacement of defective components, adjustment of service controls, or the like.

3.2.2 *Operational Maintenance* is any maintenance activity, other than *corrective maintenance*, intended to be performed by the operator and which is required in order for the equipment to serve its intended purpose.

Such activities typically include the correcting of "zero" on a panel instrument, changing charts, making records, adding ink, or the like.

3.3 *Make/Break Components* are components having contacts which can interrupt a circuit (even if the interruption is transient in nature). Examples of *make/break components* are relays, circuit breakers, servo potentiometers, adjustable resistors, switches, connectors, and motor brushes.

3.4* *Nonincendive Circuit* is a circuit in which any arc or thermal effect produced under normal operating conditions of the equipment is not capable, under the

---

[1]National Electrical Code, NFPA 70-1984, Articles 500 and 501, National Fire Protection Agency, Quincy, MA, 1984, pp. 407-425; or the Canadian Electrical Code (Part I), C22.1, Section 18, Canadian Standards Association, Rexdale, Ontario, Canada, 1982.

[2]ANSI C39.5-1974, "Safety Requirements for Electrical and Electronic Measuring and Controlling Instrumentation," American National Standards Institute, New York, New York; or CSA Standard C22.2, No. 142-M1983, "Process Control Equipment," Canadian Standards Association, Rexdale, Ontario, Canada, 1983.

[3]NFPA Standard 496, "Purged and Pressurized Enclosures for Electrical Equipment in Hazardous (Classified) Locations," National Fire Protection Agency, Quincy, MA, 40 p.

*NOTE: An asterisk following the subsection number signifies that explanatory material on that paragraph appears in Appendix A.

conditions prescribed in this standard, of igniting the specified flammable gas or vapor-in-air mixture.

3.5* *Nonincendive Circuit Field Wiring* is wiring which enters or leaves the equipment enclosure and which, under normal operating conditions of the equipment, is not capable, due to arcing or thermal effects, of igniting the specified flammable gas or vapor-in-air mixture by opening, shorting, or grounding the field wiring.

3.6 *Nonincendive Component* is a component with contacts for making or breaking a specified incendive circuit where either the contacting mechanism or the enclosure in which the contacts are housed is so constructed that the component is not capable of propagating ignition of the specified flammable gas or vapor-in-air mixture when tested according to Section 9. The housing of a nonincendive component is not intended to exclude the flammable atmosphere.

3.7 *Normal Operational Conditions* - Equipment is in *normal operational conditions* when it conforms electrically and mechanically with its design specifications and is used within the limits specified by the manufacturer. This includes:

(1) Supply voltage, current, and frequency

(2) Environmental conditions (including process interface)

(3) All tool-removable parts in place, (e.g., covers)

(4) All operator-accessible adjustments at their most unfavorable settings, and

(5) Opening, shorting, or grounding of nonincendive field wiring

3.8 *Sealed Device* is a device which is so constructed that it cannot be opened during *normal operational conditions* or operational maintenance; it has a free internal volume less than 100 cm³ (6.1 in³) and is sealed to restrict entry of an external atmosphere. It may contain normally arcing parts or internal hot surfaces.

4 GENERAL REQUIREMENTS

4.1 Requirements for equipment intended for use in Class I, Division 2 locations are established on the basis that the equipment in its normal operational condition is incapable of causing ignition of a specified flammable gas or vapor-in-air mixture. The tolerances associated with the components of the equipment must be considered.

Protection shall be provided according to Subsections 4.1.1 and 4.1.2 or by a combination of these methods that ensures that under normal operational conditions such equipment is not capable of igniting the specified flammable gas or vapor-in-air mixture.

Subsequent arcs or thermal effects within the equipment, resulting from opening, shorting, or grounding of the nonincendive circuit field wiring, shall be taken into consideration as they affect the suitability of the equipment for use in Division 2 locations.

4.1.1* Any make/break component shall be:

(1) A normal nonarcing component which meets the requirements of Section 5, or

(2) Used in a nonincendive circuit which meets the requirements of Section 8, or

(3) A nonincendive component which meets the requirements of Section 9, or

(4) A sealed device which meets the requirements of Section 10

4.1.2 Equipment with surface temperatures in excess of 100°C (212°F) shall comply with requirements of Section 6.

4.2 Equipment enclosures shall provide protection to prevent deterioration which would adversely affect the suitability of the equipment for use in Division 2 locations. While general purpose enclosures will normally suffice, particular attention should be given to the possible need for weatherproofing and general protection from corrosion, and preventive maintenance.

4.3 Equipment shall be capable of being installed according to the requirements of the applicable National Codes. (See Subsection 2.1.)

4.4* Fuses may be housed in general purpose enclosures only when used for overcurrent protection of circuits which are not subject to overloading in normal use.

Fuses used in circuits which are subject to overloading in normal use must be of a type suitable for use in Division 2 locations or housed in an enclosure rated for Division 1 locations.

NOTE: This subsection precludes a fuse housed in a general purpose enclosure from being used in a motor circuit where a possibility of a stalled motor opening the fuse exists, or where there is a possibility of an overload not caused by a fault in the circuit.

4.4.1 If a fuse is provided, a switch must also be provided to remove power from the fuse and equipment. The switch need not be integral to the equipment if the equipment installation instructions indicate the need for such a switch complying with Subsection 4.1.1.

Exception No. 1: Fuses in nonincendive circuits meeting the requirements of Section 8 need not be in a switched circuit.

---

*NOTE: An asterisk following the subsection number signifies that explanatory material on that paragraph appears in Appendix A.

Exception No. 2: Fuseholder assemblies determined to be a nonincendive component need not be in a switched circuit.

5 NORMALLY NONARCING COMPONENTS

5.1 Make/break components which are to be considered nonarcing in normal operation shall comply with the requirements of Subsections 5.2 through 5.5 where applicable.

5.2* Connectors used in incendive circuits and incorporated within equipment shall be considered normally nonarcing if disconnection is not required under operational maintenance conditions and if they are secured (for example, if a separating force of at least 15 N (3.4 pounds) is required for loosening), or if they are mechanically prevented from separating. If accessible during operational maintenance, connectors in an incendive circuit shall be provided with a cautionary marking according to Subsection 7.2.

5.3 A fuse in a fuseholder in an incendive circuit, accessible during operational maintenance, shall be provided with a cautionary marking (according to Subsection 7.2) located adjacent to the fuseholder.

Exception: A fuseholder assembly determined to be a nonincendive component need not comply with this requirement.

5.4 A lamp in a lampholder in an incendive circuit, accessible during operational maintenance, shall be provided with a cautionary marking (according to Subsection 7.2) located adjacent to the lampholder.

Exception: A lampholder assembly determined to be a nonincendive component need not comply with this requirement.

5.5 Plugs and receptacles used for connection to associated nonincendive field circuits shall be noninterchangeable.

Exception: In circuits other than branch circuits, where interchange does not affect nonincendive circuits, or plugs and receptacles are so identified that interchange is unlikely, noninterchangeable plugs and receptacles are not required.

6 SURFACE TEMPERATURE REQUIREMENTS

6.1* The maximum temperature of an external or internal surface to which a surrounding specified flammable gas or vapor-in-air mixture has access shall be determined under normal operational conditions. Such measurements need not be made on the internal parts of sealed devices. Measurements shall be made at any con-

---

TABLE 1
MAXIMUM SURFACE TEMPERATURE

Degrees C	Degrees F	Temperature Code*
450	842	T1
300	572	T2
280	536	T2A
260	500	T2B
230	446	T2C
215	419	T2D
200	392	T3
180	356	T3A
165	329	T3B
160	320	T3C
135	275	T4
120	248	T4A
100	212	T5
85	185	T6

*Underlined Temperature Codes are preferred and agree with the temperature identification system specified by the International Electrotechnical Commission (IEC 79 Series Publications).

---

venient ambient temperature normally encountered in a laboratory, corrected linearly to 40°C (104°F).

6.2* Equipment which attains temperatures higher than 100°C (212°F) based on a 40°C (104°F) ambient shall be marked either by the temperature code as given in Table 1 or by the specific temperature as measured according to Subsection 6.1, corrected linearly to a 40°C (104°F) ambient.

NOTE: Component surface temperature may exceed the marked temperature rating if it can be demonstrated by testing that no hazard exists.

7* MARKING

7.1 Each piece of equipment shall be marked with the following information:

7.1.1 Manufacturer's name, trademark, trade name, or other recognized symbol identifying the manufacturer.

7.1.2 Catalog, style, model, or other unique designation.

7.1.3 Electrical rating; volts, frequency, and amperes or volt-amperes.

7.1.4* Hazardous location suitability; Class I, Division 2, and Group(s), or, in lieu of Group(s) a specific gas according to Subsection 8.3.4.

7.1.5* Temperature marking according to Subsection 6.2.

7.1.6* Any other markings or cautions necessary for the installation and operation of the equipment.

The international symbol ⚠ may be used to refer the operator to an explanation in the equipment instructions.

---

*NOTE: An asterisk following the subsection number signifies that explanatory material on that paragraph appears in Appendix A.

7.2 Connectors, fuseholders, and lampholders, required to be marked according to Subsections 5.2, 5.3, and 5.4, shall be marked with the following warning or equivalent:

WARNING - DO NOT DISCONNECT WHILE CIRCUIT IS LIVE UNLESS AREA IS KNOWN TO BE NONHAZARDOUS

This marking shall be on or adjacent to the component. If that is not practical, this marking may be prominently displayed on the enclosure.

7.3* Nonincendive Circuit Field Wiring

7.3.1 Connections for nonincendive circuit field wiring shall be clearly identified.

Exception: Identification of terminals is not required in a device marked nonincendive for Class I, Group . . . ., Division 2 and having only one set of field wiring connections.

NOTE: Equipment located in a nonhazardous location which has nonincendive circuit field wiring connections need only comply with marking according to this section.

7.3.2* In addition to the marking required according to Subsection 7.3.1, equipment supplying energy to nonincendive circuit(s) may be marked to show:

(1) V_{oc} the maximum open-circuit voltage, and

(2) I_{sc} the maximum short-circuit current, and

(3) C_a the maximum allowable connected cable capacitance, and

(4) L_a the maximum allowable connected cable inductance, and

(5) C_n the maximum allowable connected capacitance based upon the normal circuit voltage (V_n)

(6) L_n the maximum allowable connected inductance based upon the normal circuit current (I_n)

7.3.3* In addition to the markings required according to Subsection 7.3.1, equipment receiving energy from a nonincendive circuit(s) may be marked to show:

(1) V_{max} the maximum voltage the equipment can receive, and

(2) I_{max} the maximum current the equipment can receive, and

(3) C_i the maximum unprotected internal capacitance, and

(4) L_i the maximum unprotected internal inductance

8* EVALUATION OF NONINCENDIVE CIRCUITS

8.1* Either of the following two methods may be employed to determine that a circuit(s) is nonincendive:

(1) Testing the circuit according to Subsections 8.2 through 8.5, or

(2) Comparing the calculated or measured values of current, voltage, and associated inductances and capacitances to the appropriate values in Figures 1 through 8[1] to establish that the current and voltage levels are below those specified in Subsection 8.1.2

8.1.1 In evaluating a circuit as nonincendive, ignition sources such as the following shall be considered under normal operational conditions:

(1) Discharge of a capacitive circuit

(2) Interruption of an inductive circuit

(3) Intermittent making and breaking of a resistive circuit, and

(4) Hot-wire fusing

8.1.2* The maximum voltage and current levels (dc or ac peak) in circuits determined to be nonincendive by the comparison methods shall, for given circuit constants, be less than:

(1) The current determined from Figures 1 through 6

(2) The voltage determined from Figures 7 and 8

Figures 1 and 2 apply only to circuits whose output voltage-current characteristic is a straight line drawn between open-circuit voltage and short-circuit current (i.e., no voltage or current regulators).

Figures 7 and 8 represent capacitor discharge only. They do not include the additional current which may be available from the power supply.

8.2* Test Apparatus

8.2.1 The spark test apparatus[2] used for performing ignition tests on circuits shall consist of an explosion chamber of about 250 cm³ volume, in which circuit-making-and-breaking sparks can be produced in the presence of the prescribed test gas.

---

[1]All figures except Figures 5 and 6 are reprinted from Certification Standard SFA 3012, 1972 Edition, with permission of the Department of Trade and Industry, British Approvals Service for Electrical Equipment in Flammable Atmospheres. Figures 5 and 6 are from "Some Aspects of the Design of Intrinsically-Safe Circuits," Research Report 256, 1968, by D.W. Widginton, Safety in Mines Research Establishment, Sheffield, England.
[2]IEC 79-3 (1972) "Part 3: Spark Test Apparatus for Intrinsically-Safe Circuits," International Electrotechnical Commission, Geneva, Switzerland; ANSI/UL/NFPA 4913-1979, "Electric Equipment," American National Standards Institute, New York, NY; and CSA Standard C22.2, No. 157M-1979, "Intrinsically Safe and Non-Incendive Equipment for Use in Hazardous Locations," Canadian Standards Association, Rexdale, Ontario, Canada.

*NOTE: An asterisk following the subsection number signifies that explanatory material on that paragraph appears in Appendix A.

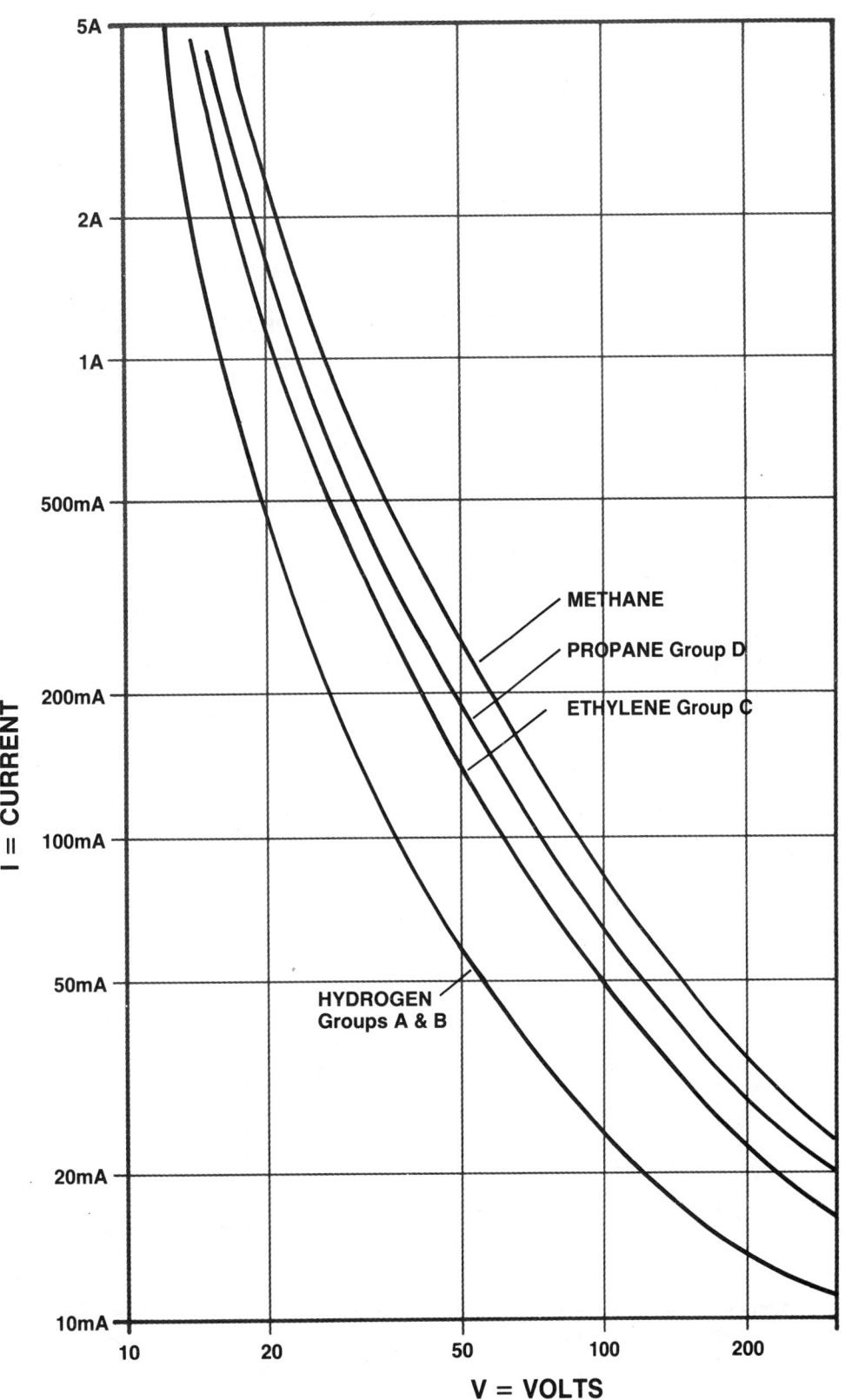

Figure 1. Resistance Circuits and Minimum Igniting Currents, Applicable to All Circuits Containing Cadmium, Zinc, or Magnesium (L < 5 microhenries)

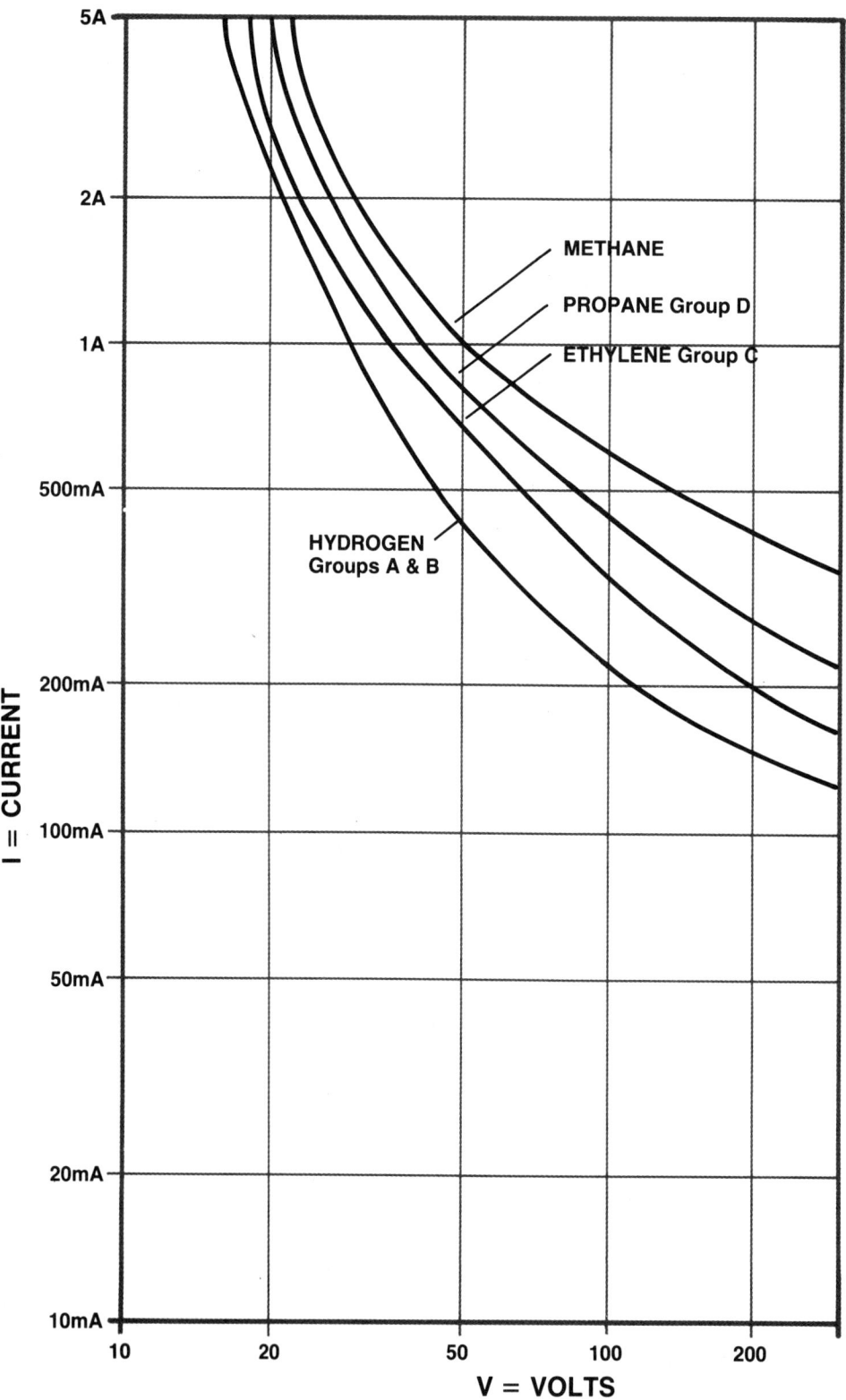

Figure 2. Resistance Circuits and Minimum Igniting Currents, Applicable to Circuits Where Cadmium, Zinc, or Magnesium Can Be Excluded (L < 5 microhenries)

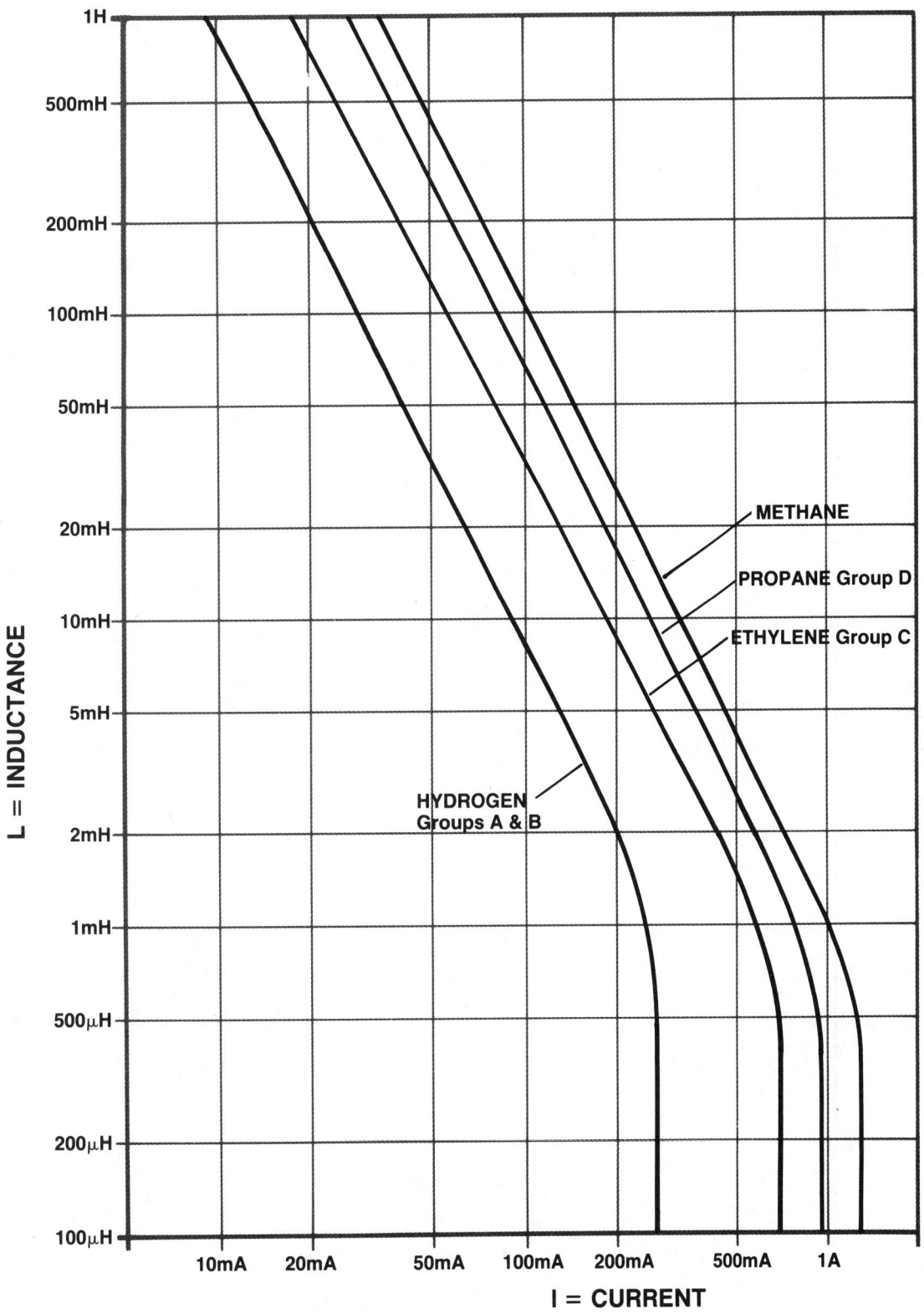

Figure 3. Inductance Circuits and Minimum Igniting Currents at 24 V, Applicable to All Circuits Containing Cadmium, Zinc, or Magnesium

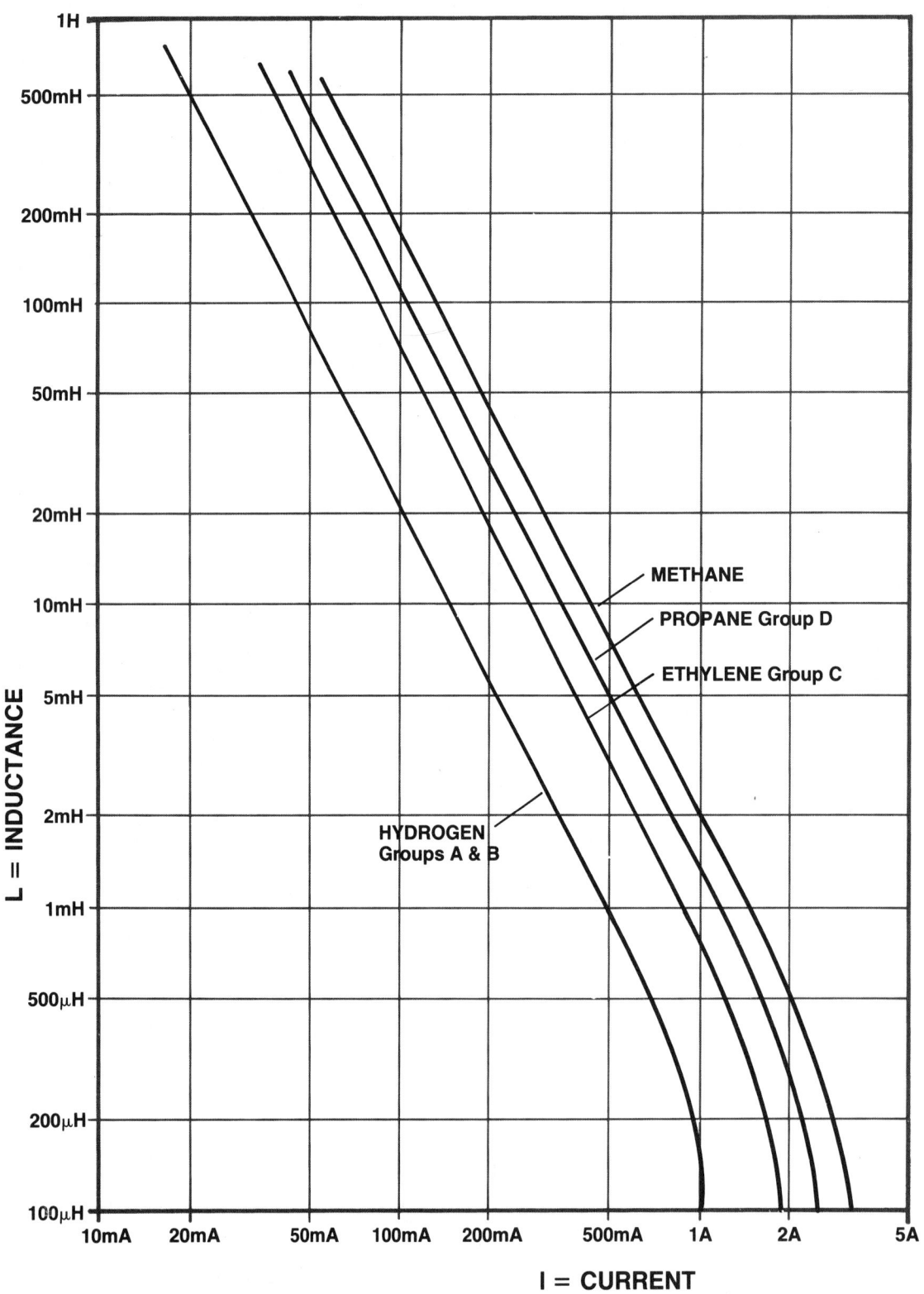

Figure 4. Inductance Circuits and Minimum Igniting Currents at 24 V, Applicable Only to Circuits Where Cadmium, Zinc, or Magnesium Can Be Excluded

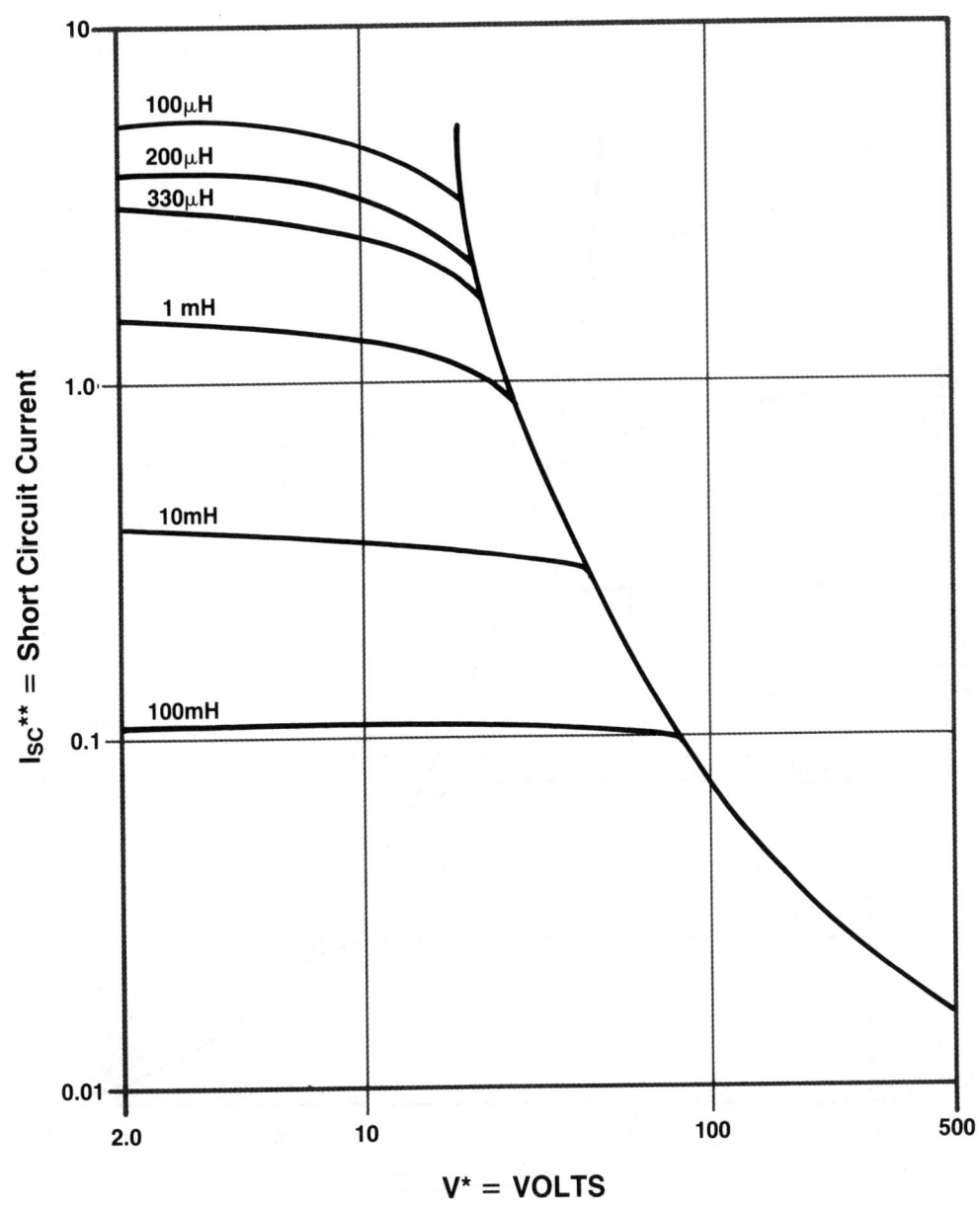

Figure 5. Inductance Circuits and Minimum Igniting Currents for Various Voltages in Methane, Applicable to All Circuits Containing Cadmium, Zinc, or Magnesium

* **Open circuit voltage V_{oc}: V**
** **Minimum value of I_{sc} for ignition: Amperes**

15

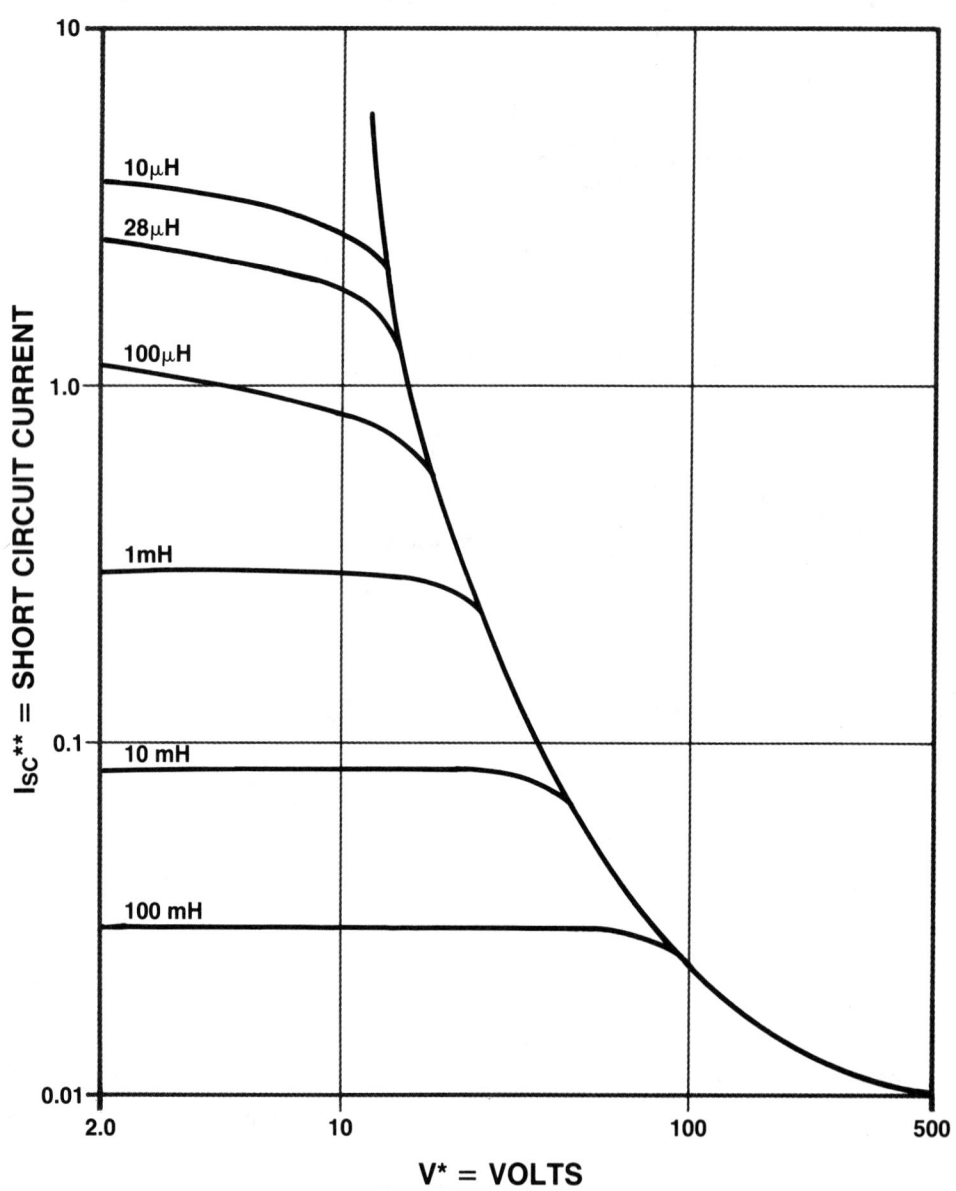

Figure 6. Inductance Circuits and Minimum Igniting Currents for Various Voltages in Group B, Applicable to All Circuits Containing Cadmium, Zinc, or Magnesium

  * Open circuit voltage V_{oc}: V
** Minimum value of I_{sc} for ignition: Amperes

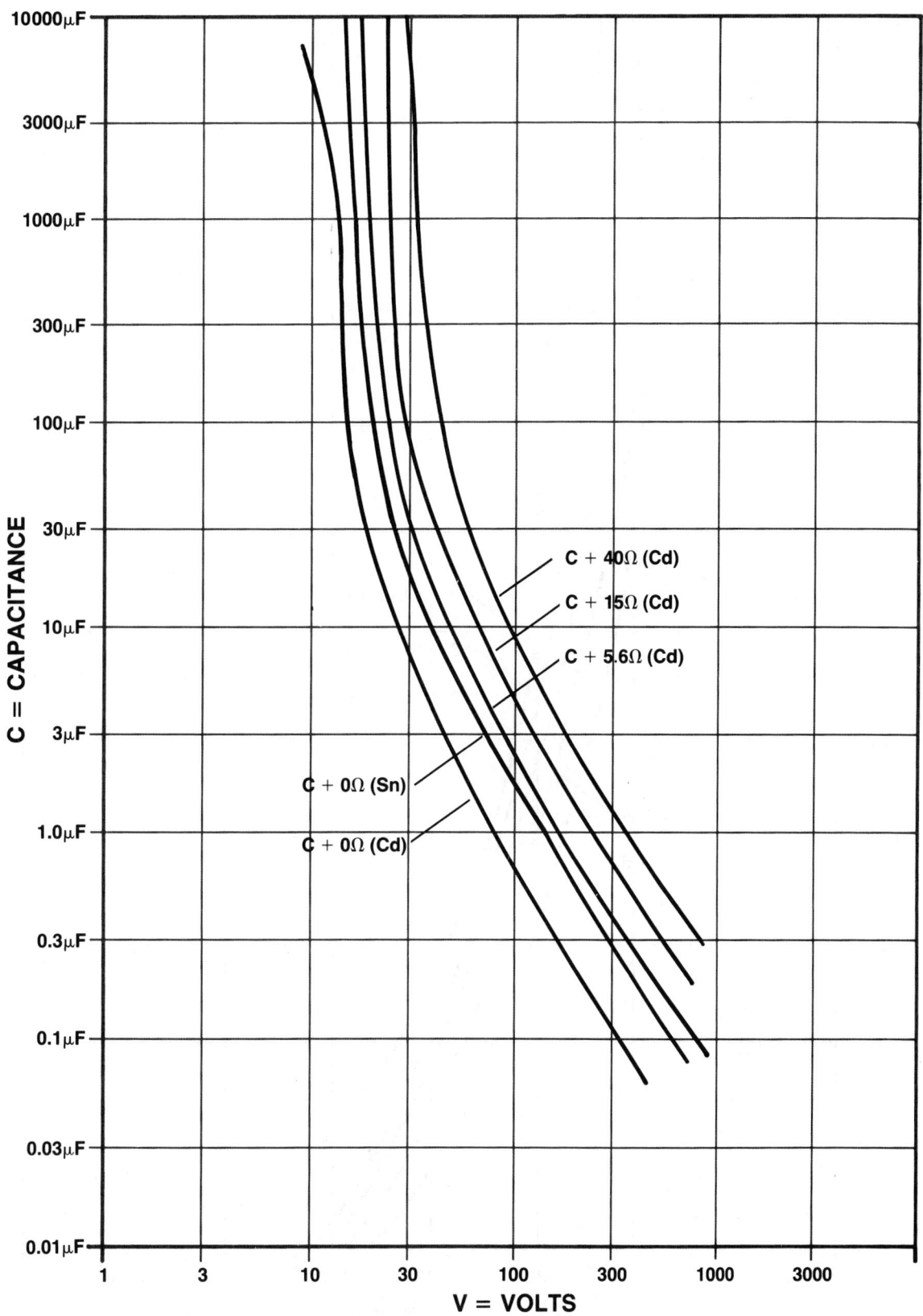

Figure 7. Capacitance Circuits and Minimum Ignition Voltage in Methane

Note: The curves corresponding to values of current limiting resistance as indicated.
The curve marked Sn is applicable only where cadmium, zinc, or magnesium can be excluded.

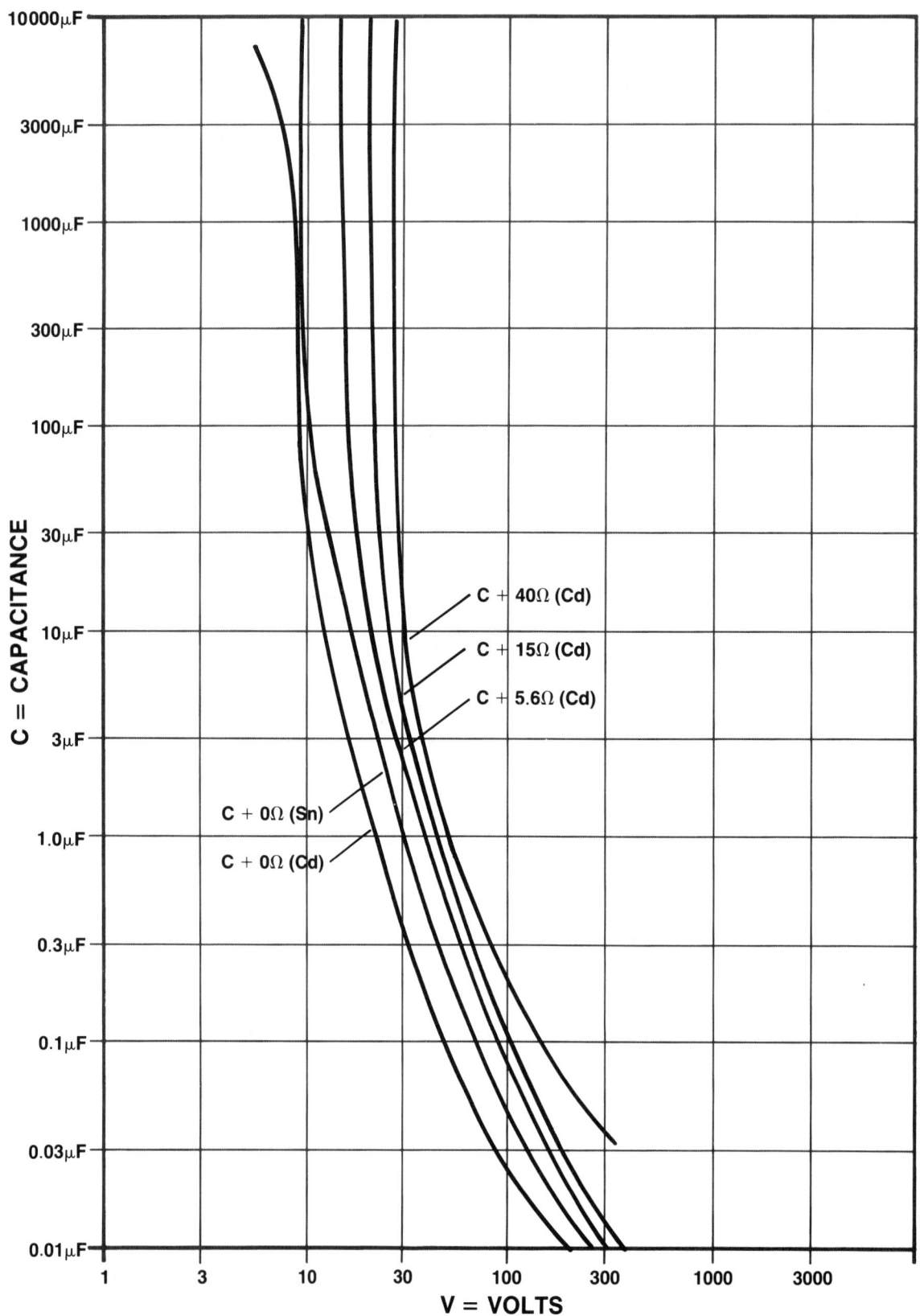

Figure 8. Capacitance Circuits and Minimum Ignition Voltage in Groups A and B

Note: The curves corresponding to values of current limiting resistance as indicated.
The curve marked Sn is applicable only where cadmium, zinc, or magnesium can be excluded.

18

8.2.2 A cadmium disk shall normally be used. When evaluating make/break components, a tin disk may be used if it can be demonstrated that cadmium, zinc, or magnesium will not be present in the equipment. It is normally assumed that cadmium or zinc is present in field wiring due to common use of these metals for corrosion protection.

8.3 Gas Mixture[1]

8.3.1 For Group D, the test mixture shall be 5.25 ± 0.25 percent propane in air.

8.3.2 For Group C, the test mixture shall be 7.8 ± 0.5 percent ethylene in air.

8.3.3 For Groups A and B, the test mixture shall be 21 ± 2 percent hydrogen in air.

Exception: Nonincendive components intended for use in Group A shall use a test mixture of 8.7 percent acetylene in air.

8.3.4 Equipment which is intended for use in a specific gas or vapor may be tested in the most easily ignitible concentration[2] of that gas or vapor-in-air mixture in lieu of the mixtures specified in Subsections 8.3.1 through 8.3.3.

8.4* Sensitivity Verification of Spark Test Apparatus

8.4.1 The sensitivity of the spark test apparatus shall be verified before and after each test series carried out in accordance with Subsection 8.5. For this purpose, the test apparatus shall be operated in one of two verification circuits — a 24 V dc circuit containing a 0.095 H air-core coil or a 24 V dc resistive circuit (inductance less than 10 μH). The currents in these circuits shall be set at the value given for the appropriate group in Tables 2 or 3.

8.4.2 The verification circuit chosen shall be that which is most appropriate to the equipment which is being tested.

8.4.3 The spark test apparatus shall be run for 400 revolutions of the tungsten wire-holder, with the holder of positive polarity, and shall be satisfactory if ignition of the test gas occurs.

8.5 Tests

TABLE 2
CURRENT IN VERIFICATION CIRCUIT
CADMIUM DISK

Group	Inductive Circuit	Resistive Circuit
D	100 mA	1000 mA
C	65 mA	700 mA
A & B	30 mA	300 mA

TABLE 3
CURRENT IN VERIFICATION CIRCUIT
TIN DISK

Group	Inductive Circuit	Resistive Circuit
D	110 mA	2500 mA
C	90 mA	2000 mA
A & B	50 mA	1500 mA

8.5.1 The spark test apparatus shall be connected in the circuit under test at each point where an interruption normally occurs, taking into account the requirements of this standard.

8.5.2 Ignition Test Conditions

There shall be no ignition of the test mixture under the following conditions:

(1) For line-connected equipment, the input voltage shall be increased to 110 percent of nominal line voltage

(2) All adjustments shall be set at their most unfavorable positions

(3) Each circuit shall be tested for the following number of revolutions of the tungsten wire-holder in the spark test apparatus:

   (a) For dc circuits, not less than 400 revolutions, 200 revolutions at each polarity

   (b) For ac circuits, not less than 1000 revolutions

8.5.3 After each circuit test, verification of the spark test apparatus shall be repeated. If the verification does not comply with Subsection 8.4, the ignition test on the circuit under investigation shall be considered invalid.

9 EVALUATION OF NONINCENDIVE COMPONENTS

9.1 Nonincendive components shall be subjected to tests according to Subsection 9.2.

9.1.1* A nonincendive component is limited in use to the rating for which it has been satisfactorily tested according to Subsection 9.2.

---

[1]The purity of commercially available gases and vapors is normally adequate for these tests, but those of purity less than 95 percent should not be used. The effect of normal variations in laboratory temperature and pressure and of the humidity of the air in the gas mixture is also likely to be small. Any significant effects of these variables will become apparent during the routine verification of the sensitivity of the spark test apparatus.

[2]The most easily ignitible concentration may not be the stoichiometric mixture.

*NOTE: An asterisk following the subsection number signifies that explanatory material on that paragraph appears in Appendix A.

9.2* Spark Ignition Test for Nonincendive Components

9.2.1 Nonincendive components shall be preconditioned by operating them 6000 times at the rate of approximately 6 times per minute carrying their normal electrical load.

9.2.2* Following the preconditioning test, the nonincendive component shall be placed in a suitable test chamber of at least 10 times the volume of the device and connected to an electrical load of 150 percent of the ac or dc current and 75 percent power factor if for ac and at maximum voltage of the circuit for which the component is being tested.

NOTE: Switches intended for use with motor loads or tungsten loads (high inrush currents) are subject to overload testing representative of actual circuit applications. Preconditioning and overload conditions shall be according to the requirements of National Standards covering these switch applications.

9.2.3* The nonincendive component shall be filled with and surrounded by a gas mixture according to Subsection 8.3.

9.2.4* The component shall be operated 50 times at not less than 10-second intervals, renewing the gas or vapor-in-air mixture after each set of 10 operations or more frequently if necessary to ensure the presence of the vapor-in-air mixture within the nonincendive component.

There shall be no ignition of the surrounding gas or vapor-in-air mixture.

10* SEALED DEVICE

10.1 This section covers the requirements for electrical equipment or parts of electrical equipment or components which contain normally arcing parts or heat-producing surfaces which by location in a sealed enclosure are intended to be made incapable of causing ignition of the specified gas or vapor-in-air mixtures at their most easily ignitible concentration.

10.1.1 Hermetically sealed devices (i.e., sealed by fusion) shall be considered to meet these requirements without test.

10.2* The free internal volume of the enclosure of other than hermetically sealed devices shall be less than 100 cm³ (6.1 in³).

10.3 Resilient gasket seals or poured seals shall be arranged so that they are not subject to mechanical damage during normal operational conditions and shall retain their sealing properties for the intended conditions of use.

10.3.1 Poured seals shall have a softening or melting point at least 20°C (68°F) higher than the maximum temperature to which the seal is exposed.

10.4 A sealed device must have structural integrity and be constructed of materials suitable for the intended environment with full consideration for atmospheric contaminants and for corrosive compounds. The enclosure shall be sufficiently rugged to withstand normal handling and assembly operations without damage to the seal.

10.5 In order to determine that damage will not occur during normal operation, precondition and testing shall be performed on sample(s) of each sealed device in the following sequence.

10.5.1 Solvent Vapor Exposure Test

Three samples shall be preconditioned by exposing them to each of the following vapors for a period of 24 hours. The specimens shall be suspended approximately 2.5 cm (1 in) above 50 ml of liquid in a flask. The flask shall be sealed to prevent loss of solvent due to evaporation. For Class I, Group C rating, diethyl ether and ethylene dichloride shall be used. For Class I, Group D rating, acetone ammonium hydroxide (20 percent by weight), benzene, ethyl acetate, n-hexane, methyl alcohol, unleaded gasoline, ASTM No. 3 oil, methyl ethyl ketone and toluene shall be used.

After preconditioning each sample shall be:

(1) Examined for visual deterioration, and

(2) Subject to an air leakage test according to Subsection 10.5.3

10.5.2 Oven Aging Test

If the device contains a gasket or seal of thermoplastic material, three samples shall be preconditioned according to the following formula:

$$t = 300e^{-(0.0693)(T-T_1)}$$

where t = time in days
T = oven aging temperature in °C
T_1 = maximum use temperature in °C
e = 2.71828

After the preconditioning, each sample shall be:

(1) Examined for visual deterioration, and

(2) Subject to an air leakage test according to Subsection 10.5.3

10.5.3 Air Leakage Test

Following the preconditioning according to Subsections 10.5.1 and 10.5.2, the samples shall pass one of the following tests:

*NOTE: An asterisk following the subsection number signifies that explanatory material on that paragraph appears in Appendix A.

20

(1) At an initial temperature of 25°C (7.7°F) the test samples shall be immersed in water at a temperature of 50°C (122°F) to a depth of 25 mm (1 in) for 1 min. If no bubbles emerged from the samples during this test, they are considered to be "sealed" for the purpose of this standard, or

(2) Be immersed to a depth of 75 mm (3 in) in water contained in an enclosure which can be partially evacuated. The air pressure within the enclosure shall then be reduced by the equivalent of 120 mm (4.7 in) of mercury. If no bubbles emerged from the samples during this test, they are considered to be "sealed" for the purpose of this standard, or

(3) Be shown to leak at a rate not greater than 10^{-5} ml of air per second at a pressure differential of 101.3 kPa (1 atmosphere) by means of suitable leak rate detector

11 MANUFACTURER'S INSTRUCTIONAL MANUAL

11.1 The manufacturer's instructional material shall include, in addition to the information required for ordinary locations, the information shown in Subsections 11.2 through 11.4 to emphasize the precautions required when operating the equipment in a Class I, Division 2 location.

11.2 The following or equivalent specification for the location of the equipment shall be included:

This equipment is suitable for use in Class I, Division 2 Groups (as applicable) or nonhazardous locations only.

11.3 The following or equivalent information for use of the equipment shall be included:

11.3.1 A list of the equipment or combination of equipment which may be connected to each nonincendive field circuit shall be provided.

Alternatively, the parameters listed under Subsections 7.3.2 and 7.3.3 may be provided.

11.3.2 Sufficient information to allow the user to correct the temperature code rating for ambient temperatures exceeding 40°C (104°F) up to maximum rated temperature if a linear correction is not considered by the manufacturer to be suitable.

11.4 The following warning or equivalent warnings for repair of the equipment shall be included:

11.4.1 If replacement of a component could ignite the flammable atmosphere: **WARNING - EXPLOSION HAZARD DO NOT REMOVE OR REPLACE LAMPS OR FUSES UNLESS POWER HAS BEEN DISCONNECTED OR THE AREA IS KNOWN TO BE NONHAZARDOUS.**

11.4.2 If disconnecting the equipment supply could ignite the flammable atmosphere: **WARNING - EXPLOSION HAZARD DO NOT DISCONNECT EQUIPMENT UNLESS POWER HAS BEEN DISCONNECTED OR THE AREA IS KNOWN TO BE NONHAZARDOUS.**

11.4.3 If components are relied upon to make the equipment suitable for Class I, Division 2 locations, then these shall be individually identified. The following is an example of identifying such components:

WARNING - SUBSTITUTION OF THE FOLLOWING COMPONENTS MAY IMPAIR SUITABILITY FOR DIVISION 2:

Reference Designation	Description	Type of Protection
K3	Relay	Sealed Contacts
OL1	Thermal Switch	Hermetically Sealed Contacts
M1	Fan	Brushless Motor
S6	Switch	Nonincendive Component

---

*NOTE: An asterisk following the subsection number signifies that explanatory material on that paragraph appears in Appendix A.

APPENDIX A

This appendix is not part of this standard, but it is included for informational purposes only.

A1.0 (Refers to Section 1.1) — **General Information**

Class I, Division 2 — A Class I, Division 2 location is a location: (1) in which volatile flammable liquids or flammable gases are handled, processed, or used, but in which the liquids, vapors, or gases will normally be confined within closed containers or closed systems from which they can escape only in case of accidental rupture or breakdown of such containers or systems, or in case of abnormal operation of equipment; or (2) in which ignitible concentrations of gases or vapors are normally prevented by positive mechanical ventilation, and which might become hazardous through failure or abnormal operation of the ventilating equipment; or (3) that is adjacent to a Class I, Division 1 location, and to which ignitible concentrations of gases or vapors might occasionally be communicated unless such communication is prevented by adequate positive-pressure ventilation from a source of clean air, and effective safeguards against ventilation failure are provided.

This classification usually includes locations where volatile flammable liquids or flammable gases or vapors are used, but which in the judgment of the authority having jurisdiction would become hazardous only in case of an accident or of some unusual operating condition. The quantity of flammable material that might escape in case of accident, the adequacy of ventilating equipment, the total area involved, and the record of the industry or business with respect to explosions or fires are all factors that merit consideration in determining the classification and extent of each location.

Piping without valves, checks, meters, and similar devices would not ordinarily introduce a hazardous condition even though used for flammable liquids or gases. Locations used for the storage of flammable liquids or of liquefied or compressed gases in sealed containers would not normally be considered hazardous unless subject to other hazardous conditions also.

Electrical conduits and their associated enclosures separated from process fluids by a single seal or barrier shall be classed as a Division 2 location if the outside of the conduit and enclosures is a nonhazardous location.

This standard was prepared to provide more detailed requirements for electrical equipment suitable for use in Class I, Division 2 hazardous locations than those identified in the National Codes.

It reinforces the practice of many years in North America of supplying general purpose electrical equipment for use in Class I, Division 2 locations which are of normal industrial quality and in which sources of electrical or thermal ignition do not exist under normal operational conditions.

The principles involved are based on the low probability of the presence of the explosive gas-air mixture occurring for a substantial period of time in an area classified as Division 2, coincident with an abnormal condition in the electrical equipment capable of igniting the gas mixture.

Reference information was obtained from CSA Standard C22.2-No. 157-M1979, "Intrinsically Safe and Non-incendive Equipment for Use in Hazardous Locations," and the proposals of IEC Technical Committee 31, Working Group 6, concerning "Apparatus with Type of Protection 'n.'" The latter document is intended to define equipment for use in IEC Zone 2 classified areas which are analogous to Division 2 locations in North America.

A1.1 (Refers to Sections as marked) — **Information Related to Specific Paragraphs in This Standard**

A1.2 (Refers to Section 2.1) — Marking exceptions may exist in National Codes, e.g., NFPA 70, Section 500-2(b).

A1.3 (Refers to Section 2.8) — The experimental data on which the requirements of this document are based were determined under normal laboratory atmospheric conditions. Ignition parameters are not easy to extrapolate from normal laboratory conditions to other conditions (such as might exist in process vessels) without careful engineering consideration. Increasing the initial temperature of a flammable or combustible mixture will decrease the amount of energy required to cause ignition so that, at the autoignition temperature of a gas or vapor, the electrical energy required for ignition will be zero. The nature of the energy variation between these limits is not well documented. Temperature variations can also change the concentrations of flammable materials in the mixture.

Oxygen enrichment decreases the energy necessary for ignition. The minimum ignition energy of mixtures of flammable materials with oxygen may be 1/100 of that required for the same material mixed with air.

As a general rule, the minimum ignition energy is inversely proportional to pressure squared. When examining a situation where the gas mixture is not at atmospheric pressure, one must consider whether a flammable mixture exists under higher pressure conditions. When the mixture is at high pressure, many flammable materials will condense.

A1.4 (Refers to Section 3.4) — The concept of a nonincendive circuit for application to equipment used in Division 2 locations was first identified in ISA Recommended Practice RP12.2.[1] This recommended practice

[1]ISA-RP12.2-1965 (Withdrawn), "Intrinsically Safe and Non-Incendive Electrical Instruments," Instrument Society of America, Research Triangle Park, NC.

covered certain aspects of equipment for use in Division 2 locations. ISA-RP12.2 was withdrawn following the availability of NFPA 493,[1] which concerns itself only with intrinsically safe equipment suitable for use in Division 1 and Division 2 locations. However, NEC Section 501-3 (b)(1)c recognizes nonincendive circuits. The ISA-S12.12 standard defines such circuits and provides a means for testing.

A1.5 (Refers to Section 3.5) — Nonincendive field circuits are recognized in the NEC, Section 501-4 (b) Exception.

A1.6 (Refers to Section 4.1.1) — It is recognized that other means of protection are acceptable. Air purging and positive pressurization are described in NFPA 496,[2] and oil immersion requirements are covered in ANSI C33.30[3] and Section 501-3 (b)(1)a and 501-6(b)(1) and (2) of the National Electrical Code and Section 18-152 of the Canadian Electrical Code (Part 1) C22.1.

A1.7 (Refers to Section 4.4) — It is unlikely some malfunction will occur causing a fuse to open concurrent with the location becoming flammable.

For "signaling," "alarm," "remote-control," and "communications systems," Section 501-14(b)(3) of the 1981 edition of the NEC permits fuses in a general-purpose enclosure without stated limits.

A1.8 (Refers to Section 5.2) — The separating force of 15 N is consistent with the IEC WG6 Type "n" draft standard.

A1.9 (Refers to Section 6.1) — Tests should be performed under worst-case heating conditions — for example, the highest or lowest extreme of specified supply voltage plus the worst-case load conditions.

A1.10 (Refers to Section 6.2) — Caution needs to be exercised in interpretation of the temperature identification code. This is based upon operation at a 40°C ambient temperature, whereas the maximum specified ambient temperature rating of the equipment may exceed 40°C.

The lowest ignition temperature of the explosive atmospheres concerned shall be above the maximum surface temperature. However, for components having a total surface area of not more than 10 cm^2 (e.g., as for transistors or resistors used in low power circuits protected by the energy limitation technique), their surface temperature may exceed that for the temperature class marked on the electrical apparatus if there is neither a direct nor an indirect risk of ignition from these components, with a safety margin of:

50 K for T1, T2, and T3 or;
25 K for T4, T5, and T6
(K = degrees Kelvin)

This safety margin shall be ensured by experience of similar components or by test of the electrical apparatus itself in representative explosive mixtures.

NOTE: During the test, the safety margin may be provided by increasing the ambient temperature.

If the temperature of a component is below the autoignition temperature (AIT) of a material, the component can be considered suitable for use in a circuit with regard to possible thermal ignition. When the temperature of a component is above the AIT, it must be determined if it will or will not thermally ignite the material. The ability of any small component to thermally ignite a material is dependent on the AIT of the material it is exposed to and the temperature, size, and shape of the heated component. A test which has been used to verify the suitability of a small component from a thermal standpoint may be conducted using a 5 percent diethyl ether in air mixture. Diethyl ether represents the material having the lowest AIT of the Class I atmospheres currently listed in Article 500 of the NEC. Various characteristics of diethyl ether are as follows:

NAME	FORMULA	LEL	UEL	AIT	SP. GR.	VAPOR DENSITY
Diethyl ether*	$C_2H_5OC_2H_5$	1.7	48	160°C (320°F)	0.715	2.55

*Synonyms: Ether, Ethyl Ether, Diethyl Oxide, Ethyl Oxide

The test apparatus consists of a test enclosure, with a gas-tight cover and viewing port. The enclosure contains a low velocity fan, diethyl ether dish, through-wall component terminal connections, and an igniter for mixture-ignition verification. The component under test is suspended within the enclosure with connection made, via the terminal connections, to the external associated component's circuitry or power source. An appropriate amount of liquid ether is placed in the dish. If a 3 L enclosure is used, 0.65 cm^3 of liquid ether will be needed for a 5 percent mixture. The cover is closed, and the fan is turned on to aid in the evaporation process and to maintain a homogeneous mixture. Power is then applied to the component until thermal equilibrium is reached, the component fails open, or the surrounding material is ignited. When the component fails to ignite the mixture, the sensitivity of the mixture shall be verified by activating the igniter circuit. A minimum of six tests shall be made.

A1.11 (Refers to Section 7) — See Appendix Clause A1.2.

A1.12 (Refers to Sections 7.1.4 through 7.1.6) — It is recognized that it may be desirable for equipment manu-

[1]NFPA Standard 493-78, "Intrinsically Safe Apparatus for Use in Division 1 Hazardous Locations," National Fire Protection Agency, Quincy, MA, 1978.
[2]See footnote 3, page 7.
[3]ANSI C33.30 was re-named ANSI 698-73, "Industrial Control Equipment for Use in Hazardous Locations, Class I, Groups A, B, C, D, and Class II, Groups E, F, and G," November 24, 1982, Underwriters Laboratories, Northbrook, IL.

facturers to (in addition to using the marking required by this standard) mark the equipment in accordance with the IEC marking practice for Zone 2 equipment.

If this marking practice is used, the following should be shown:

(a) The symbol "Ex n," followed by

(b) The group symbol "IIA," "IIB," or "IIC," whichever represents the applicable gas group, followed by

(c) Temperature class or maximum surface temperature according to Subsection 7.1.5 of this standard, followed by

(d) The symbol "/X," if there are any special conditions of use relevant to the safety of the equipment which are covered in the instructions.

Example: Ex n IIA T4/X

It is recognized that some equipment may be suitable for more than one class or division, and it may not be practical to provide all the necessary marking on the equipment. The complexity of such marking may require the use of a reference document. In these instances, reference should be made to a system or equipment control drawing, e.g., for hazardous location suitability reference Dwg. No. XXX.

A1.13 (Refers to Section 7.3) — The exception to Section 501-4(b) of NFPA 70[1] states: "Wiring, which under normal conditions cannot release sufficient energy to ignite a specific ignitible atmospheric mixture by opening, shorting or grounding, shall be permitted using any of the methods suitable for wiring in ordinary locations." This exception is intended to permit what has been termed "nonincendive field circuits."

A problem which faces both manufacturers and users in applying the nonincendive field wiring concept is the ability to interconnect different manufacturers' equipment with nonincendive field circuit connections and be assured the combination will still provide a nonincendive field circuit. In order to facilitate this interconnection, the marking method covered in this section provides a convenient way to assess the compatibility of different manufacturers' equipment with respect to nonincendive field circuits. The criteria for the comparison are that the voltage (V_{max}) and current (I_{max}) which the load device can receive must be equal to or greater than the normal circuit voltage (V_n) and normal circuit current (I_n) which can be delivered by the source device. In addition, the maximum capacitance (C_i) and inductance (L_i) of the load which is not prevented by circuit components from providing a stored energy charge to the field wiring (e.g., diode across a winding to clamp an inductive discharge) must be equal to or less than the capacitance

(C_n) or inductance (L_n) which can be driven by the source device. The capacitance and inductance of the interconnecting wiring must be equal to or less than the capacitance (C_a) or the inductance (L_a) which can be driven by the source.

When cable parameters are known, the system designer adds the parameters for lengths of cable to be used to the known parameters of the load device and compares the sum to the allowable parameter values for the source device. If the electrical parameters of the cable are unknown, the following values may be used:

Capacitance: 200 pF/m (60 pF/ft)
Inductance: 0.66 μH/m (0.20 μH/ft)

A1.14 (Refers to Section 7.3.2, Items (3) and (4)) — After determining the maximum open-circuit voltage (V_{oc}) and maximum short-circuit current (I_{sc}) of the source device according to Subsection 8.1.2, determine the appropriate values of cable capacitance (C_a) and cable inductance (L_a) from Figures 3 through 8.

A1.15 (Refers to Section 7.3.2, Items (5) and (6)) — After determining the normal circuit voltage (V_n) and normal circuit current (I_n) of the source device according to Subsection 8.1.2, determine the appropriate values of capacitance (C_n) and inductance (L_n) from Figures 3 through 8.

Marking Example:

$$V_{oc} = 40 \text{ V}$$
$$I_{sc} = 64 \text{ mA}$$
$$C_a = 0.08 \text{ μf}$$
$$L_a = 11 \text{ mH}$$
$$C_n = 0.7 \text{ μf at } V_n = 24 \text{ V dc}$$
$$L_n = 190 \text{ mH at } I_n = 20 \text{ mA}$$

A1.16 (Refers to Section 7.3.3)
(1) Determine V_{max} from the appropriate curve based on the maximum unprotected capacitance at the field wiring terminals of the device.

(2) Determine I_{max} from the appropriate curve based on the maximum unprotected inductance at the field wiring terminals.

(3) Determine analytically or experimentally, for the values of V_{max} and I_{max} determined in 1 and 2, that the device will not produce an ignition-capable temperature rise of components for that combination. Verify also that some lower current will not cause ignition-capable hot spots. This step can be time-consuming. Alternatively a manufacturer may arbitrarily specify a voltage lower than V_{max} as determined in 1.

A1.17 (Refers to Section 8) — Considerations of nonincendive circuits in the IEC TC31 WG 6 proposal require special concern for nonobvious faults of certain

[1]See footnote 1, page 7 for reference citation.

TABLE A1[1]
(Refers to Table 2)
CURRENT IN CALIBRATION CIRCUIT FOR CADMIUM DISK

Group	Inductive Circuit		Resistive Circuit	
	Must Not Ignite	Must Ignite	Must Not Ignite	Must Ignite
D	71　mA	100 mA	735 mA	1000 mA
C	49　mA	65 mA	540 mA	700 mA
A & B	25.5 mA	30 mA	209 mA	300 mA

TABLE A2[2]
(Refers to Table 3)
CURRENT IN CALIBRATION CIRCUIT FOR TIN DISK

Group	Inductive Circuit		Resistive Circuit	
	Must Not Ignite	Must Ignite	Must Not Ignite	Must Ignite
D	98 mA	110 mA	2150 mA	2500 mA
C	77 mA	90 mA	1590 mA	2000 mA
A & B	41 mA	50 mA	1305 mA	1500 mA

safety components and for redundancy of components. This standard does not include such considerations, as it is felt that the components used for this purpose are not stressed in normal operation.

A1.18 (Refers to Section 8.1) — Figures 1 through 8 represent circuits of a very simple nature. Unless the circuit can clearly be identified as conforming to the special conditions from which the curves are derived, analysis may prove invalid. Testing may be required in such cases. As an illustration, the capacitance discharge curves do not include the effect of the charging circuit.

A1.19 (Refers to Section 8.1.2) — It is recognized that the real margin of safety lies in the use of a test apparatus more sensitive than any probable ignition condition and the use of an ideal gas mixture.

A1.20 (Refers to Section 8.2) — The test apparatus using a 0.2 mm-diameter of fine tungsten wires yields ignition currents which are low because of the heating of the tungsten wires when the test currents approach 3 A. For testing higher current, wires of a different material (such as copper) or a different type of apparatus may be needed. The apparatus is suitable for testing circuits up to 300 V. For tests of capacitor circuits, modified apparatus (such as removing one or more of the tungsten wires) may be needed to allow adequate charging time.

A1.21 (Refers to Section 8.4) — Tables A1 and A2 tabulate the various currents at which ignition must and must not occur in order to verify that the spark test apparatus is working properly.

A1.22 (Refers to Section 9.1.1) — The maximum rat-

ings from the IEC TC31 WG 6 proposal are 250 V and 15 A. Their derivation may be arbitrary, but it is felt that these are reasonable limits for the equipment covered by this standard.

A1.23 (Refers to Section 9.2) — These requirements were obtained from the CSA and IEC reference documents.

A1.24 (Refers to Section 9.2.2) — A test chamber may be a plastic bag. The test factor of 150 percent of rated load is greater than the applied load in the IEC document. It is identical to the CSA proposal.

A1.25 (Refers to Section 9.2.3) — The CSA standard specifically requires that a nonincendive component be evacuated prior to filling it with the mixture. This may be only one of several ways to ensure that the component is filled with the test mixture. Therefore, this method is not specifically required in the standard.

A1.26 (Refers to Section 9.2.4) — Depending upon the method used to fill the nonincendive component, more frequent renewing of the gas or vapor-in-air mixture may be required.

A1.27 (Refers to Section 10) — The principle applied to "sealed devices" is not one which absolutely prevents entry of the external atmosphere, but restricts it to a degree commensurate with the probabilities which relate to the presence of the explosive gas-air mixture in the area in a Division 2 location. Sealed devices are covered in both CSA Standard C22.2 No. 157-M1979 and IEC proposal TC/31 WG 6.

A1.28 (Refers to Section 10.2) — The 100 cm³ limitation is consistent with the IEC and CSA referenced documents. (See General Information, Appendix Item A1.0.)

[1,2]Tables A1 and A2 provide additional information related to Tables 2 and 3, Section 8.4.

INSTRUMENT SOCIETY of AMERICA
Research Triangle Park, North Carolina

ANSI/ISA–S12.13, Part I–1986
Approved December 31, 1986

American National Standard

Performance Requirements, Combustible Gas Detectors

Instrument Society of America

ISBN 1-55617-015-7

ISA–S12.13, Part I, Performance Requirements, Combustible Gas Detectors

INSTRUMENT SOCIETY OF AMERICA
67 Alexander Drive
P.O. Box 12277
Research Triangle Park, North Carolina 27709

PREFACE

This preface is included for informational purposes and is not part of ISA-S12.13, Part I.

This standard has been prepared as part of the service of the Instrument Society of America (ISA) toward a goal of uniformity in the field of instrumentation. To be of real value, this document should not be static, but should be subject to periodic review. Toward this end, the Society welcomes all comments and criticisms, and asks that they be addressed to the Secretary, Standards and Practices Board, Instrument Society of America, 67 Alexander Drive, P.O. Box 12277, Research Triangle Park, NC 27709, Telephone (919) 549-8411.

The ISA Standards and Practices Department is aware of the growing need for attention to the metric system of units in general, and the International System of Units (SI) in particular, in the preparation of instrumentation standards. The Department is further aware of the benefits to U.S.A. users of ISA standards of incorporating suitable references to the SI (and the metric system) in their business and professional dealings with other countries. Toward this end, this Department will endeavor to introduce SI-acceptable metric units in all new and revised standards to the greatest extent possible. The *Metric Practice Guide,* which has been published by the Institute of Electrical and Electronics Engineers as ANSI/IEEE Std. 268-1982, and future revisions will be the reference guide for definitions, symbols, abbreviations, and conversion factors.

It is the policy of the Instrument Society of America to encourage and welcome the participation of all concerned individuals and interests in the development of ISA standards. Participation in the ISA standards-making process by an individual in no way constitutes endorsement by the employer of that individual, of the Instrument Society of America, or of any of the standards that ISA develops.

The information contained in the preface, footnotes, and appendices is included for information only and is not part of the standard.

The following people served as members of ISA Committee SP12.13, Part I, which prepared this standard:

NAME	COMPANY
D. N. Bishop, Chairman	Chevron U.S.A. Inc.
L. Aynardi	Monsanto Company
A. A. Bartkus	Underwriters Laboratories Inc.
A. F. Cohen	U.S. Bureau of Mines
H. G. Conner	E. I. du Pont de Nemours & Company
P. F. Crowley	Factory Mutual Research Corporation
J. A. Davenport	Industrial Risk Insurers
S. D. Delaune	MSA International
J. R. Dolphin	TOTCO*
G. L. Frago/T. Dickey	U.S. Coast Guard
G. Lobay	Canadian Explosive Atmospheres Laboratories
H. H. Marshall, Secretary	Mine Safety Appliances Company**
R. P. Merritt/R. P. Street	General Monitors, Inc.
J. Miante/P. L. Randall	U.S. Coast Guard
D. M. Nelson	Bacharach Instrument Company
E. M. Nesvig	ERDCO Engineering Corporation
J. R. Steegstra/B. J. Northam	Factory Mutual Research Corporation
D. B. Wechsler	Union Carbide Corporation

*Formerly supported by Underwriters Laboratories, Inc.
**Employer during the standard's development period.

The following people served as members of ISA Committee SP12:

NAME	COMPANY
E. M. Nesvig, Chairman	ERDCO Engineering Corporation
A. B. Anselmo	R. Stahl Inc.
A. A. Bartkus	Underwriters Laboratories Inc.
D. N. Bishop	Chevron U.S.A. Inc.
J. A. Bossert	Energy Mines and Resources Canada
R. Buschart	Monsanto Company
K. M. Collins/W. W. Shao*	Canadian Standards Association
H. G. Conner	E. I. du Pont de Nemours & Company
J. D. Cospolich	Waldemar S. Nelson & Company, Inc.
J. R. Dolphin/J. Farrell*	TOTCO
U. Dugar	Polysar Ltd.
J. C. Garrigus	Bristol Babcock Inc.
F. Kent	Fischer & Porter Company
E. C. Magison	Honeywell Inc.
F. L. Maltby/E. Cranch*	Drexelbrook Engineering
H. H. Marshall	Mine Safety Appliances Company
R. C. Masek	Bailey Controls
F. J. McGowan	The Foxboro Company
E. W. Olson	3M Company
D. L. Plahn	Fisher Controls International, Inc.
P. L. Randall	U.S. Coast Guard
J. Rennie, Secretary	Factory Mutual Research Corporation
S. A. Stilwell	Greytop Technics A/S
M. Stroup	U.S. Air Force
R. L. Swift	Consultant
R. K. Weinzler/N. Abbatiello*	Eastman Kodak Company
Z. Zborovszky	U.S. Bureau of Mines

*One vote

This standard was approved for publication by the ISA Standards and Practices Board in October 1986.

NAME	COMPANY
N. Conger, Chairman	Fisher Controls International, Inc.
D. N. Bishop	Chevron U.S.A. Inc.
W. Calder III	The Foxboro Company
R. S. Crowder	Ship Star Associates
C. R. Gross	Eagle Technology
H. S. Hopkins	Utility Products of Arizona
J. L. Howard	Boeing Aerospace Company
R. T. Jones	Philadelphia Electric Company
R. Keller	The Boeing Company
O. P. Lovett, Jr.	ISIS Corporation
E. C. Magison	Honeywell, Inc.
A. P. McCauley	Chagrin Valley Controls, Inc.
J. W. Mock	The Bechtel Group, Inc.
E. M. Nesvig	ERDCO Engineering Corporation
R. Prescott	Moore Products Company
D. E. Rapley	Rapley Engineering Services
C. W. Reimann	National Bureau of Standards
R. H. Reimer	Allen Bradley Company
J. Rennie	Factory Mutual Research Corporation
W. C. Weidman	Gilbert/Commonwealth, Inc.
K. Whitman	Fairleigh Dickinson University
*P. Bliss	Consultant
*B. A. Christensen	Continental Oil Company
*L. N. Combs	Consultant
*R. L. Galley	Consultant
*T. J. Harrison	IBM Corporation
*R. G. Marvin	Roy G. Marvin Company
*W. B. Miller	Moore Products Company
*G. Platt	Consultant
*J. R. Williams	Stearns Catalytic Corporation

*Director Emeritus

FOREWORD

This standard is based primarily on the initiative of the Canadian Standards Association, which resulted in their Standard C22.2 No. 152-1976.* The preface to that document was written by James E. Duncan of the Canadian Standards Association and is reprinted below for clarification of the concepts involved.**

This standard was initiated through industry effort in order to uphold and improve the level of electrical safety and safety-oriented performance of combustible gas detection instruments.

Throughout the development of this Standard the basic Subcommittee was augmented at several of its meetings by a substantial number of expert representatives from the oil and gas producing and transporting industries, the oil refining industry, the petro-chemical industry, the mining industry and other similar interests, representing ''users'' of this type of equipment. This included people responsible for the selection, application, installation and maintenance of combustible gas detection instruments, as well as people responsible for enforcement of safety regulations in the industrial situation. The requirements in this Standard have been designed to answer many of the expressed needs of such ''users'' based on their collective experience in the application and use of such instruments in actual service in the Canadian industrial environment.

This Standard establishes a minimum level of safety-oriented performance capability by requiring that:
    a. All such equipment pass certain tests, devised to measure the instrument's capability to properly respond to a variety of exposures simulating conditions and occurrences which may reasonably be expected to occur in actual service;
    b. All such equipment be provided with certain minimum features of construction deemed necessary to permit proper safe operation, ease of necessary adjustments and reliability;
    c. Each unit be provided with a comprehensive set of installation and operating instructions to permit the user to properly install, operate and adjust the equipment, and to properly acquaint him with the capabilities and limitations of the equipment.

Moreover, this Standard provides that, if claims are made for performance levels or construction features in excess of the minimum levels in the Standard's basic requirements, the testing procedures and requirements be modified to verify all such superior claims. This is intended to encourage further advancements in the state of the art, and to give free rein to natural competitive development in this field, while still ensuring the user that any instrument he obtains is capable of performing at least to this minimum level.

Instrument capability alone cannot ensure that the use of these instruments will properly safeguard areas or locations where combustible gases or vapours may be or are present. The level of safety obtained depends heavily not only upon the user having a full knowledge of the limitations of the instrument, but also upon proper selection for the particular situation, proper installation and especially upon frequent periodic maintenance and adjustment.

To properly safeguard any particular area, the user must also have a basic knowledge of gas/vapour properties and phenomena (how gases and vapours propagate, whether the gases are heavier or lighter than air, etc.), and the conditions which may prevail in the areas being protected (direction and velocity of gas movement, humidity and temperature variation, presence of particulates or detrimental contaminants, ease of gaining access for periodic maintenance and adjustment, etc.), in order to select an instrument suitable for these conditions and locate it where it will be effective for its intended function.

---

   *This standard was revised in 1984. See Appendix 1 for reference information.
   **Reprinted by permission of the Canadian Standards Association.

Anyone using this Standard as an aid in selecting or applying combustible gas detection instruments should review the conditions described for the various tests (both in the body of the Standard and in the Appendix A*) and compare them to the actual conditions which may prevail in the locations to be protected. If actual conditions are likely to be more severe than the minimum test conditions in any respect, the user should seek an instrument having superior properties.

It is stressed that the final and long-term effectiveness of any combustible gas detection equipment depends heavily upon the user himself, who must be responsible for its proper application, installation, proper use and regular maintenance.

The reader's attention is particularly directed to the fact that, in developing the test requirements and acceptance criteria contained in this Standard, the Subcommittee which took part in its preparation had in mind the specific test apparatus and test technique then in use or being developed by one specific testing agency, as described in Appendix A* of this Standard.

It is most important, therefore, in relating the test requirements of this Standard to conditions likely to arise in actual service, or in attempting to establish correlations with tests performed by different testing agencies, to realize and appreciate that the significance of the test results obtained is subject to interpretation based on the actual test apparatus and technique used. To ensure that the original intent of this Standard is met, therefore, the test technique and apparatus used must be as nearly as possible identical to that described in Appendix A.*

---

*The following ISA Standard does not include the appendix from the Canadian document; therefore, references to Appendix A are not applicable.

TABLE OF CONTENTS
Title

1 PURPOSE

1.1 Part I of this standard provides minimum requirements for the performance of combustible gas detection instruments and therefore enhances the safety of operations by employing these instruments.

1.2 Part II of this standard* establishes user criteria for the installation, operation, and maintenance of combustible gas detection instruments.

2 SCOPE

2.1 This standard covers the details of construction, performance, and testing of portable, mobile, and stationary electrical instruments used for sensing the presence of combustible gas or vapor concentrations in ambient air; parts of such instruments may be installed or operated in Class I hazardous locations and gaseous mines in accordance with Codes specified by authorities having jurisdiction. (See Appendix 1.)

2.2 This standard applies to line-voltage-operated instruments rated at 600 V nominal or less, and to portable, mobile, or stationary-type instruments utilizing a battery of a nonrechargeable (primary)-type or a rechargeable (secondary)-type.

2.3 This standard covers combustible gas detection instruments intended to provide an indication and/or alarm, the purpose of which is to give a warning of potential hazard.

2.4 This standard does *not* cover gas detection instruments of the laboratory- or scientific-type used for analysis or measurement, instruments used for process control and process monitoring purposes, or instruments used for residential purposes.

2.5 This standard is written for gas detection instruments which are intended to measure gas concentrations in air in the range from zero up to the lower flammable or explosive limit.

3 DEFINITIONS

Alarm Set Point is the selected gas concentration level(s) at which an indication, alarm, or other output function is initiated.

Calibration is the capability to adjust the instrument to "zero" and to set the desired "span."

---

*Part II is a separate document entitled "Installation, Operation, and Maintenance of Combustible Gas Detection Instruments." It is still in draft form and has not yet been formally approved by ISA.

Clean Air is air that is free of combustible gases and contaminating substances.

Combustible Gas is any flammable or combustible gas or vapor that can, in sufficient concentration by volume in air, become the fuel for an explosion or fire. Materials which cannot produce sufficient gas or vapor to form a flammable mixture at ambient or operating temperatures and mists formed by the mechanical atomization of combustible liquids are **NOT** considered to be combustible gases.

NOTE: For convenience, the shorter term "gas" may be used as an abbreviation within this standard.

Control Unit means that portion of a multipart gas detection instrument which is not directly responsive to the combustible gas but which responds to the electrical signal obtained from one or more detector heads to produce an indication, alarm, or other output function if gas is present at the detector head location.

Detector Head means that gas-responsive portion of a multipart gas detection instrument that is located in the area where sensing the presence of combustible gases is desired. Its location may be integral with or remote from its control unit.

NOTE: The detector head may incorporate additional circuitry such as signal-processing or -amplifying components or circuits in addition to the gas-sensing element in the same housing.

Diffusion is a method by which the atmosphere being monitored gains access to the gas-sensing element by natural molecular movement.

Flammable Range means the range of flammable vapor concentrations or gas-air mixtures in which propagation of flame will occur on contact with a source of ignition.

For purposes of this standard, the terms "lower flammable limit (LFL)" and "lower explosive limit (LEL)" are deemed to be synonymous, and likewise the terms "upper flammable limit (UFL)" and "upper explosive limit (UEL)" are deemed to be synonymous. For ease of reference, the two abbreviations "LFL" and "UFL" may be used hereafter to denote these two sets of terms. It should be recognized that particular authorities having jurisdiction may have overriding requirements that dictate the use of one of these sets of terms and not the other.

Full-Scale Gas Concentration means 100 percent of the actual marked full-scale concentration value.

NOTE: For purposes of this standard, the actual gas concentration corresponding to the lower and upper flammable limits shall be the values shown in the latest edition of nationally recognized documents (see Appendix 1).

Gas Detection Instrument means an assembly of electrical and mechanical components (either a single integrated unit, or a system comprising two or more physically separate but interconnected component parts) which senses the presence of such combustible gas and provides an indication, alarm, or other output function.

NOTE: For convenience, the shorter term "instrument" may be used as an abbreviation within this standard.

Gas-Sensing Element (Sensor) is the primary element in the gas detection system which responds to the presence of a combustible gas—including any reference or compensating unit, where applicable.

Mobile refers to a continuous-monitoring instrument mounted on a vehicle such as, but not limited to, a mining machine or industrial truck.

Portable, Continuous-Duty means a battery-operated portable or transportable instrument of a type intended to operate continuously for 8 hours (h) or more.

Portable, Continuous-Duty—Personal means a battery-operated, alarm-only portable instrument intended to be operator-worn, and to operate continuously for 8 h or more.

Portable, Intermittent-Duty means a battery-operated portable or transportable instrument of a type intended for operation for periods of only a few minutes at irregular intervals.

Sample Draw is a method to cause flow of the atmosphere being monitored to be directed to a gas-sensing element.

Stationary refers to a gas detection instrument intended for permanent installation in a fixed location.

4 GENERAL REQUIREMENTS

4.1 Gas detection instruments shall meet the applicable electrical and electronic measuring instrument requirements as they apply to personnel or property protection. (See Appendix 1.)

4.2 Any portion of a gas detection instrument which is intended for installation or use in a location where gas or vapor concentration may be present shall be suitable for use in Class I, Division 1 hazardous locations or gaseous mines and in accordance with the group classification of the gas, and shall be marked accordingly.

4.3 All gas detection instruments shall meet the minimum construction and test requirements contained in this standard. If the manufacturer makes performance claims that exceed these requirements, all such claims shall be verified to the satisfaction of parties having jurisdiction.

5 CONSTRUCTION

5.1 General

5.1.1 Gas detection instruments, their components, or remote detector heads specifically intended for use in the presence of corrosive vapors or gases, or which may produce corrosive by-products as a result of the catalytic oxidation or other chemical process, shall be constructed of materials resistant to corrosion or of materials suitably protected against corrosion. For additional information, consult Appendix 1.

5.1.2 Portable instruments of the sample-draw-type shall include the necessary sample-pumping mechanism.

5.2 Meters and Indicators

5.2.1 Stationary and continuous-duty portable gas detection instruments having an integral meter to indicate gas concentrations shall employ a meter having sufficient resolution to permit measurement with the precision required for performing the tests of Section 7.

5.2.2 Operational characteristics of nonlinear meters or indicators, when used, shall be stated in the Instruction Manual.

5.2.3 Continuous-duty portable instruments of the sample-draw-type shall incorporate a device to indicate adequate flow, except that such an indicating device may be omitted provided that the Instruction Manual contains detailed instructions as required by Clause 6.3.1(i).

5.2.4 For portable gas detection instruments which employ digital display, a means shall be provided to alert the user that a gas concentration in excess of the measuring range of the instrument has been detected.

5.3 Alarm or Output Function

5.3.1 Alarm devices, output contacts, or signal outputs (if provided as part of stationary instruments or continuous-duty portable instruments and intended to indicate a potentially hazardous gas concentration) shall be of a latching-type requiring a deliberate manual action to reset. If two or more set or alarm positions are provided, the lower may be nonlatching—based on user preference.

5.3.2 Alarm devices or signals provided as part of intermittent-duty or continuous-duty personal-type portable instruments, if of the nonadjustable alarm set point-type, shall be set to operate at a gas concentration not higher than 60 percent of the lower flammable limit, or if of the adjustable alarm set point-type, the means for adjustment shall not be capable of being set higher than 60 percent of the lower flammable limit.

5.3.3 The alarm set point adjustment(s) shall be inside the enclosure(s) and shall require a tool for adjustment.

5.4 Trouble Signals

5.4.1 A stationary or mobile gas detection instrument shall provide for a signal transfer or contact transfer to produce a trouble signal if any of the following conditions occur: instrument power failure; loss of continuity in any one or more wires to any remote detector head; loss of continuity of any gas-sensing element; or downscale indication (below zero) equivalent to 10 percent nominal LFL or more. Such signal or contact transfer shall be independent of any other alarm or shutdown signal or contact transfer.

5.4.2 Stationary and mobile sample-draw-type gas detection instruments shall be provided with flow-proving devices (either integral or nonintegral) which shall produce a trouble signal in the form of a contact transfer or signal transfer if a loss of flow occurs.

5.4.3 Continuous-duty, portable gas detection instruments shall be provided with an audible or visible indication of low battery condition, and the nature and purpose of either shall be clearly explained in the Instruction Manual.

5.5 Controls and Adjustments

5.5.1 All portable gas detection instruments shall be provided with means for facilitating calibration checks and adjustments as required.

5.5.2 Calibration and alarm(s) setting shall require a key or tool for any adjustment.

5.5.3 Fixed instruments housed in explosion proof or purged (see Appendix 1) enclosures shall have all controls for normal operation accessible from outside the enclosures. The controls for routine calibration may be inside the enclosure, provided all of the following requirements are met:

a. The enclosure is of the type with a cover which can be readily opened or closed and which does not require removal and replacement of bolts or other securing devices in order to open and reclose the enclosure.

b. All adjustments, switches, or controls which may be deliberately or accidentally operated during the calibration procedure shall involve only circuits meeting requirements for nonincendive circuits.

c. All circuits exceeding 30 V RMS or 42.4 V peak and all circuits 30 V RMS or less which are not Class 2 power-limited (as defined by Article 725 of the National Electrical Code) shall be protected from accidental contact through appropriate mechanical guards or barriers or be of an intrinsically safe design.

d. The enclosure shall be marked as described in detail in Section 6.2.4.

5.6 Batteries

5.6.1 Continuous-duty portable gas detection instruments with fresh or fully charged batteries shall be capable of continuous nonalarming operation for a period of at least 8 h, without replacement or recharge of batteries.

5.6.2 Intermittent-duty portable gas detection instruments with fresh or fully charged batteries shall be capable of nonalarming operation at a duty cycle of 10 minutes (min) "On" and 10 min "Off" for a period of 8 h (total cumulative "On" time of 4 h) without replacement or recharge of the batteries. If the instrument is provided with a switch which must be manually held in the "On" position for the duration of the measurement, the duty cycle may be reduced to 2 min "On" and 18 min "Off" for a period for 8 h (total accumulative "On" time of 48 min).

6 MARKING AND INSTRUMENT MANUFACTURERS' RESPONSIBILITY

6.1 General

The marking required in Clauses 6.2 and 6.3 is in addition to the marking requirement contained in Clause 4.2.

6.2 Marking on Instruments

6.2.1 The marking required by Clauses 6.2.2 and 6.2.3 shall appear in a clearly legible, visible, and permanent manner on each gas detection instrument in the following manner, as applicable:

a. For portable instruments the marking shall appear both on the outside surface of the instrument and its carrying case, if the latter obscures the marking as required by Clauses 6.2.2 and 6.2.3.

b. For stationary instruments the marking required by Clause 6.2.2 shall appear in a location where it will be visible after installation and in direct sight during the routine periodic recalibration and adjustment of set point(s).

NOTES:

1. For gas detection instruments which comprise a control unit and a remote detector head, it is sufficient that this marking appear on the control unit only, except that if routine recalibration can be accomplished entirely by adjustments at the remote detector locations alone, this marking is to appear both on the control unit and on the remote detector head.

2. For modular control units comprising one or more control modules in a common enclosure or mounting assembly,

the marking need not be repeated on each module, but may appear as a single marking on the common portion of the assembly.

3. Where the design of a stationary control unit is such that there is insufficient space for this marking to appear on the portion of the unit which is visible after installation (e.g., compact designs for close panel mounting, etc.), the marking required by Clause 6.2.2 may be permitted to appear elsewhere on the control unit, provided that a second duplicate label (of an acceptable adhesive-type) bearing such marking is supplied with each such control unit (or assembly of control units), together with the instructions that it is to be attached by the user in a conspicuous location after installation, as close as possible to the control unit.

6.2.2 All gas detection instruments shall be marked **"CAUTION—READ AND UNDERSTAND IN-STRUCTION MANUAL BEFORE OPERATING OR SERVICING."** The word **"CAUTION"** of the foregoing is to be in capital letters at least 3.0 mm high. The balance of the wording is to be in capital letters at least 2.5 mm high.

6.2.3 Portable gas detection instruments which employ analog meters having scales which indicate gas concentrations only below the flammable range shall be marked **"CAUTION—OFF-SCALE READINGS MAY INDI-CATE EXPLOSIVE CONCENTRATION."** The word **"CAUTION"** of the preceding marking shall be in capital letters at least 3.0 mm high. The balance of the wording is to be in capital letters at least 2.5 mm high.

6.2.4 Instruments of the type referred to in Clause 5.5.3 which are not intrinsically safe shall be marked **"CAUTION—THIS AREA MUST BE KNOWN TO BE NONHAZARDOUS PRIOR TO OPENING THE EN-CLOSURE"** in capital letters at least 5.0 mm high and marked in a permanent manner. The location is to be conspicuously visible prior to removal of the cover.

6.2.5 Where the design of special features of the instrument requires additional markings or a change in marking requirements, the additions or revisions are allowed, but the safety and instructional intent of Clause 6.2 must be met.

6.3 Instruction Manual

6.3.1 Each gas detection instrument shall be provided with an Instruction Manual, furnished by the manufacturer, which shall contain at least the following information:

a. Include a list of desensitizing or contaminating gases or substances known to the instrument manufacturer which may adversely affect proper operation of the instrument, including those gases which would not be accurately detected because of adverse reactions with the reference or compensating elements. Warning of the effects of oxygen-enriched or -deficient atmospheres must be included.

b. Include instructions for checking and calibration, both on a routine basis and following exposure to any of the contaminants referred to in (a) above and following exposure to concentrations causing operation of any alarm.

c. Include complete installation and initial start-up instructions.

d. Include operating adjustments and instructions (e.g., set point, zero, and balance adjustments).

e. Include details of operational limitations (e.g., ambient temperature limits for all parts of the instrument, humidity range, voltage range, maximum loop resistance, and minimum wire size for wiring between control unit and remote detector head[s], need for shielding of wiring, battery life and temperature limitations, maximum and minimum storage temperature limits, pressure limits, and sample velocity, as applicable).

f. For intermittent- and continuous-duty portables include a caution statement that electromagnetic interference (EMI) signals may cause incorrect operation.

g. Include wording to clearly indicate that suitable flow-monitoring devices must be provided at the time of installation, if applicable (see Clause 5.4.2).

h. Include wording to clearly indicate the nature and significance of all alarms, trouble signals, and any provisions which may be made for silencing or resetting of such alarms, etc., as applicable.

i. For continuous-duty portable instruments of the sample-draw-type which are not provided with an integral flow-indicating device, include detailed instructions regarding one or more suitable techniques which the user may employ which do not require special instruments or tools (to ensure that sample lines are intact and proper flow is established). (See Clause 5.2.3).

j. For stationary or continuous-duty portable instruments of the sample-draw-type, include wording to indicate the minimum and/or maximum flow rate or range of flow rates, pressure, and tubing size for proper operation.

k. For intermittent-duty portable gas detection instruments, include the word **"CAUTION,"** followed by wording such as: "Any rapid upscale reading followed by a declining or erratic reading may indicate a gas concentration beyond the upper scale limit. This may be hazardous."

l. For instruments provided with meters, include the word **"CAUTION,"** followed by wording such as: "Off-scale readings may indicate a flammable concentration."

m. Include an operational review to determine possible sources of a malfunction and the corrective procedures.

n. Include a listing of consumable or replacement components and recommendations for the type of storage of each item, with emphasis on the sensor.

o. Where more than one type of gas-sensing element is supplied by the manufacturer, include a list stating the specific gas or family of chemically similar gases for each sensor.

6.3.2 The Instruction Manual shall contain complete and accurate instructions for safe and proper operation, installation, and periodic servicing of the instrument. Instructions shall be consistent with the markings as required in Clause 6.2. Where the design or special nature of the instrument requires additional instruction and/or special information which are in contradiction to, or in addition to, the requirements of Clause 6.3.1, this consideration shall take precedence over the requirements of Clause 6.3.1.

6.3.3 The manufacturer shall include information concerning effects of externally generated electromagnetic interference (EMI) on instrument performance. Likewise, if his instrument is prone to generating EMI which could be detrimental to other nearby instrumentation, this information shall also be included.

6.3.4 The manufacturer shall specify the type of calibration gas or vapor mixture to be used. When the calibration gas is other than that which the instrument is to detect, the manufacturer shall supply the proper relative calibration or response comparisons.

7 PERFORMANCE TESTS

7.1 General

The tests required in this section are in *addition* to the requirements referred to in Section 4. The instrument tested shall be fully representative of instruments intended for commercial production. Unwarranted or false alarms shall be considered failure of the tests described below.

7.2 Sequence

The same instrument shall be subjected to all tests applicable to that type of instrument described in Clauses 7.6 through 7.18. The sequence of tests shall correspond to the order of these clauses.

EXCEPTION: The tests described by Clauses 7.9 through 7.15 (i.e., temperature, step change, humidity, air velocity, supply voltage, vibration, and EMI) may be done at any time after the test described by Clause 7.8, but before the test described by Clause 7.16.

NOTE: For stationary or continuous-duty instruments of the sample-draw-type, air velocity variation testing may not be applicable.

7.3 Preparation of Instrument

The instrument shall be prepared as if for actual service, in accordance with the manufacturer's instruction manual.

NOTE: For instruments having remote detector heads, all tests shall be performed with resistances (with temperature coefficients similar to those of the recommended interconnecting wire) connected in the detector circuit to simulate the maximum line resistance specified by the instrument manufacturer, except where minimum line resistance offers a more stringent test in the judgment of the parties having jurisdiction.

7.4 Conditions for Test and Test Area

7.4.1 Voltage—Except as otherwise indicated herein, all tests shall be performed at the nominal system voltage and frequency marked on the equipment, or as applicable with fresh or fully charged batteries. (See Appendix 1.)

7.4.2 Ambient Temperature—Except as otherwise indicated herein, tests may be performed at conveniently available room ambient temperatures in the range of 18–30°C.

7.4.3 Humidity—Except as otherwise indicated herein, tests may be performed in ambient air having a relative humidity of any convenient value in the range of 30–70 percent.

7.4.4 Room Air Circulation—Except as otherwise indicated herein, tests are to be performed in relatively still air (not more than 1 meter per second [m/s]) other than for those currents which may be induced by convection due to the natural heating of the equipment under test or caused by air-moving devices which are part of the equipment under test. It should be noted that the output indication may differ under conditions of a stagnant sample (sample velocities of 0.1 m/s or less).

7.4.5 Removal of Parts—For purposes of the tests in Clauses 7.7 through 7.17, where reference is made to exposing the sensing head to specified gas mixtures or to other specified conditions, in the case of remote detector heads, the entire head—including all normally attached diffusion devices or protective mechanical parts—shall be so exposed.

7.4.6 Multiple Detector Heads—For stationary, mobile, or continuous-duty portable gas detection instruments intended to be used with more than one remote detector head, for purposes of tests which call for the exposure of the remote detector head to a specified test gas or other specified set of conditions, only one detector head shall be so exposed. Dummy electrical loads (e.g., fixed resistors) may be substituted for additional heads; but if additional heads are used, all other heads shall be exposed to normal, clean air and normal conditions for tests as described in Clauses 7.4.2 through 7.4.4.

7.4.7 Recalibration or Adjustment—The instrument under test may be adjusted or recalibrated prior to the start of each of the tests described in Clauses 7.8 through 7.17; however, no further adjustments or recalibration shall be carried out for the duration of that test except where specifically permitted by the particular test procedure.

7.4.8 Stabilization Time—For purposes of the tests in Clauses 7.9, 7.11, 7.12, and 7.13, in each instance where the instrument is subjected to a different test condition the instrument shall be allowed to stabilize under these new conditions (see note below) before measurements are taken for comparison purposes.

NOTE: An instrument shall be considered to be stabilized when three successive observations of the indication taken at 5-min intervals indicate no significant change.

7.4.9 Instruments Having Alarms Only—These instruments do not have any meter or other output indication which can be compared before and after the tests described in the following clauses of Section 7. A tolerance of ±5 percent LFL shall apply to the alarm set point(s) for all tests. However, in all cases the alarm must operate at 60 percent (or less) of the LFL of the specific gas for which the instrument is calibrated. The Instruction Manual shall clearly state these limits unless the manufacturer wishes to specify closer tolerances to which the instrument shall be tested.

7.5 Selectable Gas/Range Instruments

7.5.1 For instruments having more than one selectable range or scale for the same gas, the tests of Clauses 7.8 through 7.17 shall be performed with the instrument operating at both the least and most sensitive ranges—except that if the most sensitive range has a full scale equal to or less than 25 percent of the LFL, the performance testing shall be that specified by the manufacturer in his Instruction Manual. If the manufacturer does not state the performance characteristics of the most sensitive scale where it is 25 percent of LFL or less, the testing shall be the same as for the least sensitive range.

7.5.2 For instruments having selectable ranges employing different detecting means, all of the tests shall be performed on each range.

7.5.3 For instruments having specific ranges or scales for different gas types, the tests of Clauses 7.9 through 7.18 shall be repeated at each selectable range for each different selectable gas type.

7.6 Unpowered Preconditioning Storage

Prior to tests of Clauses 7.7 through 7.18, all parts of the combustible gas detection instrument shall be exposed sequentially to the following conditions:

a. Temperature of −35°C for 24 h

b. Ambient temperature and humidity for at least 24 h

c. Temperature of +55°C for 24 h

d. Ambient temperature and humidity for at least 24 h

7.7 Calibration

The instrument shall be calibrated for testing in accordance with this standard using manufacturer's calibration fixture and specified calibration procedures. The indicator shall be calibrated to read the exact percent LFL of the known test gas applied to the sensor. This gas shall be a nominal 50 percent of the instrument's full-scale gas concentration. For continuous-duty portable instruments of the personal-type incorporating alarms only, the alarm set point shall be adjusted to 20 percent LFL. The combustible gas to be used shall be as follows:

a. Methane for instruments intended for sensing methane specifically, or intended for general-purpose combustible gas detection (including detection of methane)

b. Propane for instruments intended for general-purpose combustible gas detection that excludes methane

c. The actual specific gas or a representative gas for instruments intended for sensing a specific combustible gas or a specific family of chemically similar combustible gases

NOTES:

1. When instruments can be used for detecting more than one combustible gas by changing only the gas-sensing element, then only those tests as described in Clauses 7.7, 7.8, 7.9.3, 7.9.4, 7.10, 7.11, 7.12, 7.14, 7.16 and 7.17 need be repeated for second and subsequent gas-detecting sensing elements (utilizing the actual, specific test gas). Selection of the gas-sensing elements tested first will be the responsibility of the parties having jurisdiction. However, the selection will be "methane" if a methane-sensing element is one of those supplied by the manufacturer.

2. Unless otherwise indicated herein, the manufacturer's calibration device is to be used to supply the gas mixture to the gas-sensing element for the tests described in the paragraphs which follow. However, the instrument's response utilizing this method and the instrument's intended method of gas monitoring, if different, shall first be established.

3. Gas mixtures having the same concentrations as those used for tests in Clause 7.7 are used for various other tests described in the paragraphs which follow. For ease of reference, such gas mixtures will hereafter be referred to simply as "the initial calibration gas mixture."

7.8 Accuracy

7.8.1 All test gas concentrations shall be known to a tolerance of ±2 percent.

7.8.2 The sensing head shall be exposed to five gas concentrations falling in each of the following ranges: 9 to 11 percent, 24 to 26 percent, 49 to 51 percent, 74 to 76 percent, and 98 to 100 percent of the full-scale gas concentration. In each case, the concentration indicated by the meter or output signal shall not vary from the known test gas concentration by more than ±3 percent of full-scale gas concentration or ±10 percent of applied gas concentration, whichever is greater.

7.8.3 For instruments having alarms only, testing shall verify that each alarm actuates on exposure to gas-air mixtures whose concentrations are at the upper tolerance limit for alarm actuation and does not actuate on exposure to mixtures whose concentrations are below the lower tolerance limit. (See paragraph 7.4.9.)

7.9 Temperature Variation

7.9.1 All gas detection instruments shall first be calibrated using the initial calibration test gas, with all parts of the instrument at ambient temperature (Clause 7.4.2); they then shall be exposed to the same initial calibration gas while in a test chamber at a temperature of first 0°C and then of 50°C, as follows:

a. For instruments having the gas-sensing element integral with or directly attached to the control unit, the entire instrument shall be placed in the test chamber containing the gas mixture, or

b. For instruments comprising a control unit and a separate remote detector head, the control unit only shall be placed in clean air in the test chamber, while the detector head shall be exposed to the calibration gas at ambient temperature.

During this test, the meter or output indication during 0°C exposure or 50°C exposure shall not vary from the indication observed during ambient exposure by more than ±5 percent of full-scale gas concentration.

For continuous-duty portable instruments of the personal-type incorporating alarms only, the alarm shall not be actuated by 14–16 percent of LFL gas but shall be actuated by 24–26 percent of LFL gas while exposed to both temperature extremes.

7.9.2 With all parts of the instrument at ambient temperature, calibrate the instrument to twice the normal sensitivity so that with 50 percent full-scale gas concentration at the sensor, the instrument will indicate 100 percent full-scale concentration. If a live-zero reading, capable of at least ±10 percent full-scale deviation is not possible, offset the zero indication to ±10 percent full scale to simulate a live zero.

a. For instruments having the gas-sensing element integral with or directly attached to the control unit, the entire instrument shall be exposed to clean air in a test chamber at a temperature of 0°C and then 50°C. During the test, the indication shall not vary from the initial live-zero indication by more than ±5 percent full scale.

b. For instruments consisting of a control unit and a separate remote detector head, the remote detection head shall be exposed to clean air at ambient temperature, with the control unit placed in the test chamber in clean air at a temperature first of 0°C and then 50°C. During the test the indication shall not vary from the initial live-zero indication by more than a ±5 percent full-scale concentration.

c. For instruments consisting of a control unit and a separate remote detector head, with the control unit at ambient temperature, the remote detector head shall be exposed to clean air in a test chamber at first 75°C and then −40°C. During this test, the indication shall not vary from the initial live-zero indication by more than ±10 percent full scale.

NOTE: For instruments incorporating alarms only, the alarm shall not be activated while the detector head is exposed to both temperature extremes.

7.9.3 For stationary gas detection systems comprising a control unit and a remote detector head, with the control unit at ambient temperature (Clause 7.4.2), the detector head shall be exposed to the initial calibration gas mixture at both −40°C and 75°C.

During this test, the meter or output indication during −40°C exposure or 75°C exposure shall not vary from the indication observed during ambient exposure by more than ±10 percent of full-scale gas concentration.

NOTES:

1. The gas mixture used during exposure to the extreme temperatures in Clause 7.9.2 is to have the same percent-by-volume-in-air concentration as the gas mixture used for initial calibration at ambient temperature.

2. Observations of the meter or output indication shall be made 1 h after stabilization at each new temperature.

3. For detection of vapors whose properties are such that the concentration (due to its vapor pressure properties), cannot be obtained, another appropriate gas may be used.

7.10 Step Change Response to 100 Percent Calibration Gas (See Appendix 2)

7.10.1 Beginning with the gas-sensing element in clean air, it shall be suddenly exposed to a prepared mixture of gas-in-air having a concentration corresponding to 100 percent of full-scale gas concentration. From the instant of exposure to this gas mixture, the instrument shall respond to provide an indication as follows:

a. Sixty percent of full-scale gas concentration within 12 seconds (s).

b. When indication has stabilized, the test gas shall be removed. The indicator shall decline below 50 percent of full scale within 20 s and below 10 percent of full scale within 45 s.

NOTE: For sample-draw-type instruments, the above times do not include the transport time required for the gas sample to reach the instrument from a remote sampling point. For manually aspirated instruments without a sampling line, the times are to be measured from the time of starting the first manual aspiration.

7.10.2 When continuous-duty portable instruments of the personal-type incorporating alarms only are tested, and when optional readout capability is not offered, the following clause shall be substituted for Section 7.10.1:

Beginning with the gas-sensing element in clean air, it shall be suddenly exposed to a 100 percent LFL gas-air mixture. An alarm set to 20 percent LFL of the mixture shall respond within 10 s of exposure to the step change.

7.11 Humidity Variation

The instrument shall first be calibrated with the sensing element exposed to the initial calibration gas mixture at a relative humidity of 50 percent for 2 h. Then the sensing element shall be exposed for 2 h to the calibration gas mixture to which water vapor has been added to raise the relative humidity to a final value of 90 percent. The sensing head shall then be exposed for 2 h to the calibration gas mixture having a relative humidity of 10 percent. The meter or output indications during these exposures shall not vary from the 50 percent relative humidity exposure indication by more than ± 10 percent of full-scale concentration.

7.11.1 For continuous-duty portable instruments of the personal-type incorporating alarms only, the alarm shall not be actuated by 14 to 16 percent LFL test gas, but shall be activated by 24 to 26 percent LFL test gas while exposed to both humidity extremes.

NOTE: Relative humidity values are to be accurate within ± 5 relative humidity percentage points.

7.12 Air Velocity Variation

7.12.1 The instrument shall be calibrated first with the detector head exposed to a static mixture of the initial calibration gas. It shall then be exposed to the initial calibration gas mixture in motion so as to impinge on the detector head with a velocity of 5 ± 0.5 m/s. The meter or output indication during exposures to the mixture in motion shall not vary from that observed during exposure to the static gas mixture by more than 10 percent or −5 percent of full-scale gas concentration in the orientation which causes the greatest deviation.

7.12.2 For continuous-duty portable instruments of the personal-type incorporating alarms only, the alarm shall not be actuated on 14–16 percent LFL test gas but shall be actuated on 24–26 percent LFL test gas while exposed to the mixture in motion (5 ± 0.5 m/s) while the sensing head is oriented and exposed in each of the three principal planes.

7.13 Supply Voltage Variation

7.13.1 For gas detection instruments intended for operation on ac power supply systems and with the gas-sensing element exposed to the initial calibration gas mixture (see Clause 7.7), the supply voltage shall first be decreased to 85 percent of nominal line voltage and then increased to 110 percent of nominal line voltage.

NOTE: The method of causing these step changes in voltage shall be such as to simulate the effect of a heavy load being added to or removed from the source of supply; i.e., there shall be no actual interruption of the voltage supply during the voltage transition.

The variation in the meter or output indication from actual concentration shall not exceed ± 2 percent of full-scale gas concentration. (See Clause 7.4.8.)

7.13.2 For ac-powered instruments, incorrect functions shall not occur when the primary power is applied or removed.

NOTE: For test purposes, gas detection instruments intended for operation from an external ac power source shall be subjected to momentary power interruptions of approximately 0.5 s and 5 s.

During this test, the detector head shall be exposed to clean air (see Clauses 7.4.2 to 7.4.4), and alarms set in the same manner as for Clause 7.13.3.

7.13.3 Gas detection instruments intended for operation from an external dc power source shall be subjected to a step-change in supply voltage from nominal to 122.5 percent of nominal voltage, and from nominal to 87.5 percent of nominal voltage. During this test, the sensing head shall be exposed to clean air (see Clauses 7.4.2 through 7.4.4). Adjustable alarms shall be set to operate 10 percent of the lower flammable limit or 10 percent of full-scale concentration, whichever is lower, or at the lowest possible setting if this is greater than either of the foregoing. As a result of this test, there shall be no instrument malfunction or actuation of the alarms which would falsely indicate the presence of combustible gas.

NOTE: The method of causing these step changes in voltage shall be such as to simulate the effect of a heavy load being added to or removed from the source of supply; i.e., there shall be no actual interruption of the voltage supply during the voltage transition.

7.13.4 For self-contained battery-operated instruments, the voltage variation shall correspond to the maximum terminal voltage of a fresh or fully charged battery(s) and the minimum recommended operating voltage of that (those) battery(s), as determined by a built-in battery-condition indicator. Intermittent-duty portable instruments having provision for adjustment to compensate for battery voltage decline may be so adjusted. (See Clause 7.4.8.)

For continuous-duty portable instruments of the personal-type incorporating alarms only, the alarm shall be actuated by 24 to 26 percent LFL test gas, but not 14 to 16 percent LFL test gas, while exposed to both voltage variation extremes.

7.14 Vibration

7.14.1 Apparatus—The vibration test machine shall consist of a vibrating table, capable of producing a vibration of variable frequency and variable constant excursion (or variable constant acceleration peak) with the instrument under test mounted in place, as required by the test procedure described below.

7.14.2 Procedures—The remote detector head, the control unit, and all portable instruments shall be mounted on the vibration test machine and vibrated successively in each of three mutually perpendicular directions, respectively parallel to the edges of the instrument. The instrument shall be mounted on the vibration table in the same manner and position as intended for service using any resilient mounts, carrier, or holding devices which are provided as a standard part of the instrument.

The instrument shall be vibrated over a frequency range of 10 to 30 Hz at a total excursion of 0.5 mm for a period of 1 h in each of three mutually perpendicular directions. The rate of change of frequency shall not exceed 100 hertz per minute (Hz/min). This test procedure shall apply to the remote detector head, the control unit, and all portable instruments.

7.14.3 Test Criteria—The instruments shall not give any false alarms; there shall be no loose components or damage to the enclosure that could cause a hazard, and when tested with clean air and the initial calibration mixture, the reading shall be accurate within ±5 percent of full-scale gas concentration after this test. In lieu of the test criteria above, instruments of the personal-type incorporating alarms only shall be actuated by the 24 to 26 percent LFL test gas mixture but not be actuated by 14 to 16 percent LFL test gas mixture after this test.

7.15 Electromagnetic Interference (EMI)

Following satisfactory completion of all the applicable tests of the preceding clauses, the stationary and mobile instrument (including sensor, electronics, and interconnecting wiring) shall be subjected (1) while in an energized (operating) mode and (2) while in the position of normal calibration, to electromagnetic energy in the frequency ranges of 150 to 170 MHz and 450 to 470 MHz, using frequency-modulated portable radio transmitters (5 W maximum input to the final amplifier) at a distance of 1 m away from the instrument (i.e., its sensor, electronics, and interconnecting wiring).

Tests shall be conducted for both items (1) and (2) above, using a randomly selected frequency within each of the two frequency ranges and shall not cause the instrument to produce output changes of more than 10 percent of full scale, and/or result in an incorrect instrument function.

Tests should also be conducted following the manufacturer's suggestions concerning wiring, shielding, and installation techniques as they pertain to electromagnetic interference.

If the equipment has successfully passed testing in accordance with SAMA PMC 33.1, the above test is not required. (See Appendix 1.)

7.16 Long-Term Stability

NOTE: Repeat applicable accuracy tests (Clause 7.8) before performing the test under this Clause.

7.16.1 Stationary Instruments—The gas-sensing element shall be consecutively subjected to the following sets of conditions for the periods stated (see Clause 7.4.7):

a. Clean air at ambient temperature and humidity for six continuous days. (See Clauses 7.4.2 and 7.4.3.)

b. At the beginning of the seventh day, expose the gas-sensing element to the initial calibration gas mixture for a period of 24 h. The indicated concentration shall be noted and shall not deviate from the actual calibration gas concentration by more than ± 10 percent of full scale within 5 min.

c. Repeat (a) and (b) a total of four consecutive times (total of 28 days). Just prior to the end of the 14th and 28th days, while the gas-sensing element is still exposed to the initial calibration gas mixture, the indicated concentration shall be noted and shall not deviate from the actual calibration gas concentration by more than ± 10 percent of full scale.

NOTE: Following the deviation check at the end of the 14th and 28th days, the instrument may be zeroed and recalibrated.

d. Expose the gas-sensing element to clean air at ambient temperature and humidity for 24 h. At the end of this period, the gas-sensing element shall be exposed to the initial calibration gas mixture, and the indicated concentration shall not deviate from the actual gas concentration by more than ± 10 percent of full scale within 5 min.

NOTE: Following the deviation check at the end of (d), the instrument may be zeroed and recalibrated prior to the final accuracy test of (e).

e. Immediately after completing (d) (total elapsed time for tests [a] through [d] is 29 days), repeat the accuracy test procedure of Clause 7.8, except that the maximum allowable deviation shall not vary from the known test gas concentration by more than ± 4 percent full-scale gas concentration or 12 percent of applied gas concentration, whichever is greater.

7.16.2 Continuous-Duty Portable Instruments—The instrument shall be consecutively subjected to the following sets of conditions for the periods stated (see Clause 7.4.7):

a. The gas-sensing element shall be exposed to clean air at ambient temperature and humidity for six cycles of operation, each cycle consisting of 8 h with the instrument "On," followed by 16 h with the instrument "Off" (total elapsed time of 24 h per cycle).

b. At the beginning of the seventh day, the gas-sensing element shall be exposed to the initial calibration gas mixture for a period of 8 h with the instrument "On," following which it shall be exposed to clean air at ambient temperature and humidity with the instrument "Off" for a period of 16 h. The indicated concentration shall be noted and shall not deviate from the actual calibration gas concentration by more than ± 10 percent of full scale within 5 min.

c. Repeat (a) and (b) a total of four consecutive times (total elapsed time of 28 days).

d. The gas-sensing element shall be subjected to clean air at normal ambient temperature and normal humidity for an additional period of 6 h with the instrument "On."

e. Immediately after completing (d) (total elapsed time for tests [a] through [d] is 28 days and 6 h), repeat accuracy test procedure of Clause 7.8, except that the maximum allowable deviation shall not vary from the known test gas concentration by more than ± 4 percent of full-scale gas concentration or 12 percent of the applied gas concentration, whichever is greater.

  (i) In lieu of the accuracy tests above, instruments of the personal-type incorporating alarms only shall not be actuated by exposure to 14 to 16 percent LFL test gas mixture and shall be actuated by the 24 to 26 percent LFL test gas mixture after completing (d).

  (ii) The test described in (e) shall be performed immediately after the 6-h test of (d) and shall be completed in not more than 2 h so that the total elapsed instrument "On" time for tests (d) and (e) does not total more than 8 h.

NOTES:

1. For instruments using replaceable (nonrechargeable) batteries, a suitable dc power supply may be used or fresh batteries may be installed at the start of each of the 8-h "On" periods, except that the actual batteries specified for the instrument shall be used for the first 8 h of (a).

2. For instruments using rechargeable batteries, the same rechargeable battery shall be used for the entire test sequence of (a). A suitable dc power supply may be used for the remainder of the test.

3. Zero and span adjustments are permitted prior to each stipulated 8-h "On" period.

7.16.3 Intermittent-Duty Portable Instruments

Intermittent-duty portable instruments provided with a switch which must be manually held in the "On" position shall be subjected to the same test sequence and evaluation as described in 7.16.2, except that each 8-h "On" period shall consist of 24 cycles of 10 min "On" and 10 min "Off."

NOTES:

1. Manually aspirated-type instruments shall be aspirated continuously during each 10-min "On" period; alternatively, a vacuum pump having an equivalent flow may be utilized.

2. Instruments having provisions for voltage adjustments to compensate for battery voltage decline shall be adjusted as necessary during the above tests, including just prior to the final evaluation.

7.17 Flooding with Undiluted Gas

7.17.1 The gas-sensing element of instruments other than the manually aspirated type shall be subjected to a step change in gas concentration from 0 percent (clean air) to 100 percent gas-by-volume. The instrument shall produce an output indication corresponding to a concentration of at least 60 percent of the lower flammable limit or to full-scale concentration, whichever is lower, within 10 s of exposure to the 100 percent gas-by-volume. (See Reference 6.3.1[b].)

7.17.2 Manually aspirated instruments shall be subjected to a test whereby, using the shortest possible sample tube, they are aspirated at the rate which is recommended by the manufacturer with the sample inlet connected to a source of 100 percent gas-by-volume. During this test, the instrument shall produce an output indication corresponding to at least 60 percent of the lower flammable limit or to full scale, whichever is lower, within 10 s.

NOTES:

1. For sample-draw instruments of other than the manually aspirated type, the times given above do not include the transport time required for the gas sample to reach the instrument from a sampling point.

2. For detection of vapors whose properties are such that the concentration, due to its vapor pressure properties, cannot be obtained, another appropriate gas may be used for the tests of Clauses 7.17.1 and 7.17.2.

7.17.3 During the tests of Clauses 7.17.1 and 7.17.2, if the instrument is provided with audible or visible alarm signal devices or alarm contacts, these shall be set to the 60 percent of lower flammable limit set point or to the highest adjustable set point, whichever is lower, and shall be actuated as a result of these tests.

7.17.4 When continuous-duty portable instruments of the personal-type incorporating alarms only (without an optional readout capability) are tested, the following shall be substituted:

a. Beginning with the gas-sensing element in clean air, it shall be subjected to a step change in gas concentration from clean air to 100 percent gas-by-volume. An alarm set to 20 percent LFL of the mixture shall respond with 10 s of exposure to the undiluted flooding condition.

7.18 Dielectric Strength

Following completion of all of the applicable tests of the preceding clauses, the equipment shall be subjected to dielectric strength tests as specified by ANSI C39.5 or other comparable recognized standards.

APPENDIX 1

REFERENCES

This appendix is included for information only and is not part of the standard.

1. ANSI C39.5-1974, "Safety Requirements for Electrical and Electronic Measuring and Controlling Instrumentation," American National Standards Institute, New York, N.Y.
2. ANSI C84.1-1984, "Voltage Ratings for Electrical Power Systems and Equipment (60 Hz)," American National Standards Institute, New York, N.Y.
3. ANSI/ISA S51.1-1979, "Process Instrumentation Terminology," Instrument Society of America, Research Triangle Park, N.C.
4. ANSI/NFPA 70-1987, Articles 500 and 501, *National Electrical Code,* National Fire Protection Association, Quincy, Mass.
5. British Standard BS6020, Parts 1, 2, and 3, "Instruments for the Detection of Combustible Gases," British Standards Institution, London, England.
6. "Corrosion Data Survey, Metals Section," National Association of Corrosion Engineers," Houston, Texas, 1985.
7. CSA Standard C22.2 No. 152, "Combustible Gas Detection Instruments," Canadian Standards Association, Toronto, Canada, 1984.
8. Hilada, C. J., "A Method for Estimating Limits of Flammability," *J. Fire and Flammability,* Vol. 6, pp. 130–39 (April 1975).
9. IEC Pub. 654-1 (1979), "Operating Conditions for Industrial-Process Measurement and Control Equipment. Part I: Temperature, Humidity and Barometric Pressure," International Electrotechnical Commission, Geneva, Switzerland.
10. ISA-S12.4-1970, "Instrument Purging for Reduction of Hazardous Area Classification," Instrument Society of America, Research Triangle Park, N.C.
11. NFPA No. 325M-1977, "Fire Hazard Properties of Flammable Liquids, Gases and Volatile Solids," National Fire Protection Association, Quincy, Mass.
12. NFPA 493P-1978, "Intrinsically Safe Apparatus for Use in Division 1 Hazardous Locations," National Fire Protection Association, Quincy, Mass.
13. NFPA 496P-1982, "Standard for Purged and Pressurized Enclosures for Electrical Equipment," National Fire Protection Association, Quincy, Mass.
14. SAMA PMC 33.1-1978, "Electromagnetic Susceptibility of Process Control Instrumentation," Scientific Apparatus Manufacturers Association, Washington, D.C.
15. Title 30, Part 18, "Electric Motor Driven Mine Equipment and Accessories," *U.S. Code of Federal Regulations.*
16. "USCG Standard for Gas Detection," *Federal Register,* Vol. 44, No. 87, Sections 154.1345 and 154.1350, pp. 26, 035–36, United States Coast Guard.

APPENDIX 2

EXPLANATORY NOTES (See Clause 7.8.2)

This appendix is included for information only and is not part of the standard.

The tolerance of 3 percent of full-scale gas concentration or 10 percent of applied gas concentration (whichever is greater) were determined on a practical basis. Consider the following:

1. In the majority of applications, the instrument will be adjusted to cause alarm or other warning action at a gas concentration of 60 percent LFL or lower. Therefore, the possibility of a meter or output signal of 90 percent LFL when the actual concentration is 100 percent LFL will have no meaning.

2. Any application that requires an alarm or other warning action at a gas concentration higher than 60 percent LFL should consider calibration at the higher alarm point setting, with resultant increases in accuracy at that higher level.

3. The user should select instrumentation which will provide a higher level of accuracy than the minimum tolerances of this standard, if required by the application risk.

ISA-RP16.1,2,3-1959

Recommended Practice

Terminology, Dimensions and Safety Practices for Indicating Variable Area Meters (Rotameters, Glass Tube, Metal Tube, Extension Type Glass Tube)

Instrument Society of America

ISBN 0-87664-342-X

ISA-RP16.1, 2, 3   Terminology, Dimensions and Safety
Practices for Indicating Variable Area
Meters (Rotameters, Glass Tube,
Metal Tube, Extension Type Glass
Tube)

INSTRUMENT SOCIETY OF AMERICA
67 Alexander Drive
P.O. Box 12277
Research Triangle Park, North Carolina 27709

TABLE OF CONTENTS

FOREWORD

This Tentative Recommended Practice has been prepared as a part of the service of the Instrument Society of America toward a goal of uniformity in the field of instrumentation. To be of real value this report should not be static, but should be subject to periodic review. Toward this end, the Society welcomes all comments and criticisms, and asks that they be addressed to the Standards and Practices Board Secretary, Instrument Society of America, 67 Alexander Drive, P.O. Box 12277, Research Triangle Park, North Carolina 27709.

This report was prepared by the Committee on Variable Area Meters 8D-RP16.

L. N. Combs, Chairman — E. I. du Pont de Nemours & Company

S. Blechman — Brooks Rotameter Company

D. N. Brooks — Brooks Rotameter Company

W. A. Crawford, Past Chairman — E. I. du Pont de Nemours & Company

W. F. Dydak — Schutte & Koerting Company

W. A. Diament — Device Engineering Company

R. W. Eberly — Schutte & Koerting Company

K. Fischer — Fischer & Porter Company

R. L. Shapcott — Fischer & Porter Company

Approved for Tentative Publication by the Standards & Practices Board — September 1958.

G. C. Gallagher, The Fluor Corporation
Standards & Practices Department Vice President

F. E. Bryan — Douglas Aircraft Company
Aeronautical Standards Division Director

A. F. Sperry — Panellit, Inc.
Intersociety Relations Division Director

F. W. Winterkamp — E. I. du Pont de Nemours & Co.
Production Processes Division Director

H. S. Kindler — ISA Director
Technical & Educational Services, Secretary

The assistance of those who aided in the preparation of this Recommended Practice by answering questionnaires, offering suggestions, and in other ways is gratefully acknowledged.

1. PURPOSE

The Recommended Practice is intended to (a) establish uniformity of connection dimensions to permit interchangeability of one manufacturer's meters with another manufacturer's meters of the same size, (b) provide a common ground of understanding of the terminology, use, and component parts and accuracies of these meters, and (c) to provide a reference for the safe working pressures of these meters.

2. HISTORY AND DEVELOPMENT

The need for this Recommended Practice was established by a survey of instrument users sent out in September, 1954 to about two hundred and fifty meter users and manufacturers. The answers indicated a need for uniform practice in several portions of the variable area meter fields. These are: (a) connection dimensions, (b) terminology, (c) safety practices, (d) maintenance methods, (e) calibration procedures, and (f) installation practices.

In August, 1955, this summary was reviewed with the Main Recommended Practices Committee, who felt that the first three considerations should be investigated by a manufacturers organization and presented the problem to SAMA. SAMA declined to undertake this activity on the basis that they did not represent the majority of manufacturers of variable area meters. In September, 1956, the Main Recommended Practices Committee suggested that a subcommittee be formed on which the major manufacturers of this type of equipment would be represented. Early in 1957, this committee was formed including representatives of three major manufacturers.

In the September 8, 1957 Recommended Practices Committee (RPC) meeting, RP 16.1 and RP 16.2 were approved for release as tentative and both were issued in March, 1958. In the September 14, 1958 RPC meeting, RP 16.3 was approved for release as tentative with the recommendation that all be combined into one publication to reduce duplication of information.

3. SCOPE

This Recommended Practice combines RP 16.1 — 10″ (250 millimeter) scale length glass tube indicating variable area meters (rotameters) with horizontal inlet and outlet, either flanged or screwed construction; RP 16.2 — 5″ (125 millimeter) scale length metal tube indicating variable area meters (rotameters) with bottom inlet and horizontal outlet of flanged construction; and RP 16.3 — 5″ (125 millimeter) scale length glass tube extension type variable area meters (rotameters) with horizontal inlet and outlet, either flanged or screwed construction. Additional auxiliary functions of the extension such as indication, pneumatic or electrical transmission, alarm, local recording, etc., is included in RP 16.4.

4. UNIFORM CONNECTION DIMENSIONS

The uniform piping dimensions listed on the following page are to permit interchangeability of meters and simplified piping designs. Dimensions are established as whole dimensions or nominal fractions for easy piping detail. No dimension "B" is included for screwed meter connections, but center to center of inlet and outlet "A" dimension applies for screwed meters as well.

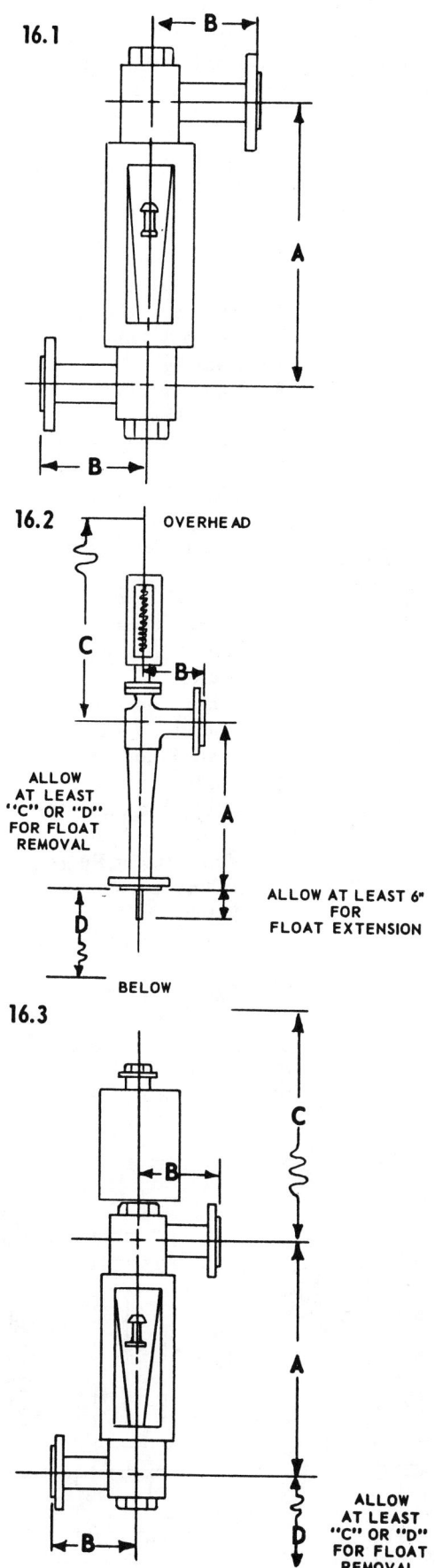

16.1

Connection Pipe Size	Center to Center Inlet to Outlet "A"	Center to Face of Flange Face "B" *	
		150# RF	300# RF
½″	16½″	3½″	4″
¾″	17½″	4″	4½″
1″	17½″	4″	4½″
1½″	20½″	5″	5½″
2″	21″	5″	5½″
3″	24″	6″	7″
4″	28″	6″	7″

* NOTE: Dimensions are for 1/16″ raised face flanges.

Safe working pressures are obtained by reference to 6.1, and not from nominal flange rating.

16.2

Connection Size Inches	Face Inlet to Center Outlet "A"	Center Meter to Face Outlet "B"*		Clearance For Float Removal	
		150# RF	300# RF	"C"	"D"
1″	12½″	3½″	4″	47″	30″
1½″	14½″	4″	4½″	51″	30″
2″	15½″	4½″	5″	51″	30″
3″	18″	5½″	6″	64″	33″
4″	24″	6½″	–	70″	39″

* NOTE: Dimension "B" in all cases is equal to dimension of ASA standard tee (centerline of run to face of side outlet.) Therefore, ASA standard tees can be used at inlet to produce a side inlet, side outlet construction. Dimensions are for 1/16″ raised face flanges.

16.3

Connection Pipe Size	Center to Center Inlet to Outlet "A"	Center to Face of Flange Face "B"*		Clearance For Float Removal	
		150# RF	300# RF	"C"	"D"
½″	11½″	3½″	4″	48″	37″
¾″	12½″	4″	4½″	48″	37″
1″	12½″	4″	4½″	48″	37″
1½″	15½″	5″	5½″	48″	39″
2″	16″	5″	5½″	50″	40″
3″	19″	6″	7″	55″	51″
4″	23″	6″	7″	60″	53″

* NOTE: Dimensions are for 1/16″ raised face flanges.

Safe working pressures are obtained by reference to 6.1, and not from nominal flange rating.

5. UNIFORM NOMENCLATURE

5.1 Listed below is a tabulation of standard terminology for component parts of the glass and metal tube meters. This portion of the Recommended Practice is intended to standardize the component terms used by each meter manufacturer (and by users) when advertising and when specifying spare parts.

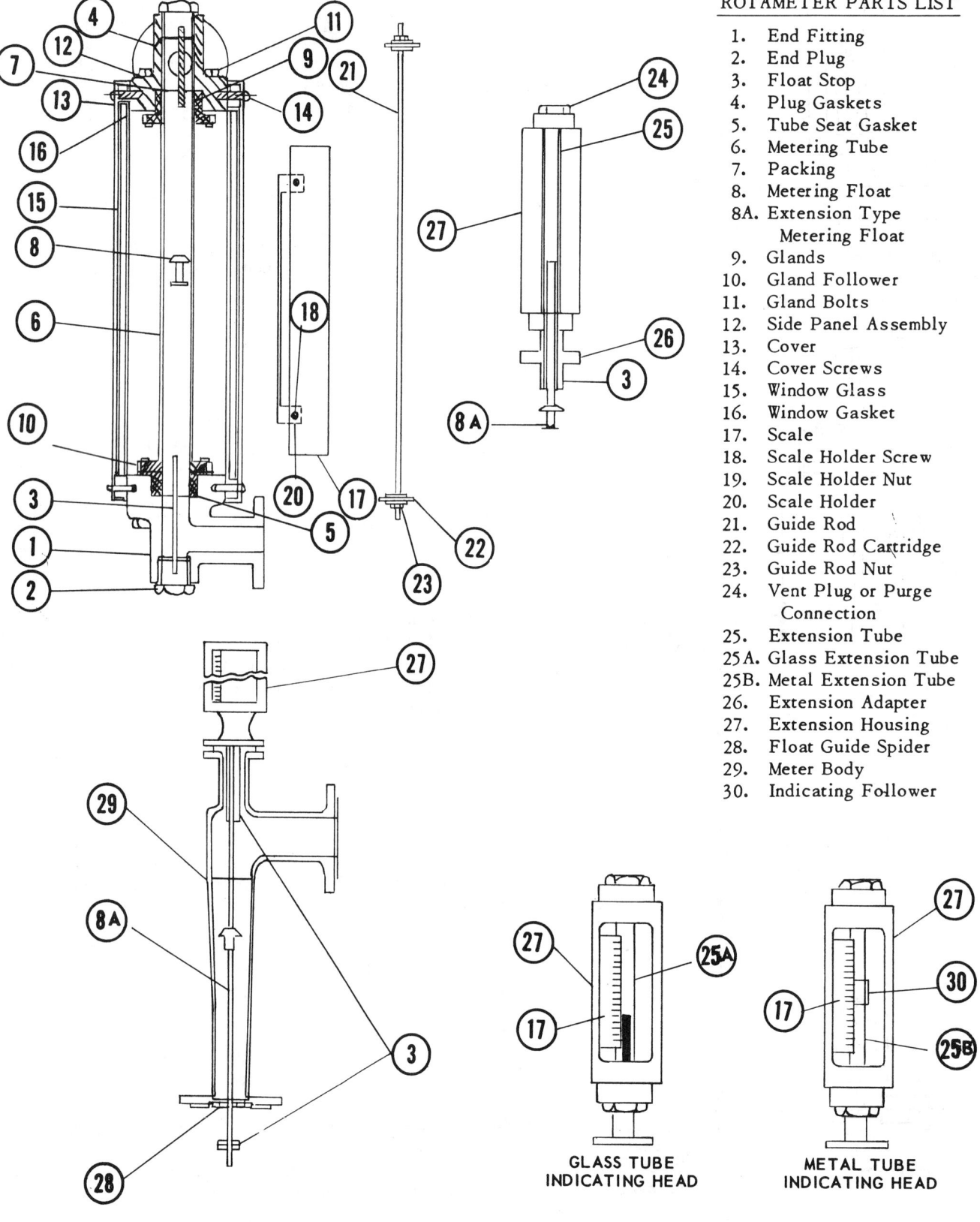

ROTAMETER PARTS LIST

1. End Fitting
2. End Plug
3. Float Stop
4. Plug Gaskets
5. Tube Seat Gasket
6. Metering Tube
7. Packing
8. Metering Float
8A. Extension Type Metering Float
9. Glands
10. Gland Follower
11. Gland Bolts
12. Side Panel Assembly
13. Cover
14. Cover Screws
15. Window Glass
16. Window Gasket
17. Scale
18. Scale Holder Screw
19. Scale Holder Nut
20. Scale Holder
21. Guide Rod
22. Guide Rod Cartridge
23. Guide Rod Nut
24. Vent Plug or Purge Connection
25. Extension Tube
25A. Glass Extension Tube
25B. Metal Extension Tube
26. Extension Adapter
27. Extension Housing
28. Float Guide Spider
29. Meter Body
30. Indicating Follower

GLASS TUBE
INDICATING HEAD

METAL TUBE
INDICATING HEAD

5.2 Uniform Terminology Concerning Meter Accuracy — A Code System is presented herewith to extablish an exact expression for Meter accuracy at specified operating conditions. Through the use of this Code System, in Specifications by both Manufacturers and Users, it is expected that there should be no misunderstanding between accuracy of maximum scale reading and accuracy of instantaneous flow reading, the limits of the scale at which the accuracy is expected, and, of course, the accuracy (maximum error) desired.

	% ACCURACY SPECIFIED	% REFERRED TO*	LOWER LIMIT (% OF SCALE) OF SPECIFIED ACCURACY
Example 1:	1	S	10

Meter specified to have accuracy of plus or minus 1% of maximum scale from 100 down to 10% of scale reading.
As written, would then be 1-S-10

Example 2:	2	R	20

Meter specified to have accuracy of plus or minus 2% of instantaneous reading from 100 down to 20% of scale reading.
As written, would then be 2-R-20.

* Note: S = full scale; R = instantaneous reading.

5.3 Uniform Tube Nomenclature — A Code System is presented to permit specifying and defining specific meter tubes.

	MANU-FACTURER (INITIALS)	INLET BORE SIZE	(DIAMETER RATIO – 1) X 100	TUBE STYLE**	TUBE LENGTH
Example 1:	F & P	1½"	27	G	10

This is a Fischer and Porter, 10", beaded guide, glass tube, 1½" size, with a 1.27 outlet to inlet diameter ratio.
As written, would then be F&P – 1½-27-G-10.

Example 2:	S & K	2"	35	P	10

This is a Schutte and Koerting, 10", plain glass tube, 2" inlet size, with 1.35 outlet to inlet diameter Ratio.
As written, would then be S&K – 2-35-P-10.

Example 3:	BR	3"	40	M	5

This is a Brooks Rotameter Company, 5" long metal tube, 3" inlet size, with 1.40 outlet to inlet diameter ratio.
As written, would then be BR – 3-40-M-5.

** Note: G = guided glass; P = plain glass; M = metal.

Please note that tube codes are included for meters not covered in the scope of this Recommended Practice. The code will apply to tubes for all styles of meters.

6. RECOMMENDED SAFE WORKING PRESSURES

6.1 RECOMMENDED SAFE WORKING PRESSURES OF BOROSILICATE GLASS TUBES

Listed below are the safe working pressures recommended by this Practice. To use meters above these pressures is considered unsafe because of the possibility of glass breakage. Particular conditions, not listed below, should be checked with the individual manufacturer.

Nominal Tube Inlet Bore	Max. Working Pressure (PSIG) Up to 200 Deg. F.	Pressure Reduction Above 200° F. PSI/Deg. F.	Max. Temp. Degrees F.
$\frac{1}{16}''$ or $\frac{1}{8}''$	550	0.75	400
$\frac{1}{4}''$	450	0.75	400
$\frac{3}{8}''$	350	0.75	400
$\frac{1}{2}''$	300	0.75	400
$\frac{3}{4}''$	240	0.60	400
1''	200	0.45	400
$1\frac{1}{2}''$	130	0.33	400
2''	100	0.25	400
3''	70	0.15	300
4''	50	0.10	300

1. Maximum working pressure ratings are for non-shock conditions. (No water hammer)

2. Recommended test pressure (static) should be 1.5 times safe working pressure.

3. Above pressure ratings apply to rib, fluted, or plain tapered tubes.

4. Maximum safe working pressures for glass tubes above 200° F. to be calculated using the pressure reduction given in table above.

5. Up to the maximum temperatures listed above, Borosilicate glass tubes are resistant to thermal shock.

6. All glass tube meters should be adequately shielded with safety glass on the reading side and amply vented on sides, back, or bottom.

6.2 RECOMMENDED SAFE WORKING PRESSURES OF METAL TUBES

These meters are suitable for operation on pressures up to that limited by the weights of the standard ASA flange connections. These metal tube meters can be tested in conformance with ASA Standard Flange Test Procedures. For specific further information, check either the Piping Code or the manufacturers.

INSTRUMENT SOCIETY of AMERICA
Research Triangle Park, North Carolina

Recommended Practice

Nomenclature and Terminology for Extension Type Variable Area Meters (Rotameters)

Instrument Society of America

ISBN 0-87664-343-8

ISA-RP16.4   Nomenclature and Terminology for
Extension Type Variable Area Meters
(Rotameters)

INSTRUMENT SOCIETY OF AMERICA
67 Alexander Drive
P.O. Box 12277
Research Triangle Park, North Carolina 27709

TABLE OF CONTENTS

FOREWORD

This Recommended Practice has been prepared as a part of the service of the Instrument Society of America toward a goal of uniformity in the field of instrumentation. To be of real value this report should not be static, but should be subject to periodic review. Toward this end the Society welcomes all comments and criticisms, and asks that they be addressed to the Standards and Practices Board Secretary, Instrument Society of America, 67 Alexander Drive, P.O. Box 12277, Research Triangle Park, North Carolina 27709.

This report was prepared by the Committee on Variable Area Meters 8D-RP16.

L. N. Combs, Chairman - E. I. du Pont de Nemours & Company
S. Blechman - Brooks Instrument Company, Inc.
W. A. Crawford, Past Chairman - E. I. du Pont de Nemours & Company
W. F. Dydak - Schutte & Koerting Company
R. W. Eberly - Schutte & Koerting Company
H. V. Mangin - Fischer & Porter Company
T. E. Quigley - E. I. du Pont de Nemours & Company
R. L. Shapcott - Weston Instruments Div., Daystrom, Inc.

Approved for Tentative Publication by the Standards & Practices Board - July 1960.

G. G. Gallagher, The Fluor Corporation
  Standards & Practices Department Vice President
F. E. Bryan - Douglas Aircraft Company
  Aero-Space Standards Division Director
R. E. Clarridge - General Electric Company
  Intersociety Standards Division Director
F. H. Winterkamp - E. I. du Pont de Nemours & Company
  Production Processes Standards Division Director
E. J. Minnar - ISA Headquarters Staff
  Standards & Practices Board Secretary

The assistance of those who aided in the preparation of this Recommended Practice by answering questionnaires, offering suggestions, and in other ways is gratefully acknowledged.

1. PURPOSE

The Recommended Practice is intended to provide a basis for nomenclature and terminology for extension type variable area meters (rotameters).

2. HISTORY AND DEVELOPMENT

The resultant of a survey sent out in 1954 to over 250 meter users and manufacturers indicated among other factors, the need for standardized terminology. Using these replies as a basis, a committee was formed and this Recommended Practice was developed.

3. SCOPE

This Recommended Practice has been prepared to define the nomenclature and terminology of various types of extensions applicable to 5 inch (125 mm) glass and metal tube variable area meters (rotameters) covered in ISA-RP 16.1.2.3.

4. DEFINITIONS AND TERMINOLOGY

An extension is a device for translating float motion into a useful secondary function, for either indicating, alarming, transmitting or other secondary functions. An extension usually consists of an extension tube, an extension housing, and the necessary adaptor to the primary rotameter, but may be any auxiliary device fixed to the rotameter which performs the functions outlined below.

4.1 Indicating Extensions

4.1.1 Magnetic

A device that provides flow rate indication by means of a magnetic coupling between the extension of the metering float and an external indicator follower surrounding the extension tube.

4.1.2 Direct

A device that provides flow rate indication by means of viewing the position of the extension of the metering float within a glass extension tube.

4.2 Transmitting Extensions

4.2.1 Pneumatic

A system that converts float position to a proportional standard pneumatic signal. A magnetic coupling connects the internal float extension with an external mechanical system linked to a pneumatic transmitter.

4.2.2 Electric or Electronic

A system that converts float position to a proportional electric signal (either AC or DC), or to a proportional shift or unbalance in impedance which is balanced by a corresponding shift in impedance in the receiving instrument.

4.3 Alarms - Extensions

4.3.1 Magnetically Actuated

A device attached to the meter body which contains an electrical switch and which is magnetically actuated by the metering float extension to signal a high or low flow. The switch is adjustable with respect to the float position over a range equal to the travel of the metering float. Standard switch ratings are usually 0.3 amperes for 110 volt, 60 cycle AC supply (five amperes or more if relays are used).

4.3.2 Electrically Operated

Usually a highly sensitive induction type device for signalling high or low flows or deviations from any set flow. The device consists of a sensing coil positioned around the extension tube of the rotameter. Movement of the metering float into the field of the coil causes a low level signal change which is usually amplified to a level suitable for performing annunciator or control functions.

4.4 Other Extensions

4.4.1 Recording

The recorder is attached directly to the meter body with the recorder pen positioned by the metering float through a magnetic coupling.

4.4.2 Integrating

In the same manner as described under 4.4.1, an integrator which derives its input from the motion of the float can be installed within the extension housing.

4.4.3 Controlling

In the same manner as described under 4.4.1, a controller which derives its input from the motion of the float can be installed within the extension housing.

INSTRUMENT SOCIETY of AMERICA
Research Triangle Park, North Carolina

ISA-RP16.5-1961

Recommended Practice

Installation, Operation, Maintenance Instructions for Glass Tube Variable Area Meters (Rotameters)

Instrument Society of America

ISBN 0-87664-344-6

ISA-RP16.5   Installation, Operation, Maintenance
             Instructions for Glass Tube Variable Area
             Meters (Rotameters)

INSTRUMENT SOCIETY OF AMERICA
67 Alexander Drive
P.O. Box 12277
Research Triangle Park, North Carolina 27709

FOREWORD

This Recommended Practice has been prepared as a part of the service of the Instrument Society of America toward a goal of uniformity in the field of instrumentation. To be of real value this report should be subject to periodic review. Toward this end the Society welcomes all comments and criticisms, and asks that they be addressed to the Standards and Practices Board Secretary, Instrument Society of America, 67 Alexander Drive, P.O. Box 12277, Research Triangle Park, North Carolina 27709.

This report was prepared by the 8D-RP16 Committee of the Production Processes Standards Division on Variable Area Meters.

L. N. Combs, Chairman - E. I. du Pont de Nemours & Company
S. Blechman - Brooks Instrument Co., Inc.
W. F. Dydak - Schutte & Koerting Company
R. W. Eberly - Schutte & Koerting Company
H. V. Mangin - Fischer & Porter Company
T. E. Quigley - E. I. du Pont de Nemours & Company
R. L. Shapcott - Daystrom, Inc., Weston Instrument Div.

The assistance of those who aided in the preparation of this Recommended Practice by answering questionnaires, offering suggestions, and in other ways is gratefully acknowledged.

The following have reviewed the Report and served as Board of Review:

B. O. Black - E. I. du Pont de Nemours & Company, Inc.
Robert Coel - The Fluor Corp., Ltd.
W. A. Crawford - E. I. du Pont de Nemours & Company, Inc.
G. N. Ehly - The Atlantic Refining Company
J. W. Percy - United States Steel Corporation
C. A. Prior - Diamond Alkali Company
J. W. Profota - Union Carbide Olefins Company
B. V. Smith - Allied Chemical Corporation
R. U. Stanley - Standard Oil Company of California

Approved for Tentative Publication by the Standards & Practices Board in June, 1961.

E. A. Adler - United Engineers & Constructors, Inc.
 Standards & Practices Department Vice President
F. E. Bryan - Douglas Aircraft Company
 Aero-Space Standards Division Director
R. E. Clarridge - General Electric Company
 Intersociety Standards Division Director
W. J. Ladniak - Minneapolis-Honeywell Reg. Co.
 Nuclear Standards Division Director
F. H. Winterkamp - E. I. du Pont de Nemours & Co., Inc.
 Production Processes Standards Division Director
E. J. Minnar - ISA Headquarters Staff
 Standards & Practices Board Secretary

1. PURPOSE

This Recommended Practice is intended to provide a useful guide for the general considerations important to the installation, operation and maintenance of glass tube variable area meters (rotameters).

2. HISTORY AND DEVELOPMENT

The need for this Recommended Practice was established by a survey of instrument users sent out initially in September, 1954, to about two-hundred-and-fifty meter users and manufacturers.

3. SCOPE

This Recommended Practice covers the general considerations important to the installation, operation and maintenance of meters to obtain the most reliable results.

Terminology, dimensions and safety practices have been covered in ISA-RP 16.1.2.3. Nomenclature and terminology for extension devices are included in ISA-RP 16.4. Calibration is included in ISA-RP 16.6.

4. UNPACKING THE INSTRUMENT

4.1 Care should be exercised when unpacking the instrument. The instrument should be carefully inspected to determine that no damage has occurred during the shipment.

Depending upon the size of rotameter, larger metering floats are packed separately and should not be discarded as packing material. Metering floats which have been packed separately should be carefully unpacked. Protective coating or tape on metering edge should not be removed until just prior to installation. Inspection for damage should be made immediately.

5. PREPARATION OF THE INSTRUMENT FOR INSTALLATION

5.1 The end fittings and metering tube should be inspected to make sure that they are free of any foreign matter and if necessary, should be cleaned with a tube brush or a soft swab on the end of a wooden dowel. Remove tape and/or protective coating from metering float, and inspect its surface for burrs or scratches. All parts of the meter should be inspected visually for assurance meter will function properly.

6. INSTALLATION

6.1 The rotameter must be installed in a vertical position with the piping aligned and supported properly to prevent any stresses being transmitted to the meter. The inlet connection is always referred to the smallest tapered end of the metering tube, or the lowest numerical scale graduation and is located at the bottom. Connecting pipelines should be the same size, but in no case, more than or less than one pipe size different than meter connection size.

The metering float should be inserted into the metering tube with the correct end up so that the float is read at the proper position with reference to the scale. Plumb-bob type floats are read at the largest uppermost diameter. Ball floats are read from mid-position of the ball. Other types of metering floats are also read at the uppermost largest diameter. In any case, the manufacturer's instructions should be checked to verify the correct position of the metering float. When inserting the float, care should be exercised to prevent tube breakage. Ball floats can be guided into the top of the metering tube by using a paper funnel. Avoid dropping metering floats other than the ball type through the outlet end fittings since this could cause tube breakage. Large metering floats usually have a tapped hole in the top to which a rod may be fitted to lower it into the metering tube. Smaller metering floats may be guided into the metering tube using the lower float stop.

The detachable scale should be aligned so that its reference mark and zero line on the metering tube coincide. If not, make the necessary adjustments. This step is not necessary if the scale is etched on the metering tube.

It is recommended that all glass tube rotameter installations incorporate the use of a by-pass arrangement for either liquid or gas service as illustrated in Fig. RP16.5.1. The use of a by-pass arrangement with unions (or flanges) and valves as illustrated in Fig. RP16.5.1 permits flushing of new lines by by-passing the rotameter, and permits the by-passing of flow when it is necessary to remove the instrument for maintenance. It also assists in maintaining stable start-up for gas service. This is described under "Start-Up" procedure.

For flashing liquids, the by-pass arrangement valving is as shown in Fig. RP16.5.1 with valve No. 1 on the downstream side of the meter being a throttling valve. Valve No. 2 (block valve) on the upstream side of the meter must be operated in the fully open position so as not to introduce excess pressure drop which may cause the liquid to flash.

For non-flashing liquids, the recommended by-pass arrangement valving is shown in Fig. RP16.5.1 with valve No. 1 in the downstream side of the meter being a throttling valve. Under certain conditions, when it is desired to protect the meter from full line pressure, or where pressure shock (water hammer) might exist, valve No. 2 could be made a throttling valve.

For gas service, the by-pass arrangement is as shown in Fig. RP16.5.1 with valve No. 1 on the downstream side of the meter being a throttling valve for stable operation. Where the by-pass arrangement cannot be incorporated for gas service, the rotameter must be installed with the throttling valve on the downstream side as shown in Fig. RP16.5.2 for stable operation. The maximum distance "L" of the throttling valve from the outlet of the rotameter can be determined from the nomograph contained in Fig. RP16.5.3. Installation according to this data will reduce or eliminate the phenomena known as "float bounce."

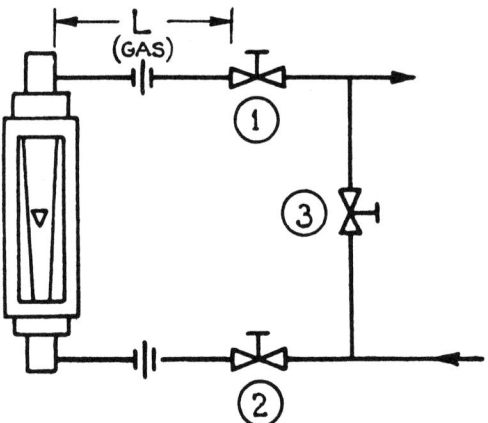

Figure RP 16.5.1

Liquid or Gas Service

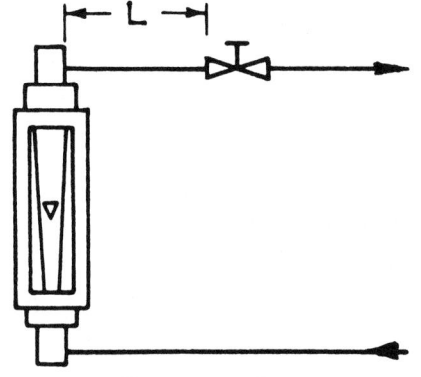

Figure RP 16.5.2

Gas Service

7. START-UP WITH BY-PASS ARRANGEMENT

7.1 In order to prevent metering tube breakage or damage to the metering float, a by-pass arrangement should be used as follows:

(1) Throttling Valve No. 1 should be closed.
(2) Block Valve No. 2 should be closed.
(3) Open Equalizing Valve No. 3 sufficiently to bring pressure up to operating pressure.
(4) Open Block Valve No. 2 gradually to full open.
(5) Open Throttling Valve No. 1 gradually to approximately 1/2 turn.
(6) Once stability is maintained, equalizing Valve No. 3 should be closed and further throttling of flow may be done by throttling Valve No. 1.

8. WATER HAMMER

In the installation of glass tube rotameters for liquid service, consideration should be given to possible damage from water hammer. Water hammer is a series of pressure shocks created by a sudden checking of the flow of liquid in a pipe, such as through the quick closing of a valve. The action, by stopping the flow, generates kinetic energy in the column of liquid by compressing the fluid and stretching the walls of the pipe. A wave of increased pressure beginning at the valve is transmitted back through the pipe at a constant magnitude and velocity. When the pressure wave has traveled upstream to a larger part of the pipe or vessel, it reverses itself and a wave of normal pressure progresses back to the valve. While the pressure surge appears first directly at the valve and lasts longest at this point, the actual pressure peak is transmitted through the entire system.

The best method of preventing water hammer is to eliminate quick-acting devices in the fluid stream, or to install surge chambers or accumulators in the system as close to the surge initiating device as possible. However, the resultant pressure peak can be calculated and selection of a rotameter capable of withstanding this pressure is possible.

The peak pressure generated by water hammer can be calculated by the following formula: [1]

$$P = P_s + 64V$$

Where —

P = Maximum surge pressure generated under condition of water hammer (p.s.i.).

P_s = Normal operating pressure (p.s.i.g.).

V = Fluid velocity at normal operating flow (feet per sec.).

Example — A rotameter system with a 6 g.p.m. rate of flow in a 3/4" (Schedule 40) pipe (fluid velocity 3.61 ft. per sec.) and a normal operating pressure of 50 p.s.i.g.

$$P = 50 + 64 \times 3.61$$
$$P = 281 \text{ p.s.i.}$$

9. MAINTENANCE

9.1 If leaking occurs at either end of the glass metering tube, the pressure in the meter should be released, then the gland bolts should be tightened uniformly. In order to prevent metering tube breakage by over-tightening gland bolts, a portable polariscope may be used to indicate excessive stresses.

A periodical visual inspection should be made of the rotameter. All metering tube packing and gaskets should be replaced if necessary. It is important that the metering tube and float be inspected for deterioration of surfaces. Refer to manufacturer for dimensional and weight tolerance of the metering float.

When it is necessary to replace metering tube or packing, reference must be made to manufacturer's maintenance instructions for correct procedure.

The following are considered normal spare parts:

(1) Metering tube
(2) Packing
(3) Gaskets
(4) Metering float
(5) Window glass

When ordering the above spare parts, complete information must be supplied, giving serial number, size of meter, metering tube, metering float, and materials of construction.

10. REFERENCE

(1) Mangin, H. V., ISA Journal, Vol. 6 No. 11 - Nov. 1959.

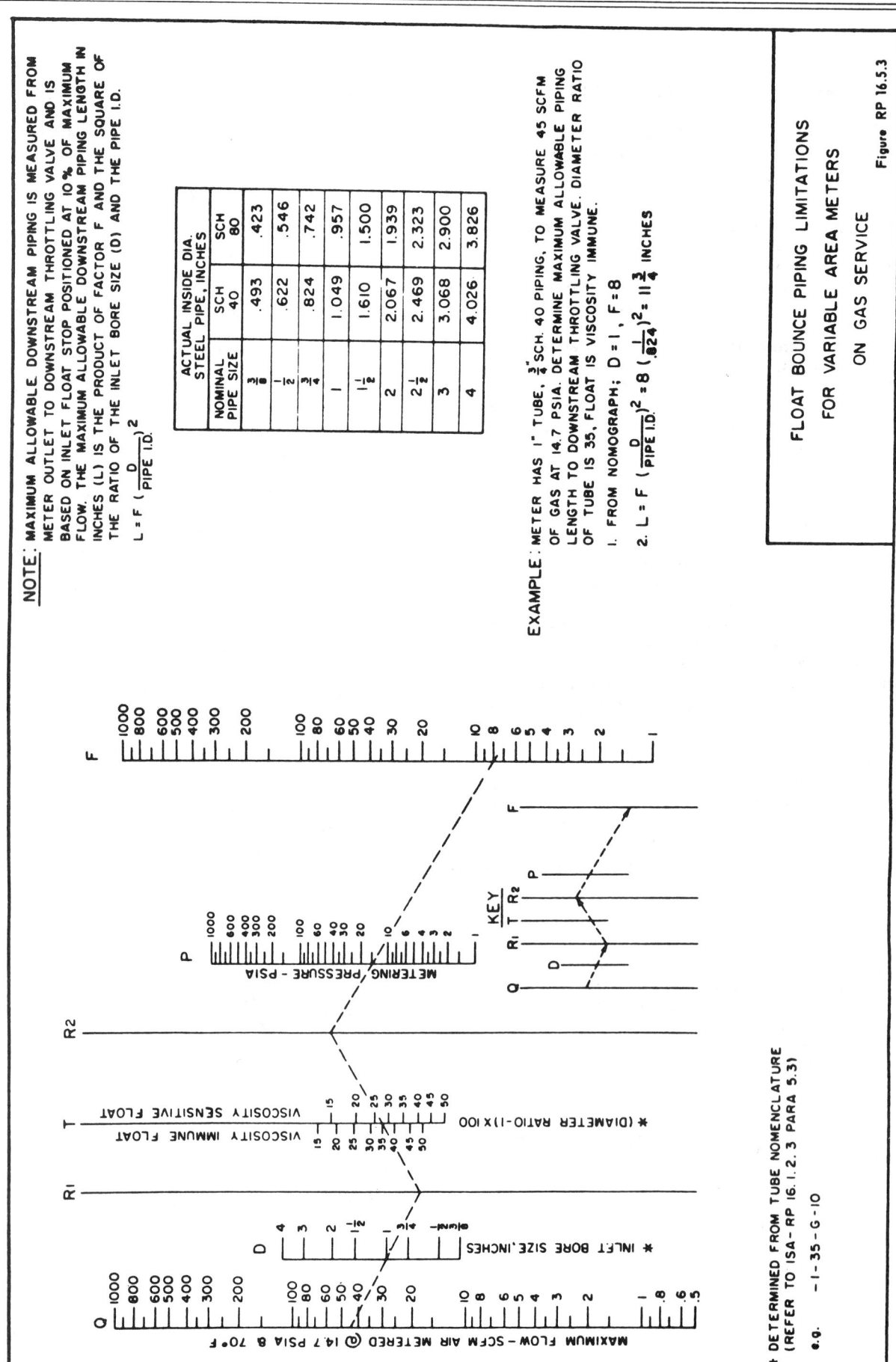

NOTE: MAXIMUM ALLOWABLE DOWNSTREAM PIPING IS MEASURED FROM METER OUTLET TO DOWNSTREAM THROTTLING VALVE AND IS BASED ON INLET FLOAT STOP POSITIONED AT 10% OF MAXIMUM FLOW. THE MAXIMUM ALLOWABLE DOWNSTREAM PIPING LENGTH IN INCHES (L) IS THE PRODUCT OF FACTOR F AND THE SQUARE OF THE RATIO OF THE INLET BORE SIZE (D) AND THE PIPE I.D.

$$L = F \left(\frac{D}{PIPE\ I.D.} \right)^2$$

| | ACTUAL INSIDE DIA. STEEL PIPE, INCHES | |
NOMINAL PIPE SIZE	SCH 40	SCH 80
$\frac{3}{8}$	.493	.423
$\frac{1}{2}$	.622	.546
$\frac{3}{4}$	.824	.742
1	1.049	.957
$1\frac{1}{2}$	1.610	1.500
2	2.067	1.939
$2\frac{1}{2}$	2.469	2.323
3	3.068	2.900
4	4.026	3.826

EXAMPLE: METER HAS 1" TUBE, $\frac{3}{4}$" SCH. 40 PIPING, TO MEASURE 45 SCFM OF GAS AT 14.7 PSIA. DETERMINE MAXIMUM ALLOWABLE PIPING LENGTH TO DOWNSTREAM THROTTLING VALVE. DIAMETER RATIO OF TUBE IS 35, FLOAT IS VISCOSITY IMMUNE.

1. FROM NOMOGRAPH; D = 1, F = 8

2. $L = F \left(\frac{D}{PIPE\ I.D.} \right)^2 = 8 \left(\frac{1}{.824} \right)^2 = 11\frac{3}{4}$ INCHES

FLOAT BOUNCE PIPING LIMITATIONS
FOR VARIABLE AREA METERS
ON GAS SERVICE

Figure RP 16.5.3

* DETERMINED FROM TUBE NOMENCLATURE
(REFER TO ISA-RP 16.1.2.3 PARA 5.3)

e.g.   -1-35-G-10

INSTRUMENT SOCIETY of AMERICA
Research Triangle Park, North Carolina

ISA-RP16.6-1961

Recommended Practice

Methods and Equipment for Calibration of Variable Area Meters (Rotameters)

Instrument Society of America

ISBN 0-87664-345-4

ISA-RP16.6   Methods and Equipment for Calibration of
Variable Area Meters (Rotameters)

INSTRUMENT SOCIETY OF AMERICA
67 Alexander Drive
P.O. Box 12277
Research Triangle Park, North Carolina 27709

TABLE OF CONTENTS

FOREWORD

This Tentative Recommended Practice has been prepared as a part of the service of the Instrument Society of America toward a goal of uniformity in the field of instrumentation. To be of real value this report should not be static, but should be subject to periodic review. Toward this end the Society welcomes all comments and criticisms, and asks that they be addressed to the Standards and Practices Board Secretary, Instrument Society of America, 67 Alexander Drive, P.O. Box 12277, Research Triangle Park, North Carolina 27709.

This report was prepared by the 8D-RP16 Committee of the Production Processes Standard Division on Variable Area Meters.

L. N. Combs, Chairman - E. I. duPont de Nemours & Company
S. Blechman - Brooks Instrument Co., Inc.
W. F. Dydak - Schutte & Koerting Company
R. W. Eberly - Schutte & Koerting Company
H. V. Mangin - Fischer & Porter Company
T. E. Quigley - E. I. duPont de Nemours & Company
R. L. Shapcott - Daystrom, Inc., Weston Instrument Div.

The assistance of those who aided in the preparation of this Recommended Practice by answering questionnaires, offering suggestions, and in other ways is gratefully acknowledged.

The following have reviewed the Report and served as Board of Review:

S. A. Cole - Wallace & Tiernan, Inc.
W. A. Crawford - E. I. duPont de Nemours & Co., Inc.
R. C. Kimball - American Viscose Corporation
H. S. Kindler - Instrument Society of America
H. L. Knight - E. I. duPont de Nemours & Co., Inc.
N. L. Massey - Badger Manufacturing Company

Approved for Tentative Publication by the Standards & Practices Board in June, 1961.

E. A. Adler - United Engineers & Constructors, Inc.
   Standards & Practices Department Vice President
F. E. Bryan - Douglas Aircraft Company
   Aero-Space Standards Division Director
R. E. Clarridge - General Electric Company
   Intersociety Standards Division Director
W. J. Ladniak - Minneapolis-Honeywell Reg. Company
   Nuclear Standards Division Director
F. H. Winterkamp - E. I. duPont de Nemours & Co., Inc.
   Production Processes Standards Division Director
E. J. Minnar - ISA Headquarters Staff
   Standards & Practices Board Secretary

1. PURPOSE

This Recommended Practice is intended to provide a useful guide for the general considerations important to the calibration of variable area meters (rotameters).

2. HISTORY AND DEVELOPMENT

The need for this Recommended Practice was established by a survey of instrument users sent out initially in September, 1954, to about two-hundred-and-fifty meter users and manufacturers.

3. SCOPE

This Recommended Practice has been prepared to describe the methods and equipment used for calibrating the glass and metal metering tube area meters (rotameters) covered in RP16.1.2.3. Nomenclature and terminology for extension devices are included in ISA-RP16.4. Installation, operation and maintenance instructions are included in ISA-RP16.5.

4. METHODS AND EQUIPMENT

When calibrating a rotameter, certain techniques for determining the proper equipment and procedures must be considered. Care should be taken in selecting equipment for controlling flows, measuring volumes and time intervals, etc., with acceptable accuracies of measurement.

The use of the rotameter dictates the accuracy of calibration required. When the rotameter is used as a flow indicating device and repeatability of readings is of paramount importance the rotameter can usually be calibrated with a reference liquid such as water and corrections for density can be made using the formulas presented in Paragraph 4.1.1.

To determine whether the viscosity of the metering liquid will permit calibration with a reference liquid, refer to Chart, Fig. RP16.6.1.

If absolute accuracy of measurement is most important, the rotameter should be calibrated with the process fluid that it will eventually handle or kinematic simulation should be employed to assure highest absolute accuracy. (Refer to Paragraph 4.2).

4.1 Three basic methods are employed in the calibration of rotameters. They are volumetric, gravimetric and comparison.

   4.1.1 Volumetric

   In these methods the volume of fluid flowing is accurately measured and timed as it passes through the rotameter into the collecting chamber at a controlled rate. Instruments in this group include gasometers, burettes, and stand pipe systems which provide measurements directly in basic volumetric terms.

   Liquids

   For liquid calibrations, the collecting chamber may vary from a small Bureau of

Standards certified burette for calibrating very small rotameters to large accurately calibrated receiving tanks for calibrating larger instruments. (Refer to Figure RP16.6.2)

Formula used with Figure RP16.6.2 — Liquid Calibration

$$Q_m = \frac{V_c}{Sec} \times 60 \times \sqrt{\frac{(\rho_f - \rho_m)\, \rho_c}{(\rho_f - \rho_c)\, \rho_m}}$$

Where:

Q_m = Volumetric flow rate of liquid to be metered in units per minute

V_c = Volume of calibrating liquid collected in units consistent with Q_m

Sec = Collection time in seconds

ρ_f = Density of metering float in grams/cc

ρ_m = Density of liquid to be metered in grams/cc

ρ_c = Density of calibrating liquid in grams/cc

Liquid to Gases

Rotameter sizes 1/2" and larger may be calibrated volumetrically using water as the flowing medium. GPM of liquid, specific gravity 1.0 times 4.15 equals SCFM gas, specific gravity 1.0, when metered at 0 psig and 70°F.

Gases

For calibrating rotameters with gases, a gasometer as illustrated in Figure RP-16.6.3 is recommended. In this method, gas at controlled flow rate, temperature, and pressure flows through the rotameter and is collected in the inverted bell.

Formulas used with Figure RP16.6.3 — Gas Calibration

To determine the calibration of the rotameter undergoing test, use the following formulas:

1. $QStd. = \dfrac{V_g}{Sec} \times \dfrac{60 \times S_c \times (P_b + P_g) \times 530}{S_o \times 14.7 \times T_g} \times \sqrt{\dfrac{V_c}{V_1}}$

2. $QAct. = \dfrac{V_g}{Sec} \times \dfrac{60 \times S_c \times (P_b + P_g) \times 530 \times V_1}{V_1 \times 14.7 \times T_g \times 13.35} \times \sqrt{\dfrac{V_c}{V_1}}$

3. $W = \dfrac{V_g}{Sec} \times \dfrac{60 \times 0.075 \times S_c \times (P_b + P_g) \times 530}{14.7 \times T_g} \times \sqrt{\dfrac{V_c}{V_1}}$

Where:

$QStd$ = Std. CFM at 14.7 psia & 70°F.

$QAct$ = Actual CFM (volume at metering temperature and pressure)

W = PPM Flow Rate

V_g = Volume gas collected in cu. ft.

Sec = Collection time, seconds

T_g = Gasometer temperature, degrees Rankine (460 + °F.)

P_b = Barometric pressure at calibration (psia)

P_g = Static pressure of Gasometer (psi) (in. H_2O x .0361)

S_c = Specific Gravity of calibrating gas at 14.7 psia and 70°F. (Air = 1.0)

S_o = Specific Gravity of gas to be metered at 14.7 psia and 70°F.

*V_c = Specific Volume of calibrating gas at calibration temperature and pressure, cu. ft./lb.

*V_1 = Specific Volume of gas to be metered at metering temperature and pressure, cu. ft./lb.

* For perfect gases

$$V_c = 13.35 \times \frac{14.7}{P_c} \times \frac{T_c}{530} \times \frac{1}{S_c}$$

Where:

P_c = metering pressure at calibration, psia

T_c = metering temperature at calibration, degrees Rankine (460 + °F.)

$$V_1 = 13.35 \times \frac{14.7}{P_1} \times \frac{T_1}{530} \times \frac{1}{S_o}$$

Where:

P_1 = operating metering pressure, psia

T = operating metering temperature, degrees Rankine (460 + °F.)

Note: Standard conditions in rotameter industry usually stated as 14.7 psia and 70°F.

4.1.2 Gravimetric

In this method the liquid at controlled flow rates passes through the rotameter under test and is collected in a receiving tank. The weight of liquid is accurately measured by a precision scale or other weighing device. Flowing time is accurately recorded and flow rate in gravimetric units computed directly by dividing scale readings by time. Automatic operation may be achieved by means of photocell or mercury contact timing devices. Equipment for this method of measurement weighs the liquid passing through the rotameter under test and provides flow rate readings in gravimetric or weight units. Unlike volumetric systems, the readings taken from the gravimetric systems are independent of density or viscosity of the fluid. (Refer to Figure RP16.6.4)

Formula used with Figure RP16.6.4 — Liquid Calibration

$$W_m = \frac{W_c \times 60}{Sec} \sqrt{\frac{(\rho_f - \rho_m)\,\rho_m}{(\rho_f - \rho_c)\,\rho_c}}$$

Where:

W_m = Mass flow rate of fluid to be metered in pounds per minute

W_c = Weight of calibrating fluid collected in pounds

Sec = Collection time in seconds

ρ_f = Density of metering float

ρ_m = Density of liquid to be metered

ρ_c = Density of calibrating liquid

4.1.3 Comparison

In this method, liquid and gas calibrations are made by comparing the rotameter under test against an accurately calibrated flow measuring instrument (See Figure RP16.6.5). Calibration accuracy will depend entirely on the accuracy of the calibrated flow measuring instrument serving as a secondary standard.

4.2 Kinematic Simulation

When it is impractical to calibrate rotameters directly because of viscosity considerations, hazards to the operator or equipment, or for economic reasons, kinematic simulation of fluid properties is employed. Through the use of a variety of fluid media and by maintaining controlled temperature conditions, viscosities and densities over a wide range can be achieved to match those physical properties of the fluid for which the rotameter is designed.

Formulas used are the same as those under 4.1.

Kinematic simulation follows the equation:

$$\text{Centistokes cal. fluid} = \text{Centistokes Metering Fluid} \times \sqrt{\frac{\text{Fluid density metering fluid}}{\text{Fluid density calibration fluid}}}$$

4.3 Field Check

Any of the above methods may be used to check rotameters. Tank gaging may be used for checking liquid handling rotameters and liquid displacement methods may be used for rotameters handling gases. Actual results will be commensurate with the care employed in the calibration and the precision of the facilities.

VISCOSITY CEILING CURVES

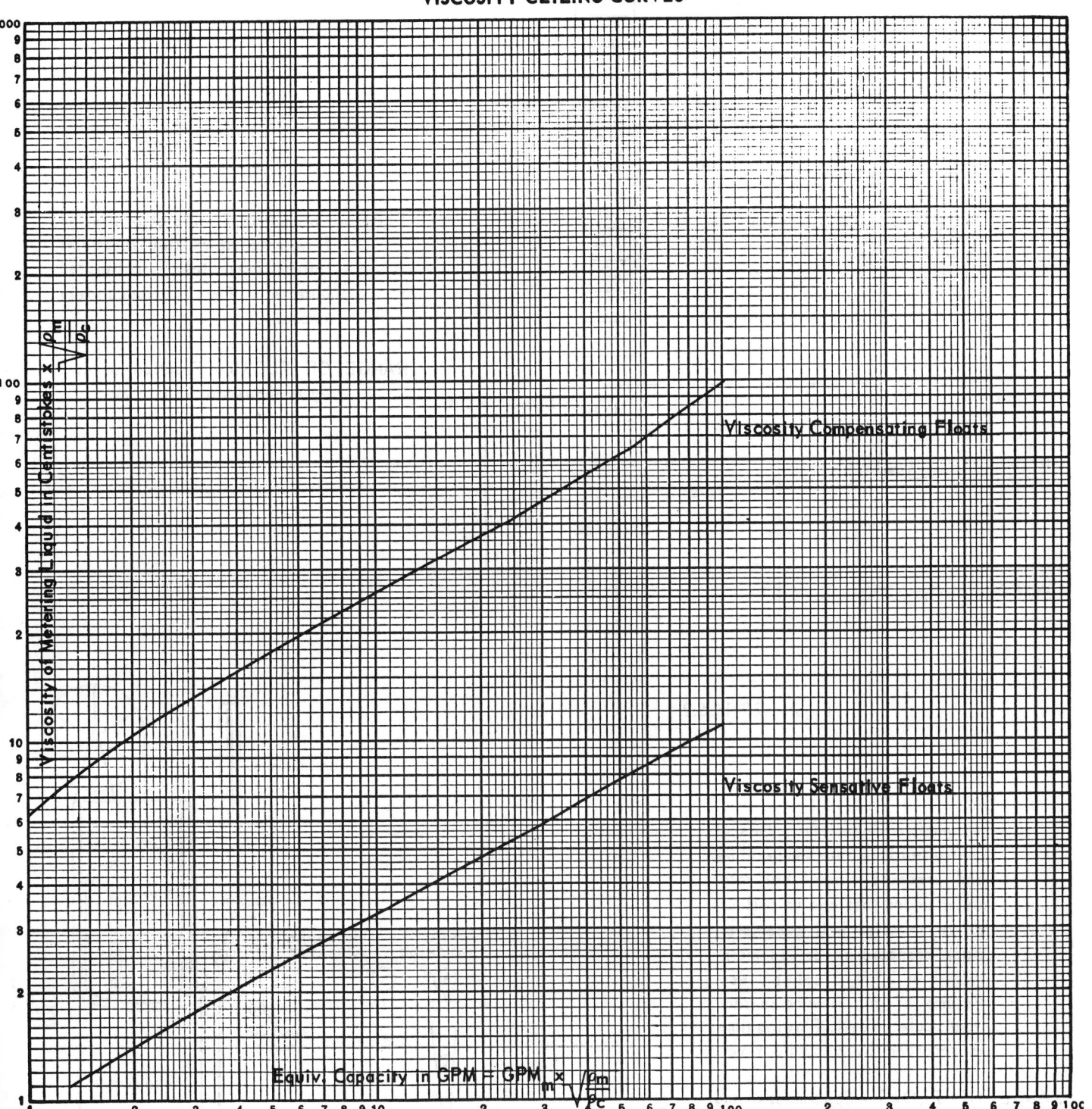

To determine whether the viscosity of the metering liquid will permit calibration with a reference liquid, calculate equivalent capacity in GPM and plot capacity vs viscosity on above curve. If this point falls below the viscosity ceiling curve for the float being used, calibration with the reference medium is permissable. If the point is above the curve, calibration by hydraulic simulation or with the actual metering liquid is required.

Figure RP 16.6.1

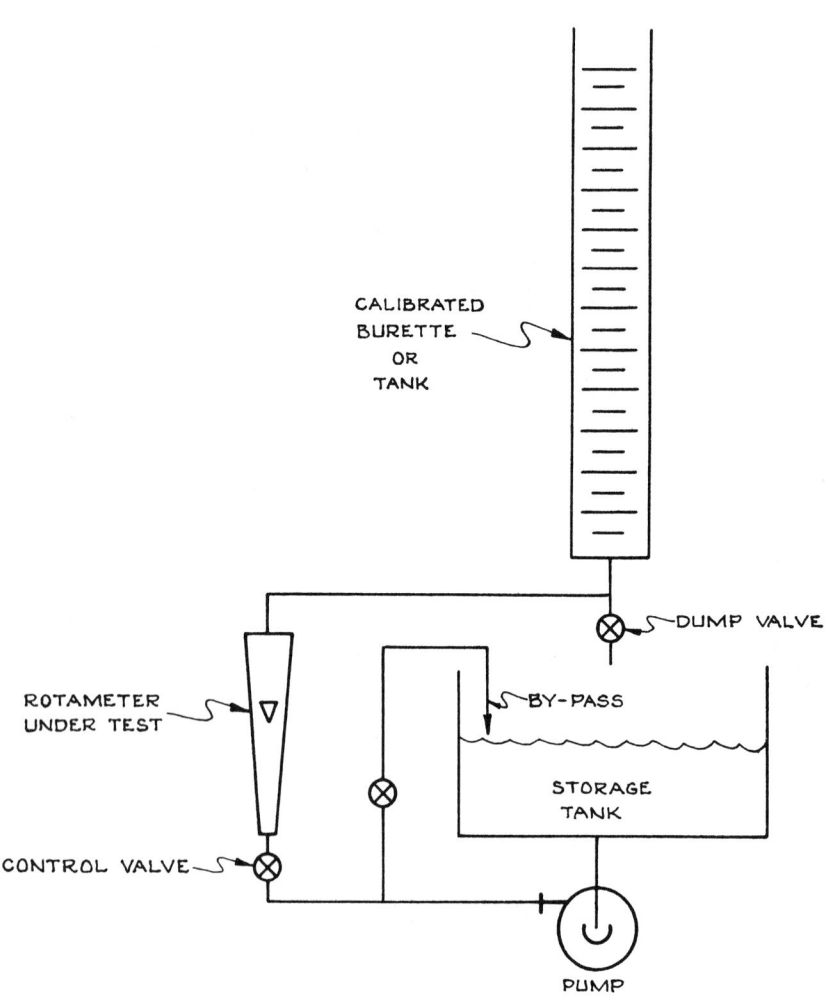

CALIBRATED
BURETTE
OR
TANK

DUMP VALVE

ROTAMETER
UNDER TEST

BY-PASS

STORAGE
TANK

CONTROL VALVE

PUMP

Figure RP 16.6.2

TYPE – Volumetric. Liquid at controlled flow rate flows thru rotameter into accurately calibrated burette. Time is accurately measured.

RANGE OF FLOWS – Fraction of c.c. Up to thousands of G.P.M.

TYPE – Volumetric Gasometer. Gas at controlled flow rate, temperature and pressure flows thru rotameter under test and is collected in inverter bell gasometer. Back pressure in bell is maintained at 2" H_2O col. or less. Volume of gas flowing is read from scale on bell. Flowing time is recorded by accurate timer.

RANGE OF FLOWS – Spirometers and gasometers are available for flow ranges from 10 cc/min to more than 200 SCFM.

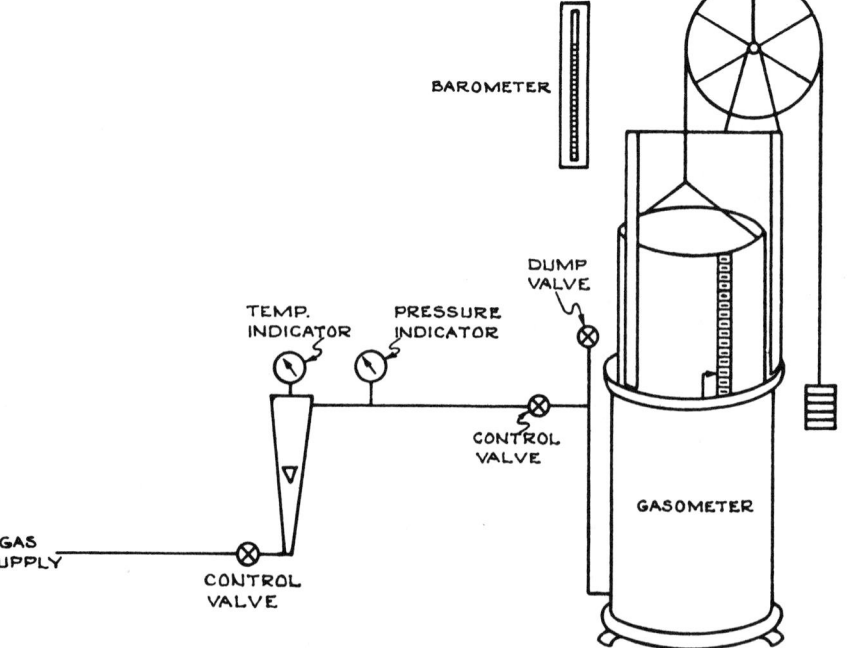

BAROMETER

DUMP
VALVE

TEMP.
INDICATOR

PRESSURE
INDICATOR

CONTROL
VALVE

GASOMETER

GAS
SUPPLY

CONTROL
VALVE

Figure RP 16.6.3

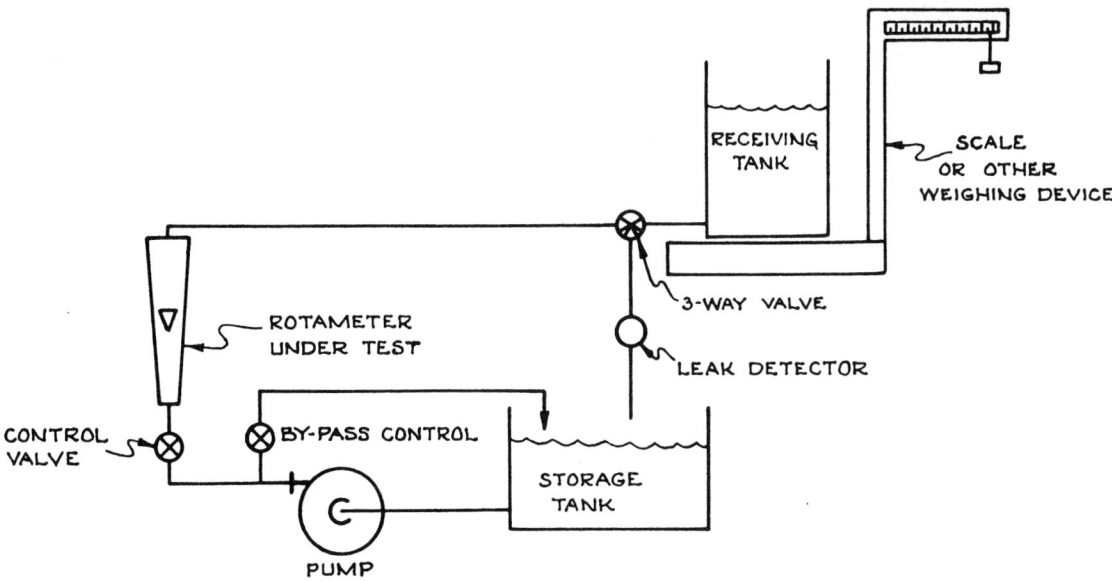

Figure RP 16.6.4

TYPE – Gravimetric. Liquid at controlled flow rate passes thru meter and is collected in receiving tank. Weight of liquid is accurately measured by scale or other weighing device flowing time is accurately recorded.

RANGE OF FLOWS – 10 P.P.H. up

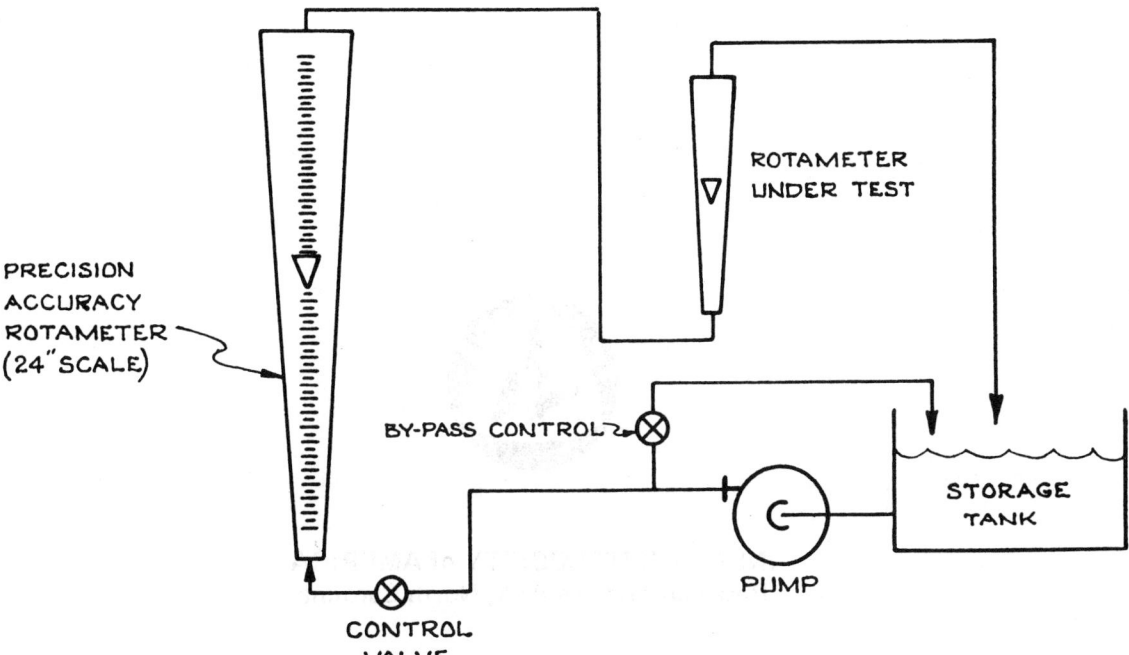

Figure RP 16.6.5

TYPE – Comparison Method. Rotameter under test is connected in series with accurately calibrated rotameter. Fluid is pumped thru system. Readings from precision rotameter and
corresponding readings from rotameter under test are recorded.

RANGE OF FLOWS – 0.1 to 150 GPM

INSTRUMENT SOCIETY of AMERICA
Research Triangle Park, North Carolina

ANSI/ISA-S18.1-1979
(R1985)
Approved March 29, 1985

American National Standard

Annunciator Sequences and Specifications

Instrument Society of America

ISBN 0-87664-346-2

ISA-S18.1   Annunciator Sequences and Specifications

INSTRUMENT SOCIETY OF AMERICA
67 Alexander Drive
P.O. Box 12277
Research Triangle Park, North Carolina 27709

PREFACE

This Preface is included for informational purposes and is not a part of Standard S18.1.

This Standard has been prepared as a part of the service of the Instrument Society of America toward a goal of uniformity in the field of instrumentation. To be of real value, this document should not be static but should be subjected to periodic review. Toward this end, the Society welcomes all comments and criticisms, and asks that they be addressed to the Standards and Practices Board Secretary, Instrument Society of America, 67 Alexander Drive, P.O. Box 12277, Research Triangle Park, North Carolina 27709.

Based on work started in 1955 by a survey committee titled Instrument Alarms and Interlocks, the 8D-RP18 Committee on Annunciator Systems of the Production Processes was formed in 1959. Tentative Recommended Practice ISA-RP18.1, titled Specifications and Guides for the Use of General Purpose Annunciators, was completed by that Committee in 1965.

The committee, reactivated as Committee SP18, Instrument Signals and Alarms, began revising ISA-RP18.1 in 1976 to reflect current industry practice for annunciators. While ISA-S18.1 is a major revision of ISA-RP18.1, the basic purpose and scope have not changed. The Standard has been updated generally, new sequence letter, option number, and first out designations have replaced the original standard sequences, and improvements have been made in the sequence presentation. Appendices provide an application guide for use when specifying annunciators and a conversion list of new sequence designations that correspond to the superseded ISA-RP18.1 sequences.

The ISA Standards and Practices Department is aware of the growing need for attention to the metric system of units in general, and the International System of Units (SI) in particular, in the preparation of instrumentation standards. The Department is further aware of the benefits to USA users of ISA Standards of incorporating suitable references to the SI (and the metric system) in their business and professional dealings with other countries. Toward this end this Department will endeavor to introduce SI and SI-acceptable metric units in all new and revised standards to the greatest extent possible. The Standard for Metric Practice, which has been published by the American Society for Testing and Materials as ANSI Z210.1 (ASTM E380-76/IEEE 268-1975), and future revisions, will be the reference guide for definitions, symbols, abbreviations, and conversion factors.

COMMITTEE 8D-RP18 (Original Committee)

Name	Company
A. P. McCauley, Chairman	The Glidden Company
L. Kurfis	Toledo-Edison Company
P. McCauchna	Mobile Oil Company
H. Wilder	Monsanto Chemical Company, Plastics Division
D. Wilhelm	Owens-Illinois Glass Company

COMMITTEE SP18 (Current Committee)

Name	Company
Richard L. Emerson, Chairman	Bechtel Power Corporation
Mitchell W. Araman	Rochester Instrument Systems, Inc.
Robert F. Barber	Ralph M. Parsons Company
James T. Campbell	Fluor Engineers & Constructors, Inc.
Claude W. Drake	Southern California Edison Company
David J. A. Hewitson	Ronan Engineering Company
Kamal S. Iskander	Los Angeles Department of Water and Power
Howard L. Joseph	Tripace Engineering Sales Company (representing Beta Products, Inc.)
C. S. Lisser	Oak Ridge National Laboratory
Frank H. Molinari	Jensen Instrument Company (representing Panalarm Division of The Riley Company)
Donald E. Swartz	Brown & Caldwell Consulting Engineers
Jonathan T. Y. Yeh	C. F. Braun & Co.

TABLE OF CONTENTS

LIST OF ILLUSTRATIONS

APPENDICES

LIST OF ILLUSTRATIONS

1 PURPOSE

The purpose of this Standard is to establish uniform annunciator terminology, sequence designations, and sequence presentation and to assist in the preparation of annunciator specifications and documentation.

This Standard is intended to improve communications among those that specify, distribute, manufacture, or use annunciators.

2 SCOPE

This Standard is primarily for use with electrical annunciators that call attention to abnormal process conditions by the use of individual illuminated visual displays and audible devices. Annunciators can range from a single annunciator cabinet, to complex annunciator systems with many lamp cabinets and remote logic cabinets.

The sequence designations provided can be used to describe basic annunciator sequences and also many sequence variations. This Standard lists types of information that should be included in annunciator specifications and types of documents that should be provided by manufacturers; however, detailed design requirements and documentation formats are beyond the scope of this Standard.

3 DEFINITION OF TERMS

The following are terms and their definitions that have special meaning in relation to annunciators. Commonly used alternate terms are shown in parentheses. Defined terms used in other definitions are in italics to provide a cross-reference.

acknowledge — the *sequence action* that indicates recognition of a new alarm.

active alarm point — see *alarm point*.

alarm — 1. an abnormal *process condition*. 2. the *sequence state* when an abnormal *process condition* occurs. 3. a device that calls attention to the existence of an abnormal *process condition*. See *annunciator*. Types of alarm include:

> **momentary** — an alarm that returns to normal before being acknowledged.

> **maintained** — an alarm that returns to normal after being acknowledged.

alarm module (point or sequence module) — a plug-in assembly containing the sequence logic circuit. Some alarm modules also contain *visual display* lamps or lamps and *windows*.

alarm point — the sequence logic circuit, *visual display*, auxiliary devices, and internal wiring related to one *visual display*. Types of alarm point include:

> **active** — an alarm point that is wired internally and completely equipped. The *window* is labeled to identify a specific monitored variable.

spare — an alarm point that is wired internally and completely equipped. The *window* is not labeled to identify a monitored variable.

future (blank) — an alarm point that is wired internally and equipped except for the plug-in *alarm module*. The *window* is not labeled to identify a monitored variable.

alert — see *process condition* and *sequence state*.

analog input point — an *alarm point* for use with an analog monitored variable signal, usually current or voltage. The logic circuit initiates an *alarm* when the analog signal is above or below a set point.

annunciator — a device or group of devices that call attention to changes in *process conditions* that have occurred. An annunciator usually calls attention to abnormal *process conditions*, but may be used also to show normal process status. Usually included are sequence logic circuits, labeled *visual displays*, *audible devices*, and manually operated *pushbuttons*.

audible device — a device that calls attention by sound to the occurrence of abnormal *process conditions*. An audible device may also call attention to return to normal conditions.

audible device follower — see *auxiliary output*.

automatic reset — see *reset*.

auxiliary contact — see *auxiliary output*.

auxiliary output (auxiliary contact) — An output signal operated by a single *alarm point* or group of points for use with a remote device. Types of auxiliary output include:

> **field contact follower** — an auxiliary output that operates while the *field contact* indicates an abnormal *process condition*.

> **lamp follower** — an auxiliary output that operates while the *visual display* lamps indicate an *alarm*, silenced, or acknowledged state.

> **audible device follower (horn relay contact)** — an auxiliary output that operates while the common alarm *audible device* operates.

> **reflash** — an auxiliary output that operates when any one of a group of *alarm points* indicates an abnormal *process condition*. The output usually returns to normal briefly when each *alarm point* changes to an abnormal *process condition* and returns to normal when all *alarm points* in the group indicate normal *process conditions*.

blank alarm point — see *alarm point*.

field contact (trouble or signal contact) — the electrical contact of the device sensing the *process condition*. The contact is either open or closed. Annunciator field contacts are identified in relation to process conditions and annunciator operation, not the disconnected position of the devices. Types of field contact include:

normally open (NO) — a field contact that is open for a normal *process condition* and closed when the *process condition* is abnormal.

normally closed (NC) — A field contact that is closed for a normal *process condition* and open when the *process condition* is abnormal.

field contact follower — see *auxiliary output*.

field contact voltage (trouble or signal contact voltage) — the voltage applied to *field contacts*.

first alert — see *first out*.

first out (first alert) — a *sequence* feature that indicates which of a group of *alarm points* operated first.

first out reset — see *reset*.

flasher — a device that causes *visual displays* to turn on and off repeatedly. Types of flashing include fast flashing, flashing, slow flashing, and intermittent flashing.

functional test — see *test*.

future alarm point — see *alarm point*.

horn relay contact — see *auxiliary output*.

integral logic annunciator — an *annunciator* that includes *visual displays* and sequence logic circuits in one assembly.

lamp cabinet — a cabinet containing *visual displays* only.

lamp follower — see *auxiliary output*.

lamp test — see *test*.

lock-in — a *sequence* feature that retains the alarm state until acknowledged when the abnormal *process condition* is *momentary*.

logic cabinet — a cabinet containing logic circuits and no *visual displays*.

maintained alarm — see *alarm*.

manual reset — see *reset*.

momentary alarm — see *alarm*.

multiple input — see *reflash*.

nameplate — see *window*.

normally closed (NC) — see *field contact*.

normally open (NO) — see *field contact*.

operational test — see *test*.

point module — see *alarm module*.

process condition — the condition of the monitored variable. The process condition is either normal or abnormal (*alarm*, alert, or off-normal).

pushbutton — A momentary manual switch that causes a change from one *sequence state* to another. Pushbutton actions include *silence*, *acknowledge*, *reset*, *first out reset*, and *test*.

reflash (multiple input) — 1. an auxiliary logic circuit that allows two or more abnormal *process conditions* to initiate or reinitiate the *alarm* state of one *alarm point* at any time. The *alarm point* cannot return to normal until all related *process conditions* return to normal. 2. one type of *auxiliary output*.

remote logic annunciator — an *annunciator* that locates *visual displays* and sequence logic circuits in separate assemblies.

reset — the *sequence action* that returns the *sequence* to the normal state. Types of reset include:

automatic — reset occurs after *acknowledge* when the *process condition* returns to normal.

manual — reset occurs after *acknowledge* when the *process condition* has returned to normal and the reset *pushbutton* is operated.

first out — reset of the *first out* indication occurs when the *acknowledge* or *first out* reset *pushbutton* is operated, whether the *process condition* has returned to normal or not, depending on the sequence.

response time — the time period between the *process condition* becoming abnormal and initiation of the *alarm* state. The minimum *momentary alarm* duration required for *annunciator* operation.

return alert — see *ringback*.

ringback (return alert) — a *sequence* feature that provides a distinct visual or audible indication or both when the *process condition* returns to normal.

sequence — the chronological series of actions and states of an *annunciator* after an abnormal *process condition* or manual *test* initiation occurs.

sequence action — a signal that causes the *sequence* to change from one *sequence state* to another. Sequence actions include *process condition* changes and manual operation of *pushbuttons*.

sequence diagram — a graphic presentation that describes *sequence actions* and *sequence states*.

sequence module — see *alarm module*.

sequence state — the condition of the *visual display* and *audible device* provided by an *annunciator* to indicate the *process condition* or *pushbutton* actions or both. Sequence states include normal, *alarm* (alert), silenced, acknowledged, and *ringback*.

sequence table — a presentation that describes *sequence actions* and *sequence states* by lines of statements arranged in columns.

signal contact — see *field contact*.

signal contact voltage — see *field contact voltage*.

silence — the *sequence action* that stops the sound of the *audible device*.

spare alarm point — see *alarm point*.

test — an *annunciator sequence* initiated by operation of the test *pushbutton* to reveal lamp or circuit failure. Types of test include:

　operational (functional) — test of the *sequence*, *visual display* lamps, *audible devices*, and *pushbuttons*.

　lamp — test of the *visual display* lamps.

trouble contact — see *field contact*.

trouble contact voltage — see *field contact voltage*.

visual display — that part of an *annunciator* or *lamp cabinet* that indicates the *sequence state*. Usually consists of an enclosure containing lamps behind a translucent *window*. The lamps can be off, flashing, or on.

window (nameplate) — a component of a *visual display* made from a translucent material that is illuminated from the rear and labeled to identify the monitored variable.

4 SEQUENCES

4.1 Operation

Annunciators usually call attention to abnormal process conditions by the use of individual illuminated visual displays and audible devices. Annunciators may also be used to show normal process status. Changes from one annunciator sequence state to another are caused by changes in process conditions and also by manual operation of pushbuttons. The new sequence state may be dependent on the process condition that exists at the time pushbuttons are operated. Process condition changes are usually sensed by field contacts.

The visual displays usually flash to indicate abnormal process conditions and change to on when alarms are acknowledged. Additional types of flashing can indicate that process conditions have returned to normal or which of a group of alarm points operated first. All of the alarm points of an annunciator usually use the same sequence; however, different sequences can be used for individual alarm points or groups of points in one annunciator.

In this Standard, sequences making use of more than one indication device as a part of each visual display to indicate the sequence state are considered to be special because of their many variations and relatively infrequent use. Examples include the use of lamps of different colors to indicate different sequence states or which is the first of a group of alarms.

4.2 Presentation

Annunciator sequence tables describe the operation of annunciators, but often do not clearly indicate all aspects of the sequences. Examples include failure to indicate the sequence actions and states when process conditions return to abnormal again before the annunciator is reset and also when pushbuttons are operated out of the normal sequence. A sequence diagram format is used in this Standard to allow annunciator sequences to be defined completely and analyzed logically. See Figures 2 to 8, pages 9 to 15.

Sequence diagrams include a block for each annunciator sequence state. The process condition, the sequence state, and the visual display and audible device conditions when in that state are indicated in each block. The blocks are arranged to describe the annunciator sequence from the normal state, through the other sequence states, and back to the normal state again. Arrows between the blocks indicate all possible sequence actions that can cause a change from one sequence state to another. Sequence actions include process condition changes and manual operation of pushbuttons.

Sequence tables are also used in this Standard since it is not always convenient to use sequence diagrams. These sequence tables are patterned after the sequence diagrams to describe all aspects of the annunciator operation. See Figures 2 to 8, pages 9 to 15.

The sequence tables include a line for the initial normal state and also a line for each possible sequence action that can cause a change from one sequence state to another. The reference line numbers in the tables are identified by suffixes A and B when the new sequence state depends on the process condition that exists at the time pushbuttons are operated. References to other lines in the table are used to avoid indicating each sequence state and the related visual display and audible device conditions more than once.

When annunciators require auxiliary outputs, the output operation should be added to sequence diagrams and sequence tables or the operation should be defined by notes. The operation of auxiliary outputs during annunciator test should be defined also.

Since most annunciators include a test pushbutton and operational test of the sequence, visual display lamps, audible devices, and pushbuttons, the sequences in this Standard include operational test as a standard feature.

4.3 Designation Method

Annunciators call attention to changes in process conditions by different visual display and audible device arrangements and by a wide variety of operating sequences. The choice depends on the requirements or preferences of the users and also on the standard or special annunciator designs that are available.

This Standard provides a sequence designation method using letters for basic sequences in common use, numbers for common sequence options, and first out designations. Combinations of letters and numbers can define many different sequence variations. Sequence designation examples and a summary of the basic sequence letters, option numbers, and first out designations are provided in Figure 1, page 8. This Standard does not designate any particular sequences as being standard.

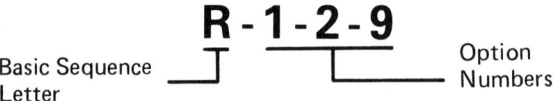

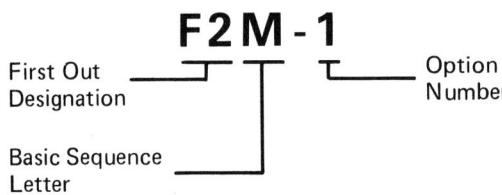

Basic Sequence Letter	Key Words
A	Automatic Reset
M	Manual Reset
R	Ringback

(See 4.4 for descriptions.)

First Out Designation	Key Words
F1	No Subsequent Alarm State
F2	No Subsequent Alarm Flashing
F3	First Out Flashing and Reset Pushbutton

(See 4.6 for descriptions.)

Option Number	Key Words
1	Silence Pushbutton
2	Silence Interlock
3	First Out Reset Interlock
4	No Lock-In
5	No Flashing
6	No Audible
7	Automatic Alarm Silence
8	Common Ringback Audible
9	Automatic Ringback Silence
10	No Ringback Audible
11	Common Ringback Visual
12	Automatic Momentary Ringback
13	Dim Lamp Monitor
14	Lamp Test

(See 4.5 for descriptions.)

Figure 1. Annunciator Sequence Designations

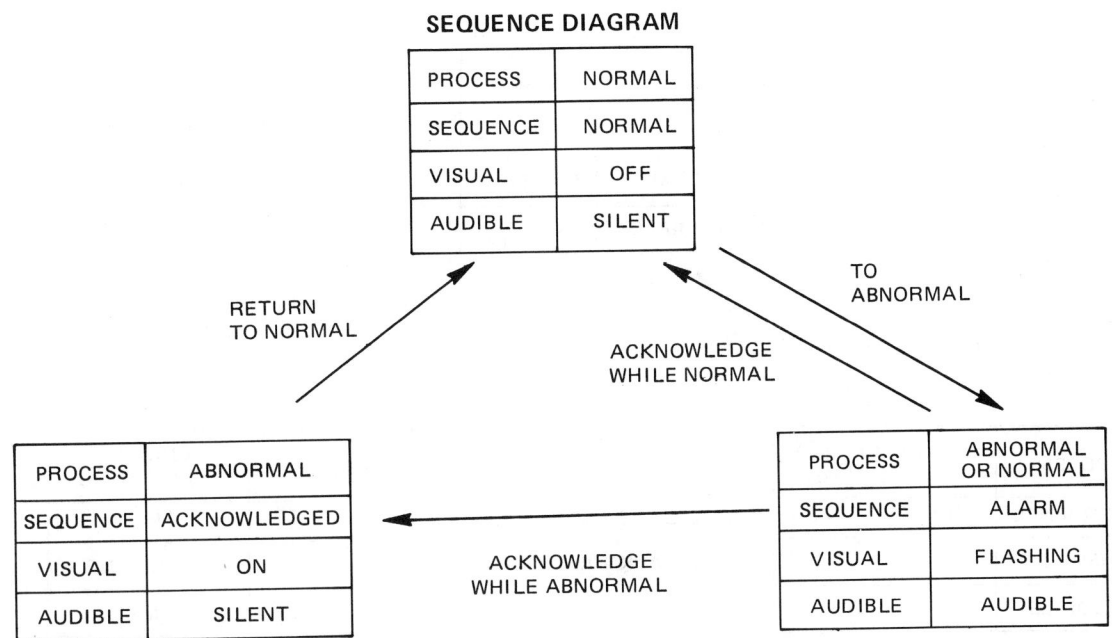

SEQUENCE DIAGRAM

PROCESS	NORMAL
SEQUENCE	NORMAL
VISUAL	OFF
AUDIBLE	SILENT

RETURN
TO NORMAL

TO
ABNORMAL

ACKNOWLEDGE
WHILE NORMAL

PROCESS	ABNORMAL
SEQUENCE	ACKNOWLEDGED
VISUAL	ON
AUDIBLE	SILENT

ACKNOWLEDGE
WHILE ABNORMAL

PROCESS	ABNORMAL OR NORMAL
SEQUENCE	ALARM
VISUAL	FLASHING
AUDIBLE	AUDIBLE

SEQUENCE TABLE

LINE	PROCESS CONDITION	PUSHBUTTON OPERATION	SEQUENCE STATE	VISUAL DISPLAY	ALARM AUDIBLE DEVICE	REMARKS
1	NORMAL	—	NORMAL	OFF	SILENT	
2	ABNORMAL	—	ALARM	FLASHING	AUDIBLE	LOCK-IN
3A	ABNORMAL	ACKNOWLEDGE	ACKNOWLEDGED	ON	SILENT	MAINTAINED ALARM
3B	NORMAL		TO LINE 4			MOMENTARY ALARM
4	NORMAL	—	NORMAL	OFF	SILENT	AUTOMATIC RESET

SEQUENCE FEATURES
1 — ACKNOWLEDGE AND TEST PUSHBUTTONS.
2 — ALARM AUDIBLE DEVICE.
3 — LOCK-IN OF MOMENTARY ALARMS UNTIL ACKNOWLEDGED.
4 — THE AUDIBLE DEVICE IS SILENCED AND FLASHING STOPS WHEN ACKNOWLEDGED.
5 — AUTOMATIC RESET OF ACKNOWLEDGED ALARM INDICATIONS WHEN PROCESS CONDITIONS RETURN TO NORMAL.
6 — OPERATIONAL TEST.

Figure 2. Sequence A, Automatic Reset

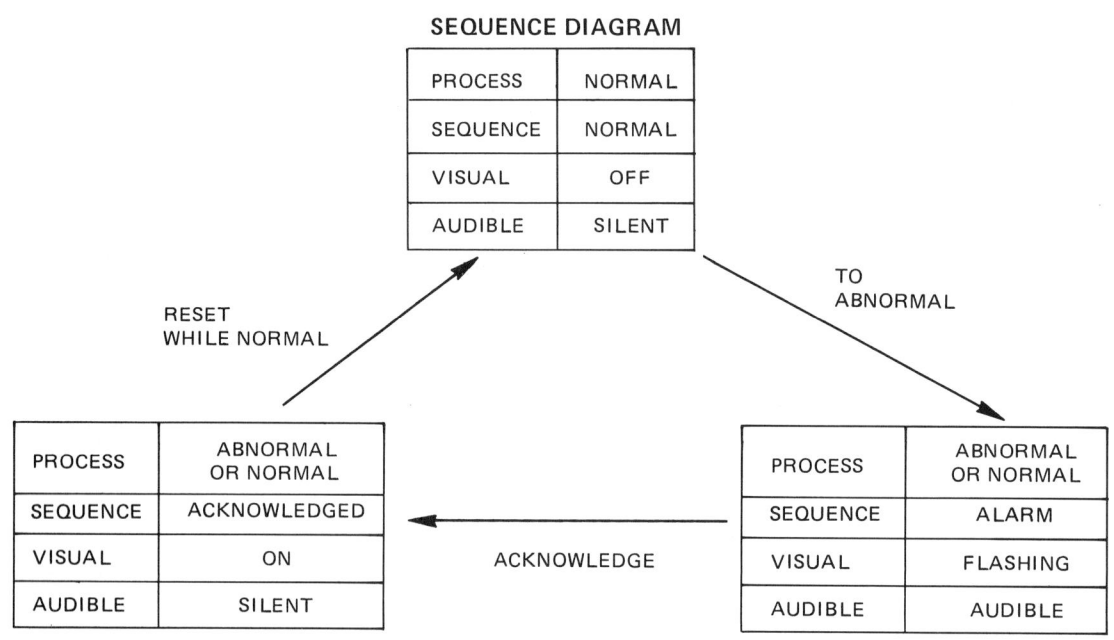

SEQUENCE DIAGRAM

PROCESS	NORMAL
SEQUENCE	NORMAL
VISUAL	OFF
AUDIBLE	SILENT

TO ABNORMAL

RESET WHILE NORMAL

PROCESS	ABNORMAL OR NORMAL
SEQUENCE	ACKNOWLEDGED
VISUAL	ON
AUDIBLE	SILENT

ACKNOWLEDGE

PROCESS	ABNORMAL OR NORMAL
SEQUENCE	ALARM
VISUAL	FLASHING
AUDIBLE	AUDIBLE

SEQUENCE TABLE

LINE	PROCESS CONDITION	PUSHBUTTON OPERATION	SEQUENCE STATE	VISUAL DISPLAY	ALARM AUDIBLE DEVICE	REMARKS
1	NORMAL	—	NORMAL	OFF	SILENT	
2	ABNORMAL	—	ALARM	FLASHING	AUDIBLE	LOCK-IN
3	ABNORMAL OR NORMAL	ACKNOWLEDGE	ACKNOWLEDGED	ON	SILENT	MANUAL RESET REQUIRED
4A	ABNORMAL	RESET	TO LINE 3			
4B	NORMAL		NORMAL	OFF	SILENT	MANUAL RESET

SEQUENCE FEATURES

1 — ACKNOWLEDGE, RESET, AND TEST PUSHBUTTONS.
2 — ALARM AUDIBLE DEVICE.
3 — LOCK-IN OF MOMENTARY ALARMS UNTIL ACKNOWLEDGED.
4 — THE AUDIBLE DEVICE IS SILENCED AND FLASHING STOPS WHEN ACKNOWLEDGED.
5 — MANUAL RESET OF ACKNOWLEDGED ALARM INDICATIONS AFTER PROCESS CONDITIONS RETURN TO NORMAL.
6 — OPERATIONAL TEST.

Figure 3. Sequence M, Manual Reset

SEQUENCE DIAGRAM

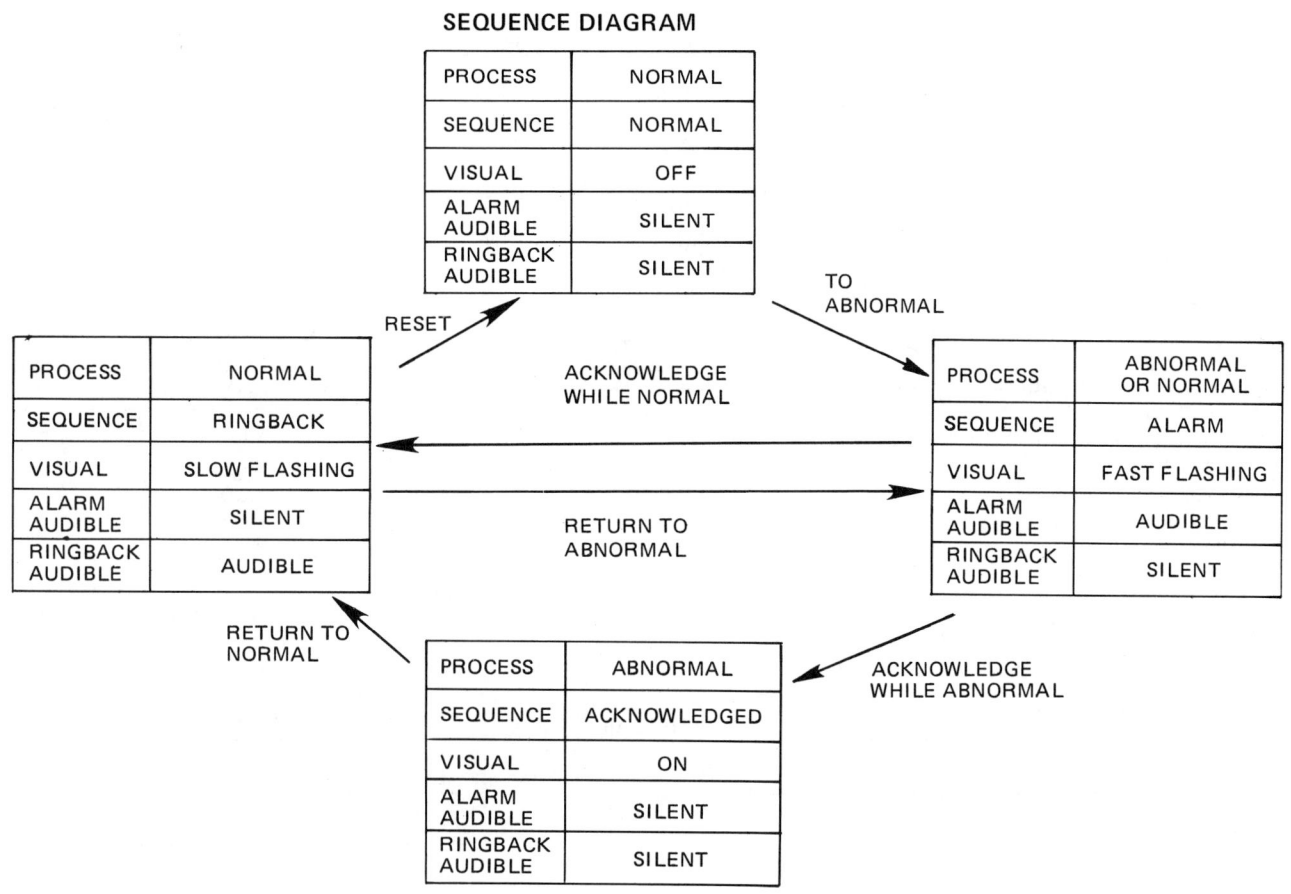

SEQUENCE TABLE

LINE	PROCESS CONDITION	PUSHBUTTON OPERATION	SEQUENCE STATE	VISUAL DISPLAY	ALARM AUDIBLE DEVICE	RINGBACK AUDIBLE DEVICE	REMARKS
1	NORMAL	—	NORMAL	OFF	SILENT	SILENT	
2	ABNORMAL	—	ALARM	FAST FLASHING	AUDIBLE	SILENT	LOCK-IN
3A	ABNORMAL	ACKNOWLEDGE	ACKNOWLEDGED	ON	SILENT	SILENT	MAINTAINED ALARM
3B	NORMAL		TO LINE 4				MONENTARY ALARM
4	NORMAL	—	RINGBACK	SLOW FLASHING	SILENT	AUDIBLE	MANUAL RESET REQUIRED
5	ABNORMAL	—	TO LINE 2				RETURN TO ABNORMAL
6	NORMAL	RESET	NORMAL	OFF	SILENT	SILENT	MANUAL RESET

SEQUENCE FEATURES

1 — ACKNOWLEDGE, RESET, AND TEST PUSHBUTTONS.
2 — ALARM AND RINGBACK AUDIBLE DEVICES.
3 — LOCK-IN OF MOMENTARY ALARMS UNTIL ACKNOWLEDGED.
4 — THE AUDIBLE DEVICE IS SILENCED AND FAST FLASHING STOPS WHEN ACKNOWLEDGED.
5 — RINGBACK VISUAL AND AUDIBLE INDICATIONS WHEN PROCESS CONDITIONS RETURN TO NORMAL.
6 — MANUAL RESET OF RINGBACK INDICATIONS.
7 — OPERATIONAL TEST.

Figure 4. Sequence R, Ringback

SEQUENCE DIAGRAM

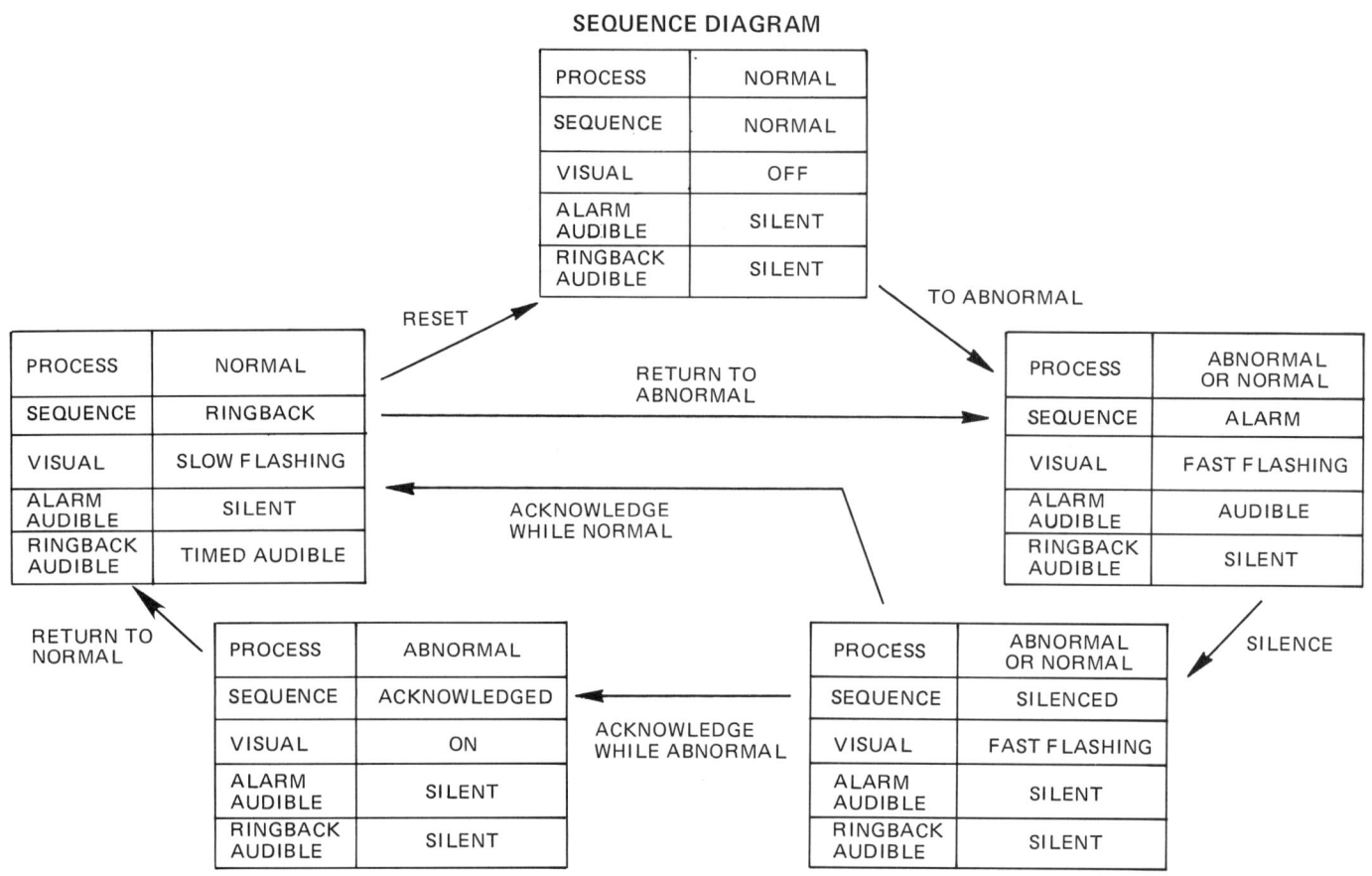

SEQUENCE TABLE

LINE	PROCESS CONDITION	PUSHBUTTON OPERATION	SEQUENCE STATE	VISUAL DISPLAY	ALARM AUDIBLE DEVICE	RINGBACK AUDIBLE DEVICE	REMARKS
1	NORMAL	—	NORMAL	OFF	SILENT	SILENT	
2	ABNORMAL	—	ALARM	FAST FLASHING	AUDIBLE	SILENT	LOCK-IN
3	ABNORMAL OR NORMAL	SILENCE	SILENCED	FAST FLASHING	SILENT	SILENT	LOCK-IN
4A	ABNORMAL	ACKNOWLEDGE	ACKNOWLEDGED	ON	SILENT	SILENT	MAINTAINED ALARM
4B	NORMAL		TO LINE 5				MOMENTARY ALARM
5	NORMAL	—	RINGBACK	SLOW FLASHING	SILENT	TIMED AUDIBLE	MANUAL RESET REQUIRED
6	ABNORMAL	—	TO LINE 2				RETURN TO ABNORMAL
7	NORMAL	RESET	NORMAL	OFF	SILENT	SILENT	MANUAL RESET

SEQUENCE FEATURES

1 — SILENCE, ACKNOWLEDGE, RESET, AND TEST PUSHBUTTONS.
2 — ALARM AND RINGBACK AUDIBLE DEVICES.
3 — LOCK-IN OF MOMENTARY ALARMS UNTIL ACKNOWLEDGED.
4 — OPTION 1 — SILENCE PUSHBUTTON TO SILENCE THE ALARM AUDIBLE DEVICE WHILE RETAINING FAST FLASHING INDICATIONS.
5 — OPTION 2 — SILENCE INTERLOCK TO REQUIRE OPERATION OF THE SILENCE PUSHBUTTON BEFORE THE ACKNOWLEDGE PUSHBUTTON.
6 — RINGBACK VISUAL AND AUDIBLE INDICATIONS WHEN PROCESS CONDITIONS RETURN TO NORMAL.
7 — OPTION 9 — AUTOMATIC RINGBACK SILENCE TO SILENCE THE RINGBACK AUDIBLE DEVICE AFTER A SET TIME.
8 — MANUAL RESET OF RINGBACK INDICATIONS.
9 — OPERATIONAL TEST.

Figure 5. Sequence R-1-2-9, Ringback with Options

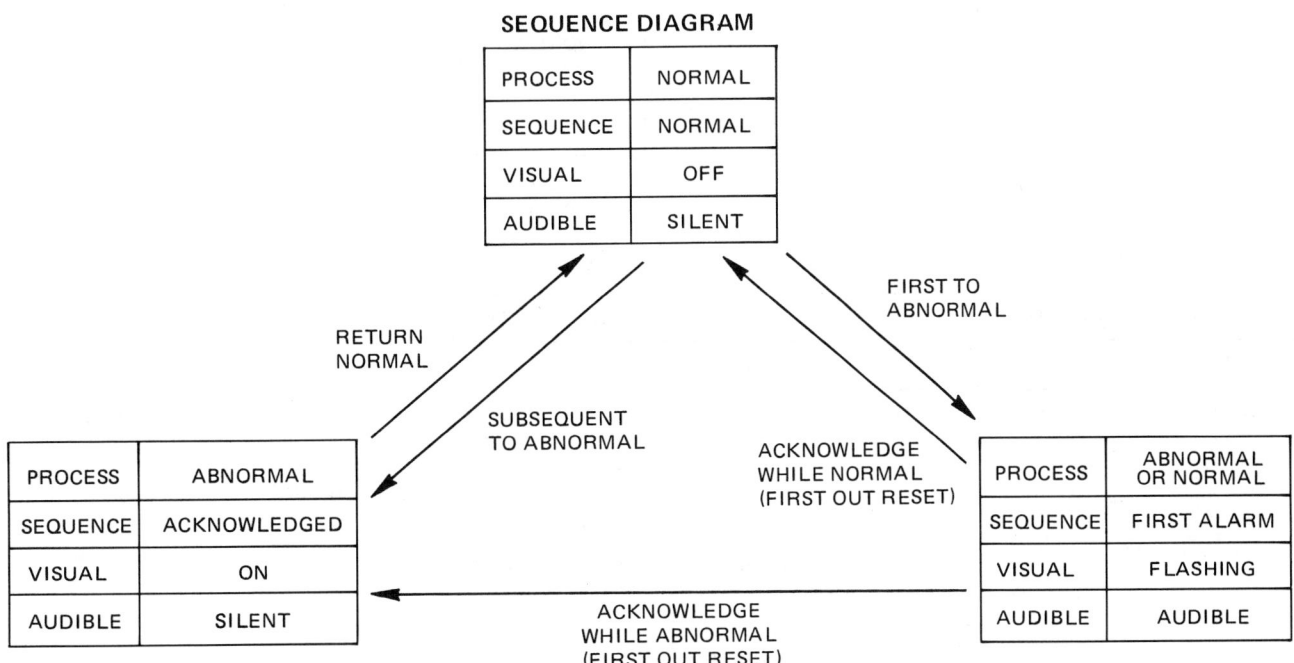

SEQUENCE TABLE

LINE	PROCESS CONDITION		PUSHBUTTON OPERATION	SEQUENCE STATE	VISUAL DISPLAY	ALARM AUDIBLE DEVICE	REMARKS
1	NORMAL		—	NORMAL	OFF	SILENT	
2	FIRST	ABNORMAL	—	FIRST ALARM	FLASHING	AUDIBLE	LOCK-IN
3	SUB.	ABNORMAL	—	ACKNOWLEDGED	ON	SILENT	NO LOCK-IN
4A	FIRST	ABNORMAL	ACKNOWLEDGE	TO LINE 3			MAINTAINED ALARM FIRST OUT RESET
4B	FIRST	NORMAL	ACKNOWLEDGE	TO LINE 5			MOMENTARY ALARM FIRST OUT RESET
5	NORMAL		—	NORMAL	OFF	SILENT	AUTOMATIC RESET

SEQUENCE FEATURES

1 — ACKNOWLEDGE AND TEST PUSHBUTTONS.
2 — ALARM AUDIBLE DEVICE.
3 — LOCK-IN OF MOMENTARY FIRST ALARM UNTIL ACKNOWLEDGED. NO LOCK-IN OF MOMENTARY SUBSEQUENT ALARMS.
4 — FLASHING AND AUDIBLE INDICATIONS FOR FIRST ALARM ONLY. NEW SUBSEQUENT ALARMS GO TO THE ACKNOWLEDGED STATE.
5 — FIRST OUT INDICATION IS RESET AND THE AUDIBLE DEVICE IS SILENCED WHEN ACKNOWLEDGED.
6 — AUTOMATIC RESET OF ACKNOWLEDGED ALARM INDICATIONS WHEN PROCESS CONDITIONS RETURN TO NORMAL.
7 — OPERATIONAL TEST.

Figure 6. Sequence F1A, Automatic Reset First Out with No Subsequent Alarm State

SEQUENCE DIAGRAM

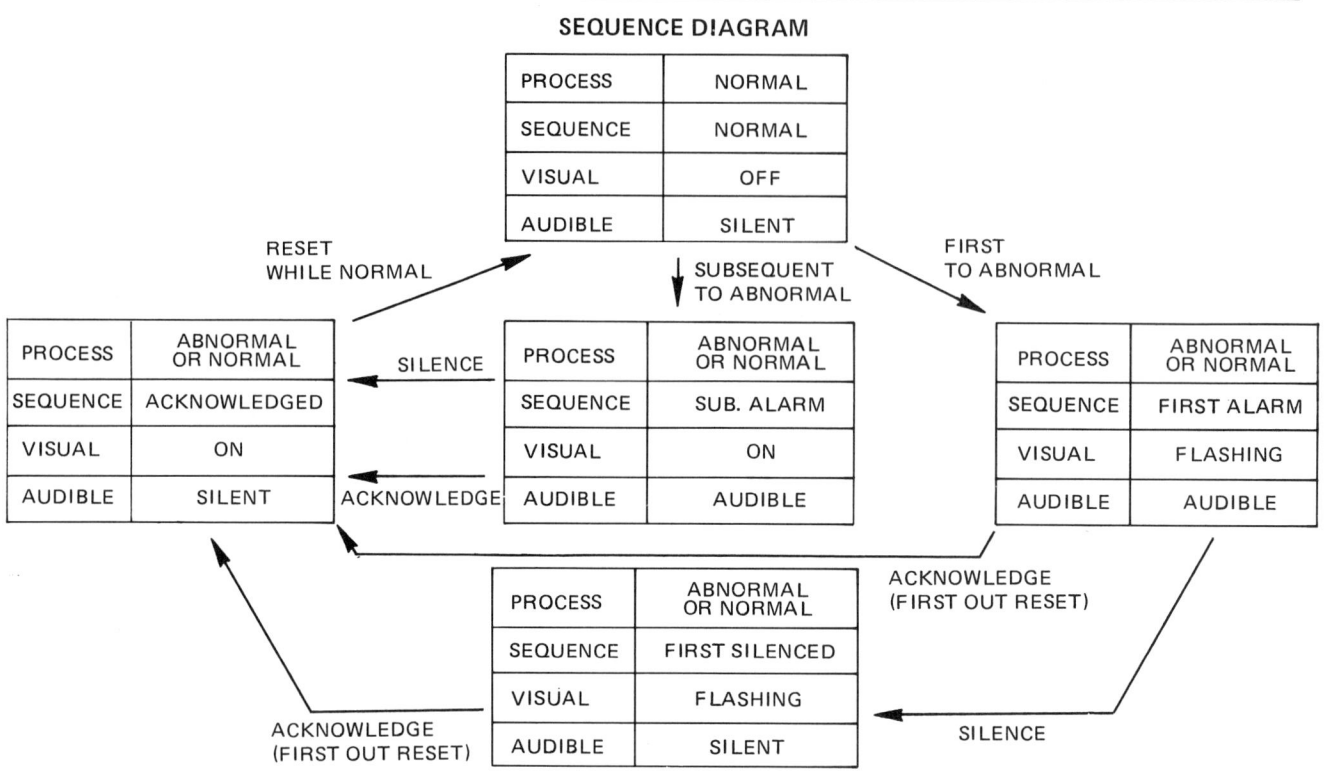

SEQUENCE TABLE

LINE	PROCESS CONDITION		PUSHBUTTON OPERATION	SEQUENCE STATE	VISUAL DISPLAY	ALARM AUDIBLE DEVICE	REMARKS
1	NORMAL		—	NORMAL	OFF	SILENT	
2	FIRST	ABNORMAL	—	FIRST ALARM	FLASHING	AUDIBLE	LOCK-IN
3	SUB.			SUB. ALARM	ON	AUDIBLE	LOCK-IN
4	FIRST	ABNORMAL OR NORMAL	ACKNOWLEDGE BEFORE SILENCE	TO LINE 7			FIRST OUT RESET
5	SUB.	ABNORMAL OR NORMAL					
6	FIRST	ABNORMAL OR NORMAL	SILENCE	FIRST SILENCED	FLASHING	SILENT	
7	SUB.	ABNORMAL OR NORMAL		ACKNOWLEDGED	ON	SILENT	MANUAL RESET REQUIRED
8	FIRST	ABNORMAL OR NORMAL	ACKNOWLEDGE AFTER SILENCE	TO LINE 7			FIRST OUT RESET
9	NORMAL		RESET	NORMAL	OFF	SILENT	MANUAL RESET

SEQUENCE FEATURES

1 — SILENCE, ACKNOWLEDGE, RESET, AND TEST PUSHBUTTONS.
2 — ALARM AUDIBLE DEVICE.
3 — LOCK-IN OF MOMENTARY ALARMS UNTIL ACKNOWLEDGED.
4 — OPTION 1 — SILENCE PUSHBUTTON TO SILENCE THE ALARM AUDIBLE DEVICE WHILE RETAINING FIRST OUT FLASHING INDICATION.
5 — FLASHING INDICATION FOR FIRST ALARM ONLY. NEW SUBSEQUENT ALARMS HAVE THE SAME VISUAL INDICATION AS ACKNOWLEDGED ALARMS.
6 — FIRST OUT INDICATION IS RESET WHEN ACKNOWLEDGED.
7 — MANUAL RESET OF ACKNOWLEDGED ALARM INDICATIONS AFTER PROCESS CONDITIONS RETURN TO NORMAL.
8 — OPERATIONAL TEST.

Figure 7. Sequence F2M-1, Manual Reset First Out with No Subsequent Alarm Flashing and Silence Pushbutton

SEQUENCE DIAGRAM

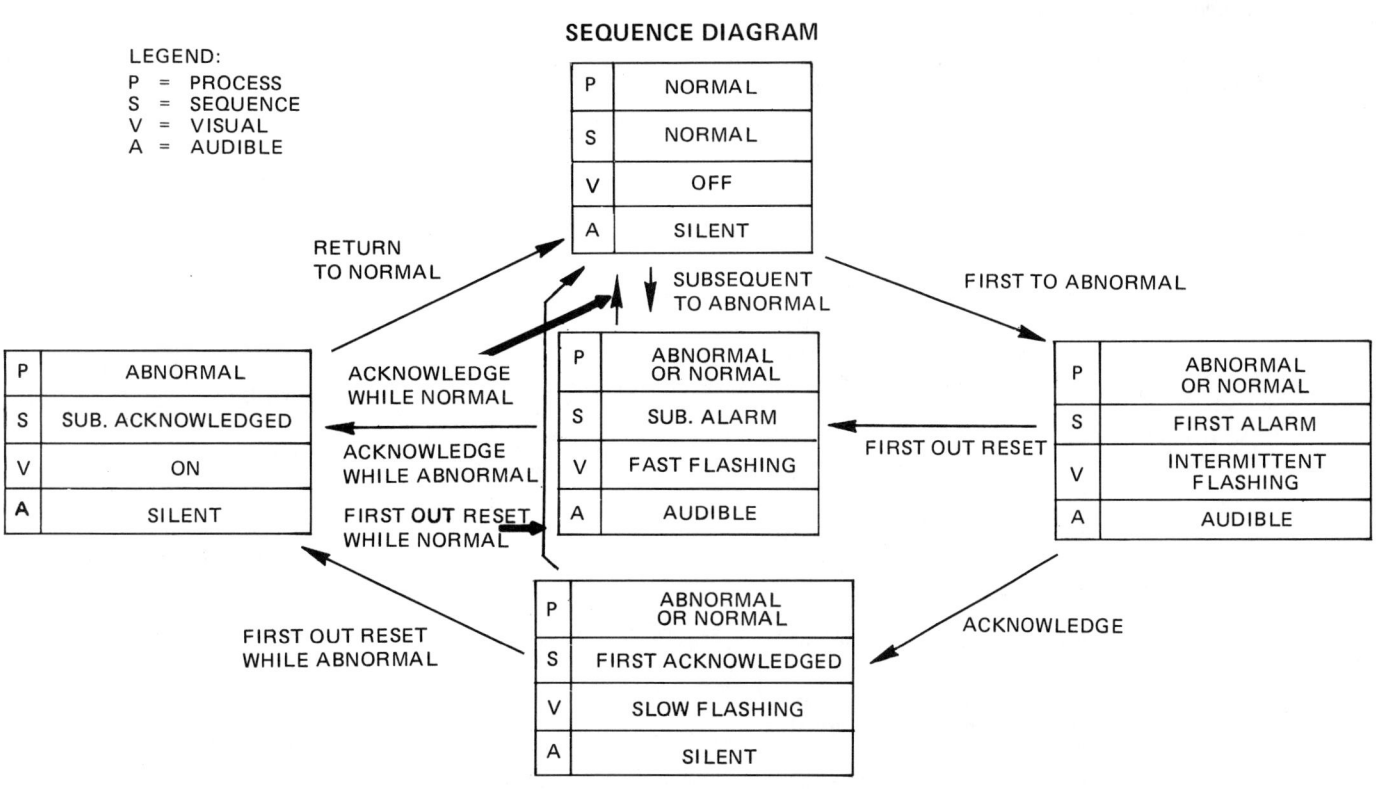

SEQUENCE TABLE

LINE	PROCESS CONDITIONS		PUSHBUTTON OPERATION	SEQUENCE STATE	VISUAL DISPLAY	ALARM AUDIBLE DEVICE	REMARKS
1	NORMAL		—	NORMAL	OFF	SILENT	
2	FIRST	ABNORMAL	—	FIRST ALARM	INTERMITTENT FLASHING	AUDIBLE	LOCK-IN
3	SUB.		—	SUB. ALARM	FAST FLASHING	AUDIBLE	LOCK-IN
4	FIRST	ABNORMAL OR NORMAL	FIRST OUT RESET BEFORE ACKNOWLEDGE	TO LINE 3			FIRST OUT RESET
5	FIRST	ABNORMAL OR NORMAL	ACKNOWLEDGE	FIRST ACKNOWLEDGED	SLOW FLASHING	SILENT	FIRST OUT RESET REQUIRED
6A	SUB.	ABNORMAL		SUB. ACKNOWLEDGED	ON	SILENT	MAINTAINED ALARM
6B		NORMAL		TO LINE 8			MOMENTARY ALARM
7A	FIRST	ABNORMAL	FIRST OUT RESET AFTER ACKNOWLEDGE	TO LINE 6A			FIRST OUT RESET
7B		NORMAL		TO LINE 8			FIRST OUT RESET
8	NORMAL		—	NORMAL	OFF	SILENT	AUTOMATIC RESET

SEQUENCE FEATURES

1 — ACKNOWLEDGE, FIRST OUT RESET, AND TEST PUSHBUTTONS.
2 — ALARM AUDIBLE DEVICE.
3 — LOCK-IN OF MOMENTARY ALARMS UNTIL ACKNOWLEDGED.
4 — FIRST OUT FLASHING DIFFERENT FROM SUBSEQUENT FLASHING.
5 — FIRST OUT RESET PUSHBUTTON TO CHANGE THE FIRST OUT VISUAL INDICATION TO BE THE SAME AS SUBSEQUENT VISUAL INDICATIONS.
6 — AUTOMATIC RESET OF ACKNOWLEDGED ALARM INDICATIONS WHEN PROCESS CONDITIONS RETURN TO NORMAL.
7 — OPERATIONAL TEST.

Figure 8. Sequence F3A, Automatic Reset First Out with First Out Flashing and Reset Pushbutton

4.4 Basic Sequence Letter Designations

Three basic types of annunciator sequence are in common use. The operation of each is different after process conditions return to normal. This Standard makes use of the following basic sequence letters to designate the three basic sequence types:

Basic Sequence Letter	Key Words	Description
A	Automatic Reset	The sequence returns to the normal state automatically after acknowledge when the process condition returns to normal.
M	Manual Reset	The sequence returns to the normal state after acknowledge when the process condition has returned to normal and the reset pushbutton is operated.
R	Ringback	The sequence provides distinct visual and audible indications when the process condition returns to normal. The sequence returns to normal after acknowledge when the process condition has returned to normal and the reset pushbutton is operated.

Sequence diagrams and sequence tables for these basic sequences are shown in Figures 2, 3, and 4, pages 9, 10, and 11.

The types of flashing shown in this Standard such as fast flashing and slow flashing are examples based on frequent use. Alternate types of flashing may be used without requiring a change in the sequence designation.

Since annunciator sequences usually include lock-in of momentary alarms, sequences in this Standard include lock-in as a standard feature. A sequence option number is provided to permit deleting the lock-in feature. Some alarm modules have provisions for deleting the lock-in feature on individual alarm points.

Variations in the basic sequences can be defined using basic sequence letter designations combined with option numbers — see 4.5, Option Number Designations.

First out sequences require a first out designation in addition to the basic sequence letter designation — see 4.6, First Out Designations.

4.5 Option Number Designations

Option numbers can be used with the basic sequence letter designations to define many different sequence variations. This Standard makes use of the following option numbers to designate many of the common sequence variations. Other sequence variations are considered to be special and should be defined by sequence diagrams, sequence tables, or notes.

An example of a sequence designation with option numbers is shown in Figure 1, page 8. Figures 5 and 7, pages 12 and 14, illustrate the use of option numbers.

Option Number	Key Words	Description
1	Silence Pushbutton	A separate pushbutton is added to allow silencing the alarm audible device without affecting the visual displays.
3	Silence Interlock	An interlock is added to require operation of the silence pushbutton before alarms can be acknowledged.
3	First Out Reset Interlock	An interlock is added to require operation of the acknowledge pushbutton before first out alarms can be reset by the first out reset pushbutton.
4	No Lock-In	The lock-in feature is deleted. Momentary alarms return to the normal sequence state without operation of the acknowledge pushbutton.
5	No Flashing	The visual display flashing feature is deleted. New alarms have the same visual display indication as acknowledged alarms.
6	No Audible	The audible device is deleted.
7	Automatic Alarm Silence	A time delay device is added to silence the alarm audible device after a set time without affecting the visual displays.
8	Common Ringback Audible	A common audible device is used to call attention to both the alarm and ringback sequence states.
9	Automatic Ringback Silence	A time delay device is added to silence the ringback audible device after a set time without affecting the visual displays.
10	No Ringback Audible	The ringback audible device is deleted.
11	Common Ringback Visual	A common type of flashing is used to indicate both the alarm and ringback sequence states.
12	Automatic Momentary Ringback	Ringback sequence momentary alarms go to the ringback sequence state without operation of the acknowledge pushbutton.

Option Number	Key Words	Description
13	Dim Lamp Monitor	The visual display indication is dim in the normal sequence state to reveal lamp failure.
14	Lamp Test	Operation of the test pushbutton tests the visual displays only.

4.6 First Out Designations

First out annunciators are used to indicate which one of a group of alarm points operated first. To accomplish this, the visual display indication for the alarm point that operates first must be different from the visual display indication for subsequent alarm points in that group. Only one first out alarm indication can exist in any one first out group.

Three methods for differentiating between first and subsequent alarms are in common use. Two make use of the usual sequence features for the first alarm and delete features for subsequent alarms. The third provides additional features to indicate first alarms. This Standard makes use of the following first out designations to designate the three methods.

First Out Designation	Key Words	Description
F1	No Subsequent Alarm State	Subsequent alarms appear in the acknowledged state. Subsequent visual displays do not flash. The audible device does not operate when subsequent alarms occur, unless still operating from the first alarm. The first out indication is reset by the acknowledge pushbutton.
F2	No Subsequent Alarm Flashing	Subsequent visual displays do not flash. The audible device operates when subsequent alarms occur. The first out indication is reset by the acknowledge pushbutton.
F3	First Out Flashing and Reset Pushbutton	Additional types of flashing are added to identify new and acknowledged first alarms. A first out reset pushbutton is added to reset the first out indication, whether the process condition has returned to normal or not.

First out sequences can be automatic reset or manual reset or can provide ringback indication when alarms return to normal. First out sequences are designated by a combination of the first out designation, the basic sequence letter designation, and option numbers. An example of a first out sequence designation is in Figure 1, page 8.

First out sequence diagrams consist of an outer loop of actions and states associated with the first alarm and an inner loop associated with subsequent alarms. The two loops have a common normal state.

Not all of the possible first out sequences are readily available. In some cases, a particular first out sequence may be a standard design for only one manufacturer. Sequence designations for a range of first out sequences are listed below. Some of these use a silence pushbutton, option number 1, to silence the audible device while retaining the visual display indications in the alarm state — see 4.5, Option Number Designations. The sequences commonly available at the time of publication are indicated. Sequence diagrams and sequence tables for three of the common first out sequences are shown in Figures 6, 7, and 8, pages 13, 14, and 15.

Key Words	Automatic Reset	Manual Reset	Ringback
No Subsequent Alarm State	F1A (Common) (Figure 6)	F1M (Common)	F1R
No Subsequent Alarm State and Silence Pushbutton	F1A-1 (Common)	F1M-1 (Common)	F1R-1
No Subsequent Alarm Flashing	F2A (Common)	F2M (Common)	F2R
No Subsequent Alarm Flashing and Silence Pushbutton	F2A-1 (Common)	F2M-1 (Common) (Figure 7)	F2R-1
First Out Flashing and Reset Pushbutton	F3A (Common) (Figure 8)	F3M	F3R

5 SPECIFICATIONS

Annunciator specifications provide manufacturers with the information necessary to prepare proposals and to design and produce the required annunciator equipment. Many details of annunciator design are relatively standard and may not need to be specified. Some annunciators require only standard specification forms[1] or brief specifications. Complex annunciator systems generally require more elaborate specifications to define the system requirements.

When preparing annunciator specifications, careful thought must be given when specifying features that are not readily available. The advantages of such features should be weighed against the disadvantages of special design.

The following types of information should be included in annunciator specifications, but other features should also be specified as required.

(1) Refer to Instrument Society of America Standard ISA-S20, "Specification Forms for Process Measurement and Control Instruments, Primary Elements, and Control Valves."

5.1 All Annunciators

(1) Logic circuit location: integral logic, remote logic cabinet

(2) Sequence: ISA designation, other identification, sequence diagram, sequence table, notes, arrangement when more than one sequence is used

(3) Number of alarm points: total, active, spare, future

(4) Power source: nominal voltage, frequency

(5) Nominal window size: dimension high, dimension wide

(6) Visual display arrangement: rows high, columns wide

(7) Cabinet mounting: flush, surface, rack

(8) Cabinet type: NEMA enclosure type or electrical classification

(9) Window engraving: legend list, lettering size, maximum number of lines, maximum characters per line (May also be provided later.)

(10) Logic circuit type: solid state, electromechanical relay

(11) Field contact operation: normally open (close to alarm), normally closed (open to alarm), mixed

(12) Field contact voltage: nominal voltage, frequency

(13) Information required with proposal: descriptions, drawings, price, delivery

(14) Documentation required: after award, before delivery, with delivery

5.2 Remote Logic Annunciators

(1) Logic cabinet mounting: surface, chassis, rack, freestanding

(2) Logic cabinet type: NEMA enclosure type or electrical classification

(3) Logic cabinet requirements: arrangement, cable entrance, color

(4) Prefabricated cables: supplier, length, conductor size, insulation type, jacket type

(5) Lamp cabinet connections: plug connectors, screw terminals

(6) Logic cabinet connections to lamp cabinet: plug connectors, screw terminals

5.3 Complex Annunciator Systems

(1) System arrangement drawing: cabinets, prefabricated cables, pushbuttons, audible devices, power sources

(2) System operation description

5.4 Annunciator Accessories and Special Features

(1) Pushbuttons: location, supplier, number, type

(2) Audible devices: location, supplier, power, type

(3) Special cabinet finish: annunciator or lamp cabinet color, logic cabinet color, materials to be used, application methods

(4) Special visual display: individual windows, graphic displays

(5) Special window or window bezel colors: colors, windows

(6) Reflash points: number of field contacts, alarm points

(7) Special field contact time delay: time delay, alarm points

(8) Analog input points: analog signals, alarm points

(9) Auxiliary outputs: function, operation on test, grouping, type, electrical rating

(10) Solid state field contacts or solid state auxiliary outputs: type, electrical rating, common potential

(11) Special power source requirements: voltage variation range, frequency variation range

(12) Backup power system: power sources, requirements

(13) Isolation from power sources: field contacts, window lamps, logic circuits

(14) Special power source for receptacles and lights: nominal voltage, frequency

(15) Power failure detector: alarm, indication

(16) Ground detector: alarm, indication, isolation switches

(17) Window legibility: lettering size, ambient light

(18) Special service conditions: temperature, humidity, ambient light, noise, hazardous atmosphere, intrinsically safe design, nonincendive design, purged cabinet, high field contact wiring impedance

(19) Special wire insulation: voltage rating, insulation type, test requirements

(20) Special field contact wiring terminations: terminal type, terminal size, wiring space

(21) Special grounding connections: terminal size, ground bus

(22) Special cabinet locks: type, keys

(23) Special factory tests: dielectric strength, functional, surge withstand capability, radio frequency interference, seismic, nuclear Class 1E

(24) Special documentation: test procedures, test reports

(25) Spare components: type, number

(26) Logic circuit tester: portable, built-in

6 DOCUMENTATION

Documentation is provided by manufacturers to describe the equipment produced to meet annunciator specifications. Before delivery of annunciators, this documentation is used to confirm that specification requirements are met and to allow design of the annunciator mounting, external wiring, and power sources. After delivery, this documentation is used by those installing, operating, and maintaining annunciators.

The following types of information should be included in the documentation provided by the manufacturer, but other documentation should be provided as required by specifications.

6.1 All Annunciators

(1) Annunciator description: appearance, standard features, special features

(2) Dimensioned drawings: outline, arrangement, enclosure type, panel cutout, mounting method, wiring entrance

(3) Sequence description: ISA designation, sequence diagram, sequence table, notes

(4) Schematic diagram drawings: logic circuit, circuit wiring, jumper and switch settings,

modules, pushbuttons, audible devices, field contact voltage, lamp voltage, power supply circuit, electrical ratings

(5) Wiring diagram drawings: external wiring, power source voltage and frequency

(6) Power requirements: normal, maximum, allowable voltage and frequency variation ranges

(7) Instruction manual: installation, operation, operating limitations, logic circuit operation, maintenance procedures, special test device operation

(8) Parts list: replacement parts, recommended spare parts, prices

6.2 Remote Logic Annunciators

(1) Logic cabinet arrangement drawings: internal devices, terminal blocks, wiring space, ground connections, receptacles, lights, fans

(2) Interconnection drawings: external wiring connections between cabinets

(3) Power supply drawings: power supply and distribution system

6.3 Complex Annunciator Systems

(1) System arrangement drawings: components, interconnections

(2) System operation description

6.4 Annunciator Accessories and Special Features

Necessary information concerning accessories and special features should be provided with the annunciator documents.

APPENDIX A

ANNUNCIATOR APPLICATION GUIDE

This Appendix is included for informational purposes and is not a part of this Standard.

A.1 Purpose

The purpose of this Appendix is to provide information to assist in specifying annunciators that will best serve the requirements of the users while making use of standard and special features that are readily available. Catalogs and other information from annunciator manufacturers should also be used because of the wide variety of features that are available.

A.2 Introduction

Annunciators are usually used to call attention to abnormal process conditions. Annunciators may also be used to show normal process status. Annunciators usually include individual illuminated visual displays that are labeled to identify the particular monitored variable that is abnormal and audible devices. Annunciators may also call attention to the return to normal of the process conditions.

Visual displays usually flash to indicate abnormal process conditions. Manual operation of pushbuttons is usually required to silence audible devices and acknowledge new alarms. Visual displays usually change from flashing to on when alarms are acknowledged. Figure A1 illustrates a typical annunciator sequence.

Additional types of flashing can indicate that process conditions have returned to normal or which of a group of alarm points operated first. Additional pushbuttons can be used to acknowledge alarms that return to normal, to reset first out indications, and to test annunciator lamps and circuits.

Annunciators are available in an almost infinite variety of physical arrangements, operating sequences, and special features. In some cases, the annunciation function is performed by computer systems using CRT (cathode ray tube) visual displays or recording annunciators — see A.5, Special Annunciators.

This Standard is primarily for use with electrical annunciators that use illuminated visual displays and audible devices. Enclosures with lamps behind labeled translucent windows are commonly used as visual displays. Annunciators usually operate from electrical contacts that are part of the devices that sense the process conditions.

In this Standard, sequences making use of more than one indication device as a part of each visual display to indicate the sequence state are considered to be special because of their many variations and relatively infrequent use. When additional wiring and sockets are provided, colored lights such as red and green may be used along with flashing to indicate sequence states or which is the first of a group of alarms. Colored lights may also be used to uniquely identify some of the alarms in an annunciator — see A.8.2, Windows.

A.3.1 Designation Method

The sequence designation method provided in this Standard uses letters for basic sequences in common use, numbers for common options, and first out designations. Combinations of letters and numbers can define many different sequence variations. Refer to Figure 1, page 8, of this Standard for sequence designation examples and a summary of the basic sequence letters, option numbers, and first out designations.

This Standard does not designate any particular sequences as being standard. Refer to Appendix B, Sequence Designation Conversion, for the ISA sequences from Instrument Society of America Recommended Practice ISA-RP18.1-1965 that are superseded and the corresponding new sequence designations.

Sequence diagrams, sequence tables, or notes should be used when sequence letter and option number designations do not adequately define required sequences.

When selecting sequence designations, careful thought must be given when including features or combinations of features that are not readily available. The advantages of such features should be weighed against the disadvantages of special design.

A.3.2 Lock-In

Lock-in is a sequence feature that retains the alarm state until acknowledged when momentary alarms occur. With lock-in, momentary alarms can be observed before the acknowledge pushbutton is operated. Without lock-in, momentary alarms may not be observed at all, even though the audible device operates briefly.

Since annunciator sequences usually include lock-in, the sequences in this Standard include lock-in as a standard feature. Option Number 4 is provided to permit deleting the lock-in feature — see 4.5, Option Number Designations. Some alarm modules include movable jumpers or switches to allow deleting the lock-in feature on individual alarm points.

When momentary alarms occur frequently enough to be a nuisance, the lock-in feature is often deleted to avoid having to acknowledge the alarms repeatedly. In such cases, a better solution may be to correct the cause of the momentary alarms or to use time delays in field contact circuits in order to alarm only abnormal process conditions that exist longer than a given time.

A.3.3 Sequence A, Automatic Reset

Sequence A is a basic annunciator sequence with automatic reset that automatically returns acknowledged alarms to normal when process conditions return to normal. A sequence diagram and sequence table for sequence A are shown in Figure 2, page 9, of this Standard.

In some applications, sequence A may have a disadvantage since new momentary alarms return to off

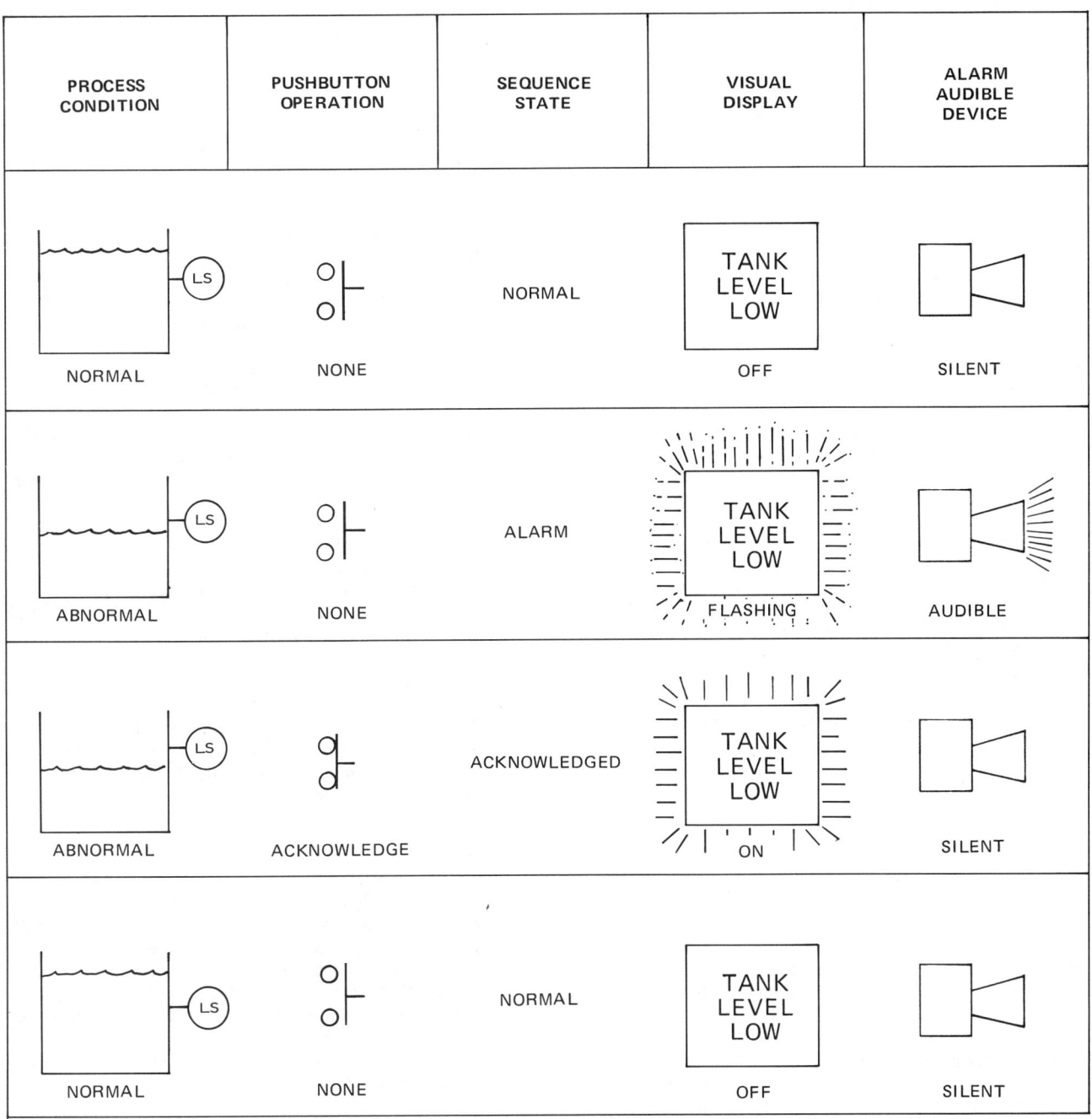

Figure A1. Typical Annunciator Sequence, Sequence A, Automatic Reset

and new maintained alarms change to on when and during the time the audible device is silenced by operation of the acknowledged pushbutton. New alarms may be lost or may be confused with existing acknowledged alarms. New alarms may have to be reviewed or logged while flashing, with the continual distraction of the audible device signal. If these features are not desirable, sequence A-1 with a silence pushbutton or sequence A-1-2 with a silence pushbutton and interlock should be used — see 4.5, Option Number Designations.

If an audible signal is required when process conditions return to normal, sequence R that includes ringback should be used — see A.3.5, Sequence R, Ringback.

A.3.4 Sequence M, Manual Reset

Sequence M is a basic annunciator sequence with manual reset that retains acknowledged alarms until the process conditions return to normal and the manual reset pushbutton is operated. A sequence diagram and sequence table for sequence M are shown in Figure 3, page 10, of this Standard.

In some applications, sequence M may have a disadvantage since new alarms change to on when and during the time the audible device is silenced by operation of the acknowledge pushbutton. New alarms may be confused with existing acknowledged alarms. New alarms may have to be reviewed or logged while flashing, with the continual distraction of the audible device signal. If these features are not desirable, sequence M-1 with a silence pushbutton or sequence M-1-2 with silence pushbutton and interlock should be used — see 4.5, Option Number Designations.

In order to reset alarms, sequence M requires that the reset pushbutton be operated repeatedly to determine if process conditions have returned to normal. When the reset pushbutton is operated, it may be difficult to observe which of a number of acknowledged alarms have returned to normal. With sequence M, it is not evident when process conditions return to normal or return again to abnormal. If these features are not desirable, sequence R that includes ringback should be used — see A.3.5, Sequence R, Ringback.

Sequence M usually cannot be used with reflash circuits or to operate from remote reflash auxiliary outputs since the sequence does not return to the alarm state when the field contact circuit returns to normal briefly. Refer to A.9.3, Shared Alarm Points, and A.9.5, Auxiliary Outputs. If this feature is required, sequence A or R should be used — see A.3.3, Sequence A, Automatic Reset and A.3.5, Sequence R, Ringback.

A.3.5 Sequence R, Ringback

Sequence R is a basic annunciator sequence with ringback that provides distinct visual and audible indications when process conditions return to normal. The ringback indications are retained until the process conditions return to normal and the manual reset pushbutton is operated. A sequence diagram and

sequence table for sequence R are shown in Figure 4, page 11, of this Standard.

In some applications, sequence R may have a disadvantage since new momentary alarms change to slow flashing and new maintained alarms change to on when and during the time the audible device is silenced by operation of the acknowledge pushbutton. New alarms may be confused with existing alarms. New alarms may have to be reviewed or logged while fast flashing, with the continual distraction of the alarm audible device signal. If these features are not desirable, sequence R-1 with a silence pushbutton or sequence R-1-2 with silence pushbutton and interlock should be used — see 4.5, Option Number Designations.

Sequence R includes different visual display indications and different audible device signals for alarm and ringback. Several variations of this arrangement can be used. Sequence R-8 uses a common audible device for both alarm and ringback and relies on the different visual display indications for differentiation. Sequence R-9 uses a time delay device to silence the ringback audible device after a set time. Sequence R-10 deletes the ringback audible device and uses only the ringback visual displays. Sequences R-9 and R-10 avoid the need for pushbutton operation to silence an audible device when process conditions return to normal. Sequence R-11 uses a common type of flashing for both alarm and ringback and relies on the different audible devices for differentiation. See 4.5, Option Number Designations.

Sequence R retains both momentary and maintained alarms in the alarm state until acknowledged. Sequence R-12 causes momentary alarms to go to the ringback sequence state as soon as process conditions return to normal. New momentary alarms are evident sooner, but may be confused with existing alarms in the ringback state.

Sequence R-1-2-9, shown as an example of the use of option number designations in Figure 5, page 12, of this Standard, includes a silence pushbutton and interlock to allow new alarms to be reviewed or logged while flashing after the audible signal has been silenced and to require operation of the silence pushbutton first. In addition, the ringback audible device is silenced after a set time to retain a ringback signal while avoiding the need for pushbutton operation to silence the ringback audible device when process conditions return to normal.

A.3.6 First Out

First out annunciators are used to indicate which one of a group of alarm points operated first. To accomplish this, the visual display indication for the alarm point that operates first must be different from the visual display indication for subsequent alarm points in that group. Only one first out alarm indication can exist in any one first out group.

When first out annunciators are used primarily to identify the first alarm, a flashing visual display can be used to indicate the first alarm and visual displays without flashing can be used to indicate subsequent alarms. With this approach, the visual display indication for subsequent alarms does not differentiate between new and acknowledged alarms. Two methods using this approach are in common use. (1) First out designation F1 designates a first out sequence with no subsequent alarm state. Subsequent alarms appear in the acknowledged state. Subsequent visual displays do not flash. The audible device does not operate when subsequent alarms occur, unless still operating from the first alarm. The first out indication is reset by the acknowledge pushbutton. It should be noted that subsequent alarms do not lock in when sequence F1A is used. (2) First out designation F2 designates a first out sequence with no flashing for subsequent alarms. The audible device operates when subsequent alarms occur. The first out indication is reset by the acknowledge pushbutton.

To allow the first out visual display indication to be reviewed or logged after silencing the audible device when using first out designations F1 and F2, a separate silence pushbutton should be used in addition to the other annunciator pushbuttons. The silence pushbutton feature is designated by option number 1 — see 4.5, Option Number Designations.

When use of the annunciator requires differentiation between new and acknowledged subsequent alarms, the first out sequence should include different types of visual display flashing to identify the first alarm while new subsequent alarms are indicated by the usual flashing visual display. One method using this approach is in common use. First out designation F3 designates a first out sequence with first out flashing to identify new and acknowledged first alarms and a first out reset pushbutton to reset the first out indication, whether the process condition was returned to normal or not. If desired, an interlock can be provided to require operation of the acknowledge pushbutton before the first out indication can be reset by the first out reset pushbutton by use of option number 3 — see 4.5, Option Number Designations.

After the first out indication is reset, that alarm point indicates the process condition in the same manner as subsequent alarms. The next alarm point to operate will display a first out indication.

First out sequences can be automatic reset or manual reset or can provide ringback indication when alarms return to normal. First out sequences are designated by a combination of the first out designation, the basic sequence letter designation, and any required option numbers. Refer to Figure 1, page 8, in this standard for a first out sequence designation example.

Sequence designations for a range of first out sequences are listed in 4.6, First Out Designations. Sequence diagrams and sequence tables for three of the common first out sequences are shown in Figures 6, 7, and 8, pages 13, 14, and 15, of this Standard. Because of the complex nature of first out annunciator sequences, use of those sequences that are readily available should be considered when making a selection.

When annunciators include both first out alarm points and alarm points without the first out feature, the

first out windows and alarm modules are usually located together for easy recognition and to facilitate first out logic bus interconnections. Colored window bezels, windows, or lamps may be used to identify first out alarm points. Several separate first out groups can be created by using several first out logic buses.

Annunciator first out windows may not be needed if a recording annunciator or computer printer is provided for sequence of events analysis.

A.3.7 Test

Annunciator test pushbuttons initiate an annunciator sequence to reveal lamp or circuit failures. Most annunciators include a test pushbutton and operational test of the sequence, visual display lamps, audible devices, and pushbuttons. The sequences in this Standard include operational test as a standard feature.

Operation of the test pushbutton simulates simultaneous abnormal process conditions on all related alarm points. Release of the test pushbutton simulates return to normal. Operation of the other annunciator pushbuttons to complete the sequence and observation of the operation of the visual displays and audible devices can reveal lamp or circuit failures. Alarm points in the acknowledged state as a result of actual abnormal process conditions usually remain in the acknowledged state during test and will be in the acknowledged state after the test sequence.

Since the test signal usually operates the sequence logic circuit in the alarm module, annunciator input auxiliary circuits such as reflash or time delay circuits and some auxiliary output circuits may not operate during the test sequence. In some cases, operation of auxiliary outputs during the test sequence may not be desirable since false alarm signals would be transmitted to the connected alarm, recording, or control circuits.

Some annunciator sequences make use of visual display indications that are dim in place of off to reveal lamp failures in addition to an operational test sequence. The dim lamp monitor feature is designated by option number 13 — see 4.5, Option Number Designations.

Test pushbuttons used with some electromechanical relay logic annunciators test only the visual display lamps. Operation of the test pushbutton turns all related visual display lamps on simultaneously. The lamp test feature is designated by option number 14 — see 4.5, Option Number Designations.

A.4 Arrangement

A.4.1 Integral Logic Annunciators

Integral logic annunciators include visual displays and sequence logic circuits in one assembly as illustrated in Figure A2. Plug-in alarm modules contain the sequence logic circuits. The visual display lamps and windows may or may not be a part of the plug-in alarm modules. Terminal blocks are provided for the

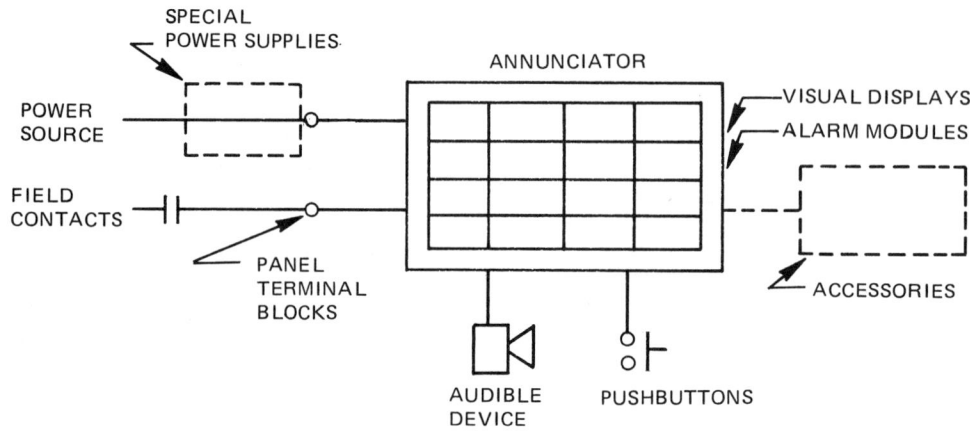

Figure A2. Integral Logic Annunciator Arrangement

field contact, pushbutton, audible device, and power source wiring.

When large annunciator windows are subdivided to form two, three, or four smaller windows, a single plug-in alarm module may serve each group of alarm points.

Common flashers, audible device drivers, and required power supplies are usually located within integral logic annunciators. Small pushbuttons and audible devices can be mounted in window positions when preferred and when annunciators are easily accessible.

Since the space in integral logic annunciator cabinets is limited, it may be necessary to locate special power supplies and accessories such as auxiliary output, reflash, ground detector, and power failure detector components in separate cabinets. The separate cabinets are usually designed for surface mounting.

Integral logic annunciators should be located so that alarm module replacement and other required maintenance can be done without excessive interference with nearby activities.

Panel wiring is often used to connect from the annunciator terminal blocks to larger panel terminal blocks where the field wiring is terminated.

A.4.2 Remote Logic Annunciators

Remote logic annunciators locate visual displays and sequence logic circuits in separate assemblies as illustrated in Figure A3. Lamp cabinets contain the visual displays and the necessary terminal blocks or plug connectors. The plug-in alarm modules, accessory components, and special power supplies can be supplied in separate assemblies for rack or cabinet mounting by the user, or else complete logic cabinets can be supplied by the manufacturer. Pushbuttons

and audible devices are usually located near the lamp cabinets.

Logic cabinets are usually arranged to suit the requirements of a particular annunciator specification. The arrangement can include space for power supplies and accessories such as auxiliary output, reflash, ground detector, and power failure detector components. A ground bus may be required and electrical receptacles or lights may be desirable for use during maintenance. Large terminal blocks and adequate space for field wiring and terminations can be provided.

When required, terminal blocks can be specified to be separated from the logic circuits to minimize circuit damage during installation or to allow installation of terminal blocks and field wiring before the logic circuits are installed.

To reduce congestion, logic cabinets are often located such that field contact wiring avoids control panel wiring areas. Special cables with large numbers of small conductors are usually used to connect between the logic cabinets and the lamp cabinets. These cables often connect directly to the lamp cabinets so that panel wiring and terminal blocks are not required. Installation time may be reduced if prefabricated lamp cabinet cables with plug connectors are provided by the manufacturer. Care must be taken when specifying the length of prefabricated cables with plug connectors on both ends. In some cases, plug connectors on one end of each cable are installed in the field after the cables are cut to the correct length.

The higher cost of remote logic annunciators is often justified for large or complex annunciator systems. Installation and maintenance costs are reduced since the logic cabinets can contain special power supplies and accessory components, field contact wiring

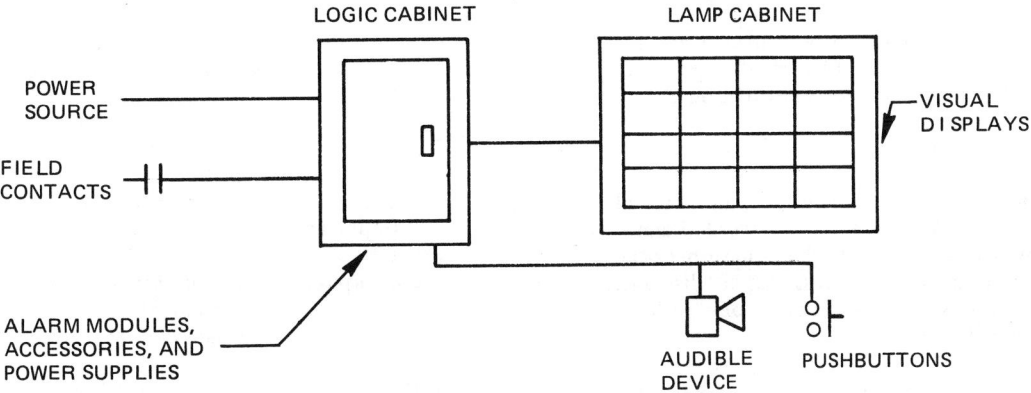

Figure A3. Remote Logic Annunciator Arrangement

terminates away from control panels, special requirements for field wiring space and terminal blocks can be met, and lamp replacement is the only annunciator maintenance required in the control panel area.

A.5 Special Annunciators

A.5.1 General

The more commonly used annunciators include groups of illuminated visual displays and sequence logic circuits that are dedicated for use with individual alarm points. The following special types of annunciators and visual displays should also be considered when their use can improve the annunciation function. The terminology, sequence designations, and sequence presentation in this Standard should be used wherever practicable when specifying these special annunciators.

A.5.2 Drop Type Annunciators

Drop type annunciators make use of inexpensive electromechanical devices that release labeled cards to fall to a position where the labels can be seen. A separate device operates the audible device. The audible device and drops can be reset independently by mechanical levers or electrical pushbuttons. Auxiliary output contacts are available.

Drop type annunciators are sometimes used at unattended locations to transmit general trouble alarms to central control room annunciators and to indicate the specific abnormal condition to maintenance personnel.

A.5.3 Individual Visual Displays

Control panel arrangements sometimes require that annunciator visual displays be located adjacent to re-

lated control and indication devices. Individual light boxes or pilot lights suitable for location on panels can be used as visual displays with remote logic annunciators. Some individual light boxes contain the sequence logic and can operate as individual annunciators.

A.5.4 Graphic Displays

Graphic displays with colored shapes to represent process equipment and interconnections are often used to enable a more clear understanding of the process operation. Individual visual displays can be included as a part of the graphic display to show the location of alarms in the process. The individual visual displays can be a part of remote logic annunciators or they can operate in parallel with the visual displays of integral logic annunciators.

A.5.5 Recording Annunciators

Recording annunciators, often called sequence of events recorders, provide a printed record of the alarm point identification and the date and time that alarms occur and return to normal. The printed identification to identify the alarms can be coded alarm point numbers or short worded alarm messages.

Recording annunciator circuits usually scan field contacts rapidly, store information when alarms occur in rapid succession, and print the information in chronological sequence. The time can be recorded to as close as the nearest millisecond to allow accurate analysis of the sequence of events. The printers are usually separate from the circuit cabinets.

Recording annunciators are sometimes used in parallel with illuminated visual display annunciators to record the operation of important alarm points. Since they cannot be read from a distance, recording annunciators can be used to both indicate and record alarms only when there is no need to react to alarms quickly or from different control locations. An alarm summary can be printed when required to indicate all points in alarm at that time.

Printers and circuit cabinets can be provided just for recording alarms, or they can be part of larger computer systems that also perform other functions.

A.5.6 CRT Visual Displays

CRTs (cathode ray tubes) can be used as alarm visual displays. Alarms can be presented by worded alarm messages or by words or symbols that appear near the related equipment on CRT process graphic displays. The alarm indications can flash or be colored or both until acknowledged.

The CRTs are usually separate from the sequence logic and CRT display control circuit cabinets.

Commonly used annunciators with conventional illuminated visual displays require additional panel space for each additional alarm point. Since CRTs usually indicate only the points that are in alarm, additional alarm points require only additional capacity in the circuit cabinets. On the other hand, alarms must be displayed on the CRT in sections when a large number of alarms exist at one time since space on a CRT is limited. When a large number of alarms is anticipated, several CRTs should be used. Each display can indicate a selected portion of the alarms.

When CRTs are used, conventional annunciator illuminated visual displays are often installed also to indicate the more important alarms. Since they are dedicated displays and can be located near the related process controls, they direct attention to the trouble area more rapidly than CRT alarm messages.

CRTs and circuit cabinets can be provided for indication of alarms only, or they can be part of larger computer systems that also perform other functions.

A.5.7 Voice Annunciators

Voice annunciators use speakers to broadcast verbal alarm messages. The words may be created by electronic synthesis or by playing messages recorded on magnetic tape.

Voice annunciators tend to command immediate response and can be projected over large control and process areas. They can also be used along with visual display annunciators to call attention to the more important alarms.

A.5.8 Combined Annunciation Systems

Annunciation systems can combine annunciators of different types in order to best serve each of the required functions. As an example, a large number of alarms could be presented in a compact control room by the combined annunciation system described below.

Important Alarms	Indicated by conventional annunciator visual displays located on the

related control panels. Reflash circuits are used to conserve panel space by indicating several related alarms on one visual display.

All Alarms	Indicated on CRTs located in several major control areas. Each alarm indicates only on the related CRT. The CRTs indicate the reflash circuit alarms individually for better alarm definition.
All Alarms and Important Process Status Inputs	Recorded by a printer for later analysis of alarms and process status and the sequence of events. The printer is located away from the control room to minimize noise in the operating area.
Audible Device and Pushbuttons	Designed to allow silencing audible devices and acknowledging alarms as a coordinated annunciation system.

A.5.9 Multiplexing

Multiplexing is the use of a single communication channel to transmit many individual signals. Multiplexing systems may include redundant circuits and communication channels for increased reliability. While multiplexing is not inherently related to alarms, it is being used with increasing frequency along with annunciators or as a part of annunciation systems to reduce the amount and the cost of field contact wiring. Multiplexing should be considered when the length of field contact wiring is greater than about 1000 feet (300 m).

A.6 Audible Devices

Annunciator audible devices call attention by sound to the occurrence of alarm and return to normal conditions. They are often located inside control panels near the annunciators or lamp cabinets. The sound can usually be heard adequately in the adjacent area. In noisy areas, openings in the panels can be provided to increase the sound or the audible devices can be mounted on top of the panels. Small audible devices can be provided on the front of annunciator cabinets. Audible devices are usually supplied with annunciators to ensure coordination with the audible device driver circuits.

Audible devices can be horns, bells, chimes, buzzers, or speakers operating from adjustable solid state tone generators. The sound can be continuous or intermittent until silenced by manual pushbutton operation or the sound can be silenced automatically after a set time. The sound can be steady or it can fluctuate or warble.

Audible devices in relatively quiet control rooms are often speakers or intermittent chimes that call attention to alarms without undue noise. Audible devices with different sounds or tones may be selected to distinguish between different functions, systems, or levels of alarm importance.

Annunciators in locations that are frequently unattended should silence audible devices automatically after a set time to avoid unnecessary noise and wear of the audible devices. Sequence option numbers 7 and 9 are provided to allow adding the automatic silence feature — see 4.5, Option Number Designations.

When annunciators are in unattended or noisy areas, attention can be drawn to alarms by large flashing lights mounted on top of the panels. These lights can be connected to the audible device driver circuits.

Audible devices should meet the environmental requirements that apply to other panel mounted devices.

A.7 Pushbuttons

Annunciator pushbuttons are momentary manual switches that cause a change from one annunciator sequence state to another. They are usually located on the panels below the annunciators or lamp cabinets. Pushbuttons can also be provided on the front of annunciator cabinets. Usually one normally open or normally closed pushbutton contact is wired to the annunciator. Pushbuttons can be supplied with annunciators or provided by others.

Miniature or heavy duty pushbuttons can be used, depending on the application and the need to coordinate with other panel devices. Since annunciator pushbuttons may be operated frequently and in a hurry when responding to alarms, they should be selected and located for convenient operation and to minimize the possibility of accidental operation of other nearby pushbuttons.

Pushbuttons should meet the environmental requirements that apply to other panel mounted devices.

An interlock is usually provided to require operation of acknowledge pushbuttons before alarms can be reset. It may also be desirable to require operation of silence and acknowledge pushbuttons or acknowledge and first out reset pushbuttons in sequence to avoid accidental loss of alarm indications. Sequence option numbers 2 and 3 are provided to add these interlocks — see 4.5, Option Number Designations.

When several annunciators or lamp cabinets in a large control room operate together as a system, the pushbuttons should be arranged and connected to allow operation from appropriate locations. As an example, the following arrangement could be used in a control room with separate supervision and control panels.

Silence Pushbuttons	Located on all supervision and control panels and connected to silence the alarm audible device for any alarm in the control room. This avoids continued noise while retaining the flashing visual displays.
Acknowledge Pushbuttons	Located only on control panels where corrective control action can be taken and connected to acknowledge new alarms related to the controls on that panel only. This encourages observation of related indicators and controls.
Reset Pushbuttons	Located on all supervision and control panels and connected to reset any

ringback alarm in the control room. Return to normal conditions usually do not require control action.

Test Pushbuttons	Located only near annunciators or lamp cabinets and connected to test the related annunciator or lamp cabinet only. This avoids disrupting the entire annunciator system by test at one time.

A.8 Visual Displays

A.8.1 General

Annunciator visual displays indicate the sequence state. They are usually enclosures containing lamps behind translucent windows. The windows are labeled to identify the monitored variables. Bullseye pilot lights with round glass lenses and adjacent labeled nameplates can also be used as annunciator visual displays. Pilot lights are sometimes better suited for use in adverse environments than are displays with translucent windows.

A.8.2 Windows

The size of annunciator windows has not been standardized among manufacturers. Windows commonly used are in the order of 2 inches (50 mm) high and 3 inches (75 mm) wide. These windows often can be subdivided to form two, three, or four smaller windows to reduce the required panel space when the viewing distance is short or a small amount of lettering is necessary. These alarm points may make use of a common window or alarm module.

Annunciator windows are usually translucent white with black letters. Clear or white lamps are usually used. When it is necessary to uniquely identify some of the alarms in an annunciator, such as more important alarms or alarms related to one process area, it is usually possible to obtained colored window bezels, windows, or lamps. Methods used and colors available vary widely. Lamps are often colored by the addition of colored boots that are slipped over the lamps. The brightness and legibility of displays are reduced when colored windows or lamps are used.

To avoid interfering with more alarms than necessary during relamping, windows or alarm modules usually can be removed individually from the front.

For best visibility, the windows should be approximately perpendicular to the line of sight of viewers. Since annunciators and lamp cabinets are often located above other devices on control panels, some windows are wedge-shaped for better visibility from below. Some control panels are designed with an inclined section at the top for improved visibility. These factors should be considered during the design of control panels.

The size of lettering used on annunciator windows should be selected to be legible from any point in the related operating area. A compromise is required, however, since larger lettering reduces the number of characters available to identify alarms. For uniform appearance, the same lettering size should be used for all windows of a given size in any one control room. Recommended reading distances for different letter sizes are included in Instrument Society of America Recommended Practice ISA-RP60.6, "Nameplates, Tags, and Labels for Control Centers," (being prepared at the time of publication).

In order to identify the monitored variable completely in the space available, abbreviations are usually required. Abbreviations should be consistent, and periods are usually omitted. For standard abbreviations, refer to American National Standards Institute Standard ANSI Y1.1, "Abbreviations for Use on Drawings and in Text." The lettering on each line is usually centered. To facilitate maintenance, windows can be marked with instrument numbers or small identification numbers to correlate with similar numbers adjacent to alarm modules. Annunciator legend lists should be prepared with care, since the legends will appear exactly as indicated.

The arrangement of information on annunciator windows should be consistent. From top to bottom, the words usually become increasingly specific. For example:

COOLING WATER	PRIMARY PMP A102
HEADER PRESS	BEARING TEMP
LOW	HIGH TRIP

A.8.3 Lamps

Annunciator visual displays usually are incandescent lamps. Two lamps are usually connected in parallel to light the windows more evenly and also to allow visual displays to function even if one lamp has failed. Lamps with bayonet bases are usually used to avoid loss of electrical contact because of vibration.

To minimize maintenance, annunciator lamps are usually operated below the rated voltage. The particular type of lamp, the rated voltage, and the applied voltage are usually selected by annunciator manufacturers. Lamps with a higher watt rating usually cannot be substituted without damage to annunciator circuits or windows.

Care should be taken so the illumination level in areas where annunciator windows are located is not so high that legibility is reduced below an acceptable level.

A.9 Logic Circuits

A.9.1 General

Annunciator logic circuits operate the visual displays and audible devices through a chronological series of states to indicate and call attention to alarm and return to normal conditions. Changes from one sequence state to another are caused by changes in process conditions and also by manual operation of pushbuttons. Annunciators usually operate from field contacts that are located in the devices that sense the process conditions.

Annunciator logic usually uses solid state circuits on plug-in printed circuit boards, although electromechanical relay circuits are also available. The widest variety of annunciator features and accessories is available when solid state circuits are used. Solid

state logic circuits usually operate on a low dc voltage, from 5 to 15 volts dc. Electromechanical relay logic circuits usually operate on the annunciator power source voltage.

Even when solid state logic circuits are used, some electromechanical relays may be provided to drive large audible devices and to provide isolation in pushbutton, auxiliary output, and reflash circuits. Miniature relays mounted on printed circuit boards are often used to provide isolation because of small size.

Alarm modules usually include movable jumpers or switches to allow operation with either normally open or normally closed field contacts. Other jumpers or switches may be provided to allow selection of the annunciator sequence.

All of the alarm points of an annunciator usually have the same operating sequence; however, different sequences can be provided for individual alarm points or groups of points in one annunciator.

All of the visual displays of an annunciator or lamp cabinet usually flash in unison to minimize confusion, particularly during test. It is usually not necessary to have all visual displays of a large system flash in unison.

A.9.2 Field Contacts

Annunciator field contacts are the electrical output contacts of the devices sensing the process conditions. In relation to annunciators, field contacts are either normally open and close to alarm or they are normally closed and open to alarm. This contact operation description is usually not adequate when specifying the devices that sense the process conditions. It is usually necessary to specify that the contacts of these devices close or open when the process conditions increase or decrease. Field contacts may be either electromechanical or solid state.

Movable jumpers or switches are usually provided as a part of alarm modules to allow sequence logic circuits to operate from either normally open or normally closed field contacts. To avoid confusion during maintenance, however, all annunciator field contacts used with one annunciator or in one installation are usually selected to be of the same type.

Many annunciators are used with normally open field contacts and many are used with normally closed contacts. Some applications and industries are fairly consistent in this selection, while others are not consistent. The selection is most often based on past practice, but differences in annunciator operation when failures occur and during maintenance can be factors. Failures in the mechanical or electrical devices sensing the process conditions affect both normally open and normally closed field contacts. Open circuits in the field contact wiring prevent alarms from normally open contacts, but cause false alarms from normally closed contacts. Short circuits in the field wiring cause just the opposite results. Disconnecting field contact wiring briefly to locate grounded wires temporarily prevents alarms from normally open contacts, but causes false alarms from normally closed contacts. Spare alarm points with normally closed field contacts require shorting jumpers on the field contact terminal blocks or some other action to avoid the continuous display of alarms.

The field contact voltage is the voltage applied to the field contacts by the annunciator circuits. It is usually from 12 to 125 volts and can be either ac or dc. Annunciators with normally open field contacts often use from 48 to 125 volts for more reliable operation when electromechanical field contacts are contaminated. Lower voltages may be used with normally closed field contacts where contamination cannot prevent an alarm and also to be compatible with other low voltage instrument and control circuits. A single voltage is used when both field contact types are used with one annunciator.

The annunciator response time is the time period between the operation of the field contacts and initiation of the alarm state. Momentary field contact operations with a duration of less than the response time do not initiate alarms. Annunciator response times are in the order of 5 to 50 milliseconds, depending on the manufacturer. Electrical transients induced in the field contact wiring may cause false annunciator operation if the response time is not long enough. Because of the filtering provided by the response time, annunciators usually do not require shielded field contact wiring.

Longer response times or special time delay circuits can usually be provided when required to prevent false operation caused by extraneous circuit pulses or to avoid nuisance alarms caused by momentary operation of field contacts.

The steady state current that passes through field contacts from annunciators is usually in the order of 1 to 10 milliamperes, depending on the manufacturer. This current is necessary to minimize false operation caused by electrical transients and to provide enough arcing to assist in cleaning electromechanical field contacts. This current is usually sufficient for proper operation of solid state field contacts.

The maximum length of field contact wiring that can be used with annunciators depends on the conductor dc resistance or ac impedance, leakage resistance, the field contact voltage, and the annunciator input circuit design. Special design may be required when field contact wiring involves unusually small conductors or long lengths.

A.9.3 Shared Alarm Points

In many cases, annunciator alarm points are connected to only one field contact. Sometimes alarm points operate from two or more contacts that are wired in a series or parallel arrangement to develop the required alarm logic or to avoid nuisance alarms during certain operating or maintenance conditions. These series or parallel contact arrangements operate together as a single field contact in relation to the annunciator alarm point.

In order to conserve space on control panels, it may be desirable to use one annunciator visual display for more than one related monitored variable. This can be done by operating alarm points from two or more normally open field contacts in parallel or normally closed field contacts in series so any field contact operation causes an alarm. When this is done, that alarm point alarms when the first abnormal process condition occurs, and can return to normal only after all process conditions have returned to normal. That alarm point cannot alarm subsequent abnormal process conditions that occur while the first abnormal process condition still exists. This method is convenient and economical because it does not require special annunciator circuits, but it should be used only when knowledge of the occurrence of subsequent abnormal process conditions is not required.

The loss of information that occurs when two or more field contacts are connected in parallel or series does not occur if reflash circuits are added to the annunciator. Reflash circuits allow two or more abnormal process conditions to initiate or reinitiate the alarm state of one annunciator alarm point at any time. Reflash alarm points cannot return to normal until all related process conditions return to normal.

Two or more field contacts are wired independently to reflash circuits. The reflash circuit outputs are connected to sequence logic circuits. The annunciator sequence operates in the usual manner when the first abnormal process condition occurs. When subsequent abnormal process conditions occur while previous abnormal process conditions still exist, the reflash circuit causes the sequence to go to the alarm state again. The sequence cannot return to the normal state until all related process conditions return to normal. Reflash circuits allow annunciators to alarm the first and all subsequent abnormal process conditions, but the identity of the individual monitored variables must be determined from nearby instruments, field instruments, or observation of process conditions.

The number of individual reflash circuits on one reflash module is determined by the manufacturer. Reflash circuit outputs are interconnected to accommodate the required number of field contacts and then wired to alarm modules. Unused individual reflash circuits on some reflash modules may not be available for use with another group, depending on the module internal connections. In some cases, annunciator circuit design limits the number of reflash circuits that can be in any one reflash group.

Reflash modules are usually located together and not intermixed with alarm modules. This allows flexibility as the annunciator requirements develop and change.

A.9.4 Analog Input Alarm Points

While annunciators usually operate from field contacts, they can operate from analog input signals also. Analog input points are for use with analog monitored variable signals. Additional logic circuits initiate alarms when the analog signals are above or below set points.

Analog input points can operate from a variety of dc currents and voltages. In some cases thermocouples, resistance temperature detectors, and ac currents and voltages can be used also. When required, terminal block points should be specified for grounding the shields of analog input point field wiring.

Solid state circuits compare the analog input signals against set points established on the printed circuit boards by adjustable potentiometers. Movable jumpers or switches can be set to determine whether alarms occur when the signals are above or below the set points. The analog signal dead band is usually adjustable.

Separate analog input modules have output contacts for connection to alarm modules. In some cases, the analog input circuits and sequence logic circuits are included in one module.

A.9.5 Auxiliary Outputs

Annunciator auxiliary outputs are operated by single alarm points or groups of points for use with remote devices. Auxiliary outputs may be contact or voltage signals that are connected to remote alarm, recording, or control devices. Auxiliary output contacts may be electromechanical or solid state. These contacts are usually isolated electrically from the annunciator and powered from the remote device. Auxiliary output contacts are usually available that are normally open and close to alarm or are normally closed and open to alarm.

The four common types of auxiliary output and the operation of a graphic display in parallel with annunciator visual displays are shown functionally in Figure A4. The type selected for a particular application depends on the required operation. The operation of auxiliary outputs should be added to sequence diagrams and sequence tables or the operation should be defined by notes.

Field contact follower auxiliary outputs operate while individual field contacts indicate abnormal process conditions. The outputs usually do not operate during annunciator test. When field contact follower operation is required, it may be possible to avoid auxiliary outputs by operating annunciators in parallel with other devices using the same field contacts. The possibility and any technical problems should be discussed with the manufacturers of the annunciator and other device.

Lamp follower auxiliary outputs operate while individual visual display lamps indicate alarm, silenced, or acknowledged states. The outputs do not change when visual display lamps flash. The outputs usually operate during operational test, but not during lamp test.

Audible device follower auxiliary outputs operate while the common alarm audible devices operate. The outputs usually operate during operational test, but not during lamp test.

Reflash auxiliary outputs operate when any one of a group of alarm points indicates an abnormal process condition. The outputs usually return to normal briefly when each alarm point changes to an abnormal process condition the outputs return to normal when all alarm points in the group indicate normal process conditions. The outputs usually operate during annunciator test.

Consideration should be given to the operation of auxiliary outputs during annunciator test. In some cases operation of auxiliary outputs can test the remote devices also; however, in other cases operation during annunciator test would be undesirable or unacceptable.

When auxiliary outputs are connected to recording devices for analysis of sequence of events, the response time of the annunciator input and auxiliary output circuits should be considered. This delay may change the recorded sequence when compared to signals that do not pass through the annunciator.

Some installations use annunciator auxiliary outputs for control and protection interlocks and others use separate field contacts. The use of auxiliary outputs is often convenient, but separate field contacts may provide improved operating reliability and avoid unexpected control actions during annunciator maintenance.

Auxiliary outputs are often used to retransmit alarms from local annunciators to annunciators in central control rooms. Local annunciator auxiliary outputs can retransmit the operation of an entire local annunciator, groups of alarm points, or individual alarm points. Outputs from local annunciator power failure detectors are often transmitted also. When a group of local annunciator alarm points is retransmitted to one alarm point in the central control room, the central annunciator indicates that a local alarm has occurred but not which alarm. The following methods are commonly used to retransmit the operation of local annunciators.

Audible Device Follower Auxiliary Outputs	The central control room annunciator operates when local alarms occur and can return to normal after they are acknowledged at local annunciators. The central control room annunciator does not indicate when local alarms return to normal or that local alarms continue to exist. Operators acknowledging alarms at local annunciators are responsible for correcting all local alarm conditions. This method is convenient, economical, and can be added easily to existing local annunciators.
Reflash Auxiliary Outputs	The central control room annunciator operates when each local alarm occurs and can return to normal after all local alarms return to normal. The central control room annunciator does not indicate when individual local alarms return to normal. This method requires reflash auxiliary outputs in local annunciators and field wires to the control room annunciator from each local annunciator.

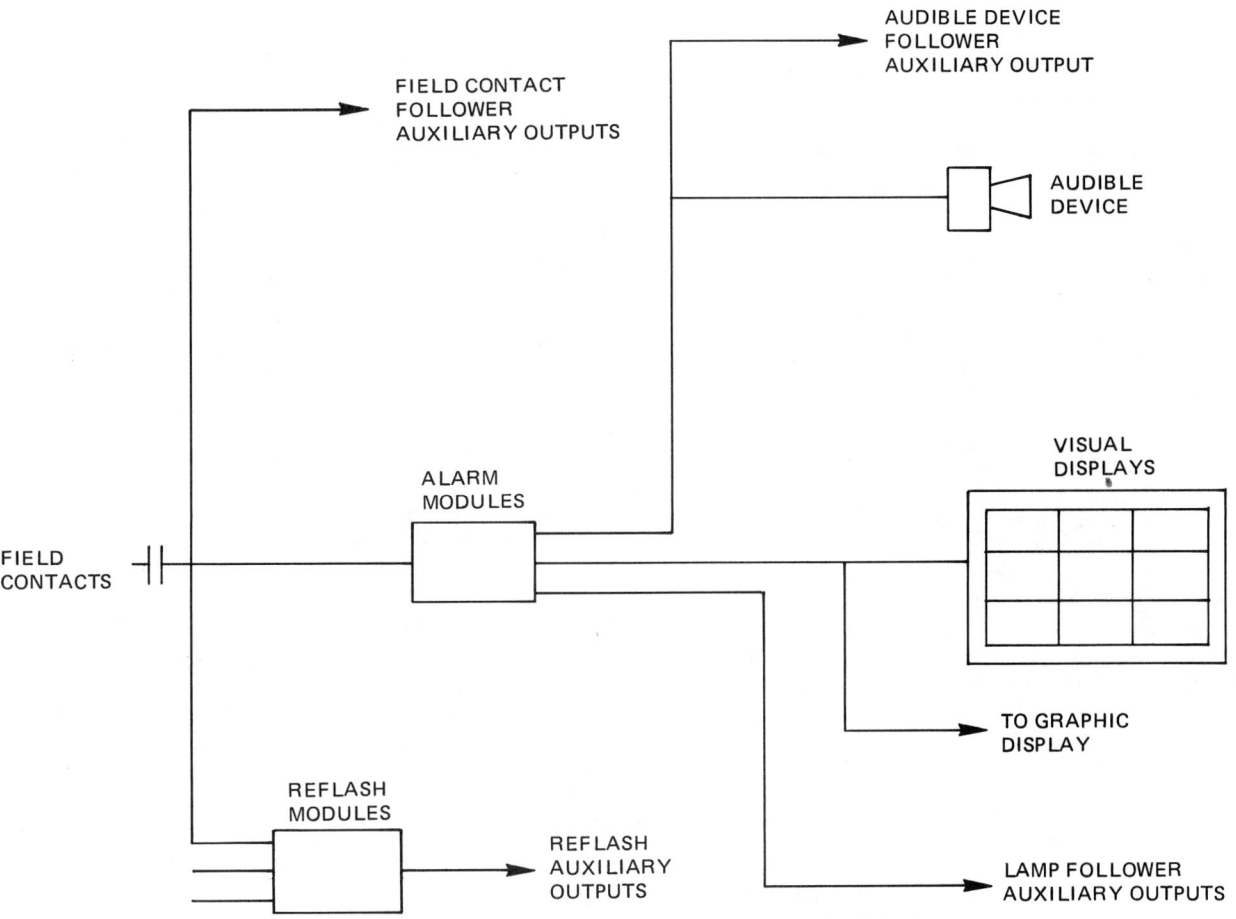

Fig. A4. Auxiliary Output Functional Diagram

Field Contact or Lamp Follower Auxiliary Outputs

Reflash alarm points in the central control room annunciator are used to indicate when each local alarm occurs and when all local alarms return to normal. The central control room annunciator does not indicate when individual local alarms return to normal. This method requires field contact or lamp follower auxiliary outputs in local annunciators, reflash alarm points in the central room annunciator, and field wires to the control room annunciator for each local annunciator alarm point.

A.10 Hazardous Atmosphere Location Annunciators

A.10.1 General

Special precautions are required when electrical equipment and circuits are installed in locations where fires or explosion hazards exist or may exist. Hazardous atmosphere locations are classified by class, division, and group. Class I locations are those in which flammable gases or vapors are or may be present in quantities sufficient to produce ignitible or explosive mixtures. Class II locations are those that are hazardous because of the presence of combustible dust. Class III locations are those that are hazardous because of the presence of easily ignitible fibers. Class I, Division 1, includes locations in which the gas or vapor hazard exists continuously or intermittently. Class I, Division 2, includes locations in which hazardous materials are normally confined but can escape when failures or abnormal conditions occur. Specific hazardous atmospheres are identified by group letters.

Special circuit or enclosure design may be avoided in Class I, Division 2, locations when current interrupting contacts are hermetically sealed against the entrance of gases and vapors.

For detailed application and installation information, refer to (1) Instrument Society of America Recommended Practice ISA-RP12.1, "Electrical Instruments in Hazardous Atmospheres," and (2) National Fire Protection Association Standard NFPA 70, "National Electrical Code," Articles 500-503.

Annunciators for use in hazardous atmosphere locations may require certification to meet the requirements of specific codes or agencies.

A.10.2 Intrinsically Safe Annunciator Systems

Intrinsically safe annunciator systems are designed and installed so that field contacts and field contact wiring are incapable of releasing sufficient electrical or thermal energy under normal or abnormal conditions to cause ignition of specific hazardous gas or vapor atmospheric mixtures in the most easily ignited concentrations. Field contacts and field contact wiring connected to intrinsically safe annunciators can be installed in many Class I, Division 1 and Division 2, hazardous locations without the use of explosion proof enclosures.

Intrinsically safe annunciator systems limit the release of energy from field contacts and field contact wiring under both normal and abnormal conditions by special power supply design, low field contact voltage, current limiting resistors, wiring barriers or separation from nonintrinsically safe circuits, and limited field contact wiring capacitance and inductance. Abnormal conditions include any two independent device or circuit failures in combination.

The annunciator cabinets, power supplies, pushbuttons, and audible devices of intrinsically safe annunciator systems are not intrinsically safe themselves. They must be installed in nonhazardous areas or must be designed for the specific hazardous atmosphere involved. Refer to A.10.3, Nonincendive Annunciators; A.10.4, Purged Cabinets; and A.10.5, Explosion Proof Cabinets.

In some applications, increased flexibility is obtained by the use of special relays and power supplies in the nonhazardous annunciator area to provide intrinsically safe barriers between the annunciator and intrinsically safe field contacts and field contact wiring. These barriers can be supplied with annunciators.

For detailed application and installation information, refer to (1) information from annunciator manufacturers, (2) Instrument Society of America Recommended Practice ISA-RP12.6, "Installation of Intrinsically Safe Instrument Systems in Class I Hazardous Locations," and (3) National Fire Protection Association Standard NFPA 493, "Intrinsically Safe Apparatus in Class I, II, and III, Division 1, Hazardous Locations."

A.10.3 Nonincendive Annunciators

Nonincendive annunciators are designed so that annunciator cabinets, power supplies, pushbuttons, and audible devices are incapable of releasing sufficient electrical or thermal energy under normal operating conditions to cause ignition of specific hazardous gas or vapor atmospheric mixtures in the most easily ignited concentrations. Nonincendive annunciators can be installed in many Class I, Division 2, hazardous locations without the use of explosion proof cabinets.

The design is not required to consider abnormal conditions such as device or circuit failure. The probability of simultaneous annunciator failure and abnormal conditions causing the escape of hazardous materials is considered negligible. No energy limit is placed on circuits or on contacts, switches, or plugs that are used only during maintenance.

Nonincendive annunciators make use of hermetically sealed contacts where current is interrupted under normal operating conditions. Annunciator cabinet covers that require the use of a tool for removal and approved warning nameplates remind the user to determine that the atmosphere is not hazardous before performing maintenance.

The annunciator field contacts and field contact wiring of nonincendive annunciators are not protected in any way by the nonincendive design. They must be designed and installed for the specific hazardous atmosphere involved. Refer to A.10.2, Intrinsically Safe Annunciator Systems.

For detailed application information, refer to (1) information from annunciator manufacturers, and (2) National Fire Protection Association Standard NFPA 70, "National Electrical Code, " Article 501.

A.10.4 Purged Cabinets

Annunciators with purged cabinets are designed so that a supply of clean air or inert gas around the electrical devices at sufficient flow reduces hazardous gas or vapors initially present to a safe level and maintains a safe level by positive pressure. Annunciators with purged cabinets can be installed in many Class I, Division 1 and Division 2, hazardous locations without the use of explosion proof cabinets. Pushbuttons and audible devices that are not in the annunciator cabinet must be designed for the specific hazardous atmosphere involved.

Purging requirements are divided into three classifications. Type X purging reduces the classification of the area within cabinets from Division 1 to nonhazardous, Type Y from Division 1 to Division 2, and Type Z from Division 2 to nonhazardous. Depending on the classification, purged cabinets require special construction, low pressure alarms, nonincendive design, power interlocks, warning nameplates, and special operating procedures.

For detailed application and installation information, refer to (1) information from annunciator manufacturers, (2) Instrument Society of America Standard ISA-S12.4, "Instrument Purging for Reduction of Hazardous Area Classification," and (3) National Fire Protection Association Standard NFPA 496, "Purged and Pressurized Enclosures for Electrical Equipment in Hazardous Locations."

A.10.5 Explosion Proof Cabinets

Annunciators with explosion proof cabinets are designed so that the cabinets are capable of withstanding explosions of a specified gas or vapor that may occur within them while preventing the ignition of specific hazardous gas or vapor atmospheric mixtures surrounding the cabinets. Annunciators with explosion proof cabinets can be installed in many hazardous locations. Pushbuttons and audible devices that are not in the annunciator cabinet must be designed for the specific hazardous atmosphere involved.

For NEMA enclosure requirements, refer to National Electrical Manufacturers Association Standard NEMA ICS 6, "Enclosures for Industrial Controls and Systems."

A.11 Cabinets

Annunciator cabinets and lamp cabinets are usually flush mounted on steel panels. The cabinet bezels and windows extend about 1 inch (25 mm) in front of the panels. Annunciator cabinets extend further behind the panels than lamp cabinets. Cabinets usually fit into rectangular cutouts in the panels and are held in place by clamps. Access for lamps and alarm modules is usually from the front. Access for wiring is usually from the rear. Openings for ventilation may be provided. Forced ventilation and high temperature alarm contacts are usually not required. Special surface mounted annunciator and lamp cabinets with access from the front are also available.

Logic circuits for use with remote logic annunciators usually can be supplied as open chassis assemblies, enclosed chassis assemblies, or 19 inch (475 mm) rack mount assemblies for mounting in logic cabinets by others. Logic circuits can also be provided in freestanding or surface mounted logic cabinets.

Annunciator cabinets are usually for general purpose indoor installation and are classified as NEMA 1. Special cabinets to meet NEMA 12 indoor driptight and dusttight requirements, and NEMA 4 indoor and outdoor watertight and dusttight requirements may also be available. For NEMA enclosure requirements, refer to National Electrical Manufacturers Association Standard NEMA ICS 6, "Enclosures for Industrial Controls and Systems." For annunciators located in hazardous atmospheres, see A.10, Hazardous Atmosphere Location Annunciators.

A.12 Power Supply Systems

A.12.1 Power Sources

Annunciators are available for use on a wide range of power source voltages, on ac at 50 or 60 hertz, and on dc. The most common power source voltage is 120 volts, 60 hertz, but other ac voltages such as 24, 48, and 230 volts are used also. Annunciators can usually operate over a range of at least 90 to 110 percent of the rated ac voltage.

Many annunciators operate on dc power sources to allow continued operation from batteries in the event ac power is lost. Larger installations usually provide 125 volt dc power for this purpose. Other dc voltages such as 24, 48, and 250 volts are used also. Annunciators can usually operate at the high equalizing charge voltage applied to some batteries during periodic maintenance and the low voltage that may exist during emergency operation. This voltage range is usually from 105 to 140 volts for annunciators designed to operate on 125 volts dc.

Annunciator electrochemical relay logic circuits usually use the power source voltage directly for the logic relays, lamps, and field contacts. Solid state logic circuits usually require power supplies to develop the dc voltages used for logic circuits, lamps, and field contacts.

A.12.2 Power Supplies

Annunciator power supplies convert ac or dc power source voltages to one or more dc voltages for use with solid state logic circuits, lamps, and field contacts. Power supplies are usually supplied with annunciators to ensure coordination with the annunciator circuits.

When dc power sources are used, special power supplies may be provided to convert dc power source voltages directly to the required dc annunciator circuit voltages, or separate inverters may be used to invert the dc to ac for connection to conventional ac to dc power supplies.

Power supplies provide electrical isolation between power sources and annunciator circuits. This isolation allows annunciator circuits to operate ungrounded. When accidental circuit grounds occur on ungrounded annunciator circuits, ground detectors operate annunciator alarms and the grounds can be located and removed without affecting the annunciator operation or the power source.

Inverters and power supplies used with annunciators should be provided with overload protection and should not require manual actions when voltage is applied or after brief overloads occur.

A.12.3 Backup Power

Annunciators are no more reliable than the power on which they operate. The power sources, power supplies, and power distribution systems for annunciators should provide the reliability required by the annunciation function.

A single power supply may provide power for only a few alarm points or for an entire annunciator. An annunciator power source may come from a 120 volt ac lighting panel or from a large battery and charger system. The failure of a single power supply or power source can disable an entire annunciator, or backup power supplies and power sources can be used so that any single failure affects no more than one alarm point. Power supply system designs should suit the power sources available and provide the required annunciator reliability.

While an annunciator powered from a dc power source provides reliable operation, a single failure in the dc distribution panel or annunciator feeder can leave the annunciator without power. In addition, dc power source buses are often ungrounded. When a ground occurs, the load circuits are switched off and on again in turn until the grounded circuit is located. When the annunciator feeder is switched off and on again, the alarm points usually assume a random sequence state and auxiliary outputs may operate incorrectly. It is then necessary for the annunciator pushbuttons to be operated before the correct alarms are displayed again.

When two independent power sources are used to power an important annunciator, the reliability becomes very high. Backup power supplies can also be used so that failure of a single power supply does not affect annunciator operation. When backup power sources or power supplies are used, power failure detectors should be provided to alarm the failure of each power source and power supply.

A.12.4 Power Failure Detectors

Power failure detectors provide outputs to alarm the loss of voltage on annunciator logic circuit, lamp, and field contact power buses. Power failure detectors cannot alarm on the failed annunciator, so another alarm device must be used.

Local annunciator power failure can be alarmed on central control room annunciators. Separate CRT visual display or recording annunciators can be used to alarm control room annunciator failure. In some cases, separate annunciator audible and visual devices powered by an independent power source are provided to alarm annunciator power failure. In these cases the independent power source must, in turn, be monitored by the annunciator.

A.12.5 Ground Detectors

Annunciator buses are often ungrounded so that accidental circuit grounds can be detected and ground alarms initiated without affecting the operation of annunciators. Grounds on the field contact wiring are the most common since this wiring is often run in areas that have an adverse environment.

Ground detectors are usually connected to alarm the presence of a ground on either conductor of an ungrounded system. The ground should be located and removed before a second ground occurs, since two or more simultaneous grounds can affect annunciator operation.

Grounds are usually located by disconnecting individual circuits in turn until the ground detector indicates that no ground exists. The ground is on the disconnected circuit.

Annunciator circuits are usually disconnected to check for grounds by briefly disconnecting field contact wires or by briefly removing individual alarm modules. Large annunciators often include power supply system disconnect switches to subdivide the system for more convenient ground isolation and maintenance. Some remote logic annunciators include field contact wiring terminal blocks that include switches or other devices to allow ground isolation without affecting the field wiring terminations.

The location of grounds is made easier if an indication of grounds is within sight of the disconnecting devices. Lights operating from ground detectors can be located in alarm module and terminal block areas for this purpose.

A.13 Factory Tests

A.13.1 General

Annunciator factory tests can include prototype, component, module, and final assembly tests. The routine tests performed by the manufacturer are often the only tests performed. In some cases, users specify specific tests to confirm that annunciators will operate as required in special applications.

Annunciators are usually tested using test procedures that apply to industrial electrical and electronic instrumentation and control equipment.

A.13.2 Dielectric Strength

Dielectric strength tests, sometimes referred to as high potential or high pot tests, are usually performed on annunciator final assemblies to confirm that insulation and clearances are adequate between circuits and cabinets and also between separate circuits. To avoid damage, solid state circuit modules are usually removed, grounded, or shorted during these tests.

Test voltages range from 1000 to 1500 volts ac and are usually applied for one minute. Dielectric strength standards are included in (1) National Electrical Manufacturers Association Standard NEMA ICS 1, Part ICS 1-109, "Industrial Controls and Systems, Tests and Test Procedures," (2) American National Standards Institute Standard ANSI C39.5, "American National Standard Safety Requirements for Electrical and Electronic Measuring and Controlling Instrumentation," and (3) Underwriters Laboratories Standard UL 508, "Industrial Control Equipment."

A.13.3 Functional

Functional tests are usually performed on annunciator modules and final assemblies to confirm that the operation is as specified. Tests on final assemblies are sometimes referred to as simulation tests.

Functional tests usually include operation in all of the normal operating modes and also in abnormal modes such as high and low power source voltage, grounds, short circuits, and unusual service conditions. Cabinet internal temperatures are often measured to confirm ventilation designs.

Annunciator modules are often subjected to a burn-in period. This reveals modules with weak components that would fail early in the annunciator life. The burn-in period varies widely, depending on the manufacturer or the specification requirements.

Logic circuit test assemblies can usually be provided to facilitate testing individual modules in the field.

A.13.4 Surge Withstand Capability

Annunciator solid state logic circuits operate at low power levels. Annunciator circuits are designed to suppress electrical surges and transients that may be induced in field wiring. Tests are usually performed on prototype modules to confirm that expected surges and transients do not cause false operation or damage.

A specific surge withstand capability test is included in Institute of Electrical and Electronics Engineers Standard IEEE 472, "IEEE Guide for Surge Withstand Capability (SWC) Tests."

A.13.5 Radio Frequency Interference

There is some possibility that false operation of annunciator solid state logic circuits can be caused by operation of nearby portable radio transmitters, especially when cabinet doors are open.

A radio frequency interference test method and a classification of signal field strengths and frequency bands are included in Scientific Apparatus Makers Association Standard SAMA PMC 33.1, "Electromagnetic Susceptibility of Process Control Instrumentation."

A.13.6 Seismic

In some applications, annunciators may be required to operate during and after specified earthquakes, or annunciator cabinets may be required only to remain intact externally so as not to endanger nearby personnel or equipment. When required, seismic tests are usually performed on prototype assemblies.

Seismic test procedures are included in Institute of Electrical and Electronics Engineers Standard IEEE 344, "IEEE Recommended Practices for Seismic Qualification of Class 1E Equipment for Nuclear Power Generating Stations."

A.13.7 Class 1E Equipment for Nuclear Generating Stations

Annunciators that perform Class 1E safety related functions in nuclear power generating stations must be qualified to confirm the adequacy of the equipment design under all expected normal and abnormal operation conditions.

Qualification procedures making use of type testing, operating experience, and analysis are included in Institute of Electrical and Electronics Engineers Standard IEEE 323, "IEEE Standard for Qualifying Class 1E Equipment for Nuclear Power Generating Stations."

APPENDIX B

SEQUENCE DESIGNATION CONVERSION

This Appendix is included for informational purposes and is not a part of this Standard.

The sequences in Instrument Society of America Recommended Practice ISA-RP18.1-1965 that are superseded and the corresponding new sequence designations used in this Standard are listed below.

ISA-RP18.1-1965 Sequences	ISA-S18.1 Sequence Designations
ISA-1	A
ISA-1A	A-5
ISA-1B	A-4
ISA-1C	A-4-5
ISA-1D	A-13
ISA-1E	Special [1]
ISA-1F	Special [1]
ISA-2	Special [1]
ISA-2A	R-8
ISA-2B	R-11
ISA-2C	M
ISA-2D	M-5
ISA-3	Special [2]
ISA-3A	Special [2]
ISA-3B	Special [2]
ISA-4	Special [1]
ISA-4A	F1A

(1) Requires two visual display indication devices.
(2) Requires three visual display indication devices.

INSTRUMENT SOCIETY OF AMERICA
Research Triangle Park, North Carolina

ISA-S20-1981

Standard

Specification Forms for Process Measurement and Control Instruments, Primary Elements and Control Valves

Instrument Society of America

ISBN 0-87664-347-0

ISA-S20 Specification Forms for Process Measurement
and Control Instruments, Primary Elements and
Control Valves

INSTRUMENT SOCIETY OF AMERICA
67 Alexander Drive
P.O. Box 12277
Research Triangle Park, North Carolina 27709

PREFACE

(This Preface is included for information purposes and is not part of S20).

This Standard has been prepared as a part of the service of the Instrument Society of America toward a goal of uniformity in the field of instrumentation. To be of real value this report should not be static, but should be subjected to periodic review. Toward this end the Society welcomes all comments and criticisms, and asks that they be addressed to the Standards and Practices Board Secretary, Instrument Society of America, 67 Alexander Drive, P.O. Box 12277, Research Triangle Park, North Carolina 27709.

This document was prepared by the Subcommittee on Instrument Specification Forms (RP20.1) and was originally published in 1956 under the direction of G. G. Gallagher of the Fluor Corporation. In 1961 additional forms were published, prepared by Committee 8D-RP20 under the direction of W. Carmack of the Fluor Corporation. This revision was prepared, with the supervision of the Chairman, R. E. Frey of Rohm and Haas Company, by the committee as listed below.

SP20 COMMITTEE

R. E. Frey (deceased)	Rohm and Haas Company
W. S. Buzzard	Fisher and Porter Company
J. G. Converse	Sun Oil Company
G. F. Erk	Sun Oil Company
J. Imber	Atlantic Richfield Company
R. D. Irwin	Honeywell, Inc.
E. F. Kremer	E. I. duPont deNemours & Co., Inc.
R. Leese	Catalytic Inc.
R. D. Prescott	Moore Products Company
A. Rosenthol	Catalytic Incorporated
F. J. Ryan, Jr.	Clifford B. Ives & Co., Inc.
W. C. Thomas	Clifford B. Ives & Co., Inc.

The assistance of those who aided in the preparation of this Standard, by their review of the draft and by offering suggestions toward its improvement, is gratefully acknowledged. The following have reviewed the report and served as Board of Review:

C. B. Anderson	Armstrong Cork Company
W. A. Bajek	UOP Process Division
A. S. Bartholomew	Reliance Electric Company
C. S. Beard	Bechtel Corporation
J. V. Becker	National Load
O. K. Booth	North American Rockwell
L. R. Brewer	Koppers Company, Inc.
A. M. Calabrese	M. W. Kellogg Company
F. I. Callisen	C. F. Braun & Company
C. P. Clark	United Technology Center
G. T. Clawson	Goddard Space Flight Center
R. Coel	The Fluor Corporation

L. Costea	Hunt-Wesson Foods
W. G. Cozart	Dow Chemical Company
E. Csaky	Dow Chemical Company
J. W. Eby	Honeywell, Inc.
M. W. Fifer	Proctor & Gamble
W. Forsyth	Honeywell, Inc.
G. G. Gallagher	The Fluor Corporation
W. J. Greter	Union Carbide Corporation
H. P. Haas	Philco-Ford
L. A. Haines	Valter, Inc.
M. G. Haines	Allied Chemical Corporation
W. Y. Harkins	Tennessee Eastman Company
E. J. Hayter	Allied Chemical Corporation
J. E. Holland	E. I. duPont deNemours & Company
E. A. Houser	Beckman Instruments
P. Kindersley	Kamyr Inc.
L. Kipnis	Instrumentation Consultant
C. D. Kolbe	NASA
W. B. Kostiw	Stone & Webster Engineering Corporation
E. R. Langston	Stearns-Roger Corporation
P. E. Larsen	Dow Chemical Company
B. G. Liptak	Crawford & Russell, Inc.
R. Loewe	Sargent & Lundy
F. Maltby	Drexelbrook Engineering Company
C. A. Master	Philco-Ford
A. F. Marks	Bechtel Corporation
M. W. Marxen	Dow Badische Company
C. McCrain	Monsanto Company
R. T. Miller	Brown & Root, Inc.
H. E. Nurmi	The Detroit Edison Company
P. L. Pettersen	Lockheed Missile & Space Company
E. Podolak	Federal Aviation Administration
C. Pray	Brown & Root, Inc.
H. D. Preszler	Johnson Service Company
C. A. Prior	E. I. duPont deNemours & Company
G. R. Rauschenberg	IBM Corporation
J. Rellford	Honeywell, Inc.
W. A. Richards	General Electric
E. J. Rogers	Bourns Life Systems
D. E. Sanchez	Jensen Instrument Company
R. H. Schipper	Jordan Valve
K. D. Smith	General Electric Company
E. F. Spalidoro	Barton Sales
W. Speight	Honeywell Controls, Ltd.
J. L. Thoma	Wyandotte Chemicals Corporation
D. J. Untener	Standard Oil Company
W. C. Voyles	Cummins Engine Company, Inc.
J. C. Walsh, Jr.	Eastman Kodak Company
N. S. Waner	Hallikainen Instruments
J. P. Wolfinger	Alpha Portland Cement Company
N. R. Zeller	Tucson Gas & Electric Company
R. Zielski	Georgia-Pacific Corporation

This Standard was approved for publication by the Standards and Practices Board on July 15, 1975.

W. B. Miller, Vice President	Moore Products Company
R. G. Hand, Secretary	Instrument Society of America
P. Bliss	Pratt Whitney Aircraft Company
L. N. Combs	retired from E. I. duPont deNemours & Company
B. A. Christensen	Continental Oil Company
R. L. Galley	Bechtel Corporation
T. J. Harrison	IBM Corporation
T. S. Imsland	Fisher Controls Company
P. S. Lederer	National Bureau of Standards
E. C. Magison	Honeywell, Inc.
J. R. Mahoney	IBM Corporation
R. L. Martin	Tex-A-Mation Engineering, Inc.
R. G. Marvin	Dow Chemical Company
A. P. McCauley	Glidden Durkee Div. SCM Corporation
T. A. Murphy	The Fluor Corporation, Ltd.
R. L. Nickens	Reynold Metals Company
G. Platt	Bechtel Corporation
A. T. Upfold	Polysar Ltd.
K. A. Whitman	Allied Chemical Corporation

Specification Forms for Process Measurement and
Control Instruments, Primary Elements and Control Valves

CONTENTS

1 PURPOSE

1.1 The purpose of this standard is to promote uniformity in instrument specifications, both in content and form. Because of the complexity of present day instruments and controls it is desirable to have some type of specification form to list pertinent details for use by all interested parties. General use of these forms by users and manufacturers offers many advantages, as listed below:

1. Assist in preparation of complete specification by listing and providing space for all principal descriptive options.

2. Promotes uniform terminology*.

3. Facilitates quoting, purchasing, receiving, accounting and ordering procedures by uniform display of information.

4. Provides a useful permanent record and means for checking the installation.

5. Improves efficiency from the initial concept to the final installation.

2 SCOPE

2.1 These forms are intended to assist the specification writer to present the basic information. In this sense they are "short-form" specifications or "check sheets" and may not include all necessary engineering data or definitions of application requirements. While the types of instruments described by these forms are more common to the process industries the forms should also prove useful in other areas if special requirements are defined elsewhere.

2.2 Some forms consist of a primary sheet and a secondary (tabulation) sheet. The primary sheet may be used by itself to specify a single instrument or to specify general requirements for a series of similar instruments which are then tabulated on the secondary sheet.

2.3 The heading used on all forms is designed to permit the user to add company name, plant location, trade mark, or specific project data.

2.4 The specification forms included in this standard are intended to cover the most commonly used instruments. The list is not a complete catalog of instruments and control valves available. It is intended that new forms shall be added with each general revision of this standard.

2.5 An instruction sheet is provided for each form to explain the terms used and the intended procedure. The instructions are keyed to the form by reference to the line numbers. The Committee has minimized dependence on the instruction sheet since the forms are frequently reprinted and used without the instructions. The explanation is omitted where the meaning is felt to be obvious.

2.6 Instrument specifications may be prepared by the use of Automatic Data Processing (ADP) techniques. The format of such specifications may be modified in order to be compatible with ADP machine capabilities. However, general consistency with this Standard shall be retained.

*Where applicable, the terminology used is in accordance with American National Standard C85.1-1963, "Terminology for Automatic Control," sponsored by the American Society of Mechanical Engineers.

		RECEIVER INSTRUMENTS				SHEET _____ OF _____	

						SHEET _____ OF _____	
		NO	BY	DATE	REVISION	SPEC. NO.	REV.
						CONTRACT	DATE
						REQ. P.O.	
						BY	CHK'D	APPR.

	1	Tag No.	Service
GENERAL	2	Function	Record ☐ Indicate ☐ Control ☐ Blind ☐ Integ ☐ Deviation ☐ Other _____
	3	Case	MFR STD ☐ Nom Size _____ Color: MFR STD ☐ Other _____
	4	Mounting	Flush ☐ Surface ☐ Rack ☐ Multi-Case ☐ Other _____ For Multiple Case, See Spec. Sheet _____
	5	Enclosure Class	General Purpose ☐ Weather Proof ☐ Explosion-Proof ☐ Class _____ For Use in Intrinsically Safe System. ☐ Other _____
	6	Power Supply	117 V 60Hz ☐ Other ac _____ dc ☐ _____ Volts
	7	Chart	____ Strip ☐ ____ Roll ☐ ____ Fold ☐ Circular _____ Time Marks ____ Range _____ Number _____
	8	Chart Drive	Speed _____ Power _____
	9	Scales	Type _____ _____ _____ _____ Range 1 _____ 2 _____ 3 _____ 4 _____
CONTROLLER	10	Control Modes	P = Prop (Gain), I = Integral (Auto Reset), D = Derivative (Rate), Sub: s = Slow, f = Fast P ☐ PI ☐ PD ☐ PID ☐ If ☐ Df ☐ Is ☐ Ds ☐ Other _____
	11	Action	On Meas. Increase Output: Increases ☐ Decreases ☐
	12	Auto-Man Switch	None ☐ MFR STD ☐ Other _____
	13	Set Point Adj.	Manual ☐ External ☐ Remote ☐ Other _____
	14	Manual Reg	None ☐ MFR STD ☐ Other _____
	15	Output	4-20 mA ☐ 10-50 mA ☐ 21-103 kPa (3-15 psig) ☐ Other _____
INPUTS	16	Input Signals	4-20 mA ☐ 10-50 mA ☐ 21-103 kPa (3-15 psig) ☐ Other _____
	17	No. of Inputs	1 ☐ 2 ☐ 3 ☐ 4 ☐
	18	Power for XMTRS	External ☐ This Inst ☐ No. of Independent Supplies _____ For Transmitters. See Spec Sheet.
ALARMS	19	Alarm Switches	Quantity _____ Form _____ Rating _____
	20	Function	Meas. Var. ☐ Deviation ☐ Contacts To _____ On Meas _____ Other _____
	21	Options	Filter-Reg ☐ Supply Gage ☐ Charts ☐ Int. Illumination ☐ Other _____ _____
	22	MFR & Model No.	_____

Notes:

ISA Form S20.1a

RECEIVER INSTRUMENTS					SHEET _____ OF _____		

	RECEIVER INSTRUMENTS				SPEC. NO.		REV.
NO	BY	DATE	REVISION		CONTRACT		DATE
					REQ. - P.O.		
					BY	CHK'D	APPR.

Rev.	Tag No.	No. of Inputs	Chart Range	Chart Number	Scale Ranges	Meas. Incr. Output	Service

Notes:

ISA Form S20.1b

RECEIVER INSTRUMENTS

Instructions for ISA Forms S20.1a and 20.1b

1. To be used for a single item. Use secondary sheet for multiple listing.

2. Check as many as apply.

3. Nominal size refers to approximate front of case dimensions; width x height.

4. It is assumed that the instrument has its own case or shelf suitable for single mounting, unless "MULTI-CASE" is checked. Shelf or separable case for multiple case mounting instrument is not included unless listed and described as an accessory.

5. Enclosure class refers to composite instrument. If electrical contacts are the case they must meet this classification inherently or by reason of the enclosure. Use NEMA identification system or ISA system RP8.1.

6. Specify electrical power to the entire instrument from an external source.

7. For multiple instruments list ranges on secondary sheet, but specify other chart options on primary sheet. Chart graduations assumed to be uniform unless otherwise noted. Circular charts assumed to have 24 hr/revolution speed; strip charts ¾ in. to 1 in. per hour.

8. Chart drive mechanism assumed to be synchronous motor operating on 117V 60 Hz and suitable for ENCLOSURE CLASS specified on line 5. If the chart drive is pneumatic so state — identify pneumatic pulser under options. Note deviations from standard (MFR) under notes i.e., dual speed or special speeds.

9. The scale type may be SEGMENTAL, VERTICAL, HORIZONTAL, DIAL (CIRCULAR) or other. Ranges 1, 2, 3 and 4 are used for multiple inputs. The first listed (No. 1) is assumed to be the controller input, if a controller is used.

10. See explanation of terminology given on specification sheet. For further definition refer to American National Standard C85.1-1963, "Terminology for Automatic Control." Specific ranges of control modes can be listed after "OTHER", if required.

11. For multiple items specify on second sheet.

12. If standard auto-manual switching is not known or not adequate, specify particular requirements, such as BUMPLESS, PROCEDURELESS, 4-POSITION, or as required.

13. Remote set point adjustment assumes full adjustment range. Specify limits if required. Under other can be noted bias or ratio.

14. Specify if applicable.

15. Specify if applicable.

16. All input signals on multi-channel instruments assumed to be the same range unless otherwise noted.

17. Specify number of inputs.

18. Check if power source for the loop is contained in this instrument or in some external instrument.

19. Form may be SPST, SPDT, DPDT or other. Rating refers to electrical rating of switch or contacts in amps.

20. Specify if alarm is actuated by measured variable or by deviation from controller set point. Give contact action if single throw form.

21. Specify required accessories and options, fill in number of charts. This is assured to be number of chart roles for strip charts.

22. After selection is made fill in manufacturer and specific model number.

SECONDARY SHEET — for listing multiple instruments. List all instruments of the same type specified on the primary sheet, with variations as shown. "Notes" refers to notes listed by number at the bottom of the sheet. Line 11 of sheet 1a is tabulated under measurement increases, output tabulate increase or decrease.

		ANNUNCIATORS				SHEET _____ OF_____		
	NO	BY	DATE	REVISION		SPEC. NO.		REV.
						CONTRACT		DATE
						REQ. - P.O.		
						BY	CHK'D	APPR.

GENERAL	1	Tag No._____ Location: _____
	2	Cabinet Size: _____ Rows High By _____ Columns Wide _____
	3	Mounting: Flush Panel □ Surface □
	4	Cabinet Style: Plug-In Light Boxes □ Swing Door □ Remote Logic Cabinet □ Watertight Door □
	5	Rating: General Purpose □ Weather proof □ Explosion proof □ Class_____ Group_____ Division _____
	6	Power Supply: 117V 60Hz □ 125 Vdc □ 12 Vdc □ 24 Vdc □ Other: _____
DISPLAY	7	Backlighted Nameplates: White Translucent □ Other: _____ Size: _____
	8	Alarm Points Per Lightbox: One □ Two □ Three □ Four □
		Lamps Per Alarm: One □ Two □ Three □ Four □
	9	Bullseye Type: Number of Lights:_____ Color:_____
	10	Other Display: _____
LOGIC	11	Logic: Electro-Mechanical Relay □ Solid-State Electronic □ Mercury Bottle □ Fluidic □
	12	In Display Cabinet □ Remote Cabinet □ Strip Chassis □
	13	General Purpose □ Weather proof □ Explosion proof □ Class_____ Group_____ Division_____ Intrinsically Safe □
	14	Field Contact Voltage: 117 Vac □ 12 Vdc □ 125 Vdc □ Other: _____
	15	On Alarm, Actuating Contacts: Open Close Field Selectable Form _____
FEATURES	16	Required Features: Lock-In of Momentary Alarms □ Auxiliary Contacts □ Sequential Alarm Circuit □
	17	Ring-Back Circuit: Via Alarm Audible Signal □ Via Other Audible Signal □
	18	Fail-Safe Circuit to Signal Own Failure □ Operational Test □ Lamp Test □
	19	Flasher: Remote □ In Cabinet □ Model No.: _____
	20	Acknowledge Common □ Unit □ Light □ Audible □ PB Location in Cabinet □ Remote □ Others □
	21	Reset Common □ Unit □ Light □ Audible □ PB Location in Cabinet □ Remote □ Others □

		STAGE	VISUAL SIGNAL	AUDIBLE SIGNAL
SEQUENCE	22	Normal		
		Alert, Initial		
		Alert, Subsequent		
		Acknowledge, Int.		
		Acknowledge, Subs.		
		Return to Normal		
		Reset		
		Test		
		ISA Sequence Number:		

OPTIONS	23	Horn:
	24	Bell:
	25	Dimmer:
	26	Color Caps:
	27	Power Supply Location
	28	
	29	Manufacturer: Model No.

Notes:

ISA Form S20.2a

	ANNUNCIATORS NAMEPLATE SCHEDULE				SHEET _____ OF _____		
		NO	BY	DATE	REVISION	SPEC. NO.	REV.
						CONTRACT	DATE
						REQ. P.O.	
						BY CHK'D APPR.	

COLUMN

COLUMN 1 2 3 4 5 6 7 8
ROW 1 2 3 4 5 6

Alarm Tag No.	Legend	Row	Column	Tag No. of Contact	Notes

ISA Form S20.2b

ANNUNCIATORS

Instructions for ISA Forms S20.2a and 20.2b

1. Write in Tag Number of entire Annunciator system.

2. Omit if single unit.

3. Specify cabinet mounting.

4. Specify type of cabinet.

5. Refers only to display and audible.

6. Specify power supply required.

7. Check WHITE TRANSLUCENT, or write in color of plate and engraving required. Specify window size in height x width.

8. Number of independent displays in one box, or position in cabinet.

9. If individual bullseyes, specify number and color required. If self-contained unit, specify number of normal and off-normal lights and color of each. (Example — two red independent off-normal and one green common normal light.)

10. Describe display if other than blacklighted nameplate or bullseye. For example; Blacklighted prism, Electrolumenescent, Two-color pneumatically operated.

11. Specify type of logic unit which operates display and audible system.

12. Check required location of logic components.

13. Check Enclosure Class of logic components and or enclosure. General purpose relays inside an explosion proof housing, or explosion proof relays will both satisfy the hazardous area classification. Use NEMA identification system or ISA system RP8.1.

14. Specify voltage across contacts which actuate alarm.

15. Give contact action.

16. Sequential Alarm refers to "First Out" system.

17. Specify type of ring back, if applicable.

18. An operational test actuates audible as well as lamps.

19. Specify flasher location and model number.

20. Specify type of Acknowledgment, and Pushbutton locations.

21. Specify reset and pushbutton location.

22. Write in ISA Sequence number as described in RP18.1, Specifications and Guides For the Use of General Purpose Annunciators, or fill in the table for the sequence required.

23. Write in the model number, or describe type, if required.

24. Write in the model number, or describe type, if required.

25. Write in the model number, or describe type, if required.

26. Specify number required, and color.

27. Specify power supply location i.e., in logic cabinet, or separate cabinet.

28. For any additional accessories required.

29. Fill in after selection is made.

©ISA S20

		NO	BY	DATE	REVISION

SHEET _____ OF _____

SPEC. NO.	REV.
CONTRACT	DATE

REQ. - P.O.

BY	CHK'D	APPR.

ISA Form S20.3a

13

		NO	BY	DATE	REVISION	SHEET _____ OF _____	
						SPEC. NO.	REV.
						CONTRACT	DATE
						REQ. - P.O.	
						BY	CHK'D	APPR.

ISA Form S20.3b

		NO	BY	DATE	REVISION	SHEET _____ OF _____	
						SPEC. NO.	REV.
						CONTRACT	DATE
						REQ. - P.O.	
						BY	CHK'D	APPR.

ISA Form S20.3c

			POTENTIOMETER INSTRUMENTS				SHEET _____ OF _____		
			NO	BY	DATE	REVISION	SPEC. NO.		REV.
							CONTRACT		DATE
							REQ. - P.O.		
							BY	CHK'D	APPR.

	1	Tag No.	Service	
GENERAL	2	Function	Record ☐ Indicate ☐ Control ☐ Blind ☐ Transmit ☐ Other _____	
	3	Type	Auto Bal. ☐ Man Bal. ☐ Galv ☐ Other_____	
	4	Case	MFR STD ☐ Nom Size _____ Color: MFR STD ☐ Other_____	
	5	Mounting	Flush ☐ Surface ☐ Rack ☐ Multi-Case ☐ Other _____ For Multiple Case Spec, See Sheet _____	
	6	Enclosure Class	Gen Purp ☐ Weather Proof ☐ Explosion-Proof ☐ Class _____ Other _____	
	7	Power Supply	117V 60 Hz ☐ Other _____	
	8	Chart	_____ Strip ☐ _____ Circ ☐ Time Marks ☐ Range _____ No _____ Chart Speed: _____ Change Gears _____	
	9	Scale	Type _____ Range 1 _____ 2 _____	
	10	Printout	No. of Points _____ Sec Per Point _____ Full Travel Speed _____ Print Character and Color _____ Point Select ☐	
	11	Selector Switches	No. and Form _____ In Case ☐ External ☐ Switch Cabinet Specs _____	
XMTR	12	Trans. Output	4-20 mA ☐ 10-50 mA ☐ 21-103 kPa (3-15 psig) ☐ Other _____ Input-Output Isolation ☐ For Receiver See Sheet _____	
CONTROLLER	13	Control Modes	P = Prop (Gain), I = Integral (Auto Reset), D = Derivative (Rate), Sub: s=Slow f=Fast If ☐ Df ☐ P ☐ PI ☐ PD ☐ PID ☐ I_s ☐ D_s ☐ Other _____	
	14	Action	On Meas. Increase Output: Increases Decreases	
	15	Auto-Man Switch	None ☐ MFR STD ☐ Specify _____	
	16	Set Point Adj.	Manual ☐ External ☐ Remote ☐ Specify_____	
	17	Manual Reg.	None ☐ MFR-STD ☐ Other _____	
	18	Output	4-20 mA ☐ 10-50 mA ☐ 21-103 kPa (3-15 psig) ☐ Other _____	
INPUT	19	Thermocouple Type	J(IC) ☐ K(CA) ☐ T(CC) ☐ E(CHR-CON) ☐ Other _____ Ref Junction Comp ☐ Lead Resistance (Galv) _____	
	20	Other Input	Resistance Temp Sensor ☐ Calibration _____ Other _____	
ALARM	21	Alarm Switches	Quantity _____ Form _____ Rating_____	
	22	Function	Meas. Var. ☐ Deviation ☐ Contacts to_____ measure _____ Other _____	
	23		Front Adj _____ Back Adj_____	
OPTIONS	24	T/C Burnout Drive	None ☐ Upscale ☐ Downscale ☐	
	25	Accessories	Case Illuminator ☐ _____ Charts _____ Filter Reg. ☐ Other_____	
	26	MFR. & Model No.	_____	

Notes:

ISA Form S20.10a

		POTENTIOMETER INSTRUMENTS					SHEET ___ OF ___			

		POTENTIOMETER INSTRUMENTS				SPEC. NO.	REV.
		NO	BY	DATE	REVISION	CONTRACT	DATE
						REQ. P.O.	
						BY	CHK'D	APPR.

Rev.	Tag No.	Range	Type Input	Scale and Chart	No. of Points	Meas. Incr., Output	Service	

Notes:

ISA Form S20.10b

POTENTIOMETER INSTRUMENTS
SPECIFICATION SHEET INSTRUCTIONS

Instructions for ISA Forms S20.10a and 20.10b

Prefix number designates line number on corresponding specification sheet.

1. To be used for single item. Use secondary sheet for multiple listing.

2. Check as many as apply.

3. Check one. Note that sheet may be used to specify galvanometric type of instrument.

4. Nominal size refers to approximate front of case dimensions; width x height.

5. It is assumed that the instrument has its own case or shelf suitable for single mounting unless "multi-case" is checked. Shelf or separable case for multiple case mounting instrument is not included in this sheet unless listed as an accessory.

6. Enclosure Class refers to composite instrument. If electrical contacts are in the case, they meet this rating inherently or by reason of the enclosure. Use NEMA identification system or ISA system presented in RP8.1.

7. Specify electrical power to entire instrument.

8. For multiple instruments list ranges on second sheet, but specify other items here.

9. Ranges 1 and 2 refer to multi-channel instruments. The first listed is assumed to be the controller input (if any).

10. For multiple items list number of points on second sheet. "Point Select" permits by-passing any or all points by a switching mechanism.

11. For multiple items show number of switches on second sheet under "No. of Points."

12. Specify if applicable.

13. See explanation of terminology given on spec. sheet. Specific ranges of control modes can be listed under "other" if required.

14. For multiple items specify on second sheet.

15. If standard auto-manual switching is not known or not adequate, specify particular requirements, such as BUMPLESS, PROCEDURELESS, 4-POSITION, or as required.

16. Remote set point adjustment assumes full adjustment range. Specify limits if required.

17. Specify if applicable.

18. Specify if applicable.

19. Check if thermocouple input applies. Lead resistance required only for galvanometer.

20. Specify any input other than thermocouple. "Calibration" refers to curve used and does not imply that element is specifically calibrated for this instrument.

21. Form may be SPST, SPDT, DPDT, etc. Rating is electrical rating of switch in amps.

22. Check if alarm is actuated by measured variable or by deviation from controller set point. Give contact action if single throw form. Specify calibrated or blind alarm index setter.

23. Specify if applicable.

24. Specify if applicable.

25. Accessories for multiple items may be covered by "notes" second sheet.

26. May be filled in after selection is made.

SECONDARY SHEET — for listing multiple instruments. List all instruments of the same type, specified on Primary Sheet, with variations as shown. "Notes" refers to notes listed by number at the bottom of the sheet. Or use Secondary Sheet to list and identify the multiple points of a single multipoint instrument.

			TEMPERATURE INSTRUMENTS (FILLED SYSTEM)				SHEET _____ OF_____		
				NO	BY	DATE	REVISION	SPEC. NO.	REV.
								CONTRACT	DATE
								REQ. - P.O.	
								BY / CHK'D / APPR.	

	1	Tag No.	Service
GENERAL	2	Function	Record ☐ Indicate ☐ Control ☐ Blind ☐ Trans ☐ Other _____
	3	Case	MFR STD ☐ Nom Size _____ Color: MFR STD ☐ Other _____
	4	Mounting	Flush ☐ Surface ☐ Yoke ☐ Other _____
	5	Enclosure Class	General Purpose ☐ Weather proof ☐ Explosion proof ☐ Class _____ For Use in Intrinsically Safe System ☐ Other_____
	6	Power Supply	117 V 60Hz ☐ Other ac _____ dc ☐ _____Volts
	7	Chart	Strip ☐ _____ Roll ☐ _____ Fold ☐ _____ Circular _____ Time Marks _____
	8	Chart Drive	Speed _____ Power _____
	9	Scales	Type _____ Range 1 _____ 2 _____ 3 _____ 4 _____
XMTR	10	Transmitter Output	4-20 mA ☐ 10-50 mA ☐ 21-103 kPa (3-15 psig) ☐ Other _____ For Receiver See Spec. Sheet _____
CONTROLLER	11	Control Modes	P=Prop (Gain), I=Integral (Auto Reset), D=Derivative (Rate), Sub: s = Slow f = Fast P ☐ PI ☐ PD ☐ PID ☐ If ☐ Df ☐ Is ☐ Ds ☐ Other _____
	12	Action	On Meas. Increase Output: Increases ☐ Decreases ☐
	13	Auto-Man Switch	None ☐ MFR STD ☐ Other _____
	14	Set Point Adj.	Manual ☐ External ☐ Remote ☐ Other _____
	15	Manual Reg.	None ☐ MFR STD ☐ Other_____
	16	Output	4-20 mA ☐ 10-50 mA ☐ 21-103 kPa (3-15 psig) ☐ Other _____
ELEMENT	17	Fill	SAMA Class _____ Compensation _____
	18	Process Data	Temp: Normal_____ Max _____ Max. Press. _____
	19	Range	Fixed ☐ Adj. Range _____ Set At _____ Overrange Protection to _____
	20	Bulb	Type _____ Mtl. _____ Extension: Length _____ Type _____ Size: Diameter _____ Length_____ Insertion _____ Conn: _____ Location_____ Ft. Above ☐ Below ☐ Instr.
	21	Capillary	MFR STD ☐ Length_____ Mtl. _____ Armor _____
	22	Well	Mtl. _____ Insertion_____ Lag Ext. _____ Conn._____ Const: Drilled ☐ Built-Up ☐ Other_____
	23	Alarm Switches	Quantity _____ Form _____ Rating _____
	24	Function	Temp ☐ Deviation ☐ Contacts To_____ On Temp. Increase
	25	Options	Filt-Reg. ☐ Sup. Gage ☐ Output Gage ☐ _____ Charts Other _____
	26	Mfr. & Model No.	_____

Notes:

ISA FORM S20.11a

			TEMPERATURE INSTRUMENTS (FILLED SYSTEMS)		SHEET _____ OF _____	

	NO	BY	DATE	REVISION	SPEC. NO.	REV.
					CONTRACT	DATE
					REQ. - P.O.	
					BY	CHK'D	APPR.

Rev.	Tag No.	Adj Range	Set Range	Well Conn.	Insert Length	Cap Length		Notes

Notes:

ISA Form S20.11b

TEMPERATURE INSTRUMENTS (Filled Systems)

Instructions for ISA Forms S20.11a and 20.11b

1. To be used for a single item. Use secondary sheet for multiple listing.

2. Check as many as apply.

3. Nominal size refers to approximate front of case dimensions; width x height.

4. Yoke refers to a bracket designed for mounting the instrument on a pipe stand.

5. Enclosure class refers to composite instrument. If electrical contacts are in the case, they must meet this classification inherently or by reason of enclosure. Use NEMA identification or ISA identification RP8.1.

6. Specify electrical power to the entire instrument from an external source.

7. Specify chart size, range and number if applicable.

8. Chart drive mechanism assumed to be synchronous motor operating in 117V 60 Hz and suitable for ENCLOSURE CLASS specified on line 5. If the chart drive is pneumatic so state — identify pneumatic pulser under options. Note deviations from standard (MFR) under notes i.e., dual speed or special speeds.

9. The scale type may be SEGMENTAL, VERTICAL, HORIZONTAL, DIAL (CIRCULAR) or other. Ranges 1, 2, 3 and 4 are used for multiple inputs. The first listed (No. 1) is assumed to be the controller input, if a controller is used.

10. Specify transmitter output if applicable.

11. See explanation of terminology given on specifications sheet. For further definition refer to American National Standard C85.1-1963, "Terminology for Automatic Control." Specific ranges of control modes can be listed after "OTHER", if required.

12. For multiple items specify on second sheet.

13. If standard auto-manual switching is not known or not adequate, specify particular requirements, such as BUMPLESS, PROCEDURELESS, 4-POSITION, or as required.

14. Remote set point adjustment assumes full adjustment range. Specify limits if required.

15. Specify if applicable.

16. Specify if applicable.

17. Filled thermal systems can be of the following SAMA classifications:

Class 1A: Liquid filled, uniform scale, fully compensated.

Class IB: Liquid filled, uniform scale, case compensated only.

Class IIA: Vapor pressure, non-linear scale with measured temperature above case and tubing temperature.

Class IIB: Vapor pressure, non-linear scale with measured temperature below case and tubing temperature.

Class IIC: Vapor pressure, non-linear scale with measured temperature above and below case and tubing temperature.

Class IIIA: Gas filled, uniform scale, fully compensated.

Class IIIB: Gas filled, uniform scale, case compensated only.

Class VA: Mercury filled, uniform scale, fully compensated.

Class VB: Mercury filled, uniform scale, case compensated only.

19. Range refers to process input span for which an output is desired. Adjustable range means that the unit can give its normal output over a range of inputs.

20. Bulb type can be plain, averaging, rigid, adjustable union connections, fixed union connection. Capillary extension length can be rigid or flexible, etc.

21. Capillary tube specifications

22. Well Specifications

23. Form may be SPST, SPDT, DPDT, etc. Rating is electrical rating of switch in volt amps.

24. Check if alarm is to be actuated by measured variable or by deviation from controller set point. Give contact action if single throw from.

	THERMOCOUPLES AND THERMOWELLS				SHEET _____ OF _____		
					SPEC. NO.		REV.
	NO	BY	DATE	REVISION	CONTRACT		DATE
					REQ.	P.O.	
					BY	CHK'D	APPR.

1. Complete Assembly ☐   Other _____
   MFR. & Model No. _____

ELEMENT

MFR. & Model No. _____
2. ISA Type _____ Wire Size _____
3. Sheathed: _____ O.D. Material _____
   Exposed ☐   Grounded ☐   Ungrounded ☐
   Enclosed ☐   Beaded Insulators ☐   Spring Loaded ☐
4. Nipple Size ____ Dimension "N" ____ Union ☐
5. Packed Connector _____

HEAD

6. Screw-Cap & Chain ☐ Other _____

7. Material _____ Conduit Conn. _____
8. Terminal Block: Single ☐   Duplex ☐

WELL OR TUBE

9. Material _____
10. Construction: Tapered ☐   Straight ☐
    Drilled ☐   Built-Up ☐   Closed End Tube ☐
11. Dimensions: MFR. STD. ☐   O.D. ____ I.D. ____
12. Connections: Process _____ INT _____
13. Other Specs.: _____

Notes:

Rev.	Tag No.	Well Dimens.		Element Length	Single Duplex	Type	Gage	Service	Notes
		"U"	"T"						

Notes:

ISA FORM S20.12a

		THERMOCOUPLES AND THERMOWELLS				SHEET _____ OF _____		
						SPEC. NO.		REV.
		NO	BY	DATE	REVISION	CONTRACT		DATE
						REQ. - P.O.		
						BY	CHK'D	APPR.

Rev.	Tag No.	Well Dimens. "U"	Well Dimens. "T"	Element Length	Single Duplex	Type	Gage	Service	Notes

Notes:

ISA FORM S20.12b

THERMOCOUPLES AND THERMOWELLS

Instructions for ISA Forms S20.12a and 20.12b

Reference: ANSI MC96.1, American National Standard
for Temperature Measurement Thermocouples.

1. Check COMPLETE ASSEMBLY, or write in ELE-
MENT ONLY, ELEMENT & HEAD, etc.

2. Specify ISA type: E Chromel/Constantan
                      J Iron/Constantan
                      K Chromel/Alumel
                      R Platinum-13 percent
                             Rhodium/Platinum
                      S Platinum-10 percent
                             Rhodium/Platinum
                      T Copper/Constantan
and wire diameter in American Wire Gage (AWG), also
known as Brown and Sharpe Gage (B & S). Thermocouple wire normally runs from AWG No. 24 (0.0201 in.
dia.) through AWG No. 8 (0.1285 in. dia.).

3. Specify required construction by filling in sheath
diameter and material, or checking BEADED INSUL-
ATORS. Check type of junction, EXPOSED, EN-
CLOSED and GROUNDED, ENCLOSED and
UNGROUNDED.

4. Specify nominal diameter of nipple, or write NONE.
Specify length N (as defined on sketch below line 8) if
appropriate.
Check UNION if required.

5. Specify connection size and material of packed
connector, and whether Fixed or Adjustable. (For
ceramic packed thermocouples only).

6. Specify general type of head.

7. Specify material of construction of head.

8. A duplex terminal block accommodates two thermocouples as listed. Refer to Notes.

9. Specify material of well or tube.

10. A built-up well has a welded tip. Check as many as
apply.

11. Give dimensions if required.

12. Process connection is external. However, INT will
cover a thread dimension if well flange is threaded.

13. Fill in any applicable company standards or specifications.

Note: For thermocouples other than arrangement shown
in sketch, space has been provided for you to draw your
own picture.

Tabulation: Fill in all applicable information. SINGLE/
DUPLEX, need only be filled in on line 8 if they are the
same for all thermocouples on the sheet.

	RESISTANCE TEMPERATURE SENSORS			SHEET _____ OF _____	
				SPEC. NO.	REV.
NO	BY	DATE	REVISION	CONTRACT	DATE
				REQ. - P.O.	
				BY · CHK'D · APPR.	

1. Complete Assembly   Other
   HEAD
2. Screwed Cover ☐   Other _____
3. Explosion Proof ☐   Class _____
4. Material _____ Cond. Conn. _____
5. Nipple Size_____ Dim. "N" _____ Union ☐
   ELEMENT
6. Platinum ☐   Nickel ☐   Other _____
7. Ice Point Resistance _____
8. Temperature Range _____
9. Leads: STD ☐   Potted ☐   Herm. Sealed ☐
10. Sheath Material _____ O.D. _____

11. Mounting Thread _____
12. Connection: 2-Wire ☐ 3-Wire ☐ 4-Wire ☐
    Lead Wires ☐   Receptacle ☐   Bayonet Lock ☐
    Other _____
    WELL OR TUBE
13. Material _____
14. Construction: Tapered ☐   Straight ☐
    Drilled ☐   Built-Up ☐   Closed-End ☐
                                              Tube
15. Dim: MFR STD ☐   O.D. _____ I.D. _____
16. Internal Thread _____
17. Process Connection _____

Rev.	Tag No.	Process Conn.	Well Dim. U	Well Dim. T	Element Length	Single or Dual	Service	Notes

Notes:

ISA FORM S20.13a

		RESISTANCE TEMPERATURE SENSORS					SHEET ____ OF____		

						SPEC. NO.		REV.

NO	BY	DATE	REVISION

CONTRACT | DATE

REQ. - P.O.

BY | CHK'D | APPR.

Rev.	Tag No.	Well Dim.		Element Length	Single or Dual	Service	Notes
		U	T				

Notes:

ISA FORM S20.13b

RESISTANCE TEMPERATURE SENSORS

Instructions for ISA Forms S20.13a and 20.13b

Refer to Scientific Apparatus Manufacturers Association (SAMA) Tentative Standard on Resistance, RC 5-10-1955.

1. Complete assembly includes head, element, and well; as shown in sketch.

5. Give size and pipe schedule of nipple. Check if union is required.

7. The ice point resistance in ohms usually defines the resistance vs. temperature curve. If not, provide additional data as an attachment.

8. Give maximum range over which the elements will be used.

9. Specify sealing of leads.

11. This thread is on the element termination, not the well.

12. It is necessary to specify the number of wires, depending on the compensation required. The other items refer to the element termination.

14. A built-up well has a welded tip and connection.

16. Internal thread of flange if well flange is threaded.

Instructions for the tabulation:

17. Process Connection is the connection on the element or well which is connected to the pipe or vessel. Well dimensions are illustrated in the sketch. It is not necessary to specify "Element Length" if well dimensions are already given. Single or Dual elements are assumed to be within the same sheath. Refer to Notes by number or letter and explain in the space at the bottom of the form.

	INDUSTRIAL BIMETAL AND GLASS THERMOMETERS				SHEET _____ OF _____	
					SPEC. NO.	REV.
	NO	BY	DATE	REVISION		
					CONTRACT	DATE
					REQ. - P.O.	
					BY CHK'D	APPR.

THERMOMETER

1. Stem: Threaded ☐ Plain ☐ Union ☐
   Material_____
2. Stem or Union Thread: ½ in. ☐ ¾ in. ☐
3. Stem Diameter: STD ☐ .250 in. ☐ .375 in. ☐
4. Case Material: STD ☐ Other _____
5. Dial Size _____ Color _____
6. Scale length _____ Color _____
7. Form: Fig. No. _____ Adjustable ☐
8. External Calibrator ☐ Herm Sealed Case ☐
9. MFR. & Model No._____

WELL

10. None ☐ Included ☐ By Others ☐
11. Material: 304SS ☐ 316SS ☐
    Other: _____
12. Construction: Drilled ☐ Built-Up ☐
    Other: _____
Well Length Must Suit Stem Length.

FIG. I BOTTOM FORM (STRAIGHT) — FIG. 2 BACK FORM (ANGLE) — STRAIGHT FIG. 3 — BACK FIG. 4-5 — LEFT RIGHT FIG. 6-7 — 135° RIGHT-LEFT FIG. 8-9

Rev.	Tag No.	Range	Operating Temp	Stem Length	Well Conn.	Lag Ext.	Service	Notes

Notes:

ISA FORM S20.14a

	BIMETAL THERMOMETERS				SHEET _____ OF _____		
		NO	BY	DATE	REVISION	SPEC. NO.	REV.
						CONTRACT	DATE
						REQ. - P.O.	
						BY	CHK'D	APPR.

Rev.	Tag No.	Range	Operating Temp	Stem Length	Well Conn.	Lag Ext.	Service	Notes

Notes:

ISA Form S20.14b

BI-METAL THERMOMETERS

Instructions for ISA Forms S20.14a and 20.14b

1. Specify mounting termination of stem and write in stem materials or "MFR. STD."

2. Select stem thread size.

3. Stem diameter standards may vary. Check specific size if this is important.

4. Write in case material if other than standard.

5. Write in nominal dial size and color.

6. Scale Length

7. The form of the thermometer is illustrated on the form. The adjustable form may be set to any angle. If a stem connection form other than shown is required, make a sketch in the space provided.

8. Check applicable options

9. List specific make and model number when selection is made.

10. Specify how well is to be furnished, if any.

11. Specify well material. If not all are the same, cover exceptions by notes in the tabulation.

12. Specify well construction. A "built-up" well has a welded tip. Special well designs should be described by a sketch in the space provided or on an attached sheet.

Tabulation:

Tag No: It is assumed that a tag number represents a single item. If multiple units have the same number, cover this with a special note.

Range: Write "F" or "C" at the top of the column. May be left blank on initial issue if Operating Temp. is specified.

Operating
Temp. Must be filled in if range is not specified.

Stem Length: Refer to illustrations on form.

Well Conn: Show thread size, such at "1 in. NPT" or flange size and rating, such as "1½ in. 150 lb". All flanges are assumed to be ANSI Standard; if not, cover by a special note.

Lag. Ext: Applies to screwed wells only.

Notes: Index notes by number or letter and specify in space below tabulation.

©ISA S20

			DIFFERENTIAL PRESSURE INSTRUMENTS				SHEET _____ OF _____	
				NO	BY	DATE	REVISION	SPEC. NO.	REV.

(Header block:)

DIFFERENTIAL PRESSURE INSTRUMENTS

NO	BY	DATE	REVISION

SHEET _____ OF _____
SPEC. NO. | REV.
CONTRACT | DATE
REQ. - P.O.
BY | CHK'D | APPR.

	1	Tag No.	Service

GENERAL

2	Function	Record ☐ Indicate ☐ Control ☐ Blind ☐ Trans ☐ Integ ☐ Other _____
3	Case	MFR STD ☐ Nom Size _____ Color: MFR STD ☐ Other _____
4	Mounting	Flush ☐ Surface ☐ Yoke ☐ Other _____
5	Enclosure Class	General Purpose ☐ Weather proof ☐ Explosion proof ☐ Class _____
		For use in Intrinsically Safe System ☐ Other _____
6	Power Supply	117V 60 Hz ☐ Other ac _____ dc ☐ _____ Volts _____
7	Chart	12 in. Circ. ☐ Other _____ Range _____ No. _____
8	Chart Drive	24 hr Other _____ Elec. ☐ Spring ☐ Other _____
9	Scale	Type _____ Range: 1 _____ 2 _____ 3 _____

XMTR

10	Transmitter Output	4-20 mA ☐ 10-50 mA ☐ 21-103 kPa (3-15 psig) ☐ Other _____
		For Receiver, See Spec Sheet _____

CONTROLLER

11	Control Modes	P=Prop (Gain), I=Integral (Auto Reset), D=Derivative (Rate) Sub: s=Slow, f=Fast If☐ Df☐ P ☐ PI ☐ PD ☐ PID ☐ I_s ☐ D_s ☐ Other _____
12	Action	On Meas. Increase Output: Increases ☐ Decreases ☐
13	Auto-Man Switch	None ☐ MFR STD ☐ Other _____
14	Set Point Adj.	Manual ☐ External ☐ Remote ☐ Other _____
15	Manual Reg.	None ☐ MFR STD ☐ Other _____
16	Output	4-20 mA ☐ 10-50 mA ☐ 21-103 kPa (3-15 psig) ☐ Other _____

UNIT

17	Service	Flow ☐ Level ☐ Diff. Pressure ☐ Other _____
18	Element Type	Diaphragm ☐ Bellows ☐ Mercury ☐ Other _____
19	Material	Body _____ Element _____
20	Rating	Overrange _____ Body Rating _____ psig
21	Diff. Range	Fixed ☐ Adj. Range _____ Set At _____
22		Elevation _____ Suppression _____
23	Process Data	Fluid _____ Max Temp. _____ Max. Press. _____
24	Process Conn.	½ in. NPT ☐ Other _____

25	Alarm Switches	Quantity _____ Form _____ Rating _____
26	Function	Meas. Var. ☐ Deviation ☐ Contacts To _____ on Inc. Meas.

27	Options	Pressure Element ☐ Range _____ Material _____ Temp. Element ☐ Range _____ Type _____ Filt Reg. ☐ Sup. Gage ☐ Output Gage ☐ _____ Charts Valve Manifold _____ Cond. Pots ☐ Adj. Damp ☐ Integral Sq. Rt. Ext. ☐ Integrator _____ Other _____

28	MFR & Model No.	_____

Notes:

ISA Form S20.20a

	DIFFERENTIAL PRESSURE INSTRUMENTS				SHEET _____ OF _____		
	NO	BY	DATE	REVISION	SPEC. NO.		REV.
					CONTRACT		DATE
					REQ. - P.O.		
					BY	CHK'D	APPR.

Rev.	Tag	Adj. Range	Set Range	Scale or Chart	Scale Factor	Service	Notes

Notes:

ISA Form S20.20b

DIFFERENTIAL PRESSURE INSTRUMENTS

Instructions for ISA Forms S20.20a and 20.20b

1. To be used for a single item. Use secondary sheet for multiple listing.

2. Check as many as apply.

3. Nominal size refers to approximate front of case dimensions; width x height.

4. Yoke refers to a bracket designed for mounting the instrument on a pipe stand.

5. Enclosure class refers to composite instrument. If electrical contacts are in the case they must meet this classification inherently or by reasons of the enclosure. Use NEMA identification system or ISA identification RP8.1.

6. Specify electrical power to the entire instrument from an external source.

7. Specify chart size, range and number if applicable.

8. "24 hr" is the time for one rotation of the chart. Other speeds should be listed in hours or days. If a spring wound clock is used fill in number of hours or days it runs between windings.

9. The scale type may be SEGMENTAL, ECCENTRIC, or DIAL (CIRCULAR). Space is provided for multiple ranges on the same scale.

10. Specify transmitter output if applicable.

11. See explanation of terminology given on specification sheet. For further definition refer to American National Standard C85-1-1963, "Terminology for Automatic Control." Specific ranges of control modes can be listed after "OTHER", if required.

12. For multiple items specify on second sheet.

13. If standard auto-manual switching is not known or not adequate, specify number of positions.

14. Remote set point adjustment assumes full adjustment range. Specify limits if required.

15. Specify if applicable.

16. Specify if applicable.

17. Specify measured variable.

18. Specify type of element or write in "MFR. STD."

19. Materials refer to wetted parts only.

20. Over-range protection refers to maximum differential pressure. The instrument can withstand without a shift in calibration.

21. Adjustable range means that the range can be changed without replacing any parts.

22. Elevation

23. Give process data affecting meter selection. Flow elements such as orifice plates are specified on separate forms.

24. Refers to connections piped to process equipment or pipe line. Special flanged connections and extended diaphragms for level applications should be described after "OTHER."

25. Form may be SPST, DPDT, or others. Rating refers to electrical rating of switch or contacts in amps.

26. Specify if alarm is actuated by measured variable or by deviation from controller set point. Give contact action if single throw form.

27. Specify required accessories. If temperature element is used, the second line is provided to specify well, length of capillary tubing and other details of the thermal system.

28. After selection is made fill in manufacturer and specific model number.

SECONDARY SHEET — for listing multiple instruments. List all instruments of the same type specified on the primary sheet, with variations as shown. "Notes" refers to notes listed by number at the bottom of the sheet.

		ORIFICE PLATES and FLANGES				SHEET _____ OF_____		
						SPEC. NO.		REV.
		NO	BY	DATE	REVISION			
						CONTRACT		DATE
						REQ. - P.O.		
						BY	CHK'D	APPR.

ORIFICE PLATES

1. Concentric ☐ Other _____
2. ISA Standard ☐ Other _____
3. Bore: Maximum Rate ☐ Nearest 1/8 in. ☐
4. Material: 304SS ☐ 316SS ☐ Other _____
5. Ring Material & Type_____
6. MFR. & Model No._____

ORIFICE FLANGES

7. Taps: Flange ☐ Vena Contracta ☐ Pipe ☐ Other _____
8. Tap Size: 1/2 in. ☐ Other_____
9. Type: Weld Neck ☐ Slip On ☐ Threaded ☐
10. Material: Steel ☐ Other _____
11. Flanges included ☐ By others ☐
12. Flange Rating _____

	13	Tag Number				
	14	Service				
	15	Line Number				
	16	Fluid				
	17	Fluid State				
	18	Maximum Flow				
	19	Normal Flow				
	20	Pressure				
	21	Temperature				
	22	Specific Gravity at Base				
	23	Operating Spec. Gravity				
FLUID DATA	24	Supercomp. Factor				
	25	Mol. Weight | Cp/Cv				
	26	Operating Viscosity				
	27	Quality % or °Superheat				
	28	Base Press. | Base Temp.				
	29	Type of Meter				
	30	Diff. Range — Dry				
	31	Seal sp. gr. at 60° F				
METER	32	Static Press. Range				
	33	Chart or Scale Range				
	34	Chart Multiplier				
	35	Beta=d/D				
	36	Orifice Bore Diameter				
PLATE &	37	Line I.D.				
FLANGE	38	Flange Rating				
	39	Vent or Drain Hole				
	40	Plate Thickness				

Notes:

ISA FORM S20-21

34

ORIFICE PLATES AND FLANGES

Instructions for ISA Form S20.21

Refer to ISA Recommended Practice RP3.2, "Flanged Mounted, Sharp Edged Orifice Plates for Flow Measurement."

1. Check if concentric bore, or write in eccentric, segmental, etc.

2. ISA Standard reference given above. This also conforms to AGA-ASME requirements.

3. Check whether plate is to be bored odd size for exact maximum rate, or to nearest 1/8 in. for approximate maximum rate.

4. Select plate material.

5. If ring joint assembly is used, give ring material and configurations.

6. Refers to plate, not flanges.

7. Select one of the standard tap locations or write in other.

8. Select tap size.

9. Select flange construction.

10. Select flange material. If stainless steel, show type; such as, "304 SS."

11. Indicate whether orifice flanges are to be included with the plate, or furnished by others.

12. Note Flange Rating.

13. Tag number or other identification No.

14. Process service.

15. Line number. Include line size.

16. List fluid, unless classified.

17. Liquid, gas, or vapor.

18. Maximum flow assumed to be meter maximum. Give flow units.

19. Figure only if units given above.

20. Upstream operating pressure and units. This is also the contract figure unless otherwise noted.

21. Operating temperature, °F or °C. See comment in 20 above.

22. Specific gravity at Base Temperature.

23. Liquid specific gravity at operating temperature given on Line 21.

24. Applies to gas, at operating pressure. Supercompressibility factor normally required for gases over 100 psig because the gas at this pressure and above does not follow the ideal gas laws.

25. Applies to vapor or gas. C_p specific heat at constant pressure, C_v specific heat a constant volumes — Ratio = K at the operating temperature.

26. Viscosity and units, at operating temperature given on line 21.

27. Applies to vapor or steam. Write "SAT" if saturated; otherwise give % quality or degrees superheat, in F or C.

28. Contract base conditions. Pressure must be given in absolute units.

29. Bellows, diaphragm, mercury, etc.

30. Set range and units.

31. Applies to wet meters.

32. Fill in if applicable.

33. Full scale range and units. See comment under 18 above.

34. Fill in if required.

35. Fill in for final records after approved bore calculation is available.

36. For final records, see comment on 35.

37. In inches; or give line size and Schedule.

38. ANSI Flange Rating, i.e., 4 in. 300 lb RF

39. If desired, state whether top or bottom.

40. Give plate thickness.

			ROTAMETERS (VARIABLE AREA FLOWMETERS)				SHEET _____ OF _____		
			NO	BY	DATE	REVISION	SPEC. NO.		REV.
							CONTRACT		DATE
							REQ. - P.O.		
							BY	CHK'D	APPR.

GENERAL	1	Tag Number					
	2	Service					
	3	Line No./Vessel No.					
	4	Function					
	5	Mounting					
	6	Power Supply					
	7	Conn. Size / Type					
	8	Inlet Dir. / Outlet Dir.					
	9	Fitting Material					
	10	Packing or O-Ring Mtl.					
	11	Enclosure Type					
METER	12	Size / Float Guide					
	13	Tube Mtl. / Float Mtl.					
	14	Meter Scale: Length & Type					
	15	Meter Scale Range					
	16	Meter Factor					
	17	Rated Accuracy					
	18	Hydraulic Calib. Required					
FLUID DATA	19	Fluid					
	20	Color or Transparency					
	21	Maximum Flow Rate					
	22	Norm Flow / Min Flow					
	23	Oper. Specific Gravity (Liq)					
	24	Max Oper. Viscosity					
	25	Oper. Press. / Oper. Temp.					
	26	Oper. Density (Gases)					
	27	Std. Density / Mol. Wgt.					
	28	Max. Allowable Press. Drop					
	29						
EXT	30	Extension Well Mtl.					
	31	Gasket Mtl.					
XMTR	32	Transmitter Output					
	33	Trans. Enclosure Class					
	34	Scale Range					
ALARM	35	No. of Contacts / Form					
	36	Rating / Housing					
	37	Action					
	38						
OPTIONS	39	Valve Size & Material					
	40	Valve Location					
	41	Const. Diff. Relay Mtl.					
	42	Purge Meter Tubing					
	43	Airset					
	43a						
	44	Manufacturer					
	45	Model Number					
	46	Tube Number					
	47	Float Number					

Notes:

ISA FORM S20.22

ROTAMETERS

Instructions for ISA Form S20.22 (Refer to ISA RP16.1,2,3,4)

1. List tag number.

2. Refers to process applications.

3. Show line number, vessel number, or line specification.

4. Give functions such as INDICATE RECORD, CONTROL TRANSMIT, INTEGRATE, etc.

5. FLUSH PANEL, FRONT PANEL, PIPE, etc.

6. Give voltage, dc or ac, and ac frequency.

7. Give nominal connection size and type such as SCREWED, 150 lb FLANGED, etc.

8. Select orientation of inlet and outlet and designated as RIGHT, LEFT, VERTICAL or REAR.

9. Select material of end fittings. Note if lining is required.

10. Select either packing or "O" ring design and note material.

11. Select type of enclosure, if any, such as SIDE PLATE, SAFETY GLASS, etc.

12. Give meter size. Note that this is not the same as connection size but refers to the nominal size of the tube and float combination.

Give the method of float guiding such as NONE, FLUTES, POLE, EXTENSIONS.

13. Select tube and float material.

14. Select type meter scale: NONE, ON GLASS, METAL STRIP. Select meter scale length.

15. Select meter scale range and flow units. Remember that rotameters' scales cannot start at zero but typically have rangeability of 10:1 or 12:1.

16. Meter factor if not direct reading.

17. Accuracy statement does not imply any specific calibration.

18. Note if hydraulic calibration is required and state required accuracy.

19. If fluid cannot be identified, state if liquid or gas.

20. Give fluid color or transparency which will affect float visibility in glass tube meters.

21. List maximum operating flow rate and units, usually the same as maximum of meter scale.

22. Show normal and minimum flow rates expected.

23. Give operating specific gravity of liquid. (Numerically equal to density in gm/cm^3).

24. Give maximum expected viscosity and units.

25. Give operating pressure and temperature, with units.

26. For gases give operating density and units, unless molecular weight is given on Line 27.

27. For gases give density at standard conditions (14.7 psia and $60°F$ unless stated otherwise, and/or molecular weight if known.

28. State maximum allowable pressure drop at full flow, if applicable.

30. If meter has an extension well, state material of well.

31. Select material of gasket on extension.

32. If meter transmits, state pneumatic or electronic output such as 21-103 kPa (3-15 psig), 4-20 mA, etc.

33. Give transmitter electrical classification such as General Purpose, Class 1, Group D, etc.

34. Give transmitter scale size and range. Note that this is not the meter scale but the scale of the attached instrument.

35. Number of alarm contacts in case.

Form of contacts: SPDT, SPST, DPDT, etc.

36. Contact electrical load rating. Contact housing — GP, Class I, GR.D, etc. Use NEMA identification.

37. HIGH, LOW, DEVIATION.

39. Specify needle valve if required.

40. Valve may be on the inlet, outlet or separately mounted. Do not list here if valve is to be furnished by others.

41. This relay may be used on purge assemblies.

44 - 47. When manufacturer is selected fill in exact model and part numbers.

			MAGNETIC FLOWMETERS	SHEET _____ OF _____		

				NO	BY	DATE	REVISION

SPEC. NO. | REV.

CONTRACT | DATE

REQ. - P.O.

BY | CHK'D | APPR.

	1		Meter Tag No.						
	2		Service						
	3		Location						
METERING ELEMENT	4	CONN'S.	Line Size, Sched.						
	5		Line Material						
	6		Connection Type						
	7		Connection Mat'ls.						
	8	METER	Tube Material						
	9		Liner Material						
	10		Electrode Type						
	11		Electrode Matl.						
	12		Meter Casing						
	13		Power Supply	Elect. Code						
	14		Grounding, Type & Matl.						
	15		Enclosure Class						
	16								
	17	FLUID	Fluid						
	18		Max. Flow, Units						
	19		Max. Velocity, Units						
	20		Norm. Flow	Min. Flow						
	21		Max. Temp.	Min. Temp.						
	22		Max. Press.	Min. Press.						
	23		Min. Fluid Conductivity						
	24		Vacuum Possibility						
	25								
ASSOCIATED INSTRUMENT	26		Instrument Tag Number						
	27		Function						
	28		Mounting						
	29		Enclosure Class						
	30		Length Signal Cable						
	31		Type Span Adjustment						
	32		Power Supply						
	33	TRANS.	Transmitter Output						
	34								
	35	DISPLAY	Scale Size	Range						
	36		Chart Drive	Speed						
	37		Chart Range	Chart No.						
	38		Integrator						
	39	CONTR.	Modes	Output						
	40		Action	Auto-Man.						
	41								
	42	ALARM	Contact No.	Form						
	43		Rating	Elec. Code						
	44		Action						
	45		Manufacturer						
	46		Meter Model Number						
	47		Instrument Model Number						

Notes:

ISA FORM S20.23

MAGNETIC FLOWMETERS

Instructions for ISA Form   S20.23

1. Tag number of meter only.

2. Refers to process application.

3. Show line number or identify associated vessel.

4. Give pipeline size and schedule. If reducers are used, so state.

5. Give material of pipe. If lined, plastic or otherwise non-conductive, so state.

6. Give connection type: FLANGED, DRESSER COUPLINGS, ETC.

7. Specify material of meter connections.

8. Select tube material. (Non-permeable material required if coils are outside tube).

9. Specify material of line.

10. Select electrode type: STD., BULLET NOSED, ULTRASONIC CLEANED, BURN OFF, etc.

11. Specify electrode material.

12. Describe casing: STD., SPASH PROOF, SUBMERSIBLE, SUBMERGED OPERATION, etc.

13. Give ac voltage and frequency, along with application NEMA identification of the electrical enclosure.

14. State means for grounding to fluid: GROUNDING RINGS, STRAPS, etc.

15. State power supply and enclosure class to meet area electrical requirements.

16.

17. State fluid by name or description.

18. Give maximum operating flow and units; usually same as maximum of instrument scale.

19. Give maximum operating velocity, usually in ft/s.

20. List normal and minimum flow rates.

21. List maximum and minimum fluid temperature °F.

22. List maximum and minimum fluid pressure.

23. List minimum (at lowest temp.) conductivity of fluid.

24. If a possibility of vacuum exists at meter, so state and give greatest value. (highest vacuum).

25.

26. List tag number of instrument used directly with meter.

27. Control loop function such as INDICATE, RECORD CONTROL, etc.

28. Mounting: FLUSH PANEL, SURFACE INTEGRAL WITH METER, etc.

29. Give NEMA identification of case type.

30. State cable length required between meter and instrument.

31. Span adjust: BLIND, ft/s DIAL, OTHER.

32. Give ac supply voltage and frequency.

33. If a transmitter, state analog output electrical or pneumatic range, or pulse train frequency for digital outputs, i.e., pulses per gallon.

34. List scale size and range.

35. List Scale Size and Range for indicating transmitter

36. Recorder chart drive — ELECT. HANDWIND, etc. and chart speed in time per revolution or inch per hour.

37. List chart range and number.

38. If integrator is used, state counts per hour, or value of smallest count; such as "10 GAL UNITS".

39. For control modes: (Per ANSI C85.1-1963, "Terminology for Automatic Control.") Write-in PI_f, I_f, PI_s, $PI_f D_f$, etc.

$$P = \text{proportional (gain)}$$
$$I = \text{integral (auto reset)}$$
$$D = \text{derivative (rate)}$$

Subscripts:

$$f = \text{fast}$$
$$s = \text{slow}$$
$$n = \text{narrow}$$

State output signal range, pneumatic or electronic.

40. Controller action in response to an increase in flowrate — INC. or DEC.

State auto-man. switch as NONE, SWITCH ONLY, BUMPLESS, etc.

42. Number of alarm lights in case. Give form of contacts; SPDT, SPST, etc.

43. Contact electrical load rating. Contact housing General Purpose, Class I, Group D, etc., if not in the same enclosure described in line 29.

44. Action of alarms: HIGH, LOW, DEVIATION, etc.

45. Fill in manufacturer and model numbers for meters
46. and
47. instrument after selection.

			TURBINE FLOWMETERS				SHEET ____ OF ____		
							SPEC. NO.		REV.
			NO	BY	DATE	REVISION			
							CONTRACT		DATE
							REQ. - P.O.		
							BY	CHK'D	APPR.

METER	1	Tag Number				
	2	Service				
	3	Meter Location				
	4	Line Size				
	5	End Connections				
	6	Body Rating				
	7	Nominal Flow Range				
	8	Accuracy				
	9	Linearity				
	10	K Factor, Cycles per Vol. Unit				
	11	Excitation				
	12	Materials: Body				
	13	Support				
	14	Shaft				
	15	Flanges				
	16	Rotor				
	17	Bearings: Type				
	18	Bearing Material				
	19	Max. Speed				
	20	Min. Output Voltage				
	21	Pickoff Type				
	22	Enclosure Class				
	23					
FLUID DATA	24	Fluid				
	25	Flow Rate: Min. Max.				
	26	Normal Flow				
	27	Operating Pressure				
	28	Back Pressure				
	29	Operating Temp. Max. Min.				
	30	Operating Specific Gravity				
	31	Viscosity Range				
	32	Percent Solids & Type				
	33					
SECONDARY INSTR.	34	Secondary Instr. Tag No.				
	35	Preamplifier				
	36	Function				
	37	Mounting				
	38	Power Supply				
	39	Scale Range				
	40	Output Range				
OPTIONS	41	Totalizer Type				
	42	Compensation				
	43	Preset Counter				
	44	Enclosure Class				
	45	Strainer Size & Mesh				
	46					
	47					
	48					
	49	Manufacturer				
	50	Meter Model No.				
	51	Secondary Instr. Model No.				

Notes:

ISA Form S20.24

40

TURBINE FLOWMETERS

Instructions for ISA Form S20.24
Refer to ISA Standard S31, "Specification, Installation, and Calibration of Turbine Flowmeters"

1. Show meter tag number. Quantity is assumed to be one unless otherwise noted.

2. Refers to process service or applications.

3. Give line number or process area.

5. Specify size and style of connections, such as "1 in. NPT", "2 in. 150 lb ANSI", etc.

6. Pressure and temperature design rating required.

7. Nominal flow range is obtained from manufacturer's data. This usually defines linear range of selected meter.

8. Turbine meter accuracy figures are in terms of percent of instantaneous flow rate.

9. Degree of linearity over nominal flow range.

10. K factor relates cycles per second to volume units. Enter this figure after selection is made.

11. Excitation modulating type only expressed as volts _____ at _____ hertz.

12 to 16. Specify materials of construction or write in "MFR STD."

17. Specify sleeve or ball bearings, or none if floating rotor design.

18. Bearing material — will be MFG STD if not stated otherwise.

19. Maximum speed or frequency which the meter can produce without physical damage.

21. Pickoff may be standard hi-temp., radio-frequency type (RF) or explosion proof. Minimum output voltage _____ volts peak to peak.

22. Specify electrical classification of enclosure such as General Purpose, Weather Proof, Class 1, Group D, etc.

23. Specify fluid data as indicated, using line 28 for additional item if required.

34. Give Tag No. of secondary instrument if different from meter Tag No.

35. Pre-amplifier if used.

36. Specify function of instrument, such as rate indicator, totalizer, or batch control.

37. Flush, surface or rack.

38. Power Supply, i.e., 117 Vac.

39. Applies to rate indicator

40. Give output range such as "40-20 mA", 21-103 kPa (3-15 psig), etc.

41. May be used for number of digits, and to state whether counter is reset or non-reset type.

42. Specify range of compensation, if required, in pressure and/or temperature units or viscosity units.

43. Pre-set counter.

44. Specify NEMA classification of enclosure.

45. Specify strainer size and mesh size. Request vendor's recommendation if not known.

50. Fill in after selection is made.

51. Fill in after selection is made.

52. Fill in after selection is made.

			POSITIVE DISPLACEMENT METERS					SHEET _____ OF_____		
			NO	BY	DATE	REVISION		SPEC. NO.		REV.
								CONTRACT		DATE
								REQ. - P.O.		
								BY	CHK'D	APPR.

	1	Tag Number				
	2	Service				
	3	Line No./Vessel No.				
METER	4	Type of Element				
	5	Size				
	6	End Connections				
	7	Temp. & Press. Rating				
	8	Flow Rate Range				
	9	Totalized Units				
	10	Enclosure Class				
	11	Power Supply				
	12	Materials: Outer Housing				
	13	Main Body Cover				
	14	Rotating Element				
	15	Shaft				
	16	Blades				
	17	Bearings: Type & Material				
	18	Packing				
	19	Type of Coupling				
	20					
COUNTER	21	Register Type				
	22	Totalizer				
	23	Reset				
	24	Capacity				
	25	Set-Stop				
	26					
FLUID DATA	27	Fluid				
	28	Flow Rate: Min. / Max.				
	29	Normal Flow				
	30	Oper. Press. / Oper. Temp.				
	31	Oper. Specific Gravity				
	32	Oper. Viscosity				
	33	Coef. of Expansion				
OPTIONS	34	Flow Units				
	35	Shut-Off Valve				
	36	Switch: Single or 2-Stage				
	37	Temp. Compensator				
	38	Transmitter Type				
	39	Transmitter Output				
	40	Air Eliminator				
	41	Strainer: Size & Mesh				
	42					
	43					
	44					
	45	Manufacturer				
	46	Model Number				

Notes:

ISA FORM S20.25

POSITIVE DISPLACEMENT METERS

Instructions for ISA Form S20.25.

1. Tag No. of instrument.

2. Process service.

3. Pipe line or vessel identification.

4. Write in type of rotating element, such as, disc, piston, vane, helical, rotors, etc.

5. Show connection pipe size.

6. Specify end connections type and ANSI rating such as 300 lb R.F.

7. Specify the manufacturer's recommended body pressure and temperature rating, such as 250 psi at 190°F.

8. Write in manufacturer's recommended normal operating range.

9. Specify smallest totalized unit, such as "Tens of Gallons", "Pounds", "Barrels".

10. Specify enclosure electrical classification, if applicable, such as "Class 1, Group D., Div. 2", "General Purpose", etc.

11. Specify power supply, if applicable.

12. Specify materials of construction. If no preference, write in, MFR. STD. (Manufacturer's Standard).

13-18. Specify materials of construction, if no preference, write in, Manufacturer's Standard (MFG-STD)

19. Specify type of coupling

20. Specify coupling such as "Magnetic", or MFR. STD.

21. Specify register type such as horizontal, vertical, inclined, inline reading, dial reading, print, etc.

22. Specify number of figures such as 6 digit, 5 digit, or 0-99, 999, etc.

23. If totalizer reset required, write in type. If reset is not required, write in "none".

24. Write in number of figures or maximum quantity (in flow units) that can be held in counter.

25. Specify by writing in "yes" if a set-stop is required to operate shutoff valve, switch, etc.

27-34. Specify fluid data as completely as possible, note at operating conditions. Be sure to note if liquid is at saturation conditions.

35. Specify by writing in "yes" if a shut-off valve is required. Valve to be manufacturer's standard construction unless otherwise noted.

36. Specify by writing in "yes" if a switch is required. Two switches are required for 2-stage shut-off control.

37. Write in "yes" if manufacturer's standard temperature compensator is required. Write in "no" if not required.

38. Specify, if transmitter is required, by writing in type such as pulse, rate of flow, etc.

39. Give transmitter output in pulse per gallon, 4-20 mA, etc.

40. Write in "yes" if air eliminator is required, otherwise write in "no".

41. Specify, if strainer is required, by writing in type such as "Y", "Basket", etc. Strainer to have same pressure and temperature rating, end connections and material as meter body unless otherwise noted.

45-46. Identify manufacturer's name and model number after selection is made.

				LEVEL INSTRUMENTS (DISPLACER or FLOAT)			SHEET _____ OF _____		
				NO	BY	DATE	REVISION	SPEC. NO.	REV.
								CONTRACT	DATE
								REQ. - P.O.	
								BY	CHK'D	APPR.

	1	Tag Number					
	2	Service					
	3	Line No./Vessel No.					
BODY/CAGE	4	Body or Cage Mtl					
		Rating					
	5	Conn Size & Location Upper					
		Type					
	6	Conn Size & Location Lower					
		Type					
	7	Case Mounting					
		Type					
	8	Rotatable Head					
	9						
	10	Orientation					
	11	Cooling Extension					
	12						
DISPLACER OR FLOAT	13	Dimensions					
	14	Insertion Depth					
	15	Displacer Extension					
	16	Disp. or Float Material					
	17	Displacer Spring/Tube Mtl					
	18						
	19						
XMTR/CONT.	20	Function					
	21	Output					
	22	Control Modes					
	23	Differential					
	24	Output Action: Level Rise					
	25	Mounting					
	26	Enclosure Class					
	27	Elec. Power or Air Supply					
	28						
SERVICE	29	Upper Liquid					
	30	Lower Liquid					
	31	sp. gr.: Upper / Lower					
	32	Press. Max. / Normal					
	33	Temp. Max. / Normal					
	34						
	35						
OPTIONS	36	Airset / Supply Gage					
	37	Gage Glass Connections					
	38	Gage Glass Model No.					
	39	Contacts: No. / Form					
	40	Contact Rating					
	41	Action of Contacts					
	42						
	43						
	44						
	45						
	46	Manufacturer					
	47	Model Number					
	48						

Notes:

ISA FORM S20.26

LEVEL INSTRUMENTS
(DISPLACER OR FLOAT)

Instructions for ISA Form S20.26.

1. Tag No. or other identification.

2. Process service.

3. Line number or vessel number on which cage or body is installed.

4. Material of chamber and/or mounting flange.

5. For float specify top or side of vessel connection. For displacer in a chamber specify upper, then lower connection; such as side-side, side-bottom, top-bottom, etc. Give flange size and rating or NPT size.

6. Same as 5.

7. Refers to position of case when viewing the front of the case relative to the chamber; the case is either to the left, right, or top.

8. On displacer instruments specify if case is to be rotatable with respect to the chamber. This only applies if there is one or more side connections.

10. Orientation of control with respect to displacer cage.

11. Cooling Extension

13. Specify float diameter or displacer length. The displacer length is also the range.

14. Insertion depth applied to ball floats. It is the mounting flange to the center of the ball.

15. The displacer extension is measured from the face of the mounting flange to the top of the displacer. This dimension is required only for top of vessel mounted instruments.

16. Includes rod.

17. Refer to MFR's standard materials or special materials.

18.

19.

20. Transmitter, controller, switch, etc.

21. Air pressure or electrical signal output of transmitter or controller.

22. P: Proportional
Pn: Narrow band proportional
PI: Proportional plus Integral (Reset).

23. Differential if controller on/off must specify differential adj. or fixed. State adjustable range or fixed amount.

24. INCREASE (Direct action) or DECREASE (Reverse Action).

25. Remote, or integral.

26. Electrical classification of housing. NEMA number.

27. Air pressure or voltage. If electronic, state whether ac or dc.

29. Used only for interface application.

30. Used for all services.

31. Specific gravities at operating temperature.

32. Operating and max. pressure, or vacuum.

33. For cryogenic service, give minimum temperature.

36. Airset assumed mounted to case.

37. Connections on chamber, give size.

38. Specify gauge glass, if required.

39. Contact form: SPST, SPDT, etc.

40. Give Volts, Amps.

41. Describe contact action with level.

47. Model number of entire assembly.

		LEVEL INSTRUMENTS (CAPACITANCE TYPE)	SHEET ___ OF ___

						SPEC. NO.		REV.
	NO	BY	DATE	REVISION		CONTRACT		DATE
						REQ. - P.O.		
						BY	CHK'D	APPR.

GENERAL	1	Tag Number				
	2	Service				
	3	Line No./Vessel No.				
	4	Application				
	5	Function				
	6	Fail-Safe				
PROBE	7	Model Number				
	8	Orientation				
	9	Style				
	10	Material				
	11	Sheath				
	12	Insertion Length				
	13	Inactive Length				
	14	Gland Size & Mat'l.				
	15					
	16	Conduit Connection				
AMPLIFIER	17	Location				
	18	Enclosure				
	19	Conduit Connection				
	20	Power Supply				
SWITCH	21	Type				
	22	Quantity and Form				
	23	Rating: Volts/Hz or dc				
	24	Amps/Watts/HP				
	25	Load Type				
	26	Contacts Open | On | Incr.				
	27	Close | Level | Decr.				
TRANS.	28	Output				
	29	Range				
	30	Enclosure Class				
OPTIONS	31	Compensation Cable				
	32	Local Indicator				
	33	I/P Transducer				
	34	Signal Lights				
	35					
SERVICE	36	Upper Fluid				
	37	Dielectric Constant				
	38	Lower Fluid				
	39	Dielectric Constant				
	40	Pressure Max. | Normal				
	41	Temp. Max. | Normal				
	42	Moisture				
	43	Material Buildup				
	44	Vibration				
	45	Manufacturer				
	46	Model Number				

Notes:

ISA Form S20.27

LEVEL INSTRUMENTS, CAPACITANCE TYPE

Specification Sheet Instructions for ISA Form S20.27

Prefix number designates line number on corresponding Specification Sheet.

1. Identification of item by tag number.

2. Process area or function.

3. Stream description and/or pipe size and number or vessel number in which probe is installed.

4. Specify solids level, liquid level, interface, foam detection, etc.

5. Specify alarm, transmit, on-off control, etc.

6. Specify high, low, none.

7. Specify probe model number if known.

8. Specify if probe axis is horizontal, vertical, etc.

9. Specify general purpose, heavy duty, knife-blade, inline plate, concentric shield, etc.

10. Specify probe material as 316 SS, etc.

11. Specify sheath, if required, as 1/4 in. Teflon, etc.

12. Specify total immersion in inches, or feet and inches.

13. Specify length of inactive extension in inches, or feet and inches.

14-15. Specify sealing gland material and size as 316 SS, 3/4 in. NPT, etc.

16. Specify conduit connection as 3/4 in. NPT hub, 7/8 in. OD knockout, etc.

17. Specify if electronics are mounted at probe or remotely located.

18. Specify general purpose, weather proof, explosion proof, etc.

19. Specify conduit connection as 3/4 in. NPT, 7/8 in. OD knockout, etc.

20. Specify power input as 115V 60 Hz, etc.

21. Specify switch type as mercury bottle, snapaction, etc.

22. Specify number of switches and contact form of each switch (SPST, SPDT, DPDT, etc.)

23. Specify switch voltage as 115V 60 Hz, 24 Vdc, etc.

24. Specify contact rating in amps, watts, or horsepower.

25. Specify load as inductive on non-inductive.

26-27. Specify if contacts open or close when the level increases or decreases.

28. Specify transmitter output as 1-5, 4-20, or 10-50 mA, 1-5 Vdc, etc.

29. Specify level range in inches or feet and inches corresponding to minimum and maximum transmitter signal.

30. Use NEMA identification numbers.

31. Specify length of special compensating cable to be furnished with probe, if required.

32. Specify size, type and range of local indicator, if required.

33. Specify if electro-pneumatic transducer 21-103 kPa (3-15 psig output) is required.

34. Specify if High, Low, HI/LO lights are required, and rating.

35. For items not covered in lines 31 through 34.

36. Specify upper fluid by name and state (liquid, vapor).

37. Specify dielectric constant of upper fluid.

38. Specify lower fluid by name and state.

39. Specify dielectric constant of lower fluid.

40. Specify maximum and normal operating pressure at probe.

41. Specify maximum and normal operating temperature at probe.

42. Specify percentage moisture content of solids.

43. Specify if material is expected to build up on probe.

44. Specify vibration environment of probe as mild, severe, etc.

45-46. Fill in manufacturer and model number after selected.

	GAGE GLASSES and COCKS				SHEET ____ OF ____	
					SPEC. NO.	REV.
	NO	BY	DATE	REVISION	CONTRACT	DATE
					REQ. - P.O.	
					BY	CHK'D	APPR.

1. Gage Column ☐   Cocks ☐
   Assembled with Nipples ☐   Unassembled ☐
   GAGE GLASSES
2. Type: Reflex ☐   Transparent ☐   Tubular ☐
   Large Chamber ☐   Weld Pad ☐
3. Conn: Size and Type_____
   Top & Bot. ☐   Side ☐   Back ☐
   Vent_____   Drain _____
4. Material_____
5. Min. Rating _____ psig at ___ °F
6. Options: Illuminator ☐   Mica Shield ☐
   Internal Tube ☐   External Jkt ☐
   Non-Frost ☐   Ext. Length _____
   Calb. Scale ☐   Other_____
7. Manufacturer & Model _____

GAGE COCKS
8. Type: Offset ☐   Straight ☐
9. Conn: Vessel _____ Gage _____ Vent/Drain _____
10. Material: Body_____ Trim _____
11. Min. Rating:_____ psig at _____ °F
12. Construction:_____
13. Type of Conn: Vessel _____
    Gage _____
14. Bonnet _____
15. Options: Ball Checks ☐   Renewable Seats ☐
    Other _____
16. Manufacturer & Model _____

Rev.	Quan.	Tag No.	Visible Glass	₵ Conn.	Model No.	Operating		Service	
						Press.	Temp.		

Notes:

ISA FORM S20.28

GAGE GLASSES AND COCKS

Instructions for ISA Form S20.28

1. Check what is to be supplied, and whether assembled or unassembled.

2. Select one type only per sheet.

3. Specify size, style and location of process connections. If side or back connections are used, vent and drain connections are available.

4. Material of gage glass chamber and connections.

5. Specify minimum rating. It is assumed that a higher rating is also acceptable.

6. This section is used only if the option applies to all items listed on the sheet. Where options apply to certain items only, use the notes column instead.

7. Use for Manufacturer and Series or Type; detailed number may be listed in the tabulation.

8. Select style of cock, if used.

9. Show connection sizes only.

10. Write in body and trim materials.

11. See Line 5 above.

12. Specify action and type of handle: plain closing or quick closing; handwheel or lever handle. This may be covered by the Model No. given on Line 17.

13. Specify type of connection on each side: plain union, spherical union, solid shank. Give flange size, rating and type, if applicable.

14. Bonnet may be screwed, union type, or bolted.

15. Options checked here apply to all items. See line 6 above. Include special packing.

16. Fill in if required, or as a final record after selection is made.

"⊄ CONN" in tabulation refers to distance between center lines of vessel connections. This figure, along with the visible glass dimension, defines the length of the column. A secondary sheet with tabulation only may be made up if required.

			TRAPS and DRAINERS				SHEET _____ OF_____		
			NO	BY	DATE	REVISION	SPEC. NO.		REV.
							CONTRACT		DATE
							REQ. - P.O.		
							BY	CHK'D	APPR.

	1	Tag Number				
	2	Service				
	3	Line No./Vessel No.				
	4					
	5	Type				
	6					
BODY	7	Material				
	8	Size: Inlet | Outlet				
	9	End Connections				
	10	Press. & Temp. Rating				
	11	Equalizing Conn. Size				
	12	Conn. Orientation				
	13					
TRIM	14	Trim Material				
	15					
OPTIONS	16	Internal Check Valve				
	17	Internal Bimetallic Vent				
	18	Thermostatic Vent | Mtl.				
	19	Gage Glass				
	20					
	21					
	22					
STRAINER	23	Internal or External				
	24	Type & Size				
	25	Body Material				
	26	Press. & Temp. Rating				
	27	End Connections				
	28	Blowoff Connections				
	29	Mesh Size & Material				
	30					
PROCESS DATA	31	Fluid				
	32	Normal Flow				
	33	Load Safety Factor				
	34	Maximum Capacity				
	35	Oper. Temp. |°Superheat				
	36	Press: In | Out				
	37	Allow Press. Diff: Max. | Normal				
	38	Oper. sp. gr. Top | Bottom				
	39					
	40					
	41	Calc. Orifice Size				
	42	Selected Orifice Size				
	43					
	44					
	45	Manufacturer				
	46	Model Number				

Notes:

ISA Form S20.29

50

TRAPS AND DRAINERS

Instructions for ISA Form S20.29

1. Identification or item number.

2. Fill in service or location.

5-6. Write in specific trap type corresponding to general classification such as, inverted bucket, float, drainer, thermodynamic, etc.

7. Specify body material required.

8. Write in inlet & outlet connection size.

9. Specify if traps are to have flanged, screwed socket welded, buttwelded end connections and specify the respective rating.

10. Write in temperature and pressure rating required.

11. Specify equalizing connection size if required (used with continuous drainers).

12. Show orientation or connections by sketch if necessary.

13. Write in any other features characteristic of the trap body.

14. Write in trim material. If to be manufacturers standard, write in "STD".

15. If specific items of trim, such as valve seats, need to be harder material than 14 above, write in material or description.

16. Indicate if internal check valve is required, state size (applies to Bucket Traps).

17. Specify if internal Bi-metallic Vent is required, (applies to Bucket Traps).

18. Indicate if thermostatic vent is required (used with Ball Floats) and specify bellows material.

19. Show if Gage Glass is required.

20. Write in any other accessory required not included in 16 through 19 above.

23. Specify if strainer is to be of internal or external variety, if to be supplied with trap. If not, write in "By others."

24. Indicate the specific type, i.e., "Y" type, Angle Type, etc., and inlet outlet connection size.

25. Write in body material.

26. Write in strainer temperature and pressure rating.

27. Specify if strainers are to be flanged or screwed and specify the respective rating.

28. Show size of Blow off connections. Also indicate if bushing or cap is required.

29. Specify mesh size and material if other than manufacturer's standard is required.

30. Write in any other strainer requirements.

31. Show fluid being handled.

32. Specify the anticipated normal flow quantity of condensate to be handled.

33. Write in the safety load factor which is added to compensate for the start-up load under reduced pressure conditions.

34. Maximum capacity of trap should always exceed normal quantity to be handled plus the load safety factor.

35. Show the steam temperature plus superheat that may be present.

36. Show the normal pressure at Trap inlet and outlet.

37. Show the allowable pressure differential across the trap or drainer.

38. Show the liquid gravity above and below the normal level being held (important for Continuous Drainers.)

41. Show the calculated orifice size.

42. Specify the orifice selected from manufacturer's charts.

45-46. Write in manufacturer and model number if desired.

		PRESSURE INSTRUMENTS			SHEET _____ OF _____	
					SPEC. NO.	REV.
		NO	BY	DATE	REVISION	CONTRACT	DATE
						REQ. P.O.	
					BY	CHK'D	APPR.

	1	Tag No.	Service
GENERAL	2	Function	Record ☐ Indicate ☐ Control ☐ Blind ☐ Trans ☐ Other _____
	3	Case	MFR STD ☐ Nom Size _____ Color: MFR STD ☐ Other _____
	4	Mounting	Flush ☐ Surface ☐ Yoke ☐ Other _____
	5	Enclosure Class	General Purpose ☐ Weather proof ☐ Explosion proof☐ Class _____ For Use In Intrin. Safe System ☐ Other _____
	6	Power Supply	117V 60Hz ☐ Other ac _____ dc _____ Volts
	7	Chart	_____ Strip ☐ _____ Roll ☐ _____ Fold ☐ Circular _____ Time Marks _____ Range _____ Number _____
	8	Chart Drive	Speed _____ Power _____
	9	Scales	Type _____ Range 1 _____ 2 _____ 3 _____ 4 _____
XMTR	10	Transmitter Output	4-20 mA ☐ 10-50 mA ☐ 21-103 kPa (3-15 psig) ☐ Other _____ For Receiver See Spec Sheet
CONTROLLER	11	Control Modes	P=Prop (Gain) I=Integral (Auto-Reset) D=Derivative (Rate) Sub: s=Slow f=Fast P ☐ PI ☐ PD ☐ PID ☐ If ☐ Df ☐ Is ☐ Ds ☐ Other _____
	12	Action	On Meas. Increase Output: Increases ☐ Decreases ☐
	13	Auto-Man Switch	None ☐ MFR STD ☐ Other _____
	14	Set Point Adj.	Manual ☐ External ☐ Remote ☐ Other _____
	15	Manual Reg.	None ☐ MFR STD ☐ Other _____
	16	Output	4-20mA ☐ 10-50mA ☐ 21-103 kPa (3-15 psig) ☐ Other _____
ELEMENT	17	Service	Gage Press. ☐ Vacuum ☐ Absolute ☐ Compound ☐
	18	Element Type	Diaphragm ☐ Helix ☐ Bourdon ☐ Bellows ☐ Other _____
	19	Material	316 SS ☐ Ber. Copper ☐ Other _____
	20	Range	Fixed ☐ Adj. Range _____ Set at _____ Overrange protection to _____
	21	Process Data	Press: Normal _____ Max _____ Element Range _____
	22	Process Conn.	¼ in. NPT ☐ ½ in. NPT ☐ Other _____ Location: Bottom ☐ Back ☐ Other _____
	23	Alarm Switches	Quantity _____ Form _____ Rating _____
	24	Function	Press ☐ Deviation ☐ Contacts To _____ on Inc Press.
OPTIONS	25	Options	Filt-Reg. ☐ Sup Gage ☐ Output Gage ☐ _____ Charts Diaph Seal ☐ Type _____ Diaph _____ Bot Bowl _____ Conn _____ Capillary: Length _____ Mtl. _____ Other _____
	26	MFR & Model No.	

Notes:

ISA Form S20.40a

				On Meas. Inc. Output		
Rev.	Tag No.	Adj. Range	Set Range		Service	Notes

PRESSURE INSTRUMENTS

SHEET _____ OF _____

SPEC. NO.		REV.
CONTRACT		DATE
REQ. - P.O.		
BY	CHK'D	APPR.

NO	BY	DATE	REVISION

Notes:

ISA Form S20.40b

PRESSURE INSTRUMENTS

Instructions for ISA Forms S20.40a and 20.40b

1. To be used for a single item. Use secondary sheet for multiple listing.

2. Check as many as apply.

3. Nominal size refers to approximate front of case dimensions; width x height.

4. Yoke refers to a bracket designed for mounting the instrument on a pipe stand.

5. Enclosure class refers to composite instrument. If electrical contacts are in the case, they must meet this classification inherently or by reason of the enclosure. Use NEMA identification or ISA identification per RP8.1.

6. Specify electrical power to the entire instrument from an external source.

7. Specify chart size, range and number if applicable.

8. Chart drive mechanism assumed to be synchronous motor operating in 117V 60 Hz and suitable for ENCLOSURE CLASS specified on line 5. If the chart drive is pneumatic so state — identify pneumatic pulser under options. Note deviations from standard (MFR) under notes, i.e., dual speed or special speeds.

9. The scale type may be SEGMENTAL, VERTICAL, HORIZONTAL, DIAL (CIRCULAR) or other. Ranges 1, 2, 3 and 4 are used for multiple inputs. The first listed (No. 1) is assumed to be the controller input, if a controller is used.

10. Specify transmitter output if applicable.

11. See explanation of terminology given on specification sheet. For further definition refer to American National Standard C85.1-1963, "Terminology for Automatic Control." Specific ranges of control modes can be listed after "OTHER" if required.

12. For multiple items specify on second sheet.

13. If standard auto-manual switching is not known or not adequate, specify particular requirements, such as BUMPLESS, PROCEDURELESS, 4-POSITION, or as required.

14. Remote set point adjustment assumes full adjustment range. Specify limits if required.

15. Specify if applicable.

16. Specify if applicable.

17. Specify pressure measurement application.

18. Specify type of pressure element.

19. Specify material of element.

20. If range is adjustable, specify range of adjustment and initial range setting.

21. Specify normal and maximum pressure.

22. Specify process connection size. If a diaphragm seal is used, connection is specified in line 26.

23. Form may be SPST, SPDT, DPDT, or other. Rating refers to electrical rating of switch or contacts in amps.

24. Specify if alarm is actuated by measured variable or by deviation from controller set point. Give contact action if single throw form.

25. Specify required accessories.

27. Use these lines to specify other options and accessories.

28. Fill in after selection is made.

	PRESSURE GAGES			SHEET _____ OF_____	

PRESSURE GAGES

NO	BY	DATE	REVISION

SHEET _____ OF_____

SPEC. NO.	REV.

CONTRACT	DATE

REQ.	P.O.

BY	CHK'D	APPR.

1. Type: Direct Rdg ☐ 3-15 lb Receiver ☐
   Other _____
2. Mounting: Surface ☐ Local ☐ Flush ☐
3. Dial: Diameter   Color
4. Case: Cast Iron ☐ Aluminum ☐ Phenol ☐
   Other _____
5. Ring: Screwed ☐ Hinged ☐ Slip ☐ Std ☐
   Other _____
6. Blow-out Protection None ☐ Back ☐ Disc ☐
   Solid Front ☐ Other _____
7. Lens: Glass ☐ Plastic ☐
8. Options: Sylphon ☐ Material
               Snubber ☐ _____
   Pressure Limit Valve ☐ _____
   Movement Damping ☐ _____
9. Nominal Accuracy Required _____

10. MFR. & Model No. _____
11. Press. Element: Bourdon ☐ Bellows ☐
    Other _____
12. Element Mtl: Bronze ☐ Steel ☐ _____ SS
    Other _____
13. Socket Mtl: Bronze ☐ Steel ☐ _____ SS
    Other _____
14. Connection-NPT: ¼ in. ☐ ½ in. ☐ Other _____
    Bottom ☐ Back ☐
15. Movement: Bronze ☐ SS ☐ Nylon ☐
    Other _____
16. Diaphragm Seal
    MFG. _____ Type _____
    Wetted Part Mtl. _____ Other Mtl. _____
    Fill Fluid _____
    Process Conn. _____ Gage Conn. _____

Rev.	Quan.	Tag No.	Range	Operating Pressure	Service

Notes:

ISA FORM S20.41a

		PRESSURE GAGES		SHEET ___ OF ___	

NO	BY	DATE	REVISION	SPEC. NO.	REV.
				CONTRACT	DATE
				REQ. - P.O.	
				BY	CHK'D	APPR.

Rev.	Quan.	Tag No.	Range	Operating Pressure	Service

ISA FORM S20.41b

PRESSURE GAGES

Instructions for ISA Forms S20.41a and 20.41b

1. When receiver gages are specified, the "Range" in the tabulation is the dial range.

2. Select mounting style.

3. Specify nominal dial diameter. Dial assumed white unless otherwise specified.

4. Select case material.

5. Specify ring style, or check "STD" if not important.

6. Specify blow-out protection. "Back" refers to a blow-out back. "Disc" refers to a blow-out disc located in the back or side of the case.

7. Specify lens material.

8. Options:

Snubber Specify type or model number.

Sylphon If sylphon required, specify material.
Material

Movement
Dampening Specify if required.

9. Specify nominal accuracy, such as "±½%".

10. Write in make and model number after selection is made.

11. Specify element type or write in "MFR. STD."

12. If stainless steel is required, write in the type; such as "316."

13. See 12.

14. Specify connection size and location.

15. Specify movement or write in "MFR. STD."

16. If Diaphram Seal is required, fill in specifications.

For convenience, write in psig or other pressure unit at the top of "Range" and "Op. Press" columns, if all are the same.

	PRESSURE SWITCHES			SHEET _____ OF_____	

<table>
<tr><td rowspan="2" colspan="2">PRESSURE SWITCHES</td><td colspan="2">SHEET _____ OF_____</td></tr>
<tr><td>SPEC. NO.</td><td>REV.</td></tr>
<tr><td>NO</td><td>BY</td><td>DATE</td><td>REVISION</td><td>CONTRACT</td><td>DATE</td></tr>
<tr><td></td><td></td><td></td><td></td><td colspan="2">REQ. - P.O.</td></tr>
<tr><td></td><td></td><td></td><td></td><td colspan="2"></td></tr>
<tr><td></td><td></td><td></td><td></td><td>BY | CHK'D</td><td>APPR.</td></tr>
</table>

GENERAL

1. Type: Press □   Vacuum □
   Comp. □   Diff. Press. □
2. Setting: Set in Field □   Factory Set □
   Internal □   External □   Dial □
3. Dead Band: Fixed □   Adj. □   Min. □

ELEMENT

4. Type: Diaphragm □   Bourdon □   Bellows □
   Other _____
5. Material: Bronze □   _____SS □   Alloy St. □
   Other _____
6. Connection: MFR STD □   Other Size_____
   Bottom □   Back □
7. Mounting: Local □   Surface □   Flush □

SWITCH

8. Type: Mercury □   Snap □
   Other _____
9. Quantity: Single □   Dual □
10. Form: SPST □   SPDT □   DPDT □
    Other _____
11. Rating: _____Amps _____V _____ Hz
    Other _____
12. Load: Inductive □   Non-Inductive □
13. Enclosure: General Purpose □   Weather proof □
    None □   Explosion proof □   Class_____
14. Conduit Connection: MFR STD □   Other_____

Manufacturer & Model No. _____

Rev.	Tag. No.	Process Condition	Adj. Range	Set Point		Operating		Service	Notes
				Process	Signal	Temp.	Press.		

Notes:

ISA FORM S20-42a

©ISA S20

				PRESSURE SWITCHES				SHEET _____ OF_____		

NO	BY	DATE	REVISION

SPEC. NO. | REV.
CONTRACT | DATE
REQ. - P.O.
BY | CHK'D | APPR.

Rev.	Tag No.	Process Condition	Adj. Range	Set Point		Operating		Service	Notes
				Process	Signal	Temp.	Press.		

Notes:

ISA FORM S20.42b

PRESSURE SWITCHES

Instructions for ISA Forms S20.42a and 20.42b

1. Specify pressure, vacuum, compound, or differential pressure.

2. Check setting in field or factory. Check internal or external setting adjustment. Check whether calibrated setting dial is required.

3. Specify fixed or adjustable dead band.

4. Specify diaphragm, bourdon, bellows, or write MFR. STD.

5. Select element material, for stainless fill in number, or write MFR. STD.

6. Specify connection size or write MFR. STD. Specify bottom or back connection.

7. Specify mounting — Local (pipe) surface or flush.

8. Check Mercury or Snap acting, or write MFR. STD.

9. Specify number of switches in common housing.

10. Specify switch form.

11. Electrical rating in amps or watts, dc, or if ac, give frequency in Hz.

12. Check inductive or non-inductive load.

13. Check one: general purpose, weather proof or explosion proof. Use NEMA identification.

14. Check MFR. STD. or specify connection size.

Tabulation:

"Process Condition" refers to process condition which actuates switch, such as "High Level." "Adj Range" refers to limits within which a set point may be established, such as "1-18#." If the pressure switch is in an instrument air line, the set point may be specified in both process and signal units. "Notes" should be indicated by a number or letter and then explained in the space below the tabulation.

PROJECT _____	DATA SHEET ___ of ___
UNIT _____	SPEC _____
P.O. _____	TAG _____
ITEM _____	DWG _____
CONTRACT _____	SERVICE _____
MFR SERIAL* _____	

1 Fluid _____ Crit Pres PC _____

		Units	Max Flow	Norm Flow	Min Flow	Shut-Off
2	Flow Rate					—
3	Inlet Pressure					
4	Outlet Pressure					
5	Inlet Temperature					
6	Spec Wt/Spec Grav/Mol Wt					—
7	Viscosity/Spec Heats Ratio					—
8	Vapor Pressure P_V					—
9	*Required C_V					—
10	*Travel	%				0
11	Allowable/*Predicted SPL	dBA	/	/	/	—
12						

(SERVICE CONDITIONS — rows 2–12)

LINE
13 Pipe Line Size In _____
14 & Schedule Out _____
15 Pipe Line Insulation _____

VALVE BODY/BONNET
16 *Type _____
17 *Size _____ ANSI Class _____
18 Max Press/Temp _____
19 *Mfr & Model _____
20 *Body/Bonnet Matl _____
21 *Liner Material/ID _____
22 End In _____
23 Connection Out _____
24 Flg Face Finish _____
25 End Ext/Matl _____
26 *Flow Direction _____
27 *Type of Bonnet _____
28 Lub & Iso Valve _____ Lube _____
29 *Packing Material _____
30 *Packing Type _____
31 _____

TRIM
32 *Type _____
33 *Size _____ Rated Travel _____
34 *Characteristic _____
35 *Balanced/Unbalanced _____
36 *Rated C_V_____ F_L_____ X_T_____
37 *Plug/Ball/Disk Material _____
38 *Seat Material _____
39 *Cage/Guide Material _____
40 *Stem Material _____
41 _____
42 _____

SPECIALS/ACCESSORIES
43 NEC Class _____ Group _____ Div _____
44 _____
45 _____
46 _____
47 _____
48 _____
49 _____
50 _____
51 _____
52 _____

ACTUATOR
53 *Type _____
54 *Mfr & Model _____
55 *Size _____ Eff Area _____
56 On/Off _____ Modulating _____
57 Spring Action Open/Close _____
58 *Max Allowable Pressure _____
59 *Min Required Pressure _____
60 Available Air Supply Pressure:
61 Max _____ Min _____
62 *Bench Range _____ / _____
63 Act Orientation _____
64 Handwheel Type _____
65 Air Failure Valve _____ Set at _____
66

POSITIONER
67 Input Signal _____
68 *Type _____
69 *Mfr & Model _____
70 *On Incr Signal Output Incr/Decr _____
71 Gauges _____ By-pass _____
72 *Cam Characteristic _____
73

SWITCHES
74 Type _____ Quantity _____
75 *Mfr & Model _____
76 Contacts/Rating _____
77 Actuation Points _____
78

AIR SET
79 *Mfr & Model _____
80 *Set Pressure _____
81 Filter _____ Gauge _____
82

TESTS
83 *Hydro Pressure _____
84 ANSI/FCI Leakage Class _____
85
86

Rev	Date	Revision	Orig	App

CONTROL VALVES
ISA Form S20.50, Rev. 1

Line	Explanation of Terms and Definitions	Examples
PROJECT	Specify project name for which control valve is intended.	XYZ Nuclear PS
UNIT	Specify unit within project.	#1
P.O.	Specify purchase order number from purchaser to control valve manufacturer.	P.O. 12345
ITEM	Specify item number of purchase order.	3
CONTRACT	Specific contract number of project for purchaser's reference.	56-V-32510
MFR SERIAL	This line may show the valve manufacturer's serial number(s) and is normally filled in at the time of shipment of the valve. Serial numbers often contain the manufacturer's shop order number.	C12650-3
DATA SHEET	Specify data sheet number. Normally assigned by purchaser.	3 of 12
SPEC	Specify number of technical specification on which valve selection is based.	FL-13265-A
TAG	Specify tag number, if any, used to designate location of valve.	FV-103
DWG	Specify piping and instrumentation diagram number, loop diagram number, engineering flow diagram number, etc.	17-453
SERVICE	Describe service of control valve and/or pipe line number.	Feedwater control Reheat spray 2″ MA 1051 WA7

NOTE: The above lines are suggested only and may be modified to fit the individual company's needs. If the provided space is insufficient, add an additional sheet and refer to it.

Line No.	Explanation of Terms and Definitions	Examples
1	Describe fluid flowing into valve and its state. Indicate corrosive or erosive service and the corrosive or erosive agents.	Superheated stream Saturated water Crude oil and natural gas
	Specify thermodynamic critical pressure of the fluid.	3206 psia
2	Specify volumetric or mass flow rate at inlet or standard conditions. Maximum flow condition, if greater than normal flow condition, is the condition for which the valve is sized.	3000 gpm 10 000 bpd 600 std. m^3/s 7500 scfm 300 kg/h

Line No.	Explanation of Terms and Definitions	Examples
3	Specify inlet pressure (gauge or absolute).	5000 psig 2000 kPa abs.
4	Specify outlet pressure (gauge or absolute).	1000 psig 400 kPa gauge
5	Specify inlet temperature in °F, °R, °C or K. Must agree with state of fluid and its inlet pressure.	750°F 200°C 815 K
6	Specify specific weight (in lb/ft^3 or kg/m^3), specific gravity, or molecular weight of fluid. Identify the appropriate term.	61.9 lb/ft^3 1.03 44.01
7	Specify viscosity in appropriate units for liquids or specific heats ratio for gases.	20 centipoise 17.8 centistokes 1.27
8	Specify vapor (saturation) pressure at inlet temperature in absolute units. Only required for liquid flow.	680 psia 46.9 bar abs.
9	Specify required C_V as calculated for each condition per ANSI/ISA S75.01-1986. No additional safety (oversize) factor should be included at this point.	260
10	Specify travel of the valve in percent of rated travel calculated from required C_V, rated C_V of the valve, trim selected, and characteristic (see lines 33, 34, and 36). 0% is full closed, 100% is full open.	78%
11	Specify laboratory-measured allowable and predicted sound pressure levels, both normally in dBA as measured per ISA-S75.07-1987.	90/87 dBA
12	Extra line for information not covered in lines 1 through 11.	Compressibility factor Z Ambient temperature Base pressure and temperature
13 & 14	Specify size and schedule (or wall thickness if nonstandard) of pipe line into which valve is installed.	8″ SCH 40 15″ OD × 0.500″ wall DN200, PN100
15	Specify pipe line insulation. This information is required for predicted sound pressure level calculations.	2″ thermal None
16	Specify type of valve body.	Globe (through, angle) Split body Double port Butterfly Ball Pinch

Line No.	Explanation of Terms and Definitions	Examples
17	Specify nominal size of valve body. Specify ANSI class in accordance with ANSI B16.34-81.	4″ 600 2500 SPECIAL
18	Specify maximum pressure and temperature of the valve.	2500 psig, 650° F
19	Specify manufacturer and model number.	XYZ Controls Model 719-2
20	Specify body and bonnet material.	Steel, ASTM A216, WCB
21	Specify body liner material, if any, and its inside diameter.	Polyurethane, 3.9″
22 & 23	Specify end connection. May be integral or welded onto body.	6″ RTJ Class 1500 flange Buttweld end 2″ FNPT
24	Specify flange face finish per ANSI B16.5-81 or special finish as required.	ANSI B16.5-81 Special finish: 32 RMS
25	Specify end extensions, if any. Normally, refers to sections of pipe or reducers welded to the body by the valve manufacturer.	6″ long, SCH 80, A106, GR. B
26	Specify direction of the flow through the body. FTO = flow-to-open, FTC = flow-to-close valve.	FTO FTC
	NOTE: The descriptors "FTO" and "FTC" refer to the direction of fluid forces on the closure member. If immaterial, leave blank. When FTO and FTC are not applicable, specify direction as appropriate.	
27	Specify type of bonnet.	Standard Cooling fin Extended
28	Specify whether a lubricator and isolation valve are required. Specify lubricant.	Yes Silicone
29	Specify packing material.	Graphite impreg. asbestos TPE Non-asbestos
30	Specify type of packing.	Braided Molded V-ring Laminated filament Pressure/Vacuum
31	Extra line for special body or bonnet not covered in lines 16 through 30.	Body drain Separable flanges Flangeless
32	Specify type of trim.	Single seat cage-guided Multi-stage Multi-hole Top- and bottom-guided Double seat
33	Specify nominal size and rated travel of installed trim.	2″ 50 mm

Line No.	Explanation of Terms and Definitions	Examples
34	Specify inherent flow characteristic of installed trim.	Linear Equal % Modified parabolic Quick-opening
35	Specify whether trim is balanced or unbalanced. Semi-balanced trim should be considered as balanced.	Balanced Unbalanced
36	Specify rated C_V, F_L, and X_T of installed trim. Refer to ANSI/ISA-S75.01-1986.	260 0.9 0.68
37	Specify closure member, i.e., plug, ball, or disk material as applicable.	17-4 PH, H-1150 316
38	Specify seat material.	420 hardened 316 hardfaced
39	Specify cage, bearing, or guide material.	410 hardened
40	Specify stem material.	17-4 PH H-1150 316
41 & 42	Extra lines for additional trim requirements not covered in lines 32 through 40.	Chrome-plate Pilot-operated
43	Specify hazardous location classification per the *National Electrical Code®*, ANSI/NFPA 70-1987.	NEC® Class 1, Div. 1, Group C
44–52	Specify special requirements and/or accessories not covered elsewhere.	Solenoid valves E/P transducer NACE MR-01-75 Seismic Net weight = 275 lb
53	Specify type of actuator.	Diaphragm, pneumatic Hydr. piston, double-acting Pneumatic rotary vane
54	Specify manufacturer and model number.	XYZ Controls P-100-160
55	Specify nominal size and effective diaphragm/piston area.	8″ 160 square inch 0.2 m²
56	Specify whether actuator is for on/off or modulating service.	Modulating On/off
57	Specify whether spring/if any, acts to open or to close valve.	Open Close None
58	Specify maximum pressure for which the actuator is designed.	100 psig 60 kPa
59	Specify minimum pressure required to fully stroke the installed valve under specified conditions.	65 psig

Line No.	Explanation of Terms and Definitions	Examples
60 & 61	Specify limits of available air or hydraulic supply pressure. If upper limit is greater than line 58, a reducing valve (air set) should be furnished. Lower limit or reducing valve setting must be higher than pressure shown on line 59.	90 psig / 70 psig
62	Specify the pressures in the actuator when valve starts travel and at its rated travel position without fluid forces acting on the valve.	8/32 psig 10/22 psig 1.2/2.1 kPa
63	Specify orientation of actuator as "VERT.UP" or "VERT.DOWN" (vertical) or "HORIZ." (horizontal). For rotary valves, also specify whether mounting is "RH" (right-hand) or "LH" (left-hand) as viewed from valve inlet, if appropriate. Specify additional information as appropriate or provide sketch.	VERT.UP HORIZ. RH LH
64	Specify type and orientation of handwheel (manual override), if any.	Top-mounted Side-mounted/LH
65	Specify if air failure valve (actuator air lock-up valve) is required and at what supply pressure it shuts.	Yes 40 psig
66	Extra line for additional actuator requirements not covered in lines 53 through 65.	Hydraulic damper Stroking speed 1″/sec. Stainless steel tubing
67	Specify input signal range for full travel.	3–15 psig 200–100 kPa 4–20 mA
68	Specify type of positioner.	None Single acting Double acting
69	Specify manufacturer and model number.	XYZ Control Co. Model AB
70	Specify whether an increasing signal increases or decreases output pressure to actuator.	Incr. Decr.

Line No.	Explanation of Terms and Definitions	Examples
71	Specify whether air pressure gauges and positioner are required.	No Yes
72	Specify cam characteristic, if positioner has a cam. Normally linear.	Linear Square root
73	Extra line for positioner requirements.	Aluminum-free
74	Specify type and quantity of limit switches.	Mech. (lever arm) Proximity Pneumatic 2
75	Specify manufacturer and model number.	ABC Electric Co. Model A20Z
76	Specify electrical rating and number of contacts and action.	10A, 600 VAC/ DPDT
77	Specify valve travel at which switches are to actuate.	Full open/full closed
78	Extra line for additional limit switch requirements not covered in lines 74 through 77.	NEMA 4 IP 65
79	Specify manufacturer and model number of air set (pressure regulator).	RBJ Co. Model R-70
80	Specify output pressure setting.	70 psig 20 psig
81	Specify whether filter and/or output pressure gauge is required.	Yes No
82	Extra line for additional air set requirements not covered in lines 79 through 81.	Mount separate from valve
83	Specify pressure of hydrostatic test. Normally per ANSI B16.37-80 or API 6A-83.	3350 psig
84	Specify leakage class per ANSI/ FCI 70-2-76.	Class IV
85 & 86	Extra lines for additional test requirements not covered in lines 83 and 84.	Hydro for 30 minutes Helium leak test Stroking time test Dead band test

			PRESSURE CONTROL VALVES PILOTS and REGULATORS					SHEET _____ OF_____		
				NO	BY	DATE	REVISION	SPEC. NO.		REV.
								CONTRACT		DATE
								REQ. P.O.		
								BY	CHK'D	APPR.

GENERAL	1.	Tag No.					
	2.	Service					
	3.	Line No./Vessel No.					
	4.	Line Size/Sched. No.					
	5.	Function					
BODY	6.	Type of Body					
	7.	Body Size | Port Size					
	8.	Guiding | No. of Ports					
	9.	End Conn. & Rating					
	10.	Body Material					
	11.	Packing Material					
	12.	Lubricator | Iso. Valve					
	13.	Seal Type					
	14.	Trim Form					
	15.	Trim Material					
	16.	Seat Material					
	17.	Required Seat Tightness					
	18.	Max. Allow Sound Level dBA					
ACTUATOR/ PILOT	19.	Type of Actuator					
	20.	Pilot					
	21.	Supply to Pilot					
	22.	Self Cont. | Ext. Conn.					
	23.	Diaphragm Material					
	24.	Diaphragm Rating					
	25.	Spring Range					
	26.	Set Point					
	27.						
ACCESSORIES	28.	Filt. Reg. | Supply Gage					
	29.	Line Strainer					
	30.	Housing Vent					
	31.	Internal Relief					
	32.						
	33.						
SERVICE	34.	FLOW UNITS	LIQUID		STEAM		GAS
	35.	Fluid					
	36.	Quant. Max. | C_v					
	37.	Quant. Oper. | C_v					
	38.	Valve C_v | Valve F_L					
	39.	Norm. Inlet Press. | ΔP					
	40.	Max. Inlet Press.					
	41.	Max. Shut Off | ΔP					
	42.	Temp. Max. | Operating					
	43.	Oper. sp. gr. | Mol. Wt.					
	44.	Oper Visc. | % Flash					
	45.	% Superheat | % Solids					
	46.	Vapor Press. | Crit. Press.					
	47.	Predicted Sound Level dBA					
	48.	Manufacturer					
	49.	Model No.					

Notes:

ISA FORM S20.51

PRESSURE CONTROL VALVES — PILOTS & REGULATORS

Instructions for ISA Form S20.51

1-4. Identification and service or location. It is assumed that each tag number is for a single valve.

5. Pressure reducing, back pressure control, or differential pressure regulator.

6. Globe, angle, or Manufacturer's Standard (MFR. STD.).

7. Body connection size and inner valve size.

8. Guiding may be top, top and bottom, skirt, or MFR. STD. Select single or double port, if applicable.

9. Specify screwed (NPT), flanged, or weld end; and flange rating, such as 150 lb ANSI.

10-11. Specify materials.

12. Write in "yes" or use check mark if required.

13. Quick open, equal percent, linear, etc.

State Characteristic:

L	=	Linear
LV	=	Linear V Port
EP	=	Equal Percentage
EPT	=	Equal Percentage Turned
EPB	=	Equal Percentage Balanced
Q	=	Quick Opening

Or use your own code and identify in notes.

14. Refers to seal between body and top works, such as diaphragm, stuffing box, etc.

15. Refers to seat, plug, stem; in general, all internal wetted parts.

16. Use only to specify soft seat, otherwise material will be same as trim specified in line 14.

17. Use if required.

18. Max allowable sound level dBA 3 ft from pipe and 3 ft downstream of the valve outlet.

19. Actuator may be spring type or springless pressure balanced.

20. The pilot is an integral or external auxiliary device which amplifies the force available through an operating medium, usually air.

21. Give pressure available and specify medium.

22. Refers to valve pressure sensing system. Specify whether controlled pressure is sensed internally or by means of an external line requiring an additional piping connection.

23-24. Specify diaphragm material and pressure or temperature limits, if applicable.

25. Range over which pressure setting can be made.

26. Specification of set pressure does not apply to factory setting. This must be called for specifically, if required.

27. Specify filter regulator, with or without gage, if required for air supply to pilot. Write "yes" or use check mark.

28. Specify if strainer is to be furnished with valve. Write "yes" to check off; or give style or model number.

30-31. Options available in gas regulators. On line 30 specify "bug-proof" if required.

34. State liquid, steam, gas units gpm, lb/hr ft^3/min. etc.

35. Name of fluid and state whether vapor or liquid if not apparent.

36. State maximum quantity required by process and corresponding C_V.

37. State operating quantity required by process and corresponding C_V.

38. The manufacturer shall fill in the valve C_V and F_L (Liquid Pressure) Recovery Factor without reducers or other accessories.

39. Operating inlet pressure and pressure differential with units (psia, psig, inches H_2O or Hg). Note at this point that one might consider how minimum conditions will fit the sizing.

40. Maximum inlet pressure if different from normal.

41. State the maximum pressure drop in shut-off position to determine proper actuator size. This is actual difference in inlet and outlet pressure stated in psi, inches of H_2O or Hg, etc.

42. State °F. or °C.

43. State operating specific gravity and molecular weight.

44. State operating viscosity and its units. State flash at valve outlet, i.e., of max flow that will be flashed to vapor because of the valve pressure drop.

45. In the case of vapors, state superheat and in the cases of liquids, state the solids, if present.

46. Note vapor pressure of fluid as well as the critical pressure.

47. Give manufacturers predicted sound level dBA.

48. Complete when available.

		SELF-ACTUATED TEMPERATURE REGULATOR				SHEET _____ OF _____		

			SELF-ACTUATED TEMPERATURE REGULATOR				SPEC. NO.	REV.
		NO	BY	DATE	REVISION	CONTRACT		DATE
						REQ. - P.O.		
						BY	CHK'D	APPR.

GENERAL	1.	Tag No.				
	2.	Service				
	3.	Line No./Vessel No.				
	4.	Line Size/Sched. No.				
	5.	Function				
VALVE	6.	Body Size / Trim Size				
	7.	Number of Ports				
	8.	End Conn. and Rating				
	9.	Body Material				
	10.	Trim Material				
	11.	Plug Form				
	12.	Seat Material				
	13.	Action On Temp. Rise				
	14.					
THERMAL SYSTEM	15.	Fill: SAMA Class				
	16.	Bulb Type				
	17.	Bulb Material				
	18.	Extension Length				
	19.	Insertion Length				
	20.	Bulb Connection				
	21.	Capillary Material				
	22.	Armor				
	23.	Capillary Length				
	24.	Well Material				
	25.	Well Connection				
	26.	"U" Dimension / "T" Dim.				
	27.	Adjustable Range				
ACC	28.					
	29.	Integral Thermometer				
	30.					
	31.					
	32.					
	33.					
SERVICE	34.	FLOW UNITS	LIQUID	STEAM	GAS	
	35.	Fluid				
	36.	Quant. Max. C_V				
	37.	Quant. Oper. C_V				
	38.	Valve C_V / Valve F_L				
	39.	Norm. Inlet Press. / ΔP				
	40.	Max. Inlet Press.				
	41.	Max. Shut Off ΔP				
	42.	Temp. Max. / Operating				
	43.	Oper. sp. gr. / Mol. Wt.				
	44.	Oper Visc. / % Flash				
	45.	% Superheat / % Solids				
	46.	Vapor Press. / Crit. Press.				
	47.	Predicted Sound Level dBA				
	48.	Manufacturer				
	49.	Model No.				

Notes:

ISA FORM S20.52

SELF-ACTUATED TEMPERATURE REGULATORS

Instructions for ISA Form S20.52

1. Identification of item by tag number.

2. Process area or function

3. Stream description and/or pipe size or vessel number with which valve is used.

4.

5. Function heating or cooling.

6. Specify nominal size of body and trim in inches.

7. 1 — single port (SP); 2 — double port (DP); 3 — three-way.

8. Specify screwed or flange rating and facing.

9. Specify material of body such as bronze, carbon steel, cast iron, etc.

10. Specify material of trim such as bronze, 316 stainless steel, etc.

11. State Characteristic:

L	= Linear		B	= Blending
LV	= Linear V Port		D	= Diverting
EP	= Equal Percentage			
EPT	= Equal Percentage Turned			
EPB	= Equal Percentage Balanced			
Q	= Quick Opening			

Or use your own code and identify in notes.

12. Specify seat material such as 316 stainless steel, Buna N, etc.

13. Specify open or close.

15. Filled thermal system instruments are classified as follows:

Class IA: Liquid filled, uniform scale, fully compensated.

Class IB: Liquid filled, uniform scale, case compensated only.

Class IIA: Vapor pressure, increasing scale, with measured temp. above case and tubing temp.

Class IIB: Vapor pressure, increasing scale, with measured temp. below case and tubing temp.

Class IIC: Vapor pressure, increasing scale, with measured temp. above and below case and tubing temp.

Class IID: Vapor pressure, increasing scale, above, at, and below case and tubing temp.

Class IIIA: Gas filled, uniform scale, fully compensated.

Class IIIB: Gas filled, uniform scale, case compensated only.

Class VA: Mercury filled, uniform scale, fully compensated.

Class VB: Mercury filled, uniform scale, case compensated only.

16. State whether plain, averaging, sanitary bulb.

17. Give material and type of bulb and extension; such as 316 SS.

18. Write in length of extension, followed by "ben" for bendable, "adj" for adjustable or "rgd" for rigid.

19. The bulb insertion length should be given if no well data are shown.

20. Specify size of jam nut or union connector; or part number.

21. Specify material of capillary tubing.

22. Specify material of armor (Bronze, 316 SS, etc.) or write "None."

23. Specify length in feet.

24. Specify well material such as bronze, 304 stainless steel, 316 stainless steel, monel, etc.

25. Specify process connection size and type, such as ¾ in. NPT, 1½ in. 150 lb RF, etc.

26. Specify "U" dimension from face of flange or bottom of thread to tip of well. Specify "T" (lagging extension) dimension in inches.

27. Note adjustable range available from the manufacturer.

29. Specify range, or write in "None."

34. State liquid, steam, gas units gpm, lb/hr, ft^3/min, etc.

35. Name of fluid and state whether vapor or liquid if not apparent.

36. State maximum quantity required by process and corresponding C_V.

37. State operating quantity required by process and corresponding C_V.

38. The manufacturer shall fill in the valve C_V and F_L (Liquid Pressure) Recovery Factor without reducers or other accessories.

39. Operating inlet pressure and pressure differential with units (psia, psig, inches H_2O or Hg). Note at this point that one might consider how minimum conditions will fit the sizing.

40. Maximum inlet pressure if differential from normal.

41. State the maximum pressure drop in shut-off position to determine proper actuator size. This is actual difference in inlet and outlet pressure stated in psi, inches of H_2O or Hg, etc.

42. State °F. or °C.

43. State operating specific gravity and molecular weight.

44. State operating viscosity and its unit. State flash at valve outlet, i.e., of max flow that will be flashed to vapor because of the valve pressure drop.

45. In the case of vapors, state superheat and in the cases of liquids, state the solids, if present.

46. Note vapor pressure of fluid as well as the critical pressure.

47. Give manufacturers predicted sound level dBA.

48. Complete when available.

©ISA S20

		PRESSURE RELIEF VALVES	SHEET ____ OF ____

				PRESSURE RELIEF VALVES		SHEET ____ OF ____		
		NO	BY	DATE	REVISION	SPEC. NO.		REV.
						CONTRACT		DATE
						REQ. - P.O.		
						BY	CHK'D	APPR.

GENERAL	1.	Tag Number				
	2.	Service				
	3.	Line No./Vessel No.				
	4.	Full Nozzle/Semi Nozzle				
	5.	Safety or Relief				
	6.	Conv., Bellows, Pilot Op.				
	7.	Bonnet Type				
CONN.	8.	Size: Inlet / Outlet				
	9.	Flange Rating or Screwed				
	10.	Type of Facing				
MATERIALS	11.	Body and Bonnet				
	12.	Seat and Disc				
	13.	Resilient Seat Seal				
	14.	Guide and Rings				
	15.	Spring				
	16.	Bellows				
	17.					
OPTIONS	18.	Cap: Screwed or Bolted				
	19.	Lever: Plain or Packed				
	20.	Test Gag				
	21.					
	22.					
	23.					
BASIS	24.	Code				
	25.	Fire				
	26.					
	27.					
FLUID DATA	28.	Fluid and State				
	29.	Required Capacity				
	30.	Mol. Wt. / Oper. sp. gr.				
	31.	Oper. Press. / Set Press.				
	32.	Oper. Temp. / Rel. Temp.				
	33.	Constant				
	34.	Back Pressure { Variable				
	35.	Total				
	36.	% Allowable Overpressure				
	37.	Overpressure Factor				
	38.	Compressibility Factor				
	39.	Latent Heat of Vaporization				
	40.	Ratio of Specific Heats				
	41.	Operating Viscosity				
	42.	Barometric Pressure				
	43.					
	44.					
	45.	Calc. Area sq. in.				
	46.	Selected Area				
	47.	Orifice Designation				
	48.	Manufacturer				
	49.	Model No.				

Notes:

ISA Form S20.53

PRESSURE RELIEF VALVES

Instructions for ISA Form S20.53

This Form is identical in content to the Pressure Relief Valve Specification Sheet of the American Petroleum Institute contained in the second edition of API Standard 526, November, 1969.

1. Where multiple valves are used, it is assumed that all have the same tag number, unless otherwise noted.

2. Process service or location designation.

3. Line number or vessel number on which valve is located.

4. Refers to valve inlet construction.

5. Specify valve classification: safety, relief, or safety-relief. These terms are defined in the American Society of Mechanical Engineers, ASME Boiler and Pressure Vessel Code, Section 1, 1968 Edition, Paragraph PG-67 (footnote), as follows:

Safety Valve: An automatic pressure relieving device actuated by the static pressure upstream of the valve and characterized by full opening pop action. It is used for gas or vapor service.

Relief Valve: An automatic pressure relieving device actuated by the static pressure upstream of the valve which opens further with the increase in pressure over the opening pressure. It is used primarily for liquid service.

Safety Relief Valve: An automatic pressure relieving device suitable for use either as a safety valve or relief valve, depending on application.

6. Specify conventional type of bellows, or pilot operated valve.

7. Bonnet may be open or closed.

8-10 Specify inlet connection in the left side and outlet connection in the right side of the spaces. Flanges assumed to be ANSI unless otherwise noted. For screwed ends, specify male or female NPT.

11-16 Specify materials of construction. If resilient seat seal is not used, write "None."

17.

19. Specify cap only if lever is not used.

20. If lifting lever is required, specify plain or packed.

21. A test gag is supplied with the safety valve, when specifically ordered, for the purpose of holding the valve closed against upstream pressure when hydrostatically testing the vessel or pipe line on which the valve is installed.

24. State applicable code, if any.

25. Check or write "yes" if selection is based on fire.

26-27. Specify other bases of selection, if applicable, such as "blocked discharge," or "thermal relief."

28. Specify whether liquid or vapor and name fluid.

29. Specify maximum quality valve will be required to pass at relief condition and give flow units.

30. For liquids, state specific gravity and for vapor or gases give molecular weight or specific gravity at $60°$F.

31. State operating pressure and the set pressure.

32. State operating temperature and relief temperature.

34-35. Back pressure conditions. State constant, variable or developed back pressure and the total.

36. Allowable overpressure is the percent increase over the set pressure permitted.

37. Overpressure factor utilized in some calculation forms i.e., 1.10 would be 10 percent allowable overpressure.

38. Compressibility Factor Z is the measure of deviation from Boyle's Law (p) obtained from gas curves.

39. Latent Heat of vaporization. The heat required to change liquid into vapor.

40. Ratio of specific heats. Cp/Cv.

41. Operating Viscosity

42. Barometric Pressure

43. Optional Information.

44. Optional Information.

45. Calculated Area.

46. Selected Area.

47. Orifice Size Designation.

48-49. Filled in after selection.

			RUPTURE DISCS				SHEET _____ OF_____		

				NO	BY	DATE	REVISION					

SPEC. NO. | REV.
CONTRACT | DATE
REQ. - P.O.
BY | CHK'D | APPR.

GENERAL		1.	Tag Number					
		2.	Service					
		3.	Line No./Vessel No.					
		4.	Line Size/Sched. No.					
SERVICE CONDITIONS		5.	Design Code					
		6.	Basis For Selection					
		7.	Primary/Secondary Relief					
		8.	Fluid					
		9.	Vapor	Pounds/Hour					
		10.	or Gas	Mol. Weight					
		11.	Liquid	gpm					
		12.		sp.gr. @ Rel. Temp.					
		13.	Corrosive Agents					
		14.	Operating Press. & Temp.					
		15.	Desired Burst Pressure					
		16.	Flowing Temperature					
		17.	Constant Back Pressure					
		18.	Vacuum: Operating	Max.					
		19.	Press: Static or Pulsating					
		20.	Bursting Pressure Range					
		21.	EST. Burst Press. @ 72° F					
		22.						
		23.						
CONSTRUCTION	**DISC**	24.	Manufact. & Model No.					
		25.	Size					
		26.	Material					
		27.	Coating: Inlet	Outlet					
		28.	Quantity per Assembly					
		29.						
	VAC. SUPP.	30.	Model No.					
		31.	Material					
		32.	Quantity per Assembly					
		33.	Attached to Disc					
	FLANGES	34.	Assembly No.					
		35.	Base Material					
		36.	Holddown Material					
		37.	I.D. of Conn. Piping					
		38.	Flange Rating & Facing					
		39.	½ in. NPT Tap in Holddown Flg.					
	OPTIONS	40.	Studs & Nuts					
		41.	Preassembly Screws					
		42.	Excess Flow Valve					
		43.	Pressure Gage					
		44.	Jackscrews					
		45.						
		46.						

Notes:

ISA FORM S20.54

RUPTURE DISCS

Instructions for ISA Form S20.54

1. Tag number of entire assembly.

2,3. Location in process equipment or pipe line.

5. Write in the Code governing the vessel or line design; ASME UPV, ASME BOILER, ANSI B9 Refrigeration, ANSI B19.1 Compressors, ANSI B31.3 Refinery Piping, API RP520, etc.

6. Specify if overpressure is caused by FIRE, BLOCKED DISCHARGE, COOLING WATER FAILURE, etc.

7. Write in PRIMARY or SECONDARY.

8-12. Fill in fluid properties under normal conditions.

13. Specify corrosive fluid and percentage if the manufacturer is to select the disc material.

14. Fill in normal conditions.

15. Fill in burst pressure at prevailing temperature.

16. Extremely high or low (cryogenic) temperature will affect the choice of material for the disc holder.

17. Write in ATMOS., or pressure of header system, if used.

18. Describe extent of vacuum, if any is possible.

19. If pressure is pulsating, specify range of pressure excursion.

20. For conventional preformed discs, a manufacturing tolerance must be applied to the desired rupture pressure. Specify MFR. STD. or write in the range required.

21. To be determined by the manufacturer.

24. Fill in after selection is made.

25. Nominal size, in inches.

26,27. List disc materials.

28. Include all spares.

31. List vacuum support material.

32. Should have one per disc, including spares.

33. Write in YES or NO.

34-37. Describe safety head or hold-down flange assembly.

38. Specify 125 lb FF, 150 lb RF, 600 lb RTJ, SCREWED, etc.

39. Write YES or NO.

40. Write YES or NO.

41. Write YES or NO.

42. Write YES or NO.

43. Write YES or NO.

44. Write YES or NO.

			SOLENOID VALVES				SHEET ____ OF ____		
							SPEC. NO.		REV.
			NO	BY	DATE	REVISION			
							CONTRACT		DATE
							REQ.	P.O.	
							BY	CHK'D	APPR.

	1.	Tag Number				
	2.	Service				
	3.	Line No./Vessel No.				
	4.	Quantity				
VALVE BODY	5.	Type				
	6.	Size – Body/Port				
	7.	Rating & Type Conn.				
	8.	Material – Body				
	9.	Material – Seat				
	10.	Material – Diaphragm				
	11.	Operation Direct/Pilot				
	12.	Packless or Type Packed				
	13.	Manual Re-Set				
	14.	Manual Operator				
	15.					
	16.					
WHEN DE-ENERGIZED	17.	2-Way Valve Opens/Close				
	18.	3-Way				
	19.	Vent Port Opens/Close				
	20.	Press Port Opens/Close				
	21.	4-Way				
	22.	Press to Cyl. I/Cyl 2				
	23.	Exh. from Cyl I/Cyl 2				
	24.					
	25.					
SOLENOID	26.	Enclosure				
	27.	Voltage/Hz				
	28.	Style of Coil				
	29.	Single or Double Coil				
	30.					
	31.					
SERVICE CONDITIONS	32.	Fluid				
	33.	Qty. Maximum				
	34.	Oper. Diff. Min/Max				
	35.	Allow. Diff. Min/Max				
	36.	Temp. Norm/Max.				
	37.	Oper. sp. gr.				
	38.	Oper. Viscosity				
	39.	Required C_V				
	40.	Valve C_V				
	41.					
	42.					
	43.					
	44.					
	45.	Manufacturer				
	46.	Model Number				

Notes:

ISA Form S20.55

SOLENOID VALVES

Instructions for ISA Form S20.55

1. Identification by tag number.

2. Process service.

3. Identification of line and vessel.

4. Number of identical valves.

5. Indicate whether 2-way, 3-way, or 4-way.

6. Specify body and port size in inches.

7. Maximum pressure rating and type of connections such as screwed or FLANGE rating.

8. Specify material such as bronze, aluminum or stainless steel.

9. Specify seat such as bronze or stainless steel, synthetic rubber, teflon, etc.

10. If diaphragm is used, specify material such as synthetic rubber, teflon.

11. Designate whether direct operated, self-pilot type or with pilot requiring auxiliary operating medium.

12. Specify packless or type packing.

13. State whether no voltage release or electrically tripped.

14. Specify if required.

15,16. Blanks for special requirements i.e., manifold valves etc.

17-23. State whether open or closed in appropriate places.

24,25. Blanks for special requirements.

26. Specify enclosure as general purpose, water tight, explosion proof.

27. State electrical characteristics voltage, ac or dc, and ac hertz.

28. Style of coil to be standard, molded, high temperature.

29. State whether single or dual coil. If dual coil, explain operation in space for notes.

30,31. Blanks for special requirements.

32. Name fluid and state whether liquid or gas if not apparent.

33. State maximum required capacity in units of flow such as gpm, lb/hr, SCFH.

34. State actual minimum and maximum differential encountered under operating conditions.

35. Vendor to state minimum operating differential required to operate valve and maximum allowable differential.

36-38. State normal operating temperature and maximum possible temperature operating, specific gravity or molecular weight and operating viscosity.

39. State calculated C_V requirement.

40. Vendor to state valve C_V.

Addendum to:

ISA Standard S20 "Specification Forms for Process Measurement and Control Instruments, Primary Elements and Control Valves."
In the ISA Standard ISA-S20-1975 the strict SI conversion, 21-103 kPa, is used for 3-15 psig. It is acceptable to round-off the 21-103 kPa to 20-100 kPa.

INSTRUMENT SOCIETY OF AMERICA 67 Alexander Drive, P.O. Box 12277
Research Triangle Park, NC 27709, Telephone (919) 549-8411

INSTRUMENT SOCIETY of AMERICA
Research Triangle Park, North Carolina

ISA-S26-1968
ANSI MC4.1-1975
Approved April 30, 1975

American National Standard

Dynamic Response Testing of Process Control Instrumentation

Instrument Society of America

ISBN 0-87664-349-7

ISA-S26   Dynamic Response Testing of
           Process Control Instrumentation

INSTRUMENT SOCIETY OF AMERICA
67 Alexander Drive
P.O. Box 12277
Research Triangle Park, North Carolina 27709

FOREWORD

This Foreword is included for informational purposes and is not part of Standard S26.

This revision to the earlier series of ISA Recommended Practices has been prepared as a continuing part of the service of the Instrument Society of America toward a goal of uniformity in the field of instrumentation. To continue to be of service to those organizations and individuals who use it, this document should not be static. The Society welcomes all comments and criticism; address letters to the Standards and Practices Board Secretary, Instrument Society of America, 67 Alexander Drive, P.O. Box 12277, Research Triangle Park, North Carolina 27709.

This document is an integration of four parts covering general recommendations for dynamic response testing, techniques for devices with pneumatic output signals, techniques for devices with electric output signals, and techniques for closed loop actuators for final control elements. The original editions were published in 1957, 1960, and 1961. The revisions providing pulse testing techniques, completed in 1966, have been added to the basic sine wave and step testing techniques. New or altered material is indicated by a vertical bar beside the text or drawing.

This composite ISA Standard was prepared by the Dynamic Response Testing Committee (SP26), established in 1956 with F. H. Winterkamp as Chairman. C. E. Ryker became Chairman in 1962. The purpose of this committee has been to establish guidelines for the successful application of dynamic response testing techniques to modern instruments and process lines. This subject has developed considerable interest among scientists and engineers concerned with accurate and efficient information from instrumental testing. The recommendations in this document are up-to-date with respect to the equipment commercially available to perform this type of testing.

COMMITTEE SP26

Name	Company
C. E. Ryker - Chairman	Cummins Engine Co. (formerly with Monsanto Co.)
P. S. Buckley	E. I. du Pont de Nemours and Co.
R. L. Flynn	Monsanto Company
J. D. Hougen	University of Texas (formerly with Monsanto Co.)
J. G. Pons	Honeywell, Inc.
K. B. Schnelle, Jr.	Vanderbilt University (formerly with ISA Headquarters)

The following individuals have served as a volunteer Board of Review for this ISA Standard:

Name	Company	Name	Company
R. P. Bigliano	E. I. du Pont de Nemours and Co.	W. H. Howe	Foxboro Company
M. Bradner	Foxboro Company	W. J. Katt	Monsanto Company
A. R. Catheron	Foxboro Company	J. T. Muller	Leslie Company
E. P. Diehl	General Electric Company	R. K. Temple	Foxboro Company
R. E. Gorton	Pratt and Whitney Aircraft	F. H. Winterkamp	E. I. du Pont de Nemours and Co.

This document was approved for publication by the ISA Standards and Practices Board on October 22, 1966:

Name	Company	Name	Company
E. C. Hutchison	Heat Technology Laboratory	L. N. Combs	E. I. du Pont de Nemours and Co., Inc.
R. E. Clarridge	International Business Machines Corp.	C. W. Dawson	Charles T. Main, Inc.
R. L. Galley	Douglas Aircraft Company	H. N. Norton	Consultant
E. J. Herbster	Mobil Oil Company	G. G. Gallagher	The Fluor Corporation, Ltd.
A. P. McCauley	The Glidden Company	J. R. Mahoney	Hoke, Inc.
C. E. Ryker	Cummins Engine Company	J. E. French	Instrument Society of America Headquarters
F. L. Maltby	Drexelbrook Engineering Company		

CONTENTS

1. PURPOSE

This ISA Standard constitutes a series recommending dynamic test procedures for measurement and control equipment for production processes.

With the continuing development and application of dynamic analysis to systems engineering, dynamic response test data are becoming an increasingly important part of overall performance data. Proper use of this standard should result in:

Data that will characterize dynamic performance in a uniform comparable manner,

data of value for control systems design as well as for performance characterization, and

a maximum amount of useful data per testing dollar.

2. SCOPE

This ISA Standard establishes the basis for dynamic response testing of measurement and control equipment with pneumatic output and electric output and for closed loop actuators for externally actuated control valves and other final control elements. General recommendations applicable to all dynamic response testing and a brief glossary, defining terms as used in this standard, are also included. Tabular format is used to simplify application in the laboratory by those familiar with dynamic response testing. A minimum of discussion and descriptive material precedes the tabulation of recommended tests. Methods for sine wave, step, and pulse-type signals are included.

Final control elements take many forms, and the actuators may be powered and signalled pneumatically, electrically, or in combination. Externally actuated valves for controlling flow are most common in the process industries, but the final control element can be a controlled volume pump, motor speed control, or many others.

It would, of course, be desirable to have a system which relates the flow response of a control valve or a variable speed pump to the input signal of the closed loop actuator. It was the committee's feeling that it would not be practical to require this.

For the purpose of this ISA Standard, all final control elements will be considered as having an input signal and an output motion or rotation and no distinction will be made between power media.

3. FACTORS TO BE CONSIDERED IN DYNAMIC RESPONSE TESTING

3.1 Use of Data

The application of dynamic response test data can be generally divided into two categories:

(1) characterization of a control or measuring device.

(2) control system design.

Non-specific designations such as *flat response* or *good response* should not to be used in interpreting dynamic response test data for measurement and control devices. Such terms, which are borrowed from audio work, are not applicable in the control field. It is recommended that data be taken and presented according to the appropriate sections of this standard and that the performance be judged only in the light of requirements for a specific application. For example, a device whose dynamic response is too slow for one application might be ideal for another where undesirable high frequency noise signals are present.

3.2 Interaction

A control system usually consists of a number of elements. Each element has one or more input and output signals. Two elements are said to be interacting when the relation between input and output of one element is affected by changing the characteristics of the other element. In studying dynamic response of a control system, elements which interact are often considered as a unit. It is important to note that interaction is not necessarily mutual.

An example of interacting elements is a pneumatic transmitter and the pneumatic tubing which is attached to it. The response of the transmitter (measured at the point where the tubing is attached) for a given input disturbance depends upon the amount of tubing attached. In the usual case the greater the amount of tubing, the slower the response. This system could be made non-interacting by placing a booster relay between the transmitter and the tubing. Then the speed of response of the transmitter would be virtually independent of the amount of tubing attached to the booster.

3.3 Nonlinearity

Nonlinearity can have a profound effect on the dynamic response of a device, particularly in causing large variations in performance with changes in output load or input signal size. Some common types of nonlinearities are dead spot, friction, hysteresis, velocity limiting, saturation, exponential measurements, and valve characteristics. Nonlinearity not only causes distortion of the signal shape, but also may result in additional phase shift and attenuation.

3.4 Power Supply to Device Being Tested

The performance of the power supply system used for the device under test can exert considerable influence on the dynamic test results. A power supply that is adequate for static calibration may be totally inadequate for the large power demands that can occur when high frequency input signals are used. It is common practice for vendors to specify the energy level to be used, but only in the electrical field is it common practice to specify a degree of regulation also.

Since these tests are to characterize devices in a uniform, comparable manner, it is recommended that the energy level specified by the vendor be used, and that the power supply be capable of regulating to $\pm 2\%$ of this level at all test conditions. It is extremely important to check this regulation immediately at the power input connection of the device being tested because in pneumatic and hydraulic work unexpected, large supply line drops can occur during high frequency tests.

3.5 Test Signals

This ISA Standard outlines methods for obtaining two types of dynamic data. The first are data that can be used for mathematical analysis, for graphical solution of control problems and for characterization of dynamic performance. These data can be obtained by use of either sine-wave or pulse-type forcing functions. The second type of dynamic data provide a qualitative evaluation of the nonlinearity of the device under test and are obtained by use of step tests.

The choice of whether to use sine-wave or pulse-type forcing functions is a matter of convenience and is left to the discretion of the person performing the test. Either type forcing function when used as outlined in this standard, will yield the same information. The committee is in agreement that the pulse method provides a marginal advantage over the sine wave method for obtaining dynamic data of instrument components. However, the pulse method does offer distinct advantages when the generation of sinusoidal input signals for a given device is difficult. The committee feels the greatest potential use of the pulse method is not on certain instrument components but on plant process equipment and processes themselves. The scope of this standard is limited to instrument components.

This Standard is organized in a manner that, it is hoped, will be convenient to the user. All references to sine wave and step tests appear before all reference to pulse tests. Redundancy has been kept to a minimum.

3.6 Qualifications of Personnel Performing Tests

This ISA Standard, as stated in the Purpose, is designed to present standardized procedures for performing dynamic tests. It is not the intent of the committee that this document be used as a textbook or instruction sheets but that it serve as a guide for use by personnel trained in the field of dynamic analysis.

4. RECOMMENDED TESTS - SINE WAVE AND STEP

4.1 Input Signals

It is recommended that sine wave and step input signals be used with various output loads.

Sine wave test data are most generally useful for mathematical analysis, for graphical solution of control problems, and for characterization of dynamic performance.

In order to arrive at a practical number of recommended tests, the number of output load and input signal configurations must be minimized. This could result in nonlinearities being unnoticed. To permit a qualitative evaluation of the nonlinearity of the device under test, step tests are recommended.

It is realized that the total data from both the recommended step and sine wave tests will not suffice to describe nonlinearity in the device being tested quantitatively. However, this standard is a practical compromise which will give actual data useful for most applications, and a qualitative indication of the effect of unusually large signals or loads.

Sine wave and step tests shown in Figure 1 below are recommended. The input span referred to here is that input signal that will drive the output signal through its full range.

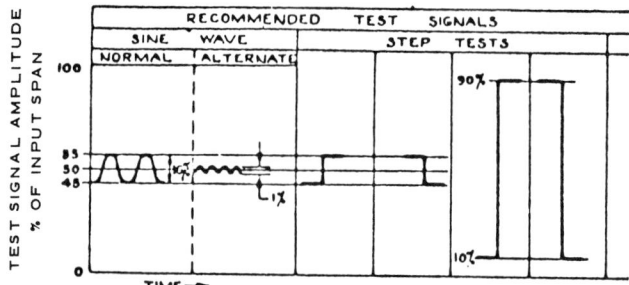

Figure 1. Recommended Test Signals

4.1.1 Sine Wave Input Signals

It is recommended that a sine wave input signal be used whose peak-to-peak magnitude is 10 percent of the actual input signal span of the device being tested. The sine wave should have the arithmetical midpoint of the input signal span as its center. The input span preferred is that span which will drive the output signal through its full range. Qualifications of the recommendation are discussed in the later documents.

The test should start at a frequency low enough that the magnitude of the output sine wave essentially equals that of the steady state output for the same input and there is essentially no phase shift between them (except if the device has pure integral or derivative functions). The frequency should be increased until the magnitude of the output sine wave is 10 percent or less of the output signal at zero frequency or the phase lag is 300°. Generally speaking, a response test should cover a frequency span of at least three decades.

Care should be taken to vary the frequency in small enough increments to describe abrupt changes in the response curve.

4.1.2 Step Input Signals

It is recommended that the following step input signals be used (in percent of actual input span):

(1) 45 to 55
(2) 55 to 45
(3) 10 to 90
(4) 90 to 10

4.2 Loading

The load presented to the output of the device has a very significant effect on dynamic performance (see Section 3.2). In selecting loadings to be used, two factors must be considered:

(1) the loadings should be as representative of actual field conditions as possible, not only to make the data useful for system design,

but to permit realistic characterization of the element.

(2) the number of tests should be kept to a minimum.

Detailed load configurations for the various tests are specified below.

4.3 Load Configurations

4.3.1 Pneumatic

Five standard load configurations are recommended in Figures 2-6. It should be noted that the recommended location of the pick-off point before the load will result in data which give the response of the device or system *alone* when coupled to the given load. To get the over-all response, it is necessary to combine the device or system response with that of the load. (See references 1-8 for dynamic response of pneumatic signal tubing.) Fewer load configurations are needed to describe typical field installations when this pick-off point location is used, than when the device or system and load are tested as a unit. For example, there can be several typical field piping configurations which appear the same to a transmitter, even though the over-all response, including load, of these installations might be quite different. The effect of the load on the response of the device or system only, when tested as recommended, reaches a limiting value as the load gets greater. For example, there is usually a difference in the system or device response if the load is changed from ten to fifty feet of one-quarter-inch tubing, but experience indicates little difference between the response curves for the device or system *alone* when the load is changed from 100 feet of one-quarter-inch tubing and a volume to 200 feet of one-quarter-inch tubing (see Section 4.2). To avoid restrictions due to flattened tubing, minimum coil diameter for one-quarter-inch copper tubing should be 16 inches; for 3/8-inch tubing, 30 inches.

The following symbols, used to illustrate typical load configurations, are consistent with ISA Standard S5.1(Y32.20).[1]

LOAD CONFIGURATIONS FOR PNEUMATIC DEVICES

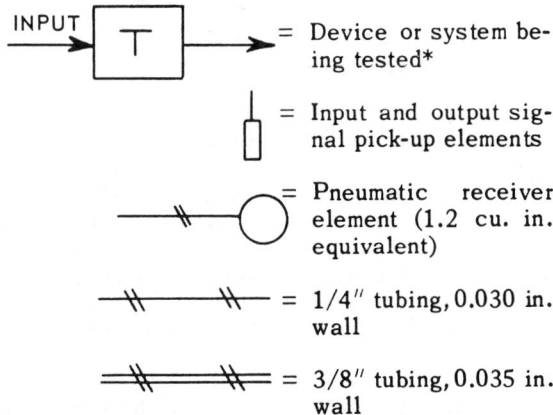

= Device or system being tested*

= Input and output signal pick-up elements

= Pneumatic receiver element (1.2 cu. in. equivalent)

= 1/4" tubing, 0.030 in. wall

= 3/8" tubing, 0.035 in. wall

*If the device being tested is a controller, the automatic-manual switch or cutoff relay should be included as part of the controller and the controller should be piped for bumpless transfer.

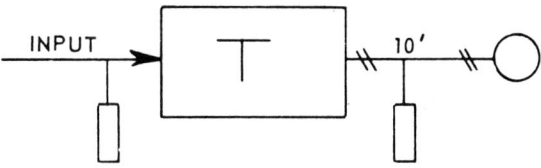

Figure 2. Configuration A

If manufacturers recommended minimum is more than 10 ft., use this minimum to prevent instability. This test will then give data for the smallest load the device or system can handle. This configuration represents a typical close-coupled pneumatic device such as a field transmitter and a controller with no long branch lines.

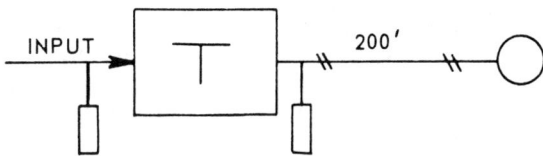

Figure 3. Configuration B

This load looks like a long transmission line to the transmitter, or a short transmission line with a long branch. The response at the transmitter is the same for either case. In using the data for a field installation, it will be necessary to add the effects of the particular load used. Experience indicates that longer transmission lines will not significantly affect the response at the transmitter.

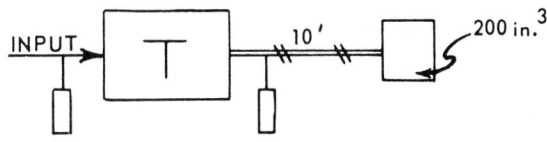

Figure 4. Configuration C

This configuration represents the case of a large load such as a diaphragm motor. It was not felt that another test with longer tubing and this large load was needed since a booster or positioner would normally be interposed if dynamic response was critical. Also, experience indicates that this load is large enough that increasing it by lengthening the tubing will not have a significant effect on the response at the instrument.

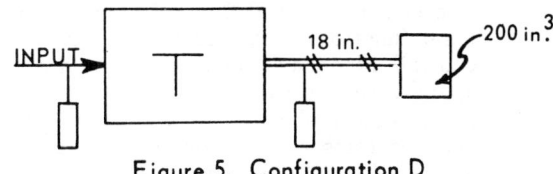

Figure 5. Configuration D

This configuration represents the case of a close-coupled large load such as would be seen by a device intended primarily to deliver pneumatic power such as a booster or positioner.

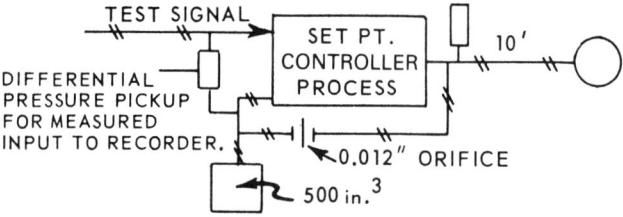

Figure 6. Configuration E

This configuration is used to test the reset function in a pneumatic controller. The controller feedback connection should be hooked up in a normal way, in addition to the connections shown above. It is possible to make dynamic tests of controller reset functions with an open loop hookup, but it requires meticulous care and very stable test signals. For this reason, the above hookup, with a feedback loop to stabilize the controller, is recommended. The restriction and volume provide sufficient time delay to permit the effects of the reset circuit to be seen. (See the section on Integral or Reset Action and reference 17.)

4.3.2 Electrical

Devices designed for a specific receiving element or load, such as commercial electronic control systems, should be tested with the actual element they would normally use as the load. Equivalent resistive loads should be used only if the element thus replaced would present a purely resistive load to the unit under test.

A special load configuration is recommended for the optional dynamic test of controller reset or integral action.

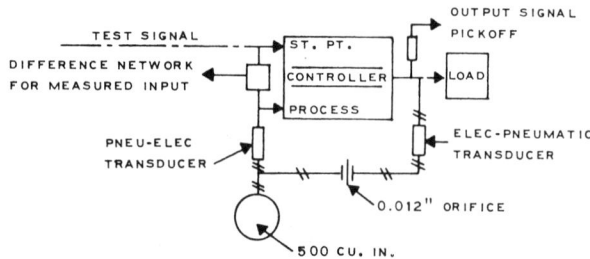

Figure 7. Test Configuration for Reset Function

The configuration of Figure 7 is used to test the reset function in an electronic controller. The controller feedback connection should be hooked up in a normal way, in addition to the connections shown above. It is possible to make dynamic tests of controller reset functions with an open loop hookup, but it requires meticulous

care and very stable test signals. For this reason, the above hookup, with a feedback loop to stabilize the controller, is recommended. The restriction and volume provide sufficient time delay to permit the effects of the reset circuit to be seen. It is easier to introduce the delay pneumatically, since many electronic controller output signals are not compatible with the input signals, and hardware of some sort is needed in any case (see Reference 18 and Section 4.5.2).

4.4 Signal Generation

The general recommendations of this ISA Standard cover signal generation for standard pneumatic pressure ranges or electrical inputs. In the case of process variable transmitters, signal generation can become a considerable problem. Where extremely high-range instruments are involved, it may be necessary to test similar instruments of lower range, because the test signals needed are too large to be practical. The following recommendations and comments on generating test signals for three of the more difficult types of measurements are intended to provide uniform and realistic tests:

4.4.1 Differential Pressure

It is recommended that dynamic tests be run on differential pressure measuring devices first with air and then with water in the measuring chamber and connecting piping. In a liquid-filled system, even a slight movement of the measuring element displaces the liquid into or from the measuring chamber through the impulse lines. The mass of liquid in the lines is moved due to the incompressibility of the liquid. Experience indicates that this can significantly affect the dynamic response of the differential pressure device. The magnitude of the effect depends on measuring element size, port size, measuring chamber geometry, and mass of liquid in the impulse lines. Figure 8 illustrates the recommended setup for the input side of the transmitter for the "wet test."

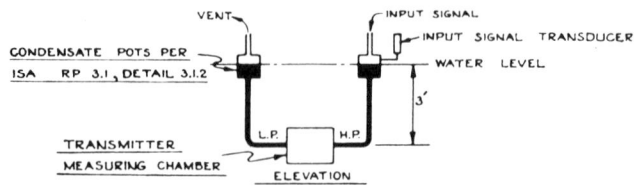

Figure 8. Recommended Wet Test Setup

Even a slight amount of unvented gas in the measuring chamber will change the test results markedly. Experience indicates that a system left standing overnight should be vented in case there are trapped bubbles coming from air that had been dissolved in the water.

In both the "wet" and "dry" tests, the low-pressure connection should be vented to atmosphere rather than held at a reference pressure level, to prevent the dynamic effects of a reference pressure system from affecting the test.

4.4.2 Temperature

Generation of sinusoidal, step, or pulse-type temperature signals which will provide useful data not only for comparison but for control system design requires that the heat transfer coefficient of the film surrounding the temperature element be known, since this is the dominant resistance to heat flow. Response in any other liquid or gas may then be computed, if the heat transfer coefficient surrounding the temperature element in the process fluid is known. Treatment of this technique is beyond the scope of this Standard (see references 12-14).

In order to obtain dynamic data with a known heat transfer coefficient, it is recommended that the sensing element be installed in a 2 inch Schedule 40 pipe tee or equivalent so that the axis of the element is parallel to the flow axis, and a flow of water at two feet/second impinges on the end of the element, well, or bulb. A bare thermocouple of #22-gauge wire with smallest feasible junction should be used to measure temperature of the stream just upstream of the element, well, or bulb. The temperature of the flowing water stream can be modulated approximately sinusoidally by switching from hot to cold sources. (Such test units are described in references 9 and 10.)

The following method has been tried and used successfully to generate sine wave and pulse input signals for the dynamic testing of temperature measuring devices:

(1) A steam water mixer was used to change the temperature of the water impinging upon the temperature measuring element. This device is designed to mix steam and water to produce relatively instantaneous hot water. The temperature of the water leaving this unit is directly proportional to both the entering steam and water pressures. The entering water pressure was held constant by a conventional pneumatic pressure indicating controller (PIC) system.[1] The pressure of the entering steam was controlled by an electric pressure indicating controller. The output of an electronic sine wave generator was cascaded into the steam PIC loop and used to obtain sinusoidal and pulse variations in water temperature.

It should be pointed out that calculation of sine wave response data from step test data or from physical dimensions is preferred by many, in the case of temperature instruments, because of the difficulties encountered in generating a signal and converting the data to the case at hand. (Such methods are described in references 11 and 15.)

However, if a step test is made by plunging the element, well, or bulb into an agitated temperature bath, it is almost impossible to determine a reasonable film coefficient value for the test conditions, hence the data are difficult to apply to process control design. It is recommended that such step tests, and calculated sine wave response data, be clearly labelled as such.

For three reasons the test should be made with the element unprotected unless this is physically impossible:

(1) The size and configuration of possible protecting tubes or wells are so varied as to make data for any specific one applicable only to a few cases.

(2) Where dynamic response is critical, a minimum mass protecting well is often built if the element cannot be inserted unprotected. The dynamic effects of such a well can be estimated fairly accurately or determined separately by a dynamic test.

(3) The data with the element unprotected are the best possible data for design or comparison purposes, since it describes the ultimate capability of the device.

It should be emphasized that this standard is intended to present an acceptable minimum number of tests, and not to exclude any other tests that might be desired for a particular case. Response data for all sine wave tests of temperature instruments should indicate whether the element was unprotected or in a thermowell.

4.4.3 Liquid Level

The geometry of possible purchased or field fabricated displacer chambers and their connecting piping and valves is so varied that no attempt is made to recommend testing in a particular chamber, although these configurations can significantly affect the overall dynamic response. These effects can be calculated or determined by tests of the particular installation. The tests recommended below will be more generally useful for performance characterization and design work. (See reference 16.)

Very few tests have been run to date in which a displacer or float was actually disturbed by a sinusoidal level change. Most tests have been made by mechanical coupling of a sinusoidal motion to the dry float or displacer. Such testing, without actual level changes, omits the resonant effects of the displacer (or float) and torque tube, which can be significant in practical process control problems.

It is possible to calculate the resonant frequency of the displacer (or float) and torque tube, and combine this data from a dry test of the remainder of the instrument. This will nor-

---

[1]*Abbreviation taken from ISA S5.1 (Y32.20), Instrumentation Symbols and Identification.*

mally result in reasonable accuracy for control design purposes. Any data so obtained should be clearly identified as such.

A proper dynamic test should include this spring-mass relationship. A chamber whose diameter is at least four times that of the float or displacer should be used for the dynamic testing and the liquid level should be varied by:

(1) raising and lowering a sinusoidally driven cylinder in the level chamber to generate a sinusoidal level change by displacement,

(2) raising and lowering the whole level chamber around the float or displacer by a mechanical drive, or

(3) coupling a reciprocating piston to a level chamber to generate sinusoidal level changes. (This might be a modified positive displacement pump piston and cylinder.)

It is known that the first scheme has been used successfully. Such a test will give the frequency response the instrument would have if the float or displacer were placed directly in the process vessel. It does not take the varying viscosities of process fluids into account, but such a test could be run for any process fluid once the apparatus is built.

Any of the three methods above could be used to generate a pulse input by removing the sinusoidal driving mechanism and actuating the cylinder, level chamber, or piston manually or with a programmed pulse source.

4.5 Parameters

The recommended tests are a compromise between the desire to reduce testing costs and keep the amount of data to a minimum and the desire for a detailed study of the effect on dynamic response of parameters such as proportional band, reset rate, derivative action, damping, and loading.

It is not intended that these recommendations be construed to exclude other tests under a greater variety of conditions. Rather, they should be considered as the minimum needed for a practical description of the dynamic response of a device.

4.5.1 Transmitting Devices

Damping adjustments on transmitting devices should be set to give minimum damping consistent with stable operation, usually 3 db down from first sign of instability. Where Figures 26 and 27 call for 100% proportional band, or gain of one, this setting should be determined by actual measurement, not by dial markings.

4.5.2 Controllers

Dynamic response testing of pneumatic or elec-

tric controllers can be divided into two parts:

(1) Finding the dynamic response of the controller with various loads, when the parameters are set at an arbitrary fixed value.

(2) Defining the dynamic response of the various controller parameters at several significant settings, with a constant load.

Item (1) is self-explanatory, but item (2) requires additional explanation. The test recommendations for this item are considered optional by the writing committee and are included to provide a uniform way of performing the tests if they are desired.

Rate or Derivative Action

Testing for dynamic response of the rate or derivative function can be performed in a relatively straight-forward manner, using an open-loop test configuration. Reset action should be minimized or removed during this test. The gain of the rate circuit will tend to cause saturation at higher test frequencies.

For this reason, the input signal magnitude should be such that at high test frequencies the output signal is 10 percent of the output span of the device. The required input signal value can be calculated by dividing 10 percent of the output signal span by the derivative or rate gain. It can also be determined by experimenting with the test setup.

Integral or Reset Action

The reset circuit of practically all process controllers has a magnitude ratio response curve as shown in Figure 9.

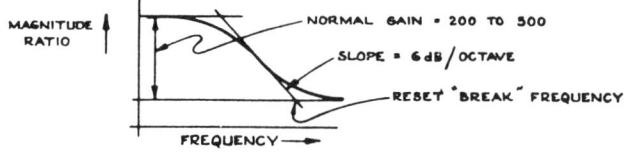

Figure 9. Magnitude Ratio Response

It is worthwhile to run step or static tests to recalibrate the reset gain and to determine the maximum dc gain of the reset circuit. Details of this procedure are not within the scope of dynamic testing. (See Reference 17 for further discussion.)

Dynamic testing of reset circuits is difficult to perform sinusoidally, especially with an open-loop design. The closed-loop test setup of Figure 6 is recommended. Dynamic testing of reset circuits is a relatively simple and straight-forward task when performed by the pulse method (see Section 7.4).

If the controller design is such that the set point is adjusted by a mechanical knob or link, the test setup of Figure 6 requires that a pneumatic-mechanical link, such as a bellows, be provided to move the set point from the test signal. The other choice for such controllers is to use an open loop test which is very difficult to perform without meticulous care and extremely stable test signals.

The time constant of the feedback network, with a 0.012 inch orifice, is such that the feedback system dynamics can be neglected for test frequencies of 0.01 cps and up. Lower test frequencies will require a smaller orifice diameter or larger tank volume. This network is included only to provide correction of any dc drift that may occur. The differential pickup measures the actual input of the reset circuit (deviation). Thus, any slight dynamic effects of the feedback circuit are eliminated by measuring the effective input signal to the controller. Sine wave tests of reset circuits require low frequency test signals and much time. Care must be taken to minimize or remove rate or derivative action.

Other Tests Further testing can be done to determine the degree of interaction between reset, derivative, and proportional settings, as desired. These are not recommended as standard tests.

4.6 Air or Power Supply and Operating Conditions

The air supply pressure, electrical power, or hydraulic pressure should be set at the level recommended by the equipment manufacturer. Where pressure or electrical settings are provided within the equipment (cushion load pressure, second-stage hydraulic booster, Wheatstone Bridge supply voltage) they should be set at the level recommended by the equipment manufacturer.

Provisions should be made to measure the power supply value immediately before it enters the equipment. The supply source should be capable of regulating to $\pm$ 2% of the desired level at all test conditions.

4.7 Final Control Elements

Final control elements may be divided into two classes, open-loop devices and closed-loop devices. For example, the pneumatic spring-opposed diaphragm-actuated control valve is an open-loop device. If a positioner is added, it becomes a closed-loop device.

The frequency response of an open-loop actuating device is principally dependent upon the dynamic characteristics and ability to deliver power of the controlling instrument as well as the impedance of the connecting tubing and parts, and therefore cannot be determined by itself.

The familiar spring and diaphragm actuator on a valve body is usually also nonlinear to some degree due to dead band caused by stuffing box or other friction forces and frequency response data may have limited significance (see Sections 3.1 and 7.1). Methods have been developed to compute the response of open-loop actuators from knowledge of physical dimensions, spring constants, and mass of plug or load. The discussion of these methods is beyond the scope of this Standard (see Reference 35). Such a computation, combined with a dynamic test of the pneumatic power source can yield reasonable results for a specific system, as can in some cases a dynamic test of the power source and actuator as one single system. This practice however, is limited to recommendations for testing closed-loop systems.

Sine wave and step tests as given in Figure 1 are recommended. The input span referred to here is that input signal that will drive the output signal through its full range.

If the recommended 10 percent peak-to-peak sine wave signal size causes excessive distortion, indicative of saturation, of the output signal before the attenuation or phase shift limits indicated below are reached, the test should be stopped and rerun with a 1 percent peak-to-peak sine wave signal. If the 10 percent signal causes excessive distortion, and the 1 percent signal is too small to be practical for the device being tested, a test may be run at an intermediate amplitude. This should be done only if the 10 percent and 1 percent signals provide results which are not usable.

The data for both tests should be plotted and clearly identified.

4.7.1 Testing of Closed-Loop Actuators Alone

The number of different loadings that might be put on a closed-loop actuator when it is used with some final element are so varied that there is little purpose to recommending an arbitrary loading when testing the actuator alone. Tests of closed-loop actuators alone should be made without a load. These data might be useful for comparison purposes and also for applications where the element being moved presents a negligible load to the actuator.

4.7.2 Testing of Closed-Loop Actuators Coupled to Final Control Elements

In the case of control valves, it is extremely desirable to test the valve and actuator assembly under flowing conditions. The infinite variety of pressure drop, damping, and coercive forces presented by varying process fluids and conditions make this impractical, in the opinion of the committee. Actuators coupled to control valve bodies should be tested under no load conditions, with the valve body drained and open to the atmosphere. The packing or stuffing box should be tightened as required to hold a hydrostatic test pressure equal to the nominal valve body pressure rating. In most cases, the response under flowing conditions will not be significantly different from the response under such a dry test (see Reference 36).

When the actuator is coupled to other elements such as variable speed drive cranks or electrical rheostats, the possible parameters should be adjusted to simulate as closely as possible the normal operation of the system. For example, in case of a variable speed drive or a metering positive displacement pump, the pump or drive should be running so that the actuator makes its excursions against the resistance it normally encounters.

It is of prime importance in all cases that the adjustments of all parameters be clearly described in the data or on the curves that are plotted from the data.

4.7.3 Test Configurations

The discussions of loading in the previous sections lead to the recommendation that in all cases the motion delivered by the actuator be measured as the output signal and the pneumatic, hydraulic, or electrical signal to the actuator as the input (see Figure 10).

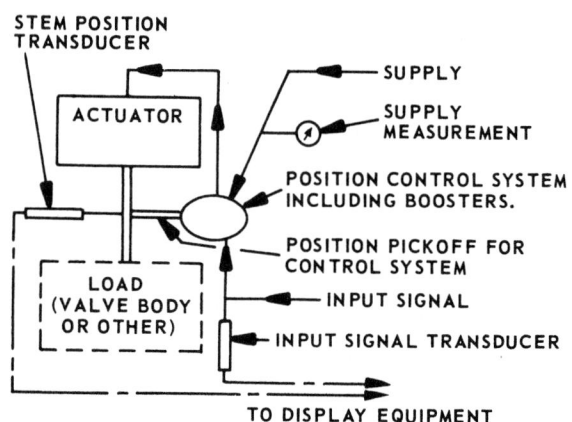

Figure 10. Recommended Loading Test Configuration

5. TEST EQUIPMENT AND PROCEDURES - SINE WAVE AND STEP

5.1 Data Required

5.1.1 Sine Wave Tests

(1) Signal frequency.

(2) Amplitude of input and output signals at each frequency.

(3) Phase relationship of input and output signals at each frequency.

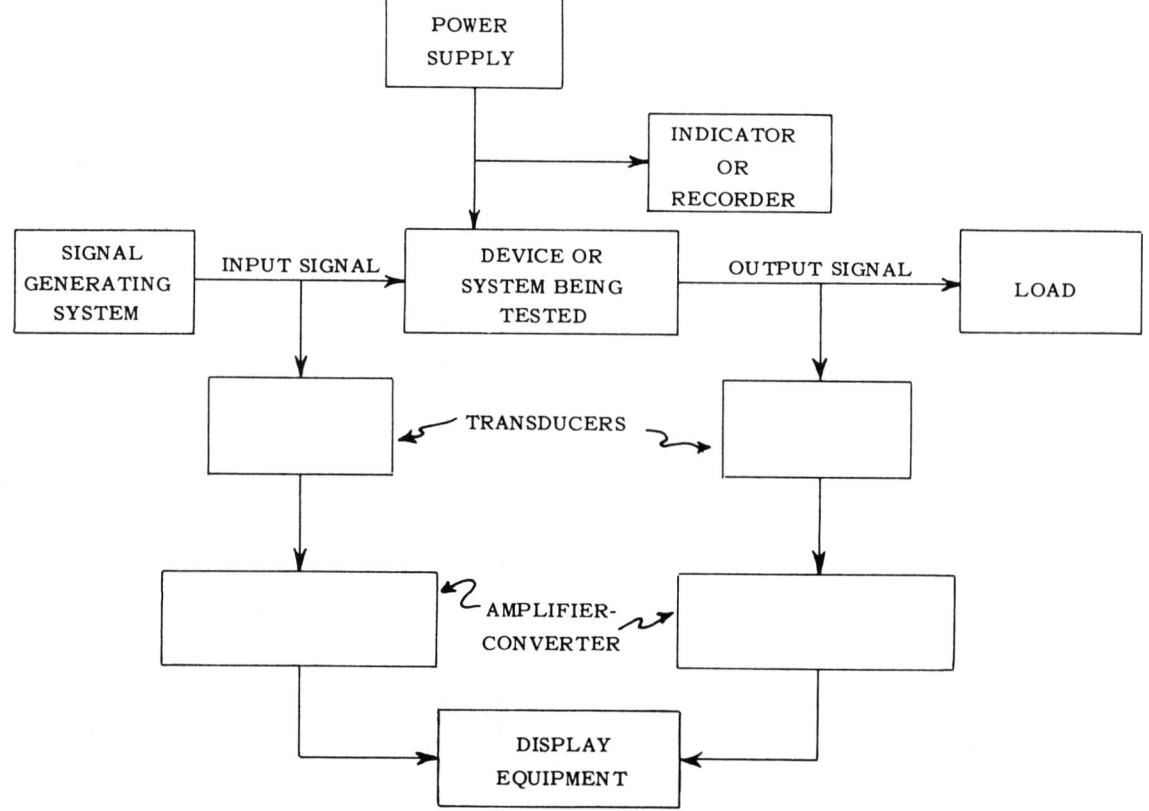

Figure 11. Generalized Dynamic Test Setup

(4) Power supply variation (to device being tested) from lowest to highest frequency at greatest load.

(5) Condition of system (adjustments, etc.).

5.1.2 Step Tests

(1) Input signal form from time zero to steady state.

(2) Output signal form from time zero to steady state.

(3) Power supply (to device being tested) variation during test.

(4) Condition of system (adjustments, etc.).

5.1.3 Supporting Data

See Section 6.3.

5.2 Generalized Test Setup

5.2.1 Block Diagram (Refer to Figure 11)

The transducers and amplifier-converters convert the input and output signals to the form required for the display equipment. With certain combinations of input signal type and display equipment (for example, electrical sine wave signals and oscilloscope) the transducers or amplifier-converters or both would not be necessary.

5.2.2 Location of Output Signal Pickoff Point

The output signal can be measured at either point ''A'' or ''B'' in Figure 12.

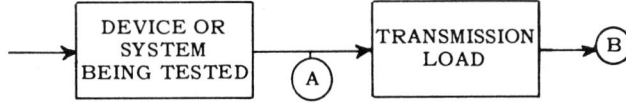

Figure 12. Measurement Points

If there is no interaction between the load and the device, it is obviously better to pickoff the output at point ''A.'' In system design, the response of the device can then be combined with the response of any load since the load does not affect the device response.

If the load will interact and affect the response of the device being tested, a case can be made for measuring the output at point ''B''. A thorough study has led to the recommendation that point ''A'' be used in all cases.

This choice does not affect the characterization function of these tests since the only requirement here is that all tests be on the same basis.

The choice of point ''A'' as the output pickoff point gives the system designer response data for the device alone (with the specific loads employed). He must then add the response of the load itself. The choice of point ''B'' would give response data for the device plus the load as a unit, for different loads.

This recommendation is based primarily on the fact that fewer tests are needed if point ''A'' is used, due to the nature of the various load configurations encountered (primarily in pneumatic work, where there is a major problem).

5.3 Test Equipment

5.3.1 Signal Generation

Sine Waves

Many devices for generating electrical sine waves are commercially available. Pneumatic and hydraulic sine wave generators are not as generally available, although some commercial units are made. One solution is to use an electrical sine wave generator with a commercial signal converting device (usually an electromechanical servo operating a pilot valve) whose output signal is pneumatic or hydraulic. Many combinations of adjustable speed sinusoidal mechanical drives coupled to a rheostat, pneumatic regulator, hydraulic regulator, or a combination of these can be constructed.

It is recommended that the sine wave generating device meet the following criteria:

(1) Although the frequency range to be covered will vary (see Paragraph 5.1.1) for process control work, the generator should cover frequencies from 0.001 cps to 20 cps.

(2) The sine wave generator must be capable of holding a given frequency without a great deal of variation.

It is recommended that acceptable performance be defined as shown in Figure 13.

At a given frequency, the period (A, B, C) of any two cycles should not vary more than ± 2%.

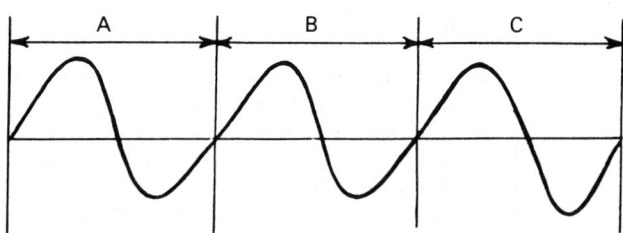

Figure 13. Definition of Acceptable Performance

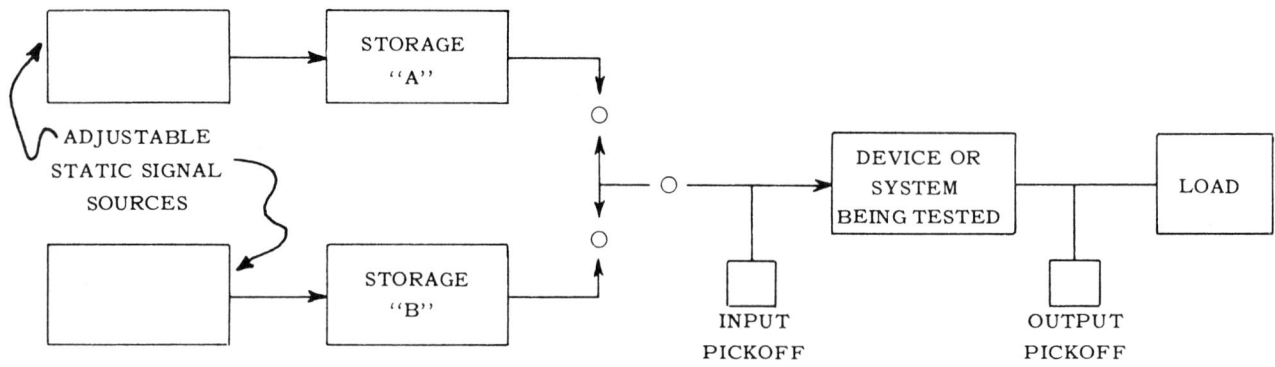

Figure 14. Step Signal Generating Circuit

Step Tests

Step signals can be generated by a signal generator-transducer combination similar to that mentioned above. A basic circuit for generating a step signal is given in Figure 14.

The following criteria should be met in constructing such a device for either electrical, pneumatic, or hydraulic testing:

(1) The connecting wiring or piping between the storage units and the device being tested (including the switch) should present as small a resistance as possible.

(2) The step input signal should rise to 95 percent of the final value in five percent or less of the time required for the output of the device under test to rise to 95 percent of the final value.

5.3.2 Display Equipment

A multi-channel high speed recorder is the fundamental tool for dynamic test work. This recorder plots both input and output sine or step signals on the same time axis. Magnitude ratio is obtained by measurements of the relative peak-to-peak distances. Phase shift is determined by measuring displacement on the time axis and converting to angular units. Step test input and output are recorded and can be read directly. Such a recorder leaves a permanent record for re-examination. This method is recommended even though the data conversion for sine wave tests can be quite time consuming.

Various combinations of types of oscilloscopes can be used to display input and output sine wave signals. Lissajous figures or calibrated phase shift networks can be used to determine phase relationship, while the magnitude ratio can be determined by peak-to-peak measurement.

Camera attachments permit a permanent record

to be made; however, these methods can cause considerable error unless used with extreme care by experienced personnel. They are not generally recommended.

5.3.3 Transducers and Amplifier-Converters

The transducer and amplifier-converter systems needed will depend upon the type of signal needed to drive the display equipment as well as the nature of the signals used for the test, i.e., pneumatic, electrical, hydraulic, or mechanical. Strain-gage transducers are commonly used for converting pressures to electrical signals. Linear variable differential transformers or multi-turn variable resistors are used for mechanical motion. The amplifier-converter units normally can best be selected by consulting the display equipment vendor, or in some cases the transducer manufacturer.

5.3.4 Test Equipment Performance Specifications

From the standpoint of results only, the following transducer-amplifier/converter display equipment system performance is recommended.

(1) Signal Amplitude-± 1 db from 0.001 cps to maximum frequency of test (at least to 20 cps).

(2) Phase shift-less than $5°$ from 0.001 cps to maximum frequency of test (at least to 20 cps(Hz).

(3) It is a general rule that the response criteria for test equipment should hold over a frequency range ten times as great as the frequency range over which the equipment being tested will meet this response criteria.

To minimize test difficulties, the following items should be carefully considered and compared before selecting test equipment:

(1) Drift, whether due to ambient temperature, supply voltage, or equipment instability,

can cause hours of extra test time. Over-all drift should be less than 5% of the full span in four hours.

(2) The chart scribing mechanism can cause difficulties at high speeds unless it is in excellent operating condition.

(3) Mechanical components (chart drives, generator linkages, etc.) should be rugged and easy to change and adjust.

5.4 Test Procedures

The following procedures are recommended as good practice and will aid in giving reliable results in minimum time.

5.4.1 Blank Run

Whenever a new test setup is completed, make a ''blank'' run, bypassing the device being tested in such a manner that the phase shift and attenuation remain at zero, as frequency increases. Points deserving particular attention are:

(1) Attenuation of one or both signals, phase shift, mechanical problems of inking and chart drive.

(2) If the equipment is new, reworked, or has not been in use, leave all equipment ''on'' in about mid-scale position, noting displayed readings and intermediate signals if possible. Recheck after four hours in ''on'' position, drift should be less than 5 percent of full scale of display equipment.

(3) Insert device to be tested and run quick check from minimum to maximum frequency, reading power supply level, at the device being tested, to check for $\pm 2\%$ specification (see Paragraph 3.4).

(4) If step tests are to be run, check test setup as specified above.

(5) Check referencing of traces of simultaneous recorder pens with respect to time. There may be an initial offset due to imperfect alignment of the pens.

5.4.2 Static Calibration

It is recommended that the overall system be statically calibrated with an independent standard (manometer or VTVM) on the actual signals. The display equipment reading is thus known accurately at the start of the test. If drift is a problem, this should be repeated periodically. Calibration by test equipment knob settings should not be substituted for the above.

5.4.3 Selecting Test Frequencies

The frequency points should be selected as the test progresses in order to provide small increments at the critical points and avoid abrupt changes in the phase or magnitude curves. A general rule is to use at least twenty points for three decades of frequency.

5.4.4 Correlating Results

Interpretation of the display equipment information is generally straightforward except when the sine wave is quite distorted. In this case, it is recommended that phase shift be determined as shown in Figure 15, assuming a multi-channel recorder is used for display.

In such a case, where precise results are desired, the magnitude ratio and phase shift may be obtained by a Fourier analysis of the output wave. Procedures for this analysis are given in many standard reference works in the electrical field. The use of Lissajous figures can be helpful in obtaining accurate phase shift data where distortion is a problem.

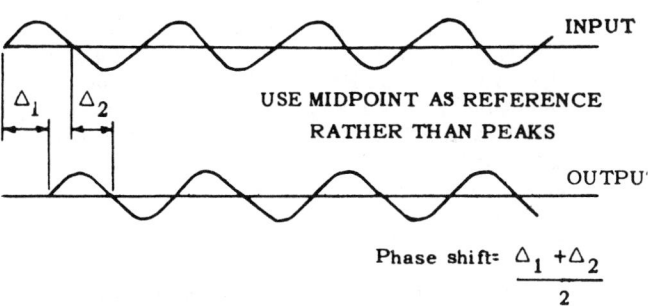

Figure 15. Determination of Phase Shift

6. DATA PRESENTATION - SINE WAVE AND STEP

6.1 Sine Wave Test Data

Data from each sine wave test should be presented as two curves. One curve should show the magnitude ratio of the output signal to the input signal as a function of frequency, the other curve should show the phase shift of the output signal from the input signal as a function of frequency. Actual data points should be shown on curves.

The choice of type and size of paper, as well as the number of tests recorded on each sheet of paper are left to the discretion and needs of the originator of the data, with the exception of the recommended coordinates and list of supporting data given below.

The recommendations below are based on the standards recommended by the ASME/IRD Dynamic Systems Committee in ASME Standard 107, ''Preferred Standards for the Presentation of Frequency Response Data.'' The recommendations in this document differ from those of the ASME/IRD only in having added some recommendations for the vertical axis of the magnitude ratio plot, and in recommending semi-log paper rather than log-log paper as a first choice for this plot.

6.1.1 Magnitude Ratio Plot

(1) The preferred recommendation is that semi-log paper be used with magnitude ratio plotted in decibels linearly on the vertical axis, which should have an absolute magnitude ratio scale superimposed. Frequency should be plotted in cycles per unit time on the horizontal, logarithmic axis.

(2) An alternate recommendation is to use log-log graph paper, with magnitude ratio plotted in absolute units on the vertical logarithmic scale, with a decibel scale superimposed. Frequency should be plotted in cycles per unit time on the horizontal logarithmic axis.

6.1.2 Phase Shift Plot

Phase shift should be plotted on semi-log paper, with phase plotted linearly on the V axis, and frequency in cycles per unit time on the horizontal logarithmic scale. Phase shift in degrees should be identified by a plus sign when the output leads the input, and by a minus sign when the output lags behind the input.

6.2 Step Test Data

Step test data should be presented on linear coordinate paper, with input and output signal magnitude in percent of its steady state change plotted on the vertical scale (which may be suppressed to bring out detail), and elapsed time in appropriate units on the horizontal scale.

In all step test plots the input signal should be plotted with a dashed or broken line, the output signal with a solid line. It is important to choose a time axis such that dead time and deviations from a true step input can be seen clearly. Since the step tests consist of a given step upset "upward" and the same size step signal "downward," these companion tests should be presented on the same graph to reduce the number of plots.

6.3 Supporting Information

The following information should be included on each plot of dynamic response data:

(1) Title of test;

(2) Description of device or system tested, including model numbers, serial numbers, and sketch of test setup if appropriate;

(3) Curve identification for each tabulation as follows:

Curve	Input Signal	Loading (Including Transmission)
Identifying code	Peak-to-peak magnitude and level for sine wave, direction and size for step test	Description or code to sketches on plot

(4) Description of test conditions.

It is suggested that for brevity's sake the statement "Tested according to ISA Standard S26" be used. Otherwise, supply regulation, details of loading circuits, pickoff points, and any other pertinent data on test conditions such as gain setting, packing friction, or reset, should be included.

(5) Supply voltage or pressure to device being tested.

(6) Date, location of test, and supervisor's initials or names.

7. RECOMMENDED TESTS - PULSE TYPE

7.1 Input Signals

Dynamic response testing by the pulse method consists of exciting a system with one pulse and obtaining, by mathematical analysis, the same data as those obtained by sinusoidal testing.

The pulse shape is not critical and does not have to follow a specific mathematical form. There are wave shapes that yield more useful information than others. The following pulse shapes are arranged in order of increasing frequency content.

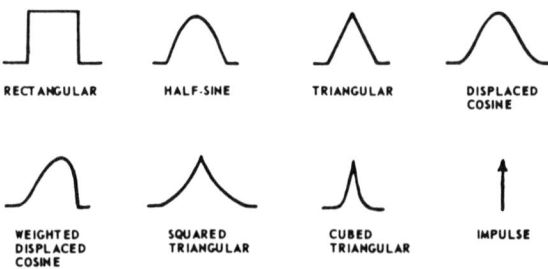

Figure 16. Pulse Input Wave Shapes

For laboratory testing of instrument components, the recommended pulse test signal is the displaced cosine. This signal shape possesses the greatest frequency content of those shapes for which signal generators are commercially available.

The recommended amplitude of the input pulse test signal is 10 percent of the actual input signal span of the device being tested. The input pulse should have the arithmetical mid-point of the input signal span as its center. The recommended input span is that span which will drive the output signal through its full range. Qualifications of this recommendation are discussed in the other documents on this series.

The degree of success achieved when a pulse-type forcing function is used to obtain dynamic data depends upon the width of the pulse used. The pulse width must be such that the system is properly excited. When the pulse width is long compared with the response of the system, the system is only moderately excited, thus suppressing high frequency information.

If the pulse width is too narrow, the device or system being tested will be forced into a region of saturation or nonlinear operation. When this happens the data obtained from the pulse test are not valid just as are the data obtained from a sine wave test at too great an amplitude or too high a frequency.

A general rule for predetermining the proper pulse width does not exist. The problem is that the shape of the output pulse does not change markedly as the device under test approaches saturation. The correct pulse width is the one that excites the device or system adequately, but does not force it into a condition of saturation. The problem of pulse width selection can be minimized by use of the guides given in this standard. Pulse widths determined by use of these guides have a reliability factor greater than 95 percent.

7.1.1 Tests When Response is Unknown

When the response of the device or system under test is completely unknown and cannot be determined even in general terms, it is recommended that a minimum of two pulse tests at different pulse widths be processed. The pulse widths should be in the general area of the width determined by Guide 1 stated in the following paragraph. The reason for running two or more tests is to ensure that the saturation or nonlinear region has not been entered.

Guides for obtaining valid tests are:

(1) The proper pulse width should possess two characteristics -

 a) the peak of the output pulse shall be well removed timewise from the peak of the input pulse, and

 b) the output pulse peak should rise to 50-75% of the input pulse peak with a system gain of unity.

(2) When input signal generating equipment that includes a selection of sine wave or

pulse-type signals is used, determine the sine wave frequency at which the output wave is attenuated to 30 percent of the input wave. Then use a pulse width which has a period of twice this frequency.

These guides are not valid for obtaining the rate or reset characteristics of controllers. These special cases are discussed in Sections 4.2 and 7.4. Examples of good and poor tests are shown in Figures 17, 18, and 19.

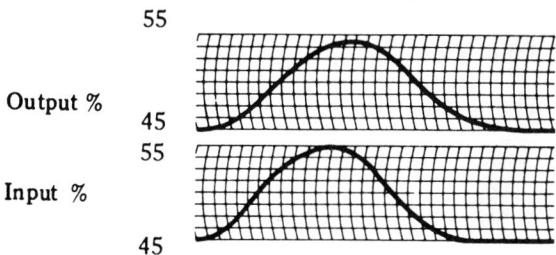

Figure 17. Poor Test, Pulse Width Too Long

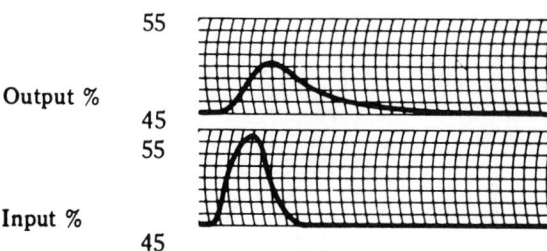

Figure 18. Good Test, Pulse Width Satisfactory

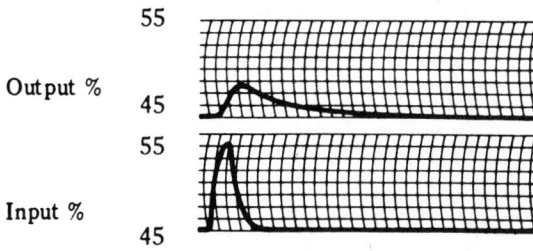

Figure 19. Poor Test, Pulse Width Too Narrow

7.1.2 Analysis of Test Data

The mathematical analysis of the input and output pulse data requires the use of either an analog or a digital computer. To perform these computations manually would be very laborious and time-consuming and outweigh any advantages gained by the use of this method. The necessity of requiring a computer to perform the mathematical computation is not considered a serious limitation. There are analog computers currently available that are especially designed to perform the necessary computation. The use of a digital computer requires a program, available from ISA for a nominal fee. Address request to Instrument Society of

America, 400 Stanwix Street, Pittsburgh, Pa., 15222. The items and figures listed under Section, 8.1 "Data Required" refer specifically to the ISA program.

7.2 Load Configurations

See Sections 4.2 and 4.3.

7.3 Signal Generation

See Section 4.4.

7.4 Parameters

See Section 4.5.

Rate or Derivative Action

See Section 4.5.2.

An example of a good pulse test for obtaining the rate characterization of a controller is shown in Figure 20.

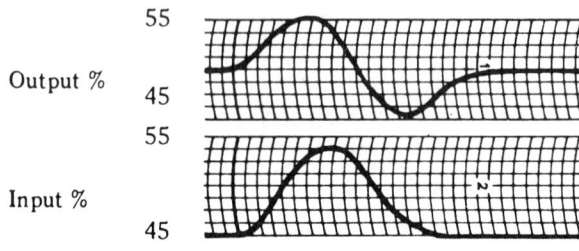

Figure 20. Good Pulse Signal for Rate Characterization Test

Integral or Reset Action

See Section 4.5.2.

The pulse method permits the procurement of the integral or reset characterization of a controller in a relatively straightforward manner, using an open-loop test configuration. A ramp-type input signal is recommended for the integral characterization test. The availability of a signal generator that can produce one triangular wave at will is desirable for this test. For reset times greater than 0.05 minute set the frequency of the triangular wave generator to $\dfrac{9}{8 \text{ (reset in seconds)}}$ cps and for reset times less than 0.05 minute set the frequency to $\dfrac{9}{4 \text{ (reset in seconds)}}$ cps. The period of the ramp signal should be of sufficient duration for the new steady state level of the output signal to increase 15 percent of full scale calibration for reset times greater than 0.05 minute and 33 percent of full scale calibration for reset times less than 0.05 minute. An example of a good test is shown in Figure 21.

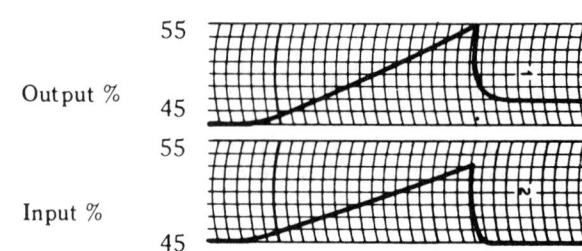

Figure 21. Good Ramp Test for Reset Action

Other Tests

Further testing can be done to determine the degree of interaction between reset, derivative, and proportional settings, as desired. These are not recommended as standard tests.

7.5 Air or Power Supply

See Section 4.6.

8. TEST EQUIPMENT AND PROCEDURES — PULSE TYPE

8.1 Data Required

8.1.1 Analog Computer

All scaling information such as transducer ranges and amplifier sensitivities should be tabulated. In addition to recording the input/output curves on paper, the curves shall be recorded in an electrical form such as on magnetic tape. The electrical recordings are used as input signals to the analog computer.

8.1.2 Digital Computer

Refer to Figure 22 for items 1-5.

(1) total number of points in each curve.

(2) the Δt values in unit time.

(3) the number of points using each Δt value.

(4) the amplitude at each point.

(5) the specific frequencies at which calculations are to be made.

(6) variations in power supply to device being tested during test.

(7) conditions of systems, including necessary adjustments.

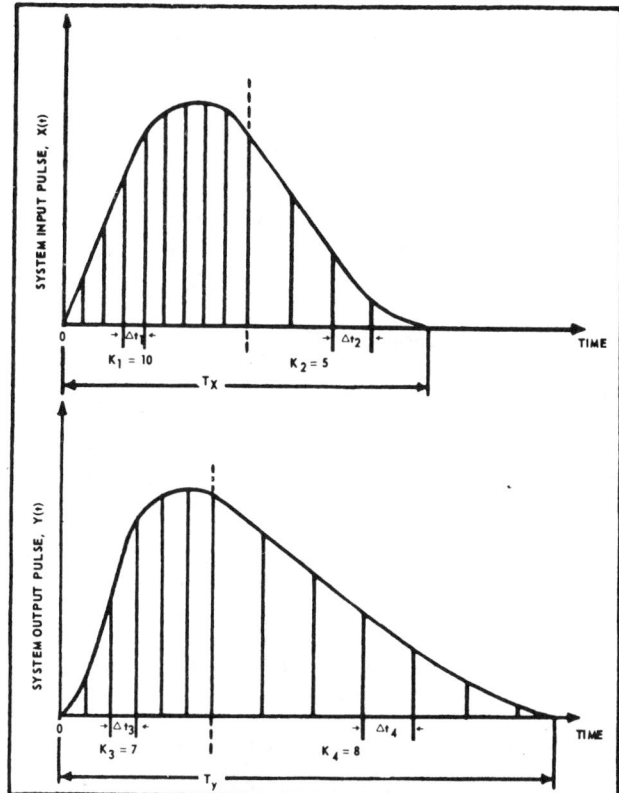

Figure 22. Typical Input and Output Pulses and Their Dissection.

In many cases, as in this illustration, one portion of the curve changes more rapidly than another portion. In such cases it is convenient to read a large number of points in the portion of fastest change, and fewer points when the change is more gradual. The digital computer program mentioned in Section 7.1.2 is written to accept two different increments of time on both the input and output pulses. Thus the program input consists of the input data points, x_0 through x_n; the output data points, y_0 through y_n; the four Δt values, two for x(t) and two for y(t); and the number of points read using Δt.

8.2 Generalized Test Setup

See Section 5.2.

8.3 Test Equipment

8.3.1 Signal Generation

Many devices for generating electrical sine waves are commercially available. A device which can be modified to generate a displaced cosine wave of one period duration at will is recommended for pulse testing. A signal converting device can then be used to obtain a pneumatic or hydraulic signal as the situation requires.

It is recommended that the sine wave generating device meet the following criteria:

(1) The frequency range to be covered will vary (see Section 7.1); however, for process control work, it is recommended that the generator cover frequencies from 0.001 to 20 cps.

(2) The sine wave generator must be capable of holding a given frequency without a great deal of variation.

It is recommended that acceptable performance be defined as shown in Figure 13.

8.3.2 Display Equipment

See Section 5.3.2.

8.3.3 Transducers and Amplifier-Converters

See Section 5.3.3.

8.3.4 Test Equipment Performance Specifications

See Section 5.3.4.

8.4 Testing Procedures

The following procedures are recommended as good practice and will aid in giving reliable results in minimum time.

8.4.1 Blank Run

See Section 5.4.1.

8.4.2 Static Calibration

See Section 5.4.2.

8.4.3 Selecting Test Frequencies

The pulse width is selected as outlined in Section 7.1. The selection of the frequencies used for calculation is arbitrary. The highest frequency used, however, should not be more than three times the frequency whose period is equal to that of the pulse width. A rule of thumb is to have 20 points for three decades of frequency.

9. DATA PRESENTATION — PULSE TYPE

9.1 Pulse Test Data

See Section 6.1.

9.2 Step Test Data

See Section 6.2.

9.3 Supporting Information

See Section 6.3.

10. GLOSSARY

Attenuation - see *Gain*.

Decibel or *dB* - a measure of magnitude ratio; magnitude ratio is dB = 20 $\log_{10}$ (magnitude ratio).

Distortion - deformation of signal shape by device or system to which it is applied.

Gain - ratio of output signal magnitude to input signal magnitude; when less than one this is usually called attenuation.

Impedance, input - the impedance presented by a device or system output element to the input.

Impedance, output - the internal impedance of an output element which limits that element's ability to deliver power.

Linearity - characteristic of a device or system which can be described by a linear differential equation with constant coefficients.

Lissajous figure - pattern on an oscilloscope screen which indicates relative phase and magnitude of sinusoidal signals.

Loading - that system connected to the output of a device, including the transmission network.

Magnitude ratio - ratio of output signal magnitude to input signal magnitude (see also *gain*).

Magnitude signal - peak-to-peak value of signal.

Octave - any group or series of eight.

Phase shift - difference between corresponding points on input and output signal wave shapes, disregarding any difference in magnitude (see Figure 23.)

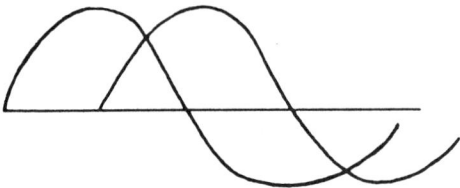

Figure 23. Phase Shift

Phase lag - phase shift when the output lags the input.

Pulse - a variation of a signal whose magnitude is normally constant; this variation is characterized by a rise and a decay, and has a finite duration.

Range - the range of an instrument is the region covered by the *span* and is expressed by stating the two end-scale values.

Response - the change in output of a device in relation to a change of input.

Span - the span of an instrument is the algebraic difference between its end-scale values.

Sine wave - a signal varying with time which can be obtained through projection of a rotating vector of constant magnitude with constant angular velocity on a linear scale (see Figure 24).

Step function signal - signal shown in Figure 25.

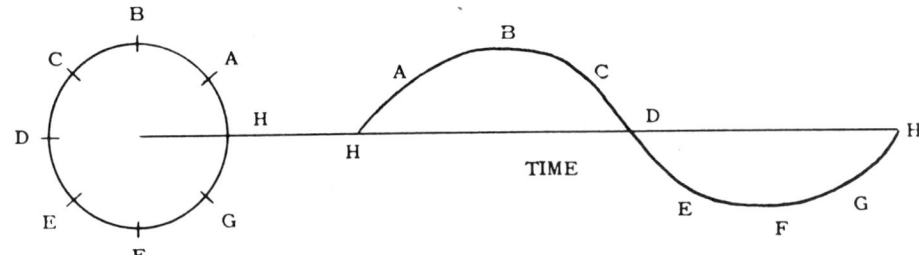

Figure 24. Definition of a Sine Wave

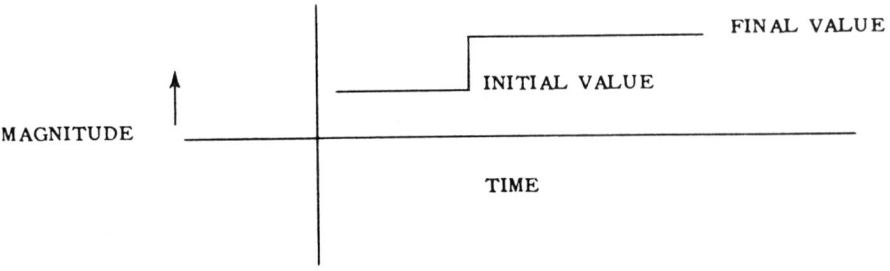

Figure 25. Definition of a Step Signal

11. RECOMMENDED TEST TABULATIONS

The following tabulations, Figures 26, 27 and 28, summarize the number of signal, loading, and parameter combinations recommended for dynamic response testing of each type of device.

Figure 26. Recommended Test Tabulation for Devices with Pneumatic Output

Component	Test No.	Test Signal*	Load** Configuration	Parameter Settings & Other Remarks
Differential Pressure Transmitter	1	Sine Wave or Pulse	A	Test with water legs on each impulse connection per Section 4.4.1. Adjust damping to minimum possible value. Set span adjustment to midpoint of the values possible for the physical components of the transmitter. Separate tests are needed for each change of measuring or feedback element size.
	2	Sine Wave or Pulse	B	
	3	Step Tests	B	
	4	Sine Wave or Pulse	A	Same as above, except dry test (no water legs), per Section 4.4.1.
	5	Sine Wave or Pulse	B	
	6	Step Tests	B	
Temperature Transmitter	1	Sine Wave or Pulse	A	Use unprotected sensing element. Test without rate (derivative) action, or with such action adjusted to give minimum effect. Separate tests are required for each change in measuring elements or internal parts such as are used for changing spans. See Section 4.4.2 for geometry of test system. Additional tests with a particular thermowell may be desirable. Data from such tests should be clearly identified.
	2	Sine Wave or Pulse	B	
	3	Step Tests	B	
Displacer or Float-Type Level Transmitters	1	Sine Wave or Pulse	A	See Section 4.4.3 for discussion of signal generation. For displacer-type instruments, adjust transmitter for 100% proportional band (gain of one) † for smallest displacer size regularly supplied. Separate tests are needed for other displacer sizes. If any other control modes are included, remove or adjust for minimum effect.
	2	Sine Wave or Pulse	B	
	3	Sine Wave or Pulse	C	
	4	Step Tests	C	
Transmitters General (Transducers)	1	Sine Wave or Pulse	A	Adjust parameters per Section 4.5.
	2	Sine Wave or Pulse	B	
	3	Step Tests	B	
Controllers	1	Sine Wave or Pulse	A	Adjust all controller modes (derivative, reset) to minimize or remove their effect. Set proportional band at 100% (gain of one) †. Automatic to manual switching devices must be in circuit for tests, per Section 4.3.
	2	Sine Wave or Pulse	B	
	3	Sine Wave or Pulse	C	
	4	Step Test	C	
	5	Sine Wave or Pulse	B	With controller modes as above, set proportional band at 50% (gain of 2) †
	6	Sine Wave or Pulse	B	With controller modes as above, set proportional band at 200% (gain of 1/2) †
				Tests 7 through 11 are intended to provide a recommended standard method for characterizing the controller parameters by dynamic tests and are optional.
Optional reset or integral action characterization tests	7	Sine Wave or Pulse	E	With 100% proportional band, and minimum or zero derivative, set reset at 1 repeat/min.
	8	Sine Wave or Pulse	E	Same as Test 7, except set reset at maximum repeats/min.
	9	Sine Wave or Pulse	E	Same as Test 7, except set reset at minimum repeats/min.
Optional derivative or rate characterization tests	10	Sine Wave or Pulse ††	B	With 100% proportional band, and minimum or no reset, set derivative for minimum effect.
	11	Sine Wave or Pulse ††	B	Same as Test 10, set derivative at 1 repeat/min.
Devices Designed for Power Output to Pneumatic Actuators	1	Sine Wave or Pulse	B	
	2	Sine Wave or Pulse	D	
	3	Step Test	D	

*As defined in Figures 1 and 10.
**See Section 4.3 for details.
†Actually measured, not based on knob setting.
††See Section 4.5.2 for methods of determining magnitude of test signals.

Figure 27. Recommended Test Tabulation for Devices with Electric Output

Component	Test No.	Test Signal per Figs. 1 and 10.	Parameter Settings & Other Remarks
Differential Pressure Transmitter	1	Sine Wave or Pulse	Test with water legs on each impulse connection per Section 4.4.1. Adjust damping to minimum possible value. Set span adjustment to midpoint of the values possible for the physical components of the transmitter. Separate tests are needed for each change of measuring or feedback element size.
	2	Step Tests	
	3	Sine Wave or Pulse	Same as above, except dry test (no water legs), per Section 4.4.1.
	4	Step Tests	
Temperature Transmitter	1	Sine Wave or Pulse	Use unprotected sensing element. Test without rate (derivative) action, or with such action adjusted to give minimum effect. Separate tests are required for each change in measuring elements or internal parts such as are used for changing spans. See Section 4.4.2 for geometry of test system. Additional tests with a particular thermowell may be desirable. Data from such tests should be clearly identified.
	2	Step Tests	
Displacer or Float-Type Level Transmitters	1	Sine Wave or Pulse	See Section 4.4.3 for discussion of signal generation. For displacer-type instruments, adjust transmitter for 100% proportional band (gain of one)* for smallest displacer size regularly supplied. Separate tests are needed for other displacer sizes. If any other control modes are included, remove or adjust for minimum effect.
	2	Step Tests	
Transmitters General (Transducers)	1	Sine Wave or Pulse	Adjust parameters per Section 4.5.
	2	Step Tests	
Controllers	1	Sine Wave or Pulse	Adjust all controller modes (derivative, reset) to minimize or remove their effect. Set proportional band at 100% (gain of one)*.
	2	Step Tests	
	3	Sine Wave or Pulse	With controller modes as above, set proportional band at 50% (gain of 2)*
	4	Sine Wave or Pulse	With controller modes as above, set proportional band at 200% (gain of 1/2)*
			Tests 5 through 9 are intended to provide a recommended standard method for characterizing the controller parameters by dynamic tests and are optional.
Optional derivative or rate characterization tests	5	Sine Wave** or Pulse	With 100% proportional band, and minimum or no reset, set derivative for minimum effect.
	6	Sine Wave** or Pulse	Ditto, set derivative knob at 1 min.
Optional reset or integral action characterization tests	7	Sine Wave or Pulse	Set up test per Fig. 10 with 100% proportional band, and minimum or zero derivative, set reset at 1 repeat/min.
	8	Sine Wave or Pulse	Ditto, set reset at maximum repeats/min.
	9	Sine Wave or Pulse	Ditto, set reset at minimum repeats/min. possible with knob setting or ability of test equipment.

* Actually measured, not by knob setting.
** See Section 4.5.2 for method of determining test sine wave magnitude.

Figure 28. Recommended Test Tabulation for Closed Loop Final Control Elements and Actuators

SYSTEM BEING TESTED	TEST SIGNALS	PARAMETER SETTINGS	REMARKS
1) Actuator Alone	a) Sine Wave	Adjust gain (or other adjustments available) to give minimum damping consistent with stable operation (usually 3 db down from first sign of instability).	Recommended Test
	b) 10% Step		
	c) 80% Step Per Section 4.7		
	d) Pulse Test Per Sections 4.2 and 4.3	(Instead of Sine Wave)	
2) Actuator on Valve Body	Same as 1)	Tighten valve packing to hold nominal body rating pressure of valve. Adjust gain or other adjustments per 1) above. Run tests with valve body open to atmosphere.	Recommended Test
3) Actuator on Other Final Element(Pump, Rheostat, etc.)	Same as 1)	Set up final element to duplicate as nearly as possible actual operating conditions. Adjust actuator system gain or other adjustments per 1) above.	Recommended Test
4) Actuator on Valve Body at Flowing Conditions	Same as 1)	Same as 2)	Optional Test

NOTES: (1) Describe loading and adjustment of parameters clearly and completely on all data or curves obtained from these tests.
(2) In all cases, measure actuator stem motion as output signal.

12. REFERENCES

1. Rohmann, C. P. and Grogan, E. C., "On the Dynamics of Pneumatic Transmission Lines," *ASME Paper No. 56-SA-1.*

2. Eckman, D. P. and Gess, L., "Pneumatic Transmission of Instrument Readings Over Long Distances," *Honeywell, Inc. Bulletin No. B59-2.*

3. Bradner, M., "Pneumatic Transmission Lag," *ISA Journal*, Vol. 4, July 1949, pp. 618-625.

4. Iberall, A. S., "Attenuation of Oscillatory Pressures in Instrument Lines," *Journal of Research, National Bureau of Standards, Research Paper 2115*, Vol. 45, July 1950.

5. Moise, J. C., "Pneumatic Transmission Lines," *ISA Journal*, Vol. 1, April 1954, pp. 35-50.

6. Humphreys, J. D., "Pressure Sensing Calculations for Aircraft and Guided Missiles," *Instrument Notes, Statham Laboratories, No. 25,* June 1953.

7. Barton, J.R., "A Note on the Evaluation of Designs of Transducers for the Measurement of Dynamic Pressures in Liquid Systems," *Instrument Notes, Statham Laboratories, No. 27,* June 1954.

8. Smith, P. H., "A Transmission Line Calculator," *Electronics*, Vol. 12, January 1939, pp. 27-32.

9. Looney, R., "A Thermal Sine Wave Generator for

Speed of Response Studies," *ASME Paper No. 54-SA-38*, 1954.

10. Higgins, S. P., "A Thermal Sine Wave Apparatus for Testing Industrial Thermometers," *ASME Paper No. 54-SA-20*, 1954.

11. "Method for Presenting the Response of Temperature-Measuring Systems," *Trans. ASME 79, 1851-6 No. 8*, November 1957.

12. Coon, G. A., "Response of Temperature-Sensing Element Analogs," *ASME Transactions*, November 1957.

13. Muller-Girard, O., "The Dynamics of Filled Temperature-Measuring Systems," *ASME Transactions*, Vol. 77, 1955, p. 591.

14. Linahan, T. C., "The Dynamic Response of Industrial Thermometers in Wells," *ASME Transactions*, Vol. 78, No. 4, May 1956, p. 759.

15. Aikman, McMillan, and Morrison, "Static and Dynamic Performance of Sheathed Industrial Thermometers," *Transactions of the Society of Instrument Technology (Great Britain)*, Vol. 5, No. 4, December 1953.

16. Grabbe, E. M., Editor, See forthcoming "Handbook of Automation, Computation, and Control, Vol. III," *John Wiley & Sons*.

17. Bigliano, R. P., "Here's A Way to Measure Pneumatic Component Dynamics," *Control Engineering*, August 1956, pp. 72-77.

18. Hougen, J. O., and R. A. Walsh, "Pulse Testing Method," *Chem. Eng. Progress*, 57, 3, p. 69 (1961).

19. Dreifke, G. E., J. O. Hougen, and G. Mesmer, "Effects of Truncation on Time to Frequency Domain Conversion," *Trans. Instr. Soc. Amer.*, 1, 4, pp. 353-368 (October, 1962).

20. Hougen, J. O., and S. Lees, "Determination of Pneumatic Controller Characteristics by Frequency Response," *Ind. Eng. Chem.*, 48, p. 1053 (1956).

21. Schuder, C. B., and R. C. Binder, "The Response of Pneumatic Transmission Lines to Step Inputs," *Trans. Am. Soc. Mech. Engrs., Jour. of Basic Engineering*, 81, p. 578 (December, 1959).

22. Hougen, J. O., O. R. Martin, and R. A. Walsh, "Dynamic Behavior of Pneumatic Transmission Lines," *Control Eng.*, 10, pp. 114-117 (1963).

23. Guisti, A. L., and J. O. Hougen, "Dynamics of pH Electrodes," *Control Eng.*, 8, pp. 136-140 (April, 1961).

24. Granton, R. L., J. O. Hougen, and G. E. Dreifke, "Steam Analyzer Dynamics," *Control Eng.*, 7, Part I, p. 113 (May, 1960); Part II, p. 104 (July, 1960).

25. Head, F. E., J. O. Hougen, and R. A. Walsh, "Determining the Properties of Continuous Flow Systems by Pulse Excitations," *Automatic and Remote Control*, 4, p. 312, Butterworths Scientific Publications, London, England (1961).

26. Vincent, G. C., J. O. Hougen, and G. E. Dreifke, "Fluid Mixing in Shell and Tube Heat Exchangers," *Chem. Engr. Progress*, 57, 7, p. 48 (July, 1961).

27. Morris, H. J., "Dynamic Response of Shell and Tube Heat Exchangers to Temperature Disturbances," *Automatic and Remote Control*, 4, p. 354, Butterworths Scientific Publications, London, England (1961).

28. Bremer, A., "Field Test Unit for Dynamic Analysis," *Instruments and Automation*, 32, p. 108, (January, 1959).

29. Cowley, P. E. A., "Application of Analog Computers to Measurement of Process Dynamics," *Trans. Am. Soc. Mech. Engrs.*, 79, p. 823 (1957).

30. Hougen, J. O., "Experiences and Experiments with Process Dynamics," *Chem. Engr. Progress, Monograph Series #4*, Vol. 60 (1964).

31. Smith, G. A., and W. C. Triplett, "Experimental Flight Methods for Evaluating Frequency-Response Characteristics of Aircraft," *Trans. Am. Soc. Mech. Engrs.*, p. 1383 (November, 1954).

32. Lees, Sidney, "Interpreting Dynamic Measurements of Physical Systems," *Trans. Amer. Soc. Mech. Engrs.*, 80, Part I, p. 833; Part II, p. 843 (May, 1958).

33. Clements, William C., Jr., and Karl B. Schnelle, Jr., "Pulse Testing for Dynamic Analysis," *I&EC Process Design and Development*, 2, pp. 94-102 (April, 1963).

34. Banham, James W., Jr., "Obtain Process Dynamics by Pulse Testing," *Control Engineering*, p. 83 (April, 1965).

35. Eckman, Donald P., "Automatic Process Control," John Wiley, 1958, pp. 197-202.

36. Harrison, J. W., Jr., "Comparison of Control Valve Frequency Response Under Wet and Dry Conditions." Presented at Sixth Annual Southeastern Conference and Exhibit of the Instrument Society of America at Pensacola, Florida, April 27-29, 1960.

13. INDEX

INSTRUMENT SOCIETY of AMERICA
Research Triangle Park, North Carolina

ANSI/ISA-RP31.1-1977
Approved 1977

Recommended Practice

Specification, Installation, and Calibration of Turbine Flowmeters

Instrument Society of America

ISBN 0-87664-371-3

ISA-RP31.1   Specification, Installation, and Calibration
                    of Turbine Flowmeters

INSTRUMENT SOCIETY OF AMERICA
67 Alexander Drive
P.O. Box 12277
Research Triangle Park, North Carolina 27709

PREFACE

(This Preface is included for information purposes and is not a part of Recommended Practice RP31.1).

This Recommended Practice has been prepared as a part of the service of the Instrument Society of America toward a goal of uniformity in the field of instrumentation. To be of real value this document should not be static but should be subjected to periodic review. Toward this end the Society welcomes all comments and criticisms, and asks that they be addressed to the Standards and Practices Board Secretary, Instrument Society of America, 67 Alexander Drive, P.O. Box 12277, Research Triangle Park, North Carolina 27709.

The ISA Standards and Practices Department is aware of the growing need for attention to the metric system of units in general, and the International System of Units (SI) in particular, in the preparation of instrumentation standards. The Department is further aware of the benefits to USA users of ISA Standards of incorporating suitable references to the SI (and the metric system) in their business and professional dealings with other countries. Towards this end this Department will endeavor to introduce SI and SI-acceptable metric units as optional alternatives to English units in all new and revised standards to the greatest extent possible. The Standard for Metric Practice, which has been published by the American Society for Testing and Materials as ANSI-Z210.1 (ASTM E 380-76; IEEE 268-1975) and future revisions, will be the reference guide for definitions, symbols, abbreviations and conversion factors.

The ISA Standards Committee on Turbine Flowmeters, SP31, operates within the ISA Standards and Practices Department, L. N. Combs, Vice-President. The men listed below serve as members of this committee:

M. J. Ford, Chairman	Pratt & Whitney Aircraft
W. J. Alspach	National Bureau of Standards
J. R. Babbitt	Daniel Industries
K. R. Benson	National Bureau of Standards
L. P. Emerson	The Foxboro Company
J. L. Hayes	Navy Plant Representative Office
J. Kopp	Fischer & Porter Company
R. McWhorter	Brooks Instrument Division, Emerson Electric
M. H. November	ITT-Barton
W. N. Seward	American Petroleum Institute
K. F. Wacker	Allison Div., General Motors Corporation

The assistance of others who aided in the preparation of RP31.1 by answering correspondence, attending many committee meetings, and in other ways is gratefully acknowledged. In particular, the assistance of Robert L. Galley, previously of General Dynamics, McDonald-Douglas, Bechtel, United Nations, and presently with Hallanger Engineers and his committee who prepared ISA Recommended Practice RP31.1, and of George J. Lyons of Pratt & Whitney Aircraft and his committee who prepared drafts of Recommended Practice RP31.3 (not published) is acknowledged.

In addition to the SP31 committee members, the following have served as a Board of Review for this Document.

D. W. Baker	National Bureau of Standards
J. W. Bauman	Dow Chemical Corp.
R. C. Butler	Bechtel Corp.
T. K. Conlan	E. I. du Pont de Nemours
S. L. Davis	E. I. du Pont de Nemours
W. J. Dreier	Appleton, Wisconsin
G. E. Dreifke	Union Electric Company
R. E. Gorton	Pratt & Whitney Aircraft
R. E. Griffin	U. S. Bureau of Mines
J. A. Harmon	White Sands Missile Range
A. E. Hayes	Consultant
T. J. Higgins	University of Wisconsin
W. R. Hytinen	Stearns Roger Corp.
T. S. Imsland	Fisher Controls Company
J. J. Jefferson	Tennessee Eastman Company
R. H. Joyce	Standard Oil Company
J. V. Kelly	LTV Aerospace Corp.
F. B. Kroeger	Ohio State University
R. Loewe	Sargent & Lundy Engineers
A. F. Marks	Bechtel Corp.
J. H. Mitchell	Cities Service Gas Company
P. A. Muller	Southern Nuclear Engineering Co.

TABLE OF CONTENTS

LIST OF FIGURES

1 PURPOSE

This document establishes the following for turbine flowmeters, especially those 2-inch diameter and smaller.

1.1 Recommended minimum information to be specified in ordering.

1.2 Recommended acceptance and qualification test methods including calibration techniques.

1.3 Uniform terminology and drawing symbols.

1.4 Recommended installation techniques.

2 SCOPE

2.1 These recommendations cover volumetric turbine flowmeters having an electrical output. The flowmeters are used to measure either flow rate or quantity of liquids, and can be either self-generating or modulating. Their use in cryogenic fluids will require special considerations, which are not fully described herein.

3 DRAWING SYMBOLS

3.1 The flow diagram symbol for a turbine flowmeter is shown below, in accordance with ISA-S5.1

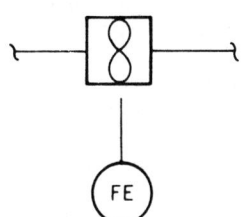

3.2 The electrical diagram symbol for a turbine flowmeter is shown below.

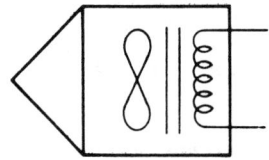

4 SPECIFICATIONS

Terminology used is defined either in ISA-S37.1 or 4.3 Specific Terminology. An asterisk appears before these terms defined in ISA-S37.1; the terms defined in 4.3 Specific Terminology appear in italics.

4.1 Design Characteristics

4.1.1 The following mechanical design characteristics shall be specified.

  (1) *Measured Fluids

    The liquids to be in contact with the wetted parts e.g. JP-4 fuel, hydrochloric acid, etc.

  (2) Configuration and Dimensions

    (a) For flared tubing (MS 33656) the nominal tube size, overall length, and equivalent fittings.

    (b) For flanged types the nominal pipe size, and length between flanges.

    (c) For male pipe-threaded flowmeters the nominal pipe size and overall length.

  (3) Mountings and Mounting Dimensions

    Unless the process connection serves as a mounting, the outline drawing shall indicate the method of mounting with hole sizes, centers and other pertinent dimensions, including thread specifications for threaded holes, if used.

  (4) The weight of the flowmeter.

  (5) The following information shall be permanently marked on the flowmeter housing.

    (a) Manufacturer
    (b) Part or Model No.
    (c) Serial No.
    (d) Flow Direction
    (e) Nominal Tube or Pipe Size

4.1.2 Supplemental Mechanical Design Characteristics

The following mechanical design characteristics shall be specified at the option of the purchaser.

  (1) Minimum and Maximum Temperatures

    The minimum and maximum temperatures of the measured liquid, and of the surrounding environment.

  (2) Pressure

    The maximum pressure of the measured liquid.

  (3) Speed

    The maximum output frequency which the flowmeter can produce without physical damage.

4.1.3 Required Electrical Design Characteristics

The following electrical characteristics shall be specified.

  (1) *Excitation (Modulating type only).

    Expressed as " _____volts at_____ hertz."

  (2) *Electrical Output

    (a) Output Voltage

    Expressed as " _____ volts" minimum peak-to-peak at any flow within the operating range.

    (b) Frequency at maximum rated flow expressed as " _____ Hz".

4.2 Performance Characteristics

The pertinent performance characteristics of turbine flowmeters should be specified in the order shown. Unless otherwise stated, they apply at *Room Conditions, as defined in ISA-S37.1 and are expressed in SI units. (Equivalent units may be substituted).

4.2.1 *Range

Expressed as "_____ to _____ meter^3 per second".

4.2.2 Linear Range

Expressed as "_____ to _____ meter^3 per second".

4.2.3 *Sensitivity

The sensitivity (K factor) is expressed as "_____ cycles per meter^3". Often used as $\overline{K}$, which is the average sensitivity across the flow range of interest to the user. Average is used here in the sense of (K_{max} + K_{min})/2.

4.2.4 *Linearity

The normal presentation of the calibration data is in cycles per meter^3 vs. frequency, which is the slope of a curve of flow vs. frequency. Linearity is then expressed as "± _____ % of the average sensitivity (K factor). (See Linearity in 4.3 Specific Terminology)

4.2.5 Pressure Drop

The pressure drop across the flowmeter at maximum rated flow, expressed as "_____ differential pascals at _____ meter^3 per second" when used with specific measured liquid.

4.3 Specific Terminology

Actual Flow—Refers to the actual volume of liquid passing through the flowmeter in a unit time as computed by applying all necessary corrections for the effects of temperature, pressure, air buoyancy, etc. to the corresponding readings indicated by the calibrator.

Air Buoyancy—The lifting effect or buoyancy of the ambient air which acts during a "weighing" procedure with open gravimetric calibrations. This is caused by displacement of air from the measuring vessel during the calibration run. The standard air (50% R.H.) for correcting to "weights" in vacuum has a density of 1.217 kg/m^3 at 288.7 K and 1.013 250E+05 Pa. When "weighings" are made against "weights", the buoyancy force on these must also be considered. For brass "weights" the net effect of air buoyancy in air at standard conditions is about 0.015%. Exact values can be determined by procedures outlined in paragraphs 3059 and 3060 of API Standard 1101, 1960 Editions, and NBS Handbook 77, Volume III, Pages 671-682.

Apparent Flow—The uncorrected volume flow as indicated by the calibrator.

Back Pressure—The absolute pressure level as measured four pipe-diameters downstream from the turbine flowmeter under operating conditions, expressed in pascals.

Calibration Curve—A graph of the performance of a turbine flowmeter, showing sensitivity as the ordinate and volume flow, flowmeter frequency, or frequency divided by kinematic viscosity as the abscissa, for a liquid of specified density, viscosity, and temperature.

Calibration System—A complete system consisting of liquid storage, pumps, and filters; flow, pressure, and temperature controls; the quantity measuring apparatus; and the associated electronic instruments used to calibrate turbine flowmeters.

Correlation Check—A procedure whereby the performance and accuracy of a calibration system is checked against another calibration system using "Master Flowmeters" as the standards.

Density—The mass of a unit volume of a liquid at a specified temperature. The units shall be stated, such as kilograms per meter^3. The form of expression shall be: Density _____ kg/m^3 at _____ Kelvin.

Dynamic Calibration—A calibration procedure in which the quantity of liquid is measured while liquid is flowing into or out of the measuring vessel.

Flow—the rate of flow of a liquid expressed in volume units per unit of time. Examples are: meter^3/second (m^3/s).

Flow Straightener—A supplementary length of straight pipe or tube, containing straightening vanes or the equivalent, which is installed directly upstream of the turbine flowmeter for the purpose of eliminating swirl from the fluid entering the flowmeter.

Gravimetric—A descriptive term used to designate an instrument or procedure in which gravitational forces are utilized. However, the results or indications of such procedures are not necessarily influenced by the magnitude of the acceleration of gravity. See discussion on mass, weight, and weighing (5.4.1 (1) Mass, Weight, and Weighing).

Linear range of a turbine flowmeter—the flow range over which the output frequency is proportional to flow (constant K factor) within the limits of linearity specified.

Linearity of a turbine flowmeter—the maximum percentage deviation from the average sensitivity ($\overline{K}$) across the linear range.

Master Flowmeter—Flowmeter used as an inter-laboratory standard in correlation checks of calibration systems.

Measuring Vessel—The container in which the liquid metered by the turbine flowmeter during calibration interval is collected and measured. In a direct-gravimetric calibration system, this is a tank on a weigh scale and the exact dimensions are not significant. In indirect-gravimetric systems and volumetric systems the cross-sectional area or actual volume, respectively, must be known to a precision compatible with the desired accuracy of calibration.

Pressure drop—the differential pressure in pascals at a maximum linear flow measured between points four pipe diameters upstream and four pipe diameters downstream from its ends, using a specified liquid, and using pipe size matching the fittings provided.

The *sensitivity of a turbine flowmeter is designated by the letter "K" and is expressed in cycles per meter^3, under the following specified conditions:

(1) **Calibration Liquid**

(a) Density (kg/m^3)
(b) Viscosity (m^2/s)
(c) Downstream Temperature (K)
(d) Back pressure (Pa abs)
(e) Flow (m^3/s)

(2) **Line Configuration**

(a) Length of straight line upstream
(b) Length of straight line downstream
(c) Configuration of flow straightener

Reference Flowmeter—Flowmeter used as a transfer standard for in-system and comparison calibrations of turbine flowmeters.

Static Calibration—A calibration procedure during which the quantity of liquid is measured while the liquid is not flowing into or out of the measuring vessel.

Swirl—A qualitative term, describing tangential motions of liquid flow in a pipe or tube.

Thermal Expansion—The increase in a volume of liquid caused by an increase in temperature.

Turbine, Turbine Supports(s), Housing and Transduction Coils—the preferred names for the major parts of a turbine flowmeter.

Turbine flowmeter with an electrical output—a flow measuring device in which the action of the entire liquid stream turns a bladed turbine at a speed nominally proportional to the volume flow, and which generates or modulates an output signal at a frequency proportional to the turbine speed.

Two-Phase—A fluid state consisting of a mixture of liquid with gas or vapors.

Vapor Pressure—The pressure of a vapor corresponding to a given temperature at which the liquid and vapor are in equilibrium. Vapor pressure increases with temperature.

Vapor Pressure, Reid—The vapor pressure of a liquid at 100°F (311K) as determined by ASTM Designation D 323-58, "Standard Method of Test for Vapor Pressure of Petroleum Products (Reid Method)."

Viscosity, Absolute—The property by which a fluid in motion offers resistance to shear. Usually expressed as newton-seconds/meter^2.

Viscosity, Kinematic—The ratio of absolute viscosity to density. The SI unit is the meter^2/s.

Weigh-Scale—A device for determining either the mass or

the weight of a body depending upon the apparatus and procedure employed.

Weighing—The process of determining either the mass or weight of a body depending upon the apparatus and procedure employed.

Weight—The force with which a body is attracted by gravity. The newton is the unit force in this Standard.

Weights—Reference units of mass such as counterpoise "weights" used with lever balances and dead "weights" used in calibrating balances, scales, and pressure gauges.

4.4 Tabulated Characteristics vs. Test Requirements

The table on the following page is intended for use as a quick reference for design and performance characteristics and tests of their proper verification as contained in this Standard.

5 INDIVIDUAL ACCEPTANCE TESTS AND CALIBRATIONS

5.1 Calibration Methods

Methods suitable for the calibration of turbine flowmeters may be classified as:

(1) Gravimetric (Direct or Indirect)
(2) Volumetric
(3) Comparison

Each of these methods has advantages and disadvantages depending upon the liquid being metered and the type of operation. The gravimetric methods require that the density of the liquid be determined accurately to provide a basis for converting mass to volume. The effect of the gas added to the "weigh" tank in closed gravimetric calibrators must also be considered. The buoyancy factor for air, in open gravimetric calibrators, as a function of liquid density is shown in Figure 1. (See also 4.3 Definitions, air buoyancy). The volumetric method is more direct in that conversions from mass to volume are not required.

The calibrator may be of either the open-type for use with low vapor pressure liquids only, or the closed-type in which a back pressure greater than atmospheric is maintained to prevent liquid loss from the measuring vessel by evaporation. Calibration methods are further classified as Static or Dynamic.

5.1.1 Static Method

In this method, "weighing" or measurement of volume occurs only while the liquid is not flowing into or out of the measurement vessel. This method is capable of high accuracy under proper conditions and should include static checks against reference units of mass or volume traceable to the National Bureau of Standards.

5.1.2 Dynamic Method

In this method, the measurement of volume or mass occurs while the liquid is flowing into or out of the measuring vessel. Although more suitable for many applications, it may involve dynamic errors which

| Characteristics | Paragraph | Design Characteristic | | Verified During | |
		Basic	Supplemental	Individual Acceptance Test	Qualification Test
Configuration, Dimensions, Mountings and Mounting Dimensions	4.1.1 (2) through 4.1.1 (3)	x		5.3.1	Special Test
Weight	4.1.1 (4)				6.2
Identification	4.1.1 (5)	x		5.3.1	
Materials in contact with measured fluid	4.1.1 (1)		x		5.3.1
Minimum & Maximum Temperatures	4.1.2 (1)		x		6.3
Pressure	4.1.2 (2)		x		6.4
Speed	4.1.2 (3)		x		6.5
Output Voltage	4.1.3 (2) (a)	x		5.3.3	
Frequency	4.1.3 (2) (b)	x		5.3.4	
Linearity	4.2.4	x		5.3.4	
Sensitivity (K factor)	4.2.3	x		5.3.4	
Linear Range	4.2.2	x		5.3.4	

cannot be detected through static checks with reference units of mass or volume. Therefore, it is important that each new dynamic calibrator of a different type or size be checked carefully by correlation or other suitable means to prove that significant dynamic errors do not exist.

Two procedures for conducting a turbine flowmeter calibration are the running start-and-stop and the standing start-and-stop. The procedure which more closely duplicates the type of service anticipated in the application of the flowmeter should be selected whenever possible.

5.1.3 Running Start-and-Stop

The running start-and-stop requires that a reasonably constant flow be maintained through the flowmeter prior to, during, and immediately after the collection of the liquid in the measuring vessel. This is accomplished by using a stream diverter, whose motion is synchronized with the starting and stopping of the electronic counter.

5.1.4 Standing Start-and-Stop

This procedure requires that a "no-flow" condition exist at the flowmeter prior to the beginning and at the end of the calibration run and that at least 95% of the total throughput be at the desired flow. This is accomplished by using solenoid valves synchronized with the action of the electronic counter.

5.2 General Precautions

General precautions which should be considered in the design of a calibration system for turbine flowmeter include the following:

5.2.1 *System Piping*

The piping between the flowmeter and the measurement vessel should be short, of small volume com-

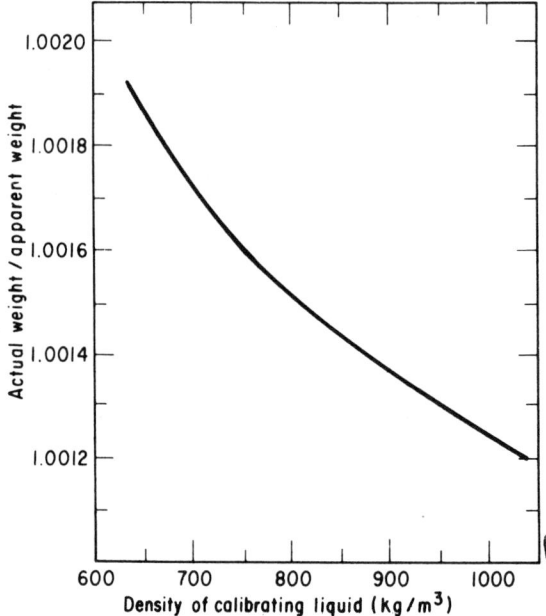

Fig. 1. Air Buoyancy Factor for Standard Air of (1,217 kg/m³)

pared to the measured volume and designed to eliminate all air, vapor, and temperature gradients. It should be constructed to assure that all of the liquid and only that liquid passing through the flowmeter is measured.

5.2.2 Throttling or Flow Control

Throttling or flow control valves should be located downstream of the flowmeter to reduce the possibility that two-phase flow may occur within the flowmeter under test. When this is not practical, a back-pressure regulator or similar device should be installed downstream to maintain the required back pressure.

5.2.3 Leak Indicators

Positive methods, visual if possible, should be provided to assure that shut-off valve action is positive and that no leakage occurs during the calibration interval.

5.2.4 Measuring Vessel Capacity

The minimum permissible capacity of the measuring vessel is dependent upon the precision required of the calibration procedure and the resolution of the readout and the flowmeter under test.

5.2.5 The liquid used in performing a calibration should be the same as that with which the flowmeter is to be used in service, and the operating conditions should be duplicated.

5.2.6 A substitute calibration medium may be employed when it is impractical to use the operating liquid. The kinematic viscosity and specific gravity of this liquid should be within 10% of those of the operating liquid. The lubricity of a liquid cannot be as well defined as viscosity and specific gravity, but this parameter should be considered when using substitutes, and should be duplicated as closely as possible.

5.2.7 Filtration shall be provided to protect the flowmeter against damage or malfunction from foreign matter. The degree of filtration required is a function of flowmeter size. A 50 micron or less filter should be employed for general usage on a calibration stand where various size flowmeters are to be calibrated.

5.2.8 Install the flowmeter as indicated by the flow direction arrow marked on the flowmeter.

5.2.9 If flow straighteners or straight sections of pipe are used as an integral part of the flowmeter service installation, the calibrations shall be performed with the same configuration. If the flowmeter must be used immediately downstream of pipe elbows or other swirl-producing pipe fittings, the calibration should be performed with the same plumbing configuration, but this arrangement does not guarantee a good calibration.

5.2.10 The flowmeter shall normally be calibrated in a horizontal position with the transduction element vertically upward since this is the usual attitude for service installations. However, when the service installation is other than horizontal, a difference in axial thrust balance may cause a change in the calibration factor. The transduction element orientation may also cause an error due to the relationship of magnetic drag and gravitational forces in some types of flowmeter. If these effects are not known to be negligible for the flowmeter being used the unit should be calibrated in the attitude in which it will be installed. When a flowmeter is equipped with two or more transduction elements, the flowmeter must be calibrated with all such elements installed.

5.3 Calibration and Test Procedure

Results obtained during the calibration and testing shall be recorded on a data sheet in 7 DATA PRESENTATION. Calibration and Testing shall be performed under *room conditions as defined in ISA-S37.1 unless otherwise specified.

The definitive paragraph under 4 SPECIFICATIONS is listed beside each of the parameters for which the test results are to be compared.

5.3.1 The flowmeter shall be inspected visually for applicable mechanical characteristics of 4.1.1 Required Mechanical Design Characteristics:

Configuration and Dimensions	4.1.1(2)
Mounting and Mounting Dimensions	4.1.1(3)
Identification	4.1.1(5)

By the use of special equipment or by formal verification of production methods and materials used, it can be determined that the materials are compatible with the measure liquids specified in 4.1.1(1).

5.3.2 The flowmeter shall be "run-in" for a period of at least five minutes at a reasonable flow prior to calibration.

5.3.3 During the run-in period, the peak voltage output shall be measured and recorded at the rated minimum and maximum flow. The wave shape of the output signal shall also be observed on a cathode-ray oscilloscope to check flowmeter for malfunctions.

5.3.4 The number of calibration points shall not be less than five and should include the minimum and maximum flow as specified by the manufacturer.

The number of runs at each calibration point shall not be less than two and shall be taken with the flow increasing and decreasing. The sensitivity, linearity, and linear range are determined from this data.

5.3.5 The back pressure shall be measured and should preclude a change in the calibration factor due to two-phase conditions. This pressure shall be measured four pipe-diameters downstream of the flowmeter. The required back pressure may be determined by specific tests, (8.3.1 Liquid Pressure) but in the absence of the exact data it may be set at a pressure equal to the sum of the vapor pressure of the liquid at the operating temperature plus three times the measured pressure drop across the flowmeter.

5.3.6 The temperature of the calibrating liquid at the

flowmeter should be measured approximately four pipe-diameters downstream. If it is necessary to install the temperature sensor upstream of the flowmeter, it should be installed one pipe-diameter upstream of the straightening vanes of a supplementary flow straightener. In all installations, the temperature sensor must be immersed to a sufficient depth to minimize thermal conduction error.

5.3.7 The total number of cycles accumulated for each calibration point is dictated by the flow measurement accuracy requirement. Since the usual electronic counter has an inherent error of $\pm$ 1 cycle, a sufficient number of cycles should be accumulated to make that error negligible.

5.3.8 The total count method does not require that flow be maintained absolutely constant when calibrating in the region in which the calibration factor is essentially independent of flow. Variations in flow of $\pm$ 4% should not introduce significant error. However, in the laminar and transition regions, the calibration factor is affected significantly by both flow and liquid viscosity. Thus, in calibrating in these regions, flow should be maintained constant to $\pm$ 1% or better; or the duration of the calibration run as well as the total pulse count should be measured, so that the exact average frequency existing during the run can be determined.

5.3.9 All gravimetric calibration methods require an accurate basis for converting mass to volume. The density of the liquid at the flowmeter temperature and pressure, should be determined to an uncertainty of $\pm$ 0.05% or less. The effect of air buoyancy must also be considered (*Air Buoyancy* in 4.3 Specific Terminology).

5.3.10 Correlation of Calibration System with NBS

(1) The National Bureau of Standards, or suitable calibration facilities which maintain current correlation checks with NBS, may be used as reference calibrating laboratories. To promote flow calibration standardization and agreement, the National Bureau of Standards maintains facilities for the calibration of flowmeters at the following maximum flows.

Hydraulic & Lubricating Oils	30 GPM	(1.9 E-03 m^3/s)
MIL-F-7024A, Type II	200 GPM	(1.3 E-02 m^3/s)
Water	9000 GPM	(5.7 E-01 m^3/s)

(2) Correlation checks shall be made using flowmeters of proven repeatability as transfer standards. These are termed "*Master Flowmeters*" as defined above, and shall have supplementary flow straighteners connected directly upstream.

(3) Identical calibration liquids and flow shall be agreed upon and used at both the reference facility and the user's calibration facility during the correlation check.

(4) To the extent that the reference facility permits, it is suggested that no less than three nor more than five flows covering the full range of each measuring vessel in the user's system be selected as calibration points.

(5) The number of test runs at each calibration point shall be 8 to 10. The allowable standard deviation of the repeat observations shall be set by agreement between the user and the reference facility.

(6) The *Master Flowmeters* may be used to check the errors of the user's calibration facility periodically as necessary and must be submitted to the reference facility for recalibration when significant shifts in their performance are suspected.

5.4 Specific Calibration Systems

5.4.1 Gravimetric Systems

(1) Mass, Weight, and "Weighing"

In the calibration and application of liquid flowmeters, it is essential that a definite distinction be made between mass and weight. This is especially true in the aeronautical and aerospace industries, with their exacting requirements, where operations are frequently conducted under conditions in which the acceleration of gravity does not have a constant value. Mass, as used herein is a direct measure of the quantity of matter. Weight will be used only as a measure of the force with which a body is attracted by the acceleration of gravity, never as a measure of the quantity of matter.

In this document, the selected unit for mass is the kilogram. This unit is used as a direct measure of the exact quantity of matter in a body. So long as no material is added to nor taken from the body its mass remains constant and is independent of all ambient conditions including the acceleration of gravity.

The selected unit of force herein is the newton. Force will only be used as a measure of pressure or weight, never as a measure of the quantity of matter. The use of the local gravitational conversion factor will be required in computing weights as:

$$\text{Weight} = kg.g = \text{newtons}$$

where kg is mass and g is the local acceleration of gravity, "Weighing" is the process of determining either the mass or weight of a body depending on the apparatus employed. Weighing on a lever balance is a comparison between an unknown mass and selected standards of mass (commonly called "weights"). Effects of variations in the acceleration of gravity are self-negating and the mass, not weight, of the unknown is determined directly. In precise work spring or load cells are usually calibrated at their place of installation with standard units of mass. Thus, these scales also read out directly in mass if they have been so calibrated. See NBS Handbook 77, Volume III, "Design and

Test of Standards of Mass," page 615-706 for a detailed discussion of Mass and Weight, and Air Buoyancy.

(2) Component Errors

The individual components which are suggested in this section are based on the premise that the uncertainty of calibration is to be no greater than 0.5% of flow. When the desired uncertainty differs materially from this, the individual components should be modified accordingly.

(a) "Weights"

Scale counterpoise "weights" for the gravimetric calibrators shall conform to the tolerances for Class C Commercial Test Weights as stated in Circular 3 appearing in Volume III, Handbook 77 of the National Bureau of Standards, the acceptance tolerances (applicable to new or newly adjusted "weights") being one-half the tabular tolerances.

(b) Weigh Scale

The weigh scale shall be installed and its ratio and sensibility reciprocal checked in accordance with NBS Handbook 44 - 2nd Edition. The ratio test shall be performed with certified test weights applied in approximately equal increments throughout the nominal capacity range of the scale under conditions of both increasing and decreasing loads. It is suggested that the performance of the scale be such that the uncertainty in weighing will not exceed ± 0.05% of the smallest net batch sample to be collected during a calibration run.

(c) Pressure Sensing Devices (Indirect Gravimetric Calibrator)

The intervals between electrical contacts in the manometer type pressure sensing instrument shall be measured by means of shadowgraph, cathetometer or other optical techniques.

Consideration must be given in the application of these techniques to the possibility of error introduced by image distortion when viewing the contacts through the transparent manometer wall. The measurements shall be made in a temperature controlled area, and at the same temperature as that to which the manometer is normally exposed. Other pressure sensing devices such as differential pressure transducers shall be calibrated at the location of the standpipe, with precision "Deadweight" testers omitting corrections for the local value of g normally applied, so that the standpipe will measure in units of mass rather than weight. Further, they shall be temperature compensated or temperature controlled. The uncertainty in pressure measurement should not exceed ± 0.05% of the smallest increment of pressure used in calibration run.

(d) Temperature Measurement

The instrument used for measurement of liquid temperature at the flowmeter shall have sufficient sensitivity and accuracy so that this temperature measurement error will not affect the density determination by more than ± 0.025%.

(e) Density Determination

The hydrometer, pycnometer, or other density determining instrument shall have sufficient sensitivity and accuracy to determine liquid density to within ± 0.05% of actual. The temperature of the liquid at the time of density determination must be precisely known so that proper corrections can be applied to obtain the density at the operating temperature of the flowmeter.

(f) Viscosity Determination

Viscosity measurement uncertainties within ± 5% of actual are adequate when calibrating in a flow region in which the calibration factor is essentially independent of flow, otherwise the viscosity should be determined to an uncertainty of ± 2% of value.

(g) Pressure Measurement

Pressure measuring instrumentation used to determine operating pressure level of the flowmeter and the pressure drop across it should be within ± 5% of actual when calibrating with conventional liquids at pressure levels where compressibility is not a factor. When compressibility is significant the pressure measurement shall be made with sufficient accuracy so that this measurement uncertainty will not affect the density determination by more than ± 0.025%.

(3) Typical Flow Diagrams

Figures 2 through 7 show the operation of typical calibration systems. Readings from all systems must be corrected for air buoyancy and the further precautions described above must be observed in each case.

In the static methods, the flow is first adjusted to a suitable value with the flow regulator (FCV). After draining the weigh tank and obtaining the tare weight, the diverter or solenoid valve is opened for an appropriate time, during which the counter records the flowmeter pulses.

In the dynamic methods, the counter is actuated by the weighing system (WI, WC), operating between selected low and high weight settings. Operation must be compensated for dynamic effects of liquid entering or leaving the weigh tank, and for any inertia in operation of weighing device. The pressurized system must also be corrected for the weight of the pressurizing gas.

In the indirect gravimetric calibrator, Figure 6, gravity corrections need not be applied providing that the mass of the measured liquid is balanced by the mass of mercury in the manometer used to

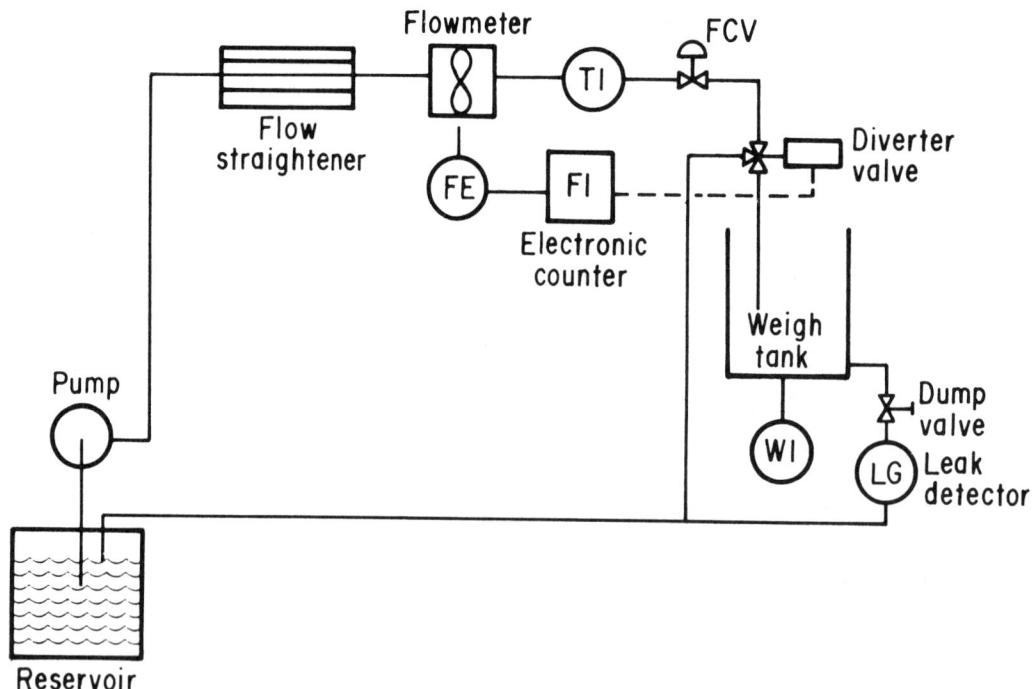

Fig. 2. Direct Gravimetric, Open, Static
Calibrator — Running Start — Stop

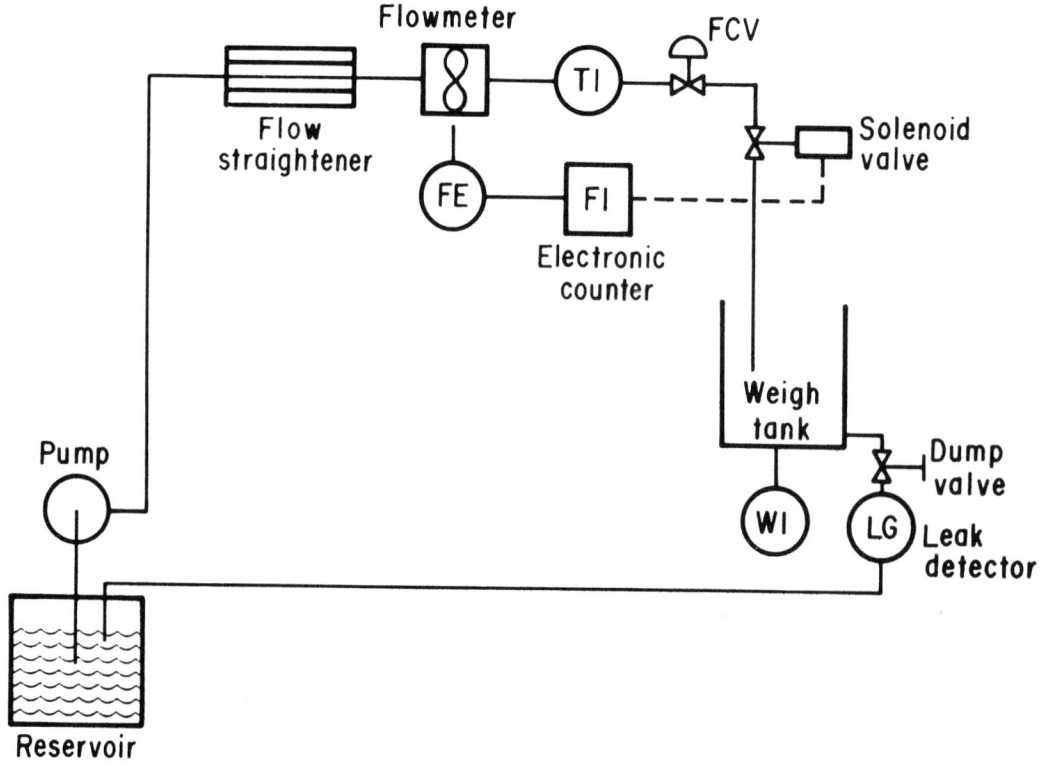

Fig. 3. Direct Gravimetric, Open, Static Calibrator
— Standing Start — Stop

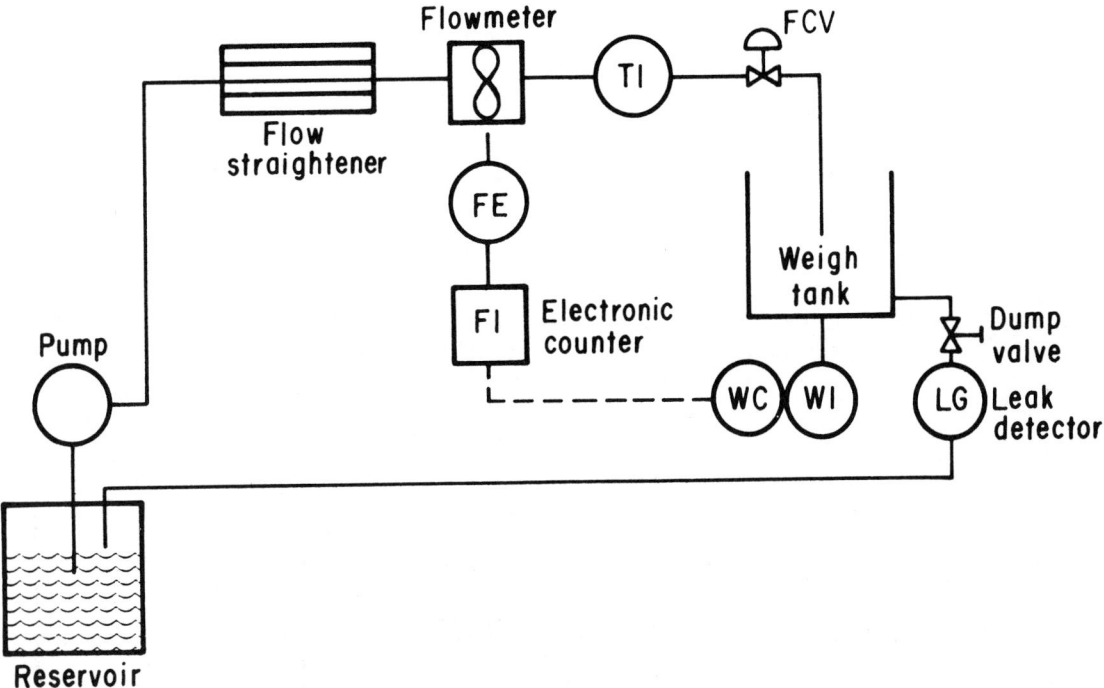

Fig. 4. Direct Gravimetric, Open,
Dynamic Calibrator

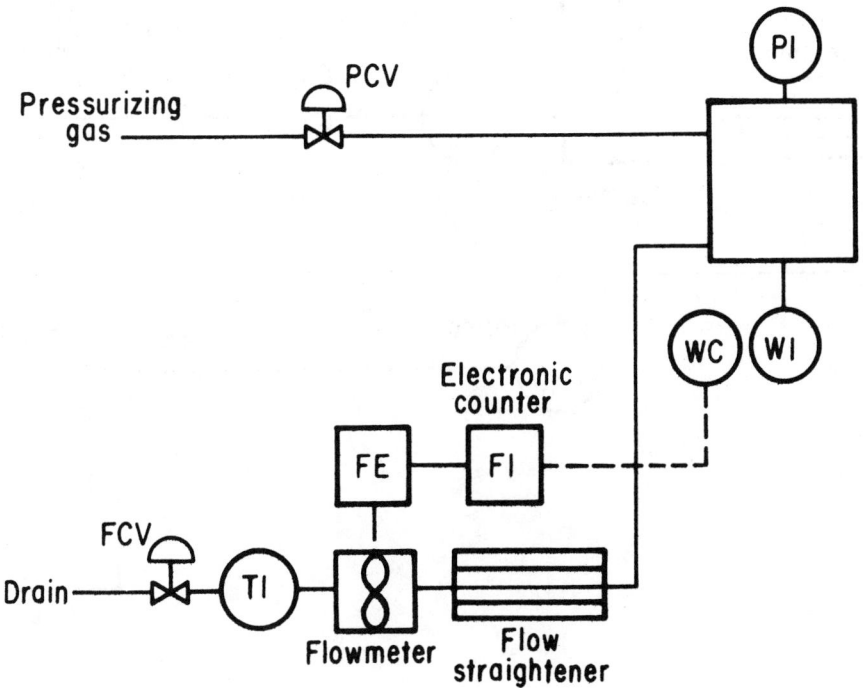

Fig. 5. Direct Gravemetric, Closed Dynamic Calibrator

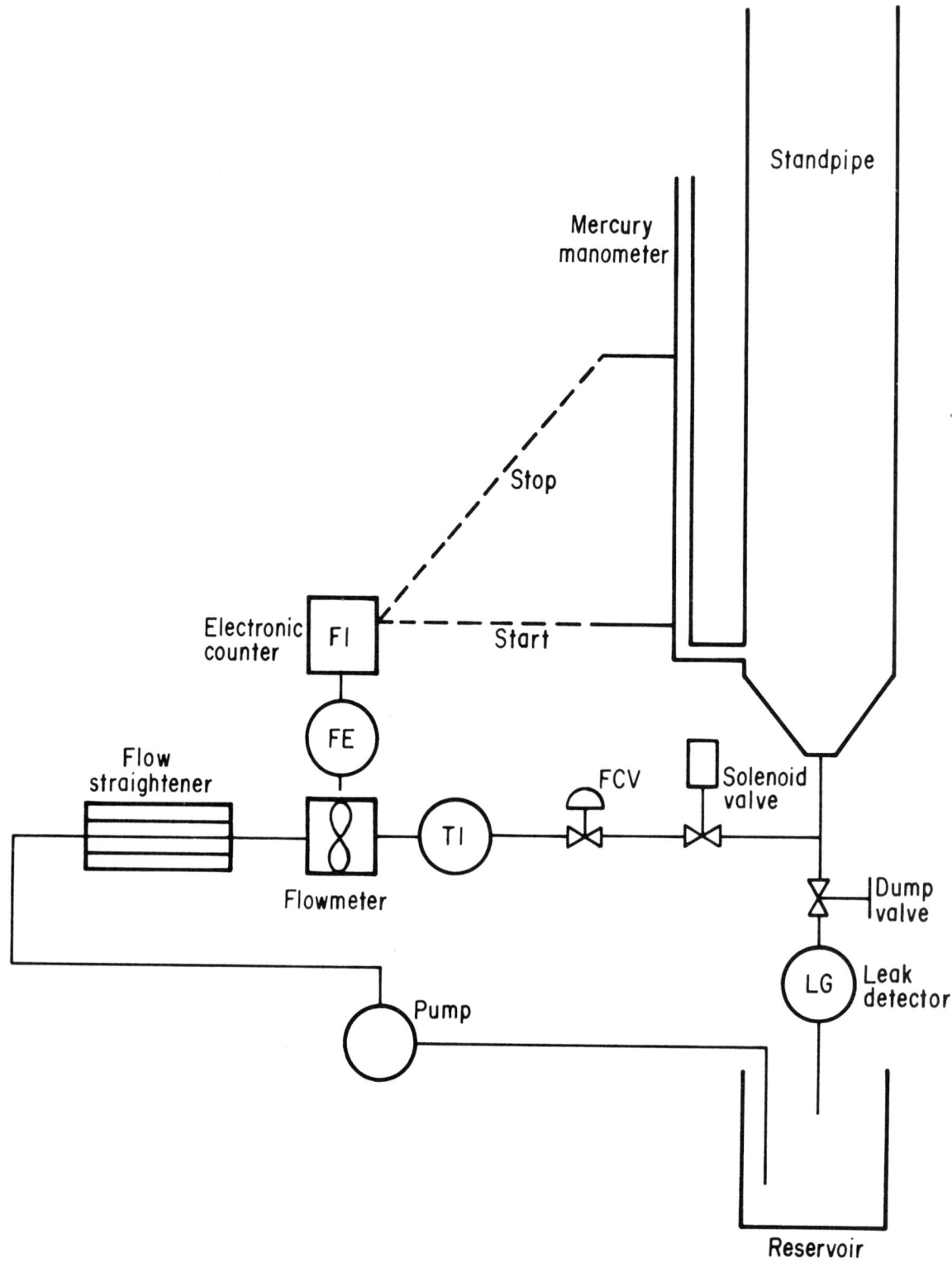

Fig. 6. Indirect, Gravimetric, Dynamic Calibrator

measure differential pressure. Contacts on the manometer start and stop the counter to establish the calibration interval. The mass of liquid passed through the flowmeter is:

$$M = k \left(\frac{\rho l}{\rho l - \rho_a} \right) A_l \times H_m$$

$$\left[\rho m \left(1 + \frac{A_t}{A_w} \right) - \rho a \left(1 + \frac{A_t}{A_l} \right) - \rho l \left(\frac{A_t}{A_w} - \frac{A_t}{A_l} \right) \right]$$

H_m = Change in elevation of manometer meniscus, i.e. distance between contacts used, in meters.

ρm = Density of manometer liquid at manometer temperature, kg/m^3.

ρ_a = Density of atmospheric air, which is assumed constant at 1.2250 kg/m^3.

ρl = Density of calibrating liquid in standpipe, kg/m^3.

A_l = Cross section area of Standpipe, in mm^2.

A_t = Cross section area of manometer tube, in mm^2.

A_w = Cross section area of manometer well, in mm^2.

k = $10^{-9} \, m^3/mm^3$

Possible sources of error with this system are: (1) the rate of rise of the liquid in the standpipe and in the manometer must be constant during the counter period; (2) the standpipe walls must be allowed to drain completely before each calibration run; and (3) correction must be made for the thermal expansion of the standpipe, the measured liquid, and the manometer liquid.

5.4.2 Volumetric Calibrators

In any of the systems described in 5.4.1 (3), a calibrated volume may be substituted for the weigh tank. In such systems, correction must be made for volume changes in the containers, and fluids with pressure and temperature.

(1) Mechanical Displacement Calibrator

The system shown in Figure 7 is also called a "ballistic" or "piston" calibrator. It is particularly suitable for use with high-vapor-pressure liquids because it is a constant volume closed system. It is the only calibrator suitable for large flow meters. The electronic counter is gated by the passage of the piston (usually a liquid-filled elastomer spheroid) across detectors at the ends of the calibrated section of the container.

This calibrator is widely used in the petroleum industry, and is fully discussed in API 2531 and APE 2534.

5.4.3 Comparison Calibration

This method requires a minimum amount of equipment and is convenient for routine calibration of turbine flowmeters. Since a secondary standard is employed in this technique, it should be recognized that the total uncertainty may increase. In operation, a reference turbine flowmeter is installed in series with the flowmeter to be calibrated and the proper steps taken to reduce pulsation and swirl at the inlet of both units. The flow points are set by the frequency output of the reference flowmeter and the flow is held constant using a frequency meter, electronic counter or cathode-ray oscilloscope for indication of flow variation. Cycles per unit volume are used as the basis for the comparison of indicated quantity as follows where:

K_1 = cycles per meter^3 of reference flowmeter

K_2 = cycles per meter^3 of flowmeter being calibrated

f_1 = cycles per time "t" of reference flowmeter

f_2 = cycles per time "t" of flowmeter being calibrated

The time base of the two counters used should be synchronous. If the temperature of the liquid is different at the two flowmeters, then a correction must be made for the difference in density.

Then $\quad K_2 = K_1 \times \dfrac{f_2}{f_1} \times \dfrac{\rho 2}{\rho 1}$

where: $\quad \rho_1$ = density of liquid at reference

ρ_2 = density of liquid at flowmeter being calibrated.

NOTE: The comparison calibration method is strictly valid only when the reference flowmeter is being operated under the identical conditions, including flow, under which it was calibrated by a primary method. It should be realized that the operation of this flowmeter at other conditions represents extrapolation of the data; the amount of the extrapolation being the extent to which the operating conditions deviate from the primary calibration conditions.

The reference flowmeter should be checked before and after a series of calibrations against either; a) a primary standard, or b) a master flowmeter to verify its repeatability.

The reference flowmeter should be calibrated in accordance with 5.3 Calibration and Test Procedure and the check calibration should not deviate from the previous calibration by more than ± 0.1%.

The reference flowmeter may be of the positive displacement type or other provided that the error requirements are met. Conversion of the output data may be necessary to effect the correlation noted above.

(1) Frequency Ratio Calibration

Better precision can be achieved in certain calibrations by the use of frequency ratio methods. Similar turbine flowmeters are plumbed together in series with the usual flow straightening sections of

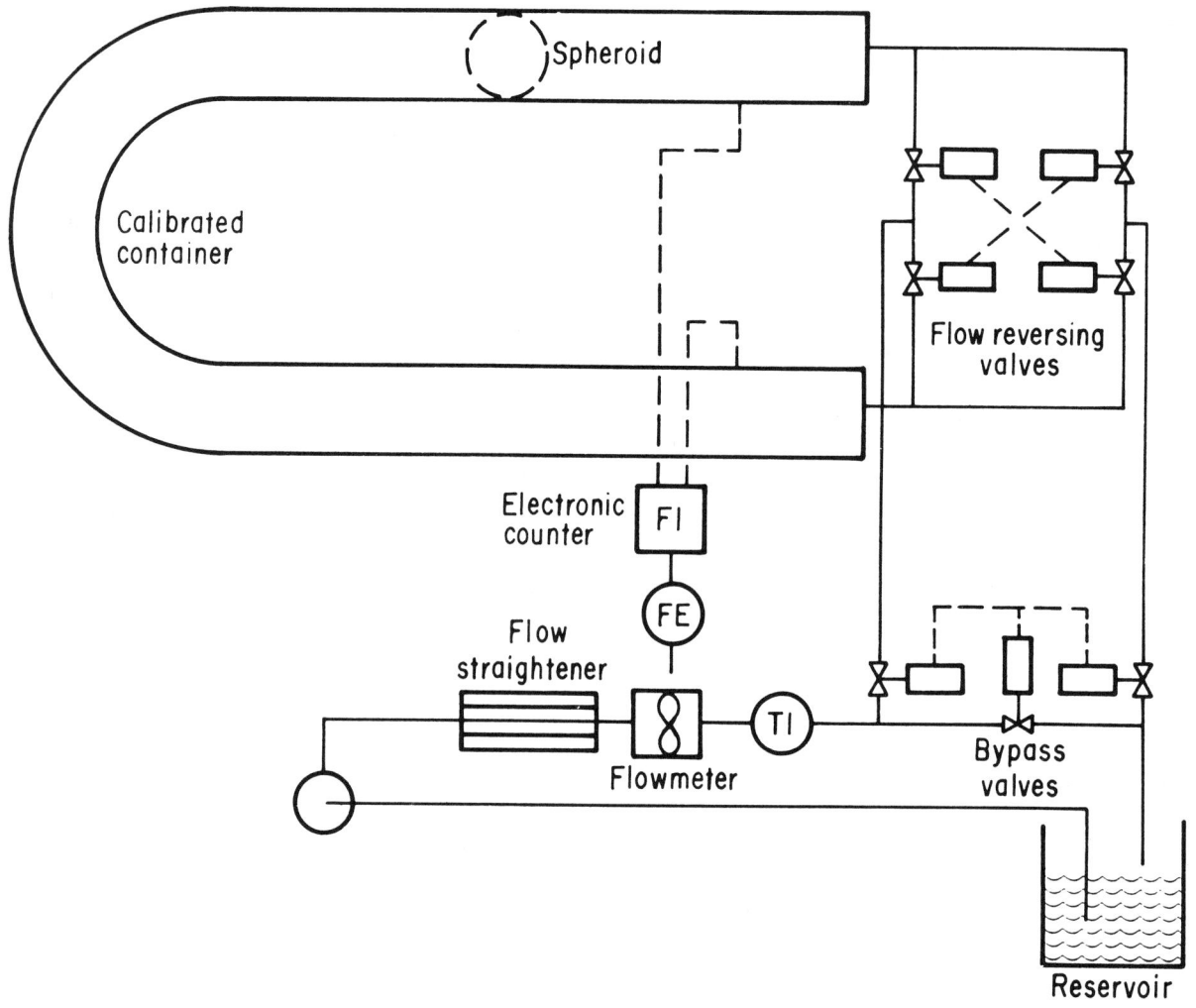

Fig. 7. Mechanical Displacement Volumetric Calibrator

pipe. These two flowmeters are calibrated simultaneously using one of the primary methods. During this calibration, the ratio of the frequencies of the two flowmeters is noted for each flow. When these flowmeters are used as reference flowmeters, if this ratio remains the same, it can be assumed that their calibrations have not changed.

In subsequent operation, these reference flowmeters and flow straighteners are installed in series with the flowmeter to be calibrated. The signals from the reference flowmeters are compared by a ratio counter that has a long counting time (20 to 30 s). The signal from the flowmeter under test is similarly compared with that from one of the reference flowmeter. In these comparisons the output of one referenced flowmeter is used as the "time base" for the other and later for the test flowmeter. This eliminates the need for a precise measurement of the counting intervals of separate calibrations. A less precise timer records the time during which the ratio count is determined. This technique is useful in several applications: two master flowmeters used as interlaboratory transfer standards can be used to check the calibration

facility, with greater confidence, or a working flowmeter can be calibrated against a reference flowmeter with better precision as explained above.

6 QUALIFICATION TEST PROCEDURES

Qualification tests shall be performed as applicable using the test forms of 7 DATA PRESENTATION as required. Upon completion of testing the form of Figure 8 shall be used to summarize all testing.

6.1 Initial Performance Tests

The tests and procedures of 5 INDIVIDUAL ACCEPTANCE TESTS AND CALIBRATIONS, individual acceptance tests and calibration, shall be run to establish reference performance during increasing and decreasing steps of 0, 20, 40, 60, 80, and 100 percent of range as a minimum.

6.2 Weight Test

The flowmeter shall be weighed on an appropriate balance or scale to establish compliance with 4.1.1.(4).

6.3 Temperature Test

The flowmeter shall be placed in a suitable temperature chamber. To establish compliance with 4.1.2 (1) Minimum and Maximum Temperatures, two calibration cycles shall be run at each combination of measured liquid temperature limits and environmental temperature limits specified therein.

NOTE: The temperature rating of the flowmeter may not in all cases be compatible with the vapor pressure of the measuring liquid used. Generally, the user is advised to test within the limitations of the intended measured liquid.

6.4 Pressure Test

During the temperature test above, at the maximum measured temperature, the liquid pressure shall be raised to the pressure limit specified in 4.1.2 (2) Pressure.

6.5 Speed Test

The flowmeter shall be operated at a flow which produces the maximum rotor speed specified in 4.1.2 (3) Speed, without regard to linearity.

7 DATA PRESENTATION

A suggested format for recording calibration data is shown on Figure 8. This format is specifically directed toward the Direct Gravimetric Static Method but with modification could be used for all types of calibrators. Since the turbine flowmeter is a volumetric device, the calibration data should be presented in volume units. The standard format should be a plot of cycles/meter^3 vs. frequency. Cycles/meter^3 is designated by the letter "K". For convenience to the user, it is also correct to present the data in a number of other ways such as: meter^3/cycle, or kilograms/cycle. When a flowmeter is calibrated with a number of liquids with various kinematic viscosities, it is sometimes useful to plot the data as cycles/meter^3 ("K") versus frequency divided by the kinematic viscosity in meter^2/s.

8 INSTALLATION

8.1 Environmental Considerations

A number of environmental conditions may affect the operation of a turbine flowmeter. Of these, moisture is the most common cause for malfunction. The electrical components should be waterproof to eliminate trouble from this source. The temperature of the flowmeter is largely determined by the temperature of the flowing liquid. However, the temperature of the transduction coil and connector may be influenced by the environment. Low temperatures do not usually cause malfunction, but high temperatures may result in failure of the insulation. Mechanical vibration will shorten the service life of a flowmeter and may also bias the data obtained. Magnetic fields in the proximity of the flowmeter or transmission cable may introduce spurious signals, if the circuit is not adequately shielded. Pulsation of the flow may produce errors or damage the flowmeter. Precautions should be taken to ensure that the operating conditions are within the limits set forth in the specification (4.1 Design Characteristics). Other special conditions such as a nuclear environment should be given consideration.

8.2 Piping Configuration

The turbine flowmeter is affected by upstream and downstream line configuration. This is caused primarily by swirl of the flowing liquid and, therefore, upstream configuration is much more influential than downstream. Rules specifying a certain number of pipe diameters of straight line upstream and downstream, which have been determined for orifice-type flowmeters, are generally conservative when applied to turbine flowmeters. However, lacking specific tests to establish the validity of less conservative configurations, these rules should be applied. These rules are well treated in API-2534, Appendix C, and in ASME, "Fluid Meters - Their Theory and Application". Without such analysis, a minimum of 20 diameters upstream and 5 diameters downstream should be used. A flow straightener is effective in eliminating swirl, and the upstream rotor support of a turbine flowmeter acts, to some extent as a straightener. The turbine flowmeter should be located with most of the available straight line upstream. Unusual downstream disturbances, such as a pump inlet, may require a straight line downstream or in extreme cases a downstream straightener may be necessary. Care should be taken not to allow piping to impose excessive stresses on the flowmeter housing.

8.2.1 Flow Direction

Install a turbine flowmeter in accordance with the flow direction arrow, or wording marked on the flowmeter.

8.2.2 Flow Straighteners

Various designs of straighteners have been proposed. Tube bundles and sections of honeycomb have been found effective. In addition to effectiveness, the pressure drop and durability should be considered.

8.2.3 Attitude of Flowmeter

Turbine flowmeters should be installed in the same position as they were calibrated, usually in a horizontal position (5.2.10). When installed in a position other than horizontal, a difference in axial thrust balance may cause a change in the calibration factor. For small flowmeters, the angular position of the transduction coil should be similarly considered.

8.2.4 Filtration

The measured liquid should not contain solid particles having a maximum dimension more than half the clearance between turbine blade tip and bore of the housing. The service life of the flowmeter will be extended by filtering the measured liquid. As a guide, the following U.S. Standard sieve mesh sizes are recommended:

Flowmeter Nominal Size	Mesh Size	Particle Size
3/8 - inch	170	88 microns
1/2 - inch	120	125 microns
3/4 - inch	45	350 microns
1 - inch	45	350 microns
1-1/2 - inch	18	1000 microns

TURBINE TRANSDUCER CALIBRATION TEST NO. _____
DIRECT-GRAVIMETRIC METHOD

Location _____ ; Date _____ ; Calibrator _____

Flowmeter Manuf. _____ ; Model _____ ; Serial # _____

Test Liquid: Type _____ ; Density _____ kg/m^2 at _____ K.

Viscosity _____ m^2/s at _____ K.

Back Pressure _____ Pascals; Pressure Drop _____ Pascals at _____ m^3/s.

Electrical Output: Max: _____ mv rms at _____ Hz.

Min: _____ mv rms at _____ Hz.

TEST MEASUREMENT DATA	Run 1	Run 2	Run 3
1. Nominal Frequency, Hz			
2. Transducer Temperature, K			
3. Density, kg/m^3			
4. Total count, cycles			
5. Time, s.			
6. Tare Reading, kg			
7. Gross Reading, kg			
8. Apparent Net, kg (7 - 6)			

CALCULATIONS

9. Air Buoyancy Factor			
10. Actual Net, kg (8 x 9)			
11. Actual Vo., m^3 (10/3)			
12. K, cycles/m^3, (4/11)			
13. Actual Freq., Hz (4/5)			
14. f/ν, (13 ÷ kinematic viscosity)			

Operator: _____

Fig. 8. Format For Recording Calibration Data

8.3 Operation Precautions

8.3.1 Liquid Pressure

A minimum back pressure for any flowmeter installation should be maintained to preclude a change in the calibration factor due to two-phase conditions. The minimum absolute back pressure is a function of the vapor pressure of the liquid and the presence of dissolved gases. The minimum back pressure can be experimentally determined under actual test conditions and is defined as the back pressure at which the calibration factor at 125% of the nominal maximum flow increases 1/2% over the corresponding calibration factor obtained at the same flow but with a 7.0E + 04 Pa higher pressure. The minimum back pressure should be measured four pipe-diameters downstream of the flowmeter.

8.3.2 Prevention of Over-Speeding

A number of conditions may cause the turbine to turn at a speed higher than its rated value. Whether such over-speeding causes damage to the flowmeter depends upon the degree and duration. Improper selection of the flowmeter or excessive flow due to failure of some component of the system can cause overspeeding. Depletion of the metered liquid in a pressurized feed system will usually cause ruinous over-speeding as the gas passes through the flowmeter. Perhaps the most frequent cause of severe over-speeding of a liquid flowmeter is gas in the line. Gas may be introduced during assembly of the system, or may result from a two-phase condition of the metered liquid. Over-speeding can be prevented by care in design and operation of the system. Proper location of the flowmeter with respect to valves can minimize risk. Finally, provision and use of means for limiting flow or bleeding gas from a liquid system is essential. Different types of over-speed protection devices are available for specific application, including mechanical and electromagnetic brakes.

8.4 Electrical Installation

8.4.1 A two-conductor or three-conductor shielded cable should be used for the electrical output. Wire size should be based on allowable signal attenuation. Avoid installation in a conduit containing power cables.

8.4.2 The cable shield is to be grounded at one point only. Normally it should be grounded at the flowmeter end, but some experimentation may be necessary in troublesome cases.

8.4.3 Excessive installation torque may damage the transduction coil.

8.5 Checkout

8.5.1 System Check

The type of check procedure will depend largely upon the flowmeter application. In some instances it may be possible and most expedient to establish test flow to check functionally the complete measuring system, with no serious consequences if proper opera-

tion is not obtained on first attempt. In other instances it may be of paramount importance that proper operation be obtained every time test flow is established so that one or more of the following pre-test checks may be justified.

8.5.2 Rotor Spin Check

The most comprehensive check that may be made of a turbine flowmeter, associated circuitry, and readout equipment, short of actually establishing flow, is to spin the rotor by tangential impingement of a jet of fluid. When such a check is desired, a pressure connection for this purpose must be provided on the housing. Any fluid that will not contaminate the system may be used. **Extreme care must be exercised in making a spin check to avoid over-speeding.** Even if excessive speeds are avoided, bearing lubrication may be inadequate so that prolonged checks should be avoided.

8.5.3 Induced Signal Check

The transduction coil, associated circuitry, and readout equipment of a self-generating flowmeter may be checked by an induced signal. A small coil, connected to an ac power source should be held near the transduction coil so as to effect an energy transfer. This functionally checks the circuit without breaking any connections and can be used.

8.5.4 Applied Signal Check

The associated circuitry and readout equipment of a self-generating flowmeter may also be checked by applying the signal directly. An oscillator can be connected in parallel with transduction coil to accomplish a check similar to that of 8.5.3 Induced Signal Check. Note, however, that the transduction coil is not checked by this method. Care must be exercised so that circuit characteristics are not altered or continuity disturbed by connection and disconnection of the oscillator.

9 SIGNAL CONDITIONING AND SYSTEMS CONSIDERATIONS

The data signal is transferred and converted in the following general manner. The electrical output from the flowmeter is sent to the signal conditioning equipment by means of a signal transmission system. After the signal has been properly conditioned, it is sent to a data presentation system, which consists of a display or recording system. The display and recording can be a combination of analog and digital readouts.

9.1 Signal Conditioning

In addition to the basic operations of amplification or attenuation, certain signal conditioning may be done to obtain data in a different form.

9.1.1 A rate-type signal conditioner counts, for repetitive short time intervals, the number of cycles of flowmeter signal produced in each of those intervals. By the choice of a suitable time interval, desired units of volumetric flow may be indicated. Flow may be indicated even in mass units if the density is constant.

9.1.2 A totalizer accumulates the number of cycles proportional to the total volumetric flow which has passed through the flowmeter.

9.1.3 An integrator provides a dc voltage level proportional to the signal frequency.

9.1.4 A scaler multiplies or divides the flowmeter output frequency by some selected factor. It is generally used to facilitate presentation and reduction of data.

9.2 Data Presentation

9.2.1 Monitor Display

Display is divided into two general categories, digital and analog. Both rate and total quantity data are easily displayed digitally, while rate can also be displayed in analog form.

9.2.2 Recorded Data

To obtain a permanent record of the data, it is usually recorded by means of an electric typewriter, counter printer, graphic recorder, oscillograph or magnetic tape system. There are numerous combinations of recording systems, the detailed discussion of which is beyond the scope of this Standard.

9.3 Supplemental Data

In many applications, the measurement of one of more additional variables may be necessary for complete interpretation of the flow data. Such supplemental data are generally limited to the following:

9.3.1 Temperature

Temperature of the measured liquid has an effect on the performance of the turbine flowmeter. Any one or more of the following effects may produce errors and must be considered. There is a mechanical effect caused by thermal expansion or contraction of the housing and turbine when the operating temperature materially differs from calibration temperature (See 9.4.1 (1) Temperature). An elevated or depressed operating temperature changes the physical properties of the liquid being measured, specifically its vapor pressure, viscosity and density. (See 9.4 Data Interpretation, 9.3.2 Viscosity, and 9.3.3 Density).

9.3.2 Viscosity

Kinematic viscosity measurement is desirable when operating over a wide temperature range or near the low end of the flowmeter's range, and when high accuracy is important. Liquid viscosity is usually determined indirectly by measuring the liquid temperature at a point four pipe diameters downstream of the flowmeter. The measured temperature information is used to obtain the liquid viscosity from known temperature vs. viscosity data for the particular liquid being measured. Available data on the viscosity-temperature relationship of the liquid being measured may be supplemented by analysis of samples from the actual batch to determine the viscosity at operating temperatures. Alternatively, a viscosimeter can be used for direct measurement.

9.3.3 Density

It is necessary to know the density of the liquid at the flowmeter when mass flow data are desired. This may be determined from a temperature measurement in the same manner as described for viscosity. Alternatively, a densitometer may be used. In some systems, the densitometer is arranged to compensate the volumetric flow measurement automatically so that flow and total quantity are presented in mass units. This approach often increases the uncertainty in the density correction.

9.3.4 Pressure

Where compressibility is an important factor, liquid pressure is measured. The pressure should be measured four pipe diameters downstream of the flowmeter.

9.4 Data Interpretation

This section deals only with the analytical factors which affect the flow data acquired. It is presumed that factors such as two-phase flow, pulsation, and electrical interference have been minimized to a degree consistent with the accuracy desired.

9.4.1 Volumetric Flow Measurement

Regardless of whether the signal conditioning equipment provides a measurement of flow rate or total quantity the readings must be divided by the calibration factor "K" (See 4.3 Specific Terminology) to obtain volumetric data. As noted in that section the factor is a function of several independent variables for which corrections may be applied according to accuracy requirements. For totalization, a mean effective calibration factor must be determined based on representative values of these variables. If the signal conditioning equipment introduces a scaling factor, it must also be applied to the data at this time. The flow through a turbine flowmeter is torsionally deflected by the rotor due to bearing friction and signal generation torque. In a properly functioning flowmeter, this effect is small and is incorporated in the calibration factor. The magnitude of the effect, however, varies inversely with the density of the flowing liquid and size of flowmeter. Consequently, if the density of the operating liquid differs from that of the calibrating liquid, an error is introduced. Even a change in density, due to temperature or pressure variation, affects the factor. Because the magnitude of this error is related to the flowmeter design, no generalized correction can be given. Fortunately, the error is small, but even so, it is desirable to calibrate with a liquid of approximately the same density as the operating liquid. Corrections, if applied at all are usually for one or more of the following:

(1) Temperature

The flowmeter temperature is usually assumed to be that of the flowing liquid. When operating temperature differs from that noted during calibration, an error is introduced by dimensional changes in the housing and turbine. A correction based on the thermal coefficient of expansion may be approximated as follows:

$$K_O = \frac{K_C}{1 + 3\,a\,(T_O - T_C)}$$

where: K_O = calibration factor at operating temperature

K_C = calibration factor at calibrating temperature

a = thermal coefficient of linear expansion

T_O = Operating temperature

T_C = Calibration temperature (in same units as T_O)

Note that K_O decreases as operating temperature increases and that correction applies only when housing and turbine have the same thermal coefficient.

(2) Flow

The sensitivity "K" is usually expressed as a constant. When a wide range of flow is to be covered, error can usually be reduced by expressing the factor as a linear function of flow, either with or without temperature correction. In some cases a non-linear expression of calibration factor may be justified.

(3) Viscosity

If viscosity of the operating liquid differs from that of the calibrating liquid, or if it varies during operation due to temperature change, it may be necessary to obtain calibration at more than one viscosity. These data are best utilized as a single relationship of K factor vs. Reynolds number if the flowmeter characteristics enable such a presentation. For simplification, the quantity f/ν (output frequency divided by kinematic viscosity) which is approximately proportional to Reynolds number, may be used instead. If the flowmeter characteristics are such that a single relationship is not obtained, the data must be utilized as a family of constant viscosity relationships of K vs. flow. In either case, these relationships may be expressed as linear or non-linear functions.

9.4.2 *Mass Flow Measurement*

When flow data obtained with a turbine flowmeter are desired in mass units, the volumetric rate or total quantity must be multiplied by the density of the metered liquid at operating conditions (See. Par. 9.3.3).

10 BIBLIOGRAPHY

1. Shafer, M. R.; "Performance Characteristics of Turbine Flowmeters"; ASME Journal of Basic Engineering, December 1962.

2. Grey, J. and Thompson, R. E.; "Turbine Flowmeter Performance Model"; Report No. AMC-3, U.S. Army Missile Command Contract DA-AH01-67-C1609, 1967.

3. ANSI MC6.1/ISA S37.1 "Standard Transducer Nomenclature and Terminology"; 1975.

4. NBS Handbook 77

5. ASTM D-323, "Standard Method of Test for Vapor Pressure of Petroleum Products (Reid Method)."

6. Shafer, M. R., and Ruegg, F. W.; "Liquid Flowmeter Calibration Techniques"; ASME Transactions, October, 1958.

7. American Petroleum Institute ANSI Z11. 299/API-2534; "Turbine Flowmeters and Application to Hydrocarbon Products."

8. Abernethy, R.B. and Thompson, J.W. Jr. "Uncertainty in Gas Turbine Measurements" AEDC TR 73-5.

9. ANSI Z11.171/API 2531. American Petroleum Institute, "Mechanical Displacement Meter Provers," 1971.

INSTRUMENT SOCIETY of AMERICA
Research Triangle Park, North Carolina

ANSI/ISA-S37.1-1975
(R1982)
Approved December 14, 1982
(Formerly ANSI MC6.1-1975)

American National Standard

Electrical Transducer Nomenclature and Terminology

Instrument Society of America

ISBN 0-87664-113-3

ISA-S37.1   Electrical Transducer
            Nomenclature and Terminology

INSTRUMENT SOCIETY OF AMERICA
67 Alexander Drive
P.O. Box 12277
Research Triangle Park, North Carolina 27709

PREFACE

This Preface is included for information purposes and is not part of S37.1.

This Standard has been prepared as a service of the Instrument Society of America toward the goal of uniformity in the field of instrumentation. To be of real value, it should not be static, but should be subject to periodic review. Toward this end, the Society welcomes all comments and criticisms and asks that they be addressed to the Standards and Practices Board Secretary, Instrument Society of America, 67 Alexander Drive, P.O. Box 12277, Research Triangle Park, North Carolina 27709.

The ISA Standards and Practices Department is aware of the growing need for attention to the metric system of units in general, and the International System of Units (SI) in particular, in the preparation of instrumentation standards. The Department is further aware of the benefits to users of ISA Standards in the USA of incorporating suitable references to the SI (and the metric system) in their business and professional dealings with other countries. Toward this end, this Department will endeavor to introduce SI and SI-acceptable metric units as optional alternatives to English units in all new and revised standards to the greatest extent possible. The Metric Practice Guide, which has been published by the American Society for Testing and Materials as ASTM E380-72 (ANSI Z210-1973), and future revisions, will be the reference guide for definitions, symbols, abbreviations and conversion factors.

It is the policy of the Instrument Society of America to encourage and welcome the participation of all concerned individuals and interests in the development of ISA Standards. Participation in the ISA standards making process by an individual in no way constitutes endorsement by the employer of that individual of the Instrument Society of American or any of the standards which ISA develops.

This Standard supersedes ISA Tentative Recommended Practice RP37.1-1963, which was develped by ISA Subcommittee 8A/RP37.1 (M. E. Binkley, H. N. Norton, T. A. Peris, and A. A. Zuehlke) between 1960 and 1963 to fill a need for standardized transducer nomenclature and specification terminology required, at that time, primarily by the aerospace industry.

As production techniques of electrical transducers advanced, associated measuring techniques and systems become more established, and as new transducer designs became more readily available, they found increasing applications in all industries and sciences in addition to those types of transducers already in widespread use. Hence, it became necessary for ISA to develop uniform transducer nomenclature and terminology for use in as many technological fields as possible.

Using RP37.1-1963 as a starting point, Committee SP37 (consisting of the chairmen of Standards Committees on individual transducer types as well as the cognizant Standards Director) created a draft version of new S37.1 which was mailed to a large review board, representing a wide variety of fields, in 1968. The results of this review indicated the general acceptability of the new Standard to most industries, sciences, and educational institutions. Numerous suggestions for improvements and clarifications were also received by the Committee. Each comment was evaluated, and suitable revisions were made with Committee concurrence.

The preparation of this Standard was coordinated with the government-sponsored Inter-Range Instrumentation Group (IRIG) as well as with ISA Committee SP51 (Measurement & Control Terminology).

The following individuals served on the 1969 SP37 committee:

NAME	COMPANY
H. N. Norton, Chairman	Jet Propulsion Laboratory
J. Z. Inskeep, Vice-Chairman	Jet Propulsion Laboratory
P. S. Lederer	National Bureau of Standards
H. C. Chandon	Rosemount Engineering Co.
L. L. Lathrop	Sandia Corp.
C. W. Silver	Revere Corp. of America
R. W. Bidstrup	Kaiser Aluminum and Chemical Co.
R. D. Bronson	General Dynamics Corp., Ft. Worth Div.

The following individuals served on the 1982 SP37 committee:

NAME	COMPANY
J. W. Mock, Chairman	Measurement Services
P. Bliss	Consultant
W. Brinkschulte	Genisco Technology Corporation
C. Kutelis	Revere Corporation of America
R. W. Lally	PCB Piezotronics
H. E. Lockery	Hottinger Baldwin Measurements, Inc.
E. W. Malone	Boeing Aerospace Co.
T. B. Miller	Lawrence Livermore Lab
K. A. Parlee	United Engineers & Constructors, Inc.
H. Pitt	
J. Powell	Rosemount Engineering
R. C. Strahm	EG & G Idaho
R. M. Whittier	ENDEVCO
N. Wilde	EG & G Idaho
E. Wong	Bell & Howell

This Standard was approved for publication by the ISA Standards and Practices Board in October 1982.

NAME	COMPANY
T. J. Harrison, Chairman	IBM Corporation
P. Bliss	Consultant
W. Calder	The Foxboro Company
N. Conger	Continental Oil Co.
B. Feikle	Bailey Controls Co.
R. T. Jones	Philadelphia Electric Co.
R. Keller	Boeing Company
O. P. Lovett, Jr.	Isis Corp.
E. C. Magison	Honeywell, Inc.
A. P. McCauley	Diamond Shamrock Corp.
J. W. Mock	Measurement Services
E. M. Nesvig	ERDCO Engineering Corp.
G. Platt	Bechtel Power Corp.
R. Prescott	Moore Products Company
W. C. Weidman	Gilbert Associates
K. A. Whitman	Allied Chemical Corp.
J. R. Williams	Stearns-Roger, Inc.
B. A. Christensen*	
L. N. Combs*	
R. L. Galley*	
R. G. Marvin*	
W. B. Miller*	Moore Products Company
R. L. Nickens*	

*Director Emeritus

ELECTRICAL TRANSDUCER NOMENCLATURE AND TERMINOLOGY

CONTENTS

1. PURPOSE

1.1 This Standard establishes:

1.1.1 Uniform nomenclature for transducers.
1.2.2 Uniform simplified terminology for transducer characteristics.

2. SCOPE

2.1 This Standard covers transducers used in electrical and electronic measuring systems.

2.2 It is realized that this Standard may not be wholly suitable for transducers used in automatic control systems and in some other specialized applications.

2.3 Emphasis on the usability of this Standard in all types of written and verbal communications has been placed in the following order of precedence:

I. Users' and manufacturers' specifications, including catalogs and advertising.
II. Calibration and test procedures and reports.
III. Technical papers, educational and reference material, and periodicals
IV. Other Communications

2.4 A recommended manner of assigning nomenclature to transducers is shown in Section 3.

2.5 Recommended terminology for transducer characteristics is shown in Section 4.

2.6 The word "simplified" (see 1.2.2) denotes the most brief, adequate definition which could be derived. The definition may be supplemented as deemed necessary by the user of the term.

3. NOMENCLATURE

3.1 NOMENCLATURE REQUIREMENTS

Nomenclature of transducers should consist of the following:

3.1.1 The noun "transducer";
3.1.2 A first modifier denoting the measurand;
3.1.3 When required, a second modifier restricting the measurand;
3.1.4 A third modifier denoting the electrical transduction principle; the adjective form should be used whenever possible.
3.1.5 An optional fourth modifier denoting the mechanical link in the transducer or any noteworthy special feature. (May be followed with the word "type").
3.1.6 When required, a modifier phrase restricting the modifier.

3.2 USAGE IN TITLES

When used in titles of drawings and specifications, headings in lists, indices, and tabulations, and when indicated by other requirements which may be applicable, the sequence shown in 3.1 should be used.

(1) Note that the standard value of the acceleration of gravity at the earth's surface — a unit of measurement — is abbreviated "g_n." A measured value of the acceleration of gravity at the earth's surface is indicated by "g." The SI mass unit "gram" is abbreviated "g".

Examples: "Transducer, Pressure, Differential, Potentiometric, 0 to 10 psid." "Transducer, Sound Pressure, Capacitive, 100 to 160 dB."
"Transducer, Acceleration, Reluctive, ± 3g."
Transducer, Pressure, Absolute, Strain Gage, Amplifying, 0 to 500 psia."

3.3 USAGE IN TEXT

For all other purposes, such as use in a sentence or in captions under pictorial representation, the exact opposite of the sequence shown in 3.1 should be used.
Examples: "A ±20g[1] piezoelectric acceleration transducer was installed on the mounting plate."
"A 0 to 300°F resistive surface temperature transducer was bonded to the tank skin."
"Hinge motion was measured with a -2 to +8 degree dc-output reluctive angular-position transducer."

3.4 OMISSION OF MODIFIERS

When generalization of transducer types or categories is desired, the omission of modifiers should proceed in the order opposite to sequence shown in 3.1 whenever possible.
Examples: "Procurement of 150 potentiometric linear-displacement transducers of various ranges has been initiated."
"Bulletin 0-0-400 describes our capacitive liquid-level transducers."
"The additional test requirements apply only to differential-pressure transducers."

3.5 NOMENCLATURE AND EXAMPLES

The construction of typical transducer nomenclature and examples of modifiers are shown in Table I.

3.6 OPTIONAL USE OF ALTERNATE NOMENCLATURE

Use of alternate nomenclature is optional in the following special cases:
3.6.1 "Accelerometer" instead of preferred "Acceleration Transducer" or "Transducer, Acceleration."
3.6.2 "Tachometer" instead of preferred "Angular Speed Transducer" or Transducer, Angular Speed."
3.6.3 "Strain Gage" instead of "Resistive Strain Transducers" or "Transducer, Strain, Resistive."
3.6.4 "Thermocouple" instead of "Thermoelectric Temperature Transducer" or "Transducer, Thermoelectric, Temperature."
3.65 "Flowmeter" instead of preferred terms "Flow Transducer" (or "Flow-Rate Transducer") or "Transducer, Flow" (or "Transducer, Flow-Rate").

3.7 NOMENCLATURE GLOSSARY

Some of the nomenclature of Table I needs clarification because of the particular meaning intended for transducers. *Italicized* words, in definitions, are defined in Section 4. *The terms in Table I with an associated asterisk (*) are defined below, for purposes of their use as modifiers in transducer nomenclature.*

3.7.1 First Modifier Definitions

Attitude — The relative orientation of a vehicle or object represented by its angles of inclination to three orthogonal reference axes.

Displacement — The change in position of a body or point with respect to a reference point.

NOTE: Position is the spatial location of a body or point with respect to a reference point.

Flow Rate — The time rate of motion of a fluid, usually contained in a pipe or duct, expressed as fluid quantity per unit time.

Heat Flux — The quantity of thermal energy transferred to a unit area per unit time.

Humidity, Absolute — The mass of water vapor present in a unit volume of air or other fluid.

Humidity, Relative — The ratio of the water vapor pressure actually present to the water vapor pressure required for saturation at a given temperature, expressed in per cent.

Jerk — The time rate of change of acceleration. Expressed in feet/s^3, cm/s^3, g$_n$/s (Refer to footnote on page 5).

Light — An electromagnetic radiation whose wavelength is between approximately 10^{-2} and 10^{-6} cm.

NOTE: By strict definition only visible radiation (4×10^{-5} to 7×10^{-5} cm) can be considered as "light."

Nuclear Radiation — The emission of charged and uncharged particles and of electromagnetic radiation from atomic nuclei.

Pressure, Absolute — The pressure measured relative to zero pressure (vacuum).

Pressure, Differential — The difference in pressure between two points of measurement.

Pressure, Gage — Pressure measured relative to *Ambient Pressure*.

TABLE I
CONSTRUCTION OF TYPICAL TRANSDUCER NOMENCLATURE
AND EXAMPLES OF MODIFIERS

Main Noun	First Modifier Measurand (Examples)	Second Modifier (Restricts Measurand) (Examples)	Third Modifier (Electrical Transduction Principle) (Examples)	Fourth Modifier[3] (Sensing Element. Special Features or Provisions) (Examples)	Range[8] (Examples)	Units[9] (Examples)
Transducer	Acceleration	Absolute	*Capacitive	AC Output	0 to 1000	A
	Air Speed	Angular	*Electromagnetic	*Amplifying	±5	°C
	*Attitude	Differential	*Inductive	*Bellows	-100 to +500	cm
	Attitude Rate	Gage	*Ionizing	*Bondable	-430 to -415	cm/s
	Current	Infrared	*Photoconductive	*Bonded		deg[10]
	*Displacement	Intensity	*Photovoltaic	*Bourdon-Tube		°F
	*Flow Rate	Linear	*Piezoelectric	*Capsule[4]		fps
	Force	Mass	*Potentiometric	*DC Output		
	*Heat Flux	Radiant	*Reluctive	*Diaphragm		Hz
	*Humidity	Relative	*Resistive	Digital-Output		ips
	*Jerk	Surface	*Strain Gage	*Discrete Increment		in.
	*Light	Total	*Thermoelectric	*Dual-Output		K
	Liquid Level	Volumetric		Exposed Element		kgf
	Mach No.			Frequency Output		lb/min.
	*Nuclear Radiation			*Gyro		m
	*Pressure			*Integrating		mmHg
	Speed[1]			*Self-Generating		N
	*Sound Pressure			*Semiconductor		% RH
	*Strain			*Servo[5] [6]		psia
	Temperature			Switch		psid
	Torque			Toothed-Rotor		psig
	Velocity[2]			Triaxial		psid
				*Turbine		psig
				*Ultrasonic		rad/s
				*Unbonded		
				Vibrating-Element[7]		
				Weldable		

(*) see Section 3.7 for definitions

(1) Scalar quantity.
(2) Vector quantity.
(3) Nomenclature may include two of these terms.
(4) Preferred to "Aneroid".
(5) Preferred to "Force Balance" or "Null Balance"
(6) When this modifier is used the third modifier (transduction principle) may be omitted.
(7) When this modifier is used together with "Frequency Output" the third modifier may be omitted.
(8) Defined in Terminology, Paragraph 4.4.
(9) Abbreviations used for units of Measurand used in specifications should generally be in accordance with ANSI Y10.19-1969 **Units Used in Science and Technology, Letter Symbols for.**
(10) Use for angular measurements.

Sound Pressure - The total instantaneous pressure at a given point in the presence of a sound wave, minus the static pressure of that point.

Strain - The deformation per unit length produced in a solid as a result of stress.

3.7.2 Third Modifier Definitions

Capacitive - Converting a change of *Measurand* into a change of capacitance.

Electromagnetic - Converting a change of *Measurand* into an *Output* induced in a conductor by a change in magnetic flux, in the absence of *Excitation.*

Inductive - Converting a change of *Measurand* into a change of the self-inductance of a single coil.

Ionizing - Converting a change of *Measurand* into a change in ionization current, such as through a gas between two electrodes.

Photoconductive - Converting a change of *Measurand* into a change in resistance or conductivity of a semiconductor material by a change in the amount of illumination incident upon the material.

Photovoltaic - Converting a change of *Measurand* into a change in the voltage generated when a junction between certain dissimilar materials is illuminated.

Piezoelectric - Converting a change of *Measurand* into a change in the electrostatic charge or voltage generated by certain materials when mechanically stressed.

Potentiometeric - Converting a change of *Measurand* into a voltage-ratio change by a change in the position of a movable contact on a resistance element across which excitation is applied.

Reluctive - Converting a change of *Measurand* into an a-c voltage change by a change in the reluctance path between two or more coils or separated portions of one coil when a-c *Excitation* is applied to the coil(s).
NOTE: Included among *Reluctive Transducers* are those employing differential-transformer, inductance-bridge, and synchro elements.

Resistive - Converting a change of *Measurand* into a change of resistance.

Strain-Gage - Converting a change of *Measurand* into a change or resistance due to strain.

Thermoelectric - Converting a change of *Measurand* into a change in the emf generated by a temperature difference between the junctions of two selected dissimilar materials.

3.7.3 Fourth Modifier Definitions

Amplifying - With integral *Output* amplifier.

Bellows - A *Pressure Sensing Element* of generally cylindrical shape whose walls contain deep convolutions, and for which the length changes when a pressure differential is applied.

Bondable - Designed to be permanently mounted to a surface by means of adhesives.

Bonded - Permanently attached over the length and width of the active element.

Bourdon Tube - A *Pressure Sensing Element* consisting of a twisted or curved tube of non-circular cross section which tends to be straightened by the application of internal pressure.

Capsule - A *Pressure Sensing Element* consisting of two metallic diaphragms joined around their peripheries.

DC Output - with integral demodulator, rectifier or frequency integrator.

Diaphragm - A *Sensing Element* consisting of a thin, usually circular, plate which is deformed by pressure differential applied across the plate.

Discrete Increment - Providing an *Output* which represents the magnitude of the *Measurand* in the form of discrete or quantized values.

Dual-Output - Providing two separate and noninteracting *Outputs* which are functions of the applied *Measurand.*

Gyro (a contraction of **gyroscope**) - A *Transducer* which makes use of a self-contained spatial directional reference.

Integrating - Providing an *Output* which is a time integral function of the *Measurand.*

Self-Generating - Providing an *Output* signal without applied *Excitation.* Examples are *Piezoelectric, Electromagnetic,* and *Thermoelectric Transducers.*

Semi-Conductor - Materials, used for *Sensing Elements* or *Transduction Elements,* whose resistivity falls between that of conductors and insulators (e.g.: germanium, silicon, etc.). Examples of useful phenomena associated with these materials are: Hall effect, temperature coefficient of resistance, photo-resistivity, photovoltaic effect, piezoresistance, etc.

Servo (a contraction of **servomechanism**) - A *Transducer* type in which the *Output* of the *Transduction Element* is amplified and fed back so as to balance the forces applied to the *Sensing Element* or its displacements. The *Output* is a function of the feedback signal.

Turbine - A bladed rotor which turns at a speed nominally proportional to the volume rate of flow.

Ultrasonic - Using frequencies above the audio-frequency range, i.e., above 20 kHz.

Unbonded - Stretched and unsupported between ends (usually refers to strain-sensitive wire).

4 TERMINOLOGY

When a term is not defined and is referenced to other terms, one of the terms referred to should be used instead.

All *italicized* terms appearing in definitions are defined in this document.

Definitions, or portions thereof, intended for use only in specifications (and their verification by testing) are preceded by "(S)".

Acceleration Error - The maximum difference, at any *Measurand* value within the specified *Range*, between *Output* readings taken with and without the application of specified constant acceleration along specified axes.

NOTE: See *Transverse Sensitivity* when applied to Acceleration *Transducer*.

Accuracy - The ratio of the *Error* to the *Full-Scale Output* or the ratio of the *Error* to the *Output*, as specified, expressed in percent.

NOTE 1: Accuracy may be expressed in terms of units of *Measurand*, or as within ± _____ per cent of *Full Scale Output*.

NOTE 2: Use of the term *Accuracy* should be limited to generalized descriptions of characteristics. It should not be used in specifications. The term *Error* is preferred in specifications and other specific descriptions of transducer performance.

Altitude - The vertical distance above a stated reference level.

NOTE: Unless otherwise specified, this reference is mean sea level.

Ambient Conditions - The conditions (pressure, temperature, etc.,) of the medium surrounding the case of the *Transducer*.

Ambient Pressure Error - The maximum change in *Output*, at any *Measurand* value within the specified *Range*, when the ambient pressure is changed between specified values.

Analog Output - *Transducer Output* which is a continuous function of the *Measurand*, except as modified by the *Resolution* of the *Transducer*.

Attitude Error - The *Error* due to the orientation of the *Transducer* relative to the direction in which gravity acts upon the *Transducer* (see *Acceleration Error*).

"Best Straight Line" - A line midway between the two parallel straight lines closest together and enclosing all *Output* vs. *Measurand* values on a *Calibration Curve*.

Breakdown Voltage Rating - (S) The dc or sinusoidal ac voltage which can be applied across specified insulated portions of a *Transducer* without causing arcing or conduction above a specified current value across the insulating material.

NOTE: Time duration of application, *Ambient Conditions,* and ac frequency must be specified

Bridge Resistance - (See *Input Impedance* and *Output Impedance*).

Burst Pressure Rating - (S) The pressure which may be applied to the *Sensing Element* or the case (as specified) of a *Transducer* without rupture of either the *Sensing Element* or *Transducer* case as specified.

NOTE: (1) Minimum number of applications and time duration of each application must be specified, (2) in the case of *Transducers* intended to measure a property of a pressurized fluid, *Burst Pressure* is applied to the portion subjected to the fluid.

Calibration - A test during which known values of *Measurand* are applied to the *Transducer* and corresponding *Output* reading are recorded under specified conditions.

Calibration Curve - A graphical representation of the *Calibration Record*.

Calibration Cycle - The application of known values of *Measurand*, and recording of corresponding *Output* readings, over the full (or specified portion of the) *Range* of a *Transducer* in an ascending and descending direction.

Calibration Record - A record (e.g., table or graph) of the measured relationship of the *Transducer Output* to the applied *Measurand* over the *Transducer Range*.

NOTE: Calibration Records may contain additional calculated points so indentified.

Calibration Simulation Provisions - Electrical connections or circuitry, contained within a *Transducer,* designed to permit the calibration of the associated measuring system by causing *Output* changes of known magnitude without varying the applied *Measurand*.

Calibration Traceability - The relation of a *Transducer Calibration,* through a specified step-by-step process, to an instrument or group of instruments calibrated by the National Bureau of Standards.

NOTE: The estimated *Error* incurred in each step must be known.

Calibration Uncertainty - The maximum calculated *Error* in the *Output* values, shown in a *Calibration Record,* due to causes not attributable to the *Transducer*.

Case Pressure - (See *Burst Pressure Rating, Proof Pressure,* or *Reference Pressure*).

Center of Seismic Mass - The point within an acceleration *Transducer* where acceleration forces are considered to be summed.

Compensation - Provision of a supplemental device,

circuit, or special materials to counteract known sources of *Error.*

Conduction Error - The *Error* in a temperature *Transducer* due to heat conduction between the *Sensing Element* and the mounting of the *Transducer.*

Conformance - (See *Accuracy* and *Error Band).*

Continuous Rating - The rating applicable to specified operation for a specified uninterrupted length of time.

Creep - A change in *Output* occurring over a specific time period while the *Measurand* and all *Environmental Conditions* are held constant.

Critical Damping - (This term is defined under "*Damping*").

Cross-Axis Acceleration - (See *Transverse Acceleration).*

Cross Sensitivity, Cross-Axis Sensitivity - (See *Transverse Sensitivity).*

Damping - The energy dissipating characteristic which, together with *Natural Frequency,* determines the limit of *Frequency Response* and the *Response-Time* characteristics of a *Transducer.*

NOTE 1: In response to a step change of *Measurand,* an underdamped (periodic) system oscillates about its final steady value before coming to rest at that value; and overdamped (aperiodic) system comes to rest without overshoot; and a critically damped system is at the point of change between the underdamped and overdamped conditions.

NOTE 2: Viscous *Damping* uses the viscosity of fluids (liquids or gases) to effect *Damping.*

NOTE 3: Magnetic *Damping* uses the current induced in electrical conductors by changes in magnetic flux to effect *Damping.*

Damping Ratio - The ratio of the actual *Damping* to the *Damping* required for *Critical Damping.*

Dead Volume - The total volume of the pressure port cavity of a *Transducer* with room barometric pressure applied.

Detector - (See *Transducer).*

Dielectric Strength - (See *Breakdown Voltage Rating* and *Insulation Resistance).*

Digital Output - *Transducer Output* that represents the magnitude of the *Measurand* in the form of a series of discrete quantities coded in a system of notation.

NOTE: Distinguished from *Analog Output.*

Directivity - The solid angle, or the angle in a specified plane, over which sound or radiant energy incident on a *Transducer* is measured within specified tolerances in a specified band of *Measurand* frequencies.

Distortion - (See *Harmonic Content).*

Dithering - The application of intermittent or oscillatory forces just sufficient to minimize static friction within the *Transducer.*

Double Amplitude - The peak-to-peak value.

Drift - An undesired change in *Output* over a period of time, which change is not a function of the *Measurand.*

Dynamic Characteristics - Those characteristics of a *Transducer* which relate to its response to variations of the *Measurand* with time.

End Device, End Instrument - (See *Transducer).*

End Points - The Outputs at the specified upper and lower limits of the *Range.*

NOTE: (S) Unless otherwise specified, *End Points* are averaged during any one *Calibration.*

End-Point Line - The straight line between the *End Points.*

Environmental Conditions - Specified external conditions (shock, vibration, temperature, etc.) to which a *Transducer* may be exposed during shipping, storage, handling, and operation.

Environmental Conditions, Operating - *Environmental Conditions* during exposure to which a *Transducer* must perform in some specified manner.

Error - The algebraic difference between the indicated value and the true value of the *Measurand.*

NOTE 1: (S) It is usually expressed in percent of the *Full Scale Output,* sometimes expressed in percent of the *Output* reading of the *Transducer.*

NOTE 2: (S) A theoretical value may be specified as true value.

Error Band - The band of maximum deviations of *Output* values from a specified reference line or curve due to those causes attributable to the *Transducer.*

NOTE 1: (S) The band of allowable deviations is usually expressed as "±——per cent of *Full Scale Output*", whereas in test and calibration reports the band of maximum actual deviations is expressed as "+ —— per cent, –——per cent of *Full Scale Output.*"

NOTE 2: (S) The *Error Band* should be specified as applicable over at least two *Calibration Cycles,* so as to include *Repeatability,* and verified accordingly.

Error Curve - A graphical representation of *Errors* obtained from a specified number of *Calibration Cycles.*

Excitation - The external electrical voltage and/or current applied to a *Transducer* for its proper operation.

NOTE 1: In the sense of a physical quantity to be measured by a *Transducer,* use *Measurand.*

NOTE 2: (S) Usually expressed as range(s) of voltage and/or current values.

NOTE 3: Also see *Maximum Excitation.*

Field of View - The solid angle, or the angle in a specified plane, over which radiant energy incident on a *Transducer* is measured within specified tolerances.

Frequency-Modulated Output - An *Output* in the form of frequency deviations from a center frequency, where the deviation is a function of the applied *Measurand.*

Frequency Output - An *Output* in the form of frequency which varies as a function of the applied *Measurand* (e.g., angular speed and flow rate).

Frequency, Natural - The frequency of free (not forced) oscillations of the *Sensing Element* of a fully assembled *Transducer.*

NOTE 1: It is also defined as the frequency of a sinusoidally applied *Measurand* at which the *Transducer Output* lags the *Measurand* by 90 degrees.

NOTE 2: (S) Applicable at *Room Temperature* unless otherwise specified.

NOTE 3: Also see *Frequency, Resonant* and *Frequency, Ringing* which are considered of more practical value than *Natural Frequency.*

Frequency, Resonant - The *Measurand* frequency at which a *Transducer* responds with maximum *Output* amplitude.

NOTE 1: (S) When major amplitude peaks occur at more than one frequency, the lowest of these frequencies is the *Resonant Frequency.*

NOTE 2. (S) A peak is considered major when it has an amplitude at least 1.3 times the amplitude of the frequency to which specified *Frequency Response* is referred.

NOTE 3: For subsidiary resonance peaks see *Resonances.*

Frequency, Ringing - The frequency of the oscillatory transient occurring in the *Transducer Output* as a result of a step change in *Measurand.*

Frequency Response - The change with frequency of the *Output/Measurand* amplitude ratio (and of the phase difference between *Output* and *Measurand*), for a sinusoidally varying *Measurand* applied to a *Transducer* within a stated range of *Measurand* frequencies.

NOTE 1: (S) It is usually specified as "within ± _____ per cent (or ± _____ db) from _____ to _____ Hertz."

NOTE 2: (S) *Frequency Response* should be referred to a frequency within the specified *Measurand* frequency range and to a specific *Measurand* value.

Frequency Response, Calculated - The *Frequency Response* of a *Transducer* calculated from its *Transient Response,* its mechanical properties, or its geometry, and so identified.

Friction - (See *Friction Error*)

Friction Error - The maximum change in *Output,* at any *Measurand* value within the specified *Range,* before and after minimizing friction within the *Transducer* by *Dithering.*

Friction-Free Error Band - The *Error Band* applicable at *Room Conditions* and with frictions within the *Transducer* minimized by *Dithering.*

Full Scale - (See *Range*).

Full-Scale Output - The algebraic difference between the *End Points.*

NOTE: (S) Sometimes expressed as "±(half the algebraic difference)" e.g., "± 2.5 volts".

Gage Factor - A measure of the ratio of the relative change of resistance to the relative change in length of a *Resistive Strain Transducer* (strain gage).

Harmonic Content - The distortion in a *Transducer's* sinusoidal *Output,* in the form of harmonics other than the fundamental component.

NOTE: (S) It is usually expressed as a percentage of rms *Output.*

Hysteresis - The maximum difference in *Output,* at any *Measurand* value within the specified Range, when the value is approached first with increasing and then with decreasing *Measurand.*

NOTE: (S) *Hysteresis* is expressed in percent of *Full Scale Output,* during any one *Calibration Cycle.* *Friction Error* is included with *Hysteresis* unless *Dithering* is specified.

Inaccuracy - (See *Error*).

Input - (See *Excitation* or *Measurand*).

Input Impedance - The impedance (presented to the *Excitation* source) measured across the *Excitation* terminals of a *Transducer.*

NOTE: (S) Unless otherwise specified, *Input Impedance* is measured at *Room Conditions,* with no *Measurand* applied, and with the *Output* terminals open-circuited.

Instability - (See *Stability*).

Insulation Resistance - (S) The resistance measured between specified insulated portions of a *Transducer* when a specified dc voltage is applied at *Room Conditions* unless otherwise stated.

Intermittent Rating - The rating applicable to specified operation over a specified number of time intervals of specified duration; the length of time between these time intervals must also be specified.

Internal Pressure - (See *Burst Pressure, Proof Pressure,* or *Reference Pressure*).

Leakage Rate - The maximum rate at which a fluid is permitted or determined to leak through a seal.

NOTE: (S) The type of fluid, the differential pressure across the seal, the direction of leakage and the location of the seal must be specified.

Least-Squares Line - The straight line for which the sum of the squares of the residuals (deviations) is minimized.

Life, Cycling - (S) The specified minimum number of full *Range* excursions or specified partial *Range* excursions over which a *Transducer* will operate as specified without changing its performance beyond specified tolerances.

Life, Operating - (S) The specified minimum length of time over which the specified *Continuous* and *Intermittent Rating* of a *Transducer* applies without change in *Transducer* performance beyond specified tolerances.

Life, Storage - (S) The specified minimum length of time over which a *Transducer* can be exposed to specified Storage Conditions without changing its performance beyond specified tolerances.

Linearity - The closeness of a *Calibration Curve* to a specified straight line.

NOTE: (S) *Linearity* is expressed as the maximum deviation of any *Calibraton* point on a specified straight line, during any one *Calibration Cycle.* It is expressed as "within ± ___ percent of *Full Scale Output.*"

Linearity, End Point - *Linearity* referred to the *End-Point Line.*

Linearity, Independent - *Linearity* referred to the "*Best Straight Line.*"

Linearity, Least Squares - *Linearity* referred to the *Least-Squares Line.*

Linearity, Terminal - *Linearity* referred to the *Terminal Line.*

Linearity, Theoretical Slope - *Linearity* referred to the *Theoretical Slope.*

Line Pressure - (See *Reference Pressure*).

Load - (See *Load Impedance*).

Load Impedance - The impedance presented to the *Output* terminals of a *Transducer* by the associated external circuitry.

Loading Error - An *Error* due to the effect of the *Load Impedance* on the *Transducer Output.*

NOTE: In the case of force *Transducers* the term "loading" has been applied to application of force.

Maximum (Minimum) Ambient Temperature - The value of the highest (lowest) ambient temperature that a *Transducer* can be exposed to, with or without *Excitation* applied, without being damaged or subsequently showing a performance degradation beyond specified tolerances.

Maximum Excitation - (S) The maximum value of *Excitation* voltage or current that can be applied to the *Transducer* at *Room Conditions* without causing damage or performance degradation beyond specified tolerances.

Maximum (Minimum) Fluid Temperature - (S) The value of the highest (lowest) *Measured-Fluid* temperature that a *Transducer* can be exposed to, with or without *Excitation* applied, without being damaged or subsequently showing a performance degradation beyond specified tolerances.

NOTE: (S) When a *Maximum* or *Minimum Fluid Temperature* is not separately specified it is intended to be the same as any specified *Maximum* or *Minimum Ambient Temperature.*

Mean Output Curve - The curve through the mean values of Output during any one *Calibration Cycle* or a different specified number of *Calibration Cycles.*

Measurand - A physical quantity, property or condition which is measured.

NOTE: The term *"Measurand"* is preferred to "input", "parameter to be measured", "physical phenomenon", "stimulus", and "variable".

Measured Fluid - The fluid which comes in contact with the *Sensing Element.*

NOTE: The chemical and/or physical properties of this fluid may be specified to insure proper *Transducer* operation.

Mounting Error - The *Error* resulting from mechanical deformation of the *Transducer* caused by mounting the *Transducer* and making all *Measurand* and electrical connections.

Natural Frequency - (See *Frequency, Natural.* See also *Frequency, Resonant.*)

Non-Linearity - (See *Linearity*).

Non-operating Conditions - (See *Environmental Conditions, Non-operating*).

Non-Repeatability - (See *Repeatability*).

Null - A condition, such as of balance, which results in a minimum absolute value of *Output*.

Operating Conditions - See *Environmental* Condition).

Output - The electrical quantity, produced by a *Transducer,* which is a function of the applied *Measurand*.

Output Impedance - The impedance across the *Output* terminals of a *Transducer* presented by the *Transducer* to the associated external circuitry.

Output Noise - The rms, peak, or peak-to-peak (as specified) a-c component of a *Transducer's* d-c *Output* in the absence of *Measurand* variations.

NOTE: (S) Unless otherwise specified, *Output Impedance* is measured at *Room Conditions* and with the *Excitation* terminals open/circuited, except that nominal *Excitation* and *Measurand* between 80 and 100 percent-of-*Span* is applied when the *Transducer* contains integral active output-conditioning circuitry.

Output Regulation - The change in *Output* due to a change in *Excitation*.

NOTE: (S) Unless otherwise specified, *Output Regulation* is measured at *Room Conditions* and with the *Measurand* applied at its upper *Range* limit.

Overload - The maximum magnitude of *Measurand* that can be applied to a *Transducer* without causing a change in performance beyond specified tolerance.

Overrange - (See *Overload*).

Overshoot - The amount of *Output* measured beyond the final steady *Output* value, in response to a step change in the *Measurand*.

NOTE: (S) Expressed in percent of the equivalent step change in *Output*.

Parameter (To be measured) - (See *Measurand*).

Peak-to-Peak - (See *Double Amplitude*).

Physical Input - (See *Measurand*).

Pickup - (See *Transducer*).

Power Input - (See *Excitation*).

Precision - (See *Repeatability* and *Stability*).

Primary Element, Primary Detector - (See *Sensing Element*).

Proof Pressure - The maximum pressure which may be applied to the *Sensing Element* of a *Transducer* without changing the *Transducer* performance beyond specified tolerances.

NOTE 1: In the case of *Transducers* intended to measure a property of pressurized fluid, proof pressure is applied to the portion subject to the fluid.

NOTE 2: (S) *Differential-Pressure Transducer* specifications should indicate whether the specified differential *Proof Pressure* is applicable at ambient or maximum specified *Reference Pressure*, or both, and whether a reverse-differential *Proof Pressure*, at ambient or maximum specified *Reference Pressure*, or both, is additionally applicable.

Range - The *Measurand* values, over which a *Transducer* is intended to measure, specified by their upper and lower limits.

Recovery Time - The time interval, after a specified event (e.g., *Overload, Excitation* transients, *Output* shortcircuiting) after which a *Transducer* again performs within its specified tolerances.

Reference Pressure - The pressure relative to which a *Differential-Pressure Transducer* measures pressure.

Reference Pressure Error - The *Error* resulting from variations of a *Differential-Pressure Transducer's Reference Pressure* within the applicable *Reference Pressure Range*.

NOTE: (S) It is usually specified as the maximum change in *Output*, at any *Measurand* value within the specified *Range*, when the *Reference Pressure* is changed from *Ambient Pressure* to the upper limit of the specified *Reference Pressure Range*.

Reference Pressure Range - (S) The range of *Reference Pressures* which can be applied without changing the *Differential-Pressure Transducer's* performance beyond specified tolerances for *Reference Pressure Error*. When no such error is specified, none is allowed.

Reference-Pressure Sensitivity Shift - The *Sensitivity Shift* resulting from variations of a *Differential-Pressure Transducer's Reference Pressure* within specified limits.

Reference-Pressure Zero Shift - The change in the *Zero-Measurand Output* of a *Differential-Pressure Transducer* resulting from variations of *Reference Pressure* (applied simultaneously to both pressure ports) within its specific limits.

Repeatability - The ability of a *Transducer* to reproduce *Output* readings when the same *Measurand* value is applied to it consecutively, under the same conditions, and in the same direction.

NOTE: (S) *Repeatability* is expressed as the maximum difference between *Output* readings; it is expressed as "within_____percent of *Full Scale Output*". Two *Calibration Cycles* are used to determine *Repeatability* unless otherwise specified.

Reproducibility - (See *Repeatability*).

Resolution - The magnitude of *Output* step changes as the *Measurand* is continuously varied over the range.

NOTE 1: This term relates primarily to *Potentiometric Transducers*.

NOTE 2: (S) Resolution is best specified as *Average* and *Maximum Resolution;* it is usually expressed in percent of *Full Scale Output.*

NOTE 3: In the sense of the smallest detectable change in *Measurand* use *Threshold.*

Resolution, Average - (S) The reciprocal of the total number of output steps over the *Range,* multiplied by 100 and expressed in percent voltage-ratio (for a *Potentiometric Transducer*) or in percent of *Full-Scale Output.*

Resolution, Maximum - (S) The magnitude of the largest of all *Output* steps over the *Range,* expressed as percent voltage-ratio (for a *Potentiometric Transducer*) or in percent of *Full-Scale Output.*

Resonances -Amplified vibrations of *Transducer* components, within narrow frequency bands, observable in the *Output,* as vibration is applied along specified *Transducer* axes.

Resonant Frequency - (See *Frequency, Resonant).*

Response Time - The length of time required for the *Output* of a *Transducer* is to rise to a specified percentage of its final value as a result of a step change of *Measurand.*

NOTE 1: (S) To indicate this percentage it can be worded so as to precede the main term, e.g., "98%-Response Time:____ milliseconds, max."

NOTE 2: Also see *Time Constant* and *Rise Time.*

Ringing Period - The period of time during which the amplitude of *Output* oscillations, excited by a step change in *Measurand,* exceed the steady-state *Output* value.

NOTE: (S) Unless otherwise specified, the *Ringing Period* is considered terminated when the *Output* oscillations no longer exceed ten percent of the subsequent steady-state *Output* value.

Rise Time - The length of time for the *Output* of a *Transducer* to rise from a small specified percentage of its final value to a large specified percentage of its final value as a result of a step-change of *Measurand.*

NOTE 1: (S) Unless otherwise specified, these percentages are assumed to be 10 and 90 percent of the final value.

NOTE 2: Also see *Time Constant.*

Room Conditions - Ambient *Environmental Conditions,* under which transducers must commonly operate, which have been established as follows:

    (a) Temperature: $25 \pm 10°$C ($77 \pm 18°$ F)

    (b) Relative Humidity: 90 percent or less.

    (c) Barometric Pressure: 26 to 32 inches Hg.

NOTE: Tolerances closer than shown above are frequently specified for transducer calibration and test environments.

Self-Heating - Internal heating resulting from electrical energy dissipated within the *Transducer.*

Sensing Element - That part of the *Transducer* which responds directly to the *Measurand.*

NOTE: This term is preferred to "Primary element", "Primary detector", "Primary detecting element".

Sensitivity - The ratio of the change in *Transducer Output* to a change in the value of the *Measurand.*

NOTE: In the sense of the smallest detectable change in *Measurand* use *Threshold.*

Sensitivity Shift - A change in the slope of the *Calibration Curve* due to a change in *Sensitivity.*

Sensor - (See *Transducer*).

Source Impedance - The impedance of the *Excitation* supply presented to the *Excitation* terminals of the *Transducer.*

Span - The algebraic difference between the limits of the *Range.*

Speed of Response - (See *Response Time, Time Constant).*

Stability - The ability of a *Transducer* to retain its performance characteristics for a relatively long period of time.

NOTE: (S) Unless otherwise stated, *Stability* is the ability of a *Transducer* to reproduce *Output* readings obtained during its original *Calibration,* at *Room Conditions,* for a specified period of time; it is then typically expressed as "within ____ percent of *Full Scale Output* for a period of____months."

Static Calibration - A *Calibration* performed under *Room Conditions* and in the absence of any vibration, shock, or acceleration (unless one of these is the *Measurand*).

Stimulus - (See *Measurand*).

Strain Error - The *Error* resulting from a strain imposed on a surface to which the *Transducer* is mounted.

NOTE 1: This term is not intended to relate to *Strain*

Transducers (strain gages).

NOTE 2: Also see *Mounting Error.*

Tapping - (See *Dithering*).

Temperature Error - The maximum change in *Output,* at any *Measurand* value within the specified *Range,* when the *Transducer* temperature is changed from *Room Temperature* to specified temperature extremes.

Temperature Error Band - The *Error Band* applicable over stated environmental temperature limits.

Temperature Gradient Error - The transient deviation in *Output* of a *Transducer* at a given *Measurand* value when the ambient temperature or the *Measured Fluid* temperature changes at a specified rate between specified magnitudes.

Temperature Range, Compensated - (See *Temperature Range, Operating*).

Temperature Range, Fluid - The range of temperature of the *Measured Fluid,* when it is not the ambient fluid, within which operation of the *Transducer* is intended.

NOTE 1: (S) Within this range of fluid temperature all tolerances specified for *Temperature Error, Temperature Error Band, Temperature Gradient Error, Thermal Zero Shift* and *Thermal Sensitivity Shift* are applicable.

NOTE 2: (S) When a *Fluid Temperature Range* is not separately specified, it is intended to be the same as the *Operating Temperature Range.*

Temperature Range, Operating - The range of ambient temperatures, given by their extremes, within which the *Transducer* is intended to operate; (S) within this range of ambient temperature all tolerances specified for *Temperature Error, Temperature Error Band, Temperature Gradient Error, Thermal Zero Shift* and *Thermal Sensitivity Shift* are applicable.

Terminal Line - A *Theoretical Slope* for which the *Theoretical End Points* are 0 and 100% of both *Measurand* and *Output.*

Theoretical Curve - The specified relationship (table, graph, or equation) of the *Transducer Output* to the applied *Measurand* over the *Range.*

Theoretical End Points - The specified points between which the *Theoretical Curve* is established and to which no *End Point* tolerances apply.

NOTE: The points can be other than 0 and 100% of both *Measurand* and *Output.*

Theoretical Slope - The straight line between the *Theoretical End Points.*

Thermal Coefficient of Resistance - The relative change in resistance of a conductor or semi-conductor per unit change in temperature over a stated range of temperature.

NOTE: (S) Expressed in ohms per ohm per degree F or C.

Thermal Compensation - (See *Compensation*).

Thermal Sensitivity Shift - (S) The *Sensitivity Shift* due to changes of the ambient temperature from *Room Temperature* to the specified limits of the *Operating Temperature Range.*

Thermal Zero Shift - (S) The *Zero Shift* due to changes of the ambient temperature from *Room Temperature* to the specified limits of the *Operating Temperature Range.*

Threshold - The smallest change in the *Measurand* that will result in a measurable change in *Transducer Output.*

NOTE: When the *Threshold* is influenced by the *Measurand* values, these values must be specified.

Time Constant - The length of time required for the *Output* of a *Transducer* to rise to 63% of its final value as a result of a step change of *Measurand.*

Total Error Band - (See *Error Band*).

Torque Error - (See *Mounting Error*).

Transducer - A device which provides a usable *Output* in response to a specified *Measurand.*

NOTE: The term *Transducer* is usually preferred to "Sensor" and "Detector" and to such terms as "Flowmeter", "Accelerometer" and "Tachometer"; it is always preferred to "Pickup", "Gage" (when not equipped with a dial-indicator), "Transmitter" (which has an entirely different meaning in telemetry technology), "Cell", and "End Instrument".

Transduction Element - The electrical portion of a *Transducer* in which the *Output* originates. (Refer to Table I, Third Modifier.)

Transient Response - The response of a *Transducer* to a step-change in *Measurand.*

NOTE: (S) *Transient Response,* as such, is not shown in a specification except as a general heading, but is defined by such characteristics as *Time Constant, Response Time, Ringing Period,* etc.

Transverse Response - (See *Transverse Sensitivity.*)

Transverse Acceleration - An acceleration perpendicular to the sensitive axis of the *Transducer.*

Transverse Sensitivity - The *Sensitivity* of a *Transducer* to *Transverse Acceleration* or other transverse *Measurand.*

NOTE: (S) It is specified as maximum *Transverse Sensitivity* when a specified value of *Measurand* is applied along the transverse plane in any direction, and is usually expressed in percent of the *Sensitivity* of the *Transducer* in its sensitive axis.

Variable - (See *Measurand*).

Vibration Error - The maximum change in *Output*, at any *Measurand* value within the specified *Range*, when vibration levels of specified amplitude and range of frequencies are applied to the *Transducer* along specified axes.

Vibration Sensitivity - (See *Vibration Error*).

Voltage Ratio - For potentiometric *Transducers*, the ratio of *Output* voltage to *Exitation* voltage, usually expressed in percent.

Warm-up Period - The period of time, starting with the application of excitation to the *Transducer*, required to assure that the *Transducer* will perform within all specified tolerances.

Zero-measurand Output - The *Output* of a *Transducer*, under *Room Conditions* unless otherwise specified, with nominal excitation and zero *Measurand* applied.

Zero Shift - A change in the *Zero-measurand Output* over a specified period of time and at *Room Conditions*.

NOTE: This *Error* is characterized by a parallel displacement of the entire *Calibration Curve*.

INSTRUMENT SOCIETY of AMERICA
Research Triangle Park, North Carolina

ISA-RP37.2-1982

Recommended Practice

Guide for Specifications and Tests for Piezoelectric Acceleration Transducers for Aero-Space Testing

Instrument Society of America

ISBN 0-87664-377-2

ISA-RP37.2   Guide for Specifications and Tests for
              Piezoelectric Acceleration Transducers
              for Aerospace Testing

INSTRUMENT SOCIETY OF AMERICA
67 Alexander Drive
P.O. Box 12277
Research Triangle Park, North Carolina 27709

PREFACE

This Preface is included for informational purposes and is not a part of ISA-RP32.1.

This Recommended Practice has been prepared as a part of the service of the Instrument Society of America toward a goal of uniformity in the field of instrumentation. To be of real value, this document should not be static, but should be subject to periodic review. Toward this end, the Society welcomes all comments and criticisms, and asks that they be addressed to the Secretary, Standards and Practices Board, Instrument Society of America, 67 Alexander Drive, P.O. Box 12277, Research Triangle Park, NC 27709, Telephone (919) 549-8411.

The ISA Standards and Practices Department is aware of the growing need for attention to the metric system of units in general, and the International System of Units (SI) in particular, in the preparation of instrumentation standards. The Department is further aware of the benefits to USA users of ISA Standards of incorporating suitable references to the SI (and the metric system) in their business and professional dealings with other countries. Towards this end, this Department will endeavor to introduce SI - acceptable metric units in all new and revised standards to the greatest extent possible. The Metric Practice Guide, which has been published by the American Society for Testing and Materials as ANSI designation Z210.1 (ASTM E380-76, IEEE Std. 268-1975), and future revisions, will be the reference guide for definitions, symbols, abbreviations, and conversion factors.

It is the policy of the Instrument Society of America to encourage and welcome the participation of all concerned individuals and interests in the development of ISA Standards. Participation in the ISA standards making process by an individual in no way constitutes endorsement by the employer of that individual of the Instrument Society of American or any of the standards which ISA develops.

The development of this Recommended Practice was initiated as a result of a survey conducted in December 1960. A total of 240 questionnaires was sent out to transducer users and manufacturers in the aerospace field. In their replies, a strong majority indicated a need for standardization of specifications and tests of several types of commonly used aerospace test transducers with electrical output.

On the basis of these replies, a Project Sub-Committee 8A-RP 37.2, Guide for Piezoelectric Acceleration Transducers for Aerospace Testing, was formed under the cognizance of Committee 8A-RP 37, Transducers for Aerospace Testing, of the Aerospace Standards Group, and this Recommended Practice was developed.

COMMITTEE 8A-RP 37 (SCOTFAST)
H. N. Norton, Chairman (General Dynamics/Astronautics)
Thomas A. Perls, Vice-Chm. (Lockheed Missiles & Space Co.)
A. A. Zuehlke, Secretary (Bourns, Inc.)
Mrs. Rose Mary Bernstein (Marketing Engineering Company)
M. E. Binkley (Lockheed Missiles & Space Company)
D. G. Egan (General Dynamics/Fort Worth)
R. G. Jewell (General Electric Company)
P. S. Lederer (National Bureau of Standards)
O. L. Smith (Marshall Space Flight Center)
Frank K. Werner (Rosemount Engineering Company)

SUB-COMMITTEE 8A-RP 37.2
Thomas A. Perls, Chm. (Lockheed Missiles & Space Company)
William L. Vandal, Sec. (McDonnell Aircraft Corporation)
Wilson Bradley, Jr. (Endevco Corporation)
Seymour Edelman (National Bureau of Standards)
E. J. Kirchman (NASA Goddard Space Flight Center)
Anthony W. Orlacchio (Gulton Industries, Inc.)

This Recommended Practice was reaffirmed in 1982 by the SP37 Committee and this reaffirmation was approved by the ISA Standards and Practices Board in October 1982.

TENTATIVE RECOMMENDED PRACTICE
GUIDE FOR SPECIFICATIONS AND TESTS
FOR PIEZOELECTRIC ACCELERATION TRANSDUCERS
FOR AERO-SPACE TESTING

CONTENTS

1. PURPOSE

This Recommended Practice establishes the following for piezoelectric acceleration transducers:

1.1 Uniform minimum general specifications for design and performance characteristics.

1.2 Uniform minimum acceptance and qualification test methods, including calibration.

1.3 Uniform presentation of minimum test data.

1.4 A drawing symbol for use in schematics.

2. SCOPE

2.1 This Recommended Practice covers piezoelectric acceleration transducers, primarily those used in aero-space test instrumentation.

2.2 Terminology used in this document follows I.S.A. RP 37.1, Nomenclature and Specification Terminology for Aerospace Test Transducers with Electrical Output, except that additional terms considered applicable to piezoelectric vibration transducers are defined in Paragraph 4.3.

3. DRAWING SYMBOL

3.1 General

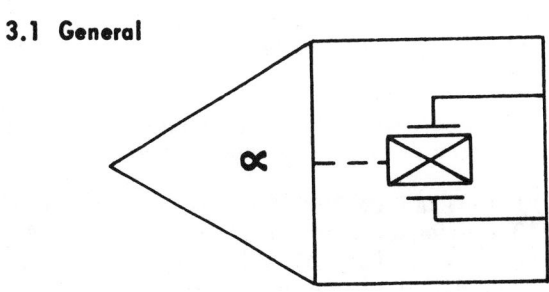

3.2 Self-checking accelerometers

3.2.1 Active

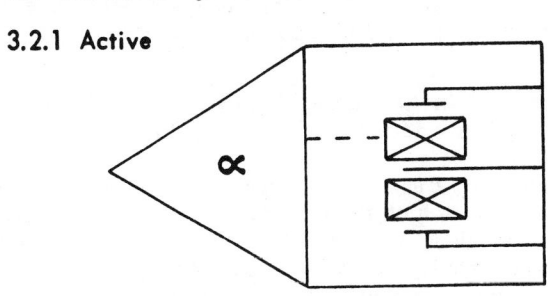

3.2.2 Passive

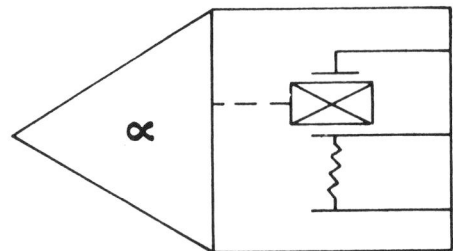

4. CHARACTERISTICS

The first two paragraphs contain, in alphabetical order, the terms to be used in manufacturers' and users' specifications for piezoelectric acceleration transducers. Some terms not already defined in ISA RP 37.1 (Nomenclature and Specification Terminology for Aero-Space Tests Transducers with Electrical Output) are marked with an asterisk (*) and are defined in Paragraph 4.3.

In Paragraphs 4.1 and 4.2 below, each characteristic listed is checked (or numbered with reference to Sections 5 or 6 below) in one or several columns to indicate the characteristics which are included in manufacturers' literature or users' specifications as follows:

BASIC - indicates characteristics which are normally specified for each transducer.

SUPPLEMENTAL - indicates additional characteristics which may be specified if desired.

The remaining three columns indicate the tests required to determine these characteristics.

INDIVIDUAL ACCEPTANCE TESTS (IAT) - tests performed on each transducer.

Within Stated Tolerances - these tests are of the "go - no go" type with respect to stated tolerances.

Measured Value - these tests provide individual measured values of the characteristic.

QUALIFICATION TESTS - tests in addition to the IAT's performed on representative samples of each transducer design. All IAT's are, of course, performed during a formal qualification test program.

Some of the BASIC and SUPPLEMENTAL characteristics are determined by the manufacturer's research and development and require no additional testing.

4.3 ADDITIONAL TERMINOLOGY
(For terms not defined here, see ISA RP 37.1.)

Acceleration Limit - the maximum vibration and shock acceleration which the transducer can accept in either direction along its sensitive axis without permanent damage, usually stated as ± ____ g's. The acceleration limits are usually much wider than the Accelera-tion Range and thereby represent a measure of the over-load capability of the transducer.

Acceleration Range - the range of accelerations over which the transducer has the specified linearity.

Acoustic Sensitivity - the output of a transducer (not due to rigid body motions) in response to a specified acoustical environment. This is sometimes expressed as the acceleration in g rms sufficient to produce the same output as induced by a specified sound pressure level spectrum having an over-all value of 140 db referred to 0.0002 dyne per sq. cm. rms.

Amplification Factor at Resonant Frequency - the ratio of the maximum sensitivity of a transducer (at its resonant frequency to its Reference Sensitivity).

Note: "Amplification factor at resonant frequency" is sometimes referred to as "Q".

Amplitude Linearity, Shock - closeness of sensitivity to reference sensitivity over a stated range of acceleration amplitudes, under shock conditions, usually specified as "within ± ____ percent for acceleration rise times longer than ____ microseconds."

Amplitude Linearity, Vibration - closeness of sensitivity over a stated range of acceleration amplitudes, at a stated fixed frequency, usually specified as "within ± ____ percent."

Electromagnetic Field Sensitivity - the maximum output of a transducer in response to a specified amplitude and frequency of magnetic field, usually expressed in gauss equivalent to a stated fraction of 1 g.

Frequency Response - the change with frequency of the sensitivity with respect to the reference sensitivity, for a sinusoidally varying acceleration applied to a transducer within a stated range of frequencies, usually specified as "within ± ____ percent of the reference sensitivity from ____ to ____ cps." The applicable total capacitance and load resistance should be stated.

Grounded or Ungrounded - refers to the presence or absence of an electrical connection between the "low" side of the transducer element and the portion of the transducer intended to be in contact with the test structure. Method of ungrounding should be stated as "internally ungrounded" or "by means of separate stud."

Markings - information shown on the transducer itself, will normally include Manufacturer, Model Number and Serial Number.

Mechanical Isolation of Transduction Element - internal construction of transducer which allows forces (particularly bending forces and external pressures) to be applied to the transducer case with negligible resulting forces on the transducer element.

Polarity - the relationship between the transducer output and the direction of the applied acceleration; taken as "standard" when a positive charge or voltage appears on the "high" side of the transducer for an ac-

(continued on page 7)

4.1 DESIGN CHARACTERISTICS

	Basic	Supple-mental	Characteristics Determined During: Individual Accep. Tests — Within Stated Tolerances	Individual Accep. Tests — Measured Value	Qual. Tests
Cable, (integral or non-integral) or Connector Type	x		5.1		
Cable, Standard, Supplied					
Type	x				
Length	x		5.3.2		
Capacitance at Room Temp.	x			5.3.2	
Temperature Range	x				6.2
Capacitance vs. Temp.		x			6.7
Noise	x		5.3.2		
Dimensions, Configuration and *Markings	x		5.1		
*Grounded or Ungrounded	x		x		
Housing Material(s)	x				
Insulation Resistance (minimum) at Maximum Rated Temperature					
Across Element	x				6.2.1
Element to Ground (if applicable)	x				6.2.1
Mounting Method (adhesive, stud, separate adaptors; state thread size, class, mounting torque, and temperature rating, if applicable.)	x				x
Temperature Range,					
Operating	x				6.2
Storage	x				
Transducer Capacitance with Stated Cable	x			5.3.3	
Transducer Seal	x		5.3.1		6.11
Transduction Element					
Material Type (proprietary name acceptable)	x				
*Sensing Mode	x				
*Mechanical Isolation		x			
Weight (state whether cable or other accessories are included)	x				x

4.2 PERFORMANCE CHARACTERISTICS

	Basic	Supple-mental	Characteristics Determined During:		
			Individual Accep. Tests		Qual. Tests
			Within Stated Tolerances	Measured Value	
*Acceleration Limit	x				
*Acceleration Range	x				6.5.1
*Amplification Factor at Resonant Frequency		x			6.3.1
Amplitude Linearity					
*Shock	x				6.5.2
*Vibration	x				6.5.1
Partial Range		x	x		
Environmental Effects					
*Acoustic Sensitivity		x			6.9
*Electromagnetic Response		x			6.10
Temperature Sensitivity Error					
-65°F to 350°F (177°C) or	x				6.7
Above 350°F.	x			5.4.1	
Below -65°F.	x			5.4.1	
*Transient Temperature Error		x			6.8
Other		x			6.12
Frequency, Resonant, Mounted; Nominal (and tolerance), or Minimum	x				6.3
*Frequency Response					
Maximum Range	x				6.4
Stated Partial Range	x			5.2.2	
Mounting Error		x			6.1
*Polarity	x		5.3.4		
*Reference Sensitivity, Charge or Voltage, Nominal (and tolerance)	x			5.2.1	
Sensitivity Stability	x				6.13
Strain Sensitivity		x			6.6
*Transverse Sensitivity					
At Stated Single Freq.	x			5.2.3	
Over Maximum Freq. Range		x			5.2.3

(continued from page 4)
celeration directed from the mounting surface into the body of the accelerometer.

Reference Sensitivity (Charge or Voltage) - the ratio of the change in charge or voltage generated by a transducer to the change in value of the acceleration that is measured under a set of defined conditions. (Amplitude, frequency, temperature, total capacitance, amplifier input resistance, mounting torque). Deviations in sensitivity should be reported as deviations from the reference sensitivity.

Sensing Mode of Transduction Element - the method used to stress the transduction element such as compression, bending or shear.

Sensitivity - the ratio of the change in transducer output to a change in the value of the acceleration.
Notes
1 Where one sensitivity under defined conditions is the basis for determining deviations in performance, use "reference sensitivity."
2 Because the use of piezoelectric acceleration transducers for the measurement of both shock and vibration, the acceleration is required to be known in either g peak or g rms. A specified sensitivity in millivolts per g is to be understood as meaning "rms millivolts per rms g" or its equivalent "peak millivolts per peak g." The use of "mixed units" such as rms millivolts per peak is to be avoided. Note, however, that an output of 10 millivolts rms is also approximately 14.1 millivolts peak, and an acceleration of 1 g rms is approximately 1.41 g peak and, therefore,

$$\frac{10mv}{g} = 10\frac{mv\,rms}{g\,rms} = 10\frac{mv\,pk}{g\,pk} = 7.07\frac{mv\,rms}{g\,pk} = 14.1\frac{mv\,pk}{g\,rms}$$

Shock - a substantial disturbance characterized by a rise and decay of acceleration from a constant value in a short period of time.

Strain Sensitivity - the sensitivity to strains applied to the base by bending, in the absence of any rigid body motion of the transducer. It is expressed as 10^{-6} times the equivalent acceleration level in g's for a strain in the plane of the base.

Temperature Range, Operating - the interval of temperatures in which the transducer is intended to be used, specified by the limits of this interval.

Temperature Sensitivity Error - the change in sensitivity of a transducer from its reference sensitivity as a result of changes in its ambient temperature over a specified operating temperature range.
Note: If changes in voltage sensitivity are specified, the total associated capacitance must be stated.

Transient Temperature Error - the output of a transducer as a result of a specified transient temperature change within a specified operating temperature range.
Note: The associated capacitance and load resistance, as well as the time, after the applied transient, at which the amplitude peak occurs must be specified.

Transverse Sensitivity - the maximum sensitivity of a uni-axial transducer to a transverse acceleration, with-

in a specified frequency range, usually expressed in percent of the reference sensitivity in the intended measuring direction.

5. INDIVIDUAL ACCEPTANCE TESTS AND CALIBRATION

Tests are listed in the order they are to be performed.

5.1 Visual Inspection

Conduct a complete visual examination for conformance to stated configuration and markings. Determine weight, dimensions, thread size, and class utilizing standard inspection instruments. Check mating of accessory cable (if any) by attaching and removing the cable. Note any discrepancy.

5.2 Initial Functional Tests

5.2.1 Reference Charge or Voltage Sensitivity

For most applications, it is recommended that the sensitivity of a transducer be determined by comparison with a standard calibrated transducer. This method is described in General Reference A.

The frequency of the driving signal should be in the range of 40 to 100 cycles per second and should be monitored continuously by a properly calibrated electronic counter.

Charge sensitivity can be determined by multiplying the voltage sensitivity in volts per g by the total capacitance of the system in picofarads, providing charge sensitivity in pico-coulombs per g.

A system consisting of a transducer, cable, and amplifier should be calibrated as a unit. The sensitivity will be that recorded at the output of the amplifier, in volts/g. Specifications should include transducer sensitivity with cable, capacitance, and amplifier gain.

If a standard transducer with calibration traceable to NBS is not available, a reasonably dependable sensitivity can be determined by methods described in General Reference A. The chatter method, described in Reference Number 5, is the simplest to use and is the most dependable for reasonable accuracy.

The temperature of the transducers should be measured and recorded.

Notes
1. The standard transducer used should have a calibration traceable (see RP 37.1) to a calibration performed at the National Bureau of Standards within the normal calibration period of one year. The transducer used as a standard should be reserved for this purpose only; it should not be exposed to large values of shock, vibration or temperature extremes; and its calibration should be checked periodically by either of the referenced methods.

2. The surface on which each transducer is mounted

and the part of the transducer base which touches that surface should be clean and flat, with a surface finish of 64 microinches or less. If oil or grease is used as a gasketing material, it should be clean, freshly applied to both surfaces just before the test, and completely removed immediately afterwards.

3. The signal used to excite the motion of the transducers should be as nearly sinusoidal as possible. The wave shape of the output signal of both transducers should be observed frequently throughout the test and no perceptible distortion should be allowed. Preferably, a distortion meter should be used, and the distortion kept below 3%.

4. Screw-attached transducers should be mounted with the torque recommended by the manufacturer, using a good grade torque wrench. A preliminary observation of the wave shape and amplitude on an oscilloscope should be made when the transducer is mounted with a torque about 10% less than recommended. Then the recommended torque should be applied. A calibration is not valid if a small increase in mounting torque changes the wave shape or amplitude of the output appreciably. Torque larger than recommended should never be used unless it is certain that no damage will result. The torque used should be reported with the results of the test.

5. The plate, on opposite sides of which the transducers are mounted, should be thick enough so that no appreciable flexure occurs and thin enough so that both pickups have the same motion.

6. The motion applied to the two transducers should be far enough above the noise level so that the system noise represents a minor error in the calibration. A signal-to-noise ratio of 40 db is desirable. Increasing the motion beyond this level is not desirable on most vibration test equipment because it increases the chance of distortion and non-axial motion. The acceleration level should be reported with the results of the test.

5.2.2 Frequency Response

Frequency response is determined by using the method described for reference sensitivity (5.2.1) at a number of frequencies over the range of interest. The frequency response of the standard in this range must be known.

For acceptance testing, ten frequencies should be chosen at which the output ratios are measured. If equally spaced on a logarithmic frequency scale, these frequency points should adequately determine the frequency response curve.

Notes
1. The plate on which the two transducers are mounted should be made of a high-modulus material such as a machinable tungsten alloy.

2. If the signal from one transducer is applied to one set of plates of an oscilloscope, and the signal from the other transducer is applied to the other set of plates, rotation of the resulting Lissajous figure as the driving frequency is increased from very low to high frequencies indicates the onset of difference in the motion of the transducers. Distortion of the figure indicates bad motion or other cause for investigation. The test should be repeated with the transducer positions interchanged. A calibration is not valid if interchanging positions changes results by more than 3%. (5% above 5,000 cps).

5.2.3 Transverse Sensitivity

Mount the transducer in a suitable test arrangement such that the known vibratory motion in a plane perpendicular to the sensing axis is at least 100 times the motion in the direction of the sensing axis. The frequency and amplitude of the motion should be stated and should be within the rated frequency and amplitude ranges of the transducer. Rotate the mounted transducer about the sensing axis through 360 degrees to determine the transverse direction of maximum sensitivity.

Express the output at this maximum sensitivity as a percentage of the output which would be obtained if the known motion were applied in the direction of the sensing axis.

5.3 Tests and Measurements

5.3.1 Transducer Seal Immersion Test

Use water at room temperature in a transparent container such as a Pyrex beaker. Heat the water to approximately 200°F. Remove detachable cables and connectors from the transducer. Immerse the transducer beneath the surface of the heated water. Any stream of air bubbles released from the transducer indicates leakage and constitutes failure. Dry the transducer without application of heat and measure the insulation resistance of the element. The minimum insulation resistance shall be met.

5.3.2 Coaxial Cable

Length and Capacitance

Measure the length of cable between the end faces of the connectors. It is acceptable, if within ± 1/4 inch or ± 2 percent of specified length, whichever is larger. For detachable cables, measure the capacitance with a capacitance bridge at 1000 cps and record.

Test for Cable Noise

The output noise of a standard length of cable (usually 3 or 4 feet) should be specified as "less than ____ mv (peak to peak)" when tested with instrumentation shown in this paragraph. (This test is adapted from Reference 20, but is somewhat more severe).

The instrumentation shall include a standard shielded capacitor (1000 pfd) connected across the cable; a weight equal to the weight of 40 feet of the test cable, in two half-cylindrical shapes that are clamped or taped to the outer jacket of the cable; a preamplifier or cathode follower with at least 10 megohms input impedance; and an oscilloscope or recorder with a capability of providing a full scale deflection from a 1 mv signal generated by the cable.

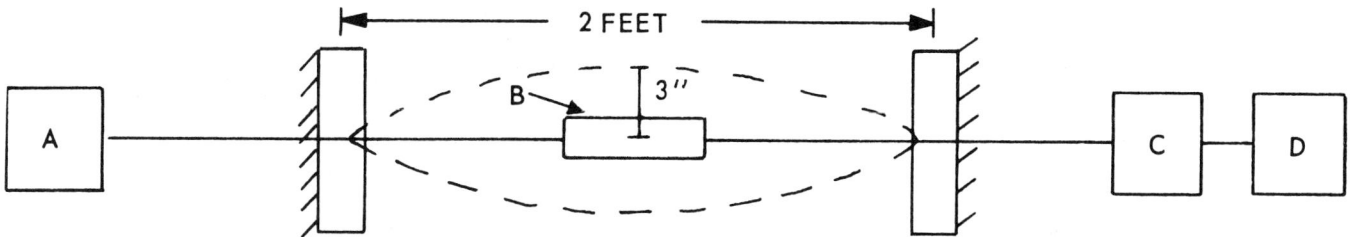

A — Shielded capacitor — 1000 pfd.
B — Cylindrical weight (2-1/2″ length). Material — brass.
C — Cathode follower or preamplifier.
D — Oscilloscope or Recorder.

Connect the cable electrically as shown in the schematic. Clamp the cable between pieces of wood to the two anchor points, allowing a 3-inch sag in the center of the cable. Clamp weight (B) to cable at center of anchored span. Raise the cable by the weight to maximum height (3 inches above its neutral position) and drop. Monitor output noise of the cable on the oscilloscope. Repeat test 3 times. Record the maximum value in mv peak to peak.

5.3.3 Transducer Capacitance

Measure capacitance of transducer with cable attached by means of a capacitance bridge at 1000 cps; measure capacitance of transducer cable at 1000 cps. Subtract cable capacitance from total capacitance to obtain transducer capacitance. Temperature should be 70°F. to 85°F.

5.3.4 Polarity (Use either the Transient Method or the Comparison Method.)

Transient Method

Check (with a battery or other DC voltage source) the transducer preamplifier and oscilloscope readout to determine the direction of scope deflection for a positive voltage input to the preamplifier.

Connect the transducer to the preamplifier and apply a transient acceleration to the transducer. The initial voltage output from the transducer should be positive for:

1. Uniaxial transducers with force applied against the base of the transducer.

2. Triaxial transducers with force applied in the direction of arrows shown on the transducer for the axes involved.

Comparison Method

Compare the test accelerometer with a standard accelerometer on an electrodynamic vibration exciter operating at a frequency below 1000 cps.

Use a phase meter or oscilloscope to indicate the phase angle between the standard and test accelerometers. Select the preamplifiers used on both the standard and test accelerometers so that their phase shifts between input and output are the same. The test accelerometer should have approximately zero phase shift.

Note: Back-to-back mounting results in opposite motions being applied to the two accelerometers at any instant.

5.4 Final Functional Tests

5.4.1 Temperature Sensitivity Error at Maximum or Minimum Rated Temperature

Determine transducer capacitance at room temperature (C_{trt}).

Mount the transducer on a calibration exciter in a temperature chamber. Determine the voltage output E_{rt} at room temperature at a frequency below 1000 cps with a known external capacitance C_{ext}. Maintain the vibration amplitude and frequency constant throughout the entire test. Stabilize the transducer at the rated temperature for 15 minutes. Measure the following:

(1) Voltage output E_{mt} at rated temperature with the known external capacitance C_{ext}.

(2) Transducer capacitance at rated temperature (C_{tmt}).

(3) Transducer resistance at rated temperature.

Cautions and Notes:

(1) The standard which is used to establish the vibration amplitude must be known to be suitable under all test conditions.

(2) Measure capacitances at 1000 cps with a capacitance bridge, without motion applied to the transducer.

(3) Measure resistance with a megohmmeter. Test voltage should be 50 vdc unless otherwise specified.

(4) Allowance may have to be made for changes in cable resistance and/or cable capacitance throughout the transducer temperature range.

(5) Monitor the output waveform of the transducer with an oscilloscope to insure that it appears sinusoidal.

(6) Do not leave the transducer in an open circuit condition while the temperature is being changed; it should be shorted or connected to a preamplifier.

Repeat the room temperature tests as above at the conclusion of the final functional tests.

If applicable, compute the percentage change in transducer voltage sensitivity at rated temperature compared with room temperature.

If applicable, compute the percentage change in transducer charge sensitivity at rated temperature compared with room temperature:

% change in charge sensitivity =

$$100 \left[\frac{(C_{tmt} + C_{ext})E_{mt} - (C_{trt} + C_{ext})E_{rt}}{(C_{trt} + C_{ext})E_{rt}} \right]$$

Low resistance or excessive sensitivity change with temperature may constitute failure. Non-repeatability of room temperature output before and after the test may constitute failure.

5.4.2 Reference Voltage or Charge Sensitivity
Repeat 5.2.1

5.4.3 Frequency Response
Repeat 5.2.2

5.4.4 Transverse Sensitivity
Repeat 5.2.3

6. QUALIFICATION TESTS

Tests which are performed on representative transducers in addition to the Individual Acceptance Tests.

6.1 Effectiveness of Mounting Technique

6.1.1 With thread-mounted transducers, determine the sensitivity at a frequency between 40 and 100 cps for various thread torques in the region of the recommended mounting torque. This determines the care which must be used in installing the transducer. Do not exceed a maximum torque rating, if such a rating exists.

6.1.2 With electrically insulated mounting studs, apply the rated maximum mounting torque to the stud several times; reject if there is visible

evidence of mechanical damage. Use the tested mounting stud if possible when conducting the final frequency response test. (See 5.4.3)

6.2 Temperature Range

6.2.1 Resistance and Capacitance at Maximum Rated Temperature (delete if rated higher than 350°F. and measured in 5.4.1).

Insert transducer with cable attached into an oven with the temperature sensing thermocouple in contact with or adjacent to the base of the transducer. Short the open end of the cable. Increase oven temperature and stabilize at the maximum rated temperature for 15 minutes. Remove short and measure the resistance of the transducer-cable combination using a 50 volt megohmmeter. Measure the capacitance at 1000 cps with a capacitance bridge. Again short the open end of the cable and return to room temperature. If the resistance is not acceptable, re-run the test with the cable alone to determine whether the cause of low resistance is within the transducer or the cable assembly.

In the same way measure the insulation resistance from both sides of the element to ground for ungrounded transducers.

Note: All connectors should be wiped with clean alcohol and a dry cloth before the test.

6.2.2 Soak Test

Soak the transducer and associated cable for one hour each at the minimum and maximum rated temperatures (the transducer should be shorted during the test). Measure the reference sensitivity before and after the test, in accordance with paragraph 5.2.1. Reject if the apparent change in reference sensitivity exceeds an allowable limit.

6.3 Mounted Resonant Frequency

The frequency response of a piezoelectric accelerometer depends on the value of the (lowest) resonant frequency of the instrument when mounted on the structure to be tested. It is shown in References 16 and 21 that this frequency is not only a function of the mass of the structure, but also of the compliance at the contact between structure and accelerometer. It does not appear possible to specify a test which will determine this frequency for all installations of a given transducer. It is therefore suggested that a test be specified which will give an indication of transducer resonant frequency under a set of standard, reproducible conditions, with the understanding that the resonant frequency in actual use will in all probability be appreciably different (generally lower, by a factor depending on the mass and compliance of the test structure).

The resonant frequency is to be determined with the transducer mounted on a small plate or anvil made of a high modulus material such as a machinable tungsten alloy. (Normally acceptable dimensions for damped natural frequencies below 50,000 cps approximate a one-inch cube. In general, the mass of the plate or anvil should be approximately ten times the active mass of the transducers.) Determine the resonant frequency by the Sinusoidal Method or the damped natural frequency by the Shock Method. State method used. (Impedance methods are applicable only to certain accelerometer designs, and it is shown in References 21 and 22 that they do not reliably establish the lowest mounted resonant frequency). Use the Sinusoidal Method, if it is known that the resonant frequency of the vibration exciter is above that of the accelerometer; otherwise use the Shock Method. It may not be practical to obtain good results from the Shock Method when the accelerometer has additional resonances near its resonant frequency. (See RP 37.1 for definitions. In practice, for transducers with low damping, the resonant, damped natural, and undamped natural frequencies are essentially the same.)

6.3.1 Sinusoidal Method

Measure the accelerometer output using the comparison method throughout the frequency range including and above the resonant frequency of the accelerometer. The resonant frequency is the frequency of maximum sensitivity. The phase angle relative to the standard accelerometer changes by almost 180 degrees in the range of frequencies near the resonant frequency of the accelerometer.

The standard accelerometer should be built into the exciter. The resonant frequency of the exciter is determined by measuring the transfer impedance between the driver coil current and the standard accelerometer output. Use a dummy mass load to simulate a transducer during the transfer impedance test. Use the exciter only throughout the range up to 95 percent of its resonant frequency. (Note: The low-frequency rigid-body rise in amplitude and change in phase may be ignored). The resonant frequency of the exciter may also be determined with an accelerometer whose damped natural frequency has been determined by the Shock Method and found to be above the exciter resonant frequency.

6.3.2 Shock Method

The damped natural frequency should be determined by mechanical excitation by a short transient impact whose pulse duration is about three times the natural period of the accelerometer. The damped natural frequency is established by the frequency of the ringing which occurs on the transducer output; the ringing should be presented on a photograph of an oscilloscope trace or on a memory oscilloscope.

6.4 Frequency Response, Maximum Range
(see 5.2.1 and 5.2.2.)

Special care and/or special techniques may be re-

quired in the frequency range of 1 to 20 cps and at frequencies exceeding 2000 cps.

6.5 Amplitude Linearity

The Vibration Linearity or Shock Linearity methods may be used. It is necessary to use the Shock Linearity Method, if the acceleration range exceeds the attainable acceleration on resonant beams or rods. Calibration errors in linearity tests often exceed 2%. In these cases, it should be stated that the test indicated no deviations from linearity greater than the calibration errors.

6.5.1 Vibration Linearity

Amplitude linearity is performed using the comparison method with a standard accelerometer. Transversely resonant beams may be used at frequencies up to 1000 cps and longitudinally resonant rods up to 10,000 cps to obtain the desired acceleration. The beams and rods are attached rigidly to the exciter and tuned to their fundamental free-free resonant transverse and longitudinal modes, respectively. When using the resonant rod, care is taken to mount both the standard and test accelerometers in close proximity on the same end and at a location far from the node point. The ratio of the two outputs is measured throughout the range of applied accelerations as determined from the standard accelerometer.

The standard accelerometer must be calibrated over the applied acceleration range by the Shock Linearity method, Section 6.5.2 or by the Reference Voltage or Charge Sensitivity method, Section 5.2.1 or by one of the absolute calibration methods listed in general Reference A.

6.5.2 Shock Linearity

Use a ballistic impact or comparison technique similar to that outlined in General Reference A, Sections 4.6 and 5, respectively. Measure and record the basic sensitivity before and after the shock linearity tests.

6.6 Strain Sensitivity

The technique used to measure strain sensitivity of an accelerometer meets the requirement of ASA Z 24.21-1957 (General Reference C) paragraph 3.1.3.7. The accelerometer is mounted on a simple cantilever beam. The radius of curvature at the point where the accelerometer is mounted is 1000 inches when the measurements are taken.

A steel beam is held as a cantilever in a vice bolted to a concrete floor. The beam is 3.0" wide by 0.5" thick and 60" long. (The free length is approximately 57 inches.) The natural frequency is very close to 5 cps. Four strain gauges are bonded to the beam adjacent to the accelerometer mounting hole (two each, top and bottom, about 1.5" from the edge of the clamp.)

A two-channel recorder is used to record the output of both the strain gauge bridge and the accelerometer under test.

The system is excited by manually deflecting the free end of the beam and allowing it to vibrate freely. The output of the accelerometer is taken from the oscillograph record at a point where the strain in the surface of the beam is 250×10^{-6} inch per inch. (This is equivalent to a radius of curvature of 1000 inches.) The strain sensitivity, in g's, for a strain of 10^{-6} inch per inch is found by dividing the above accelerometer output by 250 times the accelerometer sensitivity in millivolts per g.

6.7 Temperature Sensitivity Error

To determine the temperature response at each temperature of interest, follow procedures given in 5.4.1 but do not repeat room temperature test.

6.8 Transient Temperature Error

Mount the transducer on a one-inch cube of aluminum. Adjust the external capacitance to be approximately equal to the capacitance of the specified cable (or use the actual cable). Connect the transducer to a DC amplifier whose input resistance is approximately 10^{8} ohms. Immerse the transducer in water whose temperature is approximately 50°F. above room temperature. Measure and record the maximum quasi-DC voltage which is generated by the transducer and the time from the start of the transient at which this maximum voltage is reached. If the voltage reverses within the first two seconds and reaches a peak in the opposite polarity, record the amplitude and the time of the peak also. Convert these voltages to equivalent g's based upon the transducer reference voltage sensitivity with the specified cable.

6.9 Acoustic Sensitivity

Place the acceleration transducer in a reverberant acoustical test chamber. The transducer shall be mounted or suspended with a system whose undamped natural frequency is 25 cps or less. Subject the transducer to a specified sound pressure level spectrum covering the frequency range from 75 cps to 9600 cps. Either a swept sinusoid or a random acoustic input may be employed. Measure the rms electrical output of the acceleration transducer (using a suitable preamplifier) and convert to equivalent rms g's based upon the reference sensitivity of the transducer and corrected for external capacitance, preamplifier gain, etc. Report the maximum transducer output in equivalent rms g's; the frequency of this maximum output, if evident; and the specified sound pressure level spectrum.

Note: The acoustic sensitivity of piezoelectric accelerometers is negligible in almost all applications except possibly where vibration is to be measured on vibration-isolated components which are subjected to high-intensity airborne noise. Even in these cases, the sound pressure reaching the transducer generally causes vibratory accelerations of the structure to which the accelerometer is attached of sufficient magnitude to make negligible any output due to pressure changes alone. Tests for acoustic sensitivity, including the one suggested here, tend to be conservative in that they yield an electrical output which is too large because it includes signals due to rigid-body motions of the accelerometer.

6.10 Electromagnetic Response

Mount the transducer on a 10 to 15 lb. plate of non-magnetic material such as lead. Place the mounted transducer in a known 60 or 400 cps magnetic field so that the sensitive axis of the transducer points toward the source of electromagnetic energy and the plate is away from the source. Rotate the transducer and plate about the sensitive axis of the transducer recording the maximum transducer electrical output. Record as equivalent g per gauss based on reference sensitivity corresponding to the external capacity used in the electromagnetic test. Specify test frequency or frequencies.

Note: Induced mechanical vibrations and electrical ground loops must be eliminated from the test set up.

6.11 Transducer Seal Immersion Test (Use on all moisture sealed or hermetically sealed transducers).

De-aerate several inches of water at 70°F. in a closed transparent container by reducing the absolute pressure to about 2 psi. Return the pressure to normal and immediately immerse the transducer beneath the surface. Reduce the absolute pressure to about 2 psi. Air leakage from the transducer constitutes failure. Let the pressure return to normal atmospheric with the transducer still submerged. Dry the transducer without the application of heat and measure the insulation resistance of the element. The minimum insulation resistance shall be met.

6.12 Other Environmental Effects

Special tests, as dictated by the user's needs, may include, but are not limited to:

> Altitude
> Explosion
> Fungus
> Humidity
> Nuclear Radiation
> Radio Interference
> Salt Fog
> Sand and Dust

6.13 Sensitivity Stability

Measure the Reference Sensitivity of three transducers as in 5.2.1. Then perform the following:

(1) Soak each transducer at maximum rated temperature for one hour. Return to room temperature by allowing to cool for 24 hours. Measure and record new reference sensitivity for each.

(2) Impact each transducer three times in succession at maximum rated acceleration. Measure and record the new reference sensitivity of each 24 hours later.

(3) Soak each transducer at minimum rated temperature for one hour. Return to room temperature by allowing to warm for 24 hours. Measure and record new reference sensitivity for each.

For each transducer, note the maximum percentage change from the original reference sensitivity anywhere during the above tests, and calculate the arithmetical average of these for the three transducers. This is a measure of the average Sensitivity Stability of the transducer.

Note: The manufacturer may have data on stability sensitivity of any transducer design based on long term measurements. These measurements are compiled over years' time by periodic (monthly) reference sensitivity measurements on transducers not otherwise in use.

Model No.		Part No.
Manufacturer		Serial No.
	INDIVIDUAL ACCEPTANCE TESTS & CALIBRATIONS FOR PIEZOELECTRIC ACCELERATION TRANSDUCERS	Transducer Element
		Case Material

1. VISUAL (5.1): Dimensions ☐ Weight ☐ Finish ☐ Markings ☐ Cable ☐ Receptacle ☐
2. ELECTRICAL: Transducer Capacitance (5.3.3) ____pf Cable Capacitance (5.3.2) ____ pf
   Cable Length (5.3.2) ____ ft. Cable Noise (5.3.2) less than _____ mv.
3. TRANSDUCER SEAL (5.3.1):
4. POLARITY (5.3.5):
5. MAXIMUM TRANSVERSE SENSITIVITY (5.2.3): _____ %
6. REFERENCE CHARGE OR VOLTAGE SENSITIVITY (5.2.1): _____ at _____ cps at _____ °F

Frequency Response (5.2.2) _____ % max. Temperature Sensitivity Error (5.4.1) at _____ cps, _____ g.

Frequency cps	Amplitude g's	Output mv or pc*	Sensi- tivity		Reference Temperature °F.	T1 °F	T2 °F	T3 °F	T4 °F
				Resistance	meg ohm				
				Transducer Capacitance	pf				
				Cable Capacitance	pf				
				System Capacitance	pf				
				Output	mv				
				% Change					

BY _____ DATE _____ APPROVED _____

NOTE: Numbers in parentheses refer to sections in ISA Recommended Practices 37.2.

APPROVED _____

STAMP

*pc stands for picocoulombs.

Model No.		Part No.
	QUALIFICATION TESTS	Serial No.
Manufacturer	FOR PIEZOELECTRIC ACCELERATION TRANSDUCERS	Transducer Element
		Case Material

REFERENCE SENSITIVITY _____ at _____ cps at _____ °F.

    Transducer capacitance _____ pf; Cable capacitance _____ pf

5.1 Mounting Technique:

    Preamplifier Model _____ Gain _____ Cable capacitance _____ pf

    Transducer:

	Torque	Sensitivity at cps
Maximum		
Rated		
75% Rated		
50% Rated		

    Stud: Maximum Torque Rating _____ Evidence of damage _____

5.2 Temperature Range:

    Resistance at top rated temperature of _____°F. across element _____meg Ω; to ground _____meg Ω.

    Capacitance at top rated temperature across element _____ pf.

Transducer: Reference Sensitivity at _____ cps:
           Reference Sensitivity after maximum temperature of _____ °F: _____
           Reference Sensitivity after minimum temperature of _____ °F: _____

Stud: Evidence of damage due to maximum and minimum temperature _____

Cable: Evidence of damage due to maximum and minimum temperature _____

5.3 Mounted Resonant Frequency:

_____ Kc Measured by _____ Method with/without mounting stud.

5.4 Frequency Response: See Individual Acceptance Test (5.2.2)

Note: Bold numbers and numbers in parentheses refer to sections in ISA Recommended Practice 37.2.

6.5 Amplitude Linearity: Preamplifier Model _____ Gain _____
Cable Capacitance _____ pf

VIBRATION				SHOCK			
Amplitude	Signal	Sensitivity	Deviation from Ref. Sens.	Amplitude	Signal	Sensitivity	Deviation from Ref. Sens.

6.6 Strain Sensitivity: ___ g x 10^{-6} for a strain in the plane of the base.

6.7 Temperature Sensitivity Error: See Individual Acceptance Test (5.4.1)

6.8 Transient Temperature Error:
DC Amplifier Model # _____ Input Resistance _____ ohms
Ambient Temperature of Aluminum Block _____ °F   Water Temperature _____ °F
Cable capacitance _____ pf   Amplifier Gain _____
Maximum DC voltage generated: _____ volts and _____ g/°F

6.9 Acoustic Sensitivity:
Preamplifier Model # _____ Gain _____ Cable Capacitance _____ pf

Frequency Band	SPL	mv (rms)	Equivalent g

6.10 Electromagnetic Response:

Preamplifier Model _____ Gain _____ Cable Capacitance _____ pf

Mounting Material _____ Weight _____

Magnetic field level _____ gauss at _____ cps

Transducer output _____ mv _____ g/gauss

6.11 Transducer Seal

Water Temperature_____°F.
Pressure_____ psia.
Air Leakage?
Insulation Resistance across Element _____ meg Ω; to ground_____ meg Ω.

6.12 Other Environmental Effects

Give details of tests and results of measurements.

6.13 Sensitivity Stability

Stability Tests: (1) Soak Temperature:_____°F; Time: ____hr; Cooling time:____ hrs.

(2) Impact Acceleration:_____ g applied _____ times.

(3) Soak Temperature: _____°F; time: ____ hr; warming time:_____ hrs.

Reference Sensitivity measured at _____ cps, _____°F.

REFERENCE SENSITIVITY IN MV/G

Serial No.	Before Tests	After (1)	24 hrs. after (2)	After (3)

	Serial No.	Before Tests	After Test (1)	24 hours After (2)	After Test (3)
TRANSDUCER CAPACITANCE (pf)					
CABLE CAPACITANCE (pf)					

BY _____ DATE _____ APPROVED _____

NOTE: Test numbers and numbers in parentheses refer to procedures in ISA RP 37.2

APPROVED_____

stamp

7. REFERENCES

General

A. *American Standard Methods for the Calibration of Shock and Vibration Pickups*, (S2.2-1959), American Standards Association

B. *Auxiliary Equipment for Shock and Vibration Measurements, Method for Specifying the Characteristics of*, (S2.4-1960); American Standards Association

C. *Specifying the Characteristics of Pickups for Shock and Vibration Measurement, Method for*, (Z24.21-1957); American Standards Association

D. Roberts H. C., *Mechanical Measurements by Electrical Methods*, Pittsburgh, Instruments Publishing, 1946, 357pp

E. Den Hartog, J. P., *Mechanical Vibrations*, New York, Mc Graw-Hill, 1947, Chapter II, Section 9.10

F. Harris, Cyril M. and Crede, Charles E., *Shock and Vibration Handbook*, 1961, 3 volumes

Specific Articles on Piezoelectric Transducers

1. Levy, S., and Kroll, W. D., "Response of Accelerometers to Transient Accelerations," *Journal of Research of the National Bureau of Standards*, Vol. 45, No. 4, October 1950, pp. 303-309, (Research Paper 2138).

2. Harrison, M., Sykes, A. O., and Marcotte, P. G., *The Reciprocity Calibration of Piezoelectric Accelerometers*, Washington, D. C., David Taylor Model Basin, March 1952, (Report R-811).

3. Unholtz, Karl., "The Calibration of Vibration Pickups to 2,000 cps," *Proceedings of the Instrument Society of America*, Vol. 7, 1952, (Paper 52-26-3).

4. Conrad, R. W., and Vigness, Irwin, "Calibration of Accelerometers by Impact Techniques," *Proceedings of the Instrument Society of America*, Vol. 8, 1953, pp. 166-170.

5. Kissinger, C. W., "Determination of Sinusoidal Acceleration at Peak Levels Near That of Gravity by the 'Chatter' Method," *Proceedings of the Instrument Society of America*, Vol. 9, Part V, 1954, (Paper 50-40-1).

6. Perls, T. A., and Kissinger, C. W., "High-g Accelerometer Calibrations by Impact Methods with Ballistic Pendulum, Air Gun, and Inclined Trough," *Proceedings of the Instrument Society of America*, Vol. 9, Part V, 1954, 6 pp. (Paper 54-40-2).

7. Perls, T. A., Kissinger, C. W., Paquette, D. R., "Steady-State Calibration of Vibration Transducers at Accelerations up to ± 4000g," *Bulletin of American Physical Society*, Vol. 30, No. 3, P. 35, Program of 1955 Washington Meeting; (Paper KA 13).

8. Levy, S., and Bouche, R. R., "Calibration of Vibration Pickups by the Reciprocity Method," *Journal of Research of the National Bureau of Standards*, Vol. 57, No. 4, October 1956, pp. 227-243, (Research Paper 2714).

9. Orlacchio, A. W., "Measurements of Shock and Vibrations Under Extreme Environmental Conditions," *Electrical Manufacturing*, Vol. 59, No. 1, January 1957, pp. 78-81.

10. Jansen, J. H., "Calibration of Small Vibration Pickups", *Acustica*, Vol. 8, pp. 179, 1958.

11. Shchedrovitskic, S. S., "Methods and Equipment for Calibration and Checking of Accelerometers," *Measurement Techniques*, Vol. 6, p. 720, 1958, (Translated from Russian.)

12. Tukker, J. C., "Report on Calibration and Use of Some Piezoelectric Accelerometers," Report No. 59061, *Technisch Physische Dieust T.N.O.*, Ent. H. Delft, Holland, 12 Dec. 1959.

13. Nisbet, J. S., Brennan, J. N., and Tarpley, H. I., "High Frequency Strain Gauge and Accelerometer Calibration," *Journal of the Acoustical Society of America*, 32, p. 71, Jan. 1960.

14. Bouche, R. R., "The Absolute Calibration of Pickups on a Drop-Ball Shock Machine of the Ballistic Type," *Proceedings of the Institute of Environmental Sciences*, p. 115, April 1961.

15. Lane, Kerwin, "Equipment and Techniques used in Calibrating Vibration Measuring Transducers to 10,000 cps", *Proceedings of the Institute of Environmental Sciences*, p. 109, April 1961.

16. Schloss, Fred, "Inherent Limitations of Accelerometers for High-Frequency Vibration Measurements," *Journal of the Acoustical Society of America*, 33, p. 539, April 1961.

17. Jones, E., Edelman, S., Sizemore, K. O., "Calibration of Vibration Pickups at Large Amplitudes," *Journal of the Acoustical Society of America*, 33, P. 1462, November 1961.

18. Schmidt, V. A., Edelman, S., Smith, E. R., and Pierce, E. T., "Modulated Photoelectric Measurement of Vibration," *Journal of the Acoustical Society of America*, 34, p. 355, April 1962.

19. Massa, F., "Vibration Measurements for the Frequency Range 10 Cycles to 20 Kilocycles," *Instruments*, November 1948.

20. Perls, T. A., "A Simple Objective Test for Cable Noise Due to Shock, Vibration or Transient Pressures," PB 121583, Office of Technical Services, U. S. Government Printing Office, 1955.

21. Lederer, P. S., "Resonant Frequencies of Piezoelectric Accelerometers and the Evaluation of an Electrical Technique for their Determination," Report 7566, National Bureau of Standards, 1962,

(also to be published in abstract in *Technical News Bulletin of National Bureau of Standards*)

22. Clements, E. W., and Stone, M. G., "Techniques for the Rapid Estimation of Accelerometer Natural Frequencies," Report No. 5681, *U. S. Naval Research Laboratory*, October 1961.

Related ISA Transducer Recommended Practices

RP31.1 — Terminology and Specifications for Turbine-Type Flow Transducers

RP37.1 — Nomenclature and Specification Terminology for Aero-Space Test Transducers

This document was approved for publication by the Standards and Practices Board on September 11, 1963.

E. A. Adler   — United Engineers and Constructors, Inc.; Director-at-Large

R. E. Clarridge   — General Electric Company; Director, Intersociety Standards and Practices

G. G. Gallagher — The Fluor Corporation, Ltd.; Director-at-Large

R. L. Galley   — North American Aviation Company, Rocketdyne Division; Director, Aerospace Standards

E. C. Hutchison — Union Carbide Nuclear Company; Director, Production Processes Standards

J. R. Mahoney   — Union Carbide Corporation; Vice-President, Standards and Practices Department

E. J. Minnar   — ISA International Headquarters; Secretary Pro Tem, Standards and Practices Board

INSTRUMENT SOCIETY of AMERICA
Research Triangle Park, North Carolina

ANSI/ISA-S37.3-1975
(R 1982)
Approved December 14, 1982
(Formerly ANSI MC 6.2-1975)

American National Standard

Specifications and Tests for Strain Gage Pressure Transducers

Instrument Society of America

ISBN 0-87664-378-0

ISA-S37.3   Specifications and Tests for
             Strain Gage Pressure Transducers

INSTRUMENT SOCIETY OF AMERICA
67 Alexander Drive
P.O. Box 12277
Research Triangle Park, North Carolina 27709

PREFACE

This Preface is included for informational purposes and is not a part of ISA-S37.3.

This Standard has been prepared as a part of the service of the Instrument Society of America toward a goal of uniformity in the field of instrumentation. To be of real value, this document should not be static, but should be subject to periodic review. Toward this end, the Society welcomes all comments and criticisms, and asks that they be addressed to the Standards and Practices Board Secretary, Instrument Society of America, 67 Alexander Drive, P.O. Box 12277, Research Triangle Park, NC 27709

The ISA Standards and Practices Department is aware of the growing need for attention to the metric system of units in general, and the International System of Units (SI) in particular, in the preparation of instrumentation standards. The Department is further aware of the benefits to USA users of ISA Standards of incorporating suitable references to the SI (and the metric system) in their business and professional dealings with other countries. Towards this end, this Department will endeavor to introduce SI - acceptable metric units in all new and revised standards to the greatest extent possible. The Metric Practice Guide, which has been published by the American Society for Testing and Materials as ANSI designation Z210.1 (ASTM E380-76) and future revisions, will be the reference guide for definitions, symbols, abbreviations, and conversion factors.

It is the policy of the Instrument Society of America to encourage and welcome the participation of all concerned individuals and interests in the development of ISA Standards. Participation in the ISA standards making process by an individual in no way constitutes endorsement by the employer of that individual of the Instrument Society of American or any of the standards which ISA develops.

The development of this Standard was initiated as the result of a survey conducted by the Survey Committee on Transducers for Aero-Space Testing (8A-RP37) in December 1960. In addition to the strong need for improved and uniform transducer nonmenclature and specification terminology, many of the people surveyed also indicated the need for standardization of performance characteristic specifications, test methods, and electrical requirements for certain classes of transducers used in Aero-Space Testing. Accordingly, five subcommittees were established initially, each to deal with one of these classes of transducers. Subcommittee 8A-RP37.3 (Sub-committee on Strain Gage Pressure Transducers, 'SCOSGAPT'') was organized on May 1, 1961, to prepare a Recommended Practice for strain gage pressure transducers. Six successive drafts were prepared and submitted for review and comments to a large number of people active in aerospace industries and sciences in which strain gage pressure transducers are used. The final document, ISA-RP37.3 (Guide for Specifications and Tests for Strain Gage Pressure Transducers for Aero-Space Testing), was published by the Instrument Society of America in April 1964. It was revised in 1970 and approved as ANSI Standard MC 6.2 in October 1975.

This Standard was prepared under the direction of Paul S. Lederer (Chairman S37.3) by members of Committee SP37 by updating and expanding the previous version of the document and by obtaining extensive reviews of drafts of the Standard by representatives of transducer users and manufacturers as well as agencies of the U.S. Government. The reviewers were selected from a board cross-section of all industries and sciences in which transducers are applied for measuring purposes.

This Standard is intended as a guide for technical personnel at user facilities as well as by manufactuers' technical and sales personnel whose duties include specifying, calibrating, testing or showing performance characteristics of potentiometric pressure transducers. By basing users' specifications as well as technical advertising and reference literature on this Standard, or by referencing portions thereof, as applicable, a clear understanding of the users' needs or of the transducers' performance capabilities, and of the methods used for evaluating or proving performance, will be provided. Adhering to the specification outline, terminology and procedures shown will not only result in simple, but also complete specifications; it will also reduce design time, procurement lead time, and labor, as well as material costs. Of major importance will be the reduction of qualification tests resulting from use of a commonly accepted test procedures and uniform data presentation.

The following individuals served on the 1975 SP37.3 committee:

NAME	COMPANY
P. S. Lederer, Chairman	National Bureau of Standards
J. F. Arbogast	Hercules Powder Company
D. M. Keast	Bolt, Beranek and Newman
D. L. Limbacher	Aerojet-General
H. E. Lockery	Baldwin-Lima-Hamilton
W. R. Myers	Consolidated Electrodynamics

The following individuals served on the 1982 SP37 committee:

NAME	COMPANY
J. W. Mock, Chairman	Measurement Services
P. Bliss	Consultant
W. Brinkschulte	Genisco Technology Corporation
C. Kutelis	Revere Corporation of America
R. W. Lally	PCB Piezotronics
H. E. Lockery	Hottinger Baldwin Measurements, Inc.
E. W. Malone	Boeing Aerospace Co.
T. B. Miller	Laurence Livermore Lab
K. A. Parlee	United Engineers & Constructors, Inc.
H. Pitt	
J. Powell	Rosemount Engineering
R. C. Strahm	EG & G Idaho
R. M. Whittier	ENDEVCO
N. Wilde	EG & G Idaho
E. Wong	Bell & Howell

This Standard was approved for publication by the ISA Standards and Practices Board in October 1982.

NAME	COMPANY
T. J. Harrison, Chairman	IBM Corporation
P. Bliss	Consultant
W. Calder	The Foxboro Company
N. Conger	Continental Oil Co.
B. Feikle	Bailey Controls Co.
R. T. Jones	Philadelphia Electric Co.
R. Keller	Boeing Company
O. P. Lovett, Jr.	Isis Corp.
E. C. Magison	Honeywell, Inc.
A. P. McCauley	Diamond Shamrock Corp.
J. W. Mock	Measurement Services
E. M. Nesvig	ERDCO Engineering Corp.
G. Platt	Bechtel Power Corp.
R. Prescott	Moore Products Company
W. C. Weidman	Gilbert Associates
K. A. Whitman	Allied Chemical Corp.
J. R. Williams	Stearns-Roger, Inc.
B. A. Christensen*	
L. N. Combs*	
R. L. Galley*	
R. G. Marvin*	
W. B. Miller*	Moore Products Company
R. L. Nickens*	

*Director Emeritus

SPECIFICATIONS AND TESTS FOR
STRAIN GAGE PRESSURE TRANSDUCERS

CONTENTS

1. PURPOSE

This Standard establishes the following for strain gage pressure transducers:

1.1 Uniform minimum specifications for design and performance characteristics.

1.2 Uniform acceptance and qualification test methods, including calibration techniques.

1.3 Uniform presentation of minimum test data.

1.4 A drawing symbol for use in electrical schematics. (See note in 3.)

2. SCOPE

2.1 This Standard covers strain gage pressure transducers, but primarily those used in measurement systems.

2.2 Included among the specific versions of strain gage pressure transducers to which this Standard are applicable are the following:

> Absolute Pressure Transducers
> Differential Pressure Transducers
> Gage Pressure Transducers
> Sealed Reference Pressure Transducers

3. DRAWING SYMBOL

The drawing symbol for measuring a transducer is a square of dimensions 2x by 2x, with an added equilateral triangle, the base of which is the left side of the square. The triangle symbolizes the sensing element. The letter "P" in the triangle designates "pressure" and the subscripts denote the second modifier.

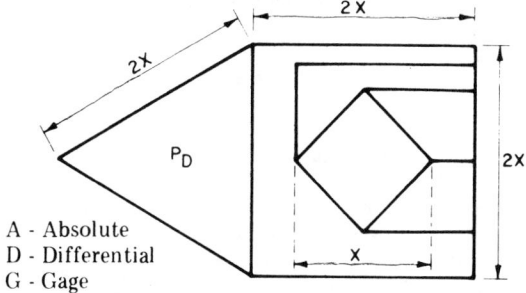

A - Absolute
D - Differential
G - Gage
S - Sealed Reference, Differential

The strain gage is symbolized by a small square, with diagonals x by x, centered in the large square. The diagonals of the small square are drawn perpendicular to the sides of the large squares. Lines from each apex of the small square projected to the right side of the large square represent the electrical leads.

Note: This symbol is not ANSI approved at this time. It has been submitted to the ANSI Y 32 Committee on Graphic Symbols for their consideration and approval.

4. SPECIFICATIONS

Terminology used in this document is defined in ISA Standard 37.1. An *asterisk* appears beside the paragraph number of terms defined in S37.1. Additional terms

considered applicable to strain gage pressure transducers are defined in Section 4.3 of this document.

4.1 DESIGN CHARACTERISTICS

4.1.1 Required Mechanical Design Characteristics

The following mechanical design characteristics shall be listed.

4.1.1.1 TYPE OF PRESSURE SENSED

> * Absolute Pressure
> * Differential Pressure, Unidirectional
> * Differential Pressure, Bidirectional
> *Differential Pressure, Sealed Reference*
> * Gage Pressure

Note: At present, no provision is made by the SI system of units for abbreviations following the pressure units to indicate the type of pressure, as is done in the U.S. customary system of units, e.g., psia for absolute pressure in psi. In the interim it is recommended that for the SI system, the type of pressure be indicated in this manner: "... An absolute pressure of _____ Pa." "... A differential pressure range of _____ kPa," etc.

Note: for differential pressure transducers, the allowable range of reference pressures shall be listed e.g., "0 to 1MPa" or "0 to 100 psi."

4.1.1.2 *Measured Fluids

The fluids in contact with pressure port(s) shall be listed, e.g., nitric acid, liquid oxygen. Requirements for and limitations on the *isolating element* (if used) shall be listed.

4.1.2.3 Materials in Contact with the Measured Fluid

The materials in contact with the measured fluid shall be listed.

Note: For differential pressure transducers, materials in both ports must be considered.

4.1.1.4 Configuration and Dimensions

The outline drawing shall show the configuration with dimensions in millimeters (inches). Unless pressure and electrical connections are specified (Reference 4.1.1.5. 4.1.3.4), the outline shall include limiting maximum dimensions for these connections.

4.1.1.5 Pressure Connection

The pressure connection(s) shall be indicated on the outline drawing. For threaded fittings, specify: Applicable Military or Industry standards or nominal size, number of threads per millimeter (threads per inch), thread series, and thread class. For hose tube fitting, specify tube size.

4.1.1.6 Mountings and Mounting Dimensions

Unless the pressure connection serves as a mounting, the outline drawing shall indicate the method of mounting with hole size, centers, and other pertinent dimensions in millimeters (inches), including thread specifications for threaded holes, if used.

4.1.1.7 Mounting Effects

The maximum mounting force or torque shall be specified if it will tend to affect transducer performance (Reference 4.2.28).

4.1.1.8 Mass

The mass of the transducer shall be specified in grams (ounces).

4.1.1.9 Case Sealing

If case sealing is necessary, the mechanism and materials used for sealing should be described. The same requirement applies to the electrical connector. The resistance of the sealing materials to cleaning solvents and commonly used measured fluids should be stated.

4.1.1.10 Identification

The following characteristics shall be permanently inscribed on the outside of the transducer case or on a suitable nameplate permanently attached to the case.

Nomenclature of transducer (acc. to ISA-S37.1, Section 3).
Name of Manufacturer, (Part number to reflect one controlled configuration), and Serial Number.
 * Range in Pa(psi) and designation of type of pressure (see 4.1.1.1.).
*Maximum *excitation.*

Identification of *Measured and Reference Ports (for differential pressure transducers).
 * Reference Pressure Range (for differential pressure transducers).
Identification of Electrical Connections.

Schematic of Electrical Connections
Nominal Bridge Resistance

Inscription of the following characteristics is optional:

> Sensitivity
> Customer Specification or Part
> Number, or Both
> Type of Electrical Connector
> (if applicable)
> Maximum Excitation
> Maximum Reference Pressure
> Maximum and Minimum
> Operating Temperature

*4.1.1.11 Maximum and Minimum Operating and Fluid Temperature

— The maximum and minimum temperature of fluids or environments which can be applied to the transducer and which will not cause permanent calibration shift shall be listed.

4.1.2 Supplemental Mechanical Design Characteristics

Listing of the following mechanical design characteristics is optional:

4.1.2.1 Case Material

4.1.2.2 Pressure Sensing Element

> Diaphragm, flat or corrugated
> Capsule
> Bellows
> Straight Tube
> Bourdon Tube, plain, spiral,
> helical, or twisted
> Liquid Filled Configuration
> (liquid shall be specified)

4.1.2.3 Type of Strain Gage Used

> Metallic; bonded or unbonded,
> wire or foil, deposited thin film
> Semiconductor; bonded, unbonded,
> or diffused

4.1.2.4 Location of Strain Gage

> Mounted directly on pressure
> sensing element
> Mounted on auxiliary member,
> and activated by pressure sensing
> element

4.1.2.5 Number of Active Strain Gage Elements

> One
> Two-arm Bridge
> Four-arm Bridge

*4.1.2.6 Dead Volume

For non-flush mounted transducers the dead volume shall be given in cubic millimeters (cubic inches). For differential pressure transducers the volume of both cavities should be listed.

4.1.2.7 Volume Change Due to Full Scale Pressure

The change in volume of the sensing element due to application of full scale pressure shall be given in cubic millimeters (cubic inches).

4.1.3 Basic Electrical Design Characteristics

The following electrical design characteristics shall be listed. They are applicable at "room conditions" according to the definition given in ISA S37.1.

*4.1.3.1 Excitation

— Expressed as " _____ volts dc" or " _____ volts rms at _____ Hz." (Preferred values are 5, 10, 20, and 28 volts) or " _____ milliamperes dc" or _____ milliamperes rms at _____ Hz."

*4.1.3.2 Maximum Excitation

— Expressed as " _____ volts dc" or " _____ volts rms at _____ Hz" or " _____ milliamperes dc" or " _____ milliamperes rms at _____ Hz."

*4.1.3.3 Input Impedance

— Expressed as " _____ ± _____ ohms at _____ ± _____ Hz." If impedance is resistive, indicate this.

Note: Output "open-circuit" for this measurement.

*4.1.3.4 Output Impedance

— Expressed as " _____ ± _____ ohms at _____ ± _____ Hz." If impedance is resistive, indicate this.

Note: If excitation terminals are "short-circuit" for this measurement, this should be indicated.

*4.1.3.5 Load Impedance — Performance characteristics values apply only for load impedance values of _____ ohms, minimum or _____ ± _____ ohms.

4.1.3.6 Electrical Connections — Whether the electrical termination is by means of a connector or a cable, the pin designations or wire color code shall conform to the following transducer wiring standard promulgated by the Western Regional Strain Gage Committee, as approved September 18, 1957, and revised May 6, 1960.

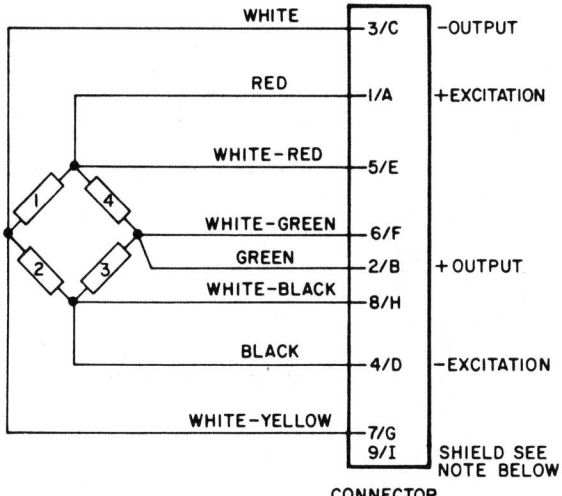

CONNECTOR

NOTES:

1. The output polarities indicated on the above wiring diagram apply when an increasing absolute pressure is applied to the pressure port (sensing end) of an absolute pressure transducer. For differential and gage pressure transducers, the indicated polarities apply when the absolute pressure at measurand port is greater than the absolute pressure at the reference pressure port.

2. The measurand (pressure) port of differential pressure transducers shall be marked "+" or optionally "high" or "meas." while the reference (pressure) port shall be marked "-" or optionally "low" or "ref."

3. The bridge elements shall be arranged so that functions producing positive output will cause increasing resistance in arms 1 and 3 of the bridge.

4. For shielded transducers, pins 5, 7, and 9 shall be shield terminals for 4, 6, and 8 wire systems respectively.

5. Position of any internal compensation network should be indicated.

*4.1.3.7 Insulation Resistance — Expressed as "_____ megohms at _____ volts dc between all terminals in parallel and the transducer case, at a temperature of _____ ± _____ °C(°F)."

4.1.4 Supplemental Electrical Design Characteristics

Listing of the following electrical design characteristic is optional:

4.1.4.1 Shunt Calibration Resistor — Expressed as "_____ ohms for _____ % of full scale output." See Section 4.3.

Note: The terminals across which this resistor is to be placed shall be specified, if the resistor is used.

4.2 Performance Characteristics

The pertinent performance characteristics of strain gage pressure transducers should be tabulated in the order shown. Unless otherwise specified, they apply at "room conditions" as defined in ISA S37.1; Temperature: 25±10°C (77±18°F); Relative Humidity: 90% maximum; Barometric Pressure: 73±7 cm of Hg (29±3 inches of Hg) and after an adequate warm-up period. (See Section 4.2.13).

Terminology used here is defined in either ISA S37.1 or Section 4.3 of this document. An asterisk appears beside the paragraph number of those terms defined in S37.1; the remaining terms are defined in Section 4.3 below.

*4.2.1 Range — Usually expressed as "_____ to _____ Pa(psia, psig)" or "± _____ Pa(psid)" or "zero to _____ Pa(psid.)"

Note: Equivalent pressure units in the SI system are expressed in Pascals.

1 psi = 6894.8 Pa
10kPa = 1.4504 psi

*4.2.2 End Points — Expressed as "_____ ± _____ mV and _____ ± _____ mV at _____ volts (milliamps) excitation."

*4.2.3 Full-Scale Output — Expressed as "_____ ± _____ mV per volt (milliamp) excitation into specified load impedance" or "_____ ± _____ mV at _____ volts (milliamp) excitation into specified load impedance."

Note: If 4.2.2 and 4.2.3 are used to specify performance characteristics, the tolerance in 4.2.3 may be omitted.

Alternately, the following may be specified (4.2.3-4.2.6):

*4.2.3 Full-Scale Output — Expressed with tolerance (see above note).

*4.2.4 Zero Measurand Output — Expressed as "_____ ± _____ mV."

*4.2.5 Zero Shift — Expressed as " ± _____ % of full scale output over a period of _____ minutes (hours, days, etc.)."

*4.2.6 Sensitivity Shift — Expressed as " ± _____ % over a period of _____ minutes (hours, days, etc.)."

*4.2.7 Linearity — Expressed as "_____ linearity within ± _____% of full scale output."

Note: The type of linearity specified shall be one of the types defined in ISA S37.1; namely, end point, independent, least squares, terminal, or theoretical slope.

*4.2.8 Hysteresis — Expressed as "_____% of full scale output." Alternately, 4.2.7 and 4.2.8 may be combined as:

4.2.9 Hysteresis and Linearity — Expressed as "combined hysteresis and _____ linearity within ± _____% of full scale output."

Note: The type of linearity shall be stated.

*4.2.10 Repeatability — Expressed as "within _____% of full scale output over a period of _____ minutes (hours, days, months)." Alternately, 4.2.7, 4.2.8, and 4.2.10 may be combined as:

*4.2.11 Static Error Band — Expressed as "± _____% of full scale output as referred to _____."

Note: The type of reference line or curve shall be stated. When an end point line is specified, the tolerances for the end points should be stated separately. When a "best straight" or terminal line or theoretical slope is specified, the static error band should also include the zero measurand output, zero shift, and sensitivity shift.

*4.2.12 Creep — Expressed as "_____ minutes for subsequent shifts in output not to exceed _____% of full scale output."

*4.2.13 Warm-up Period — Expressed as "_____ minutes for subsequent shifts in sensitivity and zero balance not to exceed _____% of full scale output."

*4.2.14 Reference Pressure Effect — Expressed as "change in zero balance not to exceed _____% of full scale for a reference pressure change of _____ Pa(psi). Sensitivity change shall not exceed_____% of full scale for a reference pressure change of _____ Pa(psi)." Alternately expressed as "operation at reference pressures from _____ Pa(psia) to _____ Pa(psia) not to cause output readings which will exceed the specified error band."

*4.2.15 Frequency Response (Amplitude) — Expressed as "within ± _____% from zero to _____ Hz."

Note: Frequency response should be referred to a frequency within the specified frequency range, preferably zero, and to a specific measurand value. Mounting conditions and measured fluid should be specified.

4.2.16 Phase Shift — Expressed as either "phase shift linear within ± _____% from zero to_____ Hz, reaching _____ degrees at_____ Hz or "phase shift less than _____ degrees between zero and_____Hz."

Alternately 4.2.15 and 4.2.16 may be replaced by:

*4.2.17 Resonant Frequency — Expressed in "hertz" or "kilohertz."

Note: If a number of resonances exist, all frequencies

should be listed. The lowest resonance frequency must be listed.

*4.2.18 Damping Ratio — Expressed as "_____% of critical damping."

Note: For any other than a second order single-degree-of-freedom system, damping ratio is not defined and ringing period, rise time and overshoot should be stated.

*4.2.19 Ringing Period — Expressed as "_____ milliseconds."

*4.2.20 Overshoot — Expressed as "_____% of applied pressure."

For transducers with relatively high damping and little overshoot, 4.2.17, 4.2.18, 4.2.19, and 4.2.20 may be replaced by 4.2.21:

*4.2.21 Rise Time — Expressed as "_____ milliseconds (microseconds) for response to rise from 10% to 90% for an applied pressure step function of _____ Pa(psi)."

Note: Existing test equipment generates ramp functions rather than step functions. Care must be taken to insure that the rise time of the generated ramp function is one-third or less of the anticipated rise time of the transducer under test.

*4.2.22 Proof Pressure — Expressed as "application of _____% of full scale for _____ minutes will not cause changes in transducer performance beyond the specified tolerances."

*4.2.23 Burst Pressure Rating — Expressed as"_____ Pa(psia, psig) (or psid) applied_____times for a period of _____ minutes each to_____." (Sensing element or case; specify.)

4.2.24 Operating Temperature Range — Expressed, as "temperatures from _____ °C(°F) to _____ °C(°F) will not cause thermal sensitivity shift of more than_____% or thermal zero shift of more than_____% of full scale."

Or, alternately, the following may be specified.

*4.2.25 Thermal Sensitivity Shift — Expressed as "_____% per °C(°F) temperature change over a temperature range from_____°C(°F) to _____°C(°F)."

*4.2.26 Thermal Zero Shift — Expressed as "_____% of full scale output per °C(°F) temperature change change over a temperature range from _____°C(°F) to _____°C(°F)."

4.2.24, 4.2.25, or 4.2.26 may be specified by:

*4.2.27 Temperature Error — Expressed as "_____% full scale output at_____ Pa(psi) for a temperature change from _____°C(°F) to _____°C(°F)."

*4.2.28 Temperature Error Band — Expressed as within ± _____% of full scale output from the straight line establishing static error band, over temperature range from _____°C(°F) to _____°C(°F)."

*4.2.29 Temperature Gradient Error — Expressed as

"less than ± ____ % of full scale output during a period of ____ minutes while subjected to a step function temperature change from ____ °C(°F) to ____ °C(°F), applied to ____ (specify particular part) of the transducer."

*4.2.30 Acceleration Error — Expressed as "less than ± ____ % of full scale output per g along ____ axis at steady acceleration level of ____ g."

Note: The error should be listed for each of the three axes or for the axis with the largest error.

Alternately 4.2.30 may be replaced by 4.2.31:

*4.2.31 Acceleration Error Band — Expressed as "within ± ____ % of full scale output for steady accelerations up to ____ g along ____ axis." See above note.

*4.2.32 Vibration Error — Expressed as "less than ± ____ % of full scale output per g along ____ axis at vibration level of ____ g peak over a frequency range from ____ Hz to ____ Hz."

Note: The error should be listed either for each of the three axes or for the axis with the largest error.
Alternately 4.2.32 may be replaced by 4.2.33:

*4.2.33 Vibration Error Band — Expressed as "within ± ____ % of full scale output for vibration level of ____ g peak over a frequency range from ____ Hz to ____ Hz along ____ axis." See above note.

*4.2.34 Life, Cycling — Expressed as "____ full scale output pressure cycles (applied at a rate of ____ hertz) over which transducer shall operate without change in characteristics beyond their specified tolerances."

*4.2.35 Mounting Error — Expressed as "within ± ____ % of full scale output," or, "within the static error band," under specified conditions of mounting force or torque.

4.2.36 Other Environmental Conditions — Other pertinent environmental conditions which shall not change transducer performance beyond specified limits should be listed; examples are:

   Shock, Triaxial
   High-Level Acoustic
    Excitation
   Humidity
   Salt Atmosphere
   Nuclear Radiation
   Magnetic Fields
   Solar (or other) Heat Radiation
   Sand and Dust
   Altitude
   Temperature Shock

4.2.37 Storage Life — Expressed as "Transducer can be exposed to Specified Environmental Storage Condition for ____ days (months, years) without changing the following performance characteristics beyond their specified tolerances."

Note: Environmental storage conditions shall be described in detail. Pertinent performance characteristics (examples: sensitivity, zero shift) shall be specified.

4.3 Additional Terminology:

Phase Shift — The amount of time by which the output of a transducer lags a sinusoidally varying measurand.

Note: Expressed as fraction of a cycle of the frequency, usually in degrees.

Sealed Reference Differential Pressure Transducer — Transducer which measures pressure difference between unknown pressure and pressure of fluid in an integral sealed reference chamber.

Shunt Calibration Resistor — A shunt resistor which, when placed across a specified element of the electrical circuit of the transducer, will electrically simulate a specified percentage of the transducer full scale output at room conditions.

5. INDIVIDUAL ACCEPTANCE TESTS AND CALIBRATIONS

5.1 Basic Equipment Necessary to Perform Individual Acceptance Tests and Calibration of Strain Gage Pressure Transducers.

The basic equipment for acceptance tests and calibrations consists of a source of pressure, a source of electrical excitation for the strain gages, and a device which measures the electrical output of the transducer. The combined errors or uncertainties of the calibration system comprising these three components should be sufficiently smaller than the permissible tolerance of the transducer performance characteristic under evaluation to result in meaningful values. (Department of Defense practice commonly uses a four-to-one ratio in calibration hierarchy.) The traceability to national standards for this measuring system should be well known.

5.1.1 Source of Pressure

A pressure medium similar to the one which the transducer is intended to measure should be used for testing. The accuracy of the pressure source should be at least five times greater than the permissible tolerance of the transducer performance characteristic under evaluation. The range of the instrument supplying or monitoring the calibration pressure should be selected to provide the necessary accuracy to 125% of the full scale range of the transducer.

The source of calibration pressure may be either continuously variable over the range of the instrument, or may be provided in discrete steps as long as the steps can be programmed in such a manner that the transition from one pressure to the next during calibration is accomplished without creating a hysteresis error in the measurement due to overshoot.

EXAMPLES OF PRESSURE SOURCES/
MONITORING EQUIPMENT

MERCURY MANOMETER (Pressure Indicating Device)

Typical Ranges

100 kPa (about 30 in. Hg) . . .
                    Accuracy ± 0.02% Full Scale

4.4 Tabulated Characteristics versus Test Requirements

This table is intended for use as quick reference for design and performance characteristics and tests of their proper verification as contained in this Standard.

Characteristic	Paragraph	Design Characteristic		Verified During	
		Basic	Supplemental	Individual Acceptance Test	Qualification Test
Type of Pressure Sensed	4.1.1.1	x		No Test	Special Test
Measured Fluids	4.1.1.2	x			
Materials in Contact with Measured Fluid	4.1.1.3		x		Special Test
Configuration, Dimensions, Mounting Pressure Connection	4.1.1.4 through 4.1.1.6	x		5.2.1	
Mounting Force or Torque	4.1.1.7	x			6.5
Weight	4.1.1.8	x			6.2
Case Sealing	4.1.1.9	x			5.2.1
Identification	4.1.1.10	x		5.2.1	
Case Material	4.1.2.1		x		5.2.1
Pressure Sensing Element	4.1.2.2				
Type of Strain Gage Used	4.1.2.3		x		5.2.1
Location of Strain Gage	4.1.2.4		x		5.2.1
Number of Active Strain Gage Elements	4.1.2.5		x		
Dead Volume	4.1.2.6		x		6.3
Volume Change Due to Full Scale Pressure	4.1.2.7		x		6.4
Maximum and Minimum Operat- and Fluid Temperature	4.1.1.11	x			
Excitation	4.1.3.1	x		5.2.6	
Input Impedance	4.1.3.3	x		5.2.11	
Output Impedance	4.1.3.4	x		5.2.11	
Load Impedance	4.1.3.5				
Electrical Connections	4.1.3.6	x		5.2.10	
Insulation Resistance	4.1.3.7		5.2.4		
Shunt Calibration Resistor	4.1.4.1	x		5.2.5 (partially)	6.7
Range	4.2.1	x		5.2.3	
End Point	4.2.2	x		5.2.3	
Full-Scale Output	4.2.3	x		5.2.3	
Zero Measurand Output	4.2.4	x		5.2.3	
Zero Shift	4.2.5	x		5.2.4	
Sensivity Shift	4.2.6	x		5.2.4	
Linearity	4.2.7	x		5.2.3	
Hysteresis	4.2.8	x		5.2.3	
Hysteresis and Linearity	4.2.9	x		5.2.3	
Repeatability	4.2.10	x		5.2.3	
Static Error Band	4.2.11	x		5.2.3	6.2
Creep	4.2.12	x		5.2.5	
Warm-up Period	4.2.13				5.2.6
Reference Pressure Error	4.2.14			5.2.7 5.2.8	
Frequency Response (Amplitude)	4.2.15				6.6
Phase Shift	4.2.16				6.6
Resonant Frequency	4.2.17	x			6.6
Damping Ratio	4.2.18	x			6.6
Ringing Period	4.2.19	x			6.6
Overshoot	4.2.20	x			6.6
Rise Time	4.2.21	x		5.2.9	
Proof Pressure	4.2.22	x			6.13
Burst Pressure Rating	4.2.23	x			6.7
Operating Temperature Range	4.2.24	x			6.7
Thermal Sensitivity Shift	4.2.25	x			6.7
Thermal Zero Shift	4.2.26	x			6.7
Temperature Error	4.2.27				
Temperature Error Band	4.2.28	x			6.7
Temperature Gradient Error	4.2.29				
Acceleration Error	4.2.30				6.8
Acceleration Error Band	4.2.31				6.8
Vibration Error	4.2.32	x			6.9
Vibration Error Band	4.2.33	x			6.9
Cycling Life	4.2.34				6.10
Mounting Error	4.2.35				6.5
Other Environmental Conditions	4.2.36				6.11
Storage Life	4.2.37				6.12

200 kPa (about 60 in. Hg) . . .
Accuracy ± 0.02% Full Scale
340 kPa (about 100 in. Hg) . . .
Accuracy ± 0.01% Full Scale

AIR PISTON (Pressure Source)

Typical Ranges

About 2 to 10 kPa (0.3 to 1.5 psi) . . .
Accuracy ± 0.15% of Reading
About 10 to 350 kPa (1.5 to 50 psi) . . .
Accuracy ± 0.015% of Reading
About 100 kPa to 1 MPa (15 to 150 psi) . . .
Accuracy ± 0.025% of Reading
About 100 kPa to 3.5 MPa (1 to 500 psi) . . .
Accuracy ± 0.025% of Reading

PRECISION DIAL GAGE (Pressure Indicating Device)

Typical Ranges

0 to 30 kPa (about 0 to 120 H_2O) . . .
Accuracy ± 0.1% Full Scale
0 to 100 kPa (about 0 to 30 in. Hg) . . .
Accuracy ± 0.1% Full Scale
0 to 100 kPa (about 0 to 100 psi) . . .
Accuracy ± 0.1% Full Scale
0 to 700 MPa (about 0 to 10 000 psi) . . .
Accuracy ± 0.1% Full Scale

Note: Pressure indicating devices generally require a supply of dry gas, e.g., dehumidified air, or nitrogen, or helium, required for reasons of safety.

OIL PISTON GAGE (Pressure Source)

Typical Ranges

About 40 kPa to 30 MPa (6 to 4000 psi) . . .
Accuracy ± 0.01% of Reading
About 400 kPa to 300 MPa (60 to 40 000 psi) . . .
Accuracy ± 0.01% of Reading
About 14 MPa to 700 MPa . . . (2000 to 100 000 psi)
Error in Piston Area Less Than ± 0.009%
About 30 MPa to 1400 MPa . . . (4000 to 200 000 psi)
Error in Piston Area Less Than 0.012%

5.1.2 Stable Source of Excitation of Accurately Known Amplitude

Commonly used sources of dc excitation are chemical batteries such as dry cells and storage batteries, or line-powered, electronically regulated, power supplies. A stable, low distortion, audio oscillator may be used to furnish ac excitation.

5.1.3 Read-out Instrument

Examples of suitable devices are:

MANUALLY BALANCED POTENTIOMETER

Typical Ranges

0 to 0.01111 volt, ± (0.008%
of reading + 0.5 microvolt);
0 to 0.1111 volt, ± (0.006%
of reading + 1 microvolt);

0 to 1.111 volt, ± (0.004%
of reading + 10 microvolts);
0 to 11.11 volts, ± (0.006%
of reading + 100 microvolts).

SELF-BALANCING POTENTIOMETER

Typical Ranges

0 to 6 millivolts, limit of
error ± 0.3%
0 to 100 millivolts, limit of
error ± 0.3%

DIGITAL ELECTRONIC VOLTMETER/RATIO METER

Typical Accuracy

± 0.01% of reading + 1 digit (4 digits display)
± 0.005% of reading + 1 digit (5 digits display)

AC RMS DIFFERENTIAL METER

Typical Accuracy

± 0.05% 10 Hz to 50 kHz
± 0.1% Hz to 10 kHz

Note: The input impedance of the readout instrument must comply with the value of load impedance specified. Unless otherwise stated, adjustments and compensation of the transducer apply with the specified load impedance across the output terminals.

5.2 Calibration and Test Procedures

Results obtained during the calibration and test procedures should be recorded on data sheets like the sample data sheet in Section 7 of this report. These procedures shall be performed under "room conditions" as defined in ISA S37.1 unless otherwise indicated.

Note: The defining paragraph under Design Characteristics (4.1) and Performance Characteristics (4.2) of this document is listed beside each of the parameters sought in the paragraphs below.

5.2.1 The transducer shall be inspected visually for mechanical defects, poor finish, and improper identification markings. The electrical connector shall also be inspected.

5.2.2 The transducer shall be connected to the pressure source and secured with the recommended force or torque. The excitation source and readout instrument shall also be connected to the transducer and turned on. Adequate warm-up time for test equipment shall be allowed before tests are conducted. The pressure source, connecting tubing, and transducer system shall have passed a prior test for leaks which would cause calibration errors.

5.2.3 Two or more complete calibration cycles shall be run consecutively. At least eleven data points shall be obtained per cycle using both ascending and descending directions. Excitation amplitude shall be monitored as required. (Time duration of calibration cycle to be stated.)

From the data obtained during these tests, the following

characteristics should be determined:

End Points	4.2.2
Full-Scale Output	4.2.3
Zero Measured Output	4.2.4
Linearity	4.2.7
Hysteresis	4.2.8
Hysteresis and Linearity	4.2.9
Repeatability	4.2.10
Static Error Band	4.2.11

5.2.4 Repeated calibration cycles over a specified period of time should establish the following characteristics for this period of time:

Zero Shift	4.2.5
Sensitivity Shift	4.2.6

Note: These tests may be abbreviated cycles with fewer data points than required in 5.2.3.

5.2.5 Application of full scale pressure to the transducer during a specified short period of time and measurement of changes in output at constant excitation during this time should establish:

Creep	4.2.12

Note: Rate of change of pressure should be as high as possible without resonant excitation of transducer.

5.2.6 By measuring zero balance and sensitivity over a period of time (one hour should suffice), starting with the application of excitation to the transducer, the following characteristic should be determined:

Warm-up Period	4.2.13

Note: It is desirable to test for these effects separately establishing the warm-up change of zero balance first.

5.2.7 From the application of the same pressure to both sides of the transducer sensing element over a range of pressures up to the maximum expected reference pressure and subsequent calibration cycles, the following should be established:

Reference Pressure Effect (zero measurand output)	4.2.14

Note: This test does not apply to absolute or fixed reference pressure transducers.

5.2.8 Application of the maximum expected reference pressure only to the low port of a differential pressure transducer and a pressure equal to the sum of the maximum expected reference pressure and the full scale pressure to the high port shall establish:

Reference Pressure Effect (sensitivity)	4.2.14

Note: Reference pressure effect on zero balance must be taken into account.

5.2.9 After application of the specifie;proof pressure a specified number of times, and in the specified direction for differential pressure transducers, at least one complete calibration cycle shall be performed to establish that the performance characteristics of the transducer are still within specifications.

Proof Pressure	4.2.22

5.2.10 Measure the insulation resistance between all terminals, or leads connected in parallel, and the case of the transducer, with a megohm meter or similar acceptable device, using a potential of 50 volts dc, unless otherwise specified. The temperature at which the insulation resistance is measured shall be specified.

Insulation Resistance	4.1.3.7

5.2.11 A wheatstone bridge (for dc) or impedance bridge shall be used to measure:

Input Impedance	4.1.3.3
Output Impedance	4.1.3.4

6. QUALIFICATION TEST PROCEDURES

Qualification Tests shall be performed as applicable using the test forms for Section 7 as required. Upon completion of all testing the form of Figure 7.6 shall be used to summarize all testing.

6.1 Initial Performance Tests (Figure 7.1)

Following a thorough inspection of the transducers, the tests and procedures of Section 5, Individual Acceptance Tests and Calibrations, shall be run to establish reference performance during increasing (and decreasing) steps of 0, 20, 40, 60, 80, and 100 percent of range as a minimum (percent of span for bidirectional transducers).

6.2 Weight Test

The transducer shall be weighed on an appropriate balance or scale. The following shall be established:

Weight	4.1.1.8

6.3 Dead Volume Test (Figure 7.6)

The pressure cavity shall be filled (both cavities for a differential transducer) with a measurable, non-corrosive fluid (under a vacuum if necessary) and the contents poured into a graduate. The following shall be established:

Dead Volume	4.1.2.6

6.4 Volume Change Test (Figure 7.6)

A fluid pressure system shall be connected to the transducer, a parallel pressure gage, and a graduated reservoir. (Provisions shal¹ be made for isolating the transducer when filled.) The pressure system shall be evacuated and filled with fluid, the valve to the transducer closed, the valve opened and the following shall be determined:

Volume Change Due to Full Scale Pressure	4.1.2.7

6.5 Mounting Test

The mounting of the actual installation shall be duplicated as closely as possible following specific instructions and one calibration run performed. The following shall be established:

Mounting Error	4.2.35
Mounting Force or Torque	4.1.1.7

6.6 Dynamic Response Test

The dynamic response characteristics of pressure transducers may be established either with transient-stimulation devices, or with sinusoidal pressure generators.

6.6.1 Transient Excitation Method

A positive step-function of pressure may be generated in gases with a shock-tube or a quick-opening valve. A hydraulic quick-opening valve is used to generate a positive pressure step function in a liquid medium. A burst diaphragm generator produces a negative pressure step in a gas medium. In all cases, the rise time of the generated step function shall be sufficiently short to shock-excite all resonances in the transducer under test. It shall also be one-third or less of the anticipated rise time of the transducer under test.

Since the tubing used to mechanically connect the transducer to the test setup will drastically affect the dynamic characteristics, it is recommended that the shortest possible tubing be installed, and that its length and diameter be stated along with the test results. Alternately the tubing used shall duplicate as closely as possible the actual installation, if this condition were specified instead of the characteristics of the transducer alone.

By applying step functions of pressure at room conditions within the full scale range of the transducer, and analyzing the electronic or electro-optical recording of the transducer output, the following can be determined: (see appropriate "notes").

Frequency Response	
Amplitude	4.2.15
Phase Shift	4.2.16
Resonant Frequency	4.2.17
Damping Ratio	4.2.18
Ringing Period	4.2.19
Overshoot	4.2.20
Rise Time	4.2.21

6.6.2 Sinusoidal Stimulation Method

Generators are now available which produce sinusoidal pressures in liquids or gases. They are generally limited to frequencies below several kilohertz and peak dynamic pressures below 10 MPa (roughly 1500 psi). These devices operate either on a piston-phone principle (such as the system used for the calibration of microphones) or by modulating fluid flow through an orifice (as exemplified by a siren).

By applying a sinusoidal pressure waveform of varying frequency and of constant and specified amplitude, the following can be obtained directly:

Frequency Response	
(Amplitude)	4.2.15
Phase Shift	4.2.16

If within the frequency range covered, the following can be established from the frequency response:

Resonant Frequency or	
Resonances	4.2.17
Damping Ratio	4.2.18

6.7 Temperature Tests

6.7.1 Steady State Temperature Test

The transducer shall be placed in a suitable temperature chamber. After allowing adequate stabilization time at a specified temperature, one or more calibration cycles shall be performed. This procedure shall be repeated at an adequate number of temperatures within the operating temperature range of the transducer, but at least at upper and lower limits of the operating temperature range. These tests should establish the following characteristics:

Thermal Sensitivity Shift	4.2.25
Thermal Zero Shift	4.2.26
Temperature Error	4.2.27
Temperature Error Band	4.2.28

These tests will also establish:

Operating Temperature	
Range	4.2.24

6.7.2 Thermal Transient Test

For a flush mounted pressure transducer, only the sensing end of the transducer is inserted rapidly from "room conditions" into a measurand fluid which is maintained at a specified temperature above or below "room conditions." And at "room" pressure the output shall be observed over a specified period of time starting from the moment of insertion.

For a cavity-type pressure transducer, fluid at a specified temperature above or below "room conditions" may be applied rapidly through the pressure port to the sensing element. The output shall be observed over the specified period of time.

Note: The type of fluid shall be specified.

These tests should establish:

Temperature Gradient Error	4.2.29

6.8 Acceleration Test

Place the transducer on a centrifuge, apply specified acceleration along specified axes, and measure changes in output. The following should be established:

Acceleration Error	4.2.30
Alternately,	
Acceleration Error Band	4.2.31

6.9 Vibration Test

Vibrate the transducer along specified axes at desired

acceleration amplitudes and over the specified frequency range with an electromagnetic or hydraulic shaker and observe or record the transducer output by means of oscilloscopes or high speed recorders. The following should be established:

Vibration Error	4.2.32
Alternately,	
Vibration Error Band	4.2.33

6.10 Life Test

After applying the specified number of full range excursions of measurand at least one complete calibration cycle shall be performed to establish minimum value of:

Cycling Life	4.2.34

6.11 Effects of Other Environments

Expose transducer to other specified environmental conditions followed in each case by at least one complete calibration cycle to test ability of transducer to perform satisfactorily after each exposure. See Section 4.2.36.

Note: In some cases, calibrations may be performed while transducer is subjected to the environment.

6.12 Storage Life Test

After storing the transducer under specified conditions (temperature, humidity, etc.) for the specified period of time, at least one complete calibration cycle shall be performed to establish:

Storage Life	4.2.37

6.13 Burst Pressure Test

The transducer shall be connected to a suitable test setup with adequate protection for equipment and personnel. The pressure shall be increased to the specified number of times and durations. The following shall be established:

Burst Pressure Rating	4.2.23

Note: If specified, burst pressure may also be applied to the inside of the case by first puncturing the sensing element.

7. TEST REPORT FORMS

7.1 The test report forms listed below are recommended for use during the testing of strain gage pressure transducers.

7.2 When using the forms all pertinent information shall be inserted in its proper place. On some forms, blank space has been provided for additional tests. Where the test is prolonged, more than one form may be required.

7.3 Individual Acceptance Tests and Calibrations (Figure 7.1) used during acceptance testing of Section 5 may also be used for initial performance tests of Section 6.1.

Environmental Test Record (Figure 7.2) used to record temperature, maximum temperature, life and other environmental tests.

Dynamic Response Tests (Figure 7.3 or Figure 7.4) used for recording test results of frequency response, resonant frequencies, damping ratio and ringing period. (Note: use 7.3 or 7.4 as applicable.)

Acceleration/Vibration Test Record (Figure 7.5) used to record acceleration and vibration test results.

Test Summary (Figure 7.6) used to compile the results of all testing.

8. BIBLIOGRAPHY

1. Beckwith, T. G. and Buck, N. L., "Mechanical Measurements" Addison-Wesley, 1961.

2. Neubert, H. K. P., "Instrument Transducers," Oxford, at the Clarendon Press, 1963.

3. Lederer, P. S., "Methods for Performance-Testing of Electromechanical Pressure Transducers," National Bureau of Standards Technical Note #411, February, 1967.

4. Schweppe, J. L., et al, "Methods for the Dynamic Calibration of Pressure Transducers," National Bureau of Standards Monograph #67, December, 1963.

5. Cross, J. L., "Reduction of Data for Piston Gage Pressure Measurements," National Bureau of Standards Monograph #65, June, 1963.

6. "Transducer Wiring Standard for Resistance Strain Gage Systems," Western Regional Strain Gage Committee, Los Angeles, California, May, 1960.

7. MIL-E-5272C (ASG) "Environmental Testing, Aircraft Electronic Equipment, General Specification for."

8. MIL-E-5400C (ASG) "Environmental Testing, Aeronautical and Associated Equipment, General Specification for."

9. MIL-STD-810 "Environmental Test Methods for Aerospace and Ground Equipment."

10. ANSI MC6.1-1975 (ISA S37.1-75) "Electrical Transducer Nomenclature and Terminology."

11. ISA S37.6 "Specifications and Tests of Potentiometric Pressure Transducers", January, 1975.

12. ISA Transducer Compendium, 2nd Edition, Part I, "Pressure, Flow and Level," Plenum Press, New York, 1969.

13. Norton, H. N., "Transducers for Electronic Measuring Systems," Prentice-Hall, 1969.

14. PMC 20.1-1973, "Process Measurement and Control Terminology", August, 1973.

15. ANSI B88.1-1971, "A Guide for the Dynamic Calibration of Pressure Transducers", ASME August, 1972.

16. ANSI Z210.1, "Standard Metric Practice Guide", ASTM (E 380-76) February, 1976.

| Vendor's Part No. | **STRAIN GAGE PRESSURE TRANSDUCER INDIVIDUAL ACCEPTANCE TESTS & CALIBRATIONS** | Customers Part No. |
| Test Facility | | Serial No. |

Ambient Conditions
Temperature_____°C(°F)
Pressure _____ mm Hg
Humidity _____%

☐ Functional Test   ☐ _____ Proof Cycle
☐ _____ Test   ☐ Calibration

Vendor

Range
_____ To _____ Pa(psi) _____

1. Visual: Mechanical ☐   Finish ☐   Nameplate ☐   Electr. Conn. ☐
2. Electrical: Input Impedance _____ohms, Output Impedance _____ohms, Ins Res. _____MΩ at _____V dc
3. Calibration and Proof Pressure Test at _____V at _____Hz Excitation, after _____minutes warm-up time
4. Load Impedance _____ohms or _____mA Excitation

Pressure Pa (psi)	Theoretical Output (mV)	Output (mV) — Run 1 Test Time: Minutes		Output (mV) — Run 2 Test Time: Minutes		Overload Output (mV)	Output (mV) — Run 3 Test Time: Minutes		Maximum Error (mV)		
		Increase	Decrease	Increase	Decrease		Increase	Decrease	+	—	—
						not applicable					
Proof Rev.		not applicable			- - - - - - -		not applicable				

STATIC ERROR BAND: + _____%, − _____% FSO (Allowed: ±_____% FSO),
Referred to _____

_____ Linearity

_____ LINEARITY: +_____, − _____% FSO (Allowed: _____% FSO)

HYSTERESIS: _____% FSO (Allowed: _____% FSO) REPEATABILITY: _____% FSO (Allowed: _____%FSO)

ZERO MEASURAND OUTPUT _____% FSO (Allowed: _____% FSO) CREEP: _____% FSO over _____minutes
(Allowed: _____% FSO)

ZERO SHIFT: _____% FSO over a period of _____ (Allowed: _____% FSO)
(number)   (units)

SENSITIVITY SHIFT: _____% over a period of _____ (Allowed: _____% FSO)
(number)   (units)

END POINTS: _____and _____mV (All'd.: _____and _____mV) BI-DIR. ONLY: ZERO-PSI PT. ____mV (All'd.____mV)

FULL SCALE OUTPUT(Run 2): _____mV (Allowed: _____mV) PROOF PRESSURE (after Run 2): _____%
Range for _____ minutes

ΔP Only: REFERENCE PRESSURE (during Run 2 and Proof Pressure Test): _____ (Pa) psia;
ZERO-OUTPUT (after Proof Pressure Test) _____ mV

ΔP Only: REVERSE ΔP (after Proof Pressure _____% Range for ____Minutes, ZERO-OUTPUT: _____mV)

Equipment Used:	Defects noted or Comments

BY: _____ DATE: _____ APPROVED BY: _____

Note: **FIGURE 7.1**

VENDOR'S PART NO.	TEST FACILITY	CUSTOMER'S PART NO.
VENDOR	**STRAIN GAGE PRESSURE TRANSDUCER**	SERIAL NO.
REPORT NO.		CUSTOMER
TYPE OF TEST	**ENVIRONMENTAL TEST RECORD**	RANGE _____ T O _____ Pa (psi) _____

Excitation _____ volts
at _____ Hz,
or _____ mA
Warmup Time _____ Minutes

Type of Environment

☐ Before
☐ During _____
☐ After   For temperature tests, temperature measured at _____

Level

Pressure	Theoretical	mV (Run 1)		mV (Run 2)		Overload	mV (Run 3)		Maximum Error mV	
Pa(psi)	mV	Increase	Decrease	Increase	Decrease	mV	Increase	Decrease	+	—
	(Pos. Proof)									
	(Neg. Proof)									

Error Band: + _____ % — _____ % FSO (Referred To _____)   Allowed: ± _____ % FSO

Proof Pressure _____ % Rated Range For _____ Minutes   Ins. Resistance: _____ megohms At _____ Vdc

For Diff. Press. Transducers Only:
Neg. Proof Pressure _____ % Rated Range For _____ Minutes   Zero Shift: _____ % FSO
                                                                Sensitivity Shift: _____ %

Ref. Press., Run 2: _____ Pa (psia)

Comments: _____
_____
_____
_____
_____
_____
_____
_____

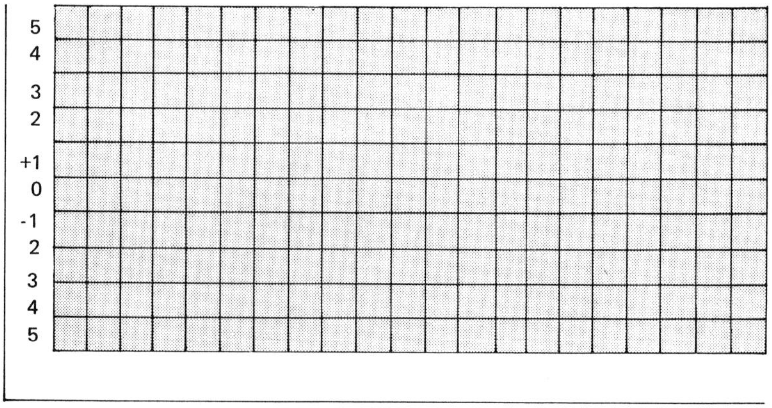

5
4
3
2
+1
0
-1
2
3
4
5

Tested By: _____   Date Test Started: _____ Date Test Finished: _____

Approved By: _____   Approved By: _____
        Title   Title

FIGURE 7.2

Transducer Type _____	DYNAMIC RESPONSE TESTS OF STRAIN GAGE PRESSURE TRANSDUCER	Purchase Order No. _____
Range _____		Serial No. _____
Vendor & Model No. _____		Part No. _____

1. Visual Inspection:
   Mechanical: _____ Finish: _____ Nameplate: _____ Connections: _____

2. Electrical: Load Impedance _____ ohms
   Input Impedance: _____ ohms, Output Impedance: _____ ohms, Insulation Resistance _____ mΩ
   Excitation _____ volts or _____ mA at _____ Hz at _____ volts

3. Ambient Conditions: Temperature _____ °C(°F); Pressure _____ cm Hg; Humidity; _____ %

4. Dynamic Response

   Step Function Generator: _____ Shock Tube, Dry Air: _____
   Mounting Location: End: _____ Side: _____

Shock No.	Initial Pressures		Shock Velocity	Step Pressure	Pronounced Resonances, Hz			Ringing Period / Rise Time
	Hi	Low						

ATTACH OSCILLOSCOPE PHOTOGRAPHS OF TRANSDUCER RESPONSES

AMPLITUDE SCALE

SHOCK 1

TIME SCALE _____

AMPLITUDE SCALE

SHOCK 3

TIME SCALE _____

AMPLITUDE SCALE

SHOCK 2

TIME SCALE _____

AMPLITUDE SCALE

SHOCK 4

TIME SCALE _____

BY: _____

APPROVED: _____

FIGURE 7.3

VENDOR'S PART NO.	TEST FACILITY	CUSTOMER'S PART NO.
VENDOR	**STRAIN GAGE PRESSURE TRANSDUCER**	SERIAL NO.
REPORT NO.	**DYNAMIC RESPONSE TESTS**	CUSTOMER
TYPE OF TEST	**(SINUSOIDAL METHOD)**	RANGE _____ To _____ (Pa) psi _____

Ambient Conditions: Temperature _____ °C(°F); Pressure _____ cm Hg; _____ Humidity _____%

Dynamic Response:
Excitation (volts or ma) _____ at _____Hz;   Load Impedance _____ohms

Sinusoidal Generator _____;   Test Fluid _____

Mounting Configuration_____

Reference Transducer_____

Test Temperature _____°C(°F);   Quiescent Static Pressure _____ Pa(psi) _____

Sinusoidal Pressure_____ Pa(psi) peak;   Port excited:_____

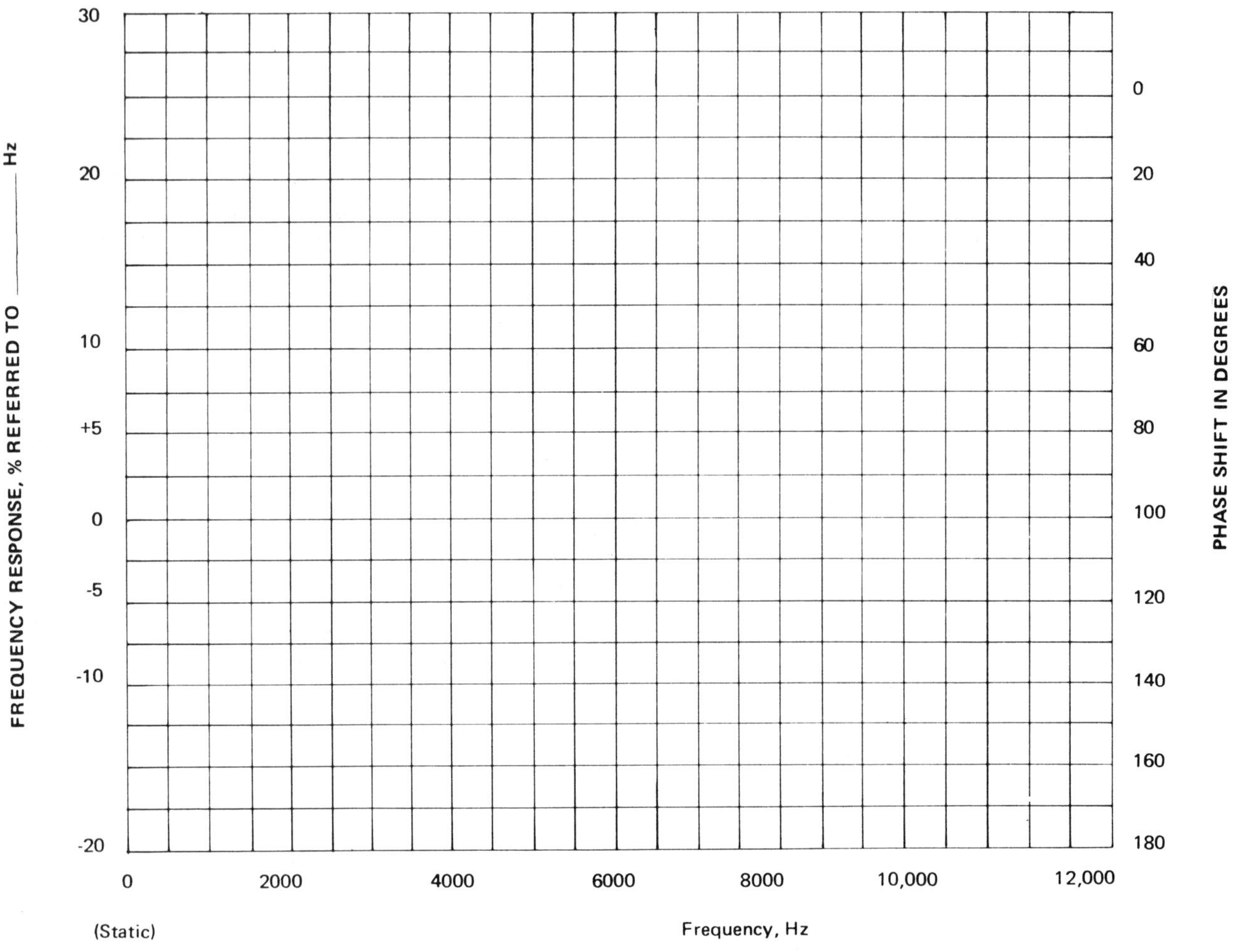

FIGURE 7.4

VENDOR'S PART NO.	TEST FACILITY	CUSTOMER'S PART NO.
VENDOR	**STRAIN GAGE PRESSURE TRANSDUCER**	SERIAL NO.
REPORT NO.		CUSTOMER
	ACCELERATION/VIBRATION TEST RECORD	
TYPE OF TEST		RANGE _____ To _____ Pa(psi) _____

SKETCH OF TRANSDUCER SHOWING AXIS ORIENTATION:

ACCELERATION TEST

AXIS	+X	-X	+Y	-Y	+Z	-Z
Output Before Accel mV						
Applied Accel. (G)						
Output During Accel. mV						
Accel. Error mV						

Excitation _____ V(mA) AT _____ Hz, Load Impedance _____ ohms

COMMENTS

Pressure Level Used: _____ Pa(psi) _____

Max. Accel. Error: + _____ , − _____ % FSO

Pre-Accel. Static Error Band: + _____ , − _____ % FSO

Accel. Error Band: + _____ , − _____ % FSO

(Allowed Accel. Error Band ± _____ % FSO

Tested By: _____ (Technician)

_____ (Test Engineer)

Date: _____ Approved By: _____

Witnessed By: _____ (_____)

Witnessed By: _____ (_____)

VIBRATION TEST

AXIS	X			Y			Z		
Pressure Level Used	_____Pa(psi)			_____Pa(psi)			_____Pa(psi)		
Output Before Vib.	mV			mV			mV		
	Freq. (Hz)	Error		Freq. (Hz)	Error		Freq. (Hz)	Error	
		Pol.	mV		Pol.	mV		Pol.	mV
Vibration Error									

Max. Vib. Error: + _____ , − _____ % FSO

Pre Vib. Static Error Band: + _____ , − _____ % FSO

Vib. Error Band: + _____ , − _____ % FSO

(Allowed Vib. Error Band: ± _____ % FSO

Tested By: _____ (Technician)

_____ (Test Engineer)

Date: _____ Approved By: _____

Witnessed By: _____ (_____)

Witnessed By: _____ (_____)

EXCITATION _____ V(mA) at _____ Hz

LOAD IMPEDANCE _____ OHMS

COMMENTS

FIGURE 7.5

VENDOR'S PART NO.	TEST FACILITY	CUSTOMER'S PART NO.
VENDOR		SERIAL NO.
REPORT NO.	**TRANSDUCER TEST REPORT**	CUSTOMER
TYPE OF TEST	**STRAIN GAGE PRESSURE TRANSDUCER**	RANGE _____ To _____ Pa(psi) _____

SUMMARY OF RESULTS

☐ Error
☐ Error Band

Test	Tested Per Proced. No. or Test Waived Per	Par. No.	Pass	Fail				+ %FSO	-% FSO
				Error	Electr.	Mech.	See Comments		
Initial P.T. (Performance Test)									
Weight									
Dead Volume								Cu	
Vol. Change over Press. Range								Cu	
Mounting									
Frequency Response								Flat (+ ___ %): ___ to ___ Hz	
Phase Shift								deg. at Hz / deg. at Hz	
Resonant Frequencies								Hz / Hz	Hz / Hz
Overshoot									
Time Constant								___ msec., ___ % Ovs.	
Low Temp _____ °C(°F)									
P.T. After Low Temp.									
Hi h Temp. + _____ °C (°F)									
Add'l. Temp. _____ °C(°F)									
P.T. After High Temp.									
_____ g Vibration									
P.T. After _____ g Vibration									
Acceleration									
P.T. After Accel.									
Thermal Gradient Error									
Life									
Burst Pressure									

Tested By: _____ Date Test Started: _____ Date Test Finished: _____

Approved By: _____ Approved By: _____
         Title   Title

FIGURE 7.6

VENDOR'S PART NO.	TEST FACILITY	CUSTOMER'S PART NO.
VENDOR		SERIAL NO.
REPORT NO.	**TRANSDUCER TEST REPORT**	CUSTOMER
TYPE OF TEST	**STRAIN GAGE PRESSURE TRANSDUCER**	RANGE _____ To _____ Pa(psi) _____

SUMMARY OF RESULTS

☐ Error
☐ Error Band

Test	Tested Per Proced. No. or Test Waived Per	Par. No.	Pass	Fail				+ %FSO	-% FSO
				Error	Electr.	Mech.	See Comments		
Initial P.T. (Performance Test)									
Weight									
Dead Volume								_____Cu_____	
Vol. Change over Press. Range								_____Cu_____	
Mounting									
Frequency Response								Flat (+ __%): ___ to ____Hz	
Phase Shift								deg. / deg.	at Hz / at Hz
Resonant Frequencies								Hz / Hz	Hz / Hz
Overshoot									
Time Constant								____msec., ____% Ovs.	
Low Temp _____ °C(°F)									
P.T. After Low Temp.									
Hi h Temp. + _____ °C (°F)									
Add'l. Temp. _____ °C(°F)									
P.T. After High Temp.									
_____ g Vibration									
P.T. After _____ g Vibration									
Acceleration									
P.T. After Accel.									
Thermal Gradient Error									
Life									
Burst Pressure									

Tested By: _____ Date Test Started: _____ Date Test Finished: _____

Approved By: _____ Approved By: _____

Title   Title

INSTRUMENT SOCIETY of AMERICA
Research Triangle Park, North Carolina

ANSI/ISA-S37.5-1975
(R 1982)
Approved December 14, 1982
(Formerly ANSI MC 6.3-1975)

American National Standard

Specifications and Tests for Strain Gage Linear Acceleration Transducers

Instrument Society of America

ISBN 0-87664-379-9

ISA-S37.5   Specifications and Tests for Strain Gage
            Linear Acceleration Transducers

INSTRUMENT SOCIETY OF AMERICA
67 Alexander Drive
P.O. Box 12277
Research Triangle Park, North Carolina 27709

PREFACE

This Preface is included for informational purposes and is not a part of ISA-S37.5.

This Standard has been prepared as a part of the service of the Instrument Society of America toward a goal of uniformity in the field of instrumentation. To be of real value, this document should not be static, but should be subject to periodic review. Toward this end, the Society welcomes all comments and critisms, and asks that they be addressed to the Secretary, Standards and Practices Board, Instrument Society of America, 67 Alexander Drive, P.O. Box 12277, Research Triangle Park, NC 27709, Telephone (919) 549-8411.

The ISA Standards and Practices Department is aware of the growing need for attention to the metric system of units in general, and the International System of Units (SI) in particular, in the preparation of instrumentation standards. The Department is further aware of the benefits to USA users of ISA Standards of incorporating suitable references to the SI (and the metric system) in their business and professional dealings with other countries. Towards this end, this Department will endeavor to introduce SI - acceptable metric units in all new and revised standards to the greatest extent possible. The Metric Practice Guide, which has been published by the American Society for Testing and Materials as ANSI designation Z210.1 (ASTM E380-76, IEEE Std. 268-1975), and future revisions, will be the reference guide for definitions, symbols, abbreviations, and conversion factors.

It is the policy of the Instrument Society of America to encourage and welcome the participation of all concerned individuals and interests in the development of ISA Standards. Participation in the ISA standards making process by an individual in no way constitutes endorsement by the employer of that individual of the Instrument Society of America or any of the standards which ISA develops.

This Standard is intended as a guide for technical personnel at user facilities as well as by manufactuers' technical and sales personnel whose duties include specifying, calibrating, testing or showing performance characteristics of strain-gage linear accelerometers. By basing users' specifications as well as technical advertising and reference literature on this Standard, or by referencing portions thereof, as applicable, a clear understanding of the users' needs or of the transducers' performance capabilities, and of the methods used for evaluating or proving performance, will be provided. Adhering to the specification outline, terminology and procedures shown will not only result in simple, but also complete specifications; it will also reduce design time, procurement lead time, and labor, as well as material costs. Of major importance will be the reduction of qualification tests resulting from use of a commonly accepted test procedure and uniform data presentation.

The development of this Standard was initiated as the result of a survey conducted in December 1960. A total of 240 questionnaires was sent out to transducer users and manufacturers. A strong majority indicated in their replies a need for transducer standardization. As strain-gage acceleration transducers were one of the types shown to be most in need of standardization, a Subcommittee, 8A-RP37.5, was formed under the former Survey Committee on Transducers for Aerospace Testing, 8A-RP37. Subcommittee 8A-RP37.5 became Standards Committee SP37.5 when the scope of the committee's work was broadened to include the applications of these transducer types by all industries and sciences. To provide a coordinated document, this committee was composed of representatives from government, user and manufacturer categories. This Standard was then processed over several mail-review and revision cycles until a consensus of reviewers was reached and it was published as ISA Standard in 1971. It was approved as ANSI Standard MC 6.3-1975 in October 1975.

The assistance of those who aided in the preparation of this document by answering questionnaires, offering suggestions, and in other ways, is gratefully achknowledged.

The following individuals served as members of the 1975 SP37.5 committee:

NAME	COMPANY
L. L. Lathrop, Chairman (1965-71)	Sandia Laboratories
R. M. Canzoneri, Secretary	Bell and Howell
G. D. Goodrich	Statham Instruments, Inc.
J. S. Hilten, Chairman (1971-)	National Bureau of Standards
G. C. Machen - Deceased	U.S. Naval Missile Center
E. D. Pettler	Consultant
D. Shannon, Alternate	Bell and Howell

The following individuals served as members of the 1982 SP37 committee:

NAME	COMPANY
J. W. Mock, Chairman	Measurement Services
P. Bliss	Consultant
W. Brinkschulte	Genisco Technology Corporation
C. Kutelis	Revere Corporation of America
R. W. Lally	PCB Piezotronics
H. E. Lockery	Hottinger Baldwin Measurements, Inc.
E. W. Malone	Boeing Aerospace Co.
T. B. Miller	Lawrence Livermore Lab
K. A. Parlee	United Engineers & Constructors, Inc.
H. Pitt	
J. Powell	Rosemount Engineering
R. C. Strahm	EG & G Idaho
R. M. Whittier	ENDEVCO
N. Wilde	EG & G Idaho
E. Wong	Bell & Howell

This Standard was approved for publication by the ISA Standards and Practices Board in October 1982.

NAME	COMPANY
T. J. Harrison, Chairman	IBM Corporation
P. Bliss	Consultant
W. Calder	The Foxboro Company
N. Conger	Continental Oil Co.
B. Feikle	Bailey Controls Co.
R. T. Jones	Philadelphia Electric Co.
R. Keller	Boeing Company
O. P. Lovett, Jr.	Isis Corp.
E. C. Magison	Honeywell, Inc.
A. P. McCauley	Diamond Shamrock Corp.
J. W. Mock	Measurement Services
E. M. Nesvig	ERDCO Engineering Corp.
G. Platt	Bechtel Power Corp.
R. Prescott	Moore Products Company
W. C. Weidman	Gilbert Associates
K. A. Whitman	Allied Chemical Corp.
J. R. Williams	Stearns-Roger, Inc.
B. A. Christensen*	
L. N. Combs*	
R. L. Galley*	
R. G. Marvin*	
W. B. Miller*	Moore Products Company
R. L. Nickens*	

*Director Emeritus

TABLE OF CONTENTS

1 PURPOSE

This Standard establishes the following for strain-gage linear acceleration transducers:

1. Uniform minimum specifications for design and performance characteristics.
2. Uniform acceptance and qualification test methods, including calibration techniques.
3. Uniform presentation of minimum test data.
4. A drawing symbol for use in electrical schematics (see Note in 3.)

2 SCOPE

2.1 This Standard covers uni-directional and bi-directional strain-gage linear acceleration transducers.

2.2 Included among the specific types of strain-gage linear acceleration transducers for which this Standard is applicable, are the following:

Bonded, unbonded, deposited metallic, or semiconductor strain gages.

2.3 Terminology used is defined in either ISA-S37.1, Electrical Transducer Nomenclature and Terminology or 4.3, Additional Terminology of this Standard. An asterisk appears before those terms defined in ISA-S37.1; a dagger (†) appears before those terms defined in this Standard.

3 UNIFORM DRAWINGS AND SYMBOLS

3.1 DRAWING SYMBOL

The electrical diagram symbol for a linear strain-gage acceleration transducer is a square of dimensions 2x by 2x, with an added equilateral triangle, the base of which is the left side of the square. The triangle symbolizes the sensing element. The letter "a" in the triangle designates linear (rectilinear) acceleration.

NOTE: Angular acceleration is designed by the Greek letter alpha (a).

The strain gage bridge is symbolized by a small square, with diagonals x by x, centered in the large square. The diagonals of the small square are drawn perpendicular to the sides of the large square. Lines from each apex of the small square projected to the right side of the large square represent the electrical leads.

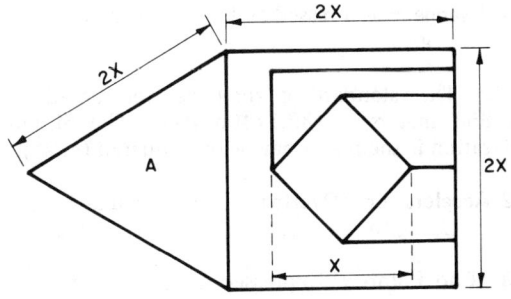

NOTE: This symbol is not ANSI approved at this time. It has been submitted to the ANSI Y32 Committee on Graphic Symbols for their consideration and approval.

3.2 Outline Drawings

Orthogrographic projection outline drawings, with tolerances should include the following information:

1. The outline dimensions.
2. The location and size of the mounting holes.
3. The indentification and location of the electrical connections; where a commercial connector is used, it and the mating connector should be identified.
4. The location of the center of the seismic mass (using the following symbol, ⊕).
5. The direction and polarity of the sensitive axis (using the following symbol, $\xrightarrow{+}$, to indicate the direction in which the case must be accelerated to produce a positive electrical output) (see 4.1.1.3, Identification, and 4.1.3.6 Polarity of Electrical Output).

3.3 Electrical Connections

Whether the electrical termination is by means of a connector or a cable, the pin designations or wire color code shall conform to the following transducer wiring standard promulgated by the Western Regional Strain Gage Committee, as approved September 18, 1957, and revised May 6, 1960.

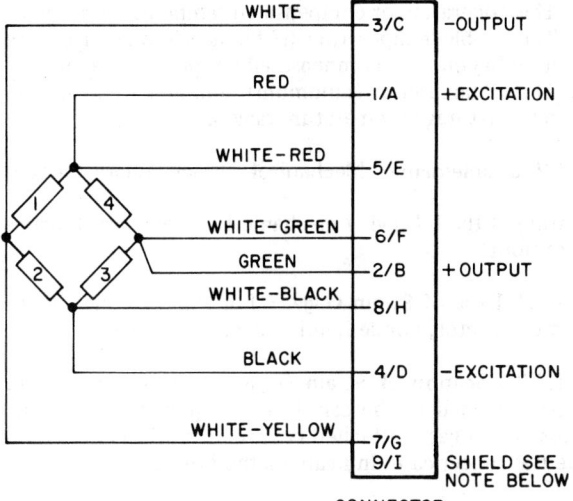

NOTES:

1. The bridge elements shall be arranged so that functions producing positive electrical output will cause increasing resistance in arms 1 and 3 of the bridge.
2. For shielded transducers, pins 5, 7, and 9 shall be shield terminals for 4, 6, and 8 wire systems respectively.

4 SPECIFICATION CHARACTERISTICS

4.1 DESIGN CHARACTERISTICS

4.1.1 Basic Mechanical Design Characteristics

4.1.1.1 Dimensions — Drawings giving dimensions (as well as other information) shall be provided as described in 3.2, Outline Drawings. Dimensions shall be given in millimeters (inches).

4.1.1.2 Mass — The mass of the transducer shall be given in grams (ounces).

4.1.1.3 Identification — The following characteristics shall be permanently inscribed on the outside of the transducer case or on a suitable nameplate permanently attached to the case:

Nomenclature of the transducer per ISA-S37.1, Electrical Transducer Nomenclature and Terminology (see 3, Nomenclature).
Name of the manufacturer
Model or part number
Serial Number
Range
Excitation
Identification of electrical connections
Direction and polarity of the sensitive axis (see 3.2, Outline Drawings).
(Optional) Customer specification or part number
(Optional) Temperature range
(Optional) Input and output impedance
(Optional) Approximate sensitivity
(Optional) Marking "delicate instrument"

4.1.1.4 Temperature Range — The following temperature ranges shall be listed in °Fahrenheit or °Celsius.

1. The *operating or compensated temperature range.
2. The usable temperature (if the accelerometer can be used beyond the compensated temperature range).
3. The *maximum (minimum) ambient temperature range (storage temperature range).

4.1.2 Supplemental Mechanical Design Characteristics

Listing of the following mechanical design characteristics is optional.

4.1.2.1 Type of Strain Gage — For example: metallic or semiconductor; bonded, unbonded, or diffused.

4.1.2.2 Location of Strain Gage — For example: strain gages attached to the cantilevers supporting the seismic mass; strain gages attached between the seismic mass and the transducer case. Indicate method of attachment.

4.1.2.3 Number of Active Strain Gage Elements — For example: two-arm bridge; four-arm bridge.

4.1.2.4 Damping — Specify type of damping (pneumatic, magnetic, or fluid); also, if fluid damping is used, specify type and characteristics of fluid.

4.1.2.5 Movement of the Seismic Mass with Acceleration — Approximate displacement in millimeters (inches) of the seismic mass due to full scale acceleration.

4.1.2.6 Mechanical Stops — The location of the stops relative to the range of the transducer.

4.1.3 Basic Electrical Design Characteristics

These characteristics are applicable at 24 ± 3°C (75 ± 5°F).

4.1.3.1 *Excitation — Expressed as _____ volts (milliamperes) dc; or _____ volts (milliamperes) ac rms at _____Hz.

4.1.3.2 *Maximum Excitation — Expressed as _____ volts (milliamperes) dc; or _____ volts (milliamperes) ac rms at _____ Hz.

4.1.3.3 *Input Impedance — Expressed as _____ ± _____ ohms at _____ ± _____Hz; or _____ ± _____ ohms (dc).

4.1.3.4 *Output Impedance — Expressed as _____ ± _____ ohms at _____ ± _____Hz; or _____ ± _____ ohms (dc).

4.1.3.5 Electrical Connections — Electrical connections shall be made as described in 3.3, Electrical Connections.

4.1.3.6 Polarity of Electrical Output — A positive output is produced by applying a positive acceleration to the case of the accelerometer. (See 3.2 Outline Drawings (5) and 3.3 Electrical Connections.)

4.1.3.7 *Insulation Resistance — Expressed as_____ megohms, minimum, at _____ volts dc between all terminals in parallel and the transducer case.

4.1.3.8 † Shunt Calibration Resistor — (Optional). Expressed as _____ ohms for _____% of full scale output.

NOTE: The circuit arrangement shall be defined.

4.1.3.9 Interference — The design characteristics incorporated in the accelerometer construction to minimize any radio frequency interference or electromagnetic interference signals being induced into the transducer or generated by the transducer by either conduction or radiation shall be described.

4.1.3.10 *Load Impedance — Performance characteristics values apply only for load impedance values of _____ ohms, minimum or _____ ± _____ohms.

4.2 Performance Characteristics

The pertinent performance characteristics of strain gage accelerometers should be tabulated in the order shown. Unless otherwise specified they apply at 24 ± 3°C (75 ± 5°F); Relative Humidity, 90% maximum; Barometric Pressure, 73 ± 7 cm Hg (29 ± 3 inches Hg).

4.2.1 *Range — Expressed as ± _____ g or _____ g to _____g.

NOTE: The standard g shall be considered to be 9.806650 meters/s^2 (32.17405 ft/s^2. The SI unit of acceleration is the meter per second squared (m/s^2).

4.2.2 Acceleration *Overload — Expressed as ± _____ g or _____ g to _____ g. (See 4.2.1, Range.)

4.2.3 *End Points — Expressed as_____ ± _____ mV and_____ ± _____mV at _____ volts (milliamperes) excitation.

4.2.4 *Full-Scale Output — Expressed as _____ ±

*Defined in ISA-S37.1
†Defined in 4.3, Additional Terminology

_____ mV per volt (milliampere) excitation; or
_____ ± _____ mV at _____ volts (milliamperes) excitation.

NOTE: At the specified load impedance.

4.2.5 *Zero-Measurand Output — Expressed as _____ ± _____ mV per volt (milliampere) excitation; or _____ ± _____ mV at _____ volts (milliamperes) excitation.

NOTE: At the specified load impedance.

4.2.6 *Linearity Expressed as _____ linearity within ± _____ % of full scale output.

NOTE: The type of linearity specified shall be one of the straight line types defined in ISA-S37.1; namely, end point, independent, least squares, terminal, or theoretical slope.

4.2.7 *Hysteresis — Expressed as _____ % of full scale output. Alternately 4.2.6 and 4.2.7 may be combined as 4.2.8.

4.2.8 *Hysteresis and Linearity — Expressed as combined hysteresis and linearity within ± _____ % of full scale output.

NOTE: The type of linearity shall be stated (see 4.2.6 Linearity).

4.2.9 *Repeatability — Expressed as within _____ % of full scale.

4.2.10 *Stability — Expressed as within _____ % of full scale output over a period of _____ (hours, days, months). Alternately sections 4.2.6, 4.2.7, 4.2.9, and 4.2.10 may be combined as 4.2.11.

4.2.11 *Static Error Band — Expressed as ± _____ % of full scale output as referred to _____ line.

NOTE: See ISA-S37.1 for listing of reference lines. The calibration cycle(s) used to establish this reference line shall be clearly identified. A least squares or end point line is preferred.

4.2.12 *Warmup Period — Expressed as _____ minutes for subsequent sensitivity shift and zero shift not to exceed _____ % of full scale output, or for the static error band not to be exceeded.

4.2.13 *Output Regulations — Expressed as sensitivity shift of ± _____ % for a change in bridge excitation of ± 10%; zero balance (mV/V) variation of ± _____ % full scale for a change in bridge excitation of ± 10%.

4.2.14 *Frequency Response — Expressed as output within ± _____ % of the output obtained in a static calibration or at a stated reference frequency over a frequency range from _____ Hz to _____ Hz at a temperature of ± _____ °C(°F).

4.2.15 *Natural Frequency and Damping Ratio (Alternate) — The natural frequency and damping ratio shall

be expressed as _____ ± _____ Hz and _____ ± _____ of critical damping respectively when the instrument temperature is _____ ± _____ °C(°F).

NOTE: At f_n a phase shift of 90 degrees will be observed between the input acceleration and the output signal.

4.2.16 † Phase Shift — Expressed as phase shift linear within ± _____ degrees from zero Hz to _____ Hz.

4.2.17 *Temperature Error — Expressed as temperature from _____ °C(°F) to _____ °C(°F) which will not cause a sensitivity shift of more than _____ % or zero shift of more than _____ % of full scale output. Alternately the following may be specified (4.2.18 and 4.2.19):

4.2.18 *Thermal Sensitivity Shift for Static Acceleration — Expressed as _____ % per _____ °C(°F) temperature change over a temperature range from _____ °C(°F) to _____ °C(°F).

4.2.19 *Thermal Zero Shift for Static Acceleration — Expressed as _____ % of full scale output _____ °C(°F) temperature change over the temperature range from _____ °C(°F) to _____ °C(°F). Alternately sections 4.2.17, 4.2.18, and 4.2.19 may be specified by:

4.2.20 *Temperature Error Band for Static Acceleration — Expressed as output values are within ± _____ % of full scale output from the straight line or curve establishing the static error band, over temperature range from _____ °C(°F) to _____ °C(°F).

4.2.21 *Temperature Gradient Error for Static Acceleration — Expressed as less than ± _____ % of full scale output while subjected to a step function temperature change from _____ °C(°F) to _____ °C(°F), lasting for _____ minutes and applied to _____ (specify particular part of the transducer).

4.2.22 † Proof Transverse Acceleration (Static) — Expressed as, shall withstand transverse static accelerations of _____ g.

4.2.23 † Proof Transverse Acceleration (Vibrational) — Expressed as, shall withstand transverse vibrational acceleration of _____ g over a frequency range of _____ to _____ Hz.

4.2.24 *Transverse Sensitivity (Static) — Expressed as a maximum of _____ % of the accelerometer sensitivity for a transverse acceleration of _____ g.

4.2.25 Transverse Sensitivity (Compound, Static) — (Optional) Expressed as a maximum of _____ % of the accelerometer sensitivity for a transverse acceleration component of _____ g and a sensitive axis acceleration component of _____ g.

4.2.26 Transverse Sensitivity (Vibrational) — Expressed as a maximum of _____ % of the accelerometer sensitivity for a transverse vibrational acceleration of _____ g and covering a frequency range of _____ Hz to _____ Hz.

4.2.27 Transverse Sensitivity (Compound, Vibrational) — (Optional) Expressed as a maximum of _____ % of

---

*Defined in ISA-S37.1
†Defined in 4.3, Additional Terminology

the accelerometer sensitivity for a transverse acceleration component of _____ g and a sensitive axis acceleration component of _____ g and covering a frequency range of _____ Hz to _____ Hz.

4.2.28 Alignment of the Sensitive Axis — Expressed as within ± _____ degrees as referenced to the mounting surface.

4.2.29 † Damping Integrity — (Optional) Expressed as no error in the predicted output greater than _____ % of full scale output (or Response Ratio) due to changes in accelerometer attitude (position relative to the field of gravity).

4.2.30 Storage Life — Expressed as _____ months (years) without changing performance characteristics beyond specified tolerances.

NOTE: Environmental storage conditions shall be described in detail.

4.2.31 *Life, Cycling — Expressed as _____ full range cycles over which the transducer shall operate without change in characteristics beyond their specified tolerances.

4.2.32 Other Conditions — Other pertinent conditions which shall not change transducer performance beyond specified limits should be listed. Examples are:

| Humidity | High Level Acoustic Excitation |
| Salt Atmosphere | Explosive Atmosphere |

*Defined in ISA-S37.1 † Defined in 4.3, Additional Terminology

Nuclear Radiation	Magnetic Fields
Shock	Sand and Dust
Over Range	Total Immersion
Fungus Resistance	(and in what medium)
Ambient Pressure (Altitude)	Solar (or other) Heat Radiation

4.3 Additional Terminology

4.3.1 Phase Shift — The phase angle by which the output of a transducer lags a sinusoidal varying measurand.

NOTE: Expressed as a fraction of a cycle of the frequency, usually in degrees, or radians.

4.3.2 Shunt Calibration Resistor — A shunt resistor which, when placed across specified points of the electrical circuit of the transducer, will electrically simulate a specified percentage of the full scale output of the transducer at room conditions.

4.3.3 Damping Integrity — The ability of the accelerometer to produce a predicted output, with no transients, during or after changes in the attitude of the transducer, due to bubbles, contamination, etc.

4.3.4 Proof Transverse Acceleration (Static) — The maximum transverse static acceleration that can be applied without causing permanent degradation in performance beyond specified tolerance.

4.3.5 Proof Transverse Acceleration (Vibrational) — The maximum transverse dynamic acceleration(s) over a specified frequency range(s) that can be applied without causing permanent degradation in performance beyond specified tolerances.

4.4 TABULATED CHARACTERISTICS VERSUS TEST REQUIREMENTS

This table is intended for use as a quick reference for design and performance characteristics and test of their proper verification as contained in this Standard.

Characteristics	Paragraph	Basic Design Characteristics	Optional Design Characteristics	Verified During Acceptance Tests	Qualification Tests
Dimensions	4.1.1.1	x		5.2.1	
Weight	4.1.1.2	x		No Test	
Identification	4.1.1.3	x		5.2.1	
Temperature Range	4.1.1.4	x		No Test	
Type of Strain Gage	4.1.2.1		x	No Test	
Location of Strain Gage	4.1.2.2		x	No Test	
Number of Active Strain Gage Elements	4.1.2.3		x	No Test	
Damping Fluid	4.1.2.4		x	No Test	
Movement of the Seismic Mass with Acceleration	4.1.2.5		x	No Test	
Mechanical Stops	4.1.2.6		x	No Test	
Excitation	4.1.3.1	x		No Test	
Maximum Excitation	4.1.3.2	x		No Test	
Input Impedance	4.1.3.3	x		5.2.2	
Output Impedance	4.1.3.4	x		5.2.2	
Electrical Connections	4.1.3.5	x		5.2.1	
Polarity of Electrical Output	4.1.3.6	x		5.2.4, 5.2.5	
Insulation Resistance	4.1.3.7	x		5.2.3	
Shunt Calibration Resistor	4.1.3.8		x	No Test	
Interference	4.1.3.9	x		No Test	
Load Impedance	4.1.3.10	x		No Test	
Range	4.2.1	x		5.2.4, 5.2.5	
Acceleration Overload	4.2.2	x			6.16
End Points	4.2.3	x		5.2.4, 5.2.5	
Full Scale Output	4.2.4	x		5.2.4, 5.2.5	
Zero Measurand	4.2.5	x		5.2.4, 5.2.5	
Linearity	4.2.6	x		5.2.4, 5.2.5	
Hysteresis	4.2.7	x		5.2.4, 5.2.5	
Hysteresis and Linearity	4.2.8	x		5.2.4, 5.2.5	
Repeatability	4.2.9	x		5.2.4, 5.2.5	
Stability	4.2.10	x		5.2.4, 5.2.5	6.17

Characteristics	Paragraph	Basic Design Characteristics	Optional Design Characteristics	Verified During Acceptance Tests	Qualification Tests
Static Error Band	4.2.11	x		5.2.4, 5.2.5	
Warmup Period	4.2.12	x			6.2
Output Regulation	4.2.13	x			6.3
Frequency Response	4.2.14	x			6.4
Natural Frequency and Damping Ratio (alt.)	4.2.15	x			6.4
Phase Shift	4.2.16	x			6.4
Temperature Error	4.2.17	x			6.5
Thermal Sensitivity Shift	4.2.18	x			6.5
Thermal Zero Shift	4.2.19	x			6.5
Temperature Error Band	4.2.20	x			6.5
Temperature Gradient Error	4.2.21	x			6.7
Proof Transverse Acceleration (Static)	4.2.22	x			6.8
Proof Transverse Acceleration (Vibrational)	4.2.23	x			6.9
Transverse Sensitivity (Static)	4.2.24	x			6.10
Transverse Sensitivity (Compound, Static) (Optional)	4.2.25		x		6.11
Transverse Sensitivity (Vibrational)	4.2.26	x			6.12
Transverse Sensitivity (Compound, Vibrational)	4.2.27		x		6.13
Alignment of the Sensitive Axis	4.2.28	x			6.14
Damping Integrity	4.2.29		x		6.15
Storage Life	4.2.30	x			6.19
Life Cycling	4.2.31	x			6.18
Other Conditions	4.2.32	x		No Test	6.20

5 INDIVIDUAL ACCEPTANCE TESTS AND CALIBRATIONS

5.1 Basic Equipment Necessary to Perform Individual Acceptance Tests and Calibrations of Strain Gage Linear Accelerometers

The basic equipment for acceptance tests and calibration consists of a source of acceleration, a monitored source of electrical excitation and a device which measures the electrical output of the transducer. The cumulative errors and uncertainties of the measuring system comprising these components should be less than 1/10, where feasible, of the permissible tolerance of the transducer performance characteristic under evaluation. The traceability to the national standards for this measuring system should be well known.

5.1.1 Source of Acceleration

The range of the instrument supplying or monitoring the calibration acceleration should be selected to provide the necessary accuracy to 125% of the full scale range of the transducer. The source of calibration signal may be either continuously variable over the range of the instrument, or may be provided in discrete steps. The steps must be programmed in such a manner that the transition from one value of acceleration to the next value of acceleration is accomplished without overshoot. Typical accelerometer calibrating devices are as follows:

5.1.1.1 Earth's Field Static Calibrator

Range 0 g to ± 1 g.
Accuracy**

5.1.1.2 Centrifuge Static Calibrator

Typical Range, 0.1 g to 1000 g.
Accuracy**

5.1.1.3 Electromagnetic Shaker Calibrator

Typical Range, up to 100 g (except as limited by displacement, velocity and table weight), 5 Hz to 10,000 Hz, 1.3 cm (0.5 inch) double amplitude.
Accuracy**

**Conservative, obtainable accuracies of the applied calibration acceleration are shown below; they are taken for illustrative purposes from the NBS Miscellaneous Publication #250, 1965, entitled "Calibration and Test Services." Static calibration in the earth's field, error no greater than 0.001 g. Static calibration on a centrifuge, error no greater than 0.2% of the applied acceleration. Dynamic calibration on an electromagnetic shaker, error no greater than 1% of the applied acceleration.

5.1.2 Stable Source of Excitation

Commonly used sources are primary and secondary batteries, such as dry cells, and storage batteries or line-powered, electronically regulated power supplies.

5.1.3 Read-out Instrument

Examples of suitable devices are:

5.1.3.1 Manually Balanced Potentiometer

Typical range: 0 to 11 mV, $\pm$ (0.008% $\pm$ 0.5μV) limit of error; 0 to 111 mV, $\pm$ (0.006% $\pm$ 1 μV) limit of error; 0 to 1.111 V, $\pm$ (0.004% $\pm$ 10 μV) limit of error; 0 to 11.110 V, $\pm$ (0.006% $\pm$ 100 μV) limit of error.

5.1.3.2 Self Balancing Potentiometer

Typical Range, 0 to 6 mV, limit of error ± 0.3%; 0 to 100 mV, limit of error ± 0.3%.

5.1.3.3 Digital Electronic Voltmeter

Typical Ranges, ± 10 mV ± 0.01% of the reading and ± 0.01% full scale; ± 100 mV or ± 1000 mV, ± 0.02% of the reading and ± 0.002% full scale.

NOTE: The input impedance of the readout instrument should be as high as possible and shall be in compliance with the load impedance specified for the transducer.

5.2 Calibration and Test Procedures

Results obtained during the calibration and test procedures should be reported on data sheets similar to the sample data sheet, Figure 9.1. These procedures shall be performed at 24 ±3°C (75± 5°F) unless otherwise indicated.

5.2.1 Visual Inspection

The transducer shall be inspected visually for mechanical defects, poor finish, improper dimensions and improper identification markings. The electrical connector shall also be inspected.

5.2.2 Impedance Measurement

A Wheatstone bridge (or other type bridge) shall be used to measure the input and output impedances of the instrument. See 4.1.3.3, Input Impedances, and 4.1.3.4, Output Impedance.

5.2.3 Insulation Resistance

Measure the insulation resistance between all transduction element terminals (or leads) connected in parallel, and the case (and ground pin) of the transducer with a megohmmeter or similar acceptable device, using a potential of 50 volts dc, unless otherwise specified.

5.2.4 Earth's Field Static Calibration

The transducer shall be attached to an earth's field static calibrator. The excitation source and readout instrument shall be connected to the transducer and turned on. Adequate warmup time for the test equipment and instrument shall be allowed before tests are conducted.

NOTE: The earth's field calibration may be waived if the range of the instrument is so high that the ± 1 g calibration will not yield useful information. Two or more calibration cycles shall be run consecutively. If the range of the instrument is greater than ±1 g the calibration cycle shall consist of at least the +1 g, 0 g, and −1 g points; if the range of the instruments is ±1 g or less the calibration cycle shall consist of a minimum of nine equally spaced data points. This calibration cycle shall include both ascending and descending directions, and, in the case of bidirectional accelerometers, both positive and negative accelerations. For' example, a ±1g instrument would be tested at the following points; 1 g, 0.5 g, 0 g, −0.5 g, −1 g, −0.5 g, 0 g, 0.5 g, and 1 g. See Reference 5.

5.2.5 Static Calibration on the Centrifuge

5.2.5.1 Incremental

The transducer shall then be attached to a centrifuge; the excitation source and readout instrument shall be connected to the transducer and turned on. Adequate warmup time for the test equipment and instrument shall be allowed before tests are conducted.

NOTE: The centrifuge calibration may be waived if the range of the instrument is ±1 g or less.

Two or more complete calibration cycles shall be run on the instrument in the centrifuge. A complete calibration shall include both ascending and descending, directions, and, in the case of bi-directional accelerometers, both positive and negative accelerations. It will include a minimum of nine equally spaced points. For example, a ±4 g instrument would be tested at the following points: 0 g, 2 g, 4 g, 2 g, 0 g, (and then reversing the instrument on the centrifuge) −2 g, −4 g, −2 g, and 0 g. The interruption of the calibration for reversing the instrument on the centrifuge should require a minimum of time.

5.2.5.2 Continuous Comparison (Alternate)

The transducer under test is mounted on the centrifuge, as is a high precision reference accelerometer having a well established low error. Both instruments are carefully positioned so as to sense the same value of acceleration. The electrical outputs of both instruments are added in opposition, using attenuation where necessary, so that the net signal represents the deviation of the test accelerometer from that of the reference accelerometer. This signal is fed into the y-axis circuit of an x-y recorder and the full output of the reference accelerometer is fed into the x-axis circuit of the recorder.

By increasing the applied acceleration from zero to full scale and then decreasing it to zero again, a complete plot of test accelerometer response deviation versus applied acceleration is obtained. For bi-directional accelerometers, the test instrument must be tested with reversed mounting position.

The rate of application of acceleration may affect the shape of the plotted characteristic; the rate should not exceed the time response capabilities of reference accelerometer, test accelerometer, or recorder.

This method may also be used in conjunction with the earth's field static calibration. This technique is rapid as well as valuable for uncovering the presence of such defects as air bubbles, foreign particles, obstructions, etc., but its validity relies entirely on the complete knowledge of the performance characteristics of the reference accelerometer and its continued stability. See Reference 6.

5.2.6 (Optional) Dynamic Calibration

The transducer shall be mounted on the moving table of an electromagnetic or similar shaker. The excitation source and readout instrument shall be connected to the transducer and turned on. Provision shall be made for determining the phase of the acceleration input (an electronic phase meter may be used). Adequate warmup time for the test equipment and instrument shall be allowed before tests are conducted. A minimum of ten frequency points, approximately equally spaced, shall be selected covering a frequency range extending to a frequency 1½ times the estimated natural frequency of the instrument. The transducer shall then be calibrated at each of the above frequencies at an acceleration level close to but not in excess of the maximum range of the instrument (care must be taken when approaching the natural frequency of high g instruments). Calibration data for the phase shift of the transducer can be taken at the same time. See Figure 9.3.

NOTE: The low frequency amplitude capability of the shaker may require some of the initial points in the ten point frequency response curve described above to be taken at a reduced input.

NOTE: Also see 4.3.3, Damping Integrity.

5.2.7 (Optional) Temperature Effects

The transducer shall be calibrated in the earth's field or on the centrifuge at a minimum of three different temperatures (including 24 ±3°C 75 ±5°F,) covering the operating temperature range of the instrument. These calibrations shall be carried out as described in 5.2.4, Earth's Field Static Calibration, or 5.2.5, Static Calibration on the Centrifuge, except that only one calibration cycle shall be run. Care shall be taken to stabilize the instrument temperature at each selected calibration temperature. See Figure 9.3.

6 QUALIFICATION TESTS

Qualification tests shall be summarized using a test form similar to that in Section 9. The sequence of the tests must be conducted in a logical order. For example, a shock acceleration may permanently distort the seismic mass suspension changing the initial alignment so the alignment test may need to be performed before and after the shock acceleration tests.

Qualification tests are to be performed on a number of representative samples to measure a transducer's performance characteristics against the values of the specification. These particular characteristics are, in general, those which are a function of basic transducer design and are not expected to vary significantly from unit to unit. However, as particular application requirements dictate, it may be deemed prudent to include certain of these tests in the acceptance tests.

6.1 Initial Performance Tests

The tests of 5, Individual Acceptance Test and Calibrations, shall be conducted to establish a reference performance; in addition, a calibration cycle shall be performed between and after the individual qualification tests and the results compared to the reference performance and the specifications.

6.2 Warmup Period

The zero balance shall be measured repeatedly over a period of at least one hour starting with the application of excitation voltage.

In a separate test the sensitivity shall be determined repeatedly over a period of at least one hour starting with the application of excitation voltage using either the earth's field or the centrifuge as an acceleration source.

NOTES: (1) It is desirable to determine the warmup characteristics of the zero balance separately from the sensitivity as this permits the series of zero measurements to be made without disturbing the accelerometer.

(2) In all tests, the equipment will have been previously warmed up.

6.3 Output Regulation

Perform one static calibration cycle at 90%, 100% and 110% of the rated excitation. Compare the values of full scale output, zero balance, linearity and hysteresis to those of the specifications and of the initial acceptance tests.

NOTE: The 110% test above shall not be performed if it exceeds the maximum allowable excitation.

6.4 Dynamic Characteristics

The transducer shall be attached to a shaker. The excitation source and readout instrument shall be connected to the transducer and turned on. Provision shall be made for determining the phase of the accelerometer output with reference to the acceleration input. An electronic phase meter may be used. Adequate warmup time for the test equipment and instrument shall be allowed before tests are conducted. Frequency points shall be selected covering a frequency range extending to a frequency 1½ times the estimated natural frequency of the transducer. These calibration points shall include at least the following (frequencies are listed as a percentage of the natural frequency): 5, 10 (these points to tie in with the static calibration), 20, 30, 40, 50, 60 (these points to cover the maximum flat region), 70, 80, 90, 100, 125, 150% (these points to establish damping and natural frequency). Calibrations shall be made at a minimum of two values of amplitude (typically 50% of full scale and 100% of full scale). The transducer phase shift shall be determined at enough of the above listed frequencies to determine the phase linearity of the instrument and the point of 90 degree phase shift.

NOTES: (1) Depending on the range of the instrument, the capability of the shaker may not permit calibration of some of the values cited above.

(2) The frequency response of the transducer is commonly very temperature dependent. See 6.6, Dynamic Characteristics.

(3) See 6.15, Damping Integrity.

6.5 Steady State Temperature Effects

A minimum of six calibrations utilizing the earth's field or centrifuge shall be performed on the transducer (as described in 5.2.4, Earth's Field Static Calibration or 5.2.5, State Calibration on the Centrifuge, except that only one calibration cycle shall be run). This series of calibrations shall begin and end with a calibration at 24 ±3°C (75 ±5°F) the remaining four, or more calibrations shall cover the operating temperature range of the accelerometer. Care should be taken to assure that the transducer is at a stabilized temperature.

6.6 Dynamic Characteristics (High and Low Temperature)

The transducer shall be calibrated as described in 5.2.6, Dynamic Calibration, and at the maximum and minimum temperatures of interest. Care should be taken that the transducer is fully stabilized at each test temperature.

6.7 Transient Thermal Effects

6.7.1 Zero

With the transducer mounted to provide a zero g input, the transducer shall be brought rapidly to a selected high or low temperature. The temperature and accelerometer output shall be continually recorded during this transient. Heating or cooling can be accomplished by conduction through the base, by directing heated or cooled air on the instrument, by placing the instrument into a temperature controlled test chamber, or by other means.

6.7.2 Sensitivity

The test of 6.7.1, Zero shall be repeated except that the earth's field or the centrifuge shall be used as an acceleration source.

6.8 Proof Transverse Acceleration (Static)

The specified maximum static transverse acceleration is applied in each of two orthogonal transverse axes.

6.9 Proof Transverse Acceleration (Vibrational)

The specified maximum transverse vibrational acceleration is imposed in each of two orthogonal transverse axes over the specified frequency range.

6.10 Transverse Sensitivity (Static)

Mount the accelerometer with its sensitive axis perpendicular to the centrifuge table. Apply the specified transverse static acceleration; keeping the location of the center of gravity of the seismic mass constant, rotate the accelerometer about its sensitive axis and take data for enough additional points to determine an approximate maximum accelerometer transverse response. Repeat the above tests with the sensitive axis parallel to the centrifuge table and perpendicular to the acceleration that will be applied by the centrifuge.

NOTE: For these tests, the sensitive axis is considered to be perpendicular (or parallel) to the accelerometer mounting surface.

6.11 (Optional) Transverse Sensitivity (Compound Static)

Mount the accelerometer on the centrifuge table with the sensitive axis of the accelerometer parallel to the centrifuge table and at a known angle with that table radius that intersects the center of gravity of the seismic mass. Apply the specified transverse static acceleration; keeping the location of the center of gravity of the seismic mass constant, rotate the accelerometer about its sensitivity axis and take data for enough additional points to determine an approximate maximum accel-erometer transverse response.

NOTE: The transverse response is determined by subtracting the computed output resulting from the sensitive axis acceleration component from the total accelerometer output. For these tests, the sensitive axis is considered to be perpendicular (or parallel) to the accelerometer mounting surface.

6.12 Transverse Sensitivity (Vibrational)

Mount the accelerometer with its sensitive axis perpendicular to the shaker motion. Apply the specified transverse vibrational acceleration over the specified frequency range; rotate the accelerometer about its sensitive axis and take enough additional points to determine the approximate maximum accelerometer transverse response.

NOTE: For these tests, the sensitive axis is considered to be perpendicular (or parallel) to the accelerometer mounting surface.

NOTE: The suspension design of some shakers creates a transverse component; care should be taken to mount the accelerometer so that its sensitive axis is perpendicular to this motion.

NOTE: For a method of improved accuracy see Reference 6.

6.13 (Optional) Transverse Sensitivity (Compound, Vibrational)

Mount the accelerometer with its sensitivity axis at a known angle to the motion that will be applied by the shaker. Apply the specified vibrational acceleration over the specified frequency range; rotate the accelerometer about its sensitive axis and take additional points to determine the approximate maximum accelerometer response.

NOTE: For these tests, the sensitive axis is considered to be perpendicular (or parallel) to the accelerometer mounting surface. The suspension design of some shakers creates a transverse component; care should be taken to mount the accelerometer so that its sensitive axis is perpendicular to this motion. The transverse response is determined by subtracting the computed output resulting from the sensitive axis acceleration component from the total accelerometer output.

NOTE: For a method of improved accuracy, see Reference 7.

6.14 Alignment

The accelerometer shall be mounted with its sensitive axis (as defined by the case and its mounting surface) perpendicular to the direction of applied acceleration. The accelerometer is then turned about its sensitive axis until the points of maximum output deviations are found. These will be two points 180 degrees apart so that one-half the difference between these points represents the value of output caused by the misalignment for the acceleration applied. From this data, the magnitude and direction of the maximum misalignment can be determined.

NOTE: The source of acceleration can be the field of gravity or a centrifuge depending on the range of the accelerometer.

6.15 Damping Integrity

Record the dynamic response of the accelerometer at a frequency approximately of 0.7 the natural frequency. Now quickly invert the accelerometer relative to the field of gravity on the vibration calibrator and, without delay, record the response at the same frequency. A significant change in response may indicate the presence of a gas bubble or foreign matter. Alternatively perform a calibration cycle on a centrifuge or in the field of gravity, then quickly invert the accelerometer and repeat the calibration.

NOTE: If the gas bubble problem is accentuated by the addition of an external vacuum environment, it is an indication of a defective case seal.

NOTE: Vacuum and/or temperature, combined with high vibration amplitudes that are within the accelerometers normal measuring range may permit cavitation and a change in frequency response even though no appreciable amount of gas is present in the damping fluid. Although liquids are incompressible at the pressures involved, sealing diaphragms, expansion chambers, and the damping chamber walls sometimes allow for an increase in volume permitting cavitation.

6.16 Acceleration Overload

The specified overload acceleration is applied in the directions and magnitudes specified; the performance during overload is monitored. Where the overload recovery time is required to be short, this time is best measured with an oscilloscope after application of a shock acceleration overload. A post overload calibration cycle is performed and compared to the specifications and the initial acceptance tests.

6.17 Stability

Calibrations shall be performed at suitable intervals of time to determine the ability of the accelerometer to reproduce the initial calibrations at room temperature. Of particular interest will be the repeatability of full scale output; additional characteristics of interest will be zero balance, linearity, hysteresis, and frequency response.

6.18 Life Test

After applying a specified number of full range excursions of measurand, at least one complete calibration cycle shall be performed to establish a minimum value of cycling life.

6.19 Storage Life Test

After storing the transducer under specified conditions for a specified period of time, one complete calibration cycle shall be performed to establish minimum storage life.

6.20 Effects of Other Environments

Expose the transducer to other specified environmental conditions followed in each case by one complete calibration cycle to test ability of the transducer to perform statisfactorily after such exposure.

7 REFERENCES

1. Transducer Wiring-Standard, Western Regional Strain Gage Committee, Los Angeles, California, May 1, 1960.

2. Instrument Notes, Statham Instruments, Inc. Numbers 2, 6, 7, 9, 12, 19, 23, 29, 32, and 33.

3. "Methods for Calibrating Motion Transducers at Low Frequencies" (0 to 20 Hz). by Otis C. Ingebritsen, NASA, Langley Research Center; ISA Paper MI8-4-MESTIND-67; September 11-14, 1967 Meeting.

4. MIL-STD-810 Environmental Test Methods

5. "Earth's Field Static Calibrator for Accelerometers" by P. S. Lederer and J. S. Hilten, National Bureau of Standards Technical Note 269; February 1, 1966.

6. "A Comparison Method to Measure Accelerometer Transverse Sensitivity," by Tom D. Finley, NASA Langley Research Center; ISA Preprint 69-666, October 27-30, 1969 ISA Annual Conference.

7. "Accelerometer Calibration with the Earth's Field Dynamic Calibrator" by J. S. Hilten, National Bureau of Standards Technical Note 517; March, 1970.

8. "Methods for the Calibration of Shock and Vibration Pickups" ANSI S2.2 — 1959 (Revised 1971).

9. "Selection of Calibrations and Tests for Electrical Transducers used for Measuring Shock and Vibration," ANSI S2.11 — 1969 (Revised 1973).

8 ISA STANDARDS COMMITTEES ON MEASUREMENT TRANSDUCERS

SP37.1 Nomenclature and Specification Terminology. Published ISA-S-37.1 "Electrical Transducer Nomenclature and Terminology" 1969, ANSI MC 6.1-1975.

SP37.2 Piezoelectric Acceleration Transducers. Published ISA-RP37.2 "Guide for Specifications and Tests for Piezoelectric Acceleration Transducers for Aero-Space Testing" 1964.

SP37.3 Strain Gage Pressure Transducers. Published ISA-S37.3 "Specifications and Tests for Strain Gage Pressure Transducers" 1970, ANSI MC 6.2-1975.

SP37.4 Resistive Temperature Transducers

SP37.5 Strain Gage Acceleration Transducers. Published ISA-S37.5 "Specifications and Tests for Strain Gage Linear Acceleration Transducers" 1971, ANSI MC 6.3-1975.

SP37.6 Potentiometric Pressure Transducers. Published ISA-S37.6 "Specifications and Tests of Potentiometric Pressure Transducers for Aerospace Testing" 1975.

SP37.7 Radiation Transducers.

SP37.8 Strain Gage Force Transducers

SP37.9 Thermoelectric Temperature Transducers

SP37.10 Piezoelectric and Sound Pressure Transducers. Published ISA-S37.10 "Specifications and Tests for Piezoelectric Pressure and Sound-Pressure Transducers" 1969, ANSI MC 6.4-1975.

SP37.11 Servo Acceleration Transducers

SP37.12 Potentiometric Displacement Transducers

SP37.13 DC Output Pressure Transducers

FIGURE 9.1 ACCEPTANCE TEST AND CALIBRATION RECORD STRAIN GAGE ACCELEROMETER

Vendor's Part No. _____

User's Part No. _____

Serial No. _____

Range _____ g

Excitation Voltage _____ volts

5.2.1 Visual Inspection:

Mechanical _____ Finish _____

Dimensions _____

Identification _____

Electrical Connectors _____

5.2.2 Input Impedance: _____ ohms

5.2.2 Output Impedance: _____ ohms

5.2.3 Insulation Resistance: _____ megohms at _____ volts dc

5.2.4 and 5.2.5 Static Calibration:

Temperature _____ °C (°F)

Acceleration Source _____ Radius _____ mm (in.)

EARTH'S FIELD				CENTRIFUGE		

Run #1

rps	Accel g	Theo. Output	+ Accel		− Accel	
			E_o mV	Dev	E_o mV	Dev
	+1					
	0					
	−1					

Run #2

rps	Accel g	Theo. Output	+ Accel		− Accel	
			E_o mV	Dev	E_o mV	Dev
	+1					
	0					
	−1					

Max Error	
+	−

Allowed

4.2.3 End Points: _____ and _____ mV (1); _____ and _____ mV (2)

4.2.4 Full Scale Output: _____ mV (1); _____ mV (2)

4.2.5 Zero Measurand Output: _____ mV (1); _____ mV (2)

4.2.6 Linearity: ± _____ % FSO (maximum)

4.2.7 Hysteresis: _____ % FSO (maximum)

4.2.9 Repeatability: _____ % FSO

4.2.11 Static Error Band: + _____ % and − _____ % FSO

Above referred to: _____ reference line or curve

FIGURE 9.2 QUALIFICATION TEST SUMMARY
STRAIN GAGE ACCELEROMETER

Report No. _____
Vendor _____
Range _____

Test Facility _____
Vendor Part No. _____
P.O. No. _____

Date Tests Started _____
Date Tests Finished _____
Test Type _____
Serial No. _____
Part No. _____

SUMMARY OF RESULTS

Test	Tested to Proc. No. Test Waived by	S37.5 Para. No.	Check if Acceptable	Check Type of Failure: Error	Electr.	Mechan.	See Comments	Error / Error Band +%FSO	−%FSO
1. Visual Inspection		5.2.1							
2. Impedance Measurement		5.2.2							
3. Insulation Resistance		5.2.3							
4. Static Calibration (1 g)		5.2.4							
5. Static Calibration (>1 g)		5.2.5							
6. Dynamic Calibration (Optional)		5.2.6							
7. Temperature Effects (Optional)		5.2.7							
8. Warmup Period		6.2							
9. Output Regulation		6.3							
10. Dynamic Characteristics		6.4							
11. Steady State Temperature Effects		6.5							
12. Dynamic Characteristics (High and Low Temp.)		6.6							
13. Transient Thermal Effects		6.7							
14. Proof Transverse Acceleration (Static)		6.8							
15. Proof Transverse Acceleration (Vibrational)		6.9							
16. Transverse Sensitivity (Static)		6.10							
17. Transverse Sensitivity (Compound, Static)		6.11							
18. Transverse Sensitivity (Vibrational)		6.12							
19. Transverse Sensitivity (Compound, Vibrational)		6.13							
20. Alignment		6.14							
21. Damping Integrity		6.15							
22. Acceleration Overload		6.16							
23. Stability		6.17							
24. Life Test		6.18							
25. Storage Life Test		6.19							

FIGURE 9.3

5.2.6 DYNAMIC CALIBRATION

Acceleration Input = _____ g (CONSTANT)

Frequency Hz	Output mV	Phase Angle Degrees (Lag)	

Temperature _____ °C(°F)
Natural Frequency _____ Hz
Damping Ratio _____ Of Critical

5.2.7 TEMPERATURE EFFECTS

Low Temperature Full Scale Output   Temperature _____ °C(°F)

Zero _____ mV
+ FS + _____ mV
− FS − _____ mV

Room Temperature Full Scale Output   Temperature _____ °C(°F)

Zero _____ mV
+ FS + _____ mV
− FS − _____ mV

Upper Temperature Full Scale Output   Temperature _____ °C(°F)

Zero _____ mV
+ FS + _____ mV
− FS − _____ mV

INSTRUMENT SOCIETY of AMERICA
Research Triangle Park, North Carolina

ANSI/ISA-S37.6-1976
(R 1982)
Approved December 14, 1982
(Formerly ANSI MC 6.5-1976)

American National Standard

Specifications and Tests of Potentiometric Pressure Transducers

Instrument Society of America

ISBN 0-87664-380-2

ISA-S37.6 Specifications and Tests of Potentiometric Pressure Transducers

INSTRUMENT SOCIETY of AMERICA
67 Alexander Drive
P.O. Box 12277
Research Triangle Park, North Carolina 27709

PREFACE

This Preface is included for information purposes and is not part of ISA S37.6.

This Standard has been prepared as a service of the Instrument Society of America toward the goal of uniformity in the field of instrumentation. To be of real value, it should not be static, but should be subject to periodic review. Toward this end, the Society welcomes all comments and criticisms and asks that they be addressed to the Standards and Practices Board Secretary, Instrument Society of America, 67 Alexander Drive, P.O. Box 12277, Research Triangle Park, North Carolina 27709, Telephone (919) 549-8411.

The ISA Standards and Practices Department is aware of the growing need for attention to the metric system of units in general, and the International System of Units (SI) in particular, in the preparation of instrumentation standards. The Department is further aware of the benefits to users of ISA Standards in the USA of incorporating suitable references to the SI (and the metric system) in their business and professional dealings with other countries. Toward this end, this Department will endeavor to introduce SI and SI-acceptable metric units as optional alternatives to English units in all new and revised standards to the greatest extent possible. The Metric Practice Guide, which has been published by the American Society for Testing and Materials as ANSI designation Z210.1 (ASTM E380-76, IEEE Std. 268-1975), and future revisions, will be the reference guide for definitions, symbols, abbreviations, and conversion factors.

It is the policy of the Instrument Society of America to encourage and welcome the participation of all concerned individuals and interests in the development of ISA Standards. Participation in the ISA standards making process by an individual in no way constitutes endorsement by the employer of that individual of the Instrument Society of American or any of the standards which ISA develops.

This Standard is intended as a guide for technical personnel at user facilities as well as by manufacturers' technical and sales personnel whose duties include specifying, calibrating, testing or showing performance characteristics of potentiometric pressure transducers. By basing users' specifications as well as technical advertising and reference literature on this Standard, or by referencing portions thereof, as applicable, a clear understanding of the users' needs or of the transducers' performance capabilities, and of the methods used for evaluating or proving performance, will be provided. Adhering to the specification outline, terminology and procedures shown will not only result in simple, but also complete specifications; it will also reduce design time, procurement lead time, and labor, as well as material costs. Of major importance will be the reduction of qualification tests resulting from use of a commonly accepted test procedure and uniform data presentation.

The development of this Standard was initiated as the result of a survey conducted in December, 1960. A total of 240 questionnaires was sent out to transducer users and manufacturers. A strong majority indicated in their replies a need for transducer standardization. As potentiometric pressure transducers were one of the types shown to be most in need of standardization, a project subcommittee, SP37.6, Potentiometric Pressure Transducers, was formed under the cognizance of Committee SP37, Transducers for Aerospace Testing, and a standard was developed and published in 1967. Subsequently, the standard was reviewed extensively, and revised in 1974. The reviewers were selected from a broad cross-section of all industries and sciences in which transducers are applied for measuring purposes.

The following individuals served on the 1969 SP37 committee:

NAME	COMPANY
P. S. Lederer, Chairman	National Bureau of Standards
J. F. Mayo-Wells	National Bureau of Standards
R. R. Bouche	ENDEVCO
H. C. Chandon	Rosemount, Inc.
J. Hilten	National Bureau of Standards
H. N. Norton	Jet Propulsion Laboratory
J. J. Elengo	Revere Corporation of America
O. C. Ingebritsen	NASA/Langley Research Center
R. D. Bronson	Lockheed Electronics Company
J. Z. Inskeep	Jet Propulsion Laboratory

The following individuals served on the 1982 SP37 committee:

NAME	COMPANY
J. W. Mock, Chairman	Measurement Services
P. Bliss	Consultant
W. Brinkschulte	Genisco Technology Corporation
C. Kutelis	Revere Corporation of America
R. W. Lally	PCB Piezotronics
H. E. Lockery	Hottinger Baldwin Measurements, Inc.
E. W. Malone	Boeing Aerospace Co.
T. B. Miller	Lawrence Livermore Lab
K. A. Parlee	United Engineers & Constructors, Inc.
H. Pitt	
J. Powell	Rosemount Engineering
R. C. Strahm	EG & G Idaho
R. M. Whittier	ENDEVCO
N. Wilde	EG & G Idaho
E. Wong	Bell & Howell

This Standard was approved for publication by the ISA Standards and Practices Board in October 1982.

NAME	COMPANY
T. J. Harrison, Chairman	IBM Corporation
P. Bliss	Consultant
W. Calder	The Foxboro Company
N. Conger	Continental Oil Co.
B. Feikle	Bailey Controls Co.
R. T. Jones	Philadelphia Electric Co.
R. Keller	Boeing Company
O. P. Lovett, Jr.	Isis Corp.
E. C. Magison	Honeywell, Inc.
A. P. McCauley	Diamond Shamrock Corp.
J. W. Mock	Measurement Services
E. M. Nesvig	ERDCO Engineering Corp.
G. Platt	Bechtel Power Corp.
R. Prescott	Moore Products Company
W. C. Weidman	Gilbert Associates
K. A. Whitman	Allied Chemical Corp.
J. R. Williams	Stearns-Roger, Inc.
B. A. Christensen*	
L. N. Combs*	
R. L. Galley*	
R. G. Marvin*	
W. B. Miller*	Moore Products Company
R. L. Nickens*	

*Director Emeritus

CONTENTS

1. PURPOSE

This Standard establishes the following for potentiometric pressure transducers:

1.1 Uniform minimum specifications for design and performance characteristics.

1.2 Uniform acceptance and qualification test methods, including calibration techniques.

1.3 Uniform presentation of minimum test data.

1.4 A drawing symbol for use in electrical schematics.

2. SCOPE

2.1 This Standard covers potentiometric pressure transducers, primarily those used in measuring systems.

2.2 Included among the specific versions of potentiometric pressure transducers to which this Standard is applicable are the following:

Absolute Pressure Transducers
Differential Pressure Transducers
Gage Pressure Transducers

2.3 Technology used in this document is defined in ISA Standard 37.1. Additional terms considered applicable to potentiometric pressure transducers are defined in section 4.3 of this document. An asterisk appears before those terms defined in S37.1. The terms defined in section 4.3 appear in italics.

3. DRAWING SYMBOL

3.1 The drawing symbol for a potentiometric pressure transducer is a square with an added equilateral triangle, the base of which is the left side of the square. The letter "P" in the triangle designates "pressure" and the subscripts denote the second modifier (the illustration shows an absolute pressure transducer, as symbolized by "P_A").

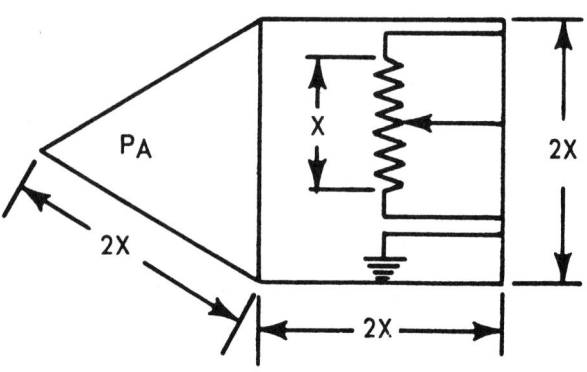

Subscripts

A=Absolute
D=Differential
G=Gage
S=Sealed Reference Differential (see 4.3)

3.2 The potentiometer is symbolized by a variable resistance of length x. The lines from it and to ground represent the electrical leads or terminations.

4. SPECIFICATION CHARACTERISTICS

4.1 Design Characteristics

4.1.1 Required Mechanical Design Characteristics

The following mechanical design characteristics shall be listed.

4.1.1.1 TYPE OF PRESSURE SENSED

* Absolute Pressure
* Differential Pressure, Unidirectional
* Differential Pressure, Bidirectional
Differential Pressure, Sealed Reference
* Gage Pressure

Note: At present, no provision is made by the SI system of units for abbreviations following the pressure units to indicate the type of pressure, as is done in the U.S. customary system of units, e.g., psia for absolute pressure in psi. In the interim it is recommended that for the SI system, the type of pressure be indicated in this manner: " . . . An absolute pressure of ____ Pa." " . . . A differential pressure range of ____ kPa," etc.

Note: For differential pressure transducers, the allowable range of reference pressures shall be listed e.g., "0 to 1MPa" or "0 to 100 psi."

4.1.1.2 * Measured Fluids

The fluids in contact with pressure port(s) shall be listed, e.g., nitric acid, liquid oxygen. Requirements for and limitations on the *isolating element* (if used) shall be listed.

4.1.1.3 Configuration and Dimensions

The outline drawing shall show the configuration with dimensions in millimeters (inches). Unless pressure and electrical connections are specified (Reference 4.1.1.5, 4.1.3.4), the outline shall include limiting maximum dimensions for these connections.

4.1.1.4 Mountings and Mounting Dimensions

Unless the pressure connection serves as a mounting, the

outline drawing shall indicate the method of mounting with hole size, centers, and other pertinent dimensions in millimeters (inches), including thread specifications for threaded holes, if used.

4.1.1.5 Pressure Connection

The pressure connection(s) shall be indicated on the outline drawing. For threaded fittings, specify: Applicable Military or Industry standards or nominal size, number of threads per millimeter (threads per inch), thread series, and thread class. For hose tube fitting, specify tube size.

4.1.1.6 Mounting Effects

The maximum mounting force or torque shall be specified if it will tend to affect transducer performance (Reference 4.2.28).

4.1.1.7 Mass

The mass of the transducer shall be specified to grams (ounces).

4.1.1.8 Case Sealing

If case sealing is necessary, the mechanism and materials used for sealing should be described. The same requirement applies to the electrical connector. The resistance of the sealing materials to cleaning solvents and commonly used measured fluids should be stated.

4.1.1.8 Identification

The following characteristics shall be permanently inscribed on the outside of the transducer case or on a suitable nameplate permanently attached to the case.

Nomenclature of transducer (acc. to ISA-S37.1, Section 3).
Name of Manufacturer, (Part number to reflect one controlled configuration), and Serial Number.
* Range in Pa(psi) and designation of type of pressure (see 4.1.1.1.). *Maximum * excitation.*
* Transduction element resistance (*Potentiometric Element*).
Identification of * Measured and Reference Ports (for differential pressure transducers).
* Reference Pressure Range (for differential pressure transducers).
Identification of Electrical Connections.

Inscription of the following characteristics is optional:

Customer Specification or Part Number or both.
Type of Electrical Connector and Mating Connector (if applicable).
Operating Temperature Range.
* Proof Pressure.

Note: Identificaton of Pressure Ports may be abbreviated MEAS & REF.

4.1.2 Supplemental Mechanical Design Characteristics

Listing of the following mechanical design characteristics is optional:

4.1.2.1 Case Material

Where applicable, state the surrounding environmental condition requirement or compatibility.

4.1.2.2 Sensing Element

The sensing element type shall be specified, e.g.,

* Diaphragm (flat or corrugated)
* Capsule
* Bellows
* Bourdon Tube-"C" or "U" shaped, spiral, helical, or twisted.

Note: Where used, an isolating element with transfer fluid shall be detailed as to composition.

4.1.2.3 Damping Fluid

Where used, the type, composition, temperature characteristics and compatibility with transducer components shall be specified.

4.1.2.4 Number of *Potentiometric Elements or Taps*

Where more than one potentiometric transducer element or a tapped element is required, they shall be specified.

4.1.2.5 Dead Volume

The dead volume shall be given in cubic millimeters (cubic inches). For differential pressure transducers the volume of both cavities should be listed.

4.1.2.6 Volume Change Due to Full Scale Pressure

The change in volume of the sensing element due to application of full scale pressure shall be given in cubic millimeters (cubic inches).

4.1.2.7 Materials in Contact with the Measured Fluid

The materials in contact with the measured fluid shall be listed.

Note: For differential pressure transducers, materials in both ports must be considered.

4.1.2.8 Gage Vent (Port)

In gage pressure transducers where the transduction element is exposed to ambient atmosphere, the allowable types and concentrations of atmospheric contaminant shall be specified.

4.1.2.9 Maximum and Minimum Temperatures

The maximum and minimum temperatures of fluids or environments which can be applied to the transducer and which will not cause permanent calibration shift shall be listed.

Note: Exposure time shall be specified, if relevant.

4.1.3 Required Electrical Design Characteristics

The following electrical design characteristics shall be listed. All are applicable at *Room Conditions.

4.1.3.1 *Excitation

Expressed as " _____ volts (milliamperes) DC" or " _____ volts (milliamperes) AC rms at ___ hertz."

4.1.3.2 Maximum Excitation

Expressed as " _____ volts (milliamperes) DC" or " _____ volts (milliamperes) AC rms at ___ hertz."

4.1.3.3 * Transduction Element Resistance

Expressed as " _____ ± _____ ohms."

4.1.3.4 Electrical Connections

Whether the electrical termination is by means of a connector or a cable, the pin designations or wire color code shall conform to the following transducer wiring standard.

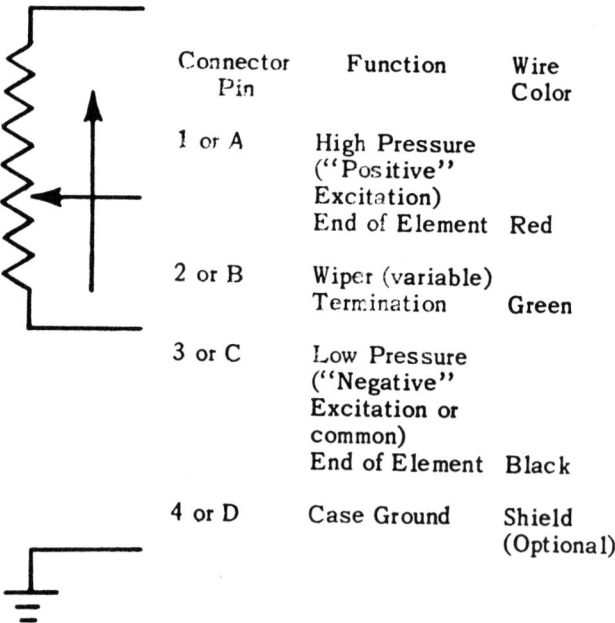

Connector Pin	Function	Wire Color
1 or A	High Pressure ("Positive" Excitation) End of Element	Red
2 or B	Wiper (variable) Termination	Green
3 or C	Low Pressure ("Negative" Excitation or common) End of Element	Black
4 or D	Case Ground	Shield (Optional)

Notes:

1. For differential pressure transducers, the arrow indicates pressure at the measurand port greater than pressure at the reference port.

2. The transduction element(s) shall be arranged with increasing "positive" voltage output as caused by increasing resistance between the wiper and low pressure (common) end of winding.

4.1.3.5 *Insulation Resistance

Expressed as " _____ megohms at _____ volts DC between all transduction element terminals in parallel and the transducer case at a temperature of _____ °C and 90% relative humidity."

4.1.3.6 *Breakdown Voltage Rating

Expressed as "Capable of withstanding _____ volts ac-rms at _____ hertz, at a temperature of _____ °C and 90% relative humidity for _____ minutes."

4.1.3.7 *Load Impedance

Expressed as " _____ ohms" (See 4.2.29).

Note: Although load impedance (the impedance presented to the output terminals of the transducer by the associated external circuitry) is not a transducer but a system characteristic, it should be specified in order to define loading error. A single, close-tolerance value of load impedance shall be specified for use during all tests where not otherwise noted.

4.2 The performance characteristics of the potentiometric pressure transducers should be tabulated in the order shown. Unless otherwise specified, they apply at Room Conditions as defined in ISA S37.1. Characteristics are usually referred to the ouput and expressed as "% VR," i.e. "percent *Voltage Ratio".

4.2.1 * Range

Expressed as " _____ to _____ Pa (psi)" or "+ _____ Pa (psid)."

Note: Equivalent pressure units in the SI system are expressed in Pascals.

$$1 \text{ psi} = 6894.8 \text{ Pa}$$
$$10 \text{KPa} = 1.4504 \text{ psi}$$

4.2.2 * End Points

Expressed as " _____ % ± _____ % and _____% ± _____ % VR."

Note: "%VR" is "percent *Voltage Ratio."

End Points shall be omitted where adequately defined using Error Band specifications.

4.2.3 * Full Scale Output

Expressed as " _____ % ± _____ % VR."

Note: Full scale output shall be omitted where adequately defined using End Points or Error Band specifications.

4.2.4 * Linearity

Expressed as " _____ linearity shall be within ± _____ % VR."

Note: The linearity modifier shall be one of those defined in ISA S37.1; namely end point, independent, least squares, terminal or theoretical slope.

4.2.5 * Hysteresis

Expressed as " _____ % VR."

Alternately, 4.2.4 and 4.2.5 may be combined as 4.2.6.

4.2.6 Combined *Hysteresis and *Linearity

Expressed as "combined hysteresis and linearity within ± ___% VR."

Note: The linearity modifier shall be stated.

4.2.7 *Friction Error

Expressed as "within ___% VR."

4.2.8 *Repeatability

Expressed as "within ___% VR over a period of __ hours."

Alternately, 4.2.2, 4.2.3, 4.2.4, 4.2.5, 4.2.7, and 4.2.8 may be combined as 4.2.9.

4.2.9 *Static Error Band

Expressed as "± ___% VR as referred to _____(curve)_____ ."

Note: The curve shall be stated as:

1) "End Point Line"—a straight line between end points.

2) "Best Straight Line"—a line midway between the two parallel lines closest together and enclosing all output versus measurand values.

3) "Least Square Line"—a straight line for which the sum of the squares of the residuals is minimized.

4) "Terminal Line"—a straight line between 0 and 100% of both measurand and output.

5) "Theoretical Slope"—a straight line between the theoretical end points.

6) Other curves shall be defined if specified, e.g., mean-output curve.

4.2.10 *Friction-Free Error Band

Expressed as "± ___% VR as, referred to _____(curve)_____ ."

Note: The reference curve shall be specified (See 4.2.9).

4.2.11 *Resolution (See also 4.3)

Expressed as "*average resolution* within ___% VR and *maximum resolution* within ___% VR."

4.2.12 Reference-Pressure Error

Expressed as "change in end points within ± ___% VR for a reference pressure change of ___ Pa (psi) over a reference pressure range of __ to ___Pa (psi)." Alternately expressed as "operation at reference pressures from ___ Pa (psi) to ___ Pa (psi) shall not cause output readings which exceed the specified error band."

4.2.13 *Frequency Response

Expressed as "within ± ___% from zero to ___hertz."

Note: Frequency response should be referred to response at a frequency within the specified frequency range, preferably zero, and to a specific static pressure. Mounting conditions and measured fluid should be specified, as should length and inside diameter of attached tubing.

Alternately, 4.2.13 may be replaced by 4.2.14, 4.2.15, and 4.2.16.

4.2.14 *Resonant Frequency

Expressed in "hertz" or "kilohertz."

Note: If a number of acoustic or mechanical, resonant frequencies exist, the lowest shall be listed and so identified.

4.2.15 *Damping Ratio

Expressed as " ___% of critical damping." Or, as " (ratio) of critical damping."

Note: For other than a single-degree-of-freedom system, the ringing period shall be stated.

4.2.16 *Ringing Period*

Expressed as " ___ milliseconds."

For transducers with relatively high damping and little overshoot, either 4.2.13 or 4.2.14, and 4.2.16 may be replaced by 4.2.17 and 4.2.18.

4.2.17 *Time Constant

Expressed as " ___ milliseconds (microseconds) for step change in measurand."

4.2.18 *Overshoot

Expressed as "maximum of ___% VR, settling within ___ cycles to ___% full range, at a frequency of ___ hertz."

4.2.19 *Proof Pressure

Expressed as " ___ Pa (psi) for ___ minutes" (will not cause changes in end points or in transducer performance characteristics beyond specified tolerances), "with output reading taken within ___ minutes after pressure removal."

4.2.20 *Burst Pressure Rating

Expressed as " ___ Pa (psi) applied ___ times for a period of ___ minutes each."

Note: Applicability to sensing element or case, or both, shall be stated.

4.2.21 *Operating Temperature Range*

Expressed as "from ___ °C to ___ °C."

4.2.22 *Temperature Error

Expressed as "within ___ % VR per ___ °C." Or "within ___% VR over the *operating temperature range.*"

Alternately 4.2.22 may be specified by 4.2.23.

4.2.23 *Temperature Error Band

Expressed as "within ± ___% VR from the reference curve established for the Static Error Band and over the *operating temperature range.*"

4.2.24 *Acceleration Error

Expressed as "within ___% VR per g_n along ___ axis at steady acceleration levels to ___g_n."

Note: The error shall be listed either for each of the three axis or for the axis with the largest error, i.e., most sensitive axis.

Alternately 4.2.24 may be replaced by 4.2.25.

4.2.25 *Acceleration Error Band

Expressed as "within ± ___% VR from the reference curve established for the Static Error Band for steady accelerations up to ___g_n along ___ axis."

Note: The error band shall be listed either for each of the three axes or for the axis with the largest error, i.e., most sensitive axis.

4.2.26 *Vibration Error

Expressed as "within ± ___% VR along ___axis over the specified vibration program." Signal "dropout" or discontinuities shall be noted.

Note: The error shall be listed either for each of the three axis or for the axis with the largest error, i.e., most sensitive axis; and the program shall be detailed, possibly by a graph.

Alternately 4.2.26 may be replaced by 4.2.27.

4.2.27 *Vibration Error Band

Expressed as "within ± ___% VR along ___axis from the reference curve established for the Static Error Band over the specified vibration program."

Note: The error band shall be listed either for each of the three axes or for the axis with the largest error, i.e., most sensitive axis; and the program shall be detailed, possibly by a graph.

4.2.28 *Mounting Error

Expressed as "within ± ____ % VR" or "within the Static Error Band."

4.2.29 *Loading Error

Expressed as "within ± ____ % VR" or "within the Static Error Band."

4.2.30 *Cycling Life

Expressed as " ____ cycles at one fourth the designated maximum operating frequency of the transducer."

4.2.31 Other Environmental Conditions

Other pertinent environmental conditions which shall not change transducer performance beyond specified limits should be listed, examples are:

*Ambient Pressure
Shock-Triaxial
High Level Acoustic Excitation
Humidity
Salt Atmosphere
Nuclear Radiation
Magnetic Fields
Sand and Dust
Total Immersion (and in what medium)
Solar (or other) Heat Radiation
Temperature Shock

4.2.32 *Storage Life

Expressed as " ____ months (years) without changing performance characteristics beyond their specified tolerances."

Note: Environmental storage conditions shall be described in detail. Where performance characteristics require additional tolerances over the storage life they shall be specified.

4.3 Additional Terminology

Average Resolution

The reciprocal of the total number of output steps over the unit range multiplied by 100 and expressed in % VR.

Damping Fluid

A fluid used to damp the single-degree-of-freedom spring/mass system, usually surrounding the reference side (transduction element side) of the sensing element.

Isolating Element

A movable membrane, usually of metal, that physically separates the measured fluid from the sensing element. Usually this membrance is considerably more flexible than the sensing element and is coupled to the sensing element using a *transfer fluid*. Its purpose is to provide material compatibility with the measured fluid while maintaining the performance integrity of the sensing element.

Maximum Excitation

The maximum allowable voltage (current) applied to the potentiometric element at Room Conditions while maintaining all other performance characteristics within their limits. (Note: The excitation value is particularly associated with temperature.)

Mounting Effects

The effects (errors) introduced into transducer performance during installation caused by fastening of the unit or its mounting hardware or by irregularities of the surface on (or to) which the transducer is mounted.

Operating Temperature Range

The range in extremes of ambient temperature within which the transducer must perform to the requirements of the "Temperature Error" or "Temperature Error Band." (See paragraphs 4.2.22 and 4.2.23, respectively.)

Potentiometric Element

The resistive part of the transduction element upon which the wiper (movable contact) slides and across which excitation is applied. It may be constructed of a continuous resistance or of small diameter wire wound on a form (mandrel).

Pressure Connection (Pressure Port)

The opening and surrounding surface of a transducer used for measured fluid access to transducer sensing element (or isolating element). This can be a standard industrial or military fitting configuration, a tube hose fitting or a hole (orifice) in a base plate. For differential pressure transducers there are two pressure connections: the measurand port and the reference.

Reference Pressure Error

The maximum change in output at specified measurand values due to a specified change in the Reference Pressure applied at both ports simultaneously.

Ringing Period

The period of time during which the amplitude of measurand step-function-excited oscillations exceed 10% of the step amplitude.

Sealed Reference Differential Pressure Transducer

A transducer which measures the pressures difference between an unknown pressure and the pressure of a gas in an integral sealed reference chamber.

Tap

A connection to a potentiometric element along its length, frequently at the element's center for use in providing bidirectional output.

Transfer Fluid

A degassed liquid used between an isolating element and a sensing element to provide hydraulic coupling of the pressure between both elements.

Worst Resolutions

The magnitude of the largest of all output steps over the unit range expresed as a percentage of VR.

4.4 Tabulated Characteristics versus Test Requirements

This table is intended for use as quick reference for design and performance characteristics and tests of their proper verification as contained in this Standard.

Characteristic	Paragraph	Basic	Supple-mental	Verified During Individual Acceptance Test	Qualifi-cation Test
Type of Pressure Sensed	4.1.1.1	x		No Test	
Measured Fluids	4.1.1.2	x			Special Test
Configuration, Dimensions, Mounting Pressure Connection	4.1.1.3 through 4.1.1.5	x		5.2.1	
Mounting Effects	4.1.1.6	x			6.6
Weight	4.1.1.7	x			6.3
Case Sealing	4.l.1.8	x			5.2.1
Identification	4.1.1.9	x		5.2.1	
Case Material	4.1.2.1		x		5.2.1
Sensing Element	4.1.2.2.		x		5.2.1
Damping Fluid	4.1.2.3		x		5.2.1
Number of Potentiometric Elements or Taps	4.1.2.4		x	5.2.2 through 5.2.6	
Dead Volume	4.1.2.5		x		6.4
Volume Change Due to Full Scale Pressure	4.1.2.6		x		6.5
Materials in Contact with Measured Fluid	4.1.2.7		x		Special Test
Gage Vent	4.1.2.8		x		Special Test
Maximum Temperature	4.1.2.9		x		Special Test
Excitation	4.1.3.1	x		5.2.6	
Maximum Excitation	4.1.3.2	x			Special Test
Transduction Element Resistance	4.1.3.3	x		5.2.2	
Electrical Connections	4.1.3.4	x		5.2.2	

| Characteristic | Paragraph | Basic | Supple-mental | Verified During | |
				Individual Acceptance Test	Qualification Test
Insulation Resistance	4.1.3.5	x		5.2.3	
Breakdown Voltage Rating	4.1.3.6	x		5.2.4	
Load Impedance	4.1.3.7	x		5.2.5 (partially)	6.7
Range	4.2.1	x		5.2.6	
End Points	4.2.2	x		5.2.6	
Full Scale Output	4.2.3	x		5.2.6	
Linearity	4.2.4	x		5.2.6	
Hysteresis	4.2.5	x		5.2.6	
Combined Hysteresis and Linearity	4.2.6	x		5.2.6	
Friction Error	4.2.7	x		5.2.6	
Repeatability	4.2.8	x		5.2.6	
Static Error Band	4.2.9	x		5.2.6	
Friction-Free Error Band	4.2.10	x		5.2.6	
Resolution	4.2.11	x			6.2
Reference-Pressure Error	4.2.12	x		5.2.7	
Frequency Response	4.2.13				6.8
Resonant Frequency	4.2.14				6.8
Damping Ratio	4.2.15				6.8
Ringing Period	4.2.16				6.8
Time Constant	4.2.17	x			6.8
Overshoot	4.2.18	x			6.8
Proof Pressure	4.2.19	x		5.2.8	
Burst Pressure Rating	4.2.20	x			6.14
Operating Temperature Range	4.2.21	x			6.9
Temperature Error	4.2.22	x			6.9
Temperaure Error Band	4.2.23	x			6.9
Acceleration Error	4.2.24	x			6.10
Acceleration Error Band	4.2.25	x			6.10
Vibration Error	4.2.26	x			6.11
Vibration Error Band	4.2.27	x			6.11
Mounting Error	4.2.28				6.6
Loading Error	4.2.29				6.7
Cycling Life	4.2.30	x			6.13
Other Environmental Conditions	4.2.31	x			6.12
Storage Life	4.2.32				Special Test (accelerated)

5. INDIVIDUAL ACCEPTANCE TESTS AND CALIBRATIONS

5.1 Basic Equipment Necessary to Perform Individual Acceptance Tests and Calibrations of Potentiometric Pressure Transducers.

The basic equipment for acceptance tests and calibrations consists of a source of pressure, a source of electrical excitation for the potentiometer, and a device which measures the electrical output of the transducer directly or as a ratio to excitation input (VR). The errors or uncertainties of the measuring system comprising these three components should be less than one-fifth of the permissible tolerance of the transducer performance characteristic under evaluations. The traceability to national standards for this measuring system shall be well known.

5.1.1 Pressure Source

A pressure medium similar to the one which the transducer is intended to measure should be used for testing. The accuracy of the pressure source should be at least five times greater than the permissible tolerance of the transducer performance characteristic under evaluation. The range of the pressure source and monitoring equipment should be selected to provide the necessary pressure, and accuracy, respectively, to 125% of the full scale of the transducer.

The pressure source may be either continuously variable over the range of the instrument, or may give discrete steps as long as the steps can be programmed in such a manner that the transition from one pressure to the next during calibration is accomplished without eliminating an existing hysteresis (or friction) error in the transducer by overshoot or fluctuation.

Note: By "similar" is meant a fluid with similar properties, bearing in mind safety and availability, i.e., H_2, N_2, O_2, silicone oils, and the like.

EXAMPLES OF PRESSURE SOURCES/MONITORING EQUIPMENT MERCURY MANOMETER
(Pressure Indicating Device)

Typical Ranges

100 KPa (about 30 in. Hg) . . .
            Accuracy ± 0.02% Full Scale
200 KPa (about 60 in. Hg) . . .
            Accuracy ± 0.02% Full Scale
340 KPa (about 100 in Hg) . . .
            Accuracy ± 0.01% Full Scale

AIR PISTON (Pressure Source)

Typical Ranges

About 2 to 10 KPa (0.3 to 1.5 psi) . . .
            Accuracy ± 0.15% of Reading
About 10 to 350 KPa (1.5 to 50 psi) . . .
            Accuracy ± 0.015% of Reading
About 100 KPa to 1 MPa (15 to 150 psi) . . .
            Accuracy ± 0.025% of Reading
About 100 KPa to 3.5 MPa (1 to 500 psi) . . .
            Accuracy ± 0.025% of Reading

PRECISION DIAL GAGE (Pressure Indicating Device)

Typical Ranges

0 to 30 KPa (about 0 to 120 Hz0) . . .
            Accuracy ± 0.1% Full Scale
0 to 100 KPa (about 0 to 30 in. Hg)
            Accuracy ± 0.1% Full Scale
0 to 100 KPa (about 0 to 100 psi) . . .
            Accuracy ± 0.1% Full Scale
0 to 700 MPa (about 0 to 10 000 psi) . . .
            Accuracy ± 0.1% Full Scale

Note: Presssure indicating devices generally require a supply of dry gas, e.g., dehumified air, or nitrogen, or helium, required for reasons of safety.

OIL PISTON GAGE (Pressure Source)

Typical Ranges

About 40 KPa to 30 MPa (6 to 4000 psi) . . .
            Accuracy ± 0.01% of Reading
About 400 KPa to 300 MPa (60 to 40 000 psi) . . .
            Accuracy ± 0.01% of Reading
About 14 MPa to 700 MPa . .
            Error in Piston Area Less Than ± 0.009%
About 30 MPa to 1400 MPa . . (400 to 200 000 psi)
            Error in Piston Area Less Than 0.012%

Note: The accuracies cited may be greater than needed for the calibration of many potentiometric pressure transducers, but may be required for the calibration of other types of pressure sensing instruments. Economic considerations suggest acquisition of minimum number of pressure sources/monitors to meet calibration needs of majority of transducers in a given installation.

5.1.2 Stable Excitation Source of accurately known amplitude (unless VR is being measured).

Commonly used sources are chemical batteries such as dry cells and storage batteries or line-powered, electronically regulated, power supplies. (A current limiting device shall be inserted in series with the transducer to preclude accidental damage of the potentiometric element.)

5.1.3 Electronic Indicating Instrument

Examples of suitable devices are:

Manually Balanced Ratiometer
Achievable Accuracy
1 part in 10 000

Self Balancing Ratiometer
Achievable Accuracy
1 part in 10 000

Digital Electronic Voltmeter/Ratiometer
Achievable Accuracy

$\pm$ 0.01% of Reading + 1 digit (4 digits display)
$\pm$ 0.005% of Reading + 1 digit (5 digits display)

Note: The input impedance of the readout instrument shall be sufficiently high to produce negligible loading error. Suggested value is 100 times the resistance of the transduction element.

5.2 Calibration and Test Procedures

Results obtained during the calibration and testing shall be recorded on a data sheet similar to the sample date sheet in Section 7, Figure 7.1 (7.7 for static error and calibration) of this standard. Calibration and testing shall be performed under Room Conditions as defined in ISA-S37.1 unless otherwise specified.

Notes:

1. The definitive paragraph under Performance Characteristics (Section 4) of this document is listed beside each of the parameters for which the test results are to be compared.

2. If more than one potentiometric element is used in the transducer, the performance of every element shall be recorded on its own form.

5.2.1 The transducer shall be inspected visually for mechanical defects, poor finish, and other applicable mechanical characteristics of 4.1.1.

Configuration and Dimensions	4.1.1.3
Mounting and Mounting Dimensions	4.1.1.4
Pressure Connection	4.1.1.5
Identification	4.1.1.9

By use of special equipment, or by formal verification of production methods and materials used, the following can be additionally determined:

Case Sealing	4.1.1.8
Case Material	4.1.2.1
Sensing Element	4.1.2.2
Damping Fluid	4.1.2.3

5.2.2 A precision resistance measuring device shall be used to measure:

Transduction Element Resistance and can be used to verify:	4.1.3.3
Number of Potentiometric Elements or Taps	4.1.2.4
Electrical Connections	4.1.3.4

5.2.3 Measure the insulation resistance between all transduction element terminals (or leads) connected in parallel and the case (and ground pin) of the transducer with a megohmmeter device, using a potential of 50 volts unless otherwise specified.

Insulation Resistance	4.1.3.5

5.2.4 Verify the Breakdown Voltage Rating, using sinusoidal ac voltage test with all transduction element terminals (or leads) paralleled, and tested to case and ground pin.

Breakdown Voltage Rating (ac-rms)	4.1.3.5

5.2.5 The transducer shall be connected to the pressure source and secured as recommended for its use. The appropriate excitation source and indicating instruments shall be properly connected to the transducer and turned on. Adequate warm-up time for indicating instruments shall be allowed before tests are conducted. The pressure source, connecting tubing, and transducer system shall pass a leak test to assure absence of calibration errors. Electrical connections shall be checked for correctness of hook-up including the appropriate load impedance (See 4.1.3.7).

5.2.6 Two or more complete calibration cycles shall be run consecutively. A minimum of eleven data points shall be obtained including both ascending and descending directions. Excitation amplitude shall be monitored as required unless VR is measured.

In order to verify performance between the discrete levels and to assure absence of noise, a full-scale X-Y plot shall be obtained, preferably inscribed diagonally across the test record form, by applying increasing then decreasing pressure to the transducer, and simultaneously to a reference transducer having continuous resolution and suitable linearity, each connected to one axis input of the plotter.

From the data obtained during these tests, the following characteristics should be determined:

End Points	4.2.2
Full Scale Output	4.2.3
Linearity	4.2.4
Hysteresis	4.2.5
(or Hysteresis and Linearity)	4.2.6
Friction Error	4.2.7
Repeatability	4.2.8

Note: To determine Friction Error or Friction-Free Error Band, at least one calibration cycle shall be run with the transducer dithered (light but sufficient vibration or shock).

5.2.7 For Differential Pressure Transducers, the performance of a three-point (e.g., 10, 50, and 90%) calibration cycle at both the minimum and maximum specified reference pressures shall establish:

| Reference Pressure Error | 4.1.12 |

5.2.8 After application of the proof pressure, at least a three-point calibration shall be performed to establish that the performance characteristics of the transducer are still within specifications. The first output reading shall be recorded within the period of time specified for this.

| Proof Pressure | 4.2.19 |

Note: For bidirectional differential transducer, proof pressure shall be applied to both ports individually. For reporting purposes these are identified as "positive" and "negative" proof pressures.

6. QUALIFICATION TEST PROCEDURES

Qualification Tests shall be performed as applicable using the test forms of Section 7 as required. Upon completion of all testing the form of Figure 7.6 shall be used to summarize all testing.

6.1 Initial Performance Tests (Figure 7.2)

The tests and procedures of Section 5, Individual Acceptance Tests and Calibrations, shall be run to establish reference performance during increasing (and decreasing) steps of 0, 20, 40, 60, 80, and 100% of range as a minimum (% of span for bidirectional transducers).

6.2 Resolution Test (Figure 7.6)

An X-Y plotter shall be connected so that the transducer output is connected to the X-Axis and a continuous-resolution reference transducer to the Y-Axis input of the plotter. As the pressure to both transducers is slowly increased (simultaneously on both transducers), the number of steps shall be recorded from 0 to 100% of the test transducer's range. The following shall be determined:

| Resolution (Average and Worst) | 4.2.11 |

6.3 Weight Test (Figure 7.6)

The transducer shall be weighed on an appropriate balance or scale. The following shall be established:

| Weight | 4.1.1.7 |

6.4 Dead Volume Test (Figure 7.6)

The pressure cavity shall be filled (both cavities for a differential transducer) with a measurable, non-corrosive liquid (under a vacuum if necessary), and the contents poured into a graduate. The following shall be established:

| Dead Volume | 4.1.2.5 |

6.5 Volume Change Test (Figure 7.6)

A liquid pressure system shall be connected to a transducer, a parallel pressure gage, and a graduated reservoir. (Provisions for isolating the transducer when filled shall be made.) The pressure system shall be evacuated and filled with liquid, the valve to the transducer closed, the pressure increased to 100% of range, the valve opened, and the following shall be determined:

| Volume Change due to Full Scale Pressure | 4.1.2.6 |

6.6 Mounting Test (Figure 7.6)

The mounting of the actual installation shall be duplicated as closely as possible following specific instructions and one calibration run performed. The following shall be established:

| Mounting Error | 4.2.28 |
| Mounting Effect | 4.1.1.6 |

6.7 Loading Test (Figure 7.6)

Approximately 66% of full range pressure shall be applied to the transducer and the total load resistance varied from the highest to the lowest ohmic value allowed. (Note: Take into account the resistance of the ratiometer or other indicating instrument.) The following shall be verified:

| Loading Error | 4.2.29 |

The loading error of a potentiometric pressure transducer is variable with wiper position ranging from zero at both extremes to a maximum value at approximately 66% of VR. As a first approximation the percentage error is equal to fifteen times the ratio of the potentiometric element resistance to the loading resistance. Unless otherwise stated, assembly adjustments of the transducer apply to the open circuit conditions at the output terminals.

6.8 Dynamic Response Test (Figure 7.4a or 7.4b as applicable)

The dynamic response characteristics of pressure transducers may be established either with transient-excitaton devices, or with sinusoidal pressure generators.

6.8.1 Transient Excitation Method

A positive step-function of pressure may be generated in gases with a shock-tube or a quick-opening valve. A hydraulic quick-opening valve is used to generate a positive pressure step function in a liquid medium. A burst diaphragm generator produces a negative pressure step in a gas medium. In all cases, the rise time of the generated step function shall be sufficiently short to shock-excite all resonances in the transducer under test. It shall also be one third or less of the anticipated rise time of the transducer under test.

Since the tubing used to mechanically connect the transducer to the test set-up may affect the dynamic characteristics, it is recommended that the shortest possible tubing be installed, or that the tubing used shall duplicate as closely as possible the actual installation, if this condition was specified instead of the characteristics of the transducer alone. Any tubing used shall be described by length, internal diameter, and curvature.

By applying step functions of pressure at Room Conditions within the full scale range of the transducer, and analyzing the electronic or electro-optical recording of the transducer output, the following can be determined.

Frequency Response (amplitude and phase)	4.2.13
Resonant Frequency	4.2.14
Damping Ratio	4.2.15
Ringing Period	4.1.16

Alternately for transducers with relatively high damping and little overshoot, the following can be determined:

Time Constant	4.2.17
Overshoot	4.2.18

6.8.2 Sinusoidal Excitation Method

For frequencies below a few kilohertz, static pressures below 100 MPa (roughly 15 000 psi) and peak dynamic pressures below 10 MPa (roughly 1500 psi), generators that produce sinusoidal pressure wave-forms in either liquids or gases are available. These sinusoidal pressure generators operate either on the pistonphone principle (which is in common use for the absolute calibration of microphones) or by modulating a fluid flow through an orifice.

A sinusoidal pressure waveform of constant amplitude and varying frequency, over a specified frequency range, shall be applied at a specified static pressure. The following shall be determined:

Frequency Response (amplitude and phase)	4.1.13

If within the frequency range covered, the following can be established from the frequency response recording by suitable calculations:

Resonant Frequency	4.2.14
Damping Ratio	4.2.15
Ringing Period	4.2.16
Time Constant	4.2.17
Overshoot	4.2.18

6.9 Temperature Tests

6.9.1 Low Temperature Test (Figure 7.3)

The transducer shall be placed in a suitable temperature chamber. The temperature of the transducer shall be stabilized for one hour at the lower specified operating temperature and two calibration cycles performed, followed by a positive-proof pressure test and a third calibration cycle. The insulating resistance shall be measured and recorded as in 5.2.3. (For differential pressure transducers only, the second calibration cycle shall be performed with maximum specified reference pressure applied, followed immediately by a negative-proof pressure test.)

These tests shall establish the following:

Temperature Error (at low temperature)	4.2.22
or	
Temperature Error Band (at low temperature)	4.2.23

6.9.2 Post Low Temperature Test (at Room Conditions) (Figure 7.3)

The transducer shall be removed from the temperature chamber and permitted to stablize for one hour at room conditions. The tests of 6.9.1 shall be repeated except that the operating temperature shall be room temperature.

6.9.3 High Temperature Test (Figure 7.3)

The tests of 6.9.1 shall be repeated except that the transducer temperature shall be stabilized for one hour at the highest specified operating temperature.

These tests shall establish the following:

Temperature Error (at high temperature)	4.2.22

or

Temperature Error Band (at
   high temperature)   4.2.23

6.9.4 Post High Temperature Test (at Room Conditions) (Figure 7.3)

The tests of 6.9.2 shall be repeated after stabilization of the transducer at room temperature for one hour.

Note: If required, thermal and post-thermal zero shift and sensitivity shift may also be calculated from the results of these tests.

6.10 Acceleration Test (Figure 7.5)

Acceleration shall be imposed on the transducer in three orthogonal directions by tilting it in the earth's gravitational field or by placing it on a centrifuge. A specific acceleration level shall be applied on specified axes, and the output measured. The following shall be established:

Acceleration Error   4.2.24
          or
Acceleration Error Band   4.2.25

6.11 Vibration Test (Figure 7.5)

With specified measurand levels applied, the transducer shall be vibrated along specified axes at specified acceleration amplitudes over the specified frequency range with an electro-magnetic or hydraulic shaker. The transducer output shall be recorded with a high-speed recorder. The following shall be established:

Vibration Error   4.2.26
         or
Vibration Error Band   4.2.27

Note: If so specified, the vibration error band can be established as the algebraic sum of maximum vibration errors and the last previously obtained static error band.

6.12 Tests For Other Environmental Conditions (Figure 7.3)

The transducer shall be exposed to other specified environmental conditions. As specified for each condition, one complete calibration cycle shall be performed during or after the test to establish the ability of the transducer to perform satisfactorily.

See Section 4.2.31

6.13 Life Test (Figure 7.3)

After applying the specified number of full range excursions of measurand, or after completion of each of several

specified portions of the total number of cycles, at least one complete calibration cycle shall be performed to establish minimum value of:

Cycling Life   4.2.30

6.14 Burst Pressure Test (Figure 7.6)

The transducer shall be connected to a suitable test setup with adequate protection for equipment and personnel. The pressure shall be increased to the specified limit and applied for the specified number of times and durations. The following shall be established :

Burst Pressure Rating   4.2.20

NOTE: If specified, burst pressure is applied to the inside of the case by first puncturing the sensing element.

7. TEST REPORT FORMS

7.1 The test report forms listed below are recommended for use during the testing of Potentiometric Pressure Transducers.

7.2 When using the forms all pertinent information shall be inserted in its proper place. On some forms, blank space has been provided for additional tests. Where the test is prolonged, e.g., Cycling Life, more than one form may be required.

7.3 Individual Acceptance Tests and Calibrations (Figure 7.1). Used during acceptance testing of Section 5.

Initial Performance Tests and Calibrations (Figure 7.2). Used for establishing the reference performance for comparison to other test results.

Environmental Test Record (Figure 7.3). Used for Temperature, Maximum Temperature, Life, and other environmental tests.

Dynamic Response Tests (Figure 7.4),. Used for recording test results of Frequency Response, Resonant Frequency, Damping Ratio, Ringing Period, Time Constant, and Overshoot. (Note: Use 7.4a or 7.4b as applicable.)

Environmental Test Record (Figure 7.5). Used to record Acceleration and Vibration Test results.

Test Summary (Figure 7.6). Used to compile the results of all testing.

Individual Acceptance Test Record (Static Error Band) (Figure 7.7). Used as an alternate for Figure 7.1 when Static Error Band Calibration is specified.

8. BIBLIOGRAPHY

1. Beckwith, T. G., and Buck, N. L.; "Mechanical Measurements"; Addison-Wesley, 1961.

2. Norton, H. N.; "Transducers for Electronic Measuring Systems"; Prentice-Hall, 1969.

3. *ISA Transducer Compendium,* 2nd Edition Part I "Pressure, Flow, and Level"; Plenum Press, New York, 1969.

4. Neubert, H.K.P.; *Instrument Transducers;* Oxford at the Clarendon Press, 1963.

5. Lederer, P.S.; "Methods for Performance Testing of Electromechanical Pressure Transducers"; National Bureau of Standards Technical Note # 411, February, 1967.

6. Cross, J. L.; "Reduction of Data for Piston Gage Pressure Measurements"; National Bureau of Standards Monogram # 65, June, 1963.

7. ISA-S37.1; "Electrical Transducer Nomenclature and Terminology"; 1969.

8. ISA-S37.3; "Specifications and Tests for Strain Gage Pressure Transducers"; 1970.

9. Schweppe et al.; "Methods for the Dynamic Calibration of Pressure Transducers"; National Bureau of Standards Monograph # 67, December, 1963.

10. MIL-E-5272C (ASG); "Environmental Testing, Aircraft Electronic Equipment, General Specification For."

11. MIL-E-5400C (ASG); "Environmental Testing, Aeronautical and Associated Equipment, General Specification For."

12. MIL-STD-810; "Environmental Test Methods for Aerospace and Ground Equipment."

13. PMC 20.1-1973; "Process Measurement and Control Terminology"; August, 1973.

14. ANSI B88.1-1971; "A Guide for the Dynamic Calibration of Pressure Transducers"; ASME, August, 1972.

15. ANSI Z210.1-1973, "Standard Metric Practice Guide"; ASTM (E 380-72E) March, 1973.

VENDOR'S PART NO.	TEST FACILITY	CUSTOMER'S PART NO.
VENDOR	**POTENTIOMETRIC PRESSURE TRANSDUCER**	SERIAL NO.
CUSTOMER	**INDIVIDUAL ACCEPTANCE TEST AND CALIBRATION RECORD**	RANGE _____ TO _____ Pa _____

Visual Inspection:

Dimensions ☐   Workmanship ☐

Finish ☐   Nameplate ☐ El. Conn. ☐

Electrical Inspection:

Element Resistance _____ ohms   Insulation Resistance _____ Megohms at _____ Vdc

Breakdown Voltage Rating @ _____ Vac, _____ Hz ☐ Z_L used _____ MΩ

Calibration @ _____ V _____ Excitation   Ambient Temperature _____ °C

Pressure (Pa)	Theor. Output (%VR)	(Undithered) First Calib. Cycle Output (%VR)		(Dithered) First Calib. Cycle Output (%VR)		(Undithered) Second Calib. Cycle Output (%VR)		(Dithered) Second Calib. Cycle Output (%VR)			(Undithered) Max. Error (%VR)		(Dithered) Max. Error (%VR)	
		Increase	Decrease	Increase	Decrease	Increase	Decrease	Increase	Decrease		+	−	+	−

* _____ Linearity : + _____ , − _____ %VR (Allowed: ± _____ %VR); * Hysteresis: _____ %VR (All'd _____ %VR))

* Hysteresis and _____ Linearity (Combined): + _____ , − _____ %VR   (Allowed: _____ ± _____ %VR)

Friction-free Error Band: + _____ , − _____ %VR (Allowed: ± _____ %VR);   Friction Error: _____ %VR (All'd _____ %VR)

Repeatability: _____ %VR (All'd: _____ %VR); * End Points: _____ and _____ %VR (All'd _____ and _____ %VR)

* Full-Scale Output: _____ %VR   Allowed _____ ± _____ %VR NOTE: * Values Determined From _____ Calib. Cycle

Pressure (Pa)	Theor. Output	P Only: Performance Test @ _____ Pa _____ Ref. Press. Output (_____)	Error (_____)		Δ P Only: Perf. Test ____ Minutes After ____ Pa ____ Neg. Proof Press. (Overload Output: _____ _____) Output (_____)	Error (_____)		Perf Test _____ Minutes After ____ Pa _____ Pos. Proof Press. (Overload Output: _____ _____) Output (_____)	Error (_____)	
			+	−		+	−		+	−
			+	−		+	−		+	−
			+	-		+	−		+	−

Full-Scale X-Y Plot: _____

Static Error Band: + _____ , − _____ %VR   Allowed: ± _____ %VR)

Error Bands Ref. To _____   Ref. Press. Error: _____ %VR

Equipment Used: _____

Tested By: _____   Date: _____

Approved By: _____

FIGURE 7.1

VENDOR'S PART NO.	TEST FACILITY		CUSTOMER'S PART NO.
VENDOR	**POTENTIOMETRIC PRESSURE TRANSDUCER**		SERIAL NO.
REPORT NO.	☐ **INITIAL PERFORMANCE TEST**		CUSTOMER
TYPE OF TEST	☐ _____ **TEST**		RANGE _____ TO _____ Pa _____

Visual Inspection:.

Dimensions ☐ Workmanship ☐

Finish ☐ Nameplate ☐ El. Conn. ☐

Electrical Inspection:

Element Resistance _____ ohms ; Insulation Resistance _____ Megohms at _____ Vdc

Breakdown Voltage Rating ℗ _____ Vac, _____ H$_z$ ☐   Z$_L$ used _____ M Ω

Calibration ℗ _____ V _____ Excitation   Ambient Temperature _____ °C

Pressure (Pa)	Theor. Output (%VR)	(Undithered) First Calib. Cycle Output (%VR)		(Dithered) First Calib. Cycle Output (%VR)		(Undithered) Second Calib. Cycle Output (%VR)		(Dithered) Second Calib. Cycle Output (%VR)			(Undithered) Max. Error (%VR)		(Dithered) Max. Error (%VR)	
		Increase	Decrease	Increase	Decrease	Increase	Decrease	Increase	Decrease		+	−	+	−

* _____ Linearity : + _____ , − _____ %VR (Allowed: ± _____ %VR); * Hysteresis: _____ %VR (All'd : _____ %VR)

* Hysteresis and _____ Linearity (Combined): + _____ , − _____ %VR   (Allowed: ± _____ %VR)

Friction-free Error Band: + _____ , − ____ %VR (Allowed): ± _____ %VR); Friction Error: _____ %VR (All'd: _____ %VR)

Repeatability: _____ %VR (All'd : _____ %VR); * End Points: _____ and _____ %VR (All'd : _____ and _____ %VR)

* Full-Scale Output: _____ %VR Allowed: _____ ± _____ %VR NOTE: * Values Determined From _____ Calib. Cycle

	△ P Only: Performance Test ℗ _____ Pa ____ Ref. Press.		△ P Only: Perf. Test __Minutes After ____ Pa __ Neg. Proof Press. (Overload Output: _____ _____)		Perf Test _____ Minutes After __ Pa __ Pos. Proof Press. (Overload Output: _____ _____)	
Pressure (Pa)	Theor. Output	Output (_____)	Error (_____)	Output (_____)	Error (_____)	Output (_____)	Error (_____)
			+ −		+ −		+ −
			+ −		+ −		+ −
			+ −		+ −		+ −

Full-Scale X-Y Plot: _____

Static Error Band: + _____ , − _____ %VR (Allowed: ± _____ %VR)

Error Bands Ref. To _____   Ref. Press. Error: _____ %VR

Equipment Used:	Defects Noted Or Comments:

Tested By: _____ Date: _____ Approved By: _____

FIGURE 7.2

VENDOR'S PART NO.	TEST FACILITY	CUSTOMER'S PART NO.
VENDOR	**POTENTIOMETRIC PRESSURE TRANSDUCERS**	SERIAL NO.
REPORT NO.		CUSTOMER
TYPE OF TEST	**ENVIRONMENTAL TEST RECORD**	RANGE _____ TO _____ Pa ____

Tested While: ☐ Undithered ☐ Dithered

☐ Before ☐ During ☐ After

Type of Environment _____

Level _____

Pressure (Pa)	Theoretical %VR	%VR (Run 1) Increase	%VR (Run 1) Decrease	%VR (Run 2) Increase	%VR (Run 2) Decrease	Overload %VR	%VR (Run 3) Increase	%VR (Run 3) Decrease	Maximum Error %VR +	Maximum Error %VR −
	(Pos. Proof)									
	(Neg. Proof)									

Error Band: + _____ % − _____ %VR (Referred To _____)   Allowed: ± _____ %VR

Proof Pressure _____ % Rated Range for _____ Minutes (POS)   Ins. Resistance: _____ Megohms at ___ Vdc

For Diff. Press. Transducers Only: Zero Shift: _____ %VR

Neg. Proof Pressure _____ %Rated Range for _____ Minutes   Sensitivity Shift: _____ %VR

Ref. Press., Run 2 : _____ Pa

Comments: _____
_____
_____
_____
_____
_____
_____
_____

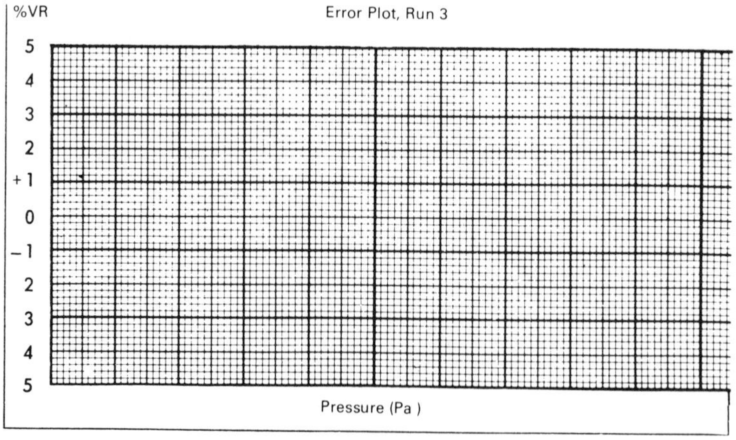

%VR   Error Plot, Run 3

5 4 3 2 +1 0 −1 2 3 4 5

Pressure (Pa)

Tested By: _____   Date Test Started: _____   Date Test Finished: _____

Approved By: _____   Approved By: _____

Title: _____   Title: _____

FIGURE 7.3

VENDOR'S PART NO.	TEST FACILITY	CUSTOMER'S PART NO.
VENDOR	**POTENTIOMETRIC PRESSURE TRANSDUCER**	SERIAL NO.
REPORT NO.		CUSTOMER
TYPE OF TEST	**DYNAMIC RESPONSE TESTS**	RANGE ___ TO ___ Pa ___

1. Ambient Conditions: Temperature: _____ °C, Pressure: _____ Pa; Humidity _____ %

2. Dynamic Response _____

    Excitation Voltage: _____

    Step Function Generator _____ Shock Tube, Dry Air: _____

    Mounting Location: End: _____ Side: _____

Shock No.	Initial Pressures		Shock Velocity	Step Pressure	Pronounced Resonances, H$_z$			Ringing Period	Time Constant
	Hi	Low							

Frequency Response _____ Hz (Allowed _____ Hz), Damping Ratio _____ (All'd _____)

Ringing Period _____ msec. (All'd _____ msec.),

Time Constant _____ msec. (All'd _____ msec.),   Overshoot _____ %VR (All'd _____ %VR)

ATTACH OSCILLOSCOPE PHOTOGRAPHS OF TRANSDUCER RESPONSES

Amplitude Scale

SHOCK 1

Time Scale

Amplitude Scale

SHOCK 3

Time Scale

Amplitude Scale

SHOCK 2

Time Scale

Amplitude Scale

SHOCK 4

Time Scale

Tested By: _____   Date Test Started: _____   Date Test Finished: _____

Approved By: _____   Approved By: _____
      Title   Title

FIGURE 7.4.a

VENDOR'S PART NO.	TEST FACILITY	CUSTOMER'S PART NO.
VENDOR	**POTENTIOMETRIC PRESSURE TRANSDUCER**	SERIAL NO.
REPORT NO.	**DYNAMIC RESPONSE TESTS**	CUSTOMER
TYPE OF TEST	**(SINUSOIDAL METHOD)**	RANGE _____ TO _____ Pa ____

Ambient Conditions: Temperature: _____ °C, Pressure: _____ Pa; _____ Humidity _____ %

Dynamic Response:

Excitation (Volts or ma) _____ ; Transducer Load _____ ohms

Sinusoidal Generator _____ ; Test Fluid _____

Mounting Configuration _____

_____

Test Temperature _____°C Quiescent Static Pressure _____ Pa _____

Sinusoidal pressure _____ Pa peak; Port excited: _____

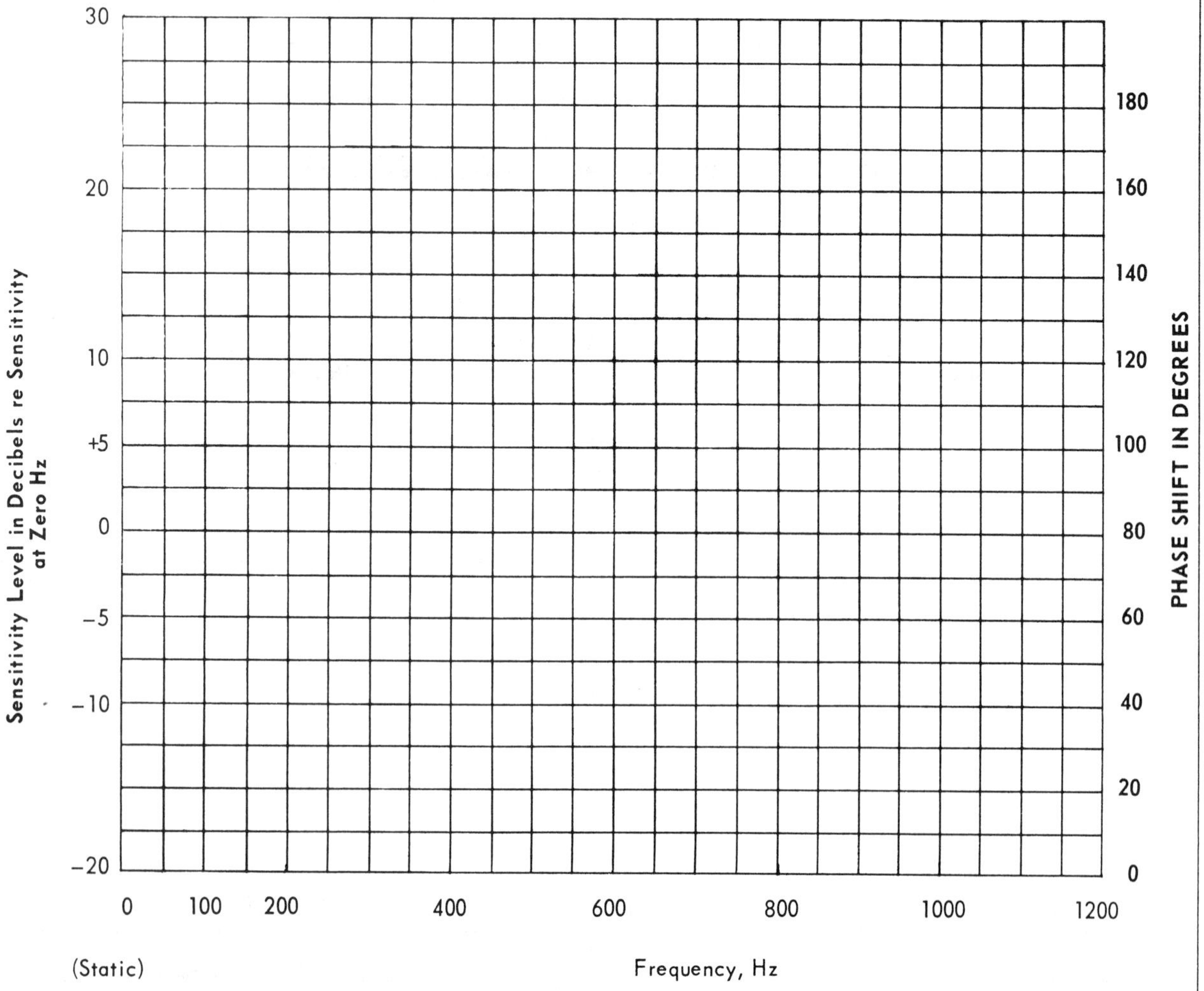

FIGURE 7.4.b

VENDOR'S PART NO.	TEST FACILITY		CUSTOMER'S PART NO.
VENDOR	**POTENTIOMETRIC PRESSURE TRANSDUCER**		SERIAL NO.
REPORT NO.			CUSTOMER
TYPE OF TEST	**ACCELERATION/VIBRATION TEST RECORD**		RANGE _____ TO _____ Pa _____

SKETCH OF TRANSDUCER SHOWING AXIS ORIENTATION:

ACCELERATION TEST

AXIS	+X	−X	+Y	−Y	+Z	−Z	
Output Before Accel (%VR)							Pressure Level Used: _____ Pa _____
Applied Accel. (G)							Max. Accel. Error: + _____ , − _____ %VR Pre-Accel. Static Error Band: + _____ , − _____ %VR
Output During Accel. (%VR)							Accel. Error Band: + _____ , − _____ %VR (Allowed Accel. Error Band ± _____ %VR)
Accel. Error (%VR)							Tested By: _____ (Technician) _____ (Test Engineer)
COMMENTS							Date: _____ Approved By: _____
							Witnessed By: _____ (_____)
							Witnessed By: _____ (_____)

VIBRATION TEST

AXIS	X			Y			Z			
Pressure Level Used	Pa			Pa			Pa			Max. Vib. Error: + _____ , − _____ %VR
Output Before Vib.	%VR			%VR			%VR			Pre-Vib. Static Error Band: + _____ , − _____ %VR
Vibration Error	Freq. (Hz)	Pol.	Error %VR	Freq. (Hz)	Pol.	Error %VR	Freq. (Hz)	Pol.	Error %VR	Vib. Error Band: + _____ , − _____ %VR (Allowed Vib. Error Band: ± _____ %VR)
										Tested By: _____ (Technician) _____ (Test Engineer)
										Date: _____ Approved By: _____
										Witnessed By: _____ (_____)
										Witnessed By: _____ (_____)
										COMMENTS

FIGURE 7.5

VENDOR'S PART NO.	TEST FACILITY	CUSTOMER'S PART NO.
VENDOR		SERIAL NO.
REPORT NO.	**TRANSDUCER TEST REPORT**	CUSTOMER
TYPE OF TEST	**POTENTIOMETRIC PRESSURE TRANSDUCER**	RANGE _____ TO _____ Pa ___

SUMMARY OF RESULTS:

☐ Error
☐ Error Band

Test	Tested Per Proced. No. or Test Waived Per	Par. No.	Pass	Fail Error	Fail Electr.	Fail Mechan.	Fail See Comments	+ %VR	− %VR
Initial P.T. (Performance Test)									
Resolution								Avg.: __ %VR Max.: __ %VR	
Weight								_____	_____
Dead Volume								_____ Cu. _____	
Vol. Change over Press. Range								_____ Cu. _____	
Mounting									
Loading Max. Z$_L$ / Min. Z$_L$									
Frequency Response								Flat (+__%): __ To __ Hz	
Response Time								_____ msec., ___ %Ovs.	
Low Temp. _____ °C									
P.T. After Low Temp.									
High Temp. + _____ °C									
Add'l. Temp. _____ °C									
P.T. After High Temp.									
_____ g$_n$ Vibration									
P.T. After _____ g$_n$ Vibr.									
Acceleration									
P.T. After Accel.									
Life									
Burst Pressure								▓▓▓▓▓▓▓	

Tested By: _____ Date Test Started: _____ Date Test Finished: _____

Approved By: _____ Approved By: _____

Title   Title

FIGURE 7.6

VENDOR'S PART NO.	TEST FACILITY		CUSTOMER'S PART NO.
VENDOR			CUSTOMER
PURCHASE ORDER NO.	**POTENTIOMETRIC PRESSURE TRANSDUCER**		SERIAL NO.
	INDIVIDUAL ACCEPTANCE TEST AND CALIBRATION RECORD (Static Error Band Calibration)		RANGE ___ TO _____ Pa ___

Visual Inspection: Dimensions ☐   Threads ☐   Finish ☐   Nameplate ☐   Receptacle (or other Electrical Conn.) ☐

Electrical Inspection: Element Resistance _____ ohms   Insulation Resistance _____ Megohms at _____ Vdc

Breakdown Voltage Rating ● _____ Vac, _____ Hz ☐   Electrical Connection ☐

Calibration   (Undithered) ● _____ V _____ Excitation Z_L used _____ Megohms

Pressure (Pa)	Theor. Output (%VR)	First Calib. Cycle (Output %VR)		Second Calib. Cycle Output (%VR)		Max. Error (%VR)		
		Increase	Decrease	Increase	Decrease	+	–	

All Error Bands Ref. To _____

	P Only: Performance Test ±● _____ Pa __ Ref. Press.		Perf. Test _____ Minutes, After___ Pa ___Neg. Proof Press. (Overload Output: _____ _____)		Perf. Test __ Minutes, After _____ Pa ____ Pos. Proof Press. (Overload Output: _____ _____)	
Pressure (Pa) / Theor. Output	Output (%VR)	Error (%VR) + / –	Output (%VR)	Error (%VR) + / –	Output (%VR)	Error (%VR) + / –
		+ / –		+ / –		+ / –
		+ / –		+ / –		+ / –
		+ / –		+ / –		+ / –
	SEB.: + _____ %, – _____ %VR		S.E.B.: + _____ %, – _____ %VR		S.E.B.: + _____ % – _____ %VR	

Full-Scale X-Y Plot: _____

Static Error Band (S.E.B.): (Calib.:)   + _____ %, – _____ %VR

S.E.B.: (All Tests) + _____ %, – _____ %VR   Allowed ± _____ %VR

Tested By: _____ Date : _____ Approved : _____

FIGURE 7.7

INSTRUMENT SOCIETY of AMERICA
Research Triangle Park, North Carolina

ANSI/ISA-S37.8-1977
(R 1982)
Approved December 14, 1982

American National Standard

Specifications and Tests for Strain Gage Force Transducers

Instrument Society of America

ISBN 0-87664-318-0

ISA-S37.8   Specifications and Tests for
Strain Gage Force Transducers

INSTRUMENT SOCIETY OF AMERICA
67 Alexander Drive
P.O. Box 12277
Research Triangle Park, North Carolina 27709

PREFACE

This Preface is included for informational purposes and is not a part of ISA-S37.8.

This Standard has been prepared as a part of the service of the Instrument Society of America toward a goal of uniformity in the field of instrumentation. To be of real value, this document should not be static, but should be subject to periodic review. Toward this end, the Society welcomes all comments and criticisms, and asks that they be addressed to the Secretary, Standards and Practices Board, Instrument Society of America, 67 Alexander Drive, P.O. Box 12277, Research Triangle Park, NC 27709, Telephone (919) 549-8411.

The ISA Standards and Practices Department is aware of the growing need for attention to the metric system of units in general, and the International System of Units (SI) in particular, in the preparation of instrumentation standards. The Department is further aware of the benefits to USA users of ISA Standards of incorporating suitable references to the SI (and the metric system) in their business and professional dealings with other countries. Towards this end, this Department will endeavor to introduce SI - acceptable metric units in all new and revised standards to the greatest extent possible. The Metric Practice Guide, which has been published by the American Society for Testing and Materials as ANSI designation Z210.1 (ASTM E380-76, IEEE Std. 268-1975), and future revisions, will be the reference guide for definitions, symbols, abbreviations,and conversion factors.

It is the policy of the Instrument Society of America to encourage and welcome the participation of all concerned individuals and interests in the development of ISA Standards. Participation in the ISA standards making process by an individual in no way constitutes endorsement by the employer of that individual of the Instrument Society of American or any of the standards which ISA develops.

This Standard is intended as a guide for technical personnel at user facilities as well as by manufacturers' technical and sales personnel whose duties specifying, calibrating, testing or showing performance characteristics of strain-gage linear accelerometers. By basing users' specifications as well as technical advertising and reference literature on this Standard, or by referencing portions thereof, as applicable, a clear understanding of the users' needs or of the transducers' performance capabilities, and of the methods used for evaluating or proving performance, will be provided. Adhering to the specification outline, terminology and procedures shown will not only result in simple, but also complete specifications; it will also reduce design time, procurement lead time, and labor, as well as material costs. Of major importance will be the reduction of qualification tests resulting from use of a commonly accepted test procedure and uniform data presentation.

The development of this Standard was initiated as the result of a survey conducted in December 1960. A total of 240 questionnaires was sent out to transducer users and manufacturers. A strong majority indicated in their replies a need for transducer standardization. As strain-gage force transducers were one of the types shown to be most in need of standardization, a Subcommittee, SP37.8, was formed. To provide a coordinated document, this committee was composed of representatives from government, user and manufacturer categories. This Standard was then processed over several mail-review and revision cycles until a consensus of reviewers was reached.

The following individuals served on the 1975 SP37.8 committee:

NAME	COMPANY
J. J. Elengo, Jr. Chairman	Revere Corporation of America
P. F. Fuselier	Lawrence Radiation Laboratory
R. E. Gorton	Pratt & Whitney Aircraft
H. E. Lockery	Consulting Engineer
H. W. Rosenburg	Naval Weapons Center
G. W. Godwin	Howe Richardson Scale Company

The following individuals served on the 1982 SP37 committee:

NAME	COMPANY
J. W. Mock, Chairman	Measurement Services
P. Bliss	Consultant
W. Brinkschulte	Genisco Technology Corporation
C. Kutelis	Revere Corporation of America
R. W. Lally	PCB Piezotronics
H. E. Lockery	Hottinger Baldwin Measurements, Inc.
E. W. Malone	Boeing Aerospace Co.
T. B. Miller	Lawrence Livermore Lab
K. A. Parlee	United Engineers & Constructors, Inc.
H. Pitt	
J. Powell	Rosemount Engineering
R. C. Strahm	EG & G Idaho
R. M. Whittier	ENDEVCO
N. Wilde	EG & G Idaho
E. Wong	Bell & Howell

This Standard was approved for publication by the ISA Standards and Practices Board in October 1982.

NAME	COMPANY
T. J. Harrison, Chairman	IBM Corporation
P. Bliss	Consultant
W. Calder	The Foxboro Company
N. Conger	Continental Oil Co.
B. Feikle	Bailey Controls Co.
R. T. Jones	Philadelphia Electric Co.
R. Keller	Boeing Company
O. P. Lovett, Jr.	Isis Corp.
E. C. Magison	Honeywell, Inc.
A. P. McCauley	Diamond Shamrock Corp.
J. W. Mock	Measurement Services
E. M. Nesvig	ERDCO Engineering Corp.
G. Platt	Bechtel Power Corp.
R. Prescott	Moore Products Company
W. C. Weidman	Gilbert Associates
K. A. Whitman	Allied Chemical Corp.
J. R. Williams	Stearns-Roger, Inc.
B. A. Christensen*	
L. N. Combs*	
R. L. Galley*	
R. G. Marvin*	
W. B. Miller*	Moore Products Company
R. L. Nickens*	

*Director Emeritus

TABLE OF CONTENTS

LIST OF FIGURES

1 PURPOSE

This standard establishes the following for strain-gage force transducers:

1.1 Uniform general specifications for design and performance characteristics.

1.2 Uniform acceptance and qualification test methods, including calibration techniques.

1.3 Uniform presentation of test data.

1.4 A drawing symbol for use in electrical schematics.

2 SCOPE

2.1 This Standard covers strain-gage force transducers, primarily those used in measurement systems.

2.2 Included among the specific versions of strain-gage force transducers, to which this Standard is applicable, are the following:

> Tension Transducers
> Compression Transducers
> Universal (Combination Compression
>   and Tension) Transducers

2.3 Terminology used in this document is defined either herein or in ISA Standard S37.1, Electrical Transducer Nomenclature and Terminology.

3 DRAWING SYMBOL

The drawing symbol for a strain-gage transducer is a square of dimensions 2x by 2x, with an added equilateral triangle, the base of which is the left side of the square. The triangle symbolizes the sensing element. The letter "F" in the triangle designates "force", and the additional sub-positioned letters denote the second modifier.

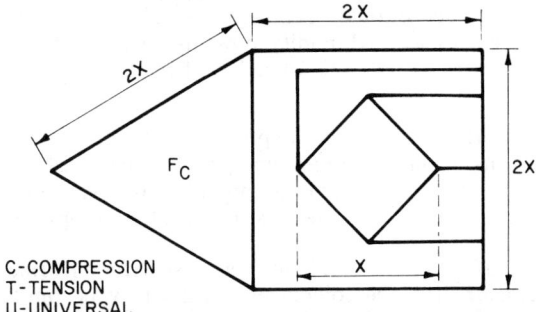

C-COMPRESSION
T-TENSION
U-UNIVERSAL

The strain-gage bridge is symbolized by a small square, with diagonals x by x, centered in the large square. The diagonals of the small square are drawn perpendicular to the sides of the large square. Lines from each apex of the small square projected to the right side of the large square represent the electrical leads.

4 CHARACTERISTICS

4.1 Design Characteristics

4.1.1 Basic Mechanical Design Characteristics

The following mechanical design characteristics shall be listed:

4.1.1.1 Type of Force Transducer — Tension, compression, or universal.

4.1.1.2 Physical Dimensions — Outline drawing to be provided with dimensions in millimeters (inches).

4.1.1.3 Force Connection — Force connections (both ends) shall be indicated on the outline drawing giving sufficient information regarding location, size, and tolerance of connection features, as well as any special considerations, to enable proper application of forces to the transducer.

4.1.1.4 Mountings and Mounting Dimensions — Outline drawing shall indicate method of mounting with dimensions in millimeters (inches).

4.1.1.5 Location of Electrical Connection — Indicate location and orientation of electrical connector or connecting wiring.

4.1.1.6 Overload Rating, Ultimate — Specify percentage of rated force that will not result in structural failure at environmental extremes.

4.1.1.7 Mounting Torque — Allowable mounting torque shall be specified if it will tend to effect transducer performance.

4.1.1.8 Weight — The weight of the transducer shall be specified in kilograms (pounds mass).

4.1.1.9 Identification — The following characteristics shall be given and preferably permanently affixed to the outside of the transducer case and indicated on the outline drawing:

Nomenclature of transducer
(per ISA S37.1, Section 3)
Manufacturer's name, part number,
and serial number
Range
Excitation
Identification of electrical connections
Bridge identification, if more than one bridge provided.

Listing of the following characteristics is optional:
Sensitivity (usually millivolts
   per volt at full scale load)
Customer's specification and/or part number
Maximum allowable force that will not
   influence prescribed performance
Maximum operating temperature range

4.1.1.10 Temperature Range, Safe — Temperature range of environment in which the transducer may be used and which will not cause permanent calibration shift or permanent change in any of its characteristics shall be listed.

4.1.2 Supplemental Mechanical Design Characteristics

Listing of the following mechanical design characteristics is optional:

Case Material
Surface Finish
Type of Strain-Gage Used — Metallic;
   bonded or unbonded, wire or foil;
   Semiconductor; bonded or unbonded
Location of Strain-Gage — Mounted
   directly on force sensing element or

mounted on auxiliary member activated by force sensing element
Number of Active Strain-Gage Bridge Arms (elements)
One, Two-arm active, Four-arm bridge
Number of Strain-Gage Bridges
Mounting Surface Requirements

4.1.3 Basic Electrical Design Characteristics

The following electrical design characteristics shall be listed. They are applicable at "ambient conditions" as specified in Section 4.2.

4.1.3.1* Excitation — Expressed as "_____volts dc" or "_____, volts rms at _____ hertz," or, expressed as "_____ milliamps dc" or "_____ milliamps rms at _____ hertz." Preferred values of voltage 5, 10, 15, 20, and 28 volts.

4.1.3.2* Maximum Excitation — Expressed as "_____ volts dc" or "_____ volts rms at _____ hertz", or, expressed as "_____ milliamps dc" or "_____ milliamps rms at _____ hertz", and defined as the maximum value of excitation voltage which will not permanently damage the transducer.

4.1.3.3* Input Impedance — Expressed as "_____ ± _____ ohms at _____ ± _____ hertz" and _____ °C (°F)." If impedance is resistive, specify "dc". Note: Output terminals are to be open-circuited for this measurement.

4.1.3.4* Output Impedance — Expressed as "_____ ± _____ohms at _____ ± _____ hertz" and _____ °C (°F)." If impedance is resistive, specify "dc". Note: If input terminals are to be short-circuited for this measurement, so specify.

4.1.3.5 Electrical Connections — Whether the electrical termination is by means of a connector or a cable, the pin designation or wire color code shall conform to the following:

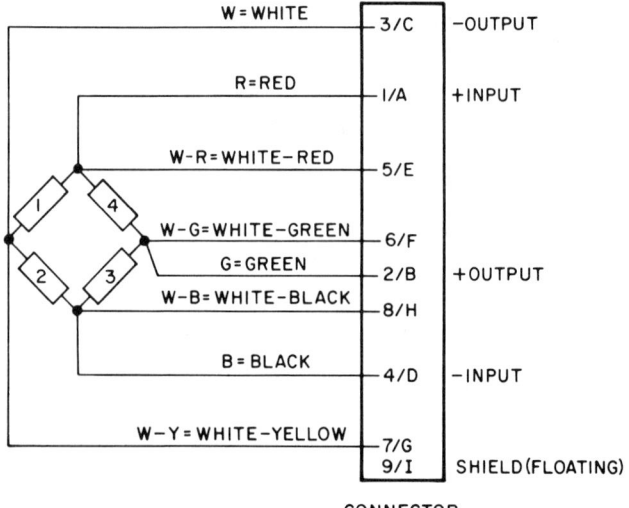

CONNECTOR

PRIMARY WIRING TERMINALS 1/A, 2/B, 3/C, 4/D
AUXILIARY WIRING TERMINALS 5/E, 6/F, 7/G, 8H
(OPTIONAL)

Notes:

1. The output polarities indicated on the above wiring diagram apply when an *increasing* force (compression or tension) is applied to the transducer. For universal force transducers, the indicated polarities apply when the tension force is applied to the transducer; a compression force will produce a negative output.

2. For shielded transducers, pins 5, 7, and 9 shall be shield terminals for 4, 6, and 8 wire systems, respectively.

3. Type connection: Solder or weld.

4.1.3.6 Insulation Resistance — Expressed as "_____ megohms at _____ volts dc at _____°C (°F) between all terminals or leads connected in parallel, and the transducer case."

4.1.4 Supplemental Electrical Design Characteristics

Listing of the following design characteristics is optional:

4.1.4.1 Shunt Calibration Resistor(s) — Expressed as "_____ ± _____ ohms for _____% ± _____% of full scale output at_____°C(°F)."

Note: The terminals across which the resistor(s) is (are) to be placed shall be specified if the resistor(s) is (are) listed.

4.2 Performance Characteristics

The pertinent performance characteristics of strain-gage force transducers shall be tabulated in the order shown. Unless otherwise specified, they apply at the following ambient conditions: Temperature 23 ±2°C (73.4°F ±3.6°F); Relative Humidity 90% maximum; Barometric Pressure 98 ±10kPa (29 ±3 inches of Hg).

4.2.1* Range — Usually expressed as "_____ to _____ newtons (pounds force) compression or tension" or "_____ to _____ newtons (pounds force) compression and _____ to _____ newtons (pounds force) tension".

Note: If 4.2.2 and 4.2.3 are used to specify performance characteristics, the tolerance in 4.2.3 may be omitted. Alternately, the following may be specified: 4.2.3 – 4.2.6.

4.2.2* End Points — Expressed as "_____ ± _____ mV and _____ ± _____ mV open circuit per volt (milliamps) excitation", or "_____ ± _____ mV and _____ ± _____ mV open circuit at _____ volts (milliamps) excitation".

4.2.3* Full Scale Output (FSO) — Expressed as "_____ ± _____ mV open circuit per volt (milliamps) excitation," or "_____ ± _____ mV open circuit at _____ volts (milliamps) excitation".

4.2.4 Zero-Measurand Output — Expressed as "± _____% of full scale output". Determined at full rated excitation, with zero measurand applied to the force transducer.

4.2.5 Zero Drift — Expressed as "± _____% of full scale output over a period of _____ (specify time) with no load applied".

4.2.6 Sensitivity Drift — Expressed as "± _____% of full scale output over a period of _____ (specify time) with _____ newtons (pounds force) applied".

4.2.7* Linearity — Expressed as "_____ linearity within ± _____% of full scale output in _____ (specify direction(s) of loading)".

Note: The type of linearity specified shall be one of the types defined in ISA S37.1; namely, end point, independent, least squares, terminal, or theoretical slope.

4.2.8 Hysteresis — Expressed as "_____% of full scale output upon application of ascending and descending forces including rated force." Alternately, 4.2.7 and 4.2.8 may be combined as:

4.2.9 Hysteresis and Linearity — Expressed as "combined hysteresis and linearity within ± _____% of full scale output upon application of ascending and descending forces including "rated force."

4.2.10* Repeatability — Expressed as "within _____% of full scale output over a period of _____ (specify time) and with _____ cycles of load application".

Alternately 4.2.7, 4.2.8, and 4.2.10 may be combined as:

4.2.11 Static Error Band — Expressed as "± _____% of full scale output as referred to _____ straight line," (see 4.2.7).
Note: The static error band includes errors due to linearity, hysteresis and repeatability.

4.2.12 Creep at Load — Expressed as "± _____% of full scale output with the transducer subjected to rated force for a period of _____ (specify time)".

4.2.13 Creep Recovery — Expressed as "± _____% of full scale output measured at no load and over a period of _____ (specify time) immediately following removal of rated force, that force having been applied for an identical period of time as specified in 4.2.12".

4.2.14* Warm-up Period — Expressed as "_____ minutes for subsequent drifts in sensitivity of zero-measurand balance not to exceed _____% of full scale output".

4.2.15 Static Spring Constant — Expressed in newtons per meter or (pounds force per inch), see Para. 6.3.

4.2.16 Equivalent Dynamic Masses — Expressed in kilograms (pounds mass), for both ends of transducer. See Para. 6.3.

4.2.17 Internal Mechanical Damping — Expressed in newtons per meter/second relative velocity (pounds force per inch/second relative velocity), between ends at a frequency of _____ Hz and a dynamic load of ± _____ newtons (pounds force).

4.2.18 Overloading Rating, Safe — Expressed as "application of _____ newtons (pounds force) for _____ minutes will not cause permanent changes in transducer performance beyond specified static error band".

4.2.19 Rated Force — Expressed as "_____ newtons (pounds force) either compression or tension." This is the maximum axial force the transducer is designed to measure within its specifications.

4.2.20* Thermal Sensitivity Shift — Expressed as "± _____% of sensitivity _____ per °C (°F) temperature change over temperature range from _____ to _____ °C (°F)".

4.2.21* Thermal Zero Shift — Expressed as "± _____% of full scale output per _____ °C (°F) temperature change over temperature range from _____ to _____ °C (°F)."

4.2.22* Temperature Error Band — Expressed as "output values are within ± _____% of full scale output from the straight line establishing static error band (as defined in 4.2.11) over temperature range from _____ to _____ °C (°F)."

4.2.23* Temperature Gradient Error — Expressed as "less than ± _____% of full scale output while at zero load and subjected to a step function temperature change from _____ to _____ °C (°F) lasting for _____ minutes and applied to _____ (specify particular part) of the transducer".

4.2.24 Cycling Life — Expressed as "_____ full scale cycles over which transducer shall operate without change in characteristics beyond its specified tolerances".

4.2.25 Other Environmental Conditions — Other pertinent environmental conditions which shall not change transducer performance beyond specified limits shall be listed. Examples are:

Shock — Triaxial
High Level Acoustic Excitation
Humidity
Salt Spray
Electromagnetic Radiation
Magnetic Fields
Nuclear Radiation

4.2.26 Storage Life — Expressed as "Transducer can be exposed to specified environmental storage condition for _____ (days, months, years) without changing the following performance characteristics beyond their specified tolerances."

Note: Environmental storage conditions shall be described in detail. Pertinent performance characteristics (examples: sensitivity, zero drift) shall be specified.

4.2.27 Abnormal Loading Effects: (Refer to Figure I on page 8.)

4.2.27.1 Concentric Angular Load Effect — Expressed as "± _____% of full scale output difference from true output (axially loaded output multiplied by cosine of angle) resulting from a load applied concentric with the primary axis at the point of application and at _____ degrees angle with respect to the primary axis".

4.2.27.2 Eccentric Angular Load Effect — Expressed as "± _____ % of full scale output difference from true output multiplied by cosine of angle) resulting from a load applied eccentric with the primary axis and at _____ degrees angle with respect to the primary axis".

4.2.27.3 Eccentric Load Effect — Expressed as "± _____ % of full scale output difference from axially loaded output resulting from a load parallel to but displaced _____ millimeters (inches) from concentricity with the primary axis".

4.3 Additional Terminology

Ambient Pressure Effects — The change in sensitivity and the change in zero-measurand output due to subjecting the transducer to a specified ambient pressure change.

Creep at Load — The change in output occurring with time under rated load and with all environmental conditions and other variables remaining constant.

Creep Recovery — The change in zero-measurand output occurring with time after removal of rated load, which had been applied for an identical period of time as employed in evaluating Creep at Load.

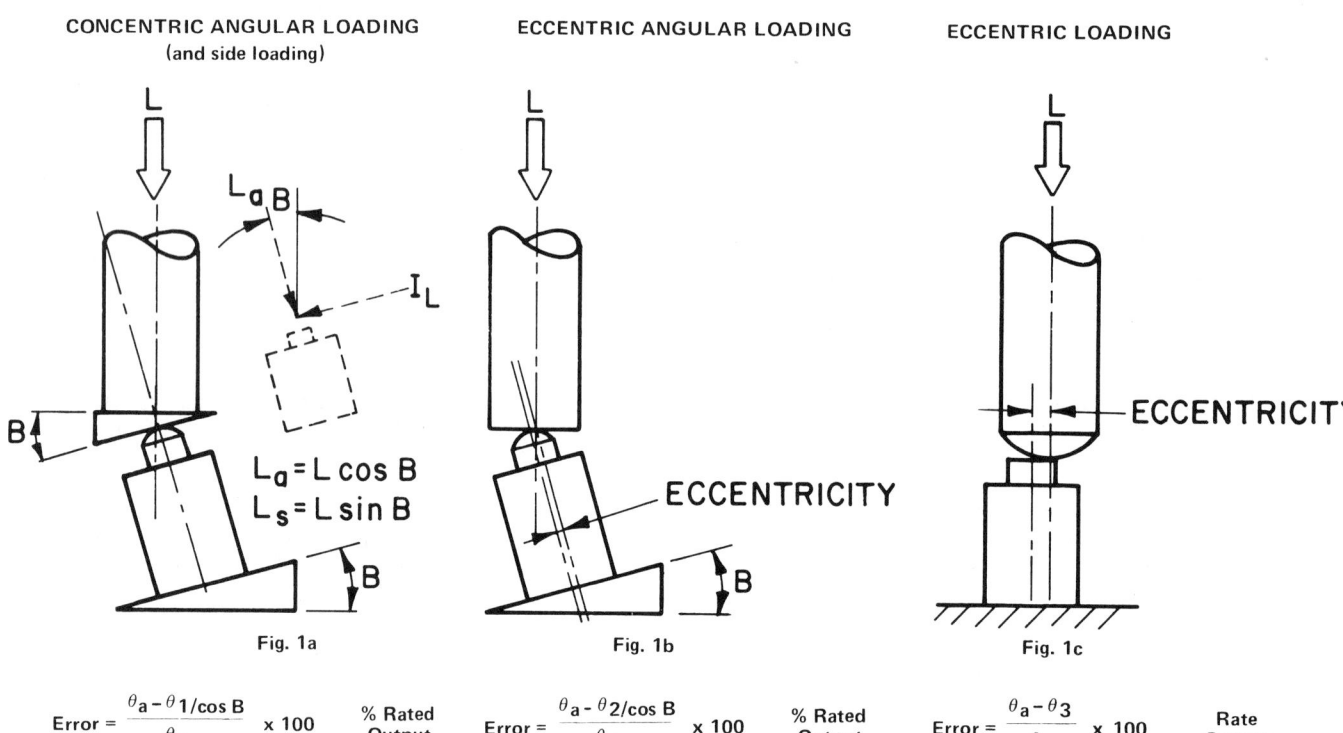

CONCENTRIC ANGULAR LOADING (and side loading)

$$L_a = L \cos B$$
$$L_s = L \sin B$$

Fig. 1a

ECCENTRIC ANGULAR LOADING

ECCENTRICITY

Fig. 1b

ECCENTRIC LOADING

ECCENTRICITY

Fig. 1c

$$\text{Error} = \frac{\theta_a - \theta_1/\cos B}{\theta_a} \times 100 \quad \% \text{ Rated Output}$$

$$\text{Error} = \frac{\theta_a - \theta_2/\cos B}{\theta_a} \times 100 \quad \% \text{ Rated Output}$$

$$\text{Error} = \frac{\theta_a - \theta_3}{\theta_a} \times 100 \quad \text{Rate Output}$$

θ_a = Rated output at rated axial loading.

$\theta_1, \theta_2, \theta_3$ = Output under any unfavorably rated loading conditions.

L = Load
L_a = Axial Load
L_s = Side Load

FIGURE 1

4.4 TABULATED CHARACTERISTICS VERSUS TEST REQUIREMENTS

Characteristic	Paragraph	Design Characteristic Basic	Supp.	Verified During Acceptance	Qual.
Type of Force Transducer	4.1.1.1	X			
Physical Dimensions	4.1.1.2	X			
Force Connection	4.1.1.3	X			
Mounting Dimensions	4.1.1.4	X			
Electrical Connection Location	4.1.1.5	X			
Overload Rating, Ultimate	4.1.1.6	X			
Mounting Force or Torque	4.1.1.7	X			
Weight	4.1.1.8	X			
Identification	4.1.1.9	X			
Temperature Range	4.1.1.10	X			
Case Material	4.1.2		X		
Surface Finish	4.1.2		X		
Type of Strain-Gage Used	4.1.2		X		
Location of Strain-Gage	4.1.2		X		
Number of Active Strain-Gage Elements	4.1.2		X		
Number of Strain-Gage Bridges	4.1.2		X		
Mounting Surface Requirements	4.1.2		X		
Excitation	4.1.3.1	X			
Maximum Excitation	4.1.3.2	X			
Input Impedance	4.1.3.3	X		5.2.9	
Output Impedance	4.1.3.4	X		5.2.9	
Electrical Connections	4.1.3.5	X			
Insulation Resistance	4.1.3.6	X		5.2.8	
Shunt Calibration Resistor	4.1.4.1		X		
Range	4.2.1	X			
End Point	4.2.2	X		5.2.3	
Full Scale Output	4.2.3	X		5.2.3	
Zero-Measurand Output	4.2.4	X		5.2.3	
Zero Drift	4.2.5	X		5.2.4	
Sensitivity Drift	4.2.6	X		5.2.4	
Linearity	4.2.7	X		5.2.3	
Hysteresis	4.2.8	X		5.2.3	
Hysteresis and Linearity	4.2.9	X		5.2.3	
Repeatability	4.2.10	X		5.2.3	
Static Error Band	4.2.11	X		5.2.3	
Creep	4.2.12	X		5.2.5	
Creep Recovery	4.2.13	X			
Warm-up Period	4.2.14	X		5.2.6	
Static Spring Constant	4.2.15	X			6.3
Equivalent Dynamic Masses	4.2.16	X			6.3
Internal Mechanical Damping	4.2.17	X			6.3
Overload Rating	4.2.18	X		5.2.7	
Rated Force	4.2.19	X			
Thermal Sensitivity Shift	4.2.20	X			6.1
Thermal Zero Shift	4.2.21	X			6.1
Temperature Error Band	4.2.22	X			6.1
Temperature Gradient Error	4.2.23	X			6.2
Cycling Life	4.2.24	X			6.4
Other Environmental Conditions	4.2.25		X		6.5
Storage Life	4.2.26	X			6.6
Abnormal Loading Effects	4.2.27	X			6.7

5 INDIVIDUAL ACCEPTANCE TESTS AND CALIBRATIONS

5.1 Basic equipment necessary to perform individual acceptance tests and calibrations of strain-gage force transducers.

The basic equipment for acceptance tests and calibration consists of a force calibrator, a source of electrical excitation for the strain-gages, and a device which measures the electrical output of the transducer. The errors or uncertainties of the measuring system comprising these three components should be less than one-third of the permissible tolerance of the transducer performance characteristic under evaluation. Traceability to the National Bureau of Standards should be established.

5.1.1 Force Calibrator — The maximum inaccuracy of the force calibrator, for ranges of 50 000 newtons and less, should be not more than one-fifth the permissible tolerance of the transducer performance characteristic under evaluation. Force ranges in excess of 50 000 newtons require force calibrator maximum inaccuracy not more than one-half the permissible tolerance of the transducer performance characteristic under evaluation. Range of the instrument supplying or monitoring the calibration force should be selected to provide the necessary accuracy to 125% of the full scale range of the transducer.

The force calibrator may be either continuously variable over the range of the instrument, or may vary in discrete steps provided that the steps can be programmed in such a manner that the transition from one force to the next during calibration is accomplished without creating a hysteresis error in the measurement due to over-shoot.

DEAD WEIGHT CALIBRATOR
Typical Ranges (Tension or Compression)

0-2000 newtons	Max. Error ±0.01% of Test Load
0-5000 newtons	Max. Error ±0.01% of Test Load
0-20 000 newtons	Max. Error ±0.01% of Test Load
0-50 000 newtons	Max. Error ±0.01% of Test Load
0-500 000 newtons	Max. Error ±0.01% of Test Load

PROVING RING CALIBRATOR
Typical Ranges (Tension or Compression)

0-200 newtons	Max. Error ±0.1% Full Scale
0-500 newtons	Max. Error ±0.1% Full Scale
0-2000 newtons	Max. Error ±0.1% Full Scale
0-5000 newtons	Max. Error ±0.1% Full Scale
0-20 000 newtons	Max. Error ±0.1% Full Scale
0-50 000 newtons	Max. Error ±0.1% Full Scale

HIGH ACCURACY REFERENCE FORCE TRANSDUCER
Typical Ranges (Tension or Compression)

0-500 newtons	Max. Error ±0.1% Full Scale
0-1000 newtons	Max. Error ±0.1% Full Scale
0-5000 newtons	Max. Error ±0.1% Full Scale
0-10 000 newtons	Max. Error ±0.1% Full Scale
0-50 000 newtons	Max. Error ±0.1% Full Scale
0-100 000 newtons	Max. Error ±0.1% Full Scale

5.1.2 Stable Source of Electrical Excitation of Accurately-Known Amplitude

For dc excitation, commonly used sources are line-powered, electronically regulated, power supplies.

For ac excitation, commonly used sources are the power line in association with a step-down transformer or an oscillator. Where radiometric measurement techniques are employed, input power shall be within both instrument and force transducer specified range.

5.1.3 Readout Instrument

Examples of suitable devices are:

MANUALLY BALANCED POTENTIOMETER
Typical Ranges

0 to 16 millivolts, Maximum error ±0.015% of reading or ±1 microvolt, whichever is greater.

0 to 160 millivolts, Maximum error ±0.015% of reading or ±3 microvolts, whichever is greater.

0 to 1.6 volts, Maximum error ±0.015% of reading or ±30 microvolts, whichever is greater.

DIGITAL ELECTRONIC VOLTMETER WITH PREAMPLIFIER

Typical Ranges	Sensitivity	Maximum Error
0 to 10 millivolts	1 microvolt	±0.02% of reading, or ±2 microvolts, whichever is greater.
0 to 100 millivolts	10 microvolts	±0.01% of reading, or ±10 microvolts, whichever is greater.

Note: The input impedance of the readout instrument should be as high as possible. Unless otherwise stated, adjustments and compensation of the transducer apply to open circuit conditions on the output terminals.

5.2 Calibration and Test Procedures

Results obtained during the calibration and test procedures should be recorded on data sheets like the sample data sheets in Section 7 of this report. These procedures shall be performed under ambient conditions as defined in Paragraph 4.2.

5.2.1 The transducer is inspected visually for mechanical defects, poor finish, and improper identification markings.

5.2.2 The transducer shall be connected to the force calibrator with axial alignment as specified by the manufacturer. The excitation source and readout instrument shall be connected to the transducer and turned on. Adequate warm-up time for test equipment shall be allowed before tests are conducted. The force calibrator and connecting hardware shall have passed a prior test for proper operation. It may be desirable prior to calibration to exercise the force transducer by applying rated load and returning to zero load, if so, the number of cycles and time duration should be noted on the data sheet.

5.2.3 Two or more complete calibration cycles (dependent on desired statistical confidence levels) are run consecutively, including five to ten points in both ascending and descending directions. Excitation amplitude shall be monitored as required.

From the data obtained during these tests, the following characteristics should be determined:

End Points	4.2.2
Full Scale Output	4.2.3
Zero Balance	4.2.4
Linearity	4.2.7
Hysteresis	4.2.8
(or Hysteresis and Linearity)	4.2.9
Repeatability	4.2.10
(or Static Error Band)	4.2.11

5.2.4 Repeated calibration cycles over a specified period of time after warmup, establish the following characteristics for that period of time:

Zero Drift	4.2.5
Sensitivity Drift	4.2.6

Note: They may be abbreviated cycles with fewer data points than required in 5.2.3.

5.2.5 Application of rated force to the transducer during a specified short period of time and measurement of changes in output at constant excitation during this time should establish:

Creep at Load	4.2.12

Note: See Figure 2 (page 12). Rate of application of force to be as high as possible without resonant excitation of transducer.

5.2.6 By measuring zero-measurand output and sensitivity over a period of time (one hour should suffice), starting with the application of excitation to the transducer, the following characteristic should be determined:

Warmup Period	4.2.14

Note: It is desirable to test for these effects separately, establishing the warmup change of zero-measurand output first.

5.2.7 After application of the specified overload a specified number of times (and in the specified duration for compression or tension), at least one complete calibration cycle shall be performed to establish that the performance characteristics of the transducer are still within specifications.

Overload Rating, Safe	4.2.18

5.2.8 Measure the insulation resistance between all terminals, or leads connected in parallel, and the case of the transducer with a megohmmeter or similar acceptable device, using a potential of 50 volts, unless otherwise specified. Insulation resistance should be measured at room temperature.

Insulation Resistance	4.1.3.6

5.2.9 A Wheatstone bridge (for dc) or impedance bridge shall be used to measure:

Input Impedance	4.1.3.3
Output Impedance	4.1.3.4

6 QUALIFICATION TESTS

6.1 Steady State Temperature Effects

The transducer shall be placed in a suitable temperature chamber. After allowing adequate stabilization time at a specified temperature, one or more calibration cycles shall be performed within the chamber. This procedure shall be repeated at an adequate number of temperatures within the operating temperature range of the transducer. These tests should establish the following characteristics:

Thermal Sensitivity Shift	4.2.20
Thermal Zero Shift	4.2.21
Temperature Error Band	4.2.22

6.2 Temperature Gradient Error

The force transducer, at "room temperature," shall be subjected to a thermal transient by immersion in a fluid which is kept at a specified temperature above or below "room conditions." With no force applied, the output is observed over a specified period of time.

Note: The type of fluid and method of application shall be specified.

These tests should establish:

Temperature Transient Error	4.2.23

6.3 Dynamic Characteristics

The dynamic response of an installed force transducer depends on the stiffness and mass distribution throughout the entire system in which the force transducer is a part. In many cases, especially those which have many springs and masses, the dynamic response of the installed force transducer may best be determined experimentally by applying a suitable time-varying force to the complete system and measuring the output vs. time response of the transducer.

Alternatively, if the distribution of stiffness and mass throughout the system is known quantitatively, the response of the system to a time-varying force may be computed analytically. This approach, however, is apt to be very cumbersome unless (1) attention is limited to frequencies from zero to a little above the lowest natural frequency and (2) the stiffness and mass system is relatively simple.

The dynamic characteristics of the force transducer itself may be determined over a wide frequency range with suitable test fixtures and instrumentation either by applying sinusoidal forces or step function forces. Generally, such a determination yields very complex results. Over a limited low frequency range, however, the dynamic response of a force transducer may be defined adequately in terms of a simple equivalent model as shown in Figure 3 (page 13).

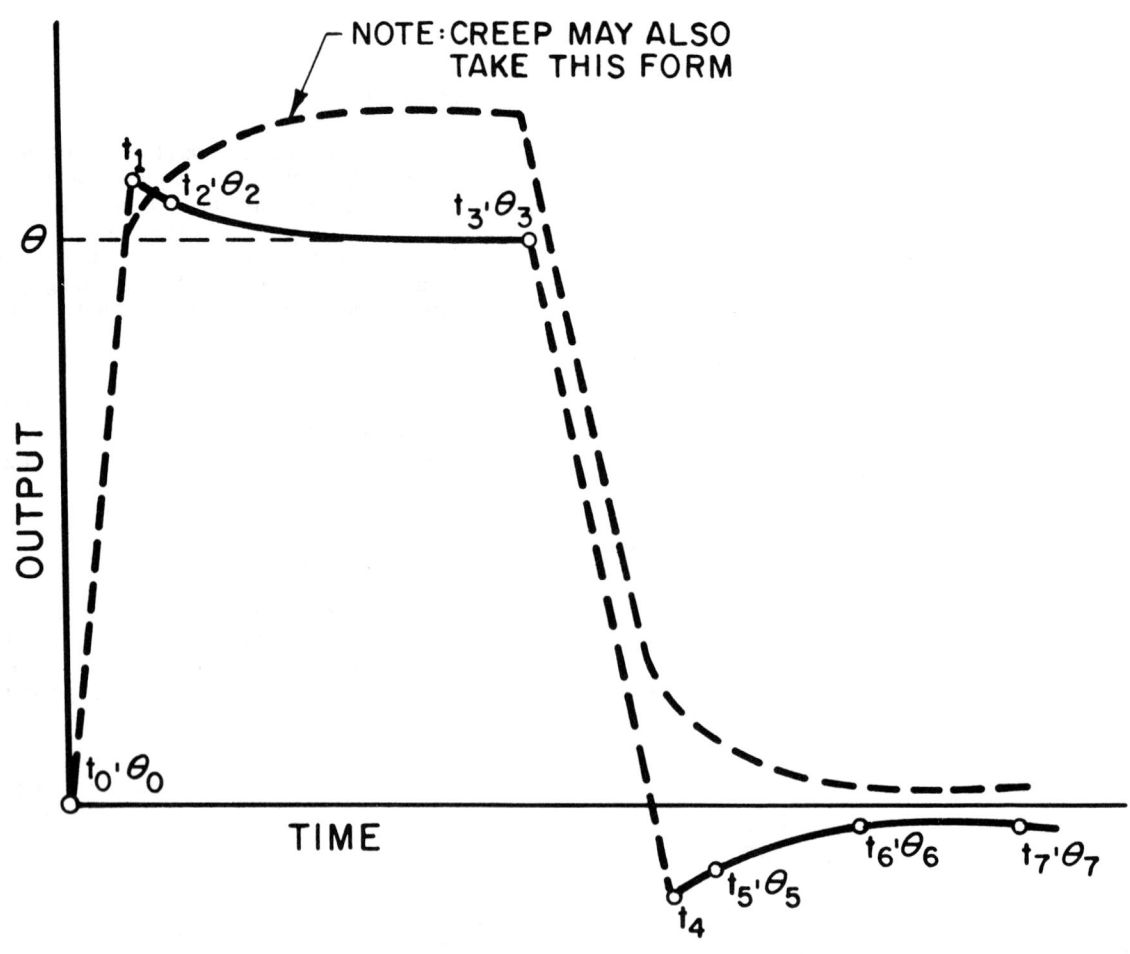

NOTE: CREEP MAY ALSO TAKE THIS FORM

$t_1 - t_0$ = Load Application. Should be carefully considered in comparative creep measurement.

$t_2 - t_1$ = Should be as short as possible. (Suggested 5-10 s.)

$t_3 - t_2$ = Creep measurement period. Suggested 3 min., for short term and 30 min. for long term.

$\dfrac{\theta_2 - \theta_3}{\theta}$ x 100 = Creep in % rated output.

$t_4 - t_3$ = Load release period. (Should equal $t_1 - t_0$).

$t_5 - t_4$ = Should be as short as possible. (Suggested 5-10 s.)

$t_6 - t_5$ = Creep recovery period. Suggested 3 min. for short term and 30 min. for long term.

$\dfrac{\theta_5 - \theta_6}{\theta}$ x 100 = Creep recovery in % rated output.

t_7 = Time at which zero return is measured.

$\dfrac{\theta_7 - \theta_0}{\theta}$ x 100 = Zero return in % rated output.

FIGURE 2

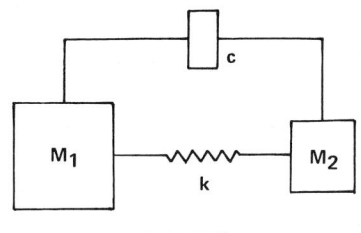

FIGURE 3

k = Static Spring Constant

c = Damping Force Parameter

M_1 = Effective Mass of "Base" of Transducer*

M_2 = Effective Mass of "Top" of Transducer*

*The "base" and "top" of the transducer are defined in the outline drawing.

M_1 and M_2 may be determined experimentally or analytically so that the calculated natural frequency of the M_1-k-M_2 model agrees with the actual first mode natural frequency of the transducer. The effective damping parameter generally varies with the force amplitude and possibly with frequency; consequently, the value of c must be determined for the particular load and frequency values of interest.

6.4 Life Test

After applying the specified number of full range excursions of force, at least one complete calibration cycle shall be performed to establish minimum value of:

Cycling Life   4.2.24

6.5 Effects of Other Environments

Expose transducer to other specified environmental conditions followed in each case by one complete calibration cycle to test ability of transducer to perform satisfactorily after such exposure.

6.6 Storage Life Test

After storing the transducer under specified conditions for the specified period of time, two complete calibration cycles shall be performed to establish:

Storage Life   4.2.26

6.7 Abnormal Loading Effects

6.7.1 For determination of the effects of concentric angular loading (and side loading), insert wedge blocks above and below the force transducer as illustrated in Figure 1a. The angle subtended by the two larger surface areas (B) of each block shall be equivalent to the angle of interest or shall result in the side load of interest.

6.7.1.1 Measure zero-measurand output.

6.7.1.2 Apply rated load and read output as soon as load has stabilized.

6.7.1.3 Remove load and record zero-measurand output after output has stabilized.

6.7.2 For determination of the effects of eccentric angular loading, remove the upper wedge block (Figure 1b.) and repeat steps 6.7.1.1 through 6.7.1.3. If eccentricities other than that obtained in the foregoing are desired, a flat load button should be used and the amount of eccentricity adjusted through placement of the force transducer under a convex loading ram surface.

6.7.3 For determination of the effects of eccentric loading, remove the lower wedge block, use a flat load button, and adjust eccentricity through placement of the force transducer under the convex loading ram surface (see Figure 1c.). Repeat steps 6.7.1.1 through 6.7.1.3.

6.7.4 The effects of the various types of loading related to axial loading conditions can be determined in accordance with the expressions included in Figure 1.

7 TEST REPORT FORMS

7.1 The test report forms listed below are recommended for use during the testing of strain-gage force transducers.

7.2 When using the forms, all pertinent information shall be inserted in its proper place. On some forms, blank space has been provided for additional tests. Where the test is prolonged, more than one form may be required.

7.3 "Individual Acceptance Tests and Calibrations" (Figure 4) used during acceptance testing of Section 5 may also be used during qualification testing of Section 6.

"Environmental Test Record" (Figure 5) used to record thermal sensitivity shift, thermal zero shift, temperature error band, temperature transient error, and other environmental tests.

8 BIBLIOGRAPHY

1. MIL-E-5272C (ASG) "*Environmental Testing, Aeronautical and Associated Equipment, General Specifications for*" (1970).

2. MIL-E-5400P, (ASG) "*Electronic Equipment, Aircraft, General Specification for*" (1974).

3. MIL-STD-810C, "*Environmental Test Methods for Aero-Space and Ground Equipment*" (1975).

4. ANSI MC6.1-1975 (ISA S37.1-75) "*Electrical Transducer Nomenclature and Terminology,*" September 10, 1975.

5. ANSI MC6.2-1975 (ISA S37.3-75) "*Specifications and Tests for Strain Gage Pressure Transducers,*" October 23, 1975.

6. "*Standard Load Cell Terminology and Definitions*", developed by the Industrial Instrument Section of the Scientific Apparatus Makers Association and published by the Scale Manufacturer's Association (1962).

7. "*Handbook of Transducers for Electronic Measuring Systems*", Norton, Harry N., Prentice-Hall, Inc. (1969).

VENDOR'S PART NO.	**STRAIN GAGE FORCE TRANSDUCER** **INDIVIDUAL ACCEPTANCE TESTS & CALIBRATIONS**	CUSTOMER'S PART NO.
TEST FACILITY		SERIAL NO. BRIDGE NO.

Ambient Conditions	☐ Calibration ☐ Concentric Angular, ____°	VENDOR
Temperature _____ °C Pressure _____ kPa Humidity _____ %	☐ _____ Overload ☐ Eccentric _____ mm ☐ Functional Test ☐ Eccentric Angular, ____° ☐ _____ Test and _____ mm	RANGE _____ TO _____ NEWTONS

1. Visual: Mechanical ☐   Finish ☐   Nameplate ☐   Electr. Conn. ☐
2. Electrical: Input Impedance _____ ohms, Output Impedance _____ ohms, Ins Res. _____ MΩ _____ Vdc
3. Calibration and Proof Load Test at _____ V at _____ Hz _____ Excitation after _____ minutes warm-up time
4. Load Impedance _____ ohms   or _____ mA Excitation
5. Exercised _____ times, to _____ newtons.

Test Load Newtons	Theoretical Output (mV)	Output (mV) - Run 1		Output (mV) - Run 2		Overload Output (mV)	Output (mV) - Run 3		Maximum Error (mV)	
		Test Time	Minutes	Test Time	Minutes		Test Time	Minutes		
			Decrease	Increase	Decrease		Increase	Decrease	+	−
						NOT APPLICABLE				
Over		NOT APPLICABLE				┄┄┄┄┄	NOT APPLICABLE			
Rev.										

STATIC ERROR BAND + _____ % _____ % FSO (Allowed: ± _____ % FSO).

Referred to _____

_____ Linearity

_____ LINEARITY: + _____ − _____ % FSO (Allowed: _____ % FSO)

HYSTERESIS: _____ % FSO (Allowed _____ % FSO) REPEATABILITY: _____ % FSO (Allowed: _____ % FSO)

ZERO-MEASURAND OUTPUT: T_____ % FSO (Allowed: _____ % FSO)   CREEP: _____ % FSO over _____ minutes

(Allowed: _____ % FSO)

ZERO SHIFT: _____ % FSO over a period of _____ (Allowed: _____ % FSO)

(number)   (units)

SENSITIVITY SHIFT: _____ % over a period of _____ (Allowed: _____ % sensitivity)

(number)   (units)

END POINTS (_____): _____ and _____ mV (Allowed: _____ and _____ mV)

FULL SCALE OUTPUT (Run 2): _____ mV (Allowed: _____ mV)   OVERLOAD (after Run 2) _____ %

Equipment Used:	Defects noted or Comments

BY: _____   DATE: _____   APPROVED BY: _____

Note:

FIGURE 4

VENDOR'S PART NO.	TEST FACILITY	CUSTOMER'S PART NO.	
VENDOR	**STRAIN GAGE FORCE TRANSDUCER**	SERIAL NO.	BRIDGE NO.
REPORT NO.		CUSTOMER	
TYPE OF TEST	**ENVIRONMENTAL TEST RECORD**	RANGE _____ TO _____ NEWTONS	

Maximum _____ volts
at _____ Hz.
or _____ mA
Warmup Time _____ Minutes

☐ Before
☐ During _____
☐ After

Type of Environment   Level

For temperature tests, temperature measured at _____

Force	Theoretical	mV (Run 1)		mV (Run 2)		Overload	mV (Run 3)		Maximum Error	
Newtons	mV	Increase	Decrease	Increase	Decrease	mV	Increase	Decrease	+	−

Error Band + _____ % − _____ % FSO (Referred To _____)   Allowed: ± _____ % FSO

Overload _____ % Rated Range for _____ Minutes

Ins. Resistance: _____ megohms At _____ Vdc

Zero Shift _____ % FSO

Sensitivity Shift _____ %

Comments : _____

_____

_____

_____

_____

_____

_____

_____

Error Plot, Run 3

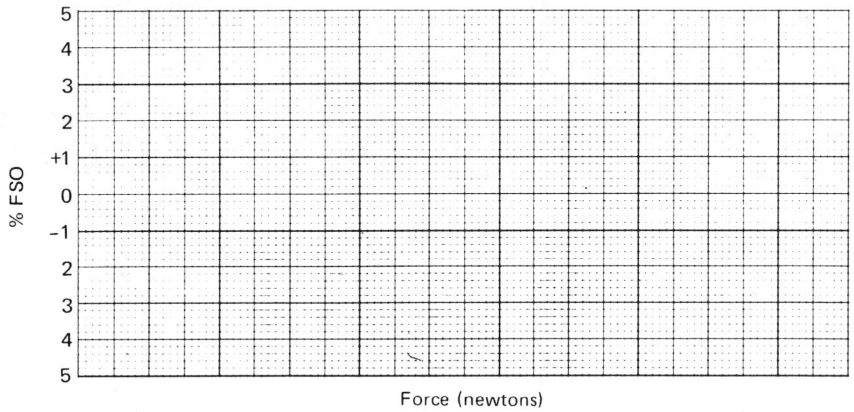

Force (newtons)

Tested By _____   Date Test Started _____   Date Test Finished _____

Approved By _____
Title

Approved By _____
Title

FIGURE 5

INSTRUMENT SOCIETY of AMERICA
Research Triangle Park, North Carolina

ANSI/ISA-S37.10-1975
(R 1982)
Approved December 14, 1982
(Formerly ANSI MC 6.4-1975)

American National Standard

Specifications and Tests for Piezoelectric Pressure and Sound-Pressure Transducers

Instrument Society of America

ISBN 0-87664-382-9

ISA-S37.10   Specifications and Tests for Piezoelectric
Pressure and Sound-Pressure Transducers

INSTRUMENT SOCIETY OF AMERICA
67 Alexander Drive
P.O. Box 12277
Research Triangle Park, North Carolina 27709

PREFACE

This Preface is included for informational purposes and is not a part of ISA-S37.10.

This Recommended Practice has been prepared as a part of the service of the Instrument Society of America toward a goal of uniformity in the field of instrumentation. To be of real value, this document should not be static, but should be subject to periodic review. Toward this end, the Society welcomes all comments and criticisms, and asks that they be addressed to the Secretary, Standards and Practices Board, Instrument Society of America, 67 Alexander Drive, P.O. Box 12277, Research Triangle Park, NC 27709, Telephone (919) 549-8411.

The ISA Standards and Practices Department is aware of the growing need for attention to the metric system of units in general, and the International System of Units (SI) in particular, in the preparation of instrumentation standards. The Department is further aware of the benefits to USA users of ISA Standards of incorporating suitable references to the SI (and the metric system) in their business and professional dealings with other countries. Towards this end, this Department will endeavor to introduce SI - acceptable metric units in all new and revised standards to the greatest extent possible. The Metric Practice Guide, which has been published by the American Society for Testing and Materials as ANSI designation Z210.1 (ASTM E380-76, IEEE Std. 268-1975), and future revisions, will be the reference guide for definitions, symbols, abbreviations,and conversion factors.

It is the policy of the Instrument Society of America to encourage and welcome the participation of all concerned individuals and interests in the development of ISA Standards. Participation in the ISA standards making process by an individual in no way constitutes endorsement by the employer of that individual of the Instrument Society of American or any of the standards which ISA develops.

This Standard was originally prepared by the SP37.10 Committee which operated under the guidance of the SP37.

The following individuals served on the original SP37.10 committee:

NAME	COMPANY
N. Keast, Chairman	Bolt, Beranek and Newman, Incorporated
G. T. Cozad	McDonnell Douglas Corporation
L. Horn	National Bureau of Standards
R. W. Lally	PCB Piezotronics
J. Rhodes	Endevco Corporation

The following individuals served on the 1982 SP37 committee:

NAME	COMPANY
J. W. Mock, Chairman	Measurement Services
P. Bliss	Consultant
W. Brinkschulte	Genisco Technology Corporation
C. Kutelis	Revere Corporation of America
R. W. Lally	PCB Piezotronics
H. E. Lockery	Hottinger Baldwin Measurements, Inc.
E. W. Malone	Boeing Aerospace Co.
T. B. Miller	Lawrence Livermore Lab
K. A. Parlee	United Engineers & Constructors, Inc.
H. Pitt	
J. Powell	Rosemount Engineering
R. C. Strahm	EG & G Idaho
R. M. Whittier	ENDEVCO
N. Wilde	EG & G Idaho
E. Wong	Bell & Howell

This Standard was approved for publication by the ISA Standards and Practices Board in October 1982.

NAME	COMPANY
T. J. Harrison, Chairman	IBM Corporation
P. Bliss	Consultant
W. Calder	The Foxboro Company
N. Conger	Continental Oil Co.
B. Feikle	Bailey Controls Co.
R. T. Jones	Philadelphia Electric Co.
R. Keller	Boeing Company
O. P. Lovett, Jr.	Isis Corp.
E. C. Magison	Honeywell, Inc.
A. P. McCauley	Diamond Shamrock Corp.
J. W. Mock	Measurement Services
E. M. Nesvig	ERDCO Engineering Corp.
G. Platt	Bechtel Power Corp.
R. Prescott	Moore Products Company
W. C. Weidman	Gilbert Associates
K. A. Whitman	Allied Chemical Corp.
J. R. Williams	Stearns-Roger, Inc.
B. A. Christensen*	
L. N. Combs*	
R. L. Galley*	
R. G. Marvin*	
W. B. Miller*	Moore Products Company
R. L. Nickens*	

*Director Emeritus

CONTENTS

1. PURPOSE

This Standard establishes the following for piezoelectric pressure and piezoelectric sound-pressure transducers:

1.1 Uniform minimum general specifications for describing design and performance characteristics.

1.2 Selected uniform acceptance and qualification test methods, including calibration techniques.

1.3 Uniform procedures for the presentation of transducer test data.

1.4 A drawing symbol for use on measurement system electrical schematics. (See Note in 3)

2. SCOPE

2.1 This standard covers piezoelectric (including ferroelectric) pressure transducers and piezoelectric sound pressure transducers. Pressure and sound-pressure types could be the same instrument differing only in the method of calibration and manner of specifying performance. With the exception of certain impedance and charge measurements, this standard is also applicable to piezoelectric transducers with built-in amplifiers.

Sound pressure transducers sense and measure the pressure oscillations within an elastic fluid medium experiencing stress-strain waves. When installed near the sound source or in the wall of a test object, the transducer behavior relates to its pressure response. When installed in a sound field, considerable interaction occurs at higher frequencies during the measuring transaction, changing the quantity being measured and relating transducer behavior to its free-field or diffuse-field response. Both aspects of transducer behavior are covered in this standard.

2.2 Included among the specific types of piezoelectric pressure transducers to which this Standard is applicable are the following:

Piezoelectric pressure transducers for transient pressure measurements.

Piezoelectric pressure transducers which, in conjunction with associated electronic equipment, have quasi-dc response to gage pressures.

Piezoelectric transducers for sound pressure levels in excess of 100 dB overall re 20 μ Pa associated with fluid-borne noise.

2.3 Terminology used in this document is defined in ISA S37.1, except that additional definitions particularly applicable to piezoelectric pressure and piezoelectric sound-pressure transducers are defined in Section 4.3 of this document.

3. DRAWING SYMBOL

The drawing symbol for a transducer is a square with an added equilateral triangle, the base of which is one side of the square. The triangle symbolizes the sensing element.

The piezoelectric element is symbolized by a small rectangle encompassing two diagonally crossing lines. Surrounding this rectangle is the electrical symbol for a capacitor. Lines from the symbolic capacitor to the right side of the large square represent the electrical leads.

NOTE: This symbol is not ANSI-approved at this time. It has been submitted to the ANSI Y32 Committee on Graphic Symbols for their consideration and approval.

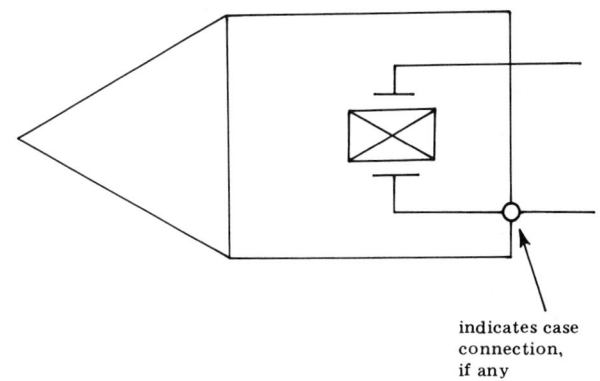

indicates case
connection,
if any

4. CHARACTERISTICS

	Summary		Individual Acceptance Tests		Qualification Tests
	Basic	Supplemental	go/no-go	measured	
Cable Characteristics (state whether integral or nonintegral)					
Type	4.1.1		5.1		
Length	4.1.1		5.8.1		
Connector(s) used	4.1.1		5.1		
Temperature range	4.1.1				

	Summary		Individual Acceptance Tests		Qualification Tests
	Basic	Supplemental	go/no-go	measured	
Capacitance at Room Temperature	4.1.3			5.8.1	
Noise Characteristics		4.1.4			6.2
Case Connector Sealing (if applicable)	4.1.1		6.1		
Dimensions, configuration and markings	4.1.1		5.1		
Connector Type	4.1.1		5.1		
Exposed Materials	4.1.1				
Transduction Element Material Type	4.1.1				
Sensing Mode	4.1.1				
Mounting Method	4.1.1		5.1		
Transducer Temperature Range (excluding cable if nonintegral) Operating	4.2.1.9				6.6
Storage		4.1.4			
Dead Volume		4.1.2			
Equivalent Volume		4.1.2			
Diaphragm Material and Thickness		4.1.2			
Vibration Isolation or Cancellation Descriptions Mechanical		4.1.2			6.4
Electrical		4.1.4			6.4
Capacitance Room Temperature	4.1.3			5.5	
Over Temperature Range		4.1.4			6.6
Grounding	4.1.3		5.7		

	Summary		Individual Acceptance Tests		Qualification Tests
	Basic	Supplemental	go/no-go	measured	
Insulation Resistance (if applicable)					
Room Temperature	4.1.3		5.5		
Over Temperature Range		4.1.4			
Shunting Resistance					
Room Temperature	4.1.3		5.5		
Over Temperature Range		4.1.4			6.6
Range	4.2.1.1			5.2	
Sensitivity					
At Room Temperature (nominal + tolerance)					
Over Temperature Range	4.2.1.7				6.6
Over Ambient Pressure Range (sound-pressure transducers only)	4.2.1.13				6.3
Stability					6.8
Frequency Response					
Pressure Transducers	4.2.1.3				5.4.1
Sound-Pressure Transducers	4.2.1.3			5.4.2	
Linearity	4.2.1.4			5.2	
Pressure-Excited Resonance					
At lowest resonant frequency	4.2.1.14			5.4.1	
At additional frequencies		4.2.2.1			5.4.1
Proof Pressure					
Pressure Transducers	4.2.1.5		5.3		
Burst Pressure Rating					
Pressure Transducers	4.2.1.6				6.9

	Summary		Individual Acceptance Tests		Qualification Tests
	Basic	Supplemental	go/no-go	measured	
Spurious Output from Temperature Gradient	4.2.1.8				6.7
Vibration Error	4.2.1.10		6.4		
Directivity	4.2.1.15				5.4.2
Susceptibility to Environments					
Mechanical Shock	4.2.1.11				
Humidity	4.2.1.11				
Salt Spray	4.2.1.11				
Nuclear Radiation	4.2.1.11				
Electromagnetic Interference	4.2.1.11				6.8
Steady-state Acceleration	4.2.1.11				6.5

4.1 Design Characteristics

4.1.1 Basic Mechanical Design Characteristics.

The following mechanical design characteristics are required to be listed:

Cable, Non-Integral. If a non-integral cable is supplied with the transducer, the type, length, connector types, and maximum operating temperature of this cable shall be specified. (Requirements for the specification of electrical cable characteristics are given in Section 4.1.3).

Case Sealing. If case sealing is employed, the mechanism and materials for sealing shall be described. The same requirement applies to the electrical connector. The resistance of the sealing materials to common and corrosive fluids shall be stated.

Connection, Pressure. The pressure connection shall be indicated on the outline drawing (see "Dimensions"). For threaded cases of fittings, indicate the nominal size, thread pitch, thread series, and thread class. For a flush-mounted transducer, indicate whether flange mounting, cemented installation, or other specified means is employed.

Connector, Electrical. The connector on the transducer shall be described. If the transducer is supplied with an integral cable, the connector at the end of the cable shall be described. The mating connector for the above connector shall also be described or designated.

Dimensions. An outline drawing of the transducer shall show its complete configuration with dimensions given in millimeters (or inches).

Identification. The following characteristics shall be permanently inscribed on the outside of the transducer case or (for very small transducers) supplied with the transducer:

    a. Nomenclature of transducer (According to ISA S37.1, Section 3).

    b. Name or trademark of manufacturer. Model (Part) Number, and Serial Number.

    c. Range (if applicable).

Materials, Housing. The case materials exposed to the environment shall be identified.

Material, Transduction Element. The type of piezoelectric material employed as the transduction element shall be identified. (A proprietary name is acceptable.) The sensing mode of this element shall also be specified.

Fluid Limitations. If specific corrosive fluids are associated with a particular transducer application, the compatibility of the transducer and its accessories with such specified fluids shall be stated.

Mounting and Mounting Dimensions. Unless the pressure connection serves as the transducer mounting, the outline drawing shall indicate the method of mounting with dimensions in millimeters or inches.

Mounting Force or Torque. Mounting force or torque shall be specified. When pressure connection is not integral with mount, pressure-connection torque shall also be specified.

Weight. The weight of the transducer shall be specified in grams (or ounces). If the transducer includes an integral cable, its weight shall also be stated.

4.1.2 Supplemental Mechanical Design Characteristics: Listing of the following mechanical design characteristics is optional.

Dead Volume. For non-flush-mounted transducers, the dead volume may be given in cubic millimeters (or cubic inches). For piezoelectric sound-pressure transducers, the *Equivalent Volume* due to the compliance of the diaphragm may be specified to assist in transducer calibrations.

Identification. The following supplemental information may be inscribed on the transducer case:

    a. Sensitivity

    b. Customer Specification, Part Number, or both

    c. Type of Electrical Connector

    d. Maximum Operating Pressure

    e. Maximum Operating Temperature

Material, Pressure Sensing. The diaphragm material and thickness may be specified, if a diaphragm is employed.

Vibration Isolation. If the transduction element is mechanically isolated in some way from the transducer mounting points (to reduce vibration sensitivity) this may be described.

4.1.3 Basic Electrical Design Characteristics: The following electrical design characteristics are required to be listed. They are applicable at "room conditions" according to the definition given in ISA S37.1.

Capacitance. The capacitance of the transducer and that of any non-integral cable shall be stated separately. Capacitances shall be expressed as "_____ picofarads."

Grounding. It shall be stated whether or not one of the transducer signal leads is internally connected to case ground electrically.

Resistance, Shunting. Expressed as "not less than _____ megohms at _____ volts dc" as applied for two minutes between the two output terminals, unless a different time is specified.

Resistance, Insulation. Insulation resistance shall be expressed as "not less than _____ megohms at _____ volts dc" as applied for two minutes between both output terminals connected in parallel and the transducer case at the mounting point. Note that this requirement is not applicable for those transducers which are internally grounded.

Temperature Range (Storage). All restrictions on the temperature at which the transducer can be safely stored shall be specified.

Load Impedance. The impedance presented by the immediately associated measuring system (cable if not integral, amplifier, etc.) to the transducer's output terminals shall be specified either as a minimum value, a range of values, or a nominal value with tolerances. All specified performance characteristics are intended to be applicable under this specified load-impedance condition.

4.1.4 Supplemental Electrical Design Characteristics. Listing of the following electrical design characteristics is optional:

Capacitance vs. Temperature. This may be given as a graph of transducer temperature. A corresponding curve of cable capacitance vs. cable temperature may also be provided.

Cable Noise. The noise produced by the transducer cable when mechanically excited in some specified way may be stated (see Section 6.2).

Insulation Resistance vs. Temperature. This may be given as a curve of the transducer insulation resistance vs. temperature.

Shunting Resistance vs. Temperature. This may be given as a curve of the shunting resistance of the transducer vs. transducer temperature.

Vibration Cancellation (Electrical). Any built-in electrical method for reducing the vibration sensitivity of the transducer may be specified.

Polarity. The positive-going output terminal for an applied increase in pressure may be specified.

4.2 Performance Characteristics

The pertinent performance characteristics of piezoelectric pressure and piezoelectric sound-pressure transducers shall be tabulated in the order shown. Unless otherwise specified, they apply at "room conditions" as defined in ISA S37.1; i.e., temperature: $25 \pm 10^\circ C$ ($77 \pm 18^\circ F$); relative humidity: not to exceed 90%; barometric pressure: 730 ± 70 millimeters Hg (29 ± 3 in. Hg).

Terminology used here is defined either in ISA S37.1 or Section 4.3 of this document. An asterisk appears beside the paragraph number of those terms defined in S37.1; the remaining terms are defined in Section 4.3 below.

It is important that all transducer performance characteristics be listed independent of the characteristics of any amplifier and/or non-integral cables supplied with the transducer. (Performance characteristics may also be supplied including the effects of amplifiers and cables, if so identified as supplemental information.) In those cases where such characteristics cannot be stated independent of amplifier properties, the pertinent amplifier and cable type, part number and properties shall also be specified.

All performance characteristics are applicable under the conditions specified for Load Impedance. Note that this practice is used in lieu of specifying "open-circuit" output characteristics (an earlier practice which did not permit verification of such characteristics since all ancillary equipment has a finite value of input impedance).

NOTE: In the following, separate statements are given for *pressure transducers* and for *piezoelectric sound-pressure transducers* in those cases where current terminologies differ for the two applications.

4.2.1 Required Performance Characteristics

*4.2.1.1 Range

Pressure Transducers. The range, usually expressed as "± _____ Pa (psi)" or "0 to _____ Pa (psi.)"

Piezoelectric Sound-Pressure Transducers. The range is usually expressed as " _____ dB Sound Pressure Level (SPL) to _____ dB SPL re 20μPa."

*4.2.1.2 Sensitivity, Charge or Voltage

Pressure Transducers. The voltage sensitivity is expressed as " _____ millivolts per Pa (psi) ± _____ %," or as " _____ ± _____ millivolts per Pa (psi)." Equivalently, charge sensitivity may be expressed as " _____ picocoulombs per Pa (psi) ± _____ %, or as " _____ ± _____ picocoulombs per Pa (psi)."

In any case, it is assumed that the pressure and electrical parameters are *both* reported as peak, average, or rms values unless stated otherwise.

Piezoelectric Sound-Pressure Transducers. The voltage sensitivity level is expressed in dB as 20 times the logarithm to the base 10 of the ratio of the sensitivity to the reference sensitivity. That is, sensitivity level S, re 1V/Pa, is

$$S = L_V - L_p = 20 \log_{10} V/p$$

where L_V is the output voltage, re 1V, produced by applied sound pressure level L_p, re 20μPa. Appropriate tolerances should be shown for a specified nominal value. Alternatively, the charge sensitivity level may be expressed as " _____ ± _____ dB re 1 picocoulomb per Pascal."

*4.2.1.3 Frequency Response

Pressure Transducers. This is expressed as "within ± _____ % of the sensitivity at _____ hertz from _____ to _____ hertz." The method for determining this frequency response should be described.

Piezoelectric Sound-Pressure Transducers. Frequency response shall be specified as " _____ (type frequency response) within ±3 dB (or, alternatively, within ±1 dB) from _____ hertz to _____ hertz." The quantity entered into the first blank shall be one of the following (as defined in Section 4.3): *calculated frequency response, pressure response, free-field grazing incidence response, free-field normal incidence response, or free-field random incidence response.

Frequency response shall be referred to a frequency within the specified frequency range of the transducer, and to a specific fluid. The methods of mounting the transducer and applying the test fluid should both be specified.

*4.2.1.4 Linearity

Pressure Transducers. Linearity is normally expressed as " _____ linearity within ± _____ % of full (or a specified partial) scale output." The type of linearity to be entered in the first blank above shall be one of the straight line types defined in ISA S37.1; namely: end point, independent, least squares, terminal, or theoretical slope.

Piezoelectric Sound-Pressure Transducers. Linearity is generally expressed as " _____ linearity within ± _____ dB." The type of linearity specified in the first blank shall be one of the straight one types defined in ISA S37.1.

*4.2.1.5 Proof Pressure

Pressure Transducers. Proof Pressure shall be expressed as (application of) " _____ Pa (psi) for _____ minutes" (will not cause changes in transducer performance which exceed its specified error limits).

*4.2.1.6 Burst Pressure Rating

Pressure Transducers. Burst Pressure Rating is stated as " _____ Pa (psi) applied _____ times for a period of _____ minutes each" (will not result in mechanical failure of the transducer housing).

*Thermal Sensitivity Shift

Pressure Transducers. Thermal Sensitivity Shift is expressed in terms of a maximum change from the (actual) room-temperature sensitivity level over the specified *operating temperature range as "_____ % maximum, from _____ °C (°F) to _____ °C (°F)."

Piezoelectric Sound-Pressure Transducers. Thermal Sensitivity Shift is expressed as "± _____ dB" (sensitivity level change over the specified *operating temperature range) "from _____ °C (°F) to _____ °C (°F)."

*4.2.1.8 Temperature Gradient Error

Expressed as "less than ± _____ millivolts output when subjected to a step-function temperature change from _____ °C (°F) to _____ °C (°F), applied to _____ (specify particular part) of the transducer" (at constant ambient pressure).

State whether Procedure I or II of Section 6.7 is to be used to verify this characteristic.

NOTE: Alternatively, the temperature-gradient error may be expressed as equivalent Pa (psi) input (for pressure transducers), or in picocoulombs output for transducers to be used with charge amplifiers, or in dB SPL.

*4.2.1.9 Maximum and Minimum Ambient Temperature

Expressed as (the transducer can be operated indefinitely at any temperature within the range from) " _____ °C (°F) to _____°C (°F)" (without incurring a permanent calibration shift).

*4.2.1.10 Vibration Error

Pressure Transducers. Vibration error limits are expressed as "less than _____ millivolts "(or alternately, picocoulombs)" rms output due to _____ g rms acceleration along any axis over a frequency range from _____ hertz to _____ hertz." The errors shall be listed for each of three mutually perpendicular axes, or for that axis expected to have the largest vibration error. State whether a swept sinusoidal or broad-band random vibration input is to be employed. In the latter case it is preferable to show a graphical representation of the vibration program.

Piezoelectric Sound-Pressure Transducers. Vibration error limits may be expressed as "less than _____ dB equivalent sound pressure level (SPL) output due to _____ g rms acceleration along any axis over a frequency range from _____ hertz to hertz." State whether a swept sinusoidal or broad-band random vibration input is to be employed.

NOTE: Alternatively, the equivalent SPL output may be expressed relative to peak acceleration, provided that this is made clear and that the type of vibration waveform is specified.

*4.2.1.11 Other *Environmental Conditions

Other Pertinent *operating or *non-operating environmental conditions which shall not affect the transducer performance beyond the specified limits shall be listed; examples are:

    a. Mechanical Shock
    b. Humidity
    c. Salt Spray
    d. Nuclear Radiation
    e. Electromagnetic Interference
    f. Acceleration (steady)
    g. Ambient Pressure

The test conditions for determining such properties shall be identified.

4.2.1.12 Sensitivity Stability

Pressure Transducers. Sensitivity stability shall be stated as: "The sensitivity shall not vary more than ± _____ % of its room-temperature value when subjected to _____ temperature cycles between _____ °C (°F) and _____ °C (°F), and to _____ pressure cycles up to _____ Pa (psi)."

Piezoelectric Sound-Pressure Transducers. Sensitivity Stability shall be stated as: "The sensitivity level shall not vary more than ± _____ dB when subjected to _____ temperature cycles between _____ °C (°F) and _____ °C (°F), and to sound pressure levels up to _____ dB re 20μPa at _____ hertz."

4.2.1.13 *Ambient-Pressure *Sensitivity Shift

Piezoelectric Sound-Pressure Transducers. The allowable sensitivity shift due to variations in ambient pressure shall be expressed as: "The microphone sensitivity level will not vary more than ± _____ dB when operated at any ambient pressure within the range from _____ Pa (psia) to _____ Pa (psia)."

4.2.1.14 Resonant Frequency Amplification Factor

Pressure Transducers. Resonant frequency amplification factor at the lowest resonant frequency shall be expressed as "the amplification factor at resonant frequency _____ hertz shall not exceed _____ ."

Piezoelectric Sound-Pressure Transducers. Resonant frequency amplification factor at the lowest resonant frequency shall be expressed as "the amplification factor at resonant frequency _____ hertz shall not exceed _____ dB."

*4.2.1.15 Directivity

Piezoelectric Sound-Pressure Transducers. The Directivity shall be specified as an allowable envelope for a specified directivity characteristic (directional response pattern, see Section 4.3).

4.2.2 Optional Performance Characteristics

4.2.2.1 Resonant Frequency Characteristics of Sensing Elements.

Amplification (damping) at additional frequencies shall be expressed as in Section 4.2.1.14.

4.3 Additional Terminology

Decibel (dB) (see Reference a) (see also Sound Pressure Level). A unit of level, where:

$$\text{Level in dB} = 10 \log_{10} \frac{P_1}{P_{ref}}$$

P_1 = a power, or, quantity directly proportional to power.

P_{ref} = a reference power, or, a corresponding reference quantity proportional to power.

Diffuse-Field Response. A frequency response of a piezoelectric sound-pressure transducer with the sound incident from random directions.

Directivity Characteristic (Directional Response Pattern) (see Reference f). A plot of the sensitivity level of piezoelectric sound-pressure transducer vs. the angle of sound incidence on its sensing element relative to the sensitivity level in a specified direction, and at a specified frequency.

Equivalent Volume (see Reference d). The volume of a gas enclosed in a rigid cavity which would give the same acoustical input impedance as that of the piezoelectrical sound-pressure transducer.

Free-Field Frequency Response (see Reference c). The free-field frequency response of a piezoelectric sound-pressure transducer is the ratio, as a function of frequency, of the transducer's output in a sound field to the free-field sound pressure existing at the transducer location in the absence of the transducer.

Free Field (Sound) (see Reference a). A free sound field is one existing in a homogeneous, isotropic medium free of any acoustically-reflecting boundaries.

Free-Field Grazing Incidence Response. A free-field frequency response of a piezoelectric sound-pressure transducer with the sound incident parallel to a specified sensing surface of the microphone.

Level (see Reference a). A measure of the logarithm of the ratio of some quantity to a reference quantity of the same kind. The reference quantity must be identified.

Pressure (Frequency) Response. The pressure frequency response (pressure response) of a piezoelectric sound-pressure transducer is the ratio, as a function of frequency, of the transducer output to a sound pressure input which is equal in phase and amplitude over the entire sensing surface of the transducer.

The pressure frequency response is generally equal to the free-field frequency response at wavelengths long compared to the maximum dimension of the piezoelectric sound-pressure transducer.

Resonant Frequency Amplification Factor. The ratio of the maximum sensitivity of a transducer at its lowest resonant frequency to its nominal sensitivity.

Sensitivity Stability. A measure of the irreversible change in transducer sensitivity level after exposure to temperature and/or pressure extremes, or with time.

Shunting Resistance. The electrical resistance observed between the two terminals of a piezoelectric transducer or its integral cable.

Sound Pressure Level (SPL) (see Reference a). Defined in decibels as:

$$\text{SPL (dB)} = 10 \log \frac{(\overline{p}^2)}{10\,(\overline{p}_{ref}{}^2)} = 20 \log \frac{p\ (\text{rms})}{10\,p_{ref}(\text{rms})}$$

Where $\overline{p}^2$ is mean square sound pressure and $\overline{p}_{ref}$ is reference pressure which shall be stated as $20\,\mu$ Pa.

NOTE: Sound level (see Reference a) is a weighted sound pressure level reading obtained by use of metering characteristics and weightings A, B, or C specified in ANSI Standard S1.4-1971. This is not to be confused with sound pressure level.

5. INDIVIDUAL ACCEPTANCE TESTS AND CALIBRATIONS

Tests are listed in the order in which they are to be performed.

Results obtained during these calibrations and test procedures shall be recorded on data sheets similar to the sample data sheet in Section 7 of this guide. These procedures shall be performed under "room conditions", as

defined in ISA S37.1, unless otherwise indicated and under specified load impedance conditions.

References to the pertinent paragraphs in Section 4.2 are given in parentheses.

5.1 Visual Inspection

Conduct a complete visual examination for conformance to stated configuration and markings. Determine dimensions, thread sizes, and thread classes utilizing standard inspection instruments. Check mating of accessory cable (if any) by attaching and removing the cable.

5.2 Voltage or Charge Sensitivity, Range, and Linearity.

For most applications, the sensitivity (level) of the transducer may be determined by comparison with a secondary standard transducer. The secondary standard transducer used shall have a calibration traceable to a calibration performed at the National Bureau of Standards (or at another suitable facility using a well-documented absolute calibration procedure) with a normal calibration period of one year. The transducer used as a standard shall be reserved for this purpose only; it shall not be exposed to large values of shock, vibration or temperature extremes; and its calibration shall be checked periodically.

In general, although the calibration of a piezoelectric pressure transducer may require the use of a connecting cable and amplifier, the effective voltage sensitivity (or equivalent charge sensitivity) of the transducer shall be determined. Statement of the calibration of a transducer, cable, and amplifier solely as a system shall be avoided.

5.2.1 Voltage and Charge Sensitivity, Range, and Linearity of Transducers with Quasi-dc Response.

A piezoelectric pressure transducer which has, in conjunction with an associated amplifier, essentially dc response may be calibrated against a standardized source of constant pressure. Typical example of such sources are the mercury manometer, air piston gauge, precision dial gauge, oil piston gauge, etc. (see Reference j). The error of the pressure source shall be one-fifth (or less) the permissible error tolerance of the transducer performance characteristics under evaluation. The range of the instruments supplying or monitoring the calibration pressure shall be selected to provide the necessary accuracy to 125% of the full-scale range of the transducer.

The source of calibration pressure may be either continuously variable over the range of the instrument, or may be provided in discrete steps, as long as the pressure is returned to zero after each step.

The transducers shall be connected to the pressure source and secured with recommended force or torque. The necessary cable, amplifier, and readout instruments shall also be connected to the transducer and turned on. Adequate warmup time shall be allowed for the test equipment before tests are conducted. The pressure source, connecting tubing, and transducer systems shall have passed a prior test for leaks which might cause calibration errors.

Two or more complete calibration cycles shall be run consecutively. At least five data points shall be obtained in both ascending and descending directions of pressure application. Amplifier characteristics shall be monitored as required.

The voltage sensitivity of the transducer is determined by the ratio (at full scale) of the transducer output in millivolts (suitably corrected for the gain of any amplifier employed, and for the effect of any cables employed) divided by the full scale static pressure applied. Charge sensitivity level can be determined by using a charge amplifier of known characteristics, or by multiplying the voltage sensitivity level in volts per Pa (psi) by the total capacitance of the system (i.e., that of the transducer and its associated cable, and the effective input capacitance of the voltage amplifier) in picofarads, yielding charge sensitivity in picocoulombs Pa (psi).

From the data obtained in these tests, the following characteristics shall be determined:

    a. Sensitivity (4.2.1.2)
    b. Range (4.2.1.1)
    c. Linearity (4.2.1.4)

5.2.2 Voltage or Charge Sensitivity, Range, and Linearity of Pressure Transducers Not Capable of Quasi-dc Response.

The sensitivity (level) of a transducer not capable of quasi-dc response may be determined by comparison with a secondary standard transducer, or by the reciprocity technique (see Reference f). The latter method is limited to use with piezoelectric sound-pressure transducers and low-range pressure transducers. The former method is described here.

It is preferable that the secondary-standard transducer be capable of dc response, so that it may be calibrated in turn in the manner described in Section 5.2.1. In order to excite the transducers, a source of transient or alternating pressure is necessary. Among the former are the shock tube, quick-opening valve devices, burst-diaphragm devices, etc. (see Reference g). Among the latter are loudspeakers, various siren-type devices and hydraulically driven actuators (see Reference g). In general, it is preferable that the transient or alternating pressure applied to the transducer be comparable to the rated full-scale pressure of the transducer under test. Frequently, however, a suitable actuating device satisfying this requirement is not available, and pressures well below full-scale must be used for calibration purposes.

The sensitivity of the transducer is computed as the ratio of the maximum instantaneous output voltage of the transducer (after suitable corrections for associated cables and amplifiers) in millivolts to the maximum instantaneous pressure applied to produce this output. Equivalently, if the pressure excitation is sinusoidal, the ratio of the rms output voltage, in millivolts, to the rms pressure applied may be determined.

Charge sensitivity can be measured with a charge calibrator (Q-step calibrator), or a charge amplifier of known characteristics, or calculated by multiplying the voltage sensitivity in volts per Pa (psi) by the total capacitance of the system in picofarads, yielding charge sensitivity in picocoulombs per psi.

By repeating the above tests at various measured amplitudes, it is possible to determine:

    a. Sensitivity (4.2.1.2)
    b. Range (4.2.1.1) (if within the capability of the pressure source)
    c. Linearity (4.2.1.4)

5.2.2.1 Voltage Sensitivity Level of Piezoelectric Sound-Pressure Transducers (4.2.1.2)

The voltage sensitivity level of microphones is expressed in dB re 1 volt per Pa. This can be computed from the voltage sensitivity measured as specified in Section 5.2.1 or 5.2.2 by employing the following equation.

Sensitivity Level =

$$20 \log_{10} \frac{\text{output in volts}}{\text{input in Pa}}$$

5.3 Proof Pressure (Pressure Transducers) (4.2.2.6)

After application of the specified proof pressure, a specified number of times, at least one complete calibration cycle shall be performed using the procedures of Sections 5.2.1 or 5.2.2, whichever is applicable, to establish that the performance characteristics of the transducer are still within specifications.

5.4 Frequency Response, Resonant Frequency and Resonant Frequency Amplification.

5.4.1 High-Range Transducers.

The dynamic response characteristics of pressure transducers may be established either with transient-excitation devices, or with sinusoidal pressure generators.

5.4.1.1 Transient Excitation Method

Transient-excitation devices generally produce step-functions of pressure. A positive step-function of pressure may be generated in gases with a shock-tube or a quick-opening valve. A hydraulic quick-opening valve is used to generate a positive pressure step function in a liquid medium. A burst-diaphragm generator is used to produce a negative pressure step in a gas medium. In all cases, the rise time of the generated step function shall be sufficiently short to shock-excite at least the lowest few resonances of the transducer under test.

Since any tubing used to mechanically connect the transducer to the test set-up will affect the dynamic characteristics, it is recommended either that the shortest possible tubing be installed, or that any tubing used shall duplicate as closely as possible the actual installation.

By applying step functions of pressure at room conditions within the full scale range of the transducer, and analyzing the electronic or electro-optical recording of the transducer output, the following can be determined:

Frequency Response (amplitude and phase) 4.2.1.3

Resonant Frequency 4.2.2.1

Note that these may be a function both of the polarity of the pressure step and of the type of test fluid.

5.4.1.2 Sinusoidal Excitation Method

Generators are available to produce sinusoidal pressure waveforms in either liquids or gases. These sinusoidal generators operate either on the "pistonphone" principle (which is in common use for the absolute calibration of microphones), by modulating a fluid flow through an orifice or by vibrating or rotating a column of liquid in a vertical plane. (See Reference l). By applying a sinusoidal pressure waveform of varying frequency at a specified static pressure, frequency response (Amplitude and Phase, Section (4.2.1.3) can be observed directly by comparison with a suitable reference transducer.

The following can be established from the frequency response:

Resonant Frequencies and their Amplification Factors

Damping Ratios

Ringing Period

Discharge Time Constant

Overshoot

5.4.2 Low-Range Pressure Transducers and Piezoelectric Sound-Pressure Transducers.

For these transducers it is generally possible to measure frequency response directly. A number of methods exist for this purpose. In general, these methods provide either a pressure response or a free-field response as defined in Section 4.3. (The pressure response is the same as the free-field response for frequencies whose wave lengths in the measure medium are large compared to the dimensions of the transducer. At higher frequencies, the pressure response differs from the free-field response because of diffraction of the sound field by the transducer and its associated mounting hardware.)

Pressure responses can be obtained with pistonphones, calibration couplers, or electrostatic actuators. Free-field responses must be measured in an anechoic space, and must be measured in the fluid medium in which the transducer will be employed (see References b, c, d, e, f).

In general, because the free-field frequency response of a transducer is affected by its geometry, this response depends upon the direction of propagation of sound waves with respect to the transducer. Therefore, it is essential that the direction of sound incidence upon the transducer be specified. Typically, free-field frequency response determinations are made for normal sound incidence, grazing sound incidence and/or random sound incidence.

To establish directivity, a set of directivity characteristics can be reported which illustrate the sensitivity of the microphone for sound incident from all possible directions. This is normally done only on a qualification basis.

From such a test, it is possible to determine, for a low-range pressure transducer:

Frequency response (4.2.1.3)
Directivity (4.2.1.15)

In some cases, the mechanical resonant frequencies (4.2.2.1) and resonant-frequency amplifications (4.2.1.5) can also be observed.

5.5 Transducer Capacitance.

Measure the capacitance of the transducer and any integral cable by means of a capacitance bridge at 1000 Hz; measure the capacitance of the transducer cable at 1000 hertz.

NOTE: When very high frequency (fast-response) pressure measurements are to be made with the transducer, capacitance measurements at 100 000 Hz should be performed additionally.

5.6 Shunting Resistance.

Measure the resistance between the two electrical leads of the transducer with an electrometer type megohmeter or similar acceptable device, using a potential of 50 volts unless otherwise specified.

5.7 Insulation Resistance (for transducers isolated from case ground).

Measure the insulation resistance between all terminals or leads connected in parallel, and the case of the transducer, with a megohmeter or similar acceptable device using a potential of 50 volts applied for two minutes unless otherwise specified.

5.8 Transducer Cable (non-integral).

5.8.1 ,Length and Capacitance.

Measure the length of any non-integral cables supplied with the transducer. It is acceptable if within ± 6 mm (±1/4 inch) or ±2% of specified length, whichever is larger, unless otherwise specified. Measure the capacitance of the non-integral cable with a capacitance bridge at 1000 Hz and record. The cable tests shall be done at room temperature as specified in RP 37.1.

NOTE: When very high frequency (fast response) pressure measurements are to be made with the transducer, capacitance measurements at 100 000 Hz should be performed additionally.

6. QUALIFICATION TESTS

These tests are performed on representative transducers in addition to the individual acceptance tests and calibrations. One transducer shall be subjected to all applicable portions of the qualification test unless otherwise specified.

6.1 Transducer Seal Test (sealed transducer only).

Use a light mineral oil (or equivalent) with a viscosity between 175 and 190 centistokes in a transparent container such as a Pyrex beaker. Heat the oil to approximately 125°C or to the maximum temperature specified for the transducer. Remove detachable cables and connectors from the transducer. Immerse the transducer beneath the surface of the heated oil. Any stream of air bubbles released from within the transducer indicates leakage and constitutes failure. Dry the transducer without application of heat (compatible solvents may be used).

6.2 Cable Noise Test.

The mechanically-induced (triboelectric) output noise of a standard length of cable that is supplied with the transducer shall be specified as "less than _____ millivolts (peak-to-peak)" when tested as described in this paragraph (adapted from Reference h).

The instrumentation shall include a standard shielded capacitor (1000 picofarads) connected across the cable; a weight equal to the weight of 12 meters (40 ft) of the cable under test in two half-cylindrical shapes that are clamped or taped to the outer jacket of the cable; a pre-amplifier or cathode follower having a specified frequency response and input impedance; and an oscilloscope or oscillograph with the capability of providing a full-scale deflection from a one millivolt signal.

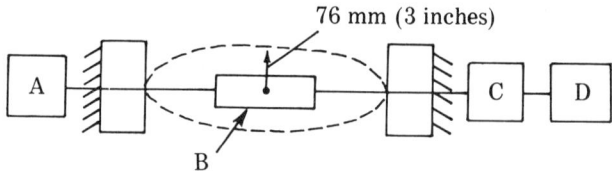

Cable Noise Test System

A   Shielded Capacitor - 1000 pfd

B   Cylindrical Weight [64 mm (2.5 in.) in length] Material - Brass

C   Cathode Follower or Pre-amplifier

D   Oscilloscope

Connect the cable electrically as shown in the above drawing. Clamp the cable between pieces of wood to the two anchor points, allowing a 76mm sag in the center of the cable. Clamp weight (B) to cable at center of anchored span. Raise the cable by the weight to a maximum height of three inches above its horizontal position and permit it to drop. Monitor the output noise of the cable on the oscilloscope. Repeat the test three times, and record the maximum output in millivolts peak-to-peak.

6.3 Ambient-Pressure Sensitivity Shift (4.2.1.12).

Certain low-range piezoelectric pressure transducers and piezoelectric sound-pressure transducers have different sensitivity levels at different ambient atmospheric pressures. In order to observe this effect, the test transducer and a secondary standard transducer (whose sensitivity as a function of ambient pressure is known) shall be inserted simultaneously or alternately into a pressure coupler calibration cavity. The static pressure in the cavity shall be established at a number of discrete points between 345 Pa (50 psi) and 0.69 Pa (0.1 psi) unless a different range is specified. At each point, the frequency of excitation of the cavity shall be varied up to the maximum frequency possible within the dimensions of the cavity while maintaining uniformity of instantaneous pressure within the cavity. The frequency response and sensitivity of the transducer shall be determined at each static pressure by comparison with the output of the secondary standard transducer.

In those cases where observations of frequency response vs. ambient pressure must be made to frequencies higher than those normally obtainable in pressure-coupler cavities, and electrostatic driving technique may sometimes be employed (see Reference f). This requires that the sensitive surface of the piezoelectric pressure transducer be electrically conductive, and that the polarizing voltage between this conductive surface and the electrostatic actuator be sufficiently small so that arcing will not occur at low ambient pressures. Under these circumstances, the transducer and electrostatic actuator combination may be inserted in an altitude chamber, and frequency response measurements made directly at various ambient pressures as described in Section 5.4.2. (The electrostatic actuator provides a pressure frequency response, as defined in Section 4.3).

6.4 Vibration Error (4.2.1.10).

Most piezoelectric pressure transducers and piezoelectric sound-pressure transducers have some electrical output due to vibration. In most test applications, it is essential that the output due to the vibration environment be insignificant when compared to the output due to the corresponding pressure environment.

Vibrate the transducer along each of the three mutually perpendicular axes, two of which lie in the mounting plane of the transducer. The acceleration amplitudes and frequencies applied shall be as specified (see Section 4.2.1.10).

Depending on the specification, either sinusoidal vibration or random vibration, or both, should be employed.

Observe the electrical output of the transducer under vibration and report the vibration error as a percent of full-scale output, or as equivalent Pa (psi) output per g.

For piezoelectric sound-pressure transducers, the equivalent SPL output due to vibration may be reported instead. This would consist of one or more curves of the equivalent SPL output of the transducer for 1 g rms (or 1 g peak, if the type of waveform is specified) excitation, plotted as a function of the frequency of excitation. Such curves can be computed from the vibration-induced output of the transducer by the application of the transducer sensitivity level calibration.

6.5 Linear-Acceleration Effects (4.2.1.11.f)

Pressure Transducers (dc-response only).

Place the transducer on a centrifuge, apply specified acceleration along specified axes, and measure changes in output. This shall establish the (linear) acceleration error of the transducer.

6.6 Thermal Sensitivity Shift at Maximum and Minimum Operating Temperature (4.2.1.7).

Determine the transducer capacitance at room temperature.

Mount the transducer in a calibration coupler at ambient temperature. Determine the output voltage of the transducer at room temperature and at a specified static pressure or at a pressure frequency within the usable range of the pressure coupler, as applicable. Then stabilize the transducer at the upper limit of the specified operating temperature range for at least 15 minutes. Measure the following:

1. Voltage output at maximum operating temperature with a known external capacitance.

2. Transducer capacitance at maximum operating temperature.

3. Transducer shunting resistance at maximum operating temperature.

NOTE 1: The standard which is used to establish the excitation amplitude must have known or reliably-calculable characteristics under all test conditions.

NOTE 2: Allowances may have to be made for changes in cable resistance and/or cable capacitance throughout the test temperature range.

NOTE 3: Do not leave the transducer in an open-circuit condition while the temperature is being changed; it should be short-circuited or connected to a preamplifier.

Allow the transducer to return again to room temperature and repeat the calibration tests of Section 5.2.2 or 5.2.3 (whichever is applicable).

Allow the transducer to stabilize at the lower limit of the specified operating temperature range and repeat the tests of Section 5.2.2 or 5.2.3 (whichever is applicable).

Allow the transducer to return again to room temperature and repeat the tests of Section 5.2.2 or 5.2.3 (whichever is applicable).

Compute the percentage change in transducer voltage sensitivity at the two temperature extremes compared to room temperature. Alternatively, compute the percentage change in transducer charge sensitivity at the operating temperature extremes compared with that at room temperature.

6.7 Temperature Gradient Error (4.2.1.8).

Procedure I: With the transducer initially at room temperature, immerse in a suitably non-conducting liquid whose temperature is at the upper limit of the operating temperature range of the transducer making sure all cavities are filled. Measure and record the maximum voltage which is generated by the transducer, and the time from the start of the transient at which this maximum voltage is reached. If the voltage reverses and reaches a peak of opposite polarity, record the amplitude and the time of that peak also. These voltages may be converted to equivalent Pa (psi) based upon the transducer voltage sensitivity with the specified cable.

Procedure II: With certain low-range pressure transducers and microphones, a qualitative test for transient temperature error may be performed as follows. Connect the transducer electrically as it would be employed for measurement purposes. Connect the output of the transducer to an oscilloscope. Approximately two feet in front of the sensitive surface of the transducer, discharge a clear No. 25 flash bulb. Observe and record the electrical output of the transducer as equivalent Pa (psi) (or dB SPL) peak.

6.8 Sensitivity Stability (4.2.1.11).

Repeat the voltage sensitivity or charge sensitivity tests for the transducer as described in 5.2.2 or 5.2.3, whichever is applicable. Then perform the following:

a. Expose the transducer to its maximum rated temperature for one hour. Return to room temperature by allowing to cool for 24 hours. Measure and record the new sensitivity.

In extreme cases for very high temperature use, the transducer may be exposed directly to the flame of a torch, and heat flux measured and recorded.

b. Subject the transducer at least ten times in succession to maximum rated pressure applied as rapidly as possible. Measure and record the new sensitivity levels 24 hours later.

c. Expose the transducer to its minimum rated temperature for one hour. Return to room temperature by allowing to stabilize for 24 hours. Measure and record the new sensitivity.

Note the maximum percentage change from the original voltage sensitivity or charge sensitivity during any portion of the above tests.

6.9 Burst Pressure (pressure transducers) (4.2.1.6)

Apply the rated burst pressure to the transducer the specified number of times and for the specified time durations. Observe safety precautions for high range transducers. Examine for mechanical damage.

7. SAMPLE DATA SHEETS

The test data sheets included in this Standard are intended to be used for the tests described in Sections 5 and 6. They consist of the following:

1. Test Summary

2. Individual Acceptance Test and Calibration

3. Piezoelectric Pressure Transducer Frequency Response Data (Transient Test and Sinusoidal Test — 2 sheets)

4. Piezoelectric Pressure Transducer Vibration Sensitivity Data

8. REFERENCES

a. "Acoustical Terminology" (S1.1-1960, R1971), American National Standards Institute.

b. "Method for the Calibration of Microphones" (S1.10-1966, R1971) American National Standards Institute.

c. "A Guide for the Dynamic Calibration of Pressure Transducers" (B88.1-1972) American National Standards Institute.

d. "Specification for Laboratory Standard Microphones" (S1.12-1967, R1972) American National Standards Institute.

e. "Calibration of Underwater Electro-Acoustic Transducers" (S1.20-1972), American National Standards Institute.

f. Beranek, L. L., Acoustic Measurements, John Wiley and Sons, New, 1949.

g. Schweppe, J. L., et al, "Methods for the Dynamic Calibration of Pressure Transducers", National Bureau of Standards Monograph 67, 12 December 1963.

h. Perls, T. A., "A Simple Objective Test for Cable Noise Due to Shock, Vibration or Transient Pressure." PB 121583, Office of Technical Services, U. S. Government Printing Office, 1955.

i. Gardner, M. F., and Barnes, J. L., Transients in Linear Systems. John Wiley and Sons, New York, 1942.

j. ANSI MC6.2-1975 (ISA 37.3) "Specifications and Tests for Strain-Gauge Pressure Transducers".

k. ANSI MC6.1-1975 (ISA 37.1) "Electrical Transducer Nomenclature and Terminology".

l. Hilten J. S., Lederer P. S. and Sethian J., "A Simple Hydraulic Sinusoidal Pressure Calibrator," National Bureau of Standards, Technical Note 720, April 1972.

TEST SUMMARY
PIEZOELECTRIC PRESSURE TRANSDUCER

Tests Started, Date _____
Tests Finished, Date _____

Report No.: _____   Linearity: _____   Test Type: _____

Vendor: _____   Vendor Part No.: _____   Serial No.: _____

Range: _____   Purchase Order No.: _____   Part No.: _____

TEST	Test Procedure	ISA S37.10	Check if Accep.	Failure Error	Failure Electr.	Failure Mechan.	Allowed Tolerances (± % FSO)
1 Visual Inspection		5.1					
2 Sensitivity Voltage		5.2					
3 Sensitivity Charge		5.2					
4 Sensitivity Level (Sound Press. Trans.)		5.2.2.1					
5 Proof Pressure		5.3					
6 Frequency Response		5.4					
7 Resonant Frequency		5.4					
8 Transducer Capacitance		5.5					
9 Shunting Resistance		5.6					
10 Insulation Resistance		5.7					
11 Transducer Cable (non-integral)		5.8	Length Capacit.				
12* Transducer Seal		6.1					
13* Cable Noise		6.2					
14* Ambient Pressure Sensitivity Shift		6.3					
15 Vibration Error		6.4					
16 Linear Acceleration Effect		6.5					
17 Thermal Sensitivity Shift		6.6					
18 Temperature Gradient Error		6.7					
19 Sensitivity Stability		6.8					
20* Burst Pressure		6.9					

Note: *Does not apply to all transducers under test.

PIEZOELECTRIC PRESSURE TRANSDUCER

Manufacturer _____

Part No. _____

Maximum Operating Pressure _____

Maximum Operating Temperature _____

┌─────────────────────────────────┐
│ INDIVIDUAL ACCEPTANCE TEST   │
│ AND CALIBRATION   │
└─────────────────────────────────┘

Sound-Pressure _____
Transducer

Pressure Transducer _____

Pressure (% FSO)	Pa(psi)	Output (Volt) Run 1	Run 2	Theo. Output	Dev. from Theo. Run 1	Run 2	Shift Run 1-2
0							
20							
0							
40							
0							
60							
0							
80							
0							
100							
0							
80							
0							
60							
0							
40							
0							
20							
0							

_____ Linearity ± % FSO

Zero Stability % FSO Over + _____ minutes after warm-up period of _____ minutes.

Thermal Calibration

Pressure Nom (% Range)	Act Pa(psi)	Output (Volt) Min __°C(°F)	Room __°C(°F)	Max __°C(°F)
0				
100				
Span				

Calibration at Room Conditions

```
D   |   +   +   +   +   +
E   |   +   +   +   +   +
V   |   +   +   +   +   +
I   |   +   +   +   +   +
A   |   +   +   +   +   +
T   |   +   +   +   +   +
I   |   +   +   +   +   +
O   |   20%  40%  60%  80%  FSO
N   |   +   +   +   +   +
    |   +   +   +   +   +
    |   +   +   +   +   +
    |   +   +   +   +   +
    |   +   +   +   +   +
    |   +   +   +   +   +
    |   +   +   +   +   +
```

Summary

CHARACTERISTIC	ALLOWED	ACTUAL	
Physical Dims.	╳	╳	
Pressure Conn.			
Electrical Conn.			
Mounting Dims.			
Identification			
Linearity			
Hysteresis			
Repeatability			
Proof Pressure Effect			
Ref. Pressure Effect			
Thermal Zero Shift			
Thermal Sens. Shift			

Rejected by _____

Approved by _____

Date _____

S37.10	SAMPLE DATA SHEET NO. 2	Sheet 1 of 2

Report No.	DYNAMIC RESPONSE TESTS OF PIEZOELECTRIC PRESSURE TRANSDUCER (Transient Test Method)	Purchase Order No.
Full Scale Range		Serial No.
Vendor & Part No.		Customer

1. Visual Inspection: before _____ after test _____

   Mechanical: _____ Finish: _____ Nameplate: _____ Connections: _____

2. Electrical: Shunting Resistance _____ Insulation Resistance: _____ MΩ

   Capacitance _____ picofarads   at _____ Volts

3. Ambient Conditions: Temperature: °C(°F); _____ Pressure: _____ mm Hg; Rel. Humidity _____ %

4. Dynamic Response: Pressure Transducer Amplifier Settings _____

   Function Generator: _____ Shock Tube, Dry Air: _____

   Tube Mounting Location: End: _____ Side: _____

Shock No.	Initial Press.		Shock Velocity	Step Pressure	Resonances, Hz	Ringing Period	Time Constant
	Hi	Low					

BY: _____   APPROVED: _____

Vendor's Part No.	Test Facility	Customer's Part No.
Vendor	PIEZOELECTRIC PRESSURE TRANSDUCER	Serial No.
Report No.	DYNAMIC RESPONSE TESTS (SINUSOIDAL METHOD)	Customer
Type of Test		Range _____ to _____ Pa(psi) _____

Ambient Conditions: Temperature _____° _____ _____   Pressure: _____ mm Hg; Humidity _____%

Dynamic Response:

     Excitation (Volts or mA) _____ ;   Transducer Load _____ ohms

     Sinusoidal Generator _____ ;   Test Fluid _____

     Mounting Configuration _____

_____

     Test Temperature _____° _____ _____ Quiescent Static Pressure _____ Pa(psi) _____

     Sinusoidal Pressure _____ Pa (psi) peak; Port Excited _____

Sensitivity Level in Decibels re Sensitivity Level at Zero Hz = $20 \log_{10} S(4)/S(0)$.

PHASE SHIFT IN DEGREES

Frequency, hertz

PIEZOELECTRIC PRESSURE TRANSDUCER VIBRATION SENSITIVITY DATA

Vendor's Part No.   Customer's Part No.

Vendor   Serial No.

Report No.   Customer

Type of Test   Range
                                          _____ to _____ Pa (psi) _____

X
Y
Z

Diaphragm   Direction of Vibration

Measured Output in Units of (check)   _____ mV _____ pC, using peak* Mag. _____
                                                    rms * Mag. _____

	*Acceleration (g)	*Measured Output	Equivalent Output			
			Output/g	Pa(psi)/g	dB@lg pk	dB@lg rms
Z Direction						
Frequency ___Hz	___	___	___	___	___	___
Frequency ___Hz	___	___	___	___	___	___
Frequency ___Hz	___	___	___	___	___	___
Random___	___	___	___	___	___	___
X Direction						
Frequency ___Hz	___	___	___	___	___	___
Frequency ___Hz	___	___	___	___	___	___
Frequency ___Hz	___	___	___	___	___	___
Random___	___	___	___	___	___	___
Y Direction						
Frequency ___Hz	___	___	___	___	___	___
Frequency ___Hz	___	___	___	___	___	___
Frequency ___Hz	___	___	___	___	___	___
Random___	___	___	___	___	___	___

*Note: Either rms or peak magnitudes may be employed if consistency is maintained including the measurement
of pressure and acceleration.

Outputs in pC/g or MV/g** are converted to outputs in Pa/g (psi/g) from knowledge of a

previous calibration in terms of mV/Pa (mV/psi)** or pC/Pa (pC/psi). Record data here if known:

Acoustic Sensitivity Level*   **Calibration Source**

—————— mV/Pa (mV/psi) as recorded by _____

—————— pC/Pa (pC/psi) as recorded by _____

Outputs in pC/g or mV/g are converted** to equivalent SPL for one g acceleration

by use of a previous calibration in terms of acoustic output at a reference SPL and

utilizing the relationship that the equivalent SPL at one g is*:

$$(\text{reference SPL in dB}) = 20 \log_{10} \frac{\text{output at reference SPL}}{\text{acceleration}}$$

Record the reference SPL here if known:

Output*   Calibration Source

_____ at _____ dB reference, as recorded by _____

*Note: Either rms or peak magnitudes may be employed *if consistency is maintained* for all quantities.

**Note: If mV/g and mV/Pa(psi) are being used, the cable capacitance should be the same in both the calibration test and the vibration test.

INSTRUMENT SOCIETY of AMERICA
Research Triangle Park, North Carolina

ANSI/ISA-S37.12-1977
(R 1982)
Approved December 14, 1982

American National Standard

Specifications and Tests for Potentiometric Displacement Transducers

Instrument Society of America

ISBN 0-87664-359-4

ISA-S37.12 Specifications and Tests for Potentiometers
Displacement Transducers

INSTRUMENT SOCIETY of AMERICA
67 Alexander Drive
P.O. Box 12277
Research Triangle Park, North Carolina 27709

PREFACE

This Preface is included for informational purposes and is not a part of ISA-S37.12.

This Standard has been prepared as a part of the service of the Instrument Society of America toward a goal of uniformity in the field of instrumentation. To be of real value, this document should not be static, but should be subject to periodic review. Toward this end, the Society welcomes all comments and critisms, and asks that they be addressed to the Secretary, Standards and Practices Board, Instrument Society of America, 67 Alexander Drive, P.O. Box 12277, Research Triangle Park, NC 27709, Telephone (919) 549-8411.

The ISA Standards and Practices Department is aware of the growing need for attention to the metric system of units in general, and the International System of Units (SI) in particular, in the preparation of instrumentation standards. The Department is further aware of the benefits to USA users of ISA Standards of incorporating suitable references to the SI (and the metric system) in their business and professional dealings with other countries. Towards this end, this Department will endeavor to introduce SI - acceptable metric units in all new and revised standards to the greatest extent possible. The Metric Practice Guide, which has been published by the American Society for Testing and Materials as ANSI designation Z210.1 (ASTM E380-76, IEEE Std. 268-1975), and future revisions, will be the reference guide for definitions, symbols, abbreviations, and conversion factors.

It is the policy of the Instrument Society of America to encourage and welcome the participation of all concerned individuals and interests in the development of ISA Standards. Participation in the ISA standards making process by an individual in no way constitutes endorsement by the employer of that individual of the Instrument Society of America or any of the standards which ISA develops.

This Standard is intended as a guide for technical personnel at user facilities as well as by manufacturers' technical and sales personnel whose duties specifying, testing or showing performance characteristics of strain-gage linear Potentiometric Displacement Transducers. By basing users' specifications as well as technical advertising and reference literature on this Standard, or by referencing portions thereof, as applicable, a clear understanding of the users' needs or of the transducers' performance capabilities, and of the methods used for evaluating or proving performance, will be provided. Adhering to the specification outline, terminology and procedures shown will not only result in simple, but also complete specifications; it will also reduce design time, procurement lead time, and labor, as well as material costs. Of major importance will be the reduction of qualification tests resulting from use of a commonly accepted test procedure and uniform data presentation.

The development of this Standard was initiated as the result of a survey conducted in December 1960. A total of 240 questionnaires was sent out to transducer users and manufacturers. A strong majority indicated in their replies a need for transducer standardization. As potentiometric displacement transducers were one of the types shown to be most in need of standardization, a project subcommittee, SP37.12. Potentiometric Displacement Transducers, was formed under the cognizance of Committee SP37, Transducers for Aerospace Testing, and a standard was drafted and reviewed extensively, and revised in 1976. The reviewers were selected from a broad cross-section of all industries and sciences in which transducers are applied for measuring purposes.

The following individuals served as members of the 1977 SP37.12 committee:

NAME	COMPANY
Robert D. Bronson (Chairman)	Lockheed Electronics Company, Inc.
Milton Brown	Bourns, Incorporated
Joseph Kauker	Binary Controls
Keith C. Posey, PE	URS/Forest & Cotton
Richard R. Richard	National Aeronautics & Space Administration
Donald Veatch	National Aeronautics & Space Administration

The following individuals served as members of the 1982 SP37 committee:

NAME	COMPANY
J. W. Mock, Chairman	Measurement Services
P. Bliss	Consultant
W. Brinkschulte	Genisco Technology Corporation
C. Kutelis	Revere Corporation of America
R. W. Lally	PCB Piezotronics
H. E. Lockery	Hottinger Baldwin Measurements, Inc.
E. W. Malone	Boeing Aerospace Co.
T. B. Miller	Lawrence Livermore Lab
K. A. Parlee	United Engineers & Constructors, Inc.
H. Pitt	
J. Powell	Rosemount Engineering
R. C. Strahm	EG & G Idaho
R. M. Whittier	ENDEVCO
N. Wilde	EG & G Idaho
E. Wong	Bell & Howell

This Standard was approved for publication by the ISA Standards and Practices Board in October 1982.

NAME	COMPANY
T. J. Harrison, Chairman	IBM Corporation
P. Bliss	Consultant
W. Calder	The Foxboro Company
N. Conger	Continental Oil Co.
B. Feikle	Bailey Controls Co.
R. T. Jones	Philadelphia Electric Co.
R. Keller	Boeing Company
O. P. Lovett, Jr.	Isis Corp.
E. C. Magison	Honeywell, Inc.
A. P. McCauley	Diamond Shamrock Corp.
J. W. Mock	Measurement Services
E. M. Nesvig	ERDCO Engineering Corp.
G. Platt	Bechtel Power Corp.
R. Prescott	Moore Products Company
W. C. Weidman	Gilbert Associates
K. A. Whitman	Allied Chemical Corp.
J. R. Williams	Stearns-Roger, Inc.
B. A. Christensen*	
L. N. Combs*	
R. L. Galley*	
R. G. Marvin*	
W. B. Miller*	Moore Products Company
R. L. Nickens*	

*Director Emeritus

CONTENTS

1. PURPOSE

This Standard establishes the following for potentiometric displacement transducers.

1.1 Uniform minimum specifications for design and performance characteristics.

1.2 Uniform acceptance and qualification test methods, including calibration techniques.

1.3 Uniform presentation of minimum test data.

1.4 A drawing symbol for use in electrical schematics.

2. SCOPE

2.1 This Standard covers potentiometric displacement transducers, primarily those used in measuring systems.

Note:
1. These specifications are not intended to cover transducers used in hazardous locations as specified in the National Electrical Code.

2. Transducers for use in nuclear power plants must conform to additional U.S. Nuclear Regulatory Commission Requirements not specifically called out in this Standard.

2.2 Included among the specific versions of potentiometric displacement transducers to which this Standard is applicable are the following:

Angular Displacement Transducers
Linear Displacement Transducers

2.3 Terminology used in this document is defined in ISA Standard 37.1. Additional terms considered applicable to potentiometric displacement transducers are defined in 4.3 of this document. An asterisk appears before those terms defined in S37.1. The terms defined in section 4.3 appear in italics.

3. DRAWING SYMBOL

3.1 The drawing symbol for a potentiometric displacement transducer is a square with an added equilateral triangle, the base of which is the left side of the square.

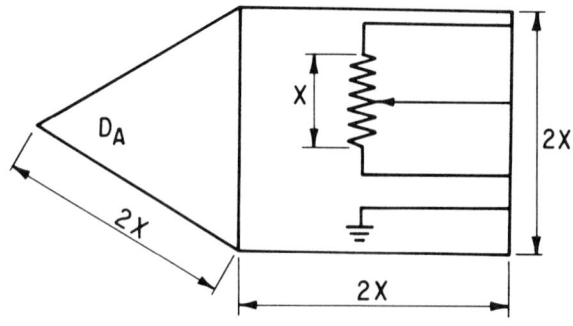

FIGURE 3.1

Subscripts
A = Angular
L = Linear

The letter "D" in the triangle designates "displacement" and the subscripts denote the second modifier. The illustration shows an angular displacement transducer, as symbolized by "D_A."

3.2 The potentiometer is symbolized by a variable resistance of length x. The lines from it and to ground represent the electrical leads or terminations.

Note:
1. This symbol is not ANSI approved at this time. It has been submitted to the ANSI Y32 Committee on Graphic Symbols for their consideration and approval.
2. For instrumentation systems use refer to ISA S5.1 "Instrumentation Symbols and Identification."

4. SPECIFICATION CHARACTERISTICS

4.1 Design Characteristics

4.1.1 Required Mechanical Design Characteristics

The following mechanical design characteristics shall be listed.

4.1.1.1 Type of Displacement Sensed

*Angular Displacement
*Linear Displacement

4.1.1.2 Configuration and Dimensions

The outline drawing shall show the configuration with dimensions in millimetres (inches). Unless electrical connections are specified (Reference 4.1.3.4), the outline shall include limiting maximum dimensions for these connections.

4.1.1.3 Mountings and Mounting Dimensions

The outline drawing shall indicate the method of mounting with hole size, centers, and other pertinent dimensions in millimetres (inches) including thread specifications for threaded holes or shafts if used.

4.1.1.4 Displacement Connection

The displacement connection(s) shall be indicated on the outline drawing. For threaded fittings, specify: Applicable military or industry standard or nominal diameter in millimetres (inches or ANSI size number) and the thread pitch in millimetres per one thread (threads per inch), and other details as necessary to define the shape of the threads.

4.1.1.5 Mounting Effects

The maximum mounting force or torque shall be specified if it will tend to affect transducer performance (Reference 4.2).

4.1.1.6 Mass

The mass of the transducer shall be specified in grams (ounces).

4.1.1.7 Case Sealing

If case sealing is necessary, the mechanism and materials used for sealing should be described. The same requirement applies to the electrical connector. The resistance of the sealing materials to cleaning solvents and commonly encountered fluids should be stated.

4.1.1.8 Dissimilar Metals

Dissimilar metals, which could cause generation of significant electrical potential, migration of metal or corrosion, shall not be used in contact with each other.

4.1.1.9 Identification

The following characteristics shall be permanently inscribed on the outside of the transducer case or on a suitable nameplate permanently attached to the case.

Nomenclature of transducer (according to ISA-S37.1, Section 3).
Name of Manufacturer, Part number (to reflect one controlled configuration), and Serial Number.
*Range in radians/millimetres (degrees/inches) and designation of type of displacement (see 4.1.1.1).
Maximum *excitation.
*Transduction Element resistance (Potentiometric Element).
Identification of Electrical Connections.

Inscription of the following characteristics is optional:

Customer Specification or Part Number or both. Type of Electrical Connector and Mating Connector (if applicable).
Operating Temperature Range.

4.1.2 Supplemental Mechanical Design Characteristics

Listing of the following mechanical design characteristics is optional:

4.1.2.1 Case Material

Where applicable, state the surrounding environmental condition requirement or compatibility.

4.1.2.2 Number of Potentiometric Elements or Taps

When more than one potentiometric transducer element or a tapped element is required, they shall be specified.

4.1.2.3 Maximum and Minimum Temperatures

The maximum and minimum temperatures of environments which can be applied to the transducer, and which will not cause permanent calibration shift, shall be listed.

Note: Exposure time shall be specified, if relevant.

4.1.3 Required Electrical Design Characteristics

The following electrical design characteristics shall be listed. All are applicable at *Room Conditions.

4.1.3.1 *Excitation

Expressed as "_____ volts (milliamperes) dc" or "_____ volts (milliamperes) ac rms at _____ hertz."

4.1.3.2 Maximum Excitation

Expressed as "_____ volts (milliamperes) dc" or "_____ volts (milliamperes) ac rms at _____ hertz, within the operating temperature range.

4.1.3.3 Resistance of *Transduction Element

Expressed as "_____ ± _____ ohms."

4.1.3.4 Electrical Connections

Whether the electrical termination is by means of a connector or a cable, the pin designations or wire color code shall conform to the following transducer wiring standard.

Connector Pin	Wire Color	Mechanical Function	Electrical Function
1 or A	red	Maximum displacement	"Positive" excitation Terminal (if dc)
2 or B	green	Variable displacement	Wiper arm terminal output
3 or C	black	Minimum displacement	"Negative" excitation (if dc), or Common, output return
4 or D	Shield (optional)	(none)	Case ground

Figure 4.1

Notes:

1. The Transduction element(s) shall be arranged to produce increasing "Positive" voltage output with increasing *wiper* displacement in direction (clockwise motion, viewed from shaft-end, for angular displacement transducers, and extending motion by linear displacement transducers shown by arrow).

2. For bidirectional range transducers, the arrow indicates increasingly positive displacement values.

3. Current flow shall be restricted to electrical elements and shall not be permitted in mechanical elements, such as shaft bushings.

4. CAUTION: A short circuit on the output or an inadvertent connection of the excitation voltage to the *wiper* may pass excessive current through the *wiper* and the Transducer Element resistance, and cause a catastrophic failure of the transducer.

5. Output Short-circuit protection (optional):

In a specialized or dedicated transducer application, output short-circuit protection may be obtained by means of an optional output current limiting resistor. The resistor is placed in series with the potentiometer wiper, preferably integral to the transducer, between the *wiper* and the output terminal (connector pin "2" or "B" or wire color "green"). The resistance value is selected such that, with the output terminals shorted, the *wiper* current is limited to a proper maximum value. For example: If the excitation voltage will always be 5 volts dc and the *wiper* current should not exceed 10 milliamperes, then the resistance value should be 500 ohms. The same resistor also protects the transducer from damage due to inadvertent connection of the 5 volt excitation to the output terminals.

4.1.3.5 *Insulation Resistance

Expressed as "_____ megohms at _____ volts dc between all transduction terminals in parallel and the transducer case at a temperature of _____°C and 90% relative humidity."

4.1.3.6 *Dielectric Withstand Voltage*

Expressed as "Capable of withstanding _____ volts ac-rms at _____ hertz, at a temperature of _____°C and 90% relative humidity for _____ minutes."

4.1.3.7 *Load Impedance

Expressed as "_____ ohms" (See 4.2.19).

Note: Although load impedance (the impedance presented to the output terminals of the transducer by the associated external circuitry) is not a transducer but a system characteristic, it affects the linearity defined in Section 4.2.4 and must be specified in order to define loading error. A single, close-tolerance value of load impedance shall be specified for use during all tests where not otherwise noted. To minimize loading error, the load impedance to transduction element resistance ratio should be as large as practicable.

4.2 Performance Characteristics

The performance characteristics of the potentiometric displacement transducers should be tabulated in the order shown. Unless otherwise specified, they apply at Room Conditions as defined in ISA S37.1. Characteristics are usually referred to the output and expressed as "% VR," i.e. "percent *Voltage Ratio."

4.2.1 *Range

Expressed as "_____ to _____ radians/millimetres (degrees/inches)."

4.2.2 *End Points

Expressed as "_____ % ± _____ % and _____ % ± _____ % VR."
End Points shall be omitted where adequately defined using Error Band specifications.

4.2.3 *Full Scale Output

Expressed as "_____ % ± _____ % VR."

Note: Full scale output shall be omitted where adequately defined using End Points or Error Band specifications.

4.2.4 *Linearity

Expressed as "_____ linearity shall be within ± _____ % VR with specified load impedance."

Note: The linearity modifier shall be one of those defined in ISA S37.1: namely end point, independent, least squares, terminal or theoretical slope. Linearity values are dependent on load impedance.

4.2.5 *Hysteresis

Expressed as "_____ % VR."

Alternately, the concepts of 4.2.4 and 4.2.5 may be replaced by 4.2.6.

4.2.6 Combined *Hysteresis and *Linearity

Expressed as "combined hysteresis and linearity within ± _____ % VR."

Note: The linearity modifier shall be stated.

4.2.7 *Repeatability

Expressed as "within _____ % VR over a period of _____ hours."

Alternately, the concepts of 4.2.2, 4.2.3, 4.2.4, 4.2.5 and 4.2.7 may be replaced by 4.2.8.

4.2.8 *Static Error Band

Expressed as "± _____ % VR as referred to
_____(curve)_____."

Note: The curve shall be stated as:
1. "End Point Line" — a straight line between end points.
2. "Best Straight Line" — a line midway between the two parallel lines closest together and enclosing all output versus measured values.
3. "Least Square Line" — a straight line for which the sum of the squares of the residuals is minimized.
4. "Terminal Line" — a straight line between 0 and 100% of both measurand and output.
5. "Theoretical Slope" — a straight line between the theoretical end points.
6. Other curves shall be defined if specified, e.g., mean-output curve.

4.2.9 *Resolution (See also 4.3.1)

Expressed as "average resolution within _____ % VR and maximum resolution within _____ % VR."

4.2.10 Frequency Response

Expressed as "within ± _____ % from zero to _____ hertz at a peak-to-peak amplitude of _____ % VR."

4.2.11 *Operating Temperature Range*

Expressed as "from _____ °C to _____ °C."

4.2.12 *Temperature Error

Expressed as "within _____ % VR per _____ °C."
Or "within _____ % VR over the operating temperature range."
Alternately, 4.2.12 may be replaced by 4.2.13.

4.2.13 *Temperature Error Band

Expressed as "within ± _____ % VR from the reference curve established for the Static Error Band, and over the operating temperature range."

4.2.14 *Acceleration Error

Expressed as "within _____ % VR per g_n along _____ axis at steady acceleration levels to _____ g_n."

Note: The error shall be listed either for each of the three axes or for the axis with the largest error, i.e., most sensitive axis.

Alternately 4.2.14 may be replaced by 4.2.15.

4.2.15 *Acceleration Error Band

Expressed as "within ± _____ % VR from the reference curve established for the Static Error Band for steady accelerations up to _____ g_n along _____ axis."

Note: The error band shall be listed either for each of the three axes or for the axis with the largest error, i.e., most sensitive axis.

4.2.16 *Vibration Error

Expressed as "within ± _____ % VR along _____ axis over the specified vibration program." Signal "dropout" or discontinuities shall be noted.

Note: The error shall be listed either for each of the three axes or for the axis with the largest error, i.e., most sensitive axis; and the program shall be detailed, preferably by a graph.

Alternately 4.2.16 may be replaced by 4.2.17.

4.2.17 *Vibration Error Bands,

Expressed as "within ± _____ % VR along _____ axis from the reference curve established for the Static Error Band over the specified vibration program."

Note: The error band shall be listed either for each of the three axes or for the axis with the largest error, i.e., most sensitive axis; and the program shall be detailed, preferably by a graph.

4.2.18 *Mounting Error*

Expressed as "within ± _____ % VR" or "within the Static Error Band."

4.2.19 *Loading Error

Expressed as "within ± _____ % VR" or "within the Static Error Band at a load impedance of _____ ohms."

4.2.20 *Cycling Life

Expressed as "_____ cycles at one fourth of the designated maximum operating frequency of the transducers."

4.2.21 Other Environmental Conditions

Pertinent environmental conditions which shall not change transducer performance beyond certain limits shall be included along with the limits beyond which the transducer performance shall not change.

Examples are:
*Ambient Pressure
  Shock-Triaxial

High Level Acoustic Excitation
Humidity
Salt Atmosphere
Nuclear Radiation
Magnetic Fields
Sand and Dust
Total Immersion (and in what medium)
Solar (or other) Heat Radiation
Temperature Shock
Electromagnetic Interference, generation or
  susceptibility
Vibration Acceleration — Triaxial

4.2.22 *Storage Life

Expressed as "_____ months (years) without
changing performance characteristics beyond their
specified tolerances."

Note: Environmental storage conditions shall be described in detail. Where performance characteristics require additional tolerances over the storage life they
shall be specified.

4.3 Additional Terminology

4.3.1 *Average Resolution*

The reciprocal of the total number of output steps over
the unit range multiplied by 100 and expressed in % VR.

4.3.2 *Maximum Excitation*

The maximum allowable voltage (current) applied to the
potentiometric element at Room Conditions while maintaining all other performance characteristics within their
limits. (Note: The excitation value is particularly associated with temperature.)

4.3.3 *Mounting Effects*

The effects (errors) introduced into transducer performance during installation caused by fastening of the unit
or its mounting hardware or by irregularities of the surface on (or to) which the transducer is mounted.

4.3.4 *Operating Temperature Range*

The range in extremes of ambient temperature within
which the transducer must perform to the requirements
of the "Temperature Error" or "Temperature Error
Band." (See paragraphs 4.2.12 and 4.2.13, respectively.)

4.3.5 *Potentiometric Element*

The resistive part of the transduction element upon
which the *wiper (movable contact)* slides and across
which excitation is applied. It may be constructed of a
continuous resistance or of small diameter wire wound
on a form (mandrel).

4.3.6 *Wiper (Movable Contact)*

That portion of the potentiometric assembly which
slides on the resistance element. It is connected to a
terminal and provides an electrical output as a function
of the shaft position relative to the body.

4.3.7 *Dielectric Withstand Voltage*

The ability of insulated portions of the transducer to
withstand a specified overvoltage for a specified time
without arcing or conduction above a specified current
value across the insulation.

4.3.8 *Tap*

A connection to a potentiometric element along its
length, frequently at the element's center for use in
providing bidirectional output.

4.3.9 *Worst Resolution*

The magnitude of the largest of all output steps over the
unit range expressed as percentage voltage ratio (%VR).

4.3.10 *Shaft*

The mechanical input element of the transducer.

4.3.11 *Shaft Position*

An indication of the position of the wiper relative to a
reference point.

4.3.12 *Noise (refer to Noise Test Circuit)*

Noise is any spurious variation in the electrical output,
not present in the input. *Noise* is defined quantitatively
in terms of an equivalent parasitic transient resistance
appearing between the *wiper* and the resistance element
while the input shaft is being moved.

The Equivalent Noise Resistance is established independently of the functional characteristics, in the *Noise* Test
Circuit. The *wiper* is required to be excited by a specified dc constant current source. The *Noise* Test Circuit
output measuring system is an oscilloscope with defined
frequency bandwidth or time constant. The magnitude
of the Equivalent Noise Resistance is measured as ohms
variation while the input shaft is moved at a specified
speed, and observed as peak-to-peak deflection on the
oscilloscope.

For example: if the constant current is one milliampere
dc and the oscilloscope deflection is 100 millivolts peak-
to-peak, then the Equivalent Noise Resistance is 100
ohms.

Note: *Noise* may be characterized as generally reproducible, exhibited as a local nonlinearity, or it may be

the classical sporadic type. Manufacturing cleanliness and improved quality control on processing may significantly reduce *noise* problems.

4.4 Supplemental Performance Characteristics

Note: In critical applications these characteristics may be of potential use in the specifications of potentiometric displacement transducers. These characteristics are typically "Qualification Test" items; however, specific test procedures are not described in this standard.

4.4.1 End Stops

The physical limits of motion provided by the transducer design which define the Total Travel of the Shaft.

Note: In order to protect the transducer from overload, the End Stops will normally be placed beyond the maximum requirements for (desired) Mechanical Travel.

4.4.2 End Point, *Shaft Position*

The shaft positions immediately before the first and after the last measureable change(s) in Output Ratio, after wiper continuity has been established, as the shaft moves in a specified direction.

4.4.3 End Point, Theoretical, *Shaft Position*

The shaft positions corresponding to the ends of the Theoretical Electrical Travel as determined from the Index Point.

The Index Point is a point of reference, fixing the relationship between a specified shaft position and the output %VR. For example; In an ideal symmetrical range bidirectional potentiometric displacement transducer, zero position input should produce 50% VR output.

4.4.4 End Resistance, *Shaft Position*

The resistance measured between the wiper terminal and end terminal with the shaft positioned at the corresponding End Point.

4.4.5 Extended Position, Linear Potentiometer

The condition when the Shaft is moved out from the transducer to either the End Stop or end of Mechanical or Electrical Travel as specified.

4.4.6 Lateral Runout, Angular Potentiometer

The perpendicularity of the mounting surface with respect to the rotational axis of the shaft, measured on the mounting surface at a specified distance from the outside edge of the mounting surface. The shaft is held fixed and the body of the potentiometer is rotated with specified loads applied radially and axially to the body of the transducer.

4.4.7 Operating Force
Non-Springloaded Designs
Breakout Force/Torque

The maximum force required to initially move the shaft due to the effects of static friction of the parts, especially in "O-ring" sealed designs. The attitude of the line of action must be specified. Angular position transducer breakout torque is defined similarly.

4.4.8 Operating Force
Non-Spring Loaded Designs
Dynamic Force/Torque

The force/torque required to continuously move the shaft after the first motion has occurred. The attitude of the line of action must be specified.

4.4.9 Springloaded Designs
Initial Force/Torque

The force/torque required to move the shaft from its normal position of rest includes the force/torque to overcome the spring forces plus the Starting Force/torque. The attitude of the line of action (direction of rotation) is specified.

4.4.10 Springloaded Designs
Final Force/Torque

The force/torque required to hold the shaft at the opposite end of the Stroke (extreme angular position of the shaft). This force/torque is measured and the attitude of the line of action (direction of rotation) must be specified.

4.4.11 Output Correlation

In a transducer which provides multiple outputs by the addition of independent resistance elements giving simultaneous reading with a common shaft motion, it is often necessary to relate the individual outputs to each other. This shall be expressed in ± % output (full scale) at a specific output reading, or number of readings, or at any specified position of the shaft measured from a given datum.

4.4.12 Overload, Shaft

The amount of force (torque) to which the Shaft may be subjected at the extremities of the Mechanical Travel (i.e., when reaching the End Stops) without damage to the transducer and without causing degradation of the specified performance characteristics.

4.4.13 Pilot Diameter Runout

The eccentricity of the pilot diameter with respect to the rotational axis of the shaft, measured on the pilot diameter. The shaft is held fixed and the body of the transducer is rotated with a specified load applied radially to the body of the transducer. This distance is measured from a specified datum.

4.4.14 Shaft Misalignment, Linear Position Potentiometer

The amount of freedom allowable or desirable of the Shaft in relation to the body of transducer when the body is held rigidly.

4.4.15 Misalignment, Body

The amount the shaft is allowed or required to move at a specified retracted position in a conical motion. This shall be measured at a specified position on the shaft.

4.4.16 Misalignment, Axial

The amount the shaft is allowed or required to deflect during all or any specified portion of the Total Travel and move in a line of parallel to the normal line of action.

4.4.17 Shaft Rotation, Linear Position Potentiometer

To be specified in rotational freedom in a direction normal to the axis of motion in radians (degrees). The definition of permissible output change, if any, in percent of Full Scale output should be included.

4.4.18 Shaft End Play, Angular Position

The total axial excursion of the shaft, measured at the end of the shaft with a specified axial load supplied alternately in opposite directions.

4.4.19 Shaft Radial Play, Angular Position

The total radial excursion of the shaft, measured at a specified distance from the front surface of the unit. A specified radial load is applied alternately in opposite directions at a specified point.

4.4.20 Shaft Runout

The eccentricity of the shaft diameter with respect to the rotational axis of the shaft, measured at a specified distance from the end of the shaft. The body of the potentiometer is held and the shaft is rotated with a specified load applied radially to the shaft.

4.4.21 Side Load

That force applied to any portion of the transducer, in any direction, which will result in bending loads and which will affect the Operating Force or Life Cycling requirements expressed in force units and direction of loading.

4.4.22 Static Stop Strength

The maximum static load that can be applied to the shaft at each mechanical stop for a specified period of time without permanent change of the stop positions greater than specified.

4.4.23 Dynamic Stop Strength

The inertial load, at a specified shaft velocity and a specified number of impacts, that can be applied to the shaft at each stop without a permanent change of the stop position greater than specified.

4.4.24 Total/Mechanical Travel, Linear Position Transducer

The physical distance the shaft may be moved from one stop to the other. This distance, that includes the Electrical Travel and the Over-Travel, may apply to either one or both ends of the electrical travel. It may be located dimensionally from some convenient output datum; e.g., ± _____ millimetres (inches) from the _____ %VR output.

4.4.25 Total/Mechanical Travel, Angular Position Transducer

The physical angle shaft may be rotated, from one stop to the other. This arc includes the Electrical Travel and the Over-Travel. It may be located dimensionally from some convenient output datum; e.g., ± _____ radians (degrees) from the _____ %VR output.

Note: Certain angular position transducers may be designed for continuous rotation.

4.4.26 Electrical Travel

That portion of the Mechanical Travel during which an output change occurs.

4.4.27 Overtravel

The differences between the required Electrical Travel and Mechanical Travel which the shaft may be moved (or rotated) beyond either or both ends of the Electrical Travel and during which no electrical·output change occurs.

TABLE 1 — TABULATED CHARACTERISTICS VERSUS TEST REQUIREMENTS

This table is intended for use as quick reference for design and performance characteristics and tests of their proper verification as contained in this Standard.

Characteristics	Paragraph	Required	Supple-mental	Individual Acceptance Test	Qualifi-cation Test
Type of Displacement Sensed	4.1.1.1	X		No Test	
Configuration, Dimensions, Mounting Displacement Connection	4.1.1.2 through 4.1.1.4	X		5.2.1	
Mounting Effects	4.1.1.5	X			6.4
Mass	4.1.1.6	X			6.3
Case Sealing	4.1.1.7	X			5.2.1
Dissimilar Metals	4.1.1.8			5.2.1	
Identification	4.1.1.9	X		5.2.1	
Case Material	4.1.2.1		X		
Number of *Potentiometeric Elements* or *Taps*	4.1.2.2		X	5.2.2 through 5.2.6	
Maximum and Minimum Temperatures	4.1.2.3		X		Special Test
Excitation	4.1.3.1	X		5.2.6	
Maximum Excitation	4.1.3.2	X			Special Test
Resistance of Transduction Element	4.1.3.3	X		5.2.2	
Electrical Connections	4.1.3.4	X		5.2.2	
Insulation Resistance	4.1.3.5	X		5.2.3	
Dielectric Withstand Voltage	4.1.3.6	X		5.2.4	
Load Impedance	4.1.3.7	X		5.2.5 (partially)	6.5
Range	4.2.1	X		5.2.6	
End Points	4.2.2	X		5.2.6	
Full Scale Output	4.2.3	X		5.2.6	
Linearity	4.2.4	X		5.2.5	
Hysteresis	4.2.5	X		5.2.6	
Combined Hysteresis and Linearity	4.2.6	X		5.2.6	
Repeatability	4.2.7	X		5.2.6	
Static Error Band	4.2.8	X		5.2.6	
Resolution	4.2.9	X			6.2
Frequency Response	4.2.10				6.7
Operating Temperature Range	4.2.11	X			6.8
Temperature Error	4.2.12	X			6.8
Temperature Error Band	4.2.13	X			6.8
Acceleration Error	4.2.14	X			6.9
Acceleration Error Band	4.2.15	X			6.9
Vibration Error	4.2.16	X			6.10
Vibration Error Bands	4.2.17	X			6.10
Mounting Error	4.2.18				6.4
Loading Error	4.2.19				6.5
Cycling Life	4.2.20	X			6.12
Other Environmental Conditions	4.2.21	X			6.11
Storage Life	4.2.22				Special Test (accelerated)
Noise	4.3.12				6.6

4.4.28 Wiper Bounce

The phenomenon of intermittent separation between the wiper and the resistive element may be caused by foreign matter, vibration, or the speed of the wiper motion. It may be measured by the value of the ratio of contact-off to contact-on time, when (a) the transducer is subjected to vibration over a specified range of frequencies and amplitudes, in the direction most likely to promote contact separation; or (b) the wiper is moved at all speeds to a specified maximum value. A convenient way to achieve this is to move the input shaft sinusoidally at specified low frequencies, with the peak-to-peak displacement approaching the range of the transducer. Steady state angular rotation may be used for Angular Displacement Transducers which do not include mechanical stops.

5. INDIVIDUAL ACCEPTANCE TEST AND CALIBRATIONS

5.1 Basic Equipment Necessary to Perform Individual Acceptance Tests and Calibrations of Potentiometric Displacement Transducers.

The basic equipment for acceptance tests and calibrations consists of a source of displacement, a source of electrical excitation for the potentiometer, and a device which measures the electrical output ratio of the transducer directly or as a ratio to excitation input (VR). The combined errors or uncertainties of the calibration system consisting of these three components should be less than one-fifth of the characteristic under evaluations. The traceability to national standards for this measuring system shall be well known.

5.1.1 Displacement

The displacement source may be either continuously variable over the range of the instrument, or may give discrete steps as long as the steps can be programmed in such a manner that the transition from one position to the next during calibration is accomplished without eliminating an existing hysteresis (or friction) error of the transducer by overshoot or fluctuation.

5.1.2 Stable Excitation Source of accurately known amplitude (unless VR is being measured).

Commonly used sources are chemical batteries such as dry cells and storage batteries or line-powered, electronically regulated, power supplies. (A current limiting device shall be inserted in series with the transducer to preclude accidental damage of the potentiometric element.)

5.1.3 Electronic Indicating Instrument

Examples of suitable devices are:

Manually Balanced Ratiometer
Achievable Accuracy
1 part in 10 000

Self Balancing Ratiometer
Achievable Accuracy
1 part in 10 000

Digital Electronic Voltmeter/Ratiometer
Achievable Accuracy

$\pm(0.01\%$ of Reading + 1 digit) (4 digit display)
$\pm(0.005\%$ of Reading + 1 digit) (5 digit display)

Note: Please refer to Section 4.1.3.7 Load Impedance.

5.2 Calibration and Test Procedures

Results obtained during the calibration and testing shall be recorded on a data sheet similar to the sample data sheet in Section 7, Figure 7.1 of this standard. Calibration and testing shall be performed under Room Conditions as defined in ISA-S37.1 unless otherwise specified.

Notes:
1. The definitive paragraph under Performance Characteristics (Section 4) of this document is listed beside each of the parameters for which the test results are to be compared.
2. If more than one potentiometric element is used in the transducer, the performance of every element shall be recorded on its own form.
3. Automatic or semi-automatic testing may be used, and should be encouraged as a means to minimize human error. The technique used should be established as satisfactory relative to the manual method.

5.2.1 The transducer shall be inspected visually for mechanical defects, poor finish, and other applicable mechanical characteristics of 4.1.1.

Configuration and Dimensions	4.1.1.2
Mounting and Mounting Dimensions	4.1.1.3
Displacement Connection	4.1.1.4
Identification	4.1.1.9

By use of special equipment, or by formal verification of production methods and materials used, the following can be additionally determined:

Case Sealing	4.1.1.7
Dissimilar Metals	4.1.1.8
Case Material	4.1.2.1

5.2.2 A precision resistance measuring device shall be used to measure:

Resistance of Transduction Element and can be used to verify:	4.1.3.3
Number of *Potentiometric Elements* or *Taps*	4.1.2.2
Electrical Connections	4.1.3.4

Note: A resistance measuring device using constant current excitation and the four-wire technique is preferred.

Caution: Care must be observed when using any resistance measurement device that the measurement current does not exceed the current rating of the transducer element. To preclude possible inadvertent damage to the transducer, the measurement current should be less than the maximum permissible transducer *wiper* current.

5.2.3 Measure the insulation resistance between all transduction element terminals (or leads) connected in parallel and the case (and ground pin) of the transducer with a megohmmeter device, using a potential of 50 volts unless otherwise specified.

Insulation Resistance   4.1.3.5

5.2.4 Verify the *Dielectric Withstand Voltage* using sinusoidal ac voltage test with all transduction element terminals (or leads) paralleled, and tested to case and ground pin.

Dielectric Withstand Voltage   4.1.3.6

5.2.5 The transducer shall be connected to the displacement source and secured as recommended for its use. The appropriate excitation source and indicating instruments shall be properly connected to the transducer and turned "on." Adequate warm-up time for indicating instruments shall be allowed before tests are conducted. Electrical connections shall be checked for correctness of hook-up including the appropriate load impedance (See 4.1.3.7).

Three complete calibration cycles shall be run consecutively. Excitation amplitude shall be monitored as required unless VR is measured.

Tapping, vibrating, or dithering the unit in any manner is not permitted unless specifically noted. Approach the points gradually, do not overshoot. Do not exceed mechanical travel limits of the unit.

Unless specified individual instructions are given dictating a unidirectional range transducer requirement, all calibrations shall be accomplished as for bidirectional range transducers. The calibration shall commence at the electrical center, 50.00 percent voltage ratio, representing zero position for plus/minus equal range units. For biased plus/minus range units, the calculated ideal electrical output voltage ratio, other than 50.00 percent voltage ratio above, shall be used as the reference zero starting point.

Set the static calibration displacement fixture to a desirable reference "zero" position. Adjust the transducer's shaft position in the fixture so that specified zero position output is obtained (this may be done with the aid of "Fine Adjustment" on the transducer if it is provided). Lock the transducer shaft in the fixture.

Calibration displacement input increments will nominal-

ly be ten percent of the transducer's full span, or ten percent of each available arc or segment for tapped potentiometers. A minimum total of 11 input points (21 individual data points: 50, 60, 70, 80, 90, 100, 90, 80, 70, 60, 50, 40, 30, 20, 10, 0, 10, 20, 30, 40, 50 % VR) must be used for each calibration run on each available arc or segment. For measurement units use "Radians" ("Degrees") for angular transducers, "Millimetres" ("Inches") for linear transducers. Do not use fractions or mixed units (e.g., degrees and minutes); use only simple decimal multiples of the above units.

Record e_o/e_i in terms of % VR.

Fill in the "Theoretical % VR" Column by calculating the e_o/e_i values in accordance with the transducer specification.

Perform three full range, increasing and decreasing calibration cycles, using selected measurand levels. Record e_o/e_i in % VR units under the "Run 1," "Run 2," and "Run 3," respectively.

Record the largest plus and minus deviations observed from the theoretical/calculated/predicted values. Record these deviations under "maximum" error." Show polarity of error for each entry.

Examine the "Maximum Error" column. Show the largest "plus" and largest "minus" error as "Error Band" in the appropriate spaces on the form. Show error band allowed by the applicable specification.

Or, in lieu of the error band concept, use the Best Straight Line.

In order to verify performance between the discrete levels and to assure absence of *noise*, a full-scale X-Y plot shall be obtained, preferably inscribed diagonally across the test record form, by applying increasing, then decreasing, displacement to the transducer, and simultaneously to a reference transducer having continuous resolution and suitable linearity, each connected to one axis input of the plotter.

From the data obtained during these tests, the following characteristics should be determined:

End Points	4.2.2
Full Scale Output	4.2.3
Linearity	4.2.4
Hysteresis	4.2.5
(or Hysteresis and Linearity)	4.2.6
Repeatability	4.2.7

6. QUALIFICATION TEST PROCEDURES

Qualification Tests shall be performed as applicable using the test forms of Section 7 as required. Upon completion of all testing the form of Figure 7.6 shall be used to summarize all testing.

6.1 Initial Performance Tests (Figure 7.2)

The tests and procedures of Section 5, Individual Acceptance Tests and Calibrations, shall be run to establish reference performance during increasing (and decreasing) steps of 50, 60, 80, 100, 80, 60, 50, 40, 20, 0, 20, 40, 50% VR minimum.

6.2 Resolution Test (Figure 7.6)

An X-Y plotter shall be connected so that the transducer output is connected to the X-Axis and a continuous-resolution reference transducer to the Y-Axis input of the plotter. As the displacement to both transducers is slowly increased (simultaneously on both transducers), the number of steps shall be recorded from 0 to 100% of the test transducer's range. The following shall be determined:

Resolution (Average and Worst)   4.2.9

6.3 Mass Test (Figure 7.6)

The transducer mass shall be determined on an appropriate balance scale. The following shall be established:

Mass   4.1.1.6

6.4 Mounting Test (Figure 7.6)

The mounting of the actual installation shall be duplicated as closely as possible following specific instructions and one calibration run performed. The following shall be established:

Mounting Error   4.2.18
Mounting Effect   4.1.1.5

6.5 Loading Test (Figure 7.6)

Approximately 67% of full range (span) displacement shall be applied to the transducer, the resultant output shall be measured open-circuited, then the specified load impedance (4.1.3.7) shall be connected across the output terminals and the output measured again. The following shall be verified:

Loading Error   4.2.19

Note: The resistance of the ratiometer or other indicating instrument must be taken into account. If the transducer is to be used with other values of load impedance, the corresponding loading error can be verified in the same manner.

The loading error of a potentiometric transducer is variable with *wiper* position, ranging from zero at both extremes to a maximum value at approximately 67% VR. As a first approximation the percentage error is equal to fifteen times the ratio of the transduction element resistance to the loading resistance. Unless otherwise stated, assembly adjustments of the transducer apply to the open circuit conditions at the output terminals.

6.6 Noise Test (Refer to Noise Test Circuit)

Using an oscilloscope and a _____ milliampere dc constant current source, the output of the Noise Test Circuit shall be less than_____ millivolts peak-to-peak, representing less than _____ ohms equivalent resistance variation in the *wiper*. This test shall be conducted at representative high and low input shaft speeds, which shall be specified.

Note: The input shaft of a potentiometric angular displacement transducer, which does not include mechanical stops, may be rotated continuously and hence easily subjected to a constant angular speed for dynamic considerations.

Most other potentiometric angular or linear displacement transducers may conveniently be driven dynamically, using a sinusoidal motion for the input stimulus.

Caution: Limiting values of frequency, velocity, and acceleration may apply to the transducer shaft (or to the motion source) for a given dynamic input displacement.

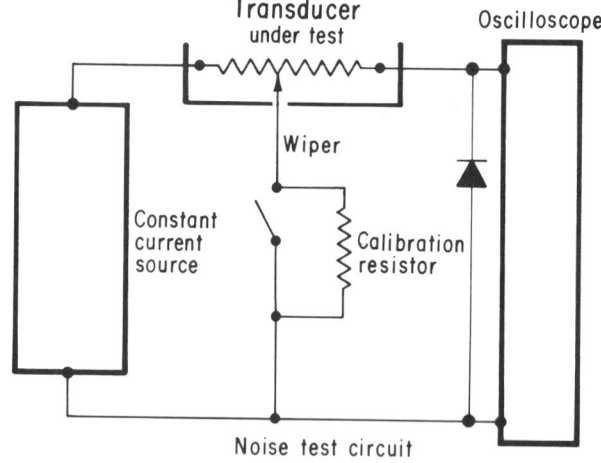

Noise test circuit

Figure 4.2

6.7 Dynamic Response Test (Figure 7.4a or 7.4b as applicable)

The dynamic response characteristics of displacement transducers may be established either with transient-excitation devices, or with sinusoidal displacement generators.

6.7.1 Transient Excitation Method

A step function of displacement may be generated with a solenoid or a spring loaded trigger mechanism.

By applying step functions of displacement at Room

Conditions within the full scale range of the transducers, and analyzing the electronic or electro-optical recording of the transducer output, the following can be determined:

Frequency Response   4.2.10

6.7.2 Sinusoidal Excitation Method

For frequencies below about 30 hertz, an oscillating mechanical table may be used.

A sinusoidal displacement waveform of constant amplitude and varying frequency, over a specified frequency range shall be applied. The following shall be determined:

Frequency Response   4.2.10

6.8 Temperature Tests

6.8.1 Low Temperature Test (Figure 7.3)

The transducer shall be placed in a suitable temperature chamber. The temperature of the transducer shall be stabilized for one hour at the lower specified operating temperature and three calibration cycles performed. The insulation resistance shall be measured and recorded as in 5.2.3.

These tests shall establish the following:

Temperature Error (at low   4.2.12
  temperature)
     or
Temperature Error Band (at low   4.2.13
  temperature)

6.8.2 Post Low Temperature Test (at Room Conditions) (Figure 7.3)

The transducer shall be removed from the temperature chamber and permitted to stabilize for one hour at room conditions. The tests of 6.7.1 shall be repeated except that the operating temperature shall be room temperature.

6.8.3 High Temperature Test (Figure 7.3)

The tests of 6.7.1 shall be repeated except that the transducer temperature shall be stabilized for one hour at the highest specified operating temperature.

These tests shall establish the following:

Temperature Error (at high   4.2.12
  temperature)
     or
Temperature Error Band (at   4.2.13
  high temperature)

6.8.4 Post High Temperature Test (at Room Conditions) (Figure 7.3)

The tests of 6.7.2 shall be repeated after stabilization of the transducer at room temperature for one hour.

Note: If required, thermal and post-thermal zero shift and sensitivity shift may also be calculated from the results of these tests.

6.9 Acceleration Test (Figure 7.5)

Acceleration shall be imposed on the transducer in three orthogonal directions by tilting it in the earth's gravitational field or by placing it on a centrifuge. A specific acceleration level shall be applied on specified axes, and the output measured. The following shall be established:

Acceleration Error   4.2.14
     or
Acceleration Error Band   4.2.15

6.10 Vibration Test (Figure 7.5)

With specified measurand levels applied, the transducer shall be vibrated along specified axes at specified acceleration amplitudes over the specified frequency range within an electro-magnetic or hydraulic shaker. The transducer output shall be recorded with a high-speed recorder. The following shall be established:

Vibration Error   4.2.16
     or
Vibration Error Band   4.2.17

Note: If so specified, the vibration error band can be established as the algebraic sum of maximum vibration errors and the last previously obtained static error band.

6.11 Tests For Other Environmental Conditions (Figure 7.3)

The transducer shall be exposed to other specified environmental conditions. As specified for each condition, one complete calibration cycle shall be performed during or after the test to establish the ability of the transducer to perform satisfactorily.

See Section 4.2.21

6.12 Life Test (Figure 7.3)

After applying the specified number of full range excursions of measurand, or after completion of each of several specified portions of the total number of cycles, at least one complete calibration cycle shall be performed to establish minimum value of:

Cycling Life   4.2.20

13

7. TEST REPORT FORMS

7.1 The test report forms listed below are recommended for use during the testing of Potentiometric Displacement Transducers.

7.2 When using the forms all pertinent information shall be inserted in its proper place. On some forms, blank space has been provided for additional tests. Where the test is prolonged, e.g., Cycling Life, more than one form may be required.

7.3 Individual Acceptance Tests and Calibrations (Figure 7.1). Used during acceptance testing of Section 5.

Initial Performance Tests and Calibrations (Figure 7.2). Used for establishing the reference performance for comparison to other test results.

Environmental Test Record (Figure 7.3). Used for Temperature, Maximum Temperature, Life and other environmental tests.

Dynamic Response Tests (Figure 7.4). Used for recording test results of Frequency Response. (Note: Use 7.4a or 7.4b as applicable.)

Acceleration/Vibration Tests (Figure 7.5). Used to record Acceleration and Vibration Test results.

Test Summary (Figure 7.6). Used to compile the results of all testing.

8. BIBLIOGRAPHY

1. Cerni, R. H. and Foster, L. E., *Instrumentation for Engineering Measurement*, John Wiley and Sons, Inc., New York, 1962.

2. Neubert, H. K. P., *Instrument Transducers*, Oxford at the Clarendon Press, 1963.

3. "Industry Standard Wirewound and Nonwirewound Precision Potentiometers," Variable Resistive Components Institute, 3525 Peterson Road, 1717 Howard St., Evanston, Illinois 60202, Revision A, March 1974.

4. MIL-E-5272C (ASG): "Environmental Testing, Aeronautical & Associated Equipment, General Specification for."

5. MIL-E-5400K (ASG): "Electronic Equipment Airborne, General Specification for."

6. MIL-STD-810B: "Environmental Test Methods"

7. ANSI MC6.1-1975 (ISA-S37.1); "Electrical Transducer Nomenclature and Terminology," 1975.

8. MIL-R-19B, "Resistor Variable, Wirebound, Precision, General Specification for."

9. MIL-R-39023, "Resistor Variable Nonwirewound, Precision, General Specification for."

10. NAS710, "Resistors; Variable, Precision."

11. MIL-STD-202D, "Test Methods for Electronic and Electrical Component Parts."

12. ANSI Z210.1-1976 (ASTM E380-76 and IEEE Std. 268-1975) "Standard for Metric Practice" February 1976.

13. ANSI MC6.2-1975 (ISA S37.3), "Specifications and Tests for Strain Gage Pressure Transducers," 1976.

14. ANSI MC6.3-1975 (ISA S37.5), "Specifications and Tests for Strain Gage Linear Acceleration Transducers," 1976.

15. "ANSI MC6.5-1976 (ISA S37.6) Specifications and Tests of Potentiometric Pressure Transducers," 1975.

16. ANSI MC6.4-1975 (ISA S37.10), "Specifications and Tests for Piezoelectric Pressure and Sound-Pressure Transducers," 1976.

17. ANSI Y32.20 (ISA S5.1), "Instrumentation Symbols and Identification," 1973.

VENDOR'S PART NO.	TEST FACILITY	USER'S PART NO.
VENDOR	**POTENTIOMETRIC DISPLACEMENT TRANSDUCER**	SERIAL NO.
LINEAR ☐ ANGULAR ☐	INDIVIDUAL ACCEPTANCE TESTS AND CALIBRATIONS	RANGE ____ TO _____

Visual Inspection:

Dimensions ☐ Workmanship ☐

Finish ☐ Nameplate ☐ El. Conn. ☐

Electrical Inspection:

Element Resistance _____ ohms   Insulation Resistance _____ Megohms at _____ Vdc

Dielectric Withstand Voltage @ _____ Vac. _____ Hz ☐ Z_L used _____ MΩ

Calibration @ _____ V ____ Excitation   Ambient Temperature _____ °C

(I = Initial, F = Final)

Input Displacement	Theor. Output (%VR)	Run 1		Run 2		Run 3				Max. Error (%VR)	
		%VR		%VR		%VR				+	−
		Increase	Decrease	Increase	Decrease	Increase	Decrease				
	100		✕		✕		✕				
	90										
	80										
	70										
	60										
	50	(I) (F)		(I) (F)		(I) (F)					
	40										
	30										
	20										
	10										
	0	✕		✕		✕					

* _____ Linearity: + ____ , − ____ %VR (Allowed: ± ____ %VR); *Hysteresis: _____ %VR (All'd ____ %VR))

* Hysteresis and _____ Linearity (Combined): + _____ , − ____ %VR   (Allowed: _____ ± _____ %VR)

Error Band: + _____ , − _____ %VR (Allowed: ± _____ %VR);

Repeatability: _____ %VR (All'd: _____ %VR);   *End Points: _____ and _____ %VR (All'd _____ and ____ %VR)

*Full-Scale Output: ____ %VR   Allowed _____ ± _____ %VR NOTE: *Values Determined From _____ Calib. Cycle

Full-Scale X-Y Pilot ____

Static Error Band: + _____ , − ____ %VR   Allowed: ± _____ %VR)

Error Bands Ref. To _____

Equipment Used:

Tested By: _____ Date _____

Approved By: _____

FIGURE 7.1

VENDOR'S PART NO.	TEST FACILITY	USER'S PART NO.
VENDOR	**POTENTIOMETRIC DISPLACEMENT TRANSDUCER**	SERIAL NO.
Linear ☐ Angular ☐	☐ INITIAL PERFORMANCE TEST	CUSTOMER
TYPE OF TEST	☐ ____________ TEST	RANGE ______ TO ______ ______

Visual Inspection:

Dimensions ☐ Workmanship ☐

Finish ☐ Nameplate ☐ El. Conn. ☐

Electrical Inspection:

Element Resistance ______ ohms; Insulation Resistance ____ Megohms at ______ Vdc

Dielectric Withstand Voltage @ ______ Vac. ______ Hz ☐ Z_L used ______ MΩ

Calibration @ ______ V ____ Excitation Ambient Temperature ______ C (I = Initial, F = Final)

Input Displacement	Theor. %VR	%VR (Run 1) Increase	%VR (Run 1) Decrease	%VR (Run 2) Increase	%VR (Run 2) Decrease		%VR Run 3 Increase	%VR Run 3 Decrease	Maximum Error %VR +	Maximum Error %VR –
	100		X		X			X		
	80									
	60									
	50	____ (I) (F)		____ (I) (F)			____ (I) (F)			
	40									
	20									
	0	X		X			X			

* ____________ Linearity: + ______ , – ______ %VR (Allowed: ± ____ %VR); *Hysteresis: ______ %VR (All'd ______ %VR)

*Hysteresis and ____________ Linearity (Combined): + ____ , – ______ %VR (Allowed: ______ ± ______ %VR)

Error Band: + ______ , – ______ %VR (Allowed: ± ____ %VR);

Repeatability: ________ %VR (All'd: ______ %VR); *End Points: ____ and ______ %VR (All'd ______ and ______ %VR)

*Full-Scale Output: ________ %VR Allowed ______ ± ______ %VR NOTE: *Values Determined From ______ Calib. Cycle

Full-Scale X-Y Plot: ________ Static Error Band: + ______ , – ______ %VR (Allowed: ± ______ %VR)

Error Bands Ref. To ____________________

Equipment Used:	Defects Noted Or Comments:

Tested By: ____________ Date: ____________ Approved By: ____________

FIGURE 7.2

VENDOR'S PART NO.	TEST FACILITY	USER'S PART NO.
VENDOR	**POTENTIOMETRIC DISPLACEMENT TRANSDUCER**	SERIAL NO.
	ENVIRONMENTAL TEST RECORD	CUSTOMER
TYPE OF TEST		RANGE ______ TO ______ ______

	Type of Environment	Level
Linear ☐ Angular ☐	☐ Before ☐ During ☐ After __________ ______	(I = Initial) ______ (F = Final)

Input Dis-place-ment	Theoretical %VR	%VR (Run 1) Increase	%VR (Run 1) Decrease	%VR (Run 2) Increase	%VR (Run 2) Decrease		%VR (Run 3) Increase	%VR (Run 3) Decrease	Maximum Error %VR +	Maximum Error %VR −
	100		X		X			X		
	90									
	80									
	70									
	60									
	50	____ (I) (F)	____ (I) (F)	____ (I) (F)			____ (I) (F)			
	40									
	30									
	20									
	10									
	0	X		X			X			

Error Band: + ______ % − ______ %VR (Referred To __________) Allowed: ± ______ %VR

Ins. Resistance: ______ Megohms at ____ Vdc

Comments: ________________

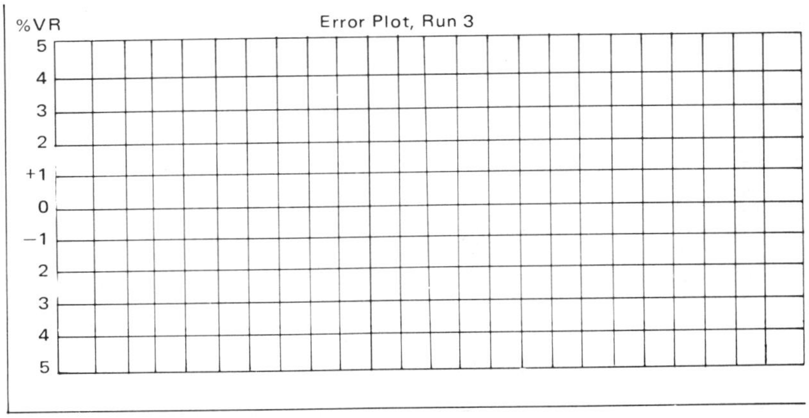
%VR
Error Plot, Run 3
5
4
3
2
+1
0
−1
2
3
4
5

Tested By: __________ Date Test Started: __________ Date Test Finished: __________

Approved By: __________ Title:

Approved By: __________ Title:

FIGURE 7.3

VENDOR'S PART NO.	TEST FACILITY	USER'S PART NO.
VENDOR	**POTENTIOMETRIC DISPLACEMENT TRANSDUCER**	SERIAL NO.
REPORT NO.		CUSTOMER
TYPE OF TEST	DYNAMIC RESPONSE TESTS	RANGE _____TO_____ ____

1. Ambient Conditions: Temperature: _____ °C; Pressure: _____ Pa; Humidity _____%
2. Dynamic Response _____

    Excitation Voltage _____

    Step Function Generator _____

    Mounting Location: End: _____

Shock No.	Hi	Low	Shock Velocity	Step	Pronounced Resonances, Hz			Remarks

Frequency Response _____ Hz (Allowed _____ Hz)

Overshoot _____ %VR (All'd _____ %VR)

ATTACH OSCILLOSCOPE PHOTOGRAPHS OF TRANSDUCER RESPONSES

Amplitude Scale

SHOCK 1

Time Scale

Amplitude Scale

SHOCK 3

Time Scale

Amplitude Scale

SHOCK 2

Time Scale

Amplitude Scale

SHOCK 4

Time Scale

Tested By: _____

Approved By: _____
             Title

Date Test Started: _____ Date Test Finished: _____

Approved By: _____
             Title

FIGURE 7.4.a

VENDOR'S PART NO.	TEST FACILITY	USER'S PART NO.
VENDOR	**POTENTIOMETRIC DISPLACEMENT TRANSDUCER**	SERIAL NO.
REPORT NO.	DYNAMIC RESPONSE TESTS	CUSTOMER
TYPE OF TEST	(SINUSOIDAL METHOD)	RANGE _____TO _____ _____

Ambient Conditions: Temperature: _____ °C, Pressure: _____ Pa; _____ Humidity _____ %

Dynamic Response:

Excitation (Volts or ma)_____ ;   Transducer Load _____ ohms

Sinusoidal Generator _____ ;   _____

Mounting Configuration _____

Test Temperature _____°C

Sinusoidal

Frequency, Hz _____

Peak to Peak, %VR _____

ATTACH OSCILLOSCOPE PHOTOGRAPH OF TRANSDUCER RESPONSE

AMPTITUDE SCALE

TIME SCALE

FIGURE 7.4.b

VENDOR'S PART NO.	TEST FACILITY		USER'S PART NO.
VENDOR	**POTENTIOMETRIC DISPLACEMENT TRANSDUCER**		SERIAL NO.
REPORT NO.			CUSTOMER
TYPE OF TEST	ACCELERATION/VIBRATION TEST RECORD		RANGE _____ TO _____ _____

SKETCH OF TRANSDUCER SHOWING AXIS ORIENTATION:

ACCELERATION TEST

AXIS	+ X	− X	+ Y	− Y	+ Z	− Z	
Output Before Accel. (%VR)							Input Displacement Used _____ _____
Applied Accel. (G)							Max. Accel. Error: + _____ , − _____ %VR
Output During Accel. (%VR)							Pre-Accel. Static Error Band: + _____ , − _____ %VR
Accel. Error (%VR)							Accel. Error Band: + _____ , − _____ %VR

Pre-Accel. Static Error Band: + _____ , − _____ %VR

Accel. Error Band: + _____ , − _____ %VR

(Allowed Accel. Error Band ± _____ %VR

Tested By: _____ (Technician)

_____ (Test Engineer)

Date: _____ Approved By: _____

Witnessed By: _____ (_____)

Witnessed By: _____ (_____)

COMMENTS

VIBRATION TEST

AXIS	X			Y			Z			
Output Before Vib.		%VR			%VR			%VR		Max. Vib. Error: + _____ , − _____ %VR
	Freq. (Hz)	Error		Freq. (Hz)	Error		Freq. (Hz)	Error		Pre-Vib. Static Error Band: + _____ , − _____ %VR
		Pol.	%VR		Pol.	%VR		Pol.	%VR	Vib. Error Band: + _____ , − _____ %VR
Vibration Error										(Allowed Vib. Error Band: ± _____ %VR

Max. Vib. Error: + _____ , − _____ %VR

Pre-Vib. Static Error Band: + _____ , − _____ %VR

Vib. Error Band: + _____ , − _____ %VR

(Allowed Vib. Error Band: ± _____ %VR

Tested By: _____ (Technician)

_____ (Test Engineer)

Date: _____ Approved By: _____

Witnessed By: _____ (_____)

Witnessed By: _____ (_____)

COMMENTS

FIGURE 7.5

VENDOR'S PART NO.	TEST FACILITY		USER'S PART NO.
VENDOR			SERIAL NO.
REPORT NO.	**TRANSDUCER TEST REPORT**		CUSTOMER
TYPE OF TEST	POTENTIOMETRIC DISPLACEMENT TRANSDUCER		RANGE _____ TO _____ ____

SUMMARY OF RESULTS:

☐ Error
☐ Error Band

Test	Tested Per Proced. No. or Test Waived Per	Par No.	Pass	Fail				+ %VR	− %VR
				Error	Electr.	Mechan.	See Cmmnts.		
Initial P.T. (Performance Test)									
Resolution								Avg.:__%VR Max.:__%VR	
Mass									
Noise									
Mounting									
Loading Max. Z_L / Min. Z_L									
Frequency Response								Flat (+__%): __To__ Hz	
								_____ms., _____%Ovs.	
Low Temp. _____ °C									
After Low Temp.									
High Temp. + _____ °C									
Add'l. Temp. _____ °C									
After High Temp.									
_____ g_n Vibration									
After _____ g_n Vibr.									
Acceleration									
After Accel.									
Life									

Tested By: _____   Date Test Started: _____ Date Test Finished: _____

Approved By: _____   Approved by: _____
                    Title   Title

FIGURE 7.6

INSTRUMENT SOCIETY of AMERICA
Research Triangle Park, North Carolina